The Earth Path

ALSO BY STARHAWK

The Spiral Dance: A Rebirth of the Ancient Religion of the Great Goddess

Dreaming the Dark: Magic, Sex and Politics

Truth or Dare: Encounters with Power, Authority and Mystery

The Fifth Sacred Thing

Walking to Mercury

The Pagan Book of Living and Dying,
with M. Macha Nightmare and the Reclaiming Collective

Circle Round: Raising Children in Goddess Tradition,
with Diane Baker and Anne Hill

The Twelve Wild Swans,
with Hilary Valentine

Webs of Power: Notes from the Global Uprising

The Earth Path

Grounding Your Spirit in the Rhythms of Nature

STARHAWK

HarperOne
An Imprint of HarperCollins*Publishers*

HarperCollins books may be purchased for educational, business, or sales promotional use. For information please write: Special Markets Department, HarperCollins Publishers, 10 East 53rd Street, New York, NY 10022.

HarperCollins Web site: http://www.harpercollins.com

FIRST HARPERCOLLINS PAPERBACK EDITION PUBLISHED IN 2005

Book design and charts by Kris Tobiassen
Illustrations by Lydia Hess

Library of Congress Cataloging-in-Publication Data is available.

ISBN: 978-0-06-000093-6

10 11 RRD(H) 10 9

Contents

List of Exercises, Meditations, and Rituals

THE SACRED

OBSERVATION

THE CIRCLE OF LIFE

AIR

FIRE

WATER

EARTH

THE CENTER

HEALING THE EARTH

Acknowledgments

This book was inspired and informed by many people. First, let me acknowledge and offer gratitude to the Pomo people, the original people of the land I live on, and to all the indigenous peoples of the earth who have been caretakers, guardians, and keepers of the earth's wisdom for millennia, and whose lands and cultures are continually under siege today.

Susan Davidson introduced me to permaculture many years ago. Penny Livingston-Stark was my first permaculture teacher, is my current teaching partner, and is a dear friend and inspiration. I am grateful also to other teachers, Blythe Reis, Patricia Michael, and Keith Johnson, and to the amazing group at Occidental Arts and Ecology Center: Brock Dolman, Dave Henson, and Adam Wolpert. Erik Ohlsen has grown from a student to a close friend and teaching partner. Abby Wing has been a coteacher and partner in actions.

Bill Mollison and David Holmgren originated the principles and practices of permaculture. Matthew Fox and Brian Swimme showed me the power of uniting science and story. Jon Young and the other teachers of the Wilderness Awareness School opened my eyes and ears and taught me to hear the language of the birds.

Evergreen Erb, Kitty Engleman, and Sunray have cotaught Earth Path workshops with me and helped develop many of the exercises and insights in this book. The many teachers I have worked with in Reclaiming over the years have cocreated and inspired much of this work.

Carol Christ and David Seaborg kindly read chapters and offered helpful feedback and clarifications. Brian Tokar attempted to awaken me to the dangers of biotech many years ago. Luke Anderson helped deepen my understanding and is a tireless activist on the issue, as is Brian.

I am grateful to the committed organizers of the Sacramento Circus, the Greenbloc, and Earth First! and to the forest defenders and all those who have put their lives on the line for the earth. And to all those who have created the many alternatives and positive solutions outlined here.

My partners in our training collective RANT (Root Activist Network of Trainers), Lisa Fithian, Hilary McQuie, Charles Williams, and Ruby Perry, have been a great support throughout the writing of this book. My friends and neighbors in the Cazadero Hills have accompanied me on walks, taught me to recognize plants and animals, shown me mushrooms, and joined me in efforts to protect our lands. Mary Dedanan and Akasha Madron have helped me hold all the complicated logistical threads of my life together, and Ken Genetti is the spider who weaves my Web page. My agent, Ken Sherman, is always a strong support and my editor, Eric Brandt, has been an understanding, flexible, and helpful reader who has sharpened the book's focus. My housemates at Black Cat and my partner, David Miller, have been loving companions on this journey. The children in my life, Kore, Aidan, Allison, Florence, Aminatou, Tijiane, Bowen, Lyra, Johanna, Casey, Emma Lee, Leif, Tashi Sophia, and Ruby, will inherit our efforts at earth-healing and the legacies of our wounds, and this book is dedicated to them.

STARHAWK
CAZADERO HILLS
JANUARY 6, 2004

ONE

Toward the Isle of Birds

On a hilltop in the coastal mountains of northern California, I meet with my neighbors just before sunset on a hot day in July to go to a fire protection ritual. All summer long, our land and homes are at risk for wildfire. In the winter, we get eighty to a hundred inches of rain in a good year, and trees and grasses and shrubs grow tall. But no rain falls from June through September, and in summer the land gets dry as tinder. A small spark from a mower, a carelessly tossed cigarette, a glass bottle full of water that acts as a magnifying lens can all be the beginning of an inferno that could claim our homes and lives.

We live with the constant risk of fire, and also with the knowledge that our land needs fire, craves fire. This land is a fire ecology. All the trees on it evolved in association with forest fires. The redwoods, with their thick, spongy bark, withstand fire. The madrones and bay laurels and tanoaks resprout from root crowns to survive fire. Fire once kept the meadows open, providing habitat for deer and their predators, coyote and cougar. Fire kept the underbrush down, favoring the big trees and reducing disease. The Pomo, the first people of this land, burned it regularly to keep it healthy. As a result, the forest floor was kept open, the fuel load was reduced, and fires were low and relatively cool. But now the woods are dense with shrubby regrowth, the grasses tall and dry. A fire today would not be cool and restorative, but a major inferno.

Below us is the small firehouse that belongs to our Volunteer Fire Department. We can look around to the far horizons and see our at-risk landscape. Deep canyons are filled with redwoods and Douglas firs, with bay laurel and madrone and vast stands of tanoak filling in the open spaces left where stands of giant conifers were logged a hundred years ago and, again, fifty years ago. The tanoaks are bushy, with multiple small stems that create a huge fire hazard. Big-leaf maples line the stream banks, and black oaks stud the open hillsides where fifty years ago sheep grazed. Tall stands of grasses in the open meadows are already dry and ready to burn. Once the meadows would have stayed green all summer with deep-rooted native bunchgrasses, but a century of grazing favored invasive European grasses that wither quickly in the summer heat. Small homes fill the wrinkles in the landscape, most built twenty years ago by back-to-the-landers out of local wood and scrounged materials. On the high ridges, we can see evidence of the latest change in land use, a proliferation of vineyards. Behind us is a huge fallen tree—a remnant of the 1978 wildfire that started just over the ridge and burned thousands of acres.

We begin by sharing some food, talking and laughing together, waiting for everyone to arrive. Then we ground, breathing deeply and with great gratitude the clean air that blows fresh from the ocean just a few ridges over. We imagine our roots going into the earth, feeling the jumble of rock formations and the volatile, shifting ground here just two ridges over from the San Andreas fault. We feel the fire of the liquid lava below our feet, and the sun's fire burning hot above our heads.

We cast our circle by describing the boundaries of the land we wish to protect—from the small town of Cazadero in the east to the *rancheria* of the Kashaya Pomo in the north; from the ocean in the west to the ridges and gulches to the south of us. We invoke the air—the actual breeze we can feel on our skin; the fire, so integral to this landscape yet so dangerous to us now; the water, the vast ocean now covered in a blanket of fog, the sweet springs that feed the land; the earth herself, these jumbled ridges and tall forests.

In the center of the circle is a small bowl. One by one, we bring water from our springs and pour it into the vessel. My neighbors know exactly where their water comes from. Each of us has spent many hours digging out springs, laying water pipes, fixing leaks.

"This is from a spring beyond that hill that flows into Camper Creek that flows into Carson Creek that flows into MacKenzie Creek that flows into Sproul Creek that flows into the South Fork of the Gualala River . . . "

We offer the combined waters to the earth with a prayer of gratitude—great gratitude that we live in one of the few places left on earth where we can drink springwater straight from the ground.

Alexandra has made our fire charm—a circle of bay laurel branches with a triangle lashed within. The triangle is the symbol of fire; the circle represents containment and also the cycle that we know someday needs to be restored. One by one, we come forward and tie on branches we have each brought from trees on our lands. Redwood, from a giant that has withstood many fires. Tanoak, suffering now from a fungal disease that fire might have cured. Madrone, of the beautiful peeling red bark, and buckeye in flower. They are as familiar as our human friends. We know them intimately, know when and how they flower and seed, have watched many individuals grow from seedlings. Some of my neighbors planted these hills after the 1978 fire, worked the creek beds to slow erosion, thinned and released the woods time after time. They know the boundaries of the soil types and the history of each patch of the woods. Ken and Alexandra bring small, uprooted firs, pulled out from a patch on their land where they grow far too thickly for any to get enough light to grow healthy and strong. Once fire would have thinned them—now people do. We add herbs and flowers from our gardens.

We pass the charm around, drumming and chanting to charge it:

Sacred fire that shapes this land,
Summer teacher, winter friend,
Protect us as we learn anew
To work, to heal, to live with you.
Green, green crown
Roots underground.
Kissed by fire,
Still growing higher.

Laughing, we dance with the charm, pass it over each other's heads and bodies. These trees and branches are part of us as we have each become part of this land. The water we have brought is our drinking water, the water that grows our gardens. We literally eat and drink the land.

When the charm has gone around, we all hold it together and chant, raising a wordless cone of power, a prayer of protection, and also a prayer for knowledge. We pray that our homes and lives can be preserved as we struggle to learn, once again, how to integrate fire with this land, how to restore the balance that has been so lost.

Then the two young girls who are with us climb the fallen tree behind us and hang the charm high on its branches, where it will overlook the land for the summer. We will see it every time we look up at Firehouse Hill. And when winter comes, and the rain returns, we will take it down and cut it apart in our rain return ritual, where we thank the rain for coming back and pray for the health of the land and the trees. We'll each take pieces of this charm to burn in our woodstoves for our winter fires, to protect our homes in the season when fire warms our hearths and cooks our food.

In thirty or more years of practicing earth-based spirituality, I've probably done thousands of rituals. Some are old and some are new; some have become traditions and some draw on ancient roots. Our fire ritual and rain return ritual are relatively young—we created them less than ten years ago. They don't correspond to the equinoxes or the major Celtic feasts or the indigenous Pomo ceremonies of this land. Yet in some ways they represent the most ancient tradition of ritual and ceremony there is: they are the rituals the land told us to do.

The fire ritual represents, for me, a shift in the way I view my own spirituality. For more than three decades, I've been a Witch, a priestess of the Goddess of birth, growth, death, and regeneration, someone who sees the sacred embodied in the natural world. I've written books, created rituals, and practiced and taught magic, "the art of changing consciousness at will."[1] I've marched, demonstrated, organized, and even gotten arrested trying to protect the integrity of the natural world. Nature has been the heart of my spirituality.

But I grew up a city girl. I didn't spend my childhood roaming the woods and splashing in pristine streams; I spent it playing handball in the parking garage of our apartment in the San Fernando Valley of L.A. There was one good climbing tree in our neighborhood, but it stood in the front yard of a woman who yelled at us to get out every time we got up into its branches. My widowed mother never took us camping, and the summer camps I went to stressed studying Hebrew and saying prayers rather than learning woodcraft. My formal education focused on art and psychology and somehow missed biology and ecology. In something like seven years of higher education, only one course, a class in botany for art majors, taught me anything about observing or interacting with the natural world.

When I began studying, teaching, and writing about Witchcraft and Goddess religion thirty years ago or more, what seemed most important to me is that Wicca (the archaic name for our tradition) valued women, the body, and the erotic. I saw magic as an ancient tradition of psychology, the understanding and training of the human mind. And those are indeed very important aspects of our tradition.

But, as I've celebrated in Pagan communities and lived in both the city and the country, as I've worked in environmental movements and other movements for social and ecological justice, I've come to feel that one aspect of our nature-based religion that too often gets neglected is our actual relationship with nature. To be a Witch, to practice magic, we can't simply honor nature's cycles *in the abstract.* We need to know them intimately and understand them in the physical as well as the psychic world. A real relationship with nature is vital for our magical and spiritual development, and our psychic and spiritual health. It is also a vital base for any work we do to heal the earth and transform the social and political systems that are assaulting her daily.

One of the most rewarding aspects of my own journey over the past decades has been a gradual process of deepening my aesthetic appreciation of nature into real knowledge and true understanding. That process became a journey that was to transform my life, my spirituality, and my understanding of the Goddess. It began my true education, and my transformation from a tourist in nature to an inhabitant—someone who not only loves trees but can plant them, prune them, and understand the complex role they play in everything from soil ecology to weather patterns. Like most eco-activists, I fully confess to being a long-term tree-hugger, and like most Witches, I've always talked to trees. But now, when they talk back, I can assess whether what I'm hearing is truly their message or my own fantasies. I've always loved birds, but now when I hear them call in the treetops around my house, I can often identify their voices and at least guess the general subject of their conversation, even if I can't translate all the details.

This journey also transformed my understanding of the Goddess. For me, now, the Goddess is the name we put on the great processes of birth, growth, death, and regeneration that underlie the living world. The Goddess is the presence of consciousness in all living beings; the Goddess is the great creative force that spun the universe out of coiled strings of probability and set the stars spinning and dancing in spirals that our entwining DNA echoes as it coils, uncoils, and evolves. The names and faces we give the Goddess, the particular aspects she takes, arise originally from the qualities of different places, different climates and ecosystems and economies. In Eleusis, once the most fertile plain in Greece, she was Demeter, Goddess of grain. Up the way, in dry, hilly Athens, she was Athena, Goddess of olives. In Hawaii, she is Pele, Goddess of the volcano. In India, each tribal village has a patron Goddess/devi of its own.

And the tradition we call Wicca arose from people who were indigenous to their own lands. In England, even until recent times, certain families passed on the tradition of "earth-walking," of knowing their own area intimately, understanding the mythological and practical significance of every hill and stream and valley, knowing the uses of the herbs and the medicinal properties of the trees and shrubs, and being responsible for the area's spiritual and ecological health.

David Clarke, in his book *Twilight of the Celtic Gods*, records the story of an informant he calls "the Guardian," who recalls his upbringing in an ancient, earth-based tradition of Yorkshire:

> I come from an old tradition, a very old tradition if the learning passed down from families is to be believed. . . . I was always told that my family and its various branches and offshoots have been in this part of the world since time began . . . and we have worked on the land as farmers, craftsmen and in related professions. . . . Yes, I suppose we are "pagans"—but only in the sense that the world of paganism originally meant the beliefs and practices of those in the countryside. . . .
>
> At the time of my "awakening," as we called it, my maternal grandmother was responsible for passing on the teachings . . . and this at first took the form of what might be called "nature walks"—remember, I was only seven at the time—in which we would walk for miles in all weathers, at all times of year and at all times of day and night. If I tried to speak or ask questions, I was hushed with a "just look and listen" or something similar. . . .
>
> My grandmother explained to me . . . that the earth was a living, breathing entity and everything was interrelated. . . . I had to learn all, and I mean all, the names—local names that is—for every single plant, tree, type of stone, animal, bird, insect, fish and so on. I had to know where they could all be found, what they looked like at any given time of year and what, if any, their uses were—practical, medical or whatever.
>
> . . . I was also eased into the fundamental belief of our tradition—that the land is sacred. And to that end we thought of ourselves as stewards, guardians of the areas where members of our family dwelt, people who could be of some use to others who had forgotten or never knew what we still held on to. . . . Farmers, stockmen, gamekeepers and many ordinary countryfolk all knew of our knowledge of plants and animals, and certain members of the family would help them with natural and herbal remedies for both animal and human problems alike. . . .
>
> The powers that we held in awe were locked inside the landscape, inherent in the power of the weather and manifest in the cycle of the changing of the seasons, and in the end they in turn ran through us.[2]

Our magical practices arose from people who were deeply connected to the natural world, and our rituals were designed to give back to that world, to help maintain its balance along with our human balance. If we leave the natural world out of our practice and rituals in any real sense, if we invoke an abstract earth but never have any real dirt under our fingernails, our spiritual, psychic, and physical health becomes devitalized and deeply unbalanced.

In one sense, this understanding of the Goddess is not new for me. More than two decades ago, I wrote about the Goddess in *The Spiral Dance:* "In the Craft, we do not *believe in* the Goddess, we connect with Her, through the moon, the stars, the ocean, the earth, through trees, animals, through other human beings, through ourselves. She is here. She is the full circle: earth, air, fire, water and essence—body, mind, spirit, emotions, change."[3]

But I understand more deeply now that what we call Goddess or God was the face and voice that people gave to the way the land spoke to them. The rituals and ceremonies and myths of the ancestors all arose from their actual relationship to a specific place on earth. And the tools of magic, that discipline of identifying and shifting consciousness, were the skills of listening to what ethnobotanist Kat Harrison calls "the great conversation,"[4] the ongoing constant communication that surrounds us.

Most of us who live in cities, who are educated to read, write, do arithmetic, and use computers, live our lives surrounded by that conversation yet are unaware of it. We may love nature, we may even profess to worship her, but most of us have barely a clue as to what she is murmuring in the night.

To be a Witch (a practitioner of the Old Religion of the Goddess) or a Pagan (someone who practices an earth-based spiritual tradition) is more than adopting a new set of terms and customs and a wardrobe of flowing gowns. It is to enter a different universe, a world that is alive and dynamic, where everything is part of an interconnected whole, where everything is always speaking to us, if only we have ears to listen. A Witch must not only be familiar with the mystic planes of existence beyond the physical realm; she should also be familiar with the trees and plants and birds and animals of her own backyard, be able to name them, know their uses and habits and what part each plays in the whole. She should understand not just the symbolic aspects of the moon's cycle, but the real functioning of the earth's water and mineral and energy cycles. She should know the importance of ritual in building human community, but also understand the function of mycorrhizal fungi and soil microorganisms in the natural community in which human community is embedded.

In fact, *everybody* should. Our culture is afflicted with a vast disconnection, an abyss of ignorance that becomes apparent whenever an issue involving the natural world arises. As a society, we are daily making decisions and setting policies that have enormous repercussions on the natural world. And those policies are being set by officials and approved by a public who are functionally eco-illiterate.

I was once giving a talk at a university about the need for earth-based spirituality, when I was stopped by a student with a question that stunned me.

"Tell me," the young man asked, "why is the earth important?"

I almost didn't know what to say. I bit back a snide retort—"What planet do you live on?"—and realized with horror that he was quite serious, that somehow all his years of higher education and graduate school had not taught him that we are utterly dependent on the earth for our lives.

"Soil bacteria—they're small things; who cares about them?" said a radio interviewer recently when I was trying to explain why we were protesting a USDA conference promoting genetic engineering to agricultural ministers of the third world. It soon became evident that neither he nor most of the audience understood the difference between genetically modifying an organism and simply breeding plants. If you, the reader, don't yet know that difference or understand why anyone who eats should care about the microorganisms in the soil, by the end of this book, you will.

To develop a real relationship with nature, we don't need to live in the country. In fact, this book and work are very much directed toward city dwellers. The vast majority of us, including the vast majority of Pagans, live in cities. It is in the cities that decisions are made that impact the health and life and balance of the natural world. If you love nature but don't really know her, if you live in the city and find yourself stunned and bewildered in the countryside, or if you perhaps know a lot intellectually about ecology but have trouble integrating your knowledge with your deepest sense of joy and connection, this book can be a guide.

Studying the language of nature can be a dangerous undertaking. For to become literate in nature's idiom, we must challenge our ordinary perceptions and change our consciousness. We must, to some extent, withdraw from many of the underlying assumptions and preoccupations of our culture.

The first set of assumptions are those about the earth and our role in it as humans. One view sees human beings as separate from and above nature. Nature exists as a resource bank that we are entitled to exploit for our own ends. She is of value only in how she can be used for our increased comfort, gain, or profit. This philosophy is held by many religions, but also by both capitalists and classical Marxists. It has resulted in unprecedented destruction of ecosystems and life-support systems all over the planet, from the clearcutting of ancient forests to the building of unsafe nuclear reactors.

But there is a counterpoint to this view, one often held by environmentalists and even some Pagans, that is more subtly destructive. That's the view that human beings are somehow worse than nature, that we are a blight on the planet and she'd be better off without us. In *Webs of Power*, I wrote about this view:

> Now, I admit that a case can be made for this view—nevertheless I think that in its own way it is just as damaging as the worldview of the active despoilers. For if we believe that we are in essence bad for nature, we are profoundly separated from the natural world. We are also subtly relieved of responsibility for listening to the great conversation, for learning to observe and interact and play an active role in nature's healing.

The humans-as-blight vision also is self-defeating in organizing around environmental issues. It's hard to get people enthused about a movement that even unconsciously envisions their extinction as a good. As long as we see humans as separate from nature, whether we place ourselves above or below, we will inevitably create false dichotomies and set up human/nature oppositions in which everyone loses.[5]

A corrective view might arise from the understanding that we are not *separate* from nature but in fact *are* nature. Penny Livingston-Stark, my teaching partner in Earth Activist Trainings that combine permaculture design training with work in earth-based spirituality and activism, often tells the story of her own evolution from believing that we must work *with* nature, to seeing us as working *within* nature, to understanding that we are nature working.

Indigenous cultures have always seen themselves as part of nature. Mabel McKay, Cache Creek Pomo healer, elder, and basketmaker, used to say, "When people don't use the plants, they get scarce. You must use them so they will come up again. All plants are like that. If they're not gathered from, or talked to and cared about, they'll die."[6]

Range management expert Allan Savory describes the vast herds of buffalo and prides of lions that stalked the land he managed in the 1950s in what is now Zambia and Zimbabwe, and he talks about how people coexisted with those creatures:

> People had lived in those areas since time immemorial in clusters of huts away from the main rivers because of the mosquitoes and wet season flooding. Near their huts they kept gardens that they protected from elephants and other raiders by beating drums throughout much of the night. . . . [T]he people hunted and trapped animals throughout the year as well.

Nevertheless, the herds remained strong and the river banks lush and well-covered with vegetation, until the government removed the people in order to make national parks.

> We replaced drum beating, gun firing, gardening and farming people with ecologists, naturalists, and tourists, under strict control to ensure that they did not disturb the animals or the vegetation. . . . Within a few decades miles of riverbank in both valleys were devoid of reeds, fig thickets and most other vegetation. With nothing but the change in behavior in one species these areas became terribly impoverished and are still deteriorating. . . . [T]he change in human behavior changed the behavior of the animals that had naturally feared them, which in turn led to the damage to soils and vegetation.[7]

The indigenous peoples of California burned the forests and grasslands to maintain a mosaic of open meadows and forest cover that was ideal for game. When they dug brodaias for food, they took the larger bulbs and scattered the smaller ones, spreading the stands and giving the young bulbs room to grow. By digging and pruning sedge roots for basketweaving, they encouraged the growth of the sedges that helped protect the soils of the riverbanks. California was a lush landscape, described by early European explorers as abundant with game, wildflowers, birds, fish, and natural beauty. Although the explorers thought they had discovered a pristine wilderness, in reality they had found a landscape so elegantly managed that they were utterly unaware of the human role in maintaining such abundance.

Some indigenous cultures have also hunted animals to extinction and turned fertile land to deserts. I don't want to romanticize other cultures, but I do think it is important to learn from them. On this continent, fire, prayer, ceremony, and myth were all ways indigenous peoples attempted to influence and understand their environment. In a world in which everything a person ate, touched, or used came from the land, humans indeed were part of the land in a deep integration we can only imagine.

Another set of assumptions we must challenge are assumptions about what constitutes knowledge. For centuries, since the start of the "scientific revolution," Western culture has pursued knowledge by breaking a subject or an object into its component parts and studying those parts. We go to doctors who specialize in one organ or one set of diseases (such as cancer or heart disease) or one technique for curing (surgery, psychiatry). We study in universities where we learn biology or chemistry or physics. We've developed a mechanistic, cause-and-effect model of the universe. Compartmentalization has taught us a lot, and produced many advances, but it is only one way of looking at the world. It doesn't allow us to look at the whole, or at the complex web of relationships and patterns that make up a whole.

Science itself has moved beyond the mechanistic model of the universe. Today science is likely to describe the world in terms of networks and probabilities and complexities, as interlocking processes and relationships. Yet our thinking and understanding as a culture does not often reflect this greater sophistication. Nor do our regulations, technologies, and practices.

Magic is, in a sense, pattern-thinking. The world is not a mechanism made up of separate parts, but a whole made up of smaller wholes. In a whole, everything is interconnected and interactive and reflective of the whole—just as in a hologram each separate bit contains an image of the whole. Astrology and Tarot, for example, work because the pattern of the stars at any given moment or the pattern the cards make when they fall reflects the whole of that moment.

Developing a deep relationship with nature means a shift in our thinking, learning to see and understand the whole and its patterns, not just the separate parts.

To really know the Goddess, we must learn to be present in and interact with the natural world that surrounds us, in the city as well as the country or wilderness. Instead of closing our eyes to meditate, we need to open our eyes and observe. Unless our spiritual practice is grounded in a real connection to the natural world, we run the risk of simply manipulating our own internal imagery and missing the real communication taking place all around us. But when we come into our senses, we can know the Goddess not just as symbol but as the physical reality of the living earth.

In developing that real relationship with the Goddess, we also need to reconcile science and spirituality. When our sense of the sacred is based not upon dogma but upon observation and wonder at what is, no contradiction exists between the theories of science and those of faith. As Connie Barlow writes in *Green Space, Green Time,*

> The more we learn about Earth and life processes, the more we are in awe and the deeper the urge to revere the evolutionary forces that give time a direction and the ecological forces that sustain our planetary home. Evolutionary biology delivers an extraordinary gift: a myth of creation and continuity appropriate for our time. . . . Finally, geophysiology, including Gaia theory, has reworked the biosphere into the most ancient and powerful of all living forms—something so much greater than the human that it can evoke a religious response.[8]

When science and spirit are reconciled, the world becomes reenchanted, full of wonder and magic. The great conversation is happening around us in many dimensions. Magic might also be called the art of opening our awareness to the consciousnesses that surround us, the art of conversing in the deep language that nature speaks. And magic teaches us also to break spells, to shatter the ensorcellment that keeps us psychologically locked away from the natural world.

To open up to the outer world, we also undergo inner changes and development. For we are part of the living earth, and to connect with her is to connect with the deepest parts of ourselves. We need the discipline of magic, of consciousness-change, in order to hear and understand what the earth is saying to us. And listening to the earth, doing the rituals the land asks us for, giving back what we are asked for, will also bring us healing, expanded awareness, and intensified life.

Opening up begins with listening. To learn to listen, however, is a long process.

Long ago I read a fairy tale about a prince who learns the language of birds. I remember only the beginning, and though I've searched many times through all my books, I've never been able to find the story again. It begins something like this:

> Once upon a time, there lived a king who had one son, whom he treasured. The king wanted to give his child every advantage, so he sent him to be educated on the sacred Isle of Birds, where he could learn the language of birds. After seven years the prince returned.
>
> "What have you learned?" the king asked.
>
> "I can hear something," the prince replied.
>
> "What! That's all? After seven years?" The king was irate. "You'd better go back and study harder."
>
> So the prince went back to the island, and, after another seven years, he returned home.
>
> "What have you learned?" the king asked again.
>
> "I can hear something, and I can understand something," the prince replied.
>
> "What! That's all? After twice seven years?" Again the king was furious, and sent his son back to the island.
>
> After another seven years, the son returned home again.
>
> "What have you learned?" the king asked, somewhat wearily this time.
>
> "Well, I can hear something, I can understand something, and I can say something," the prince replied.
>
> Angered beyond words, the king threw his son out into the wide world, and the prince was forced to make his way alone.

The story continues, but that opening has much to teach us. To begin with, it implies that true education is about learning the language of nature, and it's a slow process. To learn the language of birds takes time. It took the prince seven years just to *hear* something. We need to slow down, to learn to see and listen, to sharpen our powers of observation. Your admission fee to the Isle of Birds is simply the willingness to set aside some time in your life to be in nature—whether that's an alpine meadow in the wilderness or a vacant lot in the inner city.

Once we have learned to hear, then we can begin to understand. And only after we understand do we begin to speak, to intervene.

The story speaks to a core principle of one of the other disciplines that has deeply informed my relationship with nature: the system of ecological design known as permaculture. Developed by Bill Mollison and David Holmgren, the term *permaculture* comes from both "permanent agriculture" and "permanent culture." It includes principles, practices, and ethics that enable us to design sustainable environments that function like natural systems—for growing food but also for growing human community. Penny Livingston-Stark and Blythe Daniels taught the design course I took in 1996, and since then Penny and I have collaborated on Earth Activist Trainings that combine a permaculture design course with training in earth-based spirituality and the skills of organizing and activism.

Permaculture teaches us to begin with long and careful observation rather than careless intervention. We begin by taking the time to hear and see something, and then look for ways to make the least possible change for the greatest effect. We make small changes first, and observe their effect.

Another discipline that has influenced my ability to hear and understand has been the training I've received from the Wilderness Awareness School, a self-study program directed by Jon Young, which teaches tracking and actually has a course in learning the language of birds. I've been an erratic student, but the routines and approaches I've learned have deeply changed my way of being in the wilderness and the world.

Both those disciplines, along with three decades or more of magical practice and teaching, inform this book. My hope is that it can be a trail that takes you out into nature, that deepens and informs your magical practice as well as your daily life, and that helps ground earth-based spirituality in real ground, real earth.

My intention in this book is to do four things. First, to suggest practices and exercises that will teach us to observe, to hear something. Second, to help us understand something about how the natural world works by engaging our hearts and spirits as well as our minds—through storytelling, myth-making, and trance journeys. Third, to help us learn to speak, to create nature-based ritual, to communicate back to the living beings around us. And finally, to help us act: to know what solutions exist, to understand the practical ways that we can transform our way of living to be more in harmony with the earth, and to have a sound basis for actions in defense of the earth.

I'm writing this book on a computer powered by the sun through the solar panels on my cabin—but I don't expect every reader and every Witch to run out and convert their electrical system tomorrow. But I do expect that every reader will come away with an expanded understanding of the palette of alternative energy options, the implications of our public policy regarding energy use for the continued health of the earth and the general state of conflict in human society—and the vital relevance of these issues to a spirituality based on nature.

I'm eating food from my garden, but I don't expect that every reader and every Witch will grow their own vegetables. I do expect that readers will come away with a deeper appreciation of the Goddess as the complex cycle of birth, growth, decay, and regeneration that makes for soil fertility; with the realization that growing food and eating food are spiritual acts; and with an understanding of how the decisions we make about our food and agricultural systems impact the viability of the earth and human society.

We are animals, evolved to live in a vibrant, thriving, diverse world. It is our birthright to know pristine old-growth forests, wildflower-studded prairies, clear streams, and skies split open by the flight of falcons. The rising of the ocean from global warming is ultimately more real, and more important to the web of life on the planet, than the rising of stock prices or profit margins. The complex exchange of nutrients in the soil is more vital to life on earth than any negotiated trade agreement.

When we learn to hear and begin to understand, then the environment becomes real to us. We can start to speak: to interact sensitively with the natural world, to create visionary solutions to our problems. With all the dire crises and potential disasters that surround us, it lies within our human power to create economies and societies that can provide for our needs sustainably, that can create shared abundance while healing and restoring the environment around us, and that can nurture human freedom and creativity along with natural diversity and health. Ultimately, the test of our education comes in our ability to work with nature to transform our world.

Close this book. Walk outside, if you can, or at least go to a window and open it. Close your eyes and sniff the air. Listen. Who do you hear calling on the wind? Are the birds chattering? Are the tree frogs chanting in chorus? Do you hear the rhythmic throb of city traffic? The cycling trill of car alarms? The cries of children at play?

Everything around us is always speaking. We can heal only by first learning to hear, to understand, and, in time, to respond. As we do, the world becomes richer, a more complex and vibrant place. Open your eyes; see the patterns of light and shadow, the play of the wind. You have already begun your education in the language of nature. You have already set foot upon the Isle of Birds, which is always right here, wherever we are.

TWO

Seeds and Weapons

How We View the World

Early in the morning of June 21, 2003, a phone call awakened those of us staying in the organizers' house for the Sacramento protests against the conference organized by U.S. Secretary of Agriculture Ann Veneman and the USDA to promote biotech and industrial agriculture to ministers from WTO countries, in the run-up to the Cancun ministerial scheduled for September.

"They're raiding the Welcome Center!" a frantic voice told us. "There are a dozen cops and a paddy wagon. Come down!"

Three of us—Lisa, Bernadette, and I—had our clothes on in minutes and were in the car, racing to downtown Sacramento. We arrived at the Welcome Center, a warehouse with a large parking lot next to it, to find masses of police and a huge paddy wagon circling. The police, it turned out, had not actually obtained a search warrant or entered the center. They were entirely occupied with the dangerous materials they found in the parking lot: a bucket of nails and two buckets of seedballs made in the permaculture workshop the day before.

Seedballs are a technique for planting on abandoned and inhospitable ground. You take a variety of seeds, designed to create a "guild" (a self-sustaining mini-community of plants), roll them up in mud containing some compost and a high degree of clay, and then strew the seedballs over the ground you want to plant. The mud and clay protect the seeds from being eaten by birds, and when the rains come, the clay helps hold moisture, enabling the seeds to germinate.

These particular seedballs had been made the day before in a workshop led by Erik Ohlsen[1] and openly attended by the public and the media. They contained a mixture of native wildflowers, legumes (members of the bean and pea family that fix nitrogen and provide fertility), along with mustards and daikon radishes (to build biomass and to put deep roots into the ground and retrieve nutrients that had leached deep below). All the seeds were organic.

Bernadette and I tried to explain this to the officers on the scene, but it was clear to me that we weren't getting through. In part, we faced the same difficulty with the police that we do with the general public around issues of biotech and agriculture: a lack of understanding of the basic principles of ecology. More than that—the whole biotech industry, and the larger system of corporate industrial agriculture that it's part of, is based on a different model of the world than the one that inspired the making of the seedballs.

Industrial agriculture comes out of a mechanistic model. A plant is seen as a product, needing specific inputs of various chemicals, and soil as a stabilizing base to hold it up. Anything in that soil that is not the desired product is seen as competition, to be eliminated. Bugs and pests and diseases should also be attacked and eliminated. It's a worldview of simple causes and effects: if Bug A eats your plant, kill it and your plant will grow. If weeds compete with your corn, kill them (and everything else in the soil) and your plant will grow better. If what you want is corn, plant as much of it as you can, choosing the one variety that will produce the highest yield, so that you can maximize your true crop—profit.

This model extends to the way we view the genetic heritage of the planet. One cause produces one effect: one gene produces one trait. Therefore, why not insert the gene from a flounder, say, into a tomato, to increase its levels of protein? Why not alter soybeans to withstand herbicides so you can plant them and conveniently kill everything else?

The mechanistic model assumes that the world is knowable and controllable. Unintended consequences of an action are seen as anomalies, not "real" consequences, and therefore often go unseen, unacknowledged, and unaccounted for. "Proof" is the drawing of a clear line of simple cause and effect. This has great advantages for corporations bent on making profit. A large corporation can clearcut a hillside and spray herbicides on the exposed ground that get into the water supply: the landslides below, the cancers that arise in the community that lives nearby, the loss of the salmon that once spawned in the stream, go unaccounted for. They are "externalities," unintended consequences. Monsanto can release genetically modified canola that pollutes an organic farmer's fields with its pollen, but Monsanto does not have to add that cost to its accounts. (In fact, Monsanto can sue the farmer for royalties!)

This model is being widely sold to us as "science." It's high-tech, it's postmodern, it's the cutting edge, it will feed the world, and anyone who objects to it is accused of clinging to some romantic past.

But in reality, this model is nineteenth-century science. Science itself began to move beyond it somewhere back in the 1920s, when Heisenberg discovered the uncertainty principle and Einstein began cooking up his theories.

Einstein's theory of relativity showed that matter and energy were one seamless whole, and Heisenberg proved that the observer inevitably affects what she observes. Linear, singular cause and effect was left behind even in the thinking of many nineteenth-century scientists, such as Darwin, whose theory of evolution dealt with complex interrelationships.

The unintended consequences of applying this model to meeting our basic needs are devastating. The "Green Revolution" of the 1970s is a prime example. By applying simplistic science, technology, industrial models, and corporate structure to the agriculture of the third world, we were told, food production would increase and starvation and poverty would end. In reality, the opposite happened. Green Revolution varieties increased yields only when used in conjunction with chemical fertilizers and pesticides that destroyed the health of the soil and the community, while yielding great profits for their manufacturers. Hundreds of local varieties of rice, wheat, and corn were replaced by one or two hybrids, and much biodiversity—the fruits of thousands of years of local selection and adaptation—was lost.

Vandana Shiva, Indian social justice activist and ecofeminist, writes about the "miracle" seeds:

> In the absence of additional inputs of fertilizers and water, the new seeds perform worse than indigenous varieties. The gain in output is insignificant compared to the increase in inputs. The measurement of output is also biased by restricting it to the marketable elements of crops. But, in a country like India, crops have traditionally been bred to produce not just food for humans, but fodder for animals and organic fertilizer for soils. In the breeding strategy for the Green Revolution, multiple uses of plant biomass seem to have been consciously sacrificed for a single use. An increase in the marketable output of grain has been achieved at the cost of a decrease in the biomass available for animals and soils from, for example, stems and leaves, and a decrease in ecosystem productivity due to the over-use of resources.[2]

The Green Revolution is one example of current agricultural practices that favor a "weaponry" approach to agriculture, killing pests with toxic chemicals, tackling weeds with herbicides, and destroying soil life with chemical fertilizers. And these practices don't work: insect damage to crops has increased by twenty percent with the introduction of chemical pesticides since the 1940s.[3] These practices have destroyed farming communities from Iowa to India, driving small farmers off the land and consolidating land and food production in corporate hands.

The model represented by the seedballs comes out of the worldview being articulated by twenty-first-century science. Systems, complexity, chaos, and Gaia theories are some of its manifestations, but it is also much older, akin to the way indigenous peoples have always experienced the earth as alive and relational. "We had so many relatives," said Mihilikawna elder Lucy Smith, "and we all had to live together; so we'd better learn how to get along together. The plants, animals, birds—everything on this earth. They are our relatives and we better know how to act around them or they'll get after us."[4]

This view sees the world as a complex and dynamic web of relationships. There are no simple causes and effects: any change in the web reverberates and affects the whole; small changes can become amplified to have large effects that cannot be predicted. This is sometimes called the "butterfly effect" of chaos theory, from the analogy that a butterfly flapping its wings in Brazil could produce a tornado in Texas.

In this model, a plant is part of a living community of relationships that includes billions of soil microorganisms, worms, insects, other plants, birds, predators, and humans, all of which interact together to create a network of dynamic interactions. A crop can't be seen in isolation—it is part of the web. So our seedballs contained not just one kind of seed, but the nucleus of a group of plants that could coexist in beneficial relationships with each other, which would also benefit the health of the soil and provide conditions for increasing diversity and complexity.

This model looks at systems, not isolated elements. If bugs are devouring your plants, it's a sign that something is out of balance in the overall community. Some predator that could eat the bugs is missing, or something is putting the plants under stress and making them more vulnerable. If your plants are diseased, look to the health of the soil.

In the dynamic web model of the world, we understand that every action or change has a myriad of effects, intended and unintended. The world is not completely knowable or controllable—it's filled with complexities that go beyond our comprehension, with wonder and mystery. And because it is complex, because causes and effects are linked in networks rather than simple lines, the same act will not always produce the same effect. In making changes, therefore, we need to be responsible for any potential reverberations and careful not to produce large-scale damaging and/or irreversible effects. We do this by starting small, by carefully monitoring the changes we produce, and by making the least change necessary to produce a result.

From the dynamic worldview, genetic engineering as currently practiced is a travesty on many counts. First, genetically modifying our food plants risks unintended and irreversible consequences on a staggering, global scale. Already in southern Mexico the wild stands of teosinte, the ancestor of corn, are polluted with bioengineered genes. We simply have no way of knowing what this might mean in the long run. A precious source of biodiversity, of potential change and evolution, has been affected irreversibly.

The wild parent plants of our food plants contain the full genetic potential, the original wild vigor, the unexpressed possibilities inherent in the species. The contamination of the ancestor of corn means that potential is now diminished or lost. It also shows that there is no effective way to quarantine genetically engineered plants that, like corn, pollinate on the wind. When we discover adverse health or environmental effects from a genetic modification, there is no practical way to recall that modification from the environment.

Moreover, the assumption that one gene controls one trait is not borne out by current research. A 2002 press release from the Center for the Biology of Natural Systems (CBNS) at Queen's College, New York, described a review of scientific literature conducted by Dr. Barry Commoner, director of the Critical Genetics Project at CBNS. That review, which was subsequently published in the February 2002 issue of *Harper's* magazine,

> cites a number of recent studies "that have broken the DNA gene's exclusive franchise on the molecular explanation of inheritance." [Commoner] warns that "experimental data, shorn of dogmatic theories, point to the irreducible complexity of the living cell, which suggests that any artificially altered genetic system must sooner or later give rise to unintended, potentially disastrous consequences."
>
> Commoner charges that the central dogma—that one gene equals one trait—a seductively simple explanation of heredity, has led most molecular geneticists to believe it was "too good not to be true." As a result, the central dogma has been immune to the revisions called for by the growing array of contradictory data, allowing the biotechnology industry to unwittingly impose massive, scientifically unsound practices on agriculture.
>
> Commoner's research sounds a public alarm concerning the processes by which agricultural biotechnology companies genetically modify food crops.

> Scientists simply assume the genes they insert into these plants always produce only the desired effect with no other impact on the plant's genetics. However, recent studies show that the plant's own genes can be disrupted in transgenic plants. Such outcomes are undetected because there is little or no governmental regulation of the industry.[5]

In June 2003 Commoner himself said,

> The living cell is not merely a sack of chemicals, but a unique network of interacting components, dynamic yet sufficiently stable to survive. The living cell is made fit to survive by evolution; the marvelously intricate behavior of the nucleoprotein site of DNA synthesis is as much a product of natural selection as the bee and the buttercup. In moving DNA from one species to another, biotechnology has broken into the harmony that evolution produces, within and among species, over many millions of years of experimentation. Genetic modification is a process of very *unnatural* selection, a way to perversely reinvent the inharmonious arrangements that evolution has long ago discarded.[6]

In a worldview of simple cause and effect, we test for "safety" by testing for the effects we can anticipate or predict. But we *can't* test for the safety of effects we haven't anticipated.

In an ominous case, a German biotech company engineered a common soil bacterium, *Klebsiella planticola*, to break down wood and plant wastes and produce ethanol. It passed all its safety tests—until Michael Holmes, a graduate student at Oregon State University, decided to test it in living soil and discovered that all the plants sprouted in that soil died. Worse, it persisted in the soil, as do other genetically modified bacteria. Had it been released for use, it might have spread and, according to geneticist David Suzuki, could conceivably have wiped out all plant life on the continent.[7]

With truly dangerous organisms such as that floating around, it was somewhat surprising to see the level of fear and alarm that our innocent organic seedballs generated in the Sacramento police. They decided, after consultation with their superiors, that we could keep our bucket of nails, since we appeared to be engaged in various building projects rather than producing bombs or planning to hijack airplanes with them. However, they insisted on confiscating the seedballs as "projectile weapons."

It was clear to me that the police basically didn't understand the seedballs, and therefore were afraid of them. They had no category in their minds for "way of planting complex community of beneficial relationships," whereas they did have a category for small, round objects that could be thrown. In fact, they were looking for weapons, eager to find something that could justify the millions of dollars and massive deployment of personnel, the collection of stun guns, tear-gas guns, pepper-spray guns, rubber-bullet guns, M16s, horses, clubs, and armored personnel carriers with which they intended to protect the city from our hordes of puppet-carriers and potentially illegal gardeners.

Looking for weapons, they found our seedballs and perceived them as such. They then spent quite a bit of the day back at the station testing their ballistic capabilities, for the evening news featured cops throwing seedballs at Styrofoam walls and commenting on how they "exploded on contact."

We, on the other hand, had clearly not thought of our seedballs as weapons, or we wouldn't have left them out in plain sight in the parking lot to dry. So in a sense the police action expanded our thinking. In permaculture, we try to get multiple uses for each element in a system. Sometimes that's difficult—a rose, for example, looks pretty and its thorns might discourage intruders from an area, but aside from that most hybrids are not greatly useful in the garden. However, if I think about them as potential weapons, their uses are myriad—the prickly stalks could be used to attack unarmed civilians, the thorns could be inserted into the tires of police cars, the hips lobbed with slingshots at the windows of McDonald's. . . . And think about the lethal potential of something bigger—say, an apple tree!

Ironically, the empty boxes the police had brought to load up the seedballs were marked "Explosives," "Pepper Spray Balls," and "Rubber Bullets." Since they had turned our seeds into weapons, I felt that it would only be fair to do the reverse. But I've tried it and it doesn't work: no matter how many pepper-spray balls you bury, you won't get a single chile pepper, and planting rubber bullets won't produce any rubber trees.

The animate model of the universe is probably the most ancient way of experiencing and being in the world. Yoruba priestess Luisah Teish describes this mode of consciousness: "Prior to the white colonization of the continent, West Africans believed in an animated universe, in the process I call 'Continuous Creation.' Continuous creation means that the generation and recycling of energy is always in effect."[8]

Okanagan artist Jeanette Armstrong says, "We know there's an old, old entity that we are all just minute parts of. We are all just disturbances on the surface of that old entity we could say is humanity. We add to that consciousness continuously."[9]

The man described in Chapter One as "the Guardian," inheritor of an ancient land-based tradition in Britain, says,

> Fundamentally, the belief that was handed down to me was this: that the world and everything in it was driven by an awesome power which could be seen—but only by its effects. This power was generally considered to be female. . . . We didn't need to make representations of her like statues and the like because she was all around, everywhere. . . . [W]hy have statues and such when the whole valley you lived in can be seen as the living body of the mother on which we lived?[10]

While indigenous cultures are all different, one thing they share in common is a perception of the world as alive and themselves as embedded in a matrix of complex relationships. Myth, ritual, ceremony, prayer, and offerings are tools cultures use to maintain a balance between the human and nonhuman communities. In many parts of the world, that view remains intact. All over the world, indigenous cultures are struggling to retain their lands and way of life in the face of an assault by cultures based on very different values. Since that assault originates in what we call Western culture, it's worth looking a bit into our own origins in the West.

Archaeologist Marija Gimbutas, in her many books and excavations, documented that the origins of European civilization, too, lay in cultures that honored the earth and valued cooperation over ruthless competition and war.[11] Their depictions of the sacred expressed in art, pottery, sculpture, and architecture were images of nature and natural cycles, plants, animals, birds, fish, and insects, and of the Goddess—the birth-giving, nurturing, and death-wielding regenerative force of life.

In *Truth or Dare: Encounters with Power, Authority, and Mystery,* I explored some of the long story of how that organic, holistic worldview was replaced by one which took war as its ruling metaphor and divided power by gender.[12] That clash, between matrifocal, woman/earth-centered culture and patriarchal, male-ruled culture, has been going on for over five thousand years in Europe and the Middle East. Much of Western culture can be seen as a dialogue between those strands. "Pagan" cultures often reflected both worldviews: the Celts, for example, were a warlike, chieftain society that retained myths and legends and rites honoring the cycles of nature and awarding a high status to women.

When Christianity came into Europe, it incorporated many of the earlier, nature-centered traditions. The Virgin Mary took the place of the Mother Goddess, churches were built on ancient sacred sites, holy wells were attributed to saints rather than Gods and Goddesses, but the old practices remained. Healing traditions that came from knowledge of the land, the plants, and their properties, and traditions of divination and prophecy, spells and charms and hexes, lingered on as they do today in Latin America, where *curanderas*, traditional spiritual healers, practice alongside the priests of the Catholic Church. Those who passed on and practiced the old ways were called Witches, from an Anglo-Saxon root meaning "to bend or twist." They were the ones who could bend fate and twist the future into favorable paths.

So remnants of indigenous traditions survived in the form of healing traditions—herbalism, which gave us many of our modern medicines, as well as naturopathic and chiropractic medicine. The old religion also remained as folk customs and beliefs, songs, dances, and stories. The fairy tales we today relegate to children were originally stories for adults, the surviving myths and wisdom teachings of earlier cultures.

In the sixteenth and seventeenth centuries in Europe, new economic stresses caused by the influx of gold from the Americas challenged the power of the old ruling classes, which was based on land. A new power began to arise, based on money, trade, and the beginnings of capitalism. With it came a new ideology, the mechanistic model of the universe, which saw the world as made up of separate objects that had no inherent life, could be viewed and examined in isolation from one another, and could be exploited without constraint.

For this new economic order to be accepted, old ideas of the dynamic interrelatedness of the universe and the sacredness of nature needed to be broken down. A new ideology was enforced, and one mechanism for effecting this mass change in consciousness was the fear and terror engendered by the Witch burnings.

The sixteenth and seventeenth centuries were the prime era of Witch persecutions, when first the Catholic and later the Protestant churches attacked all that remained of the old traditions of healing and magic, and the earlier understanding of the world as alive, animate, and speaking.

Today there is much debate about exactly how many Witches were killed, and it is likely that the numbers are far below the nine million that we once postulated. But the impact of the Witch trials was nonetheless enormous.

Anyone could be suspected of being a Witch, and once accused, people found it difficult or impossible to prove the contrary. The Church, both Catholic and Protestant, defined Witchcraft as traffic with the devil, and clerics and their minions tortured suspects. In England, where the use of torture was limited, sleep deprivation, starvation, and rape were employed. On the Continent, human ingenuity was horrifically twisted to invent new ways to deliver pain: the rack, the thumbscrew, the bastinado, and other creative implements of torture were applied. People were fed suggestions of what they had done, and forced to confess. They were tortured until they implicated others, so that no one in a community was safe. The persecutions tended to focus on the peasant and working classes, however, dying away once they reached the upper echelons of society. Two-thirds of the victims were women.

Most of the victims were not actually Witches—that is, were not practitioners of the remnants of the pre-Christian, earth-based religions and healing traditions. Most were simply unfortunates, targeted because of some quarrel with a neighbor or because they perhaps owned a bit of land someone else coveted.

The Witch persecutions did not do away with the old beliefs and practices. Many still survive, even today. In rural France many villages still have a traditional healer, each of whom might specialize in a different malady. For a stiff neck, you might go to one village; for a stomachache, to another. When my friend Rose fell off a ladder in a small village in the Lot region of southern France, we took her to a *roboteuse* who manipulated her neck and cured her. These traditions are passed down in families, alternating genders in each generation.

My friend Ellen Marit was a traditional Sami shaman, from the north of Norway. Her people are called Lapps by outsiders, but they call themselves Sami. She had learned her healing traditions from her father and was passing them on to one of her sons. Their traditions included drumming, trancework, visits to special places of power, and energetic healing. Tragically, her son was murdered while still learning ancient skills, and she herself died a few years later, of a combination of grief and stomach cancer caused by the 1986 nuclear accident in Chernobyl, which strongly affected the reindeer that are the traditional food of the Sami.

Marija Gimbutas came from Lithuania, the land last to be Christianized in western Europe, where ancient Pagan traditions survive to this day, and were strongly alive in the 1920s and 1930s of her childhood. She spoke of seeing peasants kiss the ground each morning, and of how they perceived "Mother Earth as lawgiver. You didn't spit on her or strike her, especially in the spring when she was pregnant," but honored her.[13] Guardian trees were protected, and sacred snakes were fed.

In Ireland, on a walk through the Burren, the bleak but beautiful granite landscape of west County Clare, the women in our tour group visiting sacred sites are impressed with the wide-ranging knowledge of our guide, who has several advanced degrees in botany and biology. At one point, one of the women suggests we do a ritual in an Iron Age ring fort on the top of the hill.

"Not if I'm with you, you don't," our guide says. "That's where the Little People live. You don't mess with the Little People. I'm a farmer, and I need the sun to shine and the rain to fall and my cattle to give milk, and someday I might marry and I need my wife to have children. I'm not messing with the Little People!"

If these beliefs and traditions still persist today, imagine how much stronger they were centuries ago. These traditions maintained an animate and dynamic worldview, and strengthened people's attachment to place and to what was left of communal and tribal attachments to the land.

In the sixteenth and seventeenth centuries, there were still areas of common land in Europe that belonged to the community rather than to individuals. While landownership was highly concentrated and enormously hierarchical, land was nevertheless not considered mere property that could be bought or sold in isolation—but rather a nexus of rights and responsibilities deeply tied to a community. Peasants might not own any land, but they might have the hereditary right to gather wood in the lord's forest or graze their pigs under his oak trees. The folk customs—the maypoles and Morris dances and fairy tales tied to specific places on the landscape—all reinforced those traditions. Again, even today in Ireland old sites are left undisturbed, certain hilltops undeveloped, and road crews detour around certain trees where the Little People are said to live. The view of the land as animated by spirits and nonhuman intelligences was a deterrent to its wholesale exploitation.

The animate worldview and the way of life it represented were targeted by the Witch persecutions, which had several key impacts. First, they broke some of those ties to the land and attacked the underlying worldview by labeling all traffic with and attunement to those other voices as devil worship. They helped pave the way for the enclosure of the commons, the privatization of what had once been collectively held—a process which continues on today through global trade agreements and development. They also undermined the solidarity of the peasant class, which had mounted a series of rebellions over centuries.

Second, they were an attack on forms of knowledge and healing that did not have the approval of the authorities. Midwives, herbalists, and traditional healers, many of whom were women, were considered suspect, and the practice of medicine became a specialized activity concentrated in the hands of male doctors. Although the herbalists of that time were more empirical and truly "scientific" than the doctors of the day (who were busy bleeding people according to their astrological signs), the doctors' knowledge was considered official and valid while the midwives' and herbalists' knowledge was seen as superstition or outright traffic with the devil.

Finally, they were an attack on women. Most of the victims were women, and the evils of the satanic worship that the Church claimed to find were directly attributed to the generally evil nature of women. This justified increased repression of women and restriction of women's roles.

I've written at length about this period in *Dreaming the Dark: Magic, Sex, and Politics*,[14] and don't want to repeat that essay here. But I do want to examine the impact of the Witch persecutions as it still affects us today.

People often ask me, even after I've spent thirty years in this field, why I use words like "Witch" and "magic." I use "Witch" to identify with the heritage outlined above, to place myself firmly in the line of outlaw healers and purveyors of unapproved wisdom. And I use the word "magic" for much the same reason. I could say "sophisticated non-mechanistic psychology," but that term lacks the same ring.

Magic is a discipline of the mind, and it begins with understanding how consciousness is shaped and how our view of reality is constructed. Since the time of the Witch persecutions, knowledge that derives from the worldview of an animate, interconnected, dynamic universe is considered suspect—either outright evil or simply woo-woo.

But whenever an area of knowledge is considered suspect, our minds are constricted. The universe is too big, too complex, too ever-changing for us to know it completely, so we choose to view it through a certain frame—one that screens out pieces of information that conflict with the categories in our minds. The narrower that frame, the more we screen out, the less we are capable of understanding or doing. The police, in the incident that begins this chapter, could not see our seedballs as anything but potential weapons, because that is the frame they were looking through.

Our culture includes some major framemakers. The media is one; academia is another. Aviv Lavi, an Israeli journalist who participated on a panel I was on in Tel Aviv, spoke about how the media determines not just the content of what is said, but the framework of what it is possible to think or talk about. He described how for many years the possibility of a unilateral Israeli withdrawal from Lebanon was not even mentioned. The pros and cons were not debated: it just wasn't a subject for serious consideration—until in May of 2000 Peres decided to do it, and suddenly everyone had been advocating it all along.

Academia does the same. Having worked for ten years on a film about Marija Gimbutas's life, I've become aware of the tremendous backlash against her work among scholars. She is criticized, sometimes by those who haven't actually read her work, for "leaping to conclusions," lacking evidence, interpreting rather than reporting, not following the rules of academic proof, which requires clear lines of cause and effect.

Marija's scholarship reflects the dynamic worldview she wrote about. She looked for patterns, not simple causes and effects, and the encyclopedic breadth of her knowledge allowed her to see wholes, not simply examine the details of an isolated part. Of course, her conclusions can be questioned, and like any human being, she made errors. But the overall whole that she uncovered and revealed is grounded on what she saw and documented of early cultures, and it expands our sense of human possibility.

When we use language that fits into the established framework of the culture, when we try to make our ideas respectable, we limit what we can say and think. But when we use a term like "magic," when we leap out of the constrictions of respectability and cease to care if people see us as woo-woo, suddenly we can think about *anything*. We expand the range of our inquiry beyond the categories already fixed in our minds.

So the Gaia theorists, wanting to be accepted in the realm of science, are always very careful to say that despite the Goddess name for their theory, they are not talking about the earth as a being with consciousness, but about the earth as a self-regulating system that functions "like" an organism. But as soon as we proclaim ourselves Witches and start talking about magic, all the serious scientists turn their backs and leave us free to contemplate Gaia's consciousness and listen for her messages.

U.S. defense minister Donald Rumsfeld, when pressured about the missing weapons of mass destruction that were the Bush administration's main pretext for attacking Iraq, responded in February 2003 by saying, "As we know, there are known knowns. There are things we know we know. We also know there are known unknowns. That is to say we know there are some things we do not know. But there are also unknown unknowns, the ones we don't know we don't know."[15]

He's been laughed at for that statement, but actually I think it's perhaps the most illuminating thing he has ever said publicly. The media and academia deal in known knowns, and are also quite comfortable with known unknowns. But the really interesting questions are the unknown unknowns, and "magic" lets us contemplate that realm.

So we decide, on that rainy hillside in the Burren, to clap our hands three times and say, "I do believe in fairies." And we don't do a ritual inside the ancient ring fort, but look for another spot away from it. Before we begin, we address the spirits of the land and the Little People with an offering, expressing our respect. Immediately the sun comes out for the first time that day, and a raven flies overhead, cawing. The sun remains out through the duration of our ritual, and as soon as we're done, the rain begins again. There's no way to account for this by any current scientific theory I know of, and no way to "prove" that it had anything to do with our decision about the ritual or our address to the spirits. But as we continue to address the land with respect on our journey, and to note that the rain stops each time we begin a ritual, we identify a pattern. Some unknown unknown has entered the picture.

Does magic work? Not by waving a wand, Harry Potter style, and muttering the right incantation for the right result. Not by any simple sense of cause and effect. But magic does work, in the terms of its own worldview. Which is to say, once we understand the universe as a dynamic whole—a whole that we, with our human minds, are part of—we also understand that any change in any aspect of the whole affects the whole. Magic, then, is the art of discerning, choosing, and attuning onself to those changes.

Going back to our ritual on the hillside, we could say that we and our guide and the hills and rocks, the ring fort, the other living communities of that hillside, the raven, the sun and rain and clouds, the history and legends, and the spirits and energies and unknown consciousnesses around us are part of a whole that is a particular moment in the universe. If our guide were not there to warn us about the ring fort, we might be in a different whole. Once our whole includes the consciousness of the Little People, it opens the possibility of many forms of communication. Or we could say that the sunshine and the weather changes and the raven are a reflection of the harmonious whole that our acts of respect helped to form.

Whenever we are able to live for a moment within that consciousness of the whole, we become more whole, more healed. How sad, how grim and tragic, to live within an awareness that can see a seedball only as a weapon, that misses the wonder of the potential for transformation and growth within it, that is forced to view the world as an arena of danger, combat, and betrayal. In today's world we are more and more pressured to live within that limited consciousness, to accept its restrictions as reality, to discount any other possibilities as fantasy, romanticism, wishful thinking. Yet it is the constriction of our imagination that produces that grim world.

Today we live in a world so devitalized, so alienated and fragmented, that many of us are hungry for magic: for a way to perceive and experience the whole, to live in a dynamic, animate universe. It's no wonder that Harry Potter is popular worldwide among adults as well as children, for a world of talking hats and whomping willows and flying broomsticks reflects that sense we have as children that everything around us is alive and has a consciousness of its own. Instead of forcing children to "outgrow" that awareness, we should cherish it as the vital understanding that can help us become healers of this wounded world. For we cannot intervene effectively, cannot say something back to the world, unless we first understand and hear something. And we cannot hear unless we open our ears and realize that the world is speaking to us.

And as soon as we do, we become more alive, more wild, more at home in a vital and dynamic universe.

THREE

The Sacred

Earth-Centered Values

One morning I was sitting on my back deck, meditating on the question of how to make change in the world. The forest was all around me and I was asking the question "Can you change a system from within? Or from without?"

"Systems don't change from within," I heard the forest say. "Systems try to maintain themselves."

I figured that the forest, being a complex system itself, ought to know. But to say, "The forest told me," is already to create a simplified frame. It's a frame I find useful: it's a way of perceiving that's comfortable for my human awareness and allows me to hear something I might otherwise miss. But it is also a simplification of a larger framework, one that might perceive me and my mind and my question and the forest around me and the moment that includes my long-term relationship with that particular spot as a whole in which my mind and the forest's mind are not separate beings talking to each other but one process that together produced that insight.

Magic is itself a framework. Indeed, human beings cannot walk around, function, and continue to tie our shoes without putting some kind of simplified frame around the overwhelming whole of the world. Perhaps only enlightened buddhas can truly remove all frames from the world and exist in ultimate reality.

What follows is my own framework, my understanding of the values and principles that derive from a Goddess-centered view of the world. But, of course, part of the essence of that view is respect for diversity, and for the spiritual authority inherent in each person. Other Witches and Goddess *thea-logians* (*thea* as in "Goddess" instead of *theos* as in "God") may frame values and issues very differently.

Magic teaches us to be aware that we are viewing the world through a frame, warns us not to confuse it with ultimate reality or mistake the map for the territory. Moreover, part of our magical discipline is to make conscious choices about which frame we adopt.

As soon as we start making choices, we have entered the realm of values. The criteria we use for choosing one frame over another come from what we ultimately value most, what we consider sacred. To consider something sacred is to say that it is profoundly important, that it has a value in and of itself that goes beyond our immediate comfort or convenience, that we don't want to see it diminished or denigrated in any way. The word "sacred" comes from the same root as "sacrifice"—because to choose any one value is to relinquish another. If something is sacred to us, we are willing to sacrifice something to protect it, willing to take a stand or to risk ourselves in its service. We don't idealize sacrifice, however. Aligning ourselves with what is truly sacred means serving those things that also feed and renew us, that give us the greatest joy and pleasure, that evoke our deepest love.

As Witches we have a huge responsibility, because we are polytheists. We see many great powers and constellations of energies in the universe that we call Goddesses and Gods, and we choose which we will worship or ally ourselves with. We do also see the underlying unity and oneness of the universe, but being a polytheist is a way of acknowledging that no one name or description or spiritual path can do justice to that immense whole. Gods and Goddesses and sacred texts and religions are all frames, descriptions, maps. No one of them is the whole landscape itself.

But if we have no sacred text, no Ten Commandments, no Ultimate Authority, how do we know what to value? If we don't see the world as a simple battle of good versus evil but as an interplay of forces and counterforces seeking a dynamic equilibrium, on what do we base our ethics?

If we see the world as a dynamic whole, then the first question we might ask when we face a choice is "How does this action or decision impact the whole?" That's not a simple question to answer, because the whole is beyond our complete knowledge, and acts have unexpected consequences.

And how do we know if an effect is beneficial or not? Since earth-based spirituality takes nature as its frame, we can look to natural systems as a model. To understand whether something is beneficial, we need to understand what constitutes health in a natural system, and to know something about how ecosystems work.

A healthy ecosystem might be one that is characterized by cooperative and interdependent relationships among its members, and that is diverse and complex enough to be resilient, to maintain itself in the face of change. Energy and resources are spread throughout the system so that diversity can thrive. No more energy or resources are used to maintain the system than come in from the sun or are generated by the life processes of the system itself. Members of the ecological community are free to express in their unique ways the great creative energies of the universe.

Although we often think of nature as "red in tooth and claw," a field of ruthless competition for survival, today's more sophisticated understanding of ecology sees an enormous amount of cooperation and interdependence. In a forest, trees grow in conjunction with mycorrhizal fungi that interpenetrate the root hairs and extend their ability to take in food and nutrients. Voles and flying squirrels eat the fungi and excrete the spores, spreading them throughout the woods. Through the network of fungi, trees can nurture their own young, and trees in the sun share nutrients with trees in the shade, even those of different species.

In a natural system, the right level of diversity and complexity increases health and resilience. A prairie, which might have hundreds of species of grasses, forbs (or broad-leafed plants), legumes, and flowers in a single square yard, is far more diverse than a field of genetically engineered corn. If a new disease arises, it might affect one or a few of the prairie plants, but hundreds of others would survive. The ground would still be covered: the billions of soil bacteria and the worms below ground would still live. But if a new disease attacks the modified corn plants, they might all die. The exposed soil would erode, with devastating consequences to the below-soil life.

But that healthy diversity lies within a certain spectrum. If we tried to increase it by planting bananas and mangos in an Iowa prairie, obviously those newcomers would die, because they require different growing conditions. Healthy diversity is the maximum diversity that can adapt to the local conditions of life. Those differences in local adaptation create the larger mosaic of biodiversity over the earth-whole.

Abundance, or the provision of resources and energy so that members of an ecological community can thrive, is also a value. Abundance is constrained by sustainability, the need for a system to be self-replenishing, to not consume more than it can create. The margin of abundance is the free gift of the sun's energy, which is constantly showered on the earth, the only true margin of profit that exists. To benefit the whole, that abundance must be spread around and shared, not concentrated so that a few elements have most or all of the resources and others lack what they need.

Freedom and creativity are, perhaps, human values, but they are also aspects of a healthy natural system. A healthy system is dynamic, not static, ever-changing and adapting and evolving. If members of an eco-community are controlled or restricted from expressing their potential or making choices, their ability to adapt is limited. Life has shown, again and again, that it is enormously creative, and alignment with that creativity is one of the marks of health.

There are other human values that we might want to include in our definition of what constitutes "benefit": love, compassion, gratitude, joy—all characteristics that arise in the presence of a healthy, vibrant whole. But love and compassion are more—we might think of them as part of the earth-whole's immune response to dis-ease. They are the emotions that mobilize us as human beings to care for and nurture something, to heal a hurt, to right a wrong.

When a system is whole and healthy, when it is based on relationships of interdependence and cooperation that further resilience, diversity, abundance, sustainability, creativity, and freedom, it exhibits that balance we humans call "justice."

Once we have a model in our minds for what health looks and feels like, we can ask ourselves, when contemplating any act or decision, "Will this create beneficial relationships?"[1]

Answering this question, like the earlier questions, is not as simple as it might seem, because to decide if a relationship is beneficial, especially to the whole, we need to understand something about how systems work. Magic has some guiding principles to offer, and so do systems theory, permaculture, and ecology. What follows is a synthesis of all of these.

The Interplay of Consciousness, Energy, and Form

The universe is a whole, made up of many smaller wholes, circles within circles. How we define those wholes and draw their boundaries profoundly affects how we perceive them and how energy moves within them.

The world is an interplay between consciousness, energy, and matter or form. We know that energy can be transformed into matter and that the atoms of matter can be split to release enormous energy. Matter certainly affects consciousness: try being happy when you don't have enough to eat, or feeling a great sense of well-being while being hit on the head. Consciousness also affects matter: some decision I've made, some image in my mind, is moving me to hit you on the head.

Magic teaches us that consciousness can direct energy in both overt and subtle ways and that energy-flows set the patterns that result in manifestation or form.

Because everything is interdependent, there are no simple, single causes and effects. Every action creates not just an equal and opposite reaction, but a web of reverberating consequences. Everything we do affects the whole.

Every whole is made up of an interplay between consciousness, energy, and form. The universe is infused with consciousness—*is* consciousness, shifting and changing and dancing.

Every consciousness is always communicating. The more we open ourselves to hear and understand that communication, the more we can begin to speak back.

The language of that communication may not be words; it may be emotions, energies, scents, images, events. In speaking back, we also need to move beyond words.

How Energy Moves

Energy moves in cycles, circles, spirals, vortexes, whirls, pulsations, waves, and rhythms—rarely if ever in simple straight lines.

Abundance in a system comes not just from how much energy or resources flow in, but how many times that energy and those resources recirculate before flowing out. If the water you use to wash your dishes is reused to water the garden, you have double the amount of effective water. In an abundant system, waste is food; pollution is an unused resource.

Some of those cycles are *self-constraining*. (In systems theory these are called "negative feedback cycles," but people who associate "negative feedback" with criticism find that term confusing.) Self-constraining or self-regulating cycles work like the temperature regulation system in your body. When you get too hot, you begin to sweat, and the evaporation of the sweat cools you down. When you get too cold, you begin to shiver to warm up. Your body generally does a good job of maintaining an equilibrium, a base temperature. Living systems are characterized by many self-constraining cycles. Gaia as a planetary organism also includes self-regulating cycles.

Other cycles are *self-reinforcing*, or *self-amplifying*. (In systems theory, these are called "positive feedback cycles," although their effects are not always positive.) Self-reinforcing cycles can work like a good composting system: I compost my garden and kitchen wastes, which produces more fertility in the garden, which produces more wastes to compost, and so on. Or they can work like an addiction: I drink too much, so I don't show up at work, so I get fired, so I feel bad about myself, so I drink more, and so on. While self-*constraining* cycles help maintain equilibrium, self-*reinforcing* cycles are driving engines of change, for better or worse. Sometimes self-reinforcing cycles continue until the system reaches a new equilibrium—for example, I reach the absolute limit of how much my garden can produce at a heightened level of fertility. Sometimes, when they have negative effects, they continue until the system crashes, having used up its available resources—for example, I run out of unemployment insurance, friends I can borrow from, couches I can stay on, and I "hit bottom" as an alcoholic.

Energy imbalances in a system create turbulence, movement in spirals and vortexes, which evens out the spread of energy throughout the system.

Form and Matter

Form reflects underlying flows of energy. Nature is full of patterns, or forms, that repeat because they reflect ways that energy flows. Spheres, circles, branches, spirals, waves, and radials are common patterns. Trees, river systems, and the blood in our veins share a branching pattern; snail shells and pea tendrils spiral; water, light, and sand dunes travel in waves. Observing, understanding, and using these patterns can help us direct energy more effectively and create healthier systems. (In Chapter Eleven, we will delve more deeply into the mystery of patterns.)

Form is more rigid, fixed, and resistant to change than energy. The health and function of a system depend not just on what is there, but on where each element is in relation to everything else, and on when each element enters and leaves the system. The right thing in the wrong place, or at the wrong time, can be devastating. When something is in the right place at the right time, it can perform more than one function. A comfrey plant in the midst of your most fertile garden bed will take over and crowd out your vegetables. But on the edge of your garden, it can serve as a barrier to encroaching grass and provide you with medicinal poultices and healing herb tea. And that's not all. It can reclaim lost nutrients from deep soil layers. Its leaves make an excellent mulch or addition to a compost pile, and are good chicken fodder; fermented, they produce a juice that can be diluted and used as a liquid fertilizer or foliar feed for plants. Comfrey flowers feed bees and beneficial insects and are a delicious addition to a salad.

In an abundant system, each element performs multiple functions. We can do more with less, as Buckminster Fuller was fond of saying.

In a secure and stable system, each necessary function is performed by more than one element. If one thing fails, a backup can perform its function.

Making Beneficial Choices

These principles may seem abstract, but throughout this book we will be seeing examples of how they manifest and how we might apply them. But for now, let's go back to our discussion of values and decisions. Knowing what we mean by "beneficial," and coming from some basic understandings about how consciousness, energy, and matter work, how might we apply these to choices we make?

We each make decisions all the time, small ones and large ones. Do I spend an extra dollar to buy the organic tomatoes? If I consider the impact on the whole, on my own health and the health of the whole system, and if I have the dollar, then yes, I do. Do I spend the time and effort to grow tomatoes of my own? If I were to pay myself for the hours I spend gardening, account for all the money and effort and thought I expend, each tomato probably costs me thirty dollars (or more if I decide to raise my hourly rate)—terrible value for the money. But if I'm looking at more than the tomatoes, at the whole of what I need and value and take pleasure in—the value of the fertile, healthy soil I cultivate in order to grow tomatoes, the seven-year-old who lives in our house and likes to pick them off the vine (and the introduction it gives her to nature and the garden), the joy the bees take in the borage that grows with the tomatoes and the fruit the bees pollinate, the positive relationship with my friend Brook, who adores the green-tomato chutney I make, the uncountable value of eating something I have a real relationship with—then growing tomatoes is obviously of great benefit to the whole.

There are many small ways we can bring our daily lives into greater balance with our earth-centered values, from recycling our garbage to growing our own herbs for rituals. The hundreds of consumer choices we make are each an opportunity for affecting the greater balance of the whole. But there are two errors we can fall into around consumer choices.

The first error is becoming obsessive purists. In an imbalanced society, there is no way any one of us can become utterly pure. Today my computer is powered by the sun. I've traded in my pickup for an older, diesel model that can run on modified vegetable oil. I compost my garbage, grow worms, and horde every drop of water in the summertime. But I also fly on airplanes enough that I consume far more than an equitable share of the world's resources. I do that because there is no other practical way I can do the work I'm called to do, and I'm arrogant enough to think that the work is important, and justifies the fuel expended.

Not everyone has the extra dollar for the organic tomatoes, or the time or space to garden. Bringing our lives into alignment with the earth should not become a burdensome, guilt-filled project, where we are constantly in an unshriven state of eco-sin. Instead, we can think of it as a gradual, joyful process, where we look for the choices we can make that will enhance our lives. If I walk to a meeting instead of driving, I can enjoy the sights along the way and my own increased health from the exercise. If I'm too tired or rushed one day to walk, I won't flagellate myself for driving. If I can't afford to replace *all* my lightbulbs with compact fluorescents, I can replace one now, and one more each time I get a little extra money.

Making small choices that align with our values is important. It helps give us a sense of integrity, and it gradually transforms the whole of our lives to be in better balance.

The second error we can make around consumer choices is believing that those individual choices are enough to change the world. We live in a system that is currently so destructive, with so many large-scale destructive self-reinforcing cycles at play, that only *collective* action to change the larger system can hope to stay the damage and restore health.

Let's look at a larger question from the perspective of our definitions and understandings. Let's take the issue of agriculture and bioengineered foods, as posed in Chapter Two. If we were to look at the question of what benefits the whole, as defined above, Monsanto's Roundup Ready seeds would be seen as a horrifying travesty. Roundup Ready seeds are genetically engineered to withstand the herbicide glyphosate (trade name, Roundup), so that that herbicide can be applied wholesale to kill everything *else* that might compete with the crop. Although glyphosate is marketed as "safe," it has been shown to cause cancer,[2] and its use destroys the living community within the soil that creates a healthy environment for growth.

If we were truly interested in benefiting the whole, we'd boycott such products and instead look at ways to further organic agriculture and local food supplies, to support small farms, to make land available to more people, to bring food production as close as possible to where food is consumed. We'd understand that the vast majority of the billions who go hungry in this world are deprived not because there isn't enough food for them, but because they lack access to it or money to buy it. Before supporting policies that concentrate wealth in the hands of the few, we'd make sure that all people have what they need to thrive.

It's likely that policies like those outlined above would set off a new self-reinforcing cycle of benefits. Overall health would improve, from better-quality food and from a diminishment in pollutants and pesticides in the food and water supplies. Corporate profits would go down, but more real wealth and quality of life would be available to more people. City environments would improve, and small towns and rural areas would be revitalized.

We'll explore the issue of action toward the end of this book, after we have a firm grounding in the practices and insights of earth-based magic. But for now, let's look back at the question that opened this chapter, the question I asked the forest: How do systems change?

Systems change in response to forces that disturb their equilibrium. External forces, changes in conditions, new energies, and new challenges can shake up self-regulating cycles. So one way to change a system is to stir it up. That's the role of protest and direct action, and it's the reason why stronger forms of action are often necessary to bring change. Sweet reason, gentle persuasion, and dialogue that doesn't challenge the functioning of the system often end up becoming incorporated in the system's own efforts to maintain equilibrium.

Change in systems often comes from the edge. The edge, or ecotone—that place where one biological system meets another—is the most dynamic, most vulnerable, and often most diverse part of a system. The rocky shore where the ocean meets the land contains many more niches for life and diverse conditions for adaptation than either the sand dunes inland or the deep sea beyond.

So another way to change a system is to confront it with a different system. The existence of a feminist and earth-based spirituality movement offering rituals, teaching, and community completely outside the bounds of Christianity and Judaism has had profound effects on those religions over the past decades, offering support for reforms, challenges to established assumptions and practices, and creative ideas that have influenced change within the major denominations.

In spite of what the forest told me—that change has to come from outside because systems by their nature try to maintain themselves—I think systems *can* to some extent change from within. I'm not suggesting that every reader quit her job and go live in the woods. We are all part of the whole of the system, and to some extent that opens communication and makes it possible for us to influence it. To change a drum rhythm in a group of drummers, you first have to match it and join with it.

But when you are within a system, part of the whole, that system is also changing you. It is difficult to maintain your own rhythm and not simply become part of what you are trying to change.

Decades ago, feminist philosopher Mary Daly suggested that the place for feminists in the academy or other institutions was on the boundaries, neither completely within nor completely without.[3] Wherever we are, we can look for those fertile edges of systems, those places where unusual niches and dynamic forces can be found, and make change there.

Donella Meadows wrote a powerful essay many years ago entitled "Places to Intervene in a System," which detailed nine "leverage points" in increasing order of effectiveness.[4] The first two places to intervene, changing amounts and changing material stocks and flows, involve change in matter or form. If the school system is dreadful, pour more money into it or build new buildings. Sometimes those changes may be just what is needed, but they don't change the basic functioning of the system itself.

Next come changes in energy flow, looking at self-regulating and self-reinforcing systems and finding ways to intervene, either to disturb a nonfunctional equilibrium or to establish a new equilibrium to avert a crash.

Then come changes that begin to move into the realm of consciousness—changes in information flow, in rules, in self-organization, and in goals. Finally, the most overarching change comes from paradigm shift, a change in the basic premises that underlie the system.

We are faced today, in a world of global crisis, with the need for overarching change that can come only from a shift in paradigms, in our basic assumptions about the world. To change a paradigm, we must be able to express clearly what the new paradigm is.

That is the work of this book: to root us so firmly in earth that we can be walking emissaries of a new whole.

A SACRED INTENTION

Sit in a quiet spot and relax. You might want to meditate on the questions that follow, or journal about them.

Ask yourself, What is sacred to me? What do I care about so strongly that I can't bear to see it compromised or destroyed? What would I take a stand for? Risk myself for?

When you know the answer, consider for a moment what the world would be like if our social, political, and economic systems all cherished what is most sacred to you. In what ways do they already? In what ways would they need to change? What would change, in your daily life? In your community? In the world around you?

Can you describe that world in a few sentences or paragraphs?

Do you want to bring that world into being? Do you feel responsible toward it? If so, that is your sacred intention.

If not, what is your intention for your life? What are your goals?

Now consider how you spend your time and energies. Are your best energies directed toward bringing about your cherished vision of the world? Toward service of what is sacred to you?

If so, congratulations. Is there anything you need, support or opportunities or luck, to help you in that work? Who can you ask for support, in the human world? In the larger realms of the universe?

If not, what is blocking you? How would your life change if you were to put your best energies toward creating a world that cherishes what is sacred to you?

What do you need to make that change? Support? Opportunity? Courage? Luck? Who can you ask for support, in the human world? In the larger realms of the universe?

If you desire that change, affirm your sacred intention. Say, "It is my sacred intention to create a world that cherishes ________________."

Your sacred intention is the heart of your work with this book. You can revisit it and revise it, let it grow and develop, write about it in your journal, and test your daily decisions against it. As we go through this book, we will refer back to it again and again, as a touchstone for the transformations we undergo.

Rereading this exercise, I realize there's a deep assumption in it: that you, the reader, do feel a sense of responsibility for creating a world in alignment with your sacred values. Whatever you cherish is *part* of the earth, and loving it, wanting it to continue, to me logically implies a need to love and cherish the *whole*. I confess that for me that assumption is the heart of earth-based spirituality in this time when every life system on the planet is under assault. What is sacred to you may be different from what is sacred to me, and the ways you choose to serve it may not be mine, nor mine yours. But no one who loves the earth can evade responsibility right now for her well-being.

That responsibility may seem overwhelming. At times we fail to rise to it because we don't know what to do, or feel powerless or inadequate. I think it's no accident that *The Lord of the Rings* has become a highly popular series of movies at this moment in time. We are all Frodo, reluctantly carrying the burden of the ring, and not knowing clearly how we will get to our destination.

But as Galadriel, the wise elf, tells Frodo, "Even the smallest person can make a difference." Each time we act in service of our sacred intention, each time we align our energies and our actions with what we most truly love, we gain in personal power and ability. The path before us becomes clearer, and the help and allies we need come to us.

So, when you think about your intention, when you feel daunted or overwhelmed or afraid, just breathe deep and ask for help. Great powers and energies are all around us, but they cannot help us unless we ask. When we do ask, however, they are present and eager to help us serve intentions that benefit life. So you might say something like the following.

OPENING TO HELP

Great powers of creation and transformation in the universe, ancestors, allies, all beings who love the diverse and beautiful dance of life, I am open to your help and I reach out to you. I thank you for the gift of life, for the help and support I have already received, and for the great opportunity of being alive at this crucial moment. I need ____________________ to serve my sacred intention of creating a world that cherishes ____________________. I give you my gratitude for the help I know is already coming. Blessed be.

FOUR

Creation

What Every Pagan Should Know About Evolution

I managed to slide through something like nineteen years of formal education learning remarkably little science. In part, I was discouraged by a ninth-grade physics teacher whose experiments never worked. If she tried to demonstrate gravity, toy cars would refuse to roll down ramps and objects would float up. In later years, I majored in art, then film and psychology (which is science of a sort but didn't demand much grounding in biology or chemistry).

Now that I'm a Witch, I regret my ignorance and am taking steps to remedy it, mostly through reading and observation. The Goddess is embodied in the natural world, and science in its truest sense is about knowing nature. Thus our thealogy needs to be empirical as well as mystical.

Our understanding of our origins—cosmic and human—shapes our relationship to the world in subtle and profound ways. So hang on to your hats as we take a journey through the wonderful world of evolution, a topic that has always had profound religious and spiritual implications.

Most of us were raised on either the biblical creation myth or Darwin's theory—or perhaps both. From the Pagan perspective, neither of these stories is wholly satisfying or "true" (in the sense of best describing the reality around us).

The biblical creation story has a (presumed) male God making the world essentially by fiat, by word alone. The process is disembodied, entirely removed from the sweaty, bloody processes by which females create life. God's law is something imposed on nature, and God's rules are imposed on us to follow. Humans are made in God's image, and a great spiritual and existential gulf separates us from the animals. Plants, animals, and human beings were created in their finished and final forms, and have remained essentially unchanged since.

Of course, this view does not do justice to the breadth and diversity of Bible-based theology. There are strands within Christianity, Judaism, Islam, and all the major religions that celebrate and honor creation and preach a relational view of the world.[1] See, for example, the work of Matthew Fox in developing creation-centered spirituality, or Arthur Waskow's work on earth-centered Jewish ritual.[2]

Evolution, of course, was in Darwin's day a shattering and heretical challenge to the simplistic, literalistic biblical view. First, the theory of evolution holds that the world is much, much older than the Bible says. Second, humans, animals, plants, bacteria, and all other creatures are a single continuum of life. Humans are not something set apart. We are animals, and we emerged from the same natural processes by which other life-forms evolved.

From the perspective of earth-based spirituality, those insights were a vast improvement over literalistic interpretations of the Bible. Evolution restored dynamism to the universe, brought it alive as a growing, changing, interacting web of relationships.

Darwin himself was a great observer, embodying the permacultural principle of "thoughtful and protracted observation" more than a century before permaculture was formulated. He looked at the plants and animals and birds around him in the far-flung places of the world as he traveled, and he let himself ask, "I wonder . . . ": "I wonder how that tortoise got to be the way it is, how differences between those similar plants arose, what forces produced the beak on that bird." He theorized that environmental pressures and constraints select the individuals most fitted to a given environment from a range of genetic variations. Those individuals succeed best in the competition for food and scarce resources. They are also most likely to reproduce, and so they pass on their adaptations. His theory of evolution and natural selection was a brilliant example of relational thinking, focusing not just on individuals or species as separate, isolated elements, but on the whole pattern of interactions, exchanges, and effects of living communities as a whole.

But at the same time that Darwin was researching and writing, industrial capitalism was growing and consolidating its power, and looking for an ideology to justify ruthless exploitation of the poor by the rich. "Social Darwinism," a simplistic reformulation of Darwin's theory, turned natural selection into "survival of the fittest." The best win out and, by extension, the "winners" must be the best—and therefore deserving of their rewards. "Losers" are by definition inferior, maladapted, and deserving of their demise. To suggest that the winners owe anything to the losers is to interfere with nature and risk weakening the race.

This misinterpretation of Darwin's theory was a secular reformulation of earlier religious doctrines of the "elect." It was also a perfect rationale for cutthroat capitalism, in both the nineteenth century and the Reagan/Bush era. Competition is the driving force of progress in nature and, by extension, human society. The more worthy win out in time, and this, in the long run, is good for the species and for the whole. Success is its own justification, and what's good for big transnationals is good for the U.S.A.

There is a different view of evolution, one that better serves the worldview of earth-based spirituality. We might call it Gaian evolution, after the Gaia theory developed by James Lovelock and Lynn Margulis.[3] Gaian evolution is not so much a counter to Darwin as a shift in focus from the individual to the ecosystem, the whole. The earth functions like a living being, and the biosphere, the world community of life-forms, changes its environment as it is changed by it. The redwood tree does not evolve as a separate species; rather, the forest as a whole evolves, the interwoven lives of redwood and tanoak, huckleberry and salal, the mycorrhizal fungi in the soil below and the lichens in the canopy where the marbled murrelets nest. None of these creatures adapts alone, in isolation from each other—they coevolve as Forest-Being, in an interdependent dance that balances competition and cooperation. Individuals and species survive when their activities benefit the whole as well as the parts. Evolution becomes the story of how the planet herself comes alive.

Of course, scientists are very careful not to imply that this living planet has consciousness or self-awareness. Consciousness is not necessary to explain this process of life and evolution, and this becomes a messy and unnecessary part of the theory. With or without attributing consciousness or awareness to Gaia, we can still approach her life story with wonder and awe. For Witches, Pagans, and the like, however, having already removed ourselves from the realms of academic respectability, there are no reputations to protect, and thus we are free to experience Gaia as more than mechanistically alive—as a conscious being, a vast ocean of awareness in which we swim, always communicating, always present.

What follows is my synthesis of the story of Gaia coming alive, with thanks to James Lovelock and Lynn Margulis, and Elisabet Sahtouris's lovely book *EarthDance: Living Systems in Evolution.*[4]

Genesis

Before the Beginning. . .

In a swirling spiral of gas, heat, and light, a tiny grain of dust that was Gaia's seed danced and swirled. Throbbing and pulsing with an electric passion, she drew to her other grains, other seeds, until together they formed a ball, spinning and dancing in the lens of radiance that was to become the sun. The dancers flung out their arms, swirled their skirts, bumped up against each other, and fused. Growing larger and larger, spinning and dancing faster and faster, they were drawn toward each other by the passionate pull of gravity, at times colliding in a fiery death, at other times in a mating union, until at last the planets congealed into their orbits, circling a fiery sun.

Gaia was hot, her surface erupting in plumes and rivers of fire, her face bombarded by missiles of rock that left her pockmarked with craters and seeded with ice and the chemical prototypes of life. Slowly, slowly, she cooled down. On her surface, packets of energy frozen into form combined and recombined. Ice melted to primordial seas that washed a rocky shore. Lightning struck. Waves rolled to shore and retreated; the soup of energy was boiled and cooled, dried and immersed, again and again. Bubbles formed thin skins that enclosed crystalline strands of frozen energy, organized in a radically new way: a way that conveyed information, that communicated instructions for reproducing itself. The double helix of DNA was life's first great creative leap, the one that allowed all others to follow. Life was born.

The Gift of the Ancestors

Life on earth was still relatively new. At first, simple, one-celled beings filled the seas, living by changing the energy patterns around them, breaking down large molecules—complex clumps of dancing energy and form—into smaller clumps, using the energy released to move and dance. They filled the seas in promiscuous abundance, constantly exchanging bits of DNA, sidling up to one another and crooning the bacterial equivalent of "Hey, hey hey, baby . . . the thought of trading genes with you drives me c-crazy!" They formed one life-whole, one global gene pool, one planetary well of information and experimentation.

But after a time, life reached a crisis point. Life began to run out of food. There weren't enough of those complex molecules for all of life to continue, and life began to starve and die.

Yet life has always been inventive, creative. Those simple, one-celled beings were already experimenting with different forms. Some were long and skinny and wriggled and swam. Some were round and fat. Some adapted to hot and some to cold. And always they were trading genes, shifting forms, changing and transforming. At that time, there wasn't yet a brain on the planet, yet life came up with something so brilliant, so amazing, that it transformed the whole nature of existence and the atmosphere itself.

Life invented a mandala. A beautiful molecule, like a patterned flower, with a magic quality. For when a photon of sunlight struck the heart of this pattern, it began to vibrate and shiver and set off a chain of reactions that harvested the sun's energy to turn carbon dioxide and water into food. Chlorophyll and the process of photosynthesis were life's next great invention, and the green things, the sunlight-harvesters, were born.

Green things filled the seas and the crevices of the shores, flung a smear of filmy life over rock and sand. Life flourished as never before.

But there was one problem. The miraculous process that used sunlight to make food gave off a waste product, a toxic gas that burned and destroyed everything it touched. And as life grew, over hundreds of millions, a billion, then two billion years, the very air became polluted by this gas, so that life could no longer avoid its touch of death.

But life continued to experiment and invent. Some of those tiny creatures dug down in the mud to avoid the toxic gas. Some clumped together for protection.

And some discovered another miracle: that by reversing one of the moves in the dance of photosynthesis, a new process could be born—one that could take the toxic gas, which we call oxygen, and use it to burn food and make energy.

And so the breathers were born, those who dine on the sunlight-harvesters, burning their bodies as fuel for life. In burning food, the breathers give off carbon dioxide, which the green things (with the help of the sun) transform to food again. And the green things give off oxygen, which the breathers use in burning food. Gaia began to breathe, passing her breath back and forth from red to green, continuing to build up oxygen, to transform herself.

And after millions of years, the breathers took the mandala of chlorophyll, switched the atom at its heart to iron, and formed the hemoglobin that swims in the cells of our red blood.

And so the cycle is complete, and the earth breathes in and out, red to green to red.

And this air we breathe is a gift of the early ancestors. With each breath in, we take in the results of their great creativity. With each breath out, we give back.

And the balance is so perfectly kept that oxygen remains at just the right amount to sustain life. For if there were only a few more percentage points of oxygen in the air, any spark would light a fire that would ignite the whole atmosphere. And if there were only a few percentage points less, no fire would ever burn and we could not live.

Cooperation and Complexity

Breathers added something new to life's dance. The sunlight-harvesters floated in a womblike sea that contained the elements they needed to make food. The energy they needed showered down from the sun. Life was easy, and they could simply be and receive.

But breathers needed to find food, from the dead bodies of the sunlight-harvesters or from living ones. They had to be more mobile, more aggressive, pursuing and engulfing and penetrating before they could digest and dissolve their prey.

Every now and then, an aggressive breather penetrated a life-form that did not dissolve. Or took in someone it could not digest. And instead of eating each other, the life-forms coalesced and supported each other.

A breather might make food more efficiently for a scavenging bacterium. A sunlight-harvester and a breather might team up, to make best use of all possible sources of energy.

A long, skinny, wriggling creature might bury its head in this new, larger cell and provide mobility in exchange for food. A hundred, a thousand, tiny creatures might team up to become one larger being, pooling their crystalline DNA library of instructions into one central core.

And a new form of life was born, still single-celled but a thousand times larger than what had gone before, and far more complex. The *eukaryotes* were born, the cells with a nucleus that are the ancestors of all larger creatures.[5]

This new collective form opened up a wide realm of possibilities for life. For two billion years, simple bacteria had been the only model of life; now life began to experiment and change.

One of the first experiments was sex. Bacteria invented a simple form of sex, trading genes like bits of gossip throughout a worldwide pool. The variations created by this process allow them to change and evolve. When they reproduce, however, the process is still simple: each cell simply replicates itself and buds off an identical copy.

The eukaryotes each had a center, a nucleus that held a library of genetic information, arranged in paired chromosomes of DNA. Now they learned to divide those pairs, to split the deck before reshuffling. And each half-set of genes could combine with the half-set from a different individual. Sexual reproduction was born. And became very popular.

These new cells began to build on their cooperation, to form colonies. Some learned to take minerals from seawater and spin elaborate spheres of intricate, crystalline forms. Some pioneered the branching patterns of roots, the flat planes of leaves. Some built the first true bodies made of many cells, linked by the communication tubes of nerves.

And 580 million years ago, life exploded in variety. Life grew legs and began to walk on the seafloor and the shoreline. Life spun shells in elaborate, ornate forms, tried out tails, fins, flippers, segments, carapaces, and antennae. And this burgeoning life grew weirder, more delightful, more strange than all of succeeding life put together.

Until disaster hit. A meteor hit, or massive volcanoes belched smoke into the air, covering the earth with a blanket of cloud that first blocked the sun, then warmed the atmosphere. Something changed the earth's climate, and, in a great extinction, 90 percent of life on earth died.

What survived was not as diverse, not as inventive. But life tried out many variations on a few basic patterns and again began to grow and evolve. The seas were filled and the land was colonized. Dinosaurs roamed great forests of ferns, and winged lizards soared through the skies.

Then, some vast time later, another meteor crashed into the earth, leaving a great crater a hundred miles wide, and the dinosaurs died. Their small descendants, lizards and birds, still prowl and soar. A small, humble, shrewlike mammal that survived the cataclysm gave rise to mice and deer, tiger and mammoth, bison, horse, ape, and us.

And here we are, with our thumbs and our big brains, inventive, creative, aggressive, aware in some ways, oblivious in others, still struggling to learn the lessons our ancestors bequeathed us:

That everything changes. That everything is interdependent. That we survive by cooperating, sharing resources, pooling information. That change can come suddenly, cataclysmically, and when it does, the small are better fit to survive than the large. That when faced with great crisis, life is capable of great invention. That we are no less creative than the crystals that invented DNA, no less artists than the bacteria that shaped chlorophyll.

* * *

This is the story I like to remember when the world seems bleak and I wonder how we will ever survive. Or when I feel as if the Goddess and all powers of hope have deserted us. At those times I remember . . .

That life by its very nature is a great power of creativity and transformation, a power that will prevail.

And when I doubt, all I need to do is take a breath, in and out, and receive the gift of the ancestors.

Lessons of Evolution

What does this story teach us?

The first lesson is that the universe is amazingly creative, responsive, and ever-changing.

The second lesson is that survival goes not to the most ruthless or competitive, but to those who most effectively cooperate, communicate, and share resources. The cells of our body are collectives. The atmosphere we breathe is a collaborative endeavor. Adaptation is not about one species triumphing over another, but about a whole system coevolving.

Competition and predation are important aspects of this system, but they are not the only driving force. Cooperation and competition exist in a dance, a yin/yang harmonic balance of variation and culling, combining and editing.

The third lesson is the importance of variation, one of Darwin's insights. Life does not evolve one best type, but a multiplicity of fit types that retain a variety of forms and potentials. This variation, along with the random mutation of genes, allows for more diversity and more resilience in the face of environmental changes. A disease that fells one member of a species, for example, may not prove lethal to another. When a new potential food source appears, individuals who share a particular variation may be better suited than others to make use of it. Variation is every species' hidden treasure and insurance policy, and that is one reason why the loss of biodiversity and variation within species causes environmentalists such grave concern.

The fourth lesson is that change is not just a gradual process, but may also be a "punctuated equilibrium," involving periods of relative stasis punctuated by crises. The river of change generally flows gradually downhill along gentle slopes of slow, incremental change. But it may also plunge suddenly over a waterfall. Both forms of change occur in evolution and in life. When we look at potential environmental changes ahead, we do not know if we will see slow variations that our society and the biological community can adapt to, or sudden crashes down a rocky dropoff that will happen so abruptly that we cannot turn aside or cushion the fall.

If we were to extrapolate these lessons to human society, they might lead to a set of values we could call "social Gaianism" as opposed to social Darwinism. Social Gaianism would acknowledge individual needs and self-interest but see them best served in systems of cooperation and mutual aid. Our representatives in Congress might invoke the lessons of social Gaianism to bolster public support for social programs, childcare, education, health care, and other means of sharing resources and support. We would understand that the system is only as healthy as its weakest member. Libraries, those repositories of knowledge and inherited instructions, would be sacred. Knowing our interdependence with nature, and knowing also that a stressed system can suddenly crash, we would move swiftly and decisively to shift to renewable sources of energy, to phase out fossil fuels, and to keep chemicals out of our living and life-created atmosphere. When international conflicts arose, we would identify our security and survival with our ability to negotiate and cooperate, not conquer.

The Gaian story of evolution shifts our focus when we look at nature. We open our eyes and look beyond each individual tree to the pattern of the whole. We honor the unseen creatures below and above as well as what we can see. We begin to look for patterns and relationships, not just isolated individuals. We know that diverse, resilient, complex systems are most likely to survive. And only what's good for the biosphere is truly good for the U.S.A.[6]

This story also gives us hope. Enormous creativity is embedded in our very cells. Resilience is the nature of living beings. We, with our complex brains, have the inherent ability to evolve in ways that can nurture and sustain the life patterns that surround us.

And if we don't use that ability wisely, there are always those gene-swapping bacteria to sip the cocktails of our wastes and, in another billion years or so, come up again with something new.

FIVE

Observation

From my journal:

Sitting still in one place doesn't come naturally to me. But with a hurt leg, now seems the time to try it. I take my morning tea out to the deck, sit still, and practice owl eyes, the widening of perspective. I open my ears to a sphere of sound. The sun is just climbing over the hill; rays of light pierce through the forest. Tendrils of sensation up my back tell me I'm cold. I explore the sensation—could be fire, could be pain, could be just stimulation. I could sink into it, expand my comfort zone—but then I'm a middle-aged woman susceptible to colds, so I go inside and get a poncho.

A loud voice—the Eight-Note Bird, eight sharp, slightly tart notes in a string. I know he's close by but I can't see him among the branches. Nearby is also a small-voiced bird that sings simply "Wheet, wheet" at intervals. Beyond I hear the wind in the branches and the trickle of water in the lower stream. Down here in the canyon it's still and windless, but I can look up and see the tops of the redwoods swaying, and I hear an occasional sharp, ominous creaking.

The light moves, the rays change their angles, and suddenly a spiderweb gleams iridescent before me, hanging in the branches of the trees, a perfect wheel. I look and see another, and another—invisible until the light strikes them at just the right angle. Then, for a few precious minutes, they glow in rainbow colors, gold within, an iridescent purple in the outer rim. The air is full of circling bugs who are also illuminated, spiraling upward toward the sun. Threads of light gleam where spiders have spun bridges between trees.

A phone call summons me indoors, and when I return everything has changed. The sun has cleared the hill and a chorus of birds begins to sing. While Wheet and Eight-Note sing the same words over and over, now I hear a bird with a song as complex as a paragraph, an essay of rhythm and melody that plays once, and then no more. Other birdsongs, more liquid and sweet, sound in the background. In the deep forest, I hear a dove call. The ravens announce their presence with a caw that is almost a honk, then cluck a few times before they fly by. A jay squawks; I hear a woodpecker shriek, shriek, shriek as it flies.

A scratching and rustling overhead—the branches of the redwoods spring back and I look up to see two gray squirrels chasing each other through the treetops. I follow their mad dash with my eyes: from directly overhead they run out along the redwood branches, leap to the tanoak, travel up and down the highway of treetops. Another phone call. This time when I'm inside I grab my binoculars, and sure enough, when I return the squirrels are back, graceful and daring as any high-wire artists. Through the binoculars, I can see one up close—his neat white belly and fluffy gray tail. He looks at me—I've noticed that when I use binoculars, animals and birds often seem to sense that they're being watched, that some boundary of distance has been crossed. He stands still on a stump of a redwood, staring at me in a posture of tension for a long time, then hears something behind him and turns.

The other night, after a forester we were consulting informed us that squirrels had caused the tops of a couple of our redwoods to die, I dreamed I was making love to a squirrel. In the dream, he looked like a man, but when I felt his face, it felt like a muzzle. "I could tell you have serious hands," he said. "We don't want that, now, do we?" Just another guy who can't make a commitment, *I thought, and then,* But what do I expect? He's a squirrel! *Now, watching the lithe bodies leap and bounce on the high branches, spying on the sleek, gray fellow, I think a squirrel lover wouldn't be so bad. Even nicer would be to run and leap among the high branches with that freedom. Rima the Bird Girl, Julia Butterfly, the Earth First! tree-sitters who play tag eighty feet above the ground—maybe they know that delight. Me, I'm earthbound, but happy watching the squirrels perform their high-wire acts for my amusement.*

The first skill that can lead us to deepened connection with the natural world is observation. Observation seems like the simplest, most natural thing in the world. We're all born observers. From the earliest moments of life, babies stare at, listen to, and taste the world. Parents of two-year-olds soon learn to watch their language, for their observant progeny will reliably mimic any favorite swearwords or verbal quirks. By observing others, we learn how to speak and what to say, how to behave and misbehave, how to experience and navigate the world we're born into.

But between infancy and adulthood, something gets in the way of our childlike, unclouded vision. Few of us can walk into a forest and simply *be* in the forest. Instead, by adulthood we are inside a story we're telling ourselves, partly about the forest, but mostly about ourselves. Sometimes it's a story about our own weakness and inadequacy: "I'm so tired. I don't feel well. I can't keep up; everyone else is in better shape than I am." Sometimes it's a story about how wonderful we are: "I'm so spiritual. I'm attuned to the trees so much more sensitively than anyone else. I'm talking to the faeries. Don't I look good here in the woods?" Sometimes it's a story about how much we have to do: "I have that phone call I forgot to return and I haven't checked email in two days and then I have to get the car fixed and the insurance premium is overdue." Sometimes it's a story about someone else: "I'm so angry! Why did she say that? I can't believe her—why did she lie about it? I'm going to tell her to her face just what I think of her. Better yet, I'm going to send an email out to the whole listserve for the group." Sometimes it's a story about fear: "What was that? I think we're lost. Was that a cougar? I've heard cougar attacks are increasing. What if we can't find our way back? What if it starts to rain? I've heard you can die of hypothermia even on a warm day. My heart is pounding. What if I have a heart attack? What if I slip and break an ankle here? What if there's a rapist lurking in these woods?"

Whatever inner dialogue we're running, it's interference. We end up walking around inside our own heads, not in the woods.

So to truly observe, we must be able to step outside our heads and walk out into the woods. We must be able to close the book on the story, turn off the dialogue (or at least turn it down), and hear what's around us.

Of course, that's easier said than done. Buddhists spend years in practice in order to achieve a quiet mind. Psychotherapists work with their patients for session after session in order to change their internal stories.

Paradoxically, observing the outer world around us requires a great deal of inner work and discipline. It becomes a deep spiritual practice that incorporates some of the aspects of Buddhist detachment and may lead us on a journey of personal healing.

Below are some of the exercises and practices I use.

OPEN-EYED GROUNDING

Grounding is the basic practice that begins every ritual, and the tool that helps us stay calm and present in any tense situation. Grounding means being relaxed but alert, energetically connected to the earth but able to move, present and aware, in a state in which we can take in information and make conscious choices about what to do.

In a ritual or meditation practice, we often close our eyes in order to ground. Closing our eyes can be a helpful way of focusing on our inner vision and shutting out distractions. But in the woods, or in truly dangerous situations, we may also want to be able to ground ourselves with our eyes open. It helps to practice.

You can do this exercise anywhere. Ideally, you should be outside, standing on the earth, in some natural place. But you can also do it indoors or in the midst of a city.

Stand in a comfortable position, with your feet about shoulder-width apart, your knees lightly bent. Stretch and release any tension. Take some long, deep breaths into your belly. Feel your feet on the ground.

Tell yourself that just by breathing and feeling your feet on the ground, you can bring yourself into a calm and grounded state.

Imagine that, like a tree, you can extend roots into the earth, from your feet and the base of your spine. With your eyes open, notice if what you see changes as you extend your roots down.

If there is anything clouding your awareness or interfering with your ability to be present, take a deep breath and imagine letting it go down through your roots into the earth, to become compost.

Feel the living fire deep in the heart of the earth. Breathe some of that energy up through your roots, into the base of your spine and your belly, up your spine as if your spine were the flexible trunk of a tree. Feel it warm your heart and throat, and reach out through your arms and hands. Let it move up through the top of your head and out like the branches and leaves of a tree. Let those branches come all the way down to touch the earth, surrounding and protecting you. Again, notice what changes.

Feel the sunlight (or moonlight or starlight) on your leaves and branches, and breathe it in. Feed yourself on the energy, just as a tree feeds on sun. Draw it down through your head, heart, hands, and belly, down through your feet into the earth.

Look around you and notice what you see, hear, smell, and feel in your grounded state.

EARTH-WALKING: MOVING WHILE GROUNDED

Now you are grounded: calm, present, and aware, with an energetic connection to the earth established. But, although we've used the image of a tree to help us ground, we don't want to be stuck to the ground or immobile. We need to be able to move while staying rooted.

So imagine that the roots in your feet are stretchable, that when you pick each foot up the connection remains. Or you might use the image of the Ents, the tree people from Tolkien's *Lord of the Rings,* who stalk the forests on their treelike feet, with their toes spreading and gripping the ground with each step. Looking up and around you, not down at your feet, begin to move, remembering to breathe.

Imagine that your feet have sensors. Pick each foot up and set it gently and slowly down, letting it tell you about the terrain beneath. Keep your ankles loose and your knees springy. Move slowly around the space, feeling how you can keep your energetic connection with each step.

Notice how quietly you can walk. Suddenly the squeak of your shoes or the swish of your clothing becomes loud. Practice moving in this quiet, grounded state.

WIDE AWARENESS

Now stop for a moment. Looking straight ahead, bring your arms out to your sides and wiggle your fingers. Slowly bring them in until you can just see the motion of your fingers with your peripheral vision. Notice how wide your field of vision can be, how far it can extend up and down as well.

Letting your arms drop down, keep your vision extended and again move through your space, earth-walking in a calm and grounded state. Notice how quietly you can move, how much you can see and be aware of, and how you feel, earth-walking in wide awareness.

The three linked exercises above—open-eyed grounding, earth-walking, and wide awareness—are the core of my own daily spiritual practice. The key is, indeed, to *practice* them—practice them daily, moment by moment, until they become second nature, in the woods or in any situation in your life. In tense or dangerous situations, you will be safest if you don't panic and instead stay grounded, calm, able to assess what's happening and make conscious choices about what to do. In fact, when I prepare activists for political actions, I teach a version of these same exercises.

For many of us, breathing and grounding requires some deep repatterning of our basic way of being in the world. Many of us have learned to breathe high in our chest and shallowly, which keeps us in a state of heightened anxiety. Our entire culture is one of distraction, keeping us focused on other people's stories—stories we see on TV or read about in the newspapers. If we have experienced trauma or violence in our lives, we may have learned to close down and draw a shell around us, to avoid seeing or feeling what is too painful to face.

Repatterning takes time. Be patient but consistent. Learning to ground does not require hours of strenuous practice every day; it can be done in minutes, in odd moments of time—while waiting for the bus or walking the dog. The best practice is to remember to do it in moments of stress, but remembering to ground in the midst of stress takes a lot of consistent work. But when you do, when you begin to automatically breathe and ground in response to tension, you will find that you can handle situations of high intensity much more effectively and easily than before. And you will be better able to be present in the woods, nurtured by their beauty.

In magical practice, we learn to awaken subtle senses that can let us perceive energies and spirits that go beyond the physical world. But too often, we focus on the nonphysical without first being fully rooted in an awakened, sensual experience of the physical world. Below is a simple exercise to help us awaken our senses and be fully present and aware in our bodies.

COMING INTO OUR SENSES

After grounding and practicing wide awareness, find yourself a safe and interesting place in the natural world—which could be a pristine spot in the wilderness, but could also be your backyard, a quiet corner of a city park, a vacant lot. Now close your eyes, just to shift your focus away from what for most of us is our dominant sense.

Sniff the air. Take some long, deep breaths through your nose, followed by some short sniffs. Become aware of what you smell. The air is full of information. Imagine for a moment that you have the nose of a dog or a wolf. What would the breeze be telling you? Can you smell the trees? The moisture on the wind? The chemical tang of polluted air?

Taste the air in the back of your throat. What are the tastes still lingering from breakfast? Does your saliva have a taste? Roll it over your tongue as if you were tasting a fine wine. What parts of your tongue come alive? What information does it give you about your state of being?

Feel the air on your skin. Become aware of the touch of the breeze, the temperature. Are you standing in the sun or the shade? Are there patches of both, and how do they feel different? Become aware of your weight, your stance, the pull of gravity on your body, your sense of balance or energy or fatigue.

Open your ears. Imagine that you have the ears of a deer, that you can shift and point in any direction. What sounds do you hear? Do you hear birds? Traffic? Voices? Insects? What do your ears tell you?

Now open your eyes. Add sight to the information you are receiving from all of your other senses. What do you see when you focus on a point or an object? What do you see when you extend your vision into wide awareness?

Earth-walk in your space, in wide awareness with your senses open. What do you smell and taste and feel and hear and see?

Being Rooted in a Place

As the storybook prince mentioned in Chapter One discovered on the Isle of Birds, learning to hear something takes time. To truly observe the patterns of the natural world, we also need to be rooted in a place. And our observations will be most clear and focused if that place is very specific, a special spot (or "home base") that we return to over and over again, ideally on a daily basis.

Some of my great teachers about observation and the natural world have been the folks at the Wilderness Awareness School, who teach tracking and nature awareness from a spiritual perspective. The core of their teaching is the importance of a "secret spot," a place that you return to regularly to look and listen, that you ultimately know intimately.

Of course it's wonderful if that spot can be out in the pristine wilderness, but because the home base needs to be somewhere that we can return to regularly, a wilderness spot isn't practical for most people. City dwellers do better to make it our backyard or the park across the street—a place we can get to easily in the course of our normal day's activities. Very few of us have hours every day to spend observing the natural world. Given the busy, stressed lives most of us lead, we're lucky if we can carve out some minutes here and there.

Cities can be fertile places for observing the natural world. I've seen a pileated woodpecker and a great snowy owl in the heart of inner-city San Francisco. Peregrine falcons nest among skyscrapers, and raccoons plunder the garbage cans in many neighborhoods. But besides the "wildlife" that exists in an urban setting, human beings are also animals and follow our own natural patterns. Economist Jane Jacobs is one of the great observers of the ecological patterns of city life. The following passage is a beautiful example of urban observation from her classic work *The Death and Life of Great American Cities*:

> The stretch of Hudson Street where I live is each day the scene of an intricate sidewalk ballet. I make my own first entrance into it a little after eight, when I put out the garbage can, surely a prosaic occupation but I enjoy my part, my

little clang, as droves of junior high school students walk by the center of the stage dropping candy wrappers. . . .

While I sweep up the wrappers I watch the other rituals of the morning: Mr. Halpert unlocking the laundry's handcart from its mooring to a cellar door, Joe Cornacchia's son-in-law stacking out the empty crates from the delicatessen, the barber bringing out his sidewalk folding chair, Mr. Goldstein arranging the coils of wire which proclaim the hardware store is open, the three-year-old with a toy mandolin on the stoop, the vantage point from which he is learning the English his mother cannot speak. Now the primary children, heading for St. Luke's, dribble through to the south; the children for St. Veronica's cross, heading to the west, and the children for P.S. 41, heading toward the east. Two new entrances are being made from the wings: well-dressed and even elegant women and men with briefcases emerge from doorways and side streets. Most of these are heading for the bus and subways, but some hover on the curbs, stopping taxis which have miraculously appeared at the right moment, for the taxis are part of a wider morning ritual: having dropped passengers from midtown in the downtown financial district, they are now bringing downtowners up to midtown. Simultaneously, numbers of women in housedresses have emerged and as they crisscross with one another they pause for quick conversations that sound with either laughter or joint indignation, never, it seems, anything in between. It is time for me to head to work, too, and I exchange my ritual farewell with Mr. Lofaro, the short, thick-bodied, white-aproned fruit man who stands outside his doorway a little up the street, his arms folded, his feet planted, looking solid as earth itself. We nod; we each glance quickly up and down the street, then look back to each other and smile. We have done this many a morning for more than ten years, and we both know what it means: All is well.[1]

A HOME BASE

Find a spot that will be your home base, where you can practice the magical disciplines of observation. Be modest and realistic in your expectations of what you can do. Is there a park or a community garden or a vacant lot full of weeds near your workplace where you can eat your lunch regularly? Is there an untended hillside you walk by when taking your child to the playground or your dog for her daily walk? Do you have a deck where you can drink your morning coffee and look out at the garden?

Once you have your spot, spend time there. Do the awareness exercises offered above, or just sit and listen. Notice how what you observe changes over time. How are the birdsongs different through the seasons? What animals or human patterns do you observe, and are they different at different times of day? In different weather? On workdays or holidays?

And how does your own mental and emotional state affect your ability to observe?

Letting the Stories Go

I am not an obvious candidate for the role of wilderness instructor. Throughout my life, I've consistently been the last person in any group going uphill, the slowest person on every hike. I was not one of those bold, physical children who are constantly testing their abilities and challenging themselves to run, jump, climb, and do daring things. I liked to read books. Left to myself, I would have spent my entire childhood curled up in my bedroom with a good fantasy story. My mother had to insist that I go out and walk at least six blocks a day in the summertime.

As a result, it's really easy for me to slip into a story in the woods about how weak and out of shape I am, especially when I'm in a group of faster, stronger people. Instead of being in the woods, I can easily be in some internal proving ground where my sense of self-worth is on the line—and not being confirmed.

I've had to learn to let that story go. It's a constant practice, and I'm never totally finished with it. In one of my novels, *Walking to Mercury*, the character Maya Greenwood learns to let her story go while hiking up the Himalayas with a bad cough. Fiction is different from autobiography, but this incident was based on a real moment for me.

> She sighed. The path continued up and up, with no relief in sight, and she really couldn't sit down on the trail and cry. She had to go on. There was no choice. . . .
>
> All right. Here she was in Buddha land. Why not try the Buddha path . . . and let go. Let her body be, with its limitations and imperfections, its wheezing lungs and its extra pounds of flesh. Stop trying to carve it and chisel it and make it into something it isn't, and love it like a pine, like a boulder. Stop blaming herself for being sick and slow and heavy. Let go of the words *sick* and *slow* and the weight of all they conveyed, and just feel the workings of her muscles against the rock. Stop wishing the uphill were downhill; stop telling herself a story about what she should feel, and let be.
>
> She stopped. Just for a moment she let herself breathe in the clean, thin air. She was surrounded by incredible beauty; white forms on the blue crowns of mountains so high they appeared just where you'd expect only birds or stars

to be. For that one moment, she was present in the beauty, no longer wishing she were somewhere else. . . . The path was still steep, her breath still rasping, her body still slow and terribly tired, she was still worried . . . but somehow none of that was between her and the mountains any longer. She was released from a glass cage full of chattering, clamoring noise, into a world where she could feel the air on her skin and hear the bell-like tones of silence. Light danced off the glaciers to caress her eyes, and she opened, letting herself be emptied, disemboweled, a conduit for wave after wave of love deep enough to match the mountains.[2]

Three years ago I broke my ankle on a hike through the woods that was part of a goodbye party I was throwing myself before going off to Europe for three months of teaching and lectures. I slipped on the gravel of a dirt road and landed wrong on it. Instead of enjoying a barbecue with all of my friends, I ended up in the hospital having surgery. I spent the next week at home, recovering, waiting for the ankle to be ready for a cast, grumbling at my partner and housemates and occasionally tossing the odd piece of furniture around the house, while snapping viciously at the parade of well-meaning friends trying to assure me that this was the Goddess's way of telling me to slow down. I pointed out to them that if I had a boyfriend who thought I should slow down, and communicated that thought by breaking my ankle, they'd be the first ones urging me to get out of that abusive relationship and report him to the relevant authorities. I give the Goddess every opportunity to communicate with me on a daily basis, I assured them, and if she wanted me to slow down, all she had to do was say so and reduce the workload a bit, not break my ankle, which in any case didn't slow me down much. I went off to Europe with crutches and a wheelchair a week later.

Throughout the trip, I could feel stories hovering. Being in a wheelchair and dependent on others for many things I could normally do myself wasn't easy. I could hear the whispers of the stories that shaped my childhood—stories my mother told herself about being a victim, being abandoned, never being taken care of, always caring for others, and never getting enough herself. Stories I had internalized, as well.

But I also became aware of something else. When I got caught in the stories, I felt abused and misused and unhappy. When I could stay out of those stories, being in a wheelchair with a broken ankle was just something new I was dealing with. It was interesting, actually—a different sort of a trip than I had planned on or hoped for, but a trip that showed me a whole new perspective on the world that I wouldn't otherwise have seen. I had a whole new relationship to sidewalks and curb cuts and bathrooms. I understood the need for laws protecting access for the disabled in a whole new, visceral way. I look back on it now as an extremely valuable experience.

How do we let the stories go? Some of us need deep emotional support to do so, in some long-term form of counseling or psychotherapy. Twelve-step programs to help people recover from addictions are another healing discipline that can help us reshape our lives as well as our ability to be present in the woods. And a supportive ritual circle or coven can also help in our personal transformation.

The exercises above and below can help sharpen our inner observation and ability to make choices, and can work independently or in conjunction with counseling or a support group. Ultimately, magical discipline is about learning to hear, understand, and "speak" to our own inner states of consciousness as well as the outer world: the goal is to become aware of our current state of awareness and to make conscious choices about what state we want to be in. And every form of personal change requires our active will and participation.

SELF-OBSERVATION

Sit in your home base or some other safe and quiet space. Begin to practice the awareness exercises. Now notice what gets in the way. Ask yourself,

What internal dialogue do I have going on in my head? What story does it represent?
What energy does it bring with it? Does it feed my energy or drain it?
What emotions am I feeling? What is my physical body telling me? What muscles are tight? How am I breathing?

WHAT CHARACTER ARE YOU?

Once you have a sense of your story, or of the emotions and energies constricting you, ask yourself, "If I could name the character I play in my own story, what would she/he be called?" Does your character come with a favorite phrase or bit of characteristic dialogue? Sometimes I'm Ilse the Fascist Healer, who says, "You *will* get better . . . or else!" One of my students came up with Bitter Betty, who says, "I take care of everyone else, but no one takes care of me!"—the exact words I heard from my own mother, over and over again. Or I might be Frantic Frannie, who says, "I have more work to do than I can possibly do in the time I have, so let me do five things at once!"

Now consider *your* character. You might want to do some writing in a journal about her/him. If you're working in a circle or support group, your friends might also help you consider some of the following questions:

In what ways is this character's experience of the world narrower or more constricted than it might be?
Are there ways in which this character expands my experience of the world, or serves me?
When I observe through this character's eyes, what do I not see or do?
How does this character influence the choices and decisions I make?
How would my experience change if I were a different character?
What do I want to do with this character? Tell it to go away? Kill it? Love it? Absorb and integrate it? Recognize and laugh at it?

Some of the worst mistakes in judgment I've ever made have been under Ilse's influence. Now, whenever I'm engaged in healing work, I take a moment to consciously acknowledge her and send her away, to make sure that my choices are coming from some other inner state. I've learned that I can do an enormous amount of work as long as I'm *not* being Frantic Frannie; but as soon as she starts in I get frazzled, fried, and exhausted, and in fact become far less effective in what I do. I miss the freeway offramp because I'm trying to talk on the cell phone while driving to save a few moments and end up going miles out of my way, getting caught in rush-hour traffic, becoming mad and frustrated, and wasting time. When I hear Bitter Betty's voice, I know that I have to stop whatever I'm doing and rest or do some self-care, or else I will get sick.

As Ilse, my ability to observe what's going on with someone who needs healing is colored by my own need to be a savior. I cannot clearly take in information. As Frantic Frannie, I can miss even clear information that's right in front of my eyes: a huge sign saying, "San Rafael Bay Bridge, Next Exit." As Bitter Betty, I can't see or take in the many ways my friends and partner are nurturing and taking care of me. Still less can I take in the healing and nurturing of nature.

A group or circle can help us to identify these characters, and this work can also be powerful and healing for the group. When we recognize our own constricting characters and introduce them to each other, we can all laugh at them together. Knowing that my circlemate sees and acknowledges her own irritating side makes me feel more forgiving and tolerant. Knowing that she sees and accepts my own annoying personality traits makes me feel freer to be who I am, and more fully accepted and deeply loved. When I feel loved, my own love for the earth can flow freely.

Anchoring

To release those constricting characters, it's vital that we know what it feels like to be without them, to be as unimpeded as an animal on its home turf, loping along a familiar trail in its accustomed gait. We want to be in a state where we are relaxed, somewhat neutral—not agitated or putting out energy, but with energy and power available to us. This state, which we call the "core self" or "baseline state," is so valuable that I teach a magical tool called "anchoring" to help us identify it and return to it quickly.[3]

Achieving the baseline state is more than just being grounded. "Grounding" means having an energetic connection to the earth, being present, aware, and in your body. Being at baseline is all of that, but it also implies being emotionally at neutral, not inflated or pumped up, not depressed, not telling yourself a story about yourself.

Neutral is not numb. Rather, it's the state of being maximally open to the information and communications coming to you. It's the place from which you can quickly shift into a different gear, if a need arises for action or speed.

"Anchoring" is a tool to help us reach a particular state of consciousness—*any* state, including being at our baseline. It means associating a particular state of consciousness with a physical touch or posture, a visual image, and a word or phrase. With practice, we can use that anchor—that combination of touch, word, and image—to bring us quickly into the desired state of awareness. With further practice, any one of the three may be enough.

We could create an anchor, for example, to help us reach that state we call being grounded—relaxed, aware, and energetically connected to the earth. We could create a different anchor to help us move quickly into a deep trance state in which our awareness is focused internally. In the exercise below, we create an anchor to help us move into our baseline, neutral state.

ANCHORING TO CORE SELF/BASELINE

Begin in your home base or another safe space. Ground and center yourself. This time close your eyes. Think of a place or time or situation in your life where you feel relaxed and at ease, where you can just be yourself. Where you don't have to impress anyone or achieve anything or exert your power, but where you have power and energy available to you if you need them. Where your focus can be on the world around you, not on yourself.

Say your name to yourself, the name you most identify with. Notice where it resonates in your body, and touch that place, or find a posture or gesture you can make that feels connected to this baseline state.

What image or symbol or picture comes to mind that can embody this state? Hold it in your mind as you also touch that place in your body or make your gesture.

Is there a word or phrase that comes to mind, a magic word you can use that calls you into this state, a phrase that can counter the phrases of your constricting character? Say it to yourself as you visualize your image and make your gesture.

When you use these three things together—the physical gesture/touch, the word or phrase, and the image—you can bring yourself instantly into this grounded, neutral, core state.

Now go through the exercise of coming into your senses, noticing what you can smell, taste, feel, and hear. Open your eyes and notice what you see.

How does your ability to observe change in this state?

Use your anchor and practice coming into this state regularly. Add it to your daily practice of grounding, and use it whenever you go to your home base to observe.

And practice using it when you get caught up in the dialogue or vision of one of your constricting characters. Notice what changes as you shift away from a constricted state into an open, grounded, neutral place.

As with grounding, the more you practice with your anchor, especially in moments of tension and stress, the more automatic it will become, until eventually for you it becomes normal in moments of stress to ground and go to your core self. From that inner place, you will be better able not just to take in information but to make conscious choices, to act instead of react.

I have several personal anchors that I use to move in and out of states of consciousness quickly. For example, one summer when I was traveling alone in southern France, visiting the ancient caves filled with art of the old Stone Age, I very much wanted to be able to meditate in those sacred places. But the only way to get in was to go on a guided tour, which did not allow for long, silent trances in the dark. I used a simple gesture so that I could drop into a state of deep meditation while walking, and could listen for the voices of the ancestors. But be careful: when you are anchored to a state of deep meditation, you are more open and vulnerable than usual. One day the ticket-taker on the tour accused me falsely of not paying for my ticket, and I was surprised at how shaken up I felt, and how much difficulty I had explaining myself and remembering my French vocabulary—until I remembered that I was in trance.

The neutral or baseline state, however, is probably the most useful state of consciousness for me. I anchor to that state when I'm facing a difficult or dangerous situation—because when I am neutral I am most capable of taking in information. I walk in the woods in my baseline state, so that less of me and my story gets in the way of what the forest is showing and telling me. I anchor to neutral when people are praising me effusively, or damning me thoroughly, so that I don't get my own core worth confused with praise or blame.

I Have To / I Choose To

The stories and characters that constrict us limit our choices. On a personal level, they keep us acting in set patterns instead of freely choosing what we want to do. We repeat the same negative patterns in relationships or work, stay stuck in bad situations, make the same mistakes over and over. On a collective level, we often rationalize and justify our perpetuation of negative patterns in the same way: we justify abuse and violence by saying, "We have no choice."

"What can we do—we have to do *something*," my neighbor says when I ask him whether he thinks bombing Afghanistan is the proper response to 9/11. "I don't want to be doing this," a soldier who is detaining civilians tells me at a checkpoint in the West Bank of occupied Palestine. "But what can we do—we have no choice." "I don't want to step on you with my horse," a policeman said to the young woman next to me when we were sitting in front of a line of mounted police in a demonstration. "But if my captain orders me to, I won't have a choice."

The essence of nonviolence is choice: acknowledging that we always do have a choice about whether or not to use violence, and posing an expanded range of choices to those in power. In fact, we had a broad range of choices we as a society could have made in response to 9/11. The soldier at the checkpoint could have chosen—and eventually did choose—to let the group of men he was detaining go home to their village. No court in the world would have convicted the police officer on the horse for refusing an order to trample a young woman—and, in fact, the order never came and the police instead withdrew.

It was clear to me, in each case above, that the authorities were saying "I have to" in order to absolve themselves of responsibility, to avoid choice. But as I thought about those incidents, I began to become uncomfortably aware of how often I said "I have to" to myself. I had to be sitting in front of those horses, because—well, because I just felt I *had* to be. If I were to be trampled or hurt—well, I had no choice. I *had* to be there. It would not be my responsibility.

But what would change, I began to wonder, if instead of "I have to be here," I told myself, "I *choose* to be here."

Saying "I have to" disclaims both responsibility and credit. It opens the door to Bitter Betty and her self-pitying, victimized sisters.

In saying "I choose to," I claim my power. Why would I choose to sit down in front of a line of horses? Because for twenty years I'd been telling people in nonviolence trainings that that's how you stop horses, and I wanted to stop that line from advancing on the crowd. Intuitively I felt that if the line kept advancing, someone would get seriously hurt, and I was willing to take a personal risk of getting hurt in order to prevent that. If I did get hurt (and fortunately I didn't), I would not be a passive victim, but a person who had made a choice and had accepted the risks that went with it.

"I choose to" also opens the door to a wider set of questions. Before I sit down, I might ask, "Am I willing to take this risk? Is this the moment to make a stand? Is it worth it? Can the horses and their riders see us? Are there constraints on the violence they might inflict upon us? Am I choosing to do this from a grounded, anchored place? Is this truly my task?"

When we take responsibility for healing the earth and restoring the balance, we are faced with many choices. Some may involve what seems like sacrifice. No, I won't buy that new SUV. I'll walk to class instead of driving. I'll take the trouble to recycle that can. Some choices may involve situations of discomfort or danger: I'll sit in front of those horses because I'm protesting an institution that sets policies that destroy the earth and human lives.

Whatever choices we face, it's important that we stay in our own power as choosers, that we don't see ourselves as martyrs or victims, but as active agents exercising our freedom and our will. Then we remain empowered, whatever consequences we face, and are able to act joyfully, courageously, and creatively.

So observe your own inner dialogue. Whenever you find yourself saying "I have to," stop, ground, and use your anchor to your core self.

Then ask, "What would change if I said 'I *choose* to'?" What new questions would arise? What would be the basis of your choice? What responsibility would you acknowledge? What greater credit might you claim?

Frantic Frannie has to go to a meeting instead of the movie she's been wanting to see, and she sits there resentful and sighing, running out at crucial moments to make a call on a cell phone, interrupting discussion to ask when the meeting will be over. She doesn't actually contribute much to or get much out of the meeting, and the tension she generates makes the meeting less productive for everyone. Rather than letting Frannie dictate my actions, I might instead choose, from my core, grounded self, to go to the meeting instead of the movie, because it's a key part of something I care deeply about. Having chosen to be there, I'll want it to be as productive and fun as possible. I might bake an apple pie to bring, or offer to facilitate. I'll be glad to see the people I know and enjoy working with, and my gladness will generate a positive, good feeling in the room. I'll listen attentively and maybe have helpful comments to make or inspiration to share.

I don't believe that *everything* in our lives is a matter of choice. In New Age circles, I often hear people say, "We create our own reality." That's a short-sighted and simplistic misunderstanding of how reality works. We don't choose all of our circumstances, or our range of choices. The poor don't generally choose to starve, nor do the oppressed choose their oppression, and the casualties of war don't choose to die.

But we can choose how we respond to the circumstances we're presented with. I didn't choose to break my ankle, but I did choose to go on with my trip anyway, to enjoy it as much as I could, to accept that I'd have moments of anger and frustration while striving to stay open to a new mode of experiencing the world.

All this may seem to have strayed far beyond the woods and the practice of observation. But it is one of the wonderful paradoxes of magic that everything works in circles. Outer work leads us around to inner work, and inner work allows us to do the outer work.

Bitter Betty can't enjoy the woods because she is never truly in them; she's walking around inside her own replay of that nasty remark that Sullen Susan made about her that she simply *has* to respond to. Frantic Frannie can't enjoy the woods because she *has* to write this article and make that phone call and answer that email, and she doesn't have time.

But I can enjoy the woods, from my grounded, core self, because I choose to make time to do so, and to be fully present and aware when I'm there.

KEEPING A JOURNAL

One of the myths of Witchcraft is that every Witch of ancient times kept a "Book of Shadows," a magical journal that recorded her spells and charms and herbal recipes. In reality, most Witches of that day were probably illiterate, but a Book of Shadows is still a good idea, especially when we begin observing and learning from nature. Taking time to record your observations will help them become more clear to you, and over the years those observations can prove an invaluable record. What did the birds sound like at this time last year? Are there really fewer of them, or does it just seem so? What were the mushrooms I found on my walk last year? Is the weather really different? Keeping notes in your journal will help you answer such questions, and provide a record of how your own abilities to hear and understand grow and develop.

NAMING AND IDENTIFYING

Learning the names of trees and birds and mushrooms may seem like a stuffy, left-brain activity for Witches, but it is extremely valuable in helping us deepen our connection to nature. First, the process of identifying a tree or a flower will make us observe it more closely and look for characteristics we might otherwise not notice. Second, knowing the actual names of things, especially their Latin names, will help us talk about them and learn from other people's experiences. Many plants have similar common names but are actually very different.

So start collecting guidebooks and using them. The most helpful books have a key, a set of simple choices that guides you through different families and species. Does the tree have leaves or needles? Are they in bunches, or set all around the twig?

Don't get overwhelmed. There are millions of beings around us to name and identify and learn. Begin with some modest goals, such as learning the key trees, birds, mammals, and insects of your area. Or make it a goal to identify one new tree or plant on each walk. Once you know the name of a plant or animal, read up on it and learn more about it. Move from observing to understanding.

NINE WAYS OF OBSERVING

The following exercises take us through nine ways of observing. They are inspired by Bill Mollison, one of the founders of permaculture, and by the lessons I've learned from the Wilderness Awareness School. Some of them will be further developed in later chapters on the elements, but taken together, they are the beginning of learning to read a landscape.

1. *I Wonder . . .*

In your home base or other natural spot, with your attention on what is around you, say to yourself, "I wonder . . . "

"I wonder why lichen is growing on that side of the tree, only?" "I wonder why the snowdrifts are piling up in this particular pattern?" "I wonder what attracts that bug to that flower?"

Don't worry about answering your questions; just notice what questions you can generate. As much as possible, keep your questions focused on physical reality. Not "I wonder how that tree likes all that snow on its branches," but "I wonder why those branches don't break under the weight of all that snow."

This is a great exercise to use with kids. You might ask them, "How many 'I wonders' can you find in five minutes?" You could follow that exercise up at home with a session with the encyclopedia, trying to answer some of the questions. But the focus here is less on answers than on learning to generate intelligent questions.

2. *Observing Energy*

Ask yourself, "How is energy coming into this system? How is it being exchanged?" There are many different sorts of energy you might observe: sunlight, heat, energy generated by motion of air or water, food, even psychic energy (but take time to focus on the physical before you jump to the psychic.) Also, you might try sketching your spot, or a plant in it, purely as a pattern of light and shadow. Don't worry about producing a "good" drawing; just let it become a meditation on how light energy is intercepted by form.

3. *Observing Flow*

In your home base, observe flows of all kinds. How does water move through this system? How do wind and airflow affect the area? What intercepts the flows? What marks do they leave of their passage? What is the source of these flows? How is that source replenished?

4. *Observing Communities*

What is growing together with what in this area? Which trees with which bushes, which groundcovers? Are there patterns you can discern? Are there sword ferns under the redwoods, and tanoaks near the clearings? What insects, birds, and animals seem to be connected with what plants? Are some plants serving as "nurses" for the young of others? Do some plants seem to stay distant from each other? Are some plants always found together? (Note: such questions can generally be answered only by many observations over time.)

5. *Observing Patterns*

What patterns can you see there in your spot? Textures, patterns of growth, distribution patterns, stress marks, all are examples of patterns. What patterns are repeated, on what scales? Can you find spirals? Pentacles? Branching patterns? Patterns based in fours or sixes? How many times does a tree branch from twig to trunk? What functions might these patterns serve? Why are certain patterns repeated over and over again in nature?

Again, you might wish to take a session to draw patterns or forms. Put your thoughts on paper without worrying about producing a work of art, but simply as a meditation to sharpen your ability to see and focus.

6. *Observing Edges*
Where does one system meet another in your spot? As we saw earlier, edges—places where forest meets meadow, or ocean meets shore—are often the most diverse and fertile parts of an ecosystem. Is that true here? How does the edge differ from the center?

7. *Observing Limits*
What limits growth here in your spot? Shade? Lack of water? Soil fertility? Other factors? How do these limiting factors make themselves evident? What is succeeding in spite of these factors? What seems held back? How have the plants and animals adapted to these limitations? What characteristics do the successful adapters have in common?

8. *Observing from Stillness*
Just sit still in your spot for at least fifteen minutes—longer is better. Notice what you can see, and how that changes over time.

9. *Observing Past and Future*
What can you observe in this spot that can tell you about its past history, and how it might have changed over time? What can you observe that tells you something about the future of this place?

Practicing the skills of observation, taking the time to ground and listen, we begin to be able to hear something. When we clear away some of our inner obstacles so that we can open up to the outer world, when we allow ourselves to be present, we can be fed and informed and delighted by the richness of life around us.

SIX

The Circle of Life

In the Goddess tradition, all ritual takes place within a magic circle. We ground and then create a sacred space, calling the four elements of air, fire, water, and earth, and that sacred transformative spirit of the center.

The circle is the pattern of the whole, the schematic diagram that lets us know if something is complete. In *The Spiral Dance*, I discussed how to create sacred space by casting a circle, invoking the elements of air, fire, water, and earth in poetic and mythic ways.[1] In this book, now, we enter the circle, to begin a journey through the elements of life. While we know that air, fire, water, and earth are not elements in the same way that hydrogen, oxygen, nitrogen, and carbon are, they each represent great cyclical processes of transformation that sustain life. The swirling cauldron of the atmosphere, the energy exchanges fueled by the sun, the cyclical journey of water from raindrop to stream to ocean to raindrop, the endogenic cycle of rock formation and plate tectonics, and the cycles of birth, growth, decay, and regeneration are some of the most basic processes of Gaia's physiology.

Many indigenous traditions use the pattern of the sacred circle with the four elements and the four directions. Not all place the same elements in the corresponding directions, because the correspondences originate from the qualities of particular places. The correspondences I use here originate from the British Isles, where the west wind brings rain, as it does in California, where I live. The west corresponds to water, to *feeling*, to emotion, to twilight and autumn—very fitting in a land where rain returns in the fall. The north is the direction of earth and the *body*, midnight and winter. The east, the place of sunrise, is air, *mind*, thought and inspiration, dawn and spring. The south, where the sun is strongest, is fire, *energy*, high noon, and summer. The center where the elements meet is the place of *spirit*, of change and transformation, the timeless place, the heart. In another place or climate, the directions for each element might change. In Australia, for example, Witches invoke fire in the north, not the south. But regardless of such changes, the circle of the elements remains an image of wholeness, and the correspondences of air/mind, fire/energy, water/emotion, earth/body, and center/spirit hold time.

More than twenty years ago, in *The Spiral Dance*, I discussed the magical correspondences of the elements. In this book, I want to take us on a journey around the magic circle, this time experiencing the great natural cycles and life processes each element reflects.[2]

The magic circle, because it represents wholeness, is a pattern we can use for examining the whole of any subject.

The Elements of Decision-Making

When we want to know if we have considered all sides of an issue, we can think about the elements and their corresponding qualities: What do I think about this particular issue? What energy do I sense around it? What do I feel? What is my body telling me? What transformation is possible?

Or: What are the rational arguments for this proposed action? What energy will it use or generate? What flow of events, materials, or processes will affect (or be affected by) it? What are the material considerations, the constraints? What are the ethical and community concerns?

When making a decision about sustainability, for example, we can ask,

How will this proposed action affect the air, the climate? The birds and insects? Will it bring inspiration and refreshment?

How much energy will this use, and where will it come from? Will it use more energy than we take in? How much human energy will it require? Will it energize or drain us?

How will this affect the water? The fish, sea-life, and water creatures? Will it use more water than we have? How do we feel about it?

How will this affect the earth? The health of the soil? The microorganisms and soil bacteria? The plants and animals? The forests?

How does this affect our human community? Will it benefit the poorest and least advantaged among us? Does this reflect and further our deepest values? Will it feed our spirit? Will it create beneficial relationships?

WORKING THE CIRCLE

Do you have a decision to make? An issue you are considering?

Take your journal to your home base, or to a quiet, safe place. Ground and come into your senses. Now consider your issue, using one of the sets of questions above. In your journal, record your inner conversation.

Then read through the next five chapters of this book. When you're done, go back to your issue and ask the questions again, once more recording your inner dialogue in your journal.

Now compare the two entries. Has anything changed? Have *you* changed?

Casting a Circle

At times in this book, I will suggest that you cast a circle before doing an exercise. To cast a circle is to create an energetic form of protection around yourself, a boundary that can keep out interference and negative forces. Before embarking on a trance journey or a deep change in consciousness, Witches cast a circle of protection. In fact, we begin all formal rituals by casting a circle and invoking the four elements of life.

Casting the circle creates an anchor, a set of physical, verbal, and visual associations that over time facilitates your transition into and out of particular states of consciousness. You can use many different methods to cast the circle. Some good examples can be found in *The Spiral Dance* and *The Twelve Wild Swans* (which I coauthored with my friend Hilary Valentine), and I offer one below.[3] But if you find one form that you like, and use it consistently, especially when working alone, the repetition will strengthen its effectiveness as an automatic trigger to consciousness change.

FERI CASTING

Casting a circle can be a very simple procedure. You can use your hand or a magical tool—an *athame* (or Witch's knife), if you have one. I generally use my garden pruners.

Here's a very simple casting I learned from Victor Anderson, my teacher in the Feri tradition of Wicca:

First, ground yourself.

Then stand in the center of the room, facing north, and say,

"By the earth that is her body . . ."

Turning to the east:

"By the air that is her breath . . ."

To the south:

"By the fire of her bright spirit . . ."

To the west:

"And by the waters of her living womb . . ."

Turn back to the north, to complete the circle. Face center and say,

"The circle is cast, the ritual is begun. We are between the worlds. What lies between the worlds can change the world."

While you are moving and speaking, visualize a circle of light, of blue flame or whatever color of light you like best, surrounding the area you are working in.

Instead of standing in the center, you might also walk around the room and tap each of the walls in turn. If you are not a strong visualizer, physically creating the circle will make it stronger.

Once the circle is cast, try not to leave it until the ritual is over. If you need to leave, respect the circle by cutting a door into it—using your tool or hand to mime cutting an opening in the energy boundary, or taking both hands and parting the energy as if you were parting draperies. Close it behind you—and don't forget to open and close it when you come back in.

Once the form of the circle is created, invoke or acknowledge the four elements and the center, and then call in any Goddesses, Gods, ancestors, or other beings you want present in the ritual, expressing your gratitude to them for their gifts.

When the ritual is done, open the circle. You can physically walk back around the circle in reverse, or simply thank all the energies you've invoked, and each direction in reverse order, and then say goodbye. You might say,

> By the earth that is her body,
> And by the waters of her living womb . . .

By the fire of her bright spirit,
And by the air that is her breath . . .
The circle is open but unbroken.
May the peace of the Goddess go in your hearts,
Merry meet and merry part.

When the circle is cast, we are ready to meet the elements of life.

INVOKING THE ELEMENTS

There are many ways to call in the elements. You can memorize a poem or simply speak from the heart, sing, dance, drum, or whisper. But one of my favorite ways, in a group, is a small ritual Kitty Engelman and I created for a weekend workshop we taught together at Diana's Grove in Missouri.

Ask everyone in the group to be still for a moment and feel which of the four directions, plus center, they are called to. Participants then move into their specific direction and are helped to ground and come into their senses.

Everyone is then sent out to observe for fifteen minutes:

The east group is asked to observe light and motion.
The south group is asked to observe fire and energy exchanges.
The west group is asked to observe water and flow.
The north group is asked to observe earth, plants, and animals.
The center group is asked to observe patterns.

When they return, the east group is asked to step into the center of the circle and hold hands. For a moment, they acknowledge each other, then turn around and face outward. Now they are asked to tell the stories of what they observed, all speaking at the same time and speaking continuously as the circle moves around the central group.

Hearing the stories is like hearing a spoken poem, taking a journey through the images and sensations of air. When the listeners have made a complete circle, the east group is asked to turn in and take hands again, and the whole circle expresses gratitude to the air and the storytellers of the east.

Repeat that step with the south, west, north, and center groups.

Another way to do the storytelling is to have the listeners stand still and the speakers face out and slowly revolve, telling their stories over and over until they return to their original spot.

Ritual Structure

For more than twenty years, I've worked at teaching and practicing ritual with a group called Reclaiming. We began as a small collective of women and men who wanted to integrate our spirituality and our political activism, and over the decades have grown into a network that extends from California to Germany to Australia and considers itself a tradition of the Craft.

In the Reclaiming tradition of the Craft, our rituals include spontaneity and creativity, but they follow a structure that begins with grounding, cleansing, and casting a circle. Cleansing can be a meditation, a bath, a plunge into the ocean, or it can be incorporated into the grounding by taking some extra deep breaths and letting tension go. Casting the circle creates a container for the energy that will be raised. When the circle is formed, we invoke the elements of the four directions, plus the center, which is spirit, and also acknowledge and invoke any Goddesses, Gods, ancestors, or other spirits we wish to call. Then we do the work of the ritual, which can be almost infinitely varied, but generally involves some shift in consciousness or some focused direction of energy toward an intention. After energy is raised, it is grounded. Then we often bless and share food and drink, thank all the energies we invoked, and open the circle.

As we journey through the elements in the chapters that follow, I will occasionally make suggestions for rituals. Generally, I will focus on the heart of the ritual and assume that you will first ground, cleanse, cast, and call the elements and the sacred in your own way, and thank them and open the circle afterward. *The Twelve Wild Swans* also includes a full discussion of ritual structure and creation.[4]

And now let's take a journey through the elements of life.

CHAPTER 7

Air

From my journal:

I'm walking in a break in the storm. For two days it's been raining, yesterday so hard that I never went out at all. Each time I thought about a walk, the rain would pelt down in sheets. But today, as soon as it lightened up, I decided to go.

I meet my neighbor Angie halfway down the hill, and together we walk up to Transmission Hill—so called because of the sculptures and tables made of old car parts left at an abandoned campsite on the top.

At first, the long view to the north is entirely hidden in fog. But as we watch, a small window opens, like a hole in the gray through which we can see the green hills opposite us. We follow the moving hole, almost like viewing through a grand telescope. And then the fog begins to break up. First the cloud layer rises, the hills remaining shrouded while vistas open up below. In the valley, we can see the fog rolling and boiling. The vapor is so thick that the air currents are made visible, marked by shreds and tatters of gray against the dark green mountains behind. We watch them shift and dance. They look like waves, cresting and spiraling in slow motion. We can see swifter currents moving quickly down the center of the valley, leaving wisps trailing behind.

We continue on our walk, heading around the hill. Today the wind is coming from the west, bringing the rain with it. Suddenly we see a crazy raven, riding the air currents as if they were his own private roller-coaster. He, too, makes the patterns visible. He glides up and down over the great waves of the air exactly as surfers ride the waves of the sea. He soars aloft on an updraft, tucks his wings and plummets, pulling up and flipping over like a fighter plane before he nears the ground. He is having a wonderful time, showing off. Now his mate joins him, doing her own dives, tucks, and rolls. The wild storm winds have become their amusement park, their courtship ground, for this fancy flying is part of their mating dance.

Air is the most ubiquitous element. It surrounds us, and we live immersed in it. In fact, we can't live for more than a few moments without it. Invisible, it is revealed to us only through its effects on other things. We can't see the wind, but we can feel it and know the immense power of a windstorm, a hurricane, a tornado.

In traditional Craft practice, air is the element of the east, associated with thought and mind, with the rising sun, with inspiration and enlightenment. Air is the breath of the Goddess, her living inspiration.

While in these chapters we are focusing on each element individually, one cannot actually be separated from another, any more than your breath can be separated from your circulation or your bones. They are all part of the whole that is the living being of the earth.

And so this wind that carries the raven is moved by the great energetic forces at play over the sphere of the earth. The earth is like a great inside-out cauldron, with the air its bubbling, spiraling brew, heated by the sun's fire, contained not by iron walls without but by the embrace of gravity from within.

The sun pours energy out on the earth as heat and light and radiation. The tropics, the band around the earth's equator, receive that gift most strongly. There, the air is heated up, and heat rises. As the hot air moves up, it pulls in colder air from below. A band of rolling air, like an elongated fountain, stretches across the belly of the earth, creating a turbulence, a complex swirling, spiraling motion, that distributes some of that heat throughout the northern and southern hemispheres. That motion cools the tropics and warms the Arctic and the Antarctic, preventing the lands along the equator from roasting in 140°F heat, and the poles from dipping to an average of –150°F. Eventually, excess heat radiates back into space.

The winds are moved, as well, by smaller variations in energy and form. In summer, when the inland valleys heat up, warm air rises and sucks in the fog from the coast—fog that often sits like a literal wet blanket atop San Francisco. In winter, cool, moisture-laden air rides in from the coast and drops its heavy burden of rain on the slopes of the mountains as it's forced to rise.

To observe the wind, to feel it on your face or to lean into its wild power in a storm, is to literally feel the energies of the planet in motion.

WIND OBSERVATION/MEDITATION

In your home base, or wherever you happen to be, ground and come into your senses. Now focus on the air and the wind. Notice what you can smell and taste on the breeze. Where has this air come from, and what does it carry with it?

Notice the feel of the air on your skin. Is it gentle or powerful, moving or still? What is the temperature?

Listen to the sound of the wind. The wind is telling you what it's moving through, and how fast. The wind has a voice that sounds different through trees or around the corners of houses. It can tell you how much moisture it carries, and whether or not there is a storm coming. What does the wind say to you now?

Look around you and notice what is responding to the wind, and how. Are branches bending and swaying? Are there trees or bushes that show signs of having been shaped and pruned by the wind as they grew?

How does the wind respond to obstacles? Does it move differently through swaying branches than it does encountering a wall?

Get up and move around now, and find the places where the wind is strong and the places that are wind-sheltered. What makes the difference? Feel what happens when the wind hits a hard surface or a soft surface.

What can the wind teach us about movement and change? About responding to obstacles, or about softening the great forces that impact us?

Is there something in your life you'd like to let go of? Imagine that you hold it in your hand. Raise up your arm, open your fist, release it, and let the wind carry it away.

FINANCIAL DISTRICT WIND OBSERVATION/MEDITATION

Go to the downtown financial district of your city, or some other area full of high-rises and hard surfaces, on a windy day. Walk around and notice where the wind is strong and where it is blocked. Feel how the hard surfaces accelerate the strength of the wind as it's channeled through narrow corridors. Are there temperature differences? Sheltered spots? If there is a park or garden, notice if the wind changes there. Does the wind behave differently here than in your home base spot? How does sound travel here? What smells does the wind carry?

Now sit in a safe place and close your eyes for a moment. Think about what you have observed. Are there parts of yourself that, like the wind, get blocked in this environment? Are there ways in which they gain strength or force in response to the blockage?

The wind is wild. Even the most urban environment cannot keep out the wind. Breathe in some of that wildness, and let it feed and strengthen the wild parts of yourself.

Microclimates

From my journal:

I stand on a ridge, a thicket of mature tanoaks standing between me and the wind. I can see their branches waving, undulating, tossed by the wind, but I feel only a cool, gentle movement of air, not so much a breeze as a quickening. The trees filter the wind. The shapes they make through space, the thousands of leaves and twisting twigs and branches, intercept the wind's energy and slow it down, trapping it in a labyrinth.

This windbreak has such a different quality from the way a solid surface interacts with the air—the difference between a seduction and a slap in the face. I'm thinking of the way the wind is funneled in San Francisco's downtown financial district or the middle of Manhattan, between tall buildings, hitting against hard edges, and how it feels to step around the corner of a building into a sharp-edged gale. The trees, with their filtering effect, slow the wind down. The sharp-edged buildings speed it up.

I continue on my walk, noticing how the rounded shapes of the mountains create wind shelters that are subtle. From the top of the ridge, I can look out for miles and miles. The air above is a layer of gray, the cloud-forms in ridges and ripples. It's like looking up from undersea and seeing the waves from below. Moisture-laden air is water, moves like water, only it's not as dense.

Walking around the bulk of a round hill, I'm wondering which direction the wind is coming from. Not from the west today. At first it seems to be coming from the east, or even up from the south; then it slacks off as the bulk of the land shelters me. Now I round the corner and suddenly the wind is in my face, coming down from the north, and I can feel how the hill has split the wind into two streams as a boulder in a river divides the current. The wind is more forceful here. This hill is bare, with low grass and few trees, logged and then grazed for a hundred years. And this south-facing slope is oak savanna, not forest. Huge black oaks and valley oaks, bare now of leaves, reveal the direction of the prevailing wind in the patterns of their branches. The wind has sculpted them into a form which feels like motion even when the air is still and which speaks of the west wind even when, as today, the wind is blowing from a different direction.

The removal of trees has changed the way air moves over this landscape. The wind is faster, harsher, more erosive. I think about the scale of deforestation happening all over the world today. How does it change the wind patterns to remove millions of acres of trees? With the buffers gone, the winds speed up, and the great worldwide currents of air may themselves be changed. No wonder the weather is always unusual these days!

To understand how air moves, picture molasses or honey oozing over an uneven surface. Cold air will flow downhill, as molasses would, but ridges or bands of trees can trap it, creating cold pools. Warm air will flow upward, warming the ridges, creating updrafts that hawks and vultures ride aloft. These small variations in temperature and wind strength create varied living and growing conditions.

A backyard, a city street, a forest is full of these areas of small differences, called "microclimates." One area of your yard might be perfect for sunflowers, while a few feet away you might be able to grow only ferns. A north-facing city park in cool, foggy San Francisco may get very little use compared to its sunny, south-facing neighbor. The daffodils on my sunny ridge may bloom a week sooner than the ones planted down in the shady canyon.

Microclimates are useful when we think about how to put things in the right place. If you are planting an apricot tree, for example—one that tends to bloom out early in response to the first spring sun and then get caught by frost—you might want to root it in a cold hollow where it won't warm up so early and thus will delay its blossoming until warm weather has firmly arrived. If you are building a house or planting a garden, you will want to carefully consider the microclimate of your location.

One cold, north-facing park in downtown San Francisco gets no lunchtime use because it's never warm. It's marked by a black stone sculpture, popularly called "The Banker's Heart," and it feels and looks cold. In our foggy city, people seek out the sun. In the blazing heat of a Texas summer day, on the other hand, shade would be an attraction.

If you are planning a ritual, you probably want to find a spot that is wind-sheltered but warm. That enticing valley in the park could be a cold sink on a May morning. That sunny hillside may be swept with winds that will tie your Maypole ribbons into Gordian knots. Consider where the sun will be at the time of day you'll be gathering. That shady glen might turn into a blistering inferno in the afternoon. If possible, always scout a location at the time of day you intend to use it before you finalize your plans.

MICROCLIMATE OBSERVATION

You can conduct a microclimate observation in your home base or in the city, but it is most interesting in a hilly area with different elevations or patches of sun and shade. Walk around the area and notice the changes in climate and temperature with changes in elevation, amount of shelter from the wind, amount of shade and sunlight. Are these changes reflected in different vegetation? Different ways humans use the area?

Where do you feel most comfortable? At ease? Where do you feel invigorated? Where would you want to be on a hot day? A winter lunch hour?

Consider some other factors of this microclimate. The same factors that amplify wind can amplify noise. A wind-sheltered area may also be a spot where smog collects.

How does the microclimate affect the more subtle energies that you feel? If you have a chance to visit a sacred site, what can you observe about its microclimate and relationship to the land?

Shelter from the Storm

From my journal:

The wind blew down the yurt we built on a sunny hillside to house friends and caretakers who could watch our garden when we were gone. A yurt is a canvas structure originally invented by the Mongolians as a nomadic shelter. Its walls are a lattice—like an expanded version of an oversized accordion-style baby gate. A cable sits atop the lattice, and the rafters slot into the cable at their tails and tuck their noses into a central ring, which in the modern version is topped with an acrylic plastic skylight. Tension and compression work in balance to keep the thing from falling down. It all gets covered with a skin and tied down firmly. In theory, it's a very wind-resistant design—and certainly the Mongolians have plenty of wind to contend with.

We had gotten as far as the "put the thing up" part (the work of many hours)—but not as far as the "tie down firmly" part. We were then gone overnight one day—a night in which fifty-mile-an-hour winds howled over the hilltop. Our yurt platform was nestled into the side of the hill—a location which, we had hoped in the planning, would give it some shelter from high winds. Not enough, evidently. The neighbors reported seeing it billowing in the wind. We returned to find it smashed flat, the skylight cracked, the roof ripped to pieces, a few rafters broken, and many pieces of lattice cracked.

A sad and discouraging climax to months of work. The deck held firm, however. The months of work had been mainly devoted to constructing the deck platform, which our neighbor Dave had designed for us with an extended concrete foundation to provide strength against the wind. Hundred-mile-an-hour winds have been known to rip across that hillside—one actually blew off the roof of a small cabin when Doug, who built it, first lived there.

I'm now amazed at our lack of clear thinking—putting up such a strong deck and not considering the strength of the yurt itself. We were lulled into a sense of false security by the yurt brochures, which advertise the structures as being designed for wind resistance. True—when they are fully tied down. Nothing was said about the wind resistance—or lack thereof—of a half-finished yurt.

So—we've ordered new parts, with the help and sympathy of the folks at Pacific Yurts, who scoured their warehouses for seconds we could buy at a discount. The yurt will rise again!

I hope most of you will never learn quite so dramatic a lesson about the power of the wind. But anyone who has ever keeled over in a sailboat, attempted to walk or cycle into the teeth of a strong wind, or been buffeted on an airplane knows how strong and sometimes terrifying the wind can be.

Our society expends a huge amount of resources and energy keeping buildings cool or warming them up. Much of this expenditure could be eliminated if we understood better how to create shelter from the wind.

Although our yurt lasted only a day before succumbing to the wind, we have a greenhouse nearby—a sixty-foot-long womb-shaped bubble of thin plastic stretched over metal rods (much lighter than the unfortunate yurt!)—and it has withstood many fierce storms. Why so hardy? It sits behind a sheltering belt of trees. I chose its location because I always noticed, when walking along that stretch of road, how the wind died away as soon as I moved behind the trees.

Understanding how to create shelter from the wind can help us understand better how to withstand any strong force, physical or emotional.

Our usual response to a force is to try to block it. To shelter a patio, we build a wall. To withstand a criticism or an attack, we go cold, shut down, turn to stone, or otherwise defend against it.

You may have noticed in your downtown observations how the hard surfaces and high walls don't so much *stop* the wind as *redirect* it. Wind is embodied energy—when it hits a hard surface it bounces off just like a basketball hitting a backboard, and it heads in a new direction. Impervious walls create eddies and new turbulences that may be just as harsh as the original ones you were trying to block, and more unpredictable.

The elements are our teachers, in many ways. Consider for a moment whether you've ever reacted defensively against a strong force in your personal relationships. Have you built a wall to keep out someone or some group you considered different or threatening? Have you tried to shut out a criticism or block an attack? What happened?

Trees don't place an impervious barrier before the wind. Instead, they absorb and transform the wind's energy. Watch the branches sway. Each one is like a springboard, set in motion by the wind, using the wind's energy to move itself in a new dance. First the branches go with the wind, bending in its direction. Because wind is moving in spirals and eddies, the branches find an ebb moment to spring back again, only to be pushed yet again by a new wave. That reverberation uses energy, and if there are enough branches, the wind's energy will be used up, or at least diminished, its force softened.

A windbreak is a useful thing to know how to construct. If I want to shelter my home, my ritual site, my yurt from the wind, I won't build a wall around it; I'll plant trees and bushes, or set up a filter with moving parts to absorb the wind's power. By doing so, I can make my home or garden more comfortable, more energy-efficient, and safer. I'll need to expend a lot less on heating and cooling and replacing broken windows.

But a windbreak is also a great teacher, for me, about nonviolence. How do we respond to strong forces—anger, rage, even physical attack—without becoming violent in return? How do we respond to what might be well-meant but harsh criticism (whether well intentioned or intentionally hurtful)?

If we become a wall, shutting out the energies coming at us, we may actually strengthen the anger of the opposition. On the other hand, if we simply brush off or bat away criticism, the opposition may expand its criticism to include our reactions.

But there's a third alternative: if we can learn from the trees, we can take in and transform the energy coming at us. We do this by staying calm and grounded and centered, by listening rather than responding, by swaying with the wind and letting it blow itself out.

The exercise below is similar to one I often use in nonviolence trainings. In including it, I don't want to suggest that this is the *right* way or the *only* way that we should respond to attacks. It is simply one of our options, and the purpose of learning it is to expand our range of choices in any situation.

WINDBREAK/NONVIOLENCE EXERCISE

Do this role-playing exercise with a partner, or if you're in a group, have people form two lines, facing each other, with partners across from each other. The ground rules are simple: interact only with your partner, stay with the interaction and don't walk away, and don't use any physical violence in the exercise.

Choose two oppositional roles that relate to a situation you might face. In preparing for political actions, for example, one side might be peace demonstrators; the other, angry supporters of an invasion of Iraq. But you can also choose everyday situations. One partner might be a dogwalker; the other, an angry homeowner accusing the walker of letting the dog foul the sidewalk. One partner might be a leader of a public ritual; the other, a police officer who's gotten a call complaining that Satanic rituals are happening. You can also use this exercise to prepare for an anticipated difficult encounter: one partner might be a newly enthusiastic Witch going home for Thanksgiving dinner; the other, her belligerent fundamentalist brother-in-law.

The partner who is the windbreak should ground and center. The attacking partner should take a moment and think about the aggressor's role. How would you feel in that role? Have you ever been angry or wanted to attack someone? What did you do? Say? What kind of energy did you feel?

At a signal, begin. Let the interaction go for two to five minutes, with the attacking partner putting into words and gestures the anger of the role, and the windbreak partner focusing on listening and on staying grounded and centered. The windbreak partner may speak or keep silent.

Then stop the interaction and debrief. Ask the attacking partner,

How did you feel?
What did your partner do? Did she or he do anything that was effective in communicating or deescalating your anger?
Did she or he do anything ineffective? Anything that made you angrier or cut off communication?

Ask the windbreak partner,

How did you feel?
What did you do that felt effective? Ineffective?

Were you able to stay grounded and centered?
What did it feel like to absorb the energy of your partner's attack?

If in the course of the exercise you took on any negative energy that wasn't yours, take a breath and consciously release it, or do an energetic cleansing. (See the exercise called "Energy Brushdown" in Chapter Eight.)

Air and Life

Air is life. Not only does most life on the planet need air, but the atmosphere itself is a creation of life. Remember the creation story we read in Chapter Four? To understand the miracle gift of the atmosphere and the incredible dance of life that created the very air we breathe, do the following meditation.

BREATH MEDITATION

Read the story of "The Gift of the Ancestors" or have someone speak it as a guided meditation. Now close your eyes for a moment and simply breathe.[1]

Breathe in with gratitude to the ancestors and with appreciation for the great creative powers of life.

Breathe out with love, as a conscious gift back to the green world.

This air is a gift of the early ancestors. What you breathe in, this moment, originated billions of years ago. This air passed through the lungs of dinosaurs and mammoths and the earliest human beings. Is there a great teacher or hera from the past that you admire? You are breathing the same air that passed through her or his lungs, sharing inspiration. Breathe in with gratitude; breathe out with love.

Is there a problem you're stuck on, an issue that seems hopeless? A place where you've given up in despair?

Just breathe in, taking in the creative power of the ancestors, of life, asking that power to infuse you and help transform your issue.

Breathe out, with love and commitment, acknowledging your grief or pain or hopelessness, not trying to change it but just making space within, to breathe in again, filling yourself with creativity and power.

Continue as long as it feels right. Don't try to solve your problem instantly; just keep breathing into it and trust that you can shift the energies around it and open the space for inspiration to come.

The Birds and the Bees: Creatures of Air

The air not only sustains our life; it is a medium through which life moves. The birds and insects can be our teachers, too.

Philosopher and student of shamanism David Abram, in *The Spell of the Sensuous*, writes of a mystical experience he had in Bali one night, watching the brilliant stars of the night sky reflected in the still waters of the rice paddies below, feeling as if he were free-falling through space:

> I might have been able to reorient myself, to regain some sense of ground and gravity, were it not for a fact that confounded my senses entirely: between the constellations below and the constellations above drifted countless fireflies, their lights flickering like the stars, some drifting up to join the clusters of stars overhead, others, like graceful meteors, slipping down to join the constellations underfoot, and all of these paths of light upward and downward were mirrored as well in the still surface of the paddies. I felt myself at times falling through space, at other moments floating and drifting. I simply could not dispel the profound vertigo and giddiness; the paths of the fireflies, and their reflections in the water's surface, held me in a sustained trance. . . .
>
> Fireflies! It was in Indonesia, you see, that I was first introduced to the world of insects, and there I learned of the great influence that insects, such diminutive creatures, could have on the human senses.[2]

Birds are easy to appreciate and love, but insects are a challenge for many people. They seem strange and creepy. Science fiction aliens and villains are often portrayed as insectlike. In the real world, some insects sting, bite, or nibble on us and our belongings in unpleasant ways. Some are poisonous and others carry diseases.

I know too many Witches who rarely go outside because they are afraid of bugs. How sad!

We can't connect with nature without connecting with bugs. After all, about half the living things in the world are insects. Most of the plant life that we love coevolved with insects. The flowers, the fruits, and the insects that pollinate them, and the birds and animals that eat the insects, make up a whole that cannot be divided.

If insects frighten or horrify you, perhaps it is because you are being a wall instead of a tree in relationship to them. Are you shrinking away from them, trying to wall yourself off from any encounter?

Try letting insects into your awareness in gentle, nonthreatening ways first. Read about them. Pick one insect to start with, and learn about it. Get to know the life cycle of a butterfly or the strange marital dynamics of a praying mantis. When you're ready, try this observation exercise.

INSECT OBSERVATION

Sit in a garden (preferably an organic garden) or other natural spot and watch the insects. How many different insects can you count in a five-minute period? The number of insect species will tell you something about the amount of life energy in this spot.

What are the bugs doing? Are they feeding? If so, on what? Are different insects feeding on different plants? Are they drinking the nectar of flowers? Are there honeybees present? If so, are they collecting pollen? (You'll be able to see the heavy pollen collected on their legs.) What plants do they seem to prefer?

Go back to this spot in different seasons, and at different times of day, and notice how the insects change.

David Abram also goes on to suggest that shamans work not so much by contacting supernatural forces as by working with the powers and awareness of the insects, birds, animals, and other nonhuman beings around us. Insects know things that we don't and perceive things that we can't. Once we develop friends and allies in the insect world, we can experiment with augmenting our awareness.

CHOOSING AN INSECT ALLY

When you are comfortable with the insect observation, choose one insect to study. Learn about its habits and life cycle. Insects undergo amazing transformations as they develop through their life cycles. Bees, ants, and termites live in complex societies, communicating with each other through scent and the amazing symbolic communication of the bee dance.

Find a spot where you can observe your chosen insect. Ground, center, and come into your senses. Watch your insect for a bit; then close your eyes. Imagine you are face-to-face with your insect and can speak to it. Ask its permission to borrow its awareness for a moment. If you sense that the answer is yes, breathe deep and imagine your insect growing larger and larger, until you can step inside it. Feel your skin transform, grow wings and antennae; let yourself see with its eyes, hear and smell and taste as it does.

At first, you may feel as if you are simply making up an experience, telling yourself a story. Don't worry. Just enjoy the experience, noticing what's different in this mode of perception without worrying about whether or not it's "real."

Begin with just a short time, and later expand it if you feel comfortable. Before you come back to yourself, thank your insect. Imagine yourself stepping out of its body, and let it shrink back to normal size. Pat your own human body; feel your skin, your human form. Use your anchor to your grounded state; open your eyes and say your human name.

When you become comfortable and familiar with this exercise, experiment with using it to get information. You might explore your garden as a bee, for example, to see whether you've provided good habitat for pollinators. I admit that I find being a bee extremely erotic. Imagine smelling something wonderful, sipping on something sweet, healthy, and nourishing, while being stroked and caressed by velvety, feathery stamens. For that matter, each of those flickering fireflies is calling out some insect version of "Hey, hey, hey, baby—I'm the guy who can show you a *real* good time!" And then there are all those damselflies flitting about, stuck together, flaunting their relationships in an utterly shameless fashion.

A little of this exercise, and you may find your horror of bugs changing to a very different emotion.

The Language of Birds

Our prince, in the story told in the first chapter of this book, takes many years to learn the language of birds. If you hope to master that language, perhaps you'd better begin right now.

The teachers at the Wilderness Awareness School say that when songbirds sing, they are giving praise to the Creator. The Mohawk people teach that the songbirds were each set in a particular place on the earth, to sing about that place and teach us something about it. That may be, but it is my personal belief that the blue jays are here to teach us that no matter how beautiful everything is, someone will always find something to complain about.

Paying attention to the birds will tell you many things about what's going on in the world, and about your own state of being. Birdsong changes according to the time of day, the state of the weather, the season of the year, and how the birds happen to be feeling. The Wilderness Awareness School identifies five key voices of the songbirds: the call, the song, the feeding plea, male-to-male aggression, and the alarm call. A real expert in bird language can tell by the birds' vocalizations what predators are moving through the forest at any moment. Scouts among the Apache could tell when a European was three miles away by the song and behavior of the birds.

Learning the language of the birds is a long and complex process, as our fairy tale suggests. If you are truly interested, I strongly suggest that you enroll in a program like that offered by the Wilderness Awareness School. (See the section on Resources in the back of the book.)

But even without that formal training, all of us can begin to hear something.

LISTENING TO THE BIRDS

In your home base, ground and come into your senses. Now focus on your sense of hearing. What birds do you hear? How many different songs or calls can you hear? Do you know what species they belong to? If not, just give them names of your own, as I did in the observation that begins Chapter Five (the Eight-Note Bird and Wheet).

Following are some suggestions for deepening your awareness of the birds. You may want to follow some or all of them.

Where are the birds singing from? Draw a rough map of your home base, and mark approximately where you hear each bird.

For a week, come back and listen, referring each time to your map. Do you often hear the same bird in the same place? Can you identify a home base for that individual, a possible nesting place?

Keep a bird log of all the different birds you hear. Again, if you don't know the species, give it a name of your own. Try logging birds on one of the following schedules:

For half an hour at the same time of day each week (and again in different seasons)

For half an hour at dawn, just after sunrise, and at midday, twilight, and darkness

On the first day of each month, over several years (to learn if the species count changes over the years)

Few of us will be quite so thorough in our observations, but just thinking about these questions will help us sharpen our awareness and take in more information.

The Five Voices

Learning to recognize the five voices of the songbirds—the call, the song, the feeding plea, male-to-male aggression, and the alarm call—takes nothing more than time, practice, and a bit of empathy. Remember that these five voices apply only to the songbirds (the passerine, or perching birds)—not to the corvids (the crows, ravens, etc.) or the jays, who are a law unto themselves, nor to the raptors, the birds of prey.

Now as you listen to the birds in your home base, you'll be able to identify the following voices.

The Call. The call is the bird's basic "Here I am" statement. It's regular, repeated, and often echoed by a mate, as if they were saying, "I'm here, sweetie; everything's fine." "I'm here, too; all's well." "I'm still here; everything's fine." "I'm here, too; all's well."

The Song. The song is generally more elaborate, sung especially in the spring. Many birds burst into song with the sun's rising, and why not believe that they are filled with joy and delight and gratitude for a new day, singing their general well-being and thanks? Western scientists, however, maintain that they are defending their turf and advertising for mates, that their song is the bird equivalent of "Single male robin with good breeding territory seeks committed relationship. If you thrill to a sunrise and enjoy a good, juicy, early worm, if you dream of sharing a nest and rearing a clutch of eggs . . . let's meet for a twilight flight and explore possibilities."

The Feeding Plea. You're likely to hear this mostly in spring and early summer. It's the begging cry of baby birds, generally high-pitched, frantic, and recognizable to any parent: "Hey Mom, I'm hungry! Dad, more worms! Where are you guys? I'm starving! Feed me! Feed me!"

Male-to-Male Aggression. This too is a spring/early summer mating and nesting thing. It can sound like an alarm, but it's limited to one species of bird. Again, it's not hard to recognize or identify with: "This is my turf!" "Yeah, well I'm taking over; move on out!" "You move, you ***!!!*%%**." "Take that, you !!***!!*%%**."

The Alarm Call. A true alarm spreads from one species to another and often moves out over the landscape, following the path of a predator. It might be a bird's normal call, speeded up or more frantic in pitch. Squirrels and jays often join in. Again, a real expert can tell exactly what is moving through the forest by the nature of the alarm calls. Personally, I'm still at the stage of "Hmmm, something is sure upsetting that chickadee. I wonder what it is." An alarm sounds like, "Look out! Look out!" "Head's up, everybody!" "Watch out! Watch out!"

An alarm in the forest tells you that some predator is on the prowl. It might be dangerous to you: a cougar. It might be a danger only to the birds: a hawk or a housecat. It might be something simply out of place: another human in a bad temper.

What the Birds Are Saying About You

Yes, the birds are talking about you. Whenever you go into the woods or out to a meadow, the birds notice you and they respond to the state of consciousness you're in. If you are crashing through the woods, talking loudly, or *thinking* loudly about how mad you are at that person in your office who sent you the nasty email, you probably won't see or hear many birds. They will simply flee, and you will walk through a silent landscape, probably not noticing the birds at all.

When you are grounded, in your senses, practicing your awareness techniques, and attempting to walk silently and respectfully, the birds will let you come closer. Eventually, they'll simply "hook"—that is, move to a minimally safe distance away. When you gain enough practice, awareness, stillness, respect, and love, they may even approach or not change their behavior at all.

The birds and animals will be our teachers if we let them. The other day I was walking in the forest, stepping as silently as I could on a bed of fallen leaves, practicing being in my senses, not my thoughts. All was going well until my mind strayed to the upcoming WTO protest in Cancun, and the likelihood of police violence there. At that moment, a squirrel leaped onto a branch above my head and began scolding me, deeply affronted. It was as if he were saying, "How dare you bring those nasty thoughts into the woods! Here you were, walking so quietly and we thought you were a nice, safe human, and then you come out with *that!* I'm shocked at you! And just when we're rearing young ones!"

I stood still and apologized, but it did no good. He continued to follow and scold me until I was out of his home territory.

On the other hand, I was on the same trail one day, passing through the same rustling dry leaves, when my mind strayed to the possible scenario for an erotic film my friend Donna and I are always threatening to make. A whole covey of quail came out of the tanoaks and continued feeding very calmly just a few yards away from me, paying me no mind whatsoever.

CONVERSATIONAL EXCHANGE

As you sit in your home base or walk through the woods, notice how the birds and animals are responding to you. Do their responses change as your awareness changes? Your emotions? The content of your thoughts? As your ability to ground and be in your senses deepens, how do the wild animals mirror that change?

Global Warming and Climate Change

So here we are, down at the bottom of this swirling cauldron, utterly dependent for our very lives on the constant re-creation by life itself of just the right amount of oxygen in the air. And what are we doing? We are daily pumping toxic chemicals into the air, filling it full of substances that destroy the ozone layer, the protective membrane in the outer atmosphere that shields us from the most powerful and dangerous radiation from the sun. We're burning the remains of ancient organic life at such a phenomenal rate that we have increased "greenhouse gases"—gases that prevent radiation from escaping back into space—and are threatening the overall balance of the earth's climate.

"Global warming" is perhaps a misleading term, because it leads us to benign fantasies of sunbathing on otherwise chilly winter beaches and growing mangoes in Kansas City. But what global warming really means is turning up the heat on the cauldron. If you build up the flames beneath a cauldron of soup, the liquid inside boils faster. If the flames are too high, the soup may spill over the sides of the pot. In terms of the earth's atmosphere, the increased energy from global warming means that all of the turbulence of the winds is moving faster, with wilder oscillations and greater extremes. In some places, that might mean warmer winters. In others, it might mean hurricanes. In still others, drought. The patterns of climate we have planned for and adapted to can no longer be counted on, and the new patterns are likely to be more fierce and more extreme.

The earth as an organism can certainly survive this overheating, but it's not clear yet that we as a species will survive—at least not with the level of comfort and abundance we would like to enjoy. And many of the living communities of the earth that we love will be threatened, in part because so many of them are already suffering from loss of habitat, diversity, and resilience. Natural communities are capable of changing—indeed, have always changed and evolved—but they are not necessarily able to adapt as quickly as we create potential disasters. And many of the mechanisms of adaptation are blocked by human activity. Animals cannot easily migrate north or south, for example, and forests cannot retreat to higher latitudes, when freeways and housing developments and cities block their way. In earth's history, there have been several periods of great extinctions, when major life communities failed to adapt to sudden changes. We may be headed for, or already in the midst of, another such period.

We now know that global warming and climate change are underway. No impartial scientist doubts the data any longer. We see their effects in changes in the weather, in floods and hurricanes and increased storm damage. But so far the impact has not seriously affected most of us who live in the country with the greatest responsibility and lack of accountability for the problem: the United States.

But the effects of global warming and climate change won't necessarily remain as mild as they have been thus far. Ecological change is not always a gradual affair. It's more like a river heading for the rapids. Until the moment you go over the falls, the water may seem slow and peaceful, the current relatively mild. Your only warning may be the dim roaring in your ears—easily drowned out by the CD player you've brought along in your canoe. When you look ahead, the rocks and rapids are below your line of sight, and everything looks calm. And then suddenly you're over the edge and there's no turning back.

What might some of those damaging effects be? To put this in perspective, it is estimated that the earth's climate may increase by two to ten degrees on average over the next fifty to one hundred years. (A five-degree *decrease* was enough to cover the major landmasses of the earth with ice during the last Ice Age.) If ocean levels rise along with temperatures (a likely consequence, given the melting of ice), flooding will become endemic and many of our major cities will be threatened. Don't think that putting sandbags along the seafront will save us. When ocean levels rise, rivers cannot drain as easily, so inland flooding increases. We've seen that already in Bangladesh and other areas of the world.

Some places may find themselves warmer, but others may suffer a reverse effect. The Gulf Stream warms Ireland, Britain, and western Europe. Climate change could divert it, leaving the Emerald Isle with the climate of Siberia.

It is a sad failure of our current political system and our international diplomacy that we seem incapable of facing this issue. As long as the river seems placid, it is all too easy to ignore the voices warning of the rocks below, especially when what they recommend might be inconvenient or threaten the immediate profit balance of major corporations. The Kyoto treaty to limit the production of greenhouse gases and forestall climate change, itself merely a Band-Aid for the issue, was first shafted by the maneuvering of the Clinton administration and then dealt a fatal blow by Bush's outright refusal to sign.

You don't have to be a black-flag-waving anarchist to be outraged by this shortsightedness. Anyone who loves capitalism should be especially maddened—because solutions and alternative sources of energy *do* exist that could enable us to transition swiftly from our fossil-fuel-based economy to one that runs on clean, renewable energy sources that don't contribute to global warming.

There are things that we can do. First, if this whole discussion is making you feel angry, afraid, frustrated, and hopeless, take a deep breath and release some of that energy. Go back to the earlier breath meditation and breathe in some inspiration from those amazingly creative simple cells that are our ancestors. If they could figure out photosynthesis, a process so complex that scientists can barely describe it, we can figure a way out of this crisis, too.

We can make personal choices that reduce the greenhouse gases we each produce. It's important that we do this not out of a sense of guilt or resentful obligation, but as an affirmative choice to more deeply integrate our values and our everyday actions. Sometimes it's hard to believe that riding the bus on a given morning instead of driving will make a difference. But it will—especially if we make that bus ride a spell, an enacted prayer, for balance.

A SPELL FOR BALANCE

When you are faced with a choice to do something small in service of the earth that requires some sacrifice on your part, stop first, breathe and ground, and say:

"I offer up this [walk, bus ride, extra dollars spent on a compact fluorescent bulb] to the greater balance and healing of the earth. Let this small act be like a ripple that grows into a wide circle as it moves outward, leading to greater change. With every breath in, as I [walk, ride, enjoy the light], I will remember my gratitude to the ancestors for the miracle gift of air. With every breath out, I will renew my love and commitment to the balance."

The personal choices that we all make are deeply important, but change is also needed on a larger scale. So consider what groups or larger communities you belong to that might be able to make collective choices that favor the earth. Can you get your workplace or school to begin conserving energy? Switch to compact fluorescents? Run its diesel buses on biodiesel? Can you advocate for more public transportation in your community? Help educate people on this issue?

Although the Bush administration has refused to sign the Kyoto treaty, many individual cities and counties have agreed to reduce consumption to the levels mandated by Kyoto. A group known as Cities for Climate Protection is coordinating these efforts.[3]

Organizing, going to meetings, doing all the unglamorous work of political change may not seem as inspiring as ritual or a wild meditation on the wilderness winds. But these tasks can be deeply spiritual acts of service that help bring us into alignment with our deepest values. Try the blessing above before making a phone call to your representative, or in the middle of a trying meeting.

Weather-Working

As you can imagine, if you've come this far, I'm not going to suggest magical techniques for controlling the weather. The great energies that move the storms and the wind are beyond our control. Long ago, one of my first teachers in the Craft said to me, "If you want to work the weather, make friends with the clouds." If you have practiced the observation techniques in this chapter, if you have been offering blessings and gratitude to the air and breathing out your love, if the birds and animals are no longer fleeing at your approach, chances are that the clouds are feeling friendly to you. Or to put it another way, you and the air and the clouds and the creatures are now a more harmonious whole, a whole that includes the possibility that the weather will to some extent reflect your state of awareness. At my friend Penny's fiftieth birthday party, the sun shone brilliantly even though her garden near the coast is often foggy. "Who would expect anything else for Penny's birthday?" one guest remarked.

I don't like to work the weather—it seems like hubris to disturb such huge forces for my own ends. But I often find that the weather works itself—as during our ritual near the Irish ring fort when the sun came out (mentioned in Chapter Two). And when there is a true need, I simply put in a request. I stop, take a breath in with gratitude, breathe out as a conscious gift, and allow my mind to contemplate the awareness of the wind and clouds, to acknowledge that those great forces also partake of consciousness, allying with it, feeling love and appreciation for it, saying something like, "Beautiful clouds, wild wind, I love your strength and energy, and the rain that you bring is life itself. But if it doesn't disturb the greater balance, it sure would be nice to have a bit of sun and no snow tomorrow for the big peace march."

And, as friends do, the clouds generally cooperate when they can.

BLESSING FOR AIR

Praise and gratitude to the air, the breath of the living earth. We give thanks to you for our lives, for our breath, for the literal inspiration that keeps us alive. Praise and gratitude for those ancient ancestors, the first magicians, that learned to use sunlight to make food, and so gave us the gift of oxygen. Praise and gratitude to those who learned to burn food for energy, and to the great exchange, the world breath that passes from the green lung to the red and back again. Praise for the sun that sets the cauldron of the winds in motion, and to the great winds that soar over the face of the earth. Praise to the storm that brings the rains, the water of life to the land. Gratitude to the creatures of the air, the birds that lift up our hearts with their songs, the insects in their erotic caress of the flowers—a caress that brings the fruit and the seed.

May our minds be as clear and open as the air; may we learn from the wild winds how to soar across barriers and sweep away obstacles. May the air and the winds of the world be cleansed. May we learn to be good guardians and friends and allies of the air that is our life; may we make the right decisions that can restore the balance. Blessed be the air.

EIGHT

Fire

From my journal:

I'm sitting in my home base, on my back deck. Around me are clumps of redwood trees, young ones regrowing in circles around the bases of old stumps logged long ago. Above my head stretches a tanoak, and across the garden and the dry stream are beautiful big-leaf maples. Flies buzz around, and in the distance a western flycatcher calls, "Too-wheet! Too-wheet!" A raven croaks high overhead, and I hear the whuff, whuff, whuff *of its wings beating through the air. I'm thinking about energy.*

The trees all around me are capturing the sun's energy. I can see how they fill space with their leaves and needles, taking up every possible plane of interception of a photon of sunlight. That light is the power, the energy, that lets them take carbon dioxide from the air and water and make wood. Something solid out of liquid, gas, and ephemeral light.

The forest, the garden area below me, are a mass of green. All around me are organisms using sunlight to make food. And buzzing around me and crawling below me are the organisms that eat that food, and the eaters of those eaters, each a stage in the great transformation.

I have been a photon of light.

I have been a green leaf on a tree.

I have been a caterpillar munching on a leaf.
I have been a songbird dining on a worm.
I have been a hawk eating a bird.

I have been a photon of light.
I have been a blade of grass.
I have been a deer grazing in a meadow.
I have been a cougar, culling the herd.

I hear the beat of the raven's wings. It has a steady rhythm, a certain pace that I recognize as the baseline of the raven. As a drummer, I could reproduce it, hold it steady. If I had a watch, I could time it. That baseline pace represents a certain energy relationship—the size and wind resistance of the raven in relationship to the efficiency of its heart and bloodstream and the conversion of its food to energy. That pace doesn't vary much from raven to raven—not as much as, say, my comfortable rate of walking varies from that of my lithe and faster friends. But then, maybe we're just at different ends of a human baseline, which may seem to vary a lot but doesn't really—not if you compare our walking gait to that of a wolf trotting or a snail inching forward. Suddenly I'm remembering a joke told to me by a Swiss friend from Bern, whose inhabitants are said to be very slow. A Berner arrives home from Zurich, looking very exhausted and out of breath. "What happened?" asks his friend. "Why do you look so tired?" "Oh, it was terrible," the Berner says. "A snail was pursuing me all the way home!"

I'm thinking about how trees seek the light. In my city garden, our big plum tree fell down this year, on May Eve. For years I'd kept it pruned back, trying to keep some sun for the rest of the garden, but each year it grew and grew, extending its branches out to the south and west, reaching for light. All this winter, I'd looked at it and planned what branches to cut, but each time I intended to get to it, I ended up organizing some peace demonstration instead, trying to prevent a different kind of fire. So on May Eve, with the ground soft from late rains, it simply gently leaned over, pulling its roots out of the ground. We probably could have saved it even then, but I was away, and when I had time weeks later to deal with it, the roots were already dry. And the space and light that opened up in the garden with it gone were, we admitted, attractive.

So now the tree has been cut up and most of it bundled away. Instead of having plums this summer, we will have wood all winter, to burn in the woodstove, returning the tree to the heat and light, the energy, which created the wood in the first place.

Energy is fire.

The Earth's Energy Flow

With a few very obscure exceptions, like thermophilic bacteria that live in hot vents of volcanoes and such, all of Gaia's life eats sunlight—or, more accurately, lives from the food created by green plants from carbon dioxide and water, using the energy of the sun. Some forms of life, the green plants and algae, use that energy directly to sustain their lives and feed their growth. Other things feed on the plants, and still others on the plant-eaters or even other animal-eaters. But in four or five steps at most, we go back to the sun. All of life is, in a sense, a transformation of energy into form. We are sunlight at a vast costume party, dressing itself up in one form after another, discarding one outfit to pick up a different one.

That sunlight is the great grace, the one free gift that allows not just sustainability, but abundance, growth, increase, more-than-enoughness. It falls on the planet daily, without our having to do anything about it, and billions of green plants are hard at work converting that energy into various other usable forms. So don't say no one ever did anything for you!

In traditional cultures, the shaman, priest/ess, Witch, or healer was responsible for making sure that the human community remained in an energetic balance with the environment around it. David Abram, in *The Spell of the Sensuous,* writes about the shamans he witnessed in Bali and Nepal:

> The traditional or tribal shaman, I came to discern, acts as an intermediary between the human community and the larger ecological field, ensuring that there is an appropriate flow of nourishment, not just from the landscape to the human inhabitants, but from the human community back to the local earth. By his constant rituals, trances, ecstasies, and "journeys," he ensures that the relation between human society and the larger society of beings is balanced and reciprocal, and that the village never takes more from the living land than it returns to it—not just materially but with prayers, propitiations, and praise.[1]

Today the world has become deeply out of balance. Part of our role as contemporary Witches and healers must be to restore that balance, and to do so, we need a deep understanding of how both physical and subtle energies work.

PHYSICAL ENERGY OBSERVATION

In your home base, breathe and ground and come into your senses. Now observe the energy exchanges around you. How is energy coming into this place? Who is its first user? What are the subsequent tiers on the cycle? Where is the energy being stored? How does it leave this place? How many levels of energy transformation, from sunlight to plant to animal to predator, can you see around you? How much of the available sunlight is being used?

Every transformation of energy and matter uses energy. When a plant transforms carbon dioxide and water into sugar, it uses energy from the sun. When a deer eats that plant, she gains energy from it but expends some in finding, grazing, and digesting the plant. Abundance, for the deer, means expending less energy to get food than she gets from the food she eats.

Because each transformation requires energy, some energy is used up as we climb each tier; less energy is available, and therefore each tier supports fewer individuals. There are far more grass plants than there are deer, for example, and far more deer than cougars and coyotes. The tiers of transformation form a pyramid, with its base the plants that directly convert sunlight, and its top the predators.

Some of the energy expenditure in the deer's grazing goes to maintaining the basic life processes and existence of the animal, beyond the energy needed at the moment to browse that particular plant. Every organism, and every thing created by human beings or other organisms, embodies a certain amount of energy.

As I look around the forest that surrounds this cabin, I see an enormous amount of embodied energy. Redwoods tower more than a hundred feet into the air, filling the space around me with wood and needles. Birds flit through, jays dive-bomb the cat food, squirrels leap from branch to branch. I'm sitting in a cabin built of wood and glass and plasterboard, with a cast-iron woodstove and a propane-powered refrigerator, and with chairs and a bed and books and a computer, all of which represent "embodied energy"—the energy and materials used to produce these things, transport them, and (in the case of machinery) get them up and running.

The energy embodied in the living things around me is called "biomass," the sheer weight and substance of biological life in a system. In this forest, the biomass extends up for a hundred to two hundred feet, and downward into the ground in the form of roots and the organic life of the soil. In an old-growth redwood forest, the biomass might extend three hundred feet above the earth.

Energy is embodied in this system, and also stored, held available for future use. I could, in theory, cut down the redwoods and burn them in the woodstove, releasing as heat and light some of the sun's energy that grew them. I won't do that—but I will thin the tanoaks for my winter fuel. I can take as much wood out of the forest as I need or want, without hurting it in the least, as long as I remain within the solar budget.

The solar budget is the amount of the sun's free energy transformed and stored by the forest each year, beyond what it needs to maintain itself and what it uses up in making the transformations. It's the only true profit margin that exists.

I can increase real abundance by increasing the amount of sunlight that is put to use. Compare an acre of redwood forest to an acre of parking lot. In the forest, trees three hundred feet tall are madly using that sun to turn air and water into needles, twigs, and wood, feeding billions of organisms. On the asphalt, all the sunlight is doing is creating heat and glare where they're not wanted. If I'm renting out the parking spaces to shoppers, the lot may produce a form of monetary abundance for me, but in an ecological accounting, paving that space is wasting a huge potential resource.

The key to sustainability and real abundance is to remain within the solar budget. As Mr. Micawber wisely says in the novel *David Copperfield*, "Annual income twenty pounds, annual expenditure nineteen nineteen six, result happiness. Annual income twenty pounds, annual expenditure twenty pounds ought and six, result misery."[2]

As soon as I exceed the solar budget, I start to degrade the system. Were I to clearcut the redwoods, for example, I'd be removing many years' worth of embodied energy. As well, the trees represent water, nutrients, and minerals from the soil. In a pristine forest, the redwood outside my window might grow for a thousand or two thousand years, but eventually it would fall and die, returning all those minerals and nutrients to the soil in a slow process of decay that might also take a thousand years.

If I remove all that biomass and stored energy, those nutrients and minerals, from this forest, it might still have enough resources to grow back over time. But in effect, I'd be using up many decades' worth of stored energy in a short time. I could certainly get away with this once, maybe even twice or three times (if I left long enough intervals in between). But over time, the forest would lose fertility, resilience, health. And indeed, that is exactly what has happened to this forest, which was logged at least twice in the past hundred years, long before I got here. The redwoods are still growing, still beautiful, but when I compare these woods to a stand of old growth, where trees the size of the stumps around me are still alive and flourishing, I can see just how much has been lost.

When we use fossil fuels, we are using up energy that was stored millions of years ago, and that cannot be replaced in anything short of geologic time. The U'Wa people who live in the cloud forests of the Colombian Andes have been resisting oil drilling on their traditional lands by a U.S. oil company, the L.A.-based Occidental Petroleum. They believe that oil is the blood of the earth and should not be disturbed.[3] Fossil fuels are an incredible gift from Gaia, a huge reservoir of potential energy, which we are squandering daily, at the cost of enormous pollution and destabilization of the world's climate. A sane energy policy would move us as quickly as possible toward reliance on renewable sources of energy. As part of our shamanic responsibility of maintaining and restoring a balance, we should be both advocating for such policies and doing what we can as individuals to reduce our expenditure of the earth's blood.

A Simple Energy System

The energy flow in living systems is huge and complex. To understand it better, it might help to look first at a relatively simple energy system—say, the one I'm using right now.

Here in my cabin, on this hot July day, my computer is powered by solar panels that directly convert the sun's energy into electricity. The solar panels, the batteries, and the whole electrical system represent a certain amount of embodied energy. But once in place, they make direct use of the sun's gift. Not only that, they are pretty reliable and relatively trouble-free, the batteries being their weak point. Batteries are like pets—they need to be kept fed (or charged with energy) and watered, and they can't be allowed to get too hot or too cold.

My solar electrical system requires some of my own energy and attention as well. It takes more observation and consciousness than the grid electricity most of us are used to. During the day when the sun is out, the panels are producing electricity and storing it in batteries. I can sit here and write all day on the sun's bonus and still fully charge the batteries that will power my lights and computer and maybe even a video or a CD at night.

But at night, I have only the energy stored in those batteries available. In fact, if I want them to last, I can't use more than the top 20 percent of the energy they store. So I must watch what I use carefully. In the city, I can leave lights on and the stereo running and the TV going in another room and it costs me just a few dollars extra at the end of the month; *here* it might cost me a $500 set of batteries. So I carefully monitor what I use and how full those batteries remain. In the fall, especially, when down here among the redwoods we don't get a full day of sun, my supply is marginal. When the batteries get too low, I must stop using power or find another source.

In the winter, when the rains come and our stream is running, we have a micro-hydro system that provides our electricity. We catch water upstream in a tiny dam not much bigger than a milk crate and run it through a "race," or pipeline, to a spot below the cabin, where it turns a wheel that generates electricity. Then the water goes back into the stream. We "borrow" the water; we do not remove it from the ecosystem.

The micro-hydro, when it's running well, puts out five or more amps of power—not a huge amount, but enough for our needs. Without going into the technicalities of electricity, on our twenty-four-volt system that's enough to run a laptop, power a few compact fluorescent lights, and play a CD or watch a video. Because the micro-hydro runs twenty-four hours a day (and has the batteries as backup), it provides more than enough electricity for our needs—so much, in fact, that the system includes a shunt to divert excess power lest it overcharge the batteries and burn them out.

You could say this system has three basic parts. First, there's the *input*—the energy that comes from the sun or the flowing water. How much of that we take in depends on how many panels we have, or on how much water is flowing in the stream, what volume of water we collect, and how much pressure the water is under when it turns the wheel. We could increase abundance, or available energy, by increasing the number of panels or the volume of water we collect.

The second part of our system is *storage*, which provides a buffer. The batteries allow us to collect excess energy when we have a lot and to make more than our input available when we need a lot. We could also increase available energy by increasing our storage capacity. In summer, when the sun is high, and winter, when rains fill the stream and our hydro system is working, we undoubtedly produce more energy than we currently store. A second set of batteries would effectively make twice as much energy available to us. But in fall, when the hydro stops running and the sun is too low to fully charge the batteries we have, more storage wouldn't help us. And because batteries need to be kept charged to maintain their lives, having more than we can keep fully fed all year might become a liability.

The third part of our system is *output*—the use the energy is put to. We can also create more abundance by reducing our output—reducing, for example, any use that doesn't directly contribute to our abundance and quality of life. Turning the lights off when we're not using them, eliminating the "phantom load" of appliances that draw small amounts of power even when off, using compact fluorescents instead of incandescent bulbs—all these steps make more energy available for what we truly need and want.

The intake, storage, and output of energy in my system are regulated by several little gadgets that include some form of self-regulating cycle. For example, when the sun has fully charged the batteries, the charger shuts off so that they don't overheat. When the batteries are drawn down, the device again allows the sun to charge them.

Not everybody is going to run out and install an alternative electrical system, although doing so might be the best investment you could make if you own your home. In California, the state has offered a rebate on systems that are connected to the grid; the rebate pays nearly half the cost of the panels. Combined with the current low interest rates in 2003, when we began this project, our collective in the city was able to refinance the house, lower our payments, install solar panels, and still have money left over to improve insulation and help reduce our energy use.

Generally, conservation (or reducing waste output) is the easiest, most efficient, and least expensive way to generate more abundance. Increasing input would require adding more solar panels or "borrowing" more water. Increasing storage would mean adding more batteries. Both involve additional embodied energy, costly to the planet and to the bank account. But decreasing waste often requires no new material or infrastructure, no more embodied energy.

We don't all own homes or have the option of installing solar panels or a hydro system. But we can all conserve energy. Often we don't because we don't think a single lightbulb makes a difference, or because we remember our parents nagging us to turn off the lights and we still feel resentful decades later, or because we are unaware and unconscious.

But as Witches, we need to be aware of the energies around us, on every level. Paying attention to the amount of electricity we consume will help us develop awareness of the more subtle energies around us. Since part of our role is to be guardians of the balance, we will have more personal power if we cherish the balance even in small ways in our own lives. For personal power derives from integrity—the unity of our actions with our values, our speech with our acts. Each lightbulb left on represents, in a sense, a small leakage of our own personal power. Every time we turn off one of those unused lights, we are building our own store of power.

Subtle Energy

From my journal:

I'm at my home base spot, trying to observe the energies, but I'm being distracted by Tigers and Bears, the left-behind cats of a neighbor who recently moved out of the area. Originally there were three of them, and because one was gold, one was tiger-striped, and one was just big and cuddly, I called them Lions and Tigers and Bears. I've been feeding them, but they want more. Tigers is butting my legs and purring until I take him up on my lap to stroke and cuddle and scratch behind the ears. I try to brush the burrs out of his fur, but he butts my hand and the brush, just wanting to be petted. The cats aren't hungry—there is food left in the bowl. Tigers just wants attention, energy, love.

Besides the physical and electrical energies that science can measure, there are other, more subtle forms of energy. Witches know that this energy is as vital to life as the more tangible forms. We all need love, attention, and energetic support to thrive.

Many other cultures have words for this energy: the Chinese call it *ch'i*, the Hindus *prana*, the Hawaiians *mana*. Many non-Western healing traditions are based on the understanding of subtle energy and how it impacts our physical and emotional health.

Witches have also worked with these energies, for healing and in magic. We can learn to observe and shift subtle energy—indeed, that is a basic core practice of magic, a part of opening to the wider perceptual world around us.

SUBTLE ENERGY OBSERVATION

Conduct this observation with a partner. You can also do it in groups, by pairing off.

Sit with your partner for a moment. Hold hands and match your breathing. Slowly let yourself become aware of how much you know about this person, just by being with her or him. What information are you receiving? How? Which senses are you using? What subtle senses or feeling beyond the physical come into play?

Now open your eyes. Gaze into your partner's eyes and become aware of all the information you are now taking in.

Close your eyes, and return your focus to yourself. What changes?

Now bring one hand up in front of your mouth and feel your breath. Bring your hand out to the edge of your breath. Become aware of how subtle the edge is where your breath becomes part of the overall air. It may be just a sense of heat or moisture.

Opening your eyes, turn your hand around to feel your partner's breath. Again, move your hand out to find that subtle edge.

When your hand is at the edge of your partner's breath, you'll be at about the edge of her/his energy field—what we call the "aura." Continue moving your hand around until you can feel that subtle edge around different parts of your partner's body. Again, what you feel will be subtle, a sense of heat or tingling or just an urge to stop at a certain place.

Take time to explore each other's auras, to notice where the energy is stronger or weaker, hotter or colder, to be aware of any emotions or images that arise for you. Share what you perceive with your partner, to check if your perceptions make sense to her/him.

When you are done, shake out your hands to release any energies that may still be clinging to you. Close your eyes, say your own name, and think of five physical differences between you and your partner.

ENERGY BRUSHDOWN

A brushdown is a very simple way to do a quick energetic cleansing, perhaps to release negative energies you have picked up from others around you.

If you are working with a partner, first chop up your partner's aura by moving your hands swiftly through it. Then use your hands to pass through it repeatedly, combing out and flicking out any energies that don't belong. Then fluff, using a motion somewhat like fluffing up a curly hairdo. Your partner can then repeat the process on you.

You can also do a brushdown on yourself, if possible over running water to carry away any negative energies. I often do it in the shower or, in a pinch, over a toilet. Shake out your own hands after.

Volumes have been written about subtle energies. When you explored your partner's aura in the subtle energy observation, you may have noticed areas of greater energy density or concentration. These energy vortexes, or chakras, are key in many systems of healing and meditation. Many good resources exist for those who want to explore them further.

But the most important magical teaching about subtle energies is that energy follows intention. To direct energy, we must know where we want it to go, what our goals and intentions are. To accomplish anything, whether it is creating a ritual or changing the world, it helps to begin with an intention, and to monitor whether the energy you are putting out actually follows that intention.

Remember your sacred intention from Chapter Two? Just by articulating that intention, you have already begun to bring your energy into alignment with it. Now let's look at that intention from an energetic point of view.

SACRED INTENTION AND ENERGY EXERCISE

If you do this exercise with a partner, one can monitor the other partner's aura as she or he goes through the exercise. Then switch roles.

In a quiet place, ground and come into your senses. Close your eyes and think about your sacred intention. As you do, notice the quality of your own energy.

Remember your sacred intention and state it to yourself.

Think about a choice or decision you made recently that was in alignment with your sacred intention. How did you feel inside? As you remember that feeling, what happens to the quality of your energy? Is there a symbol or image or word or phrase that you can identify with that quality, to anchor and recognize it?

What was the result of that choice or decision?

Now think about a choice or decision you made recently that was *not* in alignment with your sacred intention. How did that feel? How did you feel inside? As you remember that feeling, what happens to the quality of your energy? How is that feeling different from the feeling that the previous choice inspired? Is there a symbol or image or word or phrase that you can identify with that quality, to anchor and recognize it?

What was the result of that choice or decision?

What options did you believe you had at the time?

Now use the magical tool of anchoring to your core self/baseline (from Chapter Four)—that combination of image, touch, and word or phrase that brings you to your neutral, baseline state. When you are grounded and anchored, think about the options that you felt were available to you. Do they look any different now?

Now think about the ways you are using energy in your life right now. Which are aligned with your sacred intention? Which are not aligned?

Is there a decision you are facing right now? Think about your options, each choice you might make. Does it feel like alignment or nonalignment? Does it evoke either of your earlier anchoring images? What happens to your energy if you make this particular choice?

Breathe deep, open your eyes, and relax. If you've been working with a partner, switch roles and go through the exercise again. Monitors should be sure to shake out their hands after monitoring, to release any energies they may have taken on. Now take some time to discuss what you felt and observed.

Use this exercise to help you monitor your use of energy and your evaluation of choices. As you bring your energies and decisions into alignment with your sacred intention, notice what shifts in your life, how much energy you have available to you, and what drops away. Keeping a journal can be very rewarding in this process.

Form Follows Energy

Another important magical teaching is that form and matter follow the flow of energy. In our materialistic culture, we're given the message over and over that our energetic and emotional state is conditioned by what things we possess. "If only I had a nicer kitchen," we think, "I'd stay home and cook dinner and we'd eat together as a family and feel closer."

Magic teaches us to use our "will"—our ability to consciously choose to act and direct our energies toward our intention. If our intention is to feel closer as a family, then we would choose to stay home and cook even in our old and grungy kitchen. Being there more, we might clean it up, repaint it. With the money we saved by cooking at home, we might be able to afford some new appliances or a window in that dark wall. The form of our kitchen would begin to reflect our intention and the flow of our energy.

Will is an empowering concept. It teaches us that we can make change in our material circumstances, if we put our will, intention, and energy into alignment. Sometimes that requires faith: we can't always foresee how material circumstances can change in the direction we want—and maybe they won't. But *we* will have changed, and be acting in closer alignment with what we truly value.

That kitchen might remain dingy, but we have a closer, more harmonious family.

Will is not willfulness, not whim, not mere intention to get our way. It's our ability to act "as if," to set out even if we can't foresee every step of a journey. The more we exercise our will, the stronger it gets; and the stronger our magical will, the more we are able to serve what is sacred to us, thereby realizing our deepest dreams and desires.

AWAKENING WILL EXERCISE

Consider again what is sacred to you, and how you are using your energy. Now think for a moment about your current wish list—those material conditions or objects that you think will improve your life.

Is there something you want or desire that you have unconsciously made a condition for aligning your energies and actions with your sacred intention?

Is there some act you could take, some shift you could make, toward your intention regardless of your current circumstances?

Breathe deep, focusing on your solar plexus, the energetic seat of will. Imagine filling that center with a beautiful golden light.

Say, "I honor my magical will, my ability to choose. My will is strong, and I choose to do ________________ to further my sacred intention."

Now *do* it. Do it regularly, even for short times and in a small way. Instead of saying, "Oh, if only I had money and didn't have to work, I'd write," write every day, even if it's only a page in your journal. Do it for at least a full moon cycle. Then see if any of the circumstances of your life have changed.

ASSESSING SUBTLE ENERGIES IN NATURE

Plants, trees, animals, and whole ecosystems have auras, or fields of subtle energy, just as people do. When you've become familiar with a human field, turn your attention to the energy around you. Take some time and feel the energy fields of trees, plants in your garden, your cat or dog. What do you notice about them?

Each time you do the meditation of coming into your senses, include your subtle senses, that ability to sense subtle energies.

How do the subtle energies differ in the city? The wilderness? The forest? By the ocean? What happens to your own energy field in these places?

Like individuals and single objects, places have energy fields as well. In your home base, close your eyes and extend your subtle awareness out, not to any specific object but to the whole of the place you are in. What does it feel like? Just as each person has a distinct odor that a dog can smell, each place has its own energy signature. Is there a symbol, a pattern, an image, a word that describes the quality of this place? How would you recognize its energy?

Open your eyes, come into your senses, and with eyes open, extend your more subtle awareness. Does anything change?

HOME ENERGY INVENTORY

Electricity and other forms of overt energy have an effect on our more subtle energies. As we develop awareness of the energies in nature, we also need to become aware of how the energy fields around us may impact us.

Begin in one room in your house, perhaps your bedroom. Look around and notice everything in the room that uses energy: an overhead light, a lamp, a heater, a clock radio, a computer. Where does that energy come from? What does it cost? In money? In environmental impact? Does the energy's production or delivery create waste products or pollution? What use are you making of the energy?

Are there appliances that stay on all night? Do they create sound or light? How much background noise do they produce? Does their sound or light impact your sleep?

Now use your ability to sense subtle energies. Explore the energy field around your computer, your clock radio, a lightbulb when it is on and off. What differences do you notice?

Turn the lights and other equipment on. Close your eyes, stand in the center of the room, and notice the quality of the energetic field these things create. Now turn them off and notice the difference.

Move on to the kitchen and repeat the questions. Have you learned to associate the hum of the refrigerator with food and home? How many lights are still on when you turn everything off? Do you have a microwave always ready for action? Move on to the living room, the home office. Notice the VCR, the stereo, the cordless phone, the fax, all the helpful electronic gadgets that never turn off. Everything that has one of those little black transformer boxes at the plug produces a phantom load.

How do the sounds, the lights, the constant energy use affect the whole field of the room? How do you feel in it?

Now turn everything off. If you can, even unplug the refrigerator for a moment. What changes? What sounds and feels different? How does this affect the subtle energy field of the room? Your own energetic field?

Consider ways to turn more things truly off when they are off. Stereos and computers are best plugged into surge protectors, and the switch can simply be turned off when you're done. If you are rewiring a home or building a new one, the electrical plugs can be wired to a master ON/OFF switch by the door.

As you become more aware and more sensitive, you may find that the continuous electrical stimulation that permeates modern life becomes first noticeable, then irritating. If you spend long hours at work in front of a computer, when you come home you may want at least one room to be a haven where you can turn everything off for a time. Your bedroom is especially important. Protect your sleep. Don't keep a clock radio right next to your bed—in fact, if you can, do away with it altogether and wake up by some other means. If you keep a telephone in your bedroom, make it a simple one, with no black box attached. Turn your computer and printer off before you go to sleep, and charge your cell phone in some other room. Be especially conscious of this if you have trouble sleeping or have an autoimmune disease or other disorder that drains your vital energy. If you think this whole idea is ridiculous, try it for a week or a month and notice what changes.

Becoming aware of energy use is the first step in reducing it. The energy we use at home rarely comes from a benign source. If you don't know how your utility company produces energy, find out. Does it come from burning coal? Damming a river? Running a nuclear power plant? If you know what you are implicated in by using energy, you may be motivated to use less.

Conservation is important, for bringing our own lives more into alignment with our values, for how it can impact the energetic fields of our own homes, and because small changes do add up and make a difference. But for real change to occur, we need changes in public policy. Our collective wouldn't have put solar panels on our house in the city were it not for a public policy that made it financially possible. The more we, as individuals, understand how energy works, and the more we know about alternatives, the stronger advocates we can be for those changes in policy that will help bring us into balance. It's worth burning that light an extra hour, or firing up that computer again, to write a letter to your representative urging more funding for solar panels and less for oil wars. And I will personally absolve you of a dozen left-on lightbulbs if you organize a group to pressure your utility to subsidize solar or other clean power.

As Witches, we are always working with both overt and subtle energies, increasing our ability to be aware of both of them, and eventually learning to shift and change them. We consciously "run energy"—move subtle energies through our own channels and energy field.

That kind of work *uses* energy too, and we need strong sources of the vitality found in nature in order to remain vitalized and healthy. Over the years, I've become concerned as I've seen many Witches, healers, therapists, and other spiritual teachers develop immune system disorders, chronic fatigue syndrome, and other illnesses. In order to work with energy as we do in the Craft or in other healing traditions that derive from indigenous and shamanic practices, we need a strong, ongoing connection to the natural world. These techniques come from people who were immersed in the natural world and attuned to other consciousnesses and communications. Early practitioners of the Craft didn't include in their instructions, "Go outside regularly, garden, and eat local food in season," because they didn't need to, any more than I would think of advising you to bathe regularly. I assume you do—since just about everyone in our culture does. Likewise, in the past just about everyone spent time outdoors in an environment that was more highly vital and less degraded than ours, grew some of their own food, and ate other food that was locally grown and in season because there weren't many other options.

But today we are often working with techniques of moving energy or awakening consciousness that have been removed from their original matrix of a vital, natural world and distanced from their original function of maintaining the balance of energy flow between the human and nonhuman worlds. As David Abram writes,

> Shamanism has . . . come to connote an alternative form of therapy; the emphasis is on personal insight and curing. These are noble aims, to be sure, yet they are secondary to, and derivative from, the primary role of the indigenous shaman, a role that cannot be fulfilled without long and sustained exposure to wild nature, to its patterns and vicissitudes. Mimicking the indigenous shaman's curative methods without his intimate knowledge of the wider natural community cannot, if I am correct, do anything more than trade certain symptoms for others, or shift the locus of *dis-ease* from place to place within the human community. For the source of stress lies in the relationship *between* the human community and the natural landscape.[4]

Abram may be a bit categorical: I actually think many of these techniques can be very useful in human insight and healing. But when they are practiced outside of a relationship with the natural world, they pose a danger of devitalizing the healers as well.

Your best protection, and an important aspect of your own healing, is to spend time in the natural world, preferably in a place that is high in both physical and subtle vitality. Those old Victorians had the right idea when they sent invalids to the seaside for a cure. The more time you can spend in nature, grounded and fully in your senses, opening up to its communications and gifts, the more vitality you will have to draw on. In my own life, consciously adopting a practice of observation and connection with nature has made an immense difference in my health and energy. I have far more energy now, at fifty-two, than I did ten or fifteen years ago, I enjoy better health and fewer bouts of the flu, and I can tolerate physical discomfort, danger, and intensity much more easily.

Besides observing the energy around me, I also consciously use it as a source and take it in. Below is a way to begin that practice.

DRAWING ON A HOME-BASE ENERGY SOURCE

In your home base spot, ground and come into your senses. Now activate your more subtle awareness. What energies can you feel around you? Are there energies that can feed you?

Close your eyes for a moment and feel your grounding cord. Ask the energies and spirits of your place for help and support, and breathe energy up through your roots, filling your body and aura.

Continue to breathe in, imagining that you can absorb some of the energies of the place around you through your aura. If you sense any weak spots or deficient areas in your own energy field, consciously fill them.

Breathe down energy from the sun, moon, or stars—whichever are out. Imagine that energy filling your aura, taking it in through the leaves of your own tree.

Thank your place and its energies, and open your eyes. Do this regularly, at least once a day.

You can also use this same practice away from your home base, in any place of high vitality or whenever you need to renew and replenish your energy.

GIVING BACK

"We take from the earth and say please. We give back to the earth and say thank you," says Julia Parker, a Kashia Pomo.[5]

Since our goal as Witches is to maintain and restore balance, we can't just take energy without returning something to the place and powers that feed us. There are many ways that we can give back.

Gratitude and Thanks. Just expressing our praise, thanks, and appreciation is a way of returning energy.

Prayer and Song. A chant, a liturgy, a song, a spoken prayer are all ways of giving back. Try making up a special song for your place and singing it each time you return, or choose a particular chant you know that seems to fit.

Physical Offerings. Many cultures give back to nature by offering something tangible: tobacco, ti leaves, milk. Try leaving a physical offering of some sort in your home base each time you go there. Be aware that the distinction between an "offering" and "garbage" is sometimes subtle. I prefer sacred water as an offering, as it leaves nothing that can be construed as a mess.

Actions. Actions and activities can be consciously dedicated as offerings. Maybe turning off that lightbulb can be an offering you make to the spirits of the land. You can dedicate an action by saying something like the following: "Spirits of this land, I am turning off this unused light out of gratitude for the gift of life—the sunlight that feeds all things. May the balance be restored. Blessed be."

Energy in Living Systems

Living organisms also store energy to be released as needed. If humans had no means of energy storage, we'd have to eat nonstop just to keep breathing.

We have self-regulating cycles that help monitor our intake and storage of energy. When our supply of food is low, our blood sugar drops and we feel hungry. When we eat, our blood sugar rises and we feel full.

So, too, the larger ecosystems of the earth self-regulate. Those tiers of transformation of energy, from sun to plant to grazer to carnivore, function as one self-regulating whole. On the pristine prairies of the pre-pioneer Midwest, an incredible diversity of plants used the sun's energy to make food. Their biomass extended deep underground, in root systems sometimes twenty feet deep. Herds of bison passed through periodically, grazing the grasslands, removing growth that would otherwise accumulate and choke new shoots, dropping their fertilizing dung, churning up the ground with their hooves so that rain could soak in and new plants could get a start on life. Wolves hunted the bison, culling the herds and taking out weak and sick animals. Fear of the wolves kept the bison bunched together and on the move. The surging throng of huge animals gave the plants and ground the periodic disturbance needed for health, moving on often enough that the bison did not overgraze and destroy the grasses.

Remove any part of this cycle, and you disturb the balance and flow of the whole. The prairie needs the buffalo, and the buffalo needs the wolf. Predator and prey are not at war with each other; they are part of the same system of relationships that are mutually beneficial, if not to individuals, then to the species and ecosystem as a whole.

Many environmentally conscious people choose to be vegetarians or vegans, to eat from the lower tiers of the energy pyramid. This is certainly a conscious and moral choice. The way most meat is raised today, in factory feedlots or horrific pig and chicken concentration camps, is a travesty on both ecological and compassionate grounds, and I try not to eat such meat.

But I do eat meat. I realize that this admission is likely to bring me more irate mail than even my stance on Palestine or the Iraq war, so let me state my case.

I eat meat because in thirty-five years of offering the Goddess every possible opportunity to tell me what to do, she has never suggested I become a vegetarian. On the occasions when for whatever reason I've eaten a vegetarian diet for any length of time, I've become devitalized and sick. Many people thrive on a vegan diet, but many others don't. Our bodies are different. Women, especially those of us who are prone to anemia and low blood sugar, may need a higher protein diet that includes meat. A vegetarian diet has been associated with chronic fatigue, especially in women. Meat is higher in what Chinese medicine calls *ch'i,* or vital life force, and those of us who are working with subtle energies in the way Witches do seem to crave it. Certainly I've been at many vegetarian Witch camps (week-long magical intensives) where those of us who were teaching went to great lengths to make sure we had our own supply of cold chicken in the back refrigerator, just to keep from keeling over with exhaustion by the end of the week.

But let me also make the environmental case. One of the arguments for a vegetarian diet is that it takes many pounds of grain or grass to produce a pound of beef or lamb or chicken. In terms of factory farming, that's certainly true, but I'm *not* defending our current industrial agriculture system.

With two billion hungry people in the world (the argument goes), shouldn't we be using the grain that farm animals eat to feed the masses rather than produce steaks for the rich?

That argument sets up a false choice. First, most of those hungry people are hungry not because food for them doesn't exist, but because the inequalities of our current economic system have left them unable to pay for it, or because the infrastructure and transportation don't exist to get that food to them.

Second, it assumes that land that is used for raising meat could be used, instead, to grow grain or soybeans. But not all land is suitable for growing crops. Drylands, wildlands, steep hillsides, and other areas that are limited in water and access, and that are easily eroded, could be devastated by attempting grain production on them. Indeed, that has happened in much of the original shortgrass prairie lands of the American West. Where I live, a soybean or wheat farm would be an environmental and financial disaster. But my neighbors Jim and Dave raise sheep and goats on a small scale with great sensitivity to the land, and I'm happy to help support their sustainability by buying and eating some of what they produce.

But, you might ask, don't sheep and cattle degrade and erode the land? They do if they are not managed well. However, with sensitive management, they can also improve the land. South Dakota rancher Jeff Mortenson, a student at one of our Earth Activist Trainings, has restored much of his ranch by mimicking with cattle the animal impact of the ancient bison herds. He grazes more cattle per acre than his neighbors, but keeps them concentrated for short times in areas and gives the plants time to recover. Native shrubs, grasses, and streamside vegetation have returned to his land, and its overall health is vastly better than that of neighboring ungrazed government lands.

Jeff follows a system detailed by Allan Savory in *Holistic Management.*[6] I strongly recommend Savory's book, not just for those who are actually interested in raising sheep or cattle, but also for anyone interested in improving overall decision-making. Savory makes a convincing case that land in climates that are "brittle"—that have dry periods when water is less available and organic materials do not break down easily—require animal impact for their health. Overgrazing will harm them, certainly; but so will lack of any grazing. But pastureland can be restored and revitalized by animal impact that mimics the effect of large herds moving in response to predators—the pattern with which grasslands coevolved.

Much of the hunger that exists today in the world is located in just such environments—brittle climates where the process of desertification is progressing rapidly. Contrary to the vegetarian argument, ending hunger might actually require meat production, but only production carried out in conjunction with nature and with respect.

Of course, many people choose to not eat meat because they don't want to kill another sentient being. I am certainly not going to argue with that choice, but my morality is different. Sheep, cows, chickens, and our other domesticated animals have now coevolved with human beings for ten thousand years. Were we to stop eating them, they would not be living happy, productive, fulfilled animal lives; they would cease to exist as species and become extinct. They could no longer survive in the wild. Most have lost the ability to protect themselves against predators and to find food without human help. Others, such as pigs that revert to a feral state, have a devastating impact on the environment. Yet our domesticated animals are also a precious source of genetic diversity, in the thousands of local breeds that have been developed for specific conditions over thousands of years. Their mass extinction would be a tragedy. Humans and our domesticated animals now form a whole, in which we function as top predators, and sheep and cows and chickens need us as we need them.

That doesn't give us the right to confine them in cages for their entire lives, or to condemn them to a miserable existence in meat factories. But if we raise them in humane conditions—conditions that give them a chance to do the things sheep and cows and chickens like to do, that integrate them into a healthy environment and help them perpetuate their kind—if we make sure that they are killed with a maximum of respect and a minimum of suffering, and if we give gratitude to their spirits for their lives, then I am happy to eat them. Death is a part of life, for all of us. It cannot be avoided, for humans or chickens or pigs. When my time comes, I would rather be eaten by a cougar or disposed of by vultures than embalmed and stuffed in a box. But however I am disposed of, something will eventually eat me. And so the cycle goes on.

Loving the Body

What we eat is tied to how we treat our bodies. The Goddess traditions celebrate the body and value life in this world. Our fleshy, living, breathing, eating, and excreting bodies are sacred, the matrix of form for the unique energies we each are. Treating our bodies well is one way we honor the Goddess and show respect and gratitude for the gift of life.

For many people, self-care around food seems to focus on what *not* to eat—meat, wheat, dairy, etc. But to increase the energy we receive from our food and strengthen our own vital energy, we need to think about what we *do* and *should* eat.

Some years ago, when a group of us had just acquired land in the coastal hills of northern California, I was meditating in our garden. I was questioning my own driving need to garden in an environment that was already beautiful and needed no improvement. The garden said to me, "Grow food. I want you to grow food—because if you eat the food that comes from the land, you will *be* the land."

When we eat something, we literally take in the minerals and energy of the place where it was grown. In an indigenous culture, almost everything people ate came from the land they lived on. Their bodies were literally made of the same stuff as their land. People downriver or over the hill would have smelled different. Myth and religion reflected this close identity. In Mayan mythology, for example, people were made of corn.

Jeanette Armstrong, Okanagan writer and teacher, writes,

> The Okanagan word for "our place on the land" and "our language" is the same. This means that the land has taught us our language. The way we survived is to speak the language that the land offered us as its teachings. . . . We also refer to the land and our bodies with the same root syllable. This means that the flesh which is our body is pieces of the land come to us through the things which the land is.[7]

To eat, then, was not just to take in a set of chemical nutrients. It was to be in profound relationship with a place—with the energies, elements, climate, and life community of that spot on the earth—to ingest the place and become it.

Everything you ate would also be something you had a relationship with, that you had yourself grown and tended, gathered, or hunted. All those activities would be sacred—that is, highly valued and marked by ritual and prayer and ceremony, by offering gratitude and respect for the life-forms you were culling. All food would be harvested, hunted, and prepared by people who were consciously putting themselves into a thankful and loving energetic state, and the food itself would carry that energy.

Today, few of us have that kind of relationship with place. The food we eat has often been transported halfway across the world. It may have been grown and picked and processed by people who were exploited and suffering. It may have been doused with poisons, grown in dead and devitalized soil, irradiated. If it is factory-farmed meat, we are eating confinement and torture with every mouthful. If it is fast food from a chain, it may have been prepared by underpaid, exploited, and resentful help under factory-like conditions. Overall, far more energy will have been consumed in its production and transportation than we will get by eating it.

As restorers of the balance, and as lovers of our own bodies, we need to be conscious of what we eat. The foods that we need to maintain our own vitality are also what will help restore the balance. Rather than thinking about what we shouldn't eat, let's think about what we can and should eat to increase our health and vitality.

First, we can eat high-vitality food. This means food that is organically grown, minimally processed, and locally produced (so that it doesn't lose its vitality on its way to you). If you think you can't afford to switch to an all-organic diet, think about what you can add rather than feeling guilty about what you don't want to give up. Can you substitute one organic fruit or vegetable a week? Can you take a bit of extra time to go to the farmer's market, where organic food is cheaper and you are also supporting local small producers? Can you join a local CSA, which stands for Community Supported Agriculture[8]—where for a low price you directly support a local farmer who will deliver a box of fresh fruits and vegetables each week?

We can primarily eat food that is in season. Such food is generally cheaper, can be grown locally, and keeps us in touch with the cycles and changes of the year. We can eat peaches from our garden or the farmer's market in summer, not Chilean peaches in the dead of winter. We can eat asparagus in the spring, and apples in the fall. Of course, if we live in an area with cold, snowbound winters, we will have to import some of our food, but we can still choose to eat locally through many seasons of the year.

We can grow some food of our own. Even if we grow only a small amount, growing and eating food from our place will establish a relationship with it. If nothing else, grow a few herbs or some mint on a windowsill, and use them in rituals and for tea. If you have never gardened or think you have a black thumb, consider that there are basically two kinds of plants in the world: the ones you can't grow and the ones you can't kill. Find some of the ones you can't kill in your area, and grow them. Many of the herbs and medicinal plants fall into this category, as they derive from weeds and by nature have high vitality and resilience. Ask some advice from your gardening friends. And if you can grow some of your own food, you'll find that the vital energy in a salad picked fresh from your garden is far greater than anything you can buy. The taste is also incomparable.

Cook and prepare food with love and gratitude, as a conscious meditation and offering. Karin, who heads the cook team in our Earth Activist Trainings and at the California Witch Camp, is a Buddhist who puts love, compassion, and joy into all the food she prepares. The kitchen is always a happy place to be, and the cooks are generally dancing to good music as they chop and stir. When we eat Karin's food, we feel nourished spiritually and emotionally as well as physically.

COOKING AND EATING WITH GRATITUDE

Cooking can be an action you offer up: "I give gratitude for the gift of life, for the lives of all the creatures in this food. I offer up this work of preparation as a gift of love and nurturing to all who will eat it, and a gift of thanks to all that has provided it. May the balance be restored."

Stop and give thanks and bless the food before you eat it: "I give gratitude for the gift of life, for the sunlight and the plants that use its energy to turn air and water into food, for all the beings who gave their lives to this food, to all who grew and tended and picked and transported and prepared and brought it to us. May the balance be restored. Blessed be."

Ground and come into your senses before you eat. Eat consciously, savoring the taste and noticing the energy of what you take in.

One of our primary energy storage systems in the body is fat. As humans, we have laden our flesh and fat with huge symbolic meaning. In many early Goddess cultures, as in Hawaii and Polynesia today, fat is seen as desirable, a symbol of power, a literal sign of stored energy. The earliest representations of the Goddess, from the old Stone Age, show fat women with huge breasts, bellies, and buttocks.

But for women today, in our fat-phobic culture, loving our bodies can be difficult. All around us are messages telling us that fat is ugly, undesirable, unhealthy, and a sign of weakness of character. The standard of beauty is lean and mean, and even our natural curves and softness are suspect.

Fat can certainly be a sign of imbalance—of a diet that is not truly nourishing, of lack of exercise and physical vitality. If you don't exercise, you should find some form of physical workout that fits your body and your routine, not to lose weight but to maintain the high level of physical vitality you need to work with more subtle energies. If you are not eating vital, nourishing food, you are putting your overall health at risk.

But fat can also simply be how your body metabolizes and stores energy. If you are vital, active, healthy, and careful about what you eat, and are still fat, that may be the way your body wants to be. Like redwoods, as we get older we get thicker. Hating our bodies, depriving ourselves, judging ourselves, trying out one diet after another—these are far worse energy drains than leaving on the lights. Instead of those negative perspectives, try the following meditation.

BODY LOVE MEDITATION

(Especially for Fat Women.)

(Okay, Fat Men Too.)

(Oh, all right, you don't have to be fat to do this.)

For this exercise, you need a full-length mirror and a bowl of salt water. Use only a few grains of salt, as you will eventually give the water back to the earth.

In a private place, ground and come into your senses. Take off your clothes, and look at your body in a mirror. Feel its curves, the fullness of your breasts, the roundness of your belly. Become aware of all the feelings that arise in you. Do you love your body? Hate it? Admire it? Are you ashamed of it? Want to change features of it?

Just acknowledge honestly all that you feel and think about your body. Ask yourself, How much of my energy is tied up in feelings or frustrations about my body?

Place the bowl of salt water at your feet. Breathing slowly and deeply, imagine all that energy, all of your thoughts and judgments, draining off of you into the bowl.

Now look at your body again. Going slowly, from your feet up to your head, look at every part that you might have felt judgment about, and bless it. See your curves as stored sunlight. Acknowledge and bless the stored energy in your body. Thank the plants and animals that transformed that energy into food for you. Thank the sun.

When you are full of gratitude, lift the bowl of salt water and breathe into it. Imagine filling it with sunlight and gratitude, transforming the energies you've poured into it.

When you are done, take some of the transformed salt water and anoint any places on your body that you need extra help in blessing. Then pour out what is left onto the earth. (If you are worried about too much salt affecting plants, dilute the solution even more before pouring it.)

Fire Ecology

In addition to the sun's fire and the subtle energies, fire itself is a purifying and renewing force—and also a destructive force. Fire is the first tool human beings used to alter the environment around them on a large scale.

Fire can be an energizing and renewing force in a landscape. Here in northern California, we live in a fire ecology. Native American tribes regularly burned the forests and grasslands, to keep them healthy and diverse.[9] Fire creates a mosaic of habitats. A wildfire burns at different intensities over different patches of ground, so when it passes through, it leaves a variety of different habitats and stages of succession intact. In southern Oregon, at an Earth First! gathering last May, we camped on the edge of a recent burn. Below us, we could see patches of scorched earth, patches of undamaged trees, lightly burned areas, and areas where all was black and dead. Each would provide slightly different conditions for plants and birds and animals, increasing the area's diversity and the extent of its "edge."

When our forests were burned regularly by Native Americans, they remained open and fires stayed low and relatively cool. In fact, regular burning prevented the enormously destructive wildfires we often experience today. Fire kept the meadows open, so the deer and elk could find good grazing. Burning also killed destructive insects and disease organisms, rejuvenated many plants (providing long, straight new shoots for basketweaving), and fertilized the soil.

But fire is destructive as well as benign. Even on pristine lands, fire lays bare the soil, increasing the likelihood of erosion. Fire can transform massive amounts of biomass into dead sticks and ash, leaving an area with a reduced capacity for transforming the sun's energy. And fire, of course, can easily get out of control.

So fire is not a tool we want to use lightly or automatically. In our part of northern California, where houses now dot the landscape and the forest is burdened with a huge fuel load, we can't easily or carelessly reintroduce regular burning. We are even careful not to use our woodstoves on a cold day in the dry season. Wildfires are a serious danger all summer long, because they ignite so readily. Fires have been started near us by a mower blade hitting a rock and setting off a spark, and by an abandoned bottle of water acting as a magnifying lens in dry grass. Fire is truly our teacher.

The Lessons of Fire

Fire teaches us awareness of what we do and what we leave around. Fire teaches us responsibility and mutual dependence. Fire teaches us nonattachment: we all know that we could lose our homes and possessions at any time, and it's difficult to get fire insurance up here.

Fire also teaches us about community. Our fire protection comes from a volunteer fire department, as it does in most rural areas in the United States. Men and women in the community volunteer many hours each week, without pay, to be on call in case of fire or medical emergency. They also devote many unpaid hours to trainings and meetings. If someone up here gets hurt, it's one of our neighbors who will respond. Those of us who aren't in the fire department also volunteer time and energy to raise money to support it.

Though fire can be wild and untamed, the hearth fire is a great symbol of home. It's where we cook our food and share our meals. It's the center of the fire circle, around which the community gathers for ritual, for council, for storytelling.

As we learn to work in community, fire also teaches us about our own passions and emotions, about sustaining anything that requires excitement and enthusiasm. To learn how to start a movement or spark a creative project, practice building a fire.

Building a Fire

A fire needs three things to burn: fuel, heat, and oxygen. And those things must be present in the right relationship to each other. The Creighton Ridge Fire of 1978 began on a day when the ambient heat was so high, the fuel load of dead grass so dry, that a spark from a mower ignited the grass. On a cool day, that would not have happened.

If you've ever hastily tried to build a campfire or light a woodstove and failed to get the fire going, you know something about what is necessary to start any project. Consider:

What is the need you are trying to address, the problem you are trying to solve?

What is your fuel? What is the community, the set of resources, the range of conditions that you start with?

What is your heat? What is the passion, the enthusiasm, the burning desire?

What is your oxygen? What is the fresh air, the idea, the insight, the clarity?

To build a fire, you begin with tinder, something that is easy to light—paper or pine needles or dry twigs. Who in the community will be first to catch fire with enthusiasm? Can you fire up a small group first, who will help spread the idea to others?

A fire that is only paper will soon burn out. To get a lasting fire going, the tinder must ignite the kindling, the smaller sticks chopped into thinner pieces. Once your small group of original enthusiasts is excited, how do you inspire more people to get involved? How do you keep the structure open enough for air—new ideas and insights—to get in?

To burn for a long time, to heat a room or provide a center for a council, a fire also needs some big logs. But throwing a big log on a fire too soon can crush it. At what point is your idea or group ready to expand to a larger base? What must already be inflamed to support a larger scale of work?

As a fire burns, it needs replenishment. As a project goes on, some of the original resources and people will burn out or move on. How are you going to replenish them? What is your source of additional fuel?

Eventually, a fire does burn out. A project runs its course, or a movement enters a different phase. Do you want to bank your fire, to keep a spark alive for another cycle? Or is it important to put it completely out? If so, how do you cool or smother what is left of it?

Good firebuilders, good organizers, know that you cannot rush this process or skip stages. The big logs won't burn unless you've carefully laid the fuel, left room for air, and gotten the tinder and kindling burning well first.

The following meditation (a version of which is described in *The Twelve Wild Swans*[10]) is designed for use with a group.

FIREBUILDING MEDITATION

Ask all the participants to bring something to make a fire, but don't specify exactly what. When the group gathers, ask each person what she or he brought.

Then ask whether the group as a whole now has what's needed to make a fire. (If you don't, that's the lesson!)

Lay the fire with what people brought, and light it—assuming someone brought matches.

Sit around the fire, and ask each person to consider whether what they brought reflects their role in the community, or (if it's a new group) the role they *generally* play. Did one person bring the tinder, to get a new idea going? Did someone else bring kindling, to nurture it along? Who brought a great big log? And who has ever killed a fire, or a project, by dumping a big idea on it too soon? Is someone the overly responsible one, who brought matches and kindling and tinder just in case no one else remembered to bring anything? Did someone bring nothing at all, expecting that others would take care of everything?

Does the group have what it needs to sustain this fire?

Silent Cheering: Feeding the Fire

George Lakey, a wonderful nonviolence trainer and activist, tells a story about a time when he was training a group of union officials to be trainers themselves. One of his trainees, Joe, had just set his group to doing an exercise. "What can you be doing while your group is doing their work?" George asked him.

"I don't know," Joe said.

"You could be silently cheering them on," George suggested.

"What do you mean?" Joe asked. He was highly skeptical that silent cheering would affect his group in any way. So George had him sit down on the couch and called the rest of the group over to silently cheer for him. He said that within a few minutes, Joe turned beet red and began to sweat, the impact was so strong.

When George told me this story, I recognized a valuable way to explain the concept of energetic support to those who don't think in terms of magical language and resist these woo-woo concepts. Even for those of us who do think in magical terms, cheering can be a useful concept. We cheer for a friend in a race, or for our kids' soccer team, to offer energetic support, a taste of that attention and love that the cat craves. It feels good to be cheered for.

Shortly after I heard this story from George, I was walking through my woods thinking about Sudden Oak Death, a fungal disease that is attacking many of our trees. I was remembering what Mabel McKay, a Pomo healer and elder, used to say—that the trees and plants needed human beings to eat them and use them, to talk to them and sing to them and praise them, or they would die.

Our afflicted tanoaks were one of the favorite foods of the Pomo, but now their acorns mostly go ungathered. Lumber companies think of the tanoaks as "weed trees," something to be gotten out of the way so they don't compete with the more valuable conifers. Because these "weeds" grow back more quickly than the redwoods or firs after a fire or a clearcut, our woods are choked with brushy, many-stemmed saplings that rarely grow into the magnificent grandmother trees the Pomo prized.

It occurred to me that maybe we needed to do a ritual for the trees, to sing to them and praise them, to collect their acorns and make food from them. To begin that process, I decided that on my walk I would try cheering for the trees.

At first I thought this would be an exhausting task. My daily walk takes around two hours, much of it through forest. How could I cheer for every tree? I'd be worn out! But I decided to try. So while coming into my senses and trying to walk silently and respectfully, I was mentally saying, "Go, Tanoaks! Yay, Madrones! Aren't you beautiful today! Let's hear it for the Live Oaks! Hurray! Give it up for the Black Oak over there beside the road." I made up little songs and sang them inside my head, and I collected a bagful of acorns, offering some waters of the world in exchange.

Much to my surprise, when I was done with my walk I was not exhausted, but invigorated. Cheering for the trees had put me in a high-energy state that replenished my vitality and made me feel good. I began to have a deeper understanding of why so many people love spectator sports.

We human beings are constantly looking to one another for energetic feedback for our ideas, visions, and enthusiasms. Cheering someone on is like fanning the flames of their creative fire. It can help that fire burn more brightly, more steadily. When we're in a group or relationship where people are cheering for one another, we become more creative, more joyful, more intelligent. If you've ever had to give a speech in front of a group of people, you may have noticed how even one supportive friend in the audience could bolster your confidence (and conversely, how hard it is to be brilliant and scintillating if people are shuffling their feet, yawning, looking bored, or walking out).

We can also douse each other's fire, either deliberately or out of carelessness. Have you ever been in a group where every idea was met with criticisms and dire predictions of failure? Where everyone who stepped forward to propose a direction got attacked?

In one of our Witch camps, we were working with the story of VasaLisa, a Russian fairy tale about a beautiful young girl who has the proverbial wicked stepmother and stepsisters. At one point, these nasty relatives put out every fire in the house so that VasaLisa would be forced to brave the fearsome Witch Baba Yaga to get fire.

In one nighttime ritual early on in that camp, when everyone was joyfully dancing around the fire, two priestesses acting as the stepsisters suddenly threw buckets of water all over the flames. As the fire hissed and steamed, and we all stood around it in shock, they cried out, "VasaLisa, you stupid girl. You've let the fire go out!"

Throughout that week, we went on to reflect on the ways in which we put out each other's fire. When we recognize subtle energies, we become responsible for the kind of energy we are putting forth in our community. The things we do and say about each other create a subtle energetic field that either supports our work and our relationships, or undermines them.

Malicious gossip, backbiting, unsupportive criticism, and mean-spiritedness douse even the stoutest fire. And because a fire takes energy to build and maintain, such negativity is wasteful of the community's resources; it's like using electricity not just to keep the radio on all the time, but to keep it tuned to an irritating and distracting station.

Anger and conflict don't necessarily douse a fire. Conflict is part of all human relationships—but we can have conflicts, arguments, and disagreements that strengthen rather than undermine the underlying relationships. When anger is directly and cleanly expressed, it can be a bright flame of its own, healing and strengthening a relationship.

But when anger festers, when we chew over our grievances like old bones without expressing them directly, when we meet others with sullenness or resentment, we douse not only their fire but our own.

FEEDING/DOUSING HASSLE LINE

In a group, form two lines facing each other. Reach across and shake hands, so that everyone knows who their partner is.

The facilitator should remind participants of one of my favorite quotes from Gandhi, that when you are trying to change the world, first people ignore you, then they ridicule you, then they attack you, then you win.

Everyone in the first line should ground and center, then think of something they're passionate about, something they'd like to persuade their partner to join with them in doing.

The second line should prepare to ignore, ridicule, and/or attack.

The ground rules are these: interact only with your partner, stay with the interaction and don't walk away, and use no physical contact. Give the group a moment or two to mentally and energetically get into their roles, then begin.

Let the interactions play out for three to five minutes; then debrief the group.

How did it feel to be ignored, ridiculed, attacked? How did it feel to be the ridiculer or attacker? Was there anything the passionate partner did that was effective in establishing real communication? Were you able to hold on to your grounding and passion in the face of your partner's indifference or hostility? What came up for you? Have you faced anything like this in real life? What happened to your energy? To the energy in the room?

Now switch roles and repeat the exercise and debriefing. Then have everyone take a step forward, consciously stepping out of their role, and change partners.

Again, the first line should ground and center, and think about their passion.

This time, the second line will silently cheer for them, lending them energetic support—but without speaking.

Give participants a minute to get into their roles; then go. Let the exercise run about three minutes. After you stop, have everyone switch roles, and then run the exercise again.

How was this different? How did it feel to be energetically supported? To cheer someone on? What happened to your passion or idea in that environment? To your energy? How did the energy in the room feel?

Follow this exercise with an energy brushdown, to release any lingering negativity.

The exercise above is a useful one to do with groups that are starting out, because it can lead to a discussion of what kind of energetic atmosphere the group wants to create. Whether you are in a circle, a coven of Witches, a work team, a political group, or a sports team, people will be more creative, intelligent, and skillful in an atmosphere of energetic support. But if the group is full of people who are ignoring, ridiculing, or attacking each other, no idea can catch fire.

Energetic support, cheering each other on, is one of the simple but powerful ways we can honor the great creative forces in the universe in our own groups. Besides the sun, our creativity is the other essentially unlimited resource on the planet. Whenever we support and feed each other's creative fire, we increase the abundance of the planet.

Rage and Anger

"If you aren't outraged, you aren't paying attention" is a slogan I've seen on bumper stickers. It describes our present moment all too well. As we open our eyes and come into awareness of the great beauty and interconnectedness around us, we can't help but become enraged at the incredible shortsightedness and destructiveness of our current system, and at the violence and injustice that maintain it.

Anger is like fire: it is a vital, life-force emotion that arises when we are endangered or attacked and increases the energy we have available. Like fire, it can cleanse, renew, and heal. But like fire, it can also destroy. And maintaining a high state of rage, over a long time, can burn out all of your resources.

Yet ignoring or suppressing rage can also be destructive. Unexpressed and unacknowledged anger can turn inward, becoming self-hatred or depression. It can escape in unproductive ways, turning us against our friends and allies, instead of serving as the energy we need to change a destructive situation.

Anger and rage can also keep us from taking in information. When we are in a "blind rage," we're no longer looking, listening, or thinking about what we want to do. We're not making conscious choices: we're simply reacting. In that state, we can easily be manipulated or knocked off our balance, physically or emotionally. We may do things that later we will deeply regret.

To be earth-healers, we need to be able to sustain huge amounts of rage and anger, without either losing control or burning out. The first step is to learn to ground even when enraged.

GROUNDING ANGER

If you've been practicing the grounding meditation given in Chapter Five, by now you should be familiar with what it feels like to be calm, energized, fully present, and open.

In a safe space, ground. Focus on your roots going into the earth. Think about something that makes you angry. If nothing in your life is making you angry, you might try this exercise in the morning after reading the newspaper. As the emotion begins to build, feel those roots. Remember that energy can flow down your roots as well as up them. Think of them as hollow tubes, and imagine some of the excess energy of that anger draining down through them, going back to feed the fires below the earth. Continue until you feel relaxed enough to complete your grounding by breathing some of that fire back up, as pure energy that you can use for your own creative endeavors, or to create change in the situation that's making you angry.

You may also need to do something physical to release some of this energy. Try roaring—not yelling or screaming in a way that tears out your throat, but bringing up a deep, solid roar from the pit of your stomach. If there is something in your household that needs to be torn up, chopped, or destroyed, ground first and attack that woodpile or trashbin as a conscious act of energy release. My former therapist used to recommend beating pillows, and while I resisted for a long time, thinking it would feel dumb or forced, I eventually tried it, and the physical act of hitting something was a tremendous release. Whacking a stick against the ground, banging the table with a cardboard tube from inside a roll of giftwrap, conducting a mock sword fight, and smashing a piñata are other ways to release anger physically and safely.

If you are in a situation where you cannot physically or loudly release energy, look for smaller and more subtle ways. For example, I have often embroidered my way through a difficult meeting: the needlework project gives me the opportunity to repeatedly stab something. Drumming for ritual is also very satisfying: it gives me something to hit.

Grounding anger is just the beginning of transforming it, but it is a *necessary* beginning.

The next step is to remember to breathe and ground in the midst of the situation that's making you angry.

IMAGINING A CAULDRON OF CREATIVE FIRE

When you are comfortable grounding your anger, move on to this next step. Feel the anger, ground it, and then bring it back up as fire. Feel it warm your roots, your energy center at the base of your spine, activating all your instincts and energies of survival.

Now imagine a cauldron in your belly, a bubbling brew of creative power. Tell yourself, "This is the fire in my cauldron, the heat of my passion." Feel the fire-energy swirl and dance within you, and build in intensity. Now begin to breathe it up through your body. If there is someplace in your body that needs healing, bathe it with some fire. Breathe the fire up to your heart, and feel it warm and bathe your heart. Breathe it up into your shoulders and out your hands, and imagine it energizing your hands and your power to act and do. Breathe it up into your throat, and imagine it strengthening your voice. Breathe it up into your third eye, and let it strengthen your intuition. Breathe it out the top of your head, like a fountain of fire. Stand for a moment, like a blazing firework, and imagine that all that energy is your creative power, available to you now, a resource for you to use in bringing about change.

More suggestions about transforming rage can be found *in The Twelve Wild Swans.*[11] Rage transformed can be a powerful source of energy, a healing and purifying fire that can challenge injustice and help us renew the world.

BLESSING FOR FIRE

We give praise and gratitude to the fire, the sun's fire that fuels all life on earth, the radiant heat and light that is the source of energy for all beings, from the tiniest alga to the towering redwood, from the grass to the buffalo, from the worm to the hawk. We give thanks for the wildfire that cleanses the land, and we acknowledge its awesome power to destroy and to renew. We ask help in living with the power of fire, that we learn once again how to be in balance with fire on the land. We give thanks for the hearth fire, which warms us in the cold and gives our homes and communities a heart-center. We ask that our hearts be open to learn the teachings of fire, that we understand how to feed the creative sparks that arise in each of us, and that we feed the flames of passion and love for the earth. May the flames teach us how to dance, how to transform our rage into radiant action. Blessed be the fire.

NINE

Water

From my journal:

Water brings the land alive. Water gives the land a voice. As I am writing this morning, I hear the occasional tap, tap of the trees still dripping from the night's storms. All night long—indeed, for days—squalls have been coming in, moving through on the wind, dropping their load of rain, and then passing on. The streams have come alive and their song is the constant backdrop, filling the nights now that the frogs are silent.

I spend a lot of the day watching the patterns in streams. Marija Gimbutas identified the V or chevron as a symbol linked to water. In my stream, I can see the water make V's, flowing fast between a stretch of relatively straight bank, the sides deflecting the currents which intersect each other in perfect series of chevrons, their points facing downstream. The water runs faster, there in the center, and I imagine it must be cutting the channel deeper. In the summer, looking at the streambed, I'll be able to imagine the winter flow.

Friction slows the water on the bottom of the streambed; the upper layers move faster, tumble over. Water moves in spirals. A river is an elongated horizontal whirlpool moving downstream. The shape of this flow is changed by the riverbed, by rocks and logs and obstacles and differences in the underlying soil. But, all else being equal, a perfectly straight, smooth river would still eventually meander, because this spiral flow is like a drill, cutting away soil from the banks on one side, depositing it on the other, creating an S curve. One meander creates another: water speeding up around an outside curve is slowed as it rounds the bed, while water from the inside, freed of its silt, speeds up, and so the curves reverse.

In my little stream, I can see the beginnings of meanders. Years ago, the previous owners had small check-dams built all along the stream, and they work well, creating a more varied bottom to the bed. They slow the stream down, make it hesitate for a moment, spilling silt behind them, and then plunging over in a mini-waterfall that digs a deep hole below. Were this a year-round stream, those holes would create deep pools where salmon or steelhead fry could survive the summer. As it is, they keep silt from moving further downstream, to spoil the gravel spawning beds.

Today, on this bright morning in 1998, some neighbors and I are on a mission to explore another aspect of water. We're trying to measure the drop of our stream to determine whether or not we have enough water pressure to put in a micro–hydro system to generate electricity in the winter. Our stream has a lot of flow when the weather is this wet, but only a very gentle drop where it runs near the cabin.

Building a hydro system isn't simply a matter of sticking a wheel in the stream—those do exist, but they need a much bigger and deeper flow. The system we're considering will collect water and send it through a race of two- or three-inch pipe into a small device called a Pelton wheel, where nozzles will direct the flow over a wheel that spins and generates the electricity. But to work, we'll need twenty-five pounds of water pressure.

Water pressure has to do with drop. There are formulas for these things, but generally, for hydro of this sort, fifty feet of drop is needed. It takes two feet of drop to produce a pound of pressure. We can try to measure the drop with sticks and levels, or we can hook up all the hoses in the world—or at least as many as we can borrow—fill them at the high point in the stream, and put a pressure gauge on the end to measure the pressure. With either method, we'll be able to see how far up we'd have to go to generate enough pressure, how much pipe we'd have to lay, and whether or not the whole thing would be worth it for a few months of power each year, when the rains hold steady. On a year like this, when we've just had to replace our backup generator and we've had rain for eighteen out of the last twenty days, it sure seems worth it.

Water pressure, drop, and flow—all of these things remind me of my junior high school physics class with the hapless teacher whose simplest experiments went wrong. I've avoided the subject ever since.

Water will flow from a high point to a lower point. Put it in a closed channel, such as a pipe, and you can run it up and down in between as long as the end pipe is lower than the beginning *pipe. That's how we can fill a water tank on a knoll and then run the water back down into the valley and up to a second tank behind our cabin.*

We opt for the borrowing-hose option to conduct our test, and my neighbor Ken ends up laying the hose right in the streambed. His son Galen and I run back and forth, feeding him more hose, bushwhacking down to the stream with hoses until we're able to get a steady flow through.

Then, exhausted and out of length, we decide to call it a day and finish up when we can borrow some more hose.

In a different season, high summer of 2003:

Today we're going up to the spring to find out why we're getting such a low flow of water. Mer *and* Ken *and I go tromping across the hills, hoping we can even* find *the spring. I haven't been up here much in a while, and the trees have grown. The land looks different. But we do find the spring, in a little draw above a big fir tree—a couple of pipes sticking out from the bank: a shallow depression above. I look for the Water God, a cement sculpture made by my friend Donald Engstrom, but it's gone. Maybe that's why we're not getting much water?*

We dig out the spring box, which is filled with gravel to filter the water. It's a redwood box, about a two-foot cube, with holes in the bottom boards that seem clogged with roots. Inside is a metal pipe with holes drilled into it, also clogged with roots. We ream them out with a coat hanger and a piece of conduit. There is still only a small pool of water in the box, however, and we realize the spring probably needs further work. We'll check the flow down below and come back.

I look at the box, the two or three inches of free water in it. It seems very little to support two households, the gardens, the greenhouse, the fruit trees. This morning we were getting less than a third of a gallon a minute. That's less than five hundred gallons a day—plenty for our conservative personal needs, but not much to support a big garden in this dry land.

And yet, what an incredible gift—clean, drinkable water, straight from the ground.

I walk back across the hills, past the old stock-watering pond that remains from when this land was a sheep ranch. The water is low—lower than I remember it being this time of year. It's clear, above plateaus and channels of mud where tadpoles are swimming. I'm surprised to see them so late in July—they must be fruits of a later mating. Dragonflies and damselflies hover above, the sticklike bluets, the bright red/orange flame-skimmers, and a pair of big black and white ones almost as large as hummingbirds. In the water, tiny snails crawl and a variety of bugs swim. I see something that looks just like a fly flying underwater, using its wings as propellers to move along. I've never seen that before!

On top of the water, the water-striders skim along, supported on the surface tension. Sedges grow along the edges of the pond, and green grasses, as well as lots of pennyroyal that smells pungent as I walk on it.

Below is the dry streambed, where in winter water tumbles down, eroding the steep hillside. When people built this pond years ago, they changed the flow of the stream, pushed it down a different side of the hill. Since then, it's been digging its own channel, pushing down silt that has filled up my check-dams further downstream and that clogs my now-finished hydro system in winter.

This stream flows into Camper Creek, which flows by my house, then turns north to flow down to join Carson Creek and MacKenzie Creek and make its way into the South Fork of the Gualala River.

On the other side of our land, the water runs down to streams that flow east and south, to join up with the South Fork of the Gualala before it makes its bend to head north, following the ridge lines thrown up by the San Andreas fault—lines that keep it from heading straight to the ocean.

Water and Awareness

Meditating by a lake one day, I heard the water say to me, "All water is one—one whole, one awareness. All water is continuously aware of all the other water in the world."

That insight profoundly changed my relationship to water. Instead of thinking of it as a physical substance, I began to perceive it as a flow of life-giving awareness, constantly cycling through the world. To be a Witch, to be someone who has consciously accepted the challenge of serving the powers of life and balance, we must bring ourselves into right relationship with that pervasive consciousness. Only through a balanced relationship with water can we have abundance and thriving life.

And water *knows*. Water spirits, water Goddesses and Gods, however we want to name that intelligence that is so different from ours—*something* knows and feels when we approach with love and respect.

Waters of the World

Many years ago, I asked my friend Luisah Teish, a Yoruba priestess, what I could bring back for her from a trip I was about to take. "Just bring me some water," she said. "I collect water."

I started collecting water for her, and as collecting often goes, once I started collecting it it began to seem valuable to me and I wanted my own. So I began bringing back a little bit of water from every place I went—some for Luisah and some for me.

At that year's ritual for Brigid, the Irish Goddess of the holy well and the sacred flame, we decided to create a holy well out of a punch bowl. I added all my waters, and others brought sacred water of their own, or simply brought the water they drink every day. We made a pledge to Brigid, and saved back some of that water to seed the next year's ritual.

Over the years, the tradition grew. I began carrying the waters of the world around with me in a small bottle, using it to make offerings to the land or to the waters I was sampling. We began asking people to bring water at the start of many rituals, not just Brigid but Witch camps and political actions and other gatherings. As I describe at the start of this book, we make an offering of water to begin our fire ritual every year.

People began sending us water, going to special places to collect it. Some of our waters of the world came from Ireland, from many of Brigid's wells. And we got waters from sacred rivers—from the Ganges and the Nile—from Chalice Well in Glastonbury, from the *pozo* in the town of Amatlan, where Quetzalcoatl was born, from every continent. Someone sent us Arctic waters, and someone else a bottle of melted ice from Antarctica.

A few drops of waters of the world turn any vessel of water into sacred water. It is endlessly replenishable, not subject to scarcity. We use it to honor the spirits of the water and the land. One taboo: we don't drink it. "Sacred" is not necessarily the same as "sanitary," and too many of the world's waters are, regrettably, polluted.

HONORING THE WATERS OF THE WORLD

To honor the waters of the world in ritual, we set a vessel of water in the center of the circle. One by one, people come up, add their water, and say where it is from. Then a priestess lifts the bowl and says whatever is in her heart. Generally I say something like

> *Spirits of the waters, spirits of the land, ancestors, spirits of this place, we bring you this water as a gift of gratitude, for this land, for letting us stand and walk and be here. We bring it as a sign of respect, a sign that we want to open our ears and listen to what the land has to say to us, that we*

want to learn to be healers and good allies of this land and its people. Water is life, and this water comes from many places, as does the blood that flows in our veins. We ask your permission to stand in this place, to root here, to move energy here and do our rituals here. We ask your help in opening to what this place has to teach us. We ask your help that someday all the waters of the world may be clean and run free. We thank you for the power and beauty of this land, and for the gift of life.

The priestess then sprinkles water in the four directions, and also above, below, and center. The water remains in the circle throughout the ritual, and when the ritual is done, some is kept back to add to the waters of the world, and the rest is poured out onto the land.

OBSERVING WATER

Observing water is a meditation in itself. Just watching a flowing river or a running stream can help us feel calm and renewed. Swimming, floating, being in or near water is one of the basic ways human beings relax and replenish our energies.

Following are three suggestions for observing water.

Observing Water's Effect

In your home base spot, ground and come into your senses. Look around at the form of the land, the plants, the shape of the hills, the creases and crevices. Become aware of the presence and traces of water, of the flows that have shaped the land, smoothed the rocks, of the water that permeates the soil, the water encompassed in the bodies of plants and animals. Observe the presence and flow and movement of water.

Observing Water in Motion

Sit beside a running stream, or a swift river, or the ocean. Watch the movement and form of the water. Notice the shapes and patterns that it makes, where it runs fast and where it slows down, where there are standing waves and where there are slow eddies. Notice the way the patterns of movement form and reflect the shapes of the land. The visible motion of water is only the surface layer of more complex movement below. What can the surface tell you about the depths?

Immersion

Get into water. Go swimming in a river or bodysurfing in the ocean. Be sure to be safe, have a buddy, and be aware of currents and undertows. Feel the force of the water on your body. Notice how you move in the water, how the waves and ripples feel. Dive down and feel the difference between the motion below and the motion above. Feel the temperature changes from the depths to the surface. Close your eyes, and observe the water with your skin, your muscles, your deep bodily senses.

Water Cycles

To come into relationship with water, we must understand how water works. We must treat it respectfully in very practical ways, and learn its cycles, in order to hear its deeper communication.

Life began in water, and water remains necessary to life. Plants may use the sun's energy for photosynthesis, but it is water and carbon dioxide that they break down and recombine into the carbohydrates of sugar and starch. And while the sun's energy is virtually unlimited, in many places in the world water is the limiting factor.

My garden is dependent on that small pool in the spring box. Our land receives more water than most places on earth—eighty to a hundred inches in a good year. But almost all that rain comes between November and May. We also have one of the longest dry seasons in the world—four to five months a year when essentially no rain falls at all. We are a land of extremes.

Because of those extremes, when I plant my garden I have to plant for the water I'll have available in July and August, not for the abundance I can expect in January. I have to consider how to store and conserve water, how to make the most of that little trickle flowing into my tank.

But where does January's abundance go? Each winter this land is drenched by rain, the equivalent of a lake five to eight feet deep covering every square inch of these hills. By midsummer, streams have dried to a trickle at best, bone dry at worst.

Some of the water, maybe 80 percent, runs off the land, in streams and creeks and rivers.

Some of it soaks into the soil and remains there, coating the particles and filling the spaces within the soil's structure, or sinks deep beneath the earth to pool when it hits an impervious layer of soil, either gathering in an aquifer or forming a spring and making its way to the surface again.

Some of it is in these trees towering hundreds of feet above me. A big redwood can cycle seven hundred gallons of water a day, more than the output of my little spring.

The redwoods drink water from the ground, but they also comb fog from the air. Their needles condense mist into droplets, creating their own "rain." We get no true rain in the summer, but the inland heat often pulls in fog from the coast. Redwoods can live only in fog zones. The big ones are actually too tall to pump water from below all the way to the top, and their crowns are sustained by the moisture they pull from the air. In the old growth, a whole world of fungi and lichens and other plants grows high in the canopy. Over a thousand different species live high above the earth, sustained by moisture in the atmosphere. Some of them don't even begin to grow until a redwood is 150 years old.

Where does the water come from? Gaia is more blue than green. There's an enormous amount of water on the planet, the vast majority of it in the oceans or locked in the ice of the poles. The rotation of the earth keeps the oceans stirred up and cycling, swirling in great currents that moderate the climate of the world and cycle nutrients around the seas. The pull of the moon's gravity shifts the seas back and forth in the great tides that swell and circle the planet twice each day. The sun's heat pulls water up from the depths, to evaporate and ride the skies as the vapor that forms the clouds. When the clouds from the ocean cool and touch these hills, they drop their rain. The rain falls, sinks into the earth, feeds the roots of these trees, or runs off into streams and rivers that eventually find their way back to sea. The cycle is complete.

Water has always been symbolically linked to our emotions. Our feelings ebb and flow, storm and subside, just as water does. Thus water can teach us something about emotion.

WATER TRANCE

Begin in a safe place where you will not be interrupted. You might choose to lie down to experience this trance, or to stand up and dance and move with it. If you are doing the trance in a group, one person can read the trance instructions for others. Better yet, that person can become familiar enough with the journey to make up her own words.

We often begin by singing and dancing ourselves into a light trance state, using the following chant:

> Born of water,
> Cleansing, powerful,

Healing, changing,
We are.

Breathe deep. As you dance and move, feel the water in your own body. We are made up of mostly water, and our bodies feel the pull of the tides and the rhythms of the moon. Our blood is seawater in a new form, and we can still feel within us the crash of the waves and the great, slow currents circling the earth.

And as you feel the water within you, gradually let your skin and bones and human self drop away. Feel yourself drifting and floating, calm and peaceful, until you become a single drop of water, floating in a pool deep beneath the earth.

Let yourself drift for awhile, feeling how calm and peaceful and still it is here, in the dark, with everything that is not you stripped away. Feel the power of water to be pure essence. Breathe deep, and take in that power.

But even in this stillness, something moves. Time passes. High up above you, the sun calls. Something shifts, and you begin to rise.

Higher and higher you rise, finding the cracks and crevices in the earth, until at last you emerge, under the sun, as a fresh spring of water.

Feel the joy of the sun on your face and the bright sky above you. Notice what comes to drink from you, what lives on your edge. Is there some place in your life where you need this power of water to emerge, to renew? Breathe deep and take it in.

And you are so filled with joy that you spill over and begin to flow. The spring becomes a stream, bubbling and dancing down the hillside, noticing how it feels to leap over rocks and sing. Is there a place in your life where you need that power of water to dance and play? Breathe deep, and take it in as you sing,

I am the laughing one.
I am the dancing one.
I whisper secrets
As I flow.

And you flow on, growing stronger as new streams join you. One stream merges with another. Think about where in your life you need that power to join with others, to merge. Breathe deep and take it in.

You grow stronger and deeper, carving the hillsides as you pass, smoothing the boulders, tumbling the rocks. Feel the power of water to shape and change what it touches. Is there a place in your life where you carve your own channel, where simply by being who and what you are you create change? Is there a place where you need that power? Breathe deep, take it in, and sing it out:

I am the shaper.
I am the changer.
I carve the mountains
As I flow.

And now you come to something that blocks your flow. Take a deep breath, and see and feel what this obstacle is. Does it remind you of anything in your life? Is there something written on it? Does it speak to you?

Water has many ways to move around obstacles. It can crash through them and wash them out. It can dissolve them slowly. It can move around or under them. It can back up behind them until it flows over them.

How do you move beyond this obstacle? Take a deep breath and think about all the powers of water and the power you have within you. Let your breath become a sound of power, a sound you can share and blend with others, a sound that can carry you past this block.

(Wait a moment while the sound dies down.)

Now breathe deep and look behind you. What has shifted? What has changed? Where are you now, and what power do you feel within you?

And you flow on, down through the hills and mountains, out onto the wide valleys, growing stronger and deeper. And now you spread out and flow as a great river, moving through the valley toward the sea. Feel what lives in you, what you carry, what grows along your banks. Know the power of water to nurture and sustain. Is there a place in your life where you need that power? Breathe deep and take it in.

As you flow, notice how your back grows warm from the sun, how the warm water rises and then is pushed aside by cold water warming from below, until you begin to turn like a spiral cycling up and down and around. That cycling motion becomes a pulse, a meander. You carve the banks of the river and drop silt in the bends. You curve and snake, refusing to flow in a straight line. Feel the power of water to meander, to flow in its own shape. Is there a place in your life that meanders, that does not move swiftly from one goal to the next but takes its time, finds it own route? Do you need that power? Breathe deep and take it in.

The river is flowing,
Flowing and growing,
The river is flowing
Down to the sea.
Mother, carry me,
Child I will always be,

Mother, carry me
Down to the sea.

And at last you reach the sea. You branch out into an estuary and seep out into the waves. Feel the power of water to spill itself out, to become something larger than itself. Is there a place in your life where you need that power? Breathe deep and take it in. Sing it out.

The ocean is the beginning of the earth.
The ocean is the beginning of the earth.
All life comes from the sea.
All life comes from the sea.

Feel the wild wind on the waves, as they roar and crash and beat down the shore. Breathe deep, and feel the power of water to rage, to tear down. Is there a place in your life where you need that power? Breathe it in.

And now feel the ocean grow calm and still. The wind dies down, and the waves lap quietly against the shore, mirroring the sky. Is there a place in your life where you need that power of water to soothe, to be tranquil? Breathe deep and take it in.

Sink down, deep down, below the realm of sunlight, down to the bottom. Feel the ocean's incredible depths. Dark, cold, mysterious, pressed down by the weight of all the water and life above, they nevertheless can be a source of the nutrients needed for life to thrive, a creative source. Feel your own depths, the creativity that surges upward from deep below the light of conscious awareness. Take in the depth and power of water.

And now feel the sun pulling you upward again. Rise, up toward the light, until you lie on the surface of the waves, dancing with the wind, getting lighter and lighter. Is there a place in your life where you need that power of water to lighten up? Breathe deep and take it in.

And eventually you become so light that you evaporate. You become vapor, a gas on the wind, flying high above the earth, dancing in the clouds.

And you look down below, and you see the earth beneath you. You see places that are green and flourishing, and places that are barren. You see places that are whole and thriving, and places that are wounded and healing.

And eventually you see a place that calls to you, that needs the nourishment and life that only you can bring.

You feel cooler, suddenly—heavier. Your place is calling. And you congeal, back into a drop of water, and fall.

You fall and fall and fall, until at last you reach the ground and sink in as a drop of rain. And you know the power of water to give itself away and let go.

And you sink into the earth, feeling the roots that drink from you, the tiny creatures that swim in you. But you continue to sink, down and down, through sand and soil and gravel, until anything that is not you is stripped away.

And at last you become simply a drop of water, floating in a deep pool beneath the earth, knowing what it is to be only the purest essence of what you are.

Now thank the water, for its teaching and its power and its journey. Feel how that cycle of water, with all the power it embodies, lives in you.

But begin to remember, now, that you are a human being, that while you are mostly water, you are not only water. Feel your bones begin to grow back, your skin again containing the water within you. Move your awareness back into your human mind, thanking the water and remembering its teachings.

Let's sing ourselves back:

Born of water,
Cleansing, powerful,
Healing, changing,
We are.

If you've been lying down, slowly sit up. Feel the edges of your skin, the solidity of your bones. Say your name, and clap your hands three times.

And that is the story of water.[1]

Water and Abundance

From my journal:

Mer and Ken dug out the spring yesterday, while I stayed home to write. Three of us can't actually work up there at the same time anyway—there's not enough room. The good news is, after they removed masses of roots that had grown into the silt and gravel around the spring box, they got down to the true flow, which is abundant, and now we're getting four times as much water as we were before! We can water the gardens. I can take a slightly longer shower. We might even be able to refill the pond in the greenhouse!

I feel rich. I feel as if someone just told me I had four times as much money in the bank as I thought. In fact, somehow the state of my water tank and the state of my finances have become psychologically linked for me. I guess both ultimately lead to food. Both require sources, reserves, and outflows, and both are capable of springing unexpected and disastrous leaks.

We still don't have water to squander, but we do have enough to meet our needs, and some left over. And that's abundance!

Water creates abundance. Water makes it possible for plants to use the sun's energy to create food. Abundance comes not just from how much water flows into a place, but how well it is used, whether it is available where and when it is needed, and how many times it can be recycled and reused before it flows away. I can create abundance by increasing my sources of water, as in the journal entry. I can increase my reserves, or store more water to last through the dry summers. Or I can work out ways to reuse the water. In fact, we've done all these things: built ponds and cisterns, added water tanks, and installed graywater systems (more on those later).

The best place to store water is in the soil. Here in northern California, the earth does that for us, at least in part, soaking up those heavy winter rains and releasing the water that percolates down through the ground as springs. Our springs are all "perched springs," meaning that they rest on an impermeable layer of clay and are fed purely by each winter's rains. Few people drill wells here—a hole poked through that impermeable layer can drain your neighbors' springs as well as your own, and there are no underground aquifers to tap.

In other areas, however, massive lakes of water exist deep underground, amassed through centuries. They can be tapped by deep wells, providing abundant water. But if they are tapped more quickly than they can be replenished by each year's rains, sooner or later they will be drained dry, just as my water tank will empty if I use more than comes in each day.

We are currently using up our stores of water at a phenomenal rate. In California, Owens Lake, once the third largest body of water in the state, has been sucked completely dry to feed thirsty Los Angeles.[2] Groundwater is being pumped faster than it can be recharged almost everywhere that humans have discovered it. The Oglala Aquifer, one of the world's largest underground water resources (which supplies the dry states between Texas and South Dakota), is already 60 percent tapped out after only a few decades of water mining.[3] Water demand in that region tripled between 1950 and 1990, and is expected to double again by 2015.[4]

Maude Barlowe, national chair of the environmental justice group Council of Canadians, writes,

> Global consumption of water is doubling every 20 years, more than twice the rate of human population growth. According to the United Nations, more than one billion people on earth already lack access to fresh drinking water. If

current trends persist, by 2025 the demand for fresh water is expected to rise to 56 percent more than the amount that is currently available.[5]

But we don't have to be draining our water reserves. There are many ways that even city dwellers can increase our abundance.

Let's consider storage first. As I've said, the best place to store water is in the soil. Living soil is like a sponge. It's porous, full of spaces that can hold water and air. Soil rich in organic matter can hold many times its weight in water.

On a small scale—say, your garden—the first way to encourage water storage is to make sure that the water you give your plants can't easily evaporate. There are two ways to do that. The first is by closely spacing your plantings, which creates a canopy of green over the garden bed. This requires a fairly high level of fertility in your soil, but has the added advantage of producing a lot in a fairly small space.

The second evaporation-discouraging method is mulch. Mulch is basically stuff you throw on the ground to cover it—ideally, made of organic materials. Mulch can be straw, dead leaves, dried grass clippings, last season's corn stalks or tomato vines, etc. Pile it on the garden, as high as you can, and plant through it or heap it around your plants, being careful not to smother their crowns.

A heavy cover of mulch has a number of advantages. Not only does it keep water from evaporating; it also softens the impact of splashing water drops, whether from rain or from that high-pressure jet your six-year-old lets loose from the hose. Mulch also feeds the life of the soil. It nurtures worms and soil bacteria, those creatures who contribute so much to fertility. Get enough worms tilling your soil, and mulch can save you the backbreaking work of digging, turning, and fertilizing your garden. By adding organic matter to the soil, you increase the soil's ability to hold water as well. You can set in motion a very beneficial self-reinforcing cycle: mulch decreases evaporation, increases the water held in the soil, and improves fertility, so your plants grow more lush and abundant; this in turn gives you flowers, fruit, and vegetables that give you not only food but also stems and stalks for mulch.

Water can also be stored in the soil by contouring the ground. On a larger scale, permaculturists dig swales—ditches with berms that run along the contour of the land—to catch running water and hold it so that it infiltrates the ground. Swales can be dug by hand (with a spade and hoe) or, if larger swales are desired, with earth-moving machinery. As water infiltrates over time, the swales build up a "lens" or micro-aquifer of stored water that can be tapped by trees, shrubs, and other deep-rooted plants. In the long run, swales often fill in and become terraces. In many dry climates, hillsides have been terraced over the centuries to catch and conserve water, changing the face of the hills themselves. In Spain, Italy, Greece, Nepal, and throughout the Middle East, sculptured hillsides proclaim the antiquity of agriculture.

If your garden is on sloping land, a swale or two might give you the outline for terraces and garden beds. Put the beds on the downside of the swale, to take advantage of the water collected, and plant drought-tolerant herbs on the upside. If your garden is flat, small mini-swales and channels can help more water infiltrate. Dig low paths between beds, and mulch the paths, and they will serve as infiltrators. (Don't, however, mulch them with slippery straw. Try dead leaves or wood chips.)

Water can be stored in ponds, too. The technical aspects of pond-building are beyond the scope of this book, but advice and instructions are easily available. A small backyard pond can easily be built in an afternoon, and will add much to the life of your garden.

Every garden should have a pond. Even a small pond can provide habitat for beneficial insects and a whole variety of pest-eaters: frogs, toads, birds, even turtles and snakes. A few minnows will keep mosquitoes from breeding. Water plants are some of the most efficient users of the sun's energy, and a pond can be a great source of nitrogen-rich materials for compost and mulch. You can grow food in your pond—water chestnuts, even fish if it's big enough. Plus it gives you a good way to use all those sacred rocks you've been collecting!

For children, a back-yard pond can be an introduction to the pleasures of watching nature. Seed it with frog eggs in the spring, and later watch the tadpoles grow into frogs. Inoculate it with water and muck from a natural pond, and see what grows. Many of the city-raised kids I know spend most of their free time inside, staring at computers or homework or TV screens. To entice them outside, we need to provide something equally fascinating, and a living pond can catch and hold their interest.

Don't ever leave very young children unsupervised near an unfenced pond, of course. Don't dam a running stream to make a pond or place one in a live waterway without long consideration and expert help, or you risk causing terrible erosion when storms occur. Always give a pond an outflow. And never put non-native water plants into a pond linked to an ecosystem's waterways. In our area, water hyacinths don't overwinter and my ponds don't link to running streams, but in many areas, water hyacinths and other water plants have become a menace. Near where I live, our area's reservoir (and main swimming hole) has become choked with elodea from somebody's dumped-out fish tank.

And, finally, water can be stored in tanks, cisterns, even simple rain barrels filled from roof runoff. We don't all have a spring, but if we're lucky we do have a roof, and that roof by its very nature intercepts the rain on its way down.

Instead of channeling that water into gutters that drain into the sewer system, sometimes overloading it in heavy storms, we could catch the water from our roofs and use it to water our gardens.

Roof catchment is the major source of water for many people who live on islands, or land where water is scarce. If you live in a climate that gets intermittent rain, catching and storing even some of your roof water might provide all the water you need for a flourishing garden, reducing your water bill and saving some of Gaia's squandered gift.

How much water are we talking about? Let's say your roof covers a modest home of 1,200 square feet, and you get 24 inches of rain a year. That's 2,400 cubic feet of water each year. A cubic foot of water is around 7.5 gallons. Taking that number times the cubic footage, that's 18,000 gallons of water a year. A typical garden of 1,000 square feet can thrive on a generous 100 gallons a day, so that much water could keep your garden green for 180 days![6]

Of course, storing that much water above ground would need an enormous amount of space: two cisterns twenty feet square and thirty feet high, for example. Few of us have that much space to spare. But in most places, rain comes intermittently enough that we don't need to store a whole year's worth of water. Even a couple of fifty-five-gallon drums at your downspouts could provide you with some extra water for emergencies, or with soft rainwater for your hair or handwashing or ritual blessings.

And much of that water can be stored in the soil. My friend Erik channels all of his roof water into the ponds and swales of his suburban backyard, building the water lens, that underground store of water below his garden, to help his plants through bad times.

We can also increase our water abundance by reusing water. The water from a shower or bath or from cleaning vegetables in your sink could be growing your food instead of running down the drain into the sewage system.

Such water is called graywater—not as dangerous to treat as raw sewage, but still not entirely safe to dump on your lettuces. Graywater carries bacteria from your skin, soil bacteria from the roots of those vegetables, grease, oil, soap residues, and whatever else gets dumped down the drain. It can be a fertile medium for bacteria and other beasties to breed in. But graywater is also easy to treat.

I remember, during my first permaculture class, when my friend Penny Livingston-Stark announced that we were going to learn how to clean water. I got very excited, and then mad. How to clean water—doesn't that seem like something everybody should know? And why, in something like eighteen years of formal education, had nobody ever taught me?

Living organisms clean water. In *Water: A Natural History*, Alice Outwater describes the many ways water was once kept clean and clear by plants and animals. She explains the roles beavers, prairie dogs, wallowing buffalo, and mussels once played in collecting, storing, infiltrating, and cleaning our waters, before trapping, hunting, plowing, and engineering destroyed these natural systems.[7]

Swamp plants remove excess nitrates from water. The bacteria that live around their roots gobble up fecal coliforms and other nasty beasties. Some species also take up minerals and even heavy metals. Constructed wetlands are capable of cleaning sewage. The city of Arcata, for example, has constructed a marsh that not only handles the city's sewage but provides its largest tourist attraction as habitat for many birds and wild animals. John and Nancy Jack Todd, of the Ocean Arks Institute, in Massachusetts, have pioneered the development of "living machines"—water-cleaning systems that work on the same principle as the wetlands, running sewage or contaminated water through a series of tanks containing different biological communities, from algae up to canna lilies and fish. Clean, drinkable water comes out the other end, at a fraction of the cost of conventional chemical treatment systems.[8]

A home graywater system can reuse water fairly simply. The very simplest might be to place a bucket with you in the shower, to catch your warm-up water and the splash from rinsing yourself, and then use that on ornamentals, not edible plants, in the garden or to flush the toilet. The next step up would be to hook your sink or washing machine into some kind of "biofilter": a holding tank filled with gravel, lava rock, or some other medium where helpful bacteria can grow. The tank can be covered, to prevent mosquitoes breeding, or left open as long as the water stays below the surface level. An open tank can grow water plants whose roots will also help the breakdown of dangerous bacteria. Water plants infuse oxygen into the water through their roots, creating a zone in which aerobic, or oxygen-breathing bacteria, can live. A tub containing gravel and water plants will provide habitat for both aerobic and anaerobic beneficial bacteria, which can chomp happily away on fecal coliforms and other potential disease carriers.

The water can then run from the tank into a sequence of tanks, or into a gravel bed, a small constructed wetland in your backyard, a pond, or a gravel leachfield under your plantings.

Any system that works with living things will require monitoring and adjustment. Penny runs her graywater into a duckpond. In its first incarnation, the ducks refused to enter the water. Surfactants, the agents in soap that break down oils, were still present and would have ruined the duck's feathers. Penny added another biofilter, ran the water through a small artificial stream to aerate it, and then through a gravel bed, and now the ducks enjoy their pond.[9]

Maybe you won't put down this book and go redo your gutters or put in a graywater system. Perhaps you live in an apartment and have no garden to water. But we can all bring ourselves into right relationship with water—if nothing else, by conserving it. We can turn off the tap while we brush our teeth or do the dishes, save our warm-up water from the shower, take showers instead of baths, and make them shorter showers or install a low-flow shower head. We can be conscious of water, express our gratitude when we do use it, avoid disposing toxic substances into it, and treat it with respect, as the sacred gift that it is.

Water Policy

The home-scale solutions I've outlined above are good models for the potential solutions we could put into place on a larger scale, by changing our water policies. As Witches, as people who believe that water is sacred, we should be advocating for large-scale changes in the way we treat this precious substance.

When so many elegant, cost-effective solutions exist, why aren't we putting them in place? Why isn't it the norm to build a roof-catchment/storage/irrigation system into every unit of new housing? Why are we still building costly, inefficient, environmentally damaging sewage systems?

There are two basic reasons. The first is that new ideas always meet resistance, and changing our way of thinking about things is sometimes harder than changing the things themselves. It's so much easier to stick with the tried and true than to risk the unknown unknowns.

The second reason is that many people have enormous interests vested in the system as it is. Construction, sewage treatment, water provision—all these are areas where enormous profits can be and are being made, and those who profit wield enormous influence over legislators, bureaucracies, and enforcers.

What can we do, as individuals? We can begin by educating ourselves and our communities, learning how water works and what the solutions to our problems are. We can try out solutions on a small scale to fine-tune them and show others how they work. We can ignore the rules that maintain the status quo and do things anyway. We can strongly oppose bad policies.

Here in northern California, our larger community defeated a scheme by a corporation to take water from the mouths of the Gualala and Albion Rivers and tug it down to San Diego in giant plastic bags. Besides the almost ludicrous nature of the proposal, and its potential impact on the ecology of our river systems, we were alarmed to discover that the corporation proposing it was strongly influential in the World Trade Organization. Under the rules of many international trade agreements, such as NAFTA (the North American Free Trade Agreement), corporations can sue governments for loss of projected profits if the governments pass laws interfering with their business operations, even if those laws are for public health, safety, or other good! Had the water-bag scheme succeeded, it would have meant that all of northern California's waters were now opened to profit-making, and laws regulating the sale and privatization of water would have been very difficult to make or enforce.

Electoral politics can be a fruitful field for intervening in issues around water. Many crucial decisions are made by water boards, who are often elected with very little opposition or competition. Relatively small investments of time or money can yield a high degree of influence over local directions. Rightwing fundamentalists have gained an enormous amount of political power by running for local school boards. Why shouldn't people who care about the earth gain power by running for water and utility boards?

Water and Scarcity

Water, it is predicted, will be the great issue of the twenty-first century, the center of resource wars and conflicts. Because we haven't yet been courageous enough to implement sane solutions, and because control of the world's water is becoming more and more concentrated while the population is growing, it is estimated that by the year 2020 two-thirds of the world will be without adequate supplies of clean water. Water has always been seen as a communal resource, something that should belong to all, and water delivery has long been a primary function of government, something we pay for and make decisions about *collectively*. Today, there is more and more pressure to privatize water, to place its ownership and control in the hands of corporations that can make a profit out of providing this basic human need.

What does water privatization mean? How many of us already filter our water, or buy bottled water to drink? When I was growing up, we assumed our tap water was drinkable—that was one of the hallmarks of development and democracy that America supposedly stood for. Today, we trust that our tap water probably won't give us typhoid or cholera, as it might in Mexico or India, but we suspect that it might give us cancer. Privatized water services in England, France, and Wales have meant increased rates and lack of access to water for many low-income users. In Bolivia, water privatization resulted in a 40 percent increase in cost and sparked an uprising in Cochabamba in 2000.

Maude Barlowe writes,

> Already, corporations have started to sue governments in order to gain access to domestic water sources. For example, Sun Belt, a California company, is suing the government of Canada under NAFTA because British Columbia banned water exports several years ago. The company claims that B.C.'s law violates several NAFTA-based investor rights and therefore is claiming US$10 billion in compensation for lost profits.[10]

Among the world's poorest people, who can least afford to pay for the basic necessities of life, water privatization is well advanced, often imposed on third world countries by the International Monetary Fund or by provisions of global or regional trade agreements. Most often, this means higher prices and reduced services. In Cochabamba, Bolivia, for example, the city's water supply was privatized in 2000, its control given to a company called Aguas de Tunis, a subsidiary of Bechtel. All sources of water, even those privately held, were covered. (If those provisions applied in my area, Bechtel could charge me for the water I take from my own spring!) Water prices tripled, and many people were paying a third or more of their income for water.

The people of Cochabamba rebelled. They staged a nonviolent uprising, blocking roads and commerce in the city for two weeks in April of 2000. The government eventually gave in and turned water delivery over to a committee of the people, called La Coordinadora.

Oscar Olivera, one of the leaders of the uprising, told a group of us when he visited San Francisco in 2002 that the organizers referred so often to La Coordinadora during the conflict that many people thought they were talking about a woman (since the term is feminine). Who was this larger-than-life woman, they wondered, who would provide water, distribute it equitably, and take care of their needs? Perhaps, without knowing it, they had invoked a new aspect of the Goddess.

The people of Cochabamba wrote the following declaration.

Cochabamba Declaration on the Right to Water

Here, in this city which has been an inspiration to the world for its retaking of that right through civil action, courage and sacrifice standing as heroes and heroines against corporate, institutional and governmental abuse, and trade agreements which destroy that right, in use of our freedom and dignity, we declare the following:

For the right to life, for the respect of nature and the uses and traditions of our ancestors and our peoples, for all time the following shall be declared as inviolable rights with regard to the uses of water given us by the earth:

1. *Water* belongs to the earth and all species and is sacred to life, therefore, the world's water must be conserved, reclaimed and protected for all future generations and its natural patterns respected.
2. *Water* is a fundamental human right and a public trust to be guarded by all levels of government, therefore, it should not be commodified, privatized or traded for commercial purposes. These rights must be enshrined at all levels of government. In particular, an international treaty must ensure these principles are noncontrovertible.
3. *Water* is best protected by local communities and citizens who must be respected as equal partners with governments in the protection and regulation of water. Peoples of the earth are the only vehicle to promote earth democracy and save water.[11]

The Living River

Inspired by the people of Cochabamba, our network of Pagan activists has participated in many actions around water, privatization, and corporate control of the environment by creating what we call the Living River.

The Living River began with the Quebec City demonstrations against the summit meeting of the Free Trade Area of the Americas (FTAA), the extension of NAFTA throughout the western hemisphere, in April of 2001. We formed a "river" of participants dressed in blue, carrying flowing blue cloth and following a giant blue river Goddess puppet. Our goal was to bring the Cochabamba declaration into the meetings, to say, "This is what we should be negotiating: the right to water and the need to preserve it, not the opportunity to privatize and profit from it." A nine-foot-high fence, several thousand riot police, and a barrage of tear gas kept us from entering the meetings, but we read the declaration at the gates and were able to focus attention on water issues. Since then, we have carried on this tradition at many demonstrations. We took the Living River, together with Oscar Olivera himself, to the doors of Bechtel

Corporation, one of the world's largest water privatizers, which is currently suing Bolivia for $40 million in loss of projected profits for declining its efforts to profit from their water.

The Well of Grief

Water is emotion, and water is also key to cleansing and healing. When we begin to open up to the natural world, when we drop our defenses and begin to hear what nature is saying to us, when we start to appreciate the incredible beauty and wonder of the world, we also become aware of how much is being destroyed. Grief and sadness may overwhelm us at such times.

Jon Young, director of the Wilderness Awareness School, talks about how difficult it is to train young people as trackers. The skills and techniques, even the stillness and consciousness, are not hard to learn. But when people open their awareness past a certain point, they hit what Jon calls the "wall of grief," an experience of being overwhelmed with sorrow at the loss and degradation of the natural world around us.

I experience it more as a *well* of grief, an upwelling of sorrow and tears that seems to come from the very heart of the earth, dark and cold and unimaginably deep.

Grief is like water: we can drown in it, but we can also drink from it and be strengthened and nourished by it. For our grief and sadness reflect our ability to feel, to love, and to mourn the loss of what we love so dearly.

One of the great gifts we can give each other is simply to listen to the grief, pain, anguish, fear, or whatever emotion someone needs to express, without trying to fix it, change it, or take control of it in any way. If someone is suffering from trauma or post-traumatic stress, she needs to tell her story, sometimes over and over again. When we are in grief, we need to share it with someone who will not himself be overwhelmed or shocked by it.

LISTENING MEDITATION

In pairs, decide which partner will speak first. Ground, and then just breathe together for a moment, matching breaths.

The first partner speaks first, telling about an intense experience. It can be a sad or a happy or a frightening experience—whatever that person feels moved to speak about. The second partner just listens, trying to listen on every level—to the words and content, to the energies, to the emotion, to the body language. Listen for at least three minutes without interrupting, asking questions, or offering comments. Then switch roles.

Afterward, talk about how it felt to listen so intensely, to be listened to so well. What happened to the energies between you? Within you? In the room? How do you feel about each other now? About the experience you talked about?

Grief Ritual

Ritual is one of the tools that can help us with grief. The key to a powerful ritual for grief is to open a space in which people can speak from their hearts, can name what they may not have been able to speak about before, with energetic support. In Chapter Seven, we learned about silent cheering. Energetic support in a grief ritual may not be so cheerful: it might come in the form of chanting, holding a low tone, or simply witnessing with deep attention.

Two summers ago, I was teaching in the British Columbia Witch Camp, together with Sunray and Culebra. We created a simple but beautiful healing ritual.

We placed a bowl of waters of the world in the center of the circle, after grounding, casting, and doing all the preliminaries. We also had a bowl of pebbles, and each person in the circle was given one.

One by one, each of us stepped forward and dropped our pebbles into the water, naming what we grieved for or what needed healing.

We sang a healing chant that spoke of going down to the water and letting our tears and fears be washed away. Singing, we carried the cauldron, now heavy with stones, all the way down through the camp, down the mountainside, and down to the lake, where we stripped off our clothes and jumped into the cold, cold water. We carried the cauldron in, and two women dumped the heavy load of stones into the water. The chant became joyful, jubilant as we felt cleansed and alive, burning with cold. Then, still singing, we marched naked back up the hill, completing the ritual by raising power, that is, by letting the chant build until its energy reached a peak of release. In the quiet that followed, I felt myself able to let go of some of the pain I was carrying from seeing people brutally beaten by police in the Genoa demonstrations against the G8 summit earlier that summer. The grief and sorrow remained, but my own energy was flowing again, no longer stuck and inward-turning.

I lead many grief rituals, for activists, and for all of us suffering the pain and loss of ordinary life in these difficult times. A communal ritual for grief can allow us to acknowledge and share deeply some of the emotions we ordinarily keep hidden. Together we can support each other through the sorrow, and come closer in the process. The love and compassion we share is the true healing.

A grief ritual should not try to artificially stir up emotions or push people to emote. The most powerful rituals happen when we simply create an opening and an atmosphere of receptivity, as in the ritual described above. I often use water in some form to symbolize this receptive state. Waters of the world are called for in the above ritual, but salt water or clear spring water would work just as well at the center of the circle. Likewise, we used small stones in the example above, but salt or any other object could be used as a symbol of release.

Grief may also be linked to rage. Sticks and something that can safely be hit, such as a big plastic barrel, can also be placed in the center.

The ritual also needs some way to allow us to use our voices and express our sorrow. Keening—the traditional Irish form of lamenting the dead—crying, even screaming, work sometimes, but people often find it hard and artificial to keen on command. Singing can open the emotions and provide a base of sound that allows people to moan, cry, or sob if they feel moved to do so under the cover of many raised voices. And music has its own healing power.

When the energy is released in a grief ritual, be sure to do something to cleanse, renew, and fill up again afterward. We bathed in the lake in the ritual example given here. At other times, I have plunged into the ocean, washed my face in a bowl of salt water, spilled the water of grieving onto the ground, or shared food and drink to symbolize nurturing. Feel free to create the ritual that will best serve your needs and community.

Grief, trauma, and fear can lead us to close down. When we shut down to negative feelings, we often shut off our ability to feel positive emotions as well, to take pleasure and joy in what remains of nature, to give and receive love. We may become angry toward our friends and lovers, cynical and bitter, difficult to be around or to live with. In extreme cases, grief and trauma can lead to true post-traumatic stress syndrome, deep depression, and even suicide.

But when we find someone willing to listen, when we find the support of a community that can hold our grief and not turn away, we can find the courage to open up again. Then we can drink from the well of *life*, not just the well of *grief*, and be healed.

BLESSING FOR WATER

Praise and gratitude to the sacred waters of the world, to the oceans, the mother of life, the womb of the plant life that freshens our air with oxygen, the brew that is stirred by sunlight and the moon's gravity into the great currents and tides that move across the earth, circulating the means of life, bringing warmth to the frozen Arctic and cool, fresh winds to the tropics. We give thanks for the blessed clouds and the rain that brings the gift of life to the land, that eases the thirst of roots, that grows the trees and sustains life even in the dry desert. We give thanks for the springs that bring life-giving water up from the ground, for the small streams and creeks, for the mighty rivers. We praise the beauty of water, the sparkle of the sunlight on a blue lake, the shimmer of moonlight on the ocean's waves, the white spray of the waterfall. We take delight in the sweet singing of the dancing stream and the roar of the river in flood.

We ask help to know within ourselves all the powers of water: to wear down and to build up, to ebb and to flow, to nurture and to destroy, to merge and to separate. We know that water has great powers of healing and cleansing, and we also know that water is vulnerable to contamination and pollution. We ask help in our work as healers, in our efforts to ensure that the waters of the world run clean and run free, that all the earth's children have the water they need to sustain abundance of life. Blessed be the water.

TEN

Earth

From my journal:

Sometimes the better part of gardening is not doing it. Today my friend Caerleon and I went for a walk. Originally it was planned as a short get-in-shape walk—but the day was perfect, a sunny winter day with the hills green and the sky blue, the sun warm but not too warm. We walked down our dirt road to the gate of the neighboring ranch and then for some reason continued on, down to the stream that forms the headwaters of the Gualala River, and back up an enticing dirt road that had always intrigued me.

This whole area is carved by the logging roads put in fifty years ago. Haunted by the ghosts of great trees that stood here before the loggers came, they determine the routes we travel, the paths of erosion. At best, they make great hiking trails. We climbed up beside the stream bank, looking down an almost vertical hill to see the enormous stumps of huge Douglas firs that are no more. The road led us to a meadow surrounded by a bowl of green hills, and then zigzagged up a steep grade to join the dirt road above the Big Barn, left over from the days when this land was a sheep ranch. We walked down, across a broad, green meadow, speculating on what the land would have looked like before it was logged. Would these open fields have been covered with trees, or were they natural meadows? Why haven't the trees come back if once they covered these fields? Grazing was the culprit, we speculated—

especially here, so close to the barn that was once the hub of the old sheep ranch. I explained about the mycorrhizal fungi, while Caerleon, who is an archaeologist, told tales of the ancient tribes that once lived here.

Back down by the stream that runs alongside the road, we climbed a blue schist boulder, a pitted rock marked by cupules laboriously pounded into its surface, now covered with lichens, scat, and moss. No one really knows who made the holes or what they were for, although there are theories that they had something to do with fertility. Women may have eaten the rock dust, or perhaps the pounding itself was a trance technique. "As someone who spends a lot of time in trance," I said, "I have to say that there are simpler and easier techniques. I suspect that if you used this one, you wanted to do something with the product, whatever it was."

The boulder was perched above the stream, and Caerleon said most of the pitted rocks that have been found are in places where you can look down on something. They were often associated with fishing, she speculated. Below us the stream ran clear. No steelhead spawn in it today, but as late as the fifties, the locals say, you could fish them out with a pitchfork.

Gone, gone, gone.

The stream would have been larger before the logging—and more trees mean more water. The grasses in the meadow would have been different, not these European annuals that go dry and brown in the summer but deep-rooted bunchgrasses that stay green all year. The road wouldn't have been there, of course, but undoubtedly a trail would have followed the stream.

Where you find sedge, Caerleon said, it was probably planted by the Pomo for basketweaving. I told her about the talk I'd had with Victoria, one of the Pomo teachers at Living History Day at Fort Ross. I had told her that I was interested in basketry plants. She'd showed me sedge in their display, and told me that it's difficult to find, as so many of the traditional gathering places have been built over or bulldozed under. But for baskets, you need to find the sedge that grows on sandy loam, where the roots can spread out straight. On our rocky land, they tend to twist themselves into knots and gnarls.

Nevertheless, I've been transplanting a little sedge into my stream, for the spirit energy if nothing else.

After the walk, I go over to Jim and Dave's to dig some chestnut seedlings. I take them some bee balm and some lamium—White Nancy—which have been spreading nicely in my garden. With all the failures and frustrations gardening involves, there's still the something-for-nothing satisfaction of plants that propagate themselves. Lamium is putting down roots from every node, so I've now transplanted it all around the central circle and hopefully will soon realize my vision of a perfect full moon of silvery plants in the center of the garden, punctuated by white foxgloves, white lilies, white daffodil, anemones, and geraniums.

I never leave Jim and Dave's empty-handed. They've lived on the land for twenty years, and their house is surrounded by things that happily grow and spread and reseed. And right now, in mid-December, we are at Prime Time for Propagating. Cuttings rooted now will have months of rain to establish their roots.

In just the last few days, I've given them lamb's ears and red and purply-red penstemon, but they've given me purple penstemon, Sonoma sage, coreopsis, red and yellow obedient plant, horehound, many different lavenders, a beautiful succulent, coyote mint, mother of thyme, and a dozen carnation poppy seedlings, as well as the chestnuts. They've loaned me a Havahart trap to attempt to catch what I think may be a wood rat gnawing its way between our alcove ceiling and the roof, and taught me how to use a hoedad—the tree-planting tool par excellence. A hoedad has a long, narrow blade attached to a stick—a cross between a pick and a narrow shovel. You swing it into the ground and then step on it to drive it in deeper with your body weight, wiggling it in as you go. It makes a deep, narrow cut, perfect for inserting a slim seedling with long roots.

Dave assured me it would be easier to plant the chestnuts with the hoedad, and he was right. I quickly planted eight or nine seedlings, tying ribbons on so that I could find them again later to fit them with deer protection in the form of a small wire fence or plastic protective tube. I planted three or four on our water tank knoll, two or three on the far side of the garden, the others on the bank below the road. I had time to return the hoedad and climb Firehouse Hill to watch the sun set over the ocean. The work went quickly because, as I admitted to Jim, I've grown jaded: I just planted them; I didn't pray over them for half an hour this time or invoke the chestnut deva (or guiding spirit). But I do believe, after last year's failure with the dozen or so seedlings I planted, I have more of an instinct for where they might thrive.

Mother Earth

The earth is our mother, we sing. Mother earth, mother nature—she is the literal womb of life, providing all that we need. Her living soil feeds us; her rocks make our bones; her minerals are in our life's blood. The very heart of Goddess spirituality and of other indigenous traditions is the recognition that the earth is sacred.

This understanding of the earth is very old and very widespread. The earliest works of human art are ancient figurines of full-bodied, big-breasted, and big-bellied women that embody the sacred quality of life-giving earth/flesh. The painted caves of southern France and northern Spain were the living, sacred wombs of the mother, generating the animal hordes painted so vividly on the walls. In Greece, Gaia was the eldest of Gods. The life-giving Regeneratrix underlies the later Goddesses and compassionate mother figures, from Isis in Egypt to Kwan Yin in China to the Virgin Mary. The earth mother is mountain: the paps or breasts of Anu in Ireland, the "sleeping lady" who is seen in the volcano of Popocatepetl in Mexico and Mount Tamalpais in the San Francisco Bay Area. She is the mountain in the Himalayas that we call Everest but whose true name, Chomolungma, means Goddess Mother of the World.

The understanding of the earth as a living, sacred being is also very long-lasting. Archaeologist Marija Gimbutas, who did some of the major work on the early Goddess cultures of ancient Europe (as noted in Chapter Two), described in an interview how the peasants of her Lithuanian childhood in the 1930s used to kiss the ground every morning. And the reverence continues among indigenous peoples and those who work with the earth. As an example, at a recent meeting in Mexico City of Via Campesina, the worldwide small farmers' organization, began with an *ofrenda*, an altar/offering made on the ground of seeds, flower petals, and fruits of the earth to honor the earth mother, and a ceremony to exchange seeds.

And yet today, we also live in a global culture that profoundly dishonors the earth. Words associated with the earth are used as insults: "low," "dirty," "soiled." The ecofeminist movement of the eighties was founded on the insight that the way our culture treats the earth and the way it treats women are linked. Both are identified with the flesh, the body, the bloody and messy processes of bringing life into the world and its inevitable end in death, decay, and rot. When that cycle is devalued, when what is sacred is abstract, removed from earth, transcending life and death without being marked by the cycles of life, the earth and women are both denigrated and both become victims of exploitation, assault, and rape. So, too, those who live and work close to the earth are devalued and exploited. The farmer, the peasant, the manual worker, all are "low-class" workers. Manual labor is "beneath" the dignity of "high-class" folks, who work with their minds—or better yet, don't work at all, but live off of that ultimate abstracted value—profit—that is accumulated by the labor of others.

Our spiritual and ideological rupture from the earth is reflected in every aspect of our culture, but perhaps most deeply and ominously in the way we grow our food. Industrialized agriculture, the Green Revolution, biotechnology that produces genetically modified food plants, these are all based on a mechanistic understanding of soil and growth. Soil is just a medium to support plants, in somewhat the same way as the woman's womb was believed by Aristotle and the medieval Church to be only a vessel in which the male seed, the true life germ, was nourished. Plants can be fed a few key nutrients and protected from the competition of weeds and the predation of bugs by a blanket killing of anything that is not the desired crop. The true purpose of corporate, industrial agriculture is not to grow food, but to grow profits.

This form of agriculture has become one of the most destructive human activities on the planet. In the United States, we lose six tons of topsoil for every ton of food produced. We use thirty calories of fossil fuel for every calorie of food produced.[1] And we pour hundreds of millions of tons of toxic chemicals into our air, soil, and water.

Corporate globalization is industrial agriculture on a worldwide scale. Its vision is a world where no one will eat the food that they produce or food that is locally grown. Instead, food will be just another commodity circulated on world markets, generating more profit each time it changes hands. In the last two decades, we've lost a third of our family farms to policies that support globalized agriculture.

From the point of view of profit-making, an apple grown on a corporate farm in New Zealand, with the use of pesticides and heavy machinery, picked green (and later gassed in order to ripen), waxed, irradiated, and shipped in cold storage to be sold in a giant supermarket chain in California, is a great profit generator. It produces revenue for the farm corporation, the makers of pesticides and farm equipment, the food irradiators and packagers, the truckers and shippers and haulers, and the supermarket chain. Of course, those who actually till the ground, prune the trees, and pick the fruit receive only a pittance. According to the Oregon Department of Agriculture, out of every dollar we spend on food in the U.S., farmers receive only twenty cents. In 1950, they received forty-one cents. As late as 1980, they received thirty-one cents. No wonder small farmers are going bankrupt! And if farmers receive so little, farmworkers and migrant laborers receive far, far less.[2]

Perhaps a mother buys that New Zealand apple as a healthier alternative to potato chips for her child's lunchbox. Unfortunately, the poor apple is devitalized from its various treatments and its intercontinental journey, and laden with the residues of pesticides, herbicides, wax, and radiation.

Sonoma County, where I live, used to have a thriving crop of Gravenstein apples, a variety well suited to local conditions. Imagine the difference in the apple if that same mother, instead of selecting a New Zealand apple at the supermarket, bought fruit grown on a neighboring farm, organically, picked at its point of ripeness, and sold at a nearby farmer's market or neighborhood store. That apple would generate a different kind of benefit—increased health, taste, vitality, joy. It would create less profit for big corporations and chemical companies, but more real abundance for small family farms and local stores. And it would conserve some of the diversity of crops that makes for real food security.

Earth-honoring agriculture would generate abundance, but its primary intention would be not to grow profits, but rather to grow soil—living, healthy, complex soil—as a fertile matrix for living, vital, health-sustaining food. To grow soil, we need to appreciate and understand that soil is a living matrix of incredible complexity, the product of immense cycles and great regenerative processes.

Soil scientist Elaine Ingham lists what we might find in live, healthy soil:

> What do we mean, organism-wise, when we talk about soil? Agricultural soil should have 600 million bacteria in a teaspoon. There should be approximately three miles of fungal hyphae in a teaspoon of soil. There should be 10,000 protozoa and 20 to 30 beneficial nematodes in a teaspoon of soil. No root-feeding nematodes. If there are root-feeding nematodes, that's an indicator of a sick soil.
>
> There should be roughly 200,000 microarthropods in a square meter of soil to a 10-inch depth. All these organisms should be there in a healthy soil. If those conditions are present in an agricultural soil, there will be adequate disease suppression so that it is not necessary to apply fungicides, bactericides, or nematicides. There should be 40 to 80% of the root system of the plants colonized by mycorrhizal fungi, which will protect those roots against disease.[3]

EARTH OBSERVATION

Do this exercise not with your eyes, but with your nose. Ground, center, and come into your senses. Now, put your face in the earth and smell. Breathe in the fragrance of the living being that she is. The earth breathes, taking in air to fill her pores. And she breathes out, exhaling bits of herself that communicate her state of health or disease, her level of vitality.

Take a walk through the woods or through a park. Periodically, bend down and smell the earth. How does the smell change under trees? In the grass? What does the earth smell like where it is hard and dry and compacted? Soft and fertile? Can you notice different soil types, different histories? How does your own garden smell?

Get in the habit of sniffing the earth wherever you are. Get low, close to the ground. Learn the smell of healthy soil. (Just be careful on lawns drenched with chemicals and in areas that may contain toxins.)

The Cycle of Rock to Life

Life is rock rearranging itself.

—*Elisabet Sahtouris*

Let's follow a molecule of calcium through perhaps the longest of the elemental cycles. Here is a big, prickly leaf of the herb comfrey in my garden, which I've pulled up and laid down for mulch. Red worms, bacteria, and fungi have begun the process of decay. The rain comes and leaches away a tiny bit of calcium. That calcium could have many fates. First, it could be taken up again by roots, incorporated into the body of the lettuce I grow. From there it could end up in my dinner (later strengthening my bones), or it could be drawn up into that lettuce and my young neighbor Angie could eat a leaf, passing the calcium through her breast milk to her daughter, to build Ruby's baby teeth. Eventually Angie and the baby and I will all die, and our bodies will go back to the earth, and that calcium will return to the ground that it came from.

But let's say that it isn't taken up into a living body, but dissolves into the soil and passes down through the soil into the groundwater, seeping under the hills to emerge from a stream, and flowing with the water all the way out to sea.

Suppose it is taken up by a tiny, one-celled alga that is eaten by a shrimp that is eaten by a fish that is eaten by a bigger fish. And if it is eaten, at last, by a salmon or a steelhead, it might circle right to its source, via the salmon's migration upstream to spawn and die in the headwaters of the stream of her birth. And a bear might come and eat the salmon, and return its elements, including that calcium, back to the soil of the forest, closing the circle.

But, alas, our Sonoma streams are degraded, our salmon are gone, no bears roam our woods, and the forest is hurting for calcium and phosphorus and the smell of rotting fish.

But let's imagine that our lucky calcium makes it to the ocean and escapes being eaten by that alga, instead becoming part of the body of a tiny radiolarian, a one-celled organism that constructs an intricate, fairylike shell that looks like a spherical snowflake. And that radiolarian escapes the jaws of whales and the thousand other hungry mouths in its vicinity, to live out its natural life and die, letting its shell drift back to the ocean floor to be buried under a constant slow rain of detritus.

Our calcium lies there for a long, long time. Shells and skeletons fall from above, and the great weight of the ocean presses down, down. Slowly, slowly, our calcium is crushed down into the rock below, merges with rock, *becomes* rock, a formation of limestone weighing down the edge of the earth's crust.

And after millions of years, that edge begins to sink, to push down at the rim of the ocean plate, to churn up rock and magma from below, like a plow churning up the ground, pushing up a swell of magma, squeezing up mountain ranges, grinding and slipping along the fault line. Until at last the fault gives way, the edge of the plate lurches, cities fall, and our calcium is pushed upward, lifted high above the churning currents of molten rock to become part of a limestone plateau.

The rain will wash over her. The wind will caress and buffet her face. Lichen and ferns will begin to grow, and all together they will weather down that new rock to soil. And then our particle of calcium can begin, again, a journey through life-forms, leaf and bone to rock, from rock to life. And the circle is complete.

Many scientists today believe that it was the presence of life, the weight of those billions of radiolarians and other tiny skeletons pressing down on the tectonic plates of the earth, that set those plates in motion and sent the continents off on their long, slow perambulation of the globe. The cycle of life to rock to life is probably one of the longest-lasting of the earth's great regenerative cycles. The bones in the hands that are typing this sentence are strengthened by the bodies of ancestors hundreds of millions of years old, and for that I am grateful.

FERTILITY AND DECAY

In a world in which the life of the soil is everywhere under assault, building soil fertility can be a profound act of worship. To enhance the life of the soil, we first need to understand it. As Witches, we will probably learn more from a guided journey than a long, technical explanation.

Find a safe space—ideally, outside under a tree, lying on the earth. But you can also do this inside, if necessary. Ground, come into your senses, and create a sacred space. One person can read the following meditation for the group, or you can tape it ahead of time. Even better, the journey leader can read it through enough to become familiar with it, then make it her own, using her own words and images.

Lie down, breathe deep, and relax. Take a few moments to relax completely, from your head down to your toes. Feel the weight of your body on the earth, and feel the earth as a living body, embracing you, holding you close. Feel the elements in your bones and flesh that come from the earth.

And when you are ready, take a deep breath, and for a moment, imagine that you are a leaf, hanging tight to a twig on a high branch, waving in the wind. Feel the wind and sun on your face; feel how when the sun hits you your very cells sing with the energy of light, and the chiming chord they make creates a sweetness that permeates your blood and feeds your tall and reaching body. And just for a moment, let yourself hang in the breeze, feeling what it's like to feed from light, effortlessly. And if there is someplace in your human life that you need that feeding, that nurturing, take a breath and take it in.

And now time passes. Imagine the first cold winds of winter beginning to blow. And you feel them touch your face, and a freeze comes into your veins, and you glow scarlet in the light. And you take a deep breath, maybe a sigh, for you know that the summer has been good, but now it is past, and the time of singing sweetness is done. And if there is something in your life that is complete, some phase that is ending, something sweet that you now need to let go of, take a deep breath and draw in that power of the leaf to let go. And you let go and fall, letting the wind take you, and you swirl and dance and spiral in the wind's eddies, always falling, falling down and down and down.

Until at last you come to rest on the earth, lying on your sister and brother leaves, piling body upon body. But that earth you rest on is no solid barrier. It is porous, like a sponge, a labyrinth of cells and spaces, alive with a billion hungry beings. And you take a deep breath and give yourself back to the earth, and she reaches up to embrace you, and a billion hungry mouths open wide to take you in.

The ants and the beetles come up from below and begin to break you down. They eat away the soft parts and take apart the veins. Thin threads of fungus hurl themselves across your face, beginning the process of dissolution. You are drawn down, down, into the spaces and the caverns far below. Parts of you are ingested, becoming ant or beetle for a little while, then released again as frass that tumbles down into the earth. Parts of you are held in the fungal threads and slowly dissolved. Parts of you are licked by filmy mouths of soil bacteria, and slowly, slowly you are brought back into your original elements, and slowly, slowly you are brought down into the earth.

You descend, through great caverns and chasms, past great suspended archways of crystalline rocks, over sharp-edged, gleaming silica boulders and round, smooth spheres of clay. And the great caverns within the earth are slick with water, and in the tiny pools that form in her crevices a billion creatures swim. This is a whole, rich, three-dimensional world, and you are just a tiny grain of life, one speck of luminous phosphorus down here below.

Now out of the dark spaces beyond comes a great, smooth, writhing being, slick and wet, opening its great mouth to take in chunks of the very rocks themselves, opening new tunnels and pathways. The great worm meets another, and they slide along each other's bodies, sharing the liquid lubricant of their sweat, drunk on each other's odor, coupling at last in a doubled mutuality of instinctual pleasure, each one both male and female, each fitting to the other in a double lock. You are taken in, you become part of that shuddering mating, and then you pass out again, in a casting rich with your brothers and sisters and fragrant with the promise of life.

And now something reaches for you. Thin, thin threads, long arms of the mycorrhizal fungi that stretch between the root hairs of the great trees. Sticky as spiderthread, they snake through the caverns of the soil, holding the archways and the boulders in place, wrapping them in a living binding.

And all around you now, the caverns are penetrated by the most delicate, pearly, iridescent tubes of the root hairs of the plants and great trees above. And each gives off its own fragrance, its sweetness, its unique taste. And colonies of soil bacteria, those dancing circles, surround each one—each root attracting its own clientele like a street full of ethnic restaurants, each patronized by its own fans. But the trees—they throw out these tendrils of fungi that have got you lassoed. And you feel how they link root to root, how they can pass you from one tree to another down this network, how the trees feed their young, how trees that grow in the sun will feed trees who grow in the shade. Passing nutrients and energies through this phosphorescent web that you are now a part of, like the nervous system that feeds your brain. And you take a long, deep breath, and you listen to the long, slow thoughts of the trees.

And you know how the trees speak to each other, deep below the earth, and how the forest is linked through this web. This is the web that supports the forest mind as your nervous system and brain support your mind. And feel how far this web once stretched, when the forests covered the land. Feel what still exists, and what is broken; what still speaks, and what has been silenced. And know that you are held in this web.

And then, taking a deep breath, feel yourself sucked into the root hairs, and up into the root, rising and rising on a current of sweet sap, rising up now and caught in a great upward tide. Higher and higher, through the channels within huge roots and living skin beneath the bark, and out into the branches, the twigs. Where you become part of a green bud that opens with the warmth of the spring sun, unfurling itself like a wing to catch the sunlight and sing it into sweetness.

And there you wave, a brave banner, a signal flag of life, through the warm days and the long nights, catching sunlight, singing sweet food out of air and water to feed the twigs, the bough, the trunk, the root. Until once again the cold of winter comes, and again you will let go and fall to earth, to be taken in, to be brought back down to your original elements, to sink, to be eaten, to feed the roots that feed the trunk and the boughs and the twigs where leaves cling for a time, singing sweet food from sunlight, until the time comes to fall to earth, to feed the roots, to grow the trunk to hold the boughs to carry the leaves to sing sweet sunlight into food until they fall again, to feed the roots, to grow the leaves, to fall, to sing, to feed, to grow, to fall, to sing, to feed . . .

And you breathe deep again. And you begin to remember that you have a human body, a human mind. And you thank the tree, the leaves and the roots, the fungi and the worms, the billion hungry mouths of the earth, for this journey, and for what you have learned about the cycle.

And slowly, slowly you begin to breathe yourself back into your human body, feeling your feet and legs and torso and arms and hands and head take shape again. Slowly begin to move and stretch, to feel the edges of your human body. Slowly begin to sit up, to open your eyes. Say your own name out loud. Clap your hands three times.

And that's the end of the story.

Decay Is Food

The story above teaches us one of the great lessons of both earth-based science and spirituality—that there is no such thing as waste. Waste is food. All fertility arises from decay. There is no life without death, and death feeds new life. Life and death, decay and regeneration, are part of the same cycle. We cannot have life without death, fertility without rot. But death need not be feared or viewed with horror. It is part of the cycle—a transformation, not an end.

According to Marija Gimbutas, the most ancient Goddess of Europe was, beyond all, the Regeneratrix, the one who brings fertility out of decay. We serve her whenever we take responsibility for our wastes, returning them to the cycle of fertility.

One of the simplest ways to do that is by making compost from our organic food wastes. My first coven was called the Compost coven, and a descendent of that group still exists. I have always considered compost to be sacred.

Lazy Compost

There are volumes written about various ways to make compost, and the sale of a variety of different designs for composters is a small industry. Generally speaking, there are two basic approaches to making compost, and these correspond very closely to the two basic approaches to magic. The first is the Ceremonial Magician/Alchemist School (of composting or of ritual), which involves lots of complex processes that must be done at exactly the right time, and many careful measurements. I am not going to discuss that school here, because it is not my form of practice. If you are temperamentally drawn to such things, however, they can certainly have enormous value. I encourage you to look into biodynamics, for which many good resources exist.[4]

The second school is the Kitchen Witch/Permaculturalist School, which uses whatever you have lying around. It's the method for lazy or busy people, which is why it appeals to me. (I'm not saying which I am!) I've always believed that if there are two ways of doing something, and one involves much less work than the other, that's the way to do it.

There are a number of different approaches to the Lazy School of compost-making, but all of them hold in common the essential insight that when it comes right down to it, things rot. In making compost, we're not only working with nature, we're hastening what nature would do anyway, whether we interfered or not.

So . . . the essence of the lazy method is to pile up your wastes and let them rot. There are, however, a few simple tricks that will help material rot gracefully, odorlessly, and reasonably quickly.

We can think of a compost pile as made up of two basic sorts of material, which for simplicity's sake we'll call "green" and "dry." Green stuff is high in nitrogen. It includes food scraps, fresh grass clippings, fresh-pulled weeds, etc. (It also includes manure, although that tends to be brown.) Dry materials are high in carbon: dry leaves, straw, newspaper—your basic brown and crunchy stuff.

A successful compost pile has a rough ratio of carbon to nitrogen that is thirty to one. That's right—thirty! The microorganisms in the soil that break down plant materials need about thirty times as much carbon as nitrogen to thrive.

Because green stuff also contains carbon, which is the basic structural element of life, in practice what you need is about half green and half dry. Many of the problems people have with their compost pile—smell, flies, not breaking down, etc.—can be solved by topping the pile with more dry stuff.

The other big secret is to build the pile *big* enough and *damp* enough. Rotting material generates heat, and a hot compost pile will kill weed seeds and undesirable microorganisms. To get hot enough to accomplish this, a compost pile needs to be at least three feet in diameter and about that high, and it needs to be damp—not too wet or too dry, but about the consistency of a damp sponge.

So the very simplest way to make compost is to collect enough wet and dry stuff to layer it up three feet high; then keep it damp and let it rot. If you build your compost pile on top of the bed you want to plant, you will even save yourself the trouble of later transferring it.

A mass of compost will rot down amazingly quickly. I once nobly took an entire pickup-load of rotting food wastes back home with me from Witch camp. With the help of the other teachers, who had accompanied me back to my land for what they thought was a retreat, we made a pile of food scraps layered with straw in a bin two feet wide, four feet long, and three feet high. Within a few days, the pile was a third that high. Within a week, it was half its original size. In a month, it had virtually disappeared.

Some people turn their compost piles, to aerate them and help them break down faster. You can do that, if you like or if you need the compost quickly. Essentially, you are trading work for time—more effort for faster results. Alternatively, you can leave it alone, to rot peacefully undisturbed, trading time for work.

If you live in the city, it's worth investing in one of those black plastic composters that will keep rats out of your pile. If you can't afford one, get three or four old tires. Put down a base of wire of small enough gauge to keep rats out, and affix it to the first tire. Stack the others, and pile the compost inside, with a board on top and a brick or large stone to hold it. If you make two compost piles, one can be "cooking" while you keep adding to the second.

COMPOST BLESSING

We offer gratitude to the great cycles of birth, growth, death, decay, and regeneration. We are grateful to all the beings who have made the great transformation, leaving the remains of their bodies here. We are grateful to all the hungry mouths that consume the dead. Blessings on the termite, the beetle, the ant, the spider, the worm. Blessings on the fungi and the bacteria, those that need the air and those that avoid it. Blessings on all the life in this pile that will transform decay to fertility, death to life. May I always remember that the cycle of life is a miracle. May I continue to feel a sense of wonder and joy in the presence of death and life. May I remember that waste is food, and may my eyes be open to opportunities to close the circle and create abundance and life.

COMPOST PILE SPELL

I find my compost pile to be a good ally in dealing with problems that feel stuck to me. Here's a simple spell (best done while the moon is waning).

Ground, center, create a circle, honor the elements, and bless your compost pile. In your circle, have a large piece of fruit or vegetable ready to be discarded, and materials to write or draw with, either on the fruit or on paper. Take a moment and think about your problem. You can write it out on the paper and stuff the paper into your fruit, if the fruit is large enough. Or you can carve or draw directly on your object. Hold your fruit and breathe into it, imagining that you are letting the stuck energy of your problem flow into its flesh.

Say, "I give this problem, this energy, in the body of this fruit [vegetable, apple, carrot, etc.] to the great cycle of birth, death, decay, and regeneration. May it decay in its present form, and be brought back to its essential elements. May I see those elements clearly. May it fertilize some new seed, some new growth. With the offering of this fruit, I give thanks to the cycle, thanks to the processes of life and growth. I give thanks for this transformation. Blessed be."

Now bury your fruit in the compost pile.

Thank all the powers you've invoked, and open the circle.

Sheet-Mulch—Even Lazier!

Permaculturalists employ an even easier method of creating fertility, called sheet-mulch. Essentially, we turn the whole garden into a compost pile, spread out horizontally. With sheet-mulch, we are also trading time for work. A sheet-mulched garden may take longer to establish itself than a carefully dug raised bed, but it will save you all that backbreaking digging. If you don't want to have to wait, you can dig a small area for a prize vegetable bed and sheet-mulch everything else.

One advantage of sheet-mulching is that you don't disturb the complex structure of the soil or interfere with all those billions of happily munching bacteria. Mulch creates habitat for worms, who will aerate the soil for you. But if your soil is seriously compacted and dead to begin with, digging once to aerate and break it up may be helpful.

The hardest part of sheet-mulching is collecting the materials—lots of them! Cardboard, newspaper, old rugs, and even old clothes can be used for sheet-mulching. You'll also need lots of both green and dry materials.

Push down the weeds in your garden bed. (You don't even need to cut them; just press them down well.) If you want to jumpstart the decay process, add some high-nitrogen material. This step can be as simple as peeing on the pressed-down greens! (Human urine is about eighteen percent nitrogen.) Then cover the greens with cardboard or thick layers of newspaper, about a section thick. Don't use high-gloss, colored parts of the paper. Today's papers are printed primarily with soy-based inks and should be safe to use. If your site slopes, work from the top down so that lower pieces of cardboard lie atop higher pieces, to catch water—a reverse shingle effect.

Then add a layer of green stuff, and cover again with a layer of dry stuff (the thicker the better). It's really as simple as that. Water it all down to dampen it and start the decay. If you want to plant right away, just punch a hole in the cardboard and plant down through the hole.

As the sheet-mulch settles in, it will attract worms and soil bacteria. Keep mulching. Tuck your food scraps under the mulch and they will compost quickly.

Chicken "Tractor"—The Laziest Method!

If you live in an area where you can keep chickens, you can dispense with compost and let the fowl do your gardening. Simply create an enclosure for the chickens that covers the area you intend to plant. Feed them your kitchen scraps and let the chickens dig, weed, and manure the area while eating the pesky bugs. At the same time, those chickens will provide you with eggs.

When the ground is prepared, move the chickens and plant the bed.

Worms

If you are a single person or a small family composting in a small, urban back yard, you may have difficulty keeping chickens or gathering enough material for a true compost pile. For you, I recommend worms.

I love worms! Worms get a bad rap: "Lowly as a worm," "Nobody likes me, everybody hates me, I'm going to eat some worms," etc. In reality, worms are the least lonely of creatures. Each is both male and female, and whenever worms bump into each other in crowds, they mate in writhing balls of slithery worm orgies, indiscriminately fertilizing and being fertilized simultaneously. No constricted gender roles for them!

Along with furnishing you with the vicarious enjoyment of their erotic exploits, worms are the great creators of fertility. They tunnel into the soil, turning and aerating it. They eat soil particles and rotting food, passing them through their gut and turning them into worm castings, an extremely valuable form of fertilizer high in nitrogen, minerals, and trace elements. They add soil bacteria to the mix as well, inoculating the garden with many helpful bacteria.

It is the surface-dwelling red worms that eat food scraps and waste, along with manure. Some garden stores or bait shops sell worms, but you can often find them in your compost pile or around half-digested wastes on the ground. To thrive, worms need food, moisture, and a temperature that's neither too hot nor too cold—50s through 70s Fahrenheit being ideal.

You can create a bin for your worms by drilling some holes in a plastic bin or a plastic garbage pail for air and drainage, then putting in some earth, some shredded newspaper, and some food scraps. Add your worms, and let the whole thing settle in for a couple of weeks. It may go through a disgusting phase, where everything molds and rots (especially if you added a lot of food at first). But eventually it will settle down, the worms will start chomping their dinner, and you can start collecting lovely black worm castings.

You can add the castings directly to the ground around your plants, if you want, but I like to make "worm tea," dissolving a handful of castings in a bucket of water and then using the water to fertilize the plants. At times, I can almost see the plants perk up and lick their lips.

My own preferred way to keep worms is in a "tower"—a masonry chimney pipe just a bit wider than the five-gallon bucket it contains. The pipe sits in the center of my garden bed. I have a worm colony in the bucket, with drainage holes in the bottom. From time to time I wet down the colony with the hose: the water drains through and inoculates the bed with worm tea, soil bacteria, and eggs. Sometimes I put another small bucket below the colony's bucket, fill it from above with water that drains through the top bucket quickly enough that the worms don't drown, then pull up both buckets and use the tea from the bottom bucket to water plants.

The masonry chimneys are also attractive to slugs and snails, who congregate inside, where they can be easily picked off and, after a short prayer to Kali, dispatched.

In a cold climate, you can take your worms into the garage or basement for the winter. My friends Lisa and Juniper keep their worms in the kitchen. They have three plastic bins, stacked at a slight angle. Each contains worms, shredded paper, and food scraps. They keep the worms moist, and the worm juice runs down into the lower part of the bins, where they can siphon it off with a turkey baster and use it to water their houseplants.

In Sebastopol, California, the RITES project (Returning Intention Toward Ecological Sanity) collects food scraps from the local restaurants and raises worms on a large scale. They are now selling their worm tea as fertilizer, to fund some of their programs.

Worms truly represent abundance. Last spring, I was redoing my garden beds and went to shovel out our compost bin. Worms had gotten into it at some point, and it had turned into a mass of worms and worm castings three feet in diameter and three feet high. I subdivided them into new worm colonies, started new buckets full to give away, and still had enough left to cover all my major garden beds with a couple of inches of worm castings. Then I went out to the garden store to pick up a few plants. I noticed that the store was selling a pint container of rather depressed, pathetic-looking worms for $13.95. I figured that at that rate I had just put about $10,000's worth of worms onto my garden, and I started to rethink my whole career!

And the garden has indeed exploded with fertility, with tomatoes and corn and flowers crowding all over each other. Whenever something starts to look a bit peaked, I just give it a dose of worm tea.

And although eating worms sounds unappealing, they are actually a good source of protein. Although I must admit I haven't eaten any yet, I derive a slight sense of security from knowing that I *could*, if the worst happened.

One caution with worms: don't dump your worms into a pristine wilderness environment. There is some evidence that exotic worms in northern forests, whose own worms went extinct during the Ice Age, can disturb the ecological balance and *destroy* fertility instead of *creating* it.

Fungi

Fungi also are not generally well looked upon. Yet more and more we are coming to understand the critical role they play in fertility and regeneration.

Some forms of fungi break down the tough chemical bonds in wood that keep it strong. Others are important in the general decay of plant material. Still others are symbiotic with living plants.

The mycorrhizal fungi that we met in our trance journey are a vitally important part of forest ecologies. Threadlike and spreading, they insert themselves into the root hairs of trees and almost all other plants (except grasses). They then extend the reach of the roots, drawing in more water and nutrients than the plant can reach alone, in exchange for sugars extruded by the roots. The network of sticky threads helps hold the soil in place and allows plants to communicate and share nutrients. Through the fungal network, trees can nurture their young. Trees in the sun will feed trees in the shade—even those of a different species. A clearcut forest, where the mycorrhizal fungi have died, will not easily regenerate.

Other sorts of fungi, including mushrooms, are also symbiotic with trees. The part of the mushroom we eat, the fruiting body, is only a small extrusion of the whole organism, which exists predominantly underground and often interpenetrates the roots of living trees. The mycelium, the mass of threadlike tissue that constitutes the main body organism, extrudes its fruiting bodies when conditions of temperature and moisture and disturbance are right. Collecting wild mushrooms is a specialized hobby, and the scope of it is beyond this book. There is no simple rule of thumb for telling poisonous from safe varieties, and a mistake with mushrooms is one you don't want to make. But with care, wild mushrooms can be harvested. Eating the mushrooms I find in the woods is, for me, truly eating the flesh of the land. Each winter I pick and dry the varieties I am sure of: king and queen boletes, chanterelles, and matsutakes.

Mushrooms are also great healers. Chinese medicine recognizes several varieties as medicinals—reishi, shitakes, and many others. In a good mushroom year, when I eat them through the winter, I get fewer colds and flu and have abundant energy. Turkey tail mushrooms, which grow on old logs, are a natural antiviral and antibiotic when steeped or chewed.

Paul Stametz, author, mycologist, and founder of the company Fungi Perfecti, is truly the wizard of mushrooms. He has shown that mushrooms can also be healers of the earth. Oyster mushrooms will break down diesel fuel and other toxins. Stopharia can cleanse fecal coliforms from water. Shitakes and other varieties can be grown on cut wood as part of a program of truly sustainable forestry.[5]

The humble fungus certainly deserves more honor and respect from those of us who honor the earth. The fungus is a bit like the beggar in fairy tales, who appears lowly and dirty but offers great gifts and wisdom to those who treat her with love and generosity.

Sacred Seed

At the Via Campesina ritual mentioned above, an indigenous healer spoke of the sacredness of the seed. "The seed is sacred, because the seed is the beginning of all life," he said. "Everything comes from the seed."

Seed is indeed a sacred trust. The seeds for all of our traditional food plants are a precious gift of the ancestors, who saved them, selected the best each year and put them by for the next year, over centuries and millennia of time breeding thousands of different varieties of food plants adapted to different conditions, climates, and soils, and offering different advantages.

Seeds also represent abundance. On my table sits a bowl of fava beans, saved from plants I grew. Each bean can potentially grow a new plant, yielding many pods, dozens more beans. Each seed contains the instructions for its own replication and multiplication.

Farmers and gardeners have always saved seeds, traded seeds, and gifted each other with seeds. Seed-saving and seed-sharing are part of a network of relationships that hold communities together just as the sticky threads of mycorrhizal fungi hold the soil.

Seeds are also libraries of genetic information. Each seed holds the whole history of evolution, the record of hundreds of thousands of choices and accidents. The DNA in a seed may express itself in traits of the plant, but each seed also holds unexpressed and dormant potentials that may sometime in the future give rise to new varieties. Each seed is a concentrated communication about how best to grow and live and die in a specific place.

SEED MEDITATION

Hold a seed in your hand. Close your eyes and breathe deep. Feel the life, the coiled potential that you hold. As you breathe, relax and let your mind become a clear pool. Ask the seed to speak to you, to show you its history and its wisdom. You may see the faces of the ancestors, one after another, reaching back through time—each woman or man who guarded this chain of life, growing the plants, selecting and saving the seeds. You may catch glimpses of other landscapes, other places.

Ask the seed what it needs and wants in order to thrive.

Thank the seed, the ancestors, the land, and the elements, and open your eyes.

SEED SPELL

If you've let your stuck projects and problems decay in your compost pile, you may be ready to start something new. Planting seeds is also a good way to plant new ideas or new projects. Do this in the garden, after preparing a bed.

Ground, center, create a circle, honor the elements, and bless your seeds (see below). Take a moment and think about your upcoming project or the new beginning you wish for. Hold a seed and breathe into it, imagining that you are filling it with the image of your new project and with all your enthusiasm and passion.

Say, "I ask help from the great cycles of renewal for this project. As this seed holds a world of potential, so too does the germ of my new project. May I see and realize its full potential. As this seed grows, puts down roots, and sends out shoots, may my project likewise grow. May it find the nutrients it needs to flourish; may it be well-watered; may it thrive. May it create true abundance and further the diversity and joy of life. As I plant and tend this seed, I give thanks to the cycles of life and to the great mysteries of birth, growth, death, and regeneration. Blessed be."

Thank all the powers you've invoked, plant the seed, and open the circle. Don't forget to water and tend your seed as you tend your new beginning.

SEED BLESSING

We give gratitude to the elements of life embodied in this seed, to the wisdom accumulated over billions of years. We thank the ancestors who made the choices that gave us this seed, and who tended the chain of life to preserve it. We thank the air, the sunlight, the water, and the earth that sustain the life of this seed. Within this seed are precious and unique instructions for growth and life. May we always treasure the wisdom of the seed, and may we have the help we need to continue to nurture this life for the future. Blessed be the seed.

Seeds in Jeopardy

Unfortunately, the precious heritage of seeds and diversity is threatened by today's agricultural and economic systems. Traditional seeds are "open pollinated." Open-pollinated seeds breed true—that is, their offspring are true to their parents, with only minor variations. Such seeds can be saved from year to year, crop to crop. Farmers and gardeners can save their own seeds and don't need to constantly buy new supplies from seed companies.

Hybrid seeds are the product of more genetically diverse parents. Generally, they do not breed true—their immediate offspring may be quite different from the parent plants. And many first-generation hybrids are infertile or less fertile. Over many generations, traits can become "fixed" in some hybrids, but this often takes time and professional expertise.

Over the past century, large seed companies took control of most of the world's seed production. They specialized in hybrid seeds, born of two genetically diverse parents that do not breed true. Many hybrid varieties have advantages over their open-pollinated ancestors, but they have one large disadvantage for farmers and gardeners: their offspring may be quite different from their parents, so new seed must be bought for the next crop. Of course, this is an advantage for seed companies, which reap more profit when farmers are forced to buy new seed each year.

Genetic engineering carries this process further. By contract, farmers who purchase GMO seed are not allowed to save it. Companies hire private security agencies and encourage neighbors to turn in violators in order to safeguard their royalties. As people turned away from gardening, and from saving and sharing seeds with their neighbors, becoming more dependent on buying seeds, much diversity has been lost. In *Seeds of Change*, Kenny Ausubel writes: "Of the cornucopia of reliable cultivated food plants available to our grandparents in 1900, today 97 percent are gone. Since the arrival of Europeans on this continent, 75 percent of native food plants have disappeared from the Americas."[6]

But in the past few decades, corporate globalization has mounted a campaign to control not just the seeds, but the underlying genetic information within them. International trade agreements and institutions such as the World Trade Organization have allowed the patenting of life-forms, and have enforced those regulations worldwide.

What does this mean? For centuries, farmers in Sinaloa, Mexico, have grown yellow beans. In recent years they have earned an income by exporting them to the U.S. Until a man named Larry Proctor took some beans, grew them out for two years, patented them, and began demanding six cents in royalties for every pound of beans sold. The importers simply stopped importing those beans, and the farmers of Sinaloa, descendents of the men and women who had developed the beans over millennia, lost 90 percent of their exports.[7]

In India, for thousands of years the neem tree has been a source of medicine, insecticide, and oil used for many products. Many small producers made a living from products made from the neem tree. Until a large corporation, W. R. Grace—together with the U.S. Department of Agriculture—patented it, forcing the small producers to close down. Thanks to the hard work of Vandana Shiva and the group Diverse Women for Diversity, the European Patent Office struck down this patent in 2000.[8]

This story, repeated over and over again in abundant variation, is now common throughout the world. It amounts to a form of biopiracy, a theft of the gifts of the ancestors and of the very genetic material that underlies life itself.

The theft of genetic material, in the context of a system that allows the patenting of life-forms, provides a basis that makes genetic engineering potentially profitable. The biotechnology of genetic engineering produces GMOs—genetically modified organisms.

Genetic modification is very different from the plant breeding that farmers have done for millennia. Traditional plant breeding means selecting parents that are similar enough that they can produce offspring, then selecting various of their offspring over many generations for desired traits. We can change the character of a plant—but slowly, over time, and within parameters that do not fall too far from the original characteristics. Any naturally occurring change involves the whole organism, and any flaws in the breeding will generally become apparent in the organism's failure to thrive. Selecting too assiduously for one trait may result in other, undesirable, traits being passed on. Roses bred for color, for example, may lose scent or vigor. Because whole organisms have to grow up to reproduce, any change is always being tested in the context of the survival of the organism and its relationship to the other organisms around it.

Genetic modification is something very different. It involves artificially inserting genes from one organism into another organism—one that may be entirely unrelated. Genes from a flounder have been put into a tomato, for example—presumably to increase its protein content. Bacterial genes have been inserted into corn to make its pollen toxic to insects. Corn plants have been modified to resist herbicides, so that they and everything around them can be blanket-sprayed.

"Genetically engineered crops represent a huge uncontrolled experiment whose outcome is inherently unpredictable," Dr. Barry Commoner stated in a report challenging the basic scientific assumptions upon which the technology is based. "The results could be catastrophic."[9]

We do not know the ultimate impact of the radical changes in organisms that genetic modification produces. And we do not know what the unintended consequences of these changes might be. When organisms are not changing as a whole in relationship to a whole, integrated community, but are being manipulated piecemeal, there is no safeguard of survival over time to ensure that changes will be beneficent. Earlier in this book, I described the near-debacle of *Klebsiella planticola*, the soil bacterium that unexpectedly prevented plants from growing. Luckily, that outcome was detected before its release. In a similar vein, pollen from insecticidal corn could kill butterflies as well as corn-borers, or trigger intense allergic reactions in humans.

Experimenting with such things in the laboratory might be of some value, but releasing these products into the environment is a huge threat to biodiversity and global food safety. Crops such as corn are wind pollinated, and there is no way to confine genetically modified pollen to one field. Genetically modified corn was introduced to Mexico in 1996, against many objections from farmers, environmentalists, and local indigenous communities. It has already contaminated corn crops in Oaxaca and Puebla, the region where corn originated and the center of corn's biodiversity. This contamination represents an irreparable loss. For when plant breeders seek to restore a lost trait or improve the vigor or health of a crop, they often cross back to varieties closer to the wild. If the biodiversity of the wild stock is compromised, we lose a huge amount of potential for developing new adaptations.

Moreover, once a crop is contaminated by patented pollen, farmers can be forced to pay the very corporations that have polluted their crops! Percy Schmeiser, a farmer from Saskatchewan, Canada, was sued by Monsanto after their genetically modified canola contaminated his crop. In fact, their pollen damaged his canola, but instead of Monsanto paying him, he was forced to pay them $19,000, and they sued him for back "royalties" for growing crops that include their genetic material![10] The Canadian Supreme Court ruled in Monsanto's favor, upholding their right to royalties on organisms that include their patented genes. However, Schmeiser did not have to pay damages, as they could not prove that he benefitted from having the genes in his crop.

The major crops that have been genetically modified—corn, canola, soybeans, and sunflowers—are all wind pollinated. They were chosen deliberately by corporate scientists precisely for their potential for contaminating other crops—for once a crop is tainted, the farmer can be forced to pay up and prevented from saving his or her own seeds.

Pharmaceutical crops are now being grown that include human genes. We do not yet know what eating these crops will do to human beings. We are only now, after decades of exposure, beginning to understand that pesticides and herbicides do cause cancer and other diseases. Human genes in foods could trigger allergic reactions or cause other unexpected health effects.

The proponents of GMOs often invoke the "free market" as a justification for their profiteering. They claim to be providing what the market wants. But, in fact, the market rejects GMOs wherever it is free to do so. Europe and many third world countries have resisted the introduction of GMOs. Mendocino County in California and many Vermont townships have also recently voted in bans. But once an area is contaminated, it loses one of its strongest arguments for banning them. "They're already here," the argument goes. "It's too late. Oh well."

The companies that produce GMOs have resisted any attempts in the U.S. to require labeling that would allow consumers to make a truly free choice about what they want to eat. A "free market" is one in which consumers have full information about the products they choose among. If we did have a choice, an informed public would be unlikely to make GMOs profitable.

The patenting of life and the creation of genetically modified foods are part of the privatization of nature, the attempt to turn nature into a commodity that can be bought, sold, and controlled. If we believe that nature is alive and sacred, beyond price, if we say that the heritage of knowledge and legacy of biodiversity created by the ancestors should be common to all people, we need to educate ourselves on these issues, and to oppose the patenting and genetic manipulation of life-forms.

Growing some of our own food, cherishing biodiversity, and saving and trading seeds are also ways we can honor the life forms and the legacy of the generations past. Supporting small-scale local and regional agriculture is a productive form of resistance to a profit-driven vision of a world in which no country eats what it produces, but gears everything to exportation.

We now know something about how to grow soil. Let's consider a few more ideas about plants.

Plant Communities

Abundance arises from complex webs of association and cooperation. In nature, no species grows entirely alone. Plants grow in community, in association with other plants. And in those communities, each fulfills certain roles.

Some provide shade and leaf litter that nurture others. Leguminous plants, those in the pea and bean family, fix nitrogen from the air into the soil, providing fertility for others. Dynamic accumulators, deep-rooted plants such as comfrey and burdock, bring minerals and trace elements up from deep in the earth, returning them to the areas near the surface where they can be used by others. Other plants, known as "insectaries," attract beneficial insects and pollinators. Among these are many common flowers, including asters, sunflowers, yarrow, thyme, and Queen Anne's lace. Still other plants concentrate different minerals in their leaves and stems and branches, returning them to the soil when they die and rot. And *every* plant's roots are a feeding station for fungi and bacteria, in ways we are only just beginning to understand. Birds and animals contribute too, bringing in manure, digging and pruning, and catching bugs.

In one of my fields stands a huge old Oregon oak tree. Beneath it are so many sprouting Douglas firs that I refer to the spot as "the conifer daycare center." Doug firs have been replanted in many areas of my land, and have come back naturally in others, but nowhere as thickly as under that tree. Clearly, something in its shadow or in the soil conditions it creates favors their growth.

"Guild" is a permaculture term for a self-sustaining mini-system of plants growing together in ways that support each other. Guilds are found in nature, or course. Across the road from my garden, for example, is a madrone/tanoak/gooseberry guild with firs coming back through the larger trees.

PLANT COMMUNITY OBSERVATION

In your home base, ground and come into your senses. Take a walk around and observe plant communities. What grows together with what? Do you have a sense of what roles each plant fulfills? Are some plants serving as nurses for others?

Are there patterns of association that repeat? Plants that seem to like each other? Dislike each other?

Is there a spot that's especially diverse? Any idea why? Are there certain conditions that favor certain plants?

You might want to make some notes for your journal on this one.

A Plant Guild

If we want the garden to be a self-sustaining system, we need to think about planting guilds, not just individual plants. There are many factors to consider. Let's take the case of my nectarine tree.

My nectarine tree grows on a windy hill with poor soil and produces small, wind-battered nectarines of incredible sweetness. It was planted by the former owner ten or twenty years ago, and grows outside the deer fence, away from the drip irrigation, in a field full of rough grasses, particularly Harding grass—a tough New Zealand native that was planted by mistake after the big Creighton Ridge Fire of 1978, when an aerial seed crew grabbed the wrong bag of seed. It is now the scourge of the fields around here.

What does the fruit tree need? In one sense, nothing—it's surviving if not thriving after years of complete neglect.

In another sense, what does *any* plant need? Sunlight, water, and various nutrients.

Sunlight is no problem for the nectarine. On a ridge with no obstructions to the south, it gets plenty of sun.

Water is abundant in winter and spring. In the summer, the nectarine's roots are evidently deep enough that the tree can survive our normal dry period. More water might produce more fruit, but that fruit might also be less distinctive, less concentrated in its taste. Trees are also a *source* of water. They concentrate water at their dripline and around their trunks, and their leaves and branches comb the fog for moisture. So the nectarine might actually provide some moisture for other plants.

My nectarine also needs nutrients. Nitrogen, phosphorus, and potassium are the three key nutrients, along with various minerals and trace elements. This tree has new growth in the winter, and it produces fruit, so it's not starving. Fruit trees in general don't need to be highly fertilized. Too much fertilizer can stimulate a lot of woody growth instead of fruit production. And its own leaves provide nutrients and mulch for other things I plant. Still, I might want to give the nectarine some companion plants to help feed it.

First, I might plant something from the legume family, to fix nitrogen—possibly lupines or a groundcover of clover. Peas or beans will simply get eaten by deer.

Then I could put in some dynamic accumulators: plants that concentrate minerals. Comfrey, borage, bracken, and other deep-rooted plants feed from a level below the roots of the fruit tree, thereby bringing lost nutrients back to the surface. In general, I like my fruit trees to have a comfrey "pet" to keep them company. When the comfrey leaves die and dry out, they also provide a good mulch.

The matted roots of grasses compete with fruit tree roots, so to keep back the grass I might plant a ring of bulbs that are drought-tolerant and deer-proof. Daffodils or irises would do nicely, and they would also help keep the soil loose and provide me with flowers in the spring—a small crop.

If I want wild bees and other pollinators around in the spring when the tree blossoms, I need to provide food for them all year round. In addition, I want to attract predatory insects that eat pests. So I might include some of the apiaceae, the family of small-flowered plants that used to be called umbels (as in umbrella, because of the form of their flower sprays). This includes Queen Anne's lace, fennel, parsley, and dill (although the deer might eat the latter). I'll also include some asteraceae, some plants from the daisy/sunflower/aster family—maybe perennial Mexican sunflowers or Michaelmas daisies.

I might also think about some night bloomers, to provide food for night-feeding insects, which in turn provide prey for bats and other birds. Soap lily is a native bulb with delicate, miniature lily sprays that open at dusk in the summertime. I could include some water catchers as well—plants with shiny leaves to condense water, or needles and filigree to comb fog. Other native plants provide habitat for insects and animals. And I might include some herbs—perhaps thyme, lavender, sage.

In putting together the guild that would support my nectarine, I have some other criteria. Everything I use must be drought-tolerant, requiring little or no supplementary water. It must also be deer-resistant, because I'm not going to fence this area. And it must be a light feeder, requiring little or no extra fertilizer, so as not to compete heavily with the tree or force me into rounds of fertilizing.

Everything in this guild should serve more than one function. I'm willing to include "looks pretty" as one function, but ideally every element in a system should serve at least three functions. So I might include a native ceanothus, which would fix nitrogen, bloom in spring for pollinators and beneficial insects, keep back the grass, provide habitat for natives, and look pretty. Thyme would feed insects, keep the grass back, and provide me with herbs to eat. Soap lilies would not only be natives, night bloomers, soil amenders, and grass repellers; their roots could be used to make a natural soap and shaped into natural brushes.

Many years ago, I did plant a guild very much like the idealized one described above. I began with a sheet-mulch around the tree to kill off the grass. I first cut the grass, then placed a layer of thick cardboard over it, and finally added a layer of soil that I dragged up from the bottom of a seasonal pond. I was worried that the usual sheet-mulch of manure and compost might be too much for this tree and for the plants I intended to use around it. Besides, I had the pond soil and didn't have the manure and mulch at the time, and one permaculture principle is to use local resources. I planted lupines, comfrey, thyme, lavender, irises, sage, dittany of Crete, and daffodils. All of them still survive, seven years later, as does my wind-blown nectarine, in spite of many changes and several summers of utter neglect. It has proved to be a self-sustaining guild.

Maybe this fall I'll take some time to fill in the circle with more of the various plants I've been considering, throw more sheet-mulch over the grass, and propagate the sage, a variety native to the Southwest, which thrives here and could provide a nice crop for smudge sticks. A garden can always use improvement.

Or maybe I'll just leave well enough alone.

PLANT ALLIES

Plants are always communicating with us, and we can learn to deepen and intensify that communication. When we make special friends with a plant or with a species, that plant becomes our ally. If we use allied herbs to heal ourselves, that alliance can intensify their curative powers. If we grow food or flowers, we can provide plants with spiritual as well as physical nutrition.

Plants like to be talked to, sung to, appreciated. Flowers are vain: they like to be admired. As Mabel McKay said, if we don't use the plants and talk to them, they'll die.

Here's one way I approach making an alliance with a plant:

In a safe space where your plant grows, ground and come into your senses. Create a magic circle, and bless the elements and the Goddess.

Sit with your plant for a bit. Take time to observe it with all your senses—to look, listen, smell, touch, maybe even taste (unless, of course, you're making an alliance with poison hemlock or something similar).

Sprinkle a few drops of waters of the world at the base of the plant as an offering.

Now sit, breathe deep, and close your eyes. Picture the plant in your mind's eye, and ask permission to enter it and make alliance with it.

If you sense that the answer is yes, imagine the plant growing larger and larger. When it's larger than you are, imagine a magic door that opens up. Step into the plant.

Take a deep breath. Turn to the east inside your plant, and notice what you see and hear and feel and sense.

Now take another deep breath. Turn to the south inside your plant, and notice what you see and hear and feel and sense.

Now take another deep breath. Turn to the west inside your plant, and notice what you see and hear and feel and sense.

Now take another deep breath. Turn to the north inside your plant, and notice what you see and hear and feel and sense.

Now take another deep breath. Turn to the center inside your plant, and notice what you see and hear and feel and sense.

Take some time to explore the world inside your plant. You may see something that looks and feels like the physical plant, or you may see or sense images, hear sounds, feel emotions, or notice smells that are associated with your plant.

Sit inside it for a moment, and ask its deva (or guiding spirit) to make itself known and speak to you. Ask the plant if it has information for you. Take the time you need to listen and to learn.

Ask if there is a way you can call back the energies and powers of this plant when you need them, an anchor, a word or phrase, an image.

Ask if there is an offering you can make or some way you can give back to the plant spirit.

When you are done, thank the spirit of the plant, and all beings you've encountered.

Turn to the center and say goodbye and thanks.

Turn to the north and say goodbye and thanks.

Turn to the west and say goodbye and thanks.

Turn to the south and say goodbye and thanks.

Turn to the east and say goodbye and thanks.

Remembering your anchor, say goodbye and thanks to the plant, and find the magic door. Walk back out, closing the door behind you. See and feel the plant become smaller and smaller, until it is back to its normal size.

Open your eyes. Breathe, stretch, say your name out loud, and clap your hands three times.

Thank all you've invoked, and open the circle.

Once you have a plant as an ally, it will tell you what you need to know about it and how to use it. You can have more than one plant ally.

You can also use a similar technique to connect with the deva of your whole garden, or of a forest, or of a section of your land. Before I dug swales and planted olives on one part of my land, I spent a long time making alliance with it, asking its advice and permission, making sure it felt okay about the changes I was going to inflict. We carefully spared patches of native vegetation and left a giant coyote bush in the center, as native habitat and to attract beneficial insects. So far, the olives have thrived under my usual regime of benign neglect. The spirit of the land seems to have welcomed them.

BLESSING FOR EARTH

We give gratitude to the earth, to the dust of stars that congealed into the body of this planet, our home, and that still gives form and solidity to our bones and flesh. We honor the rocks, our sisters and brothers, and their long, slow cycles of transformation into life and back to seabed, mountain, stone. We give thanks to the living soil, the mother's flesh, and the billion creatures that haunt her caves and pores and chasms, to the beetles and the ants and the termites, to the soil bacteria swimming in the slick of water that clings to her mineral archways, to the worms, wriggling, eating, coupling, and transforming within her. We bless the plants, the roots and stems and boughs, the great trees reaching upward and the deep-rooted herbs pushing down, all who contribute to the cycles of birth and growth and death and decay that lead to fertility and new growth. For all that feeds and sustains life, for all that grows, runs, leaps, and flies, we give thanks. Blessed be the earth.

ELEVEN

The Center
The Sacred Pattern

From my journal:

I am flying over the Southwest on a clear and beautiful day. For once, I have a seat in the front of the plane and can see more than a wing. Below, the hills ripple and undulate in tones of gold and sand. They are cut by the lines of streams and rivers that carve out their canyons and valleys.

I'm looking at the land, the patterns of light and shadow, noticing how the little streams flow into bigger streams in a branching pattern. There is a regularity to the shapes of those branches, just as there is a regularity to the pattern and size and number of veins in a leaf. As if I were looking down at the skeletons of invisible leaves laid over the land, each ridge with its streams seems cut in regular intervals, each larger river fed by so many streams.

And—it's hard to describe this, but suddenly I can feel the rhythm of the land, as if the streams and rivers were notations for a drum rhythm, marking off space in exactly the same way the beats of a rhythm mark off regular intervals of time. I could play the ridges and the rivulets on my drum, and I understand that those rhythms, those relationships and ratios, repeat themselves in some way throughout nature. I don't know much about sacred geometry, except that it exists, but I'm sure that this insight is what Pythagoras and Co. were all about. The earth is alive, and there is a structure to her body, a unifying heartbeat expressed in the shape of the land and the patterns of flow just as our veins and arteries are related to our heartbeat and carry our pulse. And our veins reflect the same pattern that I see before me on the land.

Pattern and Relationship

We've taken a journey around the circle of the elements. We've been introduced to each of them in turn and have learned something about the cycles of life that each represents. But the elements do not exist in isolation. They are always in relationship to one another.

Symbolically, the center is the point where the elements connect and transform, where that ethereal fifth sacred thing we call "spirit" arises. To understand how the elements of life interact, we need to look not at the isolated elements, but at the patterns around us. Magical consciousness is pattern-thinking, thinking that can comprehend not just separate parts, but wholes in relation to other wholes.

Much scientific and academic thinking is like a focused laser, beaming intensely at one aspect of a subject. Such vision might focus in detail on a needle of a redwood tree, but it could never show us the whole tree itself, let alone the forest. A thousand directed beams might reveal a thousand needles or patches of bark or single buds, but to see the tree we need to step back, broaden our view, and look in a different way. And in fact the most brilliant and creative scientists do just that. Science is full of great discoveries that came in a dream, or to a person lying under a tree, in a moment of relaxation when the laser beam was turned off and a more diffused gaze could take in the sunlight and the stars.

Those who are wedded to their laser beams defend their preferred form of perception as "objective" and "real." Narrower academics will often attack someone whose perception is grounded in the larger patterns. As I've been working on this book, I've also been involved in making a documentary about the life of Marija Gimbutas, the archaeologist who did so much groundbreaking work on the early European Goddess-centered cultures. I've become aware of the tremendous backlash against her work in academia. There are many reasons for the backlash—to assert that women once held power, that war is not inevitable, and that early cultures were based on cooperation, not violence and competition, threatens some of our most basic assumptions about human nature and culture. But part of that backlash also comes because Marija was a pattern-thinker. When Marija was doing her later work, archaeology as a discipline had turned away from attempts to interpret or find meaning in ancient artifacts, and had restricted itself to description. Much of the criticism of her work takes an isolated object from ancient Europe, examines it in a laser-focused lens, and concludes that one cannot "prove" that it represents a Goddess or anything else.

But Marija was not looking at isolated objects; she was looking at patterns. She had an incredible breadth of knowledge and experience. She read fifteen languages and had spent decades examining archaeological reports from all over Europe, including many from eastern Europe that most of her colleagues in the U.S. could not read. She examined thousands of artifacts in museums, in the field, and in the five major excavations she directed. Her interpretations came from the patterns that emerged as themes and images and forms repeated themselves again and again. The evidence she amassed was of a different kind than that of her colleagues, one that did not fit the narrow definitions of the academics.[1]

A pattern is a form or a set of actions, a way of organizing energies, that repeats. Nature is full of patterns—indeed, she seems to enjoy certain patterns and uses them over and over again. Understanding those patterns will help us to a deeper appreciation of how things work in the natural world, and how these patterns impact human culture.

And we use these patterns in our energetic and ritual work, as well. Included below are suggestions for using the basic patterns of nature in ritual and spellwork. You might draw them, paint them, bring in objects that embody them for an altar. You might include actions in a ritual based on one of the patterns, or build a ritual around them. You might simply visualize them, knowing that patterns shift and channel energies, and energy generates form.

When we can observe and truly understand some of the basic patterns in nature, we can learn not just to speak back to her, but to sing with the music of the spheres.

GENERAL PATTERN OBSERVATION

In your home base, ground and come into your senses. Take some time to observe patterns. What patterns do you see in the earth? The water? The living things around you? What forms repeat?

If you are doing this alone, you might want to take a sketchbook with you. Drawing the patterns will help you experience them in a kinesthetic way.

If you are doing this exercise in a group, have people observe on their own and come back and draw the patterns they saw on butcher paper or a blackboard.

Then share your patterns. Did several of you notice the same pattern? Do the patterns fall into groups? Why do you think they repeat? What do these patterns do?

Patterns direct and channel various energies. They repeat because they *do* something. As we go more deeply into specific patterns, each section will begin with an observation exercise. I strongly recommend that you do the observation before going on to read the rest of the section. You can cheat, of course, but here's what will happen if you do: your vision will already be shaped by expectations, and your observations will not be as open and fresh as they would be if you observed first.

Branching Patterns

BRANCHING PATTERN OBSERVATION

In your home base, ground and come into your sense. Begin to observe the branching patterns around you. Are there trees? Other plants? People or animals? Look at a tree and count the number of "orders" (or levels) of branching, from leaf to twig to branch to limb to trunk. How many times does the tree branch? Observe different trees. Is there a change in thickness at each order? What angles do the branches make?

Branching is one of nature's core patterns. Branches are patterns of flow, of collection, concentration, and dispersal. They reflect the physics of water, since most often what living organisms collect and disperse is carried in a fluid medium. And they are two-way flow patterns. Each rootlet of a tree draws up water, which is carried to a large root, up into the trunk, and out through branches and twigs to the leaves; each leaf collects energy from sunlight that is used to make sugars, which travel the reverse route down to the roots. Every root or leaf has a direct route to the central trunk—but two leaves on opposite sides of the tree have no direct link with each other.

Branching patterns have orders of branching, from capillary to vein, from twig to branch. Rarely do they include more than seven orders of branching, however—whether they are river systems or oak trees. A seven-order oak is probably an ancient and venerable tree.

The branches themselves increase in size as we move up the orders, and this, too, is relatively consistent across different systems. The ratio of your capillaries to your veins is about the same as the ratio of a twig to its next-order branch—about one to three. Why these things should be similar is somewhat of a mystery, but it is likely that it has something to do with the physics of moving water.

Branches are usually set at an angle that facilitates movement. Moving bodies lose energy if they stop, make sharp turns, or double back on themselves. Most hardwood trees branch at a roughly thirty-degree angle. Conifers are the exception: their branches head out at a ninety-degree angle from the trunk (although gravity eventually pulls them down). Highway planners apply the same principle when designing freeway onramps and cloverleafs.

Branching patterns are also the model for many human organizations, from corporations to the military. They are the model of hierarchies, of systems that collect something—generally labor and product—from a broad base, concentrate it and disperse it, and return something else, perhaps pay and directions. The branching pattern is generally thought of as embodying efficiency, and for many things it is indeed very useful.

In a tree, the hierarchy of twig to branch to trunk does not reflect a hierarchy of value; all parts of the tree are vital to its functioning. The cells of the trunk don't sit around congratulating themselves on their superiority, constructing ideologies that explain why they deserve that concentration of sap, or putting down rebellions of the disgruntled root cells ready to throw off their oppression and rise up from the underground.

But in human systems, hierarchies often reflect an unequal distribution of power and value, which quickly becomes an unequal distribution of wealth and well-being. Hierarchies can be useful—in human as well as natural systems. I certainly wouldn't want to try to persuade a tree to grow in some other form. But when hierarchies become unjust, when some members are relegated to positions of low status and value and others elevated, we need to find other forms of organizing.

Understanding how branching patterns work can give us insight into how to take them down when they become dysfunctional. Getting those lowly rootlets to communicate with each other, to trade energy and goods directly instead of sending them up the pipeline, is the time-honored way of challenging an unjust hierarchy. Organizing the workers, the students, the ordinary voters—those who form the base of the tree and provide the nutrients—reminding them that they ultimately hold the power to determine the tree's survival, begins to transform this pattern into something more equitable.

If we are designing something that involves movement and flow, perhaps a path through the garden or a drip irrigation system, we need to consider the lessons of the branching pattern. We need to think about the angles where paths join, and shape them to facilitate movement. We can't have too many orders, and we need to widen the arterial paths.

The tree of life is a powerful symbol in many traditions. The early Norse saw it as Ygdrasil, the world tree, a huge ash with a cave at its roots where the Norns, the Fates, dwelled. High in its branches lived an eagle, and a squirrel ran up and down the trunk, communicating between earth and heaven.

Goddesses and Gods were often associated with sacred trees. Athena was identified with the olive tree; Zeus, with the oak. Inanna, in ancient Sumer, was originally the Goddess of the date palm. Asherah, ancient Goddess of the Canaanites, was worshiped in sacred groves and symbolized by a pillar which actually stood in the temple of Solomon for much of its existence.

In ritual, we use the pattern of the tree for grounding, putting our roots into the earth to disperse excess energy, and we use it to gather vitality, concentrating earth energies into the "trunkline" that moves up our spine, extending an energy field out from the top of our head to catch the vital energies of the sun, moon, and stars. The two-way flow of the branching pattern lets us draw earth energies up and move sky energies down, simultaneously. When we are rooted and grounded in the earth, any energetic lightning bolts we encounter will sink harmlessly into the earth and not burn us out. And we are linked to a source of virtually endless vitality.

Circle/Sphere Patterns

CIRCLE/SPHERE PATTERN OBSERVATION

In your home base, breathe, ground, and come into your senses. Now look around you, and begin to observe circles and spheres. Where do you find round forms? How did they get that way? What function do they serve?

The circle is the symbol of wholeness, of completion, of equality, with no head or tail, no top or bottom. As a pure form, a circle encloses more area, and a sphere more volume, with less surface than any other form. That makes them protective shapes. Eggs and seeds, which risk losing vitality if they are exposed to too much heat or cold or other impacts from their environment, are often roughly spherical, to expose less surface area to weather and rain. Pebbles in streams and the ocean, rolled and tumbled and bashed around until all their edges are smoothed, eventually take on a spherical form.

In *Metapatterns*, Tyler Volk writes, "Whenever life needs to close, contain and separate by minimizing area of contact with the environment, a sphere is often the answer. Because the emphasis is on protected, internal development, animal eggs, seeds and buds of flowers and leaves do well as spheres."[2]

When we want to symbolize equality in a group, we sit in a circle. In a circle, each person can see every other person. We cast a circle to symbolize protection, and do our rituals in circles, which contain and focus the energy we raise. If we are drawing or writing out an intention or a spell, we often put a circle around it to symbolize its completion.

Spiral Patterns

SPIRAL PATTERN OBSERVATION

In your home base, breathe, ground, and come into your senses. Now look around you for spiral forms. You may find snail shells or, if you're lucky, coiled snakes. But look also at the trees. Do some of them wind upward like an elongated barber pole? Stand at their base, and look up into the branches. Is there a spiral pattern to the way they attach to the trunk? Look down into the head of a sunflower, or look at the way peas reach out for support. Where else do you see spirals?

A spiral is a dynamic form of a circle. It comes back on itself, but always with a difference. It moves somewhere. The spiral is a pattern of growth and an ancient symbol of regeneration and renewal.

The nautilus spiral is a perfect exemplar of one of the classic forms of sacred geometry. It embodies what is called a Fibonacci series, a term for a mathematical formula that says, "What is, plus what was, is what will be."

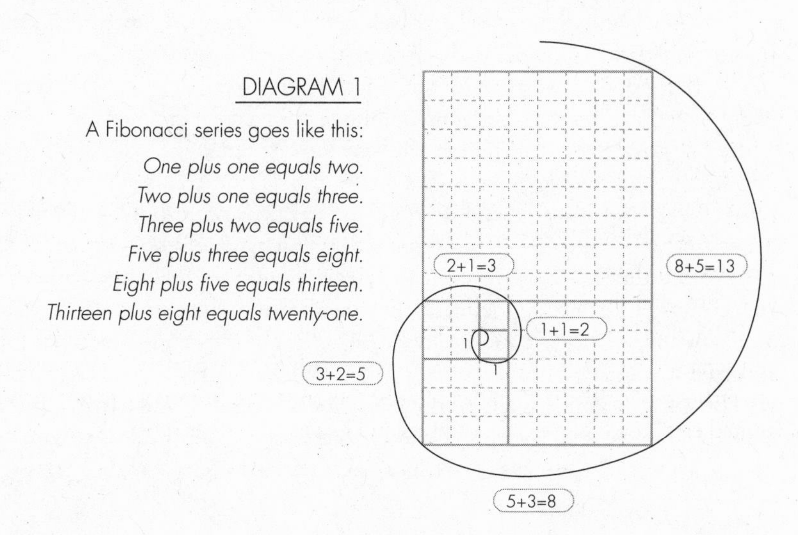

DIAGRAM 1
A Fibonacci series goes like this:
One plus one equals two.
Two plus one equals three.
Three plus two equals five.
Five plus three equals eight.
Eight plus five equals thirteen.
Thirteen plus eight equals twenty-one.
2+1=3
8+5=13
1+1=2
1
1
3+2=5
5+3=8

Draw this out, as in Diagram 1, and you have a nautilus spiral. The ratios of the numbers within each Fibonacci pairing give the golden mean, the basis of the sacred architecture of the Greeks. The Parthenon is the best-known example. Spaces based on these ratios feel intuitively right to our bodies. In fact, our bodies embody a rough Fibonacci series, in the ratios of finger joints to finger, fingers to hands, hands to arms, etc.

Tree trunks embody a different spiral, the elongated barber pole. The spiral twist gives strength to the trunk. The elongated spiral is also the form of the drill, a form for penetrating. Some seed pods take this form, drilling into your socks (or elsewhere) to implant themselves in a new environment. Water moves in spirals. A flowing river is an elongated vortex, drilling its banks, as Alice Outwater notes,

> Water that flows in a river moves like a corkscrew, twisting in on itself. The water on the river bottom is slowed by friction as it moves over obstructions on the riverbed, while the water on the surface of the river flows more quickly. Where the river bends, the faster flow at the surface pushes against the bends' outer bank and erodes it, while the slower, siltier water slips to the inner bank and drops some of its load of sand or gravel. This process creates the meanders found on lowland rivers.[3]

A spiral can stretch space. One classic permaculture design is the herb spiral, a round bed built up into a three-dimensional spiral planting surface, which adds planting space and creates interesting edges and small variations in light, shade, and temperature to please different plants.[4]

Time is a spiral, cycling back on itself but at the same time moving ahead. In the spiral dance, we coil in on ourselves, then turn outward to face each person in the group as we pass. When we wind the spiral in, we concentrate energy, eventually releasing it as an upward-spiraling cone of power.

In the northern hemisphere, the sun moves clockwise across the sky, and water forms a clockwise vortex when it drains down a hole. When we raise power for a positive end, to draw in energies and resources or to create something, we move in a clockwise or sunwise direction—or *deosil*, to use the old term. When we want to release or undo something, we move *widdershins*, or counterclockwise. In the southern hemisphere, the sun and whirlpools move counterclockwise, and our magical directions are reversed as well.

The spiral or vortex is a powerful magical form. It's the brew stirring in the cauldron. I might spin a vortex of energy counterclockwise to release obstacles to a project or plan, then spin it clockwise to draw in resources and power. I work with the form of the spiral whenever I want to create transformation and regeneration.

The spiral is one of the most widespread and long-lasting sacred symbols. Spirals are carved on the megaliths of Newgrange in Ireland, dating to the third millenium B.C.E., and decorate pottery and sculpture throughout Old Europe. The snake, which sleeps coiled in spirals and sheds its skin to emerge renewed, was an ancient symbol of the Goddess in Europe. Transformed into the dragon, it represented luck and renewal in ancient China. Spirals grace the pottery of the Hopi and the carvings of the Australian aborigines. Throughout the world, the spiral has been recognized as one of the basic patterns of life and growth.

THE SPIRAL DANCE

In ritual, we often dance the spiral. The spiral dance symbolizes regeneration. It allows everyone to come face-to-face with everyone else in the circle, and to look into one another's eyes for a moment, a moving meditation. And it can wind up a vortex of energy, serving as the base for raising a cone of power.

The spiral dance works best with forty to three hundred people, although I have done it successfully with close to two thousand. With fewer people, you will end up with more of a double line than a full spiral, and the dance will not last long enough to build a lot of power. However, you can spiral in and out several times to extend it, and still have a lot of fun.

Start in a circle. To end up dancing *deosil*—that is, clockwise—the leader frees her left hand and begins spiraling inward. When the spiral is about three lines thick, or when the center begins to look small, she turns to face the person following her, and keeps going (see Diagram 2).

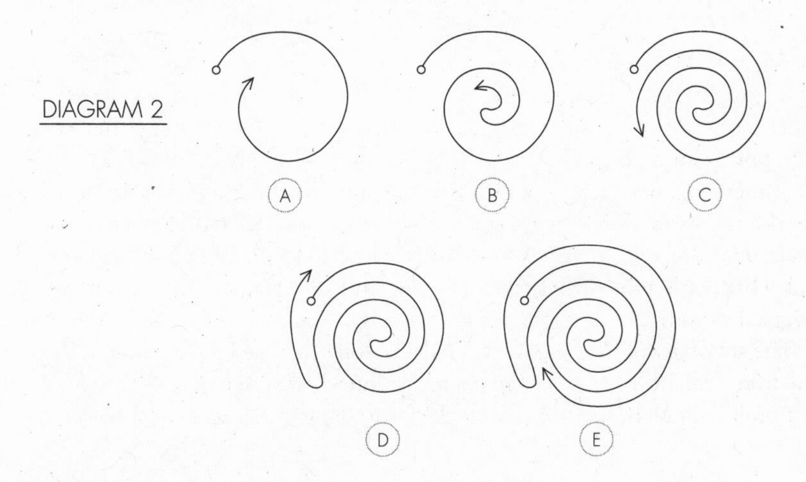

DIAGRAM 2

She continues around the inside of the circle, winding between the lines. Her path gradually takes her (and the line she's leading) out. The line following her will be passing in front of others, and people can look into each other's eyes.

When she passes the tail of the spiral, she should keep going about a third of the way around the circle. Then she again turns to face the person following her. Now she has made a loop in the line, and is moving around the outside of the circle. Again, everyone will have a chance to look into each other's eyes.

When she completes a circle and returns to the loop, she goes inside the loop. If by chance she should encounter the tail of the spiral, she must stay outside the tail. If the line unpeels and the loop disappears, she should hold the form of the spiral in her mind and simply spiral in again.

When she returns to the center, she should try not to let the center get too tight and closed. One trick is to spot a person in the line behind and keep the circle turning at a pace that keeps her even with that person as the back portion of the spiral continues to unwind. This sometimes takes some subtle communication with elbows and butt, as people tend to stop and crowd in when they reach the center, forgetting to be aware of the rest of the spiral behind them.

When the tail has unwound, and the energy is strong, the group can begin to focus on raising a cone of power, an energy vortex centered around an intention embodied in a chant, an image, or an object in the center. After a time, the chant turns into a wordless sound, like an open "Om." Participants can use their hands and bodies to direct energy, generally throwing their hands up into the air at the end and sounding together in a powerful harmonic tone. That's the moment to visualize your magical image and hold your intention strong. The energy will eventually die down, and participants can put their hands on the earth, crouching or lying down, to ground the energy and feed the earth.

The spiral dance does not need to end in a cone of power. It can wind and unwind back into a circle, or stretch into a line or snake dance.

Radial Patterns

RADIAL PATTERN OBSERVATION

In your home base, breathe and come into your senses. Now observe the radial patterns around you. Are there flowers in bloom that exhibit this pattern? Dandelions about to blow? Where else do you find this pattern?

The radial pattern is, on the one hand, the target, the pattern that says, "Come here! Come to this center!" Flowers use it to attract pollinators. Spiders incorporate it into the spokes of their web.

But a radial pattern is also the pattern of an explosion, radiating from the center, dispersing elements equally in every direction. It's the starburst, the sunburst, the seed pod exploding or dandelion blowing to send its seeds out into the world.

If I am doing a ritual or a spell to draw in some positive influence, I might create a radial pattern—draw an image, maybe, or place flowers on my altar.

Radial patterns and sun forms are also ancient sacred symbols, found in the earliest cave art, in pictographs from Norway to the Sahara. Sun Gods and Goddesses abound in mythology: Gods such as Apollo of the Greeks, Ra of the Egyptians, and Goddesses such as Sulis of the Lithuanians, Amaterasu in Japan. The Goddess is also the Star Goddess of Wicca, Inanna, Ishtar, Astarte, Aphrodite, and Venus, all love Goddesses associated with the morning and evening star.

Random/Scatter Pattern

RANDOM/SCATTER PATTERN OBSERVATION

In your home base, ground and come into your senses. Now begin to look for randomness, for what has no fixed pattern. The scattering of leaves on the ground, of stones in a stream. Where is there true randomness, and where do you sense an underlying order?

Randomness in nature is what allows freedom, movement, change. Randomness dissipates energy and breaks down form. A random scattering of rocks below a waterfall will help break the force of the water. A scatter of leaves on the ground will break down into compost.

Many cultures have honored this principle as the Trickster or Fool. The Raven/Trickster of the North American Northwest Coast First Nations was also the Creator, who brought fire to the world. Coyote, in the cultures of the West Coast, was the Trickster, whose practical jokes and bawdy humor helped keep the world in balance. Alegba, in the Yoruba tradition of West Africa, is the Trickster/Communicator, who translates messages between the world of humans and the realm of the ancestors and orishas, the greater powers beyond. The Fool in the Tarot, depicted with his knapsack and dog, about to step off a cliff, represents the freedom to take a leap, to step into the future without knowing what the outcome will be, to take a fall and land on your feet.

When energies are stuck, when forms are too rigid, introducing some randomness can help bring freedom and creativity back into a system.

Packing and Cracking/Honeycomb Pattern

PACKING AND CRACKING/ HONEYCOMB PATTERN OBSERVATION

In your home base, ground and come into your senses. Now begin to look at the different patterns of bark on trees. What are the different shapes and textures, forms and lines? Where else can you see these patterns? Look at your own skin. Observe the cracked mud in a dried-up mud puddle.

The patterns of bark and skin are patterns of expansion and contraction. A tree expands as it grows. The gray, smooth skin of a young Douglas fir grows thicker and deeply grooved as the tree ages. Our own skin expands and contracts, wearing lines into its smooth surface over time. As mud dries, it shrinks, leaving cracks and fissures.

This packing and cracking pattern is also a pattern of drainage. Channels in tree bark collect water, concentrate moisture, and move it off the surface of the tree (where it could cause rot) and down toward the roots. One way farmers in very dry lands conserve moisture is by erasing the cracks that form after rain or irrigation water evaporates, thereby preventing the remaining moisture from being channeled away. Someone who needed to drain marshy or waterlogged land might employ this pattern in digging a netlike system of water channels.

This is also a pattern of packing. Take a bunch of small, soft globes and attempt to push them into the smallest possible space: they will crunch into cells like the cells of a beehive or the geometric seeds of a sunflower. A beehive's six-sided cells are stronger than cubic cells of the same volume would be, and they pack better than spherical cells.

The amazing bee, who constructs her elaborate hives and gathers and stores honey there, who lives in a complex social system and carries out complex tasks, nurtures and feeds her young, and communicates through scent and dance the location of distant food sources, has been a sacred symbol in many cultures. Melissa, the Greek Bee Goddess, would be a fitting patroness of this pattern.

Web Pattern

WEB PATTERN OBSERVATION

In your home base, ground and come into your senses. Now begin to look for webs. Are there spiderwebs around? If so, what shapes do they take? Where do you find them? When are they visible, invisible? How do they fill space? Catch light? Transfer tension and compression? What elements are connected to which other elements? What other patterns do they incorporate? Where else does the pattern of the web repeat?

The web is one pattern in nature that we use as a magical symbol again and again. The spider's ability to spin, weave, and construct her elaborate creations has always fascinated people. It seems to hint at some larger, creative power at work in the universe, spinning reality out of her own body, weaving the tapestry of life. The Weaver, the Spider Goddess, has been sacred to many cultures. In the Southwest, she is Spider Woman of the Pueblo people, who wove the universe out of thought. In ancient Greece, she was Arachne. The Fates of the Greeks spun out a person's lifespan, measured the thread, and cut the cord. The Norns of Scandinavia were also spinners and weavers of fate.

Sitting in my home base spot one morning, I noticed three different kinds of spiderwebs within a few feet of where I sat. The first was a classic web, a two-dimensional spiral overlaid on a radial pattern. The rays of this sort of spiderweb are the anchors, giving it strength and support. Those delicate strands are stronger for their thickness than steel cable. The spiral that winds around them is the sticky trap. Movement at any point on the web sets up a vibration that alerts the spider that something has been caught. These webs are virtually invisible until the sun hits them at just the right angle, whereupon they glow like a brilliant idea, their bright, iridescent design shimmering for a short time and then disappearing.

As I looked out at the world from my home base that morning, I saw that another spider had built a domed web, an awesome engineering feat of suspension and tension that hung in the trees. This sort of web is likewise invisible until the sunlight illuminates its hanging fairy domes and palaces.

And a third web within my field of vision that day seemed almost entirely random, a wild zigzagging of sharp angles that filled space with a postmodern disdain for the usual forms.

The web is also one of the key metaphors of our age, with the emergence of the World Wide Web as a vital form of global connectedness.

If the branching pattern of hierarchy is the underlying form of our current political/economic system, the web is the pattern that seems to be emerging as a counter. Direct democratic organizing creates weblike networks of small circles linked to other small circles through spokescouncils (see Diagram 3). The difference between a web and a tree, in organizational structure, is that in a web all the parts can communicate directly with all other parts, if desired. Centralization can be useful. For example, if you want to know what programs

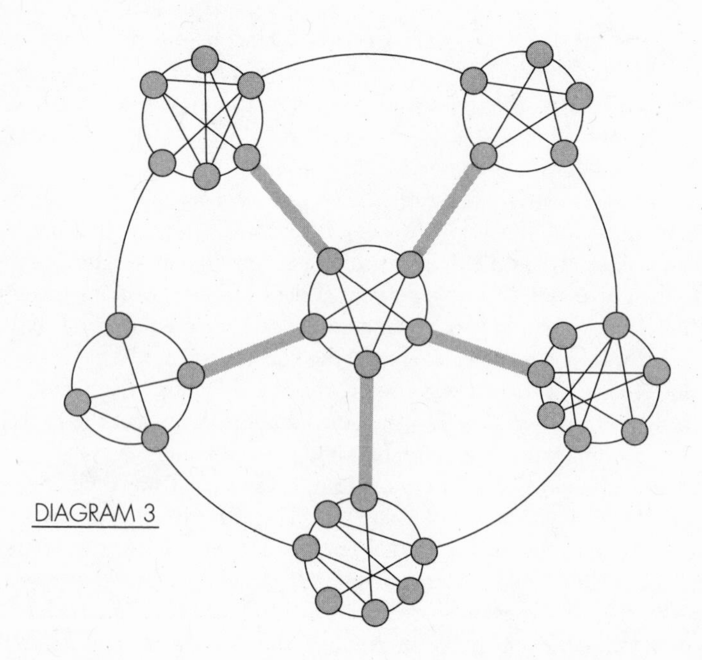

Reclaiming groups are running this summer, it's easier to go to one Web site than to check ten; if you want to find out what peace groups are operating in your area, it's much easier to go to unitedforpeace.org than to surf the Internet or thumb through the phone book or drive around town looking for them. But if the Marin Peace and Justice Center and the Santa Rosa Peace and Justice Center want to collaborate on a forum, they can talk directly to each other about it and make their own decisions. They do not need permission from some central authority.

Groups, spaces, and communities need a center that can further connectedness, a place or time when people gather for unplanned, spontaneous interactions that are often the matrix of new ideas and creativity.[5] In a home, people often gather in the kitchen, or around meals. A village has a marketplace or central square. A gathering or encampment needs a daily circle, or at least a time for announcements. Without a center, a web-based structure may lack a sense of identity and cohesion.

But centralization also means vulnerability. The key Web site can be hacked; the village market, commandeered by tanks. Whenever a structure becomes centralized, it always needs a backup. One of the permaculture principles is redundancy: we should always have more than one way of providing for every function.

WEBS IN RITUAL AND ACTION

Magically, the web is used to symbolize interconnection. Witches have many songs and chants that invoke the web and the weaver. We might sing,

> Weave and spin, weave and spin,
> This is how the work begins.
> Mend and heal, mend and heal,
> Take the dream and make it real.
>
> Or:
>
> Breath by breath, thread by thread,
> Conjure justice, weave our web.

To weave a web in ritual, provide the group with balls of yarn about the size of a tennis ball. Distribute the balls around the circle. All participants hold the end of their thread, or attach it to themselves, then toss the ball across the circle to someone else. As the balls fly back and forth, the web is woven. Often it's useful to have a "spider" or two willing to crawl beneath the web to retrieve balls that fall short. People might call out qualities they wish to weave into the web as they toss their balls.

When the web has been created, the group can circle with it, dance with it, and raise power to charge it. When the ritual is done, carefully take the web apart, perhaps saying something like "As we break this physical symbol of the web we have woven, we ask that the web of connections remain strong and whole. We will each take a thread to keep us linked to this web."

Each person can take a thread of yarn home with them. People can tie the yarn around each other's wrists while making a pledge or naming ways in which they will stay connected.

A web can also be useful in nonviolent direct action. It makes an excellent soft blockade, one that can quickly fill an intersection or block an entrance, delaying the authorities without heightening the level of tension. Chainsaws cannot cut through yarn, so forest defenders have used webs to slow down logging crews. When the yarn is first charged in ritual and then used in an action, it retains great power. I remember one blockade in Washington, D.C., when we were protesting a meeting of the International Monetary Fund and the World Bank. Our affinity group was in an intersection that had been filled with protestors before we even got there with a giant web, using yarn we'd all charged in ritual the night before. The police never tried to clear us away. Late in the day, one of our friends came back from a walk smiling. She had passed by a group of police cars and heard them discussing whether or not to clear our intersection. "No, too much yarn to deal with," was their conclusion.

Patterns for Meditation

There are some patterns that have been used for meditation and insight in many different spiritual traditions. The mandala and the labyrinth are complex patterns that can help us shift our consciousness into pattern-thinking.

The Mandala

The mandala is a quartered circle, a visual representation of the four elements within the circle of the whole. Tibetan Buddhists people their mandalas with deities and a complex pantheon of mythological beings inhabiting the four quarters, making the mandalas maps to the various realms of the spirit world. Psychologist Carl Jung recommended drawing and painting mandalas as a way to integrate one's own psyche.

But perhaps the most elegant, and certainly most vitally important, mandala is the one created by life itself to perform the miracle transformation of photosynthesis: the mandala of chlorophyll.

In *Gaia's Body*, Tyler Volk writes,

> Arguably the most important molecule of the biosphere, chlorophyll would be a perfect icon of a science-based, earth-centered religion. In the form of a molecular model, its head and tail might easily replace, for example, the Catholic chalice. A nature priestess could hold the iconic molecule by its tail and lift its illuminated, green head, truly the bringer of light to life, glittering high in the air before a reverent congregation. That might even lure me back to church![6]

MAKING A CHLOROPHYLL MOLECULE

If we had a Gaian Goddess temple, the chlorophyll molecule would make a lovely stained-glass window or floor mosaic. But, in the spirit of not taking ourselves too seriously, here's a story and directions for making a chlorophyll molecule—an enterprise that can be done with a minimum of twenty-five children or childlike adults:

The group facilitator begins by telling the following story:

Once upon a time, a long time ago . . .

A very young Gaia was amusing herself, watching the first life, the billions of simple, one-celled fermenting bacteria chase each other around her seas, gobbling complex molecules to fuel themselves, promiscuously trading genes. They were increasing at such a rate that she began to worry.

"How will they feed themselves when those complex sugars run out?" she wondered. What would they do? Would they starve? Would the seas empty of the life that she found so wriggly and entertaining?

As she mused, she began idly arranging and rearranging molecules in pretty designs.

Then the facilitator says to the group,

Now you be the molecules.

Someone jump in the center and be magnesium.

Four of you stand around her in the four directions and be molecules of nitrogen. Nitrogens—face out and hold out your hands.

Now four others of you go to each of the four nitrogens and make a little circle of five. You are carbons.

So now we have a center of magnesium, surrounded by a circle or pentacle in each of the four directions, with a nitrogen at its head and four carbons.

Now four more of you go and be a link between each circle. You are also carbons.

And all the rest of you link up and make a long, carbon tail.

DIAGRAM 4

Core Mandala within the Chlorophyll Molecule

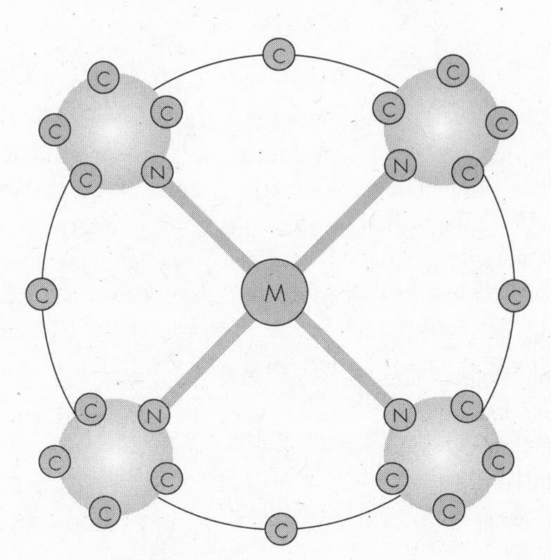

When the group members have carried out these directions (as shown in Diagram 4), the facilitator continues:

And now we've made the chlorophyll molecule, the beautiful mandala that Gaia designed. And when the tail holds the mandala to face the sun, and a photon of light strikes the magnesium in the center, she begins to vibrate. And that vibration is the energy that begins the complex, miraculous process of photosynthesis—the process that created the air that we breathe, that feeds just about all of life today.

In conclusion, the group can sing the chlorophyll song (to the tune of "O Tannenbaum"):

O chlorophyll, o chlorophyll,
You make the green world really real.
O chlorophyll, o chlorophyll,
If we don't love you, no one will.
O chlorophyll, o chlorophyll,
In parsley, sage, in thyme or dill,
O chlorophyll, o chlorophyll,
Make oxygen for lung and gill.

—*Words by Starhawk and Evergreen Erb*

The Labyrinth

The labyrinth is another sacred pattern that is very ancient. Labyrinth designs have been found on standing stones and in pottery designs from third millennium B.C.E. Crete to the contemporary American Southwest tribes. The labyrinth figures in myth: in ancient Crete it was the secret maze below the palace of King Minos where Theseus faced the minotaur, the monster with the head of a bull.

A true labyrinth, however, is not a maze, not a puzzle or trap. A labyrinth has a single pathway through it, not many confusing alternatives and dead-ends, and leads to a center that's a spot for insight and meditation. There are many forms of the labyrinth, from the classic Minoan seven-path labyrinth to the more complex pattern of the labyrinth found in Chartres cathedral. Today, many churches and healing centers are adding labyrinths as places to do a walking meditation. One of the beauties of a labyrinth is seeing many people walking and meditating together, each on her or his own unique journey, but sharing a sacred space in common.

MAKING A LABYRINTH

A labyrinth can be made simply, drawn in chalk on a floor or pavement, laid out with masking tape on a rug, mowed into a lawn or field, or laid in stone, paving, or gravel amidst plantings. For a walkable labyrinth, a space at least twenty feet on each side is needed.

First, before tackling your actual labyrinth medium, *draw* a labyrinth. (Diagram 5 illustrates the Minoan seven-path labyrinth.) Take a sheet of paper and make a cross (A). Put a small, right-angled arrow tip with its point toward the center in each quadrant, and a dot in each corner to make a square (B).

DIAGRAM 5

We can start on either side, but let's start on the top right. Connect the top of the center line to the top of the first arrow tip on the right with a curved loop (C). Now move counterclockwise and connect the top of the left arrow's tip to the top of the first dot on the right (D). Move counterclockwise again, and connect the dot on the left to the tip of the sidebar of the arrow tip on the right (E). Continue around, connecting the elements, until the labyrinth is complete.

What you've drawn above are the *walls* of a labyrinth. When you're comfortable drawing that form, experiment with drawing the path *between* the walls instead (see Diagram 6). When you mow a labyrinth, you mow the path, not the walls.

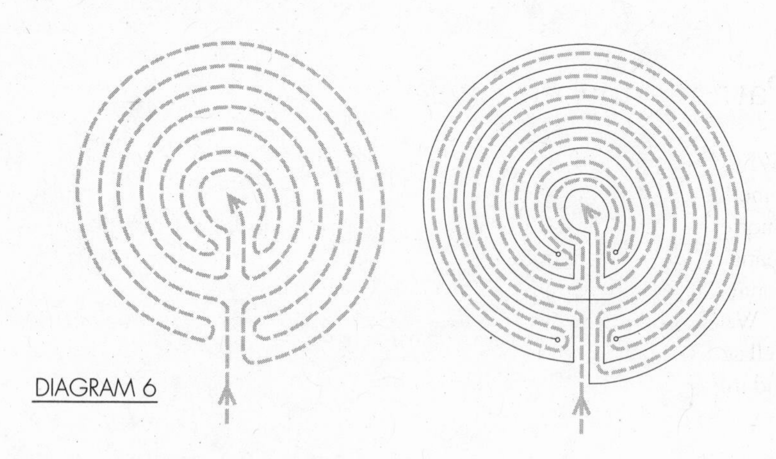

Practice chalking labyrinths or laying them out with straw or tape or some impermanent material. When you are comfortable with the form, you can easily begin to create more permanent labyrinths in the landscape.

There are many ways to use the labyrinth as a meditation tool.[7] For me, perhaps the simplest is to hold an issue, problem, or question in my mind and walk into the labyrinth slowly, examining the issue from all angles, changing my focus each time I switch directions. I might ask myself, for example, how I think or feel about it, what material factors are involved, what the heart of the matter is, what spiritual energies I can call on to help me, what my vision is, and how it might manifest.

When I reach the center, I stand in open, receptive attention, noticing what insights or visions come.

As I walk out, I focus on how I might use those insights or manifest those visions.

Before entering and on exiting, I leave an offering of waters of the world.

My favorite labyrinth is on the headlands above a beach in Sonoma County. It's a simple stone labyrinth laid on the low scrub, with a view beyond of the waves and cliffs. A path is worn into the ground from the many feet that have walked it, and in the center is a simple altar of stones, which is always graced by many offerings. The rumor is that a local woman and her friends built it to celebrate her fiftieth birthday. It's not on any map, but many people have discovered it. I stop by often when I need inspiration or grounding, and I especially love the sense that I am sharing this sacred place with many unknown others. The offerings are always changing, and sometimes stones are added to the path or taken away, but the labyrinth remains, a beautiful gift of the spirit.

Patterns in Our Lives

"Why do I always fall in love with men who abuse me?" "Why do I always do something to undermine myself just when success is within reach?" Much of emotional healing work and psychotherapy revolves around identifying and changing our own internal patterns, especially the destructive and dysfunctional actions and relationships that we tend to repeat.

We can apply the skills of observation and awareness to our own patterns as well as to the patterns of nature around us. Awareness is the first step to insight, and insight is the first step to change.

INNER PATTERN OBSERVATION

The magic circle of the elements can help us look for our own internal patterns. You can work with the questions below in several different ways. Alone, you can meditate on them, write in your journal about them, or take each for a period of days or a week to reflect upon and keep notes on.

With a friend, partner, or in a healing or mentoring relationship, you could share with each other your reflections and give each other feedback and insights.

In a group or circle with a high degree of intimacy or trust, you could go around the circle, giving each person time to respond to each question without being interrupted or challenged. At the end, you could discuss what is similar or different about your conclusions. In group work, I would suggest taking one element each session, and devoting five or six sessions to this work.

Air

Are there patterns you can identify in your thoughts? Particular phrases or words you say to yourself again and again, or snatches of inner dialogue that repeat? Names you call yourself? Images you hold? Fantasies?

How do these affect your perception of yourself? Others? How do they impact your energy? Emotions? Body? Spirit? The choices you make? The possibilities you see around you? How do these patterns restrict or harm you? How do they serve you? What do they do?

Fire

Are there patterns you can identify around your energy level and how you use your energies? Cycles or repeating ebbs and flows? Ways you dissipate or squander energy? Patterns of eating or drinking or sleeping? Repetitive ways you build up energy or stoke your fires?

How do these affect your perception of yourself? Others? How do they impact your energy? Emotions? Body? Spirit? The choices you make? The possibilities you see around you? How do these patterns restrict or harm you? How do they serve you? What do they do?

Water

Are there emotional patterns that you can identify? Cycles of feeling? Patterns in love, patterns in relationships? Patterns in the way you respond to fear? Hope? Anger? Attack? Loss?

How do these affect your perception of yourself? Others? How do they impact your energy? Emotions? Body? Spirit? The choices you make? The possibilities you see around you? How do these patterns restrict or harm you? How do they serve you? What do they do?

Earth

Are there physical patterns you can identify? Patterns in your health, fitness, muscle tone, and flexibility? Do you get sick in response to other patterns? Cyclically?

Are there patterns you follow around money or other material resources? Around providing for yourself and others? Around shopping or spending?

Do you notice patterns around your ability to set boundaries, or your encounters with others' boundaries?

How do these affect your perception of yourself? Others? How do they impact your energy? Emotions? Body? Spirit? The choices you make? The possibilities you see around you? How do these patterns restrict or harm you? How do they serve you? What do they do?

Spirit

Are there spiritual patterns that you can identify? Patterns around consciousness-change, intoxication? Addictions? Patterns in communication and connection with others?

How do these affect your perception of yourself? Others? How do they impact your energy? Emotions? Body? Spirit? The choices you make? The possibilities you see around you? How do these patterns restrict or harm you? How do they serve you? What do they do?

GROUP PATTERN OBSERVATION

Groups, too, fall into patterns. Below are some questions about the group's own processes that you might wish to consider.

Are there repeating patterns that emerge around the group's perception of itself or others? Around who has influence, whose voice is listened to by others? Around the group's energy level, the distribution of work and responsibility? Around who holds power? Patterns of attack or judgment? Emotional patterns? Patterns around acquiring, using, and distributing money or other material resources? Setting boundaries and encountering boundaries? Patterns of inclusion or exclusion? Patterns in communication, in connection or lack of connection?

Changing Patterns

A pattern may serve us or restrict us, further our growth and development or truncate it. But if a pattern is not moving us forward, we can change it.

Changing patterns is not easy, however, especially if they are ingrained or rigid, as addictions are. To change a pattern, we may not be able to simply *stop* it; instead, we may need to *replace* it with a new, consciously chosen, pattern. Twelve-step programs such as Alcoholics Anonymous replace the patterns of addiction with meetings that themselves follow a set pattern, and with a program of support and self-reflection. Many forms of counseling and psychotherapy exist that can be very helpful in identifying and changing personal patterns.

Ritual can also be an important tool for changing patterns. Ritual itself follows a pattern, and we can use our understanding of patterns (and what they do) to help us construct a ritual or spell that can move us from insight toward change.

One common pattern in ritual could be called releasing/calling in. Within the body of the ritual, we go through four steps:

We consciously acknowledge and name some pattern, force, or energy that we want to release or transform.

We do something symbolically to release it.

We identify some positive pattern, force, or energy we want to replace it with.

We do some act to symbolize that replacement.

The pattern we are releasing might be negative, but it might also simply be something that has outgrown its current form and needs to change in order to grow.

So to work a changing-pattern ritual, we must first identify the pattern we want to replace and find some object or action that will symbolize that pattern. Let's look at an example.

At one point in the growth of Reclaiming, our tradition of the Craft, we had seven or eight different Witch camps around the United States and Canada, and we called our first-ever retreat for teachers and organizers. At that time, all the decisions about teaching and programming were made by the San Francisco group that had started the camps, and we wanted to broaden the structure and empower more groups. Those of us in that San Francisco group spent many hours in meetings, and I became more and more aware of how many threads of the organization I carried personally. I taught at all the camps, and I knew everyone, all the teachers and organizers. They all knew me, but many had not met each other before that weekend. We wanted to create a web of autonomous camps, but I was still the central hub, and I wanted out of that position.

As part of our closing ritual at that retreat, I stepped forward and said that I felt as if all the energetic threads of all the camps were held in my belly. I said that I felt that I had held them well and nurtured the camps into growth, and that I was ready to let go of them. The image of myself as a spider with all the threads of the camp extending from my body symbolized the pattern we wanted to transform.

I then asked people to come forward and pull the threads out of my belly. As they did so, I felt a tremendous sense of release and relief and lightness, as if letting go of a tremendous burden. Patti Martin, one of our teachers, then reminded me that it was important for me to fill up with some positive energy to replace what I had let go of. I consciously breathed in energy from the earth, and opened to the love and support of the community around me. And, in fact, letting go of some of the work of the Witch camps opened up space and time for me to learn more about permaculture, spend more time in my garden, and (later) focus more on activism and global issues of earth-healing.

To symbolize the new pattern we wanted to create, we wove a web of yarn that represented a more open field of connection. Other challenges arose, both in the ritual and afterward, but the basic transition was made.

RITUAL FOR CHANGING PATTERNS

Before you begin, decide how you want to symbolize the pattern that you want to change as well as the new pattern you hope to adopt, and track down the necessary supplies.

There are many ways to symbolize change. You could find an object that represents the old pattern and draw on it, burn it, dissolve it, or bury it and let it compost. You could draw, sculpt, or paint a new symbol or image for the new pattern, or you could plant a seed, an herb, or a tree, weave or spin or sew a garment, dance a dance, or simply find something appropriate and breathe and pour energy into it. The discussion above of pattern can give you ideas of what to draw or how to symbolize the change.

Once you've decided on your symbols, ground and create sacred space.

Create your symbol for the old pattern. Raise energy with a dance, chant, or action to release it.

Create your symbol for the new pattern and charge it with energy by chanting, dancing, breathing, making sounds, or symbolic action. Ground the energy into the earth, and keep your symbol on your altar or in a special place where you will see it often.

After enacting your transformation, you might offer thanks, bless food and drink, and then say goodbye and thanks to any powers or energies you have invoked.

The power of the ritual is not just in the actions or words, but in all the work you do to prepare for, plan, and carry out the ritual; in the strength of your attention and focus; and in the witnessing of your community. A ritual might not instantly bring about the change you desire, but it might start you on the road to that transformation.

Patterns of Time

The Cycle of the Moon

MOON CYCLE OBSERVATION

For a full cycle, try to actually see the moon each night, even if only for a brief time. Notice when it rises and where, or when it appears in the sky. What time do you have to get up in the night to catch the waning moon? How early does the waxing moon set? Do your energies change with the moon?

In Wicca, our core religious imagery centers around the cycles of the moon and sun, which are two mythological representations of the great cycles of birth, growth, death, and regeneration throughout nature. Because I and others have written so extensively about these cycles, I will be fairly brief here, but much material can be found in *The Spiral Dance* and *Circle Round*.[8]

The moon cycle, from dark to waxing crescent to full, then to waning crescent and back to dark, swells the ocean's tides, awakens the powers of growth in plants, and strengthens human and magical energies. Farmers plant by the moon because they know that seeds will sprout and take root best when they can draw on the energies of growth and increase that the waxing moon provides. Spells and rituals for increase and growth should also be done while the moon is growing. Symbolically, the new moon is the Maiden, the young Goddess, wild and free, infused with the energies of inspiration and beginning.

The full moon is the culmination of power, the flood tide, the peak of magical and subtle energies. It's the time for rituals of fulfillment, empowerment, and transformation. The full moon is Mother, not just of children but of all creative enterprises, the nurturer, she who sustains us. And she is Lover, the erotic Goddess, riding on the flood of sexual energies that are aroused with the moon.

And finally, the waning moon is the transformation of power, the turning inward of the energies of growth, a time for letting go, releasing, going within. The waning moon is the time to plant root vegetables and those that store their nutrients underground. It is the time for divination and meditation, for looking within. Mythically, the waning moon is the Crone, the wise woman, the elder who knows the secrets of life.

When we become aware of the moon's cycle, when we craft rituals that follow her ebb and flow and when we are aware of the flux of her energies, we become embedded in the cycles and rhythms of life. Life ceases to be just a linear progression from birth to death and becomes part of a great round of constant regeneration. The moon is our monthly proof that darkness gives way to light, endings to new beginnings.

The Cycle of the Sun: The Wheel of the Year

At one point in my life, I lived a few blocks from the ocean, and I became obsessed by sunsets. Every night, I had to walk down to the beach to watch the sun set. I saw how it moved northward as the year progressed, until instead of setting over the waves it dropped below the hills that cradled Santa Monica Bay. After the summer solstice, it began to move south again, until in midwinter it lit the crests of the waves with red and purple fire.

The cycles of the sun, its daily rising and setting and the yearly cycle of the seasons, are also core cycles of birth, death, and regeneration. In Wicca, we celebrate eight major seasonal holidays, or Sabbats, evenly spaced around the year, which constitute the Wheel of the Year. They come from the Celtic traditions and are attuned to the climate cycles of the British Isles and similar temperate regions, but because they are evenly spaced, they are adaptable to many variations. Each holiday begins with sunset, not sunrise, so they span two days.

These celebrations include the solstices, the shortest and longest days of the year, and the equinoxes, the two days (signaling the beginning of spring and fall) when day and night are equal. The other holidays are the cross-quarter dates, which fall halfway in between the solstices and equinoxes. These cross-quarter dates were the key markers of the ancient Celtic year.

The names for these eight holidays, listed below, derive from Celtic or archaic English names, and have become traditional in many branches of the Craft.

Samhain, or Hallowe'en, October 31 / November 1

Yule, or Winter Solstice, December 20–23
(The exact dates of the solstices and equinoxes vary from year to year and thus must be checked with an astrological calendar.)

Brigid, February 1/2

Eostar, or Spring Equinox, March 20–23

Beltane, May Eve / Mayday, April 30 / May 1

Litha, or Summer Solstice, June 20–23

Lughnasad or Lammas, July 31 / August 1

Mabon, or Fall Equinox, September 20–23

The following text offers one way to view the Wheel of the Year as an ongoing mythology that ties into our personal growth and development. In this version the forces of change are personified and gendered female and male, but versions could be made that are very different, using multigendered or nonhuman imagery.

The year begins at *Samhain* (pronounced *sow* [as in female pig]-in). Although we are moving into the darkest and coldest time of year at this holiday, we celebrate Samhain as our New Year. The old crops are gathered in; the land is resting, waiting for new growth to begin. The Goddess is the Crone, the old wise one, she who receives the dead and comforts them. The God is the Horned God, the animal who gives away his life so that others can live. The veil is thin between the worlds of the living and the dead at this time, and we can visit our beloved dead and receive help from the ancestors. Theirs are the loving arms that gather us in when life is done, cradle us, and bring us back to rebirth. We honor our ancestors and our beloved dead.

At *Yule*, the Winter Solstice, the year is reborn. The sun, who has grown old and tired, goes to sleep in the arms of Mother Night, and is reborn at dawn. The Goddess is the Dark Mother, the giver of gifts and the teacher of lessons. Her love is the first gift given to all of her children. The God is the reborn year, all that is new, growing and possible.

The year grows, and the sun gets stronger. The days begin to lengthen. At *Brigid*, the year is beginning to grow up. The teacher of lessons gives us challenges and receives our pledges. She is Brigid, Goddess of Fire and Water, the ancient Goddess of poetry, the forge, and healing. The God becomes the poet, magician, teacher—the keeper of the mysteries. We honor all teachers, stepparents, foster parents, aunties and uncles who teach and care for us.

At *Eostar*, the Spring Equinox, day and night are equal and balanced. The Mother steps back, and the Daughter comes forth. She is life itself, who has been sleeping in the dark of winter. Now the sun wakes her up, as seeds awaken and growth begins in the warmth and rains of spring. If the Winter Solstice is the birthday of the sun, the Spring Equinox is the birthday of the earth. The Goddess is the Maiden who returns, bringing Spring, and she is the magic hare who lays the egg of life. The God is the waxing sun, and the Trickster, power of change, freedom and chance, B'rer Rabbit, Raven, Coyote.

At *Beltane*, the earth is fully awake and everything is blossoming. Just as Samhain was a time to connect with the dead, Beltane is the holiday that celebrates life. The Goddess and God become the lovers of all living things, and bless all forms of love. The gates between the worlds are open, and we can connect with the life-spirits of plants, animals, fairies, and the Mysterious Ones. We honor all mothers who bring life into the world.

At *Summer Solstice*, the sun reaches its peak and begins to decline. The Goddess becomes the lover of all things that fade and die. She is the mother of abundance and fruitfulness, who loves and feeds her children. The God, at the height of his power, begins to transform, to go into the Otherworld, carrying our messages and hopes. He represents all beings who sacrifice, who give of themselves so that life may go on. We honor fathers, who plant the seeds of life.

At *Lughnasad* (pronounced *Loo*-na-sa), the days are already growing shorter, although we are at the warmest time of year. The Goddess becomes the Harvest Mother, as we gather in the first of the crops. The God gives us the gifts of the Otherworld: life in the form of the food we eat and the skills and arts of human life. She is the Gatherer; he is the grain cut down to be planted and grow again. He is Lugh (pronounced Loo) of the Long Hand, the Shining-Faced One, and he is Lugh the Many-Skilled, God of the arts and knowledge that allow human beings to live together. We celebrate his wake at this time of hope and fear. We honor all teachers who share their arts and skills.

At *Mabon*, the Fall Equinox, day and night are equal again. We give thanks for all the gifts we have received, for everything we have harvested. The earth begins to prepare for her winter's sleep; the Goddess and God grow old and wise. They come to us in our dreams and guide us into the Otherworld.

To root our practice of ritual in the reality of the earth and her cycles in the particular place where we live, we need to do our own observation and adapt the myth to our own climate, plant and animal communities, and ecosystems.

SUN CYCLE OBSERVATION

Start a journal for a year of seasonal observations. On each Sabbat, note what the weather is like, what is growing or dying, what leaves are turning or falling, what needs to be done in the garden, what bulbs are blooming, what is budding, sprouting, blossoming, ripening, overripe, rotting, rutting, mating, birthing.

Now write your own Wheel of the Year, your own myth for the cycle of seasons as they unfold where you live.

Better yet, do this for three years, or five, or fifty, letting your own myth evolve.

This is the Wheel of the Year that I wrote for my home in the Cazadero Hills.

Wheel for the Coastal Hills

At Samhain, the year begins as the autumn rains return and renew the land. Deer are rutting, apples are ripe, and the year is getting cold and dark. The sun is low, but the streams are not yet full enough to use our hydro system, and we are reduced to running our generators on fossil fuels. We celebrate the rain's return, praying for a year of enough rain, falling gently and spaced evenly enough so that we don't have floods.

At Yule, the year is dark and cold but the rains are, with luck, abundant. The sun begins to return; our micro-hydro generators rev up and light our homes with energy produced by rushing waters. The coyote bush blooms, and so does the rosemary, and the very first narcissus puts out buds to celebrate the rebirth of the year-child.

At Brigid, the land is green, the daffodils are blooming, and the fruit trees begin to blossom. This is pruning time, time to cut away the dead wood and the overgrown branches, and planting time, the tail end of bare-root season, time to take cuttings and plant the last of the perennials and celebrate the birth of lambs and fawns. We plant, now, not just for the year but for the ages.

At Eostar, the land is at the peak of its green beauty. The soil is warming, and bulbs and fruit trees are in high blossom. We can plant peas, lettuce, and arugula, and greet the emergence of the first wild iris down on the coast, and the awakening of the wildflowers.

At Beltane, the wildflowers cover the land in gold and blue. The cool-weather crops have begun to sprout in the garden, and the collards and kale are renewed. Roses are just beginning to open in a good year, and all the land seems alive with erotic, sensual joy. We celebrate our creativity and the joys of all forms of love. We celebrate the year-child now grown into the fullness of power of the lover.

At Summer Solstice, the sun is warm and the warm-weather crops are in the ground and growing. The springs are still full of water, but the rains have stopped and the hills are beginning to turn golden. It is high summer, a time of growth but also, for us, a time of death and dormancy as the land dries up and the year-child passes into the dreamrealm.

At Lammas the land is dry and the springs are beginning to lessen in their flow. We become aware of fire, vigilant to its unexpected appearance. One spark could set this whole land ablaze. We begin to harvest squash and the first tomatoes. The sun is high, lighting our homes, but the streams are dry. We celebrate the fire ritual, honoring fire and asking for protection as we learn again to live with fire's lessons.

At Mabon, the air is cooler. The apples are ripening on the trees, and the harvest is coming in while the sun begins to decline. We ask help to get us through the last, dry portion of the year even as the very first rains begin to fall. The year-child becomes the dreamer, even as we bottle the sweet tastes of summer into jam and hide its fruits away on our shelves.

And the wheel of the year is complete.

BLESSING FOR THE CENTER

We give thanks and gratitude for the center, the heart and the hearth, and for the complex and beautiful patterns of life. We give thanks for the flow of sap through branches, water through rivers, blood through our veins. Thanks for the spiral vine and the sunflower, the daisy and the dandelion, the sunburst and the star. We bless the random element that brings freedom into the pattern, and thank the weaver of the web of life for teaching us that all is interconnected. May we be aware of our patterns, able to change those that do not serve us and to cherish and strengthen those that do. May the cycles of the moon's and the sun's journeys through the year help us remember that light arises from dark, rebirth from death. Blessings on the cycles, the patterns, and the center.

TWELVE

Healing the Earth

I am sitting in a cave in central Mexico, with a group of people who have come together for a week to share Mexican-rooted and European-rooted traditions of earth spirituality. We have made a beautiful *ofrenda*, an altar/offering of seeds and flower petals laid in patterns, to honor Tonantzin, the Aztec earth Goddess, and we have all squeezed into this cave to listen to the voice of the earth. After a time of silence, we sing and chant together, in English and Spanish,

> The earth is our mother, we must take care of her . . .
> *La tierra es nuestra madre, debemose cuidarla* . . . [1]

We raise power, an echoing, resonant tone that rings through the cave. When we ground, we lay our hands on the earth. In my mind, I hear the earth sigh with pleasure, drinking in our energy. "Do this," she says. "Feed me. Tell people to feed me, to consciously feed energy into my energy body. For I am getting weakened, and I need it."

Then suddenly I saw the earth as a great battleground. There were huge forces of destruction like thunderclouds massing, and also enormous forces of love. I resisted the vision, because I don't like to think in terms of dualities and wars of good and evil. But I heard clearly, "The forces contesting for the earth are so strong, so nearly matched, the battle so intense, that she could break apart under the strain. Feed the earth."

If you have come this far in this book, and done even a few of the practices suggested, you will have begun to experience the earth as a living being who is aware and speaking to us all the time. You will have glimpsed the miracle of her immense and intricate cycles, the tides and currents of atmosphere and ocean, her breath and blood, the passing of energy and nutrients through cycles of birth and death, decay and regeneration, the patterns that give form and substance to her flows and that form her body.

And if you have opened your awareness, you also know that she is hurting. The great cycles of the elements, the back-and-forth breath of green and red that infuses the atmosphere with oxygen, the diversity that has taken billions of years to evolve, are everywhere under assault. And we know that change may not be a gradual, slowly flowing river, but rather a stream that suddenly plunges down a rocky waterfall. At the moment, we can see and feel the earth's degradation, the unusual heat of this summer, the high rates of cancer, the loss of species—but her basic life-support systems are still functioning for us, and it's easy to close off awareness of the damage. We're like boaters drifting on a slow stretch of the stream, aware that we are descending but telling ourselves it's not so bad. Everything is normal, calm; there's nothing to get too worried about. Yet just ahead is the waterfall. We may not see it until we reach it, too late to turn back, with only time enough to cry out, "Why didn't we change course earlier?" as we plunge over the rapids.

The earth has great powers of resilience, but she is also fragile. The cycles we've learned about, the life-support systems, can adjust to immense impacts, but if they crash, they can start a cascading collapse that could have consequences devastating to much of the earth's life.

At the same time, great forces of love and healing also are growing in the world. We now have technologies that, should we actually use them, allow us to live lightly on the earth with comfort and security. We have knowledge and wisdom if we choose to apply them, about how to provide for human needs in ways that respect and enhance the balance of life. And we have a growing, global community of people committed to balanced ways of living.

In this crucial time, we are called to be healers—of the earth, of the human community, of each other and ourselves. We speak of "healing the earth," but in reality, what needs healing is our human relationship to the earth.

Healing begins with listening. Many years ago, in what now seems like a former lifetime, I was trained as a psychotherapist. Therapy/counseling is a discipline of listening. We open ourselves to our client's pains, fears, hopes, and experiences, and by listening and witnessing, we establish a healing relationship.

Earth-healing is a similar process. By following the practices of observation recommended here, by maintaining a personal practice of listening to the earth, we create a new relationship that is healing, not just for the earth, but for ourselves. For when we are out of communication with the elements and energies and processes that sustain our lives, we cannot be healthy or whole.

Listening is also a practical discipline. If I have a piece of land, a sick tree, a community to heal, I begin by listening in a particular way the Reclaiming community calls dropped and open attention.[2]

DROPPED AND OPEN ATTENTION: A HEALING MEDITATION

In the place that needs healing, ground and come into your senses. You might want to cast a simple circle of protection.

Take some time to observe this place, to notice the signs of ill health or of need, to also notice and give gratitude for the life, beauty, and diversity that exist.

Now sit comfortably or lie down, and close your eyes.

Imagine all your thoughts as a cloud of massed string around your head, and slowly begin winding them into a ball at the center of your head. Take your time, and when you are ready, notice how it feels to have your awareness focused and concentrated and centered in your head. Let the ball compress down to a point of light.

Now, breathing slowly, let that point of light, of awareness, begin to slowly drop down your spine. It floats down, breath by breath, coming to rest for a moment in your heart. Breathe deep and notice how it feels to have your awareness centered in your heart.

Now continue to breathe and let that point of light float down, down, on each breath. At last it comes to rest in your belly, just two inches below your navel. Notice how it feels to have your awareness centered in your gut-level intuition.

And now, as you breathe, let that point of light begin to expand, into a disk of light, an open awareness. Let it expand until it encompasses the edges of your physical body, and notice how it feels to let your physical reality come into your awareness.

And now let that plane of awareness expand until it encompasses the edges of your energy body, your aura, and notice how it feels to let your energetic body come into your awareness.

And now let it expand until it encompasses this place you sit in that needs healing. Take a moment, breathe deep, and notice how you feel.

Ask for information to come to you about what is needed here for healing. Stay open and breathing, and allow images and words and messages to form. You might meet someone here who can share wisdom or tell you what is needed.

After a time of openness, consider some of the following questions:

Is something lacking here?
Is there too much of something?
Is there something attacking or assaulting this system?
Is there something that needs to be released or gotten rid of?
Is there something that needs to be brought in?
Is there a healing image, sound, word, or symbol that can help strengthen the forces of life and vitality here?
Where can we find the resources needed for healing?
Are there helpers we can ask to support this healing work?

Sit with your answers, and when you are ready, thank any beings or sources of information you have encountered, and begin to draw your awareness in.

As you draw your awareness into your own energy body, consciously let go of any energies that are not yours. Imagine your aura as a filter that can exclude any toxic energies and anything that does not belong to you.

As you draw your awareness back to your physical body, again consciously let go of any toxic energies or potential physical manifestations of energies you have encountered. Consciously draw in energies of health and life.

Breathing deep, draw your awareness back into your belly until it becomes again a point of light. Now take a deep breath and let it go to wherever you want your awareness centered at this moment.

Visions of Health

My land in the coastal mountains is beautiful. Compared to the degraded landscapes most city dwellers are used to, it seems wild and natural and vibrant with energy. There are towering redwoods in the streambeds, resprouted from the stumps of giants logged decades ago. There are rolling hills that in the winter rains turn emerald green.

But if I travel a few miles down the road to the state park that preserves some stands of old-growth redwood, I can see that my land is impoverished, in a state of recovery after a century of abuse. The old trees dwarf my young redwoods. Their undergrowth is richer, denser. The streams are flowing with water and graced by a riparian vegetation that is varied and complex.

Without the old growth, I can have no picture in my mind of what true health would look like for this land. If we are to become earth-healers—creators of cultures of beauty, balance, and delight, as my friend Donald Engstrom says—we need some glimpses of what that state of health might look like.

My friend Penny Livingston-Stark and I teach courses in permaculture design and earth-healing together. Her acre of land in Point Reyes, California, serves for me as a vision of what health can be.

Penny and her partner, James, have transformed their modest ranch house. The interior is now covered in natural plasters, made of clay, sand, pigment, and flour paste, that glow in rich jewel tones. Just outside, lettuce and salad greens are growing near the door, along with herbs for cooking. Graywater from the house runs out into a biofilter of gravel planted with papyrus and cattails, and from there into a small, constructed stream that empties into a duckpond.

The ducks help keep down bugs and slugs in the garden. An arbor of natural poles supports grapes and provides shade beside the pond. The earth that was dug from the pond was used to build a cob and straw-bale office, a small building of undulating forms that looks over the pond. (Cob is a mixture of clay, straw, and sand that can be shaped into organic, flowing forms and can also be built into the strong walls of long-lasting structures.) The office walls curve into a sitting bench, where a clay oven in the form of a dragon can be fired up to bake pizza, bread, or salmon.

Further down in the garden, a fence of espaliered apples produces many varieties of fruit through the fall and summer. Chickens weed and fertilize the garden beds and provide eggs. An open-air, thatched bedstead made of Balinese bamboo invites visitors to lie down and rest. Two other small structures made of straw-bale and bamboo and covered with natural plasters serve as guest quarters.

Tucked around the edges of the garden are tomato beds, more fruit trees, potatoes growing out of brush piles, more ponds full of water plants and fish, and many sensual surprises. To visit her garden is truly to stroll through paradise, with fruit, flowers, food, and inspiration abounding everywhere.

City Repair

Last spring, my friend Delight and I drove up the Oregon Coast after priestessing a healing ritual for forest defenders in northern California who had been facing horrific violence from logging companies. Before our drive we had watched videos of young activists being beaten up in the trees; they were hogtied and dangled by ropes 180 feet up in the air, and literally tortured in an attempt to remove them from the platforms where they sat to deter the cutting of some of the last remaining old-growth redwoods in the area. We drove along a route I had bicycled twenty years before, appalled now at seeing clearcut after clearcut, barren hillsides and denuded streambeds.

We reached Portland the next morning, heartsick and weary, and drove toward where we would be staying. As we pulled up, we found ourselves stopped at an intersection beside a lush and flourishing garden that seemed to have spilled out onto all the edges of the pavement. The center of the intersection was painted in a huge, colorful mandala that filled the street, its painted radial lines marking the four directions and drawing in energy. On one street corner, a group of people were happily working together to sculpt a bench with angel wings out of cob.

The bench stood beside a small, covered kiosk where a thermos of hot water stood for making tea, with mugs and teabags provided for anyone in the neighborhood who might feel thirsty. This echoes back to the revolutionary tradition of the Boston Tea Party, and is a living example of a Goddess/service gift economy.

On another street corner of this intersection, a small hut of branches that wrapped around a living tree made a playhouse for the neighborhood children. On a third, a sculpted stand made a beautiful giveaway box for neighbors to leave things they no longer needed or books they'd already read, alongside a bulletin board for posting notices of neighborhood events. On the fourth corner, a second cob bench was under construction.

When we stopped to take a closer look, we were warmly greeted by Janelle Kapoor, who was one of the many leaders overseeing construction at sites around the city as part of the Natural Building Convergence we had come to join. She took us over to a neighboring house, where a bread oven sculpted of cob was under construction at the edge of the driveway. The woman who owns the house is a healer, and she told us the oven was pointed to face the Goddess statue in her front yard, and was oriented to the street so that all the neighbors could use it.

At that point, I sat down and burst into tears. I was overwhelmed at the generosity of spirit I could feel embodied in all these projects, and that generosity was such a contrast to the greed and violence we had witnessed in the forest. This intersection in the outskirts of Portland had become a village center, a meeting ground for a community, and by its physical existence had called that sense of community into being.

The intersection was one of the first projects of a group called City Repair, whose goal it is to repair the broken links of community that underlie our alienated cities. Mark Lakeman, one of the group's founders, is a true magician—someone able to manifest ideas and principles in physical reality. He described to me their efforts.

> The whole transformed intersection is a self-service piazza, requiring no monetary economy to run. It is grown from the gift-economy model of the universe, and so will expand and reverse the conventional paradigm directly, and through stories that inspire, and also through replication, because people intuitively and consciously recognize it as habitat in the midst of confusion.

The grid pattern of our cities, Mark explained, is really a military pattern/colonial device, designed to allow maximum control of space and easy movement through it. But it is not designed for human interactions. "The grid also makes the earth measurable as a commodity unit," Mark said. "The grid regiments space in order to homogenize experience and imagination. Indeed, it is designed to discourage and eliminate interaction, so that 'Love your neighbor' remains a high ideal rather than a daily reality."

A village, in contrast, always has at least one central meeting ground, a place where people gather informally to share news and gossip, drink tea (or something stronger), fall in love, gather to make decisions, settle a quarrel, and so on.

"Remember that freedom of assembly requires a place to assemble," Mark continued. "Villages also have many outer, supporting places which invite and encourage social interaction. . . . Such environments never need to have anyone 'build community,' because the fabric supports interaction as a matter of the flow of life."

City Repair decided to create such meeting grounds in the city. Their first project was a teahouse-treehouse, shaped in the form of an enormous, translucent womb and built around a huge conifer. (The local indigenous trees of life are conifers.) The teahouse was constructed of Mark's lifelong collection of doors and windows, which he had amassed, as builders do, knowing that someday he would use them for something.

City Repair began holding gatherings in the teahouse, free of money as a means of exchange, on Monday evenings. Monday—Moon-day—was chosen in honor of the Goddess, to reclaim her day and make it sacred. The gatherings grew and grew until finally the city got word of this unpermitted structure and ordered it dismantled.

"It was a trap," Mark said with glee. "We lured them in to attack us, which charged up the 'hood. . . . And then the 'hood was ready to do more."

The group decided to escalate by taking over an intersection, holding a gathering, closing the streets, and beginning construction of a new project. The teahouse (locally called the T-Hows) transformed into a teahorse (or T-horse). They took a used pickup, built collapsible wings of bamboo and recycled plastic that unfolded into the fans/canopies of a gathering shelter, and began taking this teahorse to public spaces to serve as a mobile meeting ground.

"We took the teahorse around and around the whole city, in a huge circular pattern," he explained, "raising the issue of the absence of crossroads places, cultural nodes, meeting places, and colonialism in general. This led to an internal movement in Portland," which is now inspiring imitators in other places.

At first, of course, the city objected to the intersection repair, but over time some magic happened, and more enlightened members of the city government recognized that the project actually solved some of the problems their own urban planners had been struggling with: improving livability, making streets safer, building local culture, increasing communication, and slowing traffic in residential areas. Intersection repair, the reclaiming and restoring of the crossroads, became legalized, and new projects were started. They require approval from all four homeowners on the corners and from eighty percent of the surrounding neighbors.

Mark began his career as a corporate architect, then became disillusioned and went traveling in Europe, Egypt, New Zealand, and throughout Mexico and Central America in search of new visions to bring back for his community. In the Lacandon rainforest of the southern Mexican region of Chiapas, he met indigenous Mayan shamans who taught him that he needed to learn to listen to the earth and to his own deeper self, to stop talking and speak through actions.

> One of the things that was said to me when I was in the rainforest in southern Mexico is that saving the world isn't going to be about words, it's going to be about actions. As I was trying to recover, coming home and being in this deep culture shock, I was trying to understand what was meant by that. And I just started to initiate these projects. How can I work out what I want to say without words? It's got to be about manifesting through action, creating examples that give people a chance to inhabit a landscape that reflects what they're already thinking. Except you don't want to try to persuade them anymore, you don't want to try to convince or debate, you just want to create it and then people can understand, viscerally; then they can know better what they have been looking to create—to manifest now.
>
> When you are doing something for the community, never ask permission of external authorities. That's what one of my rainforest teachers taught me. You go to that edge where you need to get permission, and then you don't. It's that old saying about knowing yourself, who are you, what is it that is your right, what is your nature, your habitat. What is my habitat? That is the place that we are able to create from, out of our nature, and nobody can give permission to do that. You shouldn't ever ask for permission, and nobody can ever give permission to build your birthright. And that's not my wisdom speaking, that's indigenous knowledge 101.

City Repair now has projects in many different parts of Portland. One intersection is graced by a giant Fibonacci sunflower mandala with an echoing archway of welded metal suspended beside a mosaic fountain. Another holds a Chartres-style labyrinth.

As we stood in that intersection, the women from two of the corner houses came out to welcome us.

"I see myself as a guardian of the labyrinth now," one told me, "in the tradition of all the ancient priestesses."

"It's been a while since we've had guardians of labyrinths in our cities," I agreed, "but I hope this is the beginning of a trend."

"Each thing that we do is a foothold that is taking the consciousness of the people participating, and the people observing, and the people hearing all the stories higher and higher, farther and farther, returning in a sense to who we are," Mark said. "Spirituality to me is simply communication with other people in community."[3]

Healing our relationship to the place we live is the beginning of creating community, and a healthy community is the ground for healing the land and our relationship to the natural communities that surround us.

Cancun

I kept thinking of Mark's words, and the advice he'd received in the rain forest, when we were in Cancun in September of 2003 as part of a mobilization protesting the World Trade Organization. The shaman's advice was echoed in the words we heard from our Mexican student allies, who had been deeply influenced by the Zapatista movement, which itself originated in the Lacandon rain forests of Chiapas. The Zapatistas represent the indigenous peoples who rebelled against the imposition of corporate globalization on their ancient cultures and livelihoods. When the North American Free Trade Agreement was passed in 1994, they formed an army of rebellion. Since then, they have moved away from armed struggle and embraced a political process of change. They now control many of the forested areas of Chiapas, and their example and thinking have inspired the global justice movement.

"The Zapatistas tell us that our power is not just in confronting the police and military power of the state," Abram, one of the students from Mexico City, explained in a planning meeting. "Our power is in claiming our own autonomous spaces, and building the world that we want."

Magic also teaches us to channel and direct energy, and to do that we need to know what we *want*, not just what we *oppose*. Confrontation is often necessary and unavoidable if we want to preserve and protect the earth, but when we bring our creativity and vision into the points of conflict, transformative moments can occur.

In Cancun, the Green Bloc that I was a part of had come down with the intention of making our encampments as much as possible models of ecological design. We included students from our Earth Activist Trainings, permaculturalists from the U.S., and allies from Mexico. We have an ongoing relationship with Tierra Viva, a group of street youth and punks from the slums of Mexico City who are using permaculture to transform their neighborhoods and are teaching gardening to children. Rodrigo Castellano, from the Mexican ecology group Biosfera, came down to help as well, bringing a bag of compost worms with him through airport security. "They're not on the list of forbidden objects," he explained to the security guard. "They have no sharp edges."

The city of Cancun was providing camping space for the students and the *campesinos*, the farmworkers and indigenous people, who were coming to march and to protest the policies of the World Trade Organization, which open the world's resources to exploitation by corporations and undermine citizens' ability to regulate environmental, health, and labor standards. But the city had no money to provide amenities, beyond rows of portable toilets.

We persuaded them to let us create a model handwashing/dishwashing stand next to the food tents. We fit flexible pipes to the edges of the canopies to serve as gutters and catch rain, which was stored in a tank, then pumped up to an elevated container with a simple hand-pump that worked by running a rope fitted with pistons through pipes and around a bicycle wheel. The water then went into faucets and down through small "sinks" made of bright orange funnels, to be channeled into a model graywater system consisting of a series of half-barrels. The first was filled with wood chips and duff, to filter grease, and the others were filled with gravel and water plants, to provide habitat for bacteria to break down disease organisms and toxins. We also created a simple shower installation that ran from an elevated tank fitted for rainwater collection. The installation was decorated with colorful flags made by the art collective for the mobilization that were printed with jaguars and Mayan Gods, and we also mounted photographs and printed up informational material on all the systems in both Spanish and English. The whole installation had a cheerful, colorful air about it, like an elaborate toy, and attracted much attention.

As with any project, we faced many challenges. Our original visions far outstripped our time and resources. Almost as soon as we set up our rain catchment, the rain stopped. (We had city water as a backup: one of the principles of permaculture is redundancy—that is, always having more than one source to meet any crucial need.) We had many moments of exhaustion, discouragement, and near-panic along the way.

But our little project was amazingly empowering and successful, often in ways we had not anticipated. It provided a living example that Mexicans and "gringos" could work together in a way that was mutually inspiring and respectful. It gave the media something positive to write about and photograph in the lead-up to the demonstrations, something that clearly embodied the principles we were fighting for. The project also demonstrated how an integrated system can work and proved that it can be made of simple, low-cost materials. It provided models that could be taken back to communities and used. Abby, Juniper, Riverwind, Eileen, and Cole demonstrated that women can build and design and make things. And, not least, the project provided a place for people to wash their hands before meals and to take showers!

There were many wonderful moments at the handwashing stand. Almost as soon as we set up our simple pump, a woman came over and studied it intently. "We have no running water in our village," she told me, "and a pump costs four thousand dollars. But *this* would work!"

I will always treasure the sight of Emilio, one of our black-clad, pierced, spiked, and studded punks, explaining the system to a group of thirty *campesinos*. Tierra Viva received invitations from many communities to come down and lead workshops in permaculture. And all the effort and worry and sweat were justified when Erik described what happened one morning as he was washing his hands: a five-year-old campesina girl came up to him and began explaining the entire system, from the rain catchment to the graywater, in complete and accurate detail. And she clearly understood it all: how the system turned a potential problem—runoff rainwater from the canopy—into a resource—clean water for handwashing and dishwashing—and how that water was then conserved and reused to grow plants and to be infiltrated into the ground (where, had this been a permanent installation, it could have been used to nurture a garden). We had created a living example of real abundance, and done it in a way that embodied cooperation and community.

There were many confrontations during the week of actions—at the fence the police erected to keep demonstrators out of the hotel zone, at the security barriers near the conference center where the WTO met. A Korean farmer/organizer, Hyung Hai Lee, stabbed himself to death on the first day of actions as an ultimate act of protest, bringing home to us all that these issues are matters of life and death for farmers and workers around the world. He was influenced by the suicide of a close friend who had lost his land because of economic policies that make it hard for small farmers to make a living.

Farmer suicide is a worldwide epidemic: Vandana Shiva spoke at the teach-in organized during the week of Cancun actions by the International Forum on Globalization and mentioned one area of India where 650 farmers had committed suicide in one month. In the face of those grim statistics of death, we were glad to be able to create something to embody hope.

In the middle of the week, we were able to stage a nonviolent blockade under the walls of the conference center itself. We spent the day dressed as tourists, filtering through the security system in ones or twos, and then converged just as the delegates were coming back from dinner. A small group of Mexican students and internationals moved out onto the road to blockade it by simply sitting down. Behind them, a group of us moved in and began a spiral dance, singing in Spanish,

Somos el viento que sopla,
Al imperio que colapsa.

And in English,

We are the rising of the moon,
We are the shifting of the ground,
We are the seed that takes root
When we bring the fortress down.

As we danced, members of the Green Bloc appeared carrying two trees and a bag of seeds. ("How did you get those trees past security?" I asked Rodrigo later. "We just carried them in. When they asked, 'What are you doing,' I just said, 'We're carrying these trees,' and walked on. People don't expect you to be carrying trees." Clearly, the man has a gift for this sort of thing!)

We danced around the trees, as Erik and John Henry spilled seeds on the ground in a spiral *ofrenda*, an altar/offering in the center of our circle, and raised power to charge our vision of a living world.

The ministerial ultimately collapsed in disagreements between the powerful countries of the north and the developing countries of the global south, who walked out under the leadership of Kenya. Delegates from the south told our friends inside that the many demonstrations inside the conference, outside the walls, and in the streets had given them the support they needed to take a strong stand and resist bullying from the U.S. Our chant was prophetic: we had indeed become the fresh wind that can blow away systems of destruction and open space for new seeds to grow.[4]

Community Vision

The struggle to reclaim and heal the earth is going on in every community. In one sense, anything we do to strengthen our communities and create networks that allow sharing, support, and connection is an act of earth-healing. But we are also often faced with stopping exploitation and degradation of our home environments.

In the Cazadero Hills, where I live, we have fought many battles against thoughtless development and the despoiling of the wild nature we love. We have formed land-use councils, advocacy groups, and a community land trust. When we learned that our county was undergoing a revision of its general plan, which it does every thirty years, we wanted to participate in the process. But the public meetings were held in Santa Rosa, an hour and a half from our homes. They were often rescheduled or canceled abruptly, and at best they allowed for very limited public comment.

Increasingly frustrated, we decided to conduct our own planning process. In the spring of 2002, we held a community vision day, inviting all the people who lived in our area to come together and discuss what they wanted. We facilitated *creative* processes, asking people to draw and write their visions, and also *cognitive* processes, organizing people into groups that considered the areas under discussion in the general plan. Using the results from that day, we began to write up a report, and we circulated a petition asking for a limit on vineyard development (which has been encroaching on the wildlands in our area). In the fall, when the hundred-page report was ready, we hosted a Wild Foods Breakfast, inviting county planners, elected supervisors, and the press, to present our report with a bit of fanfare that we hoped would get it noticed.

Our process inspired many other groups in the county that work on land issues, and it got good coverage in all the small local papers and even some local TV. (We were unable to persuade the major newspaper, the *Santa Rosa Press Democrat,* to cover it, however. Communities attempting to control their own destiny and resources are not "news." We finally got a reporter to come by telling him we would present the report in the nude—however, he left quickly when we appeared fully clothed.)

Our community action was not as dramatic as the frontline actions in Cancun, but it, too, was powerful and important. Articulating our community vision was empowering for everyone who took part in it. Writing it up and putting it out as a document made it part of the public record and gave it a new status in the dialogue. We had responded to an undemocratic, exclusionary process by creating our own autonomous response, and in so doing had carved out a new political space. We are not indigenous to this land, but we like to think that we are becoming indigenous, by loving the land, learning its plants and trees and animals and needs, taking responsibility for its health, and safeguarding its future.

Feeding the Earth

There are many ways to give energy and nurturing back to the earth. When we raise energy in ritual, we can consciously return it to earth with the intention of strengthening her life force and resilience. When we compost our food wastes, build soil in our gardens, heal the erosion surrounding a stream, or tend a small patch of our own ground, we are helping her overall healing.

We, too, are part of the earth, so healing ourselves and our communities is part of healing the earth. But spending time in nature, developing our relationship with her, and developing our own practices of observation and gratitude will also help heal our own wounds and imbalances.

Bringing alive visions of health and balance, carrying vision and creativity even into those battles where we stand up against the destruction of the earth, can help us to sustain our energy and can empower us with a knowledge of what we are fighting *for*, not just *against*.

Hope and Courage

When I was nearly finished with this book, I took a weekend off to join with Pagans and forest defenders in the Cascadia bioregion of Oregon, to share magic and healing. We camped near a beautiful grove of old-growth cedars and Douglas firs, and I was able to spend a long time lying among the roots of the trees, just looking and listening.

"The forces of greed are strong," the forest said to me. "But don't forget that you have immense forces working with you for the healing of the earth."

At times the process of destruction seems so advanced that we may find it hard not to sink into despair. "Action is the antidote to despair" is an Earth First! slogan, but nonstop action can be draining and disempowering if it is not thoughtful, strategic, and effective. And yet we do have great forces working with us, those same creative powers that arranged the chlorophyll mandala and learned to use sunlight to make food, that have traded genes and information for billions of years, that grew the redwood and the cedar from instructions encoded in the microscopic double-spiral DNA crystal, that move the great currents of the air and the tides of the ocean.

The earth is alive, and I don't believe that she is suicidal. She took a great gamble with this big-brain experiment, but I do believe that the consciousness we need to temper our destructive potential is arising, now, and is ultimately far, far stronger than greed. Moreover, I believe that the earth wants us to play the role we have evolved to play, a role as important as that of any worm or soil bacterium: to be her consciousness, her mirror, her great admirer and appreciator, to cheer her on and to use our specifically human abilities to help restore and sustain her balance.

When we are working in service of the earth, we can ask those powers to be with us. Indeed, they want to be asked, need to be asked. So here is a prayer for help. You may use it as written or let it inspire your own words:

PRAYER FOR HELP IN EARTH-HEALING

Great forces of creativity, growth, and love, great inventive imagination that has grown the diversity of life, mystery of the unfolding of form from energy, great powers of the trees and the grasses, the sunlight and the rain, great currents that move the continents and tides of the ocean, I send you love and gratitude and ask for your help. I am about to do ________________. It seems impossible. It seems beyond my human strength. Please lend me some of your power. Open my eyes and ears to inspiration. Release the obstacles that confront me, and draw in the resources, the luck, the energies that I need. Help me to succeed beyond my expectations, for the healing of the earth. Blessed be.

The forces of love and greed are indeed contesting with one another, and both are immensely strong—so strong that almost anything could tip the balance. Everything we do right now is vitally important. Each act of healing is a weight on the side of life. The drama moves toward its climax, and any one of us could be the small stone that starts the avalanche.

What a great time to be alive!

BLESSING FOR EARTH-HEALERS

We give thanks for all those who are moved, in their lives, to heal and protect the earth, in small ways and in large. Blessings on the composters, the gardeners, the breeders of worms and mushrooms, the soil-builders, those who cleanse the waters and purify the air, all those who clean up the messes others have made. Blessings on those who defend trees and who plant trees, who guard the forests and who renew the forests. Blessings on those who learn to heal the grasslands and renew the streams, on those who prevent erosion, who restore the salmon and the fisheries, who guard the healing herbs and who know the lore of the wild plants. Blessings on those who heal the cities and bring them alive again with excitement and creativity and love. Gratitude and blessings to all who stand against greed, who risk themselves, to those who have bled and been wounded, and to those who have given their lives in service of the earth.

May all the healers of the earth find their own healing. May they be fueled by passionate love for the earth. May they know their fear but not be stopped by fear. May they feel their anger and yet not be ruled by rage. May they honor their grief but not be paralyzed by sorrow. May they transform fear, rage, and grief into compassion and the inspiration to act in service of what they love. May they find the help, the resources, the courage, the luck, the strength, the love, the health, the joy that they need to do the work. May they be in the right place, at the right time, in the right way. May they bring alive a great awakening, open a listening ear to hear the earth's voice, transform imbalance to balance, hate and greed to love. Blessed be the healers of the earth.

Notes

CHAPTER ONE

1. Dion Fortune, an occultist and author of the nineteenth and early twentieth centuries, originated this definition. Although I've been quoting it for twenty years or more, I've never been able to track the exact reference down.

2. David Clarke with Andy Roberts, *Twilight of the Celtic Gods: An Exploration of Britain's Hidden Pagan Traditions* (London: Blandford, Cassell, 1996), pp. 22–24.

3. Starhawk, *The Spiral Dance: A Rebirth of the Ancient Religion of the Great Goddess* (San Francisco: HarperSanFrancisco, 1979, 1999), p. 103.

4. Kat Harrison, personal communication, 1994.

5. Starhawk, *Webs of Power: Notes from the Global Uprising* (Gabriola Island, B.C.: New Society Publishers, 2002), pp. 161–162.

6. Quoted by Bev Ortiz, "Contemporary California Indian Basketweavers and the Environment," in Thomas C. Blackburn and Kat Anderson, eds., *Before the Wilderness: Environmental Management by Native Californians* (Menlo Park, CA: Ballena Press, 1993), p. 199.

7. Allan Savory, *Holistic Management: A New Framework for Decision Making* (Washington, D.C., and Covelo, CA: Island Press, 1999), pp. 20–21.

8. Connie Barlow, *Green Space, Green Time: The Way of Science* (New York: Springer-Verlag, 1997), pp. 14–15.

CHAPTER TWO

1. Erik Ohlsen, who coteaches Earth Activist Trainings with me and Penny Livingston-Stark, led permaculture workshops at the Welcome Center and later blockaded with the Green Bloc who occupied the community garden.

2. Vandana Shiva, "The Green Revolution in the Punjab," *Ecologist* 21, no. 2 (March–April 1991). The article, extracted from *The Violence of the Green Revolution:*

Ecological Degradation and Political Conflict in Punjab (Dehra Dun, India: Vandana Shiva, 1989), can also be found at http://livingheritage.org/green-revolution.htm.

3. Kenny Ausubel, *Seeds of Change* (San Francisco: HarperSanFrancisco, 1994), p. 74.

4. Quoted by Bev Ortiz in Blackburn and Anderson, eds., *Before the Wilderness*, pp. 195–196.

5. Press release, "New Report Challenges Fundamentals of Genetic Engineering: Study Questions Safety of Genetically Engineered Foods," Center for the Biology of Natural Systems, http://cbns.qc.edu/harperspressrelease.pdf, Jan. 15, 2002.

6. Barry Commoner, "Unraveling the Secret of Life: DNA Self-Duplication, the Basic Precept of Biotechnology, Is Denied," *Gene Watch* (May–June 2003). See also http://www.criticalgenetics.org/gene_watch_article.htm.

7. John Robbins, "A Biological Apocalypse Averted," *Earth Island Journal* 16, no. 4 (Winter 2001–2002). See also http://www.mindfully.org/GE/GE3/Apocalypse-Averted-Robbins.htm. The study upon which this article was based is M. T. Holmes, E. R. Ingham, J. D. Doyle, and C. S. Hendricks, "Effects of *Klebsiella planticola* SDF20 on Soil Biota and Wheat Growth in Sandy Soil," *Applied Soil Ecology* 11 (1999): 67–78. See also Elaine Ingham, "Ecological Balance and Biological Integrity: Good Intentions and Engineering Organisms That Kill Wheat," www.soilfoodweb.com or http://www.organicconsumers.org/ge/klebsiella.cfm. This article is adapted from a presentation the author gave on July 18, 1998, at the First Grassroots Gathering on Biodevastation: Genetic Engineering, in St. Louis, Missouri.

8. Luisah Teish, *Jambalaya: The Natural Woman's Book of Personal Charms and Practical Rituals* (San Francisco: HarperSanFrancisco, 1985), p. 54.

9. Jeannette Armstrong, interviewed in Derrick Jensen, *Listening to the Land: Conversations About Nature, Culture, and Eros* (San Francisco: Sierra Club Books), p. 295.

10. Clarke, with Roberts, *Twilight of the Celtic Gods*, p. 38.

11. Marija Gimbutas's major works on the Goddesses of Old Europe are:

The Gods and Goddesses of Old Europe (London: Thames & Hudson; Berkeley: Univ. of California Press, 1974).

The Language of the Goddess (San Francisco: HarperSanFrancisco, 1989).

The Civilization of the Goddess: The World of Old Europe (San Francisco: HarperSanFrancisco, 1991).

The Living Goddesses, edited and supplemented by Miriam Dexter Robbins (Berkeley and Los Angeles: Univ. of California Press, 1999).

12. Starhawk, "The Dismembering of the World," in *Truth or Dare: Encounters with Power, Authority, and Mystery* (San Francisco: HarperSanFrancisco, 1987), pp. 32–70.

13. Quoted from an interview with Marija Gimbutas in the video *Signs Out of Time: The Story of Archaeologist Marija Gimbutas*, by Donna Read and Starhawk (Belili Productions, 2003). See also www.belili.org.

14. Starhawk, *Dreaming the Dark: Magic, Sex, and Politics* (Boston: Beacon Press, 1982).

15. Donald Rumsfeld at a Department of Defense news briefing, Feb. 12, 2003. For a poetic setting of this quote, see http://pages.zdnet.com/sartre65/wrack/id34.html.

CHAPTER THREE

1. Permaculturalist Patrick Whitefield defined permaculture as the "art of creating beneficial relationships" in a guest lecture at an Earth Activist Training taught by Penny Livingston-Stark and me at Ragman's Lane Farm, Gloucestershire, England, in August 2002. Patrick is the author of *How to Make a Forest Garden* (Little Clyden Lane, Clanfield, Hampshire: Permanent Publications, 2000) and *Permaculture in a Nutshell* (Little Clyden Lane, Clanfield, Hampshire: Permanent Publications, 2000).

2. For a study linking glyphosate exposure to lymphoma, see Lennart Hardell and Mikael Eriksson, "A Case Control Study pf Non-Hodgkin Lymphoma and Exposure to Pesticides," *Cancer* Vol. 85, no. 6 (March 15, 1999). See also http://www.gene.ch/genet/1999/Jun/msg00012.html.

3. Mary Daly, *Beyond God the Father* (Boston: Beacon Press, 1973). See the discussion on page 40.

4. Donella Meadows, "Places to Intervene in a System," *Whole Earth* 91 (Winter 1997): 78–84.

CHAPTER FOUR

1. A good resource for ecological awareness in all the major religions is the Religion and Ecology Web site: http://hollys7.tripod.com/religionandecology/index.html. A bibliography on Judaism and ecology can be found at http://www.coejl.org/learn/bib_basic.shtml. See also Richard C. Foltz, Frederick M. Denny, and Azizan Baharuddin, eds., *Islam and Ecology: Bestowed Trust* (Cambridge, MA: Harvard Univ. Press, 2003).

2. For a complete listing of Matthew Fox's books, see: http://www.matthewfoxfcs.org/sys-tmpl/tipstricks/. A few of his relevant works are:

The Coming of the Cosmic Christ: The Healing of Mother Earth and the Birth of a Global Renaissance (New York: Harper & Row, 1980).

Creativity: Where the Divine and the Human Meet (New York: Tarcher, 2002).

Creation Spirituality: Liberating Gifts for the Peoples of the Earth (San Francisco: HarperSanFrancisco, 1991).

Original Blessing: A Primer in Creation Spirituality (New York: Tarcher, 2002).

Passion for Creation: The Earth-Honoring Spirituality of Meister Eckhart (Burlington, VT: Inner Traditions, 1980, 2000).

Wrestling with the Prophets: Essays on Creation Spirituality and Everyday Life (New York: Tarcher, 2000).

See also *Arthur Waskow, Seasons of Our Joy: A Handbook of Jewish Festivals* (Boston: Beacon Press, 1982).

3. James Lovelock, *The Ages of Gaia: A Biography of Our Living Earth* (New York: Bantam, 1990).

4. Elisabet Sahtouris, *Earthdance: Living Systems in Evolution* (Alameda, CA: Metalog Books, 1996); and *Gaia: The Human Journey from Chaos to Cosmos* (New York:

Pocket Books, 1989). See also Timothy Ferris, *The Whole Shebang* (New York: Touchstone, 1997).

I thank Brian Swimme for the many inspirational stories I've heard him tell in many years of association with Matthew Fox's Institute for Culture and Creation Spirituality.

5. It was Lynn Margulis's groundbreaking insight that *eukaryotes* evolved as symbiotic communities of simpler life-forms.

6. For his reading of and helpful comments on this chapter, I thank David Seaborg, evolutionary biologist, environmental activist, and founder and president of the World Rainforest Fund, dedicated to saving the rain forest.

CHAPTER FIVE

1. Jane Jacobs, *The Death and Life of Great American Cities* (New York: Vintage, 1992), pp. 50–51.

2. Starhawk, *Walking to Mercury* (New York: Bantam, 1994), pp. 304–305.

3. See also the discussion of anchoring in Starhawk and Hilary Valentine, *The Twelve Wild Swans: A Journey to the Realm of Magic, Healing, and Action* (San Francisco: HarperSanFrancisco, 2000), pp. 41–43.

CHAPTER SIX

1. Starhawk, *The Spiral Dance*, pp. 80–83.

2. Starhawk, *The Spiral Dance*, pp. 87–101.

3. Starhawk, *The Spiral Dance*, pp. 87–101; Starhawk and Valentine, *Twelve Wild Swans*, pp. 80, 167–171, and 240–241.

4. Starhawk and Valentine, *Twelve Wild Swans*, pp. 167–171.

CHAPTER SEVEN

1. Teaching with activist and deep ecologist John Seed and his partner, Ruth Rosenhak, recently in Australia, I discovered that they use their own version of this same story and a very similar meditation—a beautiful example of parallel evolution at work! If Gaia is whispering the same thing into the ears of two sources on opposite sides of the world, she must really want us to do this one.

2. David Abram, *The Spell of the Sensuous: Perception and Language in a More-Than-Human-World* (New York: Pantheon Books, 1996), p. 4.

3. Cities for Climate Protection (at http://www.iclei.org/co2/) is a campaign of the International Council on Local Environmental Initiatives: http://www.iclei.org/. It offers a framework and assistance for local governments to reduce greenhouse gas emissions and achieve sustainability.

CHAPTER EIGHT

1. Abram, *Spell of the Sensuous*, p. 7.

2. Charles Dickens, *David Copperfield* (New York: The Paddington Corporation, 1965), chap. 12, p. 185. First published 1849–1850.

3. Rainforest Action Network's U'Wa campaign home page is http://www.ran.org/ran_campaigns/beyond_oil/oxy/.

4. Abram, *Spell of the Sensuous*, p. 21.

5. Quoted by Bev Ortiz in Blackburn and Anderson, eds., *Before the Wilderness*, p. 196.

6. Allan Savory, *Holistic Management*.

7. Jeanette Armstrong, "Keepers of the Earth," in Theodore Roszak, Mary E. Gomes, and Allen D. Kanner, eds., *Ecopsychology: Restoring the Earth, Healing the Mind* (San Francisco: Sierra Club Books, 1995), p. 323.

8. To find a CSA near you, check the Community Supported Agriculture farms database at http://www.nal.usda.gov/afsic/csa/csastate.htm.

9. Henry T. Lewis, "Patterns of Indian Burning in California: Ecology and Ethnohistory," in Blackburn and Anderson, eds., *Before the Wilderness*, pp. 55–116.

10. Starhawk and Valentine, *Twelve Wild Swans*, 301–302.

11. Starhawk and Valentine, *Twelve Wild Swans*; see "Anger Ritual," p. 102, and "Rage Ritual," p. 138.

CHAPTER NINE

1. Note on the chants used here: "Born of water" was created in a workshop I did with Kate Kaufman sometime in the early 1980s in Madison, Wisconsin. "I am the laughing one" and "I am the shaper" are variations that someone began singing in the middle of the trance at the British Columbia Witch Camp sometime in the late 1980s. I'm sorry I don't know the names of their creators. "The river is flowing" is by Adele Getty. "The ocean is the beginning of the earth" was made up by Delaney Johnson and me when he was six years old and I was thirty-five, and we sat overlooking the water on the Mendocino headlands.

2. Paul Simon, *Tapped Out: The Coming World Crisis in Water and What We Can Do About It* (New York: Welcome Rain, 1998), p. 21.

3. Marq de Villiers, *Water: The Fate of Our Most Precious Resource* (Boston and New York: Houghton Mifflin, 2001), p. 44.

4. Villiers, *Water*, p. 37.

5. Maude Barlowe, "The Global Water Crisis and the Commodification of the World's Water Supply," Spring 2001, http://www.canadians.org/display_document.htm?COC_token=COC_token&id=245&isdoc=1&catid=78.

6. Toby Hemenway, *Gaia's Garden: A Guide to Home-Scale Permaculture* (White River Junction, VT: Chelsea Green Publishing, 2001), pp. 89–90.

7. Alice Outwater, *Water: A Natural History* (New York: Basic Books, 1996).
8. See the Web site for Ocean Arks International at http://www.oceanarks.org/.
9. See Penny Livingston-Stark's Web site at http://www.permacultureinstitute.com.
10. Barlowe, "The Global Water Crisis."
11. The Cochabamba Declaration can be found at http://www.nadir.org/nadir/initiativ/agp/free/imf/bolivia/cochabamba.htm.

CHAPTER TEN

1. Gretel Erlich, quoted in Robert Clark, ed., *Our Sustainable Table* (San Francisco: Northpoint Press, 1990); see also http://www.public.iastate.edu/~vwindsor/Cross.html.
2. See Oregon Department of Agriculture statistics at http://www.oda.state.or.us/information/news/Food_spending.html.
3. Elaine Ingham, "Ecological Balance and Biological Integrity," www.soilfoodweb.com.
4. See, for example, the Web site of the Biodynamic Farming and Gardening Association: http://www.biodynamics.com/index.html.
5. Paul Stamets's Web site is http://www.fungiperfecti.org.
6. Kenny Ausubel, *Seeds of Change*, p. 64.
7. For more information about this, see Danielle Goldberg, "Jack and the Enola Bean," TED (Trade Environment Database) Case Study Number XXX 2003; www.american.edu/TED/enola-bean.htm.
8. "Free Tree, Free Tree," *Hindustan* (India) *Times*, June 9, 2000; at http://www1.hindustantimes.com/nonfram/090600/detOPI01.htm.
9. Press release, "New Report Challenges Fundamentals of Genetic Engineering: Study Questions Safety of Genetically Engineered Foods," Center for the Biology of Natural Systems, at http://cbns.qc.edu/harperspressrelease.pdf, Jan. 15, 2002.
10. See at http://www.percyschmeiser.com/.

CHAPTER ELEVEN

1. For information on the video and a complete bibliography of Gimbutas's work, see www.gimbutas.org/.

See the discussion of her work in J. Marler, ed., *From the Realm of the Ancestors: An Anthology in Honor of Marija Gimbutas* (Manchester, CT: Knowledge, Ideas, and Trends, 1997).

See also Carol Christ, *Rebirth of the Goddess: Finding Meaning in Feminist Spirituality* (Reading, MA: Addison-Wesley, 1997), pp. 70–88, and David Miller, *I Didn't Know God Made Honky Tonk Communists* (Berkeley, CA: Regent Press, 2002), pp. 141–155.

2. Tyler Volk, *Metapatterns: Across Space, Time, and Mind* (New York: Columbia Univ. Press, 1995), pp. 12–13.
3. Alice Outwater, *Water*, p. 57.

4. For directions on building an herb spiral, see Bill Mollison, *Introduction to Permaculture* (Tyalgum, Australia: Tagari, 1991), p. 96; and Toby Hemenway, *Gaia's Garden*, pp. 48–49.

5. Christopher Alexander, Sara Ishikawa, and Murray Silverstein, with Max Jacobson, Ingrid Fiksdahl-King, and Shlomo Angel, *A Pattern Language: Towns, Buildings, Construction* (New York: Oxford Univ. Press, 1977), pp. 618–621.

6. Tyler Volk, *Gaia's Body: Toward a Physiology of Earth* (New York: Copernicus, 1988), p. 128.

7. For examples and much more information on labyrinths, see Lauren Artress, *Walking a Sacred Path: Exploring the Labyrinth as a Spiritual Tool* (New York: Riverhead Books, 1996); and Sig Lonegren, *Labyrinths: Ancient Myths and Modern Uses* (Glastonbury: Gothic Image Publications, 1991, 1996).

8. Diane Baker, Anne Hill, and Starhawk, *Circle Round: Raising Children in Goddess Tradition* (New York: Bantam, 1998); and Starhawk, *The Spiral Dance*, pp. 193–213.

CHAPTER TWELVE

1. This chant was written by Adele Getty.

2. I credit Cybele for introducing me and the broader Reclaiming community to the practice of dropped and open attention, and she credits Wendy Palmer, *The Intuitive Body: Aikido as a Clairsentient Practice* (Berkeley, CA: North Atlantic Books, 2000).

3. Mark Lakeman's statements come from a personal interview with him I conducted in August of 2003. The City Repair Web site is www.cityrepair.org.

4. My daily updates from Cancun are posted on my Web site, www.starhawk.org. The chant was written by me and Rodrigo Castellano.

Select Bibliography

Works I've consulted in writing this book, and a small sampling of the many works that may be helpful to the reader wanting to explore these issues further.

David Abram. *The Spell of the Sensuous: Perception and Language in a More-Than-Human World*. New York: Pantheon Books, 1996.

Christopher Alexander, Sara Ishikawa, and Murray Silverstein, with Max Jacobson, Ingrid Fiksdahl-King, and Shlomo Angel. *A Pattern Language: Towns, Buildings, Construction*. New York: Oxford Univ. Press, 1977.

Luke Anderson. *Genetic Engineering, Food, and Our Environment*. White River Junction, VT: Chelsea Green Publishing, 1999.

Lauren Artress. *Walking a Sacred Path: Exploring the Labyrinth as a Spiritual Tool*. New York: Riverhead Books, 1996.

Kenny Ausubel. *Restoring the Earth: Visionary Solutions from the Bioneers*. Tiburon: H. J. Kramer, 1997.

Kenny Ausubel. *Seeds of Change*. San Francisco: HarperSanFrancisco, 1994.

Albert-Laszlo Barabasi. *Linked: The New Science of Networks*. Cambridge, MA: Perseus Publishing, 2002.

Michael Barbour, Bruce Pavlik, Frank Drysdale, and Susan Lindstrom. *California's Changing Landscapes: Diversity and Conservation of California Vegetation*. Sacramento: California Native Plant Society, 1993.

Connie Barlow. *Green Space, Green Time: The Way of Science*. New York: Springer-Verlag, 1997.

Maude Barlowe and Tony Clarke. *Blue Gold: The Battle Against Corporate Theft of the World's Water*. Toronto: Stoddart, 2002.

Maude Barlowe and Tony Clarke. *Global Showdown*. Toronto: Stoddart, 2001.

John J. Berger, ed. *Environmental Restoration: Science and Strategies for Restoring the Earth*. Washington, D.C.: Island Press, 1990.

Thomas C. Blackburn and Kat Anderson, eds. *Before the Wilderness: Environmental Management by Native Californians*. Menlo Park, CA: Ballena Press, 1993.

Carol Christ. *Rebirth of the Goddess: Finding Meaning in Feminist Spirituality*. Reading, MA: Addison Wesley, 1997.

Diana Leafe Christian. *Creating a Life Together: Practical Tools to Grow Ecovillages and Intentional Communities*. Gabriola Island, B.C.: New Society Publishers, 2003.

David Clarke, with Andy Roberts. *Twilight of the Celtic Gods: An Exploration of Britain's Hidden Pagan Traditions*. London: Blandford, 1996.

Ronnie Cummins and Ben Lilliston. *Genetically Engineered Food: A Self-Defense Guide for Consumers*. New York: Marlowe & Co., 2000.

Mary Daly. *Beyond God the Father*. Boston: Beacon Press, 1973.

Richard Dawkins. *Climbing Mount Improbable*. New York: Norton, 1996.

Irene Diamond and Gloria Orenstein, eds. *Reweaving the World: The Emergence of Ecofeminism*. San Francisco: Sierra Club Books, 1990.

Charles Dickens, *David Copperfield*. New York: The Paddington Corporation, 1965; first published 1849–1850.

Alan Drengson and Duncan Taylor. *Ecoforestry: The Art and Science of Sustainable Forest Use*. Gabriola Island, B.C.: New Society Publishers, 1997.

David Duhon. *One Circle: How to Grow a Complete Diet in Less than 1000 Square Feet*. Willits, CA: Ecology Action, 1995.

Joan Dunning. *From the Redwood Forest: Ancient Trees and the Bottom Line—A Headwaters Journey*. White River Junction, VT: Chelsea Green Publishing, 1998.

Timothy Ferris. *The Whole Shebang*. New York: Touchstone, 1997.

Richard C. Foltz, Frederick M. Denny, and Azizan Baharuddin, eds. *Islam and Ecology: Bestowed Trust*. Cambridge, MA: Harvard Univ. Press, 2003.

Matthew Fox. *The Coming of the Cosmic Christ: The Healing of Mother Earth and the Birth of a Global Renaissance*. Scranton, PA: Harper & Row, 1980.

Matthew Fox. *Creation Spirituality: Liberating Gifts for the Peoples of the Earth*. San Francisco: HarperSanFrancisco, 1991.

Matthew Fox. *Creativity: Where the Divine and the Human Meet*. New York: Jeremy P. Tarcher, Inc. 2002.

Matthew Fox. *Original Blessing: A Primer in Creation Spirituality*. New York: Jeremy P. Tarcher, Inc. 2002.

Matthew Fox. *Passion for Creation: The Earth-Honoring Spirituality of Meister Eckhart*. Vermont: Inner Traditions, 1980, 2000.

Matthew Fox. *Wrestling with the Prophets: Essays on Creation Spirituality and Everyday*. New York: Jeremy P. Tarcher, 2000.

Marija Gimbutas. *The Civilization of the Goddess: The World of Old Europe*. San Francisco: HarperSanFrancisco, 1991.

Marija Gimbutas. *The Gods and Goddesses of Old Europe*. London: Thames & Hudson; Berkeley, CA: Univ. of California Press, 1974.

Marija Gimbutas. *The Language of the Goddess*. San Francisco: Harper & Row, 1989.

Marija Gimbutas (edited and supplemented by Miriam Dexter Robbins). *The Living Goddesses*. Berkeley and Los Angeles: Univ. of California Press, 1999.

Brian Greene. *The Elegant Universe: Superstrings, Hidden Dimensions, and the Quest for the Ultimate Theory*. New York: W.W. Norton, 1999.

Jesse Wolf Hardin. *Gaia Eros: Reconnecting to the Spirit of Nature*. Reserve, NM: Earthen Spirituality Project and Sweet Medicine Women's Center, 2003; at earthway@concentric.net.

Richard Heinberg. *The Party's Over: Oil, War, and the Fate of Industrial Society*. Gabriola Island, B.C.: New Society Publishers, 2003.

Toby Hemenway. *Gaia's Garden: A Guide to Home-Scale Permaculture*. White River Junction, VT: Chelsea Green Publishing, 2001

Erich Hoyt and Ted Schultz, eds. *Insect Lives: Stories of Mystery and Romance from a Hidden World*. New York: Wiley, 1999.

Jane Jacobs. *The Death and Life of Great American Cities*. New York: Vintage, 1992.

Jane Jacobs. *The Nature of Economies*. New York: Modern Library, 2000.

John Jeavons. *How to Grow More Vegetables*. Berkeley, CA: Ten Speed Press, 1979.

Steven Johnson. *Emergence: The Connected Lives of Ants, Brains, Cities, and Software*. New York: Simon & Schuster, 2001.

David Korten. *The Post-Corporate World: Life After Capitalism*. San Francisco: Berrett-Koehler; West Hartford, CT: Kumarian, 1999.

David Korten. *When Corporations Rule the World*. San Francisco: Berrett-Koehler; West Hartford, CT: Kumarian, 1995.

Robert Kourik. *Designing and Maintaining Your Edible Landscape Naturally*. Santa Rosa, CA: Metamorphic Press, 1986.

Aldo Leopold. *A Sand County Almanac*. New York: Oxford Univ. Press, 1966.

Sig Lonegren. *Labyrinths: Ancient Myths and Modern Uses*. Glastonbury: Gothic Image, 1991, 1996.

Joanna Macy. *Thinking Like a Mountain: Toward a Council of All Beings*. Philadelphia, PA: New Society Publishers, 1988.

Malcolm Margolin. *The Way We Lived: California Indian Reminiscences, Stories, and Songs*. Berkeley, CA: Heyday Books, 1981.

Alastair McIntosh, *Soil and Soul: People Versus Corporate Power*. London: Aurum Press, 2001.

David Miller. *I Didn't Know God Made Honky Tonk Communists*. Berkeley, CA: Regent Press, 2002.

Bill Mollison. *Introduction to Permaculture*. Tyalgum (Australia): Tagari, 1991.

Bill Mollison. *Permaculture: A Designer's Manual*. Tyalgum (Australia): Tagari, 1988, 1992.

Bill Mollison, with Reny Mia Slay. *Introduction to Permaculture*. Tyalgum (Australia): Tagari Press, 1995.

Bill Moyer. *Doing Democracy: The MAP Model for Organizing Social Movements*. Gabriola Island, BC: New Society Publishers, 2001.

Beverly R. Ortiz, as told by Julia F. Parker. *It Will Live Forever: Traditional Indian Acorn Preparation*. Berkeley, CA: Heyday Books, 1991.

Alice Outwater. *Water: A Natural History*. New York: Basic Books, 1996.
Wendy Palmer. *The Intuitive Body: Aikido as a Clairsentient Practice*. Berkeley, CA: North Atlantic Books, 2000.
Marc Reisner. *Cadillac Desert: The American West and Its Disappearing Water*. New York: Penguin, 1986.
Jeremy Rifkin. *The Hydrogen Economy*. New York: Jeremy P. Tarcher, 2002.
Carolyn Roberts. *House of Straw: A Natural Building Odyssey*. White River Junction, VT: Chelsea Green Publishing, 2002.
Theodore Roszak, Mary Gomes, and Allen D. Kanner, eds. *Ecopsychology: Restoring the Earth, Healing the Mind*. San Francisco: Sierra Club Books, 1995.
Elisabet Sahtouris. *Earthdance: Living Systems in Evolution*. Alameda, CA: Metalog Books, 1996.
Elisabet Sahtouris. *Gaia: The Human Journey from Chaos to Cosmos*. New York: Pocket Books, 1989.
Greg Sarris. *Mabel McKay: Weaving the Dream*. Berkeley and Los Angeles: Univ. of California Press, 1994.
Allan Savory. *Holistic Management: A New Framework for Decision Making*. Washington, D.C./Covelo, CA: Island Press, 1999.
Randy Shaw. *The Activist's Handbook*. Berkeley and Los Angeles: Univ. of California Press, 1996.
Rupert Sheldrake. *The Presence of the Past*. London: Collins, 1988.
Rupert Sheldrake. *The Rebirth of Nature: The Greening of Science and God*. New York: Bantam, 1991.
Alix Kates Shulman. *Drinking the Rain*. New York: Farrar, Straus, Giroux, 1995.
Michael G. Smith. *The Cobber's Companion: How to Build Your Own Earthen Home*. Cottage Grove, OR: The Cob Cottage, 1998.
Starhawk. *Dreaming the Dark: Magic, Sex, and Politics*. Boston: Beacon Press, 1982.
Starhawk. *The Fifth Sacred Thing*. New York: Bantam, 1992.
Starhawk. *The Spiral Dance: A Rebirth of the Ancient Religion of the Great Goddess*. San Francisco: HarperSanFrancisco, 1979, 1999.
Starhawk. *Truth or Dare: Encounters with Power, Authority, and Mystery*. San Francisco: HarperSanFrancisco, 1987.
Starhawk. *Walking to Mercury*. New York: Bantam, 1994.
Starhawk. *Webs of Power: Notes from the Global Uprising*. Gabriola Island, B.C.: New Society Publishers, 2002.
Starhawk, Diane Baker, and Anne Hill. *Circle Round: Raising Children in Goddess Tradition*. New York: Bantam, 1998.
Starhawk and M. Macha Nightmare. *The Pagan Book of Living and Dying*. San Francisco: HarperSanFrancisco, 1997.
Starhawk and Hilary Valentine. *The Twelve Wild Swans: A Journey to the Realm of Magic, Healing, and Action*. San Francisco: HarperSanFrancisco, 2000.
Sara Stein. *Noah's Garden: Restoring the Ecology of Our Own Back Yards*. Boston: Houghton Mifflin, 1993.

Sara Stein. *Planting Noah's Garden: Further Adventures in Backyard Ecology*. Boston: Houghton Mifflin, 1997.

Brian Swimme. *The Hidden Heart of the Cosmos*. Maryknoll, NY: Orbis Books, 1999.

Brian Swimme and Thomas Berry. *The Universe Story*. San Francisco: Harper-SanFrancisco, 1994.

Luisah Teish: *Jambalaya: The Natural Woman's Book of Personal Charms and Practical Rituals*. San Francisco: Harper & Row, 1985.

Nancy Jack Todd and John Todd. *From Eco-Cities to Living Machines: Principles of Ecological Design*. Berkeley, CA: North Atlantic Books, 1994.

Brian Tokar, ed. *Redesigning Life: The Worldwide Challenge to Genetic Engineering*. London: Zed Books, 2001.

Sim Van der Ruyn and Stuart Cowan. *Ecological Design*. Washington D.C.: Island Press, 1996.

Sim Van der Ruyn. *The Toilet Papers: Recycling Waste and Conserving Water*. Sausalito, CA: Ecological Design Press, 1978.

Tyler Volk. *Gaia's Body: Toward a Physiology of Earth*. New York: Copernicus, 1998.

Tyler Volk. *Metapatterns: Across Space, Time, and Mind*. New York: Columbia Univ. Press, 1995.

Alan Weisman. *Gaviotas: A Village to Reinvent the World*. White River Junction, VT: Chelsea Green Publishing, 1995.

Patrick Whitefield. *How to Make a Forest Garden*. Little Clyden Lane, Clanfield, Hampshire: Permanent Publications, 2000.

Patrick Whitefield. *Permaculture in a Nutshell*. Little Clyden Lane, Clanfield, Hampshire: Permanent Publications, 2000.

Edward O. Wilson. *The Diversity of Life*. New York: Norton, 1992.

Linda Woodrow. *The Permaculture Home Garden*. Ringwood, Victoria (Australia): Viking Penguin, 1996.

Resources

Starhawk's Projects

STARHAWK'S WEB SITE:
www.starhawk.org

ROOT ACTIVIST NETWORK OF TRAINERS
www.rantcollective.org
(Starhawk is a member of this collective, which offers trainings in organizing and nonviolent direct action.)

EARTH ACTIVIST TRAINING
21 Fort Ross Way
Cazadero, CA 95421
707-583-2300, ext. 119
www.earthactivisttraining.org/
(These trainings combine a permaculture design course with activist and organizing skills and earth-based spirituality.)

BELILI PRODUCTIONS
www.belili.org
(Starhawk and Donna Read produce documentaries on issues related to earth-based spirituality, women, and social change. Their first project is *Signs Out of Time: The Story of Archaeologist Marija Gimbutas*. They are currently working on a documentary about permaculture.)

Pagan Resources

RECLAIMING
P.O. Box 14404
San Francisco, CA 94114
www.reclaiming.org
(Reclaiming is the group Starhawk cofounded over twenty years ago. It offers training in Goddess-based and earth-based spirituality integrated with social change. Reclaiming groups in North America, Europe, and elsewhere offer a variety of resources, including public rituals, courses, Witch camps, and gatherings.)

THE WITCHES' VOICE
www.witchvox.com
(This is the most comprehensive of the hundreds of Pagan Web sites.)

Permaculture Resources

CITY REPAIR
1237 SE Stark
P.O. Box 42615
Portland, OR 97242
503-235-1046
thecircle@cityrepair.org
www.cityrepair.org

PERMACULTURE INSTITUTE OF NORTHERN CALIFORNIA
P.O. Box 341
Point Reyes Station, CA 94956
415-663-9090
www.permacultureinstitute.com
(Penny Livingston-Stark, Starhawk's teaching partner, is codirector of PINC.)

THE PERMACULTURE ACTIVIST
P.O. Box 1209
Black Mountain, NC 28711
voicemail: 828-669-6336; fax: 828-669-5068
www.permacultureactivist.net
(This is the central clearinghouse for permaculture information and contacts in North America, and publisher of the *Permaculture Activist* magazine.)

PERMACULTURE CREDIT UNION
4250 Cerrillos Road
P.O. Box 29300
Santa Fe, NM 87592-9300
505-954-3479; toll-free: 866-954-3479
fax: 505-424-1624
www.pcuonline.org/

Wilderness Awareness Resources

WILDERNESS AWARENESS SCHOOL
www.natureoutlet.com

JON YOUNG'S TRACKING SCHOOL
www.shikari.org

TOM BROWN'S TRACKING SCHOOL
www.trackerschool.com

Acclaim for DAVID MAMET'S

Bambi VS. **GODZILLA**

"Entertaining. . . . The lack of caution in *Bambi vs. Godzilla* is bracing. . . . Filled with eccentric, keen, uncompromising insights." —*The Wall Street Journal*

"No other director has written about the movies with such a fearless mixture of amusement, anger, frustration, and rueful love." —Roger Ebert

"Feisty. . . . It may be his most confrontational [book] yet." —*Time Out New York*

"Helpful [and] pragmatic. . . . [Mamet's] style distinguishes these essays. . . . [He] is a clear, exacting thinker." —Walter Kirn, *The New York Times Book Review*

"Provocative and highly quotable. . . . [Mamet] is at the top of his game." —*The Globe and Mail* (Toronto)

"What fun to dive into this book of Mamet musings and words of wisdom! But be warned: like munching popcorn (or Raisinets) at the movies, once you get started it's hard to stop." —Leonard Maltin

"Provocative. . . . *Bambi vs. Godzilla* is always entertaining and frequently spot-on." —*The Onion*

"Piercing. . . . [He] offers a running commentary on the movie business from the inside, with anger but no artifice." —*San Jose Mercury News*

"Hilarious, pungent. . . . [It] bristles with equal parts love and hate for the movie industry." —*Pages*

"Corrosive, funny, surprising, and bracingly lucid."
—*The Post and Courier* (Charleston)

"Strong, stinging writing. . . . Mamet brings more experience—hands on and balls out—knot-head passion, and grim wit to the task than anyone since John Gregory Dunne wrote *Monster* a decade ago. He's got a great, gravelly voice, and it's worth heeding—not that anyone in Hollywood ever will." —Richard Schickel, *Film Comment*

"David Mamet is supremely talented. He is a gifted writer and observer of society and its characters. I'm sure he will be able to find work somewhere, somehow, just no longer in the movie business." —Steve Martin

"*Bambi vs. Godzilla* is far and away the best commentary on how movies are made thus far written by an American. . . . Citing everyone from Aristotle to Preston Sturges's *The Lady Eve*, Mamet demonstrates what works and what doesn't in a movie narrative, while noting what does not work: statistically, in 1958, Hollywood turned out 2000 films which listed in their credits 230 producers, while in 2003 Hollywood produced 240 films with 1200 producers listed.

"Happily, Mamet keeps on in theater and film pretty much on his own terms, and now with *Bambi vs. Godzilla*, like his great predecessor George Bernard Shaw, he can illuminate as a critic-practitioner the not-always-friendly Darwinian world he has been obliged to flourish in."
—Gore Vidal

DAVID MAMET

Bambi vs. GODZILLA

David Mamet is an Academy Award–nominated screenwriter and a Pulitzer Prize–winning playwright as well as a director, novelist, poet, and essayist. He has written the screenplays for more than twenty films, including *Heist*, *Spartan*, *House of Games*, *The Spanish Prisoner*, *The Winslow Boy*, *Wag the Dog*, and the Oscar-nominated *The Verdict*. His more than twenty plays include *Oleanna*, *The Cryptogram*, *Speed-the-Plow*, *American Buffalo*, *Sexual Perversity in Chicago*, and the Pulitzer Prize–winning *Glengarry Glen Ross*. Born in Chicago in 1947, Mamet has taught at the Yale School of Drama, New York University, and Goddard College, and he lectures at the Atlantic Theater Company, of which he is a founding member. He lives in Santa Monica, California.

ALSO BY DAVID MAMET

PLAYS

The Voysey Inheritance (adaptation)
Faustus
Boston Marriage
The Old Neighborhood
The Cryptogram
Oleanna
Speed-the-Plow
Bobby Gould in Hell
The Woods
The Shawl and *Prairie du Chien*
Reunion and *Dark Pony* and *The Sanctity of Marriage*
The Poet and the Rent
Lakeboat
Goldberg Street
Glengarry Glen Ross
The Frog Prince
The Water Engine and *Mr. Happiness*
Edmond
American Buffalo
A Life in the Theater
Sexual Perversity in Chicago and *The Duck Variations*

FICTION

The Village
The Old Religion
Wilson

NONFICTION

Jafsie and John Henry
True and False
The Cabin
On Directing Film
Some Freaks
Make-Believe Town
Writing in Restaurants
Three Uses of the Knife
South of the Northeast Kingdom
Five Cities of Refuge (with Rabbi Lawrence Kushner)
The Wicked Son

SCREENPLAYS

Oleanna
Edmond
Glengarry Glen Ross
We're No Angels
Things Change (with Shel Silverstein)
Hoffa
The Untouchables
The Postman Always Rings Twice
The Verdict
House of Games
Homicide
Wag the Dog
The Edge
The Spanish Prisoner
The Winslow Boy
State and Main
Heist
Spartan

Bambi vs. GODZILLA

DAVID MAMET

Bambi vs. GODZILLA

ON THE NATURE, PURPOSE, AND PRACTICE OF THE MOVIE BUSINESS

Vintage Books
A Division of Random House, Inc.
New York

FIRST VINTAGE BOOKS EDITION, FEBRUARY 2008

Some pieces in this book were originally published in slightly different form in *The Guardian*. Portions of this book are based on the article "Bambi v. Godzilla," which originally appeared in *Harper's* in June 2005.

The Library of Congress has cataloged the Pantheon edition as follows:
Mamet, David.
Bambi vs. Godzilla: on the nature, purpose, and practice of the movie business / David Mamet.
p. cm.
Includes index.
1. Motion picture industry—United States. I. Title.
PN1993.5.U6M244 2006 384'.80973—dc22
2006020018

Vintage ISBN: 978-1-4000-3444-4

Book design by M. Kristen Bearse

www.vintagebooks.com

Printed in the United States of America
10 9 8 7 6 5 4 3 2 1

THIS BOOK IS DEDICATED TO BARBARA TULLIVER

Douglas Fairbanks received me immediately and within a few minutes I was in his Turkish bath. This was the sort of club for the male members of high Hollywood society . . . it was a place where one lounged and steamed and heard the gossip. That day, besides ourselves there was Jack Pickford, Mary's brother, pale and slightly puffy but otherwise unmistakably a Pickford, a strange reputed Red-Indian being called Chief Longlance, and a number of the great moguls who shall be nameless because they were unbeautiful. In fact their sedentary and successful lives had made them old and fat as I am now.

—IVOR MONTAGU,
With Eisenstein in Hollywood

CONTENTS

TECHNIQUE

SOME PRINCIPLES

GENRE

PASSING JUDGMENT

CRIMES AND MISDEMEANORS

INTRODUCTION

All the rivers flow into the sea. Yet the sea is not full. Films, which began as carnival entertainments merchandising novelty, seem to have come full circle. The day of the dramatic script is ending. In its place we find a premise, upon which the various gags may be hung. These events, once but ornaments in an actual story, are now, fairly exclusively, the film's reason for being. In the thriller these events are stunts and explosions; in the horror film, dismemberments; in the crime and war films, shootouts and demolition. The film existing merely for its "high spots" has, for its provenance, the skin flick.

This deconstruction of the film as drama is the reverse slope, of which the ascendant was the genre picture. The genre film meant reassurance to the audience. They knew what they were going to get. They went into the theater, thus, to see Bette Davis, Joan Crawford, Dirty Harry, James Bond, John Wayne, Sylvester Stallone much as they might have gone to a pornographic film or, for that matter, to a stock car race.

Today, studios bet their all upon the big-tent franchise film, which is to say, upon appeal to a self-selected, preexisting audience. It is increasingly difficult to market the nonquantifiable film, as the franchise model continues its advance toward total control of the studio's budget and, thus, of the market. For all industries migrate toward monopoly, and decrease in competition inevitably results in decrease in quality.

Phenomena, when they can't get bigger, get smaller. There are wheels within wheels, and the big wheel runs on faith, and as we know, the smaller runs by the grace of God. The hucksters who invented the nickelodeon and the hucksters who administer its present incarnation are sisters beneath the skin; we, the artists and craftspersons, then as now, tear our flesh and rend our garments, bemoaning the tin ear and the monstrous cupidity of those same hucksters.

After the conflagration, in the final years of humankind, the artists will, once again, be found painting the ceilings of the caves, and the middlemen will, as always, be trying to talk the honest hunters out of their kill. And it may or may not then be remembered, or indeed believed, that there was once a time when the two groups were inextricably linked.

THE GOOD PEOPLE OF HOLLYWOOD

HARD WORK

Billy Wilder said it: you know you're done directing when your legs go. So I reflect at the end of a rather challenging shoot.

The shoot included about five weeks of nights, and I have only myself to blame, as I wrote the damn thing.

Directing a film, especially during night shooting, has to do, in the main, with the management of fatigue. The body doesn't want to get up, having had so little sleep; the body doesn't want to shut down and go to sleep at ten o'clock in the morning.

So one spends a portion of each day looking forward to the advent of one's little friends: caffeine, alcohol, the occasional sleeping pill.

The sleeping pill is occasional rather than regular, as one does not wish to leave the shoot addicted. So one recalls Nietzsche: "The thought of suicide is a great comforter. Many a man has spent a sleepless night with it."

One also gets through the day or night through a sense of responsibility to, and through a terror of failing, the workers around one.

For folks on a movie set work their butts off.

Does no one complain? No one on the crew.

The star actor may complain and often does. He is pampered, indulged, and encouraged (indeed paid) to cultivate his lack of impulse control. When the star throws a fit, the crew, ever well-mannered, reacts as does the good parent in the supermarket when the child of another, in the next aisle over, melts down.

The crew turns impassive, and the director, myself, views their extraordinary self-control, and thinks, "Thank you, Lord, for the lesson."

The director, the star players, the producer, and the writer are *above the line;* everyone else is *below.*

There is a two-tier system in the movies, just as there is in the military. Those above the line are deemed to contribute to the fundability or the potential income of the film by orders of magnitude greater than the "workers"—that is, the craftspersons—on the set, in the office, or in the labs.

On the set, the male director is traditionally addressed as "sir." This can be an expression of respect. It can also be a linguistic nicety—a film worker once explained to me he'd been taught early on that "sir" means "asshole." And, indeed, the opportunities for tolerated execrable behavior on the set abound.

I was speaking, some films back, with the prop master about bad behavior. He told me he'd been on a film with an ill-behaved star who, to lighten the mood or in a transport of jollity, took to dancing in combat boots on the roof of the prop master's brand-new Mercedes. "He did about ten thousand dollars' worth of damage," he said, "and this kind of hurt, as I'd given up my day off, unpaid, to go searching for a prop."

There exists in some stars not only a belligerence but also a litigious bent. I have seen a man take a tape measure to his trailer, as he suspected that it was *not quite* perfectly equal (as per his contract) in length to that of his fellow player.

Meanwhile, back at the ranch, the prop master is giving up his day off to ensure that the wallet or knife or briefcase or wristwatch is perfect on Monday.

This is not a picayune instance but, in my experience, the industry norm. While the star is late coming out of the trailer, while the producer is screaming obscenities on the cell phone at his assistant regarding, most likely, a botched lunch reservation, the folks on the set are doing their utmost to make a perfect movie.

I do not believe I overstate the case.

Nevil Shute wrote a rather odd book called *Round the Bend.*

Its hero is an Indonesian aircraft mechanic. He is so dedicated to both his job and the ideal of aircraft maintenance that a cult springs up around him. He is taken as an example as a teacher and then as the avatar of a new religion. In the practice of machine maintenance, he has found (and Shute closes with the notion that he may have *become*) God.

Some business people feel that they can craft a perfect (that is, financially successful) film *in general,* absent reverence, skill, or humility, and inspired and supported but by the love of gold.

But the worker is actually involved, as Leo the Lion says, in *ars gratia artis* and takes pride in working toward perfection through *the accomplishment of small and specific tasks perfectly.* Like Shute's hero.

Is the actor's hair the correct length? (The two scenes are viewed by the audience seconds apart but were shot months apart. If the hair does not match, the audience will be jolted out of the story.) Are the villain's eyes shadowed perfectly? Does the knife show just the right amount of wear?

I recall the homily of old, that thousands worked over years to build the cathedrals, and no one put his name on a single one of them.

We, of course, enjoy films because of the work of the identifiable, the actors, but *could* not enjoy them but for the work of the anonymous, the crew.

The crew is working in the service of an ideal. Faced, as they often are, with intransigence, malfeasance, bad manners, and just plain stupidity on the part of the above-the-line, they react with impassivity.

This might be taken for stolidity by the unobservant or self-involved. It is, in effect, pity.

I was taught early on that the dark secret of the movie business is this: *All* films make money. Their income, indeed, flows from on high, and the closer one is to the height of land, the more one gets. The farther from the source, the poorer. This is the meaning of the term of art "net profits," which may be loosely translated as "ha, ha."*

And just as there is gold in them thar hills (proximity to the source of the income stream), there is gold in the reduction of hard costs. This reduction includes legitimate business oversight, and may even extend, I have been told, to actual malversation of funds.

Also, we know of Pharaoh that he taxed the Israelites with harsh and unremitting labor, having them make bricks to build his palaces. He then decreed that they must gather their own straw. As did the Reagan administration when it killed the American labor movement.

The guilds and unions in the American film industry retain some strength and have the clout (at least in theory) to protect their workers against the depredations of management in that constant calculus of terror: Management: Submit or I will make all films in Hungary. Labor: Submit or we shall strike.

For any business folk in any business would be glad to take the workers' work for nothing—they, in fact, consider it their right. They would, in American films, as in hard industry, be

* Q. From whence does the money originally come? A. We recall the ancient Jewish wisdom, "If you look hard enough, *everything's treif.*"

right chuffed to see the workers race each other to the bottom, and then, having impoverished them, take the work out of the country. (As, in fact, the studios do now, shooting, I believe, the majority of American films elsewhere.)

The unions, in addition to protecting their membership against the money, must also protect them against their own love of the job. For in the practice of the movie crafts, we see the rampant American love of workmanship—and just as the true actor loves to act, the true carpenter or seamstress loves that perfect corner.

The American icon, for me, is *Rosie the Riveter.* Norman Rockwell's wartime masterpiece shows a young aircraft worker in her coveralls eating lunch. Her scuffed penny loafers rest on a copy of *Mein Kampf.*

Rosie the Riveter beat Hitler. Or, to be a little less high-flown—and in deference to the British, who were, as everyone knows, also involved in that late unpleasantness—there is a true and admirable American instinct of "getting it right."

As I was musing on the same, pondering the star, paid twenty million dollars and ruining the roof of a car, and the prop master, paid twenty thousand and giving up his one day off for the beauty of the thing, I believe I actually began to understand Marx's theory of surplus value: Q. Whom is the film "by"? Spend a day on the set and you learn. It is by everyone who worked on it.

PRODUCERS

My father was a negotiator. He opined that to conserve good feelings at the bargaining table, one should, if possible, express a negative concept in a positive form: "not meaningful" rather than "meaningless." I agree and will, therefore, now refer to contemporary movie executives as running around "like chickens without their heads cut on."

So much for humor. Now for the weather.

Life in Hollywood seems to have ground to a standstill. We have fewer and fewer films, and these are of diminishing worth and ever-inflated production costs. It is enough to drive one to the fainting couch.

There, whilst recruiting myself, I recur to my all-time favorite, Gérard de Nerval, who walked a lobster on a leash through the Tuileries. Oh, better than your pale shade Christo, I reflect, you have died too soon, a life dedicated to the exquisite remembered but in the admiration of those with much, much too little to do. Drift, then, my hand, from the scented kerchief on my forehead, to the low and accommodating bookcase but nearby. Roam, fingers, down those spines, so worn, so warm, the comfort of my grizzled age.

What have we here?

It is a novel by Captain Frederick Marryat: *The King's Own* (1831). See how the page, absent intention, falls open to a passage that may both mimic and direct my thoughts:

"Since the World Began, history is but the narrative of kingdoms and states progressing to maturity or decay. Man himself is but an epitome of the nations of men. In youth all energy, in the prime of life all enterprise and vigor; in senility, all weakness and second childhood. Then, England, learn thy fate from the unvarying page of time."

And there we have it, Spengler's two turgid volumes reduced to a mere paragraph.

Substitute "the movie business" for "England," and the thing is clear: that energy of youth, that cunning of age, must and will decay. Vigor itself will bring about death, for healthy life both breeds competition and attracts dependents—the necessity to still warring and to support legitimate claims distracts energy from the original healthy task of growth, and the walls come tumbling down.

The movie business, originally the *cosa nostra* of arcade hustlers, grew into fierce, healthily warring factions who now compete and now collude in a war fought not in the theaters but in the boardrooms.

I pass a poster for the current film and count eighteen names of producers.

On the poster?

Note that the poster is traditionally a way to attract the eye, and so the mind, to a novelty. The producers may in fact have contributed something to the film, but who in the world has *ever* gone to a film because of the identity of a producer? *No one.*

Then why list eighteen?

And here we have, to the physician, the unfortunate, inescapable, symptom—here is the sunken cheek, the dark hollow neath the eye, the foul breath and thready pulse, the herald of death: the film, perhaps, is being made no longer to attract the audience but to buttress or advance the position of the executive.

King Lear (read: Harry Cohn, B. P. Schulberg, Louis B. Mayer, Irving Thalberg) has gone to his reward, and his absence has been noted and acted upon by the canny. It is not that the fox has taken over the henhouse but, if I may, that the doorman has taken over the bordello.

In the golden days of the madam (Harry Cohn et al.), the lives of the girls may not have been better, but the lives of the customers were. Why? Because the owner-proprietor knew that her job was simply and finally *to please the customer.*

Moviemaking is an appallingly simple process. One needs a camera, film, and an idea (optional). The business of the movies, similarly, is simple hucksterism: find an attraction, present it as engagingly as possible, take the money, and guess again.

Just as the making of the film requires little more than someone to hold and someone to stand in front of the camera, so the business end requires nothing other than someone to make decisions. These decisions require intuition and/or courage, for the desires of the audience cannot be quantified; they may only, finally, be guessed at. The owner-proprietor, betting his own money, realizes that he is only rolling the dice. He stands to gain much, he stands to lose much, but he *has* to put his money on the street—that is, he has to place his bet. To that end, he strives to keep costs down, thus enhancing his chance at a greater time at the tables.

Current executives, however, have incentive to *inflate costs as much as possible,* thus necessarily minimizing the number of films produced (bets placed).

No one would think that way if he were gambling with his own money. But this new breed is not. They have tied their fortunes not to the success of a film, or of films, but of their superiors. They know that their superiors, the studio heads, will each, in time, be fired when the banks or the megaliths tire of their eventual failure. (Each must eventually fail, like the gambler who is shackled to the gaming table. *Eventually* the odds, in favor of the house, will clean him out. And for the studios the odds are, at the end of the day, in favor of the *audience.* The whims of the audience, that is, will and must break the studios, *if* the studios keep doubling their bets.)

The wise player of days gone by took his winnings and retired to a teenage wife and tennis in Malibu. Today's moguls, though, like the besotted gambler, keep doubling up. That is, they keep increasing their production costs. These increased costs propitiate not, primarily, the audience but the studio owners. "How can I take the rap for the failure of this film when I got the most expensive stars and sets and director possible?" is the exculpatory chant of the executive. "Let me try again."

The folly of this course is like that of the loser at roulette. He plays black, black loses, he doubles the bet and plays black again, black loses, he doubles his bet and plays black again. It loses. He reasons that soon black *must* come up—that the odds of red appearing five, six, seven times in a row are astronomical. Black *must* come up soon.

In his panic, however, he forgets two things:

1. His bankroll is not inexhaustible. Yes, the little ball must eventually land on black, but it might not do so until one spin after the player has gone broke.
2. The roulette wheel (like the audience) has no memory: the wheel is not aware that red has come up twelve times in a row. The odds of its appearance on the next spin are still 50-50.

The play rushes toward its dénouement. Films cost more, their increased costs attracting bevies of the sychophantic, for the larger the budget, the greater the possibility—indeed, the necessity—of waste. (The Pentagon could not, year after year, keep increasing its bloat if it did not, year after year, exhaust its stipend.) These executives scheme to hire, each and logically, his own retinue of supporters. And the gold-encrusted howdah must eventually drag down the mighty elephant.

VICTIMS AND VILLAINS

Are there *good* producers? Yes.

I met Otto Preminger in his office on Fifth Avenue. The setup was everything one might wish for in a meeting with a potentate-producer. The room was huge, and the desk might have been fashioned from the flight deck of a small carrier. He offered me a cigar. We chatted about this and that; he took me to lunch at "his" place on Fifty-fifth Street. At the end of the lunch he looked intently at me and asked if I had ever done any acting. I allowed that, yes, in my youth. . . . He nodded. "I'm looking for a young man to play an Israeli officer in my next film," he said. "Yes . . . might you consider it . . . ?" I was, of course, flattered beyond measure—that this great producer had seen into my inner soul, had seen my innate valor, strength, and capacity for both self-sacrifice and leadership. . . . "And there's something else I'd like you to consider," he said, and the something else was, of course, the kicker.

I have forgotten what service he was trying to extort out of me with his flattery, but I do remember he came damned close to doing it.

It was at this same lunch that he told me how he shot the vast crowd scene in *Exodus*. The scene is the proclamation, in Independence Square in Jerusalem, of the state of Israel.

Preminger required a packed square, some ten thousand extras. He could not pay for them.

"What did you do?" I asked.

"I charged them," he said. He papered the town with posters: BE IN A MOVIE, TEN SHEKELS. That's what I call a producer.

Preminger's dexterity in working with a limited budget puts me in mind of a situation we faced when I was directing my first film, *House of Games* (1984). Its producer was my friend and mentor, Mike Hausman.

We did the film for no money and, after a search for a suitably run-down inexpensive venue, chose Seattle. Seattle at that time was an other-than-affluent city and boasted a spectacularly seedy skid-row district, where we had hoped to shoot the film.

Here was the problem. Seattle was cheap to shoot in, but as it was virtually undiscovered as a location, all the equipment had to be schlepped up from L.A. The Teamsters Union had a rule that a film company would pay their members the going rates and benefits of that venue from which their drivers drove the equipment trucks. Their local rates in Seattle were just within the film's budget, but those of L.A. would have made filming financially impossible.

Mike scratched his head for a while, put the equipment trucks on railroad cars, and shipped them to Seattle.

We started filming, the Teamsters said, "Wait a second . . . ," and by the time the thing was adjudicated we had the film in the can. Mike and the Teamsters parted friends, and mutual respect, one pirate for another, was the order of the day.

Another great producer, Sarah Green, made several films with me in the Boston area. People bitched and moaned about the locale's intractable Teamsters, but, again, we all got along just fine.

I made a film *about* the Teamsters (*Hoffa,* screenplay by me, directed by Danny DeVito), and the film, at the time of its release in 1992, got slammed by the press for, as near as I can interpret it, a pro-labor stance. I was accused, as writer, of being an "apologist" for the Teamsters.

Funnily enough, I didn't think the Teamsters needed an apologist, as they had a *union* (a much better idea).

The money people have been bitching about the Teamsters since the Lumière Brothers schlepped out the first camera, but as a member of various unions and guilds accustomed to handwringing, I admire the Teamsters' pluck.

Capital, if it cannot call Labor "Reds" *will* call it "Thugs." Some unions, Actors' Equity, SAG, and, to a somewhat lesser extent, the Writers Guild, are addicted to going "Wee wee wee all the way home."

The Money suggests: "Give in: I can get *anybody* to act, you have no power," and the artistic guilds seem to perceive some truth in this suggestion, thus weakening their position.

They might reflect that anyone can drive a truck.

But back to Otto Preminger.

I was watching Preminger's *Exodus* the other night. There we have Paul Newman, who has my vote for the most beautiful man ever to grace the screen. He is fighting for the rights, for the lives, of Jewish refugees from Europe. Man. I love Paul Newman in this movie. He "just don't care." He doesn't care for the good opinions of the other players or, as per the script, of "the world." He endeavors to teach Eva Marie Saint, the love interest, a goodwilled American Christian woman, that nobody but the Jews cares about the Jews. This lesson is not only dramatically interesting but, in a rare coincidence, true. I sit there nodding.

Eva Marie Saint is terrific. She is, as above, goodwilled, sincere, and incredibly naïve. Quintessentially American, her answer seems to be, to everyone, "Is there not good and bad in all peoples?" Can't we just all "like each other"?

Paul's girlfriend, before the story began, got kidnapped from her kibbutz. She was, we are told, tortured, blinded, raped, her hands and feet cut off, and deposited back with her people, where she died. They name the kibbutz after her. Life goes on, and Paul has a one-night stand with Eva Marie. The next time he sees her, she reports that she regrets their fling; she is just a tourist and took a "wrong turn that night." Why? Because the young woman refugee she has grown fond of has elected to stay on the kibbutz rather than return to America as Eva's adopted daughter. Eva, in effect, "just doesn't understand these Jews." How, in effect, can they prefer each other to the company of an actual American?

The answer is, of course, that to Otto Preminger, to Leon Uris, who wrote the book, to Paul Newman, it is a fact of nature that the Americans "go home": that it is at this that we excel. Eva's foray into global politics is a prototypical example of what has come to be known as "adventure tourism." She will, pardon my French, get laid, "do good," become disenchanted, and go home (cf. Vietnam, Kuwait, and, watch-this-space, Iraq).

Exodus was a big hit. The theme music was played at every bar and bat mitzvah and Jewish wedding of my youth. They were playing our song. And lo, coincidentally, there we had this actual new country, scant time zones away, the first Jewish state in two thousand years, to go with the song and the film.

The music from *The Godfather* has also become an American racial staple.

The Godfather is ostensibly about a bunch of murderous thugs. Operationally, though, it is our American House of

Atreus. It is the story of an American family. It has gods, demigods, fates, furies, clowns, just like your family and mine. The family in question happen to be criminals. This is not only dramatically acceptable but also historically approved convention. The mafiosi are merely the Plantagenets of our day: removed, exalted, unbound by law.

Kay, played by Diane Keaton, is in love with Michael Corleone, heir to the mafia crown. He tells her he is going straight, and she responds that if he thinks she will believe him, he is naïve.

"We're just like presidents and senators," he tells her.

"Presidents and senators don't kill people," she says.

"*Now* who's being naïve?" he responds.

In both *The Godfather* and *Exodus,* the majority culture is represented by a legitimately nice, indeed, a lovely and good-willed, Protestant woman: "Why can't you" (Michael Corleone, Ari Ben Canaan) "just be like *me*?" the woman asks. The answer, in both instances, is that the hero is fighting for his life and for the life of his people, and the woman is not.

Now, *The Godfather* is, of course, the better film. Even its theme song is (marginally but nonetheless) better than that of *Exodus,* and to retire the trophy, the last time the Jews and the Italians clashed (Masada, 73 CE), the Italians won that, too.

Masada, I seem to recall, was a miniseries some years or decades back. Like *Exodus,* like *The Diary of Anne Frank,* like *Playing for Time, Schindler's List, Sophie's Choice,* it held the "magic feather." You will remember that Dumbo, the elephant, was taught by Timothy, the mouse, that he could fly if he held the magic feather. The magic feather in film is bathos: the kitten and the dog who must find their way home, the crippled child, Jews dying. *Exodus* bridged the gap. There Jews fight to *stop* dying and to start living. The message, in 1960, contained both the requisite bathos and *novelty.* Today, any potential treatment of the new state contains neither and so is dramatically unacceptable.

But Hollywood never discards the once useful. What, then, shall one do with the Jews? If we, like Ari Ben Canaan, refuse to be victims, perhaps we can be made to serve as villains. Both, we note, identify the subject as "other," and each contains the salutary potential for violence.

The film *The Sum of All Fears* discreetly brings the world to the brink of disaster because the Israelis have thoughtlessly misplaced one of their nuclear bombs. Not only have they forgotten where they put it, they also stole it in the first place, from the U.S.A., which, in the security of its own nuclear programs, was more considerate of the wishes of others.

One might think that this largely narrative off-screen identification of Israelis as villains might be an anomaly, had one not seen it reiterated daily in much of the Western press. (I will make bold to state that some readers of this piece may, in fact, "root" for the Palestinians. I take no issue with their right to that view. I merely state that, irrespective of any reader's assessment of its rectitude, much of the Western press portrays Israelis as monsters.) I will leave the press to chew its own incomprehensible cud and address myself only to the movies.

I predict a growth of the Jew as monster in the next few years' films. Well, why not? Alfonso Bedoya and John Huston inaugurated a few decades of the vicious Mexican ("We don't need no badges! I don't have to show you any stinking badges!"); Jeremy Kemp et al. made the British accent the tocsin of evil quite effectively for quite a while. So I shall naively opine that perhaps turnabout is fair play, and it is merely the Jews' turn in the barrel.

In 1960, Otto Preminger could think of no more magnificent icon than the Israeli officer. Tom Clancy, in a new day, finds them, dramatically, employed better otherwise.

JEWS IN SHOW BUSINESS

> These false Jews promote the filth of Hollywood that is seeding the American people and the people of the world and bringing you down in moral strength. . . . It's the wicked Jews, the false Jews . . .
>
> —Louis Farrakhan, 2006

Let me see if I can offend several well-meaning groups at once. I will address myself particularly to the racially punctilious and to the goodwilled but otherwise uninvolved champions of the developmentally challenged: I think it is not impossible that Asperger's syndrome helped make the movies.

The symptoms of this developmental disorder include early precocity, a great ability to maintain masses of information, a lack of ability to mix with groups in age-appropriate ways, ignorance of or indifference to social norms, high intelligence, and difficulty with transitions, married to a preternatural ability to concentrate on the minutia of the task at hand.

This sounds to me like a job description for a movie director. Let me note also that Asperger's syndrome has its highest prevalence among Ashkenazi Jews and their descendants. For those who have not been paying attention, this group constitutes, and has constituted since its earliest days, the bulk of America's movie directors and studio heads.

Neal Gabler, in his *An Empire of Their Own,* points out that the men who made the movies—Goldwyn, Mayer, Schenck, Laemmle, Fox—all came from a circle with Warsaw at its center, its radius a mere two hundred miles. (I will here proudly insert that my four grandparents came from that circle.)

Widening our circle to all of Eastern European Jewry (the Ashkenazim), we find a list of directors beginning with Joe Sternberg's class and continuing strong through Steven Spielberg's and the youth of today.

(A president of Harvard was, in the seventies, defending himself. The admissions policies, theretofore uninterested, started taking cognizance of the place of residence of the applicant. The president called the new program Geographical Diversity or some such and pointed out that in prepolicy, unenlightened days, a statistically anomalous percentage of the student body had come from "the doughnuts surrounding the cities." An alert number of the student body responded, "Those aren't doughnuts, they're bagels.")

As is the movie community.

There was a lot of moosh written in the last two decades about the "blank slate," the idea that since theoretically each child is equal under the law, each must, by extension, be equal in all things and that such a possibility could not obtain unless each child was, from birth, equally capable—environmental influences aside—of succeeding in all things.

This is a magnificent and majestic theory and would be borne out by all save those who had ever had, observed, or seriously thought about children.

Races, as Steven Pinker wrote in his refutational *The Blank Slate,* are just rather large families; families share genes and, thus, genetic dispositions. Such may influence the gene holders (or individuals) much, some, or not at all. The possibility exists, however, that a family passing down the gene for great hand-eye coordination is likely to turn out more athletes than that without. The family possessing the genes for visual acuity will most likely produce good hunters, whose skill will provide nourishment. The families of the good hunters will prosper and intermarry, thus strengthening the genetic disposition in visual acuity.

Among the sons of Ashkenazi families, nothing was more prized than genius at study and explication.

Prodigious students were identified early and nurtured—the gifted child of the poor was adopted by a rich family, which thus gained status and served the community, the religion, and the race.

These boys grew and regularly married into the family or extended family of the wealthy. The precocious ate better, and thus lived longer, and so were more likely to mate and pass on their genes.

These students grew into acclaimed rabbis and Hasidic masters, and founded generations of rabbis; the progeny of these rabbinic courts intermarried, as does any royalty, and that is my amateur Mendelian explication of the prevalence of Asperger's syndrome in the Ashkenazi.

What were the traits indicating the nascent prodigy? Ability to retain and correlate vast amounts of information, a lack of desire (or ability) for normal social interaction, idiosyncrasy, preternatural ability for immersion in minutiae; ecco, six hundred years of Polish rabbis and one hundred of their genetic descendants, American film directors.

Please note that I do not claim for myself and my extended family the *yichus* of descent from the rabbis. My own family history, and, I believe, that of most of the film directors I know (Jewish and otherwise), is firmly that of the ne'er-do-well. I suggest, however, a collateral benefit to the Ashkenazi populace-at-large of the more culturally limited inbreeding of one of its constituent portions.

One does not, of course, have to be Ashkenazi or, indeed, Jewish to succeed as a film director; my genetic divertimento may point out, however, one desideratum of the filmmaker (it need not be hereditary but had better find itself on the CV): experience as a ne'er-do-well.

Proverbs tell us that the stone the builders rejected has become the cornerstone. So it is with anyone in show business, and particularly so of the director. For how could this position, requiring a depraved generalist, attract anyone who had succeeded or was apt to succeed in a specific field?

Just as U. S. Grant failed at everything save preserving the Union, the director is probably one who, by birth, training, or disposition, is gifted and/or driven either to make order out of chaos or to reverse the process.

Loki, Raven, the Fonz, Falstaff, and Larry David are examples of this archetype, the trickster—characters who express or intuit the propensity to upset and so reorder the world on a different level of abstraction, which is the job of moviemaker.

How might one train to tell a story in pictures; assign various crafts and departments their tasks; manage and direct several hundred artists, craftspersons, and administrators; and inspire to meet the exigencies of a grueling production schedule in spite of weather, human nature, chance, et cetera?

It is a job that attracts those who thrive on challenge, chaos, uncertainty, human interaction; who love improvisation; who would rather die than revert to the general population, et cetera—in effect, quasi-criminals.

Note:

1. Yes, there are a lot of Jews in the movie business.
2. No, we did not kill Christ.

THE DEVELOPMENT PROCESS;
OR, LEARNING TO MAKE NOTHING AT ALL!

The artist is, in effect, a sort of gangster. He hitches up his trousers and goes into the guarded bank of the unconscious in an attempt to steal the gold of inspiration. The producer is like the getaway driver who sells the getaway car and waits outside the bank grinning about what a great deal he's made.

There are, admittedly, some good producers. But we must remember that even Diaghilev went into the ballet because he wanted to screw Nijinsky.

What on earth do these producers *do*?

A few are entrepreneurs, raising money for a project under their control; a few are what the shtetl knew as *shtadlans,* that is, intermediaries between the powerless (in this case, the filmmaker) and the state (or studio); the rest are clerks or clerk-sycophants.

And we have the sycophants full stop. For the powerful producer, who can drive but one Mercedes himself, can employ or cause to be employed many to drive Their Own Mercedes in his livery.

But some Greek said, or should have said, "If it exists, it probably has a cause."

So let us assume somebody's brother-in-law showed up one day in the palmy presound days of Hollywood, and *his* brother-in-law, a power on the lot or on the set, hoping to avoid a "touch," said, "People, this is Bob, and he is a producer." Bob was then entitled, under the family flag, to all the sex, drugs, and fun he could wrangle and to whatever he could hypothecate.*

Time went by, and Bob stayed on. He, or another of his ilk, caught, stole, or otherwise achieved power in some niche in the industry and, having learned a good trick, one day appointed footmen of his own.

These folk, with nothing much to do, and in the manner of functionaries down through time, schemed all their waking hours to increase and consolidate power.

The filmmakers were busy on the lot or on location, but our producers, like Jacob, stayed in the tents, free to wheedle, convince, and extort position from and in the studio system. Soon all films had a producer, then two, and, today, count 'em, an average of seven in the head titles.

Just as the royalty, another entertainment industry, has keepers of the keys, gentlemen of the bedchamber, and so on, our American charade has coexecutives, executives, and supervising producers ad infinitum.

And, just as with you in your particular racket, our ceremonial positions hide from the uninitiated gaze the empty throne.

And so the producers stay, and live long, why should they not, God bless them. They watch while the lowly make bricks and suggest, at regular intervals, that the brick makers begin to gather their own straw.

* The producer's shenanigans were, and are, regulated by unwritten, elastic but generally understood conventions. As such, they are not without precedent. In the Age of Sail, the Bosun was understood to have as his vail or perquisite a certain extent of the ship's stores. The odd damaged crate, oar, cordage, or spar might be sold over the side, and he might pocket the money. Things being what they are, some, naturally, extended the privilege, endangering the health of the ship, and were fed to the sharks. But we live in a more enlightened time.

And these producers propound heresy.

They sell all parts of the pig but the squeal. And then they sell the squeal.

I have, with my own eyes, seen the following: a sign on the Craft Service (snack) table, near the end of filming: GUM IS FOR PRINCIPAL CAST MEMBERS ONLY.

I have seen the clothing worn by the various cast members auctioned off on eBay, its value increased by the branding of the clothing with the actor's name: "Buy Ricky Jay's pants." "Buy Rebecca Pidgeon's blue jeans." The proceeds go into the producers' purse.

I have seen producers bill the movie salaries for their mistresses, for their absent yes-man, for travel and lodging never used, for services never proffered, for inedible cast and crew meals charged off as gourmet fare, for imaginary bank fees interest, and so on, and so on.

It is one of the staples of talk between actual moviemakers (cast, crew, crafts): "Do you know what that son of a bitch did?" And one comes to tire even of the introduction, "I thought I had seen it all, but . . ." Which is lassitude, indeed.

I've seen, as have we all, theft, fraud, intimidation, malversation, and seen it with such regularity that its absence provokes not comment but wonder.

My favorite offense against the gods, however, is curiously benign and is propounded, usually, not by the savage producer but by the obtuse. To wit:

"We are going about this the wrong way. Why don't we just go in a room, analyze the most successful movies of all time, and then make *that*?"

This, to the filmmaker, is enormity on the order of the resident of wartime Munich suggesting that he thought they were summer camps.

What, the filmmaker wonders, did or do you think we were

doing all of these years, all of these films, whilst you rested in the tents—can you not have seen that we, the filmmakers, were working like mad to solve, film by film, shot by shot, line by line, the problem you suggest a rational person could vanquish by simply "going in a room"?

For this desire to "go in a room" is, to the artist, heresy. It is the reductio ad absurdum of "reality" programming: having determined that it's not necessary to pay either actors or writers, the deluded additionally discover that it is not necessary even to fee the gods—that insight, idiosyncrasy, inspiration, patience, and effort are the concerns of the weak and misguided craftsperson and artist.

No, the exhortation to "go in a room" is not mere crime but blasphemy. It is not sufficient to shake one's head; one must lower the eyes.

The American educational process prepares those with second-rate intellects to thrive in a bureaucratic environment. Obedience, rote memorization, and neatness are enshrined as intellectual achievements. Just as the SAT measures the ability of the applicant to *take that test,* the bureaucratic rigors of the studio system probe the neophyte's threshold for boredom, repetition, sychophancy, and nonsense.

Actual life on the set, where the films are actually being made, is somewhat different.

On the set, a random sampling finds among the workers an actor who won the Academy Award for a short film he wrote, the world's number-two champion arm wrestler, a martial-arts master, a leatherworker, a woodworker, and so on.

The anarchistic nature of a movie set attracts individualists, autodidacts—the interested, the enthused, such as occur only among the truly self-educated.

Eric Hoffer wrote that the mark of a good and healthy society is the ability to function without good leaders. Much work on the set, while apparently hierarchical, is curiously self-directed.

The director in films is an Olympian figure, usually removed from the life of the set, his wishes or whims relayed to the workers by functionaries. In television the post seems largely ceremonial—the film *is* going to get in the can. (The ancient television phrase has it: features in the morning, documentaries in the afternoon.) How does the film get into the can?

The ostensibly hierarchical arrangement—director, department head, worker—is a chain of information that relies for execution on the man or woman on the set. The necessity for improvisation is so great that the strictly hierarchical worker could not and cannot survive. The directions given may be more or less succinct, but the possibility of their execution will almost always require intuition, improvisation—in short, art.

The art of the worker has been of the essence in the *artistic* success of films: Zoltan Korda's painted shadows in the boat-deck scene between Mary Astor and Walter Huston in *Dodsworth,* Ken Adam's code machine in the B-52 in *Dr. Strangelove,* the skill of the grips moving the dolly in the calliope shot in *The Rules of the Game,* the timing of the thrown snow in W. C. Fields's *The Fatal Glass of Beer.*

The variety and flat-out humor, the sordid tragedy, the perversity of human existence are explorable only by those experiencing them: the artists and the workers.

On the other hand, the young bureaucrat-in-training, as he progresses in the bureaucratic hierarchy, will discover—some quickly; others, their eventual lackeys, with less speed—that success comes not from pleasing the audience but from placating his superiors until that time it is reasoned effective to betray them.

He learns, in short, to bide his time.

And as time goes by, this suborned young person becomes

each day less capable of first uttering and then framing a non-bureaucratic thought.

Impulses of joy, of wonder, indeed, of rage and grief are repressed until they are no longer consciously felt.

This is called "growing savvy."

This person, like a member of a sexless marriage, ceases to feel affection, lust, desire for the permitted object, and, as in that marriage, this energy is diverted into (inter alia) depression, abuse, and treachery.

The successfully matriculated executive, marginally concerned with art and diminishingly concerned even with "product," devotes his new wisdom and increased leisure to opportunities for trickery, greed, stock manipulation, and merger, as in any business.

In the film business, one department of this glee is called "the development process."

This is the fig leaf of propriety covering, if I may, not the genitals of artistic potential but the empty space where they once lay.

Karen Horney wrote of a neurotic who could never complete anything. She was impeded now by this, now by that mischance and was eternally blighted and blocked, just on the edge of the creation of a great work. Ever saddened but still valiant, she pressed on, content in her own untested but undoubted abilities.

On completion, any creation is torn from the prospective fantasy of creative potential and consigned to the world, there to brave and bear the opinion not only of others but also of its creator.

But not the film in development.

This film need never cross that divide.

This project is, in fact, a film only by courtesy.

It functions, as do the endless and proliferating committees of Government, as a repository of bureaucratic power. This power exists, and can exist, only in *potential*—for should the committee ever come to conclusions, its task, and so its operation as a bureaucratic fiefdom, would cease. So the bureaucrat, studio or otherwise, learns not only of the inadvisability of any test or completion but also of such conclusions' absolute foolishness.

This lesson, we see, was learned very well by the folks at Enron et al.—executives who saw that the power to grow wealthy stemmed from the brave decision to stop making anything at all.

THE REPRESSIVE MECHANISM

A DARK COMEDY

America is a self-created society.

That a self-created group could rule itself, absent religious or hereditary oversight, subject only to the dictates of reason, is all very well. The body politic, however, constantly strives to re-create for itself that same irrational, repressive (that is, effective) governing authority otherwise vested in tradition, popes, and kings.

Such authority may be called patriotism and, as such, may helpfully identify convenient enemies foreign and domestic; or we may create a civil version of religion and call it "family values," liberalism, Americanism, or the American way. We may, also, uniquely in history, name this ad hoc authority "entertainment."

The absence of a historical and universally acknowledged authority to which one may pledge fealty and against which one may rebel creates factionalism: the right moves toward fascism, the left toward chaos. Democracy—in extremis—seems capable of devolving to either tyranny or civil war, and America, maddened by unimaginable prosperity and safety, incomprehensibly powerful, and bereft of threats, splits down the middle on the issue of *definition.*

Is the good person one who will not tolerate a president's lies about sex or one who will not tolerate a president's lies about war? Neither of these outrages is without precedent, but the country, in its anomie of wealth and faced with the unbearable reality of its own self-government, is reduced to what is essentially a neurosis. That is, "I know something is wrong, but I cannot identify it." The resulting free-floating anxiety is exchanged, in a good trade, for an obsession—in this case, the search for an internal enemy.

Two basic strategies, per party, are: on the right, "people are swine—endorse my position, or join them in my estimation"; and on the left, "people are basically good at heart, can't you see that, you sick fool?" One strategy to combat in the awkward burden of democracy is the establishment of aristocracies: the Kennedys, the Bushes, Frank Sinatra Jr., et al. This, of course, never answers, for the palace, the manufactory, the final clubs of Harvard, and the stage have bequeathed us little other than attenuated genes.

The continent was paved and wired by the common people free of authority; now, as we lift our heads from our labors or our depredations, the absence of authority is driving us mad. (How, we wonder, could that which was won by unchecked aggression be held by any other strategies—must not those similarly avaricious see in us that same rich continent, ripe for plunder, which our forefathers saw?) In effect, the miser cannot sleep. And in our madness, we have pressed entertainment, which is to say tincture of art, into the service of the repressive mechanism.

In the 1960s, philosophers of communications observed that television viewers sometimes did not laugh at gags that, in the studio, seemed quite hilarious. Those scientists opined that the home viewer was inhibited by the absence of the herd—that the individual, all alone, may sit unmoved at the same jokes that would convulse a full auditorium; that there was ineluctably something of the *communal* in the impulse of laughter, the viewer delighting not only in the *gag* but also in sharing with his like the notion, "Isn't life like that?"

Home viewers, then, who would not laugh alone, were freed from their inertia by the addition of the laugh track. It was later noted, however, that given this effective tool, the quality of the joke itself became moot.

One sees this brave aperçu applied in the pronouncements of our contemporary government. Periodic declarations assuage fully half of the populace absent any correspondence between that verbiage and the government's actions or, indeed, between them and the plausible. The administration, like the sitcom of old, has discovered that, given an immobilized audience, a presentation of the form itself is sufficient entertainment.

Big and bad films, summer films, blockbusters have similarly become the laugh track to our national experiment. As with the Defense Department, we are reassured by their presence rather than their content or operations. As examples of waste they appeal to our need—not for entertainment but for security.

The very vacuousness of these films is reassuring, for they ratify for the viewer the presence of a repressive mechanism and offer momentary reprieve from anxiety with this thought: "Enough money spent can cure anything. You are a member of a country, a part of a system capable of wasting two hundred million dollars on an hour and a half of garbage. You must be *somebody.*"

The same mechanism operates equally in the Defense Department.

"Soft on defense" is a war cry directed not against the country's enemies but at those who are soft on defense; the right understands (consciously or unconsciously) that a conveniently frightened populace wants security and that not *defense* but *defense spending* absolutely (if momentarily) supplies it.

This addiction is conveniently both self-ratifying and self-perpetuating, for if the country is *not* at risk after spending, the spending, now proved effective, must, logically, be continued. And if the country is *still* at risk after increased spending, the only cure must be further expenditure.

As with defense, the paradigm in blockbuster entertainment is "more will cure it." But the natural desire for true entertainment can be fulfilled only by the truly entertaining. The truly entertained audience member leaves the theater chuckling or shaking his head sagely—in either case, *fulfilled*. Those, on the other hand, who have been treated to an outing of *waste* are reenrolled in the service of the repressive mechanism, have experienced no fulfillment, and are driven to increase their expenditure and repeat their error. (Vide the hypnotized televiewer glued for five irreplaceable hours each evening in front of the set.)

The winning gambler slogs back, soon or late, into the casino, for he craves not riches but, *ultimately,* the thrill offered by his addiction. He returns to the toxic environment and its whore's promise of joy, driven mad by his own irrational actions. He tells himself he plays to win, but when he wins, he plays until he loses and, losing, plays in an effort to restore his winnings.

For America to engage in pointless, destructive, irrational foreign enterprise is, essentially, for the electorate to endorse a bad, empty, addictive entertainment.

Perhaps the success of Michael Moore's film *Fahrenheit 9/11* is due to its excellence not as a documentary but as a *comedy.* A comedy is the form in which the unsayable is said and that, thus, for a moment, breaks the corrosive cycle of repression.

AN AMERICAN TRAGEDY

My title refers not to the 1931 film with Phillips Holmes and Sylvia Sidney, nor to the story's treatment as *A Place in the Sun* (1951), nor to Dreiser's 1925 novel on which the two are based, but to another quintessential American film document, *The Jolson Story.*

This 1946 film is a pretty good musical, and musical document, of the career of Al Jolson, who started out as a busker in saloons and ended as one of the first media superstars on stage, on the radio, and in motion pictures.

The film and its attendant meditation also offer a good clinical study of American confusion.

A. J. Liebling wrote of two boxers in the ring "working out their tight little problem." That is what we have here. And the tight little problem is race. Young Asa Yoelson is the son of a cantor. We see him first singing the liturgy in shul, but he discovers vaudeville and wants to get into show business. His straightlaced and religious father forbids it. So Asa runs away. He is caught by the cops and taken for safekeeping to a Catholic church. When his dad, the cantor, catches up with him, Asa has been pressed into the choir and is singing a solo in "Ave Maria." His father, the cantor, duly impressed by that talent that can be revealed only through ecumenicism, relents, and the son goes into show business.

He leaves the shul, he leaves his home, he changes his name to Al Jolson, and there you go.

Now this newcomer pines for a break, as one does, and one night, Tom Baron, a blackface performer, gets drunk and can't go on. Al puts him to bed, blacks up, and does his act. Great impresarios are, by chance, in the audience, and Al's talent is recognized.

Not to put too fine a point upon it, but Al, a poor Jewish immigrant kid, gets his first break from impersonating a Christian (the European solution) and his second from impersonating a black man (the more American choice). He is now on his way, taken up by Lew Dockstader of Dockstader's Minstrels. These minstrels, an American tradition, are white performers in blackface, presenting a traditional evening of "darky" songs.

Al's job, for three years, is to sing, "I want a girl just like the girl who married dear old Dad." This work is steady but other than rewarding. And then one night in New Orleans, he wanders, musing and unquiet, into Storyville and hears jazz. His life is changed. He quits the minstrels and retires in meditation to reinvent himself.

As one might, at this point, he returns home. His family has not seen him for three years. He explains he is on the track of something new—he wants to bring jazz, black music, to the world at large—that is, to the white world.

Now let us become psychoanalytical. He is ushered to the family table; he sits, ready to eat; and his mother sneaks him a yarmulke. He is a Jew, his father is a member of the clergy; they are, of course, observant; one cannot eat without first blessing the meal; and a man cannot say the blessing with an uncovered head. He puts on the yarmulke, and his canny father says "Asa, did you wear the cap while you were on the road?" "No," he responds sheepishly. "Not all the time."

"Well, then," says his father, "you don't have to wear the cap for me. . . ."

Who ever referred to a yarmulke as a "cap"? Whose feelings are being spared, whose lack of intellect is being considered, in referring to a yarmulke as a cap?

Further, why would his father, a cantor, endorse (accept, perhaps, but not endorse) Al's irreligious behavior?

But he does, and all are happy to have reached this particularly American détente—or repression as the cost of assimilation—at which exact point the phone rings.

Now, in psychoanalysis, there is no such thing as accident, no such thing as coincidence or mere happenstance. Neither is there in dramaturgy. The phone rings *because* Al has set aside his race and religion.

And who is on the phone? It is, of course, Tom Baron, now a successful theater manager, offering Al a slot in a new and important Broadway show.

Yes, Al says, of course I will do it. But I have to sing "this new music"? Well, okay.

Opening night at the Winter Garden, Al is due to do a solo, but the show is running late, and Oscar Hammerstein tells the stage manager to bring down the curtain before Al's turn. Al rushes onstage and demands the orchestra play his song. The song is "Mammy," and Al brings down the house, and his career is assured. He becomes a cultural icon.

He stars in the first talking picture, *The Jazz Singer,* 1927.

In it he plays a nice Jewish boy, son of a cantor, who runs away from home to join show business. In the film's climactic moment, the cantor becomes ill, and the boy must forsake his Broadway show to rush back to the shul and sing Kol Nidre.

The lead in our film, *The Jolson Story,* is played by Larry Parks.

This is one of the most remarkable performances I've

ever seen on film. Parks lip-synchs some twenty Jolson songs, *inhabiting* them. The voice is actually Al Jolson's, the fervor, the grace, and humor are Parks's in a spectacular display of commitment, love, and skill.

What a beginning to a career. Parks went on to repeat his success in the sequel *Jolson Sings Again* and thence to the ministrations of the House Un-American Activities Committee (1951), which pilloried him for membership in the American Communist Party, at which point he was blacklisted and run out of show business. The McCarthy era ran quite a bit of the show business out of show business, and we were left with *Pillow Talk*.

But the America films, *the* mass medium, have always, and lovingly, expressed ethnocentrism. Vide Robert Alda helping Louis Armstrong and Billie Holiday to find the true meaning of jazz in *New Orleans* (1947); *Dances With Wolves,* in which Kevin Costner teaches the Lakota Sioux to hunt buffalo; *Homo sapiens'* Rae Dawn Chong in *Quest for Fire* (1981), bringing the benefits of the missionary position to the Neanderthals.

For in American film the whites teach the blacks to play jazz; Gregory Peck, a Christian, impersonates a Jew (*Gentleman's Agreement,* 1947) and lectures his Jewish secretary on her lack of racial pride; Oskar Schindler, a Christian, saves the Jews, as, in myriad films of the Pacific war, strapping GIs teach their little brown brothers, the Filipinos, how to defeat the Japs.

A tight little problem.

Samson Raphaelson wrote *The Jazz Singer.* I wrote my first screenplay, *The Postman Always Rings Twice,* for his nephew, Bob Rafelson. One day, during a script conference, Bob's uncle called, and Bob put him on the line, and Sam gave me notes on my screenplay. I thought his notes quite wrong, and I was right to think it. For who rides, decides.

One needs a vast amount of self-confidence to make movies.

When wrong-footed by a recalcitrant audience or, indeed, by time, this decisiveness is known as arrogance.

But which of us knows his time, and how many lunatic or vile creations of our day are labeled good clean fun?

Is it the place of films to address social issues, or are they merely *son et lumière,* signifying nothing?

So we see *The Jolson Story.* It is a fascinating document, a stunning performance by the subsequent nonperson, Parks, and the great singing of Jolie himself. Note also a long shot during the number "Swanee," in which, I believe, it may be Jolson himself, onstage, doing a buck-and-wing.

AN UNDERSTANDING AND A MISUNDERSTANDING OF THE REPRESSIVE MECHANISM

I once did an action picture. One sequence was supposed to take place on a rooftop in Dubai. It was shot on a rooftop in Los Angeles. The Dubai background was added by a computer. The computer background was as real, photographically, as the actors in the foreground, but the composite shot, to me, always looked false. After the film was released, it occurred to me: the foreground and the background could not *both* be in focus. Whether or not the viewer was aware of this consciously, the shot, to the unconscious, *had* to look wrong.

(I gained this insight from watching some excellent feature cartoons, or animated features, as I believe they are now called. Some brilliant animators have begun to throw the background out of focus, mimicking film.)

The movie business humbles the overly theoretical financially as well.

Every studio pays myriads of number crunchers, market analysts, and various other experts to predict and strategize. The breakaway hits, however, have usually been films that were originally discarded as "too."

"Too" what? What matter? Too original, too predictable, too mature, too infantile, too genre, not sufficiently genre, et cetera.

Harry Cohn famously commented that he knew when a

film was doing well by the feeling in his ass. I'm with him. For, finally, the decision to green-light a film or to pass on it is made by some man or woman who is sitting where the buck stops and guessing.

Napoléon frowned on councils of war, as he vowed never "to take counsel of his fears."

Executives, coming now as they do in the main from the ranks of businesspeople rather than show people, have never had the opportunity to learn how to rely on their instincts. So the film business is currently plagued by audience research.

What is wrong with audience research? It doesn't work. *If* it worked, there would be no flops.

But wait—is it not common sense to ask a potential viewer if she would see such-and-such a film, to ask a preview viewer what he would like to change? It may be common sense, but it is useless. Why?

Consider the difference between the barbershop and the jury room.

In the barbershop, beauty parlor, subway, and so on, we gossip. There is much enjoyment in knowing better than the principals, in realizing the error of the prosecution, the defense, the Defense Department, the indicted captains of industry and their mouthpieces. We form and express our vehement opinions based on information that is incomplete and, most probably, skewed or, indeed, manufactured.

Why not? That is the purpose and the joy of gossip—to strengthen community norms through essentially dramatic discourse.

In the jury room, however, we are sworn. We struggle, individually and as a group, to put *aside* prejudice, to put aside the pleasures of gossip, the proxy exercise of power, vicarious revenge, et cetera, and to act according to a set of rules.

The jury is continually taught and admonished to use

reason, as the stakes—the fate or condition of another human being—demand it.

In audience testing, the situation is reversed. Appreciation of drama, an endeavor that has been *correctly* and *necessarily* consecrated to a form of gossip, has been degraded into a mock trial. The tester insists that we put aside our *not only personal but also necessarily inchoate* reactions to a drama and apply an idealized norm of human behavior.

This norm is idealized both in the projection of a putative imaginary viewer (over whom we are to exercise responsible control) and in our self-idealization. For the questioned viewer asks himself not only "Is this the sort of movie I like?" and "Is this the sort of movie 'someone like me' might like?" but also, most corrosively, "Is this the sort of movie someone like me would *proclaim* to like?"

At this point any subjective experience of the film is banished by reason. What remains? The power to teach or admonish—both of which are death to any art.

The filmgoer has been turned into Babbitt, responsible for the film rather than a member of the audience. As a newly responsible member of a jury he will, of course, take the safest course.

What is the safest course? To rationally exclude that which may not be explained. This is much the wisest course for the surveyed, which is why the executive has enlisted him. His refusal to be moved by a film, his characterization of the disturbing or unusual as anathema, has relieved from the troubled mind of the studio bureaucrat the responsibility of *taste,* which is to say, of *choice.*

To succeed, a film must treat the audience member *as* an audience member, not as a commissar of culture. The commissar gets her thrill not from the film but from the power to admonish. (That's why moviegoers fill out the cards after a screening, engaging in a process that would be recognized as an imposition were it not for the honor of the thing.)

But the real filmmakers have to listen to the lessons of their ass.

Will they fail? Certainly. Both artistically and commercially. But (a) they have no other choice and (b) *realizing* that their final choices must be essentially subjective, they may learn to trust their instincts. Also (c) they'll have more fun.

Is it not necessary to gauge the audience? Sure thing. The way to do it is to sit in the back of the theater while the film is being screened and watch their reactions *when their attention is off themselves;* that's the way to see if the film, and any section of it, works or fails.

For that is the state the eventual viewer of the films will be in: disbelief suspended, attention on the screen—wanting to be thrilled, pleased, and diverted, hoping along with the hero, and fearing the villain; and to lead the moviegoer to that state, one cannot ask for his opinion but must pay attention to his actions.

CORRUPTION

Must all human conglomerations become corrupt? Past a certain point they seem to—that point beyond which each person in the group no longer knows the names of all the others.

It requires a genius of morality—in effect, a hero—to remain pure while involved in the conflicting rewards and temptations of power, to avoid arrogance and despair in the face of human corruptibility.

Many movie stars, directors, and producers exhibit the manners, literally, of a two-year-old—a being imagining itself to have vast power, and ignorant of responsibility, enraged by the least human noncompliance as with the broken top that refuses to spin.

The stock-in-trade, however, of the moviemaker, and, most especially, of the actor, is to show how the hero behaves in extremity—fear, greed, lust, hatred, injustice—and to inspire a like stoicism: to show the hero's valiant effort to overcome internal and external evil, to embody the villain's simple conviction of his own rectitude, and, therefore, to inculcate in the viewer a horror of sin.

Perhaps the greatest dramatic portrayal in film is Vittorio De Sica in *General della Rovere.*

Roberto Rossellini's 1959 film has De Sica as a confidence man in wartime Naples. He makes his living bilking the families of those detained by the Nazis. He sells false hope of his influence and uses his stolen gains to gamble.

His simple, kind, and completely believable appeals to his victims are terrifying. He is a monster without shame who justifies all his depredations, who sees extenuation for his incalculable viciousness. Much like you and me.

A Nazi colonel discovers him and offers a bargain. He may go free if he agrees to spy on the Partisans.

The Partisans' General della Rovere has been snuck into Italy to oppose the Nazis. He was to have made contact with the extant Partisan cell in Naples; he has, however, been shot by a Nazi sentry. The Partisans are unaware of his death. One of their cells has been apprehended. The Nazi colonel needs to know which one is the Partisan leader, and he asks De Sica to impersonate General della Rovere to gain the confidence of the group and to determine the identity of its head. In return, De Sica will go free.

De Sica, as the general, is greeted in prison with limitless respect. He hears of "his" wife, returning from safety in Switzerland to see him once more before he dies; he sees a prison mate tortured to death rather than reveal names. The colonel presses him to find out who is the head of the group. De Sica is reluctant.

Finally, the colonel sentences the entire group to death. In the long night before the execution, the head of the Partisan group reveals himself to De Sica.

In the morning, the Nazi colonel asks De Sica one last time: reveal the name, which it is now evident you know, and go free. De Sica refuses. He pens a note of courage to his (General della Rovere's) wife, he exhorts his fellow condemned to courage, and he is shot.

A thoroughly bad man has been humbled sufficiently to choose death rather than deny the voice of God.

It is a magnificent story—and portrayal—of martyrdom. One leaves the film wanting to kill not the Jews but the evil in oneself. One is humbled, as was De Sica's character, by the very distance he had fallen from good and by the extraordinary, miraculous power of awakened consciousness. Having sunk lower than imaginable in disgrace, he is given the opportunity and chooses grace and glory for himself, and we, the audience, are moved and strengthened by his sacrifice.

Movies are a potentially great art. Like any human endeavor, like you and me, they have inevitably been exposed to and have, in the main, submitted to the power of self-corruption, of self-righteousness, to the abuse of power. But like General della Rovere, like you and me, like the studio executives, they possess the possibility of beauty and, hence, for human transformation: not as preaching, not as instruction, not as doctrine—all of which, finally, are out of place in the cinema and can awaken, at best, but self-righteousness. Movies possess the power to speak to the human soul, to free us from the weight of repression.

What is repressed? Our knowledge of our own worthlessness.

The truth cleanses, but the truth hurts—everywhere but in the drama, where, in comedy or tragedy, the truth restores through art.

The audience has a right to these dramas, and the filmmaker and the studios have a responsibility to attempt them.

THE SCREENPLAY

HOW TO WRITE A SCREENPLAY

> As an American occupation, screenwriting has replaced knitting which it, in some ways, resembles: the rules for both are simple, and both involve sheep.
>
> —Richard Weisz

An expert in probability calculated that the odds against winning the lottery are so high that one does not appreciably lower them by neglecting to buy a ticket.

One might, he wrote, stand an equal chance of success by occasionally casting one's glance at the ground in hope of finding that winning ticket some careless soul had dropped.

So it is with the writing of the screenplay.

So many are written, so few are made, and the majority of those are written by that tiny coterie of Mamelukes, harem detainees, or house slaves who constitute the chosen among the Hollywood faithful.

These fattened cattle, myself included, are preserved not to write but to provide the unwary starry-eyed aspirants a further goad to their unpaid efforts.

This is the American way at its finest, which is to say, most operational, pitch—that one, if dedicated, hardworking, vicious, and blind to the blandishments of conscience, common sense, or good taste, might, through chance and/or devotion, rise to that pinnacle where he was licensed to oppress the latecomers.

We have spoken of chance.

What of devotion?

To what rule or force might one devote oneself to ensure success as a screenwriter?

Schools and individual buccaneers spring up to instruct the gullible and hopeful: "Learn the Secrets of the Second Act," "Learn How to Harness the Secret Power of Chaos," "Earn an MFA in Screenwriting."

These educational entities appeal to the common sense of the ignorant.

That is, they present a syllogism that is inherently logical.

It is, however, false.

One can study marching, the entry-level skill of the military, until one shines at it as has none other. This will not, however, make it more likely that one will be tapped to be the Secretary of the Army.

But the *true* entry-level skill of the scenarist, he discovers, is not blind obedience to authority but loathing and distrust of the same. For if the Authority—the agencies, the studios, the producers—knew what they were doing, they would all be peaceful, content, happy, and benevolent instead of caught in a constant, never-abating struggle to the knife.

What are these folks struggling for? For power.

Where does the power come from? From money, access to same, and access to material (scripts) and to stars.

One gains access to stars either through having been both poor and honest with them, or through having made them money, or through presenting the ability to do so.

One gains access to money through having been poor and dishonest along with fellow producers or agents, getting lucky, and/or getting tough. But *anyone* can gain access to a script. Because anyone can *write* a script.

Just as the young and sexually vulnerable present themselves in these western precincts as victims, so the mantra of the neophyte screenwriter is: I am young and stupid. Please abuse me.

Let us consider the screeenwriter's process structurally, as the drama that it is.

Act 1:

These just-off-the-bus newcomers arrive enthused—as is the way of the young—with the idea that they are the first to have had the idea.

The idea is: There is nothing I will not do to get ahead.

Oh, really.

But just as with the fresh-faced bimbo or gigolo in bed, there are, finally, only so many things the acolyte screenwriter can *do:*

He can do exactly as he is told, which puts him in the same applicant pool as everyone else who just got off the bus, or he can break new ground.

But the ersatz producers, the bottom feeders who hound the bus station looking for prey, are not interested in breaking new ground with film. They are not interested in film at all. They have no time. They are, like the other bus riders, possessed only with the big idea—in their case: I will take your something for nothing and sell it to my betters. For as Neal Stephenson informed us, Hollywood is just a bank.

These ersatz producers have no dramatic sense, such would, did they possess it, nix whatever chance they might have of putting the focaccia on the table—they observe that which has been successful and attempt to duplicate it. They want, in effect, to find the script for the hit of last year.

But—and here's the twist—in the *new* script, instead of being a cod, it will be a mackerel.

Yes, yes, yes, say these entrepreneurs—yes. A mackerel. Great, let me run with it.

Now we have act 2. My, what will happen? Oh no. The producer is returning to the starry-eyed, unpaid, and biddable screenwriter. Can it be good news?

Well, it could be worse, the producer reports: I, or my advisors, or that big independent money person (that gunrunning, drug money launderer) from (fill in the country) with whom I have been talking, am interested in your script: I just want you to do one thing.

Could the mackerel live in a tank in a restaurant rather than in the Indian Ocean?

Why not, one responds, what a *great* idea, thank you for the suggestion.

As this act progresses, we see the producer coming back to the writer with more and more "ideas" from his real and imaginary sources, supporters, and advisors:

Instead of a restaurant, could it be Mars, and instead of a mackerel, could it be Woodrow Wilson, et cetera.

And now what happens?

Just as he does with the nubile offerings of fresh young things, the producer gets tired of being offered "anything you want."

"Jeez," he thinks, "doesn't this person have any self-respect? I am ashamed to *know* someone who would stoop to such degrady." And the producer moves on.

Here, the producer, about to decamp, realizes that the writer brings him nothing special, that if the *writer* doesn't know good from bad, and is endlessly biddable, how can the producer trust him. He cannot.

The now-mangled script has become the fruit of the poison tree and must be discarded as unclean. Which it is. And the producer moves on in the hope that *one* of these free scripts just might, magically, attract the favorable notice of someone with a bit more power.

Let us suppose that the screenwriter has, over a period of time, paid his dues (i.e., been seduced and abandoned sufficiently to tire of it).

Here, at, as we say, the break of act 3, the sadder-but-wiser screenwriter is free to examine the tropism that got him into this mess in the first place. Why, oh why, he muses, did I come to Hollywood? To write screenplays/to find success as a screenwriter/to have an adventure.

Well, one may *still* write screenplays.

In many arts, we are told, the proselyte presents himself as food for the gods, a sacrifice of pleasing odor, saying, "Bid me."

We know of the young dancer, athlete, musician, martial artist who wants no reward greater or other than the opportunity to serve the muse. So, then, to the abused and failed immigrant, should that be your happy desire, what hinders you?

What, however, if one still craves success?

In that case, perhaps the protagonist possesses sufficient honesty to avow that his efforts to "be just like everybody else, except *more so*" have not availed and that the bloom is off the rose. Perhaps a healthy self-evaluation would lead to the conclusion that one, in fact, finds the perpetual abuse degrading and that no possible success is worth the shame and rejection. Perhaps one harbors the residual belief that learning to write a screenplay "better" might improve one's chances of selling it. I don't think so. I, however, sadly harbor the same delusion, so as a continuation of the abuse you have already taken, I will end my play on an inconclusive note (the writer's suicide is too easy—his return to a cabin in Vermont unbelievable) and tell you my advice on the subject of screenwriting.

CHARACTER, PLOT, DIALOGUE, CAMERA ANGLES, ADVICE TO THE EDITOR

Let's examine a perfect movie: *The Lady Eve,* written and directed by Preston Sturges.

His work is, to me, irrefutable proof of an afterlife, for it is impossible to make films that sweet and not go to heaven.

Here we have Barbara Stanwyck and her father, Charles Coburn. They are cardsharps and confidence tricksters plying the liners.

Here comes Henry Fonda, an amateur naturalist and the filthy rich son of Eugene Palette. He's been up the Amazon for a year and is going home.

Everyone on the liner is angling for his notice or favor. Stanwyck, of course, wins out. And she and her father set out to fell Fonda.

This is called a premise.

Stanwyck, however, makes the mistake of actually falling in love with him.

This is called a complication.

Her love is reciprocated, and Fonda proposes marriage.

But wait—before she can accept, she must confess to her past life of sin, and before she gets to do so, the ship's purser warns Fonda that she is a criminal.

He is heartbroken and tells her that he knew it all along and was just stringing her along for the entertainment value. She, now, is also heartbroken.

And now we have Billy Wilder's famous dictum posed as a Talmudic question, in re love stories: *What keeps them apart*?

Aha. The lovers are now kept apart by loathing on the part of Fonda and, upon the part of Stanwyck, by a desire for revenge.

Enter act 2.

She decides to impersonate a wealthy British countess or something, get introduced into Fonda's family's rich Connecticut set, and win him *all over again*.

She, of course, does so, and they get married.

We now have act 3. They are together, but the notional forces have not been propitiated. He has been won not through love but through actual chicanery (the very method she disdained in act 1), and Stanwyck must have her revenge.

They proceed on their honeymoon. About to consummate the marriage, she confesses first to one and then to a very lengthy run of sexual encounters, and he dumps her.

She has had her revenge, his family proposes a fat settlement, and she turns it down. All she wants is for her husband to ask outright for his release.

Note, she has won both her prize of the first act (money) and that of the second (regard) but finds that revenge is empty—that she has, in fact, gone too far. She has heaped Ossa on Pelion and now has nothing.

Fonda, she learns, is going back down the Amazon.

In a fit of inspiration, she boards his boat in her old persona as the rejected con artist.

He is overjoyed to meet her again and calls her to his bosom. Great story. And we may reflect that its description contains none of what the ignorant refer to as "characterization," nor does it contain any of their beloved "backstory."

These, and the attendant filth of authorial narrative (he comes in the room, and we see that the place to which he's going is going to be far more interesting than the place to which he's been), are the screenwriting equivalent of HIV: they spread like mad, they corrupt everything they touch, and there is no known cure.

A director could (indeed, did) *shoot* the story above. It was simple and straightforward enough to allow him to make simple choices about clothes, costumes, camera angles, music, and so on.

Actors could *act* upon those directions he gave them based upon the script.

The resultant film, though made by a master, would probably have been watchable if made by a journeyman. Why? Because we, the audience (those in their seats at the cinema and you, gentle reader, no less), wanted to know what happened next. That is more or less the total art of the film dramatist: to make the audience want to know what is going to happen next.

The garbage of exposition, backstory, narrative, and characterization spot-welds the reader into interest in what is happening *now.* It literally stops the show.

A wiser man than I might advise you, gentle reader, to write—if you must write, and if you must write for Hollywood—*two* scripts: one to appeal to those who man the conference rooms of the Valley; the other, once you've obtained their imprimatur, an actual document capable of being designed, filmed, and acted.

I can't do so. For it is one thing to take candy from a baby, another to take candy from a baby and give the baby typhus, and this two-scripts idea (though practicable in the extreme) is, to me, a heresy.

Samson (from the Hebrew for *sun*) eventually gives in to Delilah (from the Aramaic for *night*) 'cause he just can't stand her noodging.

He winds up eyeless in Gaza, and we all might just take a lesson from it. He got what he wanted (a little nookie), and he overpaid for it.

One may similarly manipulate the story, but the deviation from the essential—what does the hero want, what prevents him from getting it?—renders the writer no different from, indeed, an adjunct (read: whore) to, the forces of commoditization.

But wait, but wait, are not these the very forces one is trying to propitiate?

Perhaps, but that pursuit is not the task of screenwriting, in which case this chapter will be of no avail.

Where did we leave our burden?

We had crafted a plot reducible to five lines on one side of one sheet of paper.

It begins with a premise: the hero wants something. His desire begins with the *beginning of the film.*

That is, he cannot *just* desire something. For the screenplay to be coherent and compelling, his desire must be awakened by a new circumstance. That circumstance is the film.

Barbara Stanwyck meets the love of her life, Henry Fonda. The film starts *because* she meets him. The progress of the film is her progress toward attainment of her goal. When she attains it (in the last ten seconds), the film, the story, is over.

Each act of the film must concern her progress toward that goal and the complications thereof.

Act 1, she wants to win his money.

Act 2, she wants revenge.

Act 3, she wants reconciliation.

Each scene, act per act, is her attempt to win from the antagonist (Fonda) the special prize that, *in that act,* signifies her possession of him: money, reconciliation, love. Then, there you go, screenwriter. You are now all set. Write down your premise, tell yourself the story act by act. The small steps in each act are called scenes.

Now write each scene such that it is essential to the hero's progress toward the goal of that act.

In order to win Fonda's money, Barbara Stanwyck must first get his attention. She trips him. He breaks the heel off her shoe. To consolidate her gain (his attention), she makes him schlep her down to her cabin to find a new pair of heels.

The act culminates as she realizes that she wants to possess not his money but his love; simultaneously, he discovers she is a con woman and insults her.

Which leads us to act 2, Revenge.

What could be simpler?

And that's how one writes a screenplay.

"Cinema, at its most effective, is one scene effectively superseded by the next. Isn't that it?" (George Stevens, 1973).

I don't think he left anything out.

HELPFUL HINTS ON SCREENWRITING

"A guy comes home from college to find his mother sleeping with his uncle, and there's a ghost running around. Write it good, it's *Hamlet;* write it bad, it's *Gilligan's Island.*"

—Lorne Michaels

The audience will undergo only that journey that the hero undergoes.

Similarly, the audience will not suffer, wonder, discover, or rejoice to any extent greater than that to which the writer has been subjected. To suggest that the writer can, through exercise of craft, evade or avoid the struggle of creation is an error congruent with confounding the study of theology with prayer.

In what could a graduate course in screenwriting consist?

There are two possibilities.

It may, perhaps, instruct in first principles (i.e., it may consist of the study of that branch of philosophy called aesthetics). These first principles were best enumerated by Aristotle in his *Poetics* (a dissection of *Oedipus Rex*), and they are famously few: unity of time, place, and action.

That is: the story should come to life before our eyes—brought into being by a unique event, as opposed to an "ongoing process" (e.g., the hero should want to raise the plague on Thebes, rather than discover the cause of evil in the world). This story should take place within the space of three days, in one place, and should consist *solely* of the attempt of the hero to solve the problem whose appearance gave rise to the play. That's it.

And like the Gospels or the Torah, child rearing, or marriage, anyone truly interested can and will have to figure out the rest.

To correctly *formulate the problem,* specifically and mechanically, and then to work out the solution (the steps the hero must undergo) is a daunting process. It calls for perseverance, honesty, and, by turns, blunt candor, invention, humor, and humility. The process is made more difficult by the opportunities for chicanery.

Faced with the dead end of the first act, the author may, in fact, revert to the formulaic. He may, in effect, forge a corrupt bond with the audience—i.e., "You and I know that, at this point, a change must take place. That the barn catches fire will serve as notice that, though I, the author, have not solved the problem, I at least recognize your, the audience's, right to a solution." It is laudable to resist this nagging invitation to sloth and predictability.

(Most studio notes on a script or a film are an insistence that the intrinsically inevitable, thought-out, and, so, surprising turns of the script be junked in favor of the formulaic.)

Again, the rules of dramaturgy are few, their application difficult, their product unusual, idiosyncratic, and surprising—that is to say, dramatic.

Are there other books one might read to better understand these simple precepts?

There are. I would name *The Uses of Enchantment* by Bruno Bettelheim, *The Hero with a Thousand Faces* by Joseph Campbell, and my own *Three Uses of the Knife*.

Those bitten by a love of the form will, of course, discover a vast assortment of texts and might, if truly enthused, ponder the psychological or, indeed, the neurological nature that gave rise to the human longing for drama.

Raymond and Lorna Coppinger, in their wonderful book *Dogs* (2001), argue that, in a Darwinian sense, the various breeds indulge in breed-specific behavior *not* to achieve a (causally related) end but because the behavior in itself is enjoyable.

They cite the sled dog, which, in their decades-long experience as sled-dog racers, runs, they have found, not through fear of the driver's whip (there is no whip), and not through subjugation to some alpha animal (there is no alpha sled dog), but from the sheer joy of running in unison.

The food doled out by the driver forms a bond between him and the dogs but could not induce an animal unfitted for the task to run all day. They run from joy.

This thesis, the Coppingers note, might seem far-fetched. But they ask the reader to consider human courtship rituals. We humans enjoy these rituals for and in themselves; we are genetically wired to do so. We do not delight in them for their ability to achieve for us a mate—such end is an ancillary (and—if I may—to men, a surprising) end to the process.

Sexual intercourse, similarly, is enjoyable in itself, irrespective of and unrelated to its biologically engineered purpose: the creation of offspring.

We human beings delight in drama. We will vote, against our own best interests, for the side presenting its case most dramatically. We will endorse war, we will *fight* in wars whose benefit, in hindsight, consisted solely in the dramatic confection of national unity.

Oscar Hammerstein II most correctly summed it up: "Fish gotta swim, birds gotta fly." And we humans need to indulge in drama. What, then, when we do not?

The terrier, as per the Coppingers, needs to hunt small rodents. This is its joy. It loves to ferret out, to dig, to nip. It continues to practice its joyful innate behaviors, even in the absence of rodents. Their practice, now, however, has a new name: it is called "behavioral difficulties."

And our need to explain the world, to understand cause and effect, that joyful, innate capacity that we call "intelligence" and that we say separates us from the animals, that capacity *will* be exercised. We may employ it at the theater, which is only an experience of the story around the campfire; in gossip, and that formalized gossip we call "journalism"; in that particular subset of gossip known as politics; and in neurotic behavior in the home, the workplace, and the community.

Each of the above is the retelling or the *acting out* (the enjoyable discharge) of a dramatic, cohesive, incited, and completed view of the world.

These are the neurologic or, if I may, the phylogenetic underpinnings of drama.

Or, perhaps, the enthused individual might be charmed by the psychoanalytic paradigm. This person might see, in the formation of neurosis or psychosis, the dramatic tendency *en petit*.

For dreams and drama are the same. The power of dreams to affright, instruct, amuse, and trouble the individual might be analyzed and the fruit of that investigation applied to the construction of the drama.

Psychoanalysis is an attempt to discern in ostensibly unconnected actions and images a simple, hidden, unifying theme and, as such, may be seen as the absolute and perfect inversion of the dramatist's work.

The dramatist begins with a theme, or *quest,* and endeavors to describe its progression in ostensibly unconnected actions and images that will, at the quest's conclusion, be revealed as unified, and that revealed unity will simply state the theme, which revelation will—just as, theoretically, with the revelation by the analysis—restore order.

The rules of drama are few, its practice difficult, its implications many and fascinating.

But apart from a presentation of these few rules, the work of half a morning, what might be the purpose or, indeed, the content of a graduate course in screenwriting?

Perhaps this graduate course is designed to prepare one for the rigors, not of creation itself but of presentation, comportment, packaging—in fine, the glad-hand, one-must-take-one's-pigs-to-market aspect of any creative art, or indeed, of any business.

Such a course of study would, of necessity, involve chicanery, euphemism, obfuscation, suppression, repacking, or rechanneling of the natural instincts of aggression, ambition, greed, and envy—the skills of business, in effect.

For one must take one's pigs to market, and any market is run by the middlemen.

These middlemen in Hollywood are bureaucrats, and they have a natural foe, and this foe is the script. For a star's grosses may be quantified, and a prediction (supportable even when proved false) may be made about his or her worth. But the worth of an unshot script is moot.

How, then, to remove the potential (not for *error* but for *recrimination*) of an unfortunate choice?

By removing the unquantifiable: the surprising, the unique, the upsetting, the off-color, the provocative; by removing *drama*. A course in the *business* of screenwriting, then, might teach how to recognize, in order to obliterate, drama.

Skill in this bureaucratic endeavor, unfortunately, will avail the practitioner little, as in shunning the original, he consigns himself to a limitless applicant pool—to a pool made up of all those capable of suppressing, or incapable of possessing, a love of drama. But, again, one would not have to sign up for X years of graduate schooling in order to learn this skill.

Of what use is this graduate film diploma, then? As evidence of the bona fides of the applicant. For someone capable of putting up with X years of the nonsense of school would be odds-on willing to submit to the sit-down-and-shut-up rigors of the bureaucratic environment.

Perhaps, then, this graduate course functions, whether through design or happy accident, not to *train* but to certify house slaves.

Or perhaps this graduate course is the modern equivalent of what we ancients know as music appreciation class, wherein the students were awarded credit for listening to select tunes and praised for repeating back the phrases they had been instructed to associate with them.

Perhaps this graduate course is a congruent Pavlovian endeavor. But I suspect it is the boot camp of that side world, "the industry."

For the privileged class, in this or any other society, must take care of its own. So just as children of privilege may not "take" but may only "experiment with" drugs, the privileged may not be "out of work" but, to the contrary, be "searching for themselves." And lest this search prove too rigorous, they might relax for several years in film school watching movies.

After which point, those sufficiently connected may wave their diplomas and proceed into the studio system.

Meanwhile, back at the ranch, let us leave our suppository graduate school and investigate actual hard-won, practicable, real-world dramatic skills.

The entire practicable sentence was, of course, not "Cut to the chase" but "When in doubt, cut to the chase." Good thinking.

Other filmmakers' pearls—seldom wrong, always instructive—follow:

"Stay with the money." The audience came to see the star. The star is the hero; the drama consists solely in the quest of the hero.

"You start with a scalpel and you end with a chainsaw." Don't be too nice about cutting the film; throw away everything that's not the story.

"In the morning you're making *Citizen Kane;* after lunch you're making *The Dukes of Hazzard.*" At some point you're going to start running out of time. Plan your time by sticking to the essential story. You're going to cut everything else anyway.

These gems and their like can be learned only on the set, or in the editing room, while actually making a film.

But, the chauvinists sigh, concealing their exasperation with my Jacobin views, you forget that one *may,* at this graduate school, *actually make films.*

Yes and no.

For the real skills of filmmaking can actually be learned and practiced only in relation to an audience.

If the audience is financially involved (the studio executives), suborned (the "invited" carded test-group screening), or hired (the university professors), one will learn nothing from their responses except obedience.

The actual dramatic audience has suspended its disbelief and has sat down in order to be thrilled (in contradiction to the groups named above, that come, each in its own way, to pursue its particular goal).

Very well, but, the chauvinists continue, may not these

actual academic student filmmakers *take* their film, escaping from the academy with this golden goose, and show it to a wider world?

Sure thing; but, then, what did the school contribute?

But what of the scholarship student, what of he who could not afford tuition and entered the school only as a way to achieve, without cost, the film cameras and lights unattainable in the lay world?

The school, then, exists as a primarily charitable institution?

Fine, and all power to the young filmmaker who managed to take advantage of it. *This* person, however, would proceed directly to defrauding the chumps who thought they were doing him a good turn and divert their right-thinking charity to his own necessary ends—just as he will have to when he has moved to the wider world.

Bravo.

And as for the graduate school, so for the undergraduate and the various limitless seminars in filmmaking dotting our coasts and increasingly making inroads upon the hinterland.

They may be a rest stop for the insufficiently aggressive (those folks will find their rest expanded into a lifetime career), they may be a film club for those who cannot take their entertainment straight, they may be a boot camp for those desirous of becoming efficient at submission, or they may be a pen to hold and fleece the children of the privileged. But I think perhaps they have little to do with the actual making of movies.

THE SCRIPT

It all comes down to the script, or as they say in the theater, "If it ain't on the page, it ain't on the stage."

A good script may shine with superb actors, and a great script can be done well with amateurs. (Note the times one has heard or said, "I saw a high school amateur community production of *Waiting for Godot, Our Town, The Winslow Boy,* et cetera, and, do you know, it was perhaps the best thing I've ever seen.")

Now, when you get a great script done with great actors, then you have a classic.

The Godfather, A Place in the Sun, Dodsworth, Galaxy Quest—these are perfect films. They start with a simple premise and proceed logically, and inevitably, toward a conclusion both surprising and inevitable. The godfather wants to protect his best-loved son from contamination by the family criminal enterprise; the son eventually becomes the godfather. An attractive, poor, and ambitious young man worships wealth and beauty; he is eventually offered both beyond his wildest dreams, but his lust, as he sought them, leads him to kill the woman who would mar his plans, and he ends, on the brink of "success," caught and executed for his crime. A washed-up bunch of television actors curse the long-gone success of their show; it has mired them in supermarket openings, portraying cutout heroes; they are given the chance to inhabit that fantasy-turned-real and discover, in themselves, real heroism.

The stories are easy to summarize, as they are easy to follow. Every event in these stories can be plotted as a point on the progression described.

That is one definition of a great screenplay.

That which enthuses the actors and director enthuses the audience. To wit, I have been told the operative premise, and I want to know what happens next.

The film's precursor is the story around the campfire. In that story we hear and we imagine; in the film we see and we imagine. The structural nature of film allows the imagination to reign.

When the film turns narrative rather than dramatic, when it stands in for the viewer's imagination, the viewer's interest is lost.

The dramatic structure relies exclusively upon the progression of incident. How do we identify an *incident*? An incident (as per Aristotle) is a necessary step from the beginning proposition (a mafia don wants to spare his son the life of a criminal) to the conclusion (the son becomes the criminal chief).

To recur to the campfire story analogy, "But then the ship struck the rock and began to sink in the cold and angry sea," is better than "But then the ship struck a large, gray, wet, and ragged rock and began to sink into a frothing, viciously cold, tumultuous sea, its riptides roaring, its calmer regions full of man-eating sharks."

The rule, then, in filmmaking, as in storytelling, as in writing, is "leave out the adjectives."

> Enter: *Rock Lindquist.* He's a funny kind of guy. He walks with a swagger, but perhaps it masks an inner insecurity. Women find this attractive. They want to mother him. *We* do, too. . . .

This trash is, unfortunately, the stuff that makes, to a script reader, a "good read."

It distracts the reader, the writer, the producer, and the actor from the only question *(what is, in fact, happening?)* that would allow them to make a rational, let alone an artistic, choice. It is the coin of the realm.

The film may, perhaps, be likened to a boxer. He is going to have to deal with all the bulk his opponent brings into the ring. Common sense should indicate he had better not bring one extra ounce of flab on him—that all the weight *he* brings into the ring had better be muscle.

The extra scene: no film can stand it.

When, again, is a scene superfluous? When it does not advance the progression given at the outset as the film's purpose. What happens during this side trip? The audience's attention wanders. They have been jolted out of participation, and the filmmaker has lost his most important ally: their uncritical, which is to say, engaged, participation.

How may one learn the absolute primacy of the script?

Shoot for show, cut for dough—Hollywood has it that one learns to shoot in the editing room. It is there that the honeymoon ceases and the marriage begins. The filmmaker, wedded to the footage he has shot, now, over the breakfast table, if you will, must evaluate his choice and figure out how to live with it.

"How could I have been so foolish" invariably refers to the script and its lack of precision; and the filmmaker may well refer, to extend the connubial analogy, to the wisdom of Berry Gordy and Smokey Robinson, in their 1960 hit "Shop Around": "Pretty girls come a dime a dozen, try to find one who's gonna give ya true lovin'."

Helpful hints to the filmmaker and the viewer: The compliments—"What visuals!" "What craft!" "What use of the camera!" and "What technique!"—all mean "the script stinks."

WOMEN, WRITING FOR

What about writing for women? Some men can, and some men can't.*

I recommend John le Carré's Connie Sachs, the retired wizard of the British Secret Service, the truest and best ally of George Smiley. *There* is a portrait for you, and you can keep Blanche DuBois.

Read John Horne Burns's *The Gallery* for his whores of World War II Naples and, most particularly, for the good girl, Guilia. The ugly American captain falls in love, and his love is reciprocated. The good girl will marry him, wait for him, do anything, in short, except have sex with him before marriage. The major is about to return to combat; he calls her a demon. She looks at him surprised. "But," she says, "*all* women are."

Read John O'Hara if you'd like portraits of women by a man.

What do these depictions have in common? They are unsentimental. An absence of sentimentality is a great thing in a writer and separates the merely good from those who actually have something to say.

For it needs no prophet to remind us of the day's received wisdom—advertising will do that job quite nicely. The true writer must write not the acceptable but the true.

* The same is true of women.

True depiction of women is, I think, rather different from political sanctimony—and the political emancipation of women has given rise to a spate of pablum.

Vide the spontaneously emerging form of the women's movie. Five old friends get together to rehash their lives. All well and good. This film, however, is a prerogative of an increasingly healthy polity, not its description.

The lower classes of late seem to delight in depictions of women as victims of sex (abuse, rape, harassment); the more elevated, as victims of marriage.*

These efforts—whether of women or men—award a cheap obeisance; they are Mariolatry in its ultimate debasement: "Women are good, women are pure, how good are we to acknowledge their suffering."

As such they are also a form of emotional pornography, for, like *The Passion of the Christ* or *The Green Mile,* they may license the viewer to enjoy a disturbing spectacle through permitting him or her to protest revulsion at it.

But true depiction of women takes into account two things: the ways in which women are similar to men and the ways in which they are different.

For to say that all people are equal is not to say they are the same, and to confound the political with the practical gave us the enormity both of school busing and of feminist literary theory.

Actresses in Hollywood complain that there are no parts for women. There *are* parts for women, but they are few and

* The first category is sufficiently widespread to have, in Hollywood, its own cognomen, fem jep (females in jeopardy), e.g., all the Halloween, Friday the 13th films, *Blair Witch, Panic Room, Flightplan,* et cetera. The second category, grown out of the mid-Victorian sensation novel, survives as *The Ice Storm, The Joy Luck Club, Thelma & Louise,* et cetera. Essentially a form of whining, this "marriage as feminine ordeal" category seems to be waning in popularity as a younger audience of women increasingly accepts many of the gains of the feminist movement as a matter of course.

tend to go to the nubile. This may be good or bad, but it is true.

Is it the job of the movies to offer a well-balanced distribution-by-gender of roles? If so, who would make the choice?

Who but the slighted, seeking, as most of us do when accepting that role, not universal justice, but reparations? Even the dramatic roles for women, when viewed not as entertainment but as, if I may, art, are drivel—*Now, Voyager; Sophie's Choice;* and *Flightplan*—treating us to the noble spectacle of women either crying or bravely not crying.

Is this "writing for women"?

Well, it is writing *about* women. Or about their simulacrum. Tennessee Williams, Truman Capote, Noël Coward wrote women characters that were fantasies of men by homosexual men. Enjoyable, indeed. But hardly accurate.

What is the truth about women?

Jacqueline Kennedy was seen, in the Zapruder film, crawling out of the back of the limousine in which her husband had just been shot. The wise and kind suggested she was going for help. Lenny Bruce accurately observed that she was "hauling ass to save her own ass," and was hounded out of show business.

But would not you, and would not I, do the same (which was his message) to escape from that fusillade? What was the hypocrisy intended to protect?

The "enlightened" Western view of women—that is, the currently politically acceptable depiction of women in art, is neurotic: Diana may be a princess or she may be a whore, but that's it. Women, in mass entertainment, may be victims or postsexual comrades—eunuchs, in effect: kind old men sitting on the porch wondering where they went wrong and emerging, somehow, better for the experience.

What about the real women, and may depictions of them be written by men?

Well, who is to say? The terrible voices of that coercion known as political correctness cry—but they cry not for parity, let alone for humanity, but for power.

Let us apply, to authors, a rational application of the rational doctrine of sexual equality: Is it less heinous to inquire the sex of an applicant for our attention than of an applicant for employment? Having settled that, let us move on—and hush up, you academics, brownnosing for tenure with your authors listed by sex, race, and geographical distribution. (Who asked you?)

The sex of the author is nobody's business—which can, politics aside, be established in fact by this simple test: How many choose films based on the sex of the creator? All right, then, let us apply the same standards to literature (where questions of taste may, *disons le mot,* be corrupted by an educationally induced hypocrisy) that we do to entertainment. In entertainment, we make our true, unfettered choices based on content and worth, not on politics.

I am reminded of the memoirs of an old sixties radical. He wrote that he had been gently pursued by a gay chum for forty years and, being straight himself, had politely declined. One day, however, it occurred to him that perhaps he was being politically incorrect, and so he went to bed with the guy. Fun's fun, but that, to me, is going too far.

The question is not can one sex write for the other—if not, are we then to have only unisexual dramas?—but can the individual *write*? That is, can he (a) see and (b) tell the truth?

Lenny Bruce's remark was unacceptable because it was true.

My play *Oleanna,* when it opened in 1992, was a *succès de scandale,* a handy French phrase meaning that everyone was so enraged by it that they all had to see it.

What about this play, a rather straightforward classical tragedy, drove people berserk?

It asserted that a person could make an accusation, the truth or supportability of which was open to debate.

One would not think this enraging, *but* the accusation was made by a young female student against a male professor, and the accusation was of rape.

The play's first audience was a group of undergraduates from Brown. They came to a dress rehearsal. The play ended and I asked the folks what they thought. "Don't you think it's politically questionable," one said, "to have the girl make a false accusation of rape?"

I, in my ignorance, was stunned. I didn't realize that it was my job to be politically acceptable. I'd always thought society employed me to be dramatic; further, I wondered what force had so perverted the young that they would think that increasing political enfranchisement of a group rendered a member of that group incapable of error—in effect, rendered her other than human.

For if the subject of art is not our maculate, fragile, and often pathetic humanity, what is the point of the exercise?

HOW SCRIPTS GOT SO BAD

Here's how they got so bad.

The entry-level position at motion picture studios is *script reader.* Young folks fresh from the rigors of the academy are permitted to beg for a job summarizing screenplays. These summaries will be employed by their betters in deliberations.

These higher-ups rarely (some, indeed, breathe the word "never") read the actual screenplay; thus, the summaries, called "coverage," become the coin of the realm.

Now, like anyone newly enrolled in a totalitarian regime, these neophytes get the two options pretty quickly—conform or die. Conformity, in this case, involves figuring out *what* the studios might like (money) and giving them the illusion that the dedicated employee, through strict adherence to the mechanical weeding process, can provide it. The script reader adopts the notion that inspiration, idiosyncrasy, and depth are all very well in their place but that their place has yet to be discovered and that he would rather die than deviate from received wisdom.

The mere act of envisioning "the public," that is, "that undifferentiated mass dumber than I," consigns the script reader to life on the industrial model. He or she now is no longer an individual but a field boss, a servant of "industry," and, as the industry in question deals with the mercantilization of myth, an adjunct of oppression. Deprived of the joys of whimsy, contemplation, and creation, they are left with prerogative. So script coverage is brutal and dismissive.*

Why would this canny employee vote for the extraordinary? The industrial model demands conformity, and the job of the script reader is not to discover the financially, and perhaps morally, questionable "new" but to excel in what, for want of a better word, one must call hypocrisy.

Oh, boo hoo.

Opposed we find the scriptwriter.

As the grosses of blockbuster movies swell, the quality falls. The viewers see this trash and, correctly, exclaim, "Well, *hell, I* could do *that!*" They then write a screenplay. These screenplays are, in effect, tickets to the lottery.

The late IPO delusion, the "new economy," has taken its place with the South Sea Bubble and the Dutch Tulip Mania. These short-lived and second-class frenzies are as nothing compared to the long-lived, indeed, unkillable fantasy that

* Though the script reader, producer, studionik may *subjectively* (and legitimately) dislike any given script on its merits, he is also not only liable but likely to dislike it because of the purely mechanical operation of the development process. The monied progenitor endorses the creation of a script based on a rose-colored prospectus, known as a "pitch." However good the commissioned script may actually *be,* it is certain to awaken in the sugar daddy or mommy a feeling of disappointment because of his or her outrage that it is not "the pitch." In this fit of pique, the executive, however, is operating under a delusion similar to that undergone by the radio listener meeting in the flesh the possessor of a beloved voice. The listener inevitably thinks, "*That's* not how I envisioned him *at all!*" not recognizing that he, the listener, never *did* envision the radio personality. He just listened to his voice, the sound of which created in the listener a feeling of familiarity *similar to* that created by cognizance of an individual's physical appearance. So the surprised radio listener thinks, "That's not how I envisioned him," rather than, "Oh. *That's* what he looks like"; and the producer thinks of the script, "That's not what I had in mind," *as if* (as is not the case) he actually had a concrete prevision of an actual forthcoming script. Which he did not. He had only a warm gooshy feeling engendered by the pitch.

each and every person born can rise to prominence and wealth in show business.

This sweet folly is a capitalist's erotic dream. In general, the theoretical limit to which wages can be reduced is that at which starvation transforms worker competition into revolution. (In plain English, the boss usually has to pay the workers *something.*) In show business, however, the bitten will not only work for nothing, they will also fight for the chance to do so.

Back at the ranch, the corrupted youth, the script readers, sit at what I will imagine as their high Victorian desks and paw through the incoming screenplays, nuzzling the earth for truffles for their masters. Yes, their muzzles are tied, but someday, someday *they,* these readers, may be elevated to the rank of executive—indeed, perhaps to the very pinnacle of studio head, where they will have the power not only to *discard* but actually to endorse. They will there be showered with perquisites, first and not least among them, that they will never again have to read another screenplay.

What, however, of the *professional* screenwriter? This person, cursed with actual dramatic sense and credentials, is, similarly, self-caught in an evil net. He differs from the amateur in that he is actually getting paid for the work. This puts hummus on the table but places him at a distinct disadvantage in the lottery of production.

There are two counts against him. First, the work—if it contains inspiration, glee, sorrow; if it is complex, actually provocative or disturbing—may not be easily condensable to those three sentences allowed the script reader. Second, the fact of his *actually getting paid* enrages those involved in the studio system. Is it not monstrous, they wonder, that one should actually *pay* for that which 90 percent of all human life would do for free? The two burdens of the actual writer, his inspiration and his bill, conjoin synergistically to end in tragedy.

Most writing assignments begin with the plea "Fireman, fireman save my child" and end "Where is that half-eaten chicken I believe I left in the icebox yesterday?"

For, in fine, most executives consider the writer a thief.

BEGGING LETTERS

The language of the modern screenplay is like that of the personals column. The descriptions of the protagonist and the lovelorn aspirant are one: beautiful, smart, funny, likes long walks and dogs, affectionate kind, honest, sexy. These descriptions, increasingly, are the content of the screenplay—replacing dialogue and camera angles, the only two aspects of a screenplay actually of use.

These new screenplays are essentially begging letters, that is: See how I paint myself in the best and most general of terms and beg that someone will recognize my abasement and meet my very human but, unfortunately, mutually exclusive needs. I beg, in effect, to be both recognized for my worthlessness and to be given love.

Note that a more potentially successful strategy in a personal ad might include a personal code: an obscure reference to the literary if one wants to entrap a reader, to the stock market to attract a financier, perhaps to the Bible to collar a person of religious beliefs.

Such are not found in the personals, as, I believe, the writers might think (correctly) that to specify actual desires or attributes might limit the applicant pool of potential respondents.

The correct place to find a perfect physique is in the gym, a religious person in shul or church, a book lover at the library.

But the writer of the personal ads appeals, in extremity, to the populace at large, throwing him- or herself on its mercy and begging for a date, with this unstated reservation: "I will figure it out later—just get me on the playing field." This appeal is addressed to the similarly hopeful and desperate: "Let us indulge in codependent behavior—we will, at the very least, have *that* to share. We each know the other is far from perfect, sexy, fun, brilliant, talented, soulful, and kind, and we will agree that this frank collusion is the magical incantation necessary to establish goodwill."

The problem lies in this: these ads establish little else, and any actual date must not end, but in fact *begin,* with a measure of disappointment.

So with the screenplay.

Here, in the character description, we are told that the heroine is various things that one might find attractive in a heroine, but should the writer's words be put upon the screen, they will be found wanting, for the hopeful description of the character will not have, magically, transformed itself to the screen. We can write, "She's the kind of girl *who* . . ." all day long, but neither the actress nor the director can actually implement it; and we find, in what practice proves is nonspecific and nonimplementable language, nothing but the desire of the writer to please.

What is wrong with trying to please?

Nothing at all. *But* the writer of the "lazy Sunday mornings" gobbledegook has worked not to please the *audience* but the executive (his codependent, desired other).

Just as the personal ad is written not to attract anyone specifically but only to avoid exclusion, the "lazy Sunday mornings" screenplay strives to appeal to all—or to those who think it might appeal to all. In this it also resembles a political speech, written to lull and, by its soporific cadence and vocabulary, to allow the listener to intuit whatever the hell she wants.

"Smash, bash, crash: the world became a steel cauldron of pain." "Yes," says the young script reader. "Yes. Hot stuff *indeed*. Boss? This is hot stuff. *This* person knows how to write action."

"Loves hazy afternoons. This well-educated beauty finds loveliness all around her. Perhaps *you* do, *too* . . . ?"

(SECRET BONUS CHAPTER)

THE THREE MAGIC QUESTIONS

Here is the long-lost secret of the Incas. Anyone who wants to know how to write drama must learn to apply these questions to *all* difficulties. It is not only unnecessary but also impossible to know the answers before setting out on the individual project in question, as there are no stock answers.

This secret of the Incas, then, is like the Torah, beloved of my people, the Jews. We read the Torah, the five books of Moses, every year, in the same order. Every year the meaning of the Torah changes, though the text remains unchanged.

As the writer changes, year to year, his or her perceptions and interests change. At twenty he is interested only in sex, at thirty in sex and money, at forty in money and sex, at sixty in money and validation, et cetera.

No one can write drama without being *immersed in* the drama. Here's what that means: the writer will and must go through *exactly* the same process as the antagonist (for what is the antagonist but a creation of the writer?).

The writer may choose to supply stock, genre, or predictable answers to the magic questions, and the drama will be predictable and boring. The writer will have saved himself the agony of indecision, self-doubt—of work, in short—and so, of course, will the protagonist. The audience will view this pseudo-drama much as the graduate views a liberal arts education:

"I don't think anything happened, but I'm told I went to college, so, perhaps, I somehow got an education."

All right, you may complain, get to the fairy dust portion of the entertainment and vouchsafe to me the secret of the Incas.

Here it is.

The filmed drama (as any drama) is a succession of scenes. Each scene must end so that the hero is thwarted in pursuit of his goal—so that he, as discussed elsewhere, is forced to go on to the next scene to get what he wants.

If he is forced, the audience, watching his progress, wonders *with* him, how he will fare in the upcoming scene, as the film is *essentially* a progression of scenes. To write a successful scene, one must stringently apply and stringently answer the following three questions:

1. Who wants what from whom?
2. What happens if they don't get it?
3. Why now?

That's it. As a writer, your *yetzer ha'ra* (evil inclination) will do everything in its vast power to dissuade you from asking these questions of your work. You will tell yourself the questions are irrelevant as the scene is "interesting," "meaningful," "revelatory of character," "deeply felt," and so on; all of these are synonyms for "it stinks in ice."

You may be able to dissuade your *yetzer ha'ra* by insisting that you were and are a viewer before you were a writer, and that as a writer, these three questions are all you want to know of a scene. (You come late to a film and ask your friend there before you, "What's going on? Who is this guy? What does he want?" and your friend will, as a good dramaturge, explain that the subject of your inquiry [the hero] is the vice president of Bolivia, and he wants to determine where his boss is, as the bad guys are going to ambush him, and if he, our hero vice president, does not extract the info from the reluctant mistress, whom the president has just thrown over, the bad guys will kill his boss and bring down the country.)

1. Who wants what from whom?
2. What happens if they don't get it?
3. Why now?

As one becomes more adept in the use of these invaluable ancient tools, one may, in fact, extend their utility to the level of the actual spoken *line* and ask of the speech, no doubt beginning, "Jim, when I was young I had a puppy. . . ." "Wait a second, how does this speech help Hernando find out where his boss, the president of Bolivia, is?" And you may, then, be so happy—not with the process but with the *results* of your assiduous application of these magic questions—that finding the puppy speech wanting in their light, you will throw it to the floor and out of the scene it was just about to ruin.

These magic questions and their worth are not known to any script reader, executive, or producer. They are known and used by few writers. They are, however, part of the unconscious and perpetual understanding of that group who *will* be judging you and by whose say-so your work will stand or fall: the audience.

TECHNIQUE

STORYTELLING:
SOME TECHNICAL ADVICE

Storytelling is like sex. We all do it naturally. Some of us are better at it than others.

One learns through experience, but basically, it is a universal human instinct, and I will prove it to you.

Explaining your particular plight to the traffic cop is creating a drama. Talking your way out of a late date, a forgotten anniversary; talking a potential partner into bed, a boss into a raise, a supplier into a discount—each of these is a drama.

The bedtime stories we have heard or told are dramas, and each partakes of the same natural form as the improvisations listed above: once upon a time, and then one day, and just when everything was going so well, and just at the last moment, and they all lived happily ever after. This is the form we learn at Mother's knee, and it is the form we apply in order to understand life. It casts us, the listener, as hero of our own personal drama, as, of course, we are, and it explains that drama to us in the way nature has fitted us to understand it: as a simple, honest attempt to achieve a worthwhile goal.

On our way to the goal (the wedding party, the discount, the weekend in Vegas), we encounter resistance, we find unforeseen reserves of strength and cunning, we are almost undone by some evil force (nature, fate, the traffic cop), and we eventually triumph by recourse to those basic precepts or powers that we were apprised of at the story's opening.

Our simplicity allows us to inform the king that he has no clothes; our inventiveness, to charm the cop; our work ethic, to deliver a superior product at an economical price and so become rich.

And that's it. That is all drama comes down to.

Now, why are some folks challenged in creating it?

There are two answers: (1) everybody can throw a ball, but not everyone can throw like Sandy Koufax; and (2) self-consciousness doth make dullards of us all.

Consider the making of a speech.

We are all, again, fairly accomplished speakers given the right setting. The speech to the wayward son comes flowing to the lips, liquid, satisfying, and pleasant. It comes extempore, it needs no preparation, and though we may be saddened by its lack of result, we are never other than enamored of its sound.

Take the same speaker, however, and tell him he is about to address the Rotary on the subject of the wayward son, and you are likely to find an endless heavy hell of bombast and cliché. What happened?

He forgot that he knew how to do it naturally.

Candidates for office in our fair land sound, for all the world, each like his dull brother or sister. Their wise wranglers have instructed them to "sound presidential" or "senatorial" or what have you, and, indeed, they do, if the holder of that office must sound like a boring dolt.

We are told of each of these bores that he or she is a load of laughs off camera, and who am I to disbelieve? But *on* camera they are a fate worse than death.

They have convinced themselves, or engaged others to convince them, to forget what they know.

None of these speakers would vote for himself if all he knew of himself was the stolid, lethal stump speech.

And yet, here is a fellow or a lass who, to reach the elevated position of his or her candidacy, has had to convince a vast number of rational people of the improbable; he talked fast and fascinatingly in the quest for the candidacy. The candidate forgot that he knows how to tell a story. As does the neophyte dramatist.

How does it go?

Once upon a time, and then one day, and just when everything was going so well, when just at the last minute, and they all lived happily ever after. Period.

Knute Rockne All American, The Greatest Story Ever Told, Hamlet, Dumbo, The Godfather all utilize the same form. (In the case of certain drama and tragedy, the happily ever after is, of course, altered, per case, to, for example, "And then they all lived sadder but wiser" [the drama] or "And then, finally realizing the essence of the human condition, they put their eyes out and wandered around for a while as a blind beggar" [tragedy].)

ONCE UPON A TIME

There was a poor but honest woman, who lived with her son, Jack, in the forest.

AND THEN ONE DAY

Their money ran out, and they were forced to sell their cow.

Jack was sent to take the cow to the fair.

On the way he met a man who offered Jack this bargain: I will trade you your cow for these five magic beans. The little boy happily made the change and came back to tell his mother the happy news.

AND JUST WHEN EVERYTHING WAS GOING SO WELL

She cursed him out for a fool, threw the beans out of the window, and retired to her bed, weeping.

The little boy went to sleep, and as he slept, the beans took root and grew, until the beanstalk reached clear to the sky.

On awakening, he climbed the beanstalk and discovered, in the clouds, a giant's castle. He entered the castle and saw inside of it treasures beyond imagining. There was a golden harp and a goose that laid golden eggs.

Thinking to redeem himself, he picked up the goose and made for the beanstalk. The goose began squawking and awoke the giant, who pursued Jack.

The giant grew closer and closer as Jack threw himself onto the beanstalk and started to descend.

The giant came on roaring, and Jack's end was at hand.

WHEN JUST AT THE LAST MINUTE

Jack reached the bottom, grabbed an axe, and cut down the beanstalk; and the giant fell to his death.

AND THEY ALL LIVED HAPPILY EVER AFTER.

We may apply the same paradigm to any drama. In some, one section will predominate over another—that is, some sections, case by case, will be lengthy and elaborate, some will be relatively short or simply indicated, but all will be present in each.

LEARNING BY DOING

Every man of my age learned how to sharpen a knife. We learned it in the Boy Scouts.

One holds the knife at a 20-degree angle to the whetstone and cuts toward the oilstone. When a good enough bevel is obtained, the blade is turned, and the other side is sharpened.

But a friend of mine, Bill Bagwell, master knifesmith and a world-class expert, says that method is wrong.

Bill makes fighting and hunting knives by hand at an outdoor forge. His knives are prized by collectors and soldiers and hunters. He says one *drags* the blade over the stone rather than cutting into it. This turns the edge of the metal onto the far side of the blade. One then turns the blade and wipes this "wire edge" off.

This method works like mad. Bill used to demonstrate his wares at knife shows by cutting through several two-by-fours with his knife, cutting many inches of free hanging hemp rope, and then shaving the hair on his arm with the same blade. He'd then dull it so that one could safely draw one's thumb across it, give it a few strokes on the whetstone in his heretical fashion, and repeat the demonstration of fine and fancy cutting. QED.

I spent many years target shooting with pistols, and I became somewhat proficient through dedication to what I understood to be the two sole but inviolable rules for pistol accuracy: concentrate on the front sight and squeeze the trigger.

Again, any and all books on marksmanship contain the same advice. All wrong. The appurtenant Galileo in this case was Eric Haney, a friend who spent two decades in Delta Force as a professional gunfighter. To shoot competitively in the real world, learn to thrust the weapon toward the target, forget about the sights, and slap the trigger. That was the method he learned, taught, and trusted his life to over countless trials, and, obviously, it worked, as he had lived to correct me.

Bishop Berkeley wrote that the test of truth is "Would you trust your life to it?"

The film business has been, for a hundred years, the bailiwick of the empiricist: Ashkenazi peddlers, my cultural and racial brothers, battled to put food on the table by exhibiting moving pictures to a paying public. As the business grew, the canny hucksters, now one, now another, got a "good idea" about how to put more asses in the seats.

As the business attracted—then as now—the ne'er-do-well, these ideas were generally efficient. (As the ancient law has it, if you want to get a difficult job done, give it to a lazy man.)

Like new parents, the first movie folk had responsibility but no instruction book. And so they made it up as they went along:

The scene is supposed to be shot in the tropics, but it's cold on the beach in Santa Monica, and the actors' breath is frosting. Have them suck on ice before the take. The star can't cry? Put glycerin on her cheeks. *High Noon* is a dud? Insert shots of a ticking clock and cut to it as often as possible.

Look at the shot in *The Godfather, Part II,* where Robert De Niro shoots the vicious mafioso don somebody-or-other. De Niro puts a gun to his head, pulls the trigger, and a vast wound opens in the don's forehead.

I asked how it was accomplished—did the gun spray the wound on? How could that be, as the wound appeared to have great depth? Here's how they did it. The makeup man built the wound into the actor's forehead, covered it with a layer of flesh, and ran an ultrathin monofilament from the flesh covering to the muzzle of De Niro's pistol. When De Niro shot, he jerked his hand back, mimicking recoil; the line tore the covering free; and the wound appeared.

There are many scenes in *A League of Their Own* where the actresses, portraying a baseball team, heave the ball with fantastic speed and accuracy.

I asked the director how the actresses gained such skill. She said she simply filmed the sequences without the ball.

That is the fascination of actual moviemaking: How do you *do* it? And one is learning, and humbled, constantly.

I spent a sleepless night thinking about the cat in *The Diary of Anne Frank*.

In one of the best suspense sequences in the movies, Joseph Schildkraut comes downstairs from the attic where the Franks hide. There is a burglar in the house, and Schildkraut has to dispatch him. The burglar goes away, but before Schildkraut can close the door, a night watchman notices it ajar and calls over a couple of patrolling Nazis.

Schildkraut retreats to the attic, and all wait breathlessly while the Nazis scout the house below. Now comes the cat. She pads along a kitchen ledge in the hidden attic, she puts her head into a funnel resting on the ledge, she pushes the funnel toward the edge. Now everyone in the world holds his or her breath. Now the funnel goes *over* the ledge. But wait—the cat's head is stuck in the funnel.

Should the funnel drop off, the Nazis will hear and discover the hidden attic and kill all the inhabitants. But continue to wait—the cat now pulls its head, the funnel still on it, back *onto* the ledge, and now draws its head off. What a great sequence. But how did they do it?

I surmise that they stuck some tuna fish inside the funnel, surrounded it with glue, and turned the camera on. That would get the cat's head into the funnel, and stuck there, but then how did they get it over the ledge?

Perhaps, I reason, monofilament line. One prop guy easing the funnel over the edge, another heaving gently on another line to get it back.

Good idea. But now what do you do on take two?

The cat ain't going to put her head back into another gluey funnel. My knowledge of actual filmmaking is sufficient to consider that they just left the cat *glued to* the funnel and called it a day. This would involve shooting the cat shot *last* in the film, but why not? (Supervision of the rights of animals—and children, for that matter—on movie sets is largely hypothetical.)

But no, but no: the cat actually got her head *out* of the funnel at the end of the shot. All right, what about *magnets*?

1. Cat, with magnet hidden (or, indeed, *implanted*) in its neck, is lured to stick its head into a funnel smeared with tuna fish.
2. Funnel contains radio-controlled electromagnet. (I'm assuming such things exist.)
3. Two prop guys hold monofilament lines, one to ootz cat and funnel over the ledge, second to ootz it back.
4. Electromagnet is turned off, and grateful cat removes head from funnel.

In this scenario, I would shoot the cat sequence *first,* that is, before "establishing" the cat.

Vide: Set up four or five ledges, four or five different cats. Whichever cat-and-team first got the shot in the can, *that* cat would be the hero cat and play in the rest of the film.

This version, of course, would require many *many* cats standing by. At least a pair for each ledge.

Why? What happens if you get your shot, you establish your hero cat, and she gets run over by a milk truck or something? You're up the creek is what.

Back to the attic in Amsterdam.

There are only two things wrong with my proposed electromagnetical solution:

1. It is too elaborate.
2. The gag is so *good,* it feels like something dreamed up on the set.

That is, here we are, filming the "Nazis almost find us" sequence, and someone, the director or AD, starts jumping up shouting "ooh ooh ooh"* and comes up with the cat. And the funnel.

Maybe not, but in my experience, the prop people and the stunt people are smarter about the gags (certainly than the writer, and regularly) than the director.

But in *this* case, the director was George Stevens. Now, George Stevens started out as assistant cameraman for Hal Roach, shooting Laurel and Hardy silents.

These, to me, are the perfection of essential moviemaking. Perfect simple plot, no distracting jabber (or "dialogue," as it is more generally known), and the only thing moving the film along is *gags*—that's all there is.

The gags, here, happen to be identical to Aristotle's "incidents," that is, *those occurrences without which the plot cannot move forward*.

Here, Mr. Stevens's great gag illustrates the conjunction of

* Those three words are inevitably the herald of film genius.

the artistic, the technical, and the administrative, which constitute the nature of the director's job.

1. Ooh ooh ooh, the cat puts his head in a funnel.
2. How do we make it work?
3. How do we arrange the schedule to make it possible, and how do we protect ourselves in the event of error or catastrophe?

I will give the cat gag one more think and then surrender.

The doctors say, "You hear hoofbeats, think horses, not zebras." So, to the filmmaker, Occam's razor is "they probably shot it backwards."

(E.g., *The Pride of the Yankees.* The story of Lou Gehrig, famous and famously left-handed hitter. Gary Cooper, playing the Iron Horse, bats righty, so they printed the Yankees logo backward and had Cooper bat right, hit, and then run to third.)

But, no, the cat can't be shot backward, because he both starts and *ends* with his head free of the funnel.

(If the cat started the shot with his head in the funnel and ended it free, one could simply film a cat free, who sticks his head in the funnel, and run the film backward.)

I start again, my reasoning:

1. There existed, in antiquity, a cat trained to stick her head in a funnel, inhale sufficiently to keep the funnel stuck onto the head, et cetera. . . .
2. I give up.

I will now proceed to that portion of the entertainment known as "duh."

I will telephone George Stevens Jr., and he will describe to me the simple, and now totally obvious, method his father used. I will slap my forehead and then share it with you.

While it remains a mystery, I will note some of my favorite effects.

In *Only Angels Have Wings,* Thomas Mitchell, ex-flyer and aircraft mechanic to flyer Cary Grant, stands on the field as Grant lands in his monoplane.

Mitchell takes a cigarette from his pack, places it in his mouth. We see Grant's plane on final, about to land. Mitchell takes a wooden match and holds it above his head. The plane—seen in the background—comes closer to touchdown. Now the plane is about to touch down, and the passage of its wing over Mitchell's upraised hand lights the match, and Mitchell lights his cigarette.

This is a conjunction of a *gag* (something done on the set) and an *effect* (something done in or using the talents of the lab). In this case, there is a rear-screen projection of the approaching plane, which dips just below frame at the last moment. A physical mock-up of the *wing* then replaces it and passes over Mitchell's head to light the match.

Another favorite: Note the shot of Munchausen's boat beaching on some island or other in Terry Gilliam's *The Adventures of Baron Munchausen.* We see the prow of the ship cutting through the water. Now the water lessens, and the boat continues cutting through the *sand*.

Note also: Julia Roberts getting into her car in *Erin Brockovich.* The car pulls out into traffic and then is demolished by an oncoming vehicle. (This is an *effect:* Julia actually swaps out her place with a stuntwoman early in the sequence, the two shots are melded in the lab, and the stuntwoman takes the hit.) Another magnificent effect is the running-upstairs shot in *Contact* (director Robert Zemekis, cinematographer Don Burgess), in which the camera pulls the hero, a little girl, down a corridor, up the steps, and toward a medicine cabinet holding the drugs that will save her father from his heart attack.

She reaches forward, and the shot of her face be

shot of a medicine cabinet. The camera, however, does not move. That is, it is as if the entire shot were, somehow, a reflection in the medicine cabinet mirror. A stunning effect.

A brilliant gag is the murder of Alex Rocco in *The Godfather.* His assassin (Franco Citti) kills Rocco (as Moe Greene) in the barber chair of the St. Regis.

Rocco is getting shaved. He senses something amiss and puts on his eyeglasses. The assassin fires, and Rocco's glasses shatter as he is shot in the eye.

How did they do that?

The eyeglass temples contained a miniscule BB apparatus and air gun. On cue, the BB was propelled forward, *from* the face, *toward* the lens.

I call George Stevens Jr., as I was saying, and plead for an answer. He laughs. This shot, it seems, is *already* a part of film lore.

Mr. Stevens tells me (to my pride) that (1) there *were* a bunch of cats, and (2) his father, the director, "just turned on the cameras and shot an *unbelievable* amount of film, waiting for *some* cat to do something 'uncatlike.' "

What a wonderful business.

The *gag,* again, is the script-in-miniature, one crystal clear idea, superseded by another, creating expectations dashed in a logical and surprising way, thus, of necessity, propelling us into the next beat.

This is the paradigm understood and practiced by commercial makers—a strict adherence to the rules of drama.

Hard enough to do in the one-gag format of the thirty-second commercial; *very* hard to do over the course of the ninety-minute film.

But the *mechanism* is identical and the key to great moviemaking.

It is an improvisation based on a conscious or intuited understanding of how an audience perceives:

a. Make them wonder.
b. Answer their question in a way both surprising and inevitable.

This is the exact same mechanism as the joke:

a. Do you wake up grumpy in the morning?
b. No, he gets up on his own.

Our delight in the joke—as in the drama—comes from our momentary triumph over that ever-vigilant repressive mechanism, that distinctive, questionable human gift that, otherwise, we are required to praise: our consciousness. Participation in the drama, as in the hunt, in sex, in war, and, curiously, at the movies, regresses us to an irreducible humanity.

IMPROVISATION

Near the end of filming a movie, I lose an important location, and several months of planning and preparation go out the window. This, however, must be viewed as a blessing, as it forces me to, once again, reduce the scene from the pictorial to the schematic.

What does this mean?

I had formulated my plan of filming in *shots*. One: shot of the wheels of the plane taxiing, hero appears behind them; two: shot of petrol dump with Arabic signs and English signs reading DANGER—INFLAMMABLE; three: Land Rover stopped by large construction barricade.

Now, with no more petrol dump and no more construction barricades, I must discard my pretty (if, as yet, only imagined) pictures and return to the theoretical. That is: *if* the hero wants to get the abducted girl home, and *if* the villain has discovered his plan and means to subvert it, what stratagems will each employ in the last reel?

I am forced back to the most simple: what does the Hero want, in this scene, and how can it be designed such that its exclusion or, indeed, replacement in the progression renders the story moot? One learns to ask, as something always will go wrong.*

* Perhaps I am being teased by providence. I wrote and directed a film called *State and Main* in which a movie company is kicked out of its small-town location

Steven Spielberg confessed, in a documentary on the making of *Jaws,* that the brilliance ascribed to him in withholding the appearance of the shark until halfway through that film must be credited, instead, to the mechanical shark, which refused to function when the cameras turned. The shark wouldn't swim, so the director had to come up with the standby plan. The revised plan showed not the shark but the *effects* of the shark or the location where one might *expect* the shark, and audiences screamed at the photograph of the water.

I once asked Bob Rafelson about what I found an interesting choice at the end of his *King of Marvin Gardens.* The girl comes into the room for a "talk" scene, hair dripping from a shower, wrapping her head in a towel.

"Oh yeah," he said, and explained: she had come to the set having unilaterally decided to chop off her long red, long-established hair. Rafelson had already shot both preceding and succeeding scenes, where she had hair down to her waist, so he threw her in the shower and told her come out and throw a towel around it.

I recall a close-up of Betty Hutton in *The Miracle of Morgan's Creek.*

Preston Sturges stages one of his beautiful walk-and-talks, pulling Betty and Eddie Bracken down the small-town street for three minutes of jabber. (Cf., by the way, Tim Holt in the pony cart in *The Magnificent Ambersons,* where Orson Welles dressed not one but *both* sides of the Main Street set—we see the shop fronts of one and, reflected in their windows, the traffic in the street and the shop fronts opposite.) Eddie Bracken and Betty walk down the street, and halfway through, their

---

two days before filming is to begin. Various gags in the first few minutes concern the absence in the new town of an old mill. The screenwriter shows up and is told, blithely, to deal with the problem and make the script conform to reality; we then discover that the film they are to shoot is *called The Old Mill.*

tracking two-shot becomes a grainy, blown-up close-up of Miss Hutton, when, as any filmmaker could explain, the lab or the director found an impossible error in the otherwise uncoveraged shot.

In English, Sturges had "shot the scene in one"—that is, his only coverage of the scene was the one shot, he had nothing he could cut *to,* and, so, when the developed film showed a scratch on the negative or a previously unsuspected light stand in the background, the director confected the close-up of Hutton by overenlarging her image from the two-shot.

I will further extend my foray into Sturgesiana by here revealing, for what *must* be the first time anywhere, and to what must be a severely restricted coterie of the interested in such minutiae, what must be Sturges's inspiration for a sequence in, and perhaps for his creation of, another of his films, *Hail the Conquering Hero.* Eddie Bracken is again Sturges's protagonist. His father was awarded the Medal of Honor in World War I, in the marines. Bracken enlists in the marines in World War II but is debarred from serving because of his hay fever. He is comforting himself in a bar when a group of combat marines, fresh from Guadalcanal, comes in. He buys them a drink and tells them his sad story—he has been writing home fictitious takes of his exploits in the corps and now cannot go home to face the lie.

The marines lend him a uniform and say they'll escort him back home, he'll get off the train, go home, kiss his mom, take off the uniform, and say he doesn't want to talk about the war. A perfect comic premise, a perfect comic film.*

* Let me add to my list of loves the opening shot of *Hail the Conquering Hero.* A sultry café singer croons "Safe in the Arms of Mother" to a bunch of drunks, but the camera does not dwell on her charms, no, but slips off to frame the enormous asses of two waiters who stroll the café with her, singing harmony—as if trying to find road room to overtake and again frame the hot tomato. This

Preston Sturges. Who turned William Demarest into a household god. To whom he gave the line, "This'll put Shakespeare back with the shipping news." Preston Sturges, who, in *The Lady Eve,* shepherded Henry Fonda through *five* pratfalls in half a minute, who—but I digress. So Mr. Bracken is escorted back to town by William Demarest and the Guadalcanal marines. His plan, to scoot from the railroad station the two blocks home, change into mufti, and that's the end of it. *But* word has gone afore, and the mayor has turned the town out. There are banners and competing brass bands. The whole town throngs the square. The train pulls up; one band plays "Hail the Conquering Hero" while the other plays "The Marine Corps Hymn," and Franklin Pangborn, avatar of the fey, tears out his remaining hair as the distraught master of ceremonies. What delight then, to discover the following:

> Away the carriage goes! With the noisy populace about the wheels. What is this?—music? Yes: two opposition bands. One is playing "See the Conquering Hero Comes" while the other exhausts itself, and gets black in the face, with the exertion necessary in doing justice to "Rule Britannia."
>
> —Mary Elizabeth Braddon,
> *The Trail of the Serpent* (1885)

Over the years various journalists and other worthy folk have asked me, "Where do you get your ideas?" To which I usually reply, "I think of them." I permit myself this jolly facetiousness as the truth is, to me, more ghastly: (1) I have no idea; (2) I have so very few of them.

For film, in addition to being a structured dialectic (as per

---

filmmaker, American, and viewer smiles dopily and thinks, "Take me now, Lord. . . ."

Sergei Eisenstein), is a corporation of Good Ideas: Eisenstein's mutinous sailors herded beneath a tarpaulin to be shot in *Battleship Potemkin,* Spielberg's absence of a shark, William Wyler's tracking shot of Mary Astor and Walter Huston *just missing* each other in the American Express office in Naples (*Dodsworth,* my vote for one of the world's ten best films), George Stevens's shot of Elizabeth Taylor at the end of *A Place in the Sun*. She sits alone, and the fire in the fireplace is superimposed on her, reflected in the window.

Where *do* our ideas come from?

Perhaps the best, the true, are the reward neither of talent nor luck but of humility.

THE SLATE PIECE

> Yet still the weak offender must beg still for leniency and trust his power to avoid the sin peculiar to his discipline.
>
> —W. H. Auden

They say you get to make a movie three times: when you write it, when you shoot it, and when you cut it.

One really doesn't start to learn how to write a script until one has been on a set—on the set one learns the difference between what is *filmable* and what is merely pretty words. ("Outside the window, *New York*—in all its vicious splendor" is charming verbiage and all that, but, script-in-hand, on location, its director is going to be hard-pressed to learn from the script where to put the camera.)

Giving the actor a meaningful pause as part of his take may seem thoughtful and sensitive during filming, but the director stuck in the cutting room, watching the same interminable take, may learn, next time around, to *pick up the pace.*

The making of movies is magnificently pragmatic. As in combat, as in sex, the theoretical is all well and good if one's a commentator, but the thing itself can actually be understood only through experience. No one on any set, or in any cutting room, knows the difference (if such there is) between realism and naturalism—they are merely "telling a story with pictures." A couple of guys in a coffee shop set out to write a gag; a couple of guys with a camera set out to film a gag; a couple of guys in an editing room set out to make sense of the trash that's been dumped on their desks. That's moviemaking in its entirety—anything else is just "the suits." Through it all the clock is ticking: so many days and they take away the camera, so many days and the studio needs to release the print.

Stuck in a scene, in the editing room, sometimes the roof falls in: an actor has not picked up his cue, and the scene stops dead—there is no cutaway (no other actor to cut to, to "pace up" the sequence), and the movie grinds to a halt.

"If only," the director or editor says, "if *only* the actor sitting there like the sphinx had looked to his *left:* if he'd looked to his *left,* instead of his right, I could intercut his close-up with a shot of the other actor and pace up the scene."

But no, the actor never looked to his left, and the scene is doomed to death. But perhaps there is one hope.

The director says, "Check the slate piece."

What is the slate piece?

Here's how it goes: When the shot is set up, the actors are called in and placed. The sound guy calls "rolling," the camera is turned on, the operator tells the camera assistant to "mark it," the assistant puts the slate board (the once actual slate with chalk markings, now electronic) in front of the lens to record on film the shot's number and take. The shot is thus "slated," the director calls "action," and the take begins.

But, we may note, there was a moment, when the camera was filming, *before* the shot was slated, when the actor was waiting for action to be called. In this moment he *may* have looked to his left, his right, up or down, frowned, or smiled or yawned or done any number of things that just might magically come to the aid of a stalled or otherwise doomed shot.

This accidental, extra, hidden piece of information is called the slate piece. And most of moviemaking, as a writer, a director, a designer, is the attempt not to *invent* but to discover that hidden information—the slate piece—that is already lurking in the film.

THE WISDOM OF THE ANCIENTS

My friend Eric, career military, told me that he figured that the men who taught him in Vietnam had been taught by those who fought in World War II; that *they,* in turn, had been taught by the soldiers of World War I, who had been taught by the soldiers of the Spanish-American War, who had been taught by the Indian fighters; the hard-won direct hands-on knowledge being transmitted, by extension, to him from Thermopylae and Sumer.

Thermopylae and Sumer, in film, are, even now, just, *just* beyond living memory.

I got notes over the phone on my first screenplay from Samson Raphaelson, who wrote the first talkie.

I once had a drink with Dorothy Gish, who starred for D. W. Griffith; I made a film with Don Ameche, who was the world's biggest star in the early talkie era; I played poker for years with Eddie Bracken; the gaffer on the last film I shot is the grandson of the gaffer on *Intolerance*.

We find ourselves, still, that close to the beginning of the first new art since cave painting.

The earliest films were pure exploitation: a railroad train rushes at the viewer, a couple kisses. The novel technology put bread in the exhibitor's lunch pail.

Exhibitors found they could charge more for longer films, and greater length necessitated a dramatic structure. ("What gimmick," they wondered, "could we use to keep the audience docile between the thrills?") This same problem plagues the porn industry, which has addressed it by increasingly shortening the intervals between sex.

Mainline films have treated the problem similarly—periods between violence and sex diminish in an asymptote, the period gradually approaching the virtually nonexistent, such "spacer" material describable dramatically as "and *now*. . . ."

Multireel films created the necessity for scenarists, folks who could fill the gap between train wrecks. Their ranks included playwrights and novelists, who found themselves, from the first, in opposition to the money folk, the exhibitors, and the banks.

The exhibitors are, of course, in it for the money; films got longer in order to allow them to charge more; persons with a dramatic bent were brought on board to allow the films to grow longer, and found, and find, themselves squared off against exhibitors, who reason: "How is the audience going to be thrilled by a shot, a scene, a sequence, in which no one is being either kissed or killed?"

But dramatic structure consists of the creation and deferment of hope. That's basically all it is. The reversals, the surprises, and the ultimate conclusion of the hero's quest please in direct proportion to the plausibility of the opponent forces. The study of same, to the dramatist, scenarist, or, in the old days, "title writer," is the essence of film. These scribes brought and bring to the movies an ancient wisdom (or an approach to same), such wisdom stemming from a conscious or intuitive attempt to understand how human beings think.

What is the ancient wisdom?

Well. I suspect that a Spartan warrior, transplanted to a rifle platoon in Iraq, would find the situation rather familiar and would adapt to the new technology pretty quickly.

(The businesspeople who started the movies thought similarly: What difference selling moving pictures of a locomotive or selling gloves? It's just *selling*.)

And the dramatist, his calling older than that of the merchant and as old as the soldier's, similarly takes his skills, along with a racial intuition or predisposition, into the endeavor.

Here he is joined by the visual artist.

For the film, it has been discovered, cannot be merely the record of a stage play. It must tell the story, but it must tell it in *pictures*.

This is the new art, the conjunction of the dramatic and the plastic.

It has been approached in stagecraft and stage design, but the analogy is insufficient. Stage design exists to frame and intensify a drama of human scale; films can juxtapose seemingly random images to tell a dramatic story.

What prodigies of personal insight or of interpersonal collaboration and strife are necessitated by this new art, and how may one master its difficulties?

I've always been a delighted devotee of film wisdom, those distilled snippets of experience passed from the now-wiser to the newcomer—the wisdom of the ancients, some of which I now addend.

Stay with the money. The audience came because you advertised the star. Shoot the star. (NB: Howard Lindsay, coauthor of the plays *Arsenic and Old Lace, Life with Father, State of the Union,* et cetera, once privately printed a small volume of stage wisdom. One of his axioms was: take the great lines from the secondary characters and give them to the lead. This works like gangbusters in film and on stage. It raises the question: Why would the dramatist *want* to give the yummy lines to the second banana—as one *does;* the temptation to do so is great. Someone of a psychologic bent might suppose that the dramatist does so out of envy of the hero and the desire to show oneself superior. In any case . . .) Stay with the money.

Burn the first reel. Almost any film can be improved by throwing out the first ten minutes. That exposition, which assuaged the script reader, the coverage writer, the studio exec, the star and her handlers puts the audience to *sleep sleep* sleep. Get right into the action, and the audience will figure it out. (Simple test, for the unbelieving: when you walk into a bar and see a drama on the television, you've missed the exposition. Do you have any trouble whatever understanding what's going on?)

If you think that perhaps you should cut, cut. A film is made, and one learns to make film, first on the set but more importantly in the cutting room. If you suspect the shot, sequence, line is unnecessary, *get rid of it.* Like the dramatist giving the punch lines to the stooge, you, the filmmaker, can't quite trust yourself. Err on the side of the audience, and get on with it. Cf. "You start with a scalpel and end with an axe."

If you laughed at the dailies, you aren't going to laugh at the picture. Too true. Equally true of tears. That scene that's got all the Teamsters weeping on the set is probably coming out of the picture. Why? *It stops the show.* We're told less is more. This is a curious oxymoron, as it means more is better. But in film, *less* is always better. Tears are like homeopathic medicine: the smaller the dose, the greater the effect. And the timing of a gag, and so of a laugh, is the *essence* of the gag. And its timing will depend *on the way it is cut.* The folks on the set and the folks screening the dailies aren't seeing the thing cut and so should mistrust their reactions.

Nothing with a quill pen in it ever made a nickel. I think this is true, but it is, at very least, provocative. I myself don't respond to the Georgian in film and will addend a condign comment attributed to Harry Cohn. When head of Columbia, Mr. Cohn screened *One Million B.C.*, a cave drama with Victor Mature and Carole Landis, and remarked, "I can't vote for a film where the guy's tits are bigger than the broad's." A third film gem is the exhibitors' "Give me Tahiti in the winter and the Arctic in the summer"; all of which wisdom literature might be subsumed under the all-encompassing "Give 'em what they want."

Get out on your biggest laugh. And the aligned "Always leave 'em laughing" and "Always leave 'em wanting more." One should leave the theater thinking, "I never wanted it to end," rather than, "Now I am *sure* I've got my money's worth!" Nightclub performers had it: "When you *come* on, *be* on; and when you're done, get *off*."

If you can't figure out what the scene's about, it's probably unnecessary. The process of organizing a film for shooting is called "boarding." A board is made up showing the scene number, the players, the physical elements (cars, stunts, effects, et cetera). The scene is identified both by number and by log line, a description of the scene just sufficient for its identification, e.g., Steve finds the map; Gramps falls downstairs.

I was preparing one movie and the unit production manager came over and politely pointed out that two of the scenes on the board had the same log line. "It isn't my place," she said, "but it occurred to me that maybe one of them was unnecessary." As, indeed, one was. A corollary of "If you can't figure it out" is "If it *is* necessary, it's necessary only once."

Always get an exit and an entrance. More wisdom for the director from the cutting room. The scene involves the hero sitting in a café. Dialogue scene, blah blah blah. Well and good, but when you *shoot* it, shoot the hero coming in and sitting down. And then, at the end, shoot him getting up and leaving. Why? Because the film *is* going to tell you various things about itself, and many of your most cherished preconceptions will prove false. The scene that works great on paper will prove a disaster. An interchange of twenty perfect lines will be found to require only two, the scene will go too long, you will discover another scene is *needed,* and you can't get the hero there if he doesn't get up from the table, et cetera.

Shoot an entrance and an exit. *It's free.*

I learned a corollary from John Sayles: at the end of the take, in a close-up or one-shot, have the speaker look left, right, up, and down. Why? Because you might just find you can get out of the scene if you can have the speaker throw the focus. To what? To an actor or insert to be shot later, or to be found in (stolen from) another scene. It's free. Shoot it, 'cause you just might need it.

I once designed a scene in a bookstore. Dialogue dialogue dialogue, blah blah blah. A later scene involved the hero coming back and reinterpreting the previous evidence he found there. On the day, I announced: "Wait a second, the *second* scene can be reduced to a mere insert" (shot of an object). I felt bold and was further emboldened by my strength in resisting all polite efforts to convince me to shoot the scene as previously written. "It's *free,*" I was told. "We're in the bookstore *anyway*" (shooting the first scene). No, I bravely announced. And then I had to come back after the film was cut and, at great expense and in healthy humiliation, shoot the scene as previously written. Why? I had offended the gods. My certainty in the face of polite reminders to accept the free shots had angered the powers that be. An attendant piece of film wisdom: "If enough people tell you you're dead, lie down."

What keeps them apart? (Billy Wilder) The engine of a love story is not what attracts them—we *know* that: they're young and pretty. The work should go into the construction of the plausible opposition to their union.

More wisdom from Billy Wilder, of the audience:

"Individually, they're idiots. Collectively, they're a genius."

Anyone who speaks of the audience's understanding as diminished has never had to make a living by appealing to them. If it's coherent, they *will* get it. The filmmaker's job is not to *pander* to them but to make his vision *coherent.*

Do not shoot the pretty girl's close-up last. For some reason, this very good advice is overlooked, and one *always* ends up shooting the pretty girl's close-up last thing in the day. This may be because the male leads are generally much more demanding, and they get the grease. Perhaps it is because the filmmaker, graciously putting himself in the same position as his audience, is in love with the pretty girl and feels he can count on her. Or maybe he'd just rather have her to look at at the end of a hard day. I don't know. One always ends up, as I have said . . .

My friend the soldier passed along a compendium of the wisdom of his racket. It, similarly, dripped with the, in his case, real rather than figurative, blood of experience: "When your plan of battle is proceeding perfectly, you have just walked into an ambush"; "All combat takes place at night, in the rain, at the intersection of four map segments"; "Never trust a recruit with a weapon or an officer with a map"; "If you can't remember which way the claymore is pointed, it is pointed at you"; "It is inadvisable to parachute into an area one has just bombed."

This mutual conjunction of the fatalistic and the mechanical, the mysterious and the mundane, confirms film as a true, if late-developing, human art. As such, its practice and contemplation connect us at once with both our most- and least-human aspects. Our ability to conceptualize about both the process and the product is accompanied, and inspired, by the pure animal joy of submersion in a mystery.

SOME PRINCIPLES

THE AUDIENCE; OR,
LESSONS FROM DUCK HUNTING

A duck decoy does not need to look like a duck.

It needs to look like a duck to a duck.

Wisdom, therefore, lies not in the phenomonological question "What does a duck look like?" but, rather, in the practical "What is a duck looking for?"

The duck, in this conceit, is our friend, the Audience, and I use the term "friend" advisedly.

Sun Tzu instructs, in his *Art of War,* to treat the opponent as if he were an employee—to ask, that is, what motivates him, and to act accordingly.

The moviemaker, similarly, must treat the audience member not as an adversary but as an associate.

The untutored deal with an adversary by stealth, cunning, bribery, misdirection, blunt application of force—the tools, in short, of crime. And these are, of course, recognizable in their application by the film bureaucracy: the audience is bribed by promises of titillation, they are bombarded with endorsements from the cajoled or suborned, they are weakened by dulling repetitive advertisements, they are promised a glimpse of their favorite if they will behave (that is, attend), and so on.

The audience, knowing itself, thus, despised, reacts to films not as a legitimate entertainment but as a contest of will (one creates enmity by applying force; one may learn to prevail through understanding rather than strength—the basic tenet of jujitsu).

In using the tools of aggression and opposition, whether in Vietnam or in the calendar section of the newspaper, the aggressor strengthens the resolve of the opponent to resist, necessitating ever-greater prodigies of waste (bombing/advertisement) on the part of the powerful. Thus are the high brought low. One does not win the hearts and minds of a population by bludgeoning them.

But what if the audience were not an opponent to be bilked but a necessary adjunct of the process of creation, an associate whose needs, conscious and unconscious, the filmmakers strove to understand?

Let us return to our friend the duck.

The wealthy hunter might bespeak a decoy realistic to the nth degree. This decoy might be correct in every particular of size, form, and color, and yet the poor hunter in the next blind down might be attracting all the ducks with his roughed-out and unpainted decoy. Now, why is that?

Well, the poor man, unhampered by the capacity to waste, was forced to employ thought, and he wondered: What does a duck like? How does a duck *see*?

We have all had the experience of saying of a statue, "How lifelike," and, of a life mask, cast from the human form and painted to perfection, "How lifeless."

What was missing in the life mask? Life.

For the actual human being and the actual duck were created by, and so contain, a *mystery*. They cannot be reduced to mere measurements, and all attempts to do so (whether through the caliper of the decoy maker or through the audience testing of the social scientist) result in lifeless parody.

Life, in the art of the drama or of the carver, cannot be aped, and the attempt to remove the element of chance must doom the project absolutely. For another name for "chance" is "mystery," and another name is "art."

The artist carver, director, writer, or actor brings a (conscious or unconscious) understanding of the mystery of human interaction to his task.* And it is this ineffable element (*not* the mechanic verisimilitude) that attracts the audience. And it is this artistry that attracts the duck.

Just as the decoy cannot be all things to an incoming duck, the film cannot be all things to an audience. Better that it should be *some* thing, that it convey the desire of the filmmaker to tell a story rather than his desire to earn a living.

Most films are bad. They are, finally, just advertisements for themselves—elongated movie trailers, envisioned and cut with less skill than the trailer itself.

A good duck decoy, on the other hand, is a work of art and may be loved and admired in itself, on a tabletop, independent of its success in bringing home the game.

The films of De Sica, of Welles, of Michael Powell and Emeric Pressburger made little money and endure as spiritual delights.

On the other hand, there are films of which we, quite literally, *applaud the grosses,* while the films themselves are unwatchable (e.g., *Titanic*).

(There is a relationship of mutual exploitation fostered by the creation and marketing of the solely mercantile film. Someone said that the genius of the American tax code was that it turned everyone into a sneak and a criminal. Mass-market exploitation of the audience makes the producer and the viewer complicit in an adoration of wealth—the producer trying to mug the viewer, and the viewer submitting for the cheap thrill of the producer's notice. In this, the viewer is in the same

* As, indeed, does the inspired entrepreneur.

position as the star at the Oscars: he agrees to fawn and pant in return for a pat on the head. This is, of course, the reason for the Oscars' success as entertainment: the audience gets to see their oppressors brought low. It is like Boxing Day, when the lords of the manor had to pretend to serve the servants.)

There is another aspect to the hunting of the duck: The hunter whose poverty debars him from belief in the magical powers of the most expensive decoy (ad campaign, audience testing campaign, trailer, et cetera) must become a natural philosopher. He must teach himself to look at the totality of the situation. He may sit longer or stiller in the blind and so cultivate habits of patience, thought, and observation, perhaps, in the process, obtaining not only wisdom but also grace, as both his success and lack thereof may inspire him to a more profound understanding, so that not only his product but also his efforts achieve a sort of beauty that might be called art.

AESTHETIC DISTANCE

Charles Henry Parkhurst was a late-Victorian reformer. Like many who preceded and many who followed, his stock-in-trade was low-cost prurience. He haunted the dens of vice of Victorian New York City and wrote, at length, of their appalling, nay, demonic conditions.

The goodwilled people of that time, much like you and me, might read along, shaking their heads as they discovered the hoped-for mention of this or that preferred vice.

"Tsk tsk," they or you or I might say, as our eyes grew wide, our heart began to beat more insistently, and so on, increasing the urge to read.

For the newspaper, whatever its flag of convenience, exists to sell sex, gore, and outrage. Much like the movies.

In each, the moralistic tone is very much likely to enfold, and, indeed, to allow the sale of that denied to the high-minded. Most antiwar films succeed through the power of this engine. We viewers are titillated by images we are assured we have come to decry.

Not a Love Story, the 1981 Canadian documentary, passed as an exposé of the smut industry, but I suggest that, absent its odor of sanctity, it was powered by sexually explicit images and enjoyed by those who thought it good to watch the same.

The Green Mile, while purporting to be an indictment of

capital punishment, was a pictorial, inventive, extensive, and very graphic description of the same.

Can or could those above-named and unfortunate subjects be treated in a truly moral way? Of course. I will suggest, as for prostitution, Silvana Mangano in De Sica's *Gold of Naples* and, as for capital punishment, Stanley Kubrick's *Paths of Glory* or Daniel Mann's *I'll Cry Tomorrow.*

Each of these takes an essentially tragic tale and investigates it with dignity. Now, *Not a Love Story* and *The Green Mile* differ in degree. The second, a straightforward mercantile venture, adopts or accepts a degree of license offered by sanctimony. Why not? The first, *Not a Love Story,* sells flesh while it sails under the banner of exposé *toute entière,* to which a critic more moralistic than I might perhaps respond, "Shame on you."

Let us discuss *aesthetic distance.*

It is the goal of the dramatist to involve the audience in the working out of a hermetic syllogism.

The goal of the hero is stated, as are the impediments to that goal. The audience, then, engages its intellectual fantasies, attempting to anticipate the hero's possible solutions. This is called "getting involved." *Because* the creators have invested time and effort, they, the audience, become emotionally involved. They root for the hero, exult at his successes, are anxious for his triumph, and suffer at his reversals.

They are permitted to do so *in the degree that* the syllogism is plausible, solvable, simple, and clear.

Hamlet wants to find out who killed the king. All right, we'll play along. If Ringo can't get the sacred ring of Kali off his finger, he will be sacrificed. Ditto.

As we have signed on for what, in Hollywood, is known as "the ride," we identify with the hero (this is what the term means: that for the length of the drama, our interests are one).

The identified-with hero becomes an object of love (how

otherwise, as it is ourself?), and we want to know more about him.

The untutored mistake effect for cause. Their logical fantasy: that in the successful drama we want to know more about the hero, *therefore* a drama can be made successful by *telling* the audience more about him.

Now, the more the audience is told about the hero—the more their legitimate, indeed, induced desire is gratified—the less they care. For they have signed on to follow his journey in anticipation, glee, and dread. When the author indulges his ability to frolic away from the described path (the path, the *sole* path, to which the audience has vouchsafed its interest), the less interested the audience becomes.

(Canny test marketers hold "focus groups" at test screenings and quiz the audience on the film they've just seen. "What scenes did you like least?" Those in which the hero was in danger. "What character/s did you like least?" The villain. Oh, sigh.)

To return: the reductio ad absurdum of "we want to know more about him" is recourse to *actual* physical or biographical aspects *of the actor,* e.g., let's show his or her actual genitals, physical deformity, tattoo, et cetera; let's make reference to actual events in the actor's life that might excite interest.

This, while perhaps exciting audience interest *in general,* does so at the expense of audience interest *in the plot.* (Again, the author has thought it good to *detour* from that service for which we have paid and pledged our attention.)

This is called *violating the aesthetic distance.**

Cheap sentiment is indeed enduring. So is cheap scent.

* Steven Schachter's *Door to Door* (2002) stars William H. Macy as a man deformed by cerebral palsy. His goal is to become a door-to-door salesman. He *refuses* to let anything—including his diability—dissuade or divert him. The audience follows him in his goal and quits this most excellent film thinking the hero not a poor man but a hero.

THE FIVE-GAG FILM

The absence of affront at the violation of the aesthetic distance may be employed as a diagnostic tool indicating that the nature of the entertainment is not, *essentially,* dramatic.

Consider the use of the on-screen telephone number. The display of an actual telephone number on screen as part of a dramatic entertainment has been ruled legally actionable, as it may constitute invasion of privacy. Film and television have adopted for display the telephone exchange 555, which is never actually assigned and thus cannot incite the litigious. The nonoffending number, as we know, appears on the screen in instances such as this:

BRENDA

Sergeant Mulchahy, we have that important telephone number.

SERGEANT MULCHAHY

Give it to me, Brenda. . . .

(Brenda passes the sergeant a piece of paper on which we read "555-3948.")

The audience is shown the number. Its display, however, does not aid but ruptures the dramatic flow; for the audience realizes that the number they have been shown, contrary to the assurances of Brenda, cannot be important either to her or to the sergeant, as it exists not to identify a telephone subscriber (suspect, victim, or indeed human being) but to dissuade vexatious litigation by viewers. This is called "violating the aesthetic distance."

Other examples of such violation follow.

An actor portrays a pianist. The actor sits down to play, and the camera moves, without a cut, to his hands, to assure us, the audience, that he is *actually playing*. The filmmakers, we see, have taken pains to show the viewers that no trickery has occurred, but in so doing, they have taught us only that *the actor portraying the part can actually play the piano*. This addresses a concern that we did not have. We never wondered if the actor could actually play the piano. We accepted the storyteller's assurances that the *character* could play the piano, as we found such acceptance naturally essential to our understanding of the story, but when the camera tilts down to the actor's actual fingers, we, in effect, experience *this:*

FILMMAKER

I'm going to tell you a story about a pianist.

AUDIENCE

Oh, *good:* I wonder what happens to her!

FILMMAKER

But first, before I *do,* I will take pains to reassure you that the actor you see portraying the hero *can actually play the piano.*

We didn't care till the filmmaker brought it up, at which point we realized that, rather than being told a story, we were being shown a demonstration. We took off our "audience" hat and put on our "judge" hat. We judged the demonstration conclusive but, in so doing, got yanked right out of the drama. The aesthetic distance had been violated.

So that's the aesthetic distance.

It is a name for that condition whose existence allows the audience to suspend its judgment (to, in effect, lower its guard) in return for receipt of a *specialized* experience.*

Deeply immersed in the drama—watching a great film, a work of art, *The Magnificent Ambersons* or *The Godfather* or, in a modern example, *Whale Rider,* the display of the telephone number 555- would be shocking indeed. How, we would rightly wonder, could such a wise storyteller make such an error? "We were *with* you . . . ," the saddened viewer might think. "Why did you trust us, and yourself, so little? Why have you ruined my ability to enjoy the fantasy?" (Jewish rabbinical tradition notes that adultery is like murder, for it is a crime that cannot be undone. Violation of the aesthetic distance is a rupture of the artist's compact with the audience, and, similarly, its rupture cannot be mended.)

Stunts in films similarly have a great capacity to rupture drama. They offer (in truth or in potential) a thrill, but that thrill may very well shatter the audience's agreement with the filmmaker.

More effective—and much more difficult—is the creation of a thrill by means that do *not* draw the audience's attention

* The psychoanalyst may, indeed, fall in love with the patient, and the two may conclude that indulgence in their overwhelming ardor excuses the doctor's violation of the role of analyst. All well and jolly, but the moment they start fooling around, the *special case* relaxation of societal norms (in this case regarding self-disclosure) that permitted the analysand to speak freely is revoked and the analysis is, effectively, over.

to but further enmesh the audience *in* the storytelling process, e.g., the surprise ending of *The Sixth Sense.* Here, the essential nature of the dramatic interchange—to engage the audience in *wondering what happens next*—is employed to lead them on to a surprising, inevitable, and, thus, thrilling conclusion.

The staged, self-contained stunt—or computer-generated event—on the other hand, offers the audience a treat, thus, as in the case of the amorous analyst and patient, gratifying their senses at the cost of destroying the special interchange.

Pornographic films, though availing themselves of the protective coloration of the drama, are merely a hemstitching together of events that, spectacularly, violate the aesthetic distance. But such violation in the porno film and in the pure "stunt" film (e.g., James Bond) neither shocks nor offends. The display of the 555- number, similarly, while it would rightly wrench us out of our immersion in an Akira Kurosawa film, goes unremarked in the summer skinflick or the effects blockbuster.

Our lack of disconnection, then, may be employed diagnostically. We may say that when the aesthetic distance has been violated but we, as viewers, feel no discomfort, the presentation we are viewing is not, strictly, a drama.

The pornographic film, we note, is not a drama. For although a pretext has been stated—a male and a female astronaut are stuck in a capsule in space and endeavor to get back to Earth—neither the characters nor the actors nor the audience cares about it. Everyone knows that they're just up there to copulate and that, after the required number of copulations, an acceptable device will be used to allow the astronauts to return to Earth, and the viewers to retain a scrap of self-respect.

The malfunction of the space capsule is not the basis of a plot. It is, in this film, a pretext. Similarly, the trend in comedy, of late, is toward the nondramatic. The early *Pink Panther* films all have a plot; *It's a Mad Mad Mad Mad World* has a plot; the Ealing comedies have a plot; the late spate of summer comedies have a pretext. They are, not unlike the porno film, a loose assemblage of (in this case) humorous effects or scenes. These hemstitched entertainments are not, per se, bad or indictable. Neither are they without precedent. Their antecedent, however, is not the drama but the circus.

The circus, the vaudeville, and, indeed, performance art please through the presentation of individually complete, intellectually empty effects (tricks, turns), such that the progress, one to the next, mimics the emotional journey undergone by the listener involved in the progression of an actual drama.

The work of arranging the circus, vaudeville, or burlesque turns in the best possible order is called "routining," a most revealing term meaning "optimally ordering the arbitrary."

Routining is, rightly, prized as a showbiz skill. One wants to close the circus with the quadruple somersault, not with the farting elephant. And much of the work of current film production—that passing as both screenwriting and directing—is essentially routining.

I believe that the enjoyment of the dramatic and the nondramatic are *physiologically* different—that is, that different parts of not only the viewer's mind but of his *brain* are engaged in the two similar yet discrete experiences.

The dramatic experience is essentially *the enjoyment of the postponement of enjoyment.* The mouth waters at the prospect of a delicious meal; the palms sweat in anticipatory delight of sex. The enjoyment of the pseudodramatic entertainment has nothing to do with anticipation. It is, not only aesthetically but physiologically, akin to actual ingestion or congress.*

* Consider the difference between enjoyment and stimulation. One leaves the ballet feeling refreshed, as a promise has been fulfilled. One quits the videogame or pornographic film feeling empty and vaguely debauched—for one has

This is all well and good—as we see in the circus and pornography. And there's nothing wrong with film as compilation (of gags, effects, slapstick). That the five-gag film has virtually replaced the dramatically structured comedy is unfortunate, however. For its success feeds the historical film-industry's loathing of any presentation necessitating delayed gratification—which is to say, of drama.

---

only been stimulated. The brain, here, craves a *repetition of the stimulation*, as with any drug. One may sit in front of the television for five hours, but after *King Lear* one goes home.

BRINGING A GUN TO A KNIFE FIGHT; OR, A SHORT TOUR OF THE CONCEPT OF SUSPENSION OF DISBELIEF

It would be considered in questionable taste to demand a free dessert at the restaurant because one's aunt is dying of cancer. One commits the same solecism, however, in filming or presenting the affliction drama.

The talentless, misguided, or exploitative have long employed supradramatic devices in the construction of the drama, enlisting patriotism (see most any war movie) and right thinking (*Guess Who's Coming to Dinner,* et cetera), as these very human virtues, practiced in the wider world, are understood (subconsciously) by the viewer to trump an interest in mere entertainment.

"Do you appreciate this film, or do you hate the deaf/gays/blacks?" This is the (again, conscious or unconscious) mechanism of the issue drama.

The affliction drama *(Children of a Lesser God, The Shadow Box, Whose Life Is It Anyway?, Angels in America)* enlists the human capacity for sympathy and asks the sympathetic viewer to weep.

There's nothing wrong with a good cry (see most any film starring Bette Davis), but any claim to actual identity as a drama must rest upon the construction of a plot independent of the assignment of affliction to the protagonist.

Such a claim, like the demand for more pie, is plain bad manners.

Imagine, similarly, a candidate for office who asks for your vote because she is blind. "Yes," one might think, "that is certainly a shame, but I, the voter, pay taxes and so am entitled to representation. I will cast or withhold my vote according to my understanding of how this candidate's views reflect or inform my own; I fail to see how her affliction enters into this equation."

But we are biddable. The suspension of disbelief necessary to the dramatic transaction opens the door to its misuse by both the criminal and the well meaning.

As we enter the cinema, we relax our guard. We do so necessarily, because to resist, to insist on reality in the drama, is to rob ourselves of joy.

For who would sit through the cartoon thinking constantly, "*Wait* a second, *elephants can't fly*!"

Politicians (notably the right, in both America and Britain) have cannily understood this suspension of disbelief and have, since World War II, staged their political campaigns as *dramas,* with themes, slogans, inflammatory appeals, and villains.

This approach has put their opponents at an unfortunate disadvantage; for while the right is staging a thriller, their opponents are stuck presenting a lecture (the preferred tool of the left).

The ancient joke has a member of the majority culture taking a shortcut home through a dark alley in which he encounters a member of a despised minority. The minority fellow threatens the other with a knife, the proposed victim produces a revolver and says, "Isn't that just like a *(insert favorite racially derogatory term)* to bring a knife to a gunfight?"

Well, hijacking of the dramatic transaction is bringing a gun to a knife fight.

We are all at risk of victimization by inappropriate application of the dramatic mechanism.

Consider the Fuller brush salesman of yore. He knocked at

the door and said, "Good morning, Madam, which would you prefer today, our free brush or our free hand lotion?" The courteous and legitimately self-interested housewife opened the door to make her choice and found that, in so doing, she had made a commitment and entered into a dramatic interchange. The salesman had cleverly applied the human tropism toward friendliness, mixed it with that of greed, and his sale was made when and as the poor woman opened her door.

He had *suspended her disbelief,* circumventing her natural wariness of strangers, moving her past the trying initial encounter to a state wherein she magically (or dramatically) believed that, because of her unnamed excellences, she was to be the recipient of something for nothing.

She was flattered, bribed, and suborned through approach to her own legitimate friendliness. Game over. As were and are various happy victims of political bilge—"Recapture the dream," "It's morning in America," and my momentary favorite, "compassionate conservatism," a dramatic phrase enabling the subscriber to feel both superior and humble before God.

"Affliction drama" similarly, appeals to two of the viewer's weak spots: a desire to be politically responsible (or fashionable) and the intention to be compassionate. Overlooked in the transaction, however, is *the imaginary nature of the presentation.* The heroes, their desires, and their afflictions *are not real.* The viewer rewards himself for his compassion for a fictional victim. His compassion has cost him nothing; to the contrary, its exercise has been enjoyable—it was an entertainment.

The viewer here self-permitted an outpouring of emotion and endorsement that (he forgets) would be more problematical in the case of actual individuals.

Actual individuals are demanding, ungrateful, difficult to characterize or stereotype, combative, touchy, and easily offended (especially by unsolicited outpourings of sympathy). The actually ill display these behaviors to an increased extent.

Thus, the affliction drama turns the viewer into a sort of compassionate conservative, allowing him to believe he feels for the mass what he cannot feel for poor Aunt Sally: unlimited compassion, patience, and understanding.

GENRE

BANG-BANG

We humans love to kill. We therefore enjoy, both as fantasies and as histories, stories of murder.

We are particularly enamored of that fictive or nonfictive exploration of a "just war." What is a just war? It does not exist. War may be justifiable, but it cannot be just.

If the violence can be construed as just, our perverse entertainment is less despicable. Our enshrinement of the Greatest Generation is an attempt to co-opt what we, their descendants, perceive as their license to kill. "Just" is not a description of a war but of our enjoyment of its contemplation.

Any actual contact with violence, however, creates an abhorrence of violence.

Ex-fighters, ex-police officers, and ex-soldiers are notable for their lack of belligerence; to the contrary, those displaying arrogance or combativeness have generally never experienced or seen actual violence—their belligerence masks their fear and displays their ignorant belief that battles are somehow won by intimidation.

Violent encounters are won only by those putting themselves at risk of violence.

Though a true hero does so, the audience does not. They, thus, enjoy what they perceive as a real thrill of victory without risk.

This is the attraction of the war movie, and its pervasive influence has infected and perverted American foreign policy (vide a noncombatant president who sends young men into combat by quoting a cinematic taunt: "Bring it on"—cf. "Make my day").

The illusion of impunity has pervaded our national conflicts since Korea—as if it were possible to prevail against a foreign country without killing and being killed. The misconceived antisepsis of the Vietnam air war, of Grenada, of Iraq I and II reveals a view of impunity like that of the moviegoer.

The viewer is presented with this paradigm: The hero (i.e., you, the viewer, whom he represents) is *good*. The hero will undergo various struggles in which you, the viewer, will be able to vicariously enjoy his stoicism while undergoing no pain. Your desire to do violence will be pandered to by an incontrovertible presentation of the justice of the hero's cause and by a (ritual) period of initial restraint on his part.

This false glow of untried and (in the case of the moviegoer) proxy triumph is the drug of the bully. It seduces the weak-minded and emboldens the arrogant.

Murder has, of course, always been a staple theme of the dramatist. Its mythologic exploration is cathartic, e.g., *The Scottish Play, The Iliad, Crime and Punishment,* and, in fact, *Paths of Glory, The Ox-Bow Incident, A Place in the Sun.* These are not advertisements for, but warnings against, violence. As such they are cleansing. They artistically exhibit, they reveal and *acknowledge,* the human capacity for evil. By so doing they strip from the viewer the burden of repression.

The fiction: "I am good, I am incapable of violence, and even if I were capable of violence, I know for a fact that my cause would be just and, further, that *as* it is just, my crime would have no psychological (let alone criminal) consequences." This is the drug offered by the violent film. It represses human feelings of rage and our shame at them. It is an opiate of which increasingly larger doses must be taken for increasingly smaller effect. Its effect is anesthetic.

The O. J. Simpson show-trial was a corruptive entertainment. The probity, the careful and decent aversion of interest necessary in a civil society, were replaced with the lurid, insatiate need for retelling what the viewing audience quickly forgot was an actual human tragedy.

Reaction to the verdict split, in the United States, largely along racial lines: the whites appalled, the blacks content.

This was a political reaction: White juries had, for centuries, dismissed open-and-shut cases of white assaults on blacks. White police, defense lawyers, and prosecutors historically colluded to enforce apartheid in the criminal courts. So blacks content with an absurd verdict, the colors of the major players switched, was and is understandable. White rage and depression at the verdict were, to a large extent, compounded not by incredulity but, in fact, by an *understanding* (conscious or unconscious) of the black point of view: white society had just been mugged by the same vicious, transparent mechanism that it had immemorially employed against blacks.

Whites had been asked: "How do you like it now?" And they were forced to answer, "Not at all."

Similarly, Americans, white and black, may get all puffed out and happy at war films—the recruiting posters of the forties, *Back to Bataan;* the rather Stalinist patriotism of the cold war, *Retreat, Hell!* and *Strategic Air Command.* But notice the discrepancy between our enjoyment of this merchandising of violence and our lack of glee at the body bags of Vietnam and, now, of Iraq.

The unfortunate and inevitable concomitant of "Bring it on" is "How do you like it now?"

THE COP MOVIE

The paramilitary fantasy, dating from *Dr. No* (1962), is one of omnipotence; the superhero is unbeatable, possessed of all skills and knowledge. He may (and in the drama, of course, must) find himself in peril at stated intervals, but this peril exists only as contrast for his eventual and inevitable triumph.

The techno-thriller reduces what is essentially an automaton even further: these books (and, to a slightly lesser extent, the movies based on them) concern the primacy of *machines*.

The human protagonists of these dramas—e.g., *The Terminator, The Matrix, The Bourne Identity*—are barely human. Having no discernible feelings or desires, they are in effect machines. As the protagonist of the drama must be, in our understanding, ourself, we see here the childhood wish for certainty in its ultimate state: a wish not to feel, to be a machine.

The idea of the affectless hero (hero-as-machine) seems to date from the close of World War II, Isaac Asimov's sad robots, the *Joe Gall* books of Philip Atlee, Donald Hamilton's *Matt Helm* series, and the James Bond books. Here we see a sort of cultural autism. Western society, overloaded by the events of the preceding forty years, reimagines itself (in its hero) as uninvolved. It is not that we are incapable of emotions but that we do not *require* them.

James Bond may have sex but not love. Clint Eastwood's character name, the Man with No Name, speaks for itself.

Dirty Harry, like most of the franchise thrillers, features a protagonist who is not only autistic but also sociopathic—that is, licensed (in James Bond's case, by the state; in that of the Man with No Name, by his personality) to commit murder. These films not only license but also laud the conscienceless state. (We see their progeny in today's computer games, which are not, traditionally, games at all, but, effectively, psychological biofeedback machines, training the mind away from inherited or acquired compunction.)

The police adventure is different. It is a treatment not of the absence of emotion but of the presence of repression.

Where the paramilitary film plays out the infantile power fantasy—the infant, denied the breast, is deprived and wants to kill the offending world—the police drama plays out the rather more elaborated manipulations of the youngster confronting society.

The adolescent or preadolescent is at war not with the world but with his parents. His mutually exclusive needs for support and for freedom seem incapable of happy resolution, and this impossibility tortures him. He, like the rogue cop, withdraws or is temporarily expelled from the police department (family)—the repressive organization. He is called back for one more case and, in attempting its solution, is forced to resolve the *underlying* cause of disruption. That cause is revealed to be not his behavior, in any given instance, and not the supposed "crime" (for, psychologically, he, the hero, and we, the viewer, must know *himself* to be the perpetrator) but his *relationship to his parents*.

Put differently, the hero cop—the adolescent—is called back not to find a murderer but to acknowledge the final necessity of a break with the juvenile figures of authority.

His feelings are eventually validated as he finds, inevitably, that the real crime is not that perpetuated by the criminal but the *corruption of his superiors,* who have colluded in its commission or in its cover-up (e.g., *Bullitt, The Detective, Training Day*). Only by discovering their corruption (that is, their frail humanity) can he cease fighting with them and so depart—which is to say, mature—in peace.

The last reel of the police drama clarifies the problem: it is not that evil exists but that the hero/viewer has an insufficiently developed mechanism for dealing with it. The world will not be *cleansed* by the hero's triumph, but he, himself, will be free. Of what? Of the constant necessity of conscious service to the repressive mechanism.

Vide: at the end of most canonical police dramas, the hero leaves his organization (e.g., *Serpico, Prince of the City, Three Days of the Condor, Spy Game*). The form of these autonomically created dramas of renunciation springs full blown from our unconscious—as does the phrase "What seems to be the trouble, Officer?"

In these, the hero, like Oedipus, called to discover the cause of the city's unrest, finds it in the astounding but inevitable place. In Oedipus it is himself (a tragedy); in the police drama it is his revered superiors (a drama). Having found it, the hero is free to retire from his vocation—in grief or sorrow—to an independent life.

Stanislavsky wrote that drama stands in the same relationship to melodrama as tragedy does to comedy—that is, that tragedy is not heightened drama but heightened comedy. Here, in the perhaps special case of the cop movie, we have drama that might be tragedy manqué. In the denouement the burden of repression is lifted (as in comedy) but only at the cost of its replacement by sorrow.

FILM NOIR AND HE-MEN

I am just returned from San Francisco.

As a besotted movie lover, I cannot, of course, drive those streets without thinking of the car chase in *Bullitt*.

I meditate on the direction in this sequence and the beautiful, protracted airport chase that closes the film.

This is tough, concise, and, if I may, butch moviemaking, and I love it.

I think of Peter Yates's other crime films, *Robbery* and *The Friends of Eddie Coyle*. Two smashing films noirs—all bad guys fighting for the swag, for their lives, in a world without rules—blunt, perfect filmmaking.

My reverie shifts to another of my favorite truly tough films, *Point Blank,* and thence to another of John Boorman's films, *Deliverance*. These films are *dark*. And, it occurs to me, they're all directed by Brits.

Is the British sensibility, I wonder, more suited to the production of film noir?

Film noir is the conjunction of violence and irony, and we Americans don't do irony very well. We are a straightforward and self-righteous people, so we are rather good at viciousness and humor but lacking in irony.

Americans Stanley Kubrick, Cy Endfield, and Jules Dassin turned out some smashingly ironic stuff *(The Killing, Dr.*

Strangelove, Zulu, Night and the City, Rififi), but they, of course, ended up—by choice or as fugitives from American fury (the House Un-American Activities Committee)—as expats in Britain. Perhaps being bombed at regular intervals throughout the twentieth century has given the British a different slant on the entertainment quotient of violence.

British war films—*In Which We Serve, The Cruel Sea, One of Our Aircraft Is Missing,* indeed, Cy Endfield's magnificent *Zulu*—stress unity rather than, as in American films, a confected competition *between* comrades under arms.

Most American World War II films feature interunit antagonisms: the drill instructor vs. the recruit *(Sands of Iwo Jima),* the commander vs. the passed-over subordinate *(Run Silent, Run Deep),* the boss vs. the flyers *(Twelve O'Clock High),* two guys battling for the same girl *(Bombardier).*

This plot can also be found in *They Gave Him a Gun; The D.I.; Retreat, Hell!; Sergeant York;* and so on. It is, schematically, the essence of most American war films. The prize attendant upon the conflict's eventual resolution is (for the protagonists and the audience) the license to fight the actual enemy—in effect, to engage in violence. (Cf. American foreign policy, wherein, at this writing, the combat-reluctant are branded by the administration as traitorous—or, at best, as misguided—and the reward for the individual's or the country's overcoming its abhorrence or reluctance to battle is a "good clean" fight.)

The cleansing (if false) reduction of the American war film is "conquer we must, if our cause is just" and that of our gangster film is "crime does not pay"—two equally debatable propositions.

Your British gangster films are more involved with how-to and, to this amateur of the ironic, therefore rather more enjoyable. They deal not with misguided souls but with actually-not-very-nice people.

This is a great film tradition, and I cite *The Blue Lamp; Robbery; Peeping Tom; The Long Good Friday; Sexy Beast; Mona Lisa; The Krays; Lock, Stock and Two Smoking Barrels; Snatch; Croupier;* and Mike Hodges's *I'll Sleep When I'm Dead.*

This last film is notable for its almost complete absence of narration—a writer's dream and a moviegoer's delight. For the absence of narration leaves only the *narrative*. We watch in order to discover who the folk are, what might be their relationship, what they want, and how they are going to go about getting it.

I particularly recommend the film's protagonist, Clive Owen, in his personification of enigma.

Speaking of casting, *Circle of Danger* (1951) is a British postwar drama. In it, Ray Milland's brother has died as a member of a Special Boat Service wartime assault. Milland cannot make sense of the reports of the circumstances of his brother's death. He suspects foul play and goes to England to investigate. He finds, one by one, the surviving members of his brother's squad.

All clues indicate that the solution to the mystery lies in one Sholto Douglas, the leader of the commando squad and, as per the interviewed squad members, "the bravest man who ever lived."

Milland finally finds an address for Douglas. He rings, and the door is opened by Marius Goring. Mr. Goring is togged out in ballet gear. He opens the door in tights, a cashmere sweater jauntily tied around his neck, his feet in a very good fourth position. In the background we see the dance rehearsal that Milland's ring has interrupted.

Mr. Goring gives Milland something of a moue and says, "Yes?"

Milland: "I'm looking for Sholto Douglas."

Goring: "I'm Sholto Douglas. Bad casting, eh?"

I love the British cinema. My idea of perfection is Roger Livesey (my favorite actor) in *The Life and Death of Colonel Blimp* (my favorite film) about to fight Anton Walbrook (my other favorite actor). In the great dueling scene between the two, Livesey is asked by the referee if he is skilled in the saber. He replies, "I think I know which end to hold."

Other examples of British restraint: I give you Celia Johnson in *In Which We Serve* making a speech to the force that robs her and her friends of their husbands, their ship ("I give you the *Torrant,*" she says, simply—no "show," just perfection); Elsa Lanchester in *Rembrandt,* facing down the bailiffs ("And *that's* the *law* in *Holland* . . . "); Alec Guinness going mad in *Tunes of Glory;* Stanley Holloway in the train station in *Brief Encounter;* the slavey in *Cavalcade* ("I know where Africa is on a *map;* where is it *really* . . . ?")

In George MacDonald Fraser's magnificent war memoir, *Quartered Safe Out Here,* he writes of an infantry assault across open country. He looks at his platoon advancing determinedly into enemy fire and says his only thought was ". . . Englishmen."

Or, in the terms of my particular métier: hold the emotion, thanks; we understand.

The attitude has been tagged as stoicism, but perhaps it's just professionalism—why not let the *audience* have the experience?

Perhaps those of us who live surrounded by emotion—doctors, police, lawyers, dramatists—are not much moved by the emotional. I know I'm not. I prefer the clean statement to the plea, the film noir to the gangster film. And, if you will grant me that segue, I will continue.

The gangster film is a sentimental look at the world of crime. There are, we are told, *good* bad criminals; there is a code of honor; there is either justice or accountability. This is a film as written by a criminal, which is to say, a sentimental, self-servicing, pleasant lie. The film noir, on the other hand, depicts a cold Darwinian zero-sum world, a world without rules and without judgment. A film, if you will, written by a cop.*

I prefer the film noir.

The French, too, have, of course, had their innings, with *Bob le Flambeur, Rififi, Wages of Fear, Daybreak* (now, there's a film that just ain't kidding); and permit me to name our American *Gun Crazy* (the original) and various other black-and-white poverty row (the historical precursor of the independent film) works: *Detour, The Narrow Margin, Dillinger, The Rise and Fall of Legs Diamond, Plunder Road, Quicksand, Kiss the Blood Off My Hands,* and *T-Men.*

The American film noir grew out of postwar despair and a lack of funding on the part of the despairing. These films featured extended sequences shot on the side of a desolate road—no sets, little lighting, just great acting and a great script.†

Aristotle cautions that it is insufficient for the hero to get the idea. Many modern moviemakers, however, act as if they hadn't read his book. Their films depict the gentle progress of the protagonist toward self-actualization—usually depicted as a slow, arms-extended twirling on a beach (as if the expression of a racial memory of our descent from the shipworm).

Not the film noir. Vide *Point Blank,* where we have Lee Marvin, at the jump, robbed, shot, and left for dead on a deserted island. This fellow doesn't want self-actualization, he wants blood. And Sterling Hayden in *The Killing* wants *the money.*

The Killing is, I believe, the world's greatest film noir. The traditional subgenre here is "one last job."

* There is a long close affinity between writers and cops; we share the sad knowledge that everyone is always lying.

† A traditional recipe for genius: inspiration, a plan, not enough time.

Kubrick's team for the one last job is the greatest compilation of tough talent known to man: Ted de Corsia as the crooked cop; Elisha Cook Jr. in his perennial (and perennially brilliant) turn as the weak link; Joe Sawyer as the bartender with the ailing wife; Kola Kwariani as the ex–wrestling champion of the world; Timothy Carey as the shooter; Marie Windsor (cf. *The Narrow Margin*) as the loose-wheel bad girl who takes down the job; Jay C. Flippen as the bankroll, in love, into the bargain, with Sterling Hayden; Vince Edwards as Marie's killer boyfriend.

None of these people would tug at a heartstring to save his life. Or, to put it differently, they can be trusted.

I got to make a heist film with Gene Hackman. Like many of the stars in the above-instanced works, he is an actual tough guy. (Lee Marvin was a marine commando in the Pacific, Hayden in the Adriatic, et cetera.) Hackman was a China marine, race-car driver, stunt pilot, deep-sea diver.

These men, and their performances, are characterized by the *absence of the desire to please*. On screen, they don't have anything to prove, and so we are *extraordinarily* drawn to them.

They are not "sensitive"; they are not antiheroes; they are, to use a historic term, "he-men." How refreshing.

There will always be the same number of movie stars.

There is a table of operations, and the places must be filled, as with politicians, irrespective of the distinction of the applicant pool.

But I vote for the tone of a less sentimental time.

Look at the photographs in the family collection, of Dad or Grandad during the war or the Depression. We see individuals captured in an actual moment of their lives, not portraying themselves for the camera. I used to look at them and think one didn't see those faces today. We saw them—briefly—on September 11.

SHADOW OF A DOUBT

I should like to add to my film noir and he-men screed *Shadow of a Doubt* by that great American Alfred Hitchcock.

This, I believe, is Hitchcock's finest film. At the risk of sounding like a film student, I will refer to the construction of the shots, than which one could find no better exemplar in all of film. We have no stone lions rearing up, nor that baby carriage found in *Battleship Potemkin* (in a prescient adumbrage to Brian De Palma's 1987 *The Untouchables*). Arguably, the feat in *Shadow* is even more praiseworthy than that of Eisenstein—for Hitchcock is shooting only that to which he referred derisively as mere "pictures of people talking."

The movie is a sick and magnificent treatment of sexual abuse. It is coded as a story of a killer returned to his sister's home.

Charles, the villain (Joe Cotten), woos, marries, and kills rich widows. He comes back to Santa Rosa, posing as a retired industrialist, to live with his sis and her two daughters. The older, Charlotte, is called Charlie in honor of her uncle.

The two Charlies have a secret. She discovers that he is not what he seems, and her threats to tell are countered by his threats, first to discredit her, then to allow her to tell and so "break her mother's heart."

He proceeds to try to kill her by various stratagems, and in the last twenty seconds, she turns the tables on him, and he falls to his death in front of the oncoming train that was to've brought her demise.

Throughout the film, Cotten is standing too close and speaking too syrupily to the nubile, inevitable Teresa Wright, America's good girl.

She is first honored, then charmed, and, then, little by little, gets hep to the jive. And Hitchcock designs each sequence magnificently. There is no "master, over, close-up" about it. Each sequence is designed around its particular theme and purpose in the unfolding story. One could easily label them, e.g., alarm, suspicion, second thoughts, challenge, remorse. It may be the world's best silent film, undiminished even by the addition of dialogue.

Why are silent films potentially better?

The perfect film is the silent film, just as the perfect sequence is the silent sequence. Dialogue is inferior to picture in telling a film story. A picture, first, as we know, is worth a thousand words; the juxtaposition of pictures is geometrically more effective. If a director or writer wants to find out if a scene works, he may remove the dialogue and see if he can still communicate the idea to the audience.

Ancient theological wisdom put it thus: "Preach Christ constantly—use words if you must."

RELIGIOUS FILMS

Religious films have as much of a chance of increasing humane behavior as *Porgy and Bess* had of ending segregation. Religious films have but two subcategories: sappy and exploitative.

In the first, good things happen to good people. The heroes in this film have gone a little bit astray: they are dismissive to their inferiors or brusque with their children. An angel shows up, and all is put aright, cf. *A Christmas Carol, It's a Wonderful Life, The Bishop's Wife, Here Comes Mr. Jordan, Miracle on 34th Street.*

I list the above as religious films, as each is driven by the engine of belief in human benignity. We leave the theater thinking, "Well, I guess maybe we humans aren't such a bad lot after all." These films employ variations on the Santa Claus myth of godhood, that is, they borrow a biblical model of apotheosis, cleanse it of terror and awe, and present it as light-hearted entertainment.

The second subcategory is the straight-up or out-of-the-closet religious film, those endeavoring to depict mythical, historical, or doctrinal aspects of a specific religion—*The Robe, Ben-Hur, The Ten Commandments, The Last Temptation of Christ, The Passion of the Christ*—attempting to awaken or reengage the enthusiasm of the faithful.

These films professing fealty to received religion are, most literally, preaching to the choir, the reductio ad absurdum of "based on a true story."

But "based on a true story" is a come-on, the aesthetic equivalent of "no loan request refused."

For, at best, the creator has fashioned a film based on his understanding, interpretation, and reduction of the report of an actual occurrence. But just as it is impossible to utter a statement without inflection, it is impossible to make a film that is free of interpretation.

To claim artistic impunity, therefore, because of the sanctity of the source material is, perhaps, disingenuous.

The audience does not care if the film is based on a true story, only if it *is* a true story.

To invoke the sacrosanct is also the best trick of our friends the politicians: "Would you want your sister to marry one?" (defense of segregation); "Marriage is the backbone of civilization" (opposition to gay marriage).

In the heat of the theatrical moment, whether in the theater or the convention center, we can become enraptured to the point where we forget that our sister is going to marry whomever she wishes. And that our uncle Fred, our mother, or, indeed, we ourselves, may be as gay as possible, and that civilization nonetheless seems to continue stumbling along.

The power of iconography is such that it can endorse, and in fact exult, psychotic savagery *(Triumph of the Will)* or, devoid of content, convince an audience into thinking they have actually seen a film *about* something *(Forrest Gump)*.

Neophyte, incompetent, or bored filmmakers *show* those things they cannot dramatize, e.g., a slow pan over the heroine's desk: "Look, here are photos of her kayaking with her children, a copy of *The Theory of the Leisure Class,* a bronzed hockey skate, and a cased butterfly. She must be an interesting person."

The dramatic biographer is given a set of facts; but he, no less than the "documentary" recorder of the desktop, is still charged with making a choice. And whether charged or not, *is* making a choice.

Just as the cataloguer must decide if the kayaking photo or the hockey skate is seen first, is better lit, is closer to the camera, or "prettier," so the biographer must not only choose the facts to depict but also order and interpret them, weighing the reliability, honesty, authenticity, and possible bias not only of himself but of the original reporter(s).

This task is somewhat complicated when the source material is, to the filmmaker, holy writ.

The job of the viewer, here, is also more complicated: To what extent does civility require a nonbeliever to withhold judgment of a statement of faith not his own? To what extent must an adherent shelve aesthetic considerations in favor of an endorsement of his faith? And what if, God forbid, a religious film were to contain material actually derogatory to another faith and/or race? Would it be sufficient for the filmmaker to point to literature sacred to him but offensive to others and shrug, "What can I do? It's holy writ"?

It is written that whoever grasps the Torah without its covering will die, which the sages interpret to mean that even, and perhaps especially, holy writ may be destructive if the individual does not recognize his responsibility for interpretation: Christian tradition renders this as "The letter kills but the spirit frees."

For, just as with the placement of the hockey skates, our human mind is incapable of uncolored perception, let alone transmission. Interpretation, in the artist and in the viewer, is always and inevitably taking place, and the more the creator is aware of this, the better able both he and the viewer will be to seek out the *essential* truth of a story.

On which note, America, ever the entertainer, is currently being roiled and churned about the handy topic of gay marriage. A Massachusetts Democrat, the first openly gay member of Congress, was speaking to his constituents in a fishing town on the coast. Newly out, he was braving the puritan wrath of those who had elected him. One outraged soul asked if he was unaware that the Bible said, *explicitly,* that it was an abomination for one man to lie with another. The congressman asked if the constituent was unaware that the Bible used the same language about the eating of shellfish.

Movies possess unlimited power to entertain. They have, however, no power whatever to teach. The audience lends its attention only for the purpose of entertainment and will deny (consciously or unconsciously) its attention to any other purpose. (The child wants his bedtime story—it is an impertinence to use it as a lecture.)

But there is little in life more entertaining than self-righteousness and self-affirmation; occasionally, therefore, we find a film claiming to advance or depict a doctrine that some may find incontestable, indeed, holy, but that to others is pernicious.

Adherents, whether they know it or not, go to see these films *to be entertained*. (The correct venue for religious devotion is the holy place, not the cinema; and communion with the divine, whether the kiddush of the Jews or the Communion of the Catholics, is better celebrated with the traditional bread and wine than with popcorn and Coca-Cola.)*

Did *Taxi Driver* inspire John Hinckley Jr. to attempt assassination?

* The power of the doctrinal film consists in the ability of the doctrine in this venue and depiction to reaffirm or, unfortunately, to enrage—both entertaining emotions and inappropriate to the confines of the cinema.

We know that *The Birth of a Nation,* in its bold statement of the "unfortunate truth" that Negroes are inferior, helped endorse the rebirth of the Ku Klux Klan, and that *Triumph of the Will* helped to enthuse the Nazis.

It is, I think, the responsibility of filmmakers and distributors to assess the potential impact of their wares on those who might just mistake entertainment for exegesis, and for license.

A helpful counterexample is Joseph Goebbels's commendation of the SS: that history will record that they were capable of doing these terrible things without losing their essential humanity.

THE SEQUEL

My eight-year-old wanted to go see the sequel, so we schlepped out of the house and over to the movie theater.

The original had by no means been bad. It was intermittently funny and generally good-natured. The sequel, however, was an obscenity. How, I wondered, was it possible to make a film that bad?*

If a comedy has one hundred gags in it, one might predict that, say, three of them might be raucous, ten more pretty funny, thirty on top of that somewhat diverting, and all at the very least recognizable as an attempt to amuse.

But this sequel, this part two of a tremendously successful comedy, was not funny at all.

Not only were there no laughs, there were no premises identifiable as *intended to produce laughs*.

* I was reminded of the studies in parapsychology conducted at Duke University some years back. These famous Rhine experiments made bold to plumb the hidden powers of the mind. The paired subjects were given various flash cards blazoned with a star, a crescent, a box, a circle, then separated and asked to concentrate on the identity of the card their opposite number held. Professor Rhine found no positive correlation between the cards and the guesses. That is, the diviner could not guess correctly a percentage of times greater than predictable as random.

But wait. On review, Rhine found several instances where the *wrong* guesses far surpassed the predictable norm.

(That is, given five icon cards, a random selection must be correct approxi-

I was put in mind of contemporary American political style, in which whether or not the will of the people is subverted, it is not even mollified.

As voters, and as cinemagoers, we have agreed, as it were, to be wooed. We know we deserve to be met at our flat, perhaps given a corsage, before we step out.

But here is a concupiscent lout at our doorstep, his member in his hand, and he proclaims, "Later on, perhaps I'll buy you a sandwich. Now hike up your skirts."

"But," I say, "*but* were we, the despoiled populace, not historically entitled to a little bit of nicety, a bit of circus with our stale bread, a fine turn of phrase, at the very least?"

How about, at least, a mouthable slogan—"Remember the *Maine,*" or "Fifty-four Forty or Fight," or "No Taxation without Representation"?

Sure, the practice of the big lie has distinguished provenance (cf. the Crusaders' cry, "Hep," an acryonym meaning *Jerusalem est perdita;* "Remember the Belgian Orphans"; and the historical romance *Wag the Dog,* to name but a few). But we ship our sons off to kill or be killed for reasons that are incapable of explanation, and we do not receive in return even an acceptable bumper sticker.

It need not even be rousing—it need but be identifiable as an *honest attempt to incite* (cf., again, "Gosh, you look pretty tonight"). The swain wants to get his leg over. The oligarchy wants to rob everyone blind. Of course they do, and of course they *shall.* But wouldn't a concern for simple good manners suggest they proceed in an approved and respectful fashion?

And the motion picture megaliths, having earned more money than they could have foreseen with the original, begin

---

mately 20 percent of the time. The greater the number of attempts, the more closely the correct guesses must approach 20 percent. If one guesses one thousand times and is correct only 10 percent of the time, some force other than the operations of chance must be at work.)

eating their own entrails in a frenzy to earn all the rest of the money with the sequel. And they collude and scheme and test and confab to make sure that *each* moment of the film is recognizable as that moment that should take place at that time, in a sequel to a film whose franchise is so important that *nothing* must be risked that might endanger its success.

Professor Rhine might watch the film, as did I, in wonder. It contains not one moment of jollity, humor, or respect for the audience that paid for their inclusion. How hard those executive, associate, cosupervising, and coexecutive producers must have labored to create a product bearing no trace whatever of the human.

They, those producers, had made themselves pure and cleansed their work of the accidental, the frivolous, the whimsical. Those qualities, in film one, had made them rich, and now those qualities, like the proverbial first wife, have gone to the wall. And like the first wife, they will not return.

Jewish law states that there are certain crimes that cannot be forgiven, as they cannot be undone. It lists murder and adultery. I add this film.

That same night my wife and daughter and I watched *I Know Where I'm Going!* This film by Powell and Pressburger (1945) is one of the world's great love stories and my wife's favorite film.

A young woman from the London beau monde is en route to a Scottish isle to marry sir somebody or other, a wealthy, et cetera. A storm strands her on Mull, some few miles' sail from her destination. She meets MacNeil, poor Lord of Killoran (her intended isle); they fall in love. They are kept apart by her indecision. They eventually brave the storm to sail to Killoran, are almost drowned, and return to Mull, where they are, thank God, united for all time.*

* Obiter dicta: (1) Pride goeth before a fall. (2) MacNeil, Lord of Killoran, was originally to be played by Laurence Olivier. He was replaced, at the last

"For all time," in this case, having lasted every moment of the more than sixty years viewing by delighted moviegoers. Pressburger, an Eastern European Jew, creates the perfect fantasy of Scotland, as Warner, Fox, Laemmle, Mayer, and Goldwyn, European Jews, created the fantasy of America. But this perhaps exceeds the brief of this essay.

---

moment, by Roger Livesey. Livesey was performing onstage during the shooting of *I Know Where I'm Going!* He commuted daily from London to Shepperton Studios, and the whole of his performance, supposedly shot on the Isle of Mull, is either on a studio set or against rear-screen projections. The long shots, taken in Scotland, feature a photo double. (3) The Rhine experiments at Duke were largely discredited when it was found that two of the prime subjects were confidence people who were cheating like mad.

PASSING JUDGMENT

REVERENCE AS OPPOSED TO LOVE

One of the great cinematic delights of the sixties was the animated short *Bambi Meets Godzilla.*

For those who have not been blessed, a young, scampering Bambi crests a hill and looks winningly to the right and left. He raises his ears, sensing danger. A huge billion-ton monster, Godzilla, comes over the hill and stomps Bambi into preserves. Roll credits.

Here is my own B vs. G story.

I was at a show business party the other day, and a friend expressed a rather egregious opinion. We all, being herd creatures, recoiled; she shrugged and said that she had grown old and one of the chief delights of aging was the ability to say whatever one thought. This was the second time in two days I had heard the phrase. The French, onetime ally of America, say, "jamais deux sans trois," so I must assume that this phrase, "age and express yourself," has been in the air around me for some time and that I have just become aware of its presence.

Why would that be? I wonder, and the answer swims to the forefront of my consciousness. I have grown old.

Having grown old, I will search for prerogatives and exercise them.

I need not believe the drivel that is spoken around me—

I feel lighter already—such wistful submission has only ever earned me increased grief, and I am free to *speak my own.* I can say whatever I want, as per the sibyl of the party. I need not tell the transparent lie to avoid the wretched dinner and so on.

There are, of course, limits. The Constitution of the United States, that lovely document, draws the line at advocating violent overthrow of the government, and the usage of the British Isles has an unwritten caveat barring criticism of Laurence Olivier.

But I just can't take it anymore, and I will, like Ayn Rand's Atlas, shrug the now intolerable burden.

I can't stand Laurence Olivier's acting. He is stiff, self-conscious, grudging, coy, and ungenerous. In *Khartoum,* the whole world, Arab and Christian, refers to *his character* as the Mahdi, while he, covered in chocolate and with a false nose the size of Dorsetshire, refers to himself gutterally, as the "MACCH-di," as if to correct the pronunciation of those, fedayeen or thespian, one would have supposed to have been his colleagues.

In *That Hamilton Woman,* he whispers and turns his face from the camera throughout; in *49th Parallel,* who knows *what* the deuce he is doing, other than turning in what I believe to be the only truly bad performance in any Powell and Pressburger film. He has a moment, as Hurstwood in *Carrie,* when Dreiser's safe swings closed, and is not bad in the musician scenes as Archie Rice in *The Entertainer.* But, in general, I'm hungry for lunch, and all he's serving is an illustrated menu.

This is not to detract from his status as the world's greatest actor. He won that position fairly, kept it honorably, and contributed to the British, and to the world, theater. And those who saw him onstage speak of him with reverence. But the good he did, I say, aside: We speak of the art and artists who move us not with reverence but with love. And I cannot love Olivier's performances.

Who, then, will I class against him? What shibboleth, you wonder, will I list to augment your umbrage?

Here I pause and imagine that you have beaten me to the punch and have correctly deduced that I am about to report my love for Tony Curtis.

You may snort with contempt and recall his Brooklynese, "Yonda stands da castle of my faddah." But I will name, in support, but two of his performances. To the mention of the first you will all smile with love; those who know the second will nod sagely in agreement:

Some Like It Hot and *The Boston Strangler.*

The first is the *perfect* comic turn. As the first beleaguered, then besotted (with Marilyn Monroe) saxophone player, he is, as we say, "as clean as a hound's tooth." He plays low comedy high as it gets, and it would have been enough. Then, in the third reel, he has to throw on drag. He joins an all-girl band to escape the wrath of Al Capone. And he does the travesty not only as well as it can be done but better than anyone has ever seen it. He plays a girl for keeps. He believes it, and we believe it. Then, dear reader, and I know you are nodding along with the report, he transforms himself into a millionaire in yachting costume and does the world's best imitation of Cary Grant.

This is a performance one wishes to hug to one's chest. It is the perfection of comic acting—idiosyncratic, loving, involved, and perfectly true.

Now we see him as *The Boston Strangler.* He plays Albert DeSalvo, the murderer. The camera follows him through various quite grisly stalkings and killings. We are shocked at the seeming reason of his motivation. These acts make perfect sense to the actor—and so we see not a monster but the human capacity (yours and mine) for monstrousness. Now DeSalvo is apprehended. A psychiatrist takes him through the crimes, of which, we discover, he was unaware. He does not remember them. And in the interrogation sessions we see DeSalvo, that is, Tony Curtis, recall, little by little, the grisly murders, and we see him, before our eyes, *disintegrate*.

These are some of the greatest moments of film acting.

We do not laud and revere Mr. Curtis's "great technique"; we merely remember the moments of his performance our entire lives.

Mike Nichols told me long ago that there is no such thing as a career—that if a person has done five great things over three decades of work she is indeed blessed. I will mention Mr. Curtis in *Sweet Smell of Success,* and as the escaping convict chained to Sidney Poitier in *The Defiant Ones,* and in *Trapeze,* and that is five. Let me also cite Robert Duvall in *Apocalypse Now,* who loved the smell of napalm in the morning; Frank Morgan, who asked Dorothy not to look behind the curtain in *The Wizard of Oz;* Hattie McDaniel in *Gone with the Wind,* who instructed us that "id just ain't fittin' "; Paul Muni, whose delivery of the tagline to *I Am a Fugitive from a Chain Gang* stopped hearts (Q. How do you live? A. I steal); Brando in *The Godfather;* Welles in *Citizen Kane;* Shelley Winters's dejection in *A Place in the Sun.* These are performances that make us smile sadly or grin. These are the work of *truly* great actors, great actors with a small *g,* of whom we remember, similarly, the noun and not the adjective.

GREAT AND ROTTEN ACTING

What an odd thing is preference. I have a preference for quiet.

There is a wonderful old weeper called *Penny Serenade* (1941). Here we have Irene Dunne and Cary Grant. Their little girl has died in the great Tokyo earthquake of nineteen twenty something, and they are, of course, bereft. They are awarded provisional custody of a young orphan and raise her for four years. Cary then loses his job—it is the Depression—and the orphanage informs him that he is therefore likely to lose custody of his daughter.

He goes to the judge and pleads. Now, *pleading* is, in my experience, the hardest thing for any actor to do. It involves, onstage or off, complete self-abasement, and (again, whether in life or onstage) it is *very* painful. Most actors, asked to plead, will counterfeit the act. This is called "indicating" and means creation of a recognizable rendition of the emotion supposedly required by the script.* Cary Grant, in a magnificent piece of acting, actually pleads. He bares his soul before the judge who holds the fate of his daughter in his hands.

The performance, however, that I count as ethereal, is the one occurring *behind* him. Beulah Bondi, playing the head of

* John Barrymore flinches to indicate surprise; Bela Lugosi narrows his eyes to indicate malevolence; Danny Kaye smiles to indicate charming harmlessness.

the orphanage, has, through the film, championed the cause of Cary and Irene. She has told them that the chances of the judge awarding the little girl to a family with no income are nil. She accompanies Cary to the chambers and sits, far off in the background, to watch the proceedings.

We know she is disposed toward the supplicant. We see that she has no wish to influence the judge. We *understand* that she feels that any emotion, utterance, any comment whatever would be detrimental to the case of the pleader and, *further,* that she believes in the system as constituted. She has intervened to what she considers the limit of the acceptable, and though it is painful, she will now withhold herself from the necessary operation of the court.

Beulah Bondi accomplishes all this through sitting and watching.*

Or look at Celia Johnson in *In Which We Serve* (1942). She plays the wife of Captain Kinross (Noël Coward). In a party scene she toasts a young woman, newly betrothed to a naval officer. She recites the hardships the young woman has in store and concludes the toast in an encomium to the officer's ship. The speech, and its Ciceronian conclusion, are delivered completely without sentimentality. They are not emotionless; to the contrary, they are filled with the truth of emotion withheld.

In aid of what has the emotion been withheld? In favor of the truth.

Like Beulah Bondi, Celia Johnson will not sully the moment, she will not patronize the other player (nor, thus, the audience) with the *performance* of emotionality.

* Beula Bondi played Jimmy Stewart's mother in four films. Late in life she retired. When she was in her nineties, Stewart reached out for her to come back and act. A search revealed that she had gone on vacation in the Sierra Nevadas. The search team located her camped out by her Airstream. "We need you to come back," they said, "to play Jimmy's mother." "Yep," she replied, packing up, "I'm his mother."

Look later in the film. Here is Noël Coward, as fine an actor as anyone could hope to see. As Captain Kinross he is bidding adieu to his shipmates. They have served together, been torpedoed together, spent days in the wreckage and in the boats of their ship, and now the company is being disbanded and dispersed.

They are in an empty warehouse. The company is at attention. He says, "Gather round," and the men all step in, and he speaks to them for a few moments. He tells them what an honor it has been to serve with them, that there is not one of them with whom he would not serve again, and then shakes their hands. The ship's company lines up, and Noël Coward says good-bye to each of the forty men. The camera is stationary, the line moves forward, and we see, in each of his farewells, the nature of his relationship with that man.*

See also Ward Bond in *Gentleman Jim* (1942). It's a boxing movie about Gentleman James J. Corbett (played by Errol Flynn) and his fight for the heavyweight title (1892) against the icon John L. Sullivan (played by Bond).

Flynn was, for my money, a rather good actor. It's easy to be taken in by his perfect looks, but film by film, scene by scene, he was always simple, always truthful, and always generous. Here, he's just conquered the unconquerable Sullivan and is making merry at the postfight bun-fight. But let's look at Bond. The room grows quiet as he enters. He walks up to Flynn and, in the most authentically humbled and humble performance, gives the new champ the title belt and wishes him well. (The scene is shot in two overs: over Bond onto Flynn, over Flynn onto Bond. But the performance is so simple and striking, the director and editor played the whole two-minute speech, unbroken, on Bond.)

* See also the congruent opening scene of *Dodsworth,* where Walter Huston, resigning from the automobile company he created twenty years ago, takes his leave, walking through the crowd of workers with whom he has built the company.

I also recommend that you see Sidney Lumet's *Fail-Safe* (1964).

Henry Fonda is the president of the United States. Missiles have been sent in error against the Soviet Russians, and Fonda retires to a bomb shelter to speak on the hotline with the Soviet premier. Fonda is accompanied by his interpreter, played by Larry Hagman. The scene is just these two men in a bare white room. Hagman is talking to the Russians and translating for Fonda. The fate of the world depends on his translation. Fonda tells him to take it easy, listen carefully, take his time. He does. In this remarkable scene, it is nearly impossible to realize that they are but two actors in an imaginary situation in a set made of four white flats.

I am reminded of Ruth Draper.

For those unacquainted with her work, let me offer you a treat. Miss Draper wrote and performed one-woman dramas from the thirties through the fifties. Though they are monologues, it is insufficient, in fact, misleading, to identify them solely as such. They are true and complete dramas. She was, in my opinion, one of the great dramatists of the twentieth century.

One of her pieces is entitled "A Scottish Immigrant at Ellis Island." The group, of which she has been a part, is in line at Ellis Island, waiting for their interviews with American immigration. The young Scottish girl (played on the recording by Miss Draper, in her seventies) chats with the friends she has made. She goes on about her new life, the fiancé who is awaiting her, and she invites her friends to come visit them, after their marriage, in New Jersey. She exchanges about five lines with a young man. She tells him that she is very glad to have met him. To his reply (unheard by us), she responds that she will not forget him either, and she says good-bye.

We understand, in these five lines of hers, that the young

man has fallen irrevocably in love with her and that he will for the rest of life regret their parting, that he may eventually marry, but he will never marry his love, the young girl he met on shipboard.

We further understand that the Scottish girl herself understands this but that she will not intrude; she will not sully the young man's good-bye with pity or sympathy; she will respond to his courage with courage.

All this in a few words spoken by an actress to a person who exists only in her imagination—and now in ours.

I recommend to you Roger Livesey in just about anything. He was, to me, the British Henry Fonda—the perfect actor, incapable of falsity. He and Anton Walbrook portray a British and a German officer from the Edwardian period through the Blitz. The film, *The Life and Death of Colonel Blimp* (1943) by Powell and Pressburger, is my favorite film. Livesey, as a young man, is dispatched on a diplomatic mission to Berlin. He insults a group of Junkers and must fight a duel. His opponent, appointed by the German army, is Walbrook. They meet for the first time in a dueling academy. As the Swedish judge instructs them in the code of the duel, they exchange looks with each other. We see that each assesses the other and, having found his opponent worthy—indeed, estimable—apologizes for the necessity of savagery and regrets that personal feelings must be subordinated to duty. All in two—silent—shots.

We all know what falsity looks like. How exhilarating to see truth.

GOOD IN THE ROOM: AUDITIONS AND THE FALLACY OF TESTING

What makes some actors, some acting, and some performances false? Some of it is due to the character of the individual, some to lack of commitment, or perhaps of talent, some to the choices of the director, and much to the audition process.

I hate the audition process. Having experienced it as an actor, I found it demeaning. As a writer and director, I find it damn near useless.

The question, for the director, should not be "Can they act the part?" but "Can they *act*?" The best way to determine the applicant's skills is to watch his work.

If the applicant actually has something to offer—as evidenced in an actual performance—it should then be up to the director to decide whether the actor's skills, personality, and sensibilities make him a good choice for the particular part being cast.

Why not just give him the script and "test" him on it (in effect, the current system)? Well, because the ability to act is not always paired with the ability to audition.

In the audition room, the actor is a supplicant. Othello says, "Take me, take a soldier; take a soldier, take a King." There we have not only the actor's fantasy but also his understanding of his life and his life's work: He is allowed, encouraged, and, if gifted, driven to cast himself in various enjoyable, demanding roles and situations. These situations may not be noble, *but the work, and the joy of exploring them, is.*

The actor's ability to immerse himself freely, spontaneously, and generously is applauded, rightly, by the audience, who came to the play to be taken out of themselves. The more free the actor plays, the greater the enjoyment the audience derives.

But those across the table in the audition room are not an audience but a jury. They come not to a place of enjoyment but of drudgery.

Why? Because they spend the day disappointing people. It is taxing, constantly, to remember that though the mass of applicants is daunting, each individual is deserving of respect. The juror's attention may very well fray, and rather than indict himself, he may transfer his irritation to the next applicant. And the next. And the next. The actor, then, confronts not only a jury but also a short-tempered one.

The jurors gaze down what now seems to be a growing list of applicants and utter this prayer: "Let the next one be so bad that I may discard him immediately, thus reducing my burden." The process drags on. The jurors forget that they have come to make a discovery. They are no longer interested in discovery but in exclusion.

Why? Because *even discovery will not put a dent in their list*. The jurors are faced with an afternoon's attempt to cast one part, meaning even the joy occasioned (and that so rarely) by a brilliant performance will be immediately superseded by reserve and a return to somnolence. For the juror thinks (a) I still have twenty people to see, how could I *bear* those hours, knowing the part had already been cast?; thus, (b) I must *pace* myself. I loved this last actor, but perhaps I was deluded. Perhaps someone *even better* will come along.

And so the afternoon creeps on. Such that, at the end of the day, the initial reaction "that person was a *genius*" has been revised into "I remember I liked him very much. But was there not that *other* person, some hours later, who *also* had good qualities?"

(And it is not, I think, too Freudian to suggest that the brilliant auditioner *might* even create in the jurors some slight subconscious animosity—for was it not his audition that rendered the rest of the afternoon so problematical?)

Fatigued, angry, self-loathing (have they not heard hours of ungrantable petitions?), and confused, the jurors sit at the close of play. Each has dealt with the trauma/unpleasantness/waste/confusion differently. And this necessary difference in personality must and will manifest itself in differences of choice.

One committee member may have held out all afternoon for the brilliant actor; one, measuring not against the needs of the script but against this performance, may choose another. That is, having sat fuming for too many hours, this second juror may simply feel the need to exercise that autonomy denied him all day and strike out for the release occasioned by a true subjective choice, and let the demands of the play be damned. A third may find happiness in adopting the role of wise negotiator and counsel reserve: perhaps even *more* applicants should be seen, for if the appearance of even that one star was insufficient to create unanimity, would not wisdom suggest *something* in the fellow was somehow lacking?

The jurors have now fallen into a fatigue that they are happy to characterize as healthy caution. They no longer remember their original task, to find an actor who will enliven the part; they remember only that their brief is to "*cast* the part." They have long ceased being an audience—that is, a group that has come to be delighted. They would like to be a group of, if I may, serious, responsible judges, but they have no idea how such a group might operate.

Subjective opinion has been tried and found untrustworthy. Its day is past. What remains but consensus? Only consensus will get them out of the room. But consensus is, of course, the dead opposite of that subjectivity that is the essence of the theatrical experience.

The choice of a date, a mate, a home, a name for the baby are all transcendently subjective. Each is an assertion of an individual's understanding of his right to joy; none, finally, is capable of analysis. By each the individual asserts that there is a mystery in life and that he is entitled to participate in it.

Consensus is all very well and good. It is the core of society: the commitment to a greater good than one's own instincts. No wonder the auditioning group defaults to it. (We may note that the group, here, has become like that small subsociety immediately formed by the twin extremities of shared hardship and isolation—e.g., the basic training platoon, the survivors of a shipwreck, et cetera. Here we may see the spontaneous appearance of the lawyer, the politician, the comedian, the peacemaker, the wise elder, the willing factotum, as each member stakes out his place in the new world, which, he sees, *demands* communal effort.)

The auditioning group, like the sequestered jury, which, in effect, it is, translates its concerns from *the originally stated task at hand* to *the welfare of the group*—i.e., "I was appointed to judge whether John Smith was guilty as charged. My task has *become* how to get myself and my group out of the jury room with some semblance of self-respect."

Psychologists Ernst Heinrich Weber and Gustav Theodor Fechner wrote on the abilities of the human mind to detect small changes: How small an alteration in musical tone, in the shade of a color, in the size of an object, et cetera, could the mind perceive? Would they had addressed, and so quantified, the capacity of the mind to retain and differentiate between a vast bunch of auditioners. One may perhaps (notes or no) retain the impressions of four to five, certainly not of twenty, as the auditioner discovers. And so, now firmly self-understood as *part of a jury,* he utters the phrase that is the foundation of society and the death of art: "What do *you* think?" Consensus, enshrined as right thinking, ensues, and the stage is set for mediocrity.

The first thing an actor does on achieving success is to stop auditioning. He broadcasts that phrase he has repressed, at some cost, during the years of his struggle: if they want to know what I do, let them view my work and then make up their *own* fool minds.

Yes, some actors audition well. Some have sufficient confidence or a sufficiently idiosyncratic makeup as to consider the audition, *in itself,* a performance (which is to say, potentially enjoyable). I have heard actors say so and been told that some soi-disant auditioning classes teach this viewing of the interview, and suggest that to accept this advice, to act upon it, would, and will, increase an actor's odds of employment. But what of the actor incapable of doing so? What of the actor who translates his affront at judgment into hatred of himself or the auditioners (such generally expressed as "nervousness")?

Well, this person is in a pickle. He is, in the cant of the trade, "not good in the room." His choices are limited to philosophy, self-help, and retreat. Many retire from the business, as they can't "get out of the room"; many resort to auditioning classes, hypnosis, mantras; and if they do not profit from these, many achieve that form of anesthesia known as "experience." Some reduce this "experience" into a set of axioms actually capable of bringing about the desired result: employment. Most fail.

And the folks on the hiring side of the table reason that the process, though harsh and though inexact, does possess the merit of winnowing out that which we, in our fatigue, may allow ourselves to think of as the weak. And yes, the show must go on, and Darwin, and so forth. But note that though occasionally a part may be cast quickly and, thus, in such a manner as to conduce to a continuance of such feelings in rehearsal and performance, the process in bulk is founded upon brutality.

See the actors auditioning for television. They read, first, for a casting agent. Normally these readings are put on tape and shown to producers and directors. Progression to the next step involves an in-person reading (of the same material from the script) with producers and directors. Those still surviving proceed to their presentation to the entity (usually a studio) funding the show. Lucky winners are now walked round to the network, the final hurdle between them and stardom.

Two, three, or four contestants for each part stand in the hall. Some go over their lines one, five, ten more last times; some perform relaxation exercises; some sit in a reverie or in contemplation. The scene resembles a locker room before a martial arts competition. Those who have trained together now perceive that their onetime comrades (those with whom they shared and share hardship, perception, and aspiration) are now their opponents. Each is alone and competes with all others for possession of that indivisible sum of success, wealth, approbation that is the part. One will prevail; the rest will be discarded.

But wait. Someone in authority is coming. How might he ameliorate this brutal situation? What does he hold in his hand?

These are contracts. For the applicants, had they misunderstood their position vis-à-vis the employer, must sign a contract *before* they are allowed into their final audition. That is, lest they consider themselves special (or, indeed, individual), they are reminded *by the force of law a contract holds* that they are replaceable, that they are interchangeable, and that they will not even get a *hearing* unless they so agree, legally and as the final act before they are ushered into the presence, that it is so.

The gladiators proclaimed before the contest, "We who are about to die salute you," thus perhaps invoking, and so perhaps participating in, the honor that, in spite of the horror, may exist in personal combat. But the actor proclaims, *legally,* that his pursuit of success, far from an example of honorable combat, is furthered only at the pleasure of an authority that subjugates all to money and that does not scruple to pit its employees against each other, financially, in a race to the bottom. (I do not know if this process of preemployment contract negotiation is legal. I suppose that it is legally defensible, which is saying little. And who among the aspiring actors would contest it? Such would, of course, be career-ending folly.)

Film—and to an exponentially greater degree, television—is just business. As film and television companies grow larger and more vertically integrated, they are increasingly ruled by businesspeople. Businesspeople are trained to quantify, and it, I am sure, makes good business sense to recast quondam aesthetical decisions as cost vs. benefit. But the bottom line, in art, in entertainment, and even in the entertainment industry, is the pleasure of the audience.

This pleasure cannot be quantified. It may be propitiated only by schooled, but equally subjective, decisions on the part of the artist.

Businesspeople have always resisted not only such decisions *but also the notion that such productive subjectivity exists.* Their inclination and training rejects it, and they recur in its supposed nonexistence not to the quantifiable (the audience's whim may not be quantified) but to the *imaginable*. That is, as they have experienced the benefits of consensus (in a committee meeting, it gets people out of the room; in a consumerate, it brings money to the purveyor), they seek to apply it, inductively and deductively, and at all times.

"See," they say, "though we thought ourselves faced with a situation that seemed to admit of neither solution nor grounds for arriving at same (casting), we have made a choice!" and, "As this magic of consensus worked in the room, must we not further reason that it will work in the auditorium? Let us, therefore, *imagine* that the audience arrives at decisions much as we do here: in committee."

Entertainment businesspeople worship this supposed tool of consensus. As well they should, for it is the only tool they have. They are lost in the wilderness and prefer, as might you or I, a broken compass to no compass at all.

They take this tool to the public in test screenings and focus groups. Here the audience is invited to replace its capacity for amusement with the right of sitting in judgment.*

These invited test screeners *never engage* that portion of

* Studio folks famously opine that testing is "just another tool" and that it would be foolish to overlook the use of any tool. This might seem, at first glance, simple common sense. But consider: When choosing an accountant it might make sense to take into account his or her age and, thus, expected working lifespan. Now, there is a medically proven correlation between longevity and percentage of body fat. Why not, then, test potential accountants for body fat to predict their potential working lifespan—why reject that tool? Because in utilizing it, one might weed from the applicant pool accountants of greater skill but higher body fat. Though there may be a correlation between body fat and expected working lifespan, there is none between body fat and skill in accountancy; thus, the utilization of this "tool," far from being helpful, is actually destructive. Similarly, while it might seem "common sense" to quiz viewers on their reactions to a film, there is *no correlation between these test results and a film's performance.* Thus, attention to these tests must and does mitigate against the release of films which, although they do not "test" well, might actually *perform* better than their more sanctioned brethren.

the human mind that loves a story; no, they have become enmeshed in a fantasy of business, and they now work to imagine (as did those folks in the committee/audition room) what some notional *other* group might just like. And they vote accordingly, thumbs-up, thumbs-down, in self-congratulation at having suspended that obviously now puerile, wide-eyed state of enjoyment of the unlicensed, unschooled, and mere "member of the audience."

This extended fallacy of consensus accounts, if not for the worthlessness, at least for the uniformity of most mass entertainment. For each human being is different. And the idiosyncrasy of the artist, this supposedly (by the executive) *divisive* tropism, is actually an ability to compel—to compel a disparate group of people not into a jury capable of consensus but into a group willing to suspend its rational capacity—into an *audience*.

We may note further that the executive, in forming a lay and random group into a committee supposedly capable of forecasting dramatic success, indicts, and in fact *unsays,* his protestation of his own possession of superior financial or mercantile powers. For if a regular person wandering in a mall somewhere may be shanghaied into watching a test screening, and if his opinion, and the opinions of his like, are the basis upon which executives determine how to place their bets, why not eliminate the executives entirely and proceed directly to the mall wanderer? Which is effectively what has happened in the casting session.

CRITICS

It is a comfort to ascribe to the process of film criticism, if not *justice,* then at least a certain symmetry.

"Well," we beleaguered say, "they shat on my beautiful work *this* time, but, sigh, last time they praised certain aspects of the film I thought I might have done better"; or, truth be told, "they trashed my film, but at least the swine also trashed the film of a hated competitor."

One reads mindless, hurried, self-serving dissections of the films of others, if not with glee, then with a certain praiseworthy equanimity; such holding of one's own feet to the fire, however, prompts a recourse to the philosophic. "Who *are* these brigands?" one wonders. "And who died and left them boss?" This, in the sufficiently irate, leads to fantasies not of outright crime but of *un revers de satire*.

One imagines the critic (in general perceived as, if not actually possessing, a second-rate intellect) in debate with the more mentally nimble filmmaker. One fantasizes about the critic himself taking to art and hazarding his good name "on his own bottom," et cetera. One may, in an extended violent reverie, wake to shame and find one's thought driven ever eastward, toward the stoicism of the Mediterranean and, in absolute extremis, toward the quest for satori of Nepal.

For critics are a plague.

Being of an especially weak nature, I find my bellicosity easily awakened in the face of injustice (to myself). Injustice in the lack of preparation, appreciation, and intelligence by the critics, or, indeed, injustice in the very fact of my being criticized. I do not think I am alone.

As ants at a picnic, however, the critics will not go away, and as I (qua writer and director) do not plan to go away, I have had to learn to deal with them.

The only effective method I have found, unfortunately, is to ignore them.

Artists loathe critics, and critics know it, and artists know they know it and know that the critics view and must view any attempt at rapprochement as meeching.

Any artist of any worth is absolutely his or her own harshest critic, and a critic's gracious studied ignorance of an artistic solecism is more likely to bring about its correction than would a snide riposte.

What of trash? Is it not the critic's job to cleanse it from the public view? No.

The critic's job in America is to sell newspapers. Newspapers are sold by gossip, and most critics write gossip—they invite the reader to find fault, licensing a vicarious superciliousness.

This practice increases not only readership but also, more importantly, advertising revenues, as the more films come and go, the more films have to pour out an initial rollout ad campaign. One is as unlikely to find a champion of the public taste in the arts section as to find a champion of actual defense in the Defense Department.

The officials of that department exist to defend their department, as do the officials of the arts section. Such are effectively not critics but censors, hired to defend the status quo.

Like censors, they may perhaps be bribed.

Spend enough money on a film and enough money on its ad campaign, and the newspaper *will* absolutely respond, if not with a guaranteed good review, then with sufficient on-and-off-the-page coverage to approximate such a review's result—an increased opening-week turnout.

Is there such a thing as a good critic? Yes. I think my work has benefited over the years from one or two such. Were such supporters of my work? Yes.

But, in the main, I don't like critics, and I like even less my personal inability to come to peace with the phenomenon. I might cite my benefits derived from certain critics and name them—but would it be possible to name them without the wish that they might read, and I might, thus, not only propitiate them but have wished to do so as well? It is a philosophic can of worms, indeed, it is.

And what of the "democratization of criticism," e.g., online critics? Well, this approaches pure gossip, and, as such, it is of necessity messier and more truthful than its licensed brother. In the end, we all have to take our knocks.

THE CRITIC AND THE CENSOR

> Who are those people by whom you wish to be admired? Are they not these about whom you are in the habit of saying they are mad?
>
> What, then? Do you wish to be admired by the mad?
>
> —Epictetus*

Mr. Kipling wrote, "Tho' he who held the longer purse might hold the longer life." Similarly, the leniency of the critics of our great newspapers is proportional to the advertising budget of the offender. And who would have it otherwise? In days of yore, and still into our day, the former gossip columnist or sports reporter, and the otherwise unassignable, along with

* A long contemplation of this stoical view has left me but little better equipped to deal with the scorn of those I consider the untutored. But, recently, I have found that an increase (marginal but real) in my practicable stoicism by reasoning thus: We, (the artist), traditionally characterize our detractors as fools; therefore, to wish their good opinion is inconsistent. They (as you or I) may or may not be fools; they are, however, incontrovertibly, merchants of derision. Bad reviews sell. But, as the critic must occasionally praise (if only to preserve the illusion of open-mindedness), and if this praise—as it seems to be—is awarded *arbitrarily,* then, to wish for the critics' praise is *to wish that his censure fall on another.* Thus, to wish for critical praise is not merely inconsistent, but immoral, as it is a wish for another's suffering. Will time prove that I have allowed this progressed understanding to increase my peace of mind, and, thus, my utility to the body politic? I doubt it.

the (misguided) seeker after distinction, found, in practice, that their sway as critics had upon it the check of a rational concern, on the part of their employers, for the bottom line. The critical system may reek, to the politically savvy, of the old-fashioned retail depredations of the ward heeler—the mere settling of grudges or the promotion of some pet theory, grudge, or paramour that has traditionally been the recompense for the critic's poor pay and boring evenings.

It can be noted that advertisers, that is, those with money to burn, would, being human, expect a little warmth in return for providing the conflagration. That warmth was and is understood to be, to put it genteelly, "the benefit of the doubt." As it has been said, "Who would want to live in a town where you can't fix a parking ticket?"

Laws of eminent domain have been used for centuries to allow the state to condemn the property of the individual so that it may be used to benefit many (e.g., for the building of a dam, a road, a school). Recently, jurists have expanded the theory to allow the condemnation of any property, domestic or otherwise, that is not, in their view, generating sufficient tax revenue.

We may thank the Lord that, as we imagine and must pray is the case, a cast-iron conscience in some members of these committees of condemnation has resisted what must have been the strong, many, and varied inducements offered by the interested to misuse their wide-ranging powers.

But imagine the corruption inherent in an unchecked power to condemn. Might not the individual property holder, pitted against the power of the state and that power ceded, or cedable, to the highest bidder, not tremble and not inquire what he, in his powerlessness, might do to avert catastrophe? And might not a writer, painter, filmmaker, dramatist, or choreographer, faced with the threat of state-sanctioned censure, school himself early on to think twice?

The sobering specter of purity-of-speech and purity-of-image laws should concern all artists. In such a day, may God forbid it, our job will be done; the critic no longer will even be persuadable, and we may well yearn for that innocent day.

CRIMES AND MISDEMEANORS

MANNERS IN HOLLYWOOD

> They who lack talent expect things to happen without effort.
>
> —Eric Hoffer

Were we to construe manners in the contemporary sense of *bienséance,* this chapter would be rather short, as such manners do not exist in Hollywood. Considered, as *moeurs,* however, a survey of Hollywood manners might offer an afternoon of happy occupation to the strong-stomached anthropologist.

The adage "Don't look at the downed man" applies equally in combat and in the trenches of film production.

The observed rule in Hollywood is: "Feel free to treat everyone like scum, for if they desire something from you, they'll just have to put up with it, and should they rise to wealth and power, any past civility shown toward them will either be forgotten or remembered as some aberrant and contemptible display of weakness."

I shall not overtax the reader's credulity with tales of film-world savagery, for such tales must impose upon either his belief or his sense of outrage.

"What?" the affronted might say. "I bought (or—may heaven forfend—borrowed) this book for entertainment, and I am asked to debase myself in attention to an unclean recital."

Well, then, may my recital (for I cannot remain continent) be entertaining.

Among my favorites is the director, flown with his wife and family on the studio jet to Hawaii, for his film's opening.

It bombed, and on application to the airport, the director was told the jet had left. He intuited this to mean that he and his family were free to find their own way back home.

This is the template for the difficult passage in Hollywood—the offending party becomes not merely persona non grata but dead. I do a film with a longtime colleague, a producer. He makes a deal with the Eurotrash money folk to sell the film out. He forms a corporate partnership with them and, halfway through the film, refuses to take my calls and those of the editorial department, whose cash has been cut off and who cannot buy supplies. I finally get him on the phone. "I'm sick of carrying water for you people," he says. A relationship of twenty years.

Before the relationship (inevitably) goes wrong, all is toast and jam—sweet but bloating; after the first contretemps, war to the knife.

The exception: that interchange wherein the parties' relative status is unclear. This awakens, in those who sit on the other side of the table, apprehension, and so its expression as anger, masked as awkwardness. The technical term for the above is, I believe, smarminess (e.g., "I'm a great fan of yours. Great, great, very great"). The newcomer to Hollywood must feel like a high-functioning autistic, forced to resort to mnemonics and cheat sheets to remember how to behave in the most simple situations.

A sample list:

Should the project go awry, you will be notified by a complete lack of contact with those in whose hands its administration has rested.

Attempts to contact them will be met by the response that "she is in conference," "on vacation," or, my favorite, "out of town"—as if film executives spent any waking moment, in whatever part of the globe, other than on the phone and constantly connected to their office.

Like that autistic, the response, "Well, she may be out of town, but surely she calls in for messages; please have her return my call" is discovered by the neophyte to be other than the thing. Silence greeting such a commonplace observation is like that vouchsafed the five-year-old at the dinner party who announces that his parents have oral sex.

Any bad news expressed, rather than left to the imagination of the offending, is delivered by the lowest-possible ranked henchperson, each placeholder above that nadir understanding the essential law of the new Medes and Persians: "I need not perform any act not immediately fungible." Manners in Hollywood, in short, stink on ice.

The Academy of Motion Picture Arts and Sciences issues, each year, the Jean Hersholt Humanitarian Award.

Jean Hersholt was a character actor in the thirties and forties, and one may note that he is remembered only as a nonentity and a nice guy, and one may draw his own conclusions.

Nelson said, "Aft the most honor, forward the better man." And I agree.

The working people in Hollywood (if you have not, by this point in the book, divined my prejudices) are the salt of the earth. The motion picture industry has repaid their devotion by moving most film production out of the country. Toronto, Montreal, Belgrade, and *London* are being shot for New York.

Any business is, if not essentially, at least potentially, pillage. The surest sign of voracity and corruption is the formation of "industry standards"—a collusion of capital to muscle labor and bilk the consumer (cf. the Hayes office, the Motion Picture Academy, those opposed to Janet Jackson's breast). The desire to cover the legitimate, messy operations of commerce with a mask of sanctity reveals what the cops call "guilty knowledge"—that is, that someone is getting screwed.

Robert Evans wrote in his book *The Kid Stays in the Picture* that the best films seem to come from the most troubled sets, but with respect to Mr. Evans, I think this is a bunch of hogwash. I think that a *producer* likes a troubled set, because it allows him to "save the day" and otherwise exert undue and unfortunate influence upon a mechanism that, had he been doing his job correctly, should have run smoothly in the first place.

Were I king, I would make all the youth fomenting in film schools study deportment; for the good producer, director, agent, manager might then couple, or indeed promote, self-interest by civility—as does a *truly* smart businessperson, advancing, as has been said, to his own goals by taking the other fellow's road.

For to live in an uncivil world is debasing. It, of course, corrupts the practitioner, and it demoralizes those practiced upon, offering them the choice of learning either character or servility, and so perhaps my argument is warped, and in the stringency of life in a boorish land, there might be found some possibility of good.

But in the words of Bosquet, "C'est magnifique, mais ce n'est pas la guerre."

THEFT

Everybody on the production end always assumes the writer is stealing their money.

On the writer's side of the table, everybody always assumes the unit production manager (the comptroller) is stealing, and everybody *knows* the producer is stealing, and the thefts of the distributor are a certainty surpassing the necessity of discussion.

But among the dramatic arts and crafts, only the writer, I believe, is assumed to be criminal.

One would not accuse an actor or director of malingering—their obvious self-interest in a merchantable product bearing their name would render absurd accusations of less-than-best effort, but almost all writing for hire I have done, and almost all of that done by my colleagues, has ended with these accusations by the producers or studios: you did not do enough, you did not do as you were asked, you spent too little time, et cetera.

This is coupled to a disposition on the part of the employer to accept *nothing* as a completed job. The writer contractually obligates himself to do a certain number of revisions. These are called, variously, "drafts," "sets" (of revisions), and "polishes" and are known collectively as "steps." However many steps the writer has signed on for—that is, however many times he has agreed, contractually, to accept corrections and implement them—someone on the other side of the table *will* either request or demand more work. This demand takes two major forms: the request, the collegial, "I know what we both agreed to, but would you mind, as a favor . . . ," and its corollary, "I know what we agreed to, but refusal would be endangering not only the project but your reputation; why would you want to shoot yourself in the foot—aren't you a team player?"

The more minatory form is an outright or implied accusation of fraud: "You have not done as asked, you sloughed it off, you intend to defraud me. . . ."

O, poor writer. For I do not find the writer any less agreeable to concerted effort than the actor, director, or designer. Yet, the thing, for the writer, usually ends in recriminations.

I know there are those homeowners who always end by suing their contractors or decorators, and those decorators, et cetera, who end every contract in court. And so it is with the writer.

His urge to see the thing through, to be agreeable, to do a good job, to be well thought of, to protect his reputation in the mercantile community is accompanied by the unpleasant certainty that *someone* will most likely finally turn combative.

In addition to the usual human desire for peace, the writer dreads the havoc that continued extracontractual and, indeed, extrarational work will produce on his script.

For any project, whether the screenplay for an epic or your wife's hairdo, can be ruined by too much freely given advice, and no advice is given more freely than that supposedly correcting the fool errors of the screenwriter.

The problem is further worsened by its (to the writer) interminability. Once he gives in to cajolement or threat, he may assume such will be continued, for he is, in accepting unmannerly (as extracontractual) corrections, encouraging their continuation. (By agreeing to agree to do more than that to which he, contractually, agreed, the writer confirms in the producer, studio head, et cetera, the truth of their original prejudice: that the writer, in effect, *is* no good, and that all his work is approximate, and its worth arguable at best.)

If one treated an architect thusly ("Would you mind moving the staircase? Thank you. Now would you mind moving the skylight?"), one would quickly come to find in his acquiescence disturbing evidence of the architect's lack of structural understanding.

So does the noodger come, by the process of his cajolement, to ratify an incipient prejudice: that the writer has no damn idea what he is talking about and that the previously undiscovered literary powers of the employer had best come to the fore fast and clean house.

In the script, change follows change until, inevitably, the producer comes to the inescapable conclusion that the writer, and his now indecipherable pile of shit, must be thrown out and a new broom must sweep clean. The concern—"How can I be sure you're not going to rob me?"—is revealed as prescient when the nervous producer finds himself left with the mud he and his colleagues' notes have wrought.

No wonder all writers want to direct: one still has to put up with a load of nonsense, but even if wearing two hats (writer and director), there is one under which one is not called a thief and then raped.

TWO GREAT AMERICAN DOCUMENTS; OR, IN THE WAKE OF THE OSCARS

In a document that purports to be the Antioch College Sexual Offense Prevention Policy (SOPP) (1994):

> The spirit of Antioch's Sexual Prevention Policy is about 'yes' . . . The spirit is about a fully affirmative YES. Not an ambiguous yes, or a well-not-really-but-ok-I-guess yes . . . This is about YES, UM HUM, ABSOLUTELY YAHOO YES! . . . Being with someone you are sure YOU REALLY WANT to be with THAT is EXCITING, is EROTIC, is DEEP, is GREAT, is YES! That is consent. That is the spirit of the policy.
>
> It is exhilarating to be part of a community that is working so hard to increase equality and mutual satisfaction and to rectify domination and oppression.

And from the Academy of Motion Picture Arts and Sciences 2003 pamphlet on Academy Standards:

> There are many within the Academy today . . . who use the phrase "Academy campaign" without irony or embarrassment. It is not at all necessary, though, that such a concept or such a thing exist.
>
> The simplest most direct path to protecting the Academy Awards Process from debasement real or suspected would

be to arrive at a point at which electioneering disappeared entirely.

I envision, thus, a perfect world in which both the randy young at Antioch and the careerist film producer find a playing field as level as that which God intended.

In this world the lion shall lie down with that lamb, and only with that lamb that has spelled out its consent, vide Antioch:

> If the level of sexual intimacy increases . . . (i.e., if two people move from kissing while fully clothed . . . to undressing for direct physical contact, which is another level), the people involved need to express their clear verbal consent before moving to that new level.

But how, in what is clearly the heat of the moment, will they remember to turn aside and consider, so as to fulfill both the letter and the spirit of the law?

Our other document may help.

> The Academy does not presume to tell its members when they may and may not invite friends to their homes . . . at the same time, the Oscar-season "parties" that in fact are heavy-handed lobbying occasions have become . . . distasteful. . . . Self-tests for whether a dinner party really is a dinner or a party, as distinct from a tactical maneuver in an Oscar campaign, include the following . . .

Which list is an aide for (and a caution to) those ethically conflicted about the true nature of their canapés.

The Academy, much like the stoics of an earlier day, suggests the conflicted consider his motives and, if still in doubt, review who is footing the bill for the entertainment. Is it the host himself or a studio or production company? Are guests invited in the hope that they have gained an appreciation of the artist and his difficulties "faced and overcome" during filming?

" 'Yes' answers indicate treacherous ethical waters."

These clarifying documents, in common with *The Rights of Man,* et cetera, exist to force one's attention to abuses to which the perpetrator or victim has become desensitized. The first step toward a recompleted humanity, then, must necessarily be shock at the gulf between our practice and right reason; e.g., "All sexual contact and conduct between any two (or more!) people must be consensual."

But there will always be collusion. Collusion is, indeed, the true and necessary state of communal endeavor. Just as it seems that money is to be made in the stock market *only* by what the deluded know by the sobriquet "insider trading," so, according to the solons of Antioch, healthy sex can be ensured among the young only by a watchfulness as that of our friends the Medes and the Persians.

For will not the young collude and conspire against law, reason, custom, and tradition in order to hide the salami? And will not the depraved fame- and fortune-starved members of the Academy serve ceviche in an other-than-disinterested manner?

In short, yes.

To speak of the movie business, which is my assigned beat: wherever there is community secrecy and the possibility of invidious gain, there must be collusion.

The human mind cannot tolerate the spectre of waste presented by the possibility of chicanery without detection. The very vehemence with which the Academy presents its good-willed and patient plea informs as to the impossibility of its implementation.

For, in each, the fault is not with the participants but with their stars.

The young will not stop copulating like rabbits, and no amount of what is, finally, voyeurism and child abuse on the part of the school administration will turn them from the course of lust. Neither will the members of AMPAS cease voting for their self-interest, which is, and must be, the nature of all awards ceremonies.

The movies are a cutthroat business, the business is entertainment, and we should be glad of the entertainment value of a bunch of pirates proclaiming their mutual goodwill and probity.

Perhaps even extending to ourselves the same licensed smirk of the randy Antioch elders and receiving the same necessary rejoinder: "Okay. You first."

CONCLUSION:
IT AIN'T OVER TILL IT'S OVER

Stanislavsky wrote that the last ninety seconds are the most important in the play. Hollywood wisdom casts it thus: Turn the thing around in the last two minutes, and you can live quite nicely. Turn it around again in the last ten seconds and you can buy a house in Bel Air.

This is old-style moviemaking, wherein the loveless child relents, the previously suspect suitor is found to be "good" and gets the girl, and the wandering minstrel is revealed as the king.

We have, also, the surprise/shock ending, e.g., *Diabolique, Witness for the Prosecution, The Sixth Sense*. He who could fool the canny moviegoer up to the last beat *should* have a house in Bel Air, or whatever else his heart desires. Does he not deserve it? Absolutely. For movies do not exist to make us better but to give us a thrill or chill on Wednesday night when we are out with our best gal.

If the shark makes us say "ooh," it has earned our few dollars. If the filmmaker can make us say "ooh" of a shot of the empty water, give him his private plane.

"There are three degrees of bliss," Mr. Kipling instructed us, "at the foot of Allah's throne/and the lowest place is his/who has saved a soul by a jest."

I offer the punch line of *Some Like It Hot* as good a way of getting offstage as one could wish:

Jack Lemmon: "I'm a man."

Joe E. Brown: "Well, nobody's perfect."

What are the great endings of art? "Reader, I married him," and "Ol' Man River, he jes keeps rollin' along." Equaled in the cinema, perhaps, by the shock/surprise endings listed above and "The sonofabitch stole my watch" from *The Front Page* and Paul Muni's tag from *I Am a Fugitive from a Chain Gang:* Q: "How do you live?" Muni's answer: "I steal."

In contrast to the perfectly acceptable "I'll never leave Kansas again" or "Tomorrow is another day," there are endings that introduce new, and transformative, information in the last seconds of the film.

The Pride of the Yankees is a rather good baseball biography. Gary Cooper, playing Lou Gehrig, the world's greatest player, is dying of a wasting disease. He is given a farewell party by his New York Yankees at Yankee Stadium. He is honored, and cheered, and declares himself "the luckiest man who ever lived." He then hobbles off the field, supported by his beloved wife, Teresa Wright, walking into a medium long shot, stadium corridor, day. And as he walks from us, as we begin to stir, knowing the movie over, we hear the umpire in the distance shouting, "Play *ball.*"

This, to me, is great filmmaking. The information introduced is new and inevitable-familiar: the filmmakers have apostrophized their potentially sententious tribute (both the stadium fete and the film) in the name of a greater ideal: the game goes on.

Sailor of the King is a World War I–World War II tale. Michael Rennie is a British naval officer. In Canada, during the first war, he meets a young woman, they become intimate, but before they can marry, Michael is called away, then peace breaks out.

One war later, he is again in command at sea. A young ensign, Jeffrey Hunter, is taught the ways of command and taken under Rennie's wing, and Rennie treats him as the son he never had. Near the end of the film they lay up in a Canadian port, and Jeffrey Hunter confesses that he himself is Canadian and, perhaps, while they are at port, Rennie would like to meet his mother. We understand that the mother will be revealed to Rennie as his inamorata of the first war—the punch line of the whole thing is that Jeffrey Hunter is in truth his son.

This, while not illuminating, is enjoyable, and what is wrong with that?

On the other hand, *Cocoon* is a pretty good film. All the old folk go up in the sky at the end. And then we gather around the grave site of Hume Cronyn, who, as I remember, didn't quite have the bottle for the trip. The fellow playing the pastor does the eulogy well and then says, "Let us pray"; we see him lift his eyes skyward. I saw it and thought: *This,* now, is bold filmmaking. It is as completive as the final chord of *Tristan* and lacks only the funny hats.

Yes. *Cocoon* is not a story about space critters come down to Earth but about old people learning to face death. It is a spiritual journey. Yes, let us pray that when our time comes we can face the end with perhaps the same good humor and philosophy that the filmmakers face at the end of *their* journey.

And then the thing went on one beat too long. We were transported from the graveside to a spaceship, where we again saw Don Ameche et al., and I found, to my chagrin, that the film, no, was in fact about space critters.

"I love you" does pretty well as the end of a speech. It generally is not advisable to follow it by "and one more thing. . . ."

We remember "Au revoir, sal Juif" at the end of *Grand Illusion,* Jean Gabin's ironic farewell to his comrade, Dalio.

And Warner Baxter sitting alone by the stagedoor fire escape at the conclusion of *42nd Street,* sitting alone and unremarked as the hit play's adoring fans praise all contributions but his own.

How might one achieve this perfect completion?

First, the problem of the play must be concise. Then, the progress toward it must be direct and all incidents essential either in advancement or disruption of that progress. Finally, the conclusion must be definite (e.g., France is freed, the couple is reunited, the treasure is returned to its rightful owner).

These three steps are difficult to accomplish. The play is a syllogism, and to function perfectly, it must be structured perfectly. If A, then B.

It can take place on any level of abstraction: In order to save France, I must discover how to land on the beaches. In order to land on the beaches, I must produce a small craft cheaply and in vast numbers; in order to do so, I must obtain an enormous amount of cypress; in order to do so, I must win the trust of the crotchety octogenarian who holds the deed to . . . , et cetera.

So, my efforts to obtain that one piece of silverware that would convince the octogenarian to deed me the cypress that would allow me to build the boats. These efforts are essential to allow me to save France—each small step is essential to the clearly formulated superobjective, and the audience will follow the story, wondering what happens next.

Q. Is it possible to engross the audience when the end of the quest is already known?

Yes. Mark Twain wrote of U. S. Grant's personal memoirs that they were so well written as to make one wonder who was going to win the Civil War.

It is *more difficult* to engross the audience in a biography, as the end is known. It calls for greater skill and imagination on the part of the writer in finding an *internal* story within the generally known historical moment.

This story is usually boy-meets-loses-gets-girl. In my previous example—the quest for a piece of silverware—the idea is the same.

The superobjective may, indeed, be *concealed* from the audience. If the progress of the incidents—war, amphibious assault boats, cypress, old lady, silverware—is direct and essential, it could, indeed, be inverted, so that what seems to be a movie about matching flatware turns out to be about saving Western civilization.

Here the audience, if sufficiently engrossed, again, scene to scene, is rewarded in the last ten seconds by the revelatory recasting of the goal. They discover, in *The Sixth Sense,* that they have not been watching Bruce Willis's compassionate efforts to help a disturbed youngster with his clairvoyance but, rather, watching the youngster help Bruce come to terms with his own death.

In *Pygmalion* we are told—and find, of course—that we *knew* Eliza Doolittle was "going to marry Freddy." And in *The Pride of the Yankees* we are put in the same position as Lou Gehrig and discover that baseball itself is deeper in our hearts than our love even of its most perfect avatar.

God bless the writer who can do this, and let him or her retire, with our blessings, to the pleasures of Bel Air, whatever they may be.

Play ball.

APPENDIX: FILMS REFERENCED

The Adventures of Baron Munchausen (1988)
Starring: John Neville, Eric Idle, Sarah Polley. Director: Terry Gilliam. Written by Charles McKeown and Terry Gilliam; 126 minutes; Allied Filmmakers. Traveling with a group of misfits, a seventeenth-century aristocrat has some truly unbelievable experiences throughout Europe.

An American Tragedy (1931)
Starring: Phillips Holmes, Sylvia Sidney, Frances Dee. Director: Josef von Sternberg. Written by Samuel Hoffenstein, adapted from the classic novel by Theodore Dreiser; 96 minutes; Paramount Pictures. A working-class man finds himself torn between his love for a rich woman and his love for a lower-class coworker.

Angels in America (2003)
Starring: Al Pacino, Meryl Streep, Emma Thompson. Director: Mike Nichols. Written by Tony Kushner; 352 minutes; Avenue Pictures Productions. Miniseries based on Tony Kushner's play series about the American AIDS epidemic during the 1980s and its effects on a closeted Mormon man, a gay couple, and the notoriously homophobic lawyer Roy Cohn.

Apocalypse Now (1979)
Starring: Marlon Brando, Martin Sheen, Robert Duvall. Director: Francis Ford Coppola. Written by John Milius, Francis Ford Coppola, and Michael Herr; 153 minutes; Zoetrope Studios. An adaptation of the classic Conrad novella *Heart of Darkness*, set in the Vietnam War, in which an Army captain goes on a dangerous mission to dethrone Kurtz, a Green Beret who has gone mad.

Arsenic and Old Lace (1944)

Starring: Cary Grant, Josephine Hull. Director: Frank Capra. Written by Julius J. and Philip G. Epstein, adapted from the play by Joseph Kesserling; 118 minutes; Warner Brothers. A young man discovers that insanity runs in his family and that his two aunts have been poisoning lonely old men and hiding their bodies in the basement.

Back to Bataan (1945)

Starring: John Wayne, Anthony Quinn. Director: Edward Dymtryk. Written by Ben Barzman and Richard H. Landau, based on the story by William Gorden and Æneas Mackenzie; 95 minutes; RKO Radio Pictures. After the Philippines fall to the Japanese in World War II, a rogue American army colonel stays behind and organizes an underground resistance movement to oppose the occupation.

Bambi Meets Godzilla (1969)

Director: Marv Newland. Written by Marv Newland; 2 minutes; Rhino Wea. In this animated short, the sympathetic deer, Bambi, meets the ultimate monster, Godzilla.

Battleship Potemkin (1925)

Starring: Aleksandr Antonov, Vladimir Barsky. Director: Sergei M. Eisenstein. Written by Sergei M. Eisenstein and Nina Agadzhanova; 75 minutes; Goskino. Classic silent film about a violent mutiny on board a Russian battleship that leads to an uprising and, later, a massacre in Odessa.

Ben-Hur (1959)

Starring: Charlton Heston, Jack Hawkins. Director: William Wyler. Written by Karl Tunberg, adapted from the novel by General Lew Wallace; 212 minutes; Metro-Goldwyn-Mayer. A Jewish prince is forced into slavery after being betrayed by a lifelong Roman friend; after his release, he seeks his revenge.

The Birth of a Nation (1915)

Starring: Lillian Gish, Mae Marsh, Henry B. Walthall. Director: D. W. Griffith. Written by Frank E. Woods, Thomas F. Dixon Jr., and D. W. Griffith; 187 minutes; David W. Griffith Corp. This silent classic looks at the aftermath of the Civil War in a deeply racist light and glorifies the Ku Klux Klan.

The Bishop's Wife (1947)

Starring: Cary Grant, Loretta Young, David Niven. Director: Henry Koster. Written by Leonardo Bercovici and Robert E. Sherwood, adapted from the book by Robert Nathan; 109 minutes; Samuel Goldwyn Company. While having difficulties executing plans to have a cathedral built, a bishop prays for help and receives counsel from an angel, but the angel advises him about a completely different problem.

The Blair Witch Project (1999)

Starring: Heather Donahue, Joshua Leonard, Michael Williams. Director: Daniel Myrick and Eduardo Sanchez. Written by Daniel Myrick and Eduardo Sanchez; 86 minutes; Haxan Films. While making a documentary in the woods, three students disappear. A year later their haunting footage is seen for the first time.

The Blue Lamp (1950)

Starring: Jack Warner, Jimmy Hanley, Dirk Bogarde. Director: Basil Dearden. Written by T.E.B. Clarke; 84 minutes; Ealing Studios. As two cops go about their day, they find it interrupted by the discovery of a grisly murder of a fellow officer.

Bob le Flambeur (1955)

Starring: Isabelle Corey, Daniel Cauchy, Roger Duchesne. Director: Jean-Pierre Melville. Written by Auguste Le Breton and Jean-Pierre Melville; 98 minutes; Organisation Générale Cinématographique. A kindly down-on-his-luck gambler decides to stage a casino robbery with a pair of street kids and steal $800 million.

Bombardier (1943)

Starring: Brigadier General Eugene L. Eubank, Pat O'Brien, Randolph Scott. Director: Richard Wallace. Written by John Twist and Martin Rackin; 99 minutes; RKO Radio Pictures. A fast-moving documentary chronicling the lives and training of World War II bombers.

The Boston Strangler (1968)

Starring: Tony Curtis, Henry Fonda, George Kennedy. Director: Richard Fleischer. Written by Edward Anhalt, adapted from the book by Gerold Frank; 116 minutes; Twentieth Century Fox. During the 1960s, the Boston Strangler terrorized women, killing old, lonely women and tying pantyhose around their necks. This film re-creates the events that led to his capture.

The Bourne Identity (2002)

Starring: Matt Damon, Franka Potente, Chris Cooper. Director: Doug Liman. Written by Tony Gilroy and W. Blake Herron, adapted from the novel by Robert Ludlum; 119 minutes; The Kennedy/Marshall Company. Jason Bourne has no memory of who he is, what he does, or where he is from. All he knows is that he speaks several languages, knows how to fight, and that lots of people want him dead.

Brief Encounter (1945)

Starring: Celia Johnson, Trevor Howard. Director: David Lean. Written by Noel Coward; 86 minutes; Cineguild. Two married strangers begin an intense love affair after a chance encounter at a train station, meeting every Thursday in an attempt to sustain their passion.

Bullitt (1968)

Starring: Steve McQueen, Robert Vaughn, Jacqueline Bisset. Director: Peter Yates. Written by Alan Trustman and Harry Kleiner, adapted from the novel by Robert L. Fish; 113 minutes; Solar Productions. After a federal witness is killed, a tough cop seeks the victim's murderer and attempts to take down a mob ring.

Carrie (1952)

Starring: Laurence Olivier, Jennifer Jones, Miriam Hopkins. Director: William Wyler. Written by Ruth Goetz and Augustus Goetz, adapted from the novel (*Sister Carrie*) by Theodore Dreiser; 118 minutes; Paramount Pictures. A farm girl moves to Chicago and falls for a married man; later, she attempts to realize her dreams of Broadway fame.

Cavalcade (1933)

Starring: Diana Wynyard, Clive Brook. Director: Frank Lloyd. Written by Reginald Berkeley, adapted from the play by Noel Coward; 110 minutes; Fox Films Corporation. In this adaptation of the Noel Coward play, two British families live through such upheavals as Queen Victoria's death, the Boer War, and World War I, never losing hope despite their constantly shifting fortunes.

Children of a Lesser God (1986)

Starring: William Hurt, Marlee Matlin, Piper Laurie. Director: Randa Haines. Written by Hesper Anderson, adapted from the play by Mark Medoff; 119 minutes; Paramount Pictures. A speech teacher falls in love with a deaf custodian at the school where he works, trying to teach her to speak despite her resistance against the callous outer world.

A Christmas Carol (1951)
Starring: Alastair Sim. Director: Brian Desmond Hurst. Written by Noel Langley, adapted from the story by Charles Dickens; 86 minutes; George Minter Productions. After being haunted by several ghosts who present to him his past mistakes, Ebenezer Scrooge transforms from a miser to a charitable man.

Circle of Danger (1951)
Starring: Ray Milland, Patricia Roc, Marius Goring. Director: Jacques Tourneur. Written by Philip MacDonald, adapted from the novel by Philip MacDonald; 86 minutes; United Artists. An American travels to England to discover the circumstances surrounding his brother's death during World War II, eventually confronting his brother's killer.

Citizen Kane (1941)
Starring: Orson Welles, William Alland. Director: Orson Welles. Written by Orson Welles and Herman J. Mankiewicz; 119 minutes; Mercury Productions Inc. Classic film in which a journalist attempts to document the rise and fall—and the unanswered questions—of a newspaper tycoon's life.

Cocoon (1985)
Starring: Don Ameche, Wilford Brimley, Hume Cronyn. Director: Ron Howard. Written by Tom Benedek and David Saperstein; 117 minutes; Twentieth Century Fox. A group of elderly people find themselves enjoying newly youthful bodies after a swim in a pool full of alien cocoons.

Contact (1997)
Starring: Jodie Foster, Matthew McConaughey. Director: Robert Zemeckis. Written by James V. Hart and Michael Goldenberg, adapted from the novel by Carl Sagan; 153 minutes; Warner Brothers Pictures. An astronomer, having been contacted by aliens through radio waves, works with the government to build a giant machine to communicate with them.

Croupier (1998)
Starring: Clive Owen, Nick Reding. Director: Mike Hodges. Written by Paul Mayersberg; 94 minutes; Channel Four Films. A novelist takes a job at a posh London gaming club that quickly becomes fodder for his writing.

The Cruel Sea (1953)

Starring: Jack Hawkins, Donald Sinden, John Stratton. Director: Charles Frend. Written by Eric Ambler, adapted from the novel by Nicholas Monsarrat; 126 minutes; Ealing Studios. The veteran of a U-boat attack must decide between destroying a German boat or saving the lives of his inexperienced crew.

Dances with Wolves (1990)

Starring: Kevin Costner, Mary McDonnell, Graham Greene. Director: Kevin Costner. Written by Michael Blake; 180 minutes; Tig Productions. A former Union soldier renounces American culture and makes a new life among nature, animals, and Indians.

Daybreak (1939)

Starring: Jean Gabin, Jules Berry, Jacqueline Laurent. Director: Marcel Carné. Written by Jacques Prévert; 96 minutes; Sigma. To defend his honor, a man must kill the man who tried to seduce his lover. Following the murder, he is forced to analyze his life and the choices that brought him to his current situation.

The Defiant Ones (1958)

Starring: Tony Curtis, Sidney Poitier, Theodore Bikel. Director: Stanley Kramer. Written by Nedrick Young and Harold Jacob Smith; 97 minutes; Curtleigh Productions Inc. In order to survive and maintain their newfound freedom, two convicts handcuffed together are forced to cooperate with each other.

Deliverance (1972)

Starring: Jon Voight, Burt Reynolds, Ned Beatty. Director: John Boorman. Written by James Dickey; 109 minutes; Warner Bros. Pictures. An adaptation of the James Dickey novel about a group of friends who take a trip down a notoriously dangerous river, encountering a group of animalistic hicks.

The Detective (1968)

Starring: Frank Sinatra, Lee Remick, Ralph Meeker. Director: Gordon Douglas. Written by Abby Mann, adapted from the novel by Roderick Thorp; 114 minutes; Twentieth Century Fox. A detective investigating the murder of a young gay man discovers evidence pointing to rampant police corruption.

Detour (1945)

Starring: Tom Neal, Ann Savage, Claudia Drake. Director: Edgar G. Ulmer. Written by Martin Goldsmith; 67 minutes; Producers Releasing Corporation. A desperate drifter hitchhiking across American and searching for his girlfriend finds himself caught in a crime of identity theft.

The D.I. (1957)

Starring: Jack Webb, Don Dubbins, Jackie Loughery. Director: Jack Webb. Written by James Lee Barrett; 106 minutes; Mark VII Ltd. A young recruit threatens the power of a tough drill sergeant.

Diabolique (1955)

Starring: Simone Signoret, Vera Clouzot, Paul Meurisse. Director: H. G. Clouzot; Written by Jérôme Géronimi, Frédéric Grendel, and René Masson; 114 minutes; Filmsonor S.A. Together, two women plot to kill a man: a headmaster who is married to one of the women and having an affair with the other.

The Diary of Anne Frank (1959)

Starring: Millie Perkins, Joseph Schildkraut, Shelley Winters. Director: George Stevens. Written by Frances Goodrich and Albert Hackett, adapted from the book by Anne Frank; 180 minutes; Twentieth Century Fox. While eluding Nazis and hiding out with her family in an attic, a young Jewish girl falls in love with a fellow fugitive and meets a tragic end.

Dillinger (1945)

Starring: Lawrence Tierney, Edmund Lowe. Director: Max Nosseck. Written by Philip Yordan; 70 minutes; King Brothers Productions. Biopic that follows the rise of a small-time criminal, John Dillinger, as he becomes the quintessential American gangster through robberies, murders, and jailbreaks.

Dirty Harry (1971)

Starring: Clint Eastwood, Andy Robinson. Director: Don Siegel. Written by Harry Julian Fink, R. M. Fink, and Dean Reisner; 102 minutes; The Malpaso Company. A detective from San Francisco takes the law into his own hands as he hunts down a serial killer nicknamed "The Scorpio."

Dodsworth (1936)

Starring: Walter Huston, Ruth Chatterton. Director: William Wyler. Written by Sidney Howard; 101 minutes; Samuel Goldwyn Company. A self-made automobile tycoon begins to distance himself from his snobby wife after she reveals her attraction to another man.

Door to Door (2002)

Starring: William H. Macy, Helen Mirren, Kyra Sedgwick. Director: Steven Schachter. Written by William H. Macy and Steven Schachter; 90 minutes; Angel/Brown Productions. Despite the odds, a man suffering from cerebral palsy refuses charity and becomes a door-to-door salesman.

Dr. No (1962)

Starring: Sean Connery, Ursula Andress, Joseph Wiseman. Director: Terence Young. Written by Richard Maibaum, Johannah Harwood, and Berkely Mather, adapted from the novel by Ian Fleming; 110 minutes; Eon Productions Ltd. In this, the first James Bond movie, the British agent is sent to Jamaica to investigate a mysterious scientist and his involvement in the murder of a fellow spy.

Dr. Strangelove (1964)

Starring: Orson Welles, George C. Scott. Director: Stanley Kubrick. Written by Stanley Kubrick, Terry Southern, and Peter George; 93 minutes; Hawk Films Ltd. A black comedy about a general who announces his plans to start a nuclear war in a room full of politicians.

The Dukes of Hazzard (2005)

Starring: Johnny Knoxville, Sean William Scott, Jessica Simpson. Director: Jay Chandrasekhar. Written by John O'Brien and Jonathan L. Davis; 106 minutes; Gerber Pictures. A family of hillbillies constantly provokes members of law enforcement, who plot their revenge by trying to take the family's farm.

Dumbo (1941)

Director: Ben Sharpsteen. Written by Otto Englander, Helen Aberson, Joe Grant, and Dick Huemer; 64 minutes; Walt Disney Pictures. Due to his enormous ears, a circus elephant becomes an outcast, but with the help of a friend he learns what he is truly capable of.

The Entertainer (1960)

Starring: Laurence Olivier, Brenda De Banzie, Roger Livesey. Director: Tony Richardson. Written by Nigel Kneale and John Osborne; 96 minutes;

Woodfall Film Prodcutions. When things in his life start to come apart, a mediocre performer has no choice but to put on a good show at the seaside resort where he works.

Erin Brockovich (2000)

Starring: Julia Roberts, Aaron Eckhart, Albert Finney. Director: Steven Soderbergh. Written by Susannah Grant; 130 minutes; Jersey Films. A single mother of three takes a job as a legal secretary. After meeting several clients suffering from cancer and other illnesses, she begins to investigate a power company for illegal pollution of water.

Exodus (1960)

Starring: Paul Newman, Eva Marie Saint. Director: Otto Preminger. Written by Dalton Trumbo; 208 minutes; Carlyle Productions. A look at the founding and early days of Israel after the end of World War II, focusing on an Israeli resistance member who attempts to relocate six hundred Jews from Cyprus to Palestine.

Fahrenheit 9/11 (2004)

Director: Michael Moore. Written by Michael Moore; 122 minutes; Lions Gate Films. Documentary filmmaker Michael Moore's look at post–9/11 America and how the Bush Administration has manipulated this national tragedy to push forward a conservative political agenda.

Fail-Safe (1964)

Starring: Dan O'Herlihy, Walter Matthau. Director: Sidney Lumet. Written by Walter Bernstein, adapted from the novel by Eugene Burdick and Harvey Wheeler; 112 minutes; Columbia Pictures Corporation. An American air squadron is mistakenly sent to bomb Russia, forcing the American president to contemplate shooting down all the pilots in an attempt to stop them.

The Fatal Glass of Beer (1933)

Starring: W. C. Fields, Rosemary Thelby, George Chandler. Director: Clyde Bruckman. Written by W. C. Fields; 21 minutes; Mack Sennett. After leaving the Yukon to go to the big city, where he was briefly imprisoned, a young man returns home to his parents.

Flightplan (2005)

Starring: Jodie Foster, Peter Sarsgaard. Director: Robert Schwentke. Written by Peter A. Dowling and Billy Ray; 98 minutes; Touchstone Pictures. Flying home after the tragic death of her husband, a woman wakes up and realizes her daughter has disappeared mid-flight.

Forrest Gump (1994)

Starring: Tom Hanks, Robin Wright, Gary Sinise. Director: Robert Zemeckis. Written by Eric Roth, adapted from the novel by Winston Groom; 142 minutes; Paramount Pictures. A man born with a low IQ and leg problems makes his mark on American history, all the while pining for his childhood love, Jenny.

49th Parallel (1941)

Starring: Richard George, Eric Portman, Raymond Lovell. Director: Michael Powell. Written by Emeric Pressburger and Rodney Ackland; 123 minutes; Ortus Films Ltd. After sinking off the coast of Canada, the crew of a German U-boat has to flee to the neutral United States.

42nd Street (1933)

Starring: Warner Baxter, Bebe Daniels. Director: Lloyd Bacon. Written by Rian James and James Seymour, adapted from the novel by Bradford Ropes; 89 minutes; Warner Bros. Pictures. This twist on the classic Cinderella story has an average understudy take the place of the star in a Broadway show.

Friday the 13th (1980)

Starring: Betsy Palmer, Kevin Bacon. Director: Sean S. Cunningham. Written by Victor Miller; 95 minutes; Paramount Pictures. At a newly reopened summer camp, several counselors are murdered in a particularly gruesome fashion.

The Friends of Eddie Coyle (1973)

Starring: Robert Mitchum, Peter Boyle, Richard Jordan. Director: Peter Yates. Written by Paul Monash, adapted from the novel by George V. Higgins; 102 minutes; Paramount Pictures. On the eve of going to jail, a gangster agrees to supply guns for a bank robbery, planning to use the profits to provide for his wife while he is in jail.

The Front Page (1931)

Starring: Adolphe Menjou, Pat O'Brien, Mary Brian. Director: Lewis Milestone. Written by Bartlett Cormack, adapted from the play by Ben Hecht and Charles MacArthur; 101 minutes; The Caddo Company. A reporter planning to marry his fiancée faces opposition from his editor, who wants him to stay on.

Galaxy Quest (1999)

Starring: Tim Allen, Sigourney Weaver, Alan Rickman. Director: Dean Parisot. Written by David Howard and Robert Gordon; 102 minutes; DreamWorks SKG. While attending a sci-fi convention, the cast of a cult TV show find themselves beamed up by aliens who mistake them for actual space explorers.

General della Rovere (1959)

Starring: Vittorio De Sica, Hannes Messemer, Vittorio Caprioli. Director: Roberto Rossellini. Written by Sergio Amidei and Diego Fabbri; 129 minutes; Société Nouvelle des Établissements Gaumont (SNEG). After being arrested by the Nazis during the occupation of Italy, a swindler must pretend to be a recently executed Italian general.

Gentleman Jim (1942)

Starring: Errol Flynn, Alexis Smith, Jack Carson. Director: Raoul Walsh. Written by Vincent Lawrence, Horace McCoy, and James J. Corbett; 104 minutes; Warner Bros. Pictures. Biopic of boxer Jim Corbett, or "Gentleman Jim," who rose from meager beginnings to become the first heavyweight champion.

Gentleman's Agreement (1947)

Starring: Gregory Peck, Dorothy McGuire, John Garfield. Director: Elia Kazan. Written by Moss Hart, adapted from the novel by Laura Z. Hobson; 118 minutes; Twentieth Century Fox Film Corporation. In order to fully uncover the extent of anti-Semitism, a journalist claims to be Jewish and then writes an article about his experiences.

The Godfather (1972)

Starring: Marlon Brando, Al Pacino, James Caan. Director: Francis Ford Coppola. Written by Mario Puzo and Francis Ford Coppola; 175 minutes; Paramount Pictures. After years as the head of an organized crime syndicate, a powerful mob mogul passes power to an unlikely and unwilling son after an attack on his life.

The Godfather, Part II (1974)

Starring: Al Pacino, Robert Duvall, Diane Keaton. Director: Francis Ford Coppola. Written by Mario Puzo and Francis Ford Coppola; 200 minutes; Paramount Pictures. In this sequel to *The Godfather*, Vito Corleone's early life and rise to power is told in detail. Meanwhile, in the present day his son increases his control over his father's dynasty.

Godzilla (1954)

Starring: Akira Takarada, Momoko Kôchi. Director: Ishirô Honda. Written by Ishirô Honda, Takeo Murata, and Shigeru Kayama; 98 minutes; Toho Film (Eiga) Co. Ltd. A huge reptilian monster, created after nuclear weapons testing, terrorizes Japan.

Gold of Naples (1954)

Starring: Silvana Mangano, Sophia Loren. Director: Vittorio De Sica. Written by Vittorio De Sica, Giuseppe Marotta, and Cesare Zavattini; 131 minutes; Pomti–De Laurentiis Cinematografica. Film composed of six vignettes highlighting contemporary life in Naples.

Gone With the Wind (1939)

Starring: Clark Gable, Vivien Leigh. Director: Victor Fleming. Written by Sidney Howard, adapted from the novel by Margaret Mitchell; 222 minutes; Selznick International Pictures. A Southern belle, desperate to get the man she wants, meets and eventually marries a charming but ungentlemanly man.

Grand Illusion (1937)

Starring: Jean Gabin, Eric von Stroheim, Pierre Fresnay. Director: Jean Renoir. Written by Jean Renoir and Charles Spaak; 114 minutes; R.A.C. (Réalisation d'art cinématographique). In one of the earliest prison-break movies, three men attempt to escape a German POW camp during World War I.

The Greatest Story Ever Told (1965)

Starring: Max von Sydow, Michael Anderson Jr., Carroll Baker. Director: George Stevens. Written by James Lee Barrett and George Stevens; 225 minutes; George Stevens Productions. Epic biopic that follows the birth, life, and death of Jesus Christ.

The Green Mile (1999)

Starring: Tom Hanks, David Morse, Michael Clarke Duncan. Director: Frank Darabont. Written by Frank Darabont, adapted from the novel by Stephen King; 188 minutes; Castle Rock Entertainment. Several prison guards are forced to reconsider their racist views after discovering an innocent man on death row has the healing touch.

Guess Who's Coming to Dinner (1967)

Starring: Spencer Tracy, Katharine Hepburn, Sidney Poitier. Director: Stanley Kramer. Written by William Rose; 108 minutes; Stanley Kramer Productions. Classic film in which an affluent white couple are shocked when they discover that their daughter plans to marry a black man.

Gun Crazy (1949)

Starring: Peggy Cummins, John Dall. Director: Joseph H. Lewis. Written by MacKinlay Kantor and Dalton Trumbo; 86 minutes; King Bros. Productions Inc. A young gun-obsessed man and a female sharpshooter fall in love and take off on a crime spree.

Hail the Conquering Hero (1944)

Starring: Eddie Bracken, Ella Raines. Director: Preston Sturges. Written by Preston Sturges; 101 minutes; Paramount Pictures. Several Marines pass off a blue-collar friend as the hero of Guadalcanal and watch as his hometown celebrates their most famous citizen.

Halloween (1978)

Starring: Donald Pleasence, Jamie Lee Curtis. Director: John Carpenter. Written by John Carpenter and Debra Hill; 91 minutes; Compass International Pictures. After breaking out of a mental hospital, an insane murderer returns to the town where he killed his sister years before.

Here Comes Mr. Jordan (1941)

Starring: Robert Montgomery, Evelyn Keyes. Director: Alexander Hall. Written by Sidney Buchman and Seton I. Miller, adapted from the play (*Heaven Can Wait*) by Harry Segall; 94 minutes; Columbia Pictures Corporations. When heaven makes a mistake and calls up boxer Joe Pendleton too early, Joe gets compensated by being reborn as a millionaire heartthrob.

High Noon (1952)

Starring: Gary Cooper, Thomas Mitchell. Director: Fred Zinneman. Written by Carl Foreman and John W. Cunningham; 85 minutes; Stanley Kramer Procutions. On his wedding day, the town sheriff discovers an outlaw's plot to attack the town during the wedding. The sheriff has no choice but to appeal to the townspeople, who are reluctant to help.

Hoffa (1992)

Starring: Jack Nicholson, Danny DeVito, Armand Assante. Director: Danny DeVito. Written by David Mamet; 140 minutes; Twentieth Century Fox. Biopic based on the life of the famous union leader Jimmy Hoffa, including his battles with the government and the mysterious circumstances surrounding his disappearance.

Hollywoodism: Jews, Movies and the American Dream (1998)

Director: Simcha Jacobovici. Written by Simcha Jacobovici, adapted from the book (*An Empire of Their Own*) by Neal Gabler; 98 minutes; Canadian Broadcasting Corporation. An analysis of how Eastern European Jewish immigrants affected Hollywood in film's early years.

House of Games (1987)

Starring: Lindsay Crouse, Joe Mantegna, Mike Nussbaum. Director: David Mamet. Written by Jonathan Katz and David Mamet; 102 minutes; Filmhaus. A psychologist, attempting to help her patient break free of gambling debt, finds herself fascinated by the culture of the gambling house and its proprietor.

I Am a Fugitive from a Chain Gang (1932)

Starring: Paul Muni, Glenda Farrell, Helen Vinson. Director: Mervyn LeRoy. Written by Howard J. Green and Brown Holmes, adapted from the autobiographical book by Robert E. Burns; 93 minutes; The Vitaphone Corporation. An innocent man is haunted by his life in the prison system and terrorized by his time on a chain gang.

I Know Where I'm Going! (1945)

Starring: Wendy Hiller, Roger Livesey. Director: Michael Powell and Emeric Pressburger. Written by Michael Powell and Emeric Pressburger; 88 minutes; Rank Organisation. On the way to her wedding, a British woman finds herself stranded on an island in the Hebrides with a naval officer.

The Ice Storm (1997)

Starring: Kevin Kline, Joan Allen, Sigourney Weaver. Director: Ang Lee. Written by James Schamus, adapted from the novel by Rick Moody; 112 minutes; Fox Searchlight Pictures. The craziness of the 1970s reaches suburbia, as a pair of middle-class families walk down the path of sex, drugs, and alcohol, only to discover that they have lost control over their children.

I'll Cry Tomorrow (1955)

Starring: Susan Hayward, Richard Conte, Eddie Albert. Director: Daniel Mann. Written by Helen Deutsch and Jay Richard Kennedy, adapted from the book by Lillian Roth, Mike Connolly, and Gerold Frank; 117 minutes; Metro-Goldwyn-Mayer (MGM). After being forced by her mother into acting, a starlet's personal life crumbles as she becomes dependent on drugs and men.

I'll Sleep When I'm Dead (2003)

Starring: Clive Owen, Charlotte Rampling, Jonathan Rhys Meyers. Director: Mike Hodges. Written by Trevor Preston; 103 minutes; Revere Pictures. A former mobster searches with his ex-girlfriend for his brother's murderer, jeopardizing his dream of a quiet, crime-free life.

In Which We Serve (1942)

Starring: Noel Coward, John Mills, Bernard Miles. Director: Noel Coward and David Lean. Written by Noel Coward; 115 minutes; Two Cities Films Ltd. Classic film depicting three crew members of a British battleship who are marooned on a life raft in the Mediterranean.

Intolerance: Love's Struggle Throughout the Ages (1916)

Starring: Mae Marsh, Robert Harron, Fred Turner. Director: D.W. Griffith. Written by D.W. Griffith; 163 minutes; Triangle Film Corporation. Silent film that explores intolerance from preclassical times to contemporary America.

It's a Mad Mad Mad Mad World (1963)

Starring: Spencer Tracy, Milton Berle, Sid Caesar. Director: Stanley Kramer. Written by William Rose and Tania Rose; 192 minutes; Casey Productions. A policeman chases after several money-hungry people, all of whom are seeking a dead thief's buried treasure.

It's a Wonderful Life (1946)

Starring: James Stewart, Donna Reed. Director: Frank Capra. Written by Frances Goodrich, Albert Hackett, and Frank Capra; 130 minutes; Liberty Films Inc. An unbelievably good man whose business is failing gets a look at what life would be like if he had never been born.

Jaws (1975)

Starring: Roy Scheider, Robert Shaw, Richard Dreyfuss. Director: Steven Spielberg. Written by Peter Benchley; 124 minutes; Universal Pictures. After a shark threatens a local beach, a police chief, a shark hunter, and an oceanographer set out on a quest to kill it.

The Jazz Singer (1927)
Starring: Al Jolson, May McAvoy, Warner Oland. Director: Alan Crosland. Written by Alfred A. Cohn, adapted from the play by Samson Raphaelson; 88 minutes; Warner Bros. Pictures. A young man fulfills his dreams of becoming a jazz singer while going against his father's wishes.

Jolson Sings Again (1949)
Starring: Larry Parks, Barbara Hale, William Demarest. Director: Henry Levin. Written by Sidney Buchman; 96 minutes; Columbia Pictures Corporation. A continuation of Al Jolson's life from *The Jolson Story*, this film shows the singer's career as he comes out of retirement.

The Jolson Story (1946)
Starring: Larry Parks, Evelyn Keyes, William Demarest. Director: Alfred E. Green. Written by Stephen Longstreet; 128 minutes; Columbia Pictures Corporation. Biopic that follows Al Jolson, a celebrated performer who experienced unparalleled success, through the ups and downs of his life.

The Joy Luck Club (1993)
Starring: Kieu Chinh, Tsai Chin, France Nuyen. Director: Wayne Wang. Written by Amy Tan and Ronald Bass; 139 minutes; Hollywood Pictures. Multigenerational story about four Chinese women who were raised in a male-dominated culture and their American daughters who have led very different lives.

Khartoum (1966)
Starring: Charlton Heston, Laurence Olivier. Director: Basil Dearden, Eliot Elisofon. Written by Robert Ardrey; 134 minutes; Julian Blaustein Productions Ltd. Set in the 1830s during the British defeat in North Africa, the film follows a Christian general who disregards orders and battles the invading Arab forces.

The Killing (1956)
Starring: Sterling Hayden, Coleen Gray, Vince Edwards. Director: Stanley Kubrick. Written by Stanley Kubrick, adapted from the novel (*Clean Break*) by Lionel White; 85 minutes; Harris-Kubrick Productions. Upon his release from jail, a criminal attempts to rob a racetrack only to have his plan foiled.

King of Marvin Gardens (1972)

Starring: Jack Nicholson, Bruce Dern, Ellen Burstyn. Director: Bob Rafelson. Written by Jacob Brackman and Bob Rafelson; 103 minutes; BBS Productions. A man goes to Atlantic City to help his brother escape from jail through a series of well-played cons.

Kiss the Blood off My Hands (1948)

Starring: Joan Fontaine, Burt Lancaster, Robert Newton. Director: Norman Foster. Written by Leonardo Bercovici; 79 minutes; Norma Productions Inc. A nurse gets the shock of her life when she discovers a fugitive hiding out in her apartment.

Knute Rockne All American (1940)

Starring: Pat O'Brien, Gale Page. Director: Lloyd Bacon. Written by Robert Buckner; 98 minutes; First National Pictures Inc. A biopic of the famous Notre Dame football coach Knute Rockne and his innovative strategies that allowed Notre Dame to dominate the sport.

The Krays (1990)

Starring: Billie Whitelaw, Tom Bell, Gary Kemp, Martin Kemp. Director: Peter Medak. Written by Philip Ridley; 119 minutes; Parkfield Entertainment. Twin brothers rule organized crime with iron fists in sixties London, but their violent ways come back to haunt them.

The Lady Eve (1941)

Starring: Barbara Stanwyck, Henry Fonda, Charles Coburn. Director: Preston Sturges. Written by Preston Sturges and Monckton Hoffe; 97 minutes; Paramount Pictures. After a failed love affair, a con artist terrorizes a dimwitted, but incredibly wealthy, man, inadvertently falling in love with him as he discovers her scheme.

The Last Temptation of Christ (1988)

Starring: Willem Dafoe, Harvey Keitel, Verna Bloom. Director: Martin Scorsese. Written by Paul Schrader, adapted from the novel by Nikos Kazantzakis; 164 minutes; Universal Pictures. In this version of the story of Jesus Christ, Jesus is portrayed as an unwilling savior who is tempted to avoid his destiny by the possibility of life with Mary Magdalene.

A League of Their Own (1992)

Starring: Tom Hanks, Geena Davis, Madonna. Director: Penny Marshall. Story written by Kim Wilson, Kelly Candaele. Screenplay by Lowell Ganz and Babaloo Mandel; 128 minutes; Columbia Pictures Corporation. Two small-town sisters become stars in a woman's baseball league created during World War II, coached by a former superstar who is deeply scarred after years of heavy drinking.

The Life and Death of Colonel Blimp (1943)

Starring: James McKechnie, Neville Mapp, Vincent Holman. Director: Michael Powell and Emeric Pressburger. Written by Michael Powell and Emeric Pressburger; 163 minutes; Independent Producers. After a lifetime in military service, a retired general looks back on four decades of fighting, three wars, and multiple romances.

Life With Father (1947)

Starring: William Powell, Irene Dunne, Elizabeth Taylor. Director: Michael Curtiz. Written by Howard Lindsay and Russel Crouse, adapted from the book by Clarence Day; 128 minutes; Warner Bros. Pictures. An adaptation of Clarence Day's nostalgic memoir about the man who shaped his adolescence: his father, who ran his family like a company.

Lock, Stock, and Two Smoking Barrels (1998)

Starring: Jason Flemyng, Jason Statham, Vinnie Jones. Director: Guy Ritchie. Written by Guy Ritchie; 108 minutes; Polygram Filmed Entertainment. Four Londoners are forced to steal from thieves after losing a high-stakes poker game to a high-ranking mobster.

The Long Good Friday (1980)

Starring: Bob Hoskins, Helen Mirren, Dave King. Director: John MacKenzie. Written by Barrie Keeffe; 109 minutes; Black Lion Films Limited. As a new generation of gangsters rises to power, an underworld boss must fight off the competition.

The Magnificent Ambersons (1942)

Starring: Joseph Cotton, Dolores Costello, Anne Baxter. Director: Orson Welles. Written by Orson Welles, adapted from the novel by Booth Tarkington; 148 minutes; Mercury Productions Inc. As his once-wealthy family declines, a man plots to marry his first love after the death of his wife, much to the chagrin of his son.

Masada (miniseries, 1981)

Starring: Peter O'Toole, Peter Strauss. Director: Boris Segal. Written by Joel Oliansky and Ernest K. Gann; 394 minutes; Arnon Milchan Productions. A Roman commander hoping to compromise with the local Jewish population finds his fortress under attack.

The Matrix (1999)

Starring: Keanu Reeves, Laurence Fishburne, Carrie-Anne Moss. Directors: Andy Wachowski and Larry Wachowski. Written by Andy Wachowski and Larry Wachowski; 136 minutes; Silver Pictures. A group of leather-clad rebels attempts to show a computer geek the true nature of the world in an effort to defeat their robotic captors.

The Miracle of Morgan's Creek (1944)

Starring: Eddie Bracken, Betty Hutton, Diana Lynn. Director: Preston Sturges. Written by Preston Sturges; 99 minutes; Paramount Pictures. A farcical story about a woman who finds herself pregnant and clueless as to who the father is after an all-night party.

Miracle on 34th Street (1947)

Starring: Maureen O'Hara, John Payne, Edmund Gwenn. Director: George Seaton. Written by George Seaton and Valentine Davies; 96 minutes; Twentieth Century Fox Film Corporation. Heartwarming tale about a young girl whose faith in the existence of Santa Claus is restored as she befriends an old man who claims to be Kris Kringle himself.

Mona Lisa (1986)

Starring: Bob Hoskins, Cathy Tyson, Michael Caine. Director: Neil Jordan. Written by Neil Jordan and David Leland; 104 minutes; Handmade Films Ltd. George, a criminal with a sense of morals searching for a job after being released from prison, gets a job as the chauffeur of a pricey call girl.

The Narrow Margin (1952)

Starring: Charles McGraw, Marie Windsor, Jacqueline White. Director: Richard Fleischer. Written by Earl Felton, Martin Goldsmith, and Jack Leonard; 71 minutes; RKO Radio Pictures. The widow of a mobster is planning to testify at a grand jury despite several death threats; a hard-nosed detective is assigned to escort her as she makes the train trip from Chicago to Los Angeles, where she will give testimony.

New Orleans (1947)

Starring: Arturo De Cordova, Dorothy Patrick, Marjorie Lord, Billie Holiday. Director: Arthur Lubin. Written by Elliot Paul, Dick Irving Hyland, and Herbert J. Biberman; 90 minutes; Majestic Productions Inc. As jazz begins to captivate America, a gambler falls for a wealthy woman who attempts to reform him into respectability.

Night and the City (1950)

Starring: Richard Widmark, Gene Tierney, Googie Withers. Director: Jules Dassin. Written by Jo Eisenger, adapted from the novel by Gerald Kirsh; 101 minutes; Twentieth Century Fox Productions Ltd. A lawyer seduces the wife of an old friend who is too busy to notice, as he is trying to scam everyone he knows.

Not a Love Story: A Film About Pornography (1981)

Director: Bonnie Sherr Klein. Written by Andrée Klein, Bonnie Sherr Klein, Irene Lilienheim Angelico, and Rose-Aimée Todd; 70 minutes; National Film Board of Canada (NFB). Documentary that examines the misogynistic practices of American pornography culture.

Now, Voyager (1942)

Starring: Bette Davis, Paul Henreid, Claude Rains. Director: Irving Rapper. Written by Casey Robinson, adapted from the novel by Olive Higgins Prouty; 117 minutes; Warner Bros. Pictures. An ugly spinster transforms into a beautiful young lady through therapy, leading her down the path toward love.

One Million B.C. (1940)

Starring: Victor Mature, Carole Landis, Lon Chaney Jr. Director: Hal Roach and Hal Roach Jr. Written by Mickell Novack, George Baker, and Joseph Frickert; 80 minutes; Hal Roach Studios Inc. After being exiled by his tribe, a prehistoric man finds a new home where he learns manners.

One of Our Aircraft Is Missing (1942)

Starring: Godfrey Tearle, Eric Portman, Hugh Williams. Director: Michael Powell and Emeric Pressburger. Written by Michael Powell and Emeric Pressburger; 103 minutes; British National Films Ltd. After being shot down over Holland, a Royal Air Force crew must escape to England before being found by the Nazis.

Only Angels Have Wings (1939)

Starring: Cary Grant, Jean Arthur. Director: Howard Hawks. Written by Jules Furthman and Howard Hawks; 121 minutes; Columbia Pictures Corporation. An air squadron commander strikes up a relationship with a glamorous woman in South America after dismissing her husband from the squad.

The Ox-Bow Incident (1943)

Starring: Henry Fonda, Dana Andrews, Mary Beth Hughes. Director: William A. Wellman. Written by Lamar Trotti, adapted from the novel by Walter Van Tilburg Clark; 75 minutes; Twentieth Century Fox Film Corporation. After discovering the innocence of three men about to be lynched, a pair of drifters attempts to stop the execution.

Panic Room (2002)

Starring: Jodie Foster, Kristen Stewart, Forest Whitaker. Director: David Fincher. Written by David Koepp; 112 minutes; Columbia Pictures Corporation. One night in a new house, a woman and her daughter are forced to hide in a "panic room" as three men search their house for money.

The Passion of the Christ (2004)

Starring: James Caviezel, Maia Morgenstern, Hristo Jivkov. Director: Mel Gibson. Written by Benedict Fitzgerald and Mel Gibson; 126 minutes; Icon Productions. A brutally detailed account of the last hours and the death of Jesus Christ.

Paths of Glory (1957)

Starring: Kirk Douglas, Ralph Meeker, Adolphe Menjou. Director: Stanley Kubrick. Written by Stanley Kubrick, Calder Willingham, and Jim Thompson, adapted from the novel by Humphrey Cobb; 86 minutes; Bryna Productions. After an officer leads his men into a battle that they have no chance of winning, he must defend several of his men against attacks of desertion and cowardice.

Peeping Tom (1960)

Starring: Carol Böhm, Moira Shearer, Anna Massey. Director: Michael Powell. Written by Leo Marks; 101 minutes; Anglo-Amalgamated Productions. An insane killer gains notoriety for photographing the last expressions of his dying victims.

Penny Serenade (1941)

Starring: Irene Dunne, Cary Grant. Director: George Stevens. Written by Morrie Ryskind and Martha Cheavens; 117 minutes; Columbia Pictures Corporation. As a woman prepares to abandon her husband, she remembers their struggles to have children and listens to the music that characterized their relationship.

Pillow Talk (1959)

Starring: Rock Hudson, Doris Day, Tony Randall. Director: Michael Gordon. Written by Maurice Richlin, Stanley Shapiro, Russell Rouse, and Clarence Greene; 98 minutes; Universal International Pictures. Two combative neighbors share a phone line; as a prank, the man begins to court the woman using a disguised voice.

The Pink Panther (1963)

Starring: Peter Sellers, Robert Wagner, David Niven. Director: Blake Edwards. Written by Maurice Richlin and Blake Edwards; 113 minutes; Geoffrey Productions Inc. An incompetent French detective hunts for a jewel thief who has stolen a famous diamond known as the Pink Panther.

A Place in the Sun (1951)

Starring: Montgomery Clift, Elizabeth Taylor, Shelley Winters. Director: George Stevens. Written by Michael Wilson and Harry Brown, adapted from the play *An American Tragedy* by Patrick Kearney and the novel of the same name by Theodore Dreiser; 122 minutes; Paramount Pictures. George, a blue-collar factory worker, begins a passionate relationship with a coworker, but after leaving her for a wealthy, beautiful woman, his spurned lover decides she wants him back.

Playing for Time (1980)

Starring: Vanessa Redgrave, Jane Alexander, Maud Adams. Director: Daniel Mann. Written by Arthur Miller, adapted from the autobiography *The Musicians of Auschwitz* by Fania Fénelon; 105 minutes; Szygzy Productions. A group of female prisoners at Auschwitz forms an orchestra and performs for the Nazis in order to save themselves from the Final Solution.

Plunder Road (1957)

Starring: Gene Raymond, Jeanne Cooper, Wayne Morris. Director: Hubert Cornfield. Written by Steven Ritch and Jack Charney; 76 minutes; Regal Films Inc. Five men attempt to flee across the country with $10 million from a train robbery that's on the front page of every newspaper.

Point Blank (1967)

Starring: Lee Marvin, Angie Dickinson, Keenan Wynn. Director: John Boorman. Written by Alexander Jacobs, David Newhouse, and Rafe Newhouse, adapted from the novel by Richard Stark; 92 minutes; Metro-Goldwyn-Mayer (MGM). A gangster is betrayed by his wife and his best friend, who leave him for dead; years later, he returns to exact revenge.

Porgy and Bess (1959)

Starring: Sidney Poitier, Dorothy Dandridge, Sammy Davis Jr. Director: Otto Preminger. Written by DuBose Heyward, Dorothy Heyward (both writers of the libretto for the play *Porgy*), and N. Richard Nash. 138 minutes; Samuel Goldwyn Company. A look at the beautiful but complicated relationship that emerges between the radiant Bess and the disabled Porgy, adapted from the Gershwin opera.

The Postman Always Rings Twice (1981)

Starring: Jack Nicholson, Jessica Lange, John Colicos. Director: Bob Rafelson. Written by David Mamet, adapted from the novel by James M. Cain; 122 minutes; CIP Filmproduktion GmbH. A young drifter begins to lust after a married woman, but their electric affair leads to murder.

The Pride of the Yankees (1942)

Starring: Gary Cooper, Teresa Wright, Babe Ruth. Director: Sam Wood. Written by Jo Swerling, Herman J. Mankiewicz, and Paul Gallico; 128 minutes; Samuel Goldwyn Company. A biopic portraying the life and decline of one of the most celebrated Yankees, Lou Gehrig, as he plays in more than two thousand consecutive games only to be felled by the illness now named after him.

Prince of the City (1981)

Starring: Treat Williams, Jerry Orbach, Richard Foronjy. Director: Sidney Lumet. Written by Jay Presson Allen and Sidney Lumet, adapted from the book by Robert Daley; 167 minutes; Orion. A New York City narcotics detective begins to work with Internal Affairs and uncovers widespread corruption in his department that implicates his friends.

Pygmalion (1938)

Starring: Leslie Howard, Wendy Hiller. Directors: Anthony Asquith, Leslie Howard. Written by W. P. Liscomb and Cecil Lewis, adapted from the play by George Bernard Shaw; 96 minutes; Gabriel Pascal Productions. A cocky speech expert bets a friend that he can make a lower-class woman into a proper lady in manner, dress, and speech.

The Quest for Fire (1981)

Starring: Everett McGill, Ron Perlman, Rae Dawn Chong. Director: Jean-Jacques Annaud. Written by Gérard Brach, adapted from the novel by J. H. Rosny Sr.; 100 minutes; Belstar Productions. A group of cavemen attempt to master the creation of fire.

Quicksand (1950)

Starring: Mickey Rooney, Jeanne Cagney, Barbara Bates. Director: Irving Pichel. Written by Robert Smith; 79 minutes; Samuel H. Stiefel Productions. After stealing twenty dollars at work, a mechanic finds himself in rapidly increasing debt as a disreputable carnival owner begin to extort him.

Rembrandt (1936)

Starring: Charles Laughton, Gertrude Lawrence, Elsa Lanchester. Director: Alexander Korda. Written by June Head, Lajos Biró, and Carl Zuckmayer; 86 minutes; London Film Productions. Biopic of the Dutch painter as he struggles through personal tragedy late in his career, finding love in the most unlikely of places.

Retreat, Hell! (1952)

Starring: Frank Lovejoy, Richard Carlson, Anita Louise. Director: Joseph H. Lewis. Written by Ted Sherdeman and Milton Sperling; 94 minutes; United States Pictures. A group of American soldiers must fight against impossible odds during the Korean War.

Rififi (1955)

Starring: Jean Servais, Carl Möhner, Robert Manuel. Director: Jules Dassin. Written by Jules Dassin, René Wheeler, and Auguste Le Breton, adapted from the novel by Auguste Le Breton; 115 minutes; Indusfilms. In a classic case of the perfect crime gone wrong, four men map out a flawless robbery only to have a woman put everything in turmoil.

The Rise and Fall of Legs Diamond (1960)

Starring: Ray Danton, Karen Steele, Elaine Steward. Director: Budd Boetticher. Written by Joseph Landon; 101 minutes; United States Pictures. Two low-life crooks rise to prominence by stealing from fellow criminals during Prohibition.

Robbery (1967)

Starring: Stanley Baker, Joanna Pettet, James Booth. Director: Peter Yates. Written by Edward Boyd, George Markstein, Gerald Wilson, and Peter Yates; 110 minutes; Oakhurst Productions. A group of men create a detailed plan to rob the British Royal Mail train.

The Robe (1953)

Starring: Richard Burton, Jean Simmons, Victor Mature. Director: Henry Koster. Written by Gina Kaus, Albert Maltz, and Philip Dunne, adapted from the novel by Lloyd C. Douglas; 133 minutes; Twentieth Century Fox Film Corporation. After the execution of Jesus Christ, members of the Roman tribunal responsible for his death begin to have hallucinations that they believe are caused by Jesus' robe.

The Rules of the Game (1939)

Starring: Nora Grégor, Paulette Dubost, Mila Parély. Director: Jean Renoir., Written by Carl Koch; 143 minutes; Nouvelle edition française. When rich aristocrats and their poor servants camp out in a castle just before the World War II, the social tension builds, culminating in a murder.

Run Silent Run Deep (1958)

Starring: Clark Gable, Burt Lancaster, Jack Warden. Director: Robert Wise. Written by John Gay, adapted from the novel by Commander Edward L. Beach; 93 minutes; Hill-Hecht-Lancaster Productions. After a demotion, a U.S. submarine commander must learn to deal with his new boss as they hunt down a Japanese cruiser during World War II.

Sailor of the King (1953)

Starring: Jeffrey Hunter, Michael Rennie, Peter van Eyck. Director: Ray Boulting. Written by Valentine Davies, adapted from the novel by C. S. Forester; 85 minutes; Twentieth Century Fox Productions Ltd. In order to save his son, a British officer must make his way through the German army during World War II.

Sands of Iwo Jima (1949)

Starring: John Wayne, John Agar, Adele Mara. Director: Allen Dwan. Written by James Edward Grant and Harry Brown; 100 minutes; Republic Pictures Corporation. A notoriously tough sergeant attempts to keep his men alive during the epic World War II Battle of Iwo Jima.

Schindler's List (1993)

Starring: Liam Neeson, Ben Kingsley, Ralph Fiennes. Director: Steven Spielberg. Written by Steven Zaillian, adapted from the book by Thomas Keneally; 195 minutes; Amblin Entertainment. This drama retells the true story of Oskar Schindler, a factory owner who helped save the lives of hundreds of Polish Jews during World War II.

Sergeant York (1941)

Starring: Gary Cooper, Walter Brennan, Joan Leslie. Director: Howard Hawks. Written by Abem Finkel, Harry Chandlee, Howard Koch, and John Huston, based on a diary by Sergeant York; 134 minutes; Warner Bros. Pictures. A deeply religious country boy refuses to join the army, only to be drafted and become a decorated war hero.

Serpico (1973)

Starring: Al Pacino, John Randolph Jr., Jack Kehoe. Director: Sidney Lumet. Written by Waldo Salt and Norman Wexler, adapted from the book by Peter Maas; 129 minutes; Artists Entertainment Complex. A by-the-book cop in New York City uncovers and reveals departmental corruption, causing his former allies to become his enemies.

Sexy Beast (2000)

Starring: Ben Kingsley, Ray Winstone, Ian McShane. Director: Jonathan Glazer. Written by Louis Mellis and David Scinto; 89 minutes; FilmFour. Two gangsters find their peaceful retirement hindered by the return of an old friend who offers one last job.

The Shadow Box (1980)

Starring: Joanne Woodward, Christopher Plummer, Valerie Harper. Director: Paul Newman. Written by Michael Christofer; 96 minutes; Shadow Box Film Company. Adaptation of the play that depicts a day in the life of several terminally ill patients as they choose to live out their final moments in an experimental community.

Shadow of a Doubt (1943)

Starring: Teresa Wright, Joseph Cotton. Director: Alfred Hitchcock. Written by Thornton Wilder, Sally Benson, Alma Reville, and Gordon McDonnell; 108 minutes; Skirball Productions. A young girl, suspicious of her distant uncle, begins to suspect him of being a serial killer and attempts to reveal his crimes without raising suspicion.

The Sixth Sense (1999)

Starring: Bruce Willis, Haley Joel Osment, Toni Collette. Director: M. Night Shyamalan. Written by M. Night Shyamalan; 107 minutes; Barry Mendel Productions. Much to his mother's shock, as well as his own horror, a young boy realizes that he can see, hear, and speak with the dead.

Snatch (2000)

Starring: Brad Pitt, Jason Statham, Stephen Graham. Director: Guy Ritchie. Written by Guy Ritchie; 104 minutes; Columbia Pictures Corporation. Several less-than-competent criminals search for the stolen diamond that will make them rich beyond their wildest dreams.

Some Like it Hot (1959)

Starring: Marilyn Monroe, Jack Lemmon, Tony Curtis. Director: Billy Wilder. Written by Billy Wilder, I.A.L. Diamond, R. Thoeren, and M. Logan; 120 minutes; Ashton Productions. After witnessing a murder, two men go into hiding as the newest members of an all-girl band.

Sophie's Choice (1982)

Starring: Meryl Streep, Kevin Kline, Peter MacNicol. Director: Alan J. Pakula. Written by Alan J. Pakula, adapted from the novel by William Styron; 150 minutes; Incorporated Television Company (ITC). A woman attempts to hide her experiences during the Holocaust from two men who are deeply in love with her.

Spy Game (2001)

Starring: Robert Redford, Brad Pitt. Director: Tony Scott. Written by Michael Frost Beckner and David Arata; 126 minutes; Beacon Communications LLC. Moments from retirement, a CIA agent must risk everything to save his protégé from captivity.

State and Main (2000)

Starring: Alec Baldwin, Sarah Jessica Parker, William H. Macy, Philip Seymour Hoffman. Director: David Mamet. Written by David Mamet; 105 minutes; Filmtown Entertainment. In a dark look at fame-obsessed culture, a movie crew shakes up life in a sleepy New England town.

State of the Union (1948)

Starring: Spencer Tracy, Katharine Hepburn, Van Johson. Director: Frank Capra. Written by Myles Connolly and Anthony Veiller, adapted from the play by Russel Crouse and Howard Lindsay; 124 minutes; Liberty Films Inc. A businessman, convinced by conniving politicians to enter a presidential race, starts speaking his own mind and ignoring the advice of his handlers.

Strategic Air Command (1955)

Starring: James Stewart, June Allyson. Director: Anthony Mann. Written by Valentine Davies and Beirne Lay Jr.; 114 minutes; Paramount Pictures. A celebrity baseball player gives it all up to serve his country.

The Sum of All Fears (2002)

Starring: Ben Affleck, Morgan Freeman, James Cromwell. Director: Phil Alden Robinson. Written by Paul Attanasio and Daniel Pyne, adapted from the novel by Tom Clancy; 124 minutes; Paramount Pictures. While the president attends the Super Bowl, a terrorist cell plans to set off a nuclear bomb at the event as a CIA agent tries to avert the crisis.

Sweet Smell of Success (1957)

Starring: Burt Lancaster, Tony Curtis, Susan Harrison. Director: Alexander Mackendrick. Written by Clifford Odets and Ernest Lehman; 96 minutes; Hill-Hecht-Lancaster Productions. A New York journalist uses a ruthless press agent in attempt to derail his sister's engagement before the wedding.

T-Men (1947)

Starring: Dennis O'Keefe, Mary Meade, Alfred Ryder. Director: Anthony Mann. Written by John C. Higgins and Virginia Kellogg; 92 minutes. Two agents from the United States Treasury infiltrate a counterfeiting ring with an unusually good talent for replicating actual currency.

Taxi Driver (1976)

Starring: Robert De Niro, Cybill Shepherd, Peter Boyle. Director: Martin Scorsese. Written by Paul Schrader; 113 minutes; Columbia Pictures Corporation. Isolated in New York City, a Vietnam vet takes it upon himself to violently liberate an adolescent prostitute from her pimp.

The Ten Commandments (1956)

Starring: Charlton Heston, Yul Brynner, Anne Baxter. Director: Cecil B. DeMille. Written by Æneas MacKenzie, Jesse Lasky Jr., Jack Gariss, and Fredric M. Frank; 220 minutes; Paramount Pictures. Based on the biblical story wherein Moses rises from slavery to lead his people to the promised land, Israel.

The Terminator (1984)

Starring: Arnold Schwarzenegger, Michael Biehn, Linda Hamilton. Director: James Cameron. Written by Gale Anne Hurd and James Cameron; 108 minutes; Hemdale Film Corporation. A futuristic man is sent from the future to save a woman and her son from a threatening, murderous cyborg.

That Hamilton Woman (1941)

Starring: Vivien Leigh, Laurence Olivier, Alan Mowbray. Director: Alexander Korda. Written by Walter Reisch and R. C. Sherriff; 128 minutes; Alexander Korda Films. A married socialite who's made the perfect match falls in love with a naval officer and becomes an outcast.

Thelma & Louise (1991)

Starring: Susan Sarandon, Geena Davis, Harvey Keitel. Director: Ridley Scott. Written by Callie Khouri; 129 minutes; Metro-Goldwyn-Mayer. Two friends, hoping to flee to Mexico after accidentally murdering an attempted rapist, bond as the authorities chase after them.

They Gave Him a Gun (1937)

Starring: Spencer Tracy, Gladys George, Franchot Tome. Director: W. S. Van Dyke II. Written by Cyril Hume, Richard Maibaum, and Maurice Rapf, adapted from the book by William Joyce Cowen; 94 minutes; Metro-Goldwyn-Mayer (MGM). During World War I, two soldiers realize they love the same woman, a nurse at the local base hospital.

Three Days of the Condor (1975)

Starring: Robert Redford, Faye Dunaway, Cliff Robertson. Director: Sydney Pollack. Written by Lorenzo Semple Jr. and David Rayfiel, adapted from the novel by James Grady; 117 minutes; Paramount Pictures. While taking part in an undercover investigation, a CIA agent finds his colleagues murdered and must run for his life.

Titanic (1997)

Starring: Leonardo DiCaprio, Kate Winslet, Billy Zane. Director: James Cameron. Written by James Cameron; 194 minutes; Twentieth Century Fox. While aboard the *Titanic* a wealthy girl engaged to a man she does not love becomes frustrated with her life and finds comfort in a lower-class man.

Training Day (2001)

Starring: Denzel Washington, Ethan Hawke, Scott Glenn. Director: Antoine Fuqua. Written by David Ayer; 120 minutes; Warner Bros. Pictures. A rookie cop discovers that the man training him is not merely an unorthodox detective but a vicious criminal.

Trapeze (1956)

Starring: Burt Lancaster, Tony Curtis, Gina Lollobrigida. Director: Carol Reed. Written by Liam O'Brien and James R. Webb, adapted from the novel (*The Killing Frost*) by Max Catto; 105 minutes; Hill-Hecht-Lancaster Productions. A former superstar acrobat who is past his prime takes on a protégé, only to have their budding friendship threatened by a woman.

The Triumph of the Will (1935)

Director: Leni Riefenstahl. Written by Leni Riefenstahl and Walter Ruttmann; 114 minutes; Leni Riefenstahl-Produktion. The infamous Nazi propaganda documentary centered on a rally in Nuremberg prior to the start of World War II.

Tunes of Glory (1960)

Starring: Alec Guinness, John Mills, Dennis Price. Director: Ronald Neame. Written by James Kennaway; 106 minutes; Knightsbridge Films. When a highly qualified general finds himself demoted and a younger man takes his position, a bitter rivalry quickly develops.

Twelve O'Clock High (1949)

Starring: Gregory Peck, Hugh Marlowe. Director: Henry King. Written by Sy Bartlett and Beirne Lay Jr.; 132 minutes; Twentieth Century Fox Film Corporation. In this Oscar-winning film set during World War II, a general takes over a British bomber division and inspires his men to victory, growing closer to them throughout their training.

The Untouchables (1987)

Starring: Kevin Costner, Sean Connery, Charles Martin Smith. Director: Brian De Palma. Written by David Mamet, adapted from the novel by Oscar Fraley and Eliot Ness; 119 minutes; Paramount Pictures. An elite squad of FBI agents attempts to arrest the legendary Chicago mob boss Al Capone.

Wag the Dog (1997)

Starring: Dustin Hoffman, Robert De Niro, Anne Heche. Director: Barry Levinson. Written by Hilary Henkin and David Mamet, adapted from the book (*American Hero*) by Larry Beinhart; 97 minutes; New Line Cinema. In order to divert the nation's attention from the president's molestation of a girl, a group of political consultants creates and televises a fake war.

The Wages of Fear (1953)

Starring: Yves Montand, Charles Vanel, Peter Van Eyck. Director: Henri-Georges Clouzot. Written by Henri-Georges Clouzot and Jérome Geronimi, adapted from the novel by Georges Arnaud; 148 minutes; CICC. In a small South American town, a crew is selected and paid to move nitroglycerine without proper safety equipment.

Whale Rider (2002)

Starring: Keisha Castle-Hughes, Rawiri Paratene, Vicky Haughton. Director: Niki Caro. Written by Niki Caro, adapted from the novel by Witi Ihimaera; 105 minutes; ApolloMedia. A young girl from a long line of whale riders breaks tradition to fulfill her dreams and become the leader of her Maori tribe.

Whose Life Is It Anyway? (1981)

Starring: Richard Dreyfuss, John Cassavetes, Christine Lahti. Director: John Badham. Written by Reginald Rose, adapted from the play by Brian Clark; 118 minutes; Metro-Goldwyn-Mayer (MGM). After a tragic accident leaves a brilliant and talented sculptor paralyzed, he argues for his right to be allowed to die.

Witness for the Prosecution (1957)

Starring: Tyrone Power, Marlene Dietrich, Charles Laughton. Director: Billy Wilder. Written by Larry Marcus, Billy Wilder, and Harry Kurnitz, adapted from the play by Agatha Christie; 116 minutes; Edward Small Productions. An aging lawyer is asked to defend a man against charges of murdering a wealthy widow.

The Wizard of Oz (1939)

Starring: Judy Garland, Frank Morgan, Ray Bolger. Director: Victor Fleming. Written by Noel Langley, Florence Ryerson, and Edgar Allan Woolf, adapted from the novel (*The Wonderful Wizard of Oz*) by L. Frank Baum; 101 minutes; Metro-Goldwyn-Mayer (MGM). While a tornado strikes, Dorothy, a teenager from Kansas, and her dog, Toto, are transported to a magical world.

Zulu (1964)

Starring: Stanley Baker, Jack Hawkins. Director: Cy Endfield. Written by John Prebble and Cy Endfield; 138 minutes; Diamond Films. In Africa during the late nineteenth century, a battalion of outnumbered English soldiers fights against the Zulus.

INDEX

ALSO BY DAVID MAMET

BOSTON MARRIAGE

In this droll comedy of errors set in a Victorian drawing room, Anna and Claire are two bantering, scheming "women of fashion" who live together on the fringes of society. Anna has just become the mistress of a wealthy man. Claire, meanwhile, is infatuated with a young girl and wants to enlist the jealous Anna's help for an assignation. As the two women exchange barbs, Claire's inamorata arrives and sets off a crisis that puts both women's futures at risk.

Drama/978-0-375-70665-3

THE CABIN
Reminiscence and Diversions

The pieces in *The Cabin* are about places and things: the suburbs of Chicago, where as a boy David Mamet helplessly watched his stepfather terrorize his sister; New York City, where as a young man he had to eat his way through a mountain of fried matzoh to earn a night of sexual bliss. They are about guns, campaign buttons, and a cabin in the Vermont woods that stinks of wood smoke and kerosene—and about their associations of pleasure, menace, and regret.

Memoir/Essays/978-0-679-74720-8

THE CRYPTOGRAM

The Cryptogram is a journey back into childhood and the moment of its vanishing—the moment when the sheltering world is suddenly revealed as a place full of dangers. On a night in 1959 a boy is waiting to go on a camping trip with his father. A family friend is trying to entertain them—or perhaps distract them. Because in the dark corners of this domestic scene, there are rustlings that none of the players want to hear. And out of things as innocuous as a shattered teapot and a ripped blanket, Mamet re-creates a child's terrifying discovery that the grown-ups are speaking in code, and that that code may never be breakable.

Drama/978-0-679-74653-9

OLEANNA

A male college instructor and his female student sit down to discuss her grades and in a short time become participants in a modern reprise of the Inquisition. Socratic dialogue gives way to heated assault. And the relationship between this somewhat fatuous teacher and his seemingly hapless pupil turns into a fiendishly accurate X-ray of the mechanisms of power, censorship, and abuse.

Drama/978-0-679-74536-5

THREE USES OF THE KNIFE
On the Nature and Purpose of Drama

With bracing directness, one of our greatest living playwrights addresses the questions: What makes good drama? And why does drama matter in an age that is awash in information and entertainment? David Mamet believes that the tendency to dramatize is essential to human nature, that we create drama out of everything from today's weather to next year's elections. With a cultural range that encompasses Shakespeare, Brecht, and Ibsen, Mamet shows us how to distinguish true drama from its false variants. The result is an electrifying treatise on the playwright's art that is also a strikingly original work of moral and aesthetic philosophy.

Drama/978-0-375-70423-9

ALSO AVAILABLE:

Faustus, 978-1-4000-7648-2

The Old Neighborhood, 978-0-679-74652-2

Romance, 978-0-307-27518-9

The Spanish Prisoner and The Winslow Boy, 978-0-375-70664-6

True and False, 978-0-679-77264-4

The Voysey Inheritance, 978-0-307-27519-6

VINTAGE BOOKS

Available at your local bookstore, or visit

www.randomhouse.com

WordPress®

ALL-IN-ONE

3rd Edition

by Lisa Sabin-Wilson

WordPress® All-in-One For Dummies®, 3rd Edition

Published by: **John Wiley & Sons, Inc.,** 111 River Street, Hoboken, NJ 07030-5774, www.wiley.com

Published simultaneously in Canada

For general information on our other products and services, please contact our Customer Care Department within the U.S. at 877-762-2974, outside the U.S. at 317-572-3993, or fax 317-572-4002. For technical support, please visit https://hub.wiley.com/community/support/dummies.

Wiley publishes in a variety of print and electronic formats and by print-on-demand. Some material included with standard print versions of this book may not be included in e-books or in print-on-demand. If this book refers to media such as a CD or DVD that is not included in the version you purchased, you may download this material at http://booksupport.wiley.com. For more information about Wiley products, visit www.wiley.com.

Library of Congress Control Number: 2017933312

ISBN: 978-1-119-32777-6 (pbk); 978-1-119-32778-3 (ebk); 978-1-119-32780-6 (ebk)

Manufactured in the United States of America

10 9 8 7 6 5 4 3 2 1

Contents at a Glance

Table of Contents

Introduction

WordPress is the most popular online content management software on the planet. Between the hosted service at WordPress.com and the self-hosted software available at WordPress.org, millions of bloggers use WordPress, and to date, WordPress powers 25 percent of the Internet. That's impressive. With WordPress, you can truly tailor a website to your own tastes and needs.

With no cost for using the benefits of the WordPress platform to publish content on the web, WordPress is as priceless as it is free. WordPress makes writing, editing, and publishing content on the Internet a delightful, fun, and relatively painless experience, whether you're a publisher, a designer, a developer, or a hobbyist blogger.

About This Book

The fact that WordPress is free and accessible to all, however, doesn't make it inherently easy for everyone to use. For some people, the technologies, terminology, and coding practices are a little intimidating or downright daunting. *WordPress All-in-One For Dummies,* 3rd Edition, eliminates any intimidation about using WordPress. With a little research, knowledge, and time, you'll soon have a blog that suits your needs and gives your readers an exciting experience that keeps them coming back for more.

WordPress All-in-One For Dummies is a complete guide to WordPress that covers the basics: installing and configuring the software, using the Dashboard, publishing content, creating themes, and developing plugins. Additionally, this book provides advanced information about security, the WordPress tools, the Multisite features, and search engine optimization.

Foolish Assumptions

I make some inescapable assumptions about you and your knowledge, including the following:

- You're comfortable using a computer, mouse, and keyboard.
- You have a good understanding of how to access the Internet, use email, and use a web browser to access web pages.
- You have a basic understanding of what a website is; perhaps you already maintain your own.
- You want to use WordPress for your online publishing, or you want to use the various WordPress features to improve your online publishing.

If you consider yourself to be an advanced user of WordPress, or if your friends refer to you as an all-knowing WordPress guru, chances are good that you'll find some of the information in this book elementary. This book is for beginner, intermediate, and advanced users; there's something here for everyone.

Icons Used in This Book

The little pictures in the margins of the book emphasize a point to remember, a danger to be aware of, or information that you may find helpful. This book uses the following icons:

TIP

Tips are little bits of information that you may find useful — procedures that aren't necessarily obvious to a casual user or beginner.

WARNING

When your mother warned you, "Don't touch that pan — it's hot!" but you touched it anyway, you discovered the meaning of "Ouch!" I use this icon for situations like that one. You may very well touch the hot pan, but you can't say that I didn't warn you!

TECHNICAL STUFF

All geeky stuff goes here. I use this icon when talking about technical information. You can skip it, but I think that you'll find some great nuggets of information next to these icons. You may even surprise yourself by enjoying them. Be careful — you may turn into a geek overnight!

REMEMBER

When you see this icon, brand the text next to it into your brain so that you remember whatever it was that I thought you should remember.

Beyond the Book

On the web, you can find some extra content that's not in this book. Go online to find

- The Cheat Sheet for this book is at www.dummies.com/cheatsheet. In the Search field, type **WordPress All-in-One For Dummies Cheat Sheet** to find the Cheat Sheet for this book.
- Updates to this book, if any, are at www.dummies.com. Search for the book's title to find the associated updates.

Where to Go from Here

From here, you can go anywhere you please! *WordPress All-in-One For Dummies* is designed so that you can read any or all of the minibooks between the front and back covers, depending on what topics interest you.

Book 1 is a great place to get a good introduction to the world of WordPress if you've never used it before and want to find out more. Book 2 gives you insight into the programming techniques and terminology involved in running a WordPress website — information that's extremely helpful when you move forward to the other minibooks.

Above all else, have fun with the information contained within these pages! Read the minibooks on topics you think you already know; you might just come across something new. Then dig into the minibooks on topics that you want to know more about.

1

WordPress Basics

Contents at a Glance

» Publishing and archiving content

» Interacting through comments

» Syndicating through RSS

» Using WordPress as a content management tool to create different types of sites

Chapter 1
Exploring Basic WordPress Concepts

Blogging provides regular, nontechnical Internet users the ability to publish content on the World Wide Web quickly and easily. Consequently, blogging became extremely popular very quickly, to the point that it's now considered mainstream. In some circles, blogging is even considered passé, as it has given way to publishing all different types of content freely, and easily, using WordPress. Regular Internet users are blogging, and Fortune 500 businesses, news organizations, and educational institutions are using WordPress to publish content on the web. Today, approximately one in four websites is powered by WordPress. That is, roughly, 25 percent of all sites on the web have WordPress behind them.

Although you can choose among several software platforms for publishing web content, for many content publishers, WordPress has the best combination of options. WordPress is unique in that it offers a variety of ways to run your website. WordPress successfully emerged as a favored blogging platform and expanded to a full-featured content management system (CMS) that includes all the tools and features you need to publish an entire website on your own without requiring a whole lot of technical expertise or understanding.

In this chapter, I introduce you to such content basics such as publishing and archiving content, interacting with readers through comments, and providing ways for readers to have access to your content through social media syndication, or RSS technologies. This chapter also helps you sort the differences between a blog and a website, and introduces how WordPress, as a CMS, can help you build an entire website. Finally, I show you some websites that you can build with the WordPress platform.

Discovering Blogging

A blog is a fabulous tool for publishing your diary of thoughts and ideas; however, blogs also serve as excellent tools for business, editorial journalism, news, and entertainment. Here are some ways that people use blogs:

- **Personal:** You're considered to be a personal blogger if you use your blog mainly to discuss topics related to you or your life: your family, your cats, your children, or your interests (such as technology, politics, sports, art, or photography). My business partner, Brad Williams, maintains a personal blog at `http://strangework.com`.
- **Business:** Blogs are very effective tools for promotion and marketing, and business blogs usually offer helpful information to readers and consumers, such as sales events and product reviews. Business blogs also let readers provide feedback and ideas, which can help a company improve its services. A good example of a business blog is the Discovery Channel corporate blog, at `https://corporate.discovery.com/blog/`.
- **Media/journalism:** Popular news outlets, such as Fox News, MSNBC, and CNN, are using blogs on their websites to provide information on current events, politics, and news on regional, national, and international levels. *Variety* magazine hosts its entire website on WordPress at `http://variety.com`.
- **Government:** Governments use blogs to post news and updates to the web quickly and to integrate social media tools as a means to interact with their citizens and representatives. openNASA (`https://open.nasa.gov`) is the official site of NASA's open data initiative, which was built to provide transparency of data to benefit society and humankind. (See Figure 1-1.) NASA staff members provide content by way of blog posts, photos, and videos, and they integrate feeds from their Twitter and Facebook accounts.
- **Citizen journalism:** Citizens are using blogs with the intention of keeping the media and politicians in check by fact-checking news stories and exposing inconsistencies. Major cable news programs interview many of these bloggers because the mainstream media recognize the importance of the citizen voice that's emerging via blogs. An example of citizen journalism is Power Line at `www.powerlineblog.com`.

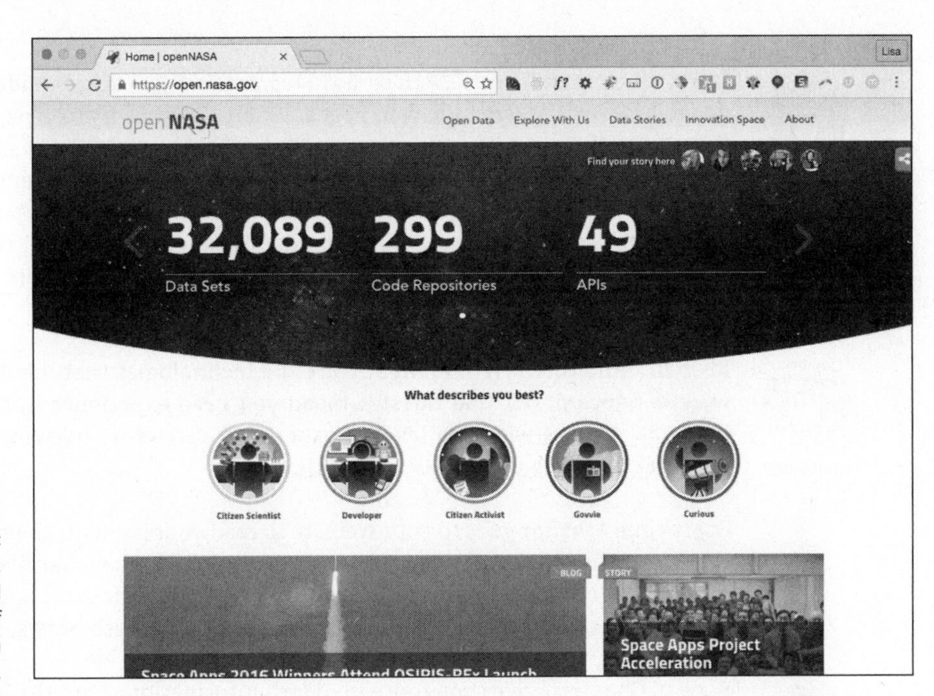

FIGURE 1-1: openNASA, the official website of NASA's open data initiative.

- **Professional:** Professional blogs typically generate revenue and provide a source of monetary income for the owner through avenues such as advertising or paid membership subscriptions. Check out Darren Rowse's ProBlogger blog at `www.problogger.net`. Darren is considered to be the grandfather of all professional bloggers.

The websites and blogs I provide in this list run on the WordPress platform. A wide variety of organizations and individuals choose WordPress to run their blogs and websites because of its popularity, ease of use, and the large and active development community.

Understanding WordPress Technologies

The WordPress software is a personal publishing system that uses a PHP-and-MySQL platform, which provides you everything you need to create your blog and publish your content dynamically without having to program the pages yourself. In short, with this platform, all your content is stored in a MySQL database in your hosting account.

PHP (which stands for *PHP Hypertext Preprocessor*) is a server-side scripting language for creating dynamic web pages. When a visitor opens a page built in PHP, the server processes the PHP commands and then sends the results to the visitor's browser. MySQL is an open-source relational database management system (RDBMS) that uses Structured Query Language (SQL), the most popular language for adding, accessing, and processing data in a database. If all that sounds like Greek to you, think of MySQL as being a big filing cabinet where all the content on your blog is stored.

Keep in mind that PHP and MySQL are the technologies that the WordPress software is built on, but that doesn't mean you need experience in these languages to use it. Anyone with any level of experience can easily use WordPress without knowing anything about PHP or MySQL.

Every time a visitor goes to your website to read your content, she makes a request that's sent to your server. The PHP programming language receives that request, obtains the requested information from the MySQL database, and then presents the requested information to your visitor through her web browser.

Book 2, Chapter 1 gives you more in-depth information about the PHP and MySQL requirements you need to run WordPress. Book 2, Chapter 3 introduces you to the basics of PHP and MySQL and provides information about how they work together with WordPress to create your blog or website.

Archiving your publishing history

Content, as it applies to the data that's stored in the MySQL database, refers to your websites posts, pages, comments, and options that you set up in the WordPress Dashboard or the control/administration panel of the WordPress software, where you manage your site settings and content (see Book 3, Chapter 2).

WordPress maintains chronological and categorized archives of your publishing history automatically. This archiving process happens with every post you publish to your blog. WordPress uses PHP and MySQL technology to organize what you publish so that you and your readers can access the information by date, category, author, tag, and so on. When you publish content on your WordPress site, you can file a post in any category you specify — a nifty archiving system that allows you and your readers to find posts in specific categories. The archives page of my business partner's blog (`http://strangework.com/archives`), for example, contains a Category section, where you find a list of categories he created for his blog posts. Clicking the Personal link below the Categories heading takes you to a listing of posts on that topic. (See Figure 1-2.)

FIGURE 1-2: A page with posts in the Personal category.

WordPress lets you create as many categories as you want for filing your content. Some sites have just one category, and others have up to 1,800 categories. When it comes to organizing your content, WordPress is all about personal preference. On the other hand, using WordPress categories is your choice. You don't have to use the category feature if you'd rather not.

TIP

When you look for a hosting service, keep an eye out for hosts that provide daily backups of your site so that your content won't be lost if a hard drive fails or someone makes a foolish mistake. Web hosting providers that offer daily backups as part of their services can save the day by restoring your site to a previous form.

REMEMBER

The theme (design) you choose for your site — whether it's the default theme, one that you create, or one that you custom-design — isn't part of the content. Those files are part of the file system and aren't stored in the database. Therefore, it's a good idea to create a backup of any theme files you're using. See Book 6 for further information on WordPress theme management.

Interacting with your readers through comments

An exciting aspect of publishing content with WordPress is receiving feedback from your readers after you publish to your site. Receiving feedback, or *comments*, is akin to having a guestbook on your site. People can leave notes for you that publish to your site, and you can respond and engage your readers in conversation. (See Figure 1-3.) These notes can expand the thoughts and ideas you present in your content by giving your readers the opportunity to add their two cents' worth.

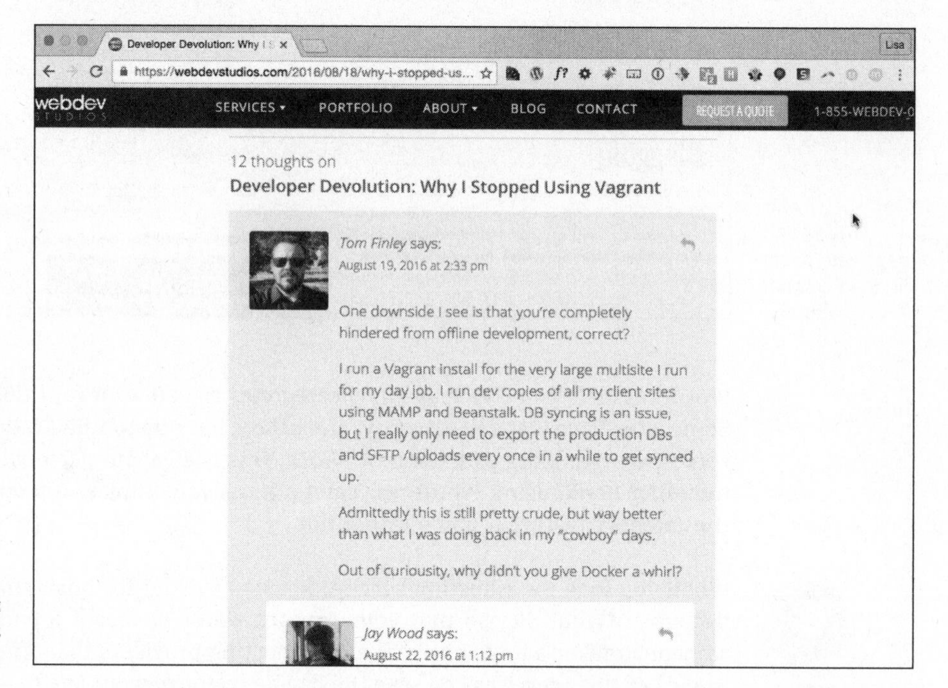

FIGURE 1-3: Blog comments and responses.

REMEMBER

The WordPress Dashboard gives you full administrative control over who can leave comments. Additionally, if someone leaves a comment with questionable content, you can edit the comment or delete it. You're also free to not allow comments on your site at all. (See Book 3, Chapter 4 for more information.)

Feeding your readers

RSS stands for *Really Simple Syndication.* An *RSS feed* is a standard feature that blog readers have come to expect. So what is RSS, really?

RSS is written to the web server in XML (Extensible Markup Language) as a small, compact file that can be read by RSS readers (as I outline in Table 1-1). Think of an RSS feed as a syndicated, or distributable, autoupdating "What's New" list for your website.

Tools like feed readers and email newsletter services can use the RSS feed from your website to consume the data and aggregate it into a syndicated list of content published on your website. Website owners allow RSS to be published to allow these tools to consume and then distribute the data in an effort to expand the reach of their publications.

Table 1-1 lists some popular tools that use RSS feeds to distribute content from websites.

TABLE 1-1 Popular RSS Feed Readers

Reader	Source	Description
Feedly	`http://feedly.com`	RSS aggregator for websites that publish an RSS feed. It compiles published stories from various user-chosen sources and allows the Feedly user to organize the stories and share the content with others.
MailChimp	`https://mailchimp.com`	MailChimp is an email newsletter service. It has an RSS-to-email service that enables you to send your recently published content to your readers via an email subscription service.
dlvr.it	`https://dlvrit.com/`	Use RSS to auto-post to Facebook, Twitter, LinkedIn, Pinterest, and other social media sites.

For your readers to stay up-to-date with the latest and greatest content you post, they can subscribe to your RSS feed. WordPress RSS feeds are *autodiscovered* by the various feed readers. The reader needs only to enter your site's URL, and the program automatically finds your RSS feed.

WordPress has RSS feeds in several formats. Because the feeds are built into the software platform, you don't need to do anything to provide your readers an RSS feed of your content.

Tracking back

The best way to understand *trackbacks* is to think of them as comments, except for one thing: Trackbacks are comments left on your site by other sites, not people. Sounds perfectly reasonable, doesn't it? After all, why wouldn't inanimate objects want to participate in your discussion?

Actually, maybe it's not so crazy after all. A trackback happens when you make a post on your site, and within the content of that post, you provide a link to a post made by another author on a different site. When you publish that post, your site sends a sort of electronic memo to the site you linked to. That site receives the memo and posts an acknowledgment of receipt in the form of a comment to the post that you linked to on the site. The information contained within the trackback includes a link back to the post on your site that contains the link to the other site — along with the date and time, as well as a short excerpt of your post. Trackbacks are displayed within the comments section of the individual posts.

The memo is sent via a *network ping* (a tool used to test, or verify, whether a link is reachable across the Internet) from your site to the site you link to. This process works as long as both sites support trackback protocol. Almost all major CMSes support the trackback protocol.

REMEMBER

Sending a trackback to a site is a nice way of telling the author that you like the information she presented in her post. Most authors appreciate trackbacks to their posts from other content publishers.

Dealing with comment and trackback spam

The absolute bane of publishing content on the Internet is comment and trackback spam. Ugh. When blogging became the "it" thing on the Internet, spammers saw an opportunity. If you've ever received spam in your email program, you know what I mean. For content publishers, the concept is similar and just as frustrating.

Spammers fill content with open comments with their links but not with any relevant conversation or interaction in the comments. The reason is simple: Websites receive higher rankings in the major search engines if they have multiple links coming in from other sites. Enter software, like WordPress, with comment and trackback technologies, and these sites become prime breeding ground for millions of spammers.

Because comments and trackbacks are published to your site publicly — and usually with a link to the commenter's website — spammers got their site links posted on millions of sites by creating programs that automatically seek websites with open commenting systems and then hammer those systems with tons of comments that contain links back to their sites.

No one likes spam. Therefore, developers of CMSes, such as WordPress, spend untold hours in the name of stopping these spammers in their tracks, and for the most part, they've been successful. Occasionally, however, spammers sneak through. Many spammers are offensive, and all of them are frustrating because they don't contribute to the conversations that occur on the websites where they publish their spam comments.

All WordPress systems have one important thing in common: Akismet, which kills spam dead. Akismet is a WordPress plugin brought to you by Automattic, the creator of the WordPress.com service. I cover the Akismet plugin, and comment spam in general, in Book 3, Chapter 4.

Using WordPress as a Content Management System

A *content management system* (CMS) is a platform that lets you run a full website on your domain. This means that WordPress enables you to create and publish all kinds of content on your site, including pages, blog posts, e-commerce pages for selling products, videos, audio files, events, and more.

A *blog* is a chronological display of content — most often, written by the blog author. The posts are published and, usually, categorized into topics and archived by date. Blog posts can have comments activated so that readers can leave their feedback and the author can respond, creating a dialogue about the blog post.

A *website* is a collection of published pages with different sections that offer the visitor different experiences. A website can incorporate a blog but usually contains other sections and features. These other features include

- **Photo galleries:** Albums of photos uploaded and collected in a specific area so that visitors can browse through and comment on them
- **E-commerce stores:** Fully integrated shopping area into which you can upload products for sale and from which your visitors can purchase them
- **Discussion forums:** Where visitors can join, create discussion threads, and respond to one another in specific threads of conversation
- **Social communities:** Where visitors can become members, create profiles, become friends with other members, create groups, and aggregate community activity
- **Portfolios:** Sections where photographers, artists, or web designers display their work
- **Feedback forms:** Contact forms that your visitors fill out with information that then gets emailed to you directly
- **Static pages (such as Bio, FAQ, or Services):** Pages that don't change as often as blog pages, which change each time you publish a new post

The preceding list isn't exhaustive; it's just a listing of some of the most common website sections.

Figure 1-4 shows what the front page of my business blog looked like at the time of this writing. Visit `https://webdevstudios.com/blog` to see that the site displays a chronological listing of the most recent blog posts.

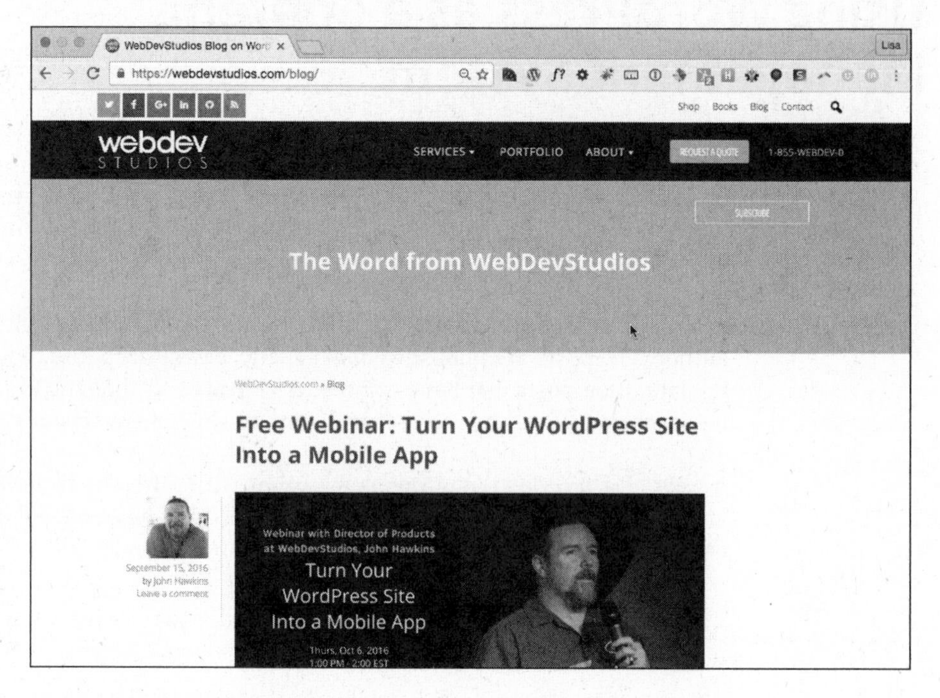

FIGURE 1-4: Visit my business blog at `http://webdevstudios.com/blog` to see an example of a chronological listing of blog posts.

My business website at `https://webdevstudios.com` also uses WordPress. This full site includes a static front page of information that acts as a portal to the rest of the site, on which you can find a blog; a portfolio of work; a contact form; and various landing pages, including service pages that outline information about the different services we offer (`https://webdevstudios.com/services`). Check out Figure 1-5 for a look at this website; it's quite different from the blog section of the site.

FIGURE 1-5: My business website uses WordPress as a CMS.

Using WordPress as a CMS means that you're creating more than just a blog; you're creating an entire website full of sections and features that offer different experiences for your visitors.

» Discovering examples of open-source projects

» Understanding WordPress licensing

» Applying WordPress licensing

Chapter 2
Exploring the World of Open-Source Software

Open-source software is a movement that started in the software industry in the 1980s. Its origins are up for debate, but most people believe that the concept came about in 1983, when a company called Netscape released its Navigator web browser source code to the public, making it freely available to anyone who wanted to dig through it, modify it, or redistribute it.

WordPress software users need a basic understanding of the open-source concept and the licensing upon which WordPress is built because WordPress's open-source policies affect you as a user — and greatly affect you if you plan to develop plugins or themes for the WordPress platforms. A basic understanding helps you conduct your practices in accordance with the license at the heart of the WordPress platform.

This chapter introduces you to open-source; the Open Source Initiative (OSI); and the GPL (General Public License), which is the specific license that WordPress is built upon (GPLv2, to be exact). You also discover how the GPL license applies to any projects you may release (if you're a developer of plugins or themes) that depend on the WordPress software and how you can avoid potential problems by abiding by the GPL as it applies to WordPress.

IANAL — *I Am Not a Lawyer* — is an acronym you often find in articles about WordPress and the GPL. I use it here because I'm not a lawyer, and the information in this chapter shouldn't be construed as legal advice. Rather, you should consider the chapter to be an introduction to the concepts of open-source and the GPL. The information presented here is meant to inform you about and introduce you to the concepts as they relate to the WordPress platform.

Defining Open-Source

A simple, watered-down definition of open-source software is software whose source code is freely available to the public and that can be modified and redistributed by anyone without restraint or consequence. An official organization called the Open Source Initiative (OSI; `https://opensource.org`), founded in 1998 to organize the open-source software movement in an official capacity, has provided a very clear and easy-to-understand definition of open-source. During the course of writing this book, I obtained permission from the OSI board to include it here.

Open-source doesn't just mean access to the source code. The distribution terms of open-source software must comply with the following criteria:

1. **Free Redistribution**

   The license shall not restrict any party from selling or giving away the software as a component of an aggregate software distribution containing programs from several different sources. The license shall not require a royalty or other fee for such sale.

2. **Source Code**

   The program must include source code, and must allow distribution in source code as well as compiled form. Where some form of a product is not distributed with source code, there must be a well-publicized means of obtaining the source code for no more than a reasonable reproduction cost preferably, downloading via the Internet without charge. The source code must be the preferred form in which a programmer would modify the program. Deliberately obfuscated source code is not allowed. Intermediate forms such as the output of a preprocessor or translator are not allowed.

3. **Derived Works**

   The license must allow modifications and derived works, and must allow them to be distributed under the same terms as the license of the original software.

4. **Integrity of the Author's Source Code**

   The license may restrict source-code from being distributed in modified form only if the license allows the distribution of "patch files" with the source code for the purpose of modifying the program at build time. The license must explicitly permit distribution of software built from modified source code. The license may require derived works to carry a different name or version number from the original software.

5. **No Discrimination Against Persons or Groups**

   The license must not discriminate against any person or group of persons.

6. **No Discrimination Against Fields of Endeavor**

   The license must not restrict anyone from making use of the program in a specific field of endeavor. For example, it may not restrict the program from being used in a business, or from being used for genetic research.

7. **Distribution of License**

   The rights attached to the program must apply to all to whom the program is redistributed without the need for execution of an additional license by those parties.

8. **License Must Not Be Specific to a Product**

   The rights attached to the program must not depend on the program's being part of a particular software distribution. If the program is extracted from that distribution and used or distributed within the terms of the program's license, all parties to whom the program is redistributed should have the same rights as those that are granted in conjunction with the original software distribution.

9. **License Must Not Restrict Other Software**

   The license must not place restrictions on other software that is distributed along with the licensed software. For example, the license must not insist that all other programs distributed on the same medium must be open-source software.

10. **License Must Be Technology-Neutral**

    No provision of the license may be predicated on any individual technology or style of interface.

The preceding items comprise the definition of open-source as provided by the Open Source Initiative. You can find this definition (see Figure 2-1) at `https://opensource.org/osd`.

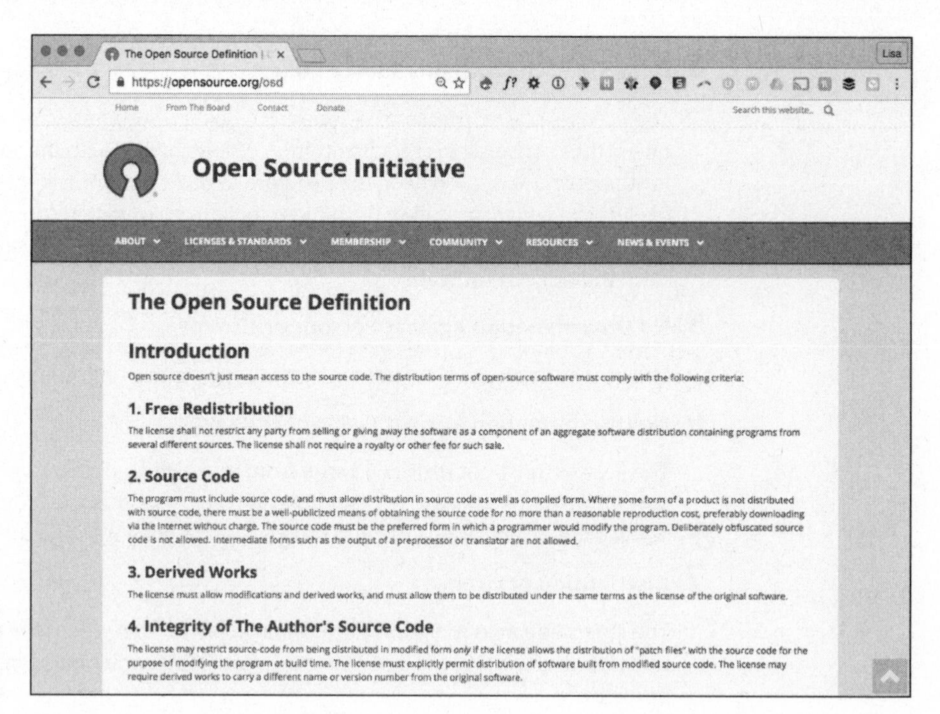

FIGURE 2-1: Definition of open-source from the OSI.

Open-source software source code must be freely available, and any licensing of the open-source software must abide by this definition. Based on the OSI definition, WordPress is an open-source software project. Its source code is accessible and publicly available for anyone to view, build on, and distribute at no cost anywhere, at any time, or for any reason.

Several examples of high-profile software enterprises, such as the ones in the following list, are also open-source. You'll recognize some of these names:

- **Mozilla (`https://www.mozilla.org`):** Community whose projects include the popular Firefox Internet browser and Thunderbird, a popular email client. All projects are open-source and considered to be public resources.
- **PHP (`http://php.net`):** An HTML-embedded scripting language that stands for PHP Hypertext Preprocessor. PHP is popular software that runs on most web servers today; its presence is required on your web server for you to run the WordPress platform successfully on your site.
- **MySQL (`www.mysql.com`):** The world's most popular open-source database. Your web server uses MySQL to store all the data from your WordPress

installation, including your posts, pages, comments, links, plugin options, theme option, widgets, and more.

- **Linux (`www.linux.org`):** An open-source operating system used by web hosting providers, among other organizations.

As open-source software, WordPress is in some fine company. Open-source itself is not a *license;* I cover licenses in the next section. Rather, open-source is a movement — some people consider it to be a philosophy — created and promoted to provide software as a public resource open to community collaboration and peer review. WordPress development is clearly community-driven and focused. You can read about the WordPress community in Book 1, Chapter 4.

Understanding WordPress Licensing

Most software projects are licensed, meaning that they have legal terms governing the use or distribution of the software. Different kinds of software licenses are in use, ranging from very restrictive to least restrictive. WordPress is licensed by the GPL (General Public License), one of the least restrictive software licenses available.

If you're bored, read the GPL text at `www.gnu.org/licenses/gpl-2.0.html`. Licensing language on any topic can be a difficult thing to navigate and understand. It's sufficient to have a basic understanding of the concept of GPL and let the lawyers sort out the rest, if necessary.

TIP

A complete copy of the GPL is included in every copy of the WordPress download package, in the `license.txt` file. The directory listing of the WordPress software files shown in Figure 2-2 lists the `license.txt` file.

Simply put, any iteration of a piece of software developed and released under the GPL must be released under the very same license in the future. Check out the nearby sidebar "The origins of WordPress," which tells the story of how the WordPress platform came into existence. Essentially, the software was *forked* — meaning that the original software (in this case, a blogging platform called b2) was abandoned by its original developer and adopted by the founders of WordPress, who took the b2 platform, called it WordPress, and began a new project with a new plan, outlook, and group of developers.

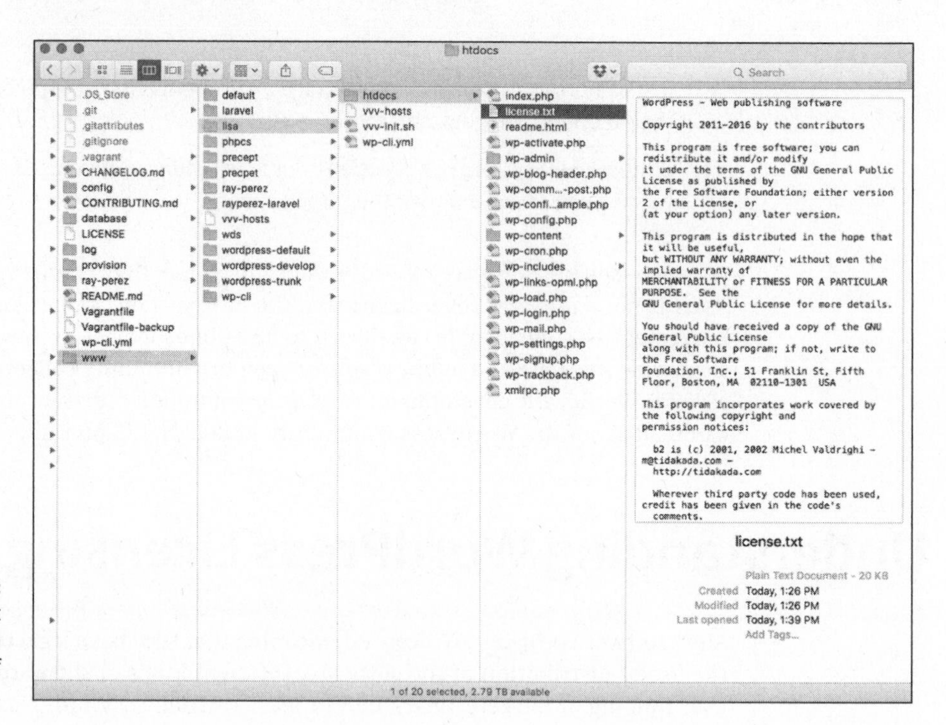

FIGURE 2-2: The GPL text is included in every copy of WordPress.

Because the b2 platform was originally developed and released under the GPL, by law, the WordPress software (all current and future iterations of the platform) must also abide by the GPL. Because of the nature of the GPL, you, your next-door neighbor, or I could do the very same thing with the WordPress platform. Nothing is stopping you, or anyone else, from taking WordPress, giving it a different name, and rereleasing it as a completely different project. Typically, open-source projects are forked when the original project development stalls or is abandoned (as was the case with b2) or (in rare cases) when the majority of the development community is at odds with the leadership of the open-source project. I'm not suggesting that you do that, though, because WordPress has one of the most active development communities of any open-source project I've come across.

THE ORIGINS OF WORDPRESS

Once upon a time, there was a simple PHP-based blogging platform called b2. This software, developed in 2001, slowly gained a bit of popularity among geek types as a way to publish content on the Internet. Its developer, Michel Valdrighi, kept development active until early 2003, when users of the software noticed that Valdrighi seemed to have disappeared. They became a little concerned about b2's future.

Somewhere deep in the heart of Texas, one young man in particular was very concerned, because b2 was his software of choice for publishing his own content on the World Wide Web. He didn't want to see his favorite publishing tool become obsolete. You can view the original post to his own blog in which he wondered what to do (`http://ma.tt/2003/01/the-blogging-software-dilemma`).

In that post, he talked briefly about some of the other software that was available at the time, and he tossed around the idea of using the b2 software to "to create a fork, integrating all the cool stuff that Michel would be working on right now if only he was around."

Create a fork he did. In the absence of b2's developer, this young man developed from the original b2 codebase a new blogging application called WordPress.

That blog post was made on January 24, 2003, and the young man's name was (and is) Matt Mullenweg. On December 26, 2003, with the assistance of a few other developers, Mullenweg announced the arrival of the first official version of the WordPress software. The rest, as they say, is history. The history of this particular piece of software surely is one for the books, as it's the most popular blogging platform available today.

Applying WordPress Licensing to Your Projects

Regular users of WordPress software need never concern themselves with the GPL of the WordPress project at all. You don't have to do anything special to abide by the GPL. You don't have to pay to use the WordPress software, and you aren't required to acknowledge that you're using the WordPress software on your site. (That said, providing on your site at least one link back to the WordPress website is common courtesy and a great way of saying thanks.)

Most people aren't even aware of the software licensing because it doesn't affect the day-to-day business of blogging and publishing sites with the platform. It's not a bad idea to educate yourself on the basics of the GPL, however. When you try to be certain that any plugins and themes you use with your WordPress installation abide by the GPL, you have peace of mind that all applications and software you're using are in compliance.

Your knowledge of the GPL must increase dramatically, though, if you develop plugins or themes for the WordPress platform. (I cover WordPress themes in Book 6 and WordPress plugins in Book 7.)

The public licensing that pertains to WordPress plugins and themes wasn't decided in a court of law. The current opinion of the best (legal) practices is just that: opinion. The opinion of the WordPress core development team, as well as the opinion of the Software Freedom Law Center (`https://www.softwarefreedom.org/services`), is that WordPress plugins and themes are derivative works of WordPress and, therefore, must abide by the GPL by releasing the development works under the same license that WordPress has.

A *derivative work*, as it relates to WordPress, is a work that contains programming whose functionality depends on the core WordPress files. Because plugins and themes contain PHP programming that call WordPress core functions, they rely on the core WordPress framework to work properly and, therefore, are extensions of the software.

TECHNICAL STUFF

The text of the opinion by James Vasile from the Software Freedom Law Center is available at `http://wordpress.org/news/2009/07/themes-are-gpl-too`.

To maintain compliance with the GPL, plugin or theme developers can't release development work under any (restrictive) license other than the GPL. Nonetheless, many plugin and theme developers have tried to release material under other licenses, and some have been successful (from a moneymaking standpoint). The WordPress community, however, generally doesn't support these developers or their plugins and themes. Additionally, the core WordPress development team considers such works to be noncompliant with the license and, therefore, with the law.

WordPress has made it publicly clear that it won't support or promote any theme or plugin that is not in 100 percent compliance with the GPL. If you're not 100 percent compliant with the GPL, you can't include your plugin or theme in the WordPress Plugin Directory hosted at `https://wordpress.org/plugins`. If you develop plugins and themes for WordPress, or if you're are considering dipping your toe into that pool, do it in accordance with the GPL so that your works are in compliance and your good standing in the WordPress community is protected.

Table 2-1 provides a brief review of what you can (and can't) do as a WordPress plugin and theme developer.

TABLE 2-1 Development Practices Compliant with GPL License

Development/Release Practice	GPL-Compliant?
Distribute to the public for free with GPL.	Yes
Distribute to the public for a cost with GPL.	Yes
Restrict the number of users of one download with GPL.	No
Split portions of your work among different licenses. (PHP files are GPL; JavaScript or CSS files are licensed with the Creative Commons license.)	Yes (but WordPress.org won't promote works that aren't 100 percent GPL across all files)
Release under a different license, such as the PHP License.	No

The one and only way to make sure that your plugin or theme is 100 percent compliant with the GPL is to do the following before you release your development work to the world:

- Include a statement in your work indicating that the work is released under the GPLv2 license in the `license.txt` file, which WordPress does. (Refer to Figure 2-2.) Alternatively, you can include this statement in the header of your plugin file:

```
<?php

 This program is free software; you can redistribute it and/or modify it under the
   terms of the GNU General Public License, version 2, as published by the Free
   Software Foundation.

 This program is distributed in the hope that it will be useful,
 but WITHOUT ANY WARRANTY; without even the implied warranty of
 MERCHANTABILITY or FITNESS FOR A PARTICULAR PURPOSE. See the
 GNU General Public License for more details.

 You should have received a copy of the GNU General Public License
 along with this program; if not, write to the Free Software
 Foundation, Inc., 51 Franklin St., Fifth Floor, Boston, MA 02110-1301 USA
*/
?>
```

- Don't restrict the use of your works by the number of users per download.
- If you charge for your work, which is compliant with the GPL, the licensing doesn't change, and users still have the freedom to modify your work and rerelease it under a different name.
- Don't split the license of other files included in your work, such as CSS or graphics. Although this practice complies with the GPL, it won't be approved for inclusion in the WordPress Plugin Directory.

» Exploring betas, release candidates, and final release versions

» Navigating WordPress release archives

» Tracking WordPress development

» Using bleeding-edge builds

Chapter 3
Understanding Development and Release Cycles

If you're planning to dip your toe into the WordPress waters (or you've already dived in and gotten completely wet), the WordPress platform's development cycle is really good to know about and understand, because it affects every WordPress user on a regular basis.

WordPress and its features form the foundation of your website. WordPress is a low-maintenance way to publish content on the web, and the software is free in terms of monetary cost. WordPress isn't 100 percent maintenance-free, however, and part of maintenance is ensuring that your WordPress software is up to date to keep your website secure and safe.

This chapter explains the development cycle for the WordPress platform and shows you how you can stay up to date and informed about what's going on. This chapter also gives you information on WordPress release cycles and shows you how you can track ongoing WordPress development on your own.

Discovering WordPress Release Cycles

Book 1, Chapter 2 introduces you to the concept of open-source software and discusses how the WordPress development community is primarily volunteer developers who donate their time and talents to the WordPress platform. The development of new WordPress releases is a collaborative effort, sometimes requiring contributions from more than 300 developers.

The public schedule for WordPress updates is roughly one new release every 120 days. As a user, you can expect a new release of the WordPress software about three times per year. The WordPress development team sticks to that schedule closely, with exceptions only here and there. When the team makes exceptions to the 120-day rule, it usually makes a public announcement so that you know what to expect and when to expect it.

Mostly, interruptions in the 120-day schedule occur because the development of WordPress occurs primarily on a volunteer basis. A few developers — employees of Automattic, the company behind WordPress.com — are paid to develop for WordPress, but most developers are volunteers. Therefore, the progress of WordPress development depends on the developers' schedules.

REMEMBER

I'm confident in telling you that you can expect to update your WordPress installation at least three, if not four, times per year.

Upgrading your WordPress experience

Don't be discouraged or frustrated by the number of times you'll upgrade your WordPress installation. The WordPress development team is constantly striving to improve the user experience and to bring exciting, fun new features to the WordPress platform. Each upgrade improves security and adds new features to enhance your (and your visitors') experience on your website. WordPress also makes the upgrades easy to perform, as I discuss in Book 2, Chapter 6.

The following list gives you some good reasons why you should upgrade your WordPress software each time a new version becomes available:

- **Security:** When WordPress versions come and go, outdated versions are no longer supported and are vulnerable to malicious attacks and hacker attempts. Most WordPress security failures occur when you're running an outdated version of WordPress on your website. To make sure that you're running the most up-to-date and secure version, upgrade to the latest release as soon as you can.

» **New features:** Major WordPress releases (I discuss the difference between major and minor, or point, releases later in the chapter) offer great new features that are fun to use, improve your experience, and boost your efficiency and productivity. Upgrading your WordPress installation ensures that you always have access to the latest, greatest tools and features that WordPress has to offer.

» **Plugins and themes:** Most plugin and theme developers work hard to make sure that their products are up to date with the latest version of WordPress. Generally, plugin and theme developers don't worry about backward compatibility, and they tend to ignore out-of-date versions of WordPress. To be sure that the plugins and themes you've chosen are current and not breaking your site, make sure that you're using the latest version of WordPress and the latest versions of your plugins and themes. (See Book 6 for information about themes and Book 7 for details about plugins.)

Understanding the cycles of a release

By the time the latest WordPress installation becomes available, that version has gone through several iterations, or *versions.* This section helps you understand what it takes to get the latest version to your website and explains some of the WordPress development terminology.

The steps and terminology involved in the release of a new version of WordPress include

» **Alpha:** This phase is the first developmental phase of a new version. Alpha typically is the "idea" phase in which developers gather ideas, including those from users and community members. During the alpha phase, developers determine which features to include in the new release and then develop an outline and a project plan. After features are decided, developers start developing and testers start testing until they reach a "feature freeze" point in the development cycle, where all new features are considered to be complete. Then development moves on to perfecting new features through user testing and bug fixes.

» **Beta:** This phase is for fixing bugs and clearing any problems that testers report. Beta cycles can last four to six weeks, if not longer. WordPress often releases several beta versions with such names as WordPress version 4.7 Beta, WordPress version 4.7 Beta 1, and so on. The beta process continues until the development team decides that the software is ready to move into the next phase in the development cycle.

» **Release candidate:** A version becomes a release candidate (RC) when the bugs from the beta versions are fixed and the version is nearly ready for final release. You sometimes see several RC iterations, referred to as RC-1, RC-2, and so on.

» **Final release:** After a version has gone through full testing in several (ideally, all) types of environments, use cases, and user experiences; any bugs from the alpha, beta, and RC phases have been squashed; and no major bugs are being reported, the development team releases the final version of the WordPress software.

After the WordPress development team issues a final release version, they start again in the alpha phase, gearing up and preparing to go through the development cycle for the next major version.

REMEMBER

Typically, a development cycle lasts 120 days, but this figure is an approximation, because any number of things can happen (from developmental problems to difficult bugs) to delay the process.

Finding WordPress release archives

WordPress keeps a historical archive of all versions it has ever released at `https://wordpress.org/download/release-archive`, as shown in Figure 3-1. On that page, you find releases dating back to version 0.17 from 2003.

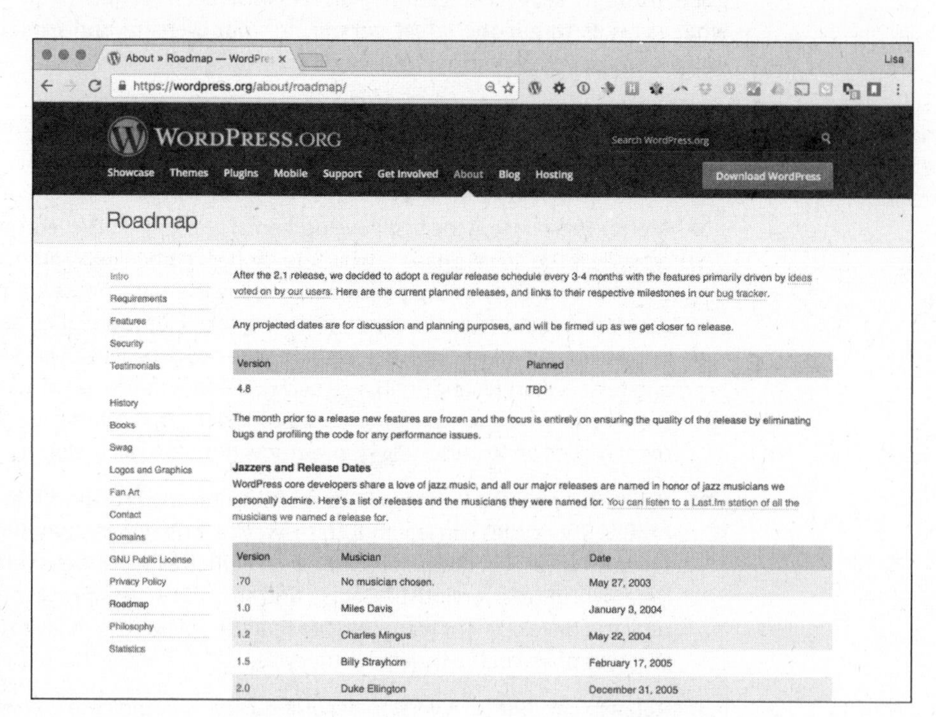

FIGURE 3-1: The archive of every WordPress release on record.

MAJOR VERSUS POINT RELEASES

You may have noticed that WordPress versions are numbered. These numbers show the progress of the development of the software, and they also tell you something else about the version you're using. *Software versioning* is a method of assigning unique numbers to each version release. Generally, the two types of versioning are

- **Point release:** Point releases usually increase the numbered version only by a decimal point or two, indicating a relatively minor release. Such releases include insignificant updates or minor bug fixes. When the version number jumps from 4.6 to 4.6.1, for example, you can be certain that the new version was released to fix minor bugs or to clean up the source code rather than to add new features.
- **Major release:** A major release most often contains new features and jumps by a more seriously incremented version number. When WordPress went from 4.6.1 to 4.7 (release 4.6 versioned into 4.6.1 before jumping to 4.7), that release was considered to be a major release because it jumped a whole number rather than a decimal point. A large jump is a sign to users that new features are included in this version, rather than just bug fixes or cleanup of code. The bigger the jump in the version number, the more major the release is. A release jumping from 4.0 to 4.5, for example, is an indication of major new features.

WARNING

None of the releases on the WordPress website is safe for you to use except the latest release in the 4.7.x series. Using an older version leaves your website open to hackers. WordPress just likes to have a recorded history of every release for posterity's sake.

Keeping Track of WordPress Development

If you know where to look, keeping track of the WordPress development cycle is easy, especially because the WordPress development team tries to make the development process as transparent as possible. You can track updates by reading about them in various spots on the Internet and by listening to conversations between developers. If you're so inclined, you can jump in and lend the developers a hand, too.

You have several ways to stay up to date on what's going on in the world of WordPress development, including blog posts, live chats, development meetings, tracking tickets, and bug reports, just to name a few. The following list gives you a solid start on where you can go to stay informed:

- **WordPress development updates (`https://make.wordpress.org/core`):** The WordPress development team's blog, Make WordPress Core, is where you can follow and keep track of the progress of the WordPress software project while it happens. (See Figure 3-2.) You find agendas, schedules, meeting minutes, and discussions surrounding the development cycles.

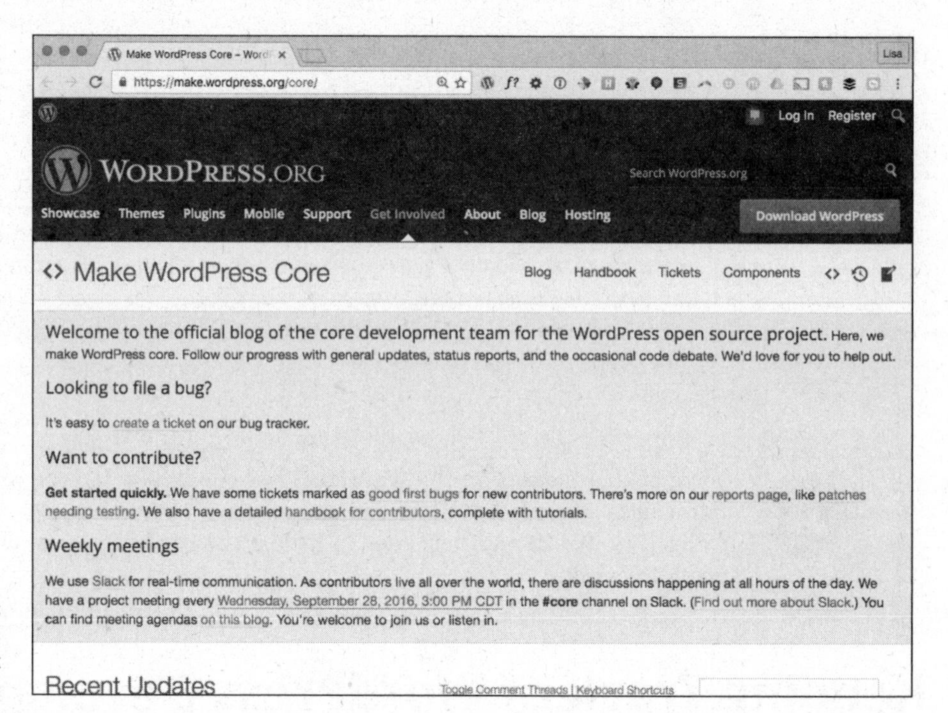

FIGURE 3-2: The WordPress development blog.

- **WordPress developers' chats (`https://make.wordpress.org/chat`):** Developers who are involved in development of WordPress core use a real-time communication platform called Slack (`https://slack.com`). You can easily participate in any of the scheduled meetings listed on `https://make.wordpress.org/core` (regular scheduled chats are listed in the right sidebar of the site).
- **WordPress Trac (`https://core.trac.wordpress.org`):** Here are ways to stay informed about the changes in WordPress development:
  - Follow the timeline: `https://core.trac.wordpress.org/timeline`
  - View the road map: `https://core.trac.wordpress.org/roadmap`

- Read reports: `https://core.trac.wordpress.org/report`
- Perform a search: `https://core.trac.wordpress.org/search`

» **WordPress mailing lists (`https://codex.wordpress.org/Mailing_Lists`):** Join mailing lists focused on different aspects of WordPress development, such as bug testing, documentation, and hacking WordPress. (For specific details about mailing lists, see Book 1, Chapter 4.)

Downloading Nightly Builds

WordPress development moves pretty fast. Often, changes in the software's development cycle occur daily. While the developers are working on alpha and beta versions and release candidates, they commit the latest core changes to the repository and make those changes available to the public to download, install, and test on individual sites. The changes are released in a full WordPress software package called a *nightly build.* This nightly build contains the latest core changes submitted to the project — changes that have not yet been released as full and final versions.

WARNING

Using nightly builds isn't a safe practice for a live site. I strongly recommend creating a test environment to test nightly builds. Many times, especially during alpha and beta phases, the core code breaks and causes problems with your existing installation. Use nightly builds in a test environment only, and leave your live site intact until the final release is available.

Hundreds of members of the WordPress community help in the development phases, even though they aren't developers or programmers. They help by downloading the nightly builds, testing them in various server environments, and reporting to the WordPress development team by way of Trac tickets (shown in Figure 3-3; check out `https://core.trac.wordpress.org/report`) any bugs and problems they find in that version of the software.

You can download the latest nightly build from the WordPress repository at `https://wordpress.org/download/nightly`. For information about installing WordPress, see Book 2, Chapter 4.

REMEMBER

Running the latest nightly build on your website is referred to as using *bleeding-edge* software because the software is an untested version, requiring you to take risks just to run it on your website.

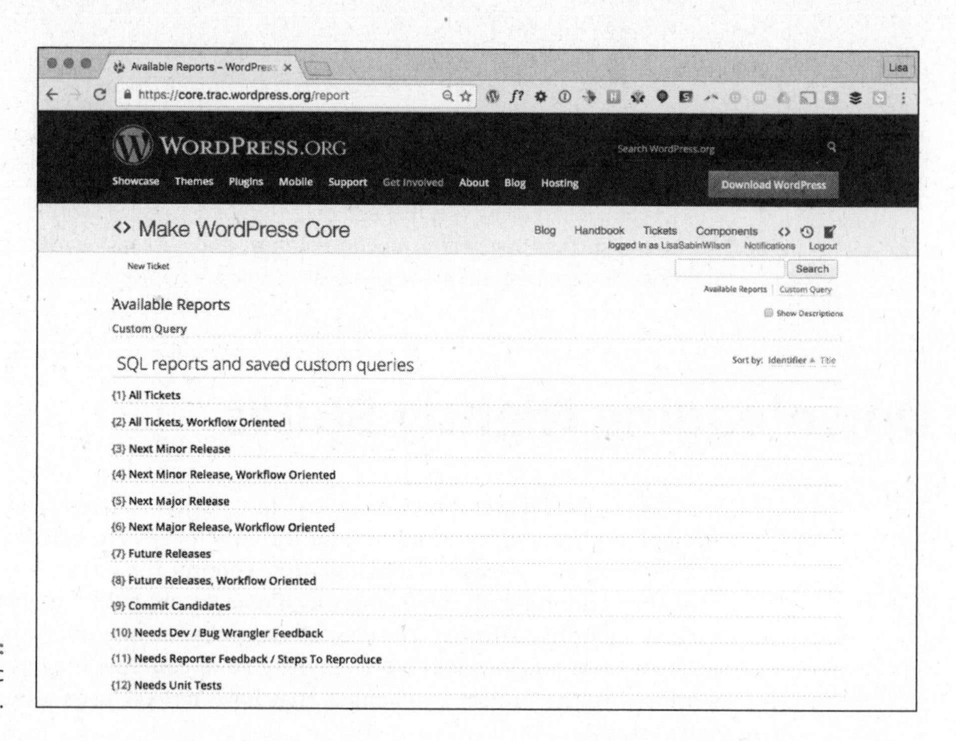

FIGURE 3-3: WordPress Trac tickets.

WordPress Beta Tester (`https://wordpress.org/plugins/wordpress-beta-tester`), by Peter Westwood, is a super plugin that enables you to use the automatic upgrade tool in your WordPress Dashboard to download the latest nightly build. For information about installing and using WordPress plugins, check out Book 7, Chapter 2.

» Locating users on social networks

» Getting help in the support forums

» Participating in testing and bug squashing

» Discovering professional consultants and services

» Attending and organizing local WordCamps

Chapter 4

Meeting the WordPress Community

Allow me to introduce you to the fiercely loyal folks who make up the WordPress user base, better known as the WordPress community. These merry ladies and gentlemen come from all around the globe, from California to Cairo, Florida to Florence, and all points in between.

Early on, in March 2005, Matt Mullenweg of WordPress proudly proclaimed that the number of WordPress downloads had reached 900,000 — an amazing landmark in the history of the software. By contrast, in 2016, the download counter for WordPress version 4.6 had exceeded 17 million. The World Wide Technology Surveys (`https://w3techs.com`) published results showing WordPress to be the most popular content management system (CMS) being used on the web today. An astounding 26.7 percent of all sites on the Internet using a CMS use WordPress. This popularity makes for a large community of users, to say the least.

This chapter introduces you to the WordPress community and the benefits of membership within that community, such as finding support forums; locating other WordPress users on various social networks; getting assistance from other users, participating in WordPress development; and hooking up with WordPress users face to face at WordPress events, such as WordCamp.

Finding Other WordPress Users

Don't let the sheer volume of users intimidate you: WordPress has bragging rights to the most helpful blogging community on the web today. Thousands of websites exist that spotlight everything, including WordPress news, resources, updates, tutorials, and training. The list is endless. Do a quick Google search for *WordPress*, and you'll get about 1.9 billion results.

My point is that WordPress users are all over the Internet, from websites to discussion forums and social networks to podcasts and more. For many people, the appeal of the WordPress platform lies not only in the platform itself, but also in its passionate community of users.

Finding WordPress news and tips on community websites

WordPress-related websites cover an array of topics related to the platform, including everything from tutorials to news and even a little gossip, if that's your flavor. The Internet has no shortage of websites related to the popular WordPress platform. Here are a few that stand out:

- **WP Tavern (`https://wptavern.com`):** A site that covers everything from soup to nuts: news, resources, tools, tutorials, and interviews with standout WordPress personalities. You can pretty much count on WP Tavern to be on top of what's new and going on in the WordPress community. WP Tavern is owned by Automattic, the parent company of WordPress.com.
- **Smashing Magazine (`https://www.smashingmagazine.com/category/wordpress/`):** A very popular and established online design magazine and resource that has dedicated a special section of its website to WordPress news, resources, tips, and tools written by various members of the WordPress community.
- **Make WordPress Core (`https://make.wordpress.org/core`):** A website that aggregates content from all the "Make WordPress" websites built and maintained by the WordPress.org community. It includes resources for contributing to WordPress core, making plugins and themes, planning WordPress events, supporting WordPress, and more.

Locating users on social networks

In addition to WordPress, many bloggers use microblogging tools such as Twitter (`https://twitter.com`) and/or social-media networks such as Facebook (`https://www.facebook.com`) to augment their online presence and market their blog, services, and products. Within these networks, you can find WordPress users, resources, and links, including the following:

- **WordPress Twitter lists:** Twitter allows users to create lists of people who have the same interests, such as WordPress. You can find a few of these lists here:
  - *Twitter:* `https://twitter.com/search?q=WordPress`
  - *Google:* `https://www.google.com/#q=WordPress+Twitter+Lists`
- **Facebook Pages on WordPress:** Facebook users create Pages and groups around their favorite topics of interest, such as WordPress. You can find some interesting WordPress Pages and groups here:
  - *WordPress.org:* `https://www.facebook.com/WordPress`
  - *Advanced WordPress:* `https://www.facebook.com/groups/advancedwp`
  - *Matt Mullenweg (founder of WordPress):* `https://www.facebook.com/matt.mullenweg`

TIP

You can include Twitter lists on your site by using the handy Twitter widget for WordPress at `https://wordpress.org/plugins/widget-twitter`.

Users Helping Users

Don't worry if you're not a member of the WordPress community. Joining is easy: Simply start your own website by using the WordPress platform. If you're already publishing on a different platform, such as Drupal or Tumblr, WordPress makes migrating your data from that platform to a new WordPress setup simple. (See Book 2, Chapter 7 for information on migrating to WordPress from a different platform.)

WordPress support forums

You can find the WordPress Forums page (shown in Figure 4-1) at `https://wordpress.org/support`. This page is where you find users helping other users in their quest to use and understand the platform.

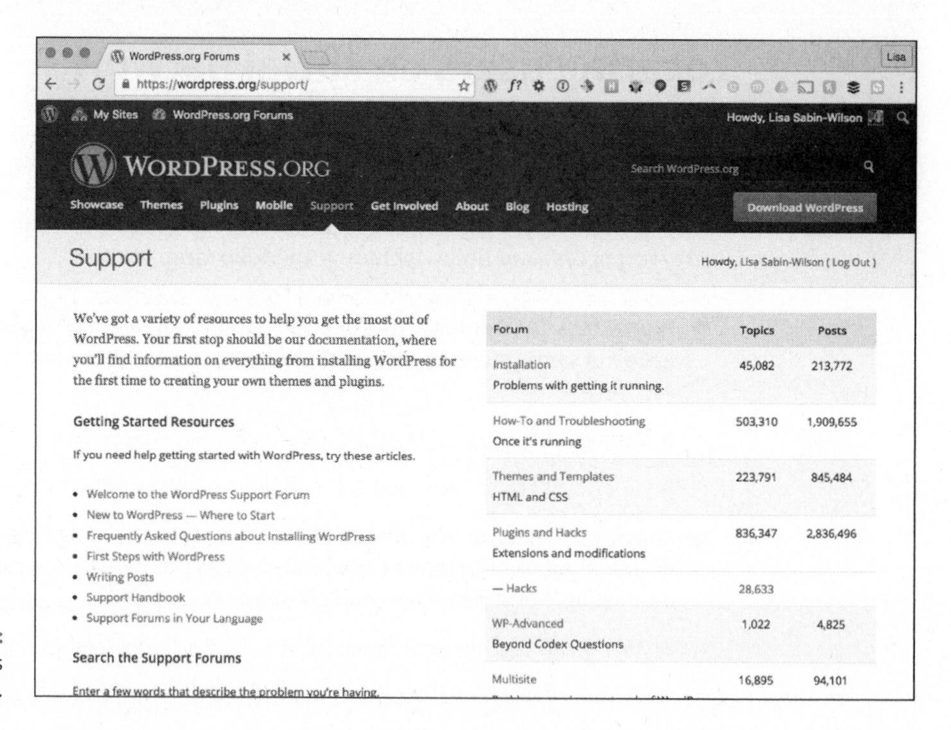

FIGURE 4-1: WordPress Forums page.

REMEMBER

The support forums are hosted on the WordPress.org website, but don't expect to find any official form of support from the WordPress developers. Instead, you find a large community of people from all walks of life seeking answers and providing solutions.

Users from beginner and novice level to the most advanced level browse the forums, providing support for one another. Each user has his or her own experiences, troubles, and knowledge level with WordPress, and the support forums are where users share those experiences and seek out the experiences of other users.

REMEMBER

It's important to keep in mind that the people you find and interact with on these official forums are offering their knowledge on a volunteer basis only, so as always, common-courtesy rules apply. "Please" and "thank you" go a long, long way in the forums.

TIP

If you find solutions and assistance in the WordPress support forums, consider browsing the forum entries to see whether you can help someone else by answering a question or two.

WordPress user manual

You can find users contributing to the very helpful WordPress Codex (a collection of how-to documents) at `https://codex.wordpress.org`. *Codex*, by the way, is Latin for *book*.

The WordPress Codex is a collaborative effort to document the use of the WordPress software. All contributors to the codex are WordPress users who donate their time as a way of giving back to the free, open-source project that has given them a dynamic piece of software for publishing freely on the web.

WordPress mailing lists

You can subscribe to various mailing lists, too. These lists offer you the opportunity to become involved in various aspects of the WordPress community as well as future development of the software. All the available WordPress mailing lists are on the Automattic website at `http://lists.automattic.com/mailman/listinfo`. The most popular ones include

- **wp-hackers (`http://lists.automattic.com/mailman/listinfo/wp-hackers`):** Subscribe to this mailing list to interact and talk to other WordPress users about *hacking* WordPress — otherwise known as altering WordPress code to make it do what you want it to do.
- **wp-testers (`http://lists.automattic.com/mailman/listinfo/wp-testers`):** This mailing list is filled with people who are testing new releases (as well as beta versions) of WordPress and reporting any bugs or problems that they find.
- **wp-edu (`http://lists.automattic.com/mailman/listinfo/wp-edu`):** This mailing list is dedicated to people in the education field who use WordPress, such as teachers and professors.

Discovering Professional WordPress Consultants and Services

You have big plans for your blog, and your time is valuable. Hiring a professional to handle the back-end design and maintenance of your blog enables you to spend your time creating the content and building your readership on the front end.

Many bloggers who decide to go the custom route by hiring a design professional do it for another reason: They want the designs/themes of their blogs to be unique. Free themes are nice, but you run the risk that your blog will look like hundreds of other blogs out there.

A *brand*, a term often used in advertising and marketing, refers to the recognizable identity of a product — in this case, your blog. Having a unique brand or design for your site sets yours apart from the rest. If your blog has a custom look, people will associate that look with you. You can accomplish branding with a single logo or an entire layout and color scheme of your choosing.

Many consultants and design professionals put themselves up for hire. Who are these people? I get to that topic in just a second. First, you want to understand what services they offer, which can help you decide whether hiring a professional is the solution for you.

Here are some of the many services available:

- Custom graphic design and CSS styling for your blog
- Custom templates
- WordPress plugin installation and integration
- Custom WordPress plugins
- WordPress software installation on your web server
- Upgrades of the WordPress software
- Web hosting and domain registration services
- Search engine optimization and site marketing

REMEMBER

Some bloggers take advantage of the full array of services provided, whereas others use only a handful. The important thing to remember is that you aren't alone. Help is available for you and your blog.

Table 4-1 pairs the three types of blog experts — designers, developers, and consultants — with the services they typically offer. Many of these folks are freelancers with self-imposed titles, but I've matched titles with typical duties. Keep in mind that some of these professionals wear all these hats; others specialize in only one area.

I wish I could tell you what you could expect to pay for any of these services, but the truth is the levels of expertise — and expense — vary wildly. Services can range from $5 per hour to $300 or more per hour. As with any purchase, do your research and make an informed decision before you buy.

TABLE 4-1 **Types of WordPress Professionals**

Title	Services
Designers	These folks excel in graphic design, CSS, and the development of custom WordPress themes.
Developers	These guys and gals are code monkeys. Some of them don't know a stitch about design, but they can provide custom code to make your blog do things you never thought possible. Usually, you'll find these people releasing plugins in their spare time for the WordPress community to use free.
Consultants	If you're blogging for a business, these folks can provide you a marketing plan for your blog or a plan for using your blog to reach clients and colleagues in your field. Many of these consultants also provide search engine optimization to help your domain reach high ranks in search engines.

Listing all the professionals who provide WordPress services is impossible, but Tables 4-2 through 4-4, later in this chapter, list some of the most popular ones. I tried to cover a diverse level of services so that you have the knowledge to make an informed decision about which professional to choose.

WordPress designers

WordPress designers can take a simple blog and turn it into something dynamic, beautiful, and exciting. These people are experts in the graphic design, CSS styling, and template tagging needed to create a unique theme for your website. Often, WordPress designers are skilled in installing and upgrading WordPress software and plugins; sometimes, they're even skilled in creating custom PHP or plugins. These folks are the ones you want to contact when you're looking for someone to create a unique design for your website that's an individual, visual extension of you or your company.

Some blog designers post their rates on their websites because they offer design *packages,* whereas other designers quote projects on a case-by-case basis because every project is unique. When you're searching for a designer, if the prices aren't displayed on the site, just drop the designer an email and ask for an estimate. Armed with this information, you can do a little comparison shopping while you search for just the right designer.

The designers and design studios listed in Table 4-2 represent a range of styles, pricing, services, and experience. All of them excel in creating custom WordPress blogs and websites. This list is by no means exhaustive, but it's a nice starting point.

TABLE 4-2 Established WordPress Designers

Who They Are	Where You Can Find Them
WebDevStudios	`https://webdevstudios.com`
10up	`https://10up.com`
Range	`http://ran.ge`

Developers

The WordPress motto sits at the bottom of the WordPress home page:

> Code is poetry.

No one knows this better than the extremely talented blog developers in the core WordPress development team. A developer can take some of the underlying code, make a little magic happen between PHP and the MySQL database that stores the content of your blog, and create a dynamic display of that content for you. Most likely, you'll contact a developer when you want to do something with your blog that's a little out of the ordinary, and you can't find a plugin that does the trick.

If you've gone through all the available WordPress plugins and still can't find the exact function that you want your WordPress blog to perform, contact one of these folks. Explain what you need. The developer can tell you whether it can be done, whether she's available to do it, and how much the job will cost. (Don't forget that last part!) You may recognize some of the names in Table 4-3 as developers/authors of some popular WordPress plugins.

TABLE 4-3 Established WordPress Developers

Who They Are	Where You Can Find Them
WebDevStudios	`https://webdevstudios.com`
eHermits, Inc.	`http://ehermitsinc.com`
Covered Web Services — Mark Jaquith	`http://coveredwebservices.com`
Voce Communications	`http://vocecommunications.com`

Consultants

Blog consultants may not be able to design or code for you, but they're probably connected to people who can. Consultants can help you achieve your goals for your blog in terms of online visibility, marketing plans, and search engine optimization. Most of these folks can help you find out how to make money with your blog and connect you with various advertising programs. Quite honestly, you can do what blog consultants do by investing just a little time and research in these areas. As with design and coding, however, figuring everything out and then implementing it takes time. Sometimes it's easier — and more cost-effective — to hire a professional than to do it yourself.

Who hires blog consultants? Typically, a business that wants to incorporate a blog into its existing website or a business that already has a blog but wants help taking it to the next level. Table 4-4 lists some people and organizations that offer this kind of consulting.

TABLE 4-4 Established Blog Consultants

Who They Are	Where You Can Find Them	Type of Consulting
Copyblogger	`www.copyblogger.com`	SEO, marketing
Convertiv	`https://www.convertiv.com`	WordPress design and development, social media consulting
WordPress 101	`https://www.wp101.com`	WordPress training

Contributing to WordPress

Contributing code to the core WordPress software is only one way of participating in the WordPress project. You don't need to be a coder or developer to contribute to WordPress — and it's easier than you might think. Here are several ways you can contribute to the project, including (but not limited to) code:

- **Code:** One of the most obvious ways you can contribute to WordPress is providing code to be used in the core files. The WordPress project has several hundred developers who contribute code at one time or another. You submit code through the WordPress Trac at `https://core.trac.wordpress.org`. Within the Trac, you can follow current development and track changes. To contribute, you can use the Trac to download and test a code patch or look at reported bugs to see whether you can offer a fix or submit a patch. Required skills include, at the very least, PHP programming, WordPress experience, and MySQL database administration. (That isn't an exhaustive list, mind you.)

- **Testing:** You can join the wp-testers mailing list (refer to "WordPress mailing lists" earlier in this chapter) to test beta versions of WordPress and report your own user experience. WordPress developers monitor this mailing list and try to fix any true bugs or problems.
- **Documentation:** Anyone can submit documentation to the WordPress Codex (the user documentation for WordPress). All you need to do is visit `https://codex.wordpress.org`, create an account, and dig in!

TIP

Be sure to check out the article titled "Codex: Contributing" (`https://codex.wordpress.org/Codex:Contributing`), which provides good tips on how to get started, including guidelines for documentation contributions.

- **Tutorials:** Do you feel that you have a few tips and tricks you want to share with other WordPress users? Take them to your blog! What better way to contribute to WordPress than sharing your knowledge with the rest of the world? Write up your how-to tutorial, publish it on your website, and then promote your tutorial on Twitter and Facebook.
- **Support forums:** Volunteer your time and knowledge on the WordPress support forums at `https://wordpress.org/support`. The involvement of the WordPress users who donate their time and talents in the support forum is an essential part of the WordPress experience.
- **Presentations:** In the next section of this chapter, I discuss live WordPress events where users meet face to face. Consider offering to speak at one of those events to share your knowledge and experience with other users — or host one in your area.

Participating in Live WordPress Events

You can not only find out about WordPress and contribute to the project online via the Internet, but also get involved in WordPress offline. Live WordPress events, called WordPress Meetups and WordCamps, are where users and fans get together to discuss, learn, and share information about their favorite platform. The two events are somewhat different:

- **WordPress Meetups:** Generally, these events involve small groups of people from the same geographical location. Typically, these speakers, organizers, and attendees enjoy gathering on a monthly or bimonthly basis.

  You can find a WordPress Meetup near your community by visiting the Meetup website at `https://www.meetup.com` or by performing a search, using the keyword *WordPress* and your city or zip code.

» **WordCamps:** These annual events are usually much larger than Meetups and are attended by people from all over the country. WordCamps are hosted in almost every major city in the United States and abroad. Usually, WordCamps cost a small amount to attend, and speakers at WordCamps are well-known personalities from the WordPress community.

  You can find a WordCamp event close to you by visiting the WordCamp website at `http://central.wordcamp.org` and browsing the upcoming WordCamps.

TIP

If there isn't a Meetup or WordCamp scheduled in your area, consider getting involved and organizing one! You can find some great tips and information about organizing WordCamps at `http://central.wordcamp.org`.

Meeting the WordPress Community

» Self-hosting with WordPress.org

» Running a network of blogs with the multisite feature

» Exploring enterprise options and VIP services

Chapter 5
Discovering Different Versions of WordPress

Website publishers have a wealth of software platforms to choose among. You want to be sure that the platform you choose has all the options you're looking for. WordPress is unique in that it offers two versions of its software. Each version is designed to meet the various needs of publishers.

One version is a hosted platform available at WordPress.com that meets your needs if you don't want to worry about installing or dealing with software; the other is the self-hosted version of the WordPress software available at `https://wordpress.org`, which offers you a bit more freedom and flexibility, as described throughout this chapter.

This chapter introduces you to both versions of the WordPress platform so you can choose which version suits your particular needs the best.

Comparing the Two Versions of WordPress

The two versions of WordPress are

- The hosted version at WordPress.com
- The self-installed and self-hosted version available at WordPress.org

Certain features are available to you in every WordPress site setup, whether you're using the self-hosted software from WordPress.org or the hosted version at WordPress.com. These features include (but aren't limited to)

- Quick and easy installation and setup
- Full-featured publishing capability, letting you publish content to the web through an easy-to-use web-based interface
- Topical archiving of your posts, using categories
- Monthly archiving of your posts, with the capability to provide a listing of those archives for easy navigation through your site
- Comment and trackback tools
- Automatic spam protection through Akismet
- Built-in gallery integration for photos and images
- Media Manager for managing video and audio files
- Great community support
- Unlimited number of static pages, letting you step out of the blog box and into the sphere of running a fully functional website
- RSS capability with RSS 2.0, RSS 1.0, and Atom support
- Tools for importing content from different content management systems (such as Blogger and Movable Type)

Table 5-1 compares the two WordPress versions.

TABLE 5-1 **Exploring the Differences between the Two Versions of WordPress**

Feature	WordPress.org	WordPress.com
Cost	Free	Free
Software download required	Yes	No
Software installation required	Yes	No
Web hosting required	Yes	No
Custom CSS control	Yes	Available in the Premium or Business plan up to $99/year
Template access	Yes	No
Sidebar widgets	Yes	Yes
RSS syndication	Yes	Yes
Access to core code	Yes	No
Ability to install plugins	Yes	No
WP themes installation	Yes	No
Multiauthor support	Yes	Yes
Unlimited number of blog setups with one account (multisite)	Yes*	Yes
Community-based support forums	Yes	Yes

**Only with the Multisite feature enabled*

Choosing the hosted version from WordPress.com

WordPress.com (see Figure 5-1) is a free service. If downloading, installing, and using software on a web server sound like Greek to you and are chores you'd rather avoid, the WordPress folks provide a solution for you at WordPress.com.

WordPress.com is a *hosted solution,* which means that it has no software requirement, no downloads, and no installation or server configurations. Everything's done for you on the back end, behind the scenes. You don't even have to worry about how the process happens; it happens quickly, and before you know it, you're making your first blog post.

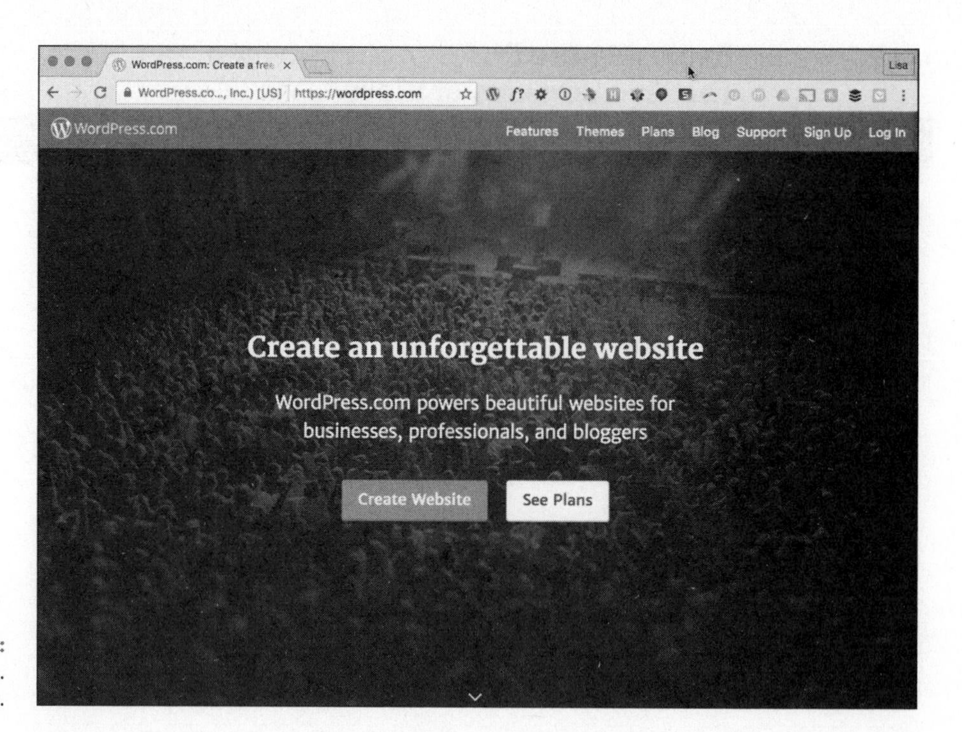

FIGURE 5-1: The WordPress.com website.

WordPress.com offers several upgrades (see Figure 5-2) to help make your publishing life easier. Here's a list of package upgrades you can purchase to enhance your WordPress.com account, with prices reflecting the annual cost:

- **Personal:** This plan allows you to add your own domain name to your WordPress.com account; see Book 2, Chapter 1. This service also provides you email and live chat support, basic design customizations, and 3GB of storage space. For the additional fee, your site also becomes ad-free. (With the free plan, WordPress.com advertisements are part of your experience.) This plan costs $2.99 per month, billed annually at $35.88 per year.
- **Premium:** This plan provides you everything included in the Personal plan and also includes more advanced theme customization (full control of the CSS), increased storage space at 13GB, the ability to monetize your site, and VideoPress support. This plan costs $8.25 per month, billed annually at $99 per year.

TIP

The VideoPress service is described in the "Discovering WordPress VIP Services" section at the end of this chapter.

- **Business:** This plan provides you everything included in both the Personal and Premium plans. In addition, you have access to premium themes, unlimited storage space, live courses, Google Analytics integration, and the removal of all WordPress.com branding — all for the cost of $24.92 per month, billed annually at $299.04 per year.

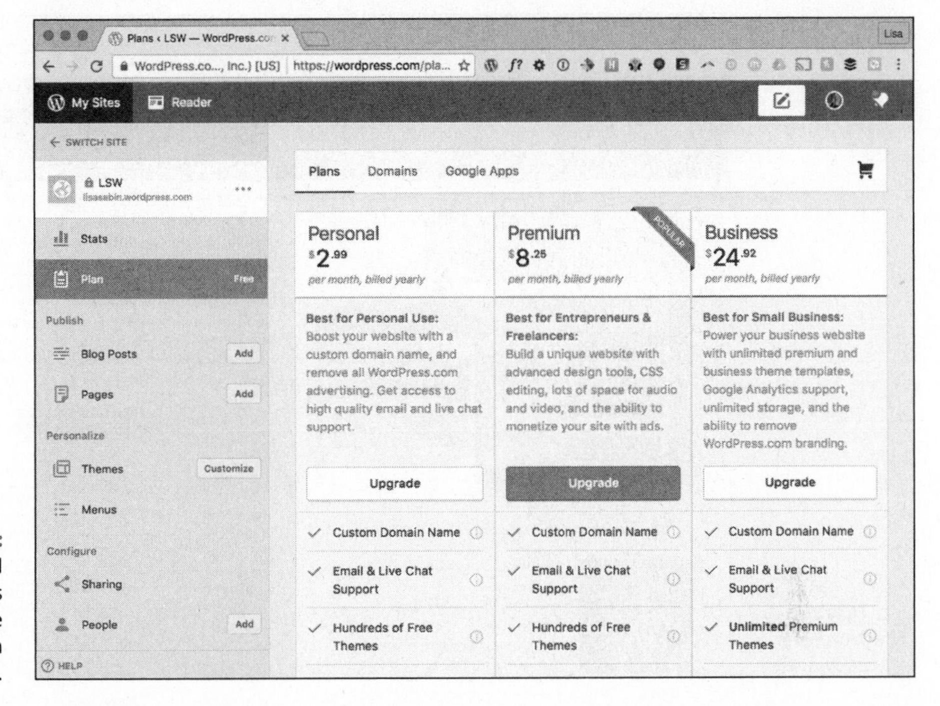

FIGURE 5-2: Several paid upgrades available on the WordPress.com free service.

TIP

WordPress.com has some limitations. You can't install plugins or custom themes, for example. Neither can you can customize the base-code files, sell advertising, or monetize your site without upgrading to a paid plan. But even with its limitations, WordPress.com is an excellent starting point if you're brand-new to Internet publishing and a little intimidated by the configuration requirements of the self-installed WordPress.org software.

The good news is that if you outgrow your WordPress.com-hosted site and want to move to the self-hosted WordPress.org software, you can. You can even take all the content from your WordPress.com-hosted site with you and easily import it into your new setup with the WordPress.org software.

Therefore, in the grand scheme of things, your options aren't really that limited.

Self-hosting with WordPress.org

The self-installed version from WordPress.org is the primary focus of *WordPress All-in-One For Dummies.* Using WordPress.org requires you to download the software from the WordPress website at `https://wordpress.org` (shown in Figure 5-3); then you need to install it on a server from which your website operates.

FIGURE 5-3: The WordPress.org website.

The WordPress.org website is an excellent repository of tools and resources for you throughout the lifespan of your WordPress-powered website, so be sure to bookmark it for future reference! Here's a list of helpful things that you can find on the website:

- **Plugins (`https://wordpress.org/plugins`):** The WordPress Plugins page houses a full directory of plugins available for WordPress. You can search for and find the plugins you need for search engine optimization (SEO) enhancement, comment management, and social media integration, among many others.
- **Themes (`https://wordpress.org/themes`):** The Theme Directory page, shown in Figure 5-4, is a repository of WordPress themes that are free for the taking. In this section of the WordPress.org website, you can browse more than 5,000 themes to use on your site to dress up your content.
- **Codex (`https://codex.wordpress.org`):** Almost every piece of software released comes with documentation and user manuals. The Support section of the WordPress.org website contains the WordPress Codex, which tries to help you answer questions about the use of WordPress and its various features and functions.

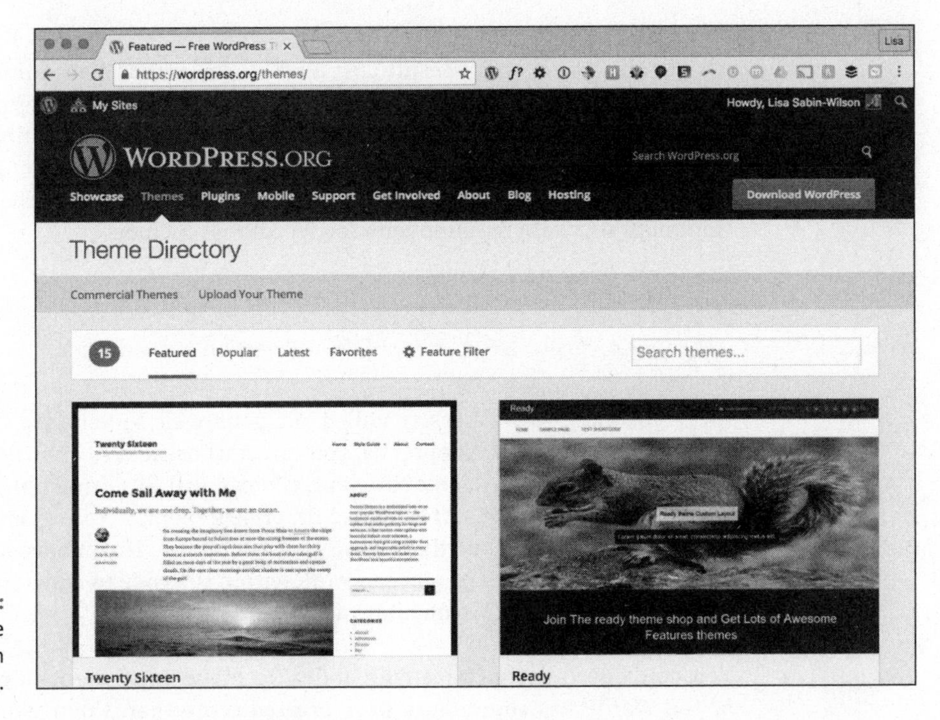

FIGURE 5-4: Theme Directory on WordPress.org.

- **Forums (`https://wordpress.org/support`):** The support forums at WordPress.org involve WordPress users from all over with one goal: finding out how to use WordPress to suit their particular needs. The support forums are very much a community of users (from beginners to experts) helping other users, and you can generally obtain a solution to your WordPress needs here from other users of the software.
- **Roadmap (`https://wordpress.org/about/roadmap`):** This section of the WordPress.org website doesn't contain support information or tools that you can download; it offers an at-a-glance peek at what's new and upcoming for WordPress. The Roadmap page gives you a pretty accurate idea of when WordPress will release the next version of its software; see Book 1, Chapter 3 for information about versions and release cycles.

TIP

Click the version number to visit the WordPress Trac and see what features developers are working on and adding.

WordPress.org is the self-installed, self-hosted software version of WordPress that you install on a web server you've set up on a domain that you've registered. Unless you own your own web server, you need to lease one. Leasing space on a web server is *web hosting*, and unless you know someone who knows someone, hosting generally isn't free.

That said, web hosting doesn't cost a whole lot, either. You can usually obtain a good, basic web hosting service for anywhere from $10 to $15 per month. (Book 2, Chapters 1 and 2 give you some great information on web hosting accounts and tools.) You need to make sure, however, that any web host you choose to work with has the required software installed on the web server. The recommended minimum software requirements for WordPress include

- PHP version 5.6 or later
- MySQL version 5.6 or later *or* MariaDB version 10.0 or later

After you have WordPress installed on your web server (see the installation instructions in Book 2, Chapter 4), you can start using it to publish to your heart's content. With the WordPress software, you can install several plugins that extend the functionality of the software, as I describe in Book 7. You also have full control of the core files and code that WordPress is built on. If you have a knack for PHP and knowledge of MySQL, you can work within the code to make changes that you think would be good for you and your website.

You don't need design or coding ability to make your site look great. Members of the WordPress community have created more than 1,600 WordPress themes (designs), and you can download them free and install them on your WordPress blog. (See Book 6, Chapter 2.) Additionally, if you're creatively inclined, like to create designs on your own, and know Cascading Style Sheets (CSS), you have full access to the template system within WordPress and can create your own custom themes. (See Book 6, Chapters 3 through 7.)

Hosting Multiple Sites with One WordPress Installation

The self-hosted WordPress.org software also lets you run an unlimited number of sites on one installation of its software platform, on one domain. When you configure the options within WordPress to enable a multisite interface, you become administrator of a network of sites. All the options remain the same, but with the multisite options configured, you can add more sites and domains, as well as allow registered users of your website to host their own sites within your network. For more information about the Multisite feature in WordPress, see Book 8.

The following types of sites use the Network options within WordPress:

- **Blog networks,** which can have more than 150 blogs. The tech giant Microsoft uses WordPress to power thousands of tech blogs at its TechNet portal: `https://blogs.technet.microsoft.com/wikininjas/2016/09/25/sunday-surprise-technet-forums`.

- **Newspapers and magazines,** such as *The New York Times,* and universities, such as Harvard Law School, use WordPress to manage the blog sections of their websites.
- **Niche-specific blog networks,** such as Edublogs.org, use WordPress to manage their full networks of free sites for teachers, educators, lecturers, librarians, and other education professionals.

TIP

Extensive information on running a network of sites by using the Multisite feature in WordPress is available in Book 8. The chapters there take you through everything: setup, maintenance, and the process of running a network of sites with one WordPress installation.

With the Multisite features enabled, users of your network can run their own sites within your installation of WordPress. They also have access to their own Dashboards with the same options and features you read about in Book 3. Heck, it probably would be a great idea to buy a copy of this book for every member within your network so everyone can become familiar with the WordPress Dashboard and features, too. At least have a copy on hand so people can borrow yours!

If you plan to run just a few of your own sites with the WordPress Multisite feature, your current hosting situation is probably well suited. (See Book 2, Chapter 1 for information on web hosting services.) If you plan to host a large network with hundreds of sites and multiple users, however, you should consider contacting your host and increasing your bandwidth and the disk space limitations on your account.

The best example of a large blog network with hundreds of blogs and users (actually, more like millions) would be the hosted service at WordPress.com, which I discuss earlier in this chapter. At WordPress.com, people are invited to sign up for an account and start a blog by using the Multisite feature within the WordPress platform on the WordPress server. When you enable this feature on your own domain and enable the user registration feature (covered later in this chapter), you invite users to

- Create an account
- Create a site on your WordPress installation (on your domain)
- Create content by publishing posts and pages
- Upload media files, such as photos, audio, and video
- Invite their friends to view their blogs or sign up for their own accounts

WARNING

In addition to the necessary security measures, time, and administrative tasks that go into running a community of sites, you have a few things to worry about. Creating a community increases the resource use, bandwidth, and disk space on your web server. In many cases, if you go over the allotted limits given to you by your web host, you incur great cost. Make sure that you anticipate your bandwidth and disk-space needs before running a large network on your website! (Don't say you weren't warned.)

REMEMBER

Many WordPress network communities start with grand dreams of being large and active. Be realistic about how your community will operate to make the right hosting choice for yourself and your community.

Small Internet communities are handled easily with a shared-server solution; larger, more active communities should consider a dedicated server solution for operation. The difference between the two lies in their names:

- **Shared-server solution:** You have one account on one server that has several other accounts on it. Think of this as apartment living. One building has several apartments under one roof.
- **Dedicated server:** You have one account on one server. The server is dedicated to your account, and your account is dedicated to the server. Think of this as owning a home where you don't share your living space with anyone else.

A dedicated-server solution is a more expensive investment for your community; a shared-server solution is more economical. Base your decision on how big and how active you estimate that your community will be. You can move from a shared-server solution to a dedicated-server solution if your community becomes larger than you expect, but starting with the right solution for your community from day one is best. For more information on hosting WordPress, see Book 2, Chapter 1.

Discovering WordPress VIP Services

The company behind the Automattic WordPress.com service is owned and operated by the WordPress cofounder, Matt Mullenweg. Although Automattic doesn't own the WordPress.org software, Automattic is a driving force behind all things WordPress.

REMEMBER

As an open-source platform, WordPress.org is owned by the community and hundreds of developers that contribute to the core code.

Have a look at the Automattic website at `https://automattic.com` (shown in Figure 5-5). The folks behind WordPress own and operate several properties and services that can extend the features of your WordPress site, including

- **WordPress.com (`https://wordpress.com`):** A hosted WordPress blogging service, discussed previously in this chapter.
- **Jetpack (`https://jetpack.com`):** A suite of plugins that can be installed on a WordPress.org self-hosted site.
- **VaultPress (`https://vaultpress.com`):** Premium backup and restoration service for your blog.
- **Akismet (`https://akismet.com`):** Spam protection for your blog. This service comes with every WordPress.org installation, but there are different levels of service, as discussed in Book 3, Chapter 4.
- **Polldaddy (`https://polldaddy.com`):** A polling and survey software that easily plugs into the WordPress platform.

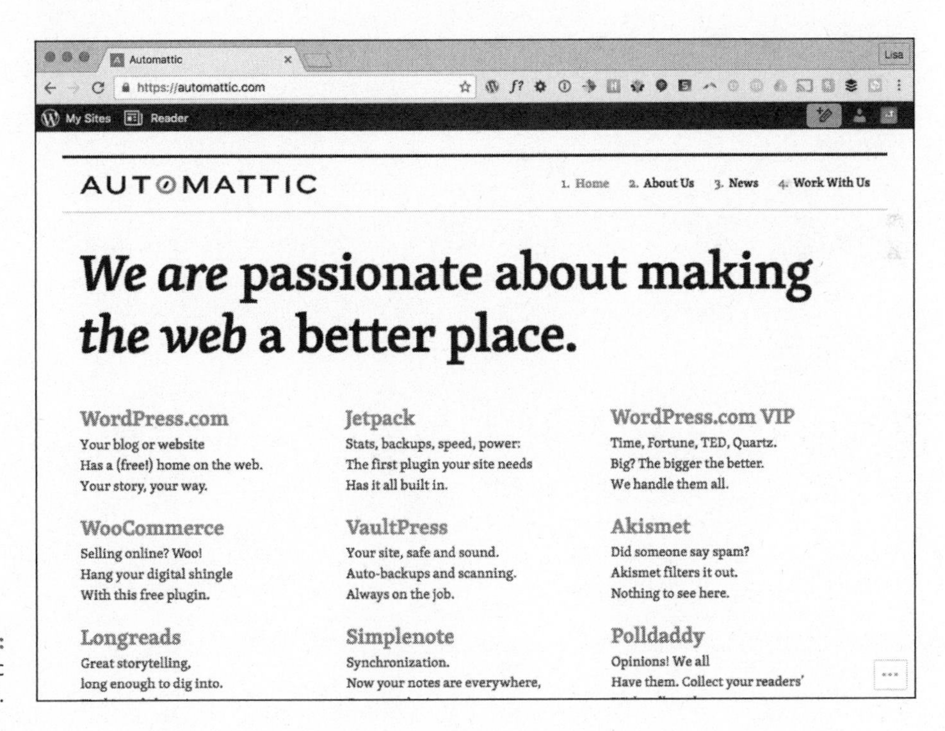

FIGURE 5-5: The Automattic website.

- **VideoPress (`https://videopress.com`):** Video hosting and sharing application for WordPress.
- **Gravatar (`https://gravatar.com`):** Photos or graphical icons for comment authors (discussed in Book 3, Chapter 2).
- **Longreads (`https://longreads.com`):** Great examples of storytelling on the Internet.
- **Simplenote (`https://simplenote.com`):** An easy way to keep notes across various iOS, Android, Mac, and Windows devices.
- **WordPress.com VIP (`https://vip.wordpress.com`):** Enterprise-level web hosting and WordPress support starting at $15,000 per year (usually reserved for heavy hitters such as CNN, BBC, and *Time* magazine, for example).
- **Cloudup (`https://cloudup.com`):** Easy sharing of media including videos, music, photos, and documents.

2

Setting Up the WordPress Software

Contents at a Glance

» Exploring web hosting environments

» Knowing the basic requirements for PHP and MySQL

» Getting web hosting recommendations for WordPress

» Understanding bandwidth and disk space needs

Chapter 1 Understanding the System Requirements

Before you can start blogging with WordPress, you have to set up your foundation. Doing so involves more than simply downloading and installing the WordPress software. You also need to establish your *domain* (your website address) and your *web hosting service* (the place that houses your website). Although you initially download your WordPress software onto your hard drive, you install it on a web hosting server.

Obtaining a web server and installing software on it is something you may already have done on your site, in which case you can move on to the next chapter. If you haven't installed WordPress, you must first consider many factors, as well as cope with a learning curve, because setting up your website through a hosting service involves using some technologies that you may not feel comfortable with. This chapter takes you through the basics of those technologies, and by the last page of this chapter, you'll have WordPress successfully installed on a web server with your own domain name.

Establishing Your Domain

You've read all the hype. You've heard all the rumors. You've seen the flashy websites powered by WordPress. But where do you start?

The first steps in installing and setting up a WordPress site are making a decision about a domain name and then purchasing the registration of that name through a domain registrar. A *domain name* is the *unique* web address that you type in a web browser's address bar to visit a website. Some examples of domain names are `WordPress.org` and `Google.com`.

REMEMBER

I emphasize *unique* because no two domain names can be the same. If someone else has registered the domain name you want, you can't have it. With that in mind, it sometimes takes a bit of time to find a domain that isn't already in use.

Understanding domain name extensions

When registering a domain name, be aware of the *extension* that you want. The `.com`, `.net`, `.org`, `.info`, or `.biz` extension that you see tagged onto the end of any domain name is the *top-level domain extension.* When you register your domain name, you're asked to choose the extension you want for your domain (as long as it's available, that is).

A word to the wise here: Just because you registered your domain as a `.com` doesn't mean that someone else doesn't, or can't, own the very same domain name with a `.net`. Therefore, if you register `MyDogHasFleas.com`, and the site becomes hugely popular among readers with dogs that have fleas, someone else can come along, register `MyDogHasFleas.net`, and run a similar site to yours in the hope of riding the coattails of your website's popularity and readership.

DOMAIN NAMES: DO YOU OWN OR RENT?

When you "buy" a domain name, you don't really own it. Rather, you're purchasing the right to use that domain name for the time specified in your order. You can register a domain name for one year or up to ten years. Be aware, however, that if you don't renew the domain name when your registration period ends, you lose it — and most often, you lose it right away to someone who preys on abandoned or expired domain names. Some people keep a close watch on expiring domain names, and as soon as the buying window opens, they snap the names up and start using them for their own websites, in the hope of taking full advantage of the popularity that the previous owners worked so hard to attain for those domains.

If you want to avert this problem, you can register your domain name with all available extensions. My personal website, for example, has the domain name `lisasabin-wilson.com`, but I also own `lisasabin-wilson.net` just in case someone else out there has the same combination of names.

Considering the cost of a domain name

Registering a domain costs you anywhere from $5 to $300 per year or more, depending on what service you use for a registrar and what options (such as storage space, bandwidth, privacy options, search engine submission services, and so on) you apply to your domain name during the registration process.

REMEMBER

When you pay the domain registration fee today, you need to pay another registration fee when the renewal date comes up again in a year, or two, or five — however many years you chose to register your domain name for. (See the nearby "Domain names: Do you own or rent?" sidebar.) Most registrars give you the option of signing up for a service called Auto Renew to automatically renew your domain name and bill the charges to the credit card you set up on that account. The registrar sends you a reminder a few months in advance, telling you that it's time to renew. If you don't have Auto Renew set up, you need to log in to your registrar account before it expires and manually renew your domain name.

TIP

When choosing a domain name for your website, you may find that the domain name you want isn't available. You know if it's available when you search for it at the domain registrar's website (listed in the next section). Have some backup domain names prepared just in case the one you want isn't available. If your chosen domain name is `cutepuppies.com`, but it's not available, you could have some variations of the domain ready to use, such as `cute-puppies.com` (notice the dash), `mycutepuppies.com`, or `reallycutepuppies.com`.

Registering your domain name

Domain registrars are certified and approved by the Internet Corporation for Assigned Names and Numbers (ICANN). Although hundreds of domain registrars exist, the ones in the following list are popular because of their longevity in the industry, competitive pricing, and variety of services they offer in addition to domain name registration (such as web hosting and website traffic builders):

- **GoDaddy:** `https://www.godaddy.com`
- **Register.com:** `https://www.register.com`
- **Network Solutions:** `https://www.networksolutions.com`
- **NamesDirect:** `http://namesdirect.com`

No matter where you choose to register your domain name, here are the steps you can take to accomplish this task:

1. **Decide on a domain name.**

   A little planning and forethought are necessary here. Many people think of a domain name as a *brand* — a way of identifying their websites or blogs. Think of potential names for your site and then proceed with your plan.

2. **Verify the domain name's availability.**

   In your web browser, enter the URL of the domain registrar of your choice. Look for the section on the registrar's website that lets you enter the domain name (typically, a short text field) to see whether it's available. If the domain name isn't available as a `.com`, try `.net` or `.info`.

3. **Purchase the domain name.**

   Follow the domain registrar's steps to purchase the name, using your credit card. After you complete the checkout process, you receive an email confirming your purchase, so be sure to use a valid email address during the registration process.

The next step is obtaining a hosting account, which the next section covers.

Some of the domain registrars have hosting services that you can sign up for, but you don't have to use those services. Often, you can find hosting services for a lower cost than most domain registrars offer. It just takes a little research.

Finding a Home for Your Blog

After you register your domain, you need to find a place for it to live: a web host. Web hosting is the second piece of the puzzle that you need to complete before you begin working with WordPress.org.

A *web host* is a business, group, or person that provides web server space and bandwidth for file transfer to website owners who don't have it. Usually, web hosting services charge a monthly or an annual fee — unless you're fortunate enough to know someone who's willing to give you server space and bandwidth free. The cost varies from host to host, but you can usually obtain quality web hosting services for $10 to $30 per month to start.

When discussing web hosting considerations, it's important to understand where your hosting account ends and WordPress begins. Support for the WordPress software may or may not be included in your hosting package.

Some web hosts consider WordPress to be a *third-party application*. This means that the host typically won't provide technical support on the use of WordPress (or any other software application) because software support generally isn't included in your hosting package. The web host supports your hosting account but typically doesn't support the software you choose to install.

On the other hand, if your web host supports the software on your account, it comes at a cost: You have to pay for that extra support. To find whether your chosen host supports WordPress, ask first. If your host doesn't offer software support, you can still find WordPress support in the support forums at `https://wordpress.org/support`, as shown in Figure 1-1.

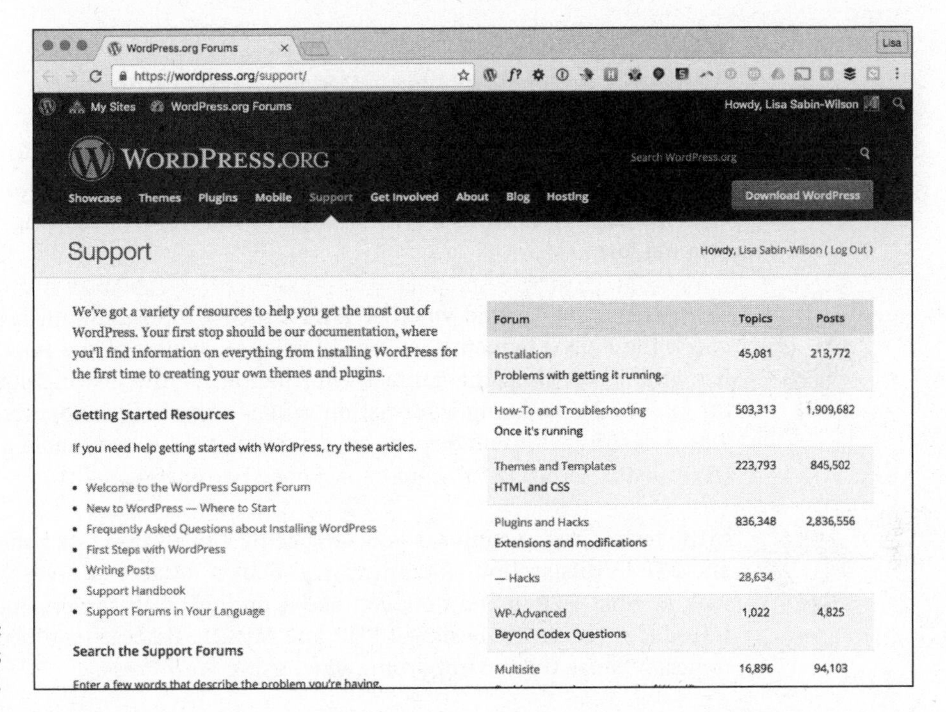

FIGURE 1-1: The WordPress support forums.

TIP

Several web hosting providers also have WordPress-related services available for additional fees. These services can include technical support, plugin installation and configuration, and theme design.

Generally, hosting services provide (at least) these services with your account:

- Hard drive space
- Bandwidth (transfer)

- Domain email with web mail access
- File Transfer Protocol (FTP) access
- Comprehensive website statistics
- MySQL database(s)
- PHP

Because you intend to run WordPress on your web server, you need to look for a host that provides the *minimum* requirements needed to run the software on your hosting account, which are

- PHP version 5.6 (or later)
- MySQL version 5.0 (or later) *or* MariaDB version 10.0 (or later)

You also want a host that provides daily backups of your site so that your content won't be lost in case something happens. Web hosting providers that offer daily backups as part of their services can save the day by restoring your site to its original form.

TIP

The easiest way to find whether a host meets the minimum requirement is to check the FAQ (Frequently Asked Questions) section of the host's website, if it has one. If not, find the contact information for the hosting company and fire off an email requesting information on exactly what it supports. Any web host worth dealing with will answer your email within a reasonable amount of time. (A response within 12 to 24 hours is a good barometer.)

TIP

If the technojargon confuses you — specifically, all that talk about PHP, MySQL, and FTP in this section — don't worry! Book 2, Chapter 2 gives you an in-depth look at what FTP is and how you use it on your web server; Book 2, Chapter 3 introduces you to the basics of PHP and MySQL. Become comfortable with these topics, because they're important when using WordPress.

Getting help with hosting WordPress

The popularity of WordPress has given birth to web services — including designers, consultants, and (yes) web hosts — that specialize in using WordPress.

Many web hosts offer a full array of WordPress features, such as an automatic WordPress installation included with your account, a library of WordPress themes, and a staff of support technicians who are very experienced in using WordPress.

Here is a list of some of those providers:

- **Pagely:** `https://pagely.com`
- **WP Engine:** `https://wpengine.com` (shown in Figure 1-2)
- **Pressable:** `https://pressable.com`

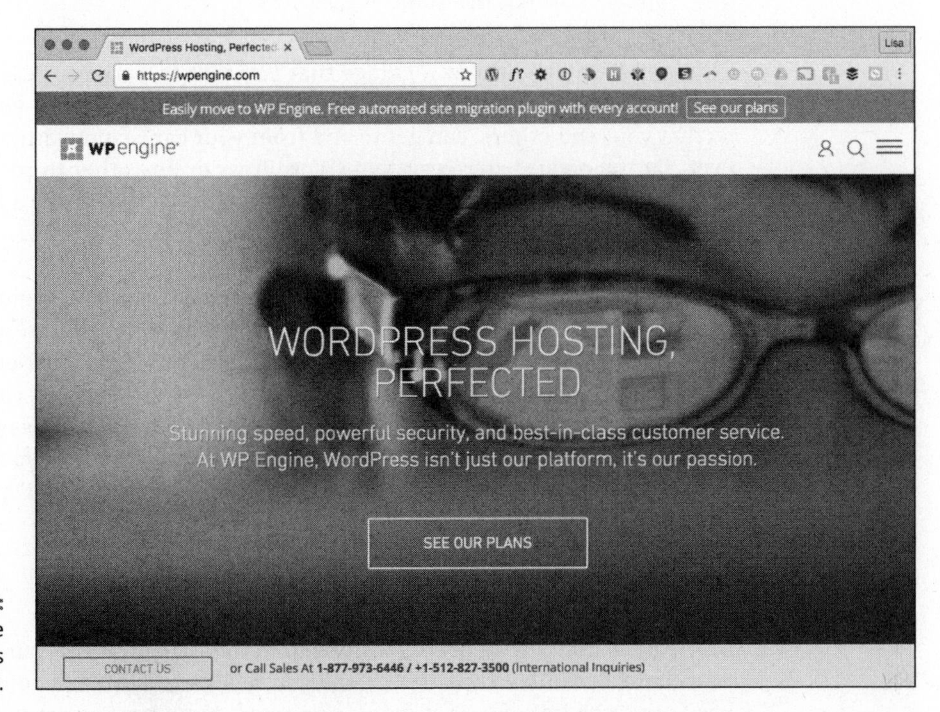

FIGURE 1-2: The WP Engine WordPress hosting provider.

WARNING

A few web hosting providers offer free domain name registration when you sign up for hosting services. Research this topic and read the terms of service because that free domain name may come with conditions. Many clients have gone this route, only to find out a few months later that the web hosting provider has full control of the domain name and the client can't move that domain off the host's servers, either for a set period (usually, a year or two) or for infinity. You need control in *your* hands, not someone else's, so stick with an independent domain registrar, such as Network Solutions.

Dealing with disk space and bandwidth

Web hosting services provide two very important things with your account:

- **Disk space:** The amount of space you can access on the web servers' hard drive, generally measured in megabytes (MB) or gigabytes (GB).
- **Bandwidth transfer:** The amount of transfer your site can do per month. Typically, traffic is measured in gigabytes.

Think of your web host as a garage that you rent to park your car in. The garage gives you the place to store your car (disk space). It even gives you the driveway so that you, and others, can get to and from your car (bandwidth). It won't, however, fix the rockin' stereo system (WordPress or any other third-party software application) that you've installed — unless you're willing to pay a few extra bucks for that service.

Most web hosting providers give you access to a hosting account manager that allows you to log in to your web hosting account to manage services. cPanel is perhaps the most popular management interface, but Plesk and NetAdmin are still widely used. These management interfaces give you access to your server logs, where you can view such things as bandwidth and hard disk usage. Get into the habit of checking those things occasionally to make sure that you stay informed about how much storage and bandwidth your site is using. Typically, I check monthly.

Managing disk space

Disk space is nothing more complicated than the hard drive on your own computer. Each hard drive has capacity, or space, for a certain amount of files. An 80GB hard drive can hold 80GB of data — no more. Your hosting account provides you a limited amount of disk space, and the same concept applies. If your web host provides you 10GB of disk space, that's the absolute limit you have. If you want more disk space, you need to upgrade your space limitations. Most web hosts have a mechanism in place for you to upgrade your allotment.

Starting with a self-hosted WordPress website doesn't take much disk space at all. A good starting point for disk space is 10GB to 20GB. If you find that you need additional space, contact your hosting provider for an upgrade.

Choosing the size of your bandwidth pipe

Bandwidth refers to the amount of data that's carried from point A to point B within a specific period (usually, only a second or two). I live out in the country — pretty much in the middle of nowhere. The water that comes to my house is provided by a private well that lies buried in the backyard somewhere. Between my house and the well are pipes that bring the water to my house. The pipes provide a free flow of water to our home so that everyone else can enjoy long, hot showers while I labor over dishes and laundry, all at the same time. Lucky me!

The very same concept applies to the bandwidth available with your hosting account. Every web hosting provider offers a variety of bandwidth limits on the accounts it offers. When I want to view your website in my browser window, the bandwidth is essentially the pipe that lets your data flow from your "well" to my computer. The bandwidth limit is similar to the pipe connected to my well: It can hold only a certain amount of water before it reaches maximum capacity and won't bring the water from the well any longer. Your bandwidth pipe size is determined by how much bandwidth your web host allows for your account. The larger the number, the bigger the pipe. A 50MB bandwidth limit makes for a smaller pipe than a 100MB limit, for example.

Web hosts are pretty generous with the amount of bandwidth they provide in their packages. Like disk space, bandwidth is measured in gigabytes. Bandwidth provision of 50GB to 100GB is generally a respectable amount to run a website with a blog.

WARNING

In my experience, I've found that if your website exceeds its allowed bandwidth, the web host won't turn off your website or limit traffic. The host will continue to allow inbound web traffic to your site but will bill you at the end of month for any bandwidth overages. Those charges can get pretty expensive, so if you find that your website is consistently exceeding the bandwidth amount every month, contact your web host to find whether if you can get an upgrade to allow for increased bandwidth.

REMEMBER

Websites that run large files — such as video, audio, or photo files — generally benefit from higher disk space compared with sites that don't involve large files. Keep this point in mind when you're signing up for your hosting account. Planning now will save you a few headaches down the road.

Be wary of hosting providers that offer things like unlimited bandwidth, domains, and disk space. Those offers are great selling points, but what the providers don't tell you outright (you may have to look into the fine print of the agreement) is that although they may not put those kinds of limits on you, they will limit your site's CPU usage.

CPU (which stands for *central processing unit*) is the part of a computer (or web server, in this case) that handles all the data processing requests sent to your web servers whenever anyone visits your site. Although you may have unlimited bandwidth to handle a large amount of traffic, if a high spike in traffic increases your site's CPU usage, your host will throttle your site because it limits the CPU use.

What do I mean by *throttle*? I mean that the host shuts down your site — turns it off. The shutdown isn't permanent, though; it lasts maybe a few minutes to an hour. The host does this to kill any connections to your web server that are causing the spike in CPU use. Your host eventually turns your site back on — but the inconvenience happens regularly with many clients across various hosting environments.

When looking into different web hosting providers, ask about their policies on CPU use and what they do to manage a spike in processing. It's better to know about it up front than to find out about it after your site's been throttled.

» Understanding secure file transfer

» Exploring easy-to-use SFTP clients

» Making sense of SFTP terminology

» Editing files by using SFTP

» Changing file permissions

Chapter 2

Using Secure File Transfer Protocol

Throughout this entire book, you run into the term *SFTP.* SFTP (Secure File Transfer Protocol) is a network protocol used to copy files from one host to another over the Internet. With SFTP, you can perform various tasks, including uploading and downloading WordPress files, editing files, and changing permissions on files.

Read this chapter to familiarize yourself with SFTP; understand what it is and how to use it; and discover some free, easy-to-use SFTP clients and programs that make your life as a WordPress website owner much easier. If you run across sections in this book that ask you to perform certain tasks by using SFTP, you can refer to this chapter to refresh your memory on how to do it, if needed.

Understanding SFTP Concepts

This section introduces you to the basic elements of SFTP, which is a method of transferring files in a secure environment. SFTP provides an additional layer of security beyond what you get with regular FTP, as it uses SSH (Secure Shell) and encrypts sensitive information, data, and passwords from being clearly transferred within the hosting network. Encrypting the data ensures that anyone monitoring the network can't read the data freely — and, therefore, can't obtain information that should be secured, such as passwords and usernames.

TIP

I highly recommend using SFTP over FTP because it's a secure connection to your web host. If your web hosting provider doesn't provide SFTP connections for you, strongly consider switching to a hosting provider that does. Almost all hosting providers these days provide SFTP as the standard protocol for transferring files.

The capability to use SFTP with your hosting account is a given for almost every web host on the market today. SFTP offers two ways of moving files from one place to another:

- **Uploading:** Transferring files from your local computer to your web server
- **Downloading:** Transferring files from your web server to your local computer

You can do several other things with SFTP, including the following, which I discuss later in this chapter:

- **View files.** After you log in via SFTP, you can see all the files that are located on your web server.
- **View date modified.** You can see the date when a file was last modified, which can be helpful when trying to troubleshoot problems.
- **View file size.** You can see the size of each file on your web server, which is helpful if you need to manage the disk space on your account.
- **Edit files.** Almost all SFTP clients allow you to open and edit files through the client interface, which is a convenient way to get the job done.
- **Change permissions.** You can control what type of read/write/execute permissions the files on your web server have. This is commonly referred to as CHMOD, which is the command that you use to change the permissions.

SFTP is a convenient utility that gives you access to the files located on your web server, which makes managing your WordPress website a bit easier.

Setting Up SFTP on Your Hosting Account

Many web hosts today offer SFTP as part of their hosting packages, so confirm that your hosting provider makes SFTP available to you for your account. In Book 2, Chapter 1, I mention the hosting account management interface called cPanel.

cPanel is by far the most popular hosting account management software used by hosts on the web, eclipsing other popular tools such as Plesk and NetAdmin. It's cPanel, or your hosting account management interface, that allows you to set up an SFTP account for your website.

TIP

In this chapter, I use cPanel as the example. If your hosting provider gives you a different interface to work with, the concepts are still the same, but you need to refer to your hosting provider for the specifics to adapt these directions to your specific environment.

Mostly, the SFTP for your hosting account is set up automatically. Figure 2-1 shows you the User Manager page in cPanel, where you set up user accounts for SFTP access.

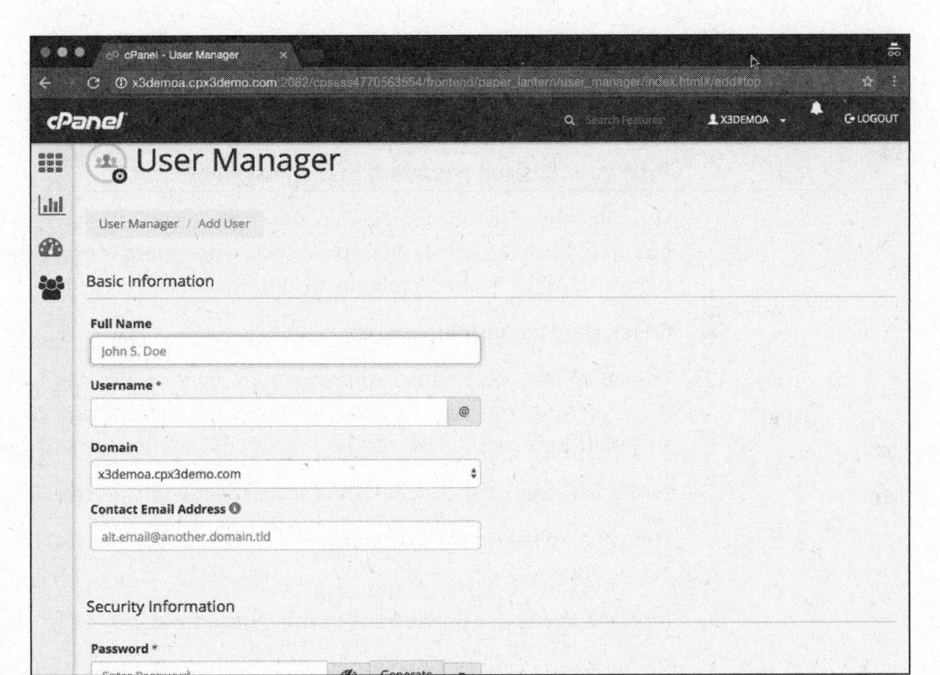

FIGURE 2-1: The User Manager page within cPanel.

Follow these steps to get to this page and set up your SFTP account:

1. **Log in to the cPanel for your hosting account.**

   Typically, you browse to `http://yourdomain.com/cpanel` to bring up the login screen for your cPanel. Enter your specific hosting account username and password in the login fields, and click OK.

2. **Browse to the User Manager page.**

   Click the Add User button in your cPanel to open the User Manager page, shown in Figure 2-1.

3. **View the existing SFTP account.**

   If your hosting provider automatically sets you up with an SFTP account, you see it listed on the User Manager page. Ninety-nine percent of the time, the default SFTP account uses the same username and password combination as your hosting account or the login information you used to log in to your cPanel in Step 1.

If the SFTP Accounts page doesn't display a default SFTP user in the User Manager page, you can create one easily in the Add SFTP Account section:

1. **Fill in the provided fields.**

   The fields of the User Manager page ask for your name, desired username, domain, and email address.

2. **Type your desired password in the Password field.**

   You can choose to type your own password or click the Password Generator button to have the server generate a secure password for you. Retype the password in the Password (Again) field to validate it.

3. **Check the Strength indicator.**

   The server tells you whether your password is Very Weak, Weak, Good, Strong, or Very Strong. You want to have a very strong password for your SFTP account so that it's very hard for hackers and malicious Internet users to guess and crack.

4. **In the Services section, click the Disabled icon within the FTP section.**

   This action changes the icon label to Enabled and enables FTP for the user you are creating.

5. **Indicate the space limitations in the Quota field.**

   Because you're the site owner, leave the radio-button selection set to Unrestricted. (In the future, if you add a new user, you can limit the amount of space, in megabytes [MB], by selecting the radio button to the left of the text field and typing the numeric amount in the text box, such as 50MB.)

6. **(Optional) Type the directory access for this user.**

   Leaving this field blank gives this new user access to the root level of your hosting account — which, as the site owner, you want, so leave this field blank. In the future, if you set up accounts for other users, you can lock down their access to your hosting directory by indicating which directory the user has access to.

7. **Click the Create button.**

   You see a new screen with a message that the account was created successfully. Additionally, you see the settings for this new user account; copy and paste them into a blank text editor window (such as Notepad for PC or TextEdit for Mac users). The settings for the user account are the details you need to connect to your web server via SFTP.

8. **Save the following settings.**

   Username, Password, and SFTP Server are specific to your domain and the information you entered in the preceding steps.

   - Username: *username@yourdomain.com*
   - Password: *yourpassword*
   - Host name: *yourdomain*.com
   - SFTP Server Port: 22
   - Quota: Unlimited MB

TIP

Ninety-nine point nine percent of the time, the SFTP Server Port will be 22. Be sure to double-check your SFTP settings to make sure that this is the case because some hosting providers have different port numbers for SFTP.

REMEMBER

At any time, you can revisit the User Accounts page to delete the user accounts you've created, change the quota, change the password, and find the connection details specific to that account.

Finding and Using Free and Easy SFTP Programs

SFTP programs are referred to as SFTP *clients* or SFTP *client software.* Whatever you decide to call it, an SFTP client is software that you use to connect to your web server to view, open, edit, and transfer files to and from your web server.

Using SFTP to transfer files requires an *SFTP* client. Many SFTP clients are available for download. Here are some good (and free) ones:

- **SmartFTP (PC):** `https://www.smartftp.com/download`
- **FileZilla (PC or Mac):** `https://sourceforge.net/projects/filezilla`

» **Transmit (Mac):** `http://panic.com/transmit`

» **FTP Explorer (PC):** `www.ftpx.com`

In Book 2, Chapter 1, you discover how to obtain a hosting account, and in the previous section of this chapter, you discover how to create an SFTP account on your web server. By following the steps in the previous section, you also have the SFTP username, password, server, and port information you need to connect your SFTP client to your web server so you can begin transferring files. In the next section, you discover how to connect to your web hosting account via SFTP.

Connecting to the web server via SFTP

For the purposes of this chapter, I use the FileZilla SFTP client (`https://sourceforge.net/projects/filezilla`) because it's easy to use and the cost is free ninety-nine (open-source geek speak for free!).

Figure 2-2 shows a FileZilla client that's not connected to a server. By default, the left side of the window displays a directory of files and folders on the local computer.

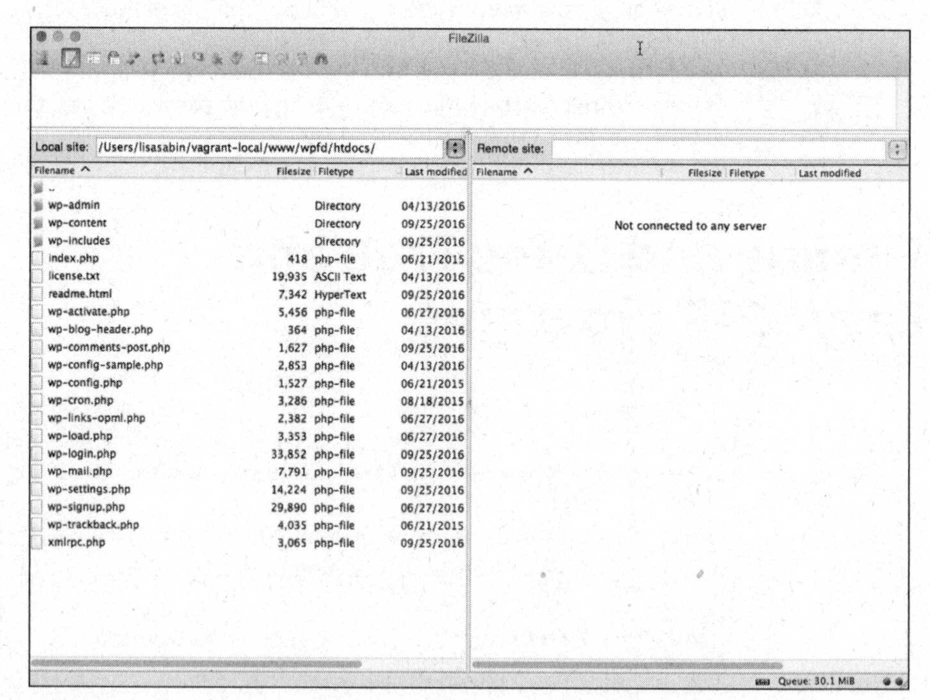

FIGURE 2-2: Mozilla FileZilla SFTP client software.

The right side of the window displays content when the FileZilla client is connected to a web server; specifically, it shows directories of the web server's folders and files.

REMEMBER

If you use different SFTP client software from FileZilla, the steps and look of the software will differ. You need to adapt your steps and practice for the specific SFTP client software you're using.

Connecting to a web server is an easy process. The SFTP settings you saved from Step 8 in "Setting Up SFTP on Your Hosting Account" section in this chapter are also the same settings you see in your cPanel User Manager page if your SFTP was set up automatically for you.

Username: *username@yourdomain.com*

Password: *yourpassword*

Server: *yourdomain*.com

SFTP Server Port: 22

Quota: Unlimited MB

This process is where you need that information. To connect to your web server via the FileZilla SFTP client, follow these few steps:

1. **Open the SFTP client software on your local computer.**

   Locate the program on your computer and click (or double-click) the program icon to launch the program.

2. **Choose File⇨Site Manager to open the Site Manager utility.**

   Site Manager appears, as shown in Figure 2-3.

3. **Click the New Site button.**

4. **Type a name for your site that helps you identify the site.**

   This site name can be anything you want it to be because it isn't part of the connection data you add in the next steps. (In Figure 2-4, you see WPFD — short for *WordPress For Dummies*.)

5. **Enter the SFTP server in the Host field.**

   Host is the same as the SFTP server information provided to you when you set up the SFTP account on your web server. In the example, the SFTP server is `wpfd.wpengine.com`, so that's entered in the Host field, as shown in Figure 2-4.

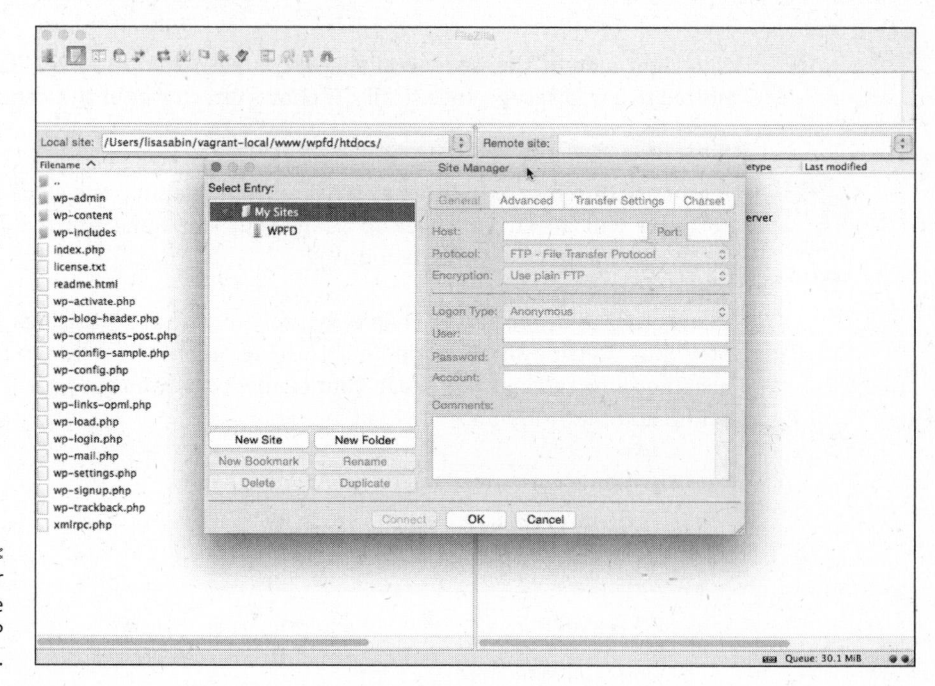

FIGURE 2-3: The Site Manager utility in the FileZilla SFTP client software.

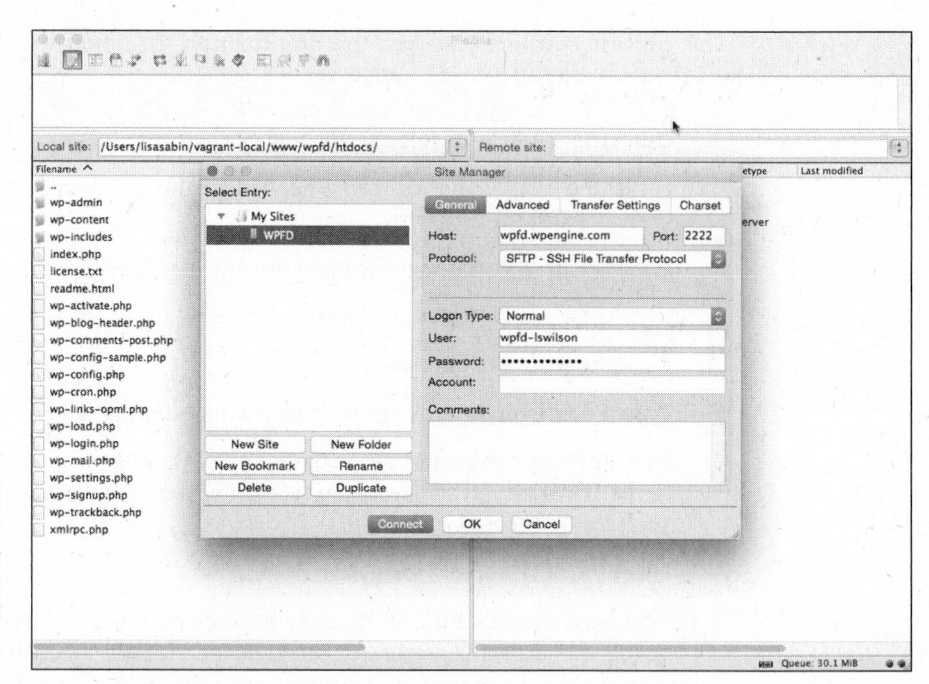

FIGURE 2-4: FileZilla Site Manager utility with SFTP account information filled in.

6. **Enter the SFTP port in the Port field.**

   Typically, in most hosting environments, SFTP uses port 22, and this setting generally never changes. My host, on the other hand, uses port 2222 for SFTP. In case your host is like mine and uses a port other than 22, double-check your port number and enter it in the Port field, as shown in Figure 2-4.

7. **Select the server type.**

   FileZilla asks you to select a server type (as do most SFTP clients). Choose SFTP – SSH File Transfer Protocol from the Protocol drop-down menu, as shown in Figure 2-4.

8. **Select the logon type.**

   FileZilla gives you several logon types to choose among (as do most SFTP clients). Choose Normal from the Logon Type drop-down menu.

9. **Enter your username in the Username field.**

   This username is given to you in the SFTP settings.

10. **Type your password in the Password field.**

    This password is given to you in the SFTP settings.

11. **Click the Connect button.**

    This step connects your computer to your web server. The directory of folders and files from your local computer displays on the left side of the FileZilla SFTP client window, and the directory of folders and files on your web server displays on the right side, as shown in Figure 2-5.

Now you can take advantage of all the tools and features SFTP has to offer you!

Transferring files from point A to point B

Now that your local computer is connected to your web server, transferring files between the two couldn't be easier. Within the SFTP client software, you can browse the directories and folders on your local computer on the left side and browse the directories and folders on your web server on the right side.

SFTP clients make it easy to transfer files from your computer to your hosting account by using a drag-and-drop method. Two methods of transferring files are

» **Uploading:** Generally, transferring files from your local computer to your web server. To upload a file from your computer to your web server, click the file you want to transfer from your local computer and then drag and drop it on the right side (the web-server side).

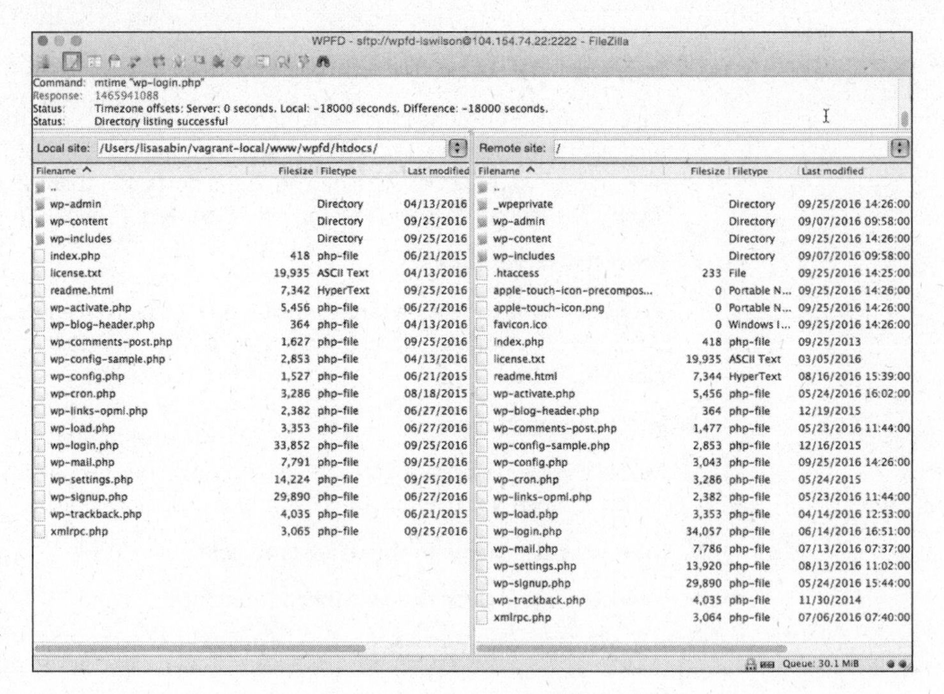

FIGURE 2-5: FileZilla displays local files on the left and server files on the right.

» **Downloading:** Transferring files from your web server to your local computer. To download a file from your web server to your local computer, click the file you want to transfer from your web server and drag and drop it on the left side (the local-computer side).

TIP

Downloading files from your web server is a very efficient, easy, and smart way of backing up files to your local computer. It's always a good idea to keep your files safe — especially things like theme files and plugins, which Books 6 and 7 cover.

Editing files by using SFTP

At times, you need to edit certain files that live on your web server. You can use the methods described in the preceding section to download a file, open it, edit it, save it, and then upload it back to your web server. Another way is to use the edit feature built into most SFTP client software by following these steps:

1. **Connect the SFTP client to your web server.**
2. **Locate the file you want to edit.**
3. **Open the file by using the internal SFTP editor.**

   Right-click the file with your mouse, and choose View/Edit from the shortcut menu. (Remember that I'm using FileZilla for these examples; your SFTP client may use different labels, such as Open or Edit.) FileZilla, like most SFTP clients, uses a program (such as Notepad for a PC or TextEdit for Mac) designated for text editing that already exists on your computer. In some rare cases, your SFTP client software may have its own internal text editor.

4. **Edit the file to your liking.**
5. **Save the changes you made.**

   Click the Save icon or choose File➪Save.
6. **Upload the file to your web server.**

   After you save the file, FileZilla alerts you that the file has changed and asks whether you want to upload the file to the server. Click the Yes button. The newly edited file replaces the old one.

That's all there is to it. Use the SFTP edit feature to edit, save, and upload files as you need to.

WARNING

When you edit files by using the SFTP edit feature, you're editing files in a "live" environment, meaning that when you save the changes and upload the file, the changes take effect immediately and affect your live website. For this reason, I strongly recommend downloading a copy of the original file to your local computer before making changes. That way, if you happen to make a typo in the saved file and your website goes haywire, you have a copy of the original to upload to restore the file to its original state.

TECHNICAL STUFF

Programmers and developers are people who generally are more technologically advanced than your average user. These folks typically don't use SFTP for editing or transferring files. Instead, they use a version-control system called Git. Git manages the files on your web server through a versioning system that has a complex set of deployment rules for transferring updated files to and from your server. Most beginners don't use such a system for this purpose, but Git *is* a system that beginners can use. If you're interested in using Git, you can find a good resource to start with at SitePoint (`https://www.sitepoint.com/git-for-beginners`).

Changing file permissions

Every file and folder on your web server has a set of assigned attributions, called *permissions*, that tells the web server three things about the folder or file. On a very simplistic level, these permissions include

- **Read:** This setting determines whether the file/folder is readable by the web server.

- **Write:** This setting determines whether the file/folder is writable by the web server.
- **Execute:** This setting determines whether the file/folder is executable by the web server.

Each set of permissions has a numeric code assigned to it, identifying what type of permissions are assigned to that file or folder. There are a lot of permissions available, so here are the most common ones that you run into when running a WordPress website:

- **644:** Files with permissions set to 644 are readable by everyone and writable only by the file/folder owner.
- **755:** Files with permissions set to 755 are readable and executable by everyone, but they're writable only by the file/folder owner.
- **777:** Files with permissions set to 777 are readable, writable, and executable by everyone. For security reasons, you shouldn't use this set of permissions on your web server unless absolutely necessary.

Typically, folders and files within your web server are assigned permissions of 644 or 755. Usually, you see PHP files — files that end with the `.php` extension — with permissions set to 644 if the web server is configured to use PHP Safe Mode.

TIP

This section gives you a very basic look at file permissions because usually, you won't need to mess with file permissions on your web server. In case you do need to dig further, you can find a great reference on file permissions from Elated.com at `www.elated.com/articles/understanding-permissions`.

You may find yourself in a situation in which you're asked to edit and change the permissions on a particular file on your web server. With WordPress sites, this situation usually happens when you're dealing with plugins or theme files that require files or folders to be writable by the web server. This practice is referred to as *CHMOD* (an acronym for Change Mode). When someone says, "You need to CHMOD that file to 755," you'll know what that person is talking about.

Here are some easy steps for using your SFTP program to CHMOD a file or edit its permissions on your web server:

1. **Connect the SFTP client to your web server.**
2. **Locate the file you want to CHMOD.**

3. Open the file attributes for the file.

Right-click the file on your web server, and choose File Permissions from the shortcut menu. (Your SFTP client, if not FileZilla, may use different terminology.)

The Change File Attributes window appears, as shown in Figure 2-6.

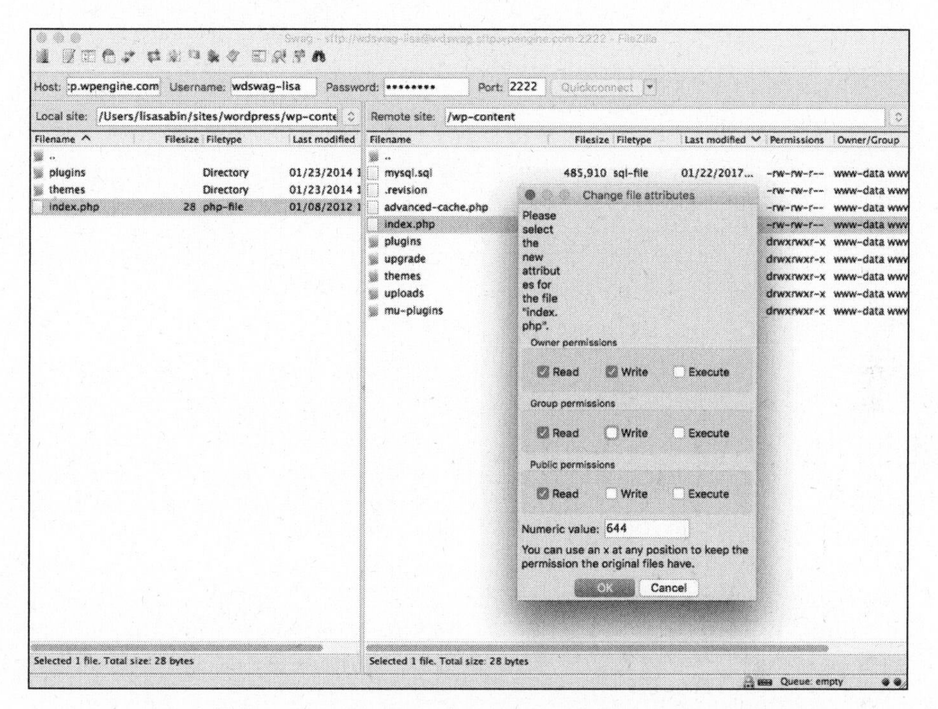

FIGURE 2-6: The Change File Attributes window in FileZilla.

4. Type the correct file permissions number in the Numeric Value field.

This number is assigned to the permissions you want to give the file. Most often, the plugin or theme developer tells you which permissions number to assign to the file or folder — typically, 644 or 755. (The permissions in Figure 2-6 are assigned the value 644.)

5. Click OK to save the file.

» Creating a page with PHP

» Managing a database with MySQL

Chapter 3
Getting to Know PHP and MySQL

In Book 6, you dig into the code necessary to create functions and features on your website. Many, if not all, of these functions and features use Hypertext Preprocessor (PHP) tags. When combined with the WordPress code, these tags make things happen (such as displaying post content, categories, archives, links, and more) on your website.

One of the reasons WordPress is the most popular content management system (CMS) is that you don't need to know PHP code to use it. That's to say, you can use WordPress easily without ever looking at any of the code or template files contained within it. If, however, you want to tweak the settings of your WordPress theme (flip to Book 6) or the code of a particular plugin (see Book 7), you need to understand the basics of how PHP works. But don't worry; you don't need to be a PHP programmer.

This chapter introduces you to the very basics of PHP and *MySQL*, which is the database system that stores your WordPress data. After you read this chapter, you'll understand how PHP and MySQL work together with the WordPress platform to serve up your website in visitors' browsers.

This book doesn't turn you into a PHP programmer or MySQL database administrator, but it gives you a glimpse of how PHP and MySQL work together to help WordPress build your website. If you're interested in finding out how to program PHP or become a MySQL database administrator, check out *PHP & MySQL For Dummies*, 4th Edition, by Janet Valade (John Wiley & Sons, Inc.).

Understanding How PHP and MySQL Work Together

WordPress uses a PHP/MySQL platform, which provides everything you need to create your own website and publish your own content dynamically without knowing how to program those pages. In short, all your content is stored in a MySQL database in your hosting account.

PHP is a server-side scripting language for creating dynamic web pages. When a visitor opens a page built in PHP, the server processes the PHP commands and then sends the results to the visitor's browser. *MySQL* is an open-source relational database management system (RDBMS) that uses *Structured Query Language* (SQL), the most popular language for adding, accessing, and processing data in a database. If all that sounds like Greek to you, just think of MySQL as a big file cabinet where all the content on your blog is stored.

Every time a visitor goes to your blog to read your content, he makes a request that's sent to a host server. The PHP programming language receives that request, makes a call to the MySQL database, obtains the requested information from the database, and then presents the requested information to your visitor through his web browser.

Here, *content* refers to the data stored in the MySQL database — that is, your blog posts, pages, comments, links, and options that you set up on the WordPress Dashboard. But the *theme* (or design) you choose to use for your blog — whether it's the default theme, one you create, or one you've custom-designed — isn't part of the content in this case. Theme files are part of the file system and aren't stored in the database. Therefore, it's a good idea to create and keep backups of any theme files that you're currently using. See Book 6 for further information on WordPress theme management.

Make sure your web host backs up your site daily so that your content (data) won't be lost in case something happens. Web hosting providers that offer daily backups as part of their services can save the day by restoring your site to its original form. Additionally, Book 2, Chapter 7 covers important information about backing up your website.

Exploring PHP Basics

WordPress requires PHP to work; therefore, your web hosting provider must have PHP enabled on your web server. If you already have WordPress up and running on your website, you know that PHP is running and working just fine. Currently, the PHP version required for WordPress is 5.6 or later.

Before you play around with template tags (covered in Book 6) in your WordPress templates or plugin functions, you need to understand what makes up a template tag and why, as well as the correct syntax, or function, for a template tag as it relates to PHP. Additionally, have a look at the WordPress files contained within the download files. Many of the files end with the .php file extension — an extension required for PHP files, which separates them from other file types, such as JavaScript (.js) and CSS (.css).

As I state earlier, WordPress is based in PHP (a scripting language for creating web pages) and uses PHP commands to pull information from the MySQL database. Every tag begins with the function to start PHP and ends with a function to stop it. In the middle of those two commands lives the request to the database that tells WordPress to grab the data and display it.

A typical template tag, or function, looks like this:

```
<?php get_info(); ?>
```

This example tells WordPress to do three things:

- **Start PHP:** <?php
- **Use PHP to get information from the MySQL database and deliver it to your blog:** get_info();
- **Stop PHP:** ?>

In this case, get_info(); represents the tag function, which grabs information from the database to deliver it to your blog. The information retrieved depends on what tag function appears between the two PHP commands.

REMEMBER

Every PHP command you start requires a stop command. For every <?php, you must include the closing ?> command somewhere later in the code. PHP commands structured improperly cause ugly errors on your site, and they've been known to send programmers, developers, and hosting providers into loud screaming fits. You find a lot of starting and stopping of PHP throughout the WordPress templates and functions. The process seems as though it would be resource-intensive, if not exhaustive, but it really isn't.

WARNING

Always, always make sure that the PHP start and stop commands are separated from the function with a single space. You must have a space after `<?php` and a space before `?>`, because if you don't, the PHP function code doesn't work. Make sure that the code looks like this

```
<?php get_info(); ?>
```

and not like this:

```
<?phpget_info();?>
```

Trying Out a Little PHP

To test some PHP code, follow these steps to create a simple HTML web page with an embedded PHP function:

1. **Open a new blank file in your default text editor — Notepad (Windows) or TextEdit (Mac) — type** <html>**, and then press Enter.**

   The `<html>` tag tells the web browser that this file is an HTML document and should be read as a web page.

2. **Type** <head> **and then press Enter.**

   The `<head>` HTML tag contains elements that tell the web browser about the document; this information is read by the browser but hidden from the web-page visitor.

3. **Type** <title>This Is a Simple PHP Page</title> **and then press Enter.**

   The `<title>` HTML tag tells the browser to display the text between two tags as the title of the document in the browser title bar.

   Note: All HTML tags need to be opened and then closed, just like the PHP tags that I describe in the preceding section. In this case, the `<title>` tag opens the command and the `</title>` tag closes it, telling the web browser that you're finished dealing with the title.

4. **Type** </head> **to close the** `<head>` **tag from Step 2 and then press Enter.**

5. **Type** <body> **to define the body of the web page and then press Enter.**

   Anything that appears after this tag displays in the web browser's window.

6. **Type** <?php **to tell the web browser to start a PHP function and then press the spacebar.**

   See "Exploring PHP Basics" earlier in this chapter for details on starting and stopping PHP functions.

7. **Type** echo '<p>Testing my new PHP function</p>'; **and then press the spacebar.**

   This function is the one that you want PHP to execute on your web page. This particular function echoes the text *Testing my new PHP function* and displays it on your website.

8. **Type** ?> **to tell the web browser to end the PHP function and then press Enter.**

9. **Type** </body> **to close the `<body>` HTML tag from Step 5 and then press Enter.**

   This tag tells the web browser that you're done with the body of the web page.

10. **Type** </html> **to close the `<html>` tag from Step 1 and then press Enter.**

    This tag tells the web browser that you're at the end of the HTML document.

When you're done with Steps 1 through 10, double-check to make sure that the code in your text editor looks like this:

```
<html>
<head>
<title>This Is a Simple PHP Page</title>
   </head>
   <body>
   <?php echo '<p>Testing my new PHP function</p>'; ?>
   </body>
   </html>
```

After you write your code, follow these steps to save and upload your file:

1. **Save the file to your local computer as `test.php`.**

2. **Upload the `test.php` file.**

   Via Secure File Transfer Protocol (SFTP), upload `test.php` to the root directory of your web server. If you need a review of how to use SFTP to transfer files to your web server, look through the information presented in Book 2, Chapter 2.

3. **Open a web browser, and type the address** `http://yourdomain.com/test.php` **in the web browser's address bar (where *yourdomain* is your actual domain name).**

   As shown in Figure 3-1, a single line of text displays: `Testing my new PHP function`.

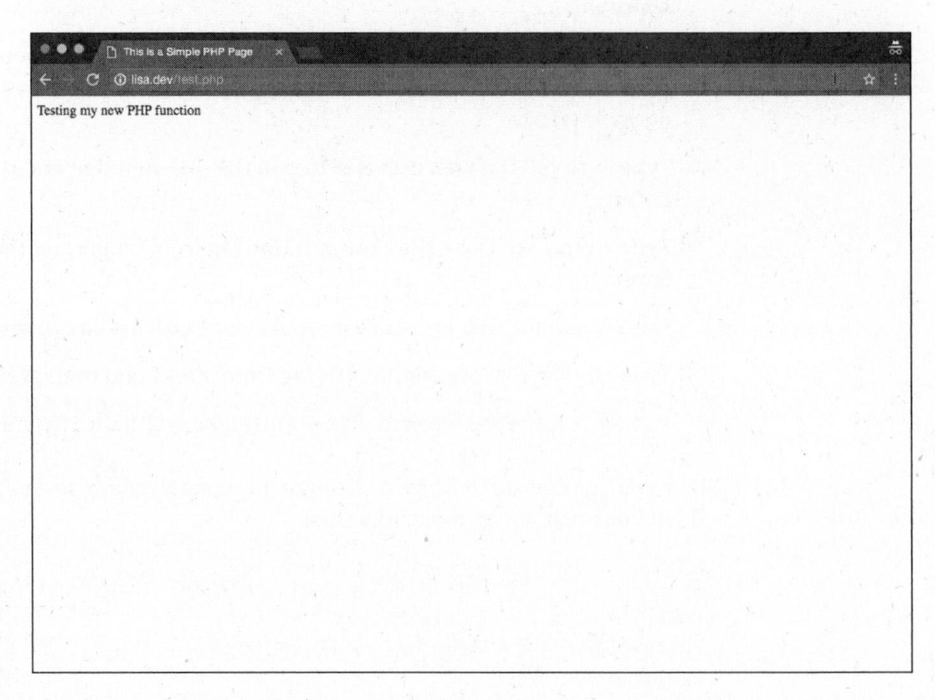

FIGURE 3-1: A basic PHP page in a browser window.

If the `test.php` file displays correctly in your browser, congratulations! You've programmed PHP to work in a web browser!

TIP

If the `test.php` file doesn't display correctly in your browser, a PHP error message gives you an indication of the errors in your code. (Usually included with the error message is the line number where the error occurs in the file.)

Managing Your MySQL Database

Many new WordPress users are intimidated by the MySQL database, perhaps because it seems to be way above their technical skills or abilities. Truth be told, regular users of WordPress — those who use it just to publish content — don't ever have to dig into the database unless they want to. You need to explore the database only if you're dealing with theme or plugin development, or if you're contributing code to the WordPress project.

This section gives you a basic overview of the WordPress database stored in MySQL so that you have an understanding of the structure and know where items are stored.

TIP

Currently, WordPress requires MySQL version 5.6 or later to work correctly. If your web hosting provider doesn't have version 5.6 or later installed on your web server, kindly ask to upgrade. WordPress can also use MariaDB version 10.0 or later, but the typical usage is MySQL.

After WordPress is installed on your server (which I discuss in Book 2, Chapter 4), the database gets populated with 12 tables that exist to store different types of data from your WordPress blog. Figure 3-2 displays the structure of the tables, as follows:

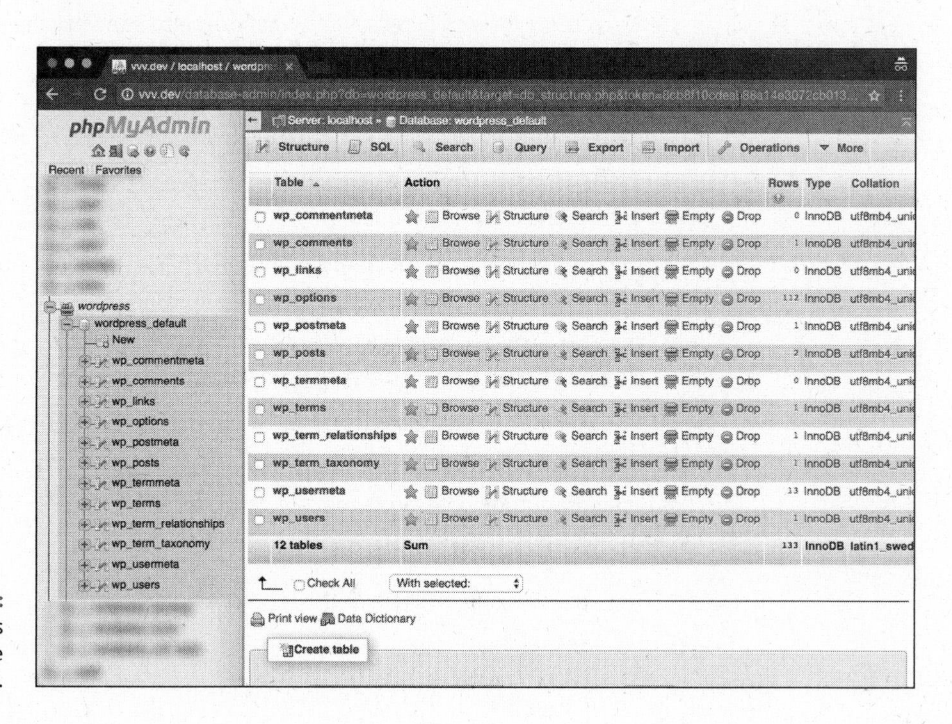

FIGURE 3-2: The WordPress database structure.

- **`wp_commentmeta`:** This table stores every comment published to your site and contains information, or *metadata,* that includes
  - A unique comment ID number
  - A comment meta key, meta value, and meta ID (unique numerical identifiers assigned to each comment left by you or visitors to your site)
- **`wp_comments`:** This table stores the body of the comments published to your site, including
  - A post ID that specifies which post the comment belongs to
  - The comment content
  - The comment author's name, URL, IP address, and email address
  - The comment date (day, month, year, and time)
  - The comment status (approved, unapproved, or spam)
- **`wp_links`:** This table stores the name, URL, and description of all links you create by using the WordPress Link Manager. It also stores all the advanced options for the links you created, if any.
- **`wp_options`:** This table stores all the option settings that you set for WordPress after you install it, including all theme and plugin option settings.
- **`wp_postmeta`:** This table includes all posts or pages published to your site and contains metadata that includes
  - The unique post ID number. (Each blog post has a unique ID number to set it apart from the others.)
  - The post meta key, meta value (unique numerical identifiers for each post created on your site), and any custom fields you've created for the post.
- **`wp_posts`:** This table features the body of any post or page you've published to your blog, including autosaved revisions and post option settings, such as
  - The post author, date, and time
  - The post title, content, and excerpt
  - The post status (published, draft, or private)
  - The post comment status (open or closed)
  - The post type (page, post, or custom post type)
  - The post comment count
- **`wp_termmeta`:** This table stores the metadata for terms (taxonomies and tags) for content.

* `wp_terms`: This table stores the categories you've created for posts and links, as well as tags that have been created for your posts.
* `wp_term_relationships`: This table stores the relationships among the posts, as well as the categories and tags that have been assigned to them.
* `wp_term_taxonomy`: WordPress has three types of taxonomies by default: category, link, and tag. This table stores the taxonomy associated for the terms stored in the `wp_terms` table.
* `wp_usermeta`: This table features metadata from every user with an account on your WordPress website. This metadata includes
  * A unique user ID
  * A user meta key, meta value, and meta ID, which are unique identifiers for users on your site
* `wp_users`: The list of users with an account on your WordPress website is maintained within this table and includes
  * The username, first name, last name, and nickname
  * The user login
  * The user password
  * The user email
  * The registration date
  * The user status and role (subscriber, contributor, author, editor, or administrator)

Most web hosting providers give you a *utility*, or an interface, to view your MySQL database; the most common one is phpMyAdmin (refer to Figure 3-2). If you're unsure how you can view your database on your hosting account, get in touch with your hosting provider to find out.

When the Multisite feature in WordPress is activated (check out Book 8 for information about the Multisite feature), WordPress adds six additional tables to the database:

* `wp_blogs`: This table stores information about each blog created in your network, including
  * A unique blog numerical ID
  * A unique site ID number (determines the ID of the site to which the blog belongs)

- The blog domain
- The blog server path
- The date the blog was registered
- The date the blog was updated
- The blog status (public, private, archived, spam; see Book 8 for more information on blog status)

» **wp_blog_versions:** This table stores general information about each network blog ID, database version, and date of last update.

» **wp_registration_log:** This table stores information about registered users, including

- Unique user numerical ID
- User email address
- User IP address
- User blog ID
- The date the user registered

» **wp_signups:** This table stores information about user sign-ups, including all the information from the wp_registration_log table, the date the user account was activated, and the unique activation key the user accessed during the sign-up process.

» **wp_site:** This table stores information about your main installation site, including the site ID, domain, and server path.

» **wp_sitemeta:** This table stores all the information about the multisite configurations set after you install the Multisite feature. (See Book 8.)

» Manually installing WordPress

» Running installation scripts

Chapter 4

Installing WordPress on Your Web Server

This chapter takes you through two installation methods for WordPress: an automatic, one-click installation with the Fantastico script installer, which is available from your web hosting provider, and a manual installation.

I also show you how to set up a MySQL database by using the cPanel web hosting management interface. By the time you're done reading this chapter, you'll be logged in to and looking at your brand-spanking-new WordPress Dashboard, ready to start publishing content right away. (If you already have WordPress installed, go ahead and skip to Book 2, Chapter 5, which contains great information about configuring WordPress for optimum performance and security.)

REMEMBER

Before you can install WordPress, you need to complete the following tasks:

- Purchase the domain-name registration for your account (Book 2, Chapter 1).
- Obtain a hosting service on a web server for your blog (Book 2, Chapter 1).
- Establish your hosting account username, password, and Secure File Transfer Protocol (SFTP) address (Book 2, Chapters 1 and 2).
- Acquire an SFTP client for transferring files to your hosting account (Book 2, Chapter 2).

If you omitted any of the preceding items, flip to the chapter listed to complete the step.

Exploring Preinstalled WordPress

The WordPress software has become such a popular publishing tool that almost all hosting providers available today provide WordPress for you in a couple of ways:

- Already installed on your hosting account when you sign up
- A user dashboard with a utility for installing WordPress from within your account management

TIP

If your hosting provider doesn't give you access to an installation utility, skip to "Installing WordPress Manually" later in this chapter for the steps to install WordPress manually via SFTP.

One of the most popular web hosts for managed WordPress hosting is a service called WP Engine, which you can find at `https://wpengine.com`. The service provides a handy, easy-to-use installation utility that's built right into your account dashboard at WP Engine to allow you to get up and running with WordPress right away.

You may not be using WP Engine, so your host may have a slightly different utility, but the basic concept is the same. Be sure to apply the same concepts to whatever kind of utility your hosting provider gives you.

To install the account dashboard of WP Engine, follow these steps:

1. **Log in to the WP Engine user dashboard.**

   a. *Browse to* `https://`my.wpengine.com *to bring up the login screen.*

   b. *Enter the email address you used to sign up, enter your password, and then click Log In. The page refreshes and displays the dashboard for your account.*

2. **Click the Add Install link.**

   The Add Install page displays in your browser window, as shown in Figure 4-1.

3. **Type the name of your new WordPress installation in the Install Name field.**

   This name is the temporary domain name of your new website. As shown in Figure 4-2, I'm using *wpfd*, which stands for *WordPress For Dummies*. This step creates the domain name `wpfd.wpengine.com`.

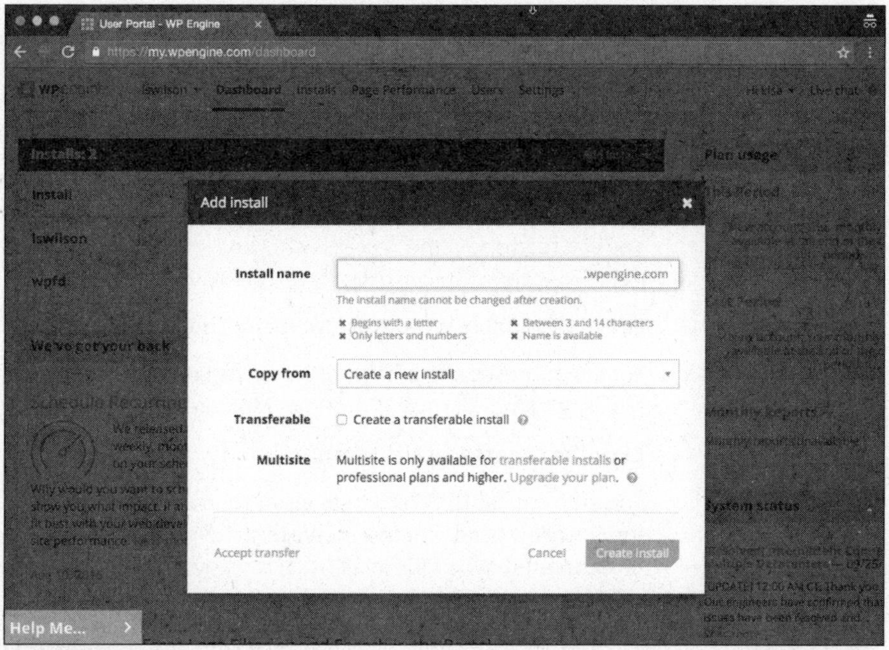

FIGURE 4-1: The Add Install module within a WP Engine account dashboard.

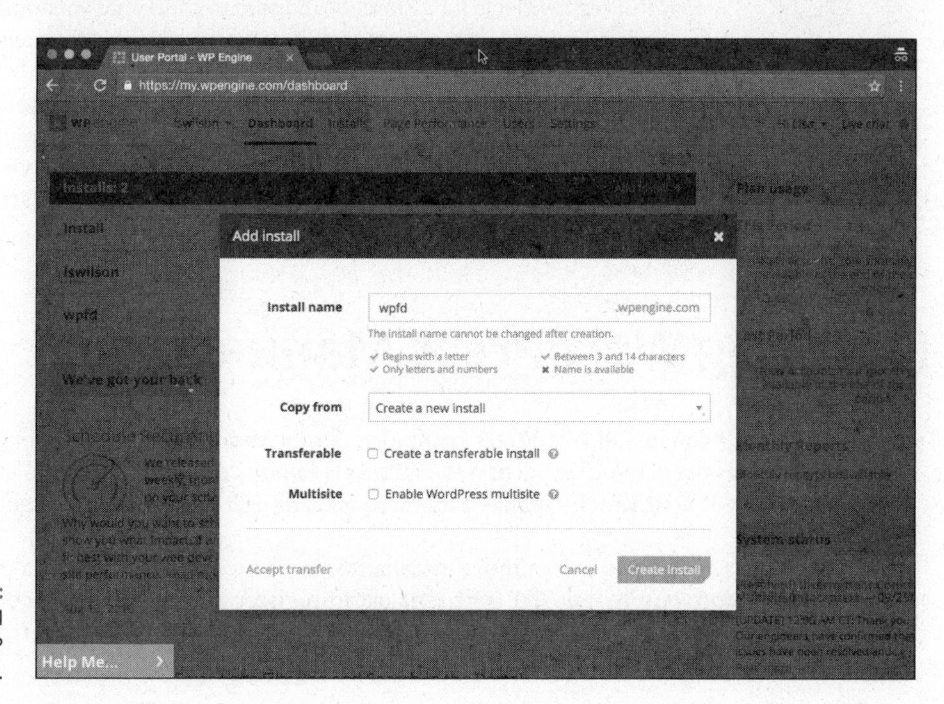

FIGURE 4-2: The Add Install page at WP Engine.

4. **Choose Create a New Install from the Copy From drop-down list.**

   This step creates a new installation of WordPress in your account.

5. **Leave the Create a Transferable Install check box unselected.**

   There may come a day where you want to create a WordPress installation that can be transferred between two WP Engine accounts, but today is not that day. You can read about the process at `https://wpengine.com/support/billing-transfer-information-for-developers`.

6. **Leave the Enable WordPress Multisite check box unselected.**

   For the purposes of basic WordPress installation and setup, don't worry about Multisite for now. I cover WordPress Multisite features in depth in Book 8.

7. **Click the Create Install button.**

   This step creates the WordPress installation in your account and takes you to the Overview page, where a message states that your WordPress installation is being created. When the installation is ready to use, you receive an email from WP Engine.

TIP

In my experience, WP Engine always has the most up-to-date version of WordPress available for installation. Be sure to check that your hosting provider is supplying the latest version of WordPress with its installation utility.

Your WordPress installation via your provider's utility is complete, and you're ready to start using WordPress on your web server. If you installed WordPress by using your provider's utility method and don't want to review the steps to install WordPress manually, flip to Book 2, Chapter 5 for the steps to optimize your WordPress installation for performance and security.

Installing WordPress Manually

If you install WordPress manually, here's where the rubber meets the road — that is, where you're putting WordPress's famous five-minute installation to the test. Set your watch, and see whether you can meet the five-minute goal.

REMEMBER

The famous five-minute installation includes only the time it takes to install the software — not the time it takes to register a domain name; obtain and set up your web hosting service; or download, install, configure, and figure out how to use the SFTP software.

Setting up the MySQL database

The WordPress software is a personal publishing system that uses a PHP/MySQL platform, which provides everything you need to create your own blog and publish your own content dynamically without knowing how to program those pages. In short, all your content (options, posts, comments, and other pertinent data) is stored in a MySQL database in your hosting account.

Every time visitors go to your blog to read your content, they make a request that's sent to your server. The PHP programming language receives that request, obtains the requested information from the MySQL database, and then presents the requested information to your visitors through their web browsers.

Every web host is different in how it gives you access to set up and manage your MySQL database(s) for your account. In this section, I use cPanel, the popular hosting interface. If your host provides a different interface, the same basic steps apply, but the setup in the interface that your web host provides may be different.

To set up the MySQL database for your WordPress site with cPanel, follow these steps:

1. **Log in to the cPanel for your hosting account:**

   a. *Browse to* `http://`yourdomain`.com/cpanel` *(where* yourdomain.com *is your actual domain name) to bring up the login screen for your cPanel.*

   b. *Enter your specific hosting account username and password in the login fields, and then click OK.*

   The page refreshes and displays the cPanel for your account.

2. **Locate the MySQL Databases icon.**

   Click the MySQL Databases icon to load the MySQL Databases page in your cPanel.

3. **Enter a name for your database in the Name text box.**

   Be sure to make note of the database name, because you need it to install WordPress.

4. **Click the Create Database button.**

   A message appears, confirming that the database was created.

5. **Click the Back button on your browser toolbar.**

   The MySQL Databases page displays in your browser window.

6. **Locate MySQL Users on the MySQL Databases page.**

   Scroll approximately to the middle of the page to locate this section.

7. **Choose a username and password for your database, enter them in the Username and Password text boxes, and then click the Create User button.**

   A confirmation message appears, stating that the username was created with the password you specified.

TIP

   For security reasons, make sure that your password isn't something that sneaky hackers can easily guess. Give your database a name that you'll remember later. This practice is especially helpful if you run more than one MySQL database in your account. If you name a database *WordPress* or *wpblog*, for example, you can be reasonably certain a year from now, when you want to access your database to make some configuration changes, that you know exactly which credentials to use.

WARNING

   Make sure that you note the database name, username, and password that you set up during this process. You need them in the section "Running the installation script" later in this chapter before officially installing WordPress on your web server. Jot these details down on a piece of paper, or copy and paste them into a text editor window; either way, make sure that you have them handy.

8. **Click the Back button on your browser toolbar.**

   The MySQL Databases page displays in your browser window.

9. **In the Add Users to Database section of the MySQL Databases page, choose the user you just set up from the User drop-down list and then choose the new database from the Database drop-down list.**

   The MySQL Account Maintenance, Manage User Privileges page appears in cPanel.

10. **Assign user privileges by selecting the All Privileges check box.**

    Because you're the *administrator* (or owner) of this database, you need to make sure that you assign all privileges to the new user you just created.

11. **Click the Make Changes button.**

    The resulting page displays a message confirming that you've added your selected user to the selected database.

12. **Click the Back button on your browser toolbar.**

    You return to the MySQL Databases page.

The MySQL database for your WordPress website is complete, and you're ready to proceed to the final step of installing the software on your web server.

Downloading the WordPress software

Without further ado, get the latest version of the WordPress software at `https://wordpress.org/download`.

TIP

WordPress gives you two compression formats for the software: `.zip` and `.tar.gz`. Use the `.zip` file because it's the most common format for compressed files and because both Windows and Mac operating systems can use the format. Generally, the `.tar.gz` file format is used for Unix operating systems.

Download the WordPress software to your computer and then *decompress* (unpack or unzip) it to a folder on your computer's hard drive. These steps begin the installation process for WordPress. Having the program on your own computer isn't enough, however. You also need to *upload* (or transfer) it to your web server account (the one discussed in Book 2, Chapter 1).

Before you install WordPress on your web server, you need to make sure that you have the MySQL database set up and ready to accept the WordPress installation. Be sure that you've followed the steps to set up your MySQL database before you proceed.

Uploading the WordPress files via FTP

To upload the WordPress files to your host, return to the `/wordpress` folder (shown in Figure 4-3) on your computer, where you unpacked the WordPress software that you downloaded earlier. If you need a review on using SFTP (Secure File Transfer Protocol) to transfer files from your computer to your web server, see Book 2, Chapter 2.

Using your SFTP client, connect to your web server and upload all these files to the root directory of your hosting account.

TIP

If you don't know what your root directory is, contact your hosting provider and ask, "What is my root directory for my account?" Every hosting provider's setup is different. The root directory is most likely the `public_html` folder, but you may find an `httpdocs` folder. The answer depends on what type of setup your hosting provider has. When in doubt, ask!

Here are a few things to keep in mind when you upload your files:

» **Upload the *contents* of the `/wordpress` folder to your web server — not the folder itself.** Most FTP client software lets you select all the files and drag and drop them to your web server. Other programs have you highlight the files and click a Transfer button.

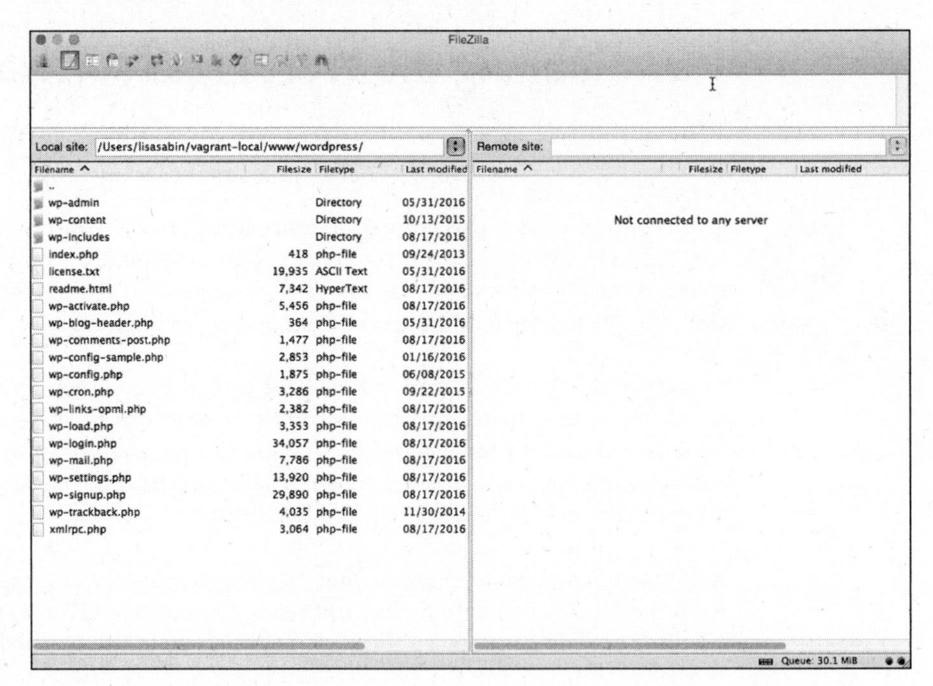

FIGURE 4-3: WordPress installation files to be uploaded to your web server.

- **Choose the correct transfer mode.** File transfers via SFTP have two forms: ASCII and binary. Most SFTP clients are configured to autodetect the transfer mode. Understanding the difference as it pertains to this WordPress installation is important so that you can troubleshoot any problems you have later:
  - *Binary transfer mode* is how images (such as JPG, GIF, BMP, and PNG files) are transferred via FTP.
  - *ASCII transfer mode* is for everything else (text files, PHP files, JavaScript, and so on).

  For the most part, it's a safe bet to make sure that the transfer mode of your SFTP client is set to autodetect. But if you experience issues with how those files load on your site, retransfer the files by using the appropriate transfer mode.
- **You can choose a different folder from the root.** You aren't required to transfer the files to the root directory of your web server. You can choose to run WordPress on a subdomain or in a different folder on your account. If you want your blog address to be `http://yourdomain.com/blog`, you transfer the WordPress files into a `/blog` folder (where *yourdomain* is your domain name).

- **Choose the right file permissions.** *File permissions* tell the web server how these files can be handled on your server — whether they're files that can be written to. Generally, PHP files need to have a permission (CHMOD is explained in Book 2, Chapter 2) of 666, whereas file folders need a permission of 755. Almost all SFTP clients let you check and change the permissions on the files, if you need to. Typically, you can find the option to change file permissions within the menu options of your FTP client.

TECHNICAL STUFF

Some hosting providers run their PHP software in a more secure format: *safe mode.* If this is the case with your host, you need to set the PHP files to 644. If you're unsure, ask your hosting provider what permissions you need to set for PHP files.

Running the installation script

The final step in the installation procedure for WordPress is connecting the WordPress software you uploaded to the MySQL database. Follow these steps:

1. **Type the URL of your website in the address bar of your web browser.**

   If you chose to install WordPress in a different folder from the root directory of your account, make sure you indicate that in the URL for the install script. If you transferred the WordPress software files to the `/blog` folder, for example, point your browser to the following URL to run the installation: `http://yourdomain.com/blog/wp-admin/install.php`. If WordPress is in the root directory, use the following URL to run the installation: `http://yourdomain.com/wp-admin/install.php` (where *yourdomain* is your domain name).

   Assuming that you did everything correctly, you should see the first step in the installation process that you see in Figure 4-4. (see Table 4-1 for help with common installation problems.)

2. **Select your preferred language from the list provided on the setup page, shown in Figure 4-4.**

   At this writing, WordPress is available in 87 languages. For these steps, I'm using English (United States).

3. **Click the Continue button.**

   You see a new page with a welcome message from WordPress and instructions that you need to gather the MySQL information you saved earlier in this chapter.

4. **Click the Let's Go button.**

   A new page loads and displays the fields you need to fill out in the next step, shown in Figure 4-5.

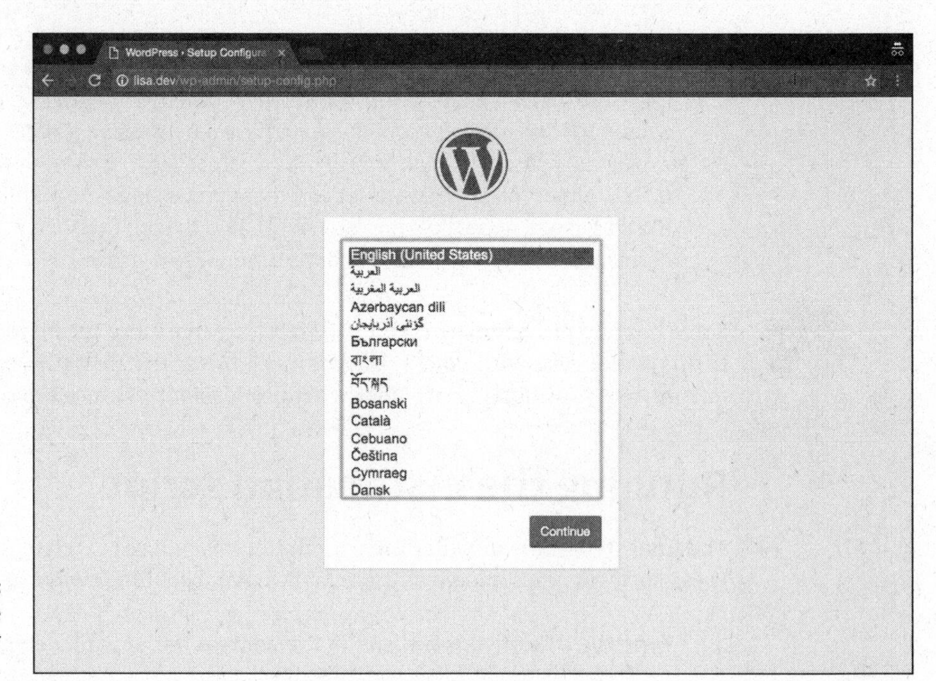

FIGURE 4-4: Choose the language for your installation.

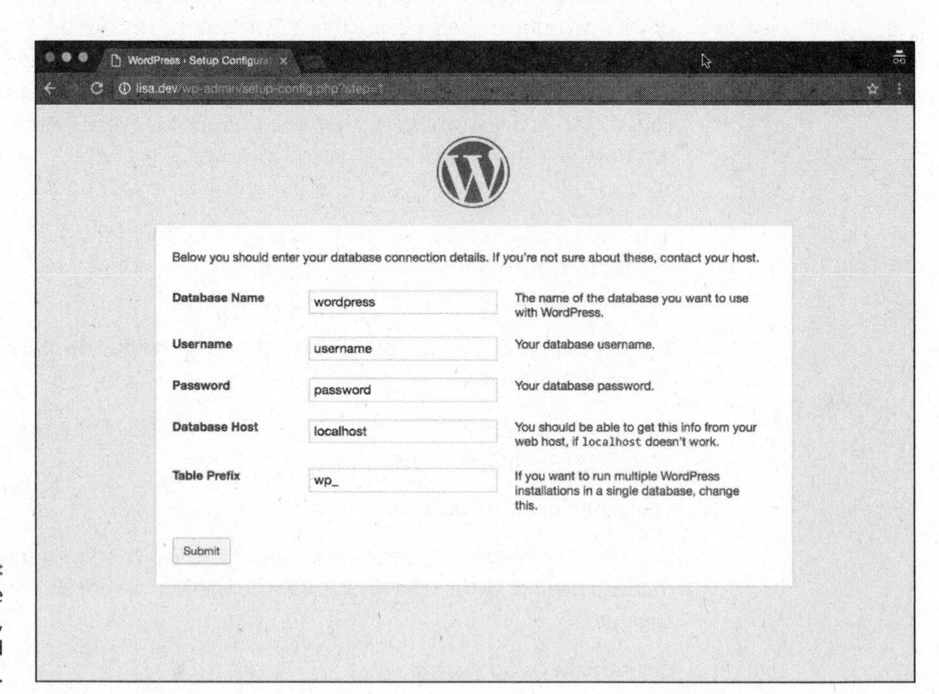

FIGURE 4-5: Entering the database name, username, and password.

5. **Dig out the database name, username, and password that you saved in the earlier section "Setting up the MySQL database," and use that information to fill in the following fields, as shown in Figure 4-5:**

   - *Database Name:* Type the database name you used when you created the MySQL database before this installation. Because hosts differ in configurations, you need to enter the database name by itself or a combination of your username and the database name, separated by an underscore (_).

     If you named your database *wordpress,* for example, you enter that in this text box. If your host requires you to append the database name with your hosting account username, you enter ***username*_wordpress**, substituting your hosting username for *username.* My username is *lisasabin,* so I enter **lisasabin_wordpress**.

   - *Username:* Type the username you used when you created the MySQL database before this installation. Depending on what your host requires, you may need to enter a combination of your hosting account username and the database username separated by an underscore (_).

   - *Password:* Type the password you used when you set up the MySQL database. You don't need to append the password to your hosting account username here.

   - *Database Host:* Ninety-nine percent of the time, you leave this field set to `localhost`. Some hosts, depending on their configurations, have different hosts set for the MySQL database server. If `localhost` doesn't work, you need to contact your hosting provider to find out the MySQL database host.

   - *Table Prefix:* Leave this field set to `wp_`.

     You can change the table prefix to create an environment that's secure against outside access. See Book 2, Chapter 5 for more information.

6. **After you fill in the MySQL database information, click the Submit button.**

   You see a message that says, `All right, sparky! You've made it through this part of the installation. WordPress can now communicate with your database. If you're ready, time to run the install!`

7. **Click the Run the Install button.**

   Another page appears, welcoming you to the famous five-minute WordPress installation process, as shown in Figure 4-6.

FIGURE 4-6: Finishing the WordPress installation.

8. **Enter the following information:**

   - *Site Title:* Enter the title you want to give your site. The title you enter isn't written in stone; you can change it later, if you like. The site title also appears on your site.
   - *Username:* Enter the name you use to log in to WordPress. By default, the username is *admin,* and you can leave it that way. For security reasons, however, I recommend that you change your username to something unique. This username is different from the one you set for the MySQL database in previous steps. You use this username when you log in to WordPress to access the Dashboard (which is covered in Book 3), so be sure to make it something you'll remember.
   - *Password:* Type your desired password in the text box. If you don't enter a password, one is generated automatically for you. For security reasons, it's a good thing to set a different password here from the one you set for your MySQL database in the previous steps; just don't get the passwords confused.

TIP

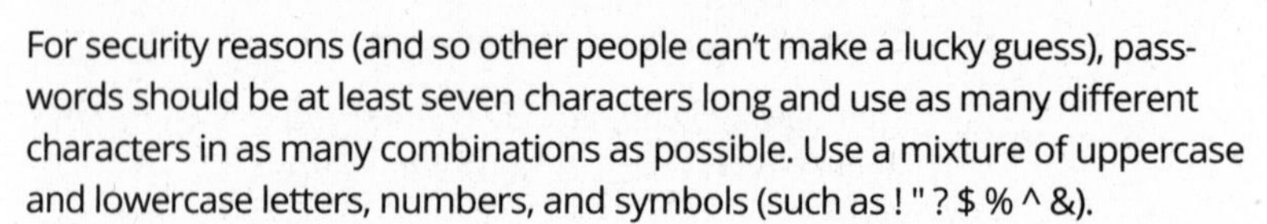

For security reasons (and so other people can't make a lucky guess), passwords should be at least seven characters long and use as many different characters in as many combinations as possible. Use a mixture of uppercase and lowercase letters, numbers, and symbols (such as ! " ? $ % ^ &).

   - *Your Email:* Enter the email address you want to use to be notified of administrative information about your blog. You can change this address later, too.

- *Allow Search Engines to Index This Site:* By default, this check box (not shown in Figure 4-6) is selected, which lets the search engines index the content of your blog and include your blog in search results. To keep your blog out of the search engines, deselect this check box. (See Book 5 for information on search engine optimization.)

9. **Click the Install WordPress button.**

   The WordPress installation machine works its magic and creates all the tables within the database that contain the default data for your blog. WordPress displays the login information you need to access the WordPress Dashboard. Make note of this username and password before you leave this page. Scribble them on a piece of paper or copy them into a text editor, such as Notepad.

REMEMBER

   After you click the Install WordPress button, you're sent an email with the login information and login URL. This information is handy if you're called away during this part of the installation process. So go ahead and let the dog out, answer the phone, brew a cup of coffee, or take a 15-minute power nap. If you somehow get distracted away from this page, the email sent to you contains the information you need to log in to your WordPress blog.

10. **Click the Log In button to log in to WordPress.**

TIP

   If you happen to lose this page before clicking the Log In button, you can always find your way to the login page by entering your domain followed by the call to the login file (such as `http://yourdomain.com/wp-login.php`, where *yourdomain* is your domain name).

You know that you're finished with the installation process when you see the login page, shown in Figure 4-7. Check out Table 4-1 if you experience any problems during this installation process; it covers some of the common problems users run into.

So do tell — how much time does your watch show for the installation? Was it five minutes? Stop by my blog sometime at `http://lisasabin-wilson.com` and let me know whether WordPress stood up to its famous five-minute installation reputation.

The good news is — you're done! Were you expecting a marching band? WordPress isn't that fancy . . . yet. Give it time, though. If anyone can produce it, the folks at WordPress can.

Let me be the first to congratulate you on your newly installed WordPress blog! When you're ready, log in and familiarize yourself with the Dashboard, which I describe in Book 3.

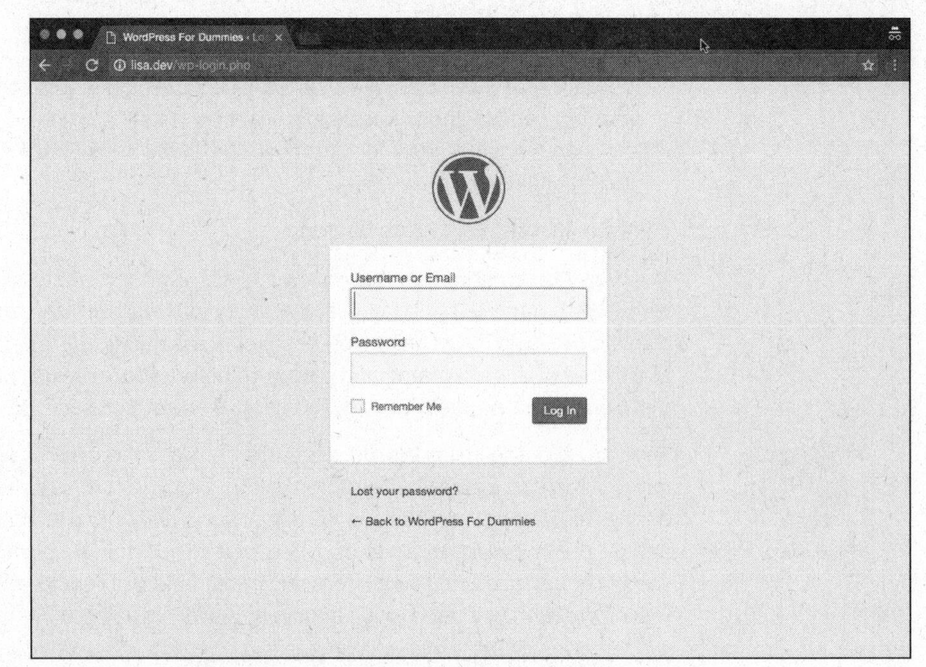

FIGURE 4-7: You know you've run a successful WordPress installation when you see the login page.

TABLE 4-1 Common WordPress Installation Problems

Error Message	Common Cause	Solution
`Error Connecting to the Database`	The database name, username, password, or host was entered incorrectly.	Revisit your MySQL database to obtain the database name, username, and password and then reenter that information.
`Headers Already Sent`	A syntax error occurred in the `wp-config.php` file.	Open the `wp-config.php` file in a text editor. The first line needs to contain only this line: <?php. The last line needs to contain only this line: ?>. Make sure that those lines contain nothing else — not even white space. Save the file changes.
`500: Internal Server Error`	Permissions on PHP files are set incorrectly.	Try setting the permissions (CHMOD) on the PHP files to 666. If that change doesn't work, set them to 644. Each web server has different settings for how it lets PHP execute on its servers.
`404: Page Not Found`	The URL for the login page is incorrect.	Double-check that the URL you're using to get to the login page is the same as the location of your WordPress installation (such as `http://yourdomain.com/wp-login.php`).
`403: Forbidden Access`	An `index.html` or `index.htm` file exists in the WordPress installation directory.	WordPress is a PHP application, so the default home page is `index.php`. Look in the WordPress installation folder on your web server. If an `index.html` or `index.htm` file is there, delete it.

» Understanding today's web threats

» Reducing the risk of attack

» Using sources you can trust

» Cleaning up to avoid a soup-kitchen server

» Hardening WordPress

Chapter 5
Configuring WordPress for Optimum Security

In this chapter, you deal with web security and how it pertains to WordPress. There are a lot of scary threats on the Internet, but with this chapter — and WordPress, of course — you'll have no problem keeping your website safe and secure.

TIP

Always have a reliable backup system in place so if something goes wrong with your website, you can reset it to the last version that you know worked. Book 2, Chapter 7 shows you how to back up your website.

Understanding the Basics of Web Security

Information security is the act of protecting information and information systems from unwanted or unauthorized use, access, modification, and disruption. Information security is built on principles of protecting confidentiality, integrity, and availability of information. The ultimate goal is managing your risk.

No silver bullet can ensure that you're never compromised. Consider your desktop: The idea of running an operating system (whether it be Windows or Mac OS X) without antivirus software is highly impractical. The same principle applies to your website. You can never reduce the percentage of risk to zero, but you can implement controls to minimize impact and to take a proactive approach to threat preparedness.

You need to be familiar with six distinct types of risk (or threats):

- **Defacements:** The motivation behind most defacements is to change the appearance of a website. Defacements are often very basic and make some kind of social stance, such as supporting a cause or bringing attention to your poor security posture. If you visit your website, and it doesn't look anything like you expect it to, contact your host to find out whether it has been defaced, and if so, ask for assistance in restoring it.
- **Search engine optimization (SEO) spam:** This kind of attack sets out to ruin your search engine results; search engines can warn viewers away from your website. The most popular one is the Pharma hack, which injects code into your website and search engine links to redirect your traffic to pharmaceutical companies and their products. If you find that your website listing disappears from major search engines, such as Google, you should be concerned that your website has been a victim of SEO spam and contact your hosting provider for assistance.
- **Malicious redirects:** Malicious-redirect attacks direct your traffic somewhere else, most likely to another website. If your domain is `http://domain.com`, for example, a malicious redirect might redirect it to `http://adifferentdomain.com`. Malicious redirects are often integrated with other attacks (SEO spam being one). If you visit your website and discover that your domain redirects to a different domain that you don't recognize, your website has been a victim of a malicious-redirect attack, and you should contact your hosting provider for assistance.
- **iFrame injections:** This kind of attack embeds a hidden iFrame in your website that loads another website onto your visitor's browser (like a pop-up ad). These embedded websites or ads can lead to malicious websites that carry a multitude of infections.
- **Phishing scams:** Phishing scams used to belong only to the world of email: You get an email from your bank asking you to confirm your login information, but if you follow the instructions, your information actually goes to the attacker's servers rather than the legitimate site.

  WordPress websites are now used for the distribution of these attacks. Attackers develop malicious files and code that look like plugins and themes and then exploit credentials on a server or WordPress site, or the attackers use a known vulnerability to infect the plugins and themes. Then they use the bait-and-hook approach through ads or emails to redirect traffic to these fake pages stored on legitimate websites. Keep an eye out for abnormal behavior on your website, such as the display of ads that you didn't insert yourself or the redirect to domains you're not familiar with. If at any time you suspect that your website and underlying files have been tampered with, contact your web hosting provider for assistance.

- **Backdoor shells:** With a backdoor shell, an attacker uploads a piece of PHP code to your website, which allows him to take control of it, download your files, and upload his own. This kind of attack is more difficult to discover because it doesn't always change the appearance of your site or your experience with it. You typically discover this kind of attack by noticing new files in your file system or notice a marked increase in your bandwidth use.

The rest of this chapter shows you how you can prevent any of these nasty attacks against your WordPress website so that you can keep yourself and your visitors safe.

REMEMBER

Part of being a website owner is keeping your website and subscribers safe from hackers.

Preventing Attacks

You can't ever be 100 percent secure. But with a WordPress website, you're in good hands. The WordPress developers understand the importance of security, and they built a highly effective system to address any vulnerabilities you'll run across.

Updating WordPress

The first way to prevent hackers is to keep your WordPress website up to date. The quick-and-easy way to do so is through the automatic update feature. Book 2, Chapter 6 takes you through the process of updating WordPress step by step.

REMEMBER

The beauty of applying updates is that they often introduce new streamlined features, improve overall usability, and work to patch and close identified or known vulnerabilities.

As technology and concepts evolve, so do attackers and their methods for finding new vulnerabilities. The farther behind you get, the harder it is to update later and the more your risk increases, which in turn affects how vulnerable you are to attacks.

Installing patches

All WordPress updates are not created equally, but you should pay special attention to a few updates of the WordPress core software.

Updates include *major releases*, which contain feature additions, interface (UI) changes, bug fixes, and security updates. You can always tell what major release you're on by the first two numbers in the version number (as in 3.4). See Book 1, Chapter 3 for more information about the difference between major and minor releases.

Then you have *point releases*, which are minor releases that can be identified by the third number in the version number (as in 4.7.1). These releases contain bug fixes and security patches but don't introduce new features.

TIP

When you see a point release, apply it. Point releases rarely cause issues with your site, and they help close off vulnerabilities in a lot of cases.

Using a firewall

A firewall builds a wall between your website and the much larger Internet; a good firewall thwarts a lot of attacks.

Your web server should also have a good firewall protecting it. Every day, countless visits, good and bad, are made to every website. Some visits are by real visitors, but many are by automated bots. A web application firewall (WAF) helps protect your WordPress installation from those bad visitors.

REMEMBER

WAFs don't offer 100 percent protection, but they're good deterrents for everyday attacks.

TECHNICAL STUFF

If you plan to manage and administer your own server, install and configure a tool such as ModSecurity (`www.modsecurity.org`), an open-source, WAF-like solution that lives at the web server level as an Apache module.

If you're using a managed hosting solution, you're probably in luck, because most of these solutions offer built-in WAF-like features.

As a user, you can also install a WordPress plugin called Cloudflare, which you can find in the official WordPress Plugin Directory at `https://wordpress.org/plugins/cloudflare`. Cloudflare (see Figure 5-1) provides the best available WAF-like features for your WordPress website on a managed hosting solution. If you'd like to use the Cloudflare plugin on your WordPress website, you need to have a Cloudflare account at `https://www.cloudflare.com`. You can open a free account or upgrade to a paid account that includes more features. After you've installed the plugin on your website, follow the instructions on the Cloudflare configuration page to connect your WordPress blog to your Cloudflare account.

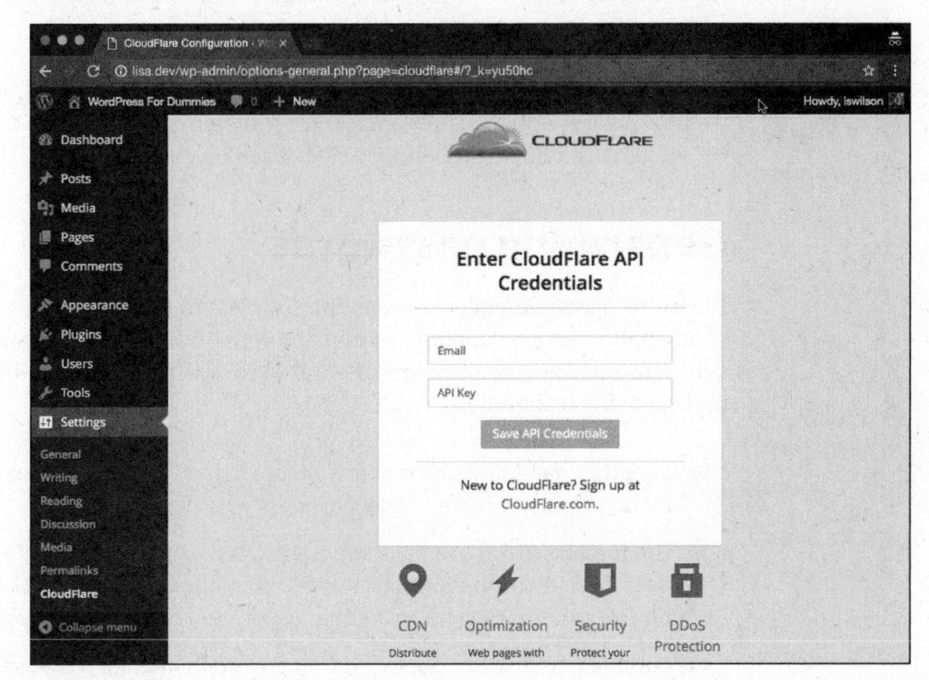

FIGURE 5-1: The Cloudflare plugin for WordPress.

Using Trusted Sources

One of the simplest things you can do to keep your website secure is vet all the people who work on your website: website administrators, website designers, developers, and web hosts. Also be sure to use trusted plugins, themes, and applications. If you're running a self-hosted WordPress website, this could be quite a few people.

If you're using themes or plugins, use the WordPress.org Theme and Plugin directories (`https://wordpress.org/themes` and `https://wordpress.org/plugins`, respectively). Each plugin and theme you find in those directories has gone through a documented review process, which reduces the risk of your downloading dangerous code.

Engage the WordPress user community. The WordPress forums (`https://wordpress.org/support`) are great places to start. Ask for community references, and identify the support mechanisms in place to support the theme or plugin over the long term.

Managing Users

The concept of *least privilege* has been in practice for ages: Give someone the required privileges for as long as he needs it to perform his job or a task. When the task is complete, reduce the privileges.

REMEMBER

Apply these safeguards not just to your WordPress Dashboard, but also to your website host's control panels and server transfer protocols. (See Book 2, Chapter 2 for information on Secure File Transfer Protocol.)

Generating passwords

Password management is perhaps the simplest of tasks, yet it's the Achilles' heel of all applications, including desktop and web-based apps. You can keep your files and data on your web server safe and secure through these simple password-management techniques:

- **Length:** Create passwords that are more than 15 characters long to make it more difficult for harmful users to guess your password.
- **Uniqueness:** Don't use the same passwords across all services. If someone does discover the password for one of your applications or services, she won't be able to use it to log in to another application or service that you manage.
- **Complexity:** A strong password contains a minimum of 8 characters and is made up of upper- and lowercase letters, numbers, and symbols, making it hard to guess.

TIP

Use password managers and generators. Two of the most popular products right now are LastPass (`https://www.lastpass.com`) and 1Password (`https://1password.com`).

Limiting built-in user roles

Not all users of your website need administrator privileges. WordPress gives you five user roles to choose among, and those roles provide sufficient flexibility for your websites.

You can find detailed information on each of the roles in Book 3, Chapter 3. You can also discover more information on users and roles in the WordPress.org codex at `https://codex.wordpress.org/Roles_and_Capabilities`.

TIP

Create a separate account with a lower role (such as Author) and use that account for everyday posting. Reserve the Administrator account purely for administration of your website.

Establishing user accountability

The use of generic accounts should be the last thing you ever consider because the more generic accounts you have, the greater your risk of being compromised. If a compromise does happen, you want to have full accountability for all users and be able to quickly answer questions like these:

- Who was logged in?
- Who made what changes?
- What did the users do while logged in?

Generic accounts preclude you from doing appropriate incident handling in the event of a compromise. In Book 3, Chapter 3, you find all the information and step-by-step details on how to create new users in your WordPress Dashboard. Keep the principles of least privilege and user accountability in mind as you're creating users.

Staying Clear of Soup-Kitchen Servers

Among the regular issues plaguing website owners are soup-kitchen servers. A *soup-kitchen server* is one that has never been maintained properly and has a combination of websites, old software, archives, unneeded files, folders, email, and so on living on its hard drive.

The real problem comes into play with the "out of sight, out of mind" phenomenon. A server owner can forget about software installations on a server that may be outdated or insecure. Over time, this forgetfulness introduces new vulnerabilities to the environment:

- Disabled installations or websites that live on the server are as accessible and susceptible to external attacks as live sites.
- When a forgotten installation or website is infected, it leads to *cross-site contamination* — a wormlike effect in which the infection can jump and replicate itself across the server.

- In many instances, these forgotten installations or websites house the backdoor and engine of the infection. This means that as you try to rigorously clean your live website, you continuously get reinfected.

Figure 5-2 demonstrates what a soup-kitchen server looks like. `$wp_version` indicates the version of WordPress that is currently installed in the directory. The many listings for `$wp_version = 2.9` (at this writing, the most recent version of WordPress is 4.7) show how many out-of-date installations of WordPress this particular soup-kitchen server has.

```
Warning: Found outdated WordPress install inside: /../httpdocs - Version: $wp_version = '2.9.1';
Warning: Found outdated WordPress install inside: /./httpdocs - Version: $wp_version = '2.9.2';
Warning: Found outdated WordPress install inside: /./httpdocs - Version: $wp_version = '3.2.1';
Warning: Found outdated WordPress install inside: /./httpdocs - Version: $wp_version = '3.3';
Warning: Found outdated WordPress install inside: /./httpdocs - Version: $wp_version = '3.0.1';
Warning: Found outdated WordPress install inside: /./httpdocs - Version: $wp_version = '2.8.6';
Warning: Found outdated WordPress install inside: /./httpdocs - Version: $wp_version = '2.8.4';
Warning: Found outdated WordPress install inside: /./httpdocs/.. - Version: $wp_version = '3.3';
Warning: Found outdated WordPress install inside: /./httpdocs - Version: $wp_version = '3.0.1';
Warning: Found outdated WordPress install inside: /./httpdocs/blog - Version: $wp_version = '2.5';
Warning: Found outdated timthumb.php version at /./httpdocs/wp-content/plugins/WordPress-popular-posts/scripts/timthumb.php (bellow 2.0). Please update asap!
Warning: Found outdated WordPress install inside: /./httpdocs - Version: $wp_version = '3.0.4';
Warning: Found outdated timthumb.php version at /./subdomains/fcdoc/httpdocs/wp-content/plugins/WordPress-gallery-plugin/timthumb.php (bellow 2.8.2). Update recommended.
Warning: Found outdated WordPress install inside: /./subdomains/fcdoc/httpdocs - Version: $wp_version = '3.2';
Warning: Found vulnerable plugin inside /./subdomains/staging/httpdocs/../wp-content/plugins/wp-spamfree . Details: http://www.exploit-db.com/exploits/17970/
Warning: Found outdated WordPress install inside: /./subdomains/staging/httpdocs/.. - Version: $wp_version = '2.5.1';
Warning: Found outdated WordPress install inside: /./subdomains/./httpdocs - Version: $wp_version = '2.8.4';
Warning: Found outdated WordPress install inside: /./subdomains/./httpdocs - Version: $wp_version = '2.9.1';
Warning: Found outdated WordPress install inside: /./subdomains/./httpdocs - Version: $wp_version = '3.0.1';
Warning: Found vulnerable plugin inside /./httpdocs/../wp-content/plugins/wp-spamfree . Details: http://www.exploit-db.com/exploits/17970/
Warning: Found outdated WordPress install inside: /./httpdocs/.. - Version: $wp_version = '2.5.1';
Warning: Found outdated WordPress install inside: /./web_users/./../.. - Version: $wp_version = '3.3';
Warning: Found outdated timthumb.php version at /./httpdocs-2-17-11/wp-content/plugins/meenews/inc/classes/timthumb.php (bellow 2.0). Please update asap!
Warning: Found outdated WordPress install inside: /./httpdocs-2-17-11 - Version: $wp_version = '3.0.4';
Warning: Found outdated timthumb.php version at /./httpdocs/wp-content/plugins/meenews/inc/classes/timthumb.php (bellow 2.0). Please update asap!
Warning: Found outdated WordPress install inside: /./httpdocs - Version: $wp_version = '3.1.4';
```

FIGURE 5-2: A file-server listing from a typical soup-kitchen server.

TIP

If you have more than one installation of WordPress on your current hosting account, try the following techniques to reduce your risk of running a soup-kitchen server:

- **Isolate each installation with its own user.** This action minimizes internal attacks that come from cross-site contamination.
- **Keep your installations up to date, and remove them when you no longer need them.** This action lessens the risk of attacks that result from outdated software on your server.

Hardening WordPress

When you *harden* (or, secure) your WordPress installation, you reduce your risk of being hacked by malicious attackers.

WEBSITE HARDENING RESOURCES

I recommend a few website resources to follow for your WordPress hardening and security needs:

- **Hardening WordPress:** `https://codex.wordpress.org/Hardening_WordPress`
- **The Definitive Guide to WordPress Security:** `https://moz.com/blog/the-definitive-guide-to-wordpress-security`
- **Sucuri:** `https://blog.sucuri.net`
- **WPsecure:** `http://wpsecure.net`

Hardening your website involves following these five steps:

1. **Enable multifactor authentication.**
2. **Limit login attempts.**
3. **Disable theme and plugin editors.**
4. **Filter by IP (Internet Protocol) address.**
5. **Kill PHP execution.**

I cover each of these steps in the following sections.

REMEMBER

Hardening your website doesn't guarantee your protection, but it definitely reduces your risk.

Enabling multifactor authentication

Authentication, in this case, refers to confirming the identity of the person who is attempting to log in and obtain access to your WordPress installation — just like when you log in to your WordPress website by using a username and password. The idea for multifactor authentication stems from the idea that one password alone isn't enough to secure access to any environment. *Multifactor authentication* (also called *strong authentication*) requires more than one user-authentication method. By default, WordPress requires only one: a username with password. Multifactor authentication adds layers of authentication measures for extra security of user logins.

To enable multifactor authentication, you can use a free plugin called Google Authenticator for WordPress, which provides two-factor user authentication through an application on your mobile or tablet device (iPhone, iPad, Droid, and so on). For this plugin to work, you need the following:

- **Google Authenticator app:** Find it at the App Store for iOS devices or the Google Play Store for Android devices.
- **Google Authenticator plugin:** You can find this plugin in the Plugin Directory at `https://wordpress.org/plugins/wp-google-authenticator`. See Book 7, Chapters 1 and 2 for details on finding, installing, and activating plugins.

Configuring Google Authenticator

When you have both of those tasks accomplished, you can configure the plugin for use on your website. Follow these steps to configure the plugin for each user on your site:

1. **Click the All Users link on the Users menu on your Dashboard.**

   The Users page opens.

2. **Select the user profile you'd like to edit by clicking the Edit link below the user's name in the Users list.**

   The Edit Users page opens.

3. **Select the Activate check box in the WP Google Authenticator Settings section, as shown in Figure 5-3.**
4. **Type a description in the Description text box.**

   This is the description you can see in the Google Authenticator application on your mobile device. I'm using my iPhone, so I typed in *Lisa's iPhone* as the description.

5. **Click the Get QR Code button.**

   In Figure 5-3, I have Google Authenticator activated, and you can see my Recovery Code, as well as a button that allows me to generate a QR code for my mobile app to use. (A *QR code* is a scannable bar code that is readable by a mobile or tablet device using the camera.)

   The QR code appears in a pop-up window.

6. **Open the Google Authenticator application on your mobile or tablet device.**

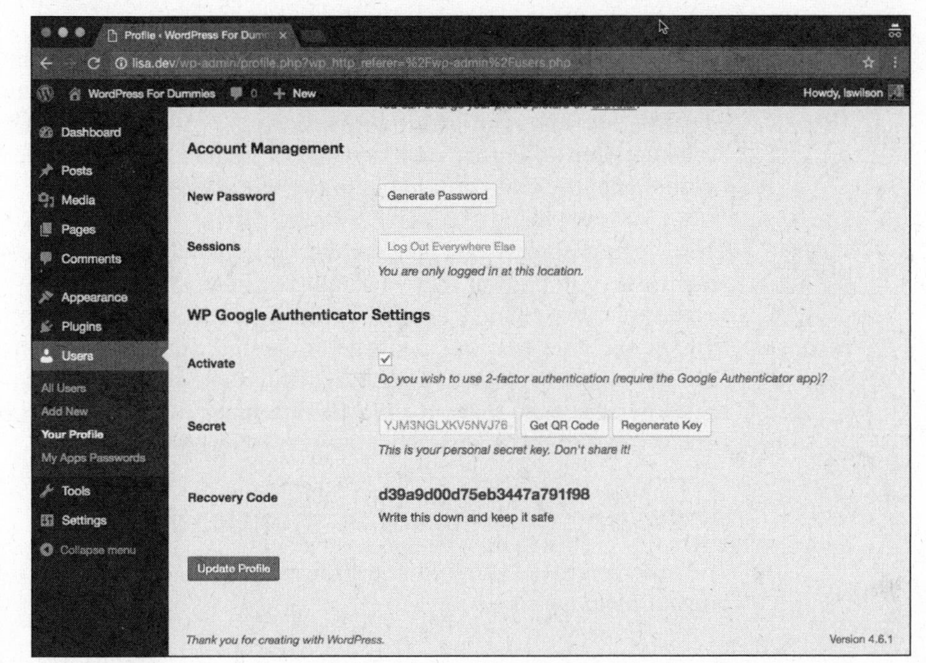

FIGURE 5-3: Google Authenticator Settings.

7. **On the Dashboard of your WordPress site, click the Regenerate Key button.**

   This step refreshes the secret key and QR code needed to connect your mobile or tablet device.

8. **In the Google Authenticator application on your mobile or tablet device, click the Scan Barcode button.**

   The camera on your device starts.

9. **Scan the bar code displayed on the Google Authenticator page of your WordPress Dashboard by taking a photo of it with your device.**

   Point your device camera at your computer screen, and line up the QR code within the camera brackets of your mobile device. The application automatically reads the QR code as soon as it's aligned correctly and displays a six-digit code identifying your blog. The six-digit code refreshes on a time-based interval. After the QR code is scanned, the user receives a message on her mobile device that contains a unique numeric code.

10. **Click the Update Profile button at the bottom of the Edit Users screen on your Dashboard.**

    This step refreshes the Edit Users page with a message at the top stating that the Google Authenticator settings have been successfully saved.

Now, with the Google Authenticator plugin in place, whenever anyone tries to log in to your WordPress Dashboard, he has to fill in his username and password, as usual, but with multifactor authentication in place, he also needs to enter the authentication code that was sent to his mobile device in Step 9. Without this unique code, the user can't log in to the WordPress Dashboard.

WARNING

The Google Authenticator application verification code is time-based, which is why it's very important that your mobile phone and your WordPress blog are set to the same time zone. If you get the message that the Google Authentication verification code you're using is invalid or expired, you need to delete the plugin and then go to your WordPress Dashboard and make sure that the time zone is set to the same time zone that your mobile or tablet device uses. See Book 3, Chapter 2 for information on time settings for your WordPress site.

Activating multifactor authentication

The following steps show you how the multifactor authentication is implemented on your blog:

1. **Log out of your WordPress Dashboard.**

   This step logs you out completely and displays the login page, shown in Figure 5-4.

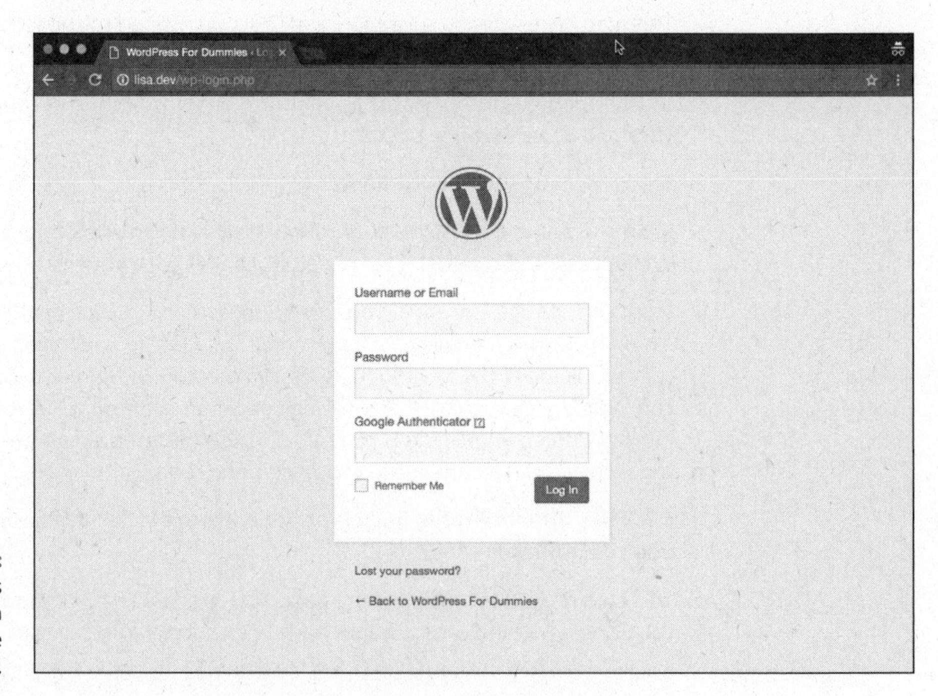

FIGURE 5-4: The WordPress login form with Google Authentication.

2. **Type your username in the Username or Email field.**

3. **Type your password in the Password field.**

   Do *not* click the Log In button yet. (If you're like me, you probably have an urge to click that button a split second after typing your password. For these steps, you have to resist that urge.)

4. **Open the Google Authenticator application on your mobile or tablet device, and locate the six-digit code assigned to your blog.**

   This six-digit code refreshes every 60 seconds. If you have more than one blog using the application, find the code that corresponds to the description you assigned to the site from Step 4 in "Configuring Google Authenticator" earlier in this chapter.

5. **Type the six-digit code in the Google Authenticator code field.**

6. **Click the Log In button.**

   You're now successfully logged in to your WordPress Dashboard via a two-factor authentication method.

The biggest shortcoming with this plugin is its inability to force all users to configure by default. For this reasons, it's important to employ the principle of least privilege on your site (see "Managing Users" earlier in this chapter). Give access only to the users who absolutely require it. In an ideal world, however, every single one of your user accounts requires two-factor authentication to log in on your site.

TIP

If you don't have access to a mobile device, WordPress has a couple of plugins you can use, including these two:

- **LaunchKey:** `https://wordpress.org/plugins/launchkey`
- **Loginizer:** `https://wordpress.org/plugins/loginizer`

Limiting login attempts

Limiting the number of times a user can attempt to log in to your WordPress site helps reduce the risk of brute-force attack. A *brute-force attack* happens when an attacker tries to gain access by guessing your username and password through the process of cycling through combinations.

To help protect against brute-force attacks, you want to limit the number of times any user can try to log in to your website. You can accomplish this task in WordPress easily enough by using the Limit Login Attempts plugin. You can find this plugin in the WordPress Plugin Directory: `https://wordpress.org/plugins/wp-limit-login-attempts/`. See Book 7, Chapters 1 and 2 for information on finding, installing, and activating it.

When you have the Limit Login Attempts plugin installed, follow these steps to configure the settings:

1. **Click the Limit Login Attempts link in the Settings menu of your Dashboard.**

   The Limit Login Attempts Settings page opens in your Dashboard, as shown in Figure 5-5.

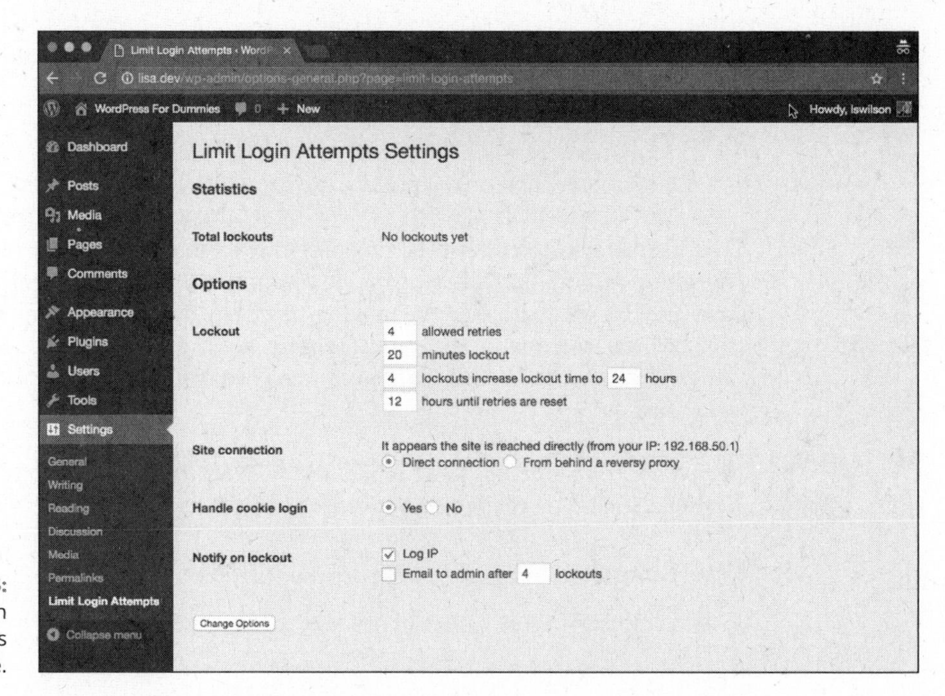

FIGURE 5-5: Limit Login Attempts Settings page.

2. **Select a configuration.**

   In the Options section, you see these four Lockout options:

   - *4 allowed retries:* This setting is the maximum number of times users are allowed to retry failed logins.
   - *20 minutes lockout:* This setting is the amount of time a user is prevented from retrying a login after she reaches the maximum allowed number.

- *4 lockouts increase lockout time to 24 hours:* If a user is locked out four times after numerous failed login attempts, he gets locked out for 24 hours.
- *12 hours until retries are reset:* This setting is the amount of time before login retries are completely reset.

3. **Select the Direct Connection option in the Site Connection section.**

   This option limits site connection to a single Internet Protocol. Alternatively, you can select this plugin to limit site connection from behind a proxy if your users are using proxy IPs to connect to the site.

4. **Select Yes in the Handle Cookie Login section.**

   This option tells WordPress to set a cookie in the user's browser for further identification. Alternatively, you can set this option to No if you're not worried about it, but having Cookie Login Handling is a good extra security measure to have in place.

5. **Select the Log IP option in the Notify on Lockout section.**

   This option notifies the site administrator via email every time a user gets locked out. Alternatively, you can select the number of lockouts that happen for a single user before the administrator is notified.

6. **Click the Change Options button at the bottom of the Limit Login Attempts Settings page.**

   This page refreshes with a message telling you that the plugin settings have been successfully saved.

TIP

If you're managing your own server, monitor your login attempts to see whether a malicious attacker is making repeated attempts to obtain passwords and usernames. Keep track of those IP addresses, and if they repeatedly attempt to log in, add them to your server firewall to prevent them from burdening your server access points.

Disabling theme and plugin editors

By default, when you log in to the WordPress Dashboard, you have the ability to edit any theme and plugin file by using the Theme Editor (click the Appearance link on the Editor menu) and the Plugin Editor (click the Plugins link on the Editor menu). The idea makes a lot of sense; it gives you the ability to do everything within your Admin panel without having to worry about logging in to your server via SFTP to edit files.

Unfortunately, having the theme and plugin editors available also gives any attacker who gains access to the Dashboard full rights to modify any theme or plugin file, which is very dangerous, because even just one line of malware code embedded within any file can grant an attacker remote access to your environment without ever having to touch your Dashboard.

You can prevent this situation by disabling the Theme Editor and Plugin Editor. To do so, add a WordPress constant (or rule) to the WordPress configuration file (`wp-config.php`), which is in the installation folder on your web server. Download the `wp-config.php` via SFTP (see Book 2, Chapter 2), and open the file in a text editor, such as Notepad (PC) or TextEdit (Mac). Look for the following line of code:

```
define('DB_COLLAT', '');
```

Add the following constant (rule) on the line directly below the preceding line:

```
define('DISALLOW_FILE_EDIT',true);
```

Although adding this constant won't prevent an attack, it helps you when it comes to reducing the impact of a compromise. You can find information about other constants you can add to the `wp-config.php` file at `https://codex.wordpress.org/Editing_wp-config.php`.

TIP

You can also disable the automatic updates in WordPress (the system by which you're allowed to automatically update WordPress core and WordPress plugins) to include the administrator. If you do, you'd have to do everything manually, via SFTP. To do this, use the following constant in your `wp-config.php` file:

```
define('DISALLOW_FILE_MODS',true);
```

Filtering by IP address

Another option is to limit access to the Dashboard to specific IP addresses. This method is also referred to as *whitelisting* (allowing) access, which complements *blacklisting* (disallowing) solutions.

Everything that touches the Internet — such as your computer, a website, or a server network — has an IP address. An IP is like your home address; it uniquely identifies you so that the Internet knows where your computer is located. An example of an IP is 12.345.67.89 — a series of numbers that uniquely identifies the physical location of a computer or network.

You can edit the `.htaccess` file on your web server so that only IPs that you approve can access your Admin Dashboard, thereby blocking everyone else from having Dashboard access.

The lines of code that define the access rules get added to the `.htaccess` file located on your web server where WordPress is installed, in a folder called `/wp-admin`. Download that file to your computer via SFTP and open it using a text editor, such as Notepad (PC) or TextEdit (Mac), and add the following lines to it:

```
order allow,deny
deny from all
allow from 12.345.67.89
```

In this example, the order defines what comes first. An IP that follows the `allow` rules is given access; any IP that doesn't follow the `allow` rules is denied access. In this example, only the IP 12.345.67.89 can access the Admin Dashboard; all other IPs are denied.

TIP

If the `/wp-admin` folder in your WordPress installation doesn't contain a file called `.htaccess`, you can easily create one using your SFTP program by opening the `/wp-admin` folder and then right-clicking to open a shortcut menu; and choose New File. Give that new file the name `.htaccess`, and make sure to add the new rules from "Disabling theme and plugin editors" earlier in this chapter.

Limiting access via IP carries the following potential negatives:

- **This technique works only with static IP addresses.** A dynamic IP changes constantly. You have ways to make this technique work with dynamic IPs, but those methods are beyond the scope of this chapter.
- **The ability to use `.htaccess` is highly dependent on a web server that's running Apache.** This technique won't do you any good if your web server is Windows-based or IIS, or if you're using the latest NGINX web server.
- **Your Apache web server needs to be configured to allow directives to be defined by `.htaccess` files.** Ask your web host about configuration.

Killing PHP execution

For most backdoor intrusion attempts to function, a PHP file has to be executed. The term *backdoor* describes ways of obtaining access to a web server through means that bypass regular authentication methods, such as file injections through programming languages such as PHP or JavaScript. Disabling PHP execution prevents an attack or compromise from taking place because PHP can't executed at all.

To disable PHP execution, add four lines of code to the `.htaccess` file on your web server:

```
<Files *.php>
Order allow,deny
Deny from all
</Files>
```

By default, you have an `.htaccess` file in the WordPress directory on your web server. You can also create an `.htaccess` file in other folders — particularly the folders in which you want to disable PHP execution.

To disable PHP execution for maximum security, create an `.htaccess` file with those four lines of code in the following folders in your WordPress installation:

- `/wp-includes`
- `/wp-content/uploads`
- `/wp-content`

This WordPress installation directory (the directory WordPress is installed in) is important because it's the only directory that has to be writeable for WordPress to work. If an image is uploaded with a modified header, or if a PHP file is uploaded and PHP execution is allowed, an attacker could exploit this weakness to create havoc in your environment. When PHP execution is disabled, however, an attacker is unable to create any havoc.

» **Backing up your database before upgrading**

» **Deactivating plugins**

» **Upgrading from the dashboard**

» **Upgrading manually**

Chapter 6
Updating WordPress

As discussed in Book 1, Chapter 3, the schedule of WordPress development and release cycles shows you that WordPress releases a new version (upgrade) of its platform roughly once every 120 days (or every 4 months). That chapter also explains why you need to keep your WordPress software up to date by using the most recent version — mostly for security purposes, but also to make sure you're taking advantage of all the latest features the WordPress developers pack within every major new release.

In this chapter, you discover the WordPress upgrade notification system and find out what to do when WordPress notifies you that a new version is available. This chapter also covers the best practices for upgrading the WordPress platform on your site to ensure the best possible outcome (that is, how not to break your website after a WordPress upgrade).

REMEMBER

The upgrade process occurs on a regular basis — at least three or four times per year. For some users, this process is a frustrating reality of using WordPress. This active development environment, however, is part of what makes WordPress the most popular platform available. Because WordPress is always adding great new features and functions to the platform, upgrading always ensures that you're on top of the game and using the latest tools and features.

Getting Notified of an Available Update

After you install WordPress and log in for the first time, you can see the version number on the WordPress Dashboard, as shown in Figure 6-1. (Note that I've scrolled down in the figure — you see the version number in the bottom-right corner.) Therefore, if anyone asks what version you're using, you know exactly where to look to find out.

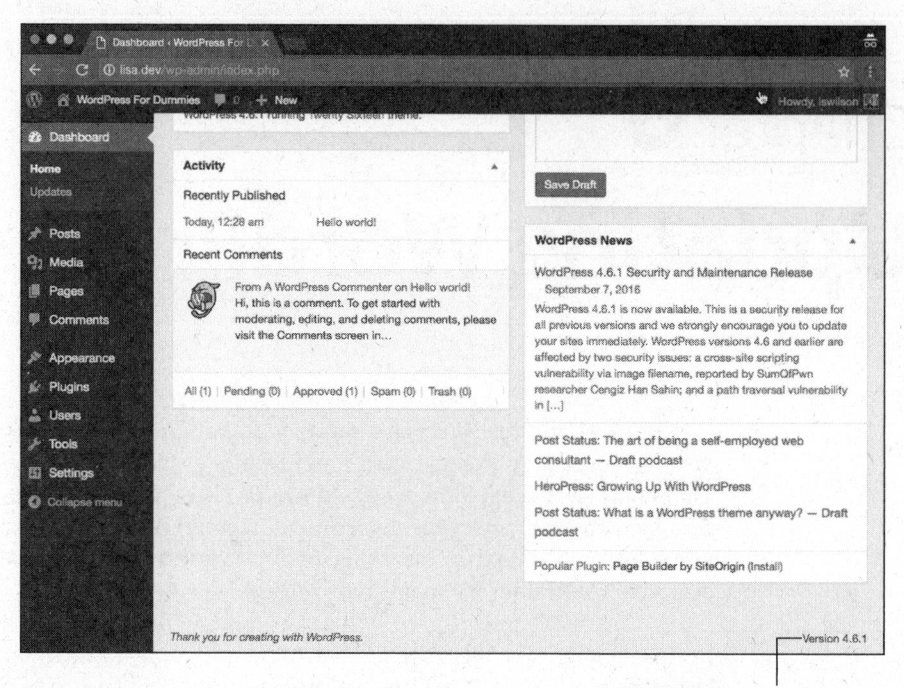

FIGURE 6-1: The WordPress version displayed in the Dashboard.

Suppose that you have WordPress installed, and you've been happily publishing content to your website with it for several weeks, maybe even months. Then one day, you log in to your Dashboard and see a message at the top of your screen you've never seen before: "WordPress X.X.X is available! Please update now." Figure 6-2 displays the update message on the Dashboard, and Figure 6-3 displays a small bubble next to the Dashboard→Updates links that indicates how many updates are currently available.

Both the message at the top of the page and the notification bubble on the Dashboard menu are visual indicators that you're using an outdated version of WordPress and that you can (and need to) upgrade the software.

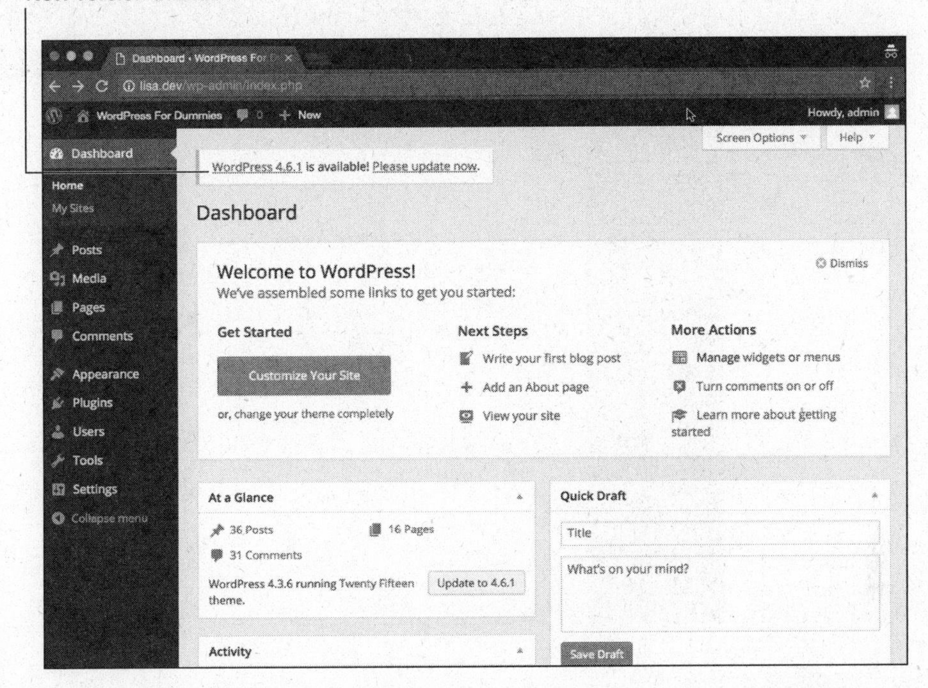

FIGURE 6-2: A Dashboard notification of an available WordPress upgrade.

The message at the top of your Dashboard includes two links that you can click for more information. (Refer to Figure 6-2.) The first is a link called WordPress 4.6.1. Clicking this link takes you to the WordPress Codex page titled Version 4.6.1, which is filled with information about the version upgrade, including

- Installation/upgrade information
- Summary of the development cycle for this version
- List of files that have been revised

The second link, Please Update Now, takes you to another page of the WordPress Dashboard: the WordPress Updates page, shown in Figure 6-3.

At the very top of the WordPress Updates page is another important message for you:

```
Important: before updating, please back up your database
    and files. For help with updates, visit the Updating
    WordPress Codex page.
```

Both links in the message take you to pages in the WordPress Codex that contain helpful information on creating backups and updating WordPress.

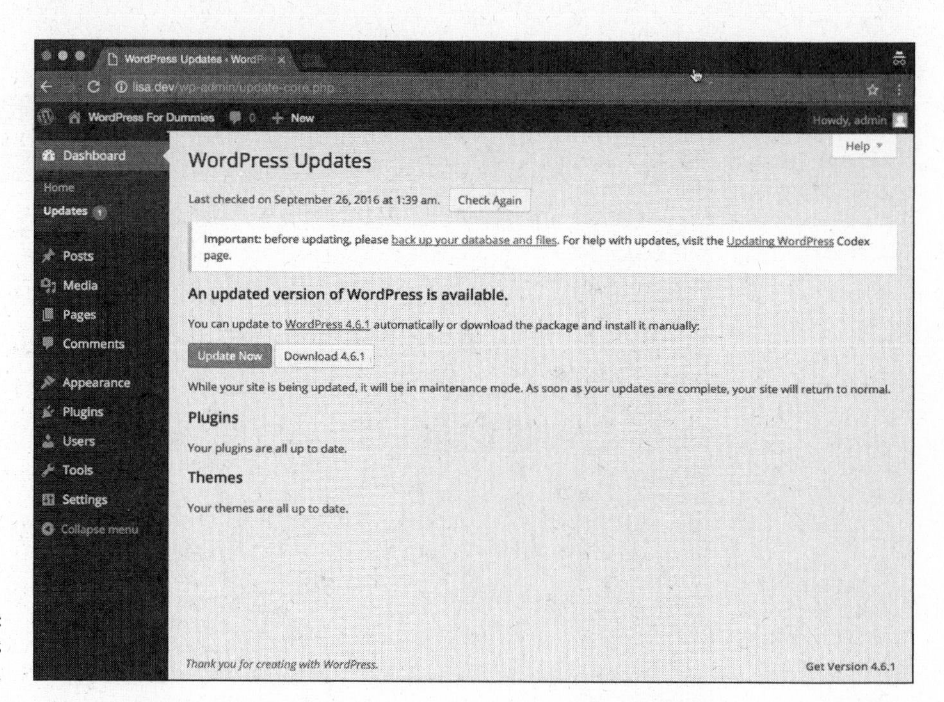

FIGURE 6-3: The WordPress Updates page.

Book 2, Chapter 7 has extensive information on how to back up your WordPress website, content, and files.

The WordPress Updates page tells you that an updated version of WordPress is available. You can update in two ways:

- Automatically, by using the built-in WordPress updater
- Manually, by downloading the files and installing them on your server

These ways to update are discussed later in the chapter.

Backing Up Your Database

Before upgrading your WordPress software installation, make sure that you back up your database. This step isn't required, of course, but it's a smart step to take to safeguard your website and ensure that you have a complete copy of your website data in the event that your upgrade goes wrong.

The best way to back up your database is to use the MySQL administration interface provided to you by your web hosting provider. (Book 2, Chapter 4 takes you through the steps of creating a new database by using the phpMyAdmin interface.)

cPanel is a web hosting interface, provided by many web hosts as a web hosting account management tool, that contains phpMyAdmin as the preferred tool for managing and administering databases. Not all web hosts use cPanel or phpMyAdmin, however, so if yours doesn't, you need to consult the user documentation for the tools that your web host provides. The instructions in this chapter use cPanel and phpMyAdmin.

Follow these steps to create a database backup by using the phpMyAdmin interface:

1. **Log in to the cPanel for your hosting account.**

   Typically, browse to `http://`*`yourdomain`*`.com/cpanel` to bring up the login screen for your cPanel. Enter your specific hosting account username and password in the login fields, and click OK to log in.

2. **Click the phpMyAdmin icon.**

   The phpMyAdmin interface opens and displays your database.

3. **Click the name of the database that you want to back up.**

   If you have more than one database in your account, the left-side menu in phpMyAdmin displays the names of all of them. Click the one you want to back up; the database loads in the main interface window.

4. **Click the Export tab at the top of the screen.**

   The page refreshes and displays the backup utility page.

5. **Select the SQL option in the Format drop-down menu.**
6. **Click the Go button.**

   A pop-up window appears, allowing you to select a location on your computer to store the database backup file.

7. **Click the Save button to download the backup file and save it to your computer.**

Book 2, Chapter 7 contains in-depth information on making a complete backup of your website, including all your files, plugins, themes, and images. For the purposes of upgrading, a database backup is sufficient, but be sure to check out that chapter for valuable information on extensive backups, including how to restore a database backup in case you ever need to go through that process.

Updating WordPress Automatically

WordPress provides you an easy, quick, and reliable method to update the core software from within your Dashboard. I recommend using this option whenever possible to make sure that you're accurately updating the WordPress software.

To update WordPress automatically, follow these steps:

1. **Back up your WordPress website.**

   Backing up your website before updating is an important step in case something goes wrong with the upgrade. Give yourself some peace of mind by knowing that you have a full copy of your website that can be restored, if needed. My advice is not to skip this step under any circumstances. If you're not sure how to back up, back up (pun intended!) to the preceding section.

2. **Deactivate all plugins.**

   This step prevents any plugin conflicts caused by the upgraded version of WordPress from affecting the upgrade process, and it ensures that your website won't break after the upgrade is completed. Find more information on working with and managing plugins in Book 7. For the purposes of this step, you can deactivate plugins by following these steps:

   a. *On the Dashboard, hover your pointer over Plugins on the navigation menu, and click the Installed Plugins link.*

      The Plugins page appears.

   b. *Select all plugins by selecting the check box to the left of the plugin names listed on that page. (See Figure 6-4.)*

   c. *From the drop-down list at the top, choose Deactivate.*

   d. *Click the Apply button.*

3. **Choose Dashboard ⇨ Updates.**

   The WordPress Updates page appears. (Refer to Figure 6-3.)

4. **Click the Update Automatically button.**

   The Update WordPress page appears with a series of messages (as shown in Figure 6-5).

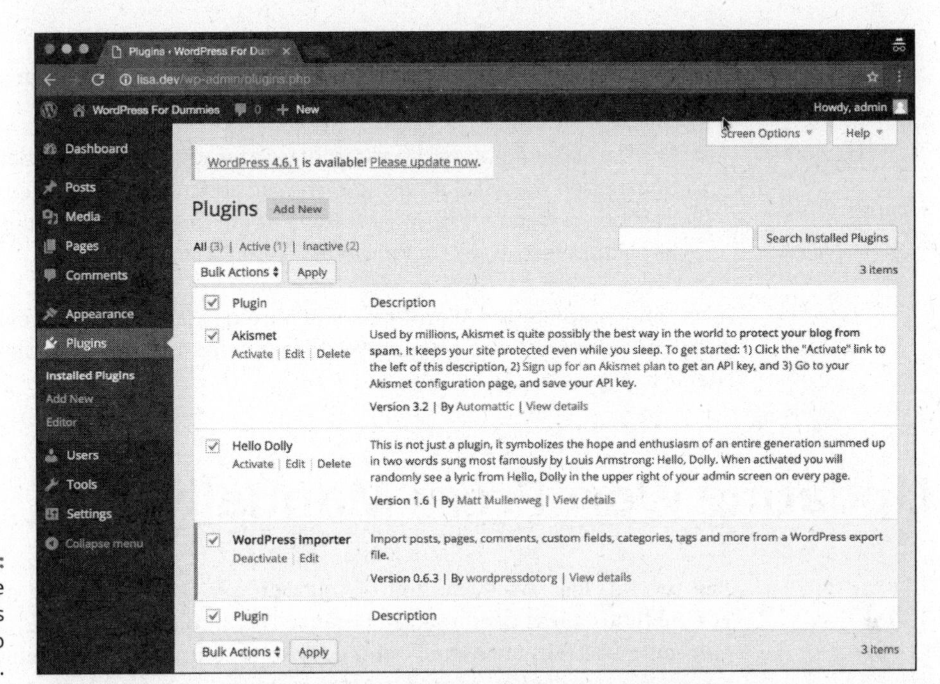

FIGURE 6-4: The Plugins page with all plugins selected, ready to deactivate.

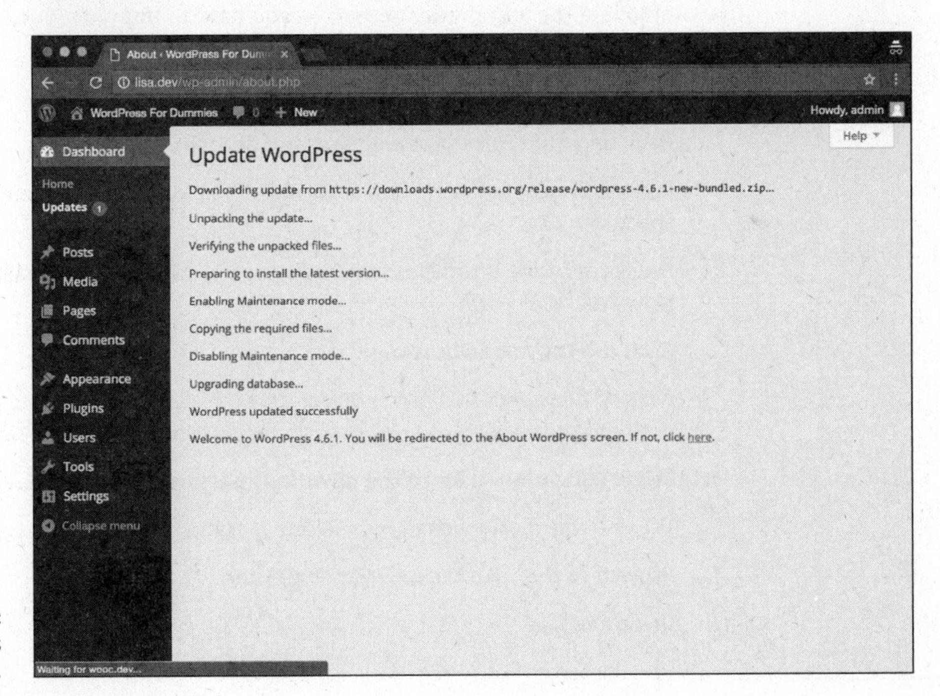

FIGURE 6-5: WordPress update messages.

5. **Wait for the Dashboard to refresh, or click the link in the last update message to visit the main Dashboard screen.**

   The Dashboard page appears in your web browser. Notice that both the update alert message at the top of the site and the notification bubble on the Dashboard menu are no longer visible. Your WordPress installation is now using the latest version of WordPress.

After you complete the WordPress software upgrade, you can revisit the Plugins page and reactivate the plugins you deactivated in Step 2 of the preceding list. (Refer to Figure 6-4.)

Updating WordPress Manually

The second, less-used method of upgrading WordPress is the manual method. The method is less used mainly because the automatic method, discussed in the preceding section, is so quick and easy to accomplish. In certain circumstances, however — probably related to the inability of your hosting environment to accommodate the automatic method — you have to upgrade WordPress manually.

To upgrade WordPress manually, take these steps:

1. **Back up your WordPress website, and deactivate all plugins.**

   Refer to Steps 1 and 2 of "Upgrading WordPress Automatically" earlier in this chapter.

2. **Navigate to the WordPress Updates page by clicking the Please Update Now link.**
3. **Click the Download button.**

   A dialog box opens that allows you to save the `.zip` file of the latest WordPress download package to your local computer, as shown in Figure 6-6.

4. **Select a location to store the download package, and click Save.**

   The `.zip` file downloads to your selected location on your computer.

5. **Browse to the `.zip` file on your computer.**
6. **Unzip the file.**

   Use a program like WinZip (`www.winzip.com`).

FIGURE 6-6: Downloading the WordPress files to your local computer.

7. Connect to your web server via SFTP.

See Book 2, Chapter 2 for details on using SFTP.

8. Delete all the files and folders in your existing WordPress installation directory *except* the following:

- `/wp-content` folder
- `.htaccess`
- `wp-config.php`

9. Upload the contents of the `/wordpress` folder to your web server — not the folder itself.

Most SFTP client software lets you select all the files to drag and drop them to your web server. Other programs have you highlight the files and click a Transfer button.

10. Navigate to the following URL on your website: `http://`*`yourdomain`*`.com/wp-admin`.

Don't panic — your database still needs to be upgraded to the latest version, so instead of seeing your website on your domain, you see a message telling you that a database update is required, as shown in Figure 6-7.

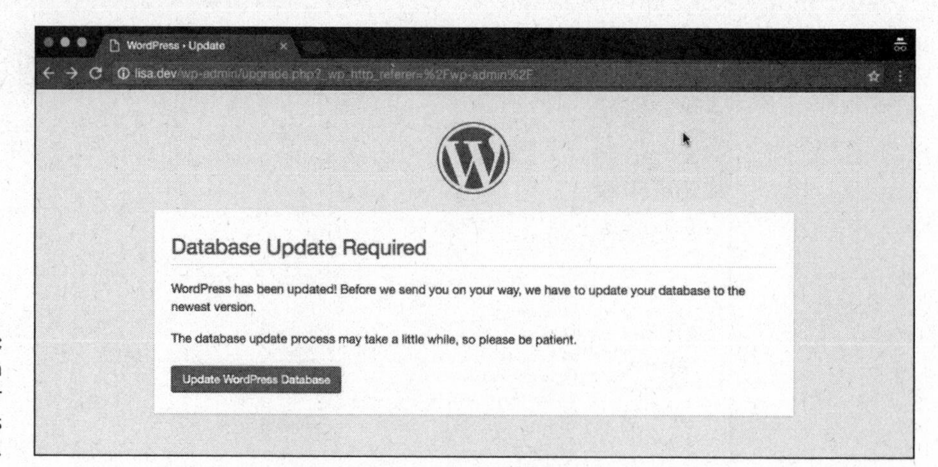

FIGURE 6-7: Click the button to update your WordPress database.

11. **Click the Update WordPress Database button.**

    WordPress initiates the upgrade of the MySQL database associated with your website. When the database upgrade is complete, the page refreshes and displays a message that the process has finished.

12. **Click the Continue button.**

    Your browser loads the WordPress login page. The upgrade is complete, and you can continue using WordPress with the newly upgraded features.

TIP

If you're uncomfortable with performing administrative tasks, such as upgrading and creating database backups, you can hire someone to perform these tasks for you — either an employee of your company (if you run a business) or a WordPress consultant who's skilled in the practice of performing these tasks. Book 1, Chapter 4 includes a listing of experienced consultants who can lend a hand.

» Handling the database backup management

» Backing up plugins and themes

» Storing images and media files

» Exporting data from WordPress

» Using plugins to make backups and moving easier

Chapter 7

Backing Up, Packing Up, and Moving to a New Host

As a WordPress website owner, you may need to move your site to a different home on the web, either to a new web host or into a different account on your current hosting account. Or maybe you're an owner who needs to move your site right now.

This chapter covers the best way to migrate a site that exists within a different platform (such as Movable Type or TypePad) to WordPress. This chapter also takes you through how to back up your WordPress files, data, and content and then move them to a new hosting provider or a different domain.

Migrating Your Existing Site to WordPress

So you have a site on a different content management system (CMS) and want to move your site to WordPress? This chapter helps you accomplish just that. WordPress makes it relatively easy to pack up your data and archives from one platform and move to a new WordPress site.

WordPress lets you move your site from such platforms as Blogger, TypePad, and Movable Type. It also gives you a nifty way to migrate from any platform via RSS feeds, as long as the platform you're importing from has an RSS feed available. Some platforms, such as Medium (`http://medium.com`), have some limitations on RSS feed availability, so be sure to check with your platform provider. In this chapter, you discover how to prepare your site for migration and how to move from the specific platforms for which WordPress provides importer plugins.

Movin' on up

TECHNICAL STUFF

For each platform, the WordPress.org platform provides a quick and easy way to install plugins so that you can import and use your content right away. The importers are packaged in a plugin format because most people use an importer just once, and some people don't use the importer tools at all. The plugins are there for you to use if you need them. WordPress.com, on the other hand, has the importers built into the software. Note the differences for the version you are using.

Website owners have a variety of reasons to migrate away from one system to WordPress:

- **Simple curiosity:** The use of WordPress — and the whole community of WordPress users — is generating a *lot* of buzz. People are naturally curious to check out something that all the cool kids are doing.
- **More control of your website:** This reason applies particularly to those who have a site on Blogger, TypePad, or any other hosted service. Hosted programs limit what you can do, create, and mess with. When it comes to plugins, add-ons, and theme creation, hosting a WordPress site on your own web server wins hands down. In addition, you have complete control of your data, archives, and backup capability when you host your site on your own server.
- **Ease of use:** Many people find the WordPress interface easier to use, more understandable, and a great deal more user-friendly than many of the other blogging platforms available today.

REMEMBER

In the WordPress software, the importers are added to the installation as plugins. The importer plugins included in the preceding list are the plugins packaged within the WordPress software; you can also find them by searching the Plugin Directory at `https://wordpress.org/plugins/tags/importer`. You can import content from several other platforms by installing other plugins that aren't available from the official WordPress Plugin Directory, but you may have to do an Internet search to find them.

Preparing for the big move

Depending on the size of your site (that is, how many posts and comments you have), the migration process can take anywhere from 5 to 30 minutes. As with any major change or update you make, no matter where your site is hosted, the very first thing you need to do is create a backup of your site. You should back up the following:

- **Archives:** Posts, comments, and trackbacks
- **Template:** Template and image files
- **Plugins:** Plugin files (do this by transferring the `/wp-content/plugins` folder from your hosting server to your local computer via SFTP)
- **Links:** Any links, banners, badges, and elements you have in your current site
- **Images:** Any images you use on your site

Table 7-1 gives you a few tips on creating the export data for your site in a few major blogging platforms. *Note:* This table assumes that you're logged in to your site software.

TABLE 7-1 Backing Up Your Website Data on Major Platforms

Blogging Platform	Backup Information
Movable Type	Click the Import/Export button on the menu of your Movable Type Dashboard and then click the Export Entries From link. When the page stops loading, save it on your computer as a `.txt` file.
TypePad	Click the name of the site you want to export and then click the Import/Export link on the Overview menu. Click the Export link at the bottom of the Import/Export page. When the page stops loading, save it on your computer as a `.txt` file.
Blogger	Back up your template by copying the text of your template to a text editor, such as Notepad. Save it on your computer as a `.txt` file.
LiveJournal	Browse to `http://livejournal.com/export.bml` and enter your information; choose XML as the format. Save this file on your computer.

(continued)

TABLE 7-1 *(continued)*

Blogging Platform	Backup Information
Tumblr	Browse to `https://www.tumblr.com/oauth/apps` and follow the directions there to create a Tumblr app. When you're done, copy the OAuth Consumer Key and Secret Key, and paste them into a text file on your computer. Use these keys to connect your WordPress site to your Tumblr account.
WordPress	Choose Tools⇨Export in the Dashboard, choose your options on the Export page, and then click the Download Export File button. Save this file on your computer.
RSS feed	Point your browser to the URL of the RSS feed you want to import. Wait until it loads fully. (You may need to set your feed to display all posts.) View the source code of the page, copy and paste that source code into a `.txt` file, and save the file on your computer.

TIP

The WordPress import script allows for a maximum file size of 128MB. If you get an "out of memory" error, try dividing the import file into pieces and uploading them separately. The import script is smart enough to ignore duplicate entries, so if you need to run the script a few times to get it to take everything, you can do so without worrying about duplicating your content. (You could also attempt to temporarily increase your PHP memory limit by making a quick edit of the `wp-config.php` file; for more information on this technique, see Book 2, Chapters 3 and 4.)

Converting templates

Every program has a unique way of delivering content and data to your site. Template tags vary from program to program; no two are the same. Also, each template file requires conversion if you want to use *your* template with your new WordPress site. In such a case, you have two options:

- **Convert the template yourself.** To accomplish this task, you need to know WordPress template tags and HTML. If you have a template that you're using on another platform and want to convert it for use with WordPress, you need to swap the original platform tags for WordPress tags. The information provided in Book 6 gives you the rundown on working with themes, as well as basic WordPress template tags; you may find that information useful if you plan to attempt a template conversion yourself.
- **Hire an experienced WordPress consultant to do the conversion for you.** See Book 1, Chapter 4 for a list of WordPress consultants.

To use your own template, make sure that you've saved *all* the template files, the images, and the stylesheet from your previous site setup. You need them to convert the template(s) for use in WordPress.

REMEMBER

Hundreds of free templates are available for use with WordPress, so it may be a lot easier to abandon the template you're currently working with and find a free WordPress template that you like. If you paid to have a custom design done for your site, contact the designer of your theme and hire him or her to perform the template conversion for you. Alternatively, you can hire several WordPress consultants to perform the conversion for you — including yours truly.

Moving your site to WordPress

You've packed all your stuff, and you have your new place prepared. Moving day has arrived!

This section takes you through the steps for moving your site from one platform to WordPress. This section assumes that you already have the WordPress software installed and configured on your own domain.

Find the import function that you need by following these steps:

1. **In the Dashboard, choose Tools⇨Import.**

   The Import page appears, listing blogging platforms such as Blogger and Movable Type from which you can import content. (See Figure 7-1.)

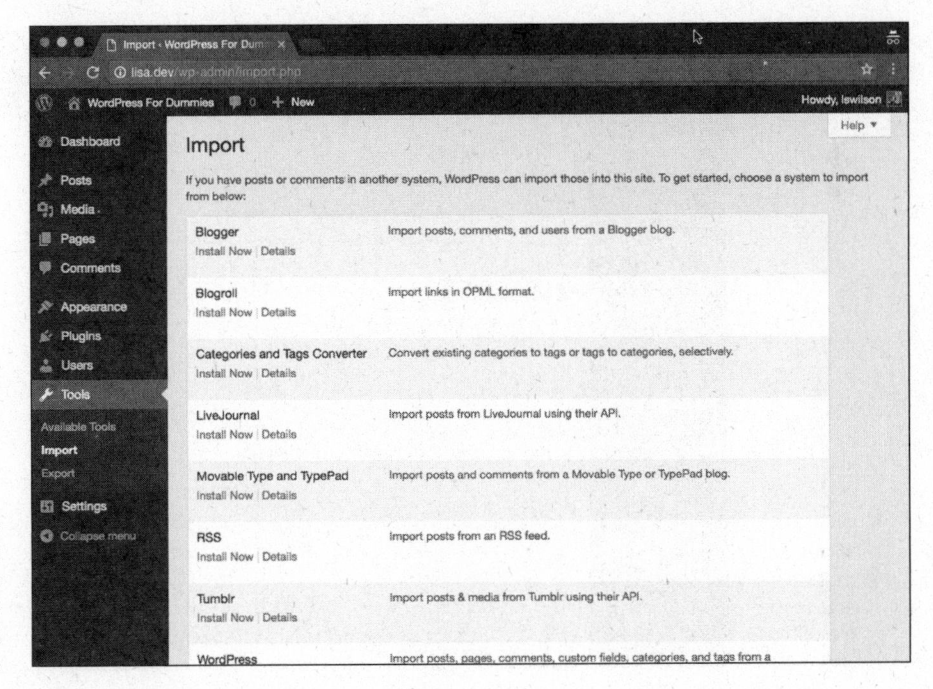

FIGURE 7-1: The Import feature of the (self-hosted) WordPress.org Dashboard.

2. **Find the blogging platform you're working with.**
3. **Click the Install Now link to install the importer plugin and begin using it.**

The following sections provide some import directions for a few of the most popular platforms (other than WordPress, that is). Each platform has its own content export methods, so be sure to check the documentation for the platform that you're using.

Importing from Blogger

Blogger (formerly called Blogspot) is the blogging application owned by Google.

To begin the import process, first complete the steps in the "Moving your site to WordPress" section, earlier in this chapter. Then follow these steps:

1. **Click the Install Now link below the Blogger heading on the Import page, and install the plugin for importing from Blogger.**
2. **Click the Run Importer link.**

   The Import Blogger page loads, with instructions for importing your file, as shown in Figure 7-2.

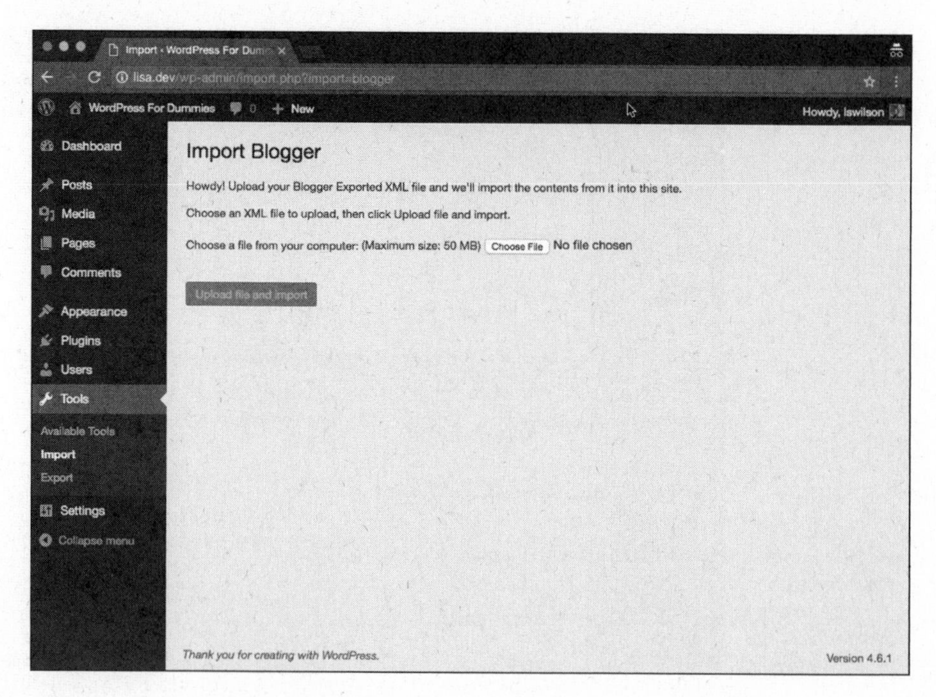

FIGURE 7-2: Import Blogger page in the WordPress Dashboard.

3. **Login to your Blogger account.**
4. **In your Blogger account, click the blog you would like to import.**
5. **In your Blogger account, select Settings⇨Other in your Blogger account.**

   This link is in the left menu.
6. **In your Blogger account, click Back Up Content⇨Save to Your Computer.**

   Save the .xml file to your local computer.
7. **In your WordPress Dashboard, on the Import Blogger page, click the Choose File button to upload the Blogger XML file.**
8. **Click the Upload and Import button.**

   This uploads the file and the screen refreshes to the Import Blogger → Assign Authors screen.
9. **Click the Set Authors button to assign the authors to the posts.**

   The Blogger username appears on the left side of the page; a drop-down menu on the right side of the page displays the WordPress login name.
10. **Assign authors by using the drop-down menu.**

    If you have only one author on each blog, the process is especially easy: Use the drop-down menu on the right to assign the WordPress login to your Blogger username. If you have multiple authors on both blogs, each Blogger username is listed on the left side with a drop-down menu to the right of each username. Select a WordPress login for each Blogger username to make the author assignments.
11. **Click Save Changes.**

    You're done!

Importing from LiveJournal

Both WordPress.com and WordPress.org offer an import script for LiveJournal users, and the process of importing from LiveJournal to WordPress is the same for each platform.

To export your site content from LiveJournal, log in to your LiveJournal site and then type `www.livejournal.com/login.bml?returnto=%2Fexport.bml` in your browser's address bar.

LiveJournal lets you export the XML files by month, so if you have a site with several months' worth of posts, be prepared to be at this process for a while. First, you have to export the entries one month at a time; then you have to import them into WordPress — yep, you guessed it — one month at a time.

TIP

To speed the process a little, you can save all the exported XML LiveJournal files in one text document by copying and pasting each month's XML file into one plain-text file (created in a text editor such as Notepad), thereby creating one long XML file with all the posts from your LiveJournal blog. Then you can save the file as an XML file to prepare it for import into your WordPress blog.

After you export the XML file from LiveJournal, return to the Import page of your WordPress Dashboard and follow these steps:

1. **Click the Install Now link below the LiveJournal heading, and install the plugin for installing from LiveJournal.**
2. **Click the Run Importer link.**

   The Import LiveJournal page loads, with instructions for importing your file, as shown in Figure 7-3.

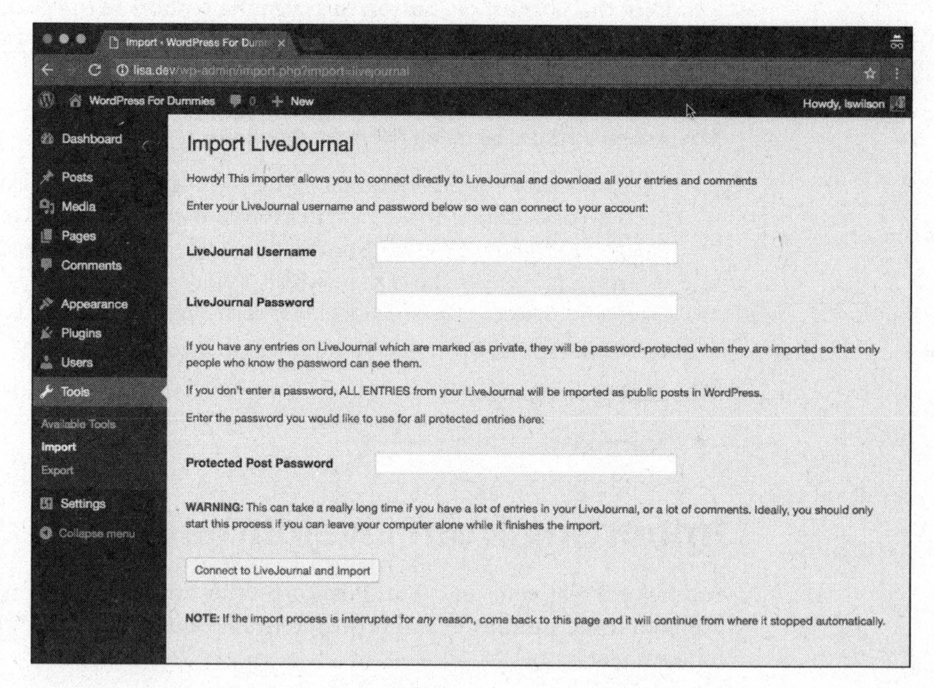

FIGURE 7-3: Import LiveJournal page on the WordPress Dashboard.

3. **In the LiveJournal Username field, type the username for your LiveJournal account.**
4. **In the LiveJournal Password field, type the password for your LiveJournal account.**

5. **In the Protected Post Password field, enter the password you want to use for all protected entries in your LiveJournal account.**

WARNING

If you don't complete this step, every entry you import into WordPress will be viewable by anyone. Be sure to complete this step if any of your entries in your LiveJournal account are password-protected (or private).

6. **Click the Connect to LiveJournal and Import button.**

This step connects your WordPress site to your LiveJournal account and automatically imports all entries from your LiveJournal into your WordPress installation. If your LiveJournal site has a lot of entries, this process could take a long time, so be patient.

Importing from Movable Type and TypePad

Six Apart created both Movable Type and TypePad. These two blogging platforms run on essentially the same code base, so the import/export procedure is the same for both. Refer to Table 7-1, earlier in this chapter, for details on how to run the export process in both Movable Type and TypePad. This import script moves all your site posts, comments, and trackbacks to your WordPress blog.

Go to the Import page of your WordPress Dashboard by following Steps 1 and 2 in the "Moving your site to WordPress" section, earlier in this chapter. Then follow these steps:

1. **Click the Install Now link below the Movable Type and TypePad heading, and install the plugin for importing from Movable Type and TypePad.**
2. **Click the Run Importer link.**

   The Import Movable Type or TypePad page, loads with instructions for importing your file, as shown in Figure 7-4.

3. **Click the Choose File button.**

   A window opens, listing your files.

4. **Double-click the name of the export file you saved from your Movable Type or TypePad blog.**
5. **Click the Upload File and Import button.**

   Sit back and let the import script do its magic. When the script finishes, it reloads the page with a message confirming that the process is complete.

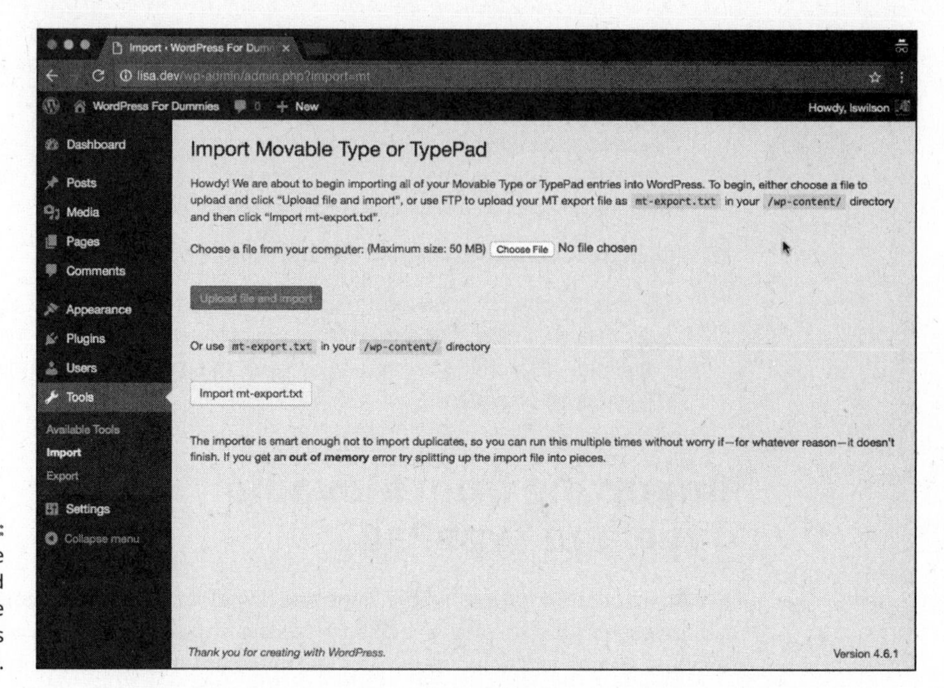

FIGURE 7-4: Import Movable Type or TypePad page on the WordPress Dashboard.

6. **When the import script finishes, assign users to the posts, matching the Movable Type or TypePad usernames with WordPress usernames.**

   If you have only one author on each blog, this process is easy; you simply assign your WordPress login to the Movable Type or TypePad username by using the drop-down menu. If you have multiple authors on both blogs, match the Movable Type or TypePad usernames with the correct WordPress login names.

7. **Click Save Changes.**

Importing from Tumblr

With the Tumblr import script for WordPress, it's easy to import the content from your Tumblr account to your WordPress blog. To complete the import, follow these steps:

1. **Go to `www.tumblr.com/oauth/apps`.**

   The Tumblr login page appears.

2. **Enter your email address and password to log in to your Tumblr account.**

   The Register Your Application page appears.

3. **Complete the Register Your Application form by filling in the following fields:**

   - *Application Name:* Type the name of your WordPress website in the text box.
   - *Application Website:* Type the URL of your WordPress website in the text box.
   - *Default Callback URL:* Type the URL of your WordPress website in the text box.

   Seven text fields are in this form, but you have to fill in these only these three fields; you can leave the rest blank.

4. **Click the Register button.**

   Make sure to select the check box that says `I'm not a robot`, to prove that you are human and not a spammer.

   The Applications page refreshes and displays your registered app information at the top.

5. **Copy the OAuth Consumer Key and paste it into a text file on your computer.**
6. **Copy the Secret Key and paste it into the same text file where you placed the OAuth Consumer Key in Step 5.**
7. **In your Dashboard, choose Tools⇨Import and then click the Tumblr link.**

   The Import Tumblr page of your Dashboard opens.

8. **Insert the OAuth Consumer Key in the indicated text box.**

   Use the OAuth Consumer Key you saved to a text file in Step 5.

9. **Insert the Secret Key in the indicated text box.**

   Use the Secret Key you saved to a text file in Step 6.

10. **Click the Connect to Tumblr button.**

    The Import Tumblr page appears with a message instructing you to authorize Tumblr.

11. **Click the Authorize the Application link.**

    The Authorization page on the Tumblr website asks you to authorize your WordPress site access to your Tumblr account.

12. **Click the Allow button.**

    The Import Tumblr page opens in your WordPress Dashboard and displays a list of your sites from Tumblr.

13. **Click the Import This Blog button in the Action/Status section.**

    The content from your Tumblr account is imported into WordPress. Depending on how much content you have on your Tumblr site, this process may take several minutes to complete. The Import Tumblr page then refreshes with a message telling you that the import is complete.

Importing from WordPress

With the WordPress import script, you can import one WordPress site into another; this is true for both the hosted and self-hosted versions of WordPress. WordPress imports all your posts, comments, custom fields, and categories into your blog. Refer to Table 7-1, earlier in this chapter, to find out how to use the export feature to obtain your site data.

When you complete the export, follow these steps:

1. **Click the Install Now link below the WordPress title on the Import page, and install the plugin to import from WordPress.**
2. **Click the Run Importer link.**

   The Import WordPress page loads, with instructions for importing your file, as shown in Figure 7-5.

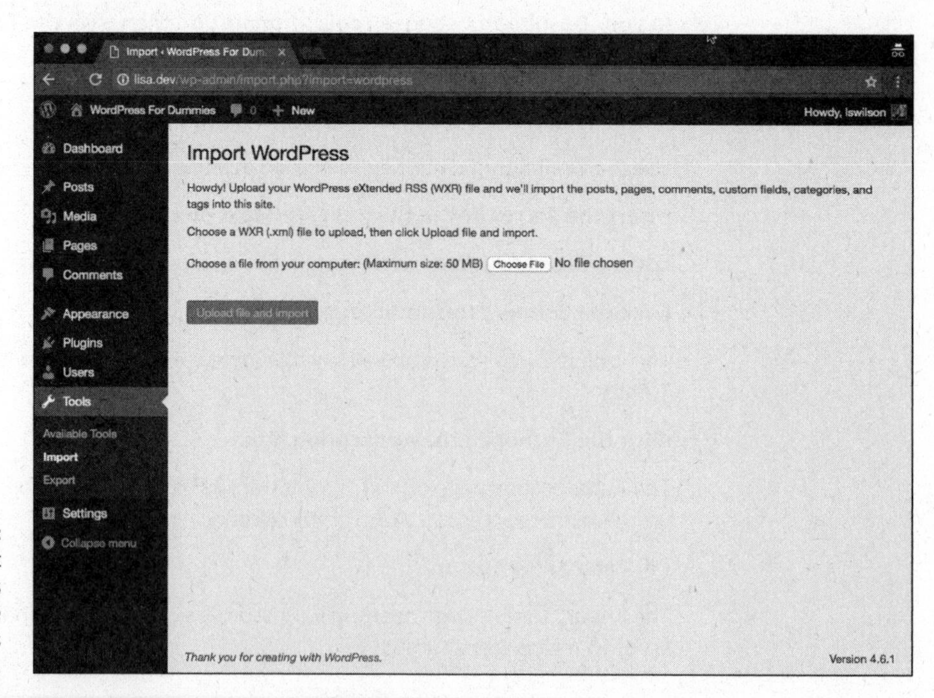

FIGURE 7-5: Import WordPress page on the WordPress Dashboard.

3. **Click the Choose File button.**

   A window opens, listing the files on your computer.

4. **Double-click the export file you saved earlier from your WordPress blog.**
5. **Click the Upload File and Import button.**

   The import script gets to work, and when it finishes, it reloads the page with a message confirming that the process is complete.

Importing from an RSS feed

If all else fails, or if WordPress doesn't provide an import script that you need for your current site platform, you can import your site data via the RSS feed for the site you want to import. With the RSS import method, you can import posts only; you can't use this method to import comments, trackbacks, categories, or users.

REMEMBER

WordPress.com currently doesn't let you import site data via an RSS feed; this function works only with the self-hosted WordPress.org platform.

Refer to Table 7-1, earlier in this chapter, for the steps to create the file you need to import via RSS. Then follow these steps:

1. **On the Import page of the WordPress Dashboard, click the Install Now link below the RSS heading, and install the plugin to import from an RSS feed.**
2. **Click the Run Importer link.**

   The Import RSS page loads, with instructions for importing your RSS file, as shown in Figure 7-6.

3. **Click the Choose File button on the Import RSS page.**

   A window opens, listing the files on your computer.

4. **Double-click the export file you saved earlier from your RSS feed.**
5. **Click the Upload File and Import button.**

   The import script does its magic and then reloads the page with a message confirming that the process is complete.

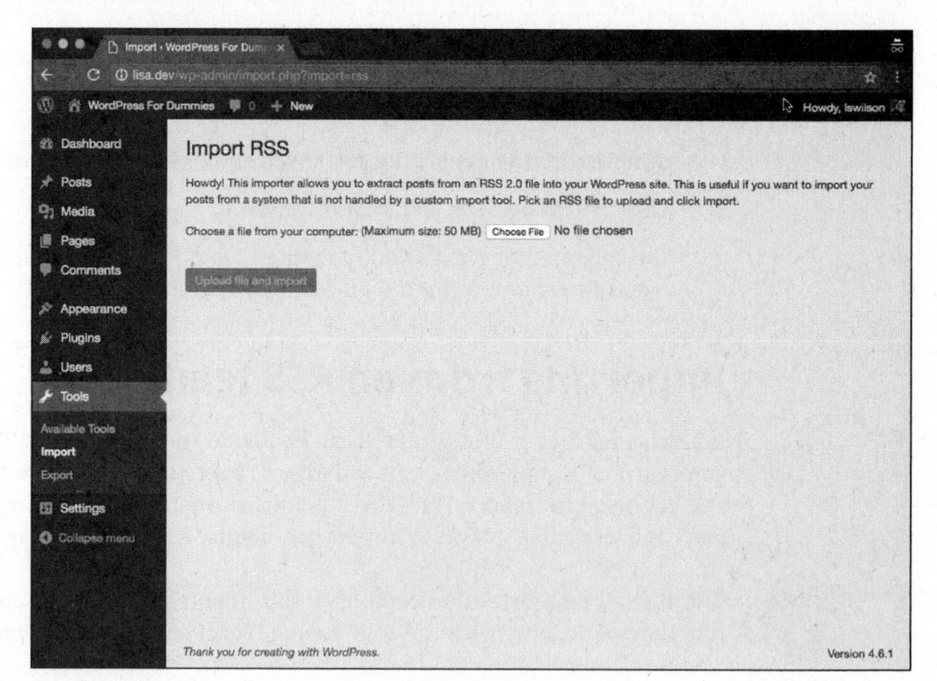

FIGURE 7-6: The Import RSS page of the WordPress Dashboard.

Finding other import resources

The WordPress Codex has a long list of other available scripts, plugins, workarounds, and outright hacks for importing from other platforms. You can find that information at `https://codex.wordpress.org/Importing_Content`.

REMEMBER

Note, however, that volunteers run the WordPress Codex. When you refer to the codex, be aware that not everything listed in it is necessarily up to date or accurate, including import information (or any other information about running your WordPress blog).

Moving Your Website to a Different Host

There may come a time that you decide that you need to switch from your current hosting provider to a new one. There are numerous reasons why you'd have to do this. Perhaps you're unhappy with your current provider and want to move to a new one, or your current provider is going out of business and you're forced to move. Yet transferring an existing website, with all its content, files, and data, from one host to another can seem to be a very daunting task. This section of the chapter should make it easier for you.

You can go about it two ways:

- Manually, by backing up your database and downloading essential files
- Using a plugin to automate as much of the process as possible

Obviously, using a tool to automate the process for you is the more desirable way to go, but just in case you need to do it manually, in the next section of this chapter, I provide the instructions for doing it both ways.

Creating a backup and moving manually

Book 2, Chapter 6 provides step-by-step instructions on how to make a backup of your database by using phpMyAdmin. Follow the steps available in that chapter, and you'll have a backup of your database with all the recent content you've published to your site — *content* being what you (or someone else) wrote or type on your site via the WordPress Dashboard, including

- Blog posts, pages, and custom post types
- Links, categories, and tags
- Post and page options, such as excerpts, time and date, custom fields, post categories, post tags, and passwords
- WordPress settings you configured on the Settings menu of the Dashboard
- All widgets that you created and configured
- All plugin options that you configured for the plugins you installed

Other elements of your website aren't stored in the database, so you need to download those elements, via SFTP, from your web server. Following is a list of those elements, with instructions on where to find them and how to download them to your local computer:

- **Media files:** Media files are the files you uploaded by using the WordPress media upload feature, including images, videos, audio files, and documents. Media files are located in the `/wp-content/uploads` folder. Connect to your web server via SFTP, and download that folder to your local computer.
- **Plugin files:** Although all the plugin settings are stored in the database, the actual plugin *files* are not. The plugin files are located in the `/wp-content/plugins` folder. Connect to your web server via SFTP, and download that folder to your local computer.

» **Theme files:** Widgets and options you've set for your current theme are stored in the database, but the physical theme template files, images, and stylesheets are not. They're stored in the `/wp-content/themes` folder. Connect to your web server via SFTP, and download that folder to your local computer.

Moving the database and files to the new host

When you have your database and WordPress files stored safely on your local computer, moving them to a new host just involves reversing the process. Follow these steps:

1. **Create a new database in your new hosting account.**

   The steps for creating a database are in Book 2, Chapter 4.

2. **Import your database backup into the new database you just created:**

   a. *Log in to the cPanel for your hosting account.*

   b. *Click the phpMyAdmin icon, and click the name of your new database in the left menu.*

   c. *Click the Import tab at the top.*

   d. *Click the Browse button, and select the database backup from your local computer.*

   e. *Click the Go button. The old database imports into the new.*

3. **Install WordPress in your new hosting account.**

   The steps for installing WordPress are in Book 2, Chapter 4.

4. **Edit the `wp-config.php` file to include your new database name, username, password, and host.**

   Information on editing the information in the `wp-config.php` file is in Book 2, Chapters 3 and 4.

5. **Upload all that you downloaded from the `/wp-content` folder to your new hosting account.**

6. **Browse to your domain in your web browser.**

   Your website should work, and you can log in to the WordPress Dashboard by using the same username and password as before, because that information is stored in the database you imported.

BACKING UP AND MOVING WITH A PLUGIN

A plugin that I use on a regular basis to move a WordPress website from one hosting environment to another is aptly named BackupBuddy. This plugin isn't free or available in the WordPress Plugin Directory; you need to pay for it. But it's worth every single penny because it takes the entire backup and migration process and makes mincemeat out of it. In other words, it's very easy to use, and you can be done in minutes instead of hours.

You can purchase the BackupBuddy plugin from iThemes at `https://ithemes.com/purchase/backupbuddy`; at this writing, pricing starts at $80 per year. After you've purchased it, you can download the plugin and install it (see plugin installation instructions in Book 7), and follow the instructions given to you in the WordPress Dashboard to make a backup copy of your website and move it to another server.

3

Exploring the WordPress Dashboard

Contents at a Glance

IN THIS CHAPTER

» Logging in to your WordPress Dashboard

» Finding your way around the Dashboard

» Configuring Dashboard modules

» Getting some help

Chapter 1
Logging In and Taking a Look Around

With WordPress successfully installed, you can explore your new publishing software. This chapter guides you through the preliminary setup of your new WordPress website by using the Dashboard. When you publish with WordPress, you spend a lot of time in the Dashboard, which is where you make all the exciting behind-the-scenes stuff happen. In this Dashboard, you can find all the settings and options that enable you to set up your site just the way you want it. (If you still need to install and configure WordPress, check out Book 2, Chapter 4.)

Feeling comfortable with the Dashboard sets you up for successful entrance into the WordPress publishing world. You'll tweak your WordPress settings several times throughout the life of your site. In this chapter, as you go through the various sections, settings, options, and configurations available to you, understand that nothing is set in stone. You can set options today and change them at any time.

Logging In to the Dashboard

I find that the direct approach (also known as *jumping in*) works best when I want to get familiar with a new software tool. To that end, follow these steps to log in to WordPress and take a look at the guts of the Dashboard:

1. **Open your web browser, and type the WordPress login-page address (or URL) in the address box.**

   The login-page address looks something like this

   ```
   http://www.yourdomain.com/wp-login.php
   ```

   TIP

   If you installed WordPress in its own folder, include that folder name in the login URL. If you installed WordPress in a folder ingeniously named `wordpress`, the login URL becomes

   ```
   http://www.yourdomain.com/wordpress/wp-login.php
   ```

2. **Type your username or email address in the Username or Email Address text box and your password in the Password text box.**

   REMEMBER

   In case you forget your password, WordPress has you covered. Click the Lost Your Password link (located near the bottom of the page), enter your username or email address, and then click the Get New Password button. WordPress resets your password and emails the new password to you.

   After you request a password, you receive two emails from your WordPress site. The first email contains a link that you click to verify that you requested the password. After you verify your intentions, you receive a second email containing your new password.

3. **Select the Remember Me check box if you want WordPress to place a cookie in your browser.**

   The cookie tells WordPress to remember your login credentials the next time you show up. The cookie set by WordPress is harmless and stores your WordPress login on your computer. Because of the cookie, WordPress remembers you the next time you visit. Because this option tells the browser to remember your login, don't select Remember Me when you're using your work computer, other devices (such as a tablet or mobile phone), or a computer at an Internet café.

   Note: Before you set this option, make sure that your browser is configured to allow cookies. (If you aren't sure how to do this, check the help documentation of the Internet browser you're using.)

4. **Click the Log In button.**

After you log in to WordPress, you see the Dashboard page.

Navigating the Dashboard

You can consider the Dashboard to be a control panel of sorts because it offers several quick links and areas that provide information about your site, starting with the actual Dashboard page shown in Figure 1-1. When you view your Dashboard for the very first time, all the modules appear in the expanded (open) position by default.

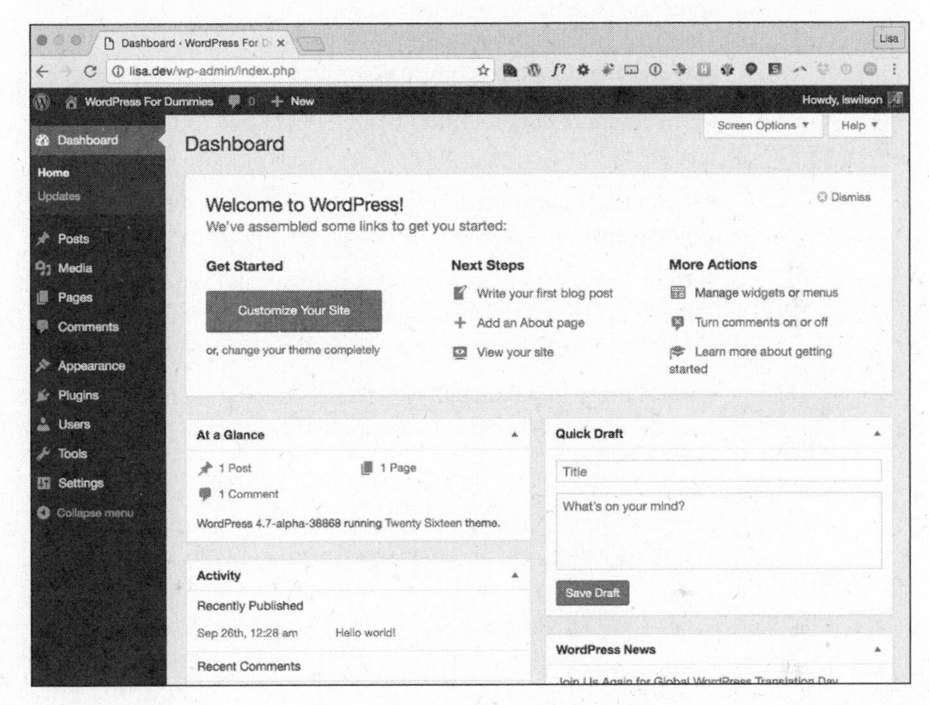

FIGURE 1-1: Log in to the WordPress Dashboard.

You can change how the WordPress Dashboard looks by modifying the order in which the modules (for example, At a Glance and Activity) appear on it. You can expand (open) and collapse (close) the individual modules by clicking anywhere within the title bar of the module. This feature is really nice because it lets you use the Dashboard for just those modules that you use regularly. The concept is easy: Keep the modules you use all the time open, and close the ones that you use only occasionally; open the latter modules only when you really need them. You save screen space by customizing your Dashboard to suit your needs.

TIP

The navigation menu of the WordPress Dashboard appears on the left side of your browser window. When you need to get back to the main Dashboard page, click the Dashboard link at the top of the navigation menu on any of the screens within your WordPress Dashboard.

DISCOVERING THE ADMIN TOOL BAR

The admin toolbar is the menu you see at the top of the Dashboard (refer to Figure 1-1). The admin toolbar appears at the top of every page on your site by default, and it appears at the top of every page of the Dashboard if you set it to do so in your profile settings (see Book 3, Chapter 3). The nice thing is that the only person who can see the admin toolbar is you, because it displays only for the user who is logged in. The admin toolbar contains shortcuts that take you to the most frequently viewed areas of your WordPress Dashboard, from left to right:

- **WordPress links:** This provides you with links to various WordPress.org sites.
- **The name of your website:** This shortcut takes you to the front page of your website.
- **Comments page:** The next link is a comment balloon icon; click it to visit the Comments page of your Dashboard.
- **New:** Hover your mouse over this shortcut, and you find links titled Post, Media, Page, and User. Click these links to go to the Add New Post, Upload New Media, Add New Page, or Add New User pages, respectively.
- **Your photo and name display:** Hover your mouse pointer over this shortcut to open a drop-down menu that provides links to two areas of your Dashboard: Edit Your Profile and Log Out.

Again, the admin toolbar is visible at the top of your site only to you, no matter what page you're on, as long as you're logged in to your WordPress site.

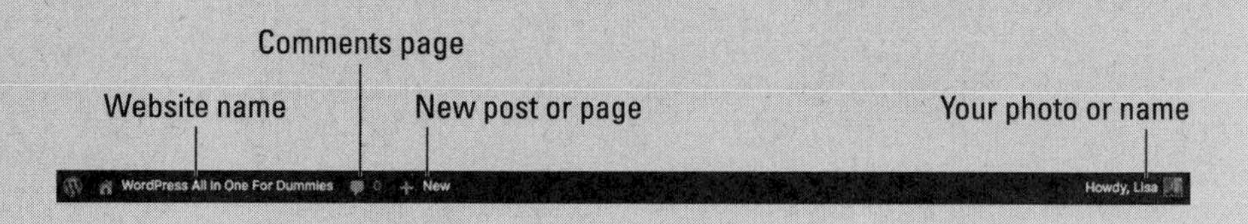

In the following sections, I cover the Dashboard page as it appears when you log in to your WordPress Dashboard for the first time. Later in this chapter, in the section "Arranging the Dashboard to Your Tastes," I show you how to configure the appearance of your Dashboard so that it best suits how you use the available modules.

Welcome to WordPress!

This module, shown in Figure 1-2, appears at the top of your Dashboard screen the first time you log in to your new WordPress installation. It can stay there, if you want it to. Also notice a small link on the right side of that module labeled Dismiss. That link allows you to remove this module if you'd rather not have it there.

The makers of the WordPress software have done extensive user testing to discover what items users want to do immediately when they log in to a new WordPress site. The result of that user testing is a group of links presented in the Welcome to WordPress! module, including

- **Get Started:** This section contains a button that, when clicked, opens the Customizer, where you can customize the active theme. Additionally, this section provides a link that takes you to the Themes page, where you can change your theme. Book 6 contains tons of information about choosing a theme, as well as customizing it to look the way you want it to.
- **Next Steps:** This section provides links to various areas within the WordPress Dashboard to get you started publishing content, including writing your first post and adding an About page. (Book 4, Chapter 1 provides information about publishing posts, and Book 4, Chapter 2 gives you information about publishing pages.) Additionally, the View Your Site link in this section opens your site, allowing you to view what it looks like to your visitors.
- **More Actions:** This section contains a few links that help you manage your site, including a link to manage widgets or menus (see Book 6, Chapter 1) and turning comments on or off (see Book 3, Chapter 4). This section also contains a link to the First Steps with WordPress link of the WordPress Codex, where you can read more information about how to start using your new WordPress site.

FIGURE 1-2: The Welcome to WordPress! module provides helpful links to get you started.

At a Glance

The At a Glance module of the Dashboard shows what's going on in your blog right now — this very second! Figure 1-3 shows the expanded At a Glance module in a brand-spanking-new WordPress blog.

FIGURE 1-3: The At a Glance module of the Dashboard, expanded so that you can see the available features.

At a Glance

2 Posts
1 Page
2 Comments

WordPress 4.8-alpha-39941 running Twenty Seventeen theme.

The At a Glance module shows the following default information:

- **The number of posts you have:** This number reflects the total number of posts you currently have in your WordPress blog. The blog in Figure 1-3, for example, has two posts. The link is blue, which means that it's clickable. When you click the link, you go to the Posts screen, where you can view and manage the posts in your blog. (Book 4, Chapter 1 covers managing posts.)
- **The number of pages:** The number of pages in your blog, which changes when you add or delete pages. (*Pages,* in this context, refers to the static pages you create in your blog.) Figure 1-3 shows that the blog has one page.

  Clicking this link takes you to the Pages screen, where you can view, edit, and delete your current pages. (Find the difference between WordPress posts and pages in Book 4, Chapter 2.)
- **The number of comments:** The number of comments on your blog. Figure 1-3 shows that this blog has two comments.

  Clicking the Comments link takes you to the Comments screen, where you can manage the comments on your blog. Book 3, Chapter 4 covers comments.

The last section of the Dashboard's At a Glance module shows the following information:

- **Which WordPress theme you're using:** Figure 1-3 shows that the blog is using the theme called Twenty Seventeen. The theme name is a link that, when clicked, takes you to the Manage Themes page, where you can view and activate themes on your blog.
- **The version of WordPress you're using:** Figure 1-3 shows that this blog is using WordPress version 4.8. This version announcement changes if you're using an earlier version of WordPress. When WordPress software is upgraded, this statement tells you that you're using an outdated version of WordPress and encourages you to upgrade to the latest version.

Activity

The module below the At a Glance module is the Activity module, shown in Figure 1-4. Within this module, you find

- **Posts most recently published:** WordPress displays a maximum five posts in this area. Each one is clickable and takes you to the Edit Post screen, where you can view and edit the post.
- **Most recent comments published to your blog:** WordPress displays a maximum of five comments in this area.
- **The author of each comment:** The name of the person who left the comment appears below it. This section also displays the author's picture (or avatar), if she has one.
- **A link to the post the comment was left on:** The post title appears to the right of the commenter's name. Click the link to go to that post in the Dashboard.
- **An excerpt of the comment:** This excerpt is a snippet of the comment this person left on your blog.
- **Comment management links:** When you hover your mouse pointer over the comment, six links appear below it. These links give you the opportunity to manage those comments right from your Dashboard: The first link is Unapprove, which appears only if you have comment moderation turned on. The other five links are Reply, Edit, Spam, Trash and View.
- **View links:** These links appear at the bottom of the Activity module. You can click All, Pending, Approved, Spam, and Trash.

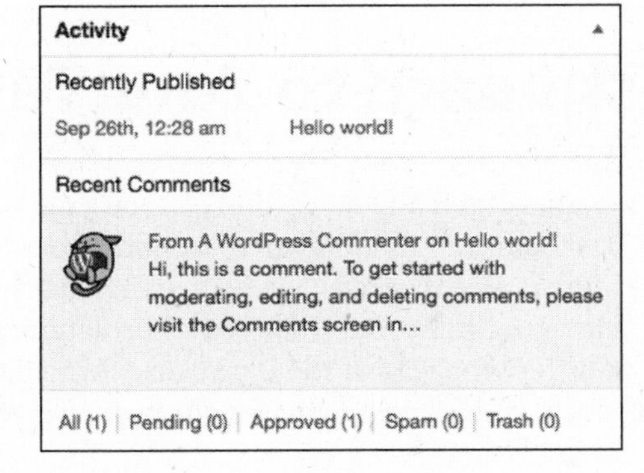

FIGURE 1-4: The Activity module of the Dashboard.

TIP

You can find even more information on managing your comments in Book 3, Chapter 4.

Quick Draft

The Quick Draft module, shown in Figure 1-5, is a handy form that allows you to write, save, and publish a blog post right from your WordPress Dashboard. The options are similar to the ones I cover in Book 4, Chapter 1.

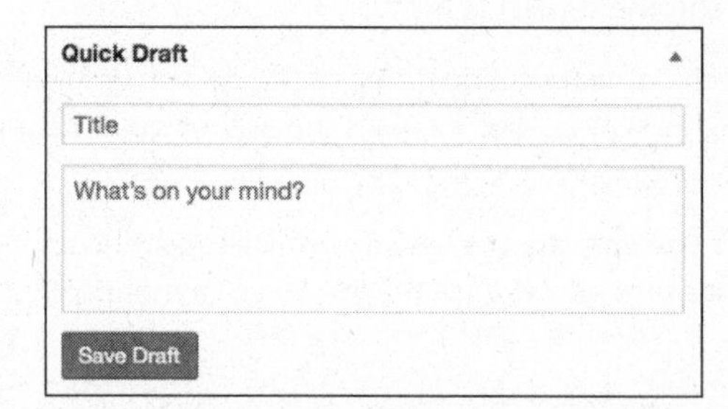

FIGURE 1-5: The Quick Draft module of the Dashboard.

WordPress News

The WordPress News module of the Dashboard pulls in posts from a site called WordPress Planet (`http://planet.wordpress.org`). By keeping the default setting in this area, you stay in touch with several posts made by folks who are involved in WordPress development, design, and troubleshooting. You can find lots of interesting and useful tidbits if you keep this area intact. Quite often, I find great information about new plugins or themes, problem areas and support, troubleshooting, and new ideas, so I tend to stick with the default setting.

Arranging the Dashboard to Your Tastes

One feature of WordPress that I'm really quite fond of allows me to create my own workspace within the Dashboard. In the following sections, you can find out how to customize your WordPress Dashboard to fit your individual needs, including modifying the layout, changing links and RSS feed information, and even rearranging the modules on different pages of the Dashboard. Armed with this information, you can open your Dashboard and create your very own workspace.

Changing the order of modules

You can arrange the order of the modules in your Dashboard to suit your tastes. WordPress places a great deal of emphasis on user experience, and a big part of that effort results in your ability to create a Dashboard that you find most useful. You can very easily change the modules to display and the order in which they display.

Follow these steps to move the At a Glance module so that it appears on the right side of your Dashboard page:

1. **Hover your mouse cursor on the title bar of the At a Glance module.**

   When the mouse is hovering over the title, your mouse cursor changes to the Move cursor (a cross with arrows).

2. **Click and hold your mouse button, and drag the At a Glance module to the right side of the screen.**

   While you drag the module, a light-gray box with a dotted border appears on the right side of your screen (see Figure 1-6). That gray box is a guide that shows you where you can drop the module.

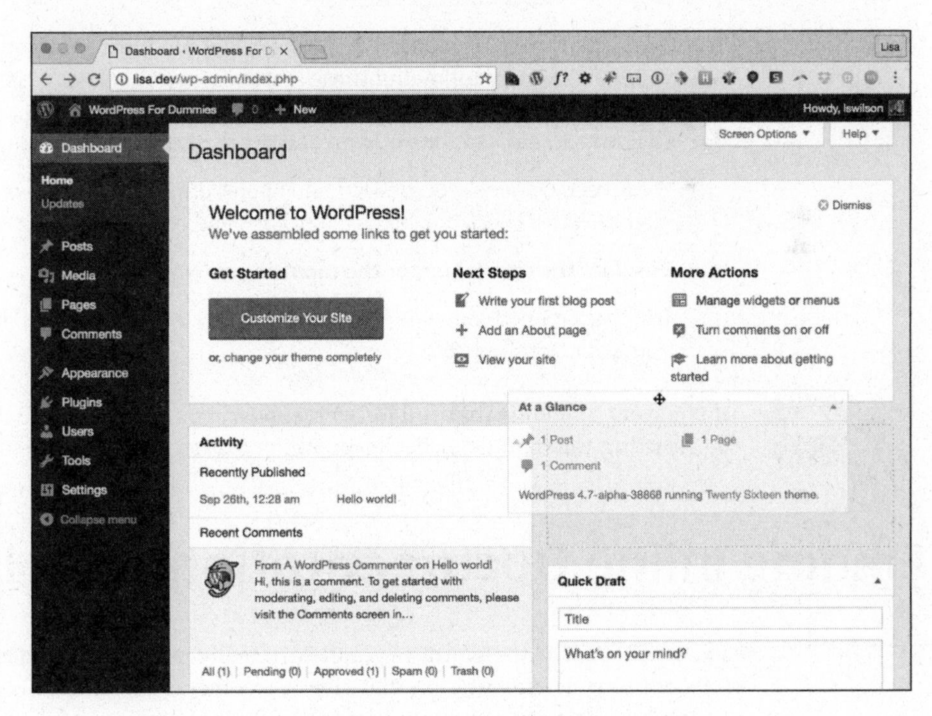

FIGURE 1-6: A light gray box appears as a guide when you drag and drop modules in the WordPress Dashboard.

3. **Release the mouse button when you have the At a Glance module in place.**

   The At a Glance module is positioned on the right side of your Dashboard page.

   The other modules on the right side of the Dashboard have shifted down, and the Activity module is the module in the top-left corner of the Dashboard page.

4. **(Optional) Click the title bar of the At a Glance module.**

   The module collapses. Click the title bar again to expand the module. You can keep that module open or closed based on your preference.

Repeat these steps with each module that you want to move on the Dashboard by dragging and dropping them so that they appear in the order you prefer.

REMEMBER

When you navigate away from the Dashboard, WordPress remembers the changes you made. When you return, you still see your customized Dashboard, and you don't need to redo these changes in the future.

Removing Dashboard modules

If you find that your Dashboard contains a few modules that you never use, you can get rid of them by following these steps:

1. **Click the Screen Options button at the top of the Dashboard.**

   The Screen Options drop-down menu opens, displaying the title of each module with a check box to the left of each title.

2. **Deselect the check box for the module you want to hide.**

   The check mark disappears from the check box, and the module disappears from your Dashboard.

TIP

If you want a module that you hid to reappear, you can simply enable that module by selecting the module's check box on the Screen Options drop-down menu.

Finding Inline Documentation and Help

The developers of the WordPress software really put in time and effort to provide tons of inline documentation that gives you several tips and hints right inside the Dashboard. You can generally find inline documentation for nearly every WordPress feature you'll use.

Inline documentation refers to those small sentences or phrases that you see alongside or below a feature in WordPress, providing a short but very helpful explanation of the feature. Figure 1-7 shows the General Settings screen, where a lot of inline documentation and guiding tips correspond with each feature. These tips can clue you into what the features are, how to use those features, and what recommended settings to use for those features.

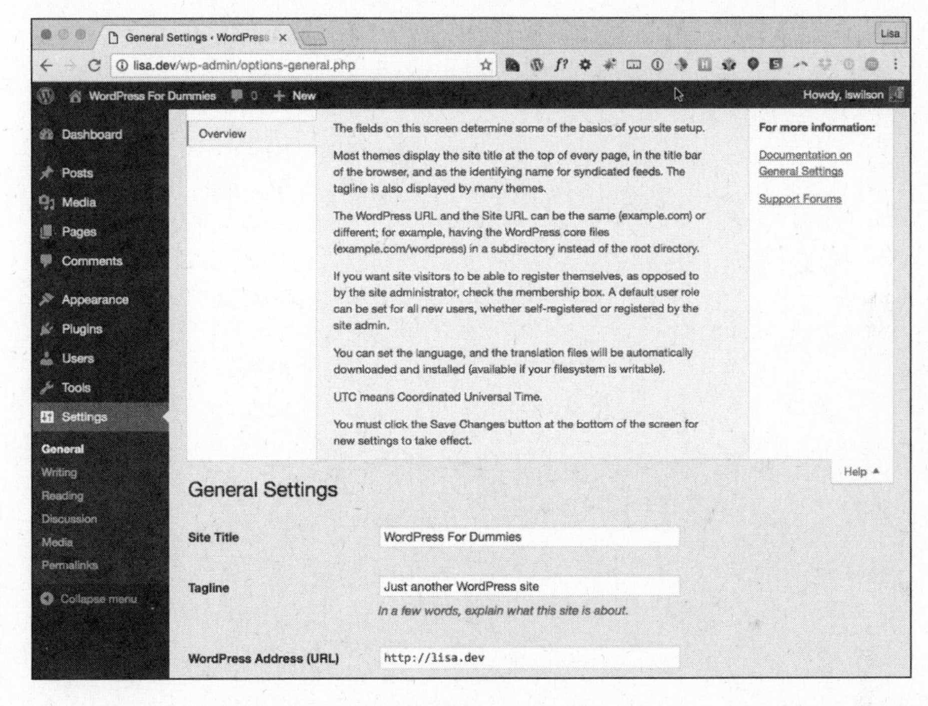

FIGURE 1-7: Inline documentation on the General Settings page in the WordPress Dashboard.

In addition to the inline documentation that you find scattered throughout the Dashboard, a useful Help tab is located in the top-right corner of your Dashboard. Click this tab to open a panel containing help text that's relevant to the screen you're currently viewing in your Dashboard. If you're viewing the General Settings screen, for example, the Help tab displays documentation relevant to the General Settings screen. Likewise, if you're viewing the Add New Post screen, the Help tab displays documentation with topics relevant to the settings and features you find on the Add New Post page of your Dashboard.

TIP

The inline documentation and the topics and text you find in the Help tab exist to assist you while you work with the WordPress platform, helping make the experience as easy to understand as possible. Another place you can visit to find help and useful support for WordPress is the WordPress Forums support page, at `http://wordpress.org/support`.

» Putting together your personal profile

» Formatting your website

Chapter 2
Exploring Tools and Settings

As exciting as it is to dig right in and start publishing right away, you should attend to a few housekeeping items first, including adjusting the settings that allow you to personalize your website. I cover these settings first in this chapter because they create your readers' experience with your website.

In this chapter, you explore the Settings menu of the WordPress Dashboard and discover how to configure items such as date and time settings, site titles, and email notification settings. This chapter also covers important aspects of your website configuration, such as permalinks, discussion options, and privacy settings.

Some of the menu items, such as those for creating and publishing new posts, are covered in detail in other chapters, but they're well worth a mention here as well.

Configuring the Settings

At the very bottom of the navigation menu, you find the Settings option. Hover over the Settings link. A menu appears that contains the following links, which I discuss in the sections that follow:

- General
- Writing
- Reading
- Discussion
- Media
- Permalinks

General

After you install the WordPress software and log in, you can put a personal stamp on your site by giving it a title and description, setting your contact email address, and identifying yourself as the author of the blog. You take care of these and other settings on the General Settings screen.

To begin personalizing your site, start with your general settings by following these steps:

1. **Choose Settings⇨General.**

   The General Settings screen appears, as shown in Figure 2-1.

2. **Enter the name of your site in the Site Title text box.**

   The title you enter here is the one that you give your blog to identify it as your own. In Figure 2-1, I gave the new blog the title *WordPress All-In-One For Dummies,* which appears on my website, as well as in the title bar of the viewer's web browser.

   TIP

   Give your website an interesting and identifiable name. You can use *Fried Green Tomatoes,* for example, if you're blogging about the topic, the book, the movie, or even anything remotely related to the lovely Southern dish.

3. **In the Tagline text box, enter a five- to ten-word phrase that describes your blog.**

   Figure 2-1 shows that the tagline is *by Lisa Sabin-Wilson.* Therefore, this blog displays the blog title, followed by the tagline: *WordPress All-in-One For Dummies by Lisa Sabin-Wilson.*

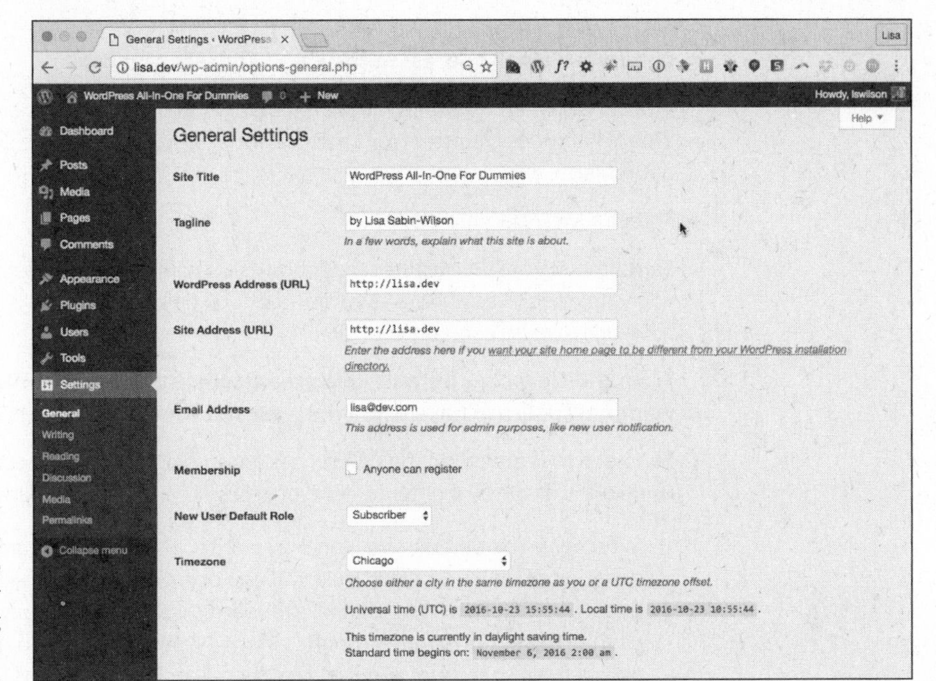

FIGURE 2-1: Personalize the settings of your WordPress blog on the General Settings screen.

REMEMBER

The general Internet-surfing public can view your site title and tagline, which various search engines (such as Google, Yahoo!, and Bing) grab for indexing, so choose your words with this fact in mind. (You can find more information about search engine optimization, or SEO, in Book 5.)

4. **In the WordPress Address (URL) text box, enter the location where you installed the WordPress software.**

   Be sure to include the `http://` portion of the URL and the entire path to your WordPress installation — for example, `http://`*`yourdomain.com`*. If you installed WordPress in a folder in your directory — in a folder called `wordpress`, for example — you need to make sure to include it here. If you installed WordPress in a folder called `wordpress`, the WordPress address would be `http://`*`yourdomain.com`*`/wordpress` (where *`yourdomain.com`* is your domain name).

5. **In the Site Address (URL) text box, enter the web address where people can find your blog by using their web browsers.**

   Typically, what you enter here is the same as your domain name (`http://`*`yourdomain.com`*). If you install WordPress in a subdirectory of your site, the WordPress installation URL is different from the Site Address (URL). If you install WordPress at `http://`*`yourdomain.com`*`/wordpress/` (WordPress Address (URL)), you need to tell WordPress that you want the blog to appear at `http://`*`yourdomain.com`* (the Site Address (URL)).

6. **Enter your email address in the Email Address text box.**

   WordPress sends messages about the details of your site to this email address. When a new user registers for your site, for example, WordPress sends you an email alert.

7. **Select a Membership option.**

   Select the *Anyone Can Register* check box if you want to keep registration on your site open to anyone. Leave the check box deselected if you'd rather not have open registration on your site.

8. **From the New User Default Role drop-down list, select the role that you want new users to have when they register for user accounts in your site.**

   You need to understand the differences among the user roles because each user role is assigned a different level of access to your site, as follows:

   - *Subscriber:* The default role. You may want to maintain this role as the one assigned to new users, particularly if you don't know who's registering. Subscribers have access to the Dashboard screen, and they can view and change the options in their profiles in the Profile screen. (They don't have access to your account settings, however — only to their own.) Each user can change his username, email address, password, bio, and other descriptors in his user profile. Subscribers' profile information is stored in the WordPress database, and your site remembers them each time they visit, so they don't have to complete the profile information each time they leave comments on your site.
   - *Contributor:* In addition to the access subscribers have, contributors can upload files and write, edit, and manage their own posts. Contributors can write posts, but they can't publish the posts; the administrator reviews all contributor posts and decides whether to publish them. This setting is a nice way to moderate content written by new authors.
   - *Author:* In addition to the access contributors have, authors can publish and edit their own posts.
   - *Editor:* In addition to the access authors have, editors can moderate comments, manage categories, manage links, edit pages, and edit other authors' posts.
   - *Administrator:* Administrators can edit all the options and settings in the WordPress site.

9. **From the Timezone drop-down list, select your UTC time.**

   This setting refers to the number of hours that your local time differs from Coordinated Universal Time (UTC). This setting ensures that all the posts and comments left on your blog are time-stamped with the correct time. If you're lucky enough, as I am, to live on the frozen tundra of Wisconsin, which is in the Central Standard Time (CST) Zone, you choose **–6** from the drop-down list because that time zone is 6 hours off UTC.

TIP

If you're unsure what your UTC time is, you can find it at the Greenwich Mean Time (`https://greenwichmeantime.com`) website. GMT is essentially the same thing as UTC. WordPress also lists some major cities in the Timezone drop-down list so that you can more easily choose your time zone if you don't know it. (Figure 2-1 displays Chicago in the drop-down menu because that's the major city closest to where I live.)

Note: The following options aren't shown in Figure 2-1; you need to scroll down to access them.

10. **For the Date Format option, select the format in which you want the date to appear in your site.**

    This setting determines the style of the date display. The default format displays time like this: January 1, 2017.

    Select a different format by selecting the radio button to the left of the option you want. You can also customize the date display by selecting the Custom option and entering your preferred format in the text box provided. If you're feeling adventurous, you can find out how to customize the date format at `https://codex.wordpress.org/Formatting_Date_and_Time`.

11. **For the Time Format option, select the format for how you want time to display on your site.**

    This setting is the style of the time display. The default format displays time like this: 12:00 a.m.

    Select a different format by selecting the radio button to the left of the option you want. You can also customize the date display by selecting the Custom option and entering your preferred format in the text box provided; find out how at `https://codex.wordpress.org/Formatting_Date_and_Time`.

TIP

You can format the time and date in several ways. Go to `http://us3.php.net/manual/en/function.date.php` to find potential formats at the PHP website.

12. **From the Week Starts On drop-down list, select the day on which the week starts in your calendar.**

    Displaying the calendar in the sidebar of your site is optional. If you choose to display the calendar, you can select the day of the week on which you want your calendar to start.

REMEMBER

Click the Save Changes button at the bottom of any page where you set new options. If you don't click Save Changes, your settings aren't saved, and WordPress reverts to the preceding options. Each time you click the Save Changes button, WordPress reloads the current screen, displaying the new options that you just set.

Writing

Choose Settings➪Writing, and the Writing Settings screen opens. (See Figure 2-2.)

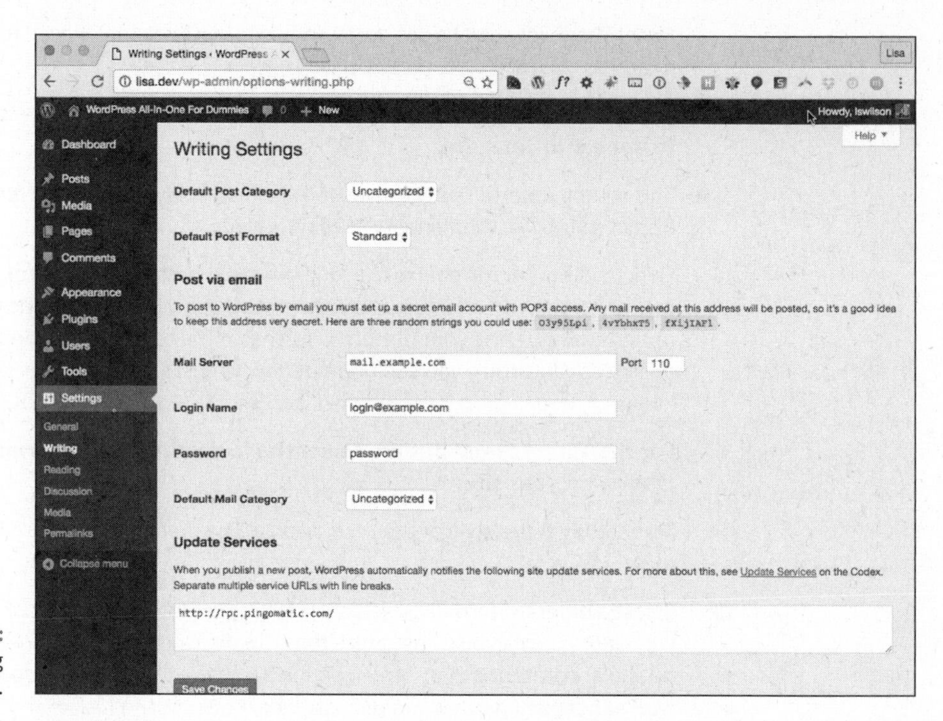

FIGURE 2-2: The Writing Settings screen.

This screen of the Dashboard lets you set some basic options for writing your content. Table 2-1 gives you some information on choosing how your content looks and how WordPress handles some specific conditions.

After you set your options, be sure to click the Save Changes button; otherwise, the changes won't take effect.

TABLE 2-1 Writing Settings Options

Option	Function	Default
Default Post Category	Select the category that WordPress defaults to any time you forget to choose a category when you publish a post.	Uncategorized
Default Post Format	Select the format that WordPress defaults to any time you create a post and don't assign a post format. (This option is theme-specific; not all themes support post formats. See Book 6, Chapter 6.)	Standard
Post via Email	Publish content from your email account by entering the email and server information for the account you'll be using to send posts to your WordPress site.	N/A
Default Mail Category	Sets the category that posts via email get submitted to when these types of posts are published.	Uncategorized
Update Services ***Note:*** This option is available only if you allow your site to be indexed by search engines (covered in the Reading settings section).	Indicates which ping service you want to use to notify the world that you've made updates, or new posts, to your blog. The default, XML-RPC (`http://rpc.pingomatic.com`), updates all the popular services simultaneously.	`http://rpc.pingomatic.com`

TIP

Go to `https://codex.wordpress.org/Update_Services` for comprehensive information on update services.

Reading

The third item in the Settings drop-down list is Reading. Choose Settings⇨Reading to open the Reading Settings screen. (See Figure 2-3.)

You can set the following options on the Reading Settings screen:

- **Front Page Displays:** Select the radio button to show a page instead of your latest posts on the front page of your site. You can find detailed information about using a static page for your front page in Book 4, Chapter 2, including information on how to set it up by using the fields in this section that appear after you select the radio button.
- **Blog Pages Show at Most:** In the text box, enter the maximum number of posts you want to appear on each blog page (default: 10).

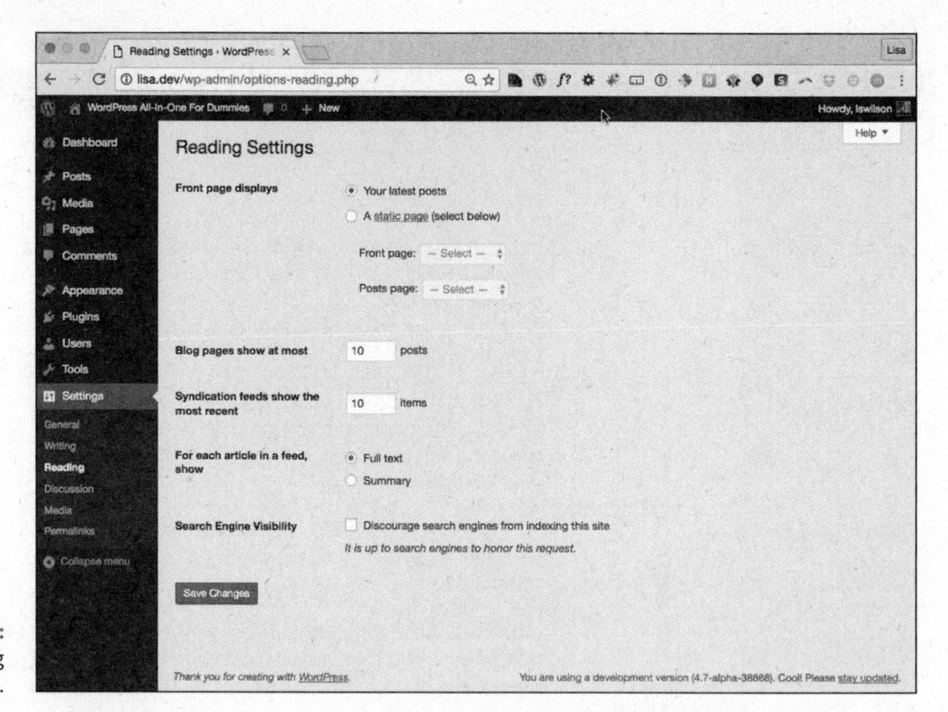

FIGURE 2-3: The Reading Settings screen.

- **Syndication Feeds Show the Most Recent:** In the text box, enter the maximum number of posts that you want to appear in your RSS feed at any time (default: 10).
- **For Each Article in a Feed, Show:** Select the Full Text or Summary radio button. Full Text publishes the entire post to your RSS feed, whereas Summary publishes an excerpt. (Check out Book 1, Chapter 1 for more information on WordPress RSS feeds.)
- **Search Engine Visibility:** By default, your website is visible to all search engines, such as Google and Yahoo!. If you don't want your site to be visible to search engines, select the check box labeled Discourage Search Engines from Indexing This Site.

TIP

Generally, you want search engines to be able to find your site. If you have special circumstances, however, you may want to enforce privacy on your site. A friend of mine has a family blog, for example, and she blocks search engine access to it because she doesn't want search engines to find it. When you have privacy enabled, search engines and other content bots can't find your website.

REMEMBER

Be sure to click the Save Changes button after you set all your options on the Reading Settings screen to make the changes take effect.

Discussion

Discussion is the fourth item on the Settings menu list; choose Settings ⇨ Discussion to open the Discussion Settings screen. (See Figure 2-4.) The sections of this screen let you set options for handling comments and publishing posts to your site.

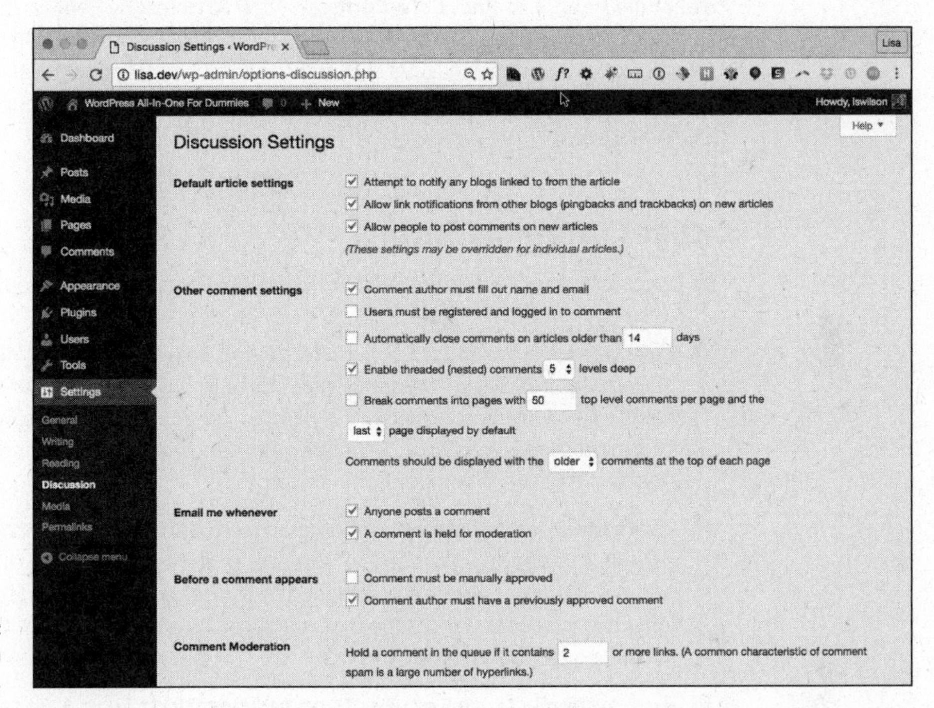

FIGURE 2-4: The Discussion Settings screen.

The following sections cover the options available to you on the Discussion Settings screen, which deals mainly with how comments and trackbacks are handled in your site.

Default Article Settings

With the Default Article Settings options, you can tell WordPress how to handle post notifications. Here are your options:

- **Attempt to Notify Any Blogs Linked to from the Article:** If you select this check box, your site sends a notification (or *ping*) to any site you've linked to in your posts. This notification is also commonly referred to as a *trackback*. (Find out more about trackbacks in Book 3, Chapter 4.) Deselect this check box if you don't want these notifications sent.

- **Allow Link Notifications from Other Blogs (Pingbacks and Trackbacks) on New Articles:** By default, this check box is selected, and your site accepts notifications via pings or trackbacks from other sites that have linked to yours. Any trackbacks or pings sent to your site appear on your site in the Comments section of the post. If you deselect this check box, your site doesn't accept pingbacks or trackbacks from other sites.
- **Allow People to Post Comments on New Articles:** By default, this check box is selected, and people can leave comments on your posts. If you deselect this check box, no one can leave comments on your content. (You can override these settings for individual articles. Find more information about this process in Book 4, Chapter 1.)

Other Comment Settings

The Other Comment Settings section tells WordPress how to handle comments:

- **Comment Author Must Fill Out Name and Email:** Enabled by default, this option requires all commenters on your site to fill in the Name and Email fields when leaving comments. This option can really help you combat comment spam. (See Book 3, Chapter 4 for information on comment spam.) Deselect this check box to disable this option.
- **Users Must Be Registered and Logged in to Comment:** Not enabled by default, this option allows you to accept comments on your site only from people who are registered and logged in as users on your site. If the user isn't logged in, she sees a message that reads `You must be logged in in order to leave a comment.`
- **Automatically Close Comments on Articles Older Than *X* Days:** Select the check box next to this option to tell WordPress you want comments on older articles to be closed automatically. Fill in the text box with the number of days you want to wait before WordPress closes comments on articles (default: 14).

REMEMBER

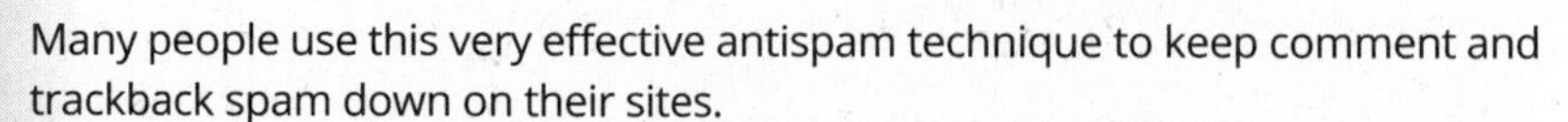
Many people use this very effective antispam technique to keep comment and trackback spam down on their sites.

- **Enable Threaded (Nested) Comments *X* Levels Deep:** From the drop-down list, you can select the level of threaded comments you want to have on your site. The default is five; you can choose up to ten levels. Instead of displaying all comments on your site in chronological order, nesting them allows you and your readers to reply to comments within the comment itself.
- **Break Comments into Pages with *X* Top Level Comments per Page and the Last/First Page Displayed by Default:** Fill in the text box with the number of comments you want to appear on one page (default: 50). This setting can really help sites that receive a large number of comments. It

provides you the ability to break the long string of comments into several pages, which makes them easier to read and helps speed the load time of your site because the page isn't loading such a large number of comments at once. If you want the last (most recent) or first page of comments to display, choose Last or First from the drop-down list.

- **Comments Should Be Displayed with the Older/Newer Comments at the Top of Each Page:** From the drop-down list, choose Older or Newer. Older displays the comments on your site in the order oldest to newest. Newer does the opposite: It displays the comments on your site in the order newest to oldest.

Email Me Whenever

The two options in the Email Me Whenever section are enabled by default:

- **Anyone Posts a Comment:** Enabling this option means that you receive an email notification whenever anyone leaves a comment on your site. Deselect the check box if you don't want to be notified by email about every new comment.
- **A Comment Is Held for Moderation:** This option lets you receive an email notification whenever a comment is awaiting your approval in the comment moderation queue. (See Book 3, Chapter 4 for more information about the comment moderation queue.) You need to deselect this option if you don't want to receive this notification.

Before a Comment Appears

The two options in the Before a Comment Appears section tell WordPress how you want WordPress to handle comments before they appear in your blog:

- **Comment Must Be Manually Approved:** Disabled by default, this option keeps every single comment left on your site in the moderation queue until you, as the administrator, log in and approve it. Select this check box to enable this option.
- **Comment Author Must Have a Previously Approved Comment:** Enabled by default, this option requires comments posted by all first-time commenters to be sent to the comment moderation queue for approval by the administrator of the site. After comment authors have been approved for the first time, they remain approved for every comment thereafter (and this setting can't be changed). WordPress stores each comment author's email address in the database, and any future comments that match any stored emails are approved automatically. This feature is another measure that WordPress has built in to combat comment spam.

Comment Moderation

In the Comment Moderation section, you can set options to specify what types of comments are held in the moderation queue to await your approval.

To prevent spammers from spamming your blog with a *ton* of links, enter a number in the Hold a Comment in the Queue If It Contains *X* or More Links text box. The default number of links allowed is 2. Try that setting, and if you find that you're getting a lot of spam comments that contain links, consider dropping that number to 1, or even 0, to prevent those comments from being published on your site. Sometimes, legitimate commenters include a link or two in the body of their comments; after a commenter is marked as approved, she's no longer affected by this method of spam protection.

The large text box in the Comment Moderation section (not shown in Figure 2-4 because it's at the bottom of the page) lets you type keywords, URLs, email addresses, and IP addresses so that if they appear in comments, you want to hold those comments in the moderation queue for your approval.

Comment Blacklist

In this section, type a list of words, URLs, email addresses, and/or IP addresses that you want to flat-out ban from your site. Items placed here don't even make it into your comment moderation queue; the WordPress system filters them as spam. I'd give examples of blacklist words, but the words I have in my blacklist aren't family-friendly and have no place in a nice book like this one.

WHAT ARE AVATARS, AND HOW DO THEY RELATE TO WORDPRESS?

An *avatar* is an online graphical representation of a person. It's a small icon that people use to represent themselves on the web in areas where they participate in conversations, such as discussion forums and blog comments.

Gravatars are globally recognized avatars; they're avatars that you can take with you wherever you go. They appear alongside comments, posts, and discussion forums as long as the site you're interacting with is Gravatar-enabled. In October 2007, Automattic, the core group behind the WordPress platform, purchased the Gravatar service and integrated it into WordPress so that everyone could enjoy and benefit from the service.

Gravatars aren't automatic; you need to sign up for an account with Gravatar so that you can receive an avatar via your email address. Find out more about Gravatar by visiting `http://gravatar.com`.

Avatars

The final section of the Discussion Settings screen is Avatars. (See the nearby sidebar "What are avatars, and how do they relate to WordPress?" for information about avatars.) In this section, you can select different settings for the use and display of avatars on your site, as follows:

1. **For the Avatar Display option (see Figure 2-5), decide how to display avatars on your site.**

   Select the Show Avatars check box to have your blog display avatars.

2. **Next to the Maximum Rating option, select the radio button for the maximum avatar rating you want to allow for the avatars that do appear on your site.**

   This feature works much like the American movie-rating system. You can select G, PG, R, and X ratings for the avatars that appear on your site, as shown in Figure 2-5. If your site is family-friendly, you probably don't want it to display R- or X-rated avatars, so select G or PG.

3. **Select the radio button for a default avatar next to the Default Avatar option (see Figure 2-5).**

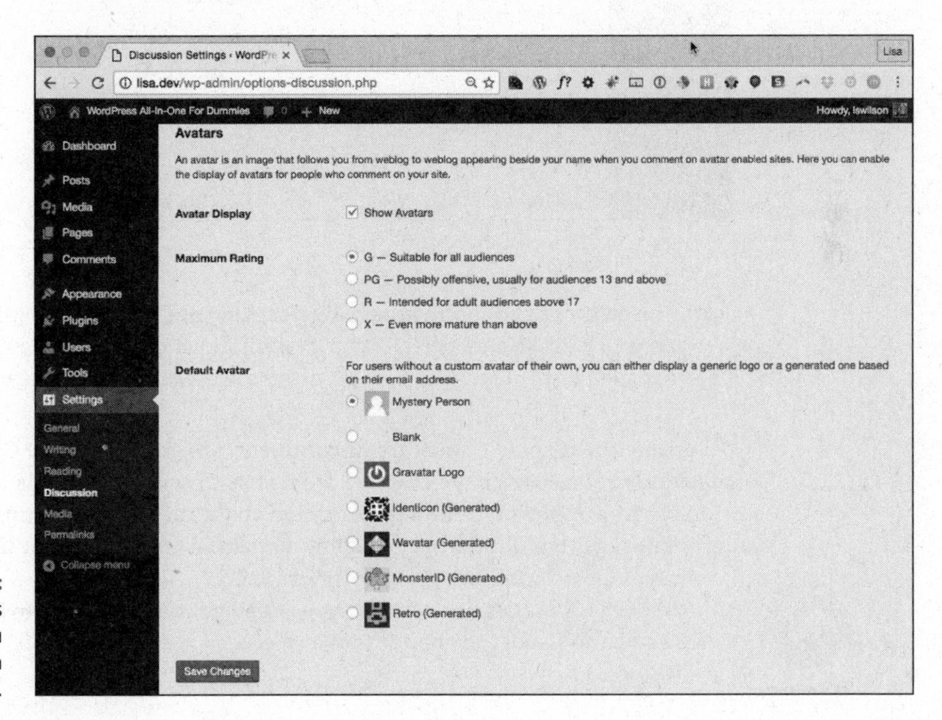

FIGURE 2-5: Default avatars that you can display in your blog.

Avatars appear in a couple of places:

» **The Comments screen of the Dashboard:** In Figure 2-6, the first two comments display either the commenter's avatar or the default avatar if the commenter hasn't created his or her own.

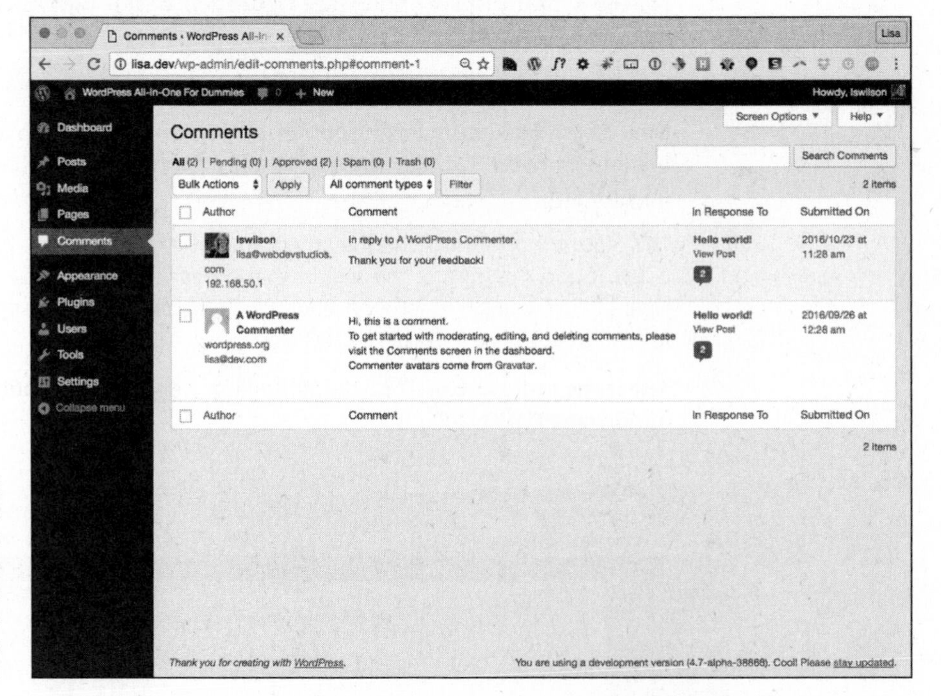

FIGURE 2-6: Comment authors' avatars appear in the Comments screen of the WordPress Dashboard.

» **The comments on individual blog posts in your blog:** Comments displayed on your website show the users' avatars. If a user doesn't have an avatar assigned from `http://gravatar.com/`, the default avatar appears.

To enable the display of avatars in comments on your blog, the Comments template (`comments.php`) in your active theme has to contain the code to display them. Hop on over to Book 6 to find information about themes and templates, including template tags that allow you to display avatars in your comment list.

REMEMBER

Click the Save Changes button after you set all your options on the Discussion Settings screen to make the changes take effect.

Media

The next item on the Settings menu is Media. Choose Settings⇨Media to make the Media Settings screen open. (See Figure 2-7.)

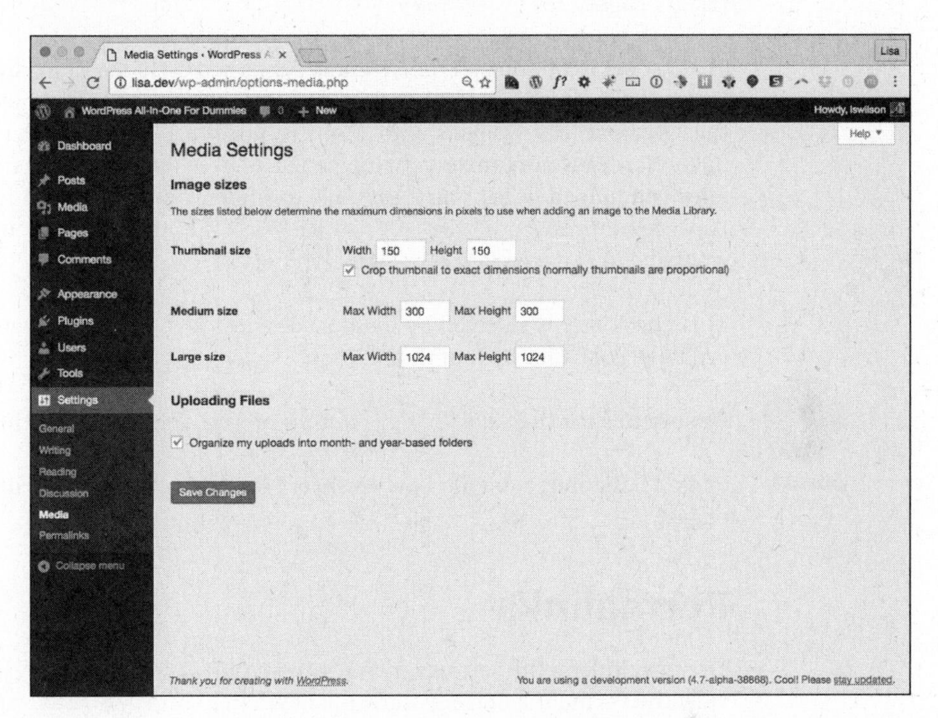

FIGURE 2-7: The Media Settings screen.

On the Media Settings screen, you can configure the options for how your image files (graphics and photos) are resized for use in your site.

The first set of options on the Media Settings page deals with images. WordPress automatically resizes your images for you in three sizes. The dimensions are referenced in pixels first by width and then by height. (The setting 150 x 150, for example, means 150 pixels wide by 150 pixels high.)

- **Thumbnail Size:** The default is 150 x 150; enter the width and height of your choice. Select the Crop Thumbnail to Exact Dimensions check box to resize the thumbnail to the exact width and height you specify. Deselect this check box to make WordPress resize the image proportionally.
- **Medium Size:** The default is 300 x 300; enter the width and height numbers of your choice.
- **Large Size:** The default is 1024 x 1024; enter the width and height numbers of your choice.

TIP

Book 4 goes into detail about WordPress themes and templates, including how you can add image sizes other than just these three. You can use these additional image sizes in and around your website. There's also a feature called Featured Image, which you can use in posts and articles that display on archive and search results pages.

The last set of options on the Media Settings screen is the Uploading Files section. Here, you can tell WordPress where to store your uploaded media files. Select the Organize My Uploads into Month- and Year-Based Folders check box to have WordPress organize your uploaded files in folders by month and by year. Files you upload in February 2017, for example, would be in the following folder: `/wp-content/uploads/2017/02/`. Likewise, files you upload in January 2017 would be in `/wp-content/uploads/2017/01/`.

This check box is selected by default; deselect it if you don't want WordPress to organize your files by month and year.

REMEMBER

Be sure to click the Save Changes button to save your configurations!

Book 4, Chapter 3 details how to insert images into your WordPress posts and pages.

Permalinks

The next link on the Settings menu is Permalinks. Choose Settings⇨Permalinks to view the Permalink Settings screen, shown in Figure 2-8.

Each WordPress post is assigned its own web page, and the address (or URL) of that page is called a *permalink.* Posts that you see in WordPress blogs usually have the post permalink in four typical areas:

- The title of the post
- The Comments link below the post
- A Permalink link that appears (in most themes) below the post
- The titles of posts appearing in a Recent Posts sidebar

Permalinks are meant to be permanent links to your posts (which is where the *perma* part of that word comes from, in case you're wondering). Ideally, the permalink of a post never changes. WordPress creates the permalink automatically when you publish a new post.

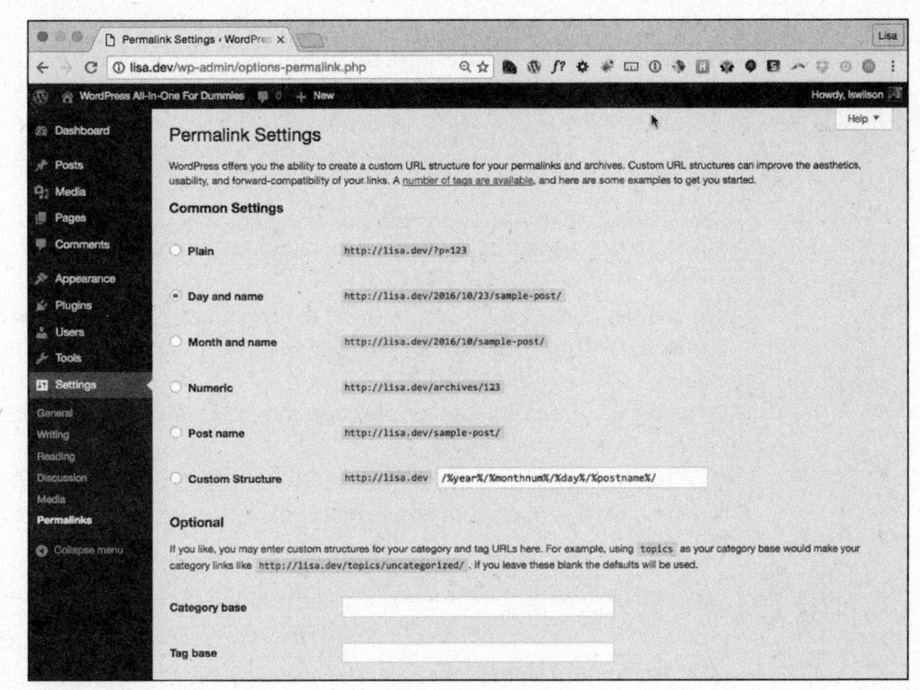

FIGURE 2-8: The Permalink Settings screen.

A plain post permalink in WordPress looks like this:

```
http://yourdomain.com/?p=100/
```

The `p` stands for *post*, and `100` is the ID assigned to the individual post. You can leave the permalinks in this format if you don't mind letting WordPress associate each post with an ID number.

WordPress, however, lets you take your permalinks to the beauty salon for a bit of makeover so that you can create pretty permalinks. You probably didn't know that permalinks could be pretty, did you?

WARNING

Changing the structure of your permalinks in the future affects the permalinks for all the posts on your site . . . new and old. Keep this fact in mind if you ever decide to change the permalink structure. An especially important reason: Search engines (such as Google and Yahoo!) index the posts on your site by their permalinks, so changing the permalink structure makes all those indexed links obsolete.

Making your post links pretty

Pretty permalinks are links that are more pleasing to the eye than standard links and, ultimately, more pleasing to search-engine spiders. (See Book 5 for an explanation of why search engines like pretty permalinks.) Pretty permalinks look something like this:

```
http://yourdomain.com/2017/01/01/pretty-permalinks
```

Break down that URL, and you see the date when the post was made, in year/month/day format. You also see the topic of the post.

To choose how your permalinks look, choose Settings⇨Permalinks in the Dashboard. The Permalink Settings screen opens. (Refer to Figure 2-8.)

On this page, you can find several options for creating permalinks:

- **Plain** (ugly permalinks): WordPress assigns an ID number to each blog post and creates the URL in this format: `http://yourdomain.com/?p=100`.
- **Day and Name** (pretty permalinks): For each post, WordPress generates a permalink URL that includes the year, month, day, and post slug/title: `http://yourdomain.com/2017/01/01/sample-post/`.
- **Month and Name** (also pretty permalinks): For each post, WordPress generates a permalink URL that includes the year, month, and post slug/title: `http://yourdomain.com/2017/01/sample-post/`.
- **Numeric** (not so pretty): WordPress assigns a numerical value to the permalink. The URL is created in this format: `http://yourdomain.com/archives/123`.
- **Post Name** (my preferred): WordPress takes the title of your post or page and generates the permalink URL from those words. If I were to create a page that contains my bibliography of books and give it the title *Books,* with the Post Name permalink structure, WordPress would create the permalink URL `http://lisasabin-wilson.com/books`. Likewise, a post titled *WordPress Is Awesome* would get a permalink URL like this `http://lisasabin-wilson.com/wordpress-is-awesome`.
- **Custom Structure:** WordPress creates permalinks in the format you choose. You can create a custom permalink structure by using tags or variables, as I discuss in the next section.

To create the pretty-permalink structure, select the Day and Name radio button; then click the Save Changes button at the bottom of the page.

Customizing your permalinks

A *custom permalink structure* is one that lets you define which variables you want to see in your permalinks by using the tags in Table 2-2.

TABLE 2-2 **Custom Permalinks**

Permalink Tag	Results
`%year%`	Four-digit year (such as 2017)
`%monthnum%`	Two-digit month (such as 01 for January)
`%day%`	Two-digit day (such as 30)
`%hour%`	Two-digit hour of the day (such as 15 for 3 p.m.)
`%minute%`	Two-digit minute (such as 45)
`%second%`	Two-digit second (such as 10)
`%postname%`	Text — usually, the post name — separated by hyphens (such as `making-pretty-permalinks`)
`%post_id%`	The unique numerical ID of the post (such as 344)
`%category%`	The text of the category name in which you filed the post (such as `books-i-read`)
`%author%`	The text of the post author's name (such as `lisa-sabin-wilson`)

If you want your permalink to show the year, month, day, category, and post name, select the Custom Structure radio button in the Customize Permalink Structure page, and type the following tags in the Custom Structure text box:

```
/%year%/%monthnum%/%day%/%category%/%postname%/
```

If you use this permalink format, a link for a post made on January 1, 2017, called WordPress All-in-One For Dummies and filed in the Books I Read category would look like this:

```
http://yourdomain.com/2017/01/01/books-i-read/wordpress-all-in-one-for-dummies/
```

REMEMBER

Be sure to include the slashes before tags, between tags, and at the very end of the string of tags. This format ensures that WordPress creates correct, working permalinks by using the correct rewrite rules located in the `.htaccess` file for your site. (See the following section for more information on rewrite rules and `.htaccess` files.)

REMEMBER

Don't forget to click the Save Changes button at the bottom of the Customize Permalink Structure screen; otherwise, your permalink changes aren't saved!

Making sure that your permalinks work with your server

After you set the format for the permalinks for your site by using any options other than the default, WordPress writes specific rules, or directives, to the `.htaccess` file on your web server. The `.htaccess` file in turn communicates to your web server how it should serve up the permalinks, according to the permalink structure you chose to use.

To use an `.htaccess` file, you need to know the answers to two questions:

- » Does your web server configuration use and give you access to the `.htaccess` file?
- » Does your web server run Apache with the `mod_rewrite` module?

If you don't know the answers, contact your hosting provider to find out.

If the answer to both questions is yes, proceed to the following section. If the answer is no, check out the "Working with servers that don't use Apache mod_rewrite" sidebar in this chapter.

Creating .htaccess files

You and WordPress work together in glorious harmony to create the `.htaccess` file that lets you use a pretty-permalink structure in your blog.

To create the `.htaccess` file, you need to be comfortable uploading files via SFTP and changing permissions. Turn to Book 2, Chapter 2 if you're unfamiliar with either of those tasks.

TIP

If `.htaccess` already exists, you can find it in the root of your directory on your web server — that is, the same directory where you find your `wp-config.php` file. If you don't see it in the root directory, try changing the options of your SFTP client to show hidden files. (Because the `.htaccess` file starts with a period [`.`], it may not be visible until you configure your SFTP client to show hidden files.)

If you don't already have an `.htaccess` file on your web server, follow these steps to create an `.htaccess` file on your web server and create a new permalink structure:

1. **Using a plain-text editor (such as Notepad for Windows or TextEdit for a Mac), create a blank file; name it `htaccess.txt` and upload it to your web server via SFTP.**
2. **After the file is uploaded to your web server, rename the file `.htaccess` (notice the period at the beginning), and make sure that it's writable by the server by changing permissions to either 755 or 777.**
3. **Create the permalink structure on the Customize Permalink Structure page of your WordPress Dashboard.**
4. **Click the Save Changes button at the bottom of the Customize Permalink Structure page.**

   WordPress inserts into the `.htaccess` file the specific rules necessary for making the permalink structure functional on your site.

If you follow the preceding steps correctly, you have an `.htaccess` file on your web server with the correct permissions set so that WordPress can write the correct rules to it. Your pretty-permalink structure works flawlessly. Kudos!

If you open the `.htaccess` file and look at it now, you see that it's no longer blank. It should have a set of code in it called *rewrite rules,* which looks something like this:

```
# BEGIN WordPress
<IfModule mod_rewrite.c>
RewriteEngine On
RewriteBase /
RewriteCond %{REQUEST_FILENAME} !-f
RewriteCond %{REQUEST_FILENAME} !-d
RewriteRule . /index.php [L]
</IfModule>

# END WordPress
```

TECHNICAL STUFF

I could delve deeply into `.htaccess` and all the things you can do with this file, but I'm restricting this chapter to how it applies to WordPress permalink structures. If you want to unlock more mysteries about `.htaccess`, check out "Comprehensive Guide to .htaccess" at `www.javascriptkit.com/howto/htaccess.shtml`.

WORKING WITH SERVERS THAT DON'T USE APACHE MOD_REWRITE

Using permalink structures requires your web hosting provider to have a specific Apache module option called `mod_rewrite` activated on its servers. If your web hosting provider doesn't have this item activated on its servers, or if you're hosting your site on a Windows server, the custom permalinks work only if you type **index.php** in front of any custom permalink tags.

Create the custom permalink tags like this:

```
/index.php/%year%/%month%/%date%/%postname%/
```

This format creates a permalink like this:

```
http://yourdomain.com/index.php/2017/01/01/wordpress-all-in-one-for-
    dummies
```

You don't need an `.htaccess` file to use this permalink structure.

Creating Your Personal Profile

To personalize your blog, visit the Profile screen of your WordPress Dashboard.

To access the Profile screen, hover over the Users link in the Dashboard navigation menu, and click the Your Profile link. The Profile screen appears, as shown in Figure 2-9.

Here are the settings on this page:

- **Personal Options:** In the Personal Options section, you can set these preferences for your site:
  - *Visual Editor:* Select this check box to indicate that you want to use the Visual Editor when writing your posts. The Visual Editor refers to the formatting options you find on the Write Post screen (discussed in detail in Book 4, Chapter 1). By default, the check box is deselected, which means that the Visual Editor is off. To turn it on, select the check box.
  - *Admin Color Scheme:* These options set the colors for your Dashboard. The default is the Gray color scheme, but you have other color options: Light, Blue, Coffee, Ectoplasm, Midnight, Ocean, and Sunrise.

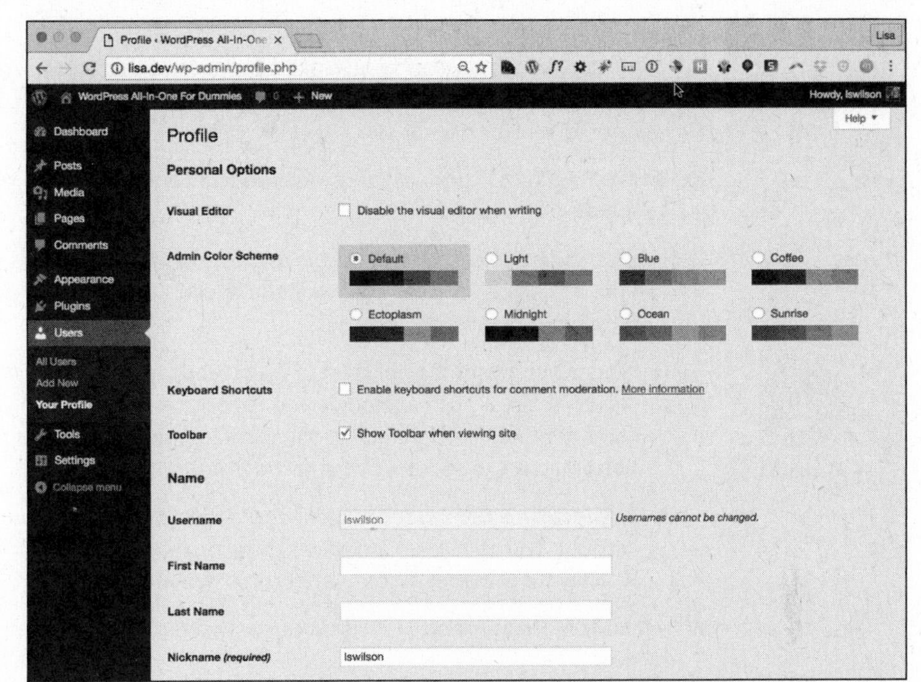

FIGURE 2-9: Establish your profile details on the Profile screen.

- *Keyboard Shortcuts:* This check box enables you to use keyboard shortcuts for comment moderation. To find out more about keyboard shortcuts, click the More Information link; you're taken to the Keyboard Shortcuts page (`https://codex.wordpress.org/Keyboard_Shortcuts`) of the WordPress codex, which offers some helpful information
- *Toolbar:* This setting allows you to control the location of the admin toolbar (see Book 3, Chapter 1) on your site. By default, the admin toolbar displays at the top of every page of your site when you're viewing it in your browser. It's important to understand that the admin toolbar appears only to users who are logged in. Regular visitors who aren't logged in to your site can't see the admin toolbar.

» **Name:** Input personal information, such as your first name, last name, and nickname, and specify how you want your name to appear publicly. Fill in the text boxes with the requested information.

The rest of the options aren't shown in Figure 2-9; you have to scroll down to see them.

» **Contact Info:** In this section, provide your email address and other contact information to tell your visitors who you are and where they can contact you. Your email address is the only required entry in this section. This address is the one WordPress uses to notify you when you have new comments or new user registrations on your blog. Make sure to use a real email address so that you get these notifications. You can also insert your website URL into the website text field.

- **About Yourself:** Provide a little bio for yourself and change the password for your blog, if you want:
  - *Biographical Info:* Type a short bio in the Biographical Info text box. This information can appear publicly if you're using a theme that displays your bio, so be creative!

    When your profile is published to your website, anyone can view it, and search engines, such as Google and Yahoo!, can pick it up. Always be careful with the information in your profile. Think hard about the information you want to share with the rest of the world!
  - *Profile Picture:* Display the current photo that you've set in your Gravatar account. You can set up a profile picture or change your existing one, within your Gravatar account at `http://gravatar.com/`.
- **Account Management:** Manage your password and user sessions, as follows:
  - *New Password:* When you want to change the password for your blog, click the Generate Password button in the New Password section. You can use the password that WordPress generates for you or type your own password in the text field that appears.

    Directly below the New Password text field is a password helper, where WordPress helps you create a secure password. It alerts you if the password you chose is too short or not secure enough by telling you that it's Weak or Very Weak. When creating a new password, use a combination of letters, numbers, and symbols to make it hard for anyone to guess (such as *b@Fmn2quDtnSLQblhml%jexA*). When you create a password that WordPress thinks is a good one, it lets you know by saying that the password is Strong.

    Change your password frequently. Some people on the Internet make it their business to attempt to hijack sites for their own malicious purposes. If you change your password monthly, you lower your risk by keeping hackers guessing.
  - *Sessions:* If you're logged into your site on several devices, you can log yourself out of those locations by clicking the Log Out Everywhere Else button. This option keeps you logged in at your current location but logs you out of any other location where you may be logged in. If you're not logged in anywhere else, the button is inactive, and a message appears that says `You are only logged in at this location.`

REMEMBER

When you finish setting all the options on the Profile screen, don't forget to click the Update Profile button to save your changes.

Setting Your Blog's Format

In addition to setting your personal settings in the Dashboard, you can manage the day-to-day maintenance of your site. The following sections take you through the links to these pages on the Dashboard navigation menus.

Posts

Hover your mouse over the Posts link on the navigation menu to reveal a submenu list with four links: All Posts, Add New, Categories, and Tags. Each link gives you the tools you need to publish content to your blog:

- **All Posts:** Opens the Posts screen, where a list of all the saved posts you've written on your site appears. On this screen, you can search for posts by date, category, or keyword. You can view all posts, only posts that have been published, or just posts that you've saved but haven't published *(drafts).* You can also edit and delete posts from this page. Check out Book 4, Chapter 1 for more information on editing posts on your site.
- **Add New:** Opens the Add New Post screen, where you can compose your posts, set the options for each post (such as assigning a post to a category, or making it a private or public post), and publish the post to your site. You can find more information on posts, post options, and publishing in Book 4, Chapter 1.

TIP

You can also get to the Add New Post screen by clicking the Add New button on the Posts screen or by clicking the +New link on the admin toolbar and selecting Post.

- **Categories:** Opens the Categories screen, where you can view, edit, add, and delete categories on your site. Find more information on creating categories in Book 3, Chapter 5.
- **Tags:** Opens the Tags screen in your WordPress Dashboard, where you can view, add, edit, and delete tags on your site. Book 3, Chapter 5 provides more information about tags and using them on your blog.

Media

Hover your mouse over the Media link on the navigation menu to reveal a submenu list with two links:

- **Library:** Opens the Media Library screen. On this screen, you can view, search, and manage all the media files you've ever uploaded to your WordPress site.
- **Add New:** Opens the Upload New Media screen, where you can use the built-in uploader to transfer media files from your computer to the media directory in WordPress. Book 4, Chapters 3 and 4 take you through the details of how to upload images, videos, and audio files by using the WordPress upload feature.

TIP

You can also get to the Upload New Media screen by clicking the Add New button on the Media Library screen or by clicking the +New link on the admin toolbar and selecting Media.

Pages

People use this feature to create pages on their sites such as About Me or Contact Me. Turn to Book 4, Chapter 2 for more information on pages. Hover your mouse over the Pages link on the navigation menu to reveal a submenu list with two links:

- **All Pages:** Opens the Pages screen, where you can search, view, edit, and delete pages in your WordPress site.
- **Add New:** Opens the Add New Page screen, where you can compose, save, and publish a new page on your site. Book 4, Chapter 2 describes the difference between a post and a page. The difference is subtle, but posts and pages are very different!

  You can also get to the Add New Page screen by clicking the Add New button on the Pages screen or by clicking the +New link on the admin toolbar and selecting Page.

Comments

Comments in the navigation menu don't have a submenu list of links. You simply click Comments to open the Comments screen, where WordPress gives you the options to view the following:

- **All:** Shows all comments that currently exist on your site, including approved, pending, and spam comments
- **Pending:** Shows comments that you haven't yet approved but are pending in the moderation queue
- **Approved:** Shows all comments that you previously approved
- **Spam:** Shows all the comments that are marked as spam
- **Trash:** Shows comments that you marked as Trash but haven't deleted permanently from your site

Book 3, Chapter 4 gives you details on how to use the Comments section of your WordPress Dashboard.

Appearance

When you hover your mouse over the Appearance link on the Dashboard navigation menu, a submenu list of links appears, displaying the following links:

- **Themes:** Opens the Themes screen, where you can manage the themes available on your site. Check out Book 6, Chapter 2 to find out about using themes on your WordPress site and managing themes on this page.
- **Customize:** Opens the Customizer screen, where you can edit various features available in the active theme on your site.
- **Widgets:** Opens the Widgets screen, where you can add, delete, edit, and manage the widgets that you use on your site.
- **Menus:** Opens the Menus screen, where you can build navigation menus that will appear on your site. Book 6, Chapter 1 provides information on creating menus by using this feature.
- **Header:** Opens the Header Image screen in the Customizer, where you can upload an image to use in the *header* (or top) of your WordPress site. This menu item and screen exist only if you're using a theme that has activated the Custom Header feature (covered in Book 6). The Twenty Seventeen theme is activated by default on all new WordPress sites, which is why I include this menu item in this list. Not all WordPress themes use the Custom Header feature, so you don't see this menu item if your theme doesn't take advantage of that feature.

» **Background:** Opens the Background Image screen in the Customizer, where you can upload an image to use as the background of your WordPress site design. Like the Custom Header option, the Custom Background option exists on the Appearances menu only if you have a theme that has activated the custom background feature (covered in Book 6).

» **Editor:** Opens the Edit Themes screen, where you can edit your theme templates. Book 6 has extensive information on themes and templates.

Uploading custom header and background images helps you individualize the visual design of your website. You can find more information on tweaking and customizing your WordPress theme in Book 6, as well as a great deal of information about how to use WordPress themes (including where to find, install, and activate them in your WordPress site) and detailed information about using WordPress widgets to display the content you want.

Book 6 provides information about WordPress themes and templates. You can dig deep into WordPress template tags and tweak an existing WordPress theme by using Cascading Style Sheets (CSS) to customize your theme a bit more to your liking.

Plugins

The next item in the navigation menu is Plugins. Hover your mouse over the Plugins link to view the submeu list:

» **Installed Plugins:** Opens the Plugins screen, where you can view all the plugins currently installed on your site. On this page, you also have the ability to activate, deactivate, and delete plugins on your blog. (Book 7 is all about plugins.)

» **Add New:** Opens the Add Plugins screen, where you can search for plugins from the official WordPress Plugin Directory by keyword, author, or tag. You can also install plugins directly to your site from the WordPress Plugin Directory; find out all about this exciting feature in Book 7, Chapter 2.

» **Editor:** Opens the Edit Plugins screen, where you can edit the plugin files in a text editor. Don't plan to edit plugin files unless you know what you're doing (meaning that you're familiar with PHP and WordPress functions). Head over to Book 7, Chapter 4 to read more information on editing plugin files.

Users

The Users submenu list has three links:

- **All Users:** Go to the Users screen, where you can view, edit, and delete users on your WordPress site. Each user has a unique login name and password, as well as an email address assigned to his account. You can view and edit a user's information on the Users page.
- **Add New:** Opens the Add New User screen, where you can add new users to your WordPress blog. Simply type the user's username, first name, last name, email (required), website, and a password in the fields provided, and click the Add User button. You can also select whether you want WordPress to send login information to the new user by email. If you want, you can also assign a new role for the new user. Turn to the section "Configuring the Settings" earlier in this chapter for more info about user roles.
- **Your Profile:** Turn to the "Creating Your Personal Profile" section earlier in this chapter for more information about creating a profile page.

Tools

The last item on the navigation menu (and subsequently in this chapter!) is Tools. Hover your mouse over the Tools link to view the submenu list of links that includes

- **Available Tools:** Opens the Tools screen on your Dashboard. WordPress comes packaged with two extra features that you can use on your site, if needed. The features are Press This and Category/Tag Converter.
- **Import:** Clicking this link opens the Import screen of your Dashboard. WordPress allows you to import from a different publishing platform. This feature is covered in depth in Book 2, Chapter 7.
- **Export:** Clicking this menu item opens the Export screen of your Dashboard. WordPress allows you to export your content from WordPress so that you can import it into a different platform or to another WordPress-powered site.

» Allowing new users to register

» Adding a new user

» Making changes to user profiles

» Using tools to manage multiauthor sites

Chapter 3
Managing Users and Multiple Authors

A *multiauthor site* involves inviting others to coauthor, or contribute articles, posts, pages, or other content to your site. You can expand the offerings on your website by using multiauthor publishing because you can have several different people writing on different topics or offering different perspectives on the same topic. Many people use this type of site to create a collaborative writing space on the web, and WordPress doesn't limit you in the number of authors you can add to your site.

Additionally, you can invite other people to register as *subscribers,* who don't contribute content but are registered members of the site, which can have benefits, too. (You could make some content available to registered users only, for example.)

This chapter takes you through the process of adding users to your site, takes the mystery out of the different user roles and capabilities, and gives you some tools for managing a multiauthor website.

Understanding User Roles and Capabilities

Before you start adding new users to your site, you need to understand the differences among the user roles, because each user role is assigned a different level of access and grouping of capabilities to your site, as follows:

- **Subscriber:** Subscriber is the default role. Maintain this role as the one assigned to new users, particularly if you don't know who's registering. Subscribers get access to the Dashboard page, and they can view and change the options in their profiles on the Profile screen. (They don't have access to your account settings, however — only to their own.) Each user can change her username, email address, password, bio, and other descriptors in her user profile. The WordPress database stores subscribers' profile information, and your site remembers them each time they visit, so they don't have to complete the profile information each time they leave comments on your site.
- **Contributor:** In addition to the access subscribers have, contributors can upload files and write, edit, and delete their own posts. Contributors can write posts, but they can't publish the posts; the administrator reviews all contributor posts and decides whether to publish them. This setting is a nice way to moderate content written by new authors.
- **Author:** In addition to the access contributors have, authors can publish and edit their own posts.
- **Editor:** In addition to the access authors have, editors can moderate comments, manage categories, manage links, edit pages, and edit other authors' posts.
- **Administrator:** Administrators can edit all the options and settings in the WordPress blog.
- **Super Admin:** This role exists only when you have the multisite feature activated in WordPress. See Book 8 for more about the multisite feature.

Table 3-1 gives you an at-a-glance reference for the basic differences in roles and capabilities for WordPress users.

TABLE 3-1 **WordPress User Roles and Capabilities**

	Super Admin	Administrator	Editor	Author	Contributor	Subscriber
Manage multisite features	Yes	No	No	No	No	No
Add/edit users	Yes	Yes	No	No	No	No
Add/edit/install plugins	Yes	Yes	No	No	No	No
Add/edit/install themes	Yes	Yes	No	No	No	No
Manage comments	Yes	Yes	Yes	No	No	No
Manage categories, tags, and links	Yes	Yes	Yes	No	No	No
Publish posts	Yes	Yes	Yes	Yes	No (moderated)	No
Edit published posts	Yes	Yes	Yes	No	No	No
Edit others' posts	Yes	Yes	Yes	No	No	No
Edit own posts	Yes	Yes	Yes	Yes	Yes	No
Publish pages	Yes	Yes	Yes	No	No	No
Read	Yes	Yes	Yes	Yes	Yes	Yes

TIP

Table 3-1 doesn't offer exhaustive information, by any means, but it covers the basic user roles and capabilities for WordPress or the most common capabilities for each user role. For a full list of user roles and capabilities, check out the WordPress Codex at `https://codex.wordpress.org/Roles_and_Capabilities`.

Allowing New User Registration

As you can see in Table 3-1, each user level has a different set of capabilities. Book 3, Chapter 2 discusses the General Settings of the WordPress Dashboard, in which you set the default role for users who register on your website. Keep the default role set to Subscriber because when you open registration to the public, you don't always know who's registering until after they register — and you don't want to arbitrarily hand out higher levels of access to the settings of your website unless you know and trust the user.

When users register on your website, you, as the Administrator, get an email notification (sent to the email address you set on the General Settings screen) so you always know when new users register, and you can then go to your Dashboard and edit the user to set his role any way you see fit.

REMEMBER

New users can register on your site only after you enable the Anyone Can Register option on the General Settings screen within your Dashboard (Book 3, Chapter 2). If you don't have this option enabled, users see a message on the Registration page telling them that registration isn't allowed, as shown in Figure 3-1.

FIGURE 3-1: The message to users that registration isn't allowed.

By the way, the direct URL for registration on a blog that has registration enabled is `http://`*`yourdomain.com`*`/wp-login.php?action=register`. With registration enabled (in the General Settings), a user sees a form inviting her to input her desired username and email address. After she does, she gets a confirmation notice in her inbox including an authorization link that she must click to authenticate her registration.

After a user has registered, you, as the site Administrator, can manage her user account and assign a user role. (Refer to Table 3-1.)

Adding New Users Manually

Allowing new users to register by using the WordPress registration interface is only one way to add users to your site. As the site Administrator, you have the ability to add new users manually by following these steps:

1. **Log in to your WordPress Dashboard by inputting your username and password in the form at** `http://`*`yourdomain.com`*`/wp-admin`**.**
2. **Click the Add New link in the Users submenu of the Dashboard.**

   The Add New User screen loads, as shown in Figure 3-2.

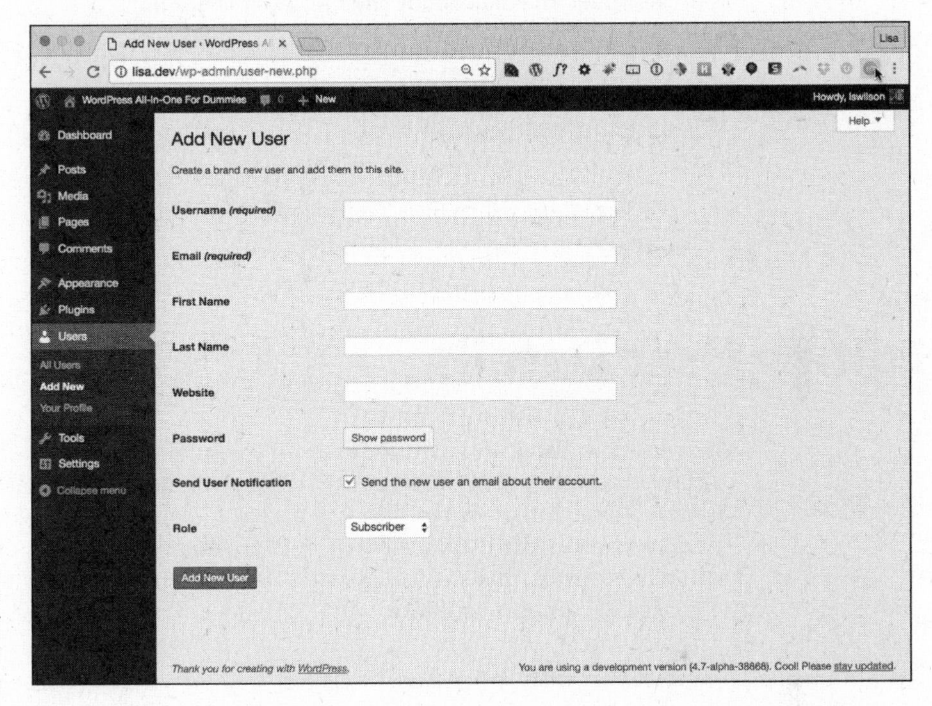

FIGURE 3-2: The Add New User screen of the WordPress Dashboard.

3. **Enter the username in the Username text box.**

   You can't skip this text box. The new user types this username when he's prompted to log in to your site.
4. **Enter the user's email address in the Email text box.**

   You can't skip this text box, either. The user receives notifications from you and your site at this email address.
5. **Enter the user's first name in the First Name text box.**
6. **Enter the user's last name in the Last Name text box.**
7. **Enter the URL of the user's website in the Website text box.**
8. **Click the Show Password button.**

   WordPress provides a random, strong password for you — or you can type your own password in the text field. WordPress provides a strength indicator that gives you an idea of how *strong,* or secure, your chosen password is. You want secure passwords so that no one can easily guess them, so make the password at least seven characters long and use a combination of letters, numbers, and symbols (such as @, #, $, and ^).

9. **If you want the user to receive his password by email, select the Send the New User an Email about Their Account check box.**

10. **From the Role drop-down list, choose Subscriber, Contributor, Author, Editor, or Administrator.**

11. **Click the Add New User button.**

    The Add New User screen loads, and the email notification is sent to the user you just added. When the screen loads, all the fields are cleared, allowing you to add another new user if you want.

Editing User Details

After users register and settle into their accounts on your site, you, as the site Administrator, have the ability to edit their accounts. You may never have to edit user accounts at all, but you have the option to do so if you need to. Most often, users can access the details of their own accounts and change email addresses, names, passwords, and so on. Following are some circumstances under which you may need to edit user accounts:

- **Edit user roles.** When a user registers, you may want to increase her role, or level of access, on your site; promote an existing user to Administrator; or demote an existing Administrator or Editor a notch or two.
- **Edit user emails.** If a user loses access to the email account that she registered with, she may ask you to change her account email address so that she can access her account notifications again.
- **Edit user passwords.** If a user loses access to the email account with which she registered, she can't use WordPress's Lost Password feature, which allows users to gain access to their account password through email recovery. In that case, a user may ask you to reset her password for her so that she can log in and access her account again.

In any of these circumstances, you can make the necessary changes by clicking the Users link on the Users menu on your WordPress Dashboard, which loads the Users screen, shown in Figure 3-3.

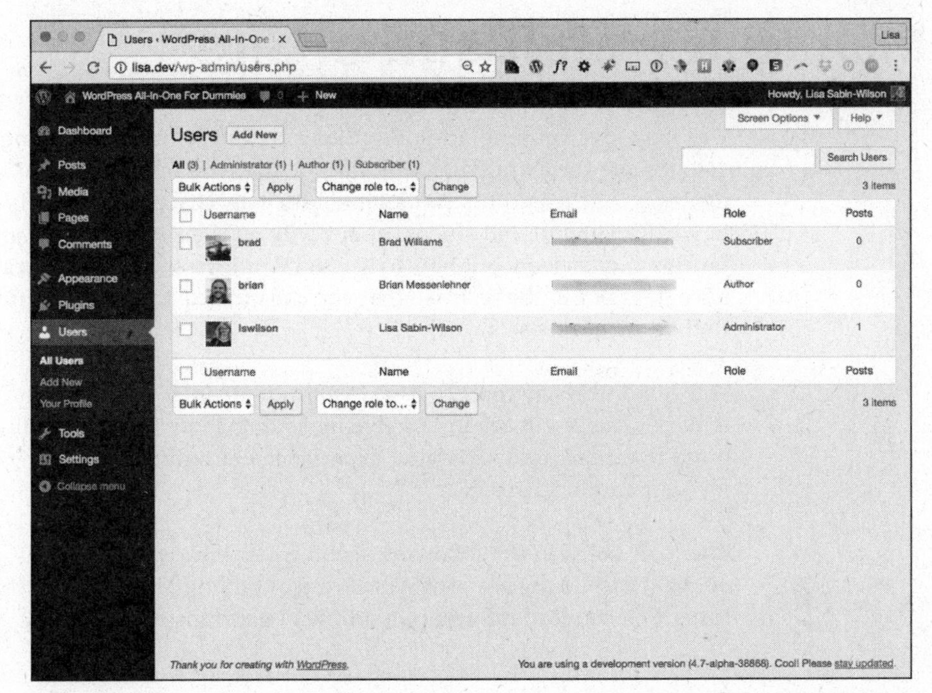

FIGURE 3-3: The Users screen lets you manage all the users on your site.

Figure 3-3 shows the Users screen on a site that has multiple users who have different levels of access, or roles. (The email addresses are blurred in Figure 3-3 to protect the users' privacy.)

When you hover your mouse over the name of a user, an Edit link appears below the user listing. Click that Edit link to access the Edit User screen, where you can edit different pieces of information for that user, including

- **Personal Options:** These options include Visual Editor, Color Scheme, Keyboard Shortcuts, and Toolbar preferences.
- **Name:** Specify a user's role, first and last names, nickname, and display name.
- **Contact Info:** These options include the user's email address and website.
- **Biographical Info:** This section provides a few lines of biographical info for the user (optional, but some WordPress themes display authors' biographies). This section also displays the user's profile picture.
- **New Password:** Here, you can change the password for the user.

The Edit User screen looks the same, and has the very same features, as the Profile screen that you deal with in Book 3, Chapter 2. Feel free to visit that chapter to get the lowdown on the options and settings on this screen.

Managing a Multiauthor Site

You may love running a multiauthor site, but it has its challenges. The minute you become the owner of a multiauthor site, you immediately assume the role of manager for the authors you invited into your space. At times, those authors look to you for support and guidance, not only on their content management, but also for tips and advice about how to use the WordPress interface. It's a good thing you have this book at the ready so that you can offer up the gems of information you're finding within these pages!

You can find many tools to assist you in managing a multiauthor site, as well as making your site more interactive by adding some features, which can make it a more rewarding and satisfying experience not only for you and your readers, but for your authors as well.

The tools listed in the following sections come by way of plugins, which are add-ons that extend the scope of WordPress by adding different functionality and features. You can find information on the use and installation of plugins in Book 7.

Tools that help authors communicate

When you're running a multiauthor site, communication is crucial for sharing information, giving and receiving inspiration, and making certain that no two authors are writing the same article (or similar articles) on your site. Use the following tools to manage the flow of communication among everyone involved:

- **Post Status Notifications:** In "Understanding User Roles and Capabilities" earlier in this chapter, I mention that the role of Contributor can write and save posts to your site, but those posts don't get published to the site until an Administrator approves them. This plugin notifies the Administrator, via email, when a new post is submitted for review. Additionally, the contributing author gets an email notification when an Administrator has published the post to the site.

  `https://wordpress.org/plugins/wpsite-post-status-notifications`

- **Editorial Calendar:** This plugin gives you an overview of scheduled posts, post authors, and the dates when you scheduled the posts to publish to your site. This plugin can help you prevent multiple author posts from publishing too close together or, in some cases, right on top of one another; you simply reschedule posts by using a drag-and-drop interface.

  `https://wordpress.org/plugins/editorial-calendar`

- **Email Users:** This plugin allows you to send emails to all registered users of your site, and users can send emails back and forth to one another by using the plugin interface of the Dashboard. This tool enables the authors and users on your multiauthor site to keep in touch and communicate with one another.

  `https://wordpress.org/plugins/email-users`

- **Author Customization:** This plugin expands author management with features such as per-post author information and rich user-biography editing.

  `https://wordpress.org/plugins/author-customization`

- **Dashboard Notepad:** This plugin gives you a widget that appears on your main Dashboard page and allows you and other users (depending on the user roles that you set in the plugin options) to leave notes for one another. You can use this plugin to ask and answer questions and to create to-do lists for your authors.

  `https://wordpress.org/plugins/dashboard-notepad`

Tools to promote author profiles

One way to operate a successful multiauthor site involves taking every opportunity to promote your authors and their information. Authors often get involved in posting content on other websites, in addition to yours, for exposure. The plugins in this list give you tools to promote authors bios, links, social network feeds, and more:

- **Authors Widget:** This plugin gives you a widget you can place on your site that displays a list of authors along with the number of posts they've published, a link to the RSS feeds for their posts, and their names and photos (see Figure 3-4).

  `https://wordpress.org/plugins/authors`

- **NS Author Widget:** This plugin provides a widget you can use to display a single author's information on the page that displays a post he has written. The widget displays the author's name, photo, and description, as well as links to all his posts. The widget also allows the author to include links to his various social media profiles (such as Twitter, Facebook, and so on).

  `https://wordpress.org/plugins/ns-author-widget`

- **Author Spotlight:** This plugin provides a widget that you can place in your sidebar, displaying the profile of the author of the post being viewed. The author information automatically appears on a single post page and displays the profile of the author of the post.

  `https://wordpress.org/plugins/author-profile`

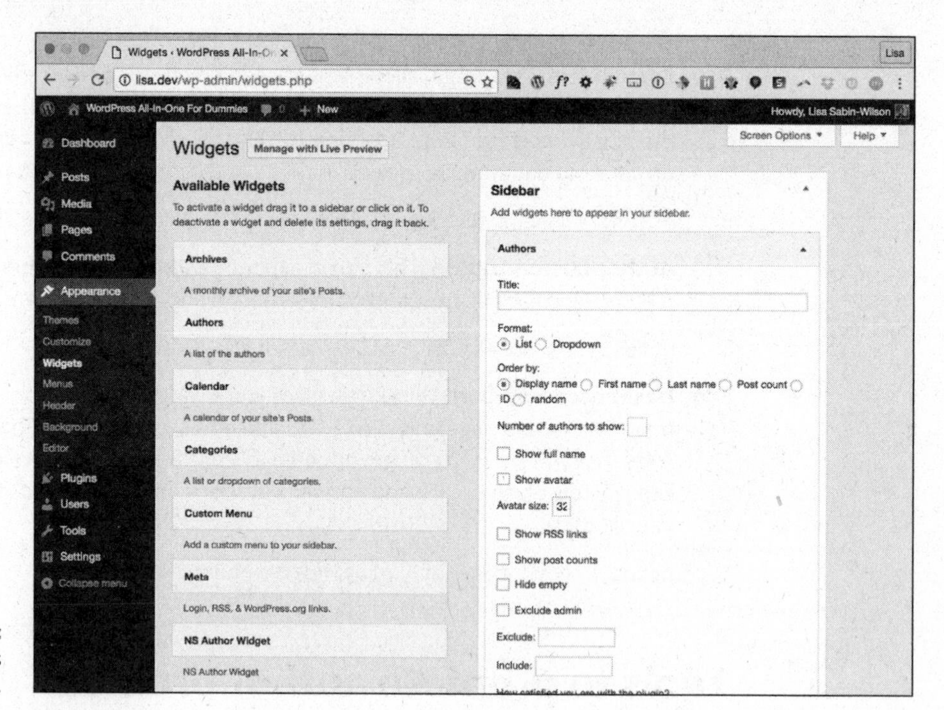

FIGURE 3-4: The Authors Widget options.

- **Author Recent Posts:** This plugin gives you a widget that allows an author to display a listing of her recent posts in the sidebar of posts she's published on your site.

  `https://wordpress.org/plugins/author-recent-posts`

Tools to manage multiauthor blog posts

The plugins listed in this section can help you, the site Administrator, manage your group of authors and registered users by giving you some tools to track users' activity, list their posts, and stay up to date and notified when your authors publish new content:

- **Co-Authors Plus:** This plugin allows you to assign multiple authors to one post, which you may find especially helpful when you have several authors collaborating on one article, allowing those authors to share the byline and credit.

  `https://wordpress.org/plugins/co-authors-plus`

- **Authors Post Widget:** This plugin provides a very easy way to show a list of authors on a site with a count of the number of posts each author has published.

  `https://wordpress.org/plugins/authors-posts-widget/`

- **Co-Authors Plus:** This plugin provides you with a handy tool that allows you to assign multiple author bylines per post. It also allows you to create guest author bylines without having to create a full user profile for guest authors.

  `https://wordpress.org/plugins/co-authors-plus/`

- **Audit Trail:** This plugin records the actions of the registered users on your site, such as when they log in or log out, when they publish posts and pages, and when they visit pages within your site. As the site Administrator, you can keep track of the actions your authors and users take on your website.

  `https://wordpress.org/plugins/audit-trail`

» Using comments to interact with readers

» Working with comments and trackbacks

» Using Akismet to help combat spam

Chapter 4

Dealing with Comments and Spam

One of the most exciting aspects of publishing with WordPress is getting feedback from your readers on articles you publish to your site. Feedback, also known as *blog comments,* is akin to having a guestbook on your site.

People leave notes for you that are published to your site, and through these notes, you can respond to and engage your readers in conversation about the topic. Having this function on your site allows you to expand the thoughts and ideas you present in your posts by giving readers the opportunity to add their two cents' worth.

In this chapter, you can decide whether to allow comments on your site, figure out how to manage those comments, use trackbacks, and discover the negative aspects of allowing comments (such as spam).

Deciding to Allow Comments on Your Site

Some publishers say that a blog without comments isn't a blog at all because the point of having a blog, in some minds, is to foster communication and interaction between the site authors and the readers. This belief is common in the publishing community because experiencing visitor feedback via comments is part of what's made Internet publishing so popular. Allowing comments is a personal choice, however, and you don't have to allow them if you don't want to.

Positive aspects of allowing comments

Allowing comments on your site lets audience members actively involve themselves in your site by creating a discussion and dialogue about your content. Mostly, readers find commenting to be a satisfying experience when they visit sites because comments make them part of the discussion.

Depending on the topic you write about, allowing comments sends the message that you, as the author/owner of the site, are open to the views and opinions of your readers. Having a comment form on your site that readers can use to leave their feedback on your articles (such as the one shown in Figure 4-1) is like having a great big Welcome to My Home sign on your site; it invites users in to share thoughts and participate in discussions.

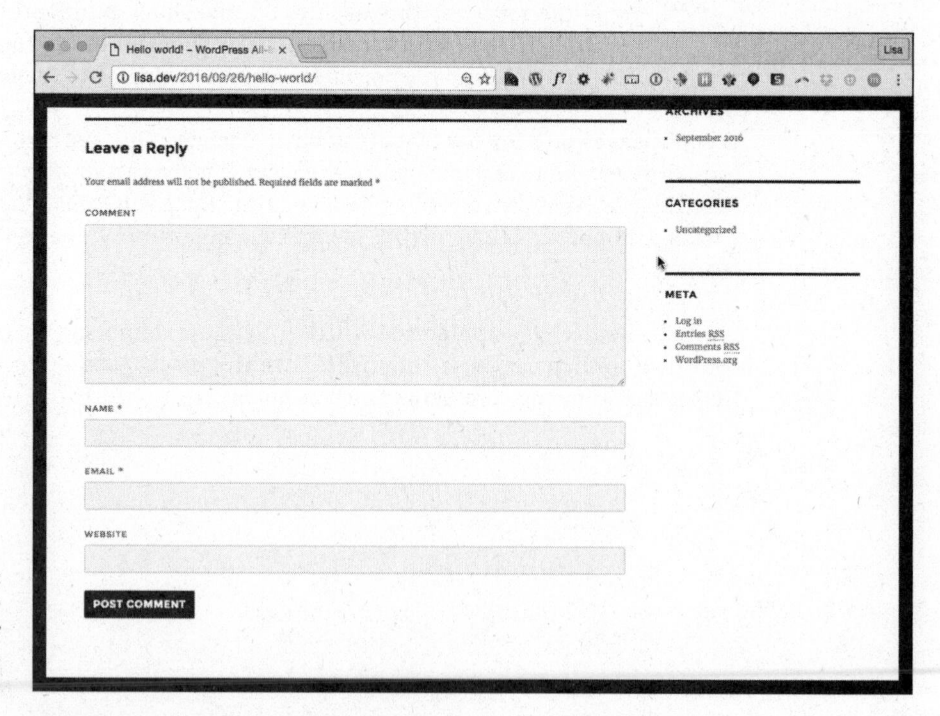

FIGURE 4-1: Readers use the Leave a Reply form to share their comments.

If you want to build a community of people who come back to your site frequently, respond to as many comments that your readers leave on your posts as possible. When people take the time to leave you a comment on your content, they like to know that you're reading it, and they appreciate hearing your feedback to them. Also, open comments keep discussions lively and active on your site. Figure 4-2 illustrates what comments look like after they're published to your site. (*Note:* The actual design and layout of the comments on sites varies from theme to theme; you can find information on theme design in Book 6.)

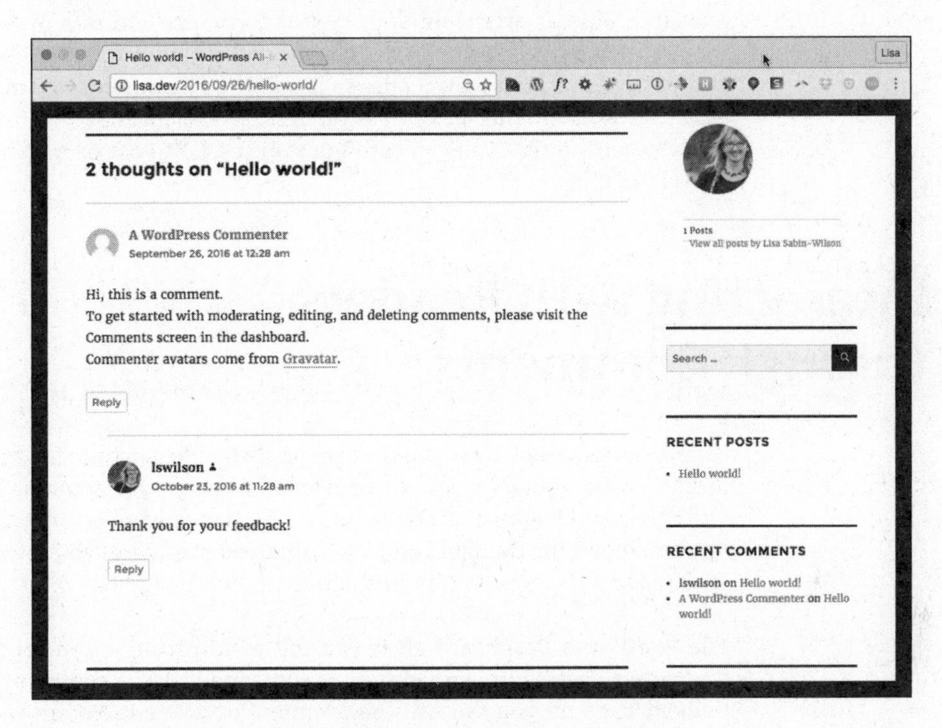

FIGURE 4-2: Visitors comment on a post.

Reasons to disallow comments

Under certain circumstances, you may not want to allow readers to leave comments freely on your site. If you wrote a post on a topic that's considered to be very controversial, for example, you may not want to invite comments, because the topic may incite flame wars or comments that are insulting to you or your readers. If you're not interested in the point of view or feedback of readers on your site, or if your content doesn't really lend itself to reader feedback, you may decide to disallow comments.

REMEMBER

In making the decision to have comments, you have to be prepared for the fact that not everyone is going to agree with what you write — especially if you're writing on a topic that invites a wide array of opinions, such as politics or religion. As a site owner, you make the decision ahead of time about whether you want readers dropping in and leaving their own views, or even disagreeing with you about yours (sometimes vehemently!).

If you're on the fence about whether to allow comments, the WordPress platform allows you to toggle that decision on a per-post basis. Therefore, each time you publish a post or article on your website, you can indicate in Post Options (on the Add New Post screen of your Dashboard) whether this particular post should allow discussion. You may choose to disallow all comments on your site (a setting that you can configure in Discussion Settings) or disallow them for only certain posts (a setting that you can configure on the Edit Post page, which I talk about in Book 4, Chapter 1).

Interacting with Readers through Comments

People can leave notes for you that are published to your site, and you can respond to and engage your readers in conversation about the topic at hand. (Refer to Figure 4-1 and Figure 4-2.) Having this function on your site creates the opportunity to expand the thoughts and ideas that you present in your blog post by giving your readers the opportunity to share their own thoughts.

The WordPress Dashboard gives you full administrative control of who can and can't leave comments. In addition, if someone leaves a comment that has questionable content, you can edit the comment or delete it. You're also free to disallow comments on your blog. The Discussion Settings screen of your Dashboard contains all the settings for allowing or disallowing comments on your site. See Book 3, Chapter 2 to dig into those settings, what they mean, and how you can use them to configure the exact interactive environment that you want for your site.

Tracking back

The best way to understand trackbacks is to think of them as comments except for one thing: *Trackbacks* are comments left on your blog by other blogs, not by actual people. Although this process may sound mysterious, it's actually perfectly reasonable.

A trackback happens when you make a post on your site and, within that post, you provide a link to a post made by another author on a different site. When you publish that post, your site sends a sort of electronic memo to the site you linked to. That site receives the memo and posts an acknowledgment of receipt in a comment within the post that you linked to on that site. Trackbacks work between most blogging platforms — between WordPress and Blogger, for example, as well as between WordPress and Typepad.

That memo is sent via a *network ping* (a tool used to test, or verify, whether a link is reachable across the Internet) from your site to the site you link to. This process works as long as both sites support trackback protocol. Trackbacks can also come to your site by way of a *pingback* — which really is the same thing as a trackback, but the terminology varies from platform to platform.

Sending a trackback to a site is a nice way of telling the author that you like the information she presented in her post. Every author appreciates the receipt of trackbacks to her posts from other authors. Figure 4-3 shows one trackback link.

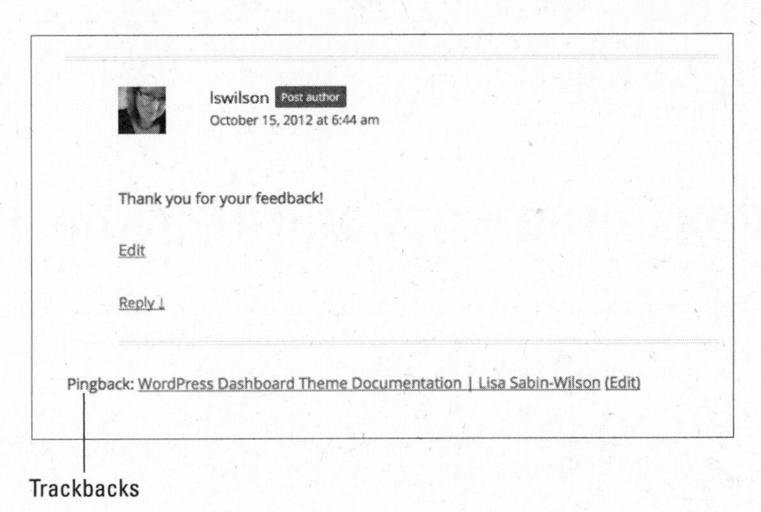

FIGURE 4-3: Trackback links on a blog.

Enabling comment and trackback display

Almost every single WordPress theme displays comments at the bottom of each post published in WordPress. You can do custom styling of the comments so that they match the design of your site by using several items:

- **WordPress template tags:** Template tags are related to the display of comments and trackbacks. For more on these tags, see Book 6, Chapter 3.

- **Basic HTML:** Using HTML markup helps you provide unique styles to display content. For information about the use of basic HTML, check out Book 6.
- **CSS:** Every WordPress theme has a Cascading Style Sheet (CSS) template called `style.css`. Within this CSS template, you define the styles and CSS markup that creates a custom look and feel for the comment and trackback display on your site. You can find more information about using CSS in Book 6.
- **Graphics:** Using graphics to enhance and define your branding, style, and visual design is an integral part of web design. Because a single chapter isn't sufficient to fully cover graphic design, you may want to check out *WordPress Web Design For Dummies,* 3rd Edition (John Wiley & Sons, Inc.), which I wrote, for great information on graphic and website design with WordPress.
- **WordPress widgets:** WordPress has a built-in widget to display the most recent comments published to your site by your visitors. You also can find several plugins that display comments in different ways, including top comments, most popular posts based on the number of comments, and comments that display the author's photo. For information about widgets and plugins for these purposes, flip to Book 6, Chapter 1 and Book 7, Chapters 1 and 2, respectively.

Managing Comments and Trackbacks

When you invite readers to comment on your site, you, as the site administrator, have full access to manage and edit those comments through the Comments page, which you can access in your WordPress Dashboard.

To find your comments, click the Comments link on the Dashboard navigation menu; the Comments screen opens. (See Figure 4-4.)

When you hover your mouse pointer over a comment, several links appear that give you the opportunity to manage the comment:

- **Unapprove:** This link appears only if you have comment moderation turned on. Also, it appears only on approved comments. The comment is placed in the moderation queue, which you can get to by clicking the Pending link that appears below the Comments page header. The moderation queue is kind of a holding area for comments that haven't yet been published to your site.

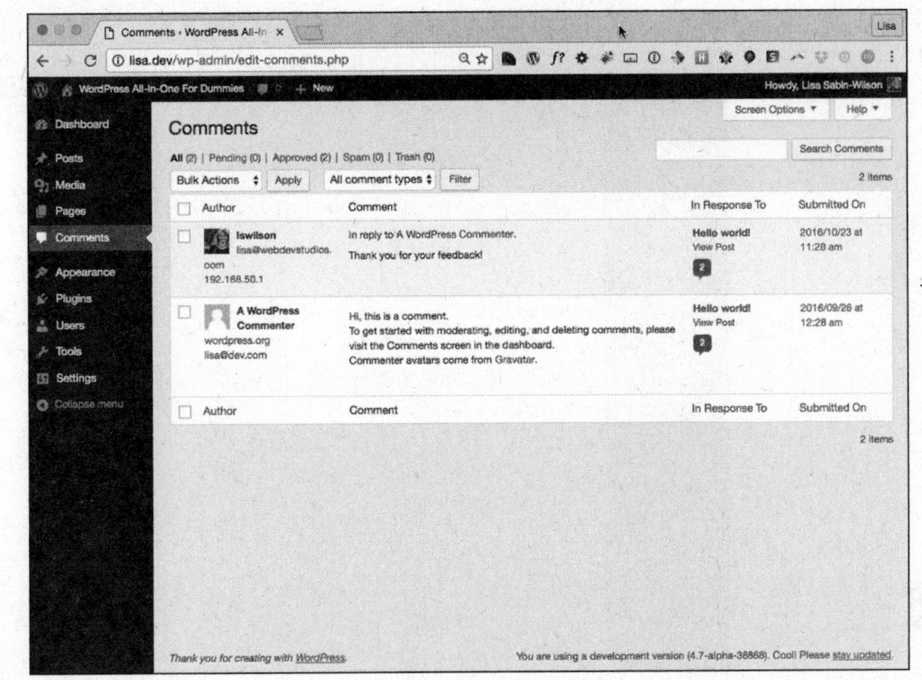

FIGURE 4-4: The Comments screen contains all the comments and trackbacks on your site.

- **Reply:** This link makes a text box visible. In the text box, you can type and submit your reply to this person. This feature eliminates the need to load your live site to reply to a comment.
- **Quick Edit:** This link opens the comment options inline without leaving the Comments page. You can configure options such as name, email address, URL, and comment content. Click the Save button to save your changes.
- **Edit:** This link opens the Edit Comment screen, where you can edit the different fields, such as name, email address, URL, and comment content. (See Figure 4-5.)
- **Spam:** This link marks the comment as spam and marks it as spam in the database, where it will never be heard from again! (Actually, it's stored in the database as spam; you just don't see it in your comments list unless you click the Spam link at the top of the Comments screen.)
- **Trash:** This link does exactly what it says: sends the comment to the trash and deletes it from your blog. You can access comments that have been sent to the trash to permanently delete them from your blog or to restore them.

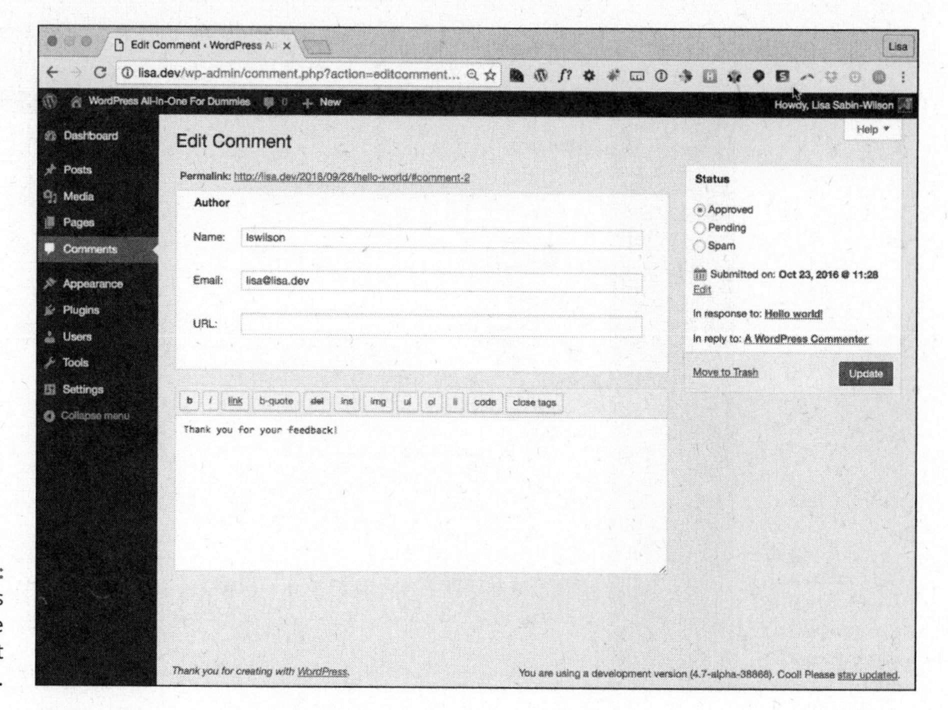

FIGURE 4-5: Edit a user's comment on the Edit Comment screen.

TIP

If you have a lot of comments listed in the Comments screen and want to edit them in bulk, select the check boxes to the left of all the comments you want to manage. Then choose one of the following from the Bulk Actions drop-down list in the upper-left corner of the page: Approve, Mark As Spam, Unapprove, or Delete.

If you have your options set so that comments aren't published to your site until you approve them, you can approve comments from the Comments screen as well. Just click the Pending link to list the comments that are pending moderation. If you have comments and/or trackbacks awaiting moderation, they appear on this page, and you can approve them, mark them as spam, or delete them.

WordPress immediately notifies you of any comments sitting in the moderation queue, awaiting your action. This notification, which appears on every single page, is a small circle, or bubble, on the left navigation menu, to the right of Comments. Figure 4-6 shows that I have three comments pending moderation.

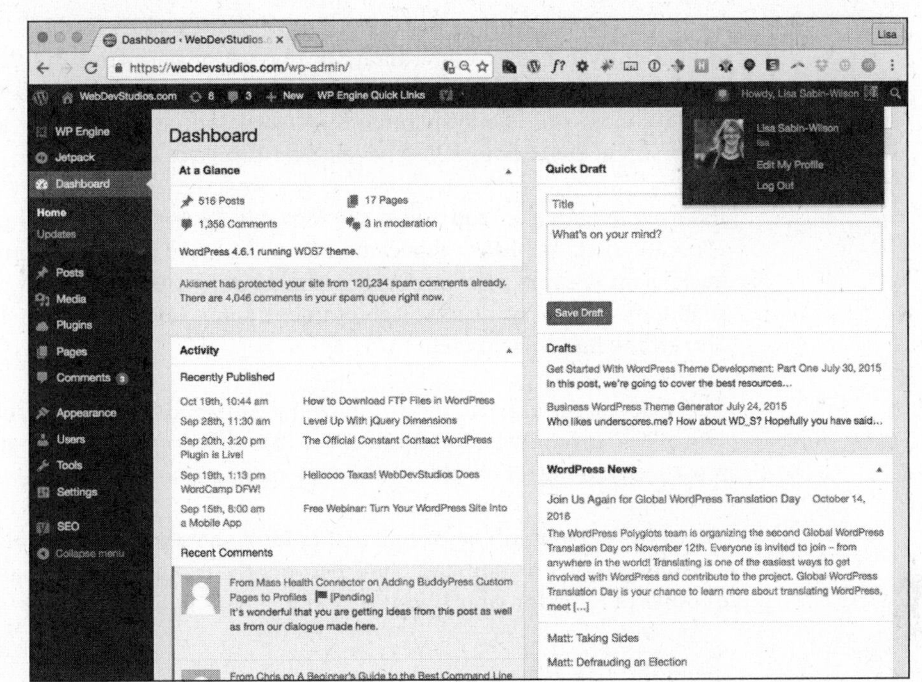

FIGURE 4-6: A small circle tells me that I have three comments pending moderation.

Tackling Spam with Akismet

No one likes spam. In fact, services such as WordPress have spent untold hours in the name of stopping spammers in their tracks, and for the most part, the services have been successful. Occasionally, however, spammers sneak through. Many spammers are offensive, and all of them are frustrating because they don't contribute to the ongoing conversations that occur in blogs. (A spammer's only goal is to generate traffic to his website.)

All WordPress installations have one significant thing in common: Akismet, a WordPress plugin. It's my humble opinion that Akismet is the mother of all plugins and that no WordPress site is complete without a fully activated version of Akismet running on it.

Apparently, WordPress agrees, because the plugin has been packaged in every WordPress software release beginning way back with version 2.0. Akismet was created by the folks at Automattic, the same folks who brought you the WordPress.com hosted version.

Akismet is the answer to combating comment and trackback spam. The Akismet website (`https://akismet.com`) explains it quite well: "Akismet is an advanced hosted anti-spam service aimed at thwarting the underbelly of the web.

It efficiently processes and analyzes masses of data from millions of sites and communities in real time. To fight the latest and dirtiest tactics embraced by the world's most proficient spammers, it learns and evolves every single second of every single day. Because you have better things to do."

I started blogging in 2002 with the Movable Type platform and moved to WordPress in 2003. As blogging became more and more popular, comment and trackback spam became more and more of a nuisance. One morning in 2004, I found that 2,300 pieces of disgusting comment spam had been published to my blog. Something had to be done!

The folks at Automattic did a fine thing with Akismet. Since the emergence of Akismet, I've barely had to think about comment or trackback spam except for the few times a month I check my Akismet spam queue.

This chapter wouldn't be complete if I didn't show you how to activate and use the Akismet plugin on your site. Book 7 covers the use, installation, and management of other plugins for your WordPress site.

Activating Akismet

Akismet is already included in every WordPress installation; you don't have to worry about downloading and installing it, because it's already there. Follow these steps to activate and begin using Akismet:

1. **Click the Plugins link on the left navigation menu of the Dashboard to load the Plugins screen.**
2. **Click the Activate link below the Akismet plugin name and description.**

   A green box appears at the top of the page, saying `Almost done- activate Akismet and say goodbye to spam`. (See Figure 4-7.)
3. **Click the Activate Your Akismet Account button in the green box to navigate to the Akismet Configuration screen.**
4. **If you already have an API key, enter it in the Manually Enter an API Key text field; then click the Use This Key button to save your changes.**

   You can stop here if you already have a key, but if you don't have an Akismet key, keep following the steps in this section.

   An *API key* is a string of numbers and letters that functions like a unique password given to you by Akismet; it's the key that allows your WordPress.org application to communicate with your Akismet account.

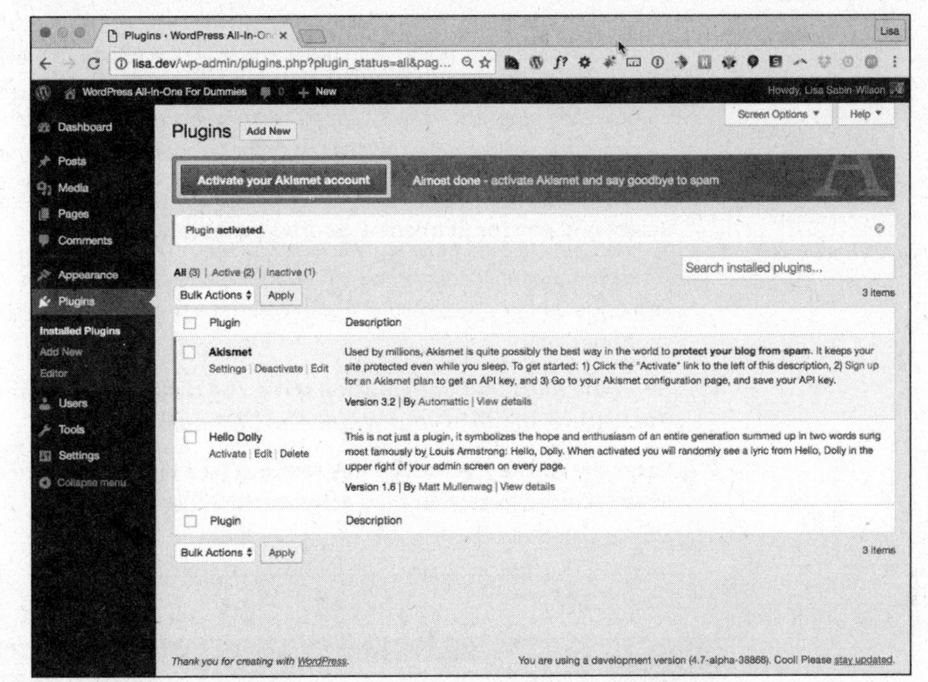

FIGURE 4-7: After you activate Akismet, WordPress tells you that the plugin isn't quite ready to use.

5. **Click the Get Your API key button on the Akismet Configuration screen.**

   The WordPress page of the Akismet website opens (`https://akismet.com/wordpress`).

6. **Click the Get an Akismet API Key button.**

   The sign-up page of the Akismet website opens (`https://signup.wordpress.com/signup`).

7. **Enter your email address, desired username, and password in the provided text fields; then click the Sign Up button.**

   The Pick Your Plan page opens.

8. **Choose among these options for obtaining an Akismet key:**

   - *Premium:* $9 per month for people who own commercial or professional sites or blogs and want additional security screening and malware protection.
   - *Plus:* $5 per month for people who own a small commercial or professional site or blog.

- *Basic:* Name your price. Type the amount you're willing to pay for the Basic plan. This option is for people who own one small, personal, WordPress-powered blog. You can choose to pay nothing ($0), but if you'd like to contribute a little cash toward the cause of combating spam, you can opt to spend up to $120 per year for your Akismet key subscription.

9. **Select and pay for (if needed) your Akismet key.**

   After you've gone through the sign-up process, Akismet provides you an API key. Copy that key by selecting it with your mouse pointer, right-clicking, and choosing Copy.

10. **Go to the Akismet Configuration screen by clicking the Akismet link on the Settings menu in your WordPress Dashboard.**

11. **Enter the API key in the API Key text box (see Figure 4-8), and click the Save Changes button to fully activate the Akismet plugin.**

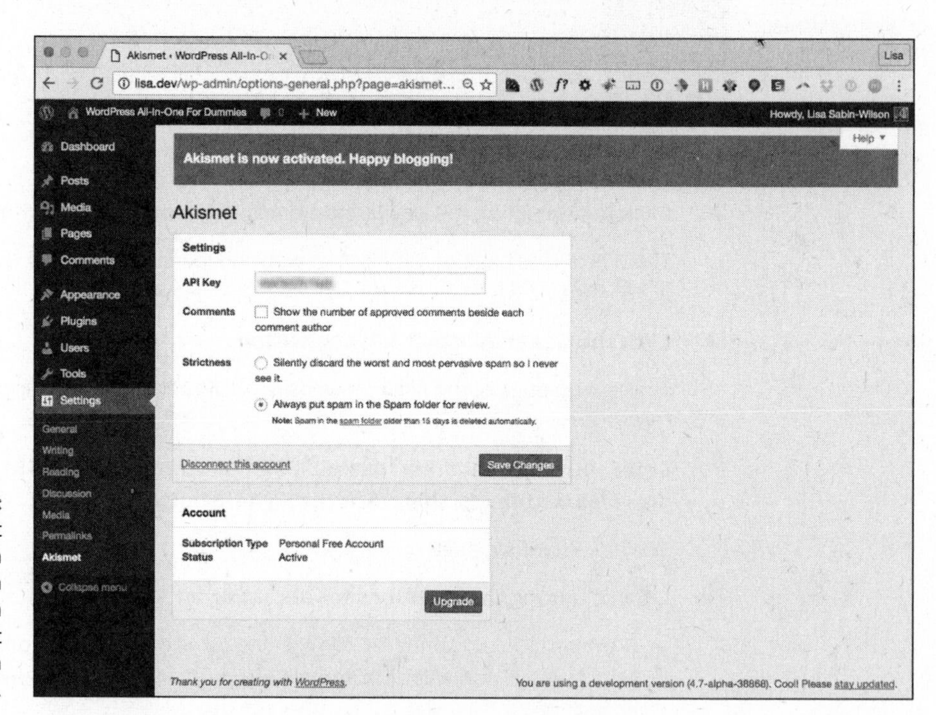

FIGURE 4-8: Akismet verification confirmation message on the Akismet Configuration screen.

Configuring Akismet

On the Akismet Configuration screen, after you've entered and saved your key, you can select two options to further configure your spam protection:

- **Comments:** Select this option to display the number of approved comments beside each comment author.
- **Strictness:** By default, Akismet puts spam in the Spam comment folder for you to review at your leisure. If you feel that this setting isn't strict enough, you can select the option to have Akismet silently delete the worst and most pervasive spam so that you never have to see it.

Akismet catches spam and throws it into a queue, holding the spam for 15 days and then deleting it from your database. It's probably worth your while to check the Akismet Spam page once a week to make sure that the plugin hasn't captured any legitimate comments or trackbacks.

Rescuing nonspam comments and trackbacks

You can rescue any nonspam captured comments and trackbacks by following these steps (after you log in to your WordPress Dashboard):

1. **Click Comments on the left navigation menu.**

   The Comments screen appears, displaying a list of the most recent comments on your blog.

2. **Click the Spam link.**

   The Comments screen displays all spam comments that the plugin caught.

3. **Browse the list of spam comments, looking for any comments or trackbacks that are legitimate.**
4. **If you locate a comment or trackback that's legitimate, click the Approve link directly below the entry.**

   The comment is marked as legitimate. In other words, WordPress recognizes that you don't consider this comment to be spam. WordPress then approves the comment and publishes it on your blog.

REMEMBER

Check your spam filter often. I just found four legitimate comments caught in my spam filter and was able to de-spam them, releasing them from the binds of Akismet and unleashing them upon the world.

» Creating and editing categories

» Creating tags and editing them

Chapter 5

Creating Categories and Tags

WordPress provides you many ways to organize, categorize, and archive content on your website. Packaged within the WordPress software is the capability to automatically maintain chronological, categorized archives of your publishing history, which provides your website visitors different ways to find your content. WordPress uses PHP and MySQL technology to sort and organize everything you publish in an order that you and your readers can access by date and category. This archiving process occurs automatically with every post you publish to your site.

In this chapter, you find out all about WordPress archiving, from categories to tags and more. You also discover how to take advantage of the category description feature to improve your search engine optimization (SEO), how to distinguish between categories and tags, and how to use categories and tags to create topical archives of your site content.

Archiving Content with WordPress

When you create a post on your WordPress site, you can file that post in a category that you specify. This feature makes for a nifty archiving system in which you and your readers can find articles/posts that you've placed within a specific category.

Articles you post are also sorted and organized by date (day/month/year) so that you can easily locate articles that you posted at a certain time. A plugin called Archive Page gives you the ability to easily create a nicely formatted site map that displays the archives of a site on a page.

Visit `http://wp-time.com/archives` to see an example of an archive page. The example page contains chronological sections by day, month, and year. It also contains the latest posts and the different categories and tags found within that site. If you click a date, tag or category on that page, you're taken to a page with a full listing of articles from that date, tag, or category, and each article title is linked to that article. (See Figure 5-1.)

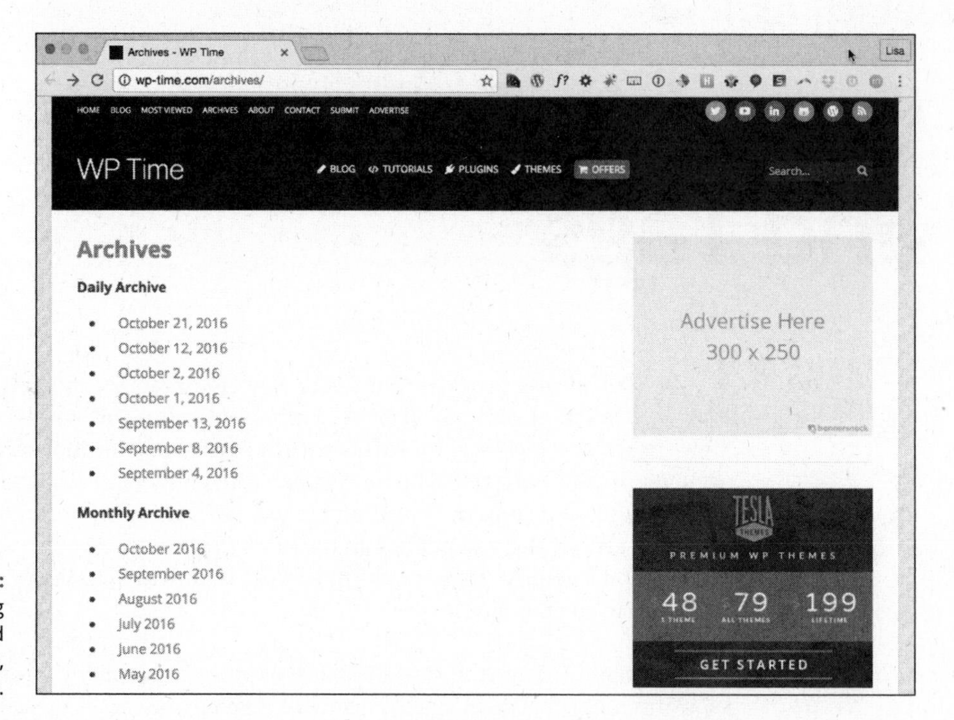

FIGURE 5-1: An archive listing of published posts by date, tag, or category.

TIP

You can easily create an archive listing like the one shown in Figure 5-1 by using a WordPress plugin called Archive Page, which you can find in the WordPress Plugin Directory at `https://wordpress.org/plugins/archive-page`. This plugin is easy to install and to use. You just need to create a page and add any of the short codes provided by the plugin to automatically build an archives page that links to all the content you've published on your site. Easy archives!

WordPress archives and organizes your content for you in more ways than by date and category. In this section, I give you an overview of the several other ways. The different types of archives and content include:

- **Categories:** Create categories of topics in which you can file your articles so that you can easily archive relevant topics. Many websites display content by category — typically referred to as a *magazine theme,* in which all content is displayed by topic rather than in a simple chronological listing. Figure 5-2 shows an example of a magazine theme. You can find out how to create one of your own by customizing your site (see Book 6). Also be sure to check out Book 6, Chapter 6 to discover how to use template tags and category templates to display category-specific content. Exciting stuff!

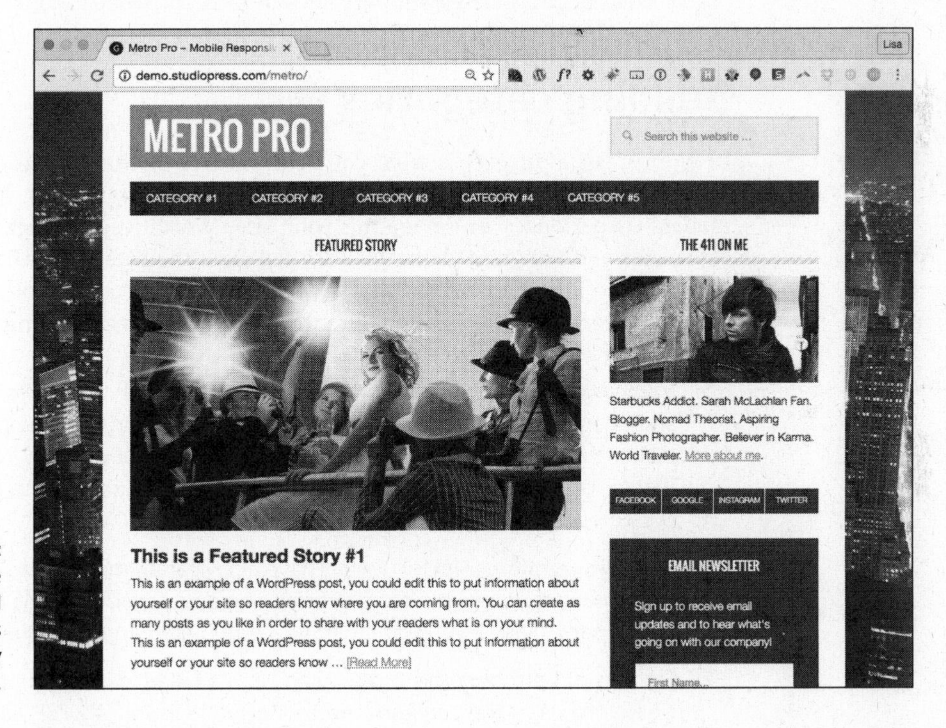

FIGURE 5-2: A magazine theme created with WordPress (Metro Pro by StudioPress).

- **Tags:** Tagging your posts with micro keywords, called *tags,* further defines related content within your site, which can improve your site for SEO purposes by helping the search engines find related and relevant content, as well as provide additional navigation to help your readers find relevant content on your site.
- **Date Based:** Your content is automatically archived by date based on the day, month, year, and time of day you publish it.

- **Author:** Content is automatically archived by author based on the author of the post and/or page. You can create an author archive if your site has multiple content contributors.
- **Keyword (or Search):** WordPress has a built-in search function that allows you and your readers to search for keywords, which presents an archive listing of content that's relevant to your chosen keywords.
- **Custom Post Types:** You can build custom post types based on the kind of content your site offers. You can find detailed information on custom post types and how to create them in Book 6, Chapter 7.
- **Attachments:** WordPress has a built-in media library where you can upload different media files, such as photos, images, documents, videos, and audio files (to name a few). You can build an archive of those files to create things such as photo galleries, eBook archives (PDFs), and video galleries.

Building categories

In WordPress, a *category* is what you determine to be the main topic of a post. By using categories, you can file your posts under topics by subject. To improve your readers' experiences in navigating your site, WordPress organizes posts by the categories you assign to them. Visitors can click the categories they're interested in to see the posts you've written on those particular topics. You can display the list of categories you set up on your site in a few places, including the following:

- **Body of the post:** In most WordPress themes, you see the title followed by a statement such as `Filed In:` *`Category 1, Category 2.`* The reader can click the category name to go to a page that lists all the posts you've made in that particular category. You can assign a single post to more than one category.
- **Navigation menu:** Almost all sites have a navigation menu that visitors can use to navigate your site. You can place links to categories on the navigation menu, particularly if you want to draw attention to particular categories.
- **Sidebar of your blog theme:** You can place a full list of category titles in the sidebar by using the Categories widget included in your WordPress installation. A reader can click any category to open a page on your site that lists the posts you made within that particular category.

Subcategories (also known as *category children*) can further refine the main category topic by listing specific topics related to the main *(parent)* category. In your WordPress Dashboard, on the Manage Categories page, subcategories appear directly below the main category. Here's an example:

Books I Enjoy (main category)

Fiction (subcategory)

Nonfiction (subcategory)

Trashy romance (subcategory)

Biographies (subcategory)

For Dummies (subcategory)

You can create as many levels of categories as you like. Biographies and *For Dummies* could be subcategories of Nonfiction, for example, which is a subcategory of the Books I Enjoy category. You aren't limited to the number of category levels you can create.

Changing the name of a category

When you install WordPress, it gives you one default category called Uncategorized. (See the Categories screen shown in Figure 5-3.) This category name is pretty generic, so you definitely want to change it to one that applies to you and your blog. (On my site, I changed it to Life in General. Although that name's still a bit on the generic side, it doesn't sound quite so . . . well, uncategorized.)

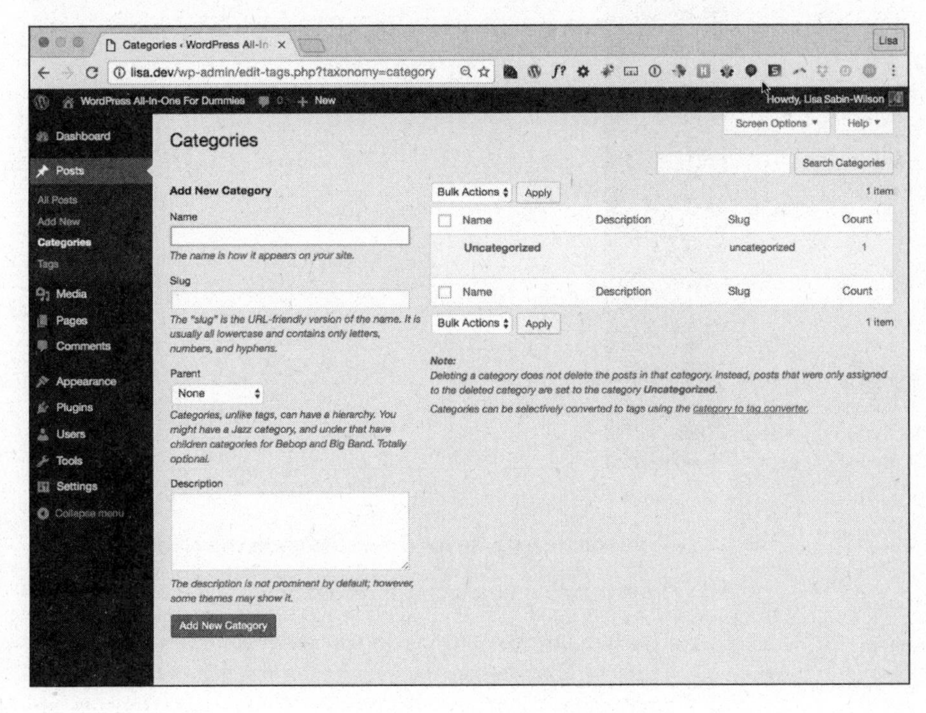

FIGURE 5-3: The Categories screen in the Dashboard of a brand-new WordPress site shows the default Uncategorized category.

REMEMBER

The default category also serves as kind of a fail-safe. If you publish a post to your site and don't assign that post to a category, the post is assigned to the default category automatically, no matter what you name the category.

So how do you change the name of that default category? When you're logged in to your WordPress Dashboard, just follow these steps:

1. **Click the Categories link in the Posts submenu of the Dashboard navigation menu.**

   The Categories screen opens, containing all the tools you need to set up and edit category titles for your site.

2. **Click the title of the category that you want to edit.**

   If you want to change the Uncategorized category, click the word *Uncategorized* to open the Edit Category screen. (See Figure 5-4.)

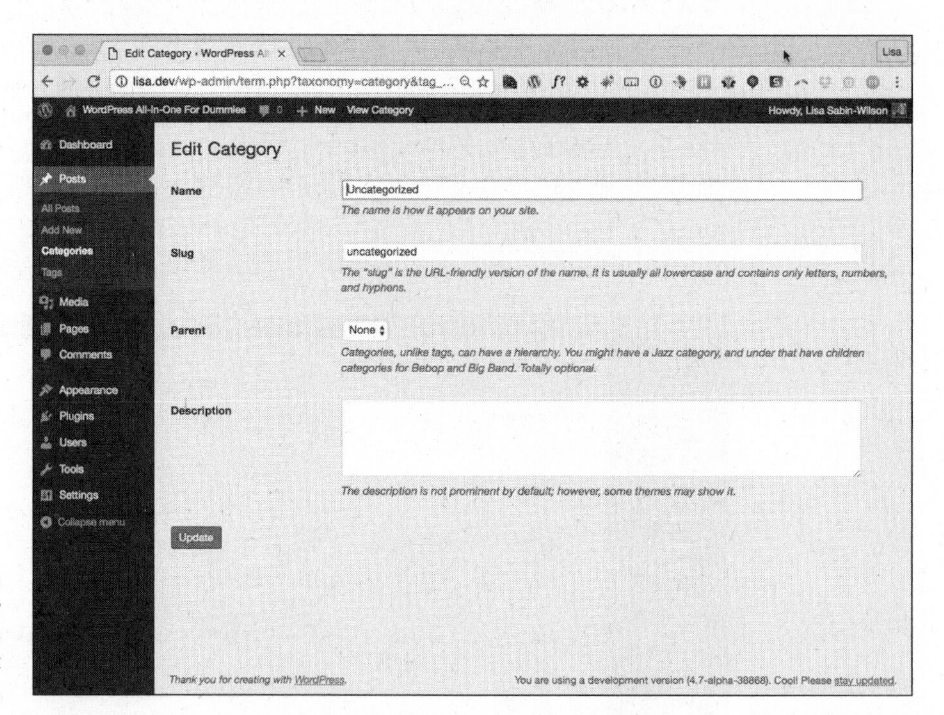

FIGURE 5-4: Editing a category in WordPress on the Edit Category screen.

3. **Type the new name for the category in the Name text box.**

4. **Type the new slug in the Slug text box.**

   The term *slug* refers to the word(s) used in the web address for the specific category. The category Books, for example, has a web address of `http://`*`yourdomain`*`.com/category/books;` if you change the Category Slug to Books

I Like, the web address is `http://`*`yourdomain.com`*`/category/books-i-like`. (WordPress automatically inserts a dash between the slug words in the web address.)

5. **Choose a parent category from the Parent drop-down list.**

   If you want this category to be a main category, not a subcategory, choose None.

6. **(Optional) Type a description of the category in the Description text box.**

   Use this description to remind yourself what your category is about. Some WordPress themes display the category description right on your site, too, which your visitors may find helpful. (See Book 6 for more about themes.) You know that your theme is coded in this way if your site displays the category description on the category page(s).

7. **Click the Update button.**

   The information you just edited is saved, and the Categories screen reloads, showing your new category name.

Creating new categories

Today, tomorrow, next month, next year — while your blog grows in size and age, continuing to add new categories further defines and archives the history of your posts. You aren't limited in the number of categories and subcategories you can create for your site.

Creating a new category is as easy as following these steps:

1. **Click the Categories link on the Posts submenu, which is on the Dashboard navigation menu.**

   The Categories screen opens. The left side of the Categories page displays the Add New Category section. (See Figure 5-5.)

2. **Type the name of your new category in the Name text box.**

   Suppose that you want to create a category in which you file all your posts about the books you read. In the Name text box, type something like **Books I Enjoy**.

3. **Type a name in the Slug text box.**

   The slug creates the link to the category page that lists all the posts you made in this category. If you leave this field blank, WordPress automatically creates a slug based on the category name: If the category is Books I Enjoy, WordPress automatically creates a category slug like `http://yourdomain.com/category/books-i-enjoy`. If you want to shorten it, however, you can! Type **books** in the Category Slug text box, and the link to the category becomes `http://yourdomain.com/category/books`.

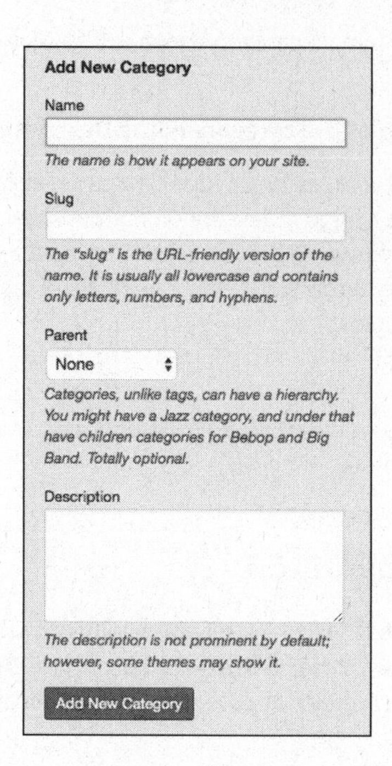

FIGURE 5-5: Create a new category on your site.

4. **Choose the category's parent from the Parent drop-down list.**

   Choose None if you want this new category to be a parent (or top-level) category. If you want to make this category a subcategory of another category, choose the category that you want to be the parent of this one.

5. **(Optional) Type a description of the category in the Description text box.**

   Some WordPress templates are set up to actually display the category description directly below the category name. (See Book 6.) Providing a description further defines the category intent for your readers. The description can be as short or as long as you want.

6. **Click the Add New Category button.**

   That's it! You've added a new category to your site. Armed with this information, you can add an unlimited number of categories to your new site.

You can delete a category by hovering your mouse pointer on the title of the category you want to delete and then clicking the Delete link that appears below the category title.

REMEMBER

Deleting a category doesn't delete the posts and links in that category. Instead, posts in the deleted category are reassigned to the Uncategorized category (or whatever you've named the default category).

TECHNICAL STUFF

If you have an established WordPress site that has categories already created, you can convert some or all of your categories to tags. To do so, look for the Category to Tag Converter link on the bottom right of the Category page of your WordPress Dashboard. Click it to convert your categories to tags. (See the nearby sidebar "What are tags, and how/why do I use them?" for more information on tags.)

Book 6, Chapter 6 shows you how to take advantage of categories in WordPress to build a dynamic theme that displays your content in a way that highlights the different topics available on your site. Book 6 describes how to use WordPress template tags to manipulate category archives for display and distribution on your website.

WHAT ARE TAGS, AND HOW/WHY DO I USE THEM?

Don't confuse tags with categories (as a lot of people do). *Tags* are clickable, comma-separated keywords that help you microcategorize a post by defining the topics in it. Unlike WordPress categories, tags don't have a hierarchy; you don't assign parent tags and child tags. If you write a post about your dog, for example, you can put that post in the Pets category — but you can also add some specific tags that let you get a whole lot more specific, such as *poodle* or *small dogs*. If someone clicks your *poodle* tag, he finds all the posts you ever made that contain the *poodle* tag.

Besides defining your post topics for easy reference, you have another reason to use tags: Search-engine spiders harvest tags when they crawl your site, so tags help other people find your site when they search for specific words.

You can manage your tags in the WordPress Administration panel by choosing Tags from the Pages drop-down list. The Tags page opens, allowing you to view, edit, delete, and add tags.

Creating and Editing Tags

In Book 4, Chapter 1, you can find out all about publishing your posts in WordPress and assigning tags to your content. This section takes you through the steps of managing tags, which is similar to the way you manage categories. To create a new tag, follow these steps:

1. **Click the Tags link on the Posts submenu, which is on the Dashboard navigation menu.**

   The Tags screen opens, as shown in Figure 5-6. The left side of the screen displays the Add New Tag section.

TIP

   Unlike what it does for categories and links, WordPress doesn't create a default tag for you, so when you visit the Tags page for the first time, no tags are listed on the right side of the page.

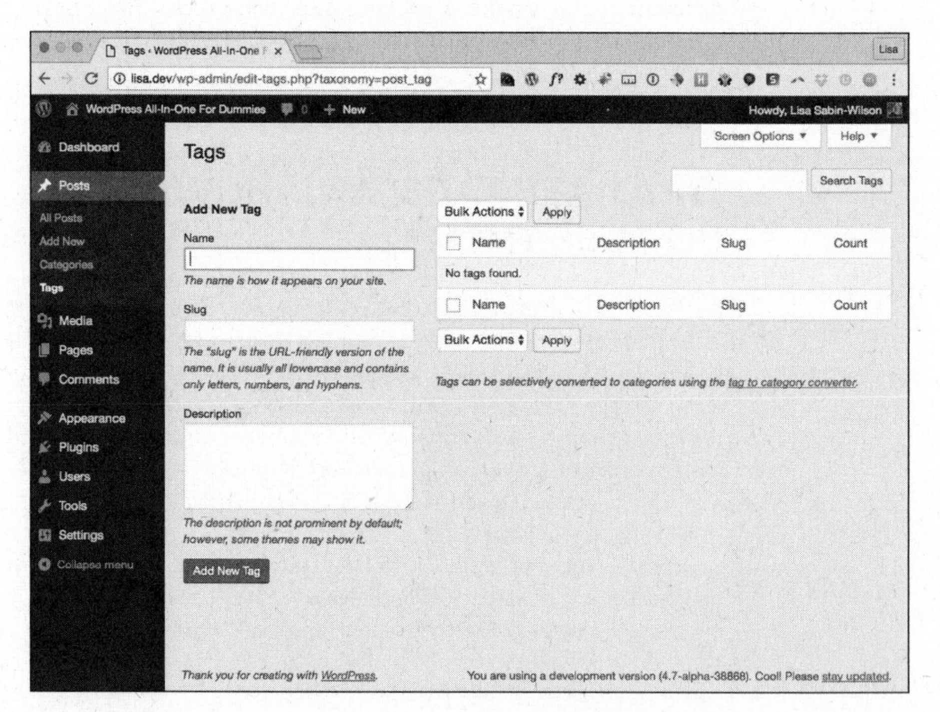

FIGURE 5-6: The Tags screen of the Dashboard.

2. **Type the name of your new tag in the Name text box.**

   Suppose that you want to create a tag in which you file all your posts about the books you read. In the Name text box, type something like Fictional Books.

3. **Type a name in the Slug text box.**

   The *slug* is the permalink of the tag and can help identify tag archives on your site by giving them their own URL, such as `http://`*`yourdomain.com`*`/tag/fictional-books`. By default, the tag slug adopts the words from the tag name.

4. **(Optional) Type a description of the tag in the Description text box.**

   Some WordPress templates are set up to actually display the tag description directly below the tag name. Providing a description further defines the category intent for your readers. The description can be as short or as long as you want.

5. **Click the Add New Tag button.**

   That's it! You've added a new tag to your site. The Add New Tag screen refreshes in your browser window with blank fields, ready for you to add another tag to your site.

6. **Repeat Steps 1 through 5 to add an unlimited number of tags to your blog.**

TIP

You use the Tags and Categories pages of your Dashboard to manage, edit, and create new tags and categories to which you assign your posts when you publish them. Book 4, Chapter 1 contains a lot of information about how to go about assigning tags and categories to your posts, as well as a few good tips on how you can create new categories and tags right on the Edit Posts screen.

4 Publishing Your Site with WordPress

Contents at a Glance

» Writing and publishing your posts

» Creating a unique work space for writing

Chapter 1
Writing Your First Post

It's time to write your first post on your new WordPress site! The topic you choose to write about and the writing techniques you use to get your message across are all up to you; I have my hands full writing this book! I *can* tell you, however, all about the techniques you'll use to write the wonderful passages that can bring you fame. Ready?

This chapter covers everything you need to know about the basics of publishing a post on your site, from writing a post to formatting, categorizing, tagging, and publishing it to your site.

Composing Your Post

Composing a post is a lot like typing an email: You give it a title, you write the message, and you click a button to send your words into the world. This section covers the steps you take to compose and publish a post on your site. By using the different options that WordPress provides — discussion options, categories, and tags, for example — you can configure each post however you like.

TIP

You can collapse or reposition all the modules on the Add New Post screen to suit your needs. The only parts of the Add New Post screen that can't be collapsed and repositioned are the actual Title and Post text boxes (where you write your post).

Follow these steps to write a basic post:

1. **Click the Add New link on the Posts menu of the Dashboard.**

   The Add New Post screen opens, as shown in Figure 1-1.

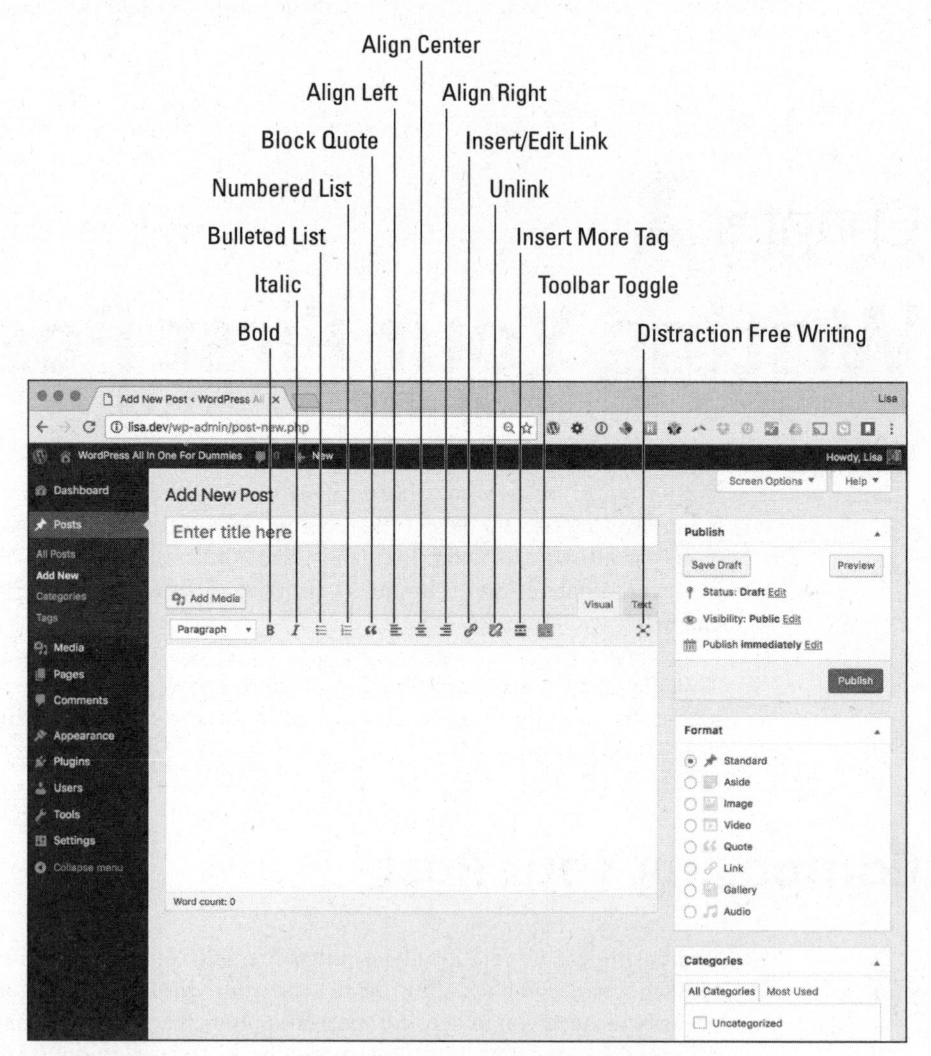

FIGURE 1-1: Give your post a title, and write your post body.

2. **Type the title of your post in the Enter Title Here text field at the top of the Add New Post screen.**

3. **Type the content of your post in the large text box below the Enter Title Here text box.**

   You can use the Visual Text Editor to format the text in your post. I explain the Visual Text Editor, and the buttons and options, later in this section.

4. **Click the Save Draft button in the Publish module, located in the top-right corner of the Add New Post screen.**

   The screen refreshes with your post title and content saved but not yet published to your site.

   At this point, you can skip to the "Publishing Your Post" section of this chapter for information on publishing your post to your site, or you can continue with the following sections to discover how to refine the options for your post.

By default, the area in which you write your post is in Visual Editing mode, as indicated by the Visual tab that appears above the text. Visual Editing mode provides WYSIWYG (What You See Is What You Get) options for formatting. Rather than having to embed HTML code in your post, you can simply type your post, highlight the text you want to format, and click the buttons (shown in Figure 1-1) that appear above the text box in which you type your post.

If you've ever used a word processing program, such as Microsoft Word, you'll recognize many of these buttons:

- **Bold:** Embeds the <strong> </strong> HTML tag to emphasize the text in bold. Example: **Bold Text.**
- **Italic:** Embeds the <em> </em> HTML tag to emphasize the text in italics. Example: *Italic Text.*
- **Strikethrough:** Embeds the <strike> </strike> HTML tag that puts a line through your text. Example: ~~Strikethrough Text~~.
- **Bulleted List:** Embeds the <ul><li> </li></ul> HTML tags that create an *unordered* (bulleted) list.
- **Numbered List:** Embeds the <ol><li> </li></ol> HTML tags that create an *ordered* (numbered) list.
- **Blockquote:** Inserts the <blockquote> </blockquote> HTML tag that indents the paragraph or section of text you've selected.
- **Horizontal Line:** Inserts the <hr> </hr> HTML tag that creates a horizontal line.
- **Align Left:** Inserts the <p align="left"> </p> HTML tag that lines up the selected text against the left margin.
- **Align Center:** Inserts the <p align="center"> </p> HTML tag that positions the selected text in the center of the page.

- **Align Right:** Inserts the `<p align="right"> </p>` HTML tag that lines up the selected text against the right margin.
- **Insert/Edit Link:** Inserts the `<a href=" "> </a>` HTML tag around the text you've selected to create a hyperlink.
- **Unlink:** Removes the hyperlink from the selected text, if it was previously linked.
- **Insert More Tag:** Inserts the `<!--more-->` tag, which lets you split the display on your blog page. It publishes the text written above this tag with a Read More link, which takes the user to a page with the full post. This feature is good for really long posts.
- **Toolbar Toggle:** Drops down additional formatting options, described in detail after this list.
- **Distraction Free Writing Mode:** Lets you focus purely on writing, without the distraction of all the other options on the page. Click this button, and the Post text box expands to fill the full height and width of your browser screen and displays only the barest essentials for writing your post. To bring the Post text box back to its normal state, click the Exit Full Screen link. Voilà — it's back to normal!

But wait — there's more! When you click the Toolbar Toggle icon, you see a lot more formatting options you can use for your posts. Click the Toolbar Toggle icon to see a new formatting list, which provides options for a veritable kitchen sink full of options, including underlining, font color, custom characters, undo, and redo:

- **Format:** This drop-down list allows you to select the different text formatting available:
  - *Paragraph:* Inserts the `<p> </p>` HTML tags around the text to indicate paragraph breaks.
  - *Headings 1, 2, 3, 4, 5, 6:* Inserts header HTML tags such as `<H1> </H1>` around text to indicate HTML headings. (H1 defines the largest, and H6 defines the smallest. Heading formats are usually defined in the CSS [see Book 6, Chapter 4] with font size and/or colors.)
  - *Preformatted:* Inserts the `<pre> </pre>` HTML tags around the text to indicate preformatted text and preserves both spaces and line breaks.
- **Underline:** Inserts the `<u> </u>` HTML tags around the text to display it as underlined.
- **Justify:** Inserts the `<p align="justify"> </p>` HTML , which aligns the selected text along both the left and right margins. Letter- and word-spacing is adjusted so each line stretches the full width of the text column.
- **Text Color:** Displays the text in the color chosen.

- **Paste As Text:** Useful if you copy text from another source. This option removes all formatting and special/hidden characters from the text and adds it to your post as unformatted text.
- **Clear Formatting:** Removes all formatting inside the post.
- **Special Character:** Opens a pop-up window appears (see Figure 1-2), offering characters such as $, %, &, and ©. In the pop-up window, click the symbol that you want to add to your post.

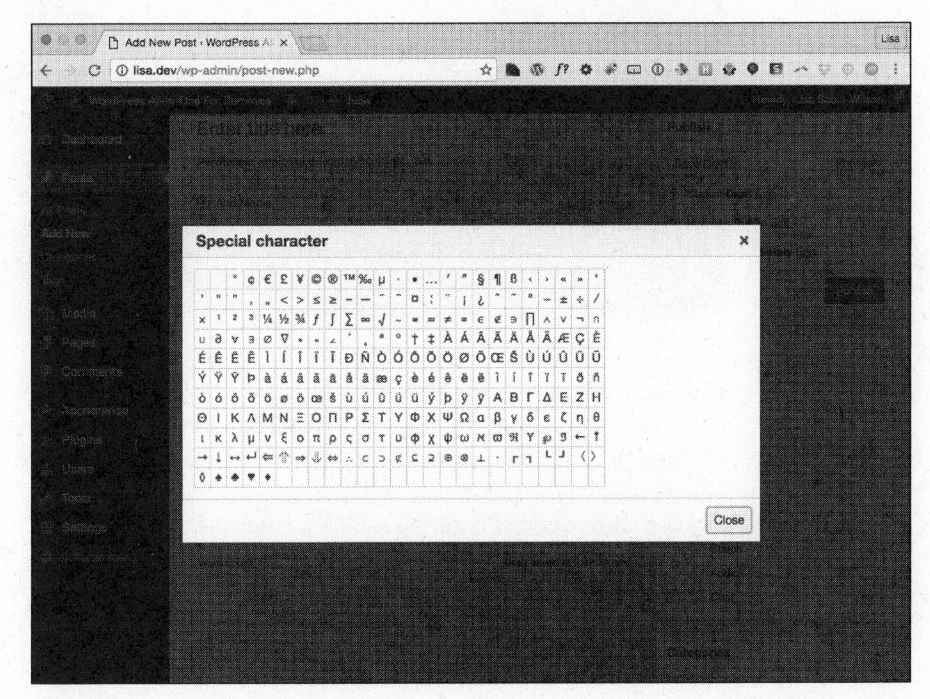

FIGURE 1-2: Insert custom characters into your post.

- **Decrease Indent:** Moves text to the left one preset level with each click.
- **Increase Indent:** Moves text to the right one preset level with each click.
- **Undo:** Undoes your last formatting action.
- **Redo:** Redoes the last formatting action that you undid.
- **Keyboard Shortcuts:** Pops open a window with helpful information about using timesaving keyboard shortcuts, as shown in Figure 1-3.

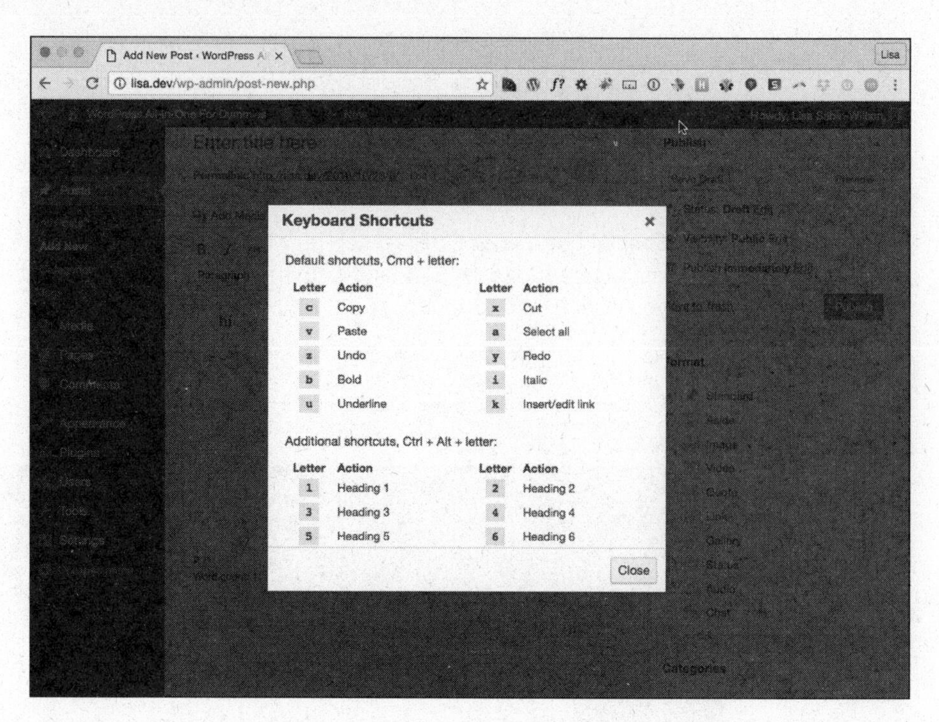

FIGURE 1-3: Keyboard shortcuts.

You can turn off the Visual Text Editor by clicking the Your Profile link on the Users menu. Select the Disable the Visual Editor When Writing check box to turn off this editor if you'd rather insert the HTML code yourself in your posts.

TECHNICAL STUFF

If you'd rather embed your own HTML code and skip the Visual Text Editor, click the Text tab that appears to the right of the Visual tab. If you're planning to type HTML code in your post — for a table or video files, for example — you have to click the Text tab before you insert that code. If you don't, the Visual Text Editor formats your code, and it most likely will look nothing like you intended it to.

REMEMBER

WordPress has a nifty built-in autosave feature that saves your work while you're typing and editing a new post. If your browser crashes or you accidentally close your browser window before you've saved your post, it will be there for you when you get back. Those WordPress folks are so thoughtful!

Directly above the Visual Text Editor row of buttons, you see the Add Media button. Click this button if you want to insert images/photos, photo galleries, videos, and audio files into your posts. WordPress has an entire Media Library capability, which you can find out about in great detail in Book 4, Chapters 3 and 4.

Refining Your Post Options

After you write the post, you can choose a few extra options before you publish it for the entire world to see. By default, the Add New Post screen of the WordPress Dashboard displays only a few options because they're the minimum options you need to publish a post. Book 3, Chapter 1 explains the Screen Options tab, which contains additional options for the screen you're currently viewing in the Dashboard. On the Add New Post screen, click the Screen Options tab (shown in Figure 1-4) to reveal additional modules that you can activate on this screen, including

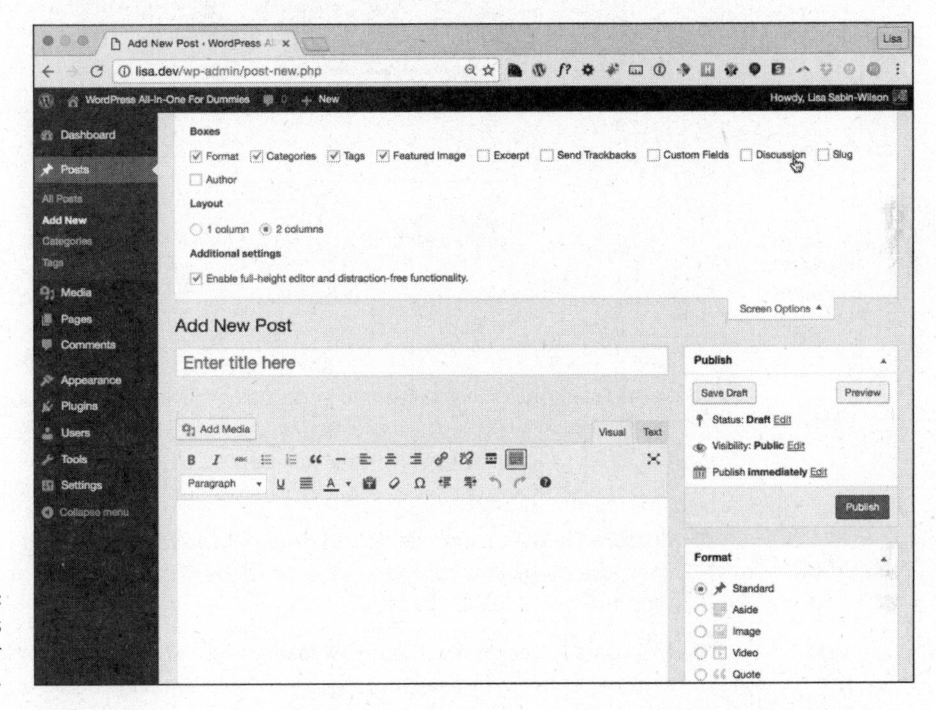

FIGURE 1-4: Several options are available for your blog posts.

- **Excerpt:** Excerpts are short summaries of your posts. Many authors use snippets to show teasers of their posts on their website, thereby encouraging the reader to click the Read More links to read the posts in their entirety. Type your short summary in the Excerpt box. Excerpts can be any length in terms of words, but the point is to keep them short and sweet to tease your readers into clicking the Read More link.

  Figure 1-5 shows a post published to my business site. It displays an excerpt of the post on the front page, requiring the reader to click the title link to view the post in its entirety. (Some blog themes include a Continue Reading link for readers to click to read the rest of the post.)

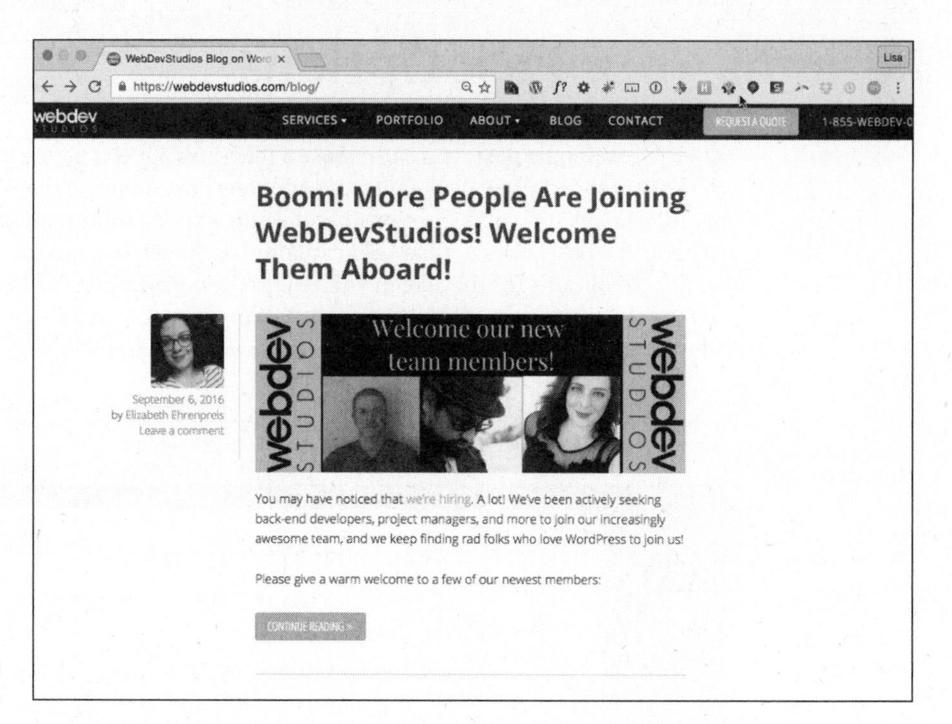

FIGURE 1-5: A post excerpt.

- **Send Trackbacks:** If you want to send a trackback to another site, enter the site's trackback URL in the Send Trackbacks To text box. You can send trackbacks to more than one blog; just be sure to separate trackback URLs with spaces. For more on trackbacks, refer to Book 3, Chapter 4.
- **Custom Fields:** Custom fields add extra data to your posts, and you can fully configure them. You can read more about the Custom Fields feature in WordPress in Book 4, Chapter 5.
- **Discussion:** Decide whether to let readers submit comments through the comment system by selecting the Allow Comments on This Post check box. By default, the box is selected; deselect it to disallow comments on this post.
- **Slug:** A slug is part of the URL of your post, added onto your domain. For example, for a post titled "WordPress tips," WordPress automatically creates a URL from that title like `http://domain.com/wordpress-tips`. The Slug option allows you to set a different slug for your post (or page) than the one WordPress automatically creates for you. So, for example, you could shorten the slug for the post title "WordPress Tips" to just "wordpress" so that the URL for the post is `http://domain.com/wordpress`.
- **Author:** If you're running a multiauthor blog, you can select the name of the author who wrote this post. By default, your own author name is selected in the Author drop-down list.

These optional modules apply to the post you're currently working on — not to any future or past posts. You can find these options below and to the right of the Post text box after they're activated in Screen Options. (Refer to Figure 1-4.) Click the title of each option to make the settings for that specific option expand or collapse.

REMEMBER

You can reposition the various post option modules on the Add New Post page to fit the way you use this page by using the drag-and-drop method.

Here are the options that appear to the right of the Post text box:

- **Publish:** These options are covered in the "Publishing Your Post" section later in this chapter.
- **Format:** This module appears only when the theme that you're using on your site supports a WordPress feature called Post Formats (which I cover in detail in Book 6, Chapter 6). In the Format module, you can select the type of format you want to use for the post you're publishing.
- **Categories:** You can file your posts in different categories to organize them by subject. (See more about organizing your posts by category in Book 3, Chapter 5.) Select the check box to the left of the category you want to use. You can toggle between listing all categories or displaying just the categories you use most by clicking the All Categories or Most Used link, respectively.

  TIP

  Don't see the category you need? Click the Add New Category link, and add a category right there on the Add New Post page.

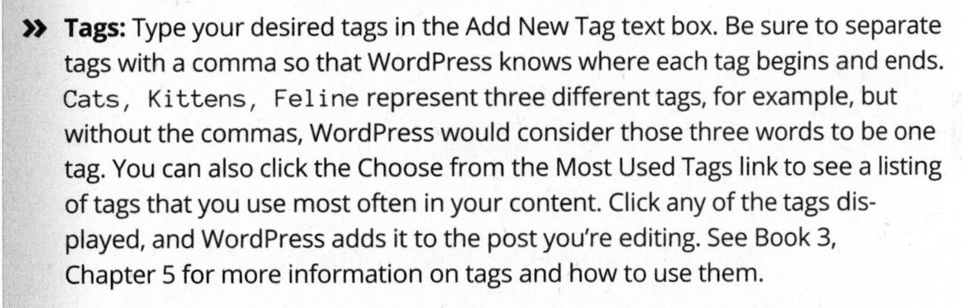

- **Tags:** Type your desired tags in the Add New Tag text box. Be sure to separate tags with a comma so that WordPress knows where each tag begins and ends. `Cats, Kittens, Feline` represent three different tags, for example, but without the commas, WordPress would consider those three words to be one tag. You can also click the Choose from the Most Used Tags link to see a listing of tags that you use most often in your content. Click any of the tags displayed, and WordPress adds it to the post you're editing. See Book 3, Chapter 5 for more information on tags and how to use them.
- **Featured Image:** Some WordPress themes are configured to use an image (photo) to represent each post that you have on your site. The image can appear on the home/front page, blog page, archives page, or anywhere within the content display on your website. If you're using a theme that has this option, you can easily define the post thumbnail by clicking Set Featured Image below the Featured Image module on the Add New Post page. You can find more information about using Featured Images in Book 6, Chapter 6.

WARNING

When you finish setting the options for your post, don't navigate away from this page; you haven't yet fully saved your options. The following section on publishing your post covers all the options you need for saving your post settings!

Publishing Your Post

You've given your new post a title and written the content of the post. Maybe you've even added an image or other type of media file to the post (see Book 4, Chapters 3 and 4), and you've definitely configured the tags, categories, and other options. Now the question is this: To publish? Or not to publish (yet)?

WordPress gives you three options for saving or publishing your post when you're done writing it. The Publish module is located on the right side of the Add New (or Edit) Post screen. Just click the title of the Publish module to expand the settings you need. Figure 1-6 shows the available options in the Publish module.

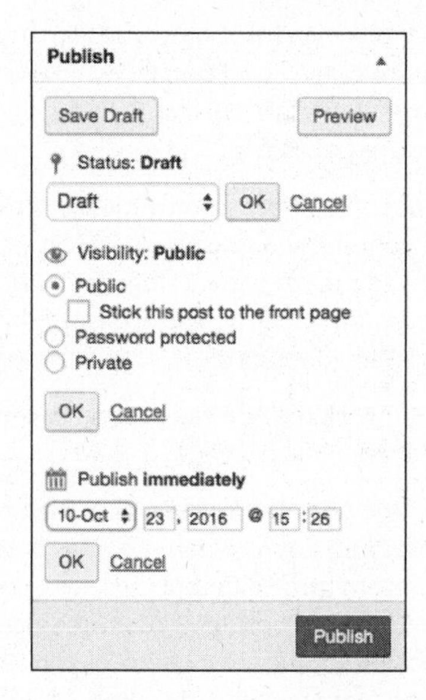

FIGURE 1-6: The publish status for your blog posts.

The Publish module has several options:

- **Save Draft:** Click this button to save your post as a draft. The Add New Post screen reloads with all your post contents and options saved. You can continue editing now, tomorrow, the next day, or next year; the post is saved as a draft until you decide to publish it or delete it. Posts saved as drafts can't be seen by visitors to your site. To access your draft posts, click the Posts link on the Posts menu.
- **Preview:** Click the Preview button to view your post in a new window, as it would appear on your live site if you'd published it. Previewing the post doesn't publish it to your site yet. Previewing simply gives you the opportunity

to view the post on your site and check it for any formatting or content changes you want to make.

- **Status:** Click the Edit link (shown in the Publish module in Figure 1-4) to open the settings for this option. A drop-down list appears, from which you can choose one of two options:
  - *Draft:* Save the post in draft form but don't publish it to your site.
  - *Pending Review:* The post shows up in your list of drafts next to a Pending Review header. This option lets the administrator of the blog know that contributors have entered posts that are waiting for administrator review and approval (helpful for sites that have multiple authors). Generally, only contributors use the Pending Review option.

  Click the OK button to save your Status setting.
- **Visibility:** Click the Edit link to open the settings for this option:
  - *Public:* Select this option to make the post viewable to everyone who visits your site. Select the Stick This Post to the Front Page check box to have WordPress publish the post to your site and keep it at the very top of all posts until you change this setting for the post.

    REMEMBER

    This option is otherwise known as a *sticky post*. Typically, posts are displayed in chronological order on your site, displaying the most recent post on top. If you make a post sticky, it remains at the very top, no matter how many other posts you make after it. When you want to unstick the post, deselect the Stick This Post to the Front Page check box.

    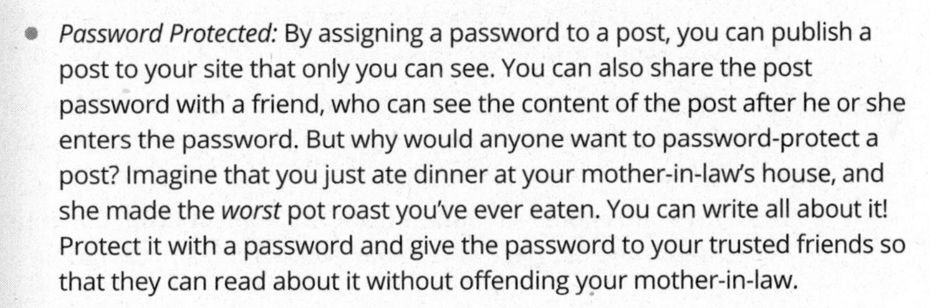
  - *Password Protected:* By assigning a password to a post, you can publish a post to your site that only you can see. You can also share the post password with a friend, who can see the content of the post after he or she enters the password. But why would anyone want to password-protect a post? Imagine that you just ate dinner at your mother-in-law's house, and she made the *worst* pot roast you've ever eaten. You can write all about it! Protect it with a password and give the password to your trusted friends so that they can read about it without offending your mother-in-law.

    Figure 1-7 shows a published post that's private; visitors see that a post exists, but they need to enter a password in the text box and then click Submit to view it.
  - *Private:* Publish this post to your blog so that only you can see it; no one else will be able to see it, ever. You may want to protect personal and private posts that you write only to yourself (if you're keeping a personal diary, for example).
- **Publish Immediately:** Click the Edit link to make the publish date options appear, where you can set the time stamp for your post. If you want the post to have the current time and date, ignore this setting.

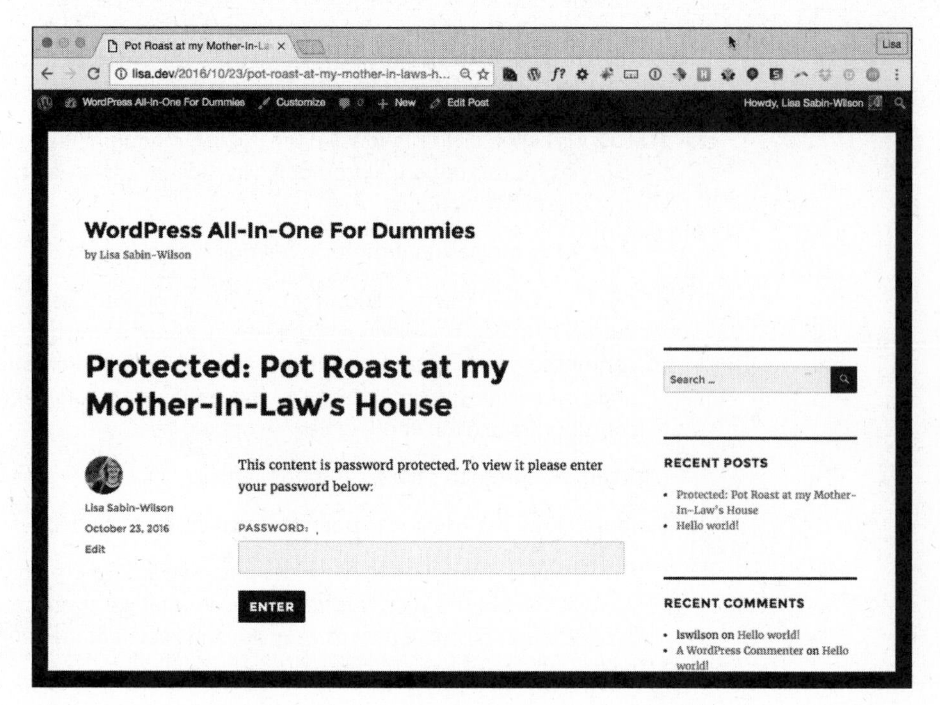

FIGURE 1-7: A password-protected post.

If you want to future-publish this post, you can set the time and date for anytime in the future. If you have a vacation planned and don't want your site to go without updates while you're gone, for example, you can write a few posts and set the date for a time in the future. Those posts are published to your site while you're somewhere tropical, diving with the fishes.

- **Publish:** This button wastes no time! It bypasses all the previous draft, pending review, and sticky settings to publish your post immediately.

After you choose an option from the Publish drop-down list, click the OK button. The Write Post page saves your publishing-status option.

TIP

If you click Publish and for some reason don't see the post on your site, you probably left the Status drop-down list set to Draft. Your new post appears in the draft posts, which you can find on the Dashboard's Posts page. Just click the Posts link on the navigation menu.

Being Your Own Editor

While I write this book, I have editors and proofreaders looking over my shoulder, making recommendations, correcting typos and grammatical errors, and telling me when I get too long-winded. You, on the other hand, are not so lucky! You're your own editor and have full control of what you write, when you write it, and how you write it.

You can always go back and edit previous posts to correct typos, grammatical errors, and other mistakes by following these steps:

1. **Find the post that you want to edit by clicking the All Posts link on the Posts menu of the Dashboard.**

   The Posts screen opens and lists the 20 most recent posts you've created.

   TIP

   To filter that listing of posts by date, choose a date from the All Dates drop-down list at the top of the Posts screen (Dashboard⇨Posts). If you choose February 2017, the Posts page reloads, displaying only those posts that were published in the month of February 2017.

   You can also filter the post listing by category. Choose your desired category from the All Categories drop-down list.

2. **When you find the post you need, click its title.**

   Alternatively, you can click the Edit link that appears below the post title.

   The Edit Post screen opens. In this screen, you can edit the post and/or any of its options.

   TIP

   If you need to edit only the post options, click the Quick Edit link. A drop-down Quick Edit menu appears, displaying the post options that you can configure, such as title, status, password, categories, tags, comments, and time stamp. Click the Save button to save your changes.

3. **Edit your post; then click the Update Post button.**

   The Edit Post window refreshes with all your changes saved.

Creating Your Own Workspace for Writing

In Book 3, Chapter 1, you discover how to organize the Dashboard to create your own customized workspace by rearranging modules and screen options. The Add New Post screen, where you write, edit, and publish your post, has the same options available, allowing you to fully control the workspace arrangement to create your own custom space that suits your writing needs.

To start customizing your workspace, open the Add New Post page by clicking the Add New link on the Posts menu of the Dashboard.

Adjusting screen options

Several items appear on the Add New Post page, as described in "Publishing Your Post," earlier in this chapter. You may not use all these items; in fact, you may find that simply removing them from the Add New Post page (and the Edit Post page) makes writing your posts easier and more efficient. To remove an item, follow these steps:

1. **Click the Screen Options tab at the top of the screen.**

   The Screen Options panel drops down. (Refer to Figure 1-4.)

2. **Select or deselect items below the Boxes heading.**

   Select an item by placing a check mark in the check box to the left of its name; deselect it by removing the check mark. Selected items appear on the page, and deselected items are removed from the page.

   TIP

   If you deselect an item that you want to include again on the Add New Post screen, it's not gone forever! Revisit the Screen Options panel and reselect its check box to make that item appear on the page once again.

3. **Select your preferred Screen Layout.**

   You can choose whether your Dashboard appears in one column or two columns (the default option).

4. **Click the Screen Options tab when you're done.**

   The Screen Options panel closes, and the options you've chosen are saved and remembered by WordPress.

Figure 1-8 shows my Add New Post screen displayed with a one-column layout because I like to have as much of the editor box showing as possible on my computer screen.

Arranging post modules

Aside from being able to make the Post text box bigger (or smaller) by using options in the Screen Options tab, you can't edit the Post text box module. You can configure all other modules on the Add New Post screen (and the Edit Post page). You can remove them (in the Screen Options panel, as I discuss in the preceding section), expand and collapse them, and drag them around to place them in different spots on your screen.

Collapse (that is, close) any of the modules by clicking the down arrow that appears to the right of the module name, as shown in Figure 1-9 for the Categories module. Likewise, you can *expand* (or open) a module by doing the same when it's collapsed.

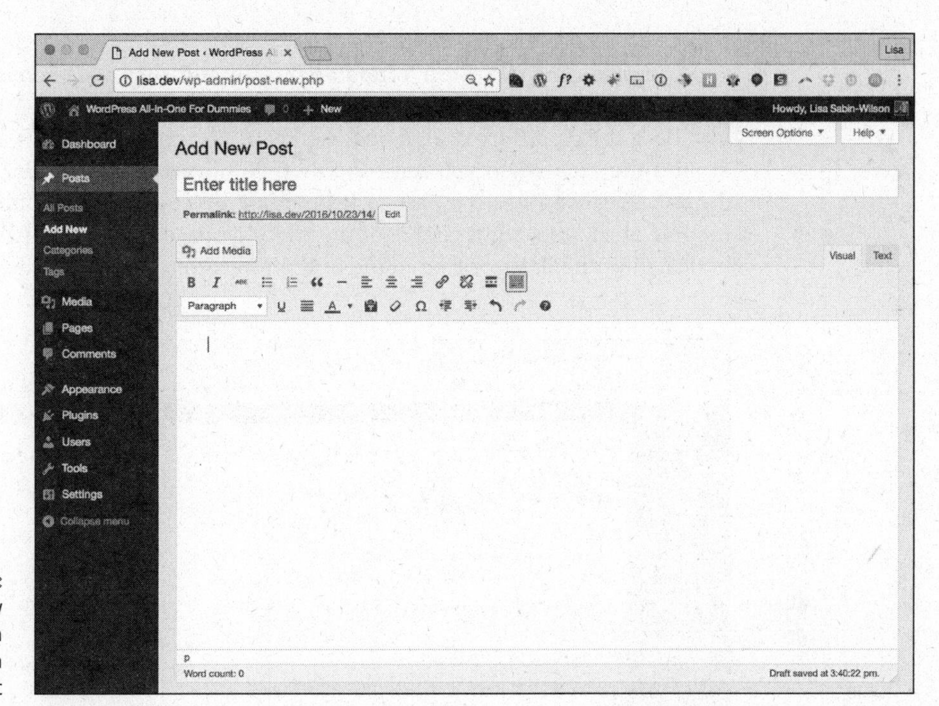

FIGURE 1-8: The Add New Post screen with a one-column format

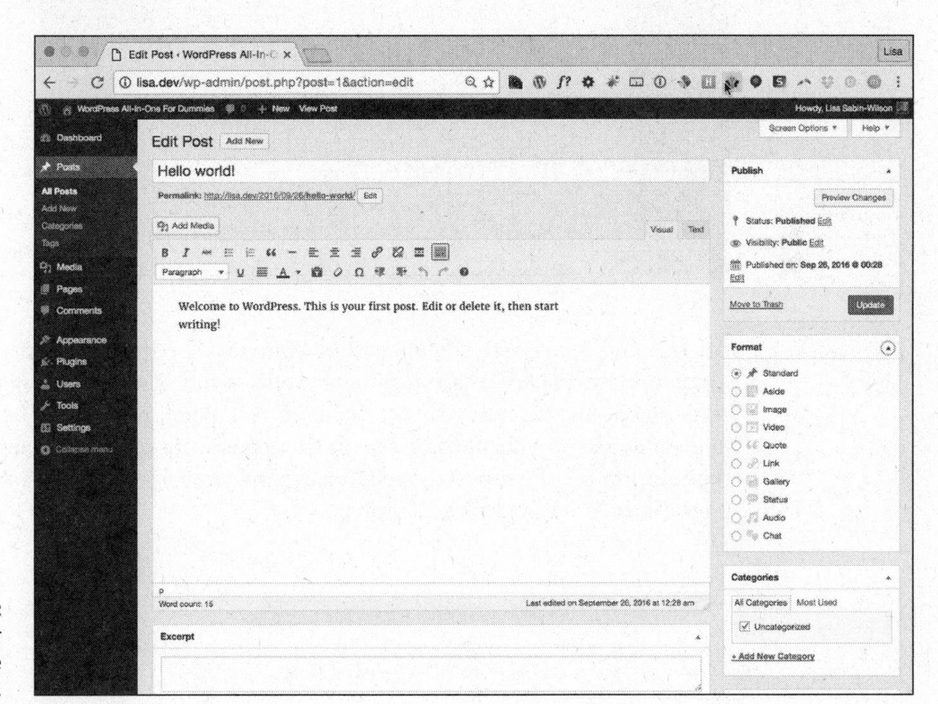

FIGURE 1-9: Expand or collapse modules.

You can also drag and drop a module on the Add New Post screen to position it wherever you want. Just click a module and, while holding down the mouse button, drag it to a different area of the screen. WordPress displays a dashed border around the area when you have the module hovering over a spot where you can drop it. Because I use the Categories module on every post I publish, I want that module at the top-right corner of my writing space. Figure 1-10 shows the action of dragging the Categories module to the top right of the Add New Post screen.

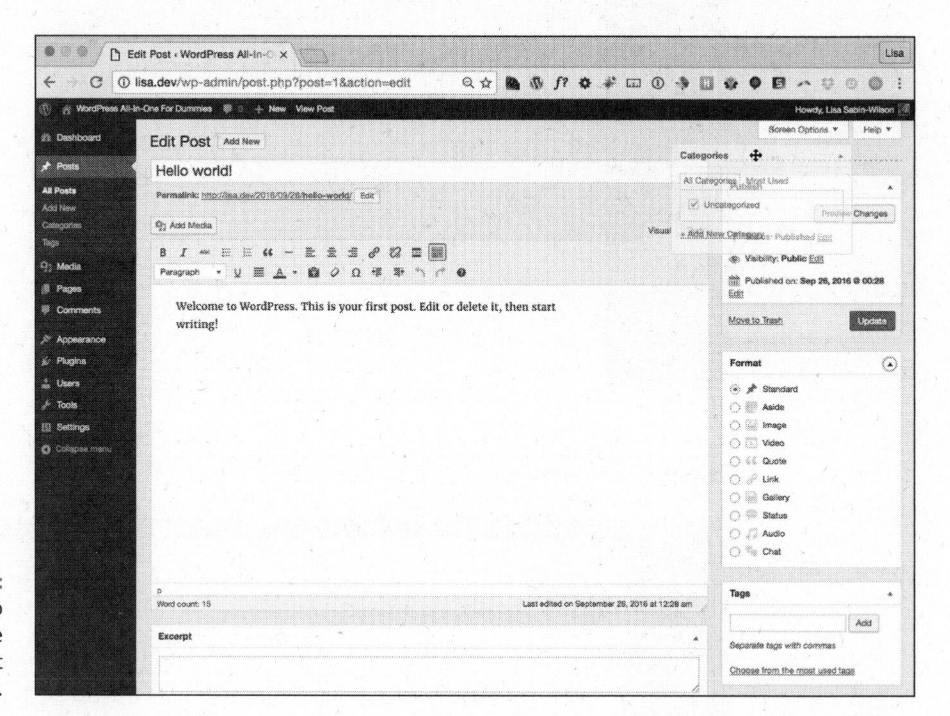

FIGURE 1-10: Drag and drop modules on the Add New Post page.

On the Add New Post screen within your Dashboard, you can really configure a custom workspace that suits your style, work habits, and needs. WordPress remembers all the changes you make on this page, including the screen options and modules, so you have to set up this page only once. You can drag and drop modules on any Dashboard page in the same way you do on the main Dashboard page, as covered in Book 3, Chapter 1.

» Creating a new static page on your website

» Adding a blog to your site

Chapter 2
Creating a Static Page

In Book 3, Chapter 5, I discuss the different ways that content gets archived by WordPress, and in Book 3, Chapter 1, I give you a very brief introduction to the concept of pages and where to find them on the WordPress Dashboard.

This chapter takes you through the full concept of pages in WordPress, including how to write and publish them. This chapter also fully explains the difference between posts and pages in WordPress so that you know which to publish for different situations.

Understanding the Difference between Posts and Pages

Pages, in WordPress, are different from posts because they don't get archived the way your posts do. They aren't categorized or tagged, don't appear in your listing of recent posts or date archives, and aren't syndicated in the RSS feeds available on your site — because content within pages generally doesn't change. (Book 3, Chapter 5 gives you all the details on how the WordPress archives work.)

REMEMBER

Use pages for static or stand-alone content that exists separately from the archived post content on your site.

With the page feature, you can create an unlimited amount of static pages separate from your posts. People commonly use this feature to create About Me or Contact Me pages, among other things. Table 2-1 illustrates the differences between posts and pages by showing you the different ways the WordPress platform handles them.

TABLE 2-1 Differences between a Post and a Page

WordPress Options	Page	Post
Appears in blog post listings	No	Yes
Appears as a static page	Yes	No
Appears in category archives	No	Yes
Appears in monthly archives	No	Yes
Appears in Recent Posts listings	No	Yes
Appears in site RSS feed	No	Yes
Appears in search results	Yes	Yes
Uses tags and/or categories	No	Yes

Creating the Front Page of Your Website

For the most part, when you visit a site powered by WordPress, the blog can appear on the main page or as a separate page of the site. My business partner, Brad Williams, keeps a personal blog at `http://strangework.com/`, powered by WordPress (of course). His blog shows his latest posts on the front page, along with links to the post archives (by month or by category) in the sidebar. This setup is typical of a site run by WordPress. (See Figure 2-1.)

But the front page of our business site at `http://webdevstudios.com`, also powered by WordPress, contains no blog and displays no blog posts. (See Figure 2-2.) Instead, it displays the contents of a static page that we created in the WordPress Dashboard. This static page serves as a portal that displays pieces of content from other sections within the entire site. The site does include a blog but also serves as a full-blown business website with all the sections I need to provide my clients the information they want.

Both of my sites are powered by the self-hosted version of WordPress.org, so how can they differ so much in what they display on the front page? The answer lies in the templates in the WordPress Dashboard.

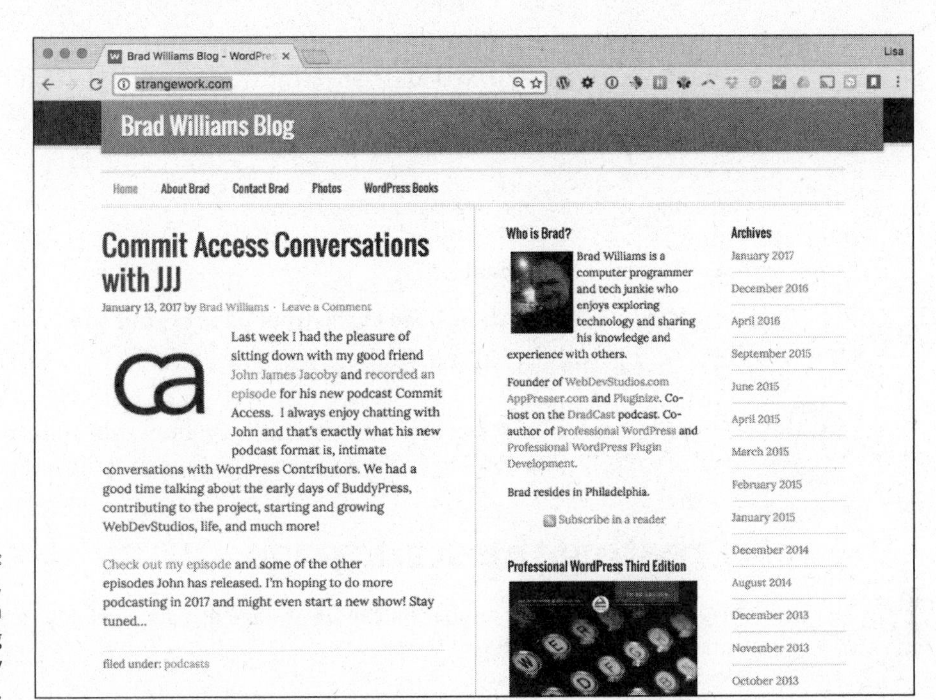

FIGURE 2-1: A personal blog, set up like a standard blog powered by WordPress.

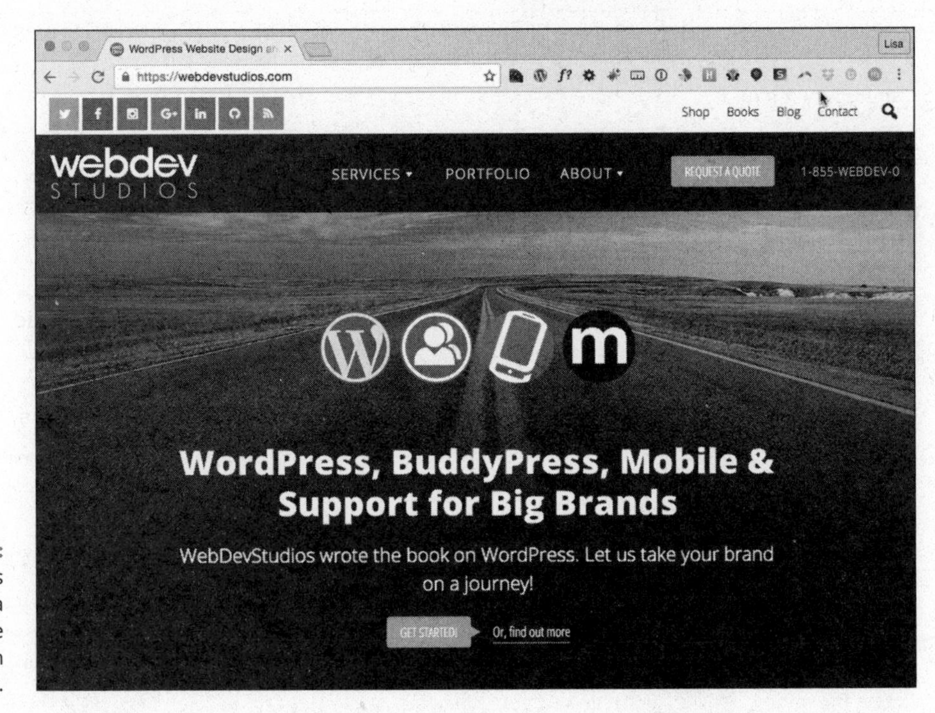

FIGURE 2-2: My business site, set up as a business website rather than a blog.

You use static pages in WordPress to create content that you don't want to appear as part of your blog but do want to appear as part of your overall site (such as a bio page, a page of services, and so on).

Creating a front page is a three-step process:

1. **Create a static page.**
2. **Designate that static page as the front page of your site.**
3. **Tweak the page to look however you like.**

REMEMBER

By using this method, you can create unlimited numbers of static pages to build an entire website. You don't even need to have a blog on this site unless you want one.

Creating the static page

To have a static page appear on the front page of your site, you need to create that page. Follow these steps:

1. **Click the Add New link on the Pages menu of the Dashboard.**

   The Add New Page screen opens, allowing you to write a new page for your WordPress site, as shown in Figure 2-3.

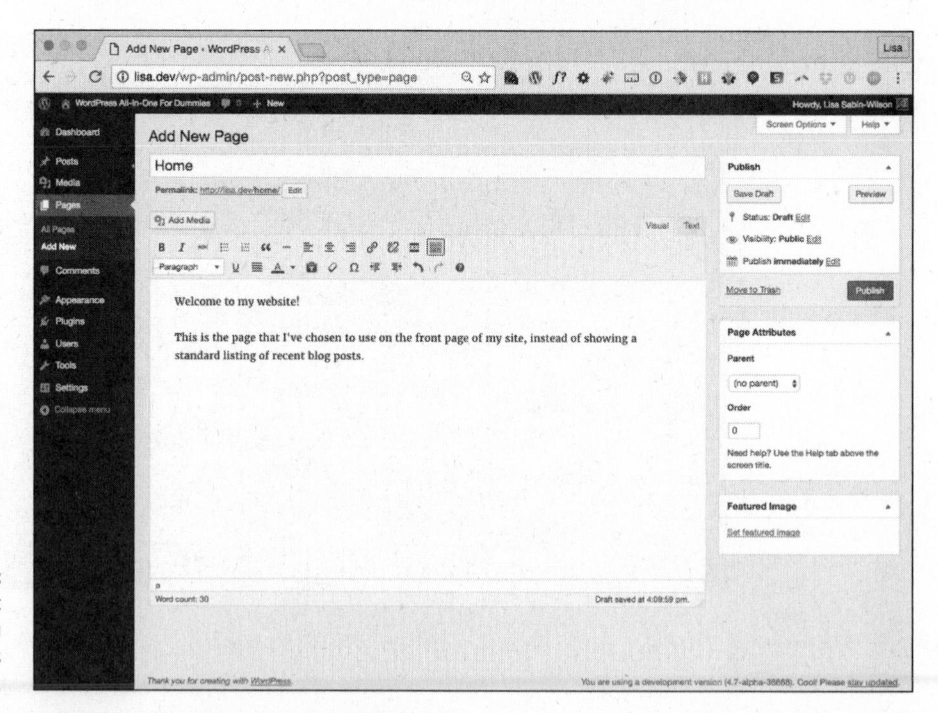

FIGURE 2-3: Create the static page that you want to use as your front page.

2. **Type a title for your page in the text box at the top.**

3. **Type the content of your page in the large text box.**

4. **Set the options for this page.**

   I explain the options on this page in the following section.

5. **Click the Publish button.**

   The page is saved to your database and published to your WordPress site with its individual URL (or *permalink*). The URL for the static page consists of your blog URL and the title of the page. If you titled your page About Me, for example, the URL of the page is `http://yourdomain.com/about-me`. (See Book 3, Chapter 2 for more information about permalinks.)

TECHNICAL STUFF

The Page Template option is set to Default Template. This setting tells WordPress that you want to use the default page template (`page.php` in your theme template files) to format the page you're creating. The default page template is the default setting for all pages you create; you can assign a different page template to pages you create if your theme has made different page templates available for use. In Book 6, Chapter 6, you can find extensive information on advanced WordPress themes, including information on page templates and how to create and use them on your site.

Setting page options

Before you publish a new page to your site, you can change options to use different features available in WordPress. These features are similar to the ones available for publishing posts, which you can read about in Book 4, Chapter 1:

- **Custom Fields:** Custom fields add extra data to your page, and you can fully configure them. You can read more about the Custom Fields feature in Book 4, Chapter 5.
- **Discussion:** Decide whether to let readers submit comments through the comment system by selecting or deselecting the Allow Comments text box. By default, the box is selected; deselect it to disallow comments on this page.

  TIP

  Typically, you don't see a lot of static pages that have the Comments feature enabled because pages offer static content that generally doesn't lend itself to a great deal of discussion. There are exceptions, however, such as a Contact page, which might use the Comments feature as a way for readers to get in touch with you through that specific page. Of course, the choice is yours to make based on the specific needs of your website.

- **Author:** If you're running a multiauthor site, you can select the name of the author you want to be attributed to this page. By default, your own author name is selected here.
- **Publish:** The publishing options for your post are covered in Book 4, Chapter 1.
- **Parent:** Select a parent for the page you're publishing. Book 3, Chapter 5 covers the different archiving options, including the ability to have a hierarchical structure for pages that create a navigation of main pages and subpages (called parent and child pages).
- **Template:** You can assign the page template if you're using a template other than the default one. (Book 6, Chapter 6 contains more information about themes and templates, including using page templates on your site.)
- **Order:** By default, this option is set to 0 (zero). You can enter a number, however, if you want this page to appear in a certain spot on the page menu of your site.

  If you're using the built-in menu feature in WordPress, you can use this option. You don't have to use it, however, because you can define the order of pages and how they appear on your menu by assigning a number to the page order. A page with the page order of 1 appears first on your navigation menu, a page with the page order of 2 appears second, and so on. Book 6, Chapter 1 covers the Menu feature in greater detail.
- **Featured Image:** Some WordPress themes are configured to use an image (photo) to represent each post that you have on your site. The image can appear on the home/front page, blog page, archives, or anywhere within the content display on your website. If you're using a theme that has this option, you can easily define a post's thumbnail by clicking the Set Featured Image link below the Featured Image module on the Add New Post screen. Then you can assign an image that you've uploaded to your site as the featured image for a particular post.

Assigning a static page as the front page

After you create the page you want to use for the front page of your website, tell WordPress that you want the static page to serve as the front page of your site. Follow these steps:

1. **Click the Reading link on the Settings menu of the Dashboard to display the Reading Settings screen.**
2. **Select the option A Static Page in the Front Page Displays section to indicate you'd like to use a static page on the front page.**
3. **From the drop-down list, choose the page you want to use for the front page of your site (see Figure 2-4).**

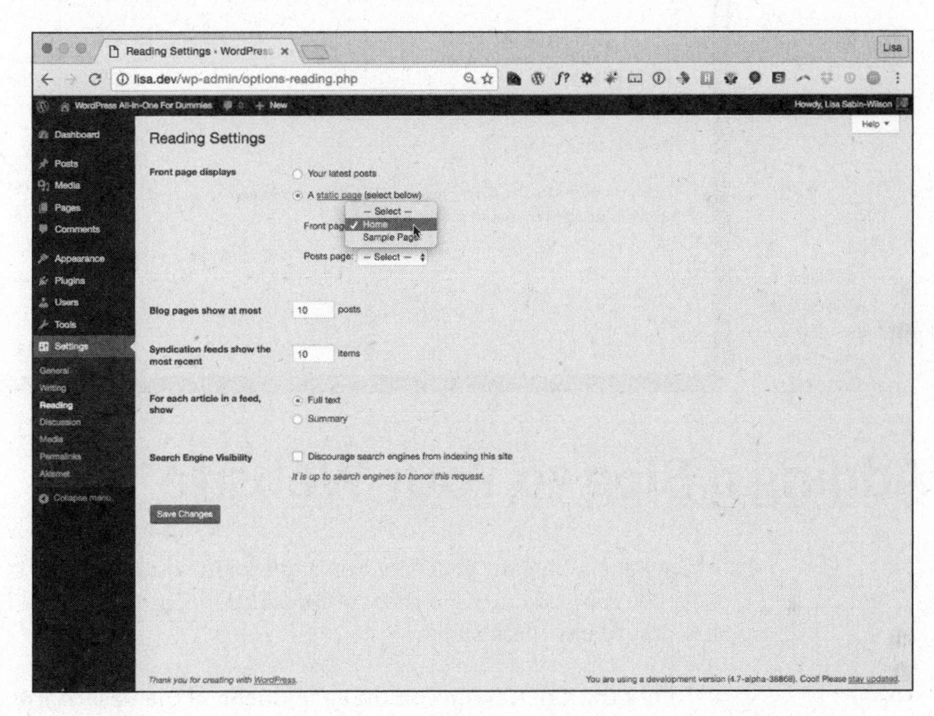

FIGURE 2-4: Choosing which page to display as the front page.

4. **Click the Save Changes button at the bottom of the Reading Settings screen.**

   WordPress displays the page you selected in Step 3 as the front page of your site. Figure 2-5 shows my site displaying a static page.

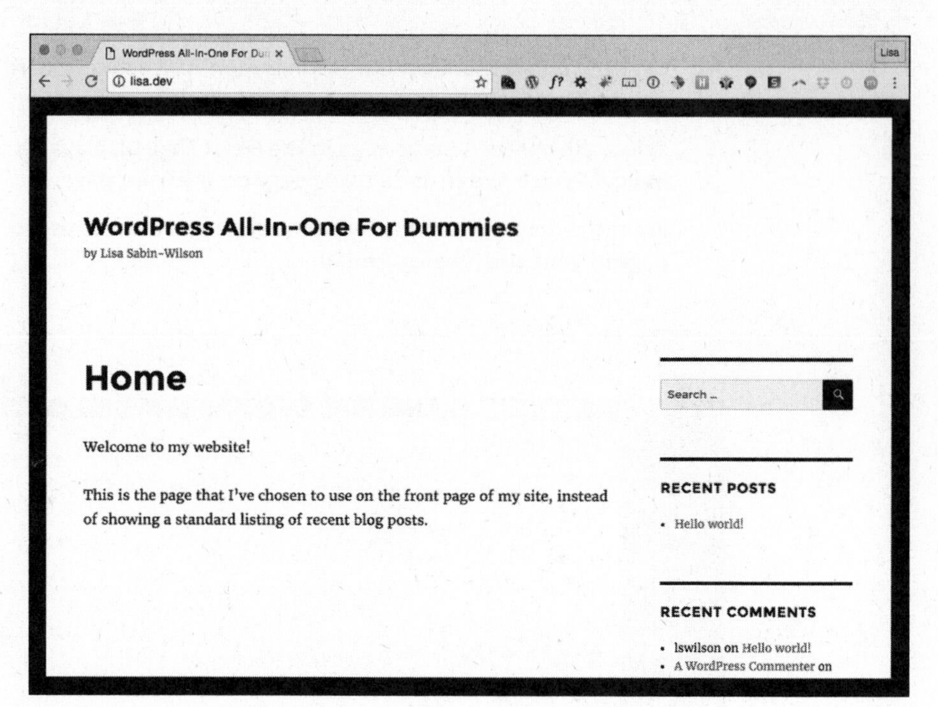

FIGURE 2-5: WordPress displays the page you selected as your front page.

Adding a Blog to Your Website

If you want a blog on your site but don't want to display the blog on the front page, you can add one from the WordPress Dashboard. To create the blog for your site, first follow these steps:

1. **Click the Add New link on the Pages menu of the Dashboard.**

   The page where you can write a new page to your WordPress site opens.

2. **Type** Blog **in the Title text box.**

   The page slug is automatically set to `/blog`. (Read more about slugs in Book 3, Chapter 5.)

3. **Leave the Page Content text box blank.**
4. **Click the Publish button.**

   The page is saved to your database and published to your WordPress site. Now you have a blank page that redirects to `http://`*`yourdomain`*`.com/blog`.

   Next, you need to assign the page you just created as your blog page.

5. **Click the Reading link on the Settings menu of the Dashboard.**

   The Reading Settings screen opens.

6. **Select the Blog page in the drop-down menu labeled Posts Page.**
7. **Click the Save Changes button.**

   The options you just set are saved, and your blog is now at `http://yourdomain.com/blog` (where *yourdomain.com* is the actual domain name of your site). When you navigate to `http://yourdomain.com/blog`, your blog appears.

REMEMBER

This method of using the `/blog` page slug works only if you're using custom permalinks with your WordPress installation. (See Book 3, Chapter 2 if you want more information about permalinks.) If you're using the default permalinks, the URL for your blog page is different; it looks something like `http://yourdomain.com/?p=4` (where 4 is the ID of the page you created for your blog).

» Inserting a photo into your post

» Creating photo galleries in WordPress

Chapter **3**

Uploading and Displaying Photos and Galleries

Adding images and photos to your posts can really dress up the content. By using images and photos, you give your content a dimension that you can't express in plain text. Through visual imagery, you can call attention to your post and add depth to it. With WordPress, you can insert single images or photographs, or you can use a few nifty plugins to turn some of the pages in your site into a full-fledged photo gallery.

In this chapter, you discover how to add some special touches to your site posts by adding images and photo galleries, all by using the built-in image-upload feature and image editor in WordPress.

Inserting Images into Your Posts

You can add images to a post pretty easily by using the WordPress image uploader. Jump right in and give it a go: From the Dashboard, click the Add New link on the Posts menu, and then, on the Add New Post screen, click the Add Media button.

The Insert Media window that appears lets you choose images from your hard drive or from a location on the web. (See Figure 3-1.)

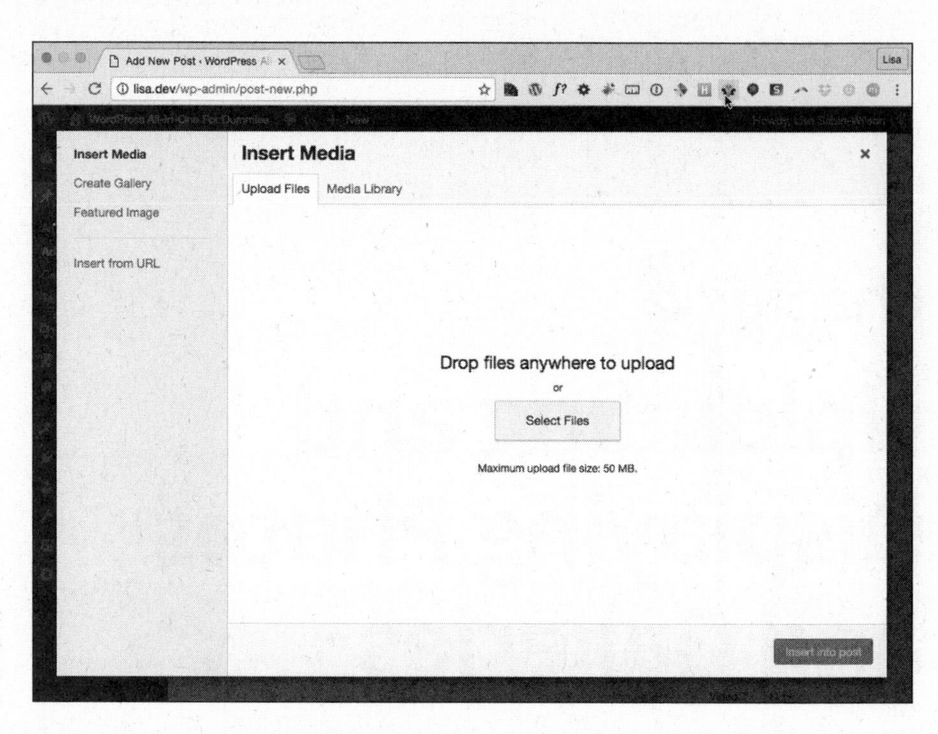

FIGURE 3-1: The WordPress Insert Media window.

Adding an image from your computer

To add an image from your own hard drive, follow these steps:

1. **Click the Add Media button on the Add New Post screen.**

   The Insert Media window appears. (Refer to Figure 3-1.)

2. **Click the Upload Files tab at the top and then click the Select Files button.**

   A dialog box, from which you can select an image (or multiple images) from your hard drive, opens.

3. **Select your image(s) and then click Open.**

   The image is uploaded from your computer to your web server, and the Insert Media window displays your uploaded image selected and ready for editing.

4. **Edit the details for the image(s) in the Attachment Details section of the Insert Media window (see Figure 3-2).**

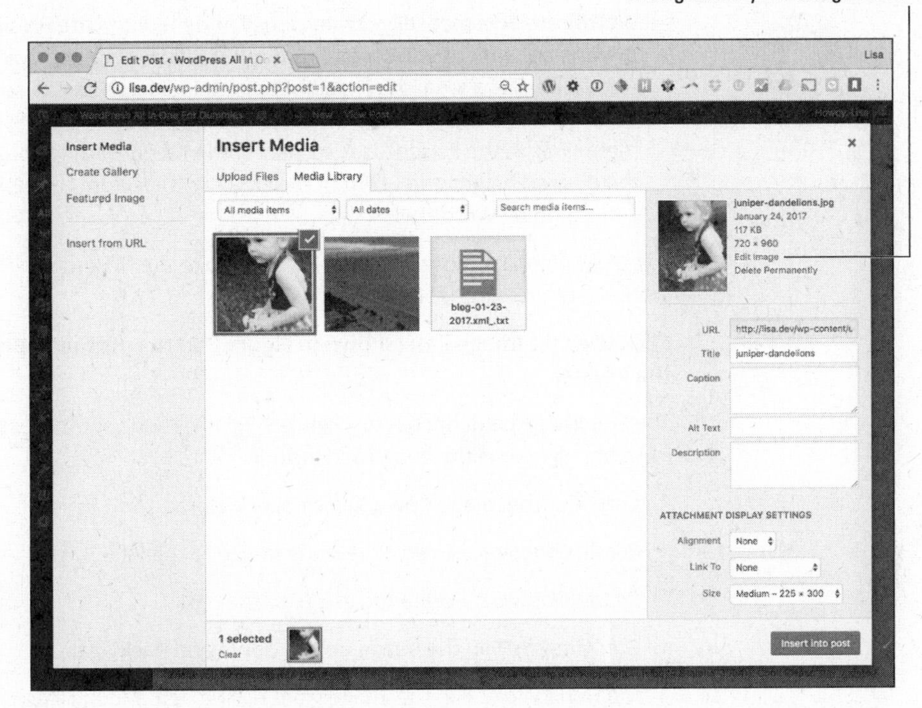

FIGURE 3-2: You can set several options for your images after you upload them.

The Attachment Details section provides several image options:

- *URL:* This isn't a configurable option; WordPress does not allow you to change the URL of uploaded media.
- *Title:* Type a title for the image.
- *Caption:* Type a caption for the image (such as **Little Miss Sunshine**).
- *Alt Text:* Type the alternative text for the image. More information on alternative text is in Book 4, Chapter 6.
- *Description:* Type a description of the image.
- *Alignment:* Select None, Left, Center, or Right. (See Table 3-1, later in this chapter, for styling information regarding image alignment.)
- *Link To:* If you want the image to be linked to a URL, type that URL in this text box. Alternatively, select the appropriate option button to determine where your readers go when they click the image you uploaded: Selecting None means the image isn't clickable, Media File URL directs readers to the image itself, Attachment URL directs readers to the post in which the image appears, and Custom URL allows you to enter whatever URL you like.
- *Size:* Select Thumbnail, Medium, or Full Size.

TIP

WordPress automatically creates small and medium-size versions of the original images you upload through the built-in image uploader. A thumbnail is a smaller version of the original file. You can edit the size of the thumbnail by clicking the Settings link and then clicking the Media menu link. In the Image Sizes section of the Media Settings page, designate the desired height and width of the small and medium thumbnail images generated by WordPress.

If you're uploading more than one image, skip to the "Inserting a Photo Gallery" section later in this chapter.

5. **Click the Edit Image link (shown in Figure 3-2) to edit the appearance of the image.**

   The Edit Image page opens (see Figure 3-3). Its options are represented by icons shown above the image and include

   - *Crop:* Cut the image down to a smaller size.
   - *Rotate Counterclockwise:* Rotate the image to the left.
   - *Rotate Clockwise:* Rotate the image to the right.
   - *Flip Vertically:* Flip the image upside down and back again.
   - *Flip Horizontally:* Flip the image from right to left and back again.
   - *Undo:* Undo any changes you made.
   - *Redo:* Redo images edits that you've undone.
   - *Scale Image:* Set a specific width and height for the image.

6. **Click the Save button on the Edit Image page when you're done editing the image.**
7. **Return to the post into which you'd like to insert the image by clicking the All Posts link on the Posts menu and then clicking the title of the post you need.**
8. **Click the Add Media button.**
9. **Click the Media Library tab in the Insert Media Window.**

   This step loads all the images you've ever uploaded to your site.

10. **Select the image you'd like to use by clicking it.**
11. **Click the Insert into Post button.**

    The Add an Image window closes, and the Add New Post screen (or the Add New Page screen, if you're writing a page) reappears. WordPress has inserted the HTML to display the image in your post, as shown in Figure 3-4. You can continue editing your post, save it, or publish it.

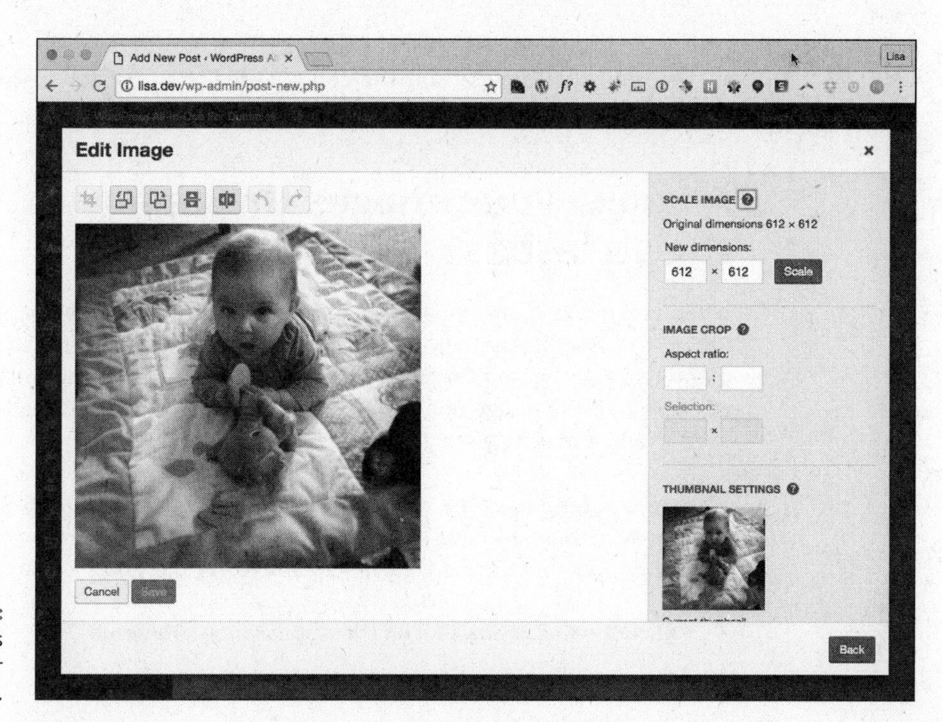

FIGURE 3-3: The WordPress image editor options.

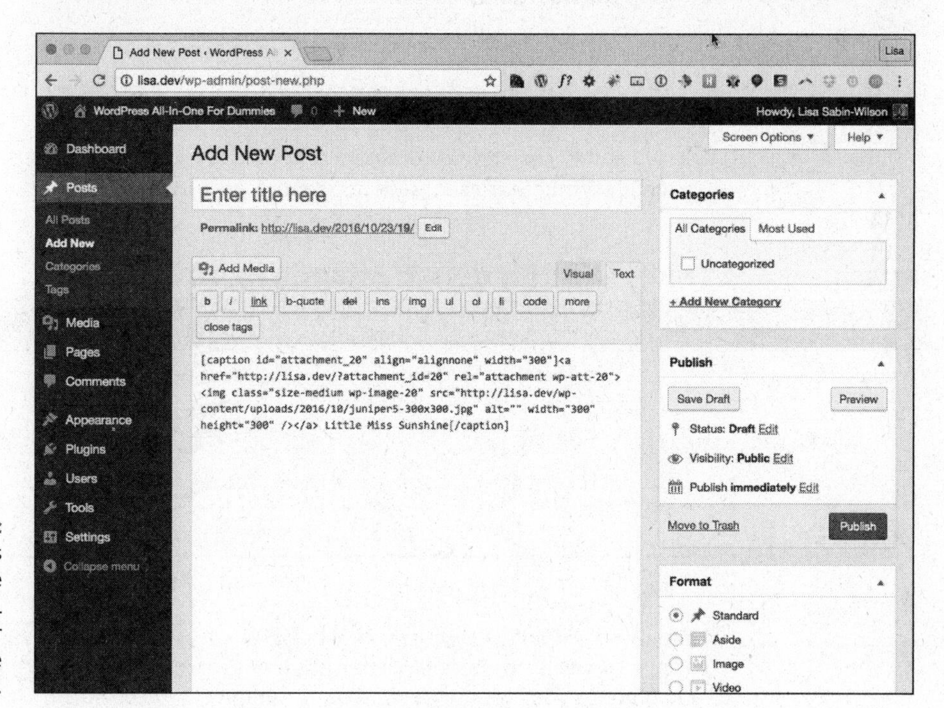

FIGURE 3-4: WordPress inserts the correct HTML code for your uploaded image into your post.

TIP

To see the actual image, not the code, click the Visual tab just above the Post text box.

Aligning your images through the stylesheet

When you upload your image, you can set its alignment as None, Left, Center, or Right. The WordPress theme you're using, however, may not have these alignment styles accounted for in its stylesheet. If you set the alignment to Left, for example, but the image on your blog doesn't appear to be aligned at all, you may need to add a few styles to your theme's stylesheet.

Themes and templates are discussed in greater detail in Book 6. For purposes of making sure that you have the correct image alignment for your newly uploaded images, follow these steps for a quick-and-dirty method:

1. **Click the Customize link on the Appearance submenu.**

   The Customizer screen opens.

2. **Click the Additional CSS link.**

   The Customizing Additional CSS screen opens.

3. **Add your desired styles to the text box.**

Table 3-1 shows the styles you can add to your stylesheet to make sure that image-alignment styling is present and accounted for in your theme.

TABLE 3-1 Styling Techniques for Image Alignment

Image Alignment	Add This to Your Stylesheet (style.css)
None	`img.alignnone {float:none; margin: 5px 0 5px 0;}`
Left	`img.alignleft {float:left; margin: 5px 10px 5px 0;}`
Center	`img.aligncenter {display:block; float:none; margin: 5px auto;}`
Right	`img.alignright {float:right; margin: 5px 0 5px 10px;}`

These styles are just examples of what you can do. Get creative with your own styling. You can find more information about using CSS (Cascading Style Sheets) to add style to your theme(s) in Book 6, Chapter 4.

Figure 3-5 displays the WordPress Customizer open with the additional CSS shown on the left side.

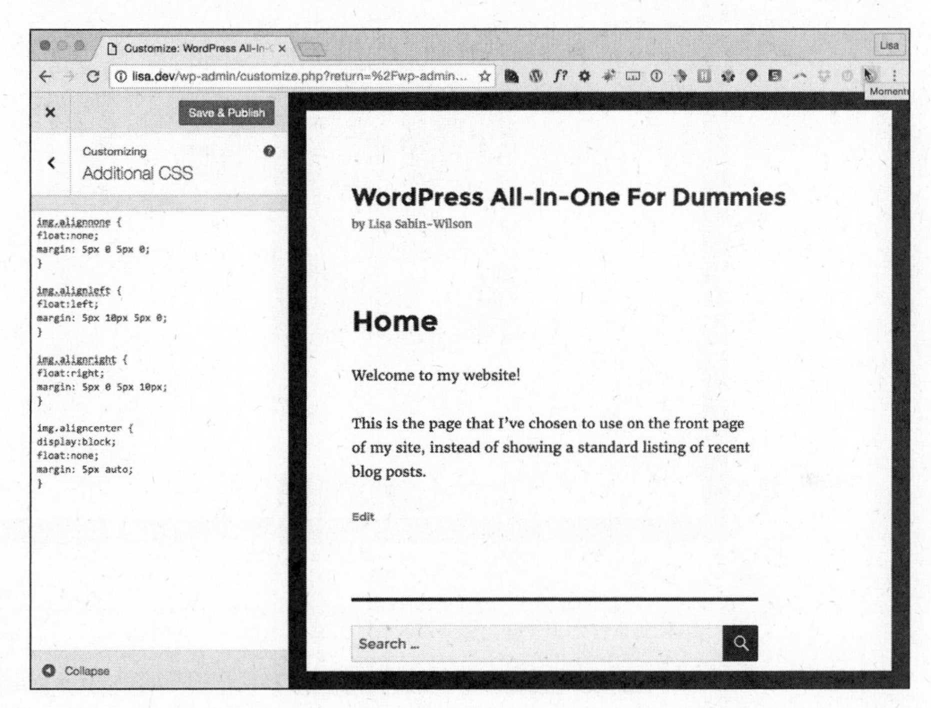

FIGURE 3-5: Additional CSS in the WordPress Customizer.

Inserting a Photo Gallery

You can also use the Insert Media window to insert a full photo gallery into your posts. Upload all your images and then, instead of clicking the Insert into Post button, click the Create Gallery link on the left side of the Insert Media window. The Create Gallery window opens and displays all the images you've uploaded to your site.

Follow these steps to insert a photo gallery into a blog post:

1. **In the Create Gallery window, select the images you want to use in your gallery.**

   Click each image once to select it for use in the gallery. A selected image displays a small check mark in its top-right corner. (See Figure 3-6.)

2. **Click the Create a New Gallery button.**

   The Edit Gallery screen opens, shown in Figure 3-7.

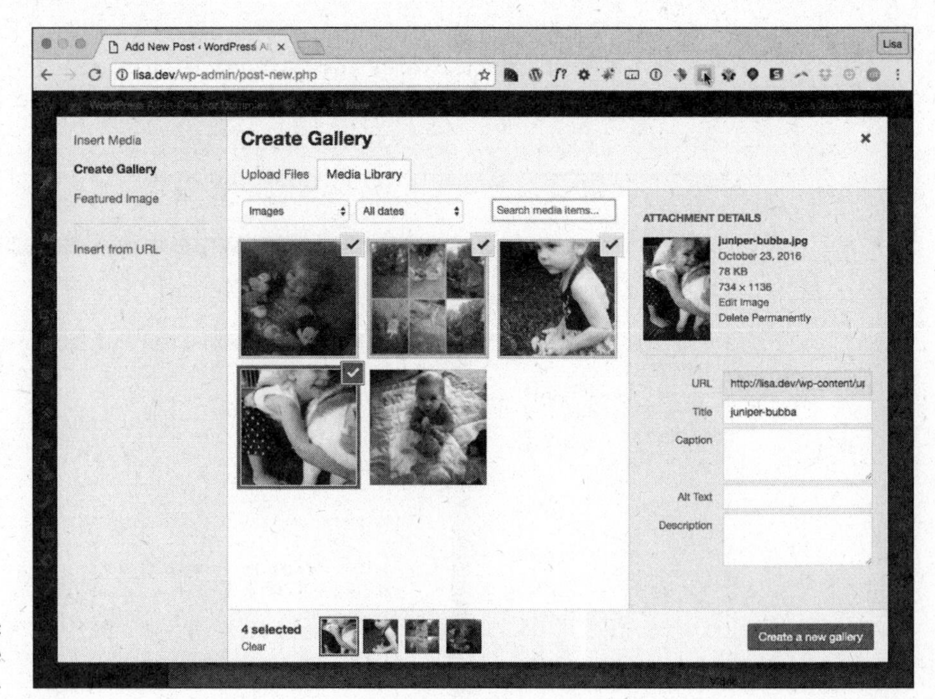

FIGURE 3-6: The Create Gallery window.

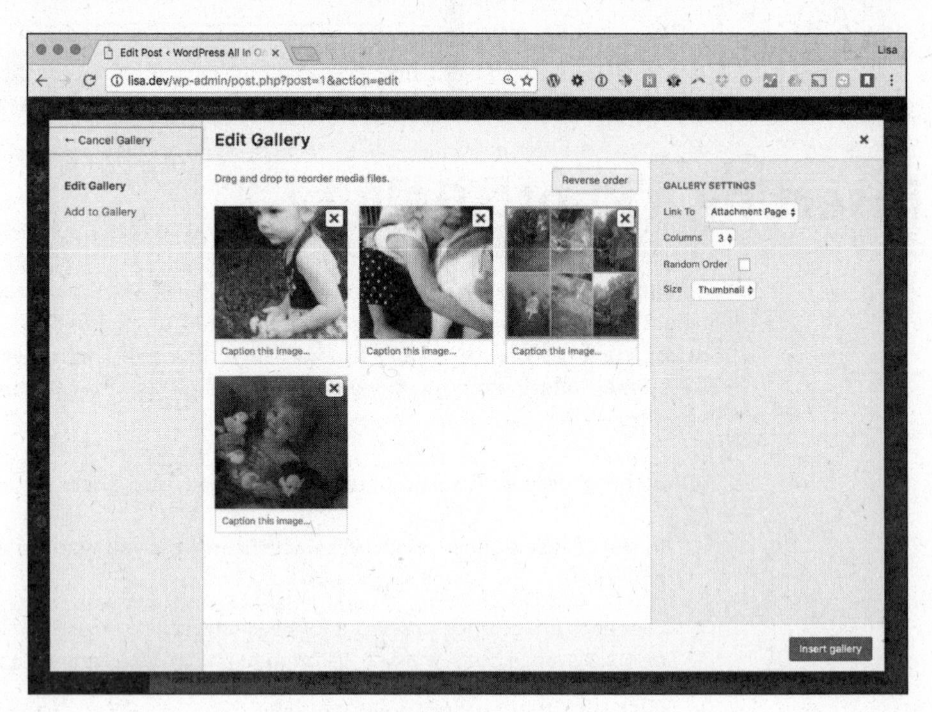

FIGURE 3-7: The Edit Gallery screen.

3. (Optional) Add a caption for each image by clicking the Caption This Image area and typing a caption or short description for the image.

4. (Optional) Set the order in which the images appear in the gallery by using the drag-and-drop option on the Edit Gallery page.

Click and then drag and drop images to change the order.

5. (Optional) Specify the following options in the Gallery Settings section on the right side of the Edit Gallery page:

- *Link To:* Select either Attachment Page, Media File, or None to tell WordPress what you'd like the images in the gallery to link to.
- *Random Order:* Select to randomize the order in which the images are displayed in the gallery.
- *Gallery Columns:* Select how many columns of images you want to appear in your gallery.

6. Click the Insert Gallery button.

WordPress inserts into your post a piece of shortcode that looks like this: `[gallery ids"1,2,3"]`. A shortcode is a small piece of WordPress-specific code that allows you to do big things with little effort, such as embed photo galleries and media files.

7. (Optional) Change the order of appearance of the images in the gallery, as well as the markup (HTML tags or CSS selectors).

Use the WordPress gallery shortcode (see Table 3-2) to change different aspects of the display of the gallery in your post:

- *`captiontag`:* Changes the markup that surrounds the image caption by altering the gallery shortcode. `[gallery captiontag="div"]`, for example, places `<div></div>` tags around the image caption. (The `<div>` tag is considered to be a block-level element and creates a separate container for the content. For more about `<div>` tags and CSS, see Book 6, Chapter 4.) If you want to have the gallery appear on a line of its own, the `[gallery captiontag="p"]` code places `<p class="gallery-caption"></p>` tags around the image caption. The default markup for the `captiontag` option is `dd`.
- *`icontag`:* Defines the HTML markup around each individual thumbnail image in your gallery. Change the markup around the `icontag` (thumbnail icon) of the image by altering the gallery shortcode to something like `[gallery icontag="p"]`, which places `<p class="gallery-icon"></p>` tags around each thumbnail icon. The default markup for `icontag` is `dt`.

- *`itemtag`:* Defines the HTML markup around each item in your gallery. Change the markup around the `itemtag` (each item) in the gallery by altering the gallery shortcode to something like `[gallery itemtag= "span"]`, which places `<span class="gallery-item"></span>` tags around each item in the gallery. The default markup for the `itemtag` is `dl`.
- *`orderby`:* Defines the order in which the images are displayed within your gallery. Change the order used to display the thumbnails in the gallery by altering the gallery shortcode to something like `[gallery orderby= "menu_order ASC"]`, which displays the thumbnails in ascending menu order. Another parameter you can use is `ID_order ASC`, which displays the thumbnails in ascending order according to their IDs.

8. **Define the style of the `<span>` tags in your CSS stylesheet.**

   The `<span>` tags create an inline element. An element contained within a `<span>` tag stays on the same line as the element before it; there's no line break. You need a little knowledge of CSS to alter the `<span>` tags. Click the Customize link on the Appearance submenu in your WordPress Dashboard and then click the Additional CSS link to add CSS to your theme. Here's an example of what you can add to the stylesheet (`style.css`) for your current theme:

```
span.gallery-icon img {
padding: 3px;
background: white;
border: 1px solid black;
margin: 0 5px;
}
```

   Placing this bit of CSS in the stylesheet (`style.css`) of your active theme automatically places a 1-pixel black border around each thumbnail, with 3 pixels of padding and a white background. The left and right margins are 5 pixels wide, creating nice spacing between images in the gallery.

9. **Click the Save and Publish button to save the CSS changes in the Customizer.**

Figure 3-8 shows my post with a photo gallery displayed, using the preceding steps and CSS example in the default WordPress theme: Twenty Seventeen. I used the gallery shortcode for the gallery shown in Figure 3-8 — `[gallery columns="2" link="file" ids="11,10,9,8" orderby="rand"]`.

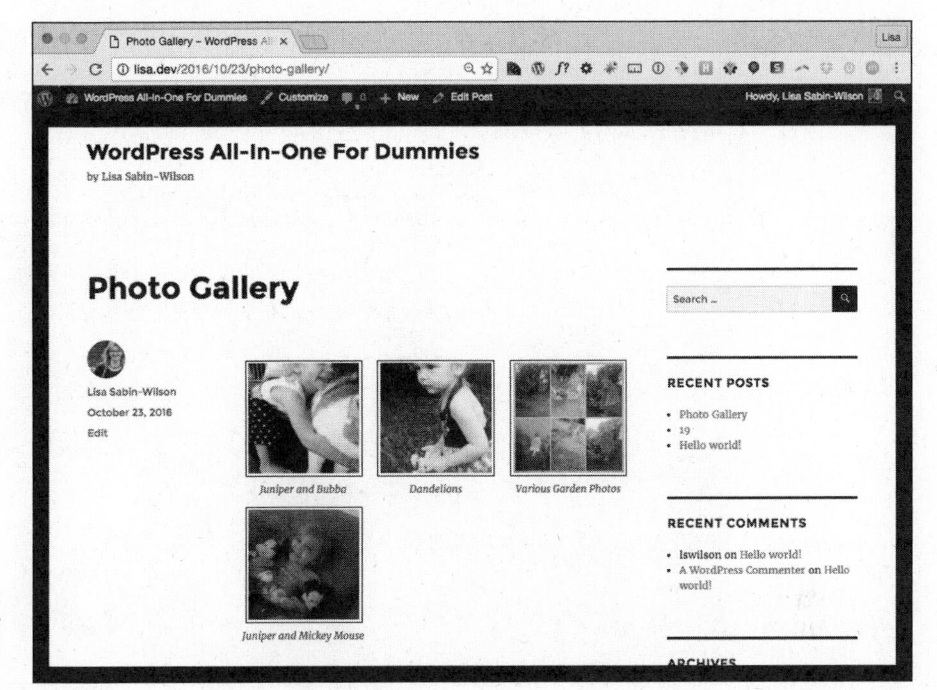

FIGURE 3-8: A photo gallery displayed in a post.

TABLE 3-2 Gallery Shortcode Examples

Gallery Shortcode	Output
`[gallery columns="4" size="medium"]`	A four-column gallery containing medium-size images
`[gallery columns="10" id="215" size="thumbnail"]`	A ten-column gallery containing thumbnail images pulled from the blog post with the ID 215
`[gallery captiontag="p" icontag="span"]`	A three-column (default) gallery in which each image is surrounded by `<span></span>` tags and the image caption is surrounded by `<p></p>` tags

TIP

Matt Mullenweg, cofounder of the WordPress platform, created a very extensive photo gallery by using the built-in gallery options in WordPress. Check out the fabulous photo gallery at `https://matt.blog`, shown in Figure 3-9.

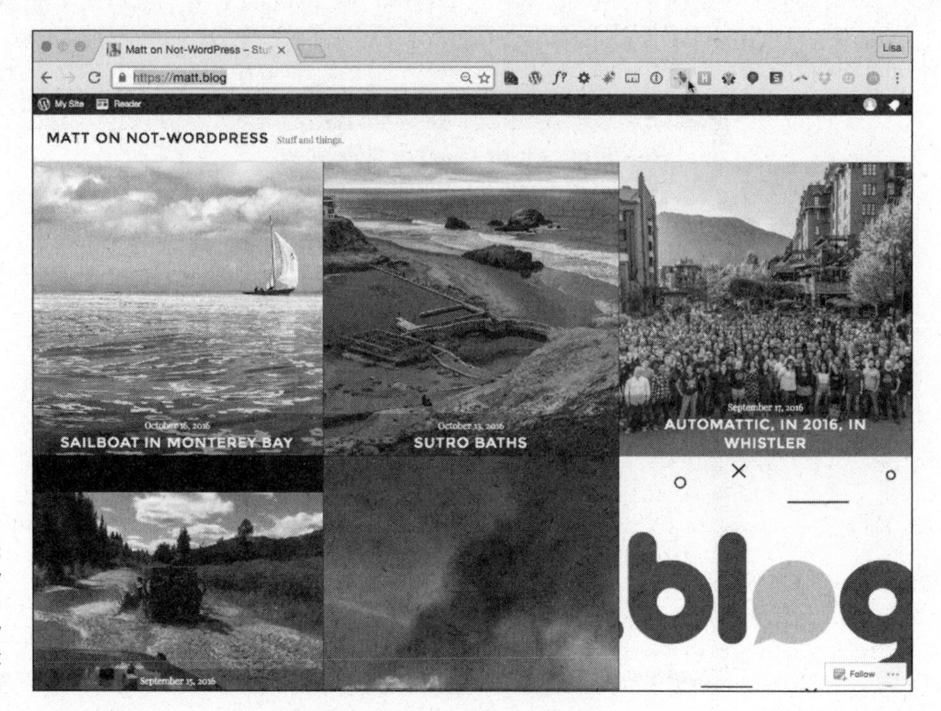

FIGURE 3-9: A photo gallery created with WordPress by founder Matt Mullenweg.

WORDPRESS GALLERY PLUGINS

Here are a few great gallery plugins:

- **NextGEN Gallery by Photocrati (`https://wordpress.org/plugins/nextgen-gallery`):** Creates sortable photo galleries and more
- **Image Widget by Modern Tribe (`https://wordpress.org/plugins/image-widget`):** Uses the WordPress media library to add image widgets to your site
- **Image Viewer Made Easy by Rakhitha Nimesh Ratnayake (`https://wordpress.org/plugins/image-viewer-made-easy`):** Creates image galleries within a few minutes while providing advanced features such as rotation, zooming, scaling, flipping, and playing as a slideshow

» Adding audio files to posts

» Exploring podcasting and vlogging

» Using the Media Library

Chapter 4

Exploring Podcasting and Video Blogging

Many website owners want to go beyond just offering written content for the consumption of their visitors by offering different types of media, including audio and video files. WordPress makes it pretty easy to include these different types of media files in your posts and pages by using the built-in file-upload feature.

The audio files you add to your site can include music or voice in formats such as `.mp3`, `.midi`, or `.wav` (to name just a few). Some website owners produce their own audio files in regular episodes, called *podcasts*, to create an Internet radio show. Often, you can find these audio files available for syndication through RSS and can subscribe to them in a variety of audio programs, such as iTunes.

You can include videos in posts or pages by embedding code offered by popular third-party video providers such as YouTube (`www.youtube.com`) and Vimeo (`www.vimeo.com`). Website owners can also produce and upload their own video shows, an activity known as *vlogging* (video blogging).

This chapter takes you through the steps to upload and embed audio and video files within your content, and it provides some tools that can help you embed those files without having to use elaborate coding techniques.

WARNING

When dealing with video and audio files on your site, remember to upload and use only media that you own or have permission to use. Copyright violation is a very serious offense, especially on the Internet, and using media that you don't have permission to use can have serious consequences, such as having your website taken down, facing heavy fines, and even going to jail. I'd really hate to see that happen to you. So play it safe, and use only those media files that you have permission to use.

Inserting Video Files into Your Content

Whether you're producing your own videos for publication or embedding other people's videos, placing a video file in a post or page has never been easier with WordPress.

TIP

Check out a good example of a video blog at `http://www.tmz.com/videos`. TMZ is a popular celebrity news website that produces and displays videos for the web and for mobile devices.

Several video galleries on the web today allow you to add videos to blog posts. Google's YouTube service (`www.youtube.com`) is a good example of a third-party video service that allows you to share its videos.

Adding a link to a video from the web

TIP

Adding a video from the web, in these steps, adds only a hyperlink to the video. Use these steps if all you want to do is link to a page that has the video on it, rather than embed the video in your blog post or page (covered in the "Adding video with Auto-Embed" section later in this chapter.)

To add a link to a video from the web, follow these steps:

1. **Click the Add Media button on the Add New Post screen to open the Insert Media window.**
2. **Click the Insert from URL link on the left side.**

   The Insert from URL page appears, as shown in Figure 4-1.
3. **Type the URL (Internet address) of the video in the text box.**

   Type the full URL, including the `http://` and `www` portions of the address. Video providers, such as YouTube, usually list the direct links for the video files on their sites; you can copy and paste one of those links into the text box.

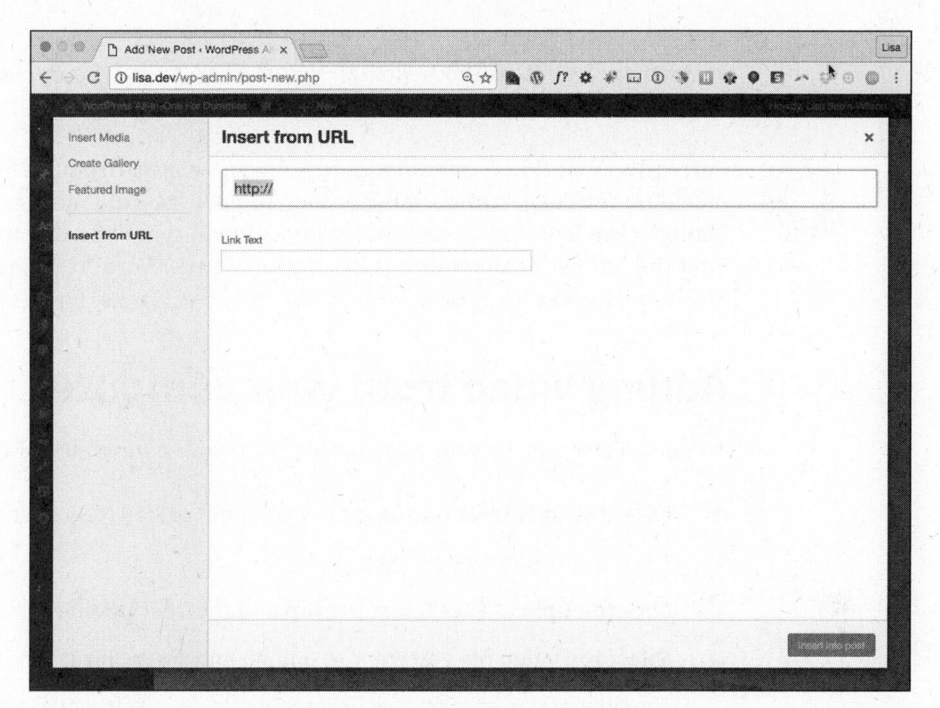

FIGURE 4-1: Add a video by linking to a URL.

4. **(Optional) Type the title of the video in the text box at the top of the Insert from URL screen.**

   Giving a title to the video allows you to provide a bit of a description. Provide a title if you can so that your readers know what the video is about.

5. **Click the Insert into Post button.**

   A link to the video is inserted into your post. WordPress doesn't embed the actual video in the post; it inserts only a link to the video. Your site visitors click the link to load another page on which the video plays.

Adding video with Auto-Embed

The preceding steps enable you to insert a hyperlink that your readers can click to view the video on another website (such as YouTube). If you use WordPress's nifty Auto-Embed feature, however, WordPress automatically embeds many of these videos within your posts and pages.

With this feature, WordPress automatically detects that a URL you typed in your post is a video (from YouTube, for example) and wraps the correct HTML embed code around that URL to make sure that the video player appears in your post (in a standard, XHTML-compliant way). The Auto-Embed feature is automatically enabled on your WordPress site; all you need to do is type the video URL within the content of your post or page.

TIP

Currently, WordPress automatically embeds videos from several sources — including YouTube, Vimeo, DailyMotion, Blip, Flickr, Hulu, Viddler, Qik, Revision3, Photobucket, and Vine— as well as VideoPress-type videos from WordPress.tv. Find the full list of supported video embeds in the WordPress Codex at `https://codex.wordpress.org/Embeds#Okay.2C_So_What_Sites_Can_I_Embed_From.3F`.

Adding video from your computer

To upload and post to your blog a video from your computer, follow these steps:

1. **Click the Add Media button on the Edit Post or Add New Post screen.**

   The Insert Media window appears.

2. **Click the Upload Files tab at the top and then click the Select Files button.**
3. **Select the video file you want to upload and then click Open.**

   The video is uploaded from your computer to your web server, and the Insert Media window displays your uploaded video selected and ready for editing.

4. **In the Attachment Details section, type a title for the file in the Title text box, a caption in the Caption text box, and a description in the Description text box.**
5. **Still in the Attachment Details section, select the Link To option.**

   You can link to a custom URL, the attachment page, or the media file, or you can link to nothing at all.

6. **Click the Insert into Post button.**

   WordPress doesn't embed a video player in the post; it inserts only a link to the video. If you have the Auto-Embed feature activated, however, WordPress attempts to embed the video within a video player. If WordPress can't embed a video player, it displays the link that your visitors can click to open the video in a new window.

TIP

I don't really recommend uploading your own videos directly to your WordPress site. Many video service providers, such as YouTube and Vimeo, give you free storage for your videos. Embedding videos in a WordPress page or post from one of those services is so easy, and by using those services, you're not using your own storage space or bandwidth limitations to provide these videos on your site.

Inserting Audio Files into Your Blog Posts

Audio files can be music files or voice recordings, such as recordings of you speaking to your readers. These files add a nice personal touch to your blog. You can easily share audio files on your blog by using the Add Media feature of WordPress. After you insert an audio file into a blog post, your readers can listen to it on their computers or download it to an MP3 player and listen to it while driving to work, if they want.

To upload an audio file to your site, follow the same steps outlined in the preceding section for uploading a video.

REMEMBER

WordPress doesn't automatically include an audio-player interface for playing your file. Instead, WordPress inserts a link that your readers can click to listen to the audio file.

Some great WordPress plugins for handling audio can enhance the functionality of the file uploader and help you manage audio files in your blog posts. Check out Book 7 for information on how to install and use WordPress plugins in your blog.

Podcasting with WordPress

When you provide regular episodes of an audio show that visitors can download to a computer and listen to on an audio player, you're *podcasting*. Think of a podcast as a weekly radio show that you tune into, except that it's hosted on the Internet rather than on a radio station.

In the sidebar "WordPress video and audio plugins" in this chapter, you can find a few plugins that allow you to easily insert audio files in your WordPress posts and pages. The plugins that are dedicated to podcasting provide features that go beyond embedding audio files in a website. Some of the most important of these features include

- **Archives:** You can create an archive of your audio podcast files so that your listeners can catch up on your show by listening to past episodes.
- **RSS Feed:** An RSS feed of your podcast show gives visitors the opportunity to subscribe to your syndicated content so that they can be notified when you publish future episodes.
- **Promotion:** A podcast isn't successful without listeners, right? You can upload your podcast to iTunes (`www.apple.com/itunes`) so that when people search iTunes for podcasts by subject, they find your podcast.

These plugins go beyond just audio-file management. They're dedicated to podcasting and all the features you need:

- **PowerPress (`https://wordpress.org/plugins/powerpress`):** PowerPress includes full iTunes support; audio players; multiple file-format support (`.mp3`, `.m4a`, `.ogg`, `.wma`, `.ra`, `.mp4a`, `.m4v`, `.mp4v`, `.mpg`, `.asf`, `.avi`, `.wmv`, `.flv`, `.swf`, `.mov`, `.divx`, `.3gp`, `.midi`, `.wav`, `.aa`, `.pdf`, `.torrent`, `.m4b`, `.m4r`, `.epub`, `.mp4`, `.oga`, `.ogv`); statistics to track the popularity of your different podcast offerings; and tagging, categorizing, and archiving of podcast files.
- **Seriously Simple Podcasts (`https://wordpress.org/plugins/seriously-simple-podcasting`):** This plugin uses the native WordPress interface with minimal settings to make it as easy as possible to podcast with WordPress. You can run multiple podcasts; obtain stats on who's listening; do both audio- and videocasting; and publish to popular services such as iTunes, Google Play, and Stitcher.

TIP

I discuss web hosting requirements in Book 2. If you're a podcaster who intends to store audio files in your web hosting account, you may need to increase the storage and bandwidth for your account so that you don't run out of space or incur higher fees from your web hosting provider. Discuss these issues with your web hosting provider to find out up front what you have to pay for increased disk space and bandwidth needs.

Keeping Media Files Organized

If you've been running your blog for any length of time, you can easily forget what files you've uploaded by using the WordPress uploader. The WordPress Media Library allows you to conveniently and easily discover which files are in your Uploads folder.

To find an image, a video, or an audio file you've already uploaded by using the file uploader and to use that file in a new post, follow these steps:

1. **Click the Add Media icon on the Add New Post screen to open the Insert Media window.**
2. **Click the Media Library tab at the top of the window.**

   All the files you've ever uploaded to your site appear because of the File Uploader feature. (See Figure 4-2.) Files you uploaded through other methods, such as SFTP, don't appear in the Media Library.

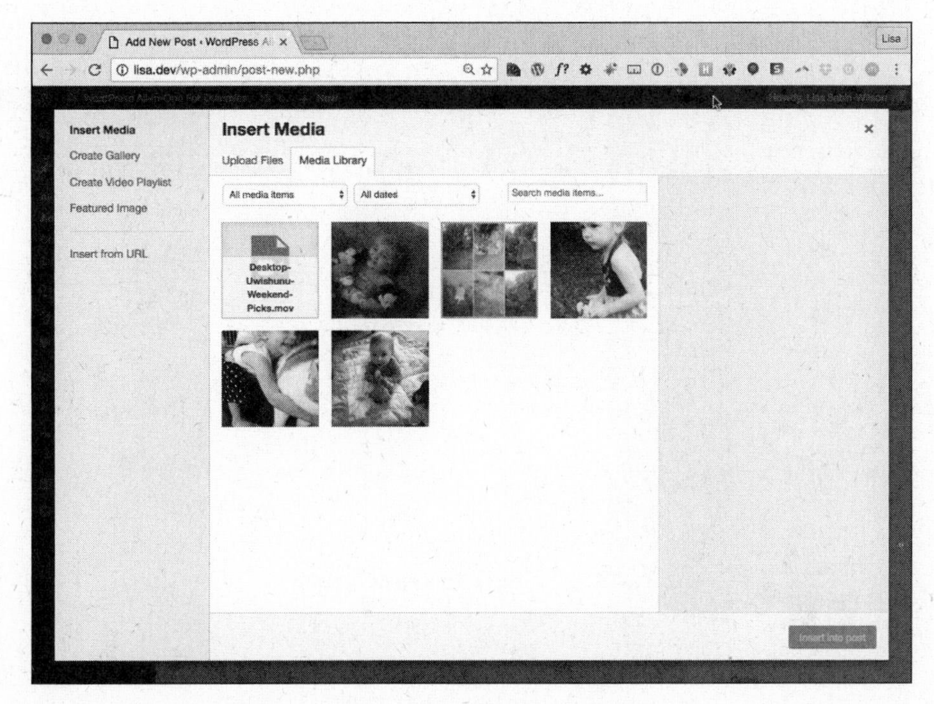

FIGURE 4-2: The Media Library shows all the files you've ever uploaded to your site.

3. **Select the file that you want to reuse.**
4. **On the settings menu that appears, set the options for that file: Title, Caption, Description, Link URL, Order, Alignment, and Size.**
5. **Click the Insert into Post button.**

   The correct HTML code is inserted into the Post text box.

TIP

If you want to view only the files you've uploaded, click the Library link on the Media menu (on the left navigation menu of the Dashboard), which opens the Media Library screen. The Media Library screen lists all the files you've ever uploaded to your WordPress site. By default, the page displays all types of files, but you can quickly navigate to see just the Images, Audio, or Video by using the All Media Items drop-down menu to specify which file type you want to see (as shown in Figure 4-3).

You can do the following tasks on the Media Library page:

- **Filter media files by date.** If you want to view all media files that were uploaded in January 2017, choose that date from the drop-down list and click the Filter button. The page reloads and displays only the media files uploaded in the month of January 2017.

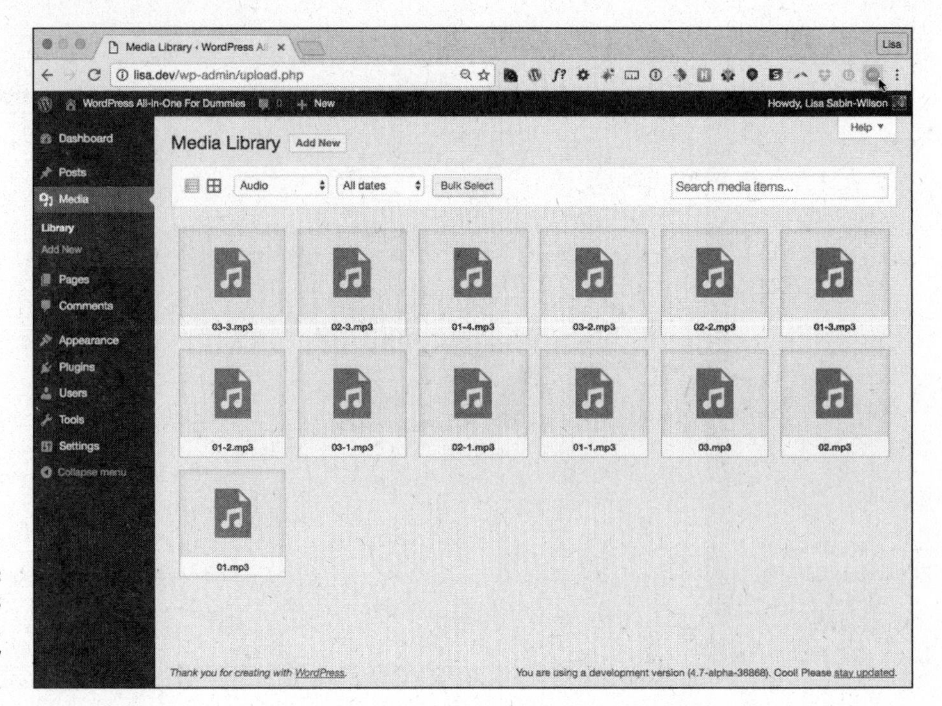

FIGURE 4-3: The WordPress Media Library, displaying only audio files.

- **Search media files by using a specific keyword.** If you want to search your Media Library for all files that reference kittens, type the word **kittens** in the Search box in the top-right corner of the Media Library screen. Then click the Search Media button. The page reloads and displays only media files that contain the keyword or tag *kittens.*
- **Delete media files.** To delete files, click the small white box that appears to the left of the file's thumbnail on the Media Library screen (on the Dashboard, hover your pointer over Media and click the Library link); then click the Delete button, which appears in the top-left corner of the page. The page reloads, and the media file you just deleted is gone.
- **View media files.** On the Media Library screen, click the thumbnail of the file you want to view. The actual file opens in your web browser. If you need the URL of the file, you can copy the permalink of the file from your browser's address bar.

WORDPRESS VIDEO AND AUDIO PLUGINS

You can find some great WordPress plugins for handling audio and video. Check out Book 7 for information on how to install and use WordPress plugins.

Here are a few great plugins for audio:

- **Simple Audio Player (`https://wordpress.org/plugins/simple-audio-player`):** Gives you a configurable audio player with several options
- **Html5 Audio Player (`https://wordpress.org/plugins/html5-audio-player`):** Integrates an HTML5 audio player into your WordPress website
- **Playlist Audio Player (`https://wordpress.org/plugins/playlist-audio-player`):** Quick and easy-to-use music player allows you to create playlists of five tracks each.

Here are a few great plugins for video:

- **Automatic Featured Images from Videos by WebDevStudios (`https://wordpress.org/plugins/automatic-featured-images-from-videos`):** Grabs an image from a YouTube or Vimeo video to create an image and set it as the featured image
- **Smart YouTube PRO by Vladimir Prelovac (`https://wordpress.org/plugins/smart-youtube`):** Inserts YouTube videos into blog posts, comments, and RSS feeds
- **Video Sidebar Widgets (`https://wordpress.org/plugins/video-sidebar-widgets`):** Embeds video from various sources

» Working with the Custom Fields interface

» Adding custom field codes to your templates

» Using custom fields in a variety of ways

Chapter 5
Working with Custom Fields

In Book 4, Chapter 1, I discuss all the different elements you can add to your blog posts and pages when you publish them. By default, WordPress allows you to give your posts and pages titles and content, to categorize and tag posts, to select a date and time for publishing, and to control the discussion options on a per-post or per-page basis.

Sometimes, however, you may want to add extra items to your posts — items you may not want to add to every post, necessarily, but that you add often enough to make manually adding them each time you publish a nuisance. These items can include a multitude of things, from telling your readers your current mood to what you're currently listening to or reading — pretty much anything you can think of.

WordPress gives you the ability to create and add *metadata* (additional data that can be added to define you and your post) to your posts by using a feature called Custom Fields. In Book 4, Chapter 2, I briefly touch on the Custom Fields interface on the Add New Post screen of the Dashboard. In this chapter, I go through Custom Fields in depth by explaining what they are and how to implement them, as well as offering some cool ideas for using Custom Fields on your site.

Understanding Custom Fields

A WordPress template contains static pieces of data that you can count on to appear on your site. These static items include elements such as the title, the content, the date, and so on. But what if you want more? Suppose that you write a weekly book-review post on your site and want to include a listing of recent reviews and accompanying thumbnails of the books. Through the use of Custom Fields, you can do all that without having to retype the list each time you do a review.

REMEMBER

You can add thousands of autoformatted pieces of data (such as book reviews or movie reviews) by adding Custom Fields to your WordPress site. Okay, thousands of Custom Fields would be pretty difficult, if not impossible, to manage; my point here is that the Custom Fields feature doesn't limit the number of fields you can add to your site.

You create Custom Fields on a per-post or per-page basis, which means that you can create an unlimited amount of them and add them only to certain posts. They help you create extra data for your posts and pages by using the Custom Fields interface, which is covered in the following section.

So what can you do with Custom Fields? Really, the only right answer is this: anything you want. Your imagination is your only limit when it comes to the different types of data you can add to your posts by using Custom Fields. Custom Fields allow you the flexibility of defining certain pieces of data for each post.

To use Custom Fields, you do need a bit of knowledge about how to navigate WordPress theme templates, because you have to insert a WordPress function tag, with specific parameters, into the body of the template file. Book 6 takes you through all the information you need to understand WordPress themes, templates, and template tags — so you may want to hit that minibook before you attempt to apply what I discuss in the rest of this chapter. If you're already comfortable and familiar with WordPress templates and tags, you probably won't have any trouble with this chapter at all.

Exploring the Custom Fields Interface

The Custom Fields module appears on both the Add New Post and Add New Page (see Book 4, Chapters 1 and 2) screens on the WordPress Dashboard, below the Post text box, as shown in Figure 5-1.

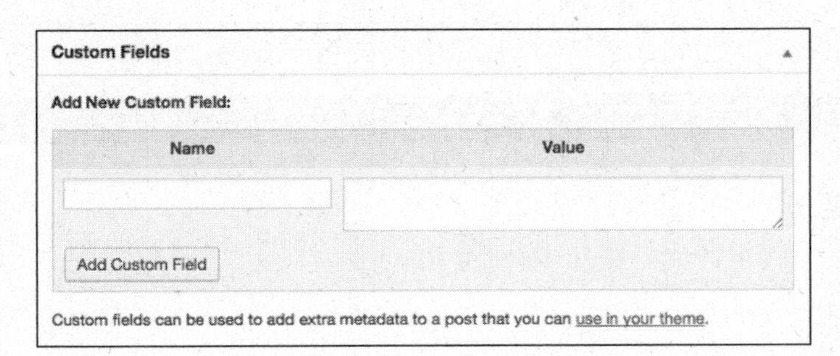

FIGURE 5-1: The Custom Fields module on the Add New Post screen of the Dashboard.

The Custom Fields module has two text boxes:

- **Name:** Also known as the Key. You give this name to the Custom Field you're planning to use. The name needs to be unique: It's used in the template tag that you can read about in the section "Adding Custom Fields to Your Template File" later in this chapter. Figure 5-2 shows a Custom Field with the name `mood`.
- **Value:** Assigned to the Custom Field name and displayed in your blog post on your site if you use the template tag that you can also read about in the section "Adding Custom Fields to Your Template File" later in this chapter. In Figure 5-2, the value assigned to the `mood` (the Custom Field name) is `Happy`.

Custom Fields
Name
Value
mood
Happy
Delete
Update
Add New Custom Field:
Name
Value
Add Custom Field
Custom fields can be used to add extra metadata to a post that you can use in your theme.

FIGURE 5-2: A Custom Field that have name and value assigned.

Simply fill out the Name and Value text boxes and then click the Add Custom Field button to add the data to your post or page. Figure 5-2 shows a Custom Field that I added to my post with the Name of `mood` and with the assigned value `Happy`. In the section "Adding Custom Fields to Your Template File" later in this chapter, I show you the template tag you need to add to your WordPress theme template to display this Custom Field, which appears in my post like this: `My Current Mood is: Happy`, shown in Figure 5-3, where the Custom Field appears at the end of my post.

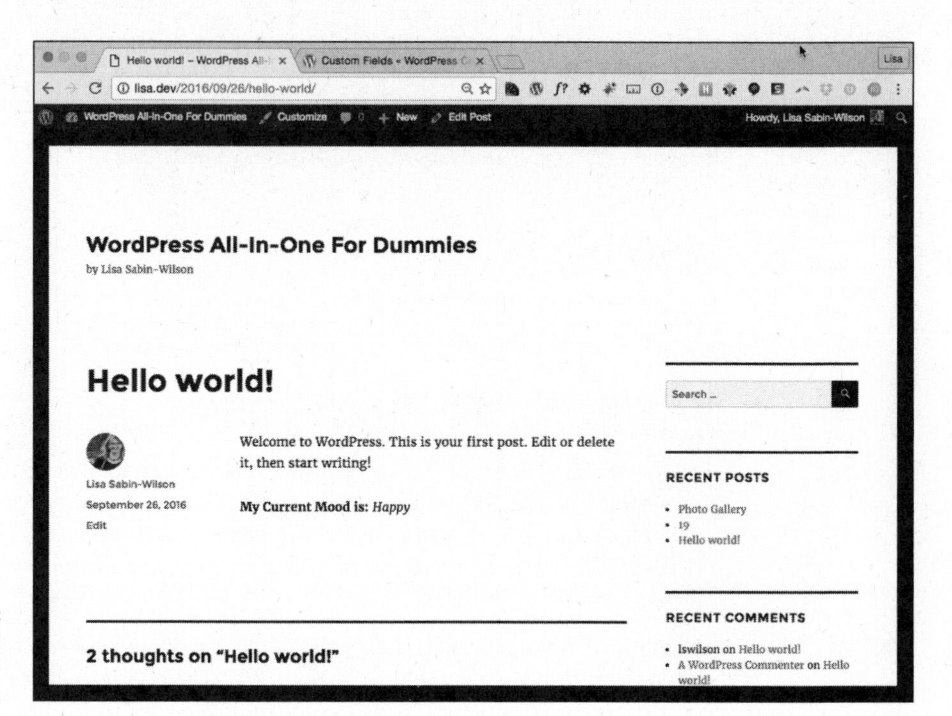

FIGURE 5-3: A Custom Field output appears in a published post.

You can add multiple Custom Fields to one post. To do so, simply add the name and the value of the Custom Field in the appropriate text boxes on the Add New Post page; then click the Add Custom Field button to assign the data to your post. Do this for each Custom Field you want to add to your post.

TIP

After you add a particular Custom Field (such as the `mood` Custom Field in Figure 5-2), you can always add it to future posts. So you can make a post tomorrow and use the `mood` Custom Field but assign a different value to it. If tomorrow, you assign the value `Sad`, your post displays `My Current Mood is: Sad`. You can easily use just that one Custom Field on subsequent posts.

You can access your Custom Fields from the drop-down list below the Name field, as shown in Figure 5-4. You can easily select it again and assign a new value to it in the future, because WordPress saves that Custom Field Key, assuming that you may want to use it again sometime in the future.

TECHNICAL STUFF

Custom Fields are considered to be extra data, separate from the post content itself, for your blog posts. WordPress refers to Custom Fields as *metadata*. The Custom Field name and value get stored in the database in the `wp_postmeta` table, which keeps track of which names and values are assigned to each post. See Book 2, Chapter 7 for more information about the WordPress database structure and organization of data.

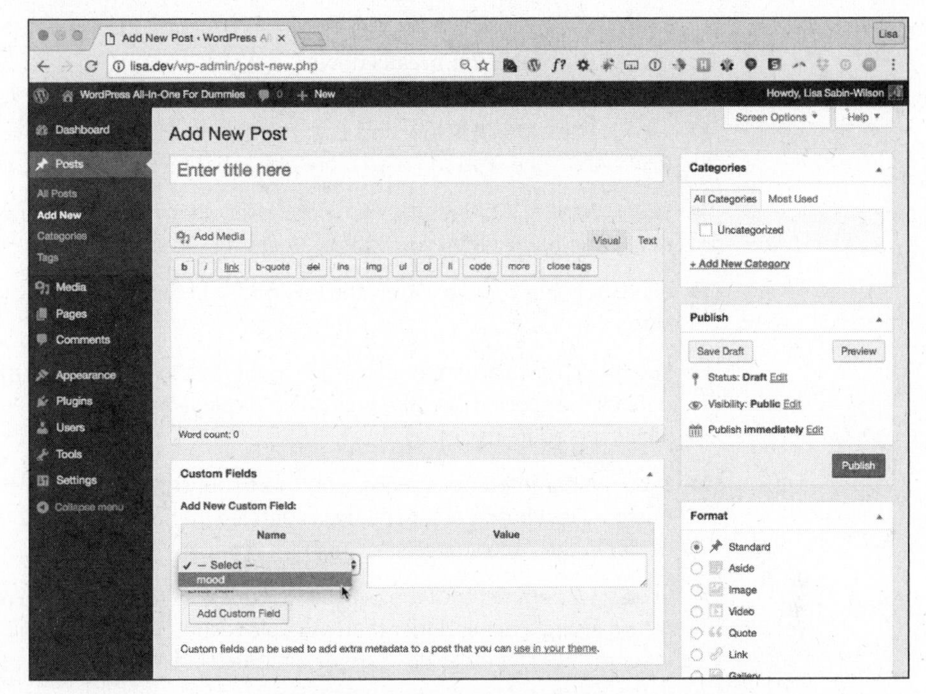

FIGURE 5-4: Custom Field names are saved and displayed in a drop-down list for future use.

REMEMBER

You can find a Custom Fields module on the Add New Page screen of the Dashboard as well, so you can add Custom Fields to either your posts or pages as needed.

Adding Custom Fields to Your Template File

If you followed along in the preceding sections and added the `mood` Custom Field to your own site, notice that the data didn't appear on your site the way it did on mine. To get the data to display properly, you must open the template files and dig into the code a little bit. If the idea of digging into the code of your template files intimidates you, you can put this section aside and read up on WordPress themes, template files, and template tags in Book 6.

You can add Custom Fields to your templates in several ways to display the output of the fields you've set. The easiest way involves using the `get_post_meta();` template tag function, which looks like this:

```
<?php $key="NAME"; echo get_post_meta($post->ID, $key, true); ?>
```

Here's how that function breaks down:

- `<?php`: Starts the PHP function. Every template tag or function needs to start PHP with `<?php`. You can read more about basic PHP in Book 2, Chapter 3.
- `$key="NAME";`: Defines the name of the key that you want to appear. You define the name when you add the Custom Field to your post.
- `echo get_post_meta`: Grabs the Custom Field data and displays it on your site.
- `$post->ID,`: A parameter of the `get_post_meta` function that dynamically defines the specific ID of the post being displayed so that WordPress knows which metadata to display.
- `$key,`: A parameter of the `get_post_meta` function that gets the value of the Custom Field based on the name, as defined in the `$key="NAME";` setting earlier in the code string.
- `true);`: A parameter of the `get_post_meta` function that tells WordPress to return a single result rather than multiple results. (By default, this parameter is set to `true`; typically, don't change it unless you're using multiple definitions in the Value setting of your Custom Field.)
- `?>`: Ends the PHP function.

Based on the preceding code, to make the `mood` Custom Field example, you define the `key` name as `mood` (replace the *NAME* in the preceding code with the word `mood`). It looks like this:

```
<?php $key="mood"; echo get_post_meta($post->ID, $key, true); ?>
```

The part of the function that says `$key="mood";` tells WordPress to return the value for the Custom Field with the Name field of `mood`.

Entering the code in the template file

So that you can see how to enter the code in your template file, I use a WordPress theme called Twenty Sixteen in this section. If you're using a different theme (thousands of WordPress themes are available), you need to adapt these instructions to your particular theme.

Follow these steps to add the template tag to your theme, along with a little HTML code to make it look nice. (These steps assume that you've already added the `mood` Custom Field to your blog post and assigned a value to it.)

1. **Log in to your WordPress Dashboard.**
2. **Click the Editor link on the Appearances menu.**

   The Edit Themes screen loads in the Dashboard, as shown in Figure 5-5.

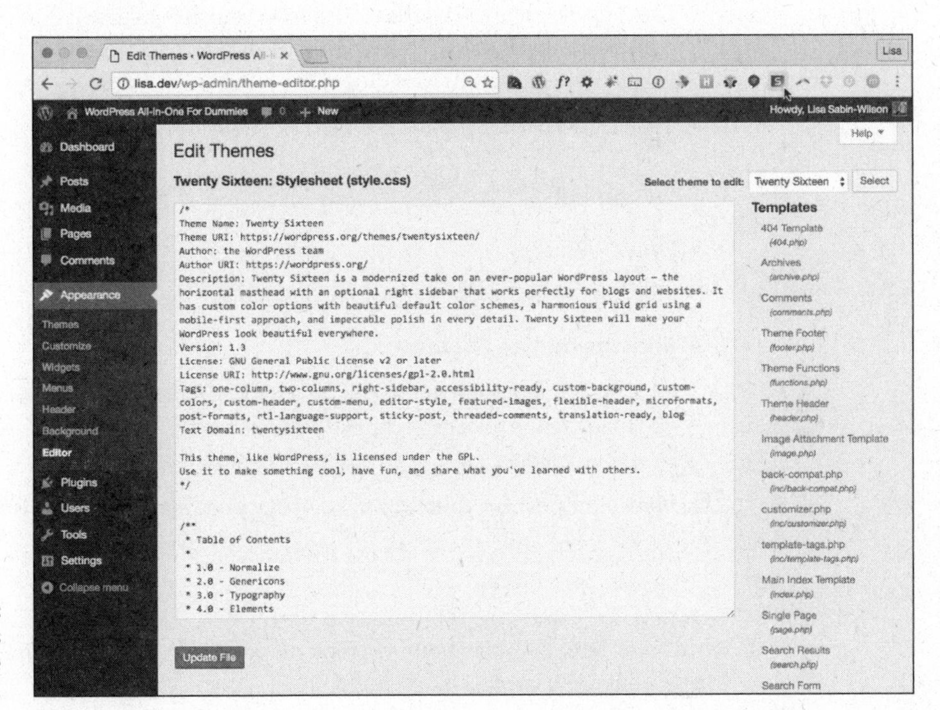

FIGURE 5-5: The Edit Themes screen of the Dashboard.

3. **Locate the template files for your theme (in this case, Twenty Twelve).**

   The available templates are listed on the right side of the Edit Themes page, as shown in Figure 5-5.

4. **Click `content-single.php` in the list of templates.**

   The `content-single.php` template opens in the text editor on the left side of the screen, where you can edit the template file.

5. **Scroll down and locate the template tag that looks like this: `</div> <!-- .entry-content -->`.**

6. **On the new line directly above the line in Step 5, type this:**

```
<p><strong>My Current Mood is:</strong><em>
```

<p> and <strong> open the HTML tags for paragraph and bold text, respectively, followed by the words to display in your template (`My Current Mood is:`). <em> opens the HTML tag for italic-style text, which gets applied to the value. (The </strong> HTML tag closes the bold text display.)

7. **Type the PHP that makes the Custom Field work:**

```
<?php $key="mood"; echo get_post_meta($post->ID, $key, true); ?>
```

8. **Type** `</em></p>`.

   This code closes the HTML tags you opened in Step 6.

9. **Click the Update File button.**

   Located at the bottom of the Edit Themes screen, this step saves the changes you made to the `content.php` file and reloads the page with a message that says your changes have been successfully saved.

10. **View your post on your site to see your Custom Field data displayed.**

   The data should look just like the `My Current Mood is: Happy` message shown earlier in Figure 5-3.

WordPress now displays your current mood at the bottom of the posts to which you've added the `mood` Custom Field.

The entire code, put together, should look like this in your template:

```
<p><strong>My Current Mood is:</strong> <em><?php $key="mood"; echo
    get_post_meta($post->ID, $key, true); ?></em></p>
```

WARNING

The code is case-sensitive, which means that the words you input for the key in your Custom Field need to match case with the `$key` in the code. If you input `mood` in the Key field, for example, the code needs to be lowercase as well: `$key="mood"`. If you attempt to change the case to `$key="Mood"`, the code won't work.

You have to add this code for the `mood` Custom Field only one time. After you add the template function code to your template for the `mood` Custom Field, you can define your current mood in every post you publish to your site by using the Custom Fields interface.

REMEMBER

This example is just one type of Custom Field that you can add to your posts.

Getting WordPress to check for your Custom Field

The previous sections show you how to add the necessary code to your template file to display your Custom Field. But what if you want to publish a post in which you *don't* want the `mood` Custom Field to appear? If you leave your template file as you set it up by following the steps in the preceding sections, even if you don't add the `mood` Custom Field, your blog post displays `My Current Mood is:` without a mood because you didn't define one.

But you can easily make WordPress check first to see whether the Custom Field is added. If it finds the Custom Field, WordPress displays your mood; if it doesn't find the Custom Field, WordPress doesn't display the Custom Field.

If you followed along in the preceding sections, the code in your template looks like this:

```
<p><strong>My Current Mood is:<strong> <em><?php $key="mood"; echo
   get_post_meta($post->ID, $key, true); ?></em></p>
```

To make WordPress check to see whether the `mood` Custom Field exists, add this code to the line above your existing code:

```
<?php if ( get_post_meta($post->ID, 'mood', true) ) : ?>
```

Then add this line of code to the line below your existing code:

```
<?php endif; ?>
```

Put together, the lines of code in your template should look like this:

```
<?php if ( get_post_meta($post->ID, 'mood', true) ) : ?>
<p><strong>My Current Mood is:</strong> <em><?php $key="mood"; echo
    get_post_meta($post->ID, $key, true); ?></em></p>
<?php endif; ?>
```

The first line is an `IF` statement that basically asks "Does the `mood` key exist for this post?" If it does, the value gets displayed. If it doesn't, WordPress skips the code, ignoring it so that nothing gets displayed for the `mood` Custom Field. The final line of code simply puts an end to the `IF` question. See the nearby "IF, ELSE" sidebar to find out about some everyday situations that explain the `IF` question. Apply this statement to the code you just added to your template, and you get this: `IF` the `mood` Custom Field exists, then WordPress will display it, or `ELSE` it won't.

IF, ELSE

In your daily life, you probably deal with `IF, ELSE` situations every day, as in these examples:

- **IF** I have a dollar, then I'll buy coffee, or **ELSE** I won't.
- **IF** it's warm outside, then I'll take a walk, or **ELSE** I won't.
- **IF** I understand this code, then I'll be happy, or **ELSE** I won't.

REMEMBER

You can find extensive information on working with WordPress template files within your theme in Book 6.

Exploring Different Uses for Custom Fields

In this chapter, I use the example of adding your current mood to your site posts by using Custom Fields. But you can use Custom Fields to define all sorts of data on your posts and pages; you're limited only by your imagination when it comes to what kind of data you want to include.

Obviously, I can't cover every possible use for Custom Fields, but I can give you some ideas that you may want to try on your own site. At the very least, you can implement some of these ideas to get yourself into the flow of using Custom Fields, and they may spark your imagination on what types of data you want to include on your site:

- **Music:** Display the music you're currently listening to. Use the same method I describe in this chapter for your current mood, except create a Custom Field named music. Use the same code template, but define the key as `$key="music";` and alter the wording from *My Current Mood is:* to *I am Currently Listening to:*.
- **Books:** Display what you're currently reading by creating a Custom Field named book, defining the key in the code as `$key="book";`, and then altering the wording from *My Current Mood is:* to *I Am Currently Reading:*.
- **Weather:** Let your readers know what the weather is like in your little corner of the world by adding your current weather conditions to your published blog posts. Create a Custom Field named `weather`, and use the same code for the template. Just define the key as `$key="weather";` and alter the wording from *My Current Mood is:* to *Current Weather Conditions:*.

If you want to get really fancy with your Custom Fields, you can also define an icon for the different metadata displays. Using the mood Custom Field, for example, you can add little *emoticons* (smiley-face icons that portray mood) after your mood statement to give a visual cue of your mood, as well as a textual one. Follow these steps to add an emoticon to the `mood` Custom Field from earlier sections of this chapter:

1. **Visit the Posts screen on the Dashboard by clicking the Posts link on the left navigation menu.**
2. **Click the title of the post that you want to edit.**
3. **Add a new Custom Field by choosing Enter New link from the drop-down list and entering** mood-icon **in the Name text box.**
4. **Click the Add Media button above the Post text box to open the Add Media window.**

   The Add Media window opens, as shown in Figure 5-6.

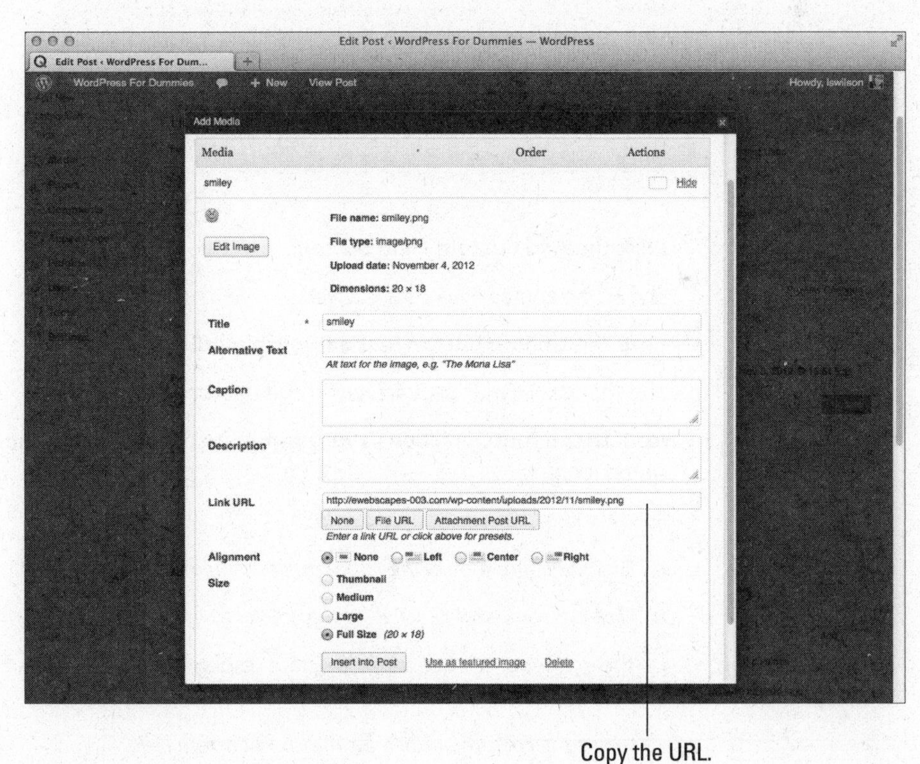

FIGURE 5-6: The Link URL in the Add Media window.

5. **Click the Select Files button, and upload an image from your computer.**

   See Book 4, Chapter 3 for information on uploading images.

6. **From the Link URL text box, copy the file URL of the image you uploaded.**
7. **Click the X in the top-right corner to close the Add Media window.**
8. **Paste the Link URL in the Value text box for the mood-icon Name (see Figure 5-7).**

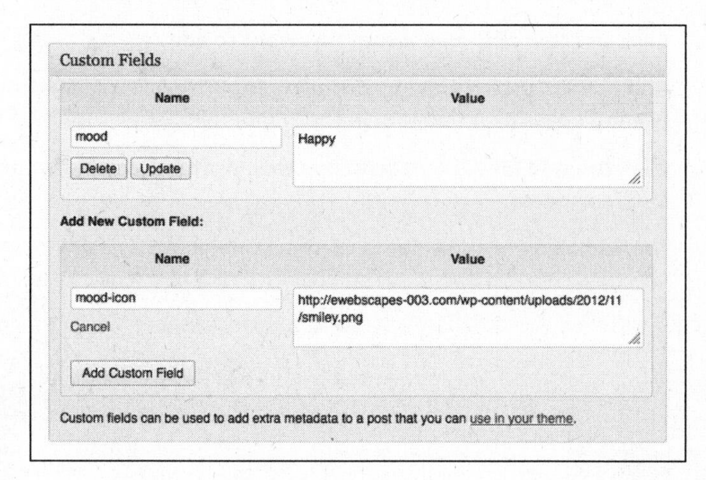

FIGURE 5-7: Adding a mood icon.

9. **Click the Add Custom Field button.**

   The Name and Key values are saved.

10. **Click the Update button in the Publish module.**

    The changes in your post are saved and updated on your site.

11. **Update the function code in your template file to include the new mood icon.**

    Follow these steps to add that code:

    a. *Click the Editor link on the Appearance menu of your Dashboard.*

    b. *Click the `content-single.php` file.*

       The `content-single.php` template displays in the text box on the left side of the page.

    c. *Locate the code you added for the `mood` Custom Field.*

d. *Before the closing </strong> HTML tag, add the following line of code:*

```
<img src="<?php $key="mood-icon"; echo get_post_meta($post->ID, $key, true); ?>"/>
```

The <img src=" that precedes the Custom Field code is the HTML tag used to display an image on your site. The " /> code that appears after the Custom Field code is part of the HTML tag, and it closes the <img src=" HTML tag. I changed the $key to indicate that I'm calling the mood-icon Custom Field.

e. *Click the Update File button to save your changes.*

f. *Visit the post on your site to view your new mood icon.*

You can see my mood icon in Figure 5-8.

The entire snippet of code you added in the preceding steps should look like this when put together (be sure to double-check your work!):

```
<?php if ( get_post_meta($post->ID, 'mood', true) ) : ?>
<p><strong>My Current Mood is:</strong> <em><?php $key="mood"; echo get_post_
   meta($post->ID, $key, true); ?></em> <img src="<?php $key="mood-icon"; echo
   get_post_meta($post->ID, $key, true); ?>"/></strong></p>
<?php endif; ?>
```

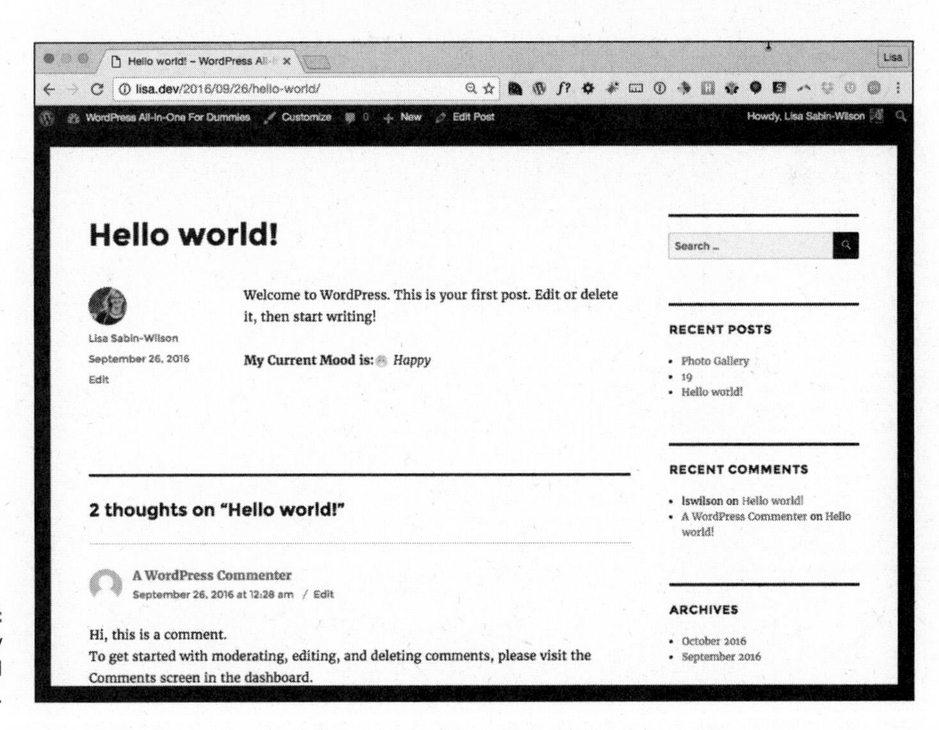

FIGURE 5-8: Displaying my current mood with a mood icon.

» Creating a template for each static page, post category, and sidebar

» Working with custom styles for sticky posts, categories, and tags

» Using Custom Post Types

» Optimizing for search engine success

Chapter 6
Using WordPress as a Content Management System

If you've avoided using WordPress as a solution for building your own website because you think it's only a blogging platform and you don't want to have a blog (not every website owner does, after all), it's time to rethink your position. WordPress is a powerful content management system that's flexible and extensible enough to run an entire website — with no blog at all, if you prefer.

A *content management system* (CMS) is a system used to create and maintain your entire site. It includes tools for publishing and editing, as well as for searching and retrieving information and content. A CMS lets you maintain your website with little or no knowledge of HTML. You can create, modify, retrieve, and update your content without ever having to touch the code required to perform those tasks.

This chapter shows you a few ways that you can use the WordPress platform to power your entire website, with or without a blog. It covers different template configurations that you can use to create separate sections of your site. This chapter also dips into a feature in WordPress called Custom Post Types, which lets you control how content is displayed on your website.

TIP

This chapter touches on working with WordPress templates and themes, a concept that's covered in depth in Book 6. If you find templates and themes intimidating, check out Book 6 first.

You can do multiple things with WordPress to extend it beyond the blog. I use the Twenty Sixteen theme to show you how to use WordPress to create a fully functional website that has a CMS platform — anything from the smallest personal site to a large business site.

Creating Different Page Views Using WordPress Templates

As I explain in Book 4, Chapter 2, a *static page* contains content that doesn't appear on the blog page, but as a separate page within your site. You can have numerous static pages on your site, and each page can have a different design, based on the template you create. (Flip to Book 6 to find out all about choosing and using templates on your site.) You can create several static-page templates and assign them to specific pages within your site by adding code to the top of the static-page templates.

Here's the code that appears at the top of the static-page template we use for our About page at `https://webdevstudios.com/about`:

```
<?php
/*
Template Name: About
*/
?>
```

Using a template on a static page is a two-step process: Upload the template and then tell WordPress to use the template by tweaking the page's code.

TIP

In Book 6, you can discover information about Custom Menus, including how to create different navigation menus for your website. You can create a menu of links that includes all the pages you created on your WordPress Dashboard. You can display that menu on your website by using the Custom Menus feature.

Uploading the template

To use a page template, you have to create one. You can create this file in a text-editor program, such as Notepad. (To see how to create a template, flip to Book 6, which gives you extensive information on WordPress templates and themes.) To create an About page, for example, you can save the template with the name `about.php`.

TIP

For beginners, the best way to get through this step is to make a copy of your theme's `page.php` file, rename the file `about.php`, and then make your edits (outlined below) in the new `about.php` file. As you gain more confidence and experience with WordPress themes and template files, you'll be able to create these files from scratch without having to copy other files to make them — or you can keep copying from other files. Why re-create the wheel?

When you have your template created, follow these steps to make it part of WordPress:

1. **Upload the template file to your WordPress theme folder.**

   You can find that folder on your web server in `/wp-content/themes`. (See Book 2, Chapter 2 for more information about SFTP.)

2. **Log in to your WordPress Dashboard, and click the Editor link on the Appearance menu.**

   The Edit Themes screen opens.

3. **Click the `about.php` template link located on the right side of the page.**

4. **Type the Template Name tag directly above the `get_header()` template tag.**

   The header tag looks like this: `get_header(); ?>`.

   If you're creating an About Page, the code to create the Template Name contains code that looks like this:

```
<?php
/*
Template Name: About
*/
get_header();
?>
```

5. **Click the Update File button.**

   The file is saved, and the page refreshes. If you created an About page template, the `about.php` template is now called About Page in the template list on the right side of the page.

Figure 6-1 shows the Page template and displays the code needed to define a specific name for the template.

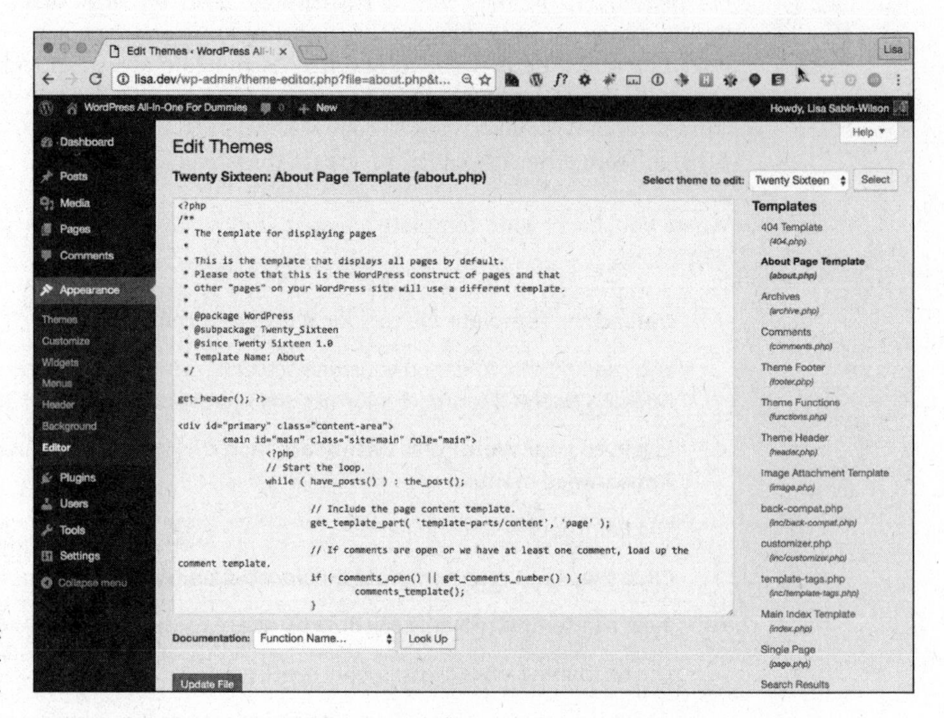

FIGURE 6-1: Naming a static-page template.

Assigning the template to a static page

After you create the template and name it the way you want, assign that template to a page by following these steps:

1. **Click the Add New link on the Pages menu of the Dashboard.**

   The Add New Page screen opens, allowing you to write a new page for your WordPress site.

2. **Type the title in the Title text box and the page content in the large text box.**

3. **Choose the page template from the Template drop-down list.**

   By default, the Template drop-down list in the Page Attributes module appears on the right side of the page. You can reposition the modules on this page; see Book 3, Chapter 1 for more information.

4. **Click the Publish button to save and publish the page to your site.**

Figure 6-2 shows the layout of my home page on my business site at `http://webdevstudios.com` and the information it contains, whereas Figure 6-3 shows the layout and information provided on the About page at `https://webdevstudios.com/about/team`. Both pages are on the same site, in the same WordPress installation, with different page templates to provide different looks, layouts, and sets of information.

FIGURE 6-2: Lisa's business home page at WebDev Studios.com.

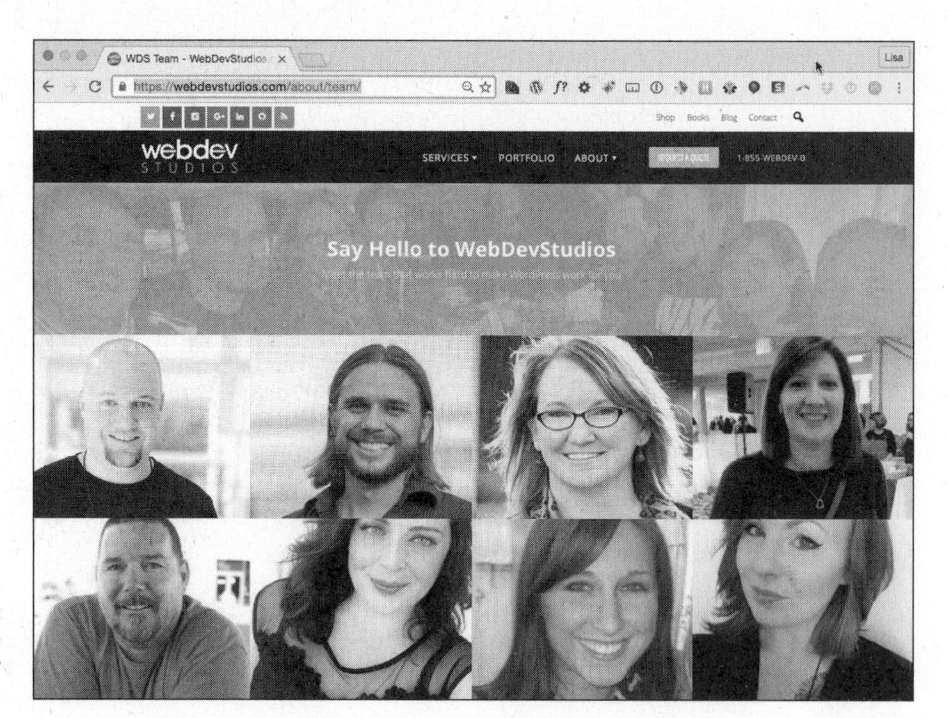

FIGURE 6-3: The About page at WebDev Studios.com.

Creating a Template for Each Post Category

You don't have to limit yourself to creating a static-page template for your site. You can use specific templates for the categories you've created on your blog (which I talk about in Book 3, Chapter 5) and create unique sections for your site.

Figure 6-4 shows my company's design portfolio. Portfolio is the name of a category that we created in the WordPress Dashboard. Instead of using a static page for the display of the portfolio, we're using a category template to handle the display of all posts made to the Portfolio category.

You can create category templates for all categories in your site simply by creating template files that have filenames that correspond to the category slug and then upload those templates to your WordPress `themes` directory via SFTP. (See Book 2, Chapter 2.) Here's the logic to creating category templates:

- A template that has the filename `category.php` is a catch-all for the display of categories.
- Add a dash and the category slug to the end of the filename (shown in Table 6-1) to specify a template for an individual category.

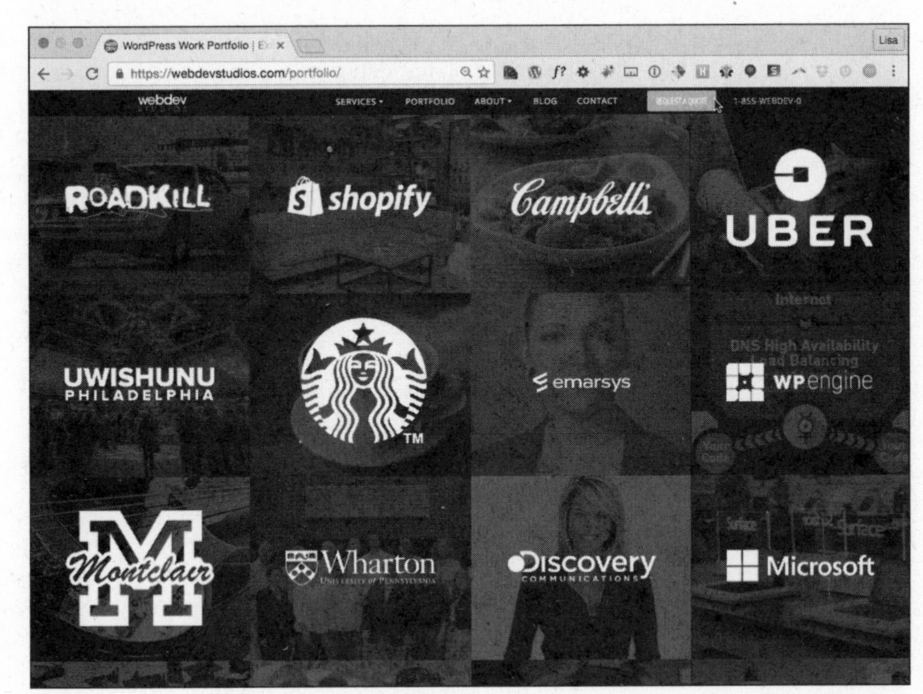

FIGURE 6-4: The Portfolio page, which uses a category template.

» If you don't have a `category.php` or `category-slug.php` file, the category display gets defined from the Main Index template (`index.php`).

Table 6-1 shows three examples of the category template naming requirements.

TABLE 6-1 WordPress Category Template Naming Conventions

If the Category Slug Is . . .	The Category Template Filename Is . . .
portfolio	`category-portfolio.php`
books	`category-books.php`
music-i-like	`category-music-i-like.php`

Pulling in Content from a Single Category

WordPress makes it possible to pull in very specific types of content on your website through the use of the `WP_Query` class. If you include `WP_Query` before The Loop (see Book 6, Chapter 3), it lets you specify which category you want to pull information from. If you have a category called *WordPress* and want to display the last three posts from that category — on your front page, in your sidebar, or somewhere else on your site — you can use this template tag.

TIP

The WP_Query class accepts several parameters that let you display different types of content, such as posts in specific categories and content from specific pages/posts or dates in your blog archives. The WP_Query class lets you pass so many variables and parameters that I just can't list all the possibilities. Instead, you can visit the WordPress Codex and read about the options available with this tag: https://codex.wordpress.org/Class_Reference/WP_Query%23Parameters.

Here are two parameters that you can use with WP_Query:

- posts_per_page=*X*: This parameter tells WordPress how many posts you want to display. If you want to display only three posts, enter **posts_per_page=3**.
- category_name=*slug*: This parameter tells WordPress that you want to pull posts from the category with a specific slug. If you want to display posts from the WordPress category, this would be category_name=wordpress.

Follow these steps to filter posts by category using WP_Query:

1. **Click the Editor link on the Appearance menu of the Dashboard.**

   The Edit Themes screen opens.

2. **Click the template in which you want to display the content.**

   If you want to display content in a sidebar, for example, choose the Sidebar template: sidebar.php.

3. **Locate the ending </aside> tag at the bottom of the template for the theme you're using.**

   In the Twenty Sixteen theme, the ending </aside> tag is the last line.

4. **Type the following code directly above the ending </aside> tag:**

```
<section id="query" class="widget widget_meta"><h2 class="widget-
   title">Category Posts</h2>

<?php $query = new WP_Query( array( 'category_name' => 'wordpress' ) ); ?>
<?php while ( $query->have_posts() ) : $query->the_post(); ?>
<ul>
<li><strong><a href="<?php the_permalink() ?>" rel="bookmark"
   title="Permanent Link to <?php the_title_attribute(); ?>"><?php
   the_title(); ?></a></strong><br/>
<?php the_excerpt(); ?> </li>
</ul>
```

```
<?php endwhile; wp_reset_postdata(); ?>
</section>
```

5. **Click the Update File button.**

   The changes you just made are saved to the `sidebar.php` template.

TECHNICAL STUFF

In past versions of WordPress, you used the `query_posts();` tag to pull content from a specific category, but the `WP_Query` class is more efficient. Although the `query_posts();` tag provides the same result, it increases the number of calls to the database and also increases page load and server resources, so please don't use `query_posts();` (no matter what you see written on the Internet!).

Using Sidebar Templates

You can create separate sidebar templates for different pages of your site by using a simple `include` statement. When you write an `include` statement, you're simply telling WordPress that you want it to include a specific file on a specific page.

The code that pulls the usual Sidebar template (`sidebar.php`) into all the other templates, such as the Main Index template (`index.php`), looks like this:

```
<?php get_sidebar(); ?>
```

What if you create a page and want to use a sidebar that has different information from what you have in the Sidebar template (`sidebar.php`)? Follow these steps:

1. **Create a new sidebar template in a text editor such as Notepad.**

   See Book 6 for information on template tags and themes. My recommendation is to make a copy of the existing `sidebar.php` file in the Twenty Sixteen theme and rename it.

2. **Save the file as `sidebar2.php`.**

   In Notepad, choose File⇨Save. When you're asked to name the file, type **sidebar2.php** and then click Save.

3. **Upload `sidebar2.php` to your Themes folder on your web server.**

   See Book 2, Chapter 2 for SFTP information, and review Book 6, Chapter 2 for information on how to locate the Themes folder.

   The template is now in your list of theme files on the Edit Themes screen. (Log in to your WordPress Dashboard and click the Editor link on the Appearance drop-down menu.)

4. **To include the `sidebar2.php` template in one of your page templates, replace `<?php get_sidebar(); />` with this code:**

   ```
   <?php get_template_part('sidebar2'); ?>
   ```

   This code calls in a template you've created within your theme.

TIP

By using that `get_template_part` function, you can include virtually any file in any of your WordPress templates. You can use this method to create footer templates for pages on your site, for example. First, create a new template that has the filename `footer2.php`. Then locate the following code in your template

```
<?php get_footer(); ?>
```

and replace it with this code:

```
<?php get_template_part('footer2'); ?>
```

Creating Custom Styles for Sticky, Category, and Tag Posts

In Book 6, you can find the method for putting together a very basic WordPress theme, which includes a Main Index template that uses the WordPress Loop. You can use a custom tag to display custom styles for sticky posts, categories, and tags on your blog. That special tag looks like this:

```
<div <?php post_class() ?> id="post-<?php the_ID(); ?>">
```

The `post_class()` section is the coolest part of the template. This template tag tells WordPress to insert specific HTML markup in your template that allows you to use CSS to make custom styles for sticky posts, categories, and tags.

REMEMBER

In Book 4, Chapter 1, I tell you all about how to publish new posts to your blog, including the different options you can set for your blog posts, such as categories, tags, and publishing settings. One of the settings is the Stick This Post to the Front Page setting. In this chapter, I show you how to custom-style those sticky posts. It's not as messy as it sounds!

Suppose that you publish a post with the following options set:

» Stick this post to the front page.

* File it in a category called WordPress.
* Tag it News.

When the `post_class()` tag is in the template, WordPress inserts HTML markup that allows you to use CSS to style *sticky posts,* or posts assigned to specific tags or categories, with different styling from the rest of your posts. WordPress inserts the following HTML markup for your post:

```
<div class="post sticky category-wordpress tag-news">
```

In Book 6, you can discover CSS selectors and HTML markup, and see how they work together to create style and format for your WordPress theme. With the `post_class()` tag in place, you can go to your CSS file and define styles for the following CSS selectors:

* `.post`: Use this tag as the generic style for all posts on your site. The CSS for this tag is

  ```
  .post {background: #ffffff; border: 1px solid silver; padding: 10px;}
  ```

  A style is created for all posts that have a white background with a thin silver border and 10 pixels of padding space between the post text and the border of the post.

* `.sticky`: You stick a post to your front page to call attention to that post, so you may want to use different CSS styling to make it stand out from the rest of the posts on your site:

  ```
  .sticky {background: #ffffff; border: 4px solid red; padding: 10px;}
  ```

  This code creates a style for all posts that have been designated as sticky in the post options on the Write Post page to appear on your site with a white background, a thick red border, and 10 pixels of padding space between the post text and border of the post.

* `.category-wordpress`: Because I blog a lot about WordPress, my readers may appreciate it if I give them a visual cue as to which posts on my blog are about that topic. I can do that through CSS by telling WordPress to display a small WordPress icon in the top-right corner of all my posts in the WordPress category:

  ```
  .category-wordpress {
  background: url(wordpress-icon.jpg) top right no-repeat;
  height: 100px; width: 100px;
  }
  ```

This code inserts a graphic — `wordpress-icon.jpg` — that's 100 pixels in height and 100 pixels in width in the top-right corner of every post I assign to the WordPress category on my site.

- `.tag-news`: I can style all posts tagged with News the same way I style the WordPress category:

```
.tag-news {
background: #f2f2f2;
border: 1px solid black;
padding: 10px;
}
```

This CSS styles all posts tagged with News with a light gray background and a thin black border with 10 pixels of padding between the post text and border of the post.

You can easily use the `post-class()` tag, combined with CSS, to create dynamic styles for the posts on your blog!

Working with Custom Post Types

A nice feature of WordPress (as of version 3.0) is Custom Post Types. This feature allows you, the site owner, to create different content types for your WordPress site that give you more creative control of how different types of content are entered, published, and displayed on your WordPress website.

REMEMBER

Personally, I wish that WordPress had called this feature Custom Content Types so that people didn't incorrectly think that Custom Post Types pertain to posts only. Custom Post Types aren't really the posts that you know as blog posts. Custom Post Types are a way of managing your blog content by defining what type of content it is, how it's displayed on your site, and how it operates — but they're not necessarily posts.

By default, WordPress already has different post types built into the software, ready for you to use. These default post types include

- Blog posts
- Pages
- Navigation menus (see Book 6, Chapter 6)

- Attachments
- Revisions

Custom Post Types give you the ability to create new and useful types of content on your website, including a smart and easy way to publish those content types to your site.

You really have endless possibilities for using Custom Post Types, but here are a few ideas that can kick-start your imagination. (They're some of the most popular and useful ideas that others have implemented on their sites.)

- Photo gallery
- Podcast or video
- Book reviews
- Coupons and special offers
- Events calendar

To create and use Custom Post Types on your site, you need to be sure that your WordPress theme contains the correct code and functions. In the following steps, I create a very basic Custom Post Type called Generic Content. Follow these steps to create the Generic Content basic Custom Post Type:

1. **Click the Editor link on the Appearance menu of the Dashboard to open the Theme Editor screen.**
2. **Click the Theme Functions template link to open the `functions.php` file in the text editor on the left side of the page.**
3. **Add the Custom Post Types code to the bottom of the Theme Functions template file.**

   Scroll down to the bottom of the `functions.php` file, and include the following code to add a Generic Content Custom Post Type to your site:

```
add_action( 'init', 'create_my_post_types' );
 function create_my_post_types() {
     register_post_type( 'generic_content', array(
        'label' => __( 'Generic Content' ),
        'singular_label' => __( 'Generic Content' ),
        'description' => __( 'Description of the Generic Content type' ),
         'public' => true,
         )
     );
 }
```

4. **Click the Update File button to save the changes you made in the `functions.php` file.**

The function `register_post_type` can accept several arguments and parameters, which are detailed in Table 6-2. You can use a variety and combination of different arguments and parameters to create a specific post type. You can find more information on Custom Post Types and using the `register_post_type` function in the official WordPress Codex at `https://codex.wordpress.org/Function_Reference/register_post_type`.

After you complete the preceding steps to add the Generic Content Custom Post Type to your site, a new post type labeled Generic appears on the left navigation menu of the Dashboard.

You can add and publish new content by using the new Custom Post Type, just as you do when you write and publish blog posts. (See Book 4, Chapter 1.) The published content isn't added to the chronological listing of blog posts; rather, it's treated like separate content from your blog (just like static pages).

View the permalink for the published content, and you see that it adopts the custom post type name Generic Content and uses it as part of the permalink structure, creating a permalink that looks like `http://`*`yourdomain.com`*`/generic-content/new-article`.

TABLE 6-2 Arguments and Parameters for register_post_type();

Parameter	Information	Default	Example
`label`	The name of the post type.	None	`'label' => __( 'Generic Content' ),`
`singular_label`	Same as label, but singular. If your label is "Movies," the singular label would be "Movie."	None	`'singular_label' => __( 'Generic Content' ),`
`description`	The description of the post type; displayed in the Dashboard to represent the post type.	None	`'description' => __( 'This is a description of the Generic Content type' ),`

Parameter	Information	Default	Example

`public`
`show_ui`
`publicly_queryable`
`exclude_from_search`
	Sets whether the post type is public.
	There are three other arguments:
	`show_ui`: whether to show admin screens.
	`publicly_queryable`: whether to query for this post type from the front end.
	`exclude_from_search`: whether to show post type in search results.
		`true` or `false`
		Default is `false`
			`'public' => true,`
			`'show_ui' => true,`
			`'publicly_queryable' => true,`
			`'exclude_from_search' => false,`

`menu_position`	Sets the position of the post type menu item on the Dashboard navigation menu.	Default: 20 By default, appears after the Comments menu on the Dashboard Set integer in intervals of 5 (5, 10, 15, 20, and so on)	`'menu_position' => 25,`
`menu_icon`	Defines a custom icon (graphic) to the post type menu item in the Dashboard navigation menu. Creates and uploads the image to the images directory of your theme folder.	None	`'menu_icon' => get_stylesheet_directory_uri() . '/images/generic-content.png',`
`hierarchical`	Tells WordPress whether to display the post type content list in a hierarchical manner.	`true` or `false` Default is `true`	`'hierarchical' => true,`

(continued)

TABLE 6-2 ***(continued)***

Parameter	Information	Default	Example
`query_var`	Controls whether this post type can be used with a query variable such as `query_posts` (see the "Adding `query_posts` Tag" section) or `WP_Query`.	`true` or `false` Default is false	`'query_var' => true,`
`capability_type`	Defines permissions for users to edit, create, or read the Custom Post Type.	`post` (default) Gives the same capabilities for those users who can edit, create, and read blog posts	`'query_var' => post,`
`supports`	Defines what meta boxes, or modules, are available for this post type in the Dashboard.	`title`: Text box for the post title `editor`: Text box for the post content `comments`: Check boxes to toggle comments on/off `trackbacks`: Check boxes to toggle trackbacks and pingbacks on/off `revisions`: Allows post revisions to be made `author`: Drop-down list to define post author `excerpt`: Text box for the post excerpt `thumbnail`: The featured image selection `custom-fields`: Custom fields input area `page-attributes`: The page parent and page template drop-down lists	`'supports' => array( 'title', 'editor', 'excerpt', 'custom-fields', 'thumbnail' ),`

Parameter	Information	Default	Example

`rewrite`	Rewrites the permalink structure for the post type.	`true` or `false`
		
		Two other arguments are available:
		
		`slug`: Permalink slug to use for your Custom Post Types
		
		`with_front`: If you've set your permalink structure with a specific prefix, such as `/blog`
			`'rewrite' => array( 'slug' => 'my-content', 'with_front' => false ),`

`taxonomies`	Uses existing WordPress taxonomies (category and tag).	Category `post_tag`	`'taxonomies' => array( 'post_tag', 'category'),`

Two helpful plugins for building Custom Post Types quickly in WordPress are

- **Custom Post Type UI:** Written by our team at WebDevStudios, this plugin (`https://wordpress.org/plugins/custom-post-type-ui`) gives you a clean interface within your WordPress Dashboard that can help you easily and quickly build Custom Post Types on your website. It eliminates the need to add the code to your `functions.php` file by giving you options and settings so that you can configure and build the Custom Post Type that you want. Figure 6-5 shows the Custom Post Type UI options page on the Dashboard.
- **Betta Boxes CMS:** Available in the WordPress Plugin Directory (`http://wordpress.org/extend/plugins/betta-boxes-cms`), this plugin provides an interface in your Dashboard that you can use to create *meta boxes,* or special Custom Fields (see Book 4, Chapter 5) for the Custom Post Types that you build. As an example, Figure 6-6 shows some custom meta boxes built by using this plugin. This website features theater productions and the Custom Post Types for those shows. On the right side of Figure 6-6, the Purchase Link box (created by using custom meta boxes) gives the website owner a quick and easy field to fill out so that he or she can include information on where to purchase show tickets in every show post published.

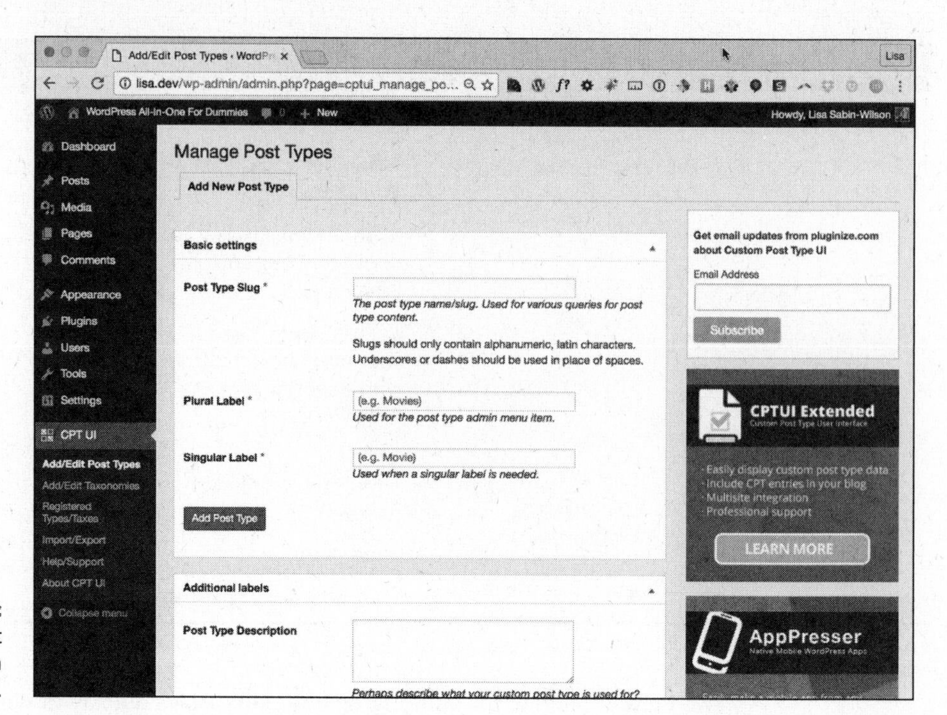

FIGURE 6-5: The Custom Post Type UI plugin options page.

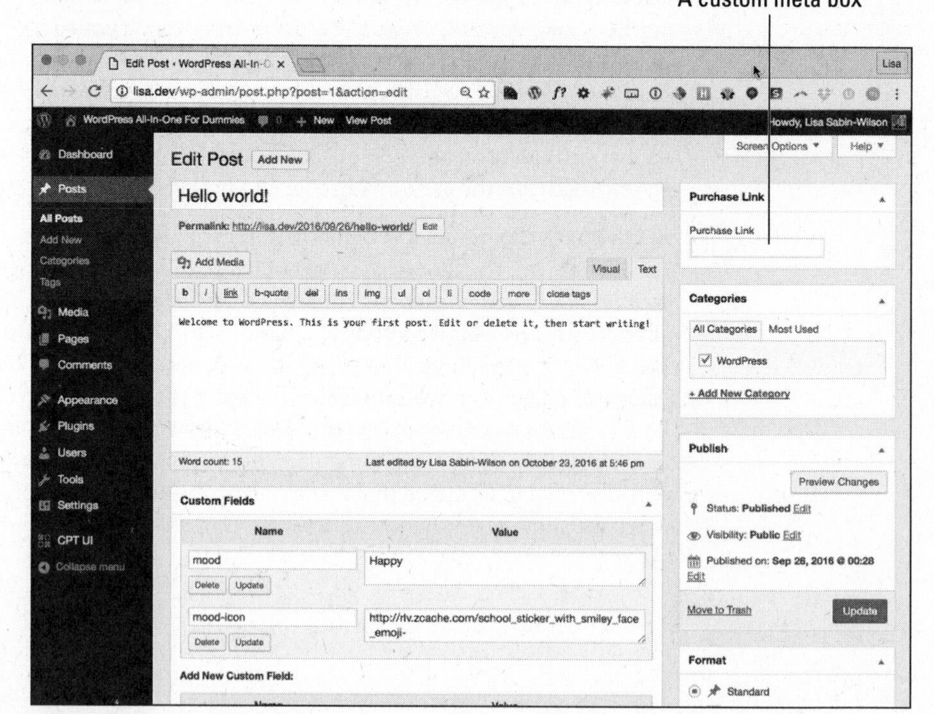

FIGURE 6-6: Purchase Link meta box created for this Custom Post Type.

Optimizing Your WordPress Site

Search engine optimization (SEO) is the practice of preparing your site to make it as easy as possible for the major search engines to crawl and cache your data in their systems so that your site appears as high as possible in the search returns. Book 5 contains more information on search engine optimization, as well as information about marketing your site and tracking its presence in search engines and social media by using analytics. This section gives you a brief introduction to SEO practices with WordPress. From here, you can move on to Book 5 to take a hard look at some of the things you can do to improve and increase traffic to your website.

If you search for the keywords *WordPress website design and development* in Google, my business site at WebDevStudios is in the top-ten search results for those keywords (at least, it is while I'm writing this chapter). Those results can change from day to day, so by the time you read this book, someone else may very well have taken over that coveted position. The reality of chasing those high-ranking search engine positions is that they're here today, gone tomorrow. The goal of SEO is to make sure that your site ranks as high as possible for the keywords that you think people will use to find your site. After you attain those high-ranking positions, the next goal is to keep them.

WordPress is equipped to create an environment that's friendly to search engines, giving them easy navigation through your archives, categories, and pages. WordPress provides this environment with a clean code base, content that's easily updated through the WordPress interface, and a solid navigation structure.

To extend SEO even further, you can tweak five elements of your WordPress posts, pages, and templates:

- **Custom permalinks:** Use custom permalinks, rather than the default WordPress permalinks, to fill your post and page URLs with valuable keywords. Check out Book 3, Chapter 2 for information on WordPress permalinks.
- **Posts and page titles:** Create descriptive titles for your posts and pages to provide rich keywords in your site.
- **Text:** Fill your posts and pages with keywords for search engines to find and index. Keeping your site updated with descriptive text and phrases helps the search engines find keywords to associate with your site.
- **Category names:** Use descriptive names for the categories you create in WordPress to place great keywords right in the URL for those category pages, if you use custom permalinks.

» **Images and <ALT> tags:** Place <ALT> tags in your images to further define and describe the images on your site. You can accomplish this task easily by using the description field in the WordPress image uploader.

Planting keywords in your website

If you're interested in a higher ranking for your site, use custom permalinks. By using custom permalinks, you're automatically inserting keywords into the URLs of your posts and pages, letting search engines include those posts and pages in their databases of information on those topics. If a provider that has the Apache `mod_rewrite` module enabled hosts your site, you can use the custom permalink structure for your WordPress-powered site.

Keywords are the first step on your journey toward great search engine results. Search engines depend on keywords, and people use keywords to look for content.

The default permalink structure in WordPress is pretty ugly. When you're looking at the default permalink for any post, you see a URL something like this:

```
http://yourdomain.com/p?=105
```

This URL contains no keywords of worth. If you change to a custom permalink structure, your post URLs automatically include the titles of your posts to provide keywords, which search engines absolutely love. A custom permalink may appear in this format:

```
http://yourdomain.com/2017/01/01/your-post-title
```

I explain setting up and using custom permalinks in full detail in Book 3, Chapter 2.

Optimizing your post titles for search engine success

Search engine optimization doesn't completely depend on how you set up your site. It also depends on you, the site owner, and how you present your content.

You can present your content in a way that lets search engines catalog your site easily by giving your blog posts and pages titles that make sense and coordinate with the actual content being presented. If you're doing a post on a certain topic, make sure that the title of the post contains at least one or two keywords about that particular topic. This practice gives the search engines even more ammunition to list your site in searches relevant to the topic of your post.

As your site's presence in the search engines grows, more people will find your site, and your readership will increase as a result.

A post with the title A Book I'm Reading doesn't tell anyone *what* book you're reading, making it difficult for people searching for information on that particular book to find the post. If you give the post the title *WordPress All-in-One For Dummies: My Review,* however, you provide keywords in the title, and (if you're using custom permalinks) WordPress automatically inserts those keywords into the URL, giving the search engines a triple keyword play:

- » Keywords exist in your blog post title.
- » Keywords exist in your blog post URL.
- » Keywords exist in the content of your post.

Writing content with readers in mind

When you write your posts and pages and want to make sure that your content appears in the first page of search results so that people will find your site, you need to keep those people in mind when you're composing the content.

When search engines visit your site to crawl through your content, they don't see how nicely you've designed your site. They're looking for words to include in their databases. You, the site owner, want to make sure that your posts and pages use the words and phrases that you want to include in search engines.

If your post is about a recipe for fried green tomatoes, for example, you need to add a keyword or phrase that you think people will use when they search for the topic. If you think people would use the phrase *recipe for fried green tomatoes* as a search term, you may want to include that phrase in the content and title of your post.

A title such as A Recipe I Like isn't as effective as a title such as A Recipe for Fried Green Tomatoes. Including a clear, specific title in your post or page content gives the search engines a double-keyword whammy.

Creating categories that attract search engines

One little-known SEO tip for WordPress users: The names you give the categories you create for your site provide rich keywords that attract search engines like honey attracts bees. Search engines also see your categories as keywords that are relevant to the content on your site. So make sure that you're giving your categories names that are relevant to the content you're providing.

If you sometimes write about your favorite recipes, you can make it easier for search engines to find your recipes if you create categories specific to the recipes you're blogging about. Instead of having one Favorite Recipes category, you can create multiple category names that correspond to the types of recipes you blog about: Casserole Recipes, Dessert Recipes, Beef Recipes, and Chicken Recipes, for example.

REMEMBER

Creating specific category titles not only helps search engines, but also helps your readers discover content that's related to topics they're interested in.

You can also consider having one category called Favorite Recipes and creating subcategories (also known as *child categories*) that give a few more details on the types of recipes you've written about. (See Book 3, Chapter 5 for information on creating categories and child categories.)

Categories use the custom permalink structure, just like posts do. So links to your WordPress categories also become keyword tools within your site to help the search engines — and, ultimately, search engine users — find the content. Using custom permalinks creates category page URLs that look something like this:

```
http://yourdomain.com/category/category_name
```

The `category_name` portion of that URL puts the keywords right into the hands of search engines.

Using the <ALT> tag for images

When you use the WordPress media uploader to include an image in your post or page, a Description text box appears; in it, you can enter a description of the image. (I cover using the WordPress image uploader in detail in Book 4, Chapter 3.) This text automatically becomes what's referred to as the `<ALT>` tag, or alternative text.

The `<ALT>` tag's real purpose is to provide a description of the image for people who, for some reason or another, can't actually see the image. In a text-based browser that doesn't display images, for example, visitors see the description, or `<ALT>` text, telling them what image would be there if they could see it. Also, the tag helps people who have impaired vision and rely on screen-reading technology because the screen reader reads the `<ALT>` text from the image. You can read more about website accessibility for people with disabilities at `https://www.w3.org/WAI/intro/people-use-web/Overview.html`.

An extra benefit of <ALT> tags is that search engines gather data from them to further classify the content of your site. The following code inserts an image with the <ALT> tag of the code in bold to demonstrate what I'm talking about:

```
<img src="http://yourdomain.com/image.jpg" alt="This is an ALT tag"/>
```

Search engines harvest those <ALT> tags as keywords. The WordPress image uploader gives you an easy way to include those <ALT> tags without having to worry about inserting them into the image code yourself. Just fill out the Description text box before you upload and add the image to your post. Book 4, Chapter 3 provides in-depth information on adding images to your site content, including how to add descriptive text for the <ALT> tag and keywords.

5

Examining SEO and Social Media

Contents at a Glance

IN THIS CHAPTER

» Making your content easy for readers to share

» Figuring out when to interact with readers

» Using Twitter through WordPress

» Connecting Facebook to your WordPress blog

Chapter 1
Exposing Your Content

After you launch your website, getting your content in front of an interested audience is one of the most important strategic decisions you make, and this chapter focuses on how to get your content in front of potential new readers. The idea that people will eventually find any content you write is a pretty big falsehood. You might have the best rock band in the world, but if you don't leave your garage and get your music in front of potential fans, you can't ever sell out arenas.

By creating good content, making it easily shareable, and then participating within groups of interested people, you can establish expertise and build a community around your content. A community is much more powerful than a bunch of empty visitors; people in a community often become advocates and cheerleaders for your site.

REMEMBER

You want to gain readers, not random visitors. There's a big difference between a reader and a visitor: *Readers* visit your site on a consistent basis, and *visitors* check out your site and then move on to the next page that grabs their attention.

Understanding the Three Cs of the Social Web

Before I dive into the technical how-to stuff, I should talk about general social media philosophy. Technical tips without philosophy are meaningless. If you don't have the general philosophy down, your results are going to be poor because your interactions are going to be very one-sided affairs.

You can concentrate your daily actions on the web on the three Cs: content, communication, and consistency. The next few sections go into detail about each topic. By applying the three Cs, you can avoid a lot of mistakes and have success with your website and blog.

Content

The first pillar of the social web is content. Although the web has seen a growing shift from content to community, content is still king. Communities based on common interests fall flat unless they have the content for people to gravitate to. Facebook groups, for example, dominate because of the wealth of content they offer: the posts, links, videos, and other media that people create within that group. Without the content, the group wouldn't exist.

Content for the sake of content isn't necessarily in your best interest, however. To ensure that you provide the best content possible, make sure that you do these things:

- **Focus your content.** People expect tailored content. If you write about just anything, people won't know what to expect and will visit less often or stop coming to your site altogether. People will come back to your site for certain reasons, and they want content tailored to what they expect. The most successful publishers have a narrow focus, and they write for a niche.

CONSIDERING SOCIAL-VOTING TOOLS

A lot of online vendors recommend that you drive as many eyeballs as possible to your site by using social-voting tools (such as Facebook's Like button and Twitter's Tweet button) and other methods. Although this strategy increases your traffic numbers and may temporarily boost your confidence, it's a short-term solution. Most of your new visitors won't have a lot of interest in your content and therefore won't return to your site.

When Problogger.net author Darren Rowse (`www.problogger.net`), an authority on professional online publishing, began blogging, he tried a wide-ranging approach but discovered it didn't work. Rowse said:

> "My blog had four main themes and different readers resonated differently with each one. A few readers shared my diverse interests in all four areas, but most came to my blog to read about one of the (or at most a couple of) topics. A number of regular loyal readers became disillusioned with my eclectic approach to blogging and gave up coming."

Stick to two or three related topics (such as WordPress and related technology topics); you can still cover and talk about a wide variety of subjects that you excel in. People will know what they're coming to your site for and what to expect from you.

- **Have a voice that people want to hear.** Some people don't necessarily care about the mechanics of your writing as much as they care about your voice. Publishers, especially ones that post large amounts of content, often have typos and errors in their posts. Tucker Max (`http://tuckermax.com`), one of the most popular comedy bloggers, switches between past and present tense often — a grammar no-no. He's aware of this problem and doesn't care, and neither do his readers.

  Max knows that he's developing his own style:

  > "I know, I know. The whole concept of tense in speech has always given me problems. In undergrad and law school, I never really took any creative writing or English courses; it was pretty much all econ, law, history, etc, so some of the basic things that most writers get right, I fail. Of course I could learn tenses, but I have never really made an effort to get it right for a reason: I want to write in my own voice, regardless of whether or not it is 'correct' grammar or not. By switching tenses, I write the way I speak, and by alternating between past and present I put the reader into the story, instead of just recounting it."

  Max says that the only time people complain about his grammar mistakes is when they want to argue about the content of his site. They use the grammar mistakes as a plank in their attack. This attempt to belittle him hasn't slowed his growth or success, however. His voice, after all, is what has made him successful.

REMEMBER

Your grammar and spelling don't always have to be perfect, but you should always ensure that your posts are readable. Just don't let perfect grammar get in the way of your individual voice.

» **Present your content well.** The actual look of your presentation matters greatly. Adding images, for example, enhances your posts in several ways, including

- Giving posts a visual point of interest
- Grabbing attention (really making your casual readers stop and read)
- Drawing people's eyes down beyond the first few lines of a post
- Illustrating examples
- Giving your site a personal touch
- Engaging the emotions and senses of readers
- Giving posts a professional feel, which can lead to an air of authority

REMEMBER

Be sure that the only images you use on your website are those images you have permission to use. The best-case scenario is that you use images that you, yourself, own the copyright to. Outside of that, be sure that you've obtained permission from the owner of the image before using it on your website. Alternatively, you can purchase images for use through reputable, commercial, stock-photography sites such as iStockPhoto (`www.istockphoto.com`) and Getty Images (`www.gettyimages.com`) or free image sources such as FreeImages (`www.freeimages.com`) or Pixabay (`https://pixabay.com`).

If you write long, poorly formatted articles, people most likely won't comment or interact with your content — not because of the length of those postings per se, but because of the way that you displayed them, as long paragraphs of endless text. Pictures, highlighted words, bullet points, and other such tricks give the reader's eye a break and can make your published content more attractive and more professional-looking.

» **Write often.** The more you write, the more people will spread the word about your writing, and you can grow your audience. Successful publishers tend to publish content multiple times per week.

All these publishing elements are extremely important on the social web. People want to read and view information that they find interesting, content that's well presented, and content that's specific to their needs. Make sure that you consider all these facets of a website when you create content for your site.

Communication

Communication is the second pillar of the social web. The more you write, the more comments you'll get, assuming that you have comments enabled on your site. Use these tips to manage communication with commenters:

» **Respond to those comments!** The whole point of the social web is communication, and people expect to engage you in a conversation. Successful publishers engage readers in the comment section and create conversations; they use articles as jumping-off points for larger discussions.

  WordPress guru Lorelle VanFossen (`http://lorelle.wordpress.com`) expresses the true value of comments and how they changed how she uses the web:

  > "Comments change how you write and what you write. I suddenly wasn't writing static information. People could question what I said. They could make me think and reconsider my point of view. They could offer more information to add value to my words. And most of all, they could inspire me to write more. Comments made writing come alive."

» **Develop a community.** When you participate in the conversation, you'll retain more readers, who times will revisit your page many during the day to see the new comments and replies in the discussion. The evolution into community discussion can result in a drastic increase in traffic and comments on your site. VanFossen writes of her site:

  > "My site isn't about 'me' or 'my opinion' any more. It's about what I have to say and you say back and I say, and then she says, and he says, and he says to her, and she reconsiders, and I jump in with my two shekels, and then he responds with another view . . . and it keeps going on. Some of these conversations never end. I'm still having discussions on topics I wrote 11 months ago."

» **Don't ignore a person's comments on multiple posts.** You can offend a commenter and lose him. Reply to most comments that your site receives, even if it's only to say thanks for the comments.

REMEMBER

Having the approach that you only want to take from the social web ultimately leaves you unsuccessful. No matter how great your content is, you need to participate and make people feel that you're communicating with them, not just speaking at them.

Consistency

The final pillar of the social web is consistency. When you produce and offer any type of content multiple times a week or on a daily basis, people begin to expect consistency. Many online publishers don't post consistently, and as a result, they frustrate their readers.

COMMENTING ON OTHER SITES

In addition to responding to comments on your site, you should go to different sites and take part in the discussions there. Choose sites that are similar to yours; you can be part of the larger publishing community beyond just your own. (Visitors of those sites will see your witty comments and most likely follow you to your site, thereby building your audience.)

Understanding the social aspect of the social web is vital to your success. People use the social web as a major mode of communication. The communication aspects of your site and others play into the overall online conversation that's going on — a conversation that can get started by an article, which a writer covers in a post on her site about that topic, which a reader comments on, which prompts another person to compose a response to those comments or that article, which gets its own set of comments. Having a grasp of this concept and seeing how it operates not only brings you better success on the social web, but also makes you a better participant.

TIP

Although consistency applies to online publishing, in general, it really matters on social networks such as Twitter and Facebook, where interconnectivity between the author and the audience reaches new heights. If you have large followings on Facebook and Twitter, and use those services as your main point of contact with your readers, be sure to post on a regular basis there as well.

Build good habits by following these consistency guidelines:

- **Set a schedule, and stick to it.** As a site owner, you have to give people a pattern to expect so that eventually, they know when to look for your posts. This idea is like knowing when a favorite TV program is on; you come to expect it and maybe even plan around it. If you miss a day on which you usually post, you just might hear from readers wondering where your post is for that day.

  If you plan to write five days a week, actually write five days a week, and try not to deviate from that schedule. If you plan to post only two to three times a week, stick to the days that you usually post (unless you want to cover some important breaking news).

WARNING

- **Don't let the increasing number of readers and comments affect your posting schedule.** The last thing you want to do is overpost. Although some people would argue that you should keep momentum on a particularly popular post, you run the risk of overexposing yourself and burning yourself out. Also, your content can quickly become watered down. The quality of the content — what the people are there for — quickly begins to erode, and you can lose the audience you've built.

By sticking with a routine and establishing consistency in your posting, you let readers know what to expect, and your site becomes a part of their routine. If you ingrain yourself in someone's life, he or she is going to return to your site frequently and become an advocate for what you're doing.

Plan. You also need to account for long breaks in your posting schedule. You can prewrite posts when you have a lot to say and save them as drafts so that you can post them at times when you aren't inspired to write.

WARNING

Some online publishers take a month off from writing or post very sporadically. But if you really want to build an audience, you can't suddenly decide to take a month off because you're tired of it. Taking a long stretch of time off can kill your site's momentum and audience.

You can explore other options instead of leaving your site dormant. If you've built an audience, you can easily find a guest author to step in for a bit to publish content on your site.

- **Keep the quality consistent.** Take pains to ensure that the quality content you produce doesn't suffer for any reason. Sites often capitalize on a popular post, gain an audience, and then become inconsistent with the quality of their content. They either shift away from their original niche or begin to post poorly thought-out or poorly put-together articles. When their content quality suffers, those sites begin to lose their audience, and sometimes, they never recover.
- **Expect some ups and downs.** You can't easily judge which articles are going to be successful and which aren't. You might write articles in five minutes that get more views and have a better reception than articles you take hours to craft. But readers can really tell when you're posting for the sake of posting. If you repeatedly have to force yourself to post, and if that goes on for too long, the quality of your content and your consistency can go out the window.

Making It Easy for Users to Share Your Content

When I was a child, I loved to go to a country store on a lake near where I lived. One time, my mother and I went to the store to pick up a few things, but my mother didn't have any cash (this was before ATMs were everywhere) and wanted to pay by check or credit card. The store owner told her that they accepted cash only. We put the items back on the shelves and headed to a large supermarket.

When we got into the car, my mother said to me, "I wanted to give them money, but they made it too hard for me to do it." That sentiment has stuck with me my entire life: Never put up barriers to actions that will ultimately benefit you. I'm sure that the store had reasons for not taking checks or credit cards, but it ultimately lost a sale and probably a customer.

Think of your site as the store and your content as the products. When people want to take your content and give it to someone, you put up a barrier if you make it hard for them to pass that content along. Make it as easy as possible for people to share your content with their friends, family members, and co-workers.

One of the best things about the social web is that you can share what you find with other people. Sharing is a basic concept — an easy, thoughtful, and fun thing to do. You find content that you like and share it with your friends on the web, who might find what you shared helpful or interesting and pass it on to their friends. But a lot of sites do a very poor job of allowing users to share content. While you set up your WordPress site, think about how you want readers to share your content.

REMEMBER

Test, test, and test some more. How to best lay out your sharing options on your site takes continual testing. You can't get it right the first time — or the first five times. Sometimes, it takes months to find the right mix.

The following sections give you some simple tips that make sharing content from your website easy.

Enable the user to share content

Enabling sharing is the first thing you want to do. If people don't have the ability to share your content, that content isn't going to go anywhere.

TIP

Sharing content doesn't mean just social media sharing; your content can get spread through other methods. Allow readers to email or print your posts. Although you may feel that email and printing are outdated features, your users may not.

Don't overwhelm the user with choices

Sites can include too many sharing options. The reader becomes overwhelmed and probably also has trouble finding the network that he or she uses.

ADDTOANY VERSUS SHARETHIS

WordPress offers a multitude of plugins (see Book 7) that blend social sharing with more traditional options, such as individual share icons, printing, and emailing. The AddToAny plugin (`https://wordpress.org/plugins/add-to-any`) provides share buttons for almost all the popular social networks (Facebook, Twitter, Google Plus, Pinterest, WhatsApp, and more than 100 other sites), as well as universal email sharing and Google Analytics integration that allows you to track your content-sharing analytics.

Other popular plugins offer similar options, with some drawbacks. The ShareThis plugin (`www.sharethis.com`), for example, provides a green button that, when clicked, expands so that users can select the networks on which they want to share your content, or print or email that content. Making users click an additional button to see their sharing options adds an extra step to the process. The AddToAny plugin (`https://www.addtoany.com/`) puts individual icons on your posts, getting rid of the extra step that users must take to share your content through ShareThis.

Just remember that when you use the ShareThis button, a reader can easily overlook it. The individual buttons are more visible and not as easily overlooked. Test both plugins and see which method gets more shares.

Pick a few sharing sites to which you want to link, test them, and cycle in new ones that people may use. Offer a low number of sharing options at a time so that people can share your content easily. Determine which of these networks your content applies to. If you write celebrity gossip, your content may do better being distributed on sites where people can share quickly with their friends, such as Facebook (`www.facebook.com`) and Twitter (`https://twitter.com`). If you write in-depth technical resources, a social bookmarking site such as Reddit (`https://www.reddit.com`) may be a better place to share your content and bring your blog additional traffic. If you write about fashion and beauty, perhaps providing a sharing button to Pinterest (`https://www.pinterest.com`) can get you traffic from people who are interested in those topics.

Make sure that the sharing options you give visitors apply to sites where your content makes sense. Don't be afraid to try different sites. Study your statistics to see where readers are discovering your content. Many of these sites allow you to search by domain, so you can check to see how often people are sharing your website and what specific content they're sharing.

Put sharing buttons in the right place

Where you present the sharing buttons really depends on the type of content you're posting and the audience reading it. If you post a picture and include a comment below it, the content could push your sharing buttons below the fold, so make sure that your major sharing options appear next to or above the content. Some of the most popular places to display sharing buttons are the top of the post, the bottom of the post, and the left and right margins of the website.

TECHNICAL STUFF

Below the fold refers to what doesn't appear in a user's web browser unless the user scrolls down to view it. The term is taken from newspaper printing, in which some items appear below the fold on the front page.

TIP

To get some ideas about how best to deploy your sharing buttons, check out sites that are similar to yours, and see where some of the most successful site publishers place their buttons.

Think about the user, not yourself

Take this major lesson away from this section. Too many times, people get excited about the latest gadget or tool for their sites. They get eager to try it out and excited to deploy it, but in the end, they aren't thinking about whether it can help the user and whether the user is going to enjoy it.

REMEMBER

How you use the web and how you navigate a site can be completely different from how most other people use it. Review button use and where people are sharing your blog posts. Also use tools such as Google Analytics to see how people interact with your page.

TIP

By using its site-overlay feature, Google Analytics (see Book 5, Chapter 3) allows you to see how often someone clicks various items on your website. You can sign up for Google Analytics for free and deploy it very easily. (You just need to paste the tracking code in your WordPress footer.)

The site-overlay feature currently works in the Google Chrome browser on either a Mac or PC. To access the site-overlay feature from your Google Analytics Dashboard, follow these steps:

1. **Go to the Page Analytics (by Google) page in the Chrome Webstore.**

   It is available at `https://chrome.google.com/webstore/detail/page-analytics-by-google/fnbdnhhicmebfgdgglcdacdapkcihcoh`.

2. **Click the Add to Chrome button.**

   This installs and adds the extension to your Chrome browser by adding a Page Analytics icon to the browser toolbar.

3. **Visit your website in your browser window.**

4. **Click the Page Analytics button in your browser toolbar.**

   The page analytics using the site-overlay feature is shown in Figure 1-1.

Now, on the home page of your site, little text boxes for the various links on your home page appear, displaying percentages. (See Figure 1-1.) The percentages within these text boxes reflect how popular the various links are within your site. If you navigate your site while using the site-overlay feature, you can see, page by page, how people are interacting with your navigation, content, sharing features, and other content.

FIGURE 1-1: The Site Overlay feature in Google Analytics.

Determining Where You Need to Participate

Communication is an important part of social media, and communication is a two-way street. In social media, communication isn't a bullhorn; you need to interact with people. If you want the rewards of participation, you need to listen as well as talk. This idea often gets lost when people start using social media to promote their content.

Determining whom you want to interact with and where to interact with them are large parts of using social media in your marketing strategy. Finding the best communities in which to participate and actively engage in conversations is the quickest way to build a loyal audience.

TIP

Although reaching out to audiences who are known to be receptive to your site's content is a good strategy, you may find that you're following a well-trod path. Other publishers may have already found success there. Don't be afraid to try areas where others who have sites similar to yours aren't participating. If you think the audience is there, go for it. Be original; trail-blaze a little.

As a writer online, you often work as the marketing person for your own site. To gain readership, you need to participate with your potential audience members in communities where they are already participating. Additionally, you can really leverage participating in these communities if you understand the authors in your niche, work with them to possibly get a guest-author slot, or even get links from them on their sites.

Taking the time to create a list of potential audiences goes a long way toward creating your own marketing strategy. Your list should include social networks and message boards where you think your content will be greeted with open arms, authors who publish content in the niche you participate in that you want to monitor, and users who have influence on other social networks (such as someone who has a large Twitter following in your niche or your particular area of interest/expertise).

The most important thing you can do while constructing a list is understand the niche in which you're building a readership. Here are some items of interest to look for when finding out about a niche:

- **Who's in the niche?** Check out the links in a post. Start with a major site in your niche, and see where the links lead, including the links to commenters' sites, and the sites they mention in their content, to get a wide view of the niche. Knowing who associates with whom and what circles people run in can help you discover a lot about a niche. You can determine who the power players are, as well as whether the niche is competitive or has a collegial atmosphere. This information helps you determine how you want to approach your outreach.
- **Is there a niche social media site or group that acts as a connecting point for the community?** Often, in various groups, you can find one or more niche social media sites that connect websites. These sites can be great resources for discovering some of the top sites, and they may help you flesh out your list of writers quickly. Additionally, see whether you can get your site listed on

these sites. Most of these kinds of sites allow free submittals and offer forms to fill out or email addresses to which you can submit your content.

This kind of online community might be a directory with social features, a Facebook community, or a group on a large social network. Whatever the case may be, you can often find large groups that have discussions within a niche. These niche sites can tell you what people in the niche you are targeting find important, what the hot topics are, and what other people are doing in this niche, such as pitches people have made to other site owners. The site TastyKitchen.com (`http://tastykitchen.com`) is a great example of a small niche community — in this case, a community of people who like to cook and share recipes.

TIP

Additionally, these sites feature the type of content that people in your niche may find interesting. Keep a Microsoft Word document open to write down content ideas based on the conversations on these sites.

- **Are common discussions occurring throughout the community?** You can often discover opportunities to get your site in front of new people or for topics to cover by looking for common threads within a niche. Maybe the community is talking about how public-relations people are pitching them, a charity cause that they all support, or an event that they regard as important. A common theme may give you information, opportunity, or direction on how you should approach this niche.
- **Do they use other media to have discussions?** Find out what other social media sites people in this niche use. Maybe they use Twitter a lot, or maybe you see high use of Facebook, LinkedIn, Pinterest, FriendFeed, or YouTube. You may find secondary ways to reach this niche where you can build a following for your site.

  It pays to determine the social media sites your niche prefers. Certain niches (such as wine sites) have taken to Twitter; others have strong ties to Facebook or other social networking sites. Make the most of these sites when you pitch your blog to people. They may prefer that method of connection.

You may think you can simply buy a list, slam together a bunch of search results into a spreadsheet, and then mass-email everyone whom you want to contact. Without studying how your niche operates, however, you can't create mutually beneficial relationships, you can't become a voice in the community, and you probably won't see a lot of success. Instead, you come off as an outsider just trying to push your message down the throats of these publishers, and your campaign will have very poor results.

Finding Influencers

After you compile lists of sites you want to target, you can begin to break the list down and determine who are the influencers in your niche, including the hidden influencers. *Hidden influencers* are people who have a large social imprint that doesn't necessarily show up on their sites. Some people don't have a lot of commenters on their articles, for example, but their Twitter feeds are followed by tens of thousands of people. Here are some ways to determine whether a blogger is an influencer:

- **Subscriber count:** A lot of sites that have large audiences display their subscriber numbers on their websites. (See Figure 1-2.)

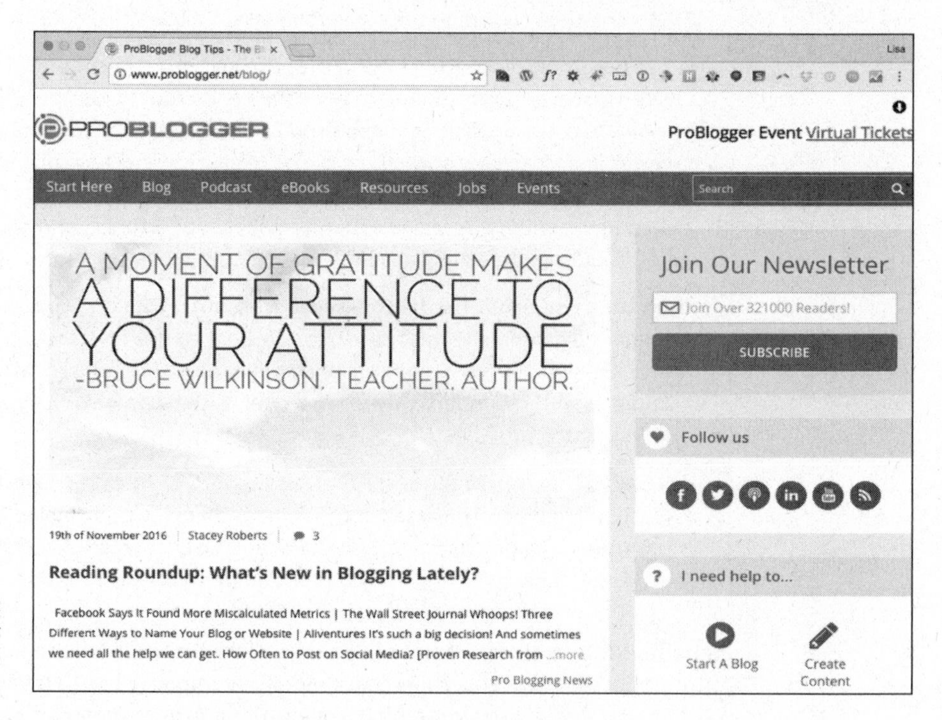

FIGURE 1-2: Subscriber count on Problogger.net.

- **Comment count:** An active community and commentary group on a blog usually shows that the blog has a large readership. Be wary of a site whose author interacts with only two or three people. When an author pays attention to only a couple of commenters, she usually has a pretty narrow vision. You want to target authors who participate with more people in their audience.

- **Alexa score:** Alexa (`www.alexa.com`) measures traffic to a site. It isn't 100 percent accurate, but it does a decent job of giving you a picture of the amount of traffic a site gets. Add a column to your list of sites where you can record the Alexa score, and see how the scores compare with yours.
- **Klout score:** Klout (`https://klout.com`) helps you evaluate the influence of Twitter, Facebook, and LinkedIn users. Sometimes, publishers have a very large reach on those social networks and are more active there than on their own sites.

After you identify the influencers, you want to attract them to your site. If influencers read your site, they may offer you guest-author spots, share your content, recommend your site to their readers, and form a mutually beneficial relationship with you.

To turn these influencers into readers, you can try multiple tactics, including the following:

- **Comment on their content.** Reading and commenting on a popular site can help you start to build your name in your niche — if you leave quality, well-thought-out comments, of course. Most sites allow your username to link back to a website; make sure that you use this link as a way for people to find your site.

  You can get the attention of a popular writer by engaging in conversation on his or her site, and also get the attention of that author's reader. If the readers and commentators like your contribution, you can get additional traffic, new readers, and even potentially high-ranking backlinks into your website, all because you left a comment on the site.
- **Email them.** Depending on the niche, top influencers may get slammed with email, so this approach may not be the best way to reach out to them. But it doesn't hurt to write a personal note that lets the author know about you and your site, and perhaps offer to guest-write if he or she ever accepts posts from other people. Make sure that the email isn't all about you, which is the quickest way to turn someone off. Talk about the person's site, and show that you have knowledge about what he's writing about. Show that you have actually read his site, and demonstrate genuine interest in what he's doing.
- **Interact with them on their platforms of choice.** Sometimes, influencers and popular writers participate in areas other than their sites. They might use message boards, forums, Twitter, Facebook, or other types of social media sites. Interacting with an author on his or her platform of choice can help you differentiate yourself from other sites.
- **Link to them.** Linking to sites in the content you create — especially if you're posting rebuttals to their posts — can really get influencers' attention.

REMEMBER

When you use any of the tactics in the preceding list, the three Cs (see "Understanding the Three Cs of the Social Web" earlier in this chapter) come into effect. When you communicate with other site authors, you need to make sure that you have consistent content on your own site. Trying to reach out to another author when you have only three posts total doesn't present the most credibility. After you've worked at it for a few months, doing outreach can provide a good way to grow your audience.

Leveraging Twitter for Social Media Success

Twitter has become one of the most effective ways for site owners to build an audience. You can use Twitter to find people who have the same interests that you do, communicate with them, and steer a ton of traffic to your site.

Building a Twitter profile into a successful tool to generate traffic is pretty straightforward. Just follow these steps in your account at `https://twitter.com`:

1. **Make sure that your profile is completely filled out, including your picture.**
2. **Follow the three Cs — content, communication, and consistency — when you post to Twitter.**

   By posting quality content consistently on Twitter, you *will* build an audience. Period. When you mix in the communication aspect and retweet the quality content of others, answer questions, and interact with other Twitter users, your profile will grow that much more.
3. **Find people who are interested in what you're writing about, and interact with them.**
4. **Use a tool such as Twitter Search (`https://twitter.com/search-home`) to search for specific keywords related to content that you're writing about — then discover the people who follow that content as well.**

   You may want to follow and interact with these people.

Building your Twitter account by using automated tools

I hesitate to include this section, because using automated tools is a fast way to get your account deleted by Twitter. Automated tools allow you to do mass additions or removals to your account. You can remove people who aren't active or who aren't following you, as well as target the friends list of other users to add them to your account. Using these mass adding-and-removal tools goes against the spirit of Twitter, where you're supposed to be discovering cool content and not just mass-promoting. So I'm warning you right now: If you go down this path, you need to see losing your account as an acceptable risk. If you use the Twitter tool that I discuss in a logical and nonaggressive way, it can help you target and build an audience quickly.

I include automated tools in my discussion of building your social media accounts because a lot of people use this technique, including people who shun them. (A lot of social media experts who deride these tools have used them to get where they are.) I don't believe in giving you half the information; you need to make this choice on your own.

WARNING

If you hyperaggressively add people and then unfollow them, Twitter probably will quickly ban you.

To target users on Twitter, here are the steps you can take:

1. **Go to Refollow (`www.re-follow.com`).**

   Refollow.com is a great service, but it isn't free. You can sign up for a free six-day trial; then, if you want to pay for the service, click the Upgrade button at the top of the site to upgrade. Prices range from $20 to $150 per month.

2. **Log in to Refollow by using your Twitter account login information.**

   This step loads the Refollow Control Panel page, shown in Figure 1-3.

3. **Click the link labeled Create New Campaign.**
4. **From the Find Users Who drop-down menu, select Follow.**
5. **Type** WordPress **in the next text field.**
6. **Select the search options of your choice, such as**
   - Language Preferences
   - People Who Have Actively Tweeted within the Past Week
   - People Who Already Follow You
   - People Who Have a Photo
7. **Click the Save Settings button.**

   The page refreshes and displays the Twitter users who follow the account you specified in Step 4. On this page, you can follow these users on Twitter or remove them from your Dashboard.

FIGURE 1-3: The Refollow Dashboard.

Don't follow more than 100 to 200 people a day. This tool allows you to follow up to 500, but if you follow that many people each day, Twitter will probably ban you after a few days.

You can use Refollow to find people who are following people within your niche and add them to your Twitter account so that they may notice your content. I advise adding people in bulk only once in a 24-hour period so you don't look like you're gaming the system. Once a week, unfollow everyone who isn't following you back to balance your following ratio. You should not be following more people than are following you.

Updating Twitter from your WordPress blog

Getting back to WordPress (that's why you bought the book, right?), you can find tons of plugins to integrate Twitter into your WordPress site. From how the tweets show up on your sidebar to how tweets are integrated into your comments, the WordPress community has tons of solutions to help you show off your Twitter account on your site.

These plugins change often, so try different ones, depending on how you want to integrate Twitter into your site. But if you want to turn your WordPress Dashboard into more of a social media command center, you can tweet right from your WordPress Dashboard.

Although tools such as TweetDeck (`https://tweetdeck.twitter.com`), Buffer (`https://buffer.com`), and Hootsuite (`https://hootsuite.com`) are better designed for an active and strategic Twitter presence, having the ability to tweet from your WordPress Dashboard allows you to update all your social media from one spot. If you're getting started in social media, this integration makes your social media use efficient and continually reminds you to participate.

One of the best WordPress integration plugins for this purpose is the Jetpack for WordPress plugin by Automattic (`https://jetpack.com`). This plugin can update your Twitter feed whenever you publish a new post. You can update Twitter about new posts by using Hootsuite, FeedBurner, and other free tools, but going with the Jetpack plugin allows you to automatically post your new content to Twitter.

Engaging with Facebook

Facebook integration is another key strategy to consider when you're setting up your website. First, integrate the Facebook-sharing feature within your website, which you can do with the ShareThis (`https://wordpress.org/plugins/share-this`) or AddThis plugin (`https://wordpress.org/plugins/addthis`). With more than 1.86 billion users, Facebook is a must-have sharing option for any website.

Next, decide how you want your website to interact with Facebook. Are you running a personal site? Then you may want to use a Facebook profile as your connecting place on Facebook. Some WordPress plugins (such as the Facebook Dashboard Widget) integrate a Facebook profile so that you can update your status right from the WordPress Dashboard.

If you don't want your Facebook account attached to your website, you may want to consider creating a Facebook Page. A Facebook Page doesn't have the Dashboard controls that a profile does, but it allows you to leverage your social media presence. By setting up a Facebook Page, you can deeply integrate the Facebook Like option, which allows users to Like your site and become followers of your page with a couple of clicks. Integrating the Like feature allows you to get exposure for your website through each of your followers' friends on Facebook.

When you have a Facebook Page, you can display a community widget on the side of your WordPress blog, letting everyone know who your followers are on Facebook. If a Facebook user clicks the Like button on your page, he can show up in this widget. Facebook offers a lot of badges and Like-button integration features in its Developers section at `https://developers.facebook.com/docs/plugins/like-button`.

In this Developers section, you can dig deep into integrating Facebook into your blog. You can display the friends of a visitor who likes your site, recommendations based on what the visitor's friends have liked, and numerous other combinations.

» Finding out which monitoring tools are right for you

» Cleaning and aggregating your monitoring data

» Turning your WordPress Dashboard into a listening post

Chapter 2
Creating a Social Media Listening Hub

This chapter focuses on the importance of listening to social media, using the free monitoring services available to you, and integrating these sources into your WordPress installation so that you can turn your run-of-the-mill WordPress installation into a social media listening hub.

A *social media listening hub* is a collection of information from several sources, including mentions of your site, keywords, or topics that you write about, and even information about competitors. You can sign up for services that monitor these topics, such as Salesforce Marketing Cloud (`https://www.salesforce.com/products/marketing-cloud/overview`) and Gigya (`www.gigya.com`). But most of these services cost money and give you another place to log in to — and you may not use these kinds of services to their full capacity. For a small business or an independent site owner, the investment (both time and financial) doesn't always make sense. By leveraging the power of the WordPress platform, you can easily cut down on both the time and financial commitment of monitoring platforms.

In this chapter, I walk you through determining what sources you should pull your data from, determining and searching for the keywords you deem important, and integrating your search results into your WordPress Dashboard. Additionally, I look at some other tools that can help you expand your monitoring practices.

Exploring Reasons for a Social Media Listening Hub

When you begin to engage in the world of social media, one of the most important things you can do is monitor what Internet users are saying about your company, your site, you, or your products. By investigating what Internet users are saying, you can find and participate in discussions about your site or company and come to an understanding about the way your community views your blog (or company). With this information, you can participate by responding to comments on other blogs, Twitter, or message boards or by creating targeted content on your own site.

The conversations happening about your area of interest or niche amount to really great intelligence. For a business, regardless of whether you participate in social media, social media users are talking about your company, so you need to be aware of what they're saying. If you're blogging about a particular topic, you can evolve your content by tracking what members of your niche are saying about it.

Eavesdropping on yourself

By monitoring your niche, you can eavesdrop on thousands of conversations daily and then choose the ones in which you want to participate. The social media listening hub you create allows you to follow various conversations going on through microblogging services such as Twitter, Facebook, blogs, news sites, message boards, and even comments on YouTube. If someone says something negative about you, you can respond quickly to fix the situation. You can attempt to step in and make sure that people are informed about what you're doing.

Keeping tabs on your brand

Think about what keywords or phrases you want to monitor. You want to monitor your name and your blog/company name, of course, as well as other keywords that are directly associated with you.

TIP

Monitor common misspellings and permutations of the name of your brand. The Bing Ads Intelligence tool (`https://advertise.bingads.microsoft.com/en-us/solutions/tools/bing-ads-intelligence`) can help you determine all the common spellings and uses of the keywords you're monitoring. You can also find common misspellings for your brand by examining some of the terms people used to find your page with Google Analytics (`https://www.google.com/analytics`) or a paid tool such as Trellian Keyword Discovery Tool (`www.keyworddiscovery.com/search.html`).

If Aaron Rodgers, Green Bay Packers quarterback, wanted to set up a monitoring service, for example, he might use the keywords *Aaron Rodgers, Arron Rodgers, Aaron Rogers, Green Bay Packers Quarterback*, and perhaps even the phrase associated with his touchdown celebration: *discount double check*. If he wanted to expand this service past direct mentions of him or his team, he could also include more general terms such as *NFL Football* or even *NFC Teams*. The general term *NFL Football* may be *too* general, though, producing too many results to monitor.

Additionally, you may want to view your site or company through the lens of your customers. What terms do they associate with your company? Looking at your site from other points of view can provide good ideas for keywords, but not always. Although you don't always want your company to be known for these terms and may not see yourself that way, getting the perspective of other people can open your eyes to how users view your website.

Don't think of this process as just pulling in keywords, either. You can pull in multiple feeds, just as you do with an RSS reader, which allows you to monitor specific sites. So if you concentrate on an industry, and a website deals specifically with your industry and has an active news flow pushed through an RSS feed, you mat want to consider adding specific websites to the mix of feeds you run through WordPress.

The setup in WordPress that I describe in this chapter gives you the convenience of having everything in one place and can help you monitor your brand, company, or blog. The limitations of the WordPress platform mean that you can monitor only five groupings, so you can't use this method as a replacement for an enterprise-monitoring tool for a large company. Additionally, if you own a restaurant, hotel, or bar and want to pick up review sites such as Yelp and Trip Advisor, these tools can't do the job. Most social media monitoring tools don't count review sites as social media. Tools such as Reputation Ranger (`http://reputationranger.com`) can monitor ratings sites for a nominal monthly fee if you want to pay attention to those types of sites.

REMEMBER

When your content changes, change what you're monitoring to match the evolution of what you're writing about.

Exploring Different Listening Tools

You can find tons of monitoring and listening tools that oversee the social media space. If you work for a large company, you can use large, paid tools such as Salesforce Marketing Cloud (`https://www.salesforce.com/products/marketing-cloud/overview`), Gigya (`www.gigya.com`), Alterian (`www.alterian.com`), and Lithium (`www.lithium.com`). Pricing for these tools runs from a few hundred dollars to tens of thousands per month. Most individuals and small businesses can't make that investment. If you're one of the smaller guys, you can create your own monitoring service right in WordPress by importing free monitoring tools into your Dashboard to create a social media listening hub.

Some monitoring tools pick up site coverage, Twitter remarks, and message board comments. Other tools pick up content created with video and pictures. Try the monitoring services mentioned in the following sections, and determine which give you the best results and which make you feel the most comfortable. Then choose the best tools to create a good monitoring mix. One solution probably can't cover everything, so experiment with different combinations of tools.

TIP

Most, but not all, of these tools use Boolean search methods, so you need to understand how to narrow your searches. If you want to combine terms, put an AND between two items (*cake AND pie*). If you use OR, you can broaden your search, such as to track common misspellings *(MacDonalds OR McDonalds).* Finally, if you want to exclude terms, you can use the NOT operator to exclude items from your search. Use NOT if you want to search for a term that could have an alternative meaning that's irrelevant to what you're actually looking for (*Afghan NOT blanket,* for example, if you're blogging about Afghanistan).

Although some of the monitoring tools in the following sections don't apply to every type of website, I'd include them in most monitoring setups.

For each search that you do on a monitoring service, you need to log the feed address. To make recording these addresses easy, open a spreadsheet or a document into which you can paste the various feeds. You can collect them in one place before you begin to splice them together (which you do in the "Creating Your Own Personal Monitoring Mix" section later in this chapter). Think of this document as a holding area.

Monitoring with Google Alerts

Most social media experts consider Google Alerts (`www.google.com/alerts`) to be a must-use monitoring source for anyone dabbling in social media. Google Alerts allows you to set up monitoring on news sites, blogs, pictures, videos, and groups. You can toggle the amount of results you see, from 20 to 50, and you can choose how often they come in (in real time, daily, or weekly). You can also have Google deliver your alerts to your email or via RSS.

Google Alerts isn't perfect, but it doesn't have many drawbacks. Some of the specialized searches (such as Boardreader, which targets message boards; see "Searching communities with Boardreader" later in this chapter) pick up more in their areas of expertise than Google Alerts does, but in general, and compared with other tools, Google Alerts covers the widest range of content.

You can easily set up Google Alerts by following these steps:

1. **Navigate to `www.google.com/alerts` in your web browser.**

   The Alerts page loads, welcoming you to the Google Alerts website.

2. **In the search text box, type the keyword or phrase that you want to monitor.**

   If you enter a phrase in which the words have to go in that particular order, put the phrase in quotation marks.

3. **From the Show Options drop-down list, choose the type of monitoring that you want to use.**

   The options send you different kinds of alerts:

   - *How Often:* Determine how often you'll receive these notices. Because you'll receive the updates via RSS and not email (which you set up in Step 4), you want the highest frequency possible, so choose the As-It-Happens option. Other options include a daily, weekly, and monthly digest.
   - *Sources:* Select the type of sources you want Google Alerts to search in (news, blogs, video, and so on).
   - *Language:* Select the language you want Google Alerts to search.
   - *Region*: Select the region you want Google Alerts to search.
   - *How Many:* Select how many results you want to receive. If you choose As-It-Happens for How Often, for example, you receive items in real time, so you don't need to specify the number of items; choose Only the Best Results or All Results.

4. **Choose your delivery type from the Deliver To drop-down list.**

   To make the delivery source an RSS feed, as opposed to an email, choose Feed.

5. **Click the Create Alert button.**

   You see your Google Alert Management screen, where you can get the RSS feeds for all your Google Alerts.

6. **To get the URL of the RSS feed, right-click the RSS Feed icon and choose Copy Link Address from the shortcut menu that appears.**

7. **Paste the copied link location into a document in which you list all the feeds that you plan to aggregate later.**

8. **Repeat Steps 2 through 7 for all the terms you want to monitor.**

TIP

Before you start importing the feed into your WordPress Dashboard, you may want to receive the update via email for a few days to test the quality of the results you're getting. If your results aren't quite right, you can always narrow your search criteria. Doing this saves you the time of parsing all your RSS feeds, blending them, and then having to go back and edit everything because your RSS feeds are set up wrong. Using email as a test is a massive timesaver.

Tracking conversations on Twitter with RSS

Tracking mentions on Twitter via RSS is relatively simple. You just need to know what you're looking for and how to build the RSS links so you can monitor them. You can look for several items to monitor your brand and reputation via the Twitter social network, including the following:

- **Username:** Monitor when your Twitter name is mentioned and by whom.
- **Hashtags:** Monitor specific Twitter hashtags (such as #wordpress).
- **Keywords:** Monitor Twitter for a specific word.

Twitter doesn't make it easy to locate the native RSS feed URL on its website, but you can find the feeds by using a URL such as this example: `http://search.twitter.com/search.atom?q=XXX`, where *XXX* is the keyword, hashtag, or username you want to monitor. The following examples can help you build various RSS links for Twitter:

- **Twitter RSS URL for a username:** `https://twitter.com/search?f=tweets&q=Lisa+Sabin-Wilson`
- **Twitter RSS URL for a keyword:** `https://twitter.com/search?f=tweets&q=wordpress`
- **Twitter RSS URL for a hashtag:** `https://twitter.com/search?f=tweets&q=%23wordpress`

Build the Twitter RSS URLs that you want to monitor and include them in the document where you're listing all the RSS feeds you want to aggregate later.

Searching communities with Boardreader

Boardreader (`http://boardreader.com`) is a must-add tool because its niche focuses on groups and message boards, where conversations have been happening much longer than on Facebook and Twitter. Many other monitoring tools overlook these areas when talking about monitoring the web, but you can find many vibrant communities that are worth being part of, in addition to monitoring what's being said about your blog or company.

To set up your Boardreader tracking, follow these steps:

1. **Navigate to `http://boardreader.com`.**
2. **In the text box, type the search term that you want to monitor; then click the Search button.**
3. **Copy the URL from your browser address bar.**
4. **Paste this URL into a document in which you list all the feeds that you plan to aggregate later.**
5. **Repeat Steps 2 through 4 to search for and monitor as many search terms as you want.**

Microblog searching with Twingly

Twingly Blog Search (`https://www.twingly.com/search`) deals with real-time search microblogging and traditional blog search. This microblog service monitors conversations only on Twitter.

TIP

Compare the results you get from Twingly with Twitter Search, just to make sure that Twingly is picking up everything on Twitter.

To monitor conversations by using Twingly, follow these steps:

1. **Navigate to `https://www.twingly.com/search`.**
2. **In the text box, type the search term that you want to monitor; then click the Search button.**

   All the relevant search results appear, as shown in Figure 2-1.
3. **On the right sidebar, right-click the Subscribe to RSS link next to the orange RSS icon, and choose Copy Link Location from the shortcut menu that appears.**
4. **Paste this link into a document in which you list all the feeds that you plan to aggregate later.**
5. **Repeat Steps 2 through 4 for all the terms you want to monitor.**

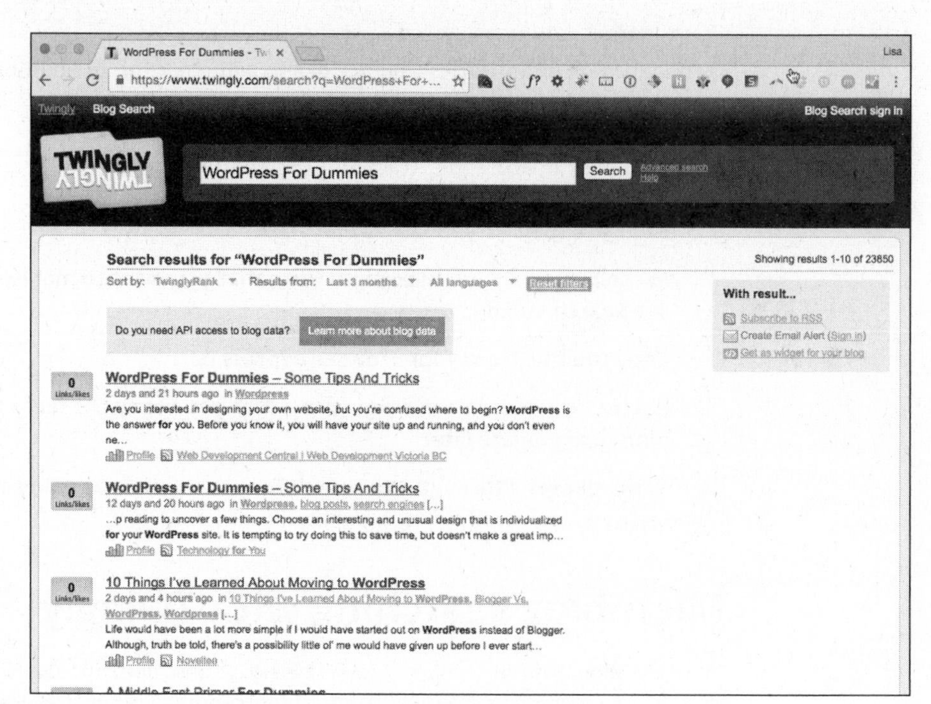

FIGURE 2-1: Searching for WordPress For Dummies on Twingly.

Creating Your Own Personal Monitoring Mix

After trying out the various monitoring services, you can create a mix of services to import into your WordPress Dashboard. You import the results of these monitoring services with the help of RSS (Really Simple Syndication). You can combine different single RSS feeds into one RSS feed and create an organized setup for all the information you have to manage. If you have various RSS feeds from different sources around the keyword *cookies*, for example, you can combine them into one RSS feed. Or if you want to combine various feeds based on sources such as all your Twitter RSS feeds, you can do that as well.

Look for the orange RSS icon that's usually located on the URL bar of your browser. Additionally, some sites offer an RSS export link on the right sidebar or the search bar. Grab the address of all these feeds by clicking the feed name and copying the feed URL from the browser.

After you copy the locations for all your RSS feeds in one document, you need to group those RSS feeds. Grouping your RSS feeds keeps your monitoring system nice and tidy, and allows you to set up the WordPress Dashboard easily. After you group these feeds, you splice them together to make one master feed per grouping (see the following section). If you're tracking a variety of keywords, you may want to put your feeds into groups. Wendy's Restaurants, for example, could make these keyword groupings:

- **Grouping 1:** Your brand, products, and other information about your company
  - Wendy's (the company name)
  - Frosty (a prominent product name)
  - Wendy Thomas (a prominent person in the company)
- **Grouping 2:** Competitors
  - McDonald's
  - Burger King
  - In-N-Out
- **Grouping 3:** Keyword-based searches (Burgers)
  - Hamburgers
  - Cheeseburgers
- **Grouping 4:** Keyword-based searches (Fast Food)
  - Fast food
  - Drive-through
- **Grouping 5:** Keyword-based searches (Chicken)
  - Chicken sandwiches
  - Chicken salad
  - Chicken nuggets

In each of these groups, you place your Google Alerts feed, Twingly feed, and whatever other feeds you feel will provide information about that subject area. You can blend each group of feeds into one master feed for that group and bring them into WordPress.

REMEMBER

WordPress limits you to five groups. Any more than five groups slows the Dashboard and is more than WordPress can handle.

Grouping all your various feeds gives you the most complete monitoring solution by covering multiple monitoring tools and blending them. You get more coverage of your brand or blog than you would by using Google Alerts alone. On the down side, you may see some duplicates because of overlaps between the services.

If you feel overwhelmed by duplicate search results, you can blend one feed that covers only your brand, or you can simplify setting up your monitoring even more by keeping one feed for each item, as follows:

- **General overview:** Google Alerts or Social Mention
- **Message boards:** Boardreader
- **Microblogging:** Twitter Search

Editing the Dashboard to Create a Listening Post

After you choose your data sources, clean up your feeds, and put them all in individual RSS feeds, you can finally bring them into WordPress and set up your social media listening hub.

You can bring these RSS feeds into your Dashboard through a plugin called Dashboard Widgets Suite, which you can find in the WordPress Plugin Directory at the following page:

```
https://wordpress.org/plugins/dashboard-widgets-suite
```

Follow these steps to set up the Dashboard Widgets Suite plugin and configure it to create a social-listening Dashboard in WordPress:

1. **From the Plugins menu on the left side of your WordPress installation, choose Add New.**

   This step takes you to the form where you can search for new plugins.

2. **In the Search text box, type** Dashboard Widgets Suite**; then click the Search Plugins button.**

   The search results page appears.

3. **Search for the Dashboard Widgets Suite plugin, and click the Install Now link, which installs the plugin on your site.**

4. **When the installation is complete, activate the plugin by clicking the Activate button on the Add Plugins screen that appears to the right of the Dashboard Widgets Suite plugin name.**

5. **Choose Dashboard Widgets from the Dashboard Settings menu.**

   The Dashboard Widgets screen appears.

6. **Ensure that the Enable the Control Panel Widget check box is selected (as it should be by default).**

7. **Leave all the rest of the default settings in place, and click the Save Changes button.**

8. **Choose Widgets from the Appearance menu.**

   The Widget Settings page appears.

9. **Add the RSS Widget to the Dashboard Widgets Suite widget area.**

   Drag the RSS Widget into the Dashboard Widgets Suite widget.

10. **Configure the RSS Widget to display information from your selected RSS feed (see Figure 2-2):**

    - Enter the RSS feed URL.
    - Optionally, give the RSS feed a title by typing a title in the text field.

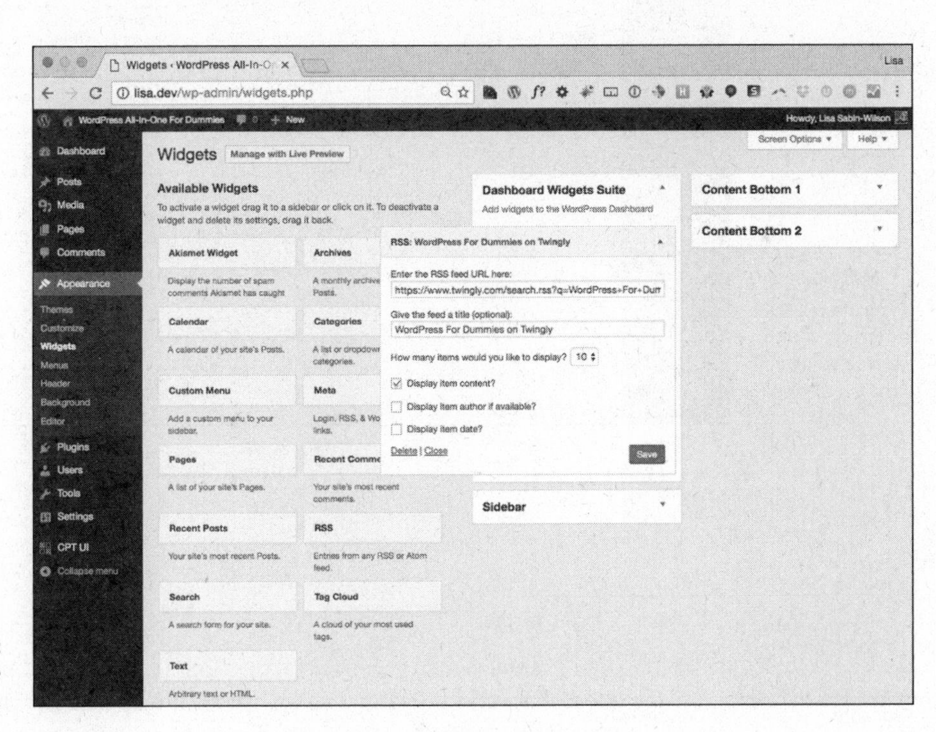

FIGURE 2-2: Configuring the Feed Box Widget.

11. **Repeat Steps 8 through 10 for the other widgets on your Dashboard by using your other selected feeds.**

    After you have your feeds set up, you can configure the appearance of your WordPress Dashboard.

12. **Drag and drop the new widget boxes where you want them on your Dashboard.**

Figure 2-3 displays my Dashboard with the RSS feed from the search I found on Twingly for the term *WordPress For Dummies*, shown in the top-right corner.

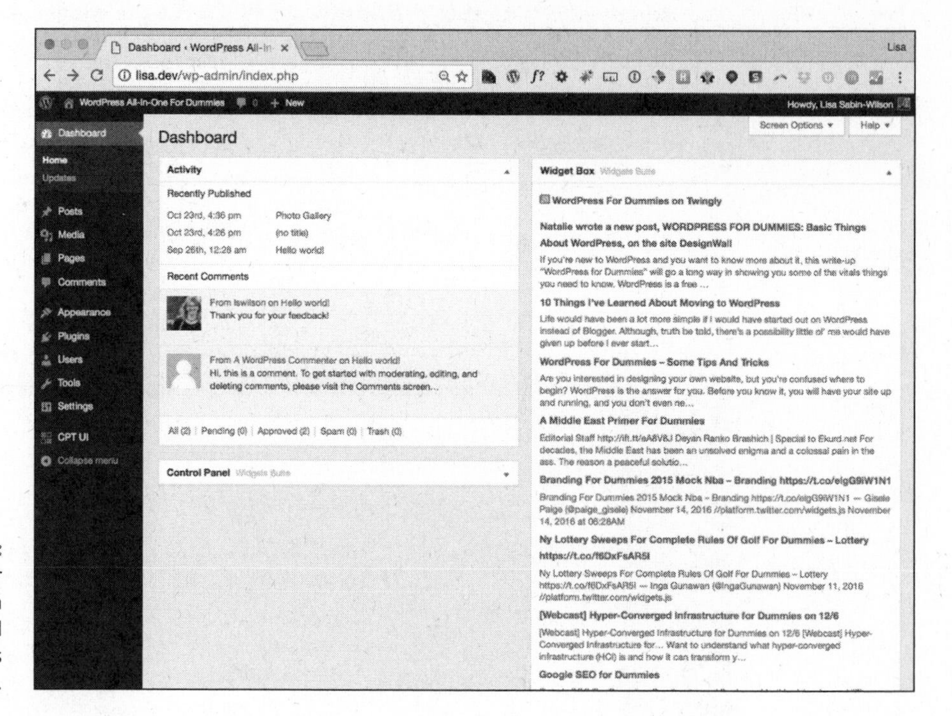

FIGURE 2-3: WordPress For Dummies search term displayed in the WordPress Dashboard.

» Choosing an analytics-tracking option

» Deciphering analytics terminology

» Working with Google Analytics

Chapter 3 Understanding Analytics

Every business on the face of the Earth needs to figure out what works and what doesn't if it wants to succeed. Site owners often know basic statistics about their sites, such as the current number on their hit counters or how many people visit their site daily or weekly. These stats may give you the big picture, but they don't really address why something is or isn't working.

You need to get at least a basic understanding of analytics if you want to make the most of your site. The data provided by free programs such as Google Analytics can really help you grow as a content publisher. In this chapter, you discover how to incorporate various data-measuring tools into your WordPress installation, decipher what the data is telling you, and determine how to act on it.

Google Analytics provides you a tremendous amount of information about your content. The goal of this chapter is to help you interpret the data, understand where your traffic is coming from, understand which of your content is the most popular among your visitors, know how to draw correlations between various data sets, and use this information to shape the content you write. This process may sound very geeky and accountant-like, but in reality, it gives you a road map that helps you improve your business.

Understanding the Importance of Analytics

Personally, I avoid math like my nephew avoids vegetables. Most people's eyes glaze over when they hear the word *analytics* followed by *stats,* any type of *percentages,* and anything that sounds like accountant-speak.

You should view analytics not as a bunch of numbers, however, but as a tool set that tells a story. It can tell you how people are finding your content, what content is most popular, and where users are sharing that content. Knowing what type of content is popular, where your site is popular (in which time zones, countries, and states, for example), and even what time of day your posts get more readers is all valuable information. Understanding your audience's interest in your content, as well as their preferences on when and how they read your content, is important.

At one point in my life, I had a pretty popular political blog. Through studying analytics and reactions to my content, I figured out that if I posted my blog between 9:30 a.m. and 10 a.m. EST, my posts garnered the most comments and got the most traffic throughout the day. When I posted after noon, my blog got about half as many comments and half as much traffic over a 24-hour period. Additionally, I saw that my site was getting shared and voted for on the social news site Reddit (`https://www.reddit.com`) more often than on Digg (`http://digg.com`), another social news site, so I replaced the Digg button with a Reddit button. This change increased the amount of traffic I received from Reddit because people had the visual reminder to share the post with their friends and vote for posts as favorites.

I was able to continue to drill down from there. Not only did I have the information on where my content was being shared, but also, I was able to garner more information for analytics. Posts that had a picture mixed in with the first three paragraphs often had a lower *bounce rate* (the interval of time it takes for a visitor to visit a site and then bounce away to a different site) than posts that had no picture at all. If I wrote the post while elevating my left leg and wearing a tinfoil helmet, I saw a 25 percent bump in traffic. (Okay, maybe that last one isn't true.)

Exploring the Options for Tracking Data

You have a lot of options when it comes to tracking data on your site. Google Analytics is the most popular tool, but several options are available. Analytics is popular because of its widespread use, the amount of content written on how to maximize it, and the fact that it's free.

Here are three popular tools:

- **StatCounter (`http://statcounter.com`):** StatCounter has both a free and a paid service. The paid service doesn't kick in until you get to 250K page views a month.

  StatCounter (shown in Figure 3-1) uses the log generated by your server and gives you the ability to configure the reports to fit your needs. If you want to use a log file, you need to have a self-hosted blog and to know where your log file is stored. StatCounter requires a little more technical knowledge than your average analytics app because you have to deal with your log file instead of cutting and pasting a line of code into your site. The main advantage of StatCounter is that it reports in real time, whereas Google Analytics always has a little bit of lag in its reporting.

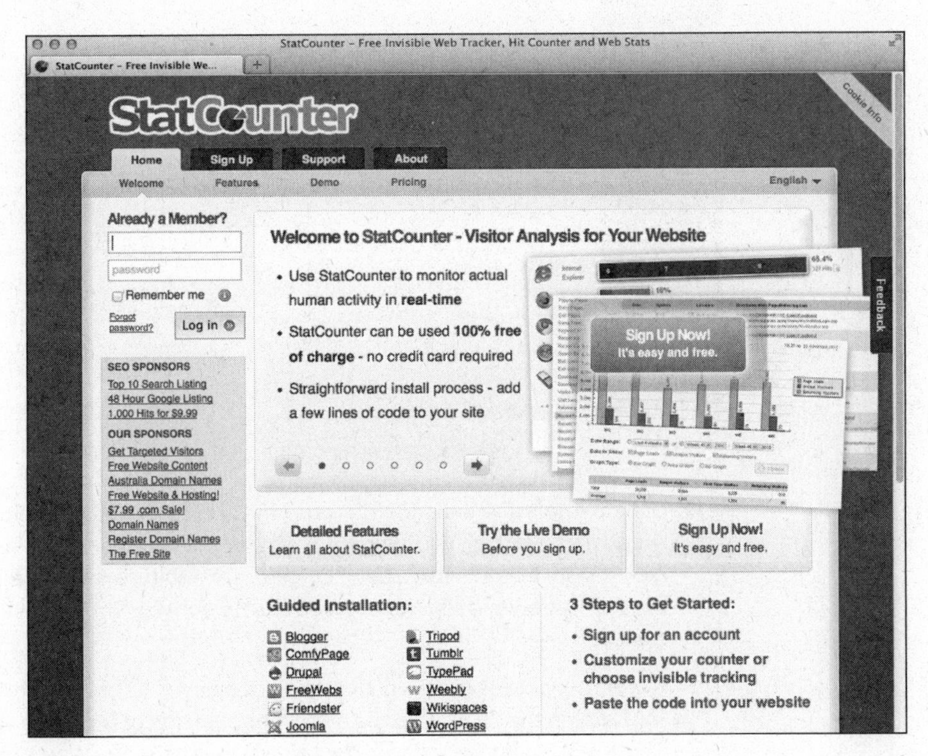

FIGURE 3-1: StatCounter offers real-time stats.

- **Jetpack (`https://wordpress.org/plugins/jetpack`):** The Jetpack plugin provides a pretty good stat package for its hosted-blog users. Shortly after launching, WordPress.com provided a WordPress Stats plugin that self-hosted users can use. (See Figure 3-2.) If you use this package, your stats appear on the WordPress Dashboard, but to drill down deeper into them, you need to access the stats on WordPress.com. The advantages of WordPress stats are that they are pretty easy to install and present a very simplified overview of your data. On the downside, they don't drill as deep as Google Analytics, and the reporting isn't as in-depth. Neither can you customize reports.

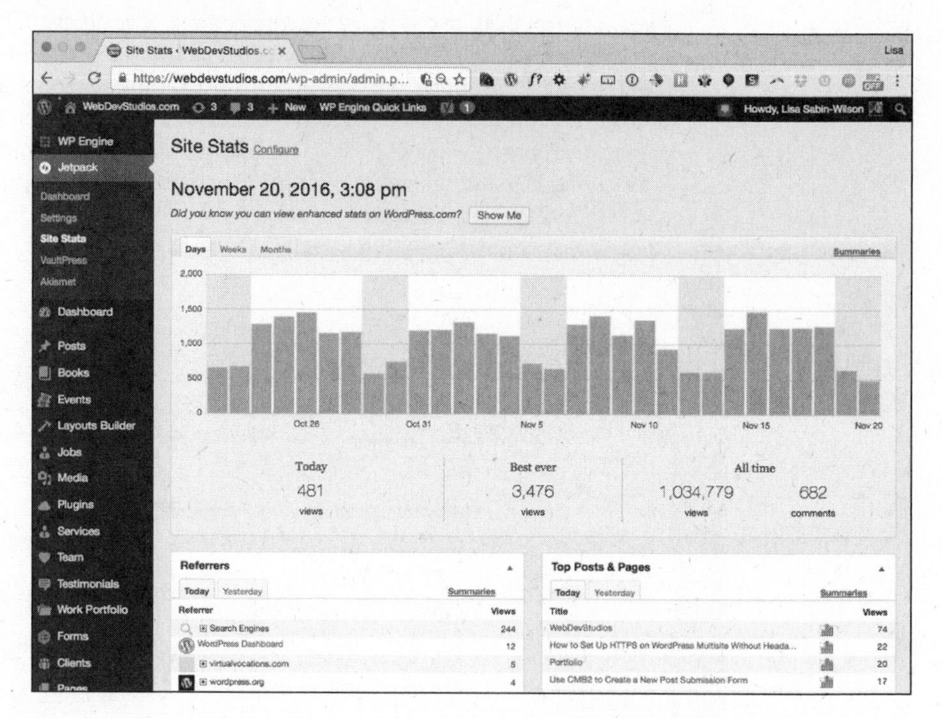

FIGURE 3-2: Try the Jetpack plugin.

» **Google Analytics (`www.google.com/analytics`):** Google Analytics can seem overwhelming when you sit in front of it for the first time, but it has the most robust stats features this side of Omniture. (Omniture is an enterprise-level stats package, which is overkill if you're a personal or small-business site owner.)

  WordPress plugins (covered in "Adding Google Analytics to Your WordPress Site" later in this chapter) bring a simplified version of Google Analytics (see Figure 3-3) to your WordPress Dashboard, much like the WordPress.com Stats plugin. If you feel overwhelmed by Google Analytics and prefer to have your stats broken down in a much more digestible fashion, this plugin is for you: It provides a good overview of analytics information, including goals that you can set up. Although the plugin doesn't offer everything that Google Analytics brings to the table, it provides more than enough so that you can see the overall health of your website and monitor where your traffic is coming from, what posts are popular, and how people are finding your website. Besides the Dashboard Stats Overview, this plugin gives you a breakdown of traffic to each post, which is a nice bonus.

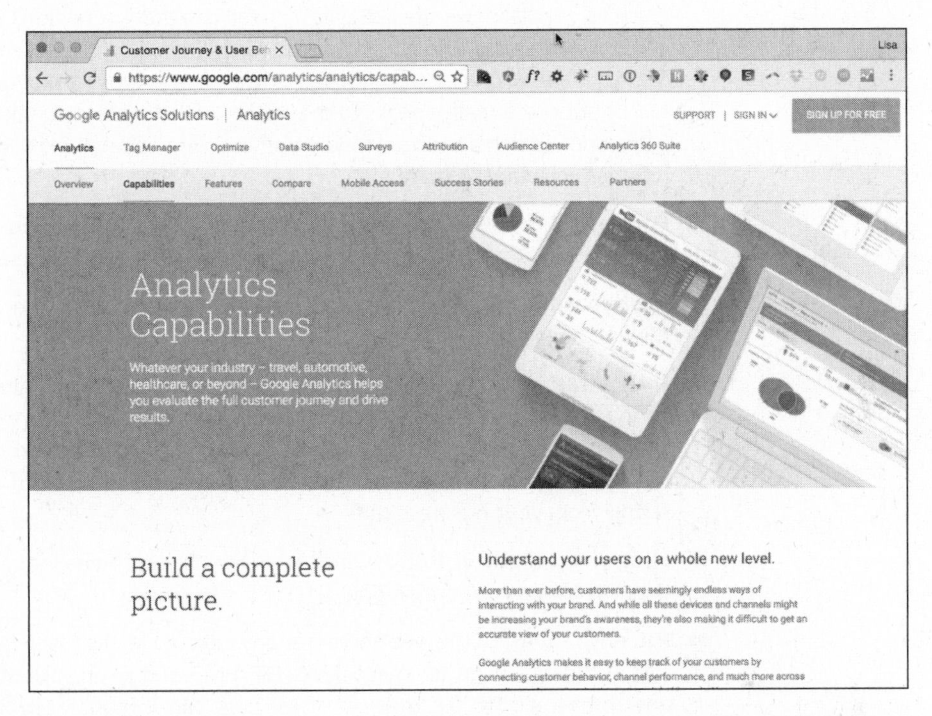

FIGURE 3-3: Google Analytics is a powerful tool.

Understanding Key Analytics Terminology

One of the reasons why people find analytics programs so overwhelming is that they use obscure terminology and jargon. Here, I've taken the time to define some of the more popular terms. (I even spent the time putting them in alphabetical order for you; you can thank me later.)

- **Bounce rate:** The percentage of single-page visits or visits in which the person leaves your site from the entrance page. This metric measures visit quality. A high bounce rate generally indicates that visitors don't find your site entrance pages relevant to them.

  The more compelling your landing pages are, the more visitors stay on your site and convert to purchasers or subscribers, or complete whatever action you want them to complete. You can minimize bounce rates by tailoring landing pages to each ad that you run (in the case of businesses) or to the audience based on the referring site (a special bio page for your Twitter profile, for example). Landing pages should provide the information and services that the ad promises.

When it comes to site content, a high bounce rate from a social media source (such as a social news site like Digg) can tell you that users didn't find the content interesting, and a high bounce rate from search engines can mean that your site isn't what users thought they were getting. In blogging, having a low bounce rate really speaks to the quality of the content on your site. If you get a lot of search and social media traffic, a bounce rate below 50 percent is a number you want to strive for.

- **Content:** The different pages within the site. (The Content menu of Google Analytics breaks down these pages so that they have their own statistics.)
- **Dashboard:** The interface with the overall summary of your analytics data. It's the first page you see when you log in to Google Analytics.
- **Direct traffic:** Traffic generated when web visitors reach your site by typing your web address directly in their browsers' address bars. (Launching a site by clicking a bookmark also falls into this category.) You can get direct-traffic visitors because of an offline promotion, repeat readers or word of mouth, or simply from your business card.
- **First-time unique visitor:** A visitor to your website who hasn't visited before the time frame you're analyzing.
- **Hit:** Any request to the web server for any type of file, not just a post on your site, including a page, an image (JPEG, GIF, PNG, and so on), a sound clip, or any of several other file types. An HTML page can account for several hits: the page itself, each image on the page, and any embedded sound or video clips. Therefore, the number of hits a website receives doesn't give you a valid popularity gauge, but indicates server use and how many files have been loaded.
- **Keyword:** A database index entry that identifies a specific record or document. (That definition sounds way fancier than a keyword actually is.) Keyword searching is the most common form of text search on the web. Most search engines do text query and retrieval by using keywords. Unless the author of the web document specifies the keywords for his or her document (which you can do by using meta tags), the search engine has to determine them. (So you can't guarantee how Google indexes the page.) Essentially, search engines pull out and index words that it determines are significant. A search engine is more likely to deem words important if those words appear toward the beginning of a document and are repeated several times throughout the document.
- **Meta tag:** A special HTML tag that provides information about a web page. Unlike normal HTML tags, meta tags don't affect how the page appears in a

user's browser. Instead, meta tags provide information such as who created the page, how often it's updated, the title of the page, a description of the page's content, and keywords that represent the page's content. Many search engines use this information when they build their indexes, although most major search engines rarely index the `keywords` meta tag anymore because it has been abused by people trying to fool search results.

- **Pageview:** Refers to the number of unique views a web page has received. A *page* is defined as any file or content delivered by a web server that would generally be considered a web document, which includes HTML pages (`.html`, `.htm`, `.shtml`), posts or pages within a WordPress installation, script-generated pages (`.cgi`, `.asp`, `.cfm`), and plain-text pages. It also includes sound files (`.wav`, `.aiff`, and so on), video files (`.mov`, `.mpeg`, and so on), and other nondocument files. Only image files (`.jpeg`, `.gif`, `.png`), JavaScript (`.js`), and Cascading Style Sheets (`.css`) are excluded from this definition. Each time a file defined as a page is served or viewed in a visitor's web browser, a *pageview* is registered by Google Analytics. The pageview statistic is more important and accurate than a hit statistic because it doesn't include images or other items that may register hits on your site.
- **Path:** A series of clicks that results in distinct pageviews. A path can't contain nonpages, such as image files.
- **Referrals:** Occur when a user clicks any hyperlink that takes him or her to a page or file in another website, which could be text, an image, or any other type of link. When a user arrives at your site from another site, the server records the referral information in the hit log for every file requested by that user. If the user found the link by using a search engine, the server records the search engine's name and any keywords used as well. Referrals give you an indication of what social-media site, as well as links from other websites, are directing traffic to your blog.
- **Referrer:** The URL of an HTML page that refers visitors to a site.
- **Traffic sources:** A metric that tells you how visitors found your website, such as via direct traffic, referring sites, or search engines.
- **Unique visitors:** The number of unduplicated (counted only once) visitors to your website over the course of a specified time period. The server determines a unique visitor by using *cookies,* which are small tracking files stored in your visitors' browsers that keep track of the number of times they visit your site.
- **Visitor:** A stat designed to come as close as possible to defining the number of distinct people who visit a website. The website, of course, can't really determine whether any one "visitor" is really two people sharing a computer, but a good visitor-tracking system can come close to the actual number. The most accurate visitor-tracking systems generally employ cookies to maintain tallies of distinct visitors.

Adding Google Analytics to Your WordPress Site

In the following sections, you sign up for Google Analytics, install it on your blog, and add the WordPress plugin to your site.

Signing up for Google Analytics

To sign up for Google Analytics, follow these steps:

1. **Go to `https://www.google.com/analytics/analytics`, and click the Sign Up For Free button, which is located in the top-right corner of the page.**

   A page where you can sign up for a Google account or sign in via an existing Google account appears. If you don't have a Google account, follow the link to sign up for one.

2. **Sign in via your Google account by entering your email address and password in the text boxes and then clicking Sign In.**

   The first of a series of walk-through pages appears.

3. **Click the Sign Up button.**
4. **On the page that appears (see Figure 3-4), fill in this information:**
   - A name to identify your account (which really doesn't matter; you can call it your website's name)
   - The URL of your website
   - The industry category of your website
   - The country and time zone you're in
5. **Click the Get Tracking ID button.**

   Scroll down to see this button.

   The Tracking Code page appears. (See Figure 3-5.)

   At the bottom of the page, Google Analytics provides your Google tracking code, as shown in Figure 3-6.

6. **Copy the tracking code by selecting it and pressing Ctrl+C on a PC or Command+C on a Mac.**

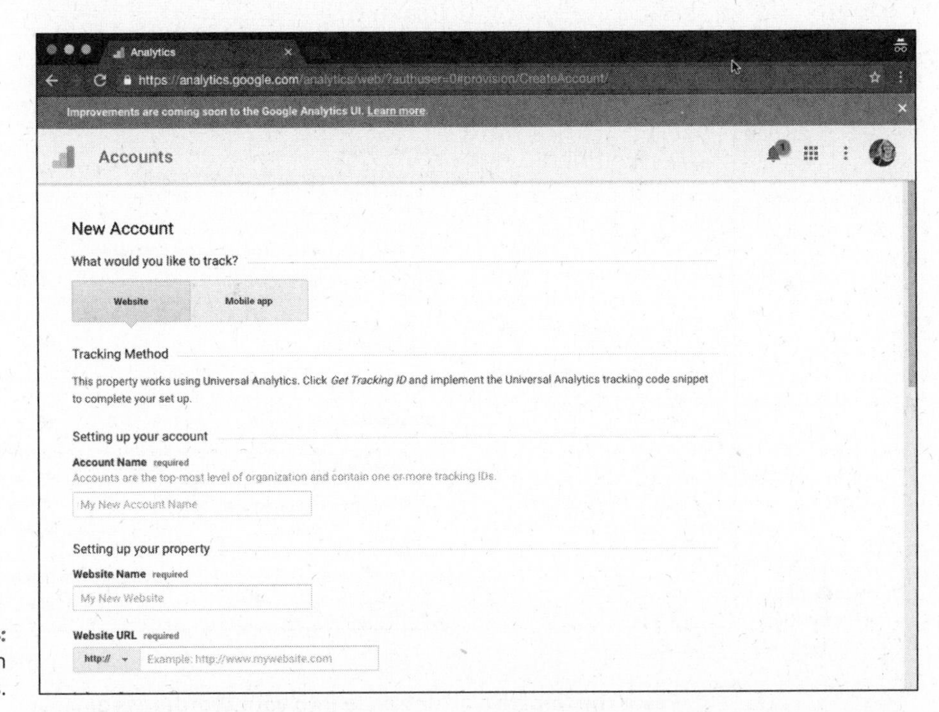

FIGURE 3-4: Accounts page in Google Analytics.

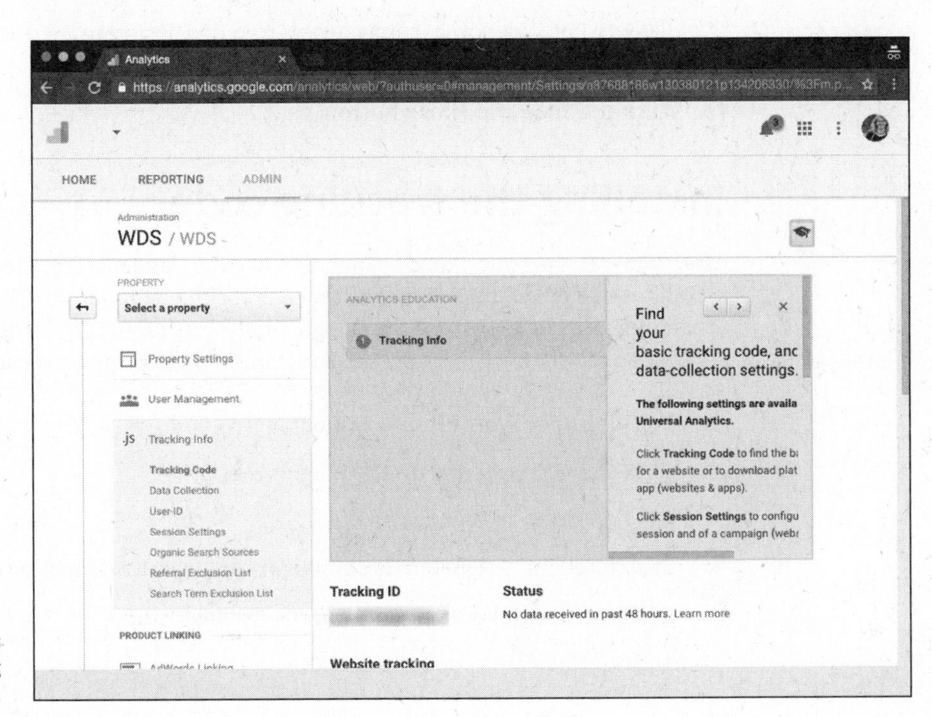

FIGURE 3-5: The Tracking Code page.

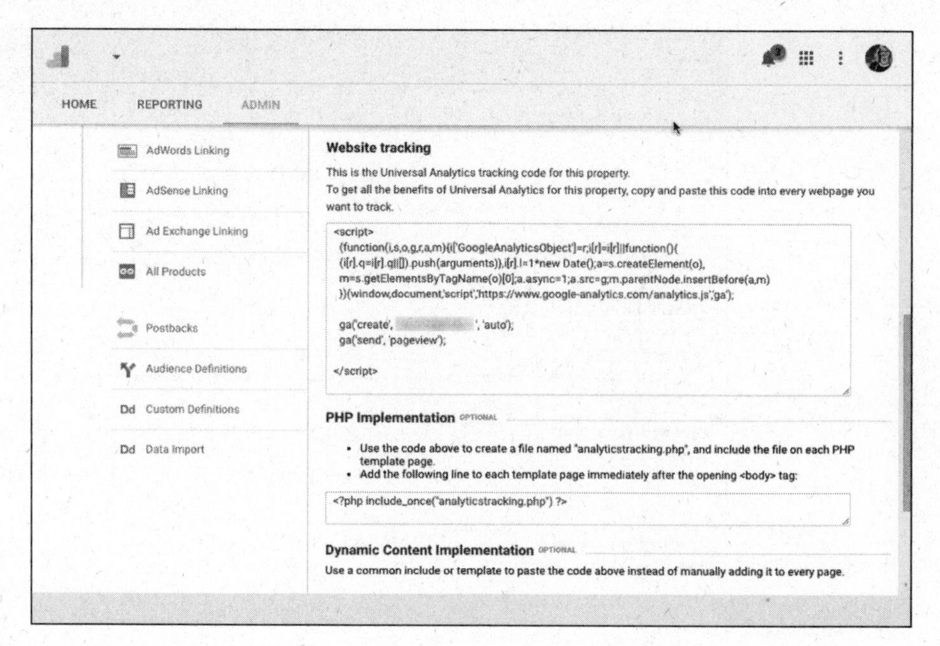
HOME
REPORTING
ADMIN
AdWords Linking
AdSense Linking
Ad Exchange Linking
All Products
Postbacks
Audience Definitions
Custom Definitions
Data Import
Website tracking
This is the Universal Analytics tracking code for this property.
To get all the benefits of Universal Analytics for this property, copy and paste this code into every webpage you want to track.
<script>
(function(i,s,o,g,r,a,m){i['GoogleAnalyticsObject']=r;i[r]=i[r]||function(){
(i[r].q=i[r].q||[]).push(arguments)},i[r].l=1*new Date();a=s.createElement(o),
m=s.getElementsByTagName(o)[0];a.async=1;a.src=g;m.parentNode.insertBefore(a,m)
})(window,document,'script','https://www.google-analytics.com/analytics.js','ga');
ga('create', ', 'auto');
ga('send', 'pageview');
</script>
PHP Implementation OPTIONAL
Use the code above to create a file named "analyticstracking.php", and include the file on each PHP template page.
Add the following line to each template page immediately after the opening <body> tag:
<?php include_once("analyticstracking.php") ?>
Dynamic Content Implementation OPTIONAL
Use a common include or template to paste the code above instead of manually adding it to every page.

FIGURE 3-6: Get your Google Analytics code.

7. **Paste the Google tracking code into your WordPress blog.**

   If you're not sure how to complete this step, see "Installing the tracking code" later in this chapter.

8. **Click the Save and Finish button.**

Installing the tracking code

After you set up your Google Analytics account and obtain the tracking ID to install in your WordPress site, you're ready for the installation. The easiest way to accomplish this task is to use the Google Analytics plugin for WordPress (`https://wordpress.org/plugins/googleanalytics`) by following these steps:

1. **Log in to your WordPress Dashboard.**
2. **Click the Add New link on the Plugins menu.**

   The Add Plugins screen appears.

3. **Search for the Google Analytics plugin, and install and activate it.**
4. **Click the Google Analytics link on the Settings menu.**

   The Google Analytics screen appears.

5. **In the Web Property ID text field, enter the tracking ID given to you in your Google Analytics account (see Figure 3-7).**

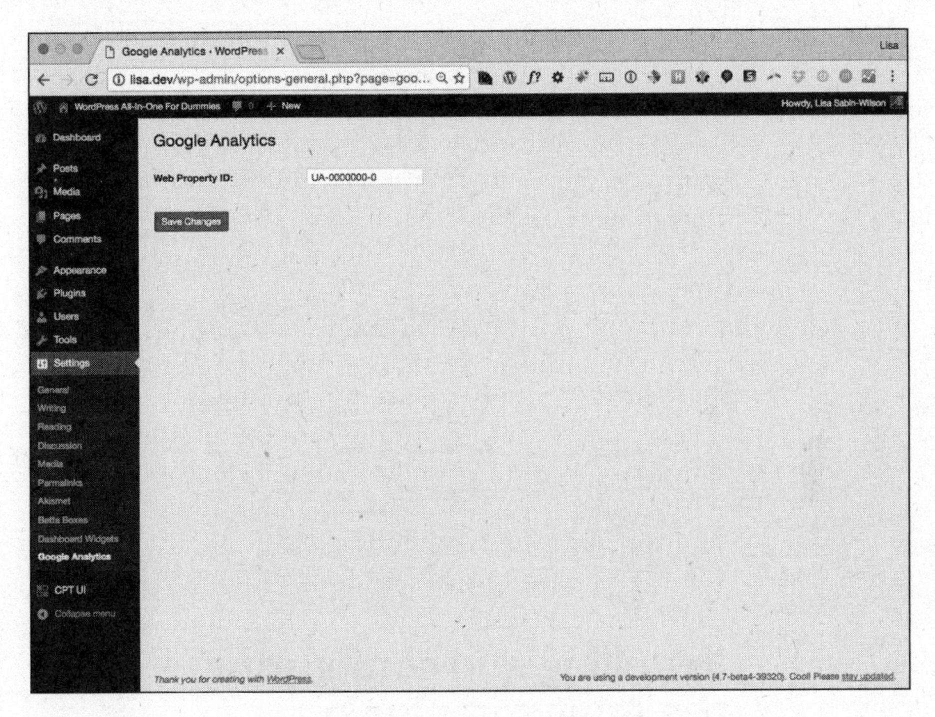

FIGURE 3-7: Google Analytics plugin screen.

6. **Click the Save Changes button.**

   This step saves your settings and inserts the correct JavaScript snippet into your site. Now Google is tracking your website analytics.

Verifying that you installed the code properly

After you install your code, check whether you installed it correctly. When you log back in to Google Analytics, your Dashboard appears. (See Figure 3-8.) The Dashboard shows the tracking data obtained from your website so far. Because your account is new, your tracking data will likely be a big fat zero, much like what you see in Figure 3-8.

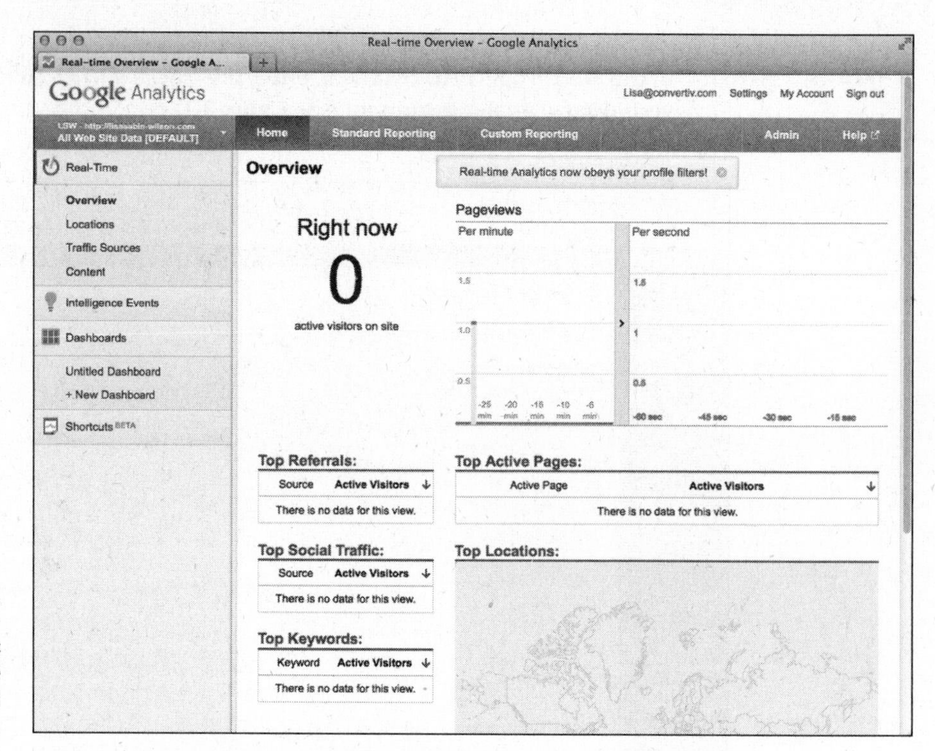

FIGURE 3-8: Tracking code in place and collecting data.

Installing and configuring the Analytics plugin

After you install Google Analytics on your WordPress site and make sure that the tracking code is working properly, you can install the plugin so that you can get a basic version of your stats right on your WordPress Dashboard. Just follow these steps:

1. **Log in to your WordPress Dashboard.**
2. **On the Plugins menu, click the Add New link.**

   A search box appears.
3. **In the Search text box, type** google analytics dashboard **and then click the Install Now link.**

   This step takes you to the Installing Plugin page.
4. **Activate the plugin by clicking the Activate button.**

5. **Click the Google Analytics link on the WordPress Dashboard menu.**

   You see the Google Analytics Settings – Plugin Authorization screen.

6. **Click the Authorize Plugin button.**

   You see a page where you log in to your Google account; then the Google Analytics Settings screen reloads.

7. **Click the Get Access Code link.**

   You see a page where Google gives you an access code; copy that code.

8. **Enter the access code from Step 7 in the Access Code text box on the Google Analytics screen in your Dashboard.**

9. **Click the Save Access Code button.**

   The Google Analytics Settings screen appears. (See Figure 3-9.)

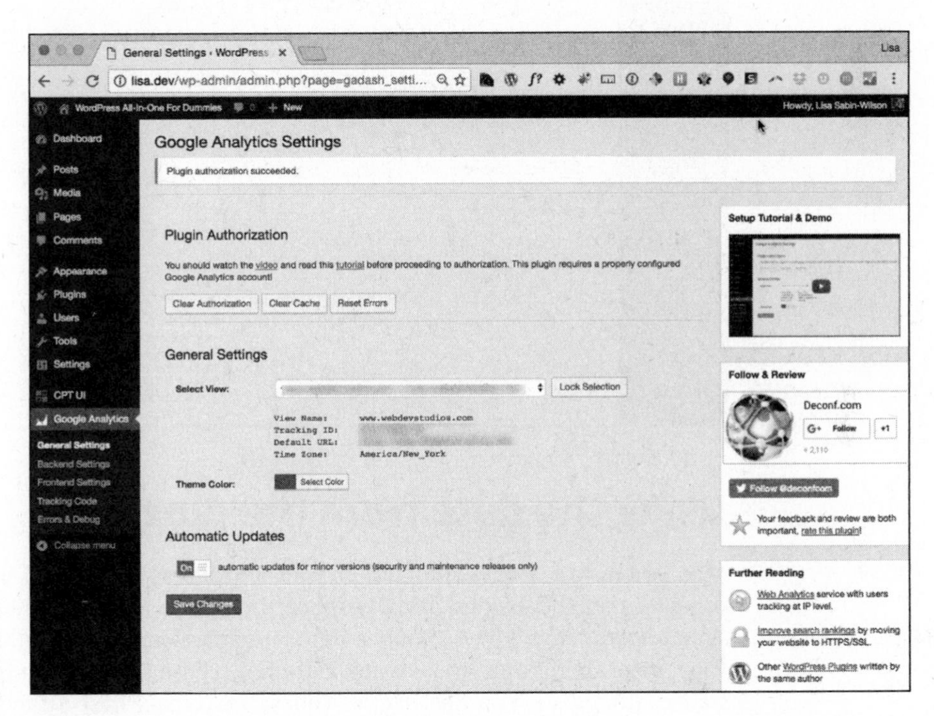

FIGURE 3-9: The Google Analytics Settings screen.

10. **From the Select View drop-down menu, choose the analytics account from which you want to pull your stats.**

11. **Click the Save Changes button.**

The plugin appears on your Dashboard.

12. **Drag and drop the plugin to the position you prefer.**

Figure 3-10 shows a WordPress Dashboard with the Google Analytics plugin displayed in the top-right corner.

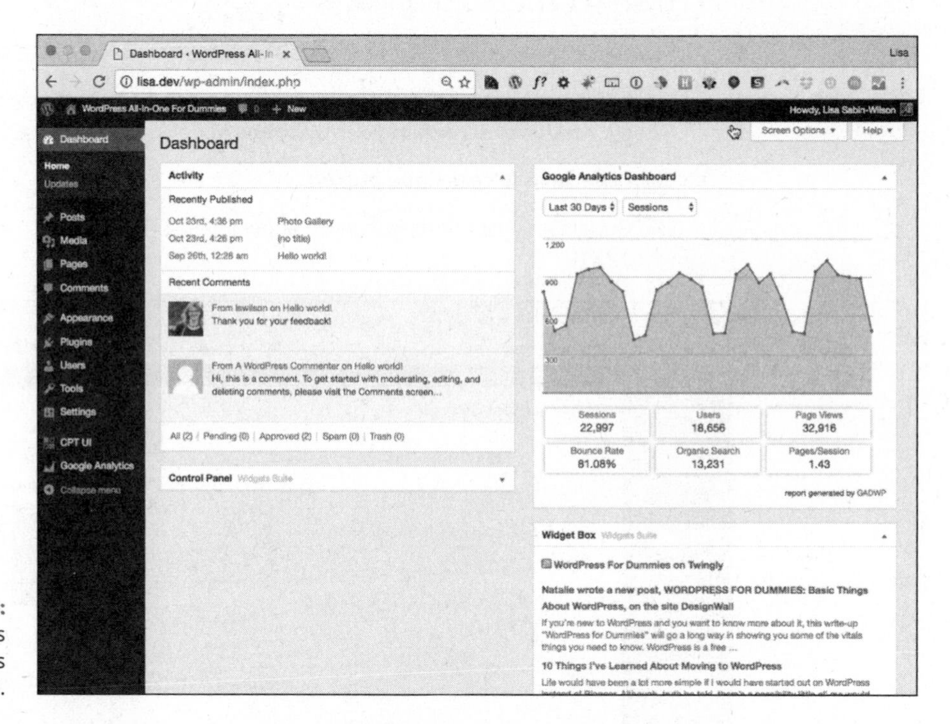

FIGURE 3-10: Google Analytics on a WordPress Dashboard.

Using the data from the plugin

After you install Google Analytics on your WordPress Dashboard, you can examine the data that it provides. Your Dashboard displays analytics stats, such as Sessions, Users, Page Views, Bounce Rate, Organic Search, Pages/Session, Time on Page, Page Load Time, and Session Duration. These stats can be filtered by time-frame options such as Real-Time, Today, Yesterday, Last 7 Days, Last 14 Days, Last 30 Days, Last 90 Days, One Year, and Three Years.

These stats show you the most popular content on your site, the ways people are finding your site, and the sources of your traffic. If you want even more detailed information on high-performing individual pages, you can configure the Dashboard widget to display the number of views on Pages (see Figure 3-11), where you can find per-page stats for on your site.

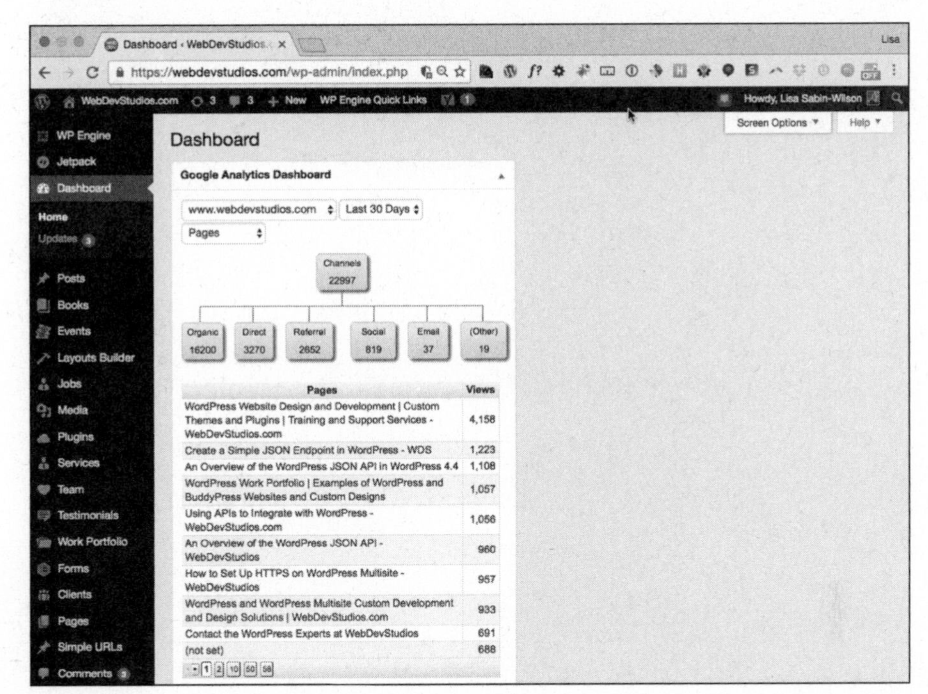

FIGURE 3-11: A per-page breakdown by Google Analytics for WordPress.

By examining the two data sets, you can get a handle on the traffic that's coming to your blog. Pay attention to the data and use it to answer the following questions:

- What posts are popular?
- Do the popular posts have a unique theme or type?
- Do long posts or short posts help increase traffic?
- Do videos, lists, or any other specific types of posts give you more traffic than the rest?

The answers to these questions can help you draw various conclusions and adapt your publishing schedule, content type, and writing style to optimize the popularity of your blog.

IN THIS CHAPTER

» Appreciating search engine optimization (SEO) benefits

» Improving your SEO with WordPress

» Getting your blog into good SEO shape

» Finding information about your niche

» Creating SEO-improvement strategies

Chapter 4
Search Engine Optimization

Google, Yahoo!, Bing, and other search engines have a massive impact on a website. Search engines can easily refer the largest amount of traffic to your site and, if dealt with properly, can help you grow a large audience over time. Often, bloggers don't discover the importance of search engine optimization (SEO) until their sites have been around for a while. By taking the time to make sure that you're following SEO best practices from the get-go, you can reap the rewards of a consistent flow of search engine traffic.

If you've been publishing content on the web for a while and haven't been following the practices in this chapter, roll up your sleeves and dive back into your site to fix some of the SEO practices that you may have overlooked (or just didn't know about) over the history of your website. If you've been publishing content on your website for only a few months, this process doesn't take long. If you have a large backlog of content . . . well, pull up a chair; this fix is going to take a while. But don't worry. You're in safe hands. This chapter helps you through the difficult task of optimizing your site for search engines.

Understanding the Importance of Search Engine Optimization

Talk about SEO usually puts most people to sleep. I'm not going to lie: Hardcore SEO is a time-consuming job that requires a strong analytical mind. Casual bloggers, or even most small-business owners, don't need to understand all the minute details that go into SEO. Everyone with a website who desires traffic, however, needs to get familiar with some of the basic concepts and best practices. Why, you ask?

One thousand pageviews. That's why.

You're not going to get 1,000 page views right off the bat by changing your SEO, of course.

SEO deals with following best practices when it comes to writing content on the web. By following these simple guidelines and by using WordPress, you can increase search engine traffic to your site. Period. To be honest, you probably won't rank number one in really tough categories just by following SEO best practices. But you definitely can increase your traffic significantly and improve your rank for some long-tail keywords. *Long-tail keywords* are keywords that aren't searched for often, but when you amass ranking for a lot of them over a period of time, the traffic adds up.

REMEMBER

You want as many search results as possible on the first two pages of Google and other search engines to be from your site(s). (Most search-engine visitors don't go past the first two pages of Google.) This search-results aim is a more reasonable goal than trying to rank number one for a highly competitive keyword.

TIP

If you really do want to rank number one in a competitive space, check out sites such as SEOBook (`www.seobook.com`) and Moz (`https://moz.com/`), which can help you achieve that difficult goal.

Outlining the Advantages That WordPress Presents for SEO

Using WordPress as your content management system comes with some advantages, including the fact that WordPress is designed to function well with search engines. Search engines can crawl the source code of a WordPress site pretty easily, which eliminates issues that a lot of web programmers face when optimizing a site. The following list outlines some of WordPress's SEO advantages:

- **Permalinks:** URLs where your content is permanently housed. As your site grows and you add more content, the items on your front page get pushed and replaced by recent content. Visitors can easily bookmark and share permalinks so that they can return to that specific post on your site, so these old articles can live on. One of the technical benefits of WordPress is that it uses the Apache `mod_rewrite` module to establish the permalink system, which allows you to create and customize your permalink structure. (See Book 3, Chapter 2 for more information on custom permalinks.)
- **Pinging:** When you post new content, WordPress has a built-in pinging system that notifies major indexes automatically so that they can come crawl your site again. This system helps speed the indexing process and keeps your search results current and relevant.
- **Plugins:** The fact that WordPress is so developer-friendly allows you to use the latest SEO plugins. Do you want to submit a site map to Google? There's a plugin for that. Do you want to edit the metadata of a post? There's a plugin for that. Do you want to alert Google News every time you post? Guess what . . . there's a plugin for that, too. More than 10,000 plugins were available at press time, which means that you can use an advanced plugin ecosystem to help power your blog. Book 5, Chapter 5 covers a few key plugins that can help you with SEO.
- **Theme construction:** SEO, social media, and design go hand in hand. You can push a ton of people to your web page by using proper SEO and robust social media profiles, but if your site has a confusing or poorly done design, visitors aren't going to stay. Likewise, a poorly designed site prevents a lot of search engines from reading your content.

  In this situation, *poorly designed* doesn't refer to aesthetics — how your site looks to the eye. Search engines ignore the style of your site and your CSS, for the most part. But the structure — the coding — of your site can affect search engines that are attempting to crawl your site. WordPress is designed to accommodate search engines: It doesn't overload pages with coding so that search engines can easily access the site. Most WordPress themes have valid code, which is code that's up to standards based on the recommendations of the World Wide Web Consortium (`www.w3.org`). Right from the start, having valid code allows search engines to access your site much more easily.

REMEMBER

When you start changing your code or adding a lot of plugins to your site, check to see whether your code validates. Validated code means that the code on your website fits a minimum standard for browsers. Otherwise, you could be preventing search engines from easily crawling your sites.

TIP

If you want to check out whether your site validates, use the free validator tool at `http://validator.w3.org`. (See Figure 4-1.)

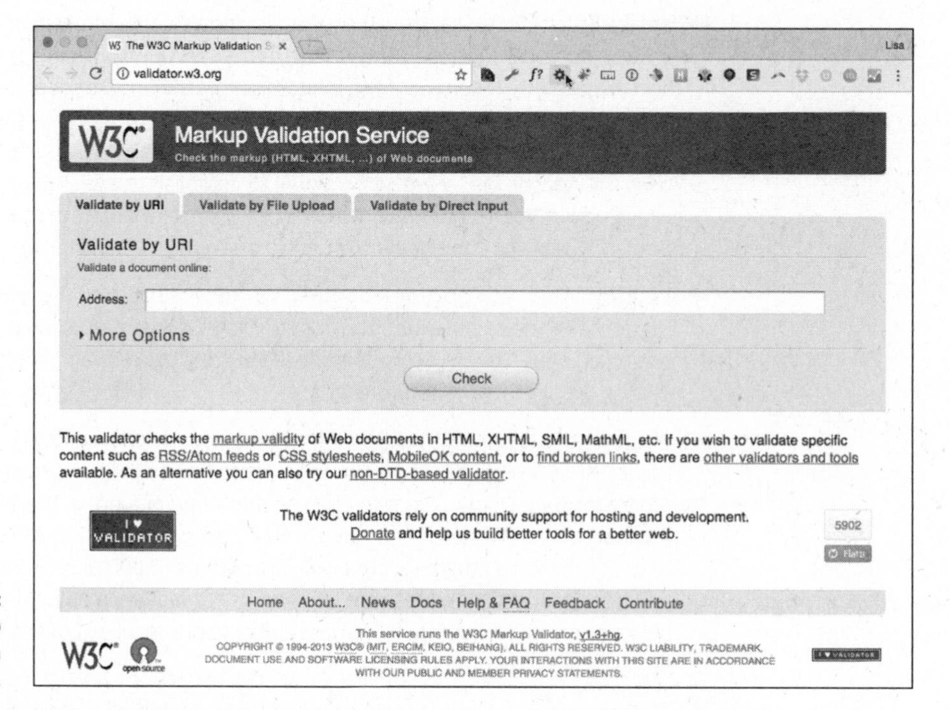

FIGURE 4-1: The W3C Markup Validation Service.

Understanding How Search Engines See Your Content

Search engines don't care what your site looks like because they can't see what your site looks like; their crawlers care only about the content. The crawlers care about the material on your site, the way it's titled, the words you use, and the way you structure those words.

You need to keep this focus in mind when you create the content of your site. Your URL structure and the keywords, post titles, and images you use in posts all have an effect on how your blog ranks. Having a basic understanding of how search engines view your content can help you write content that's more attractive to search engines. Here are a few key areas to think about when you craft your content:

- **Keywords in content:** Search engines take an intense look at the keywords or combination of keywords you use. Keywords are often compared with the words found within links that guide people to the post and in the title of the post itself to see whether they match. The better these keywords align, the better ranking you get from the search engine.

- **Post title:** Search engines analyze the title of your post or page for keyword content. If you're targeting a specific keyword in your content, and that keyword is mentioned throughout the post, mention it in the post title as well. Also, both people and search engines place a lot of value on the early words of a title.
- **URL structure:** One of the coolest things about WordPress is the way it allows you to edit permalinks for a post or page. (See Figure 4-2.) You can always edit the URL to be slightly different from the automated post title so that it contains relevant keywords for search terms, especially if you write a cute title for the post.

  Suppose that you write a post about reviewing Facebook applications and title it "So Many Facebook Applications, So Little Time." You can change the URL structure to something much more keyword-based — perhaps something like `facebook-applications-review`. This reworking removes a lot of the fluff words from the URL and goes right after keywords you want to target.

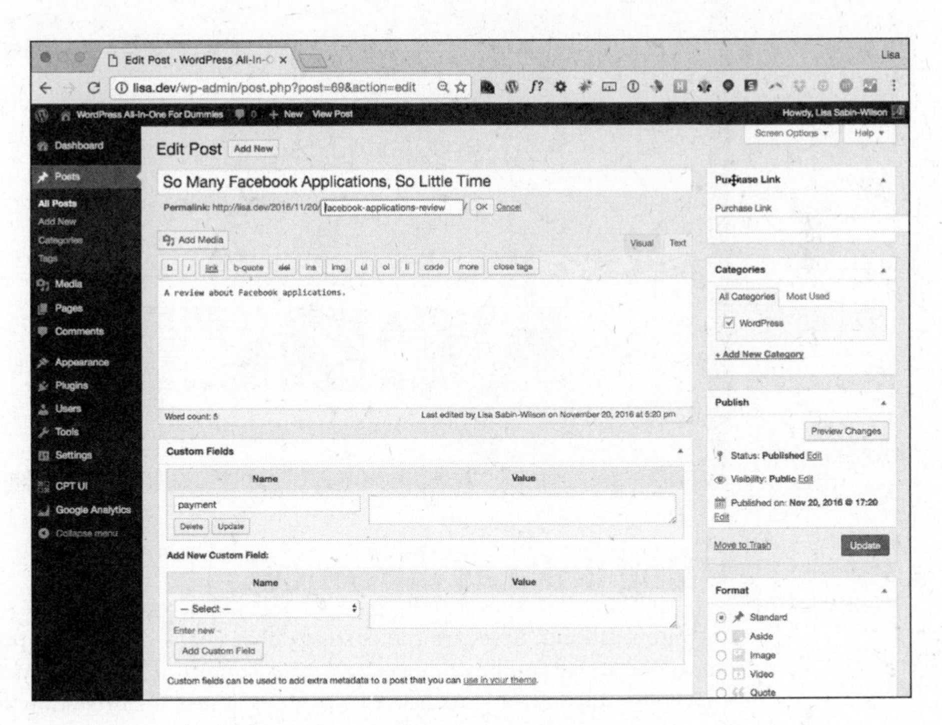

FIGURE 4-2: Editing a permalink.

» **Image titles and other image information:** This item is probably the most-missed item when it comes to SEO. You need to fill out the image information for your posts because this information is a powerful way for people to discover your content and an additional piece of content that can tie keywords to your posts. (See Figure 4-3.) This information includes the filename of your image. Saving an image file to your site as `DS-039.jpg`, for example, offers nothing for readers or search engines and thus has no value to search engines or for you because it doesn't contain a real keyword. Name a picture of a Facebook application, for example, as `Facebook-application.jpg`. Leverage the keyword title and alt tags (alternative text added to the image within the HTML markup that tells search engines what the picture is) because they provide extra content for the search engines to see, and using them can help you get a little more keyword saturation within your posts.

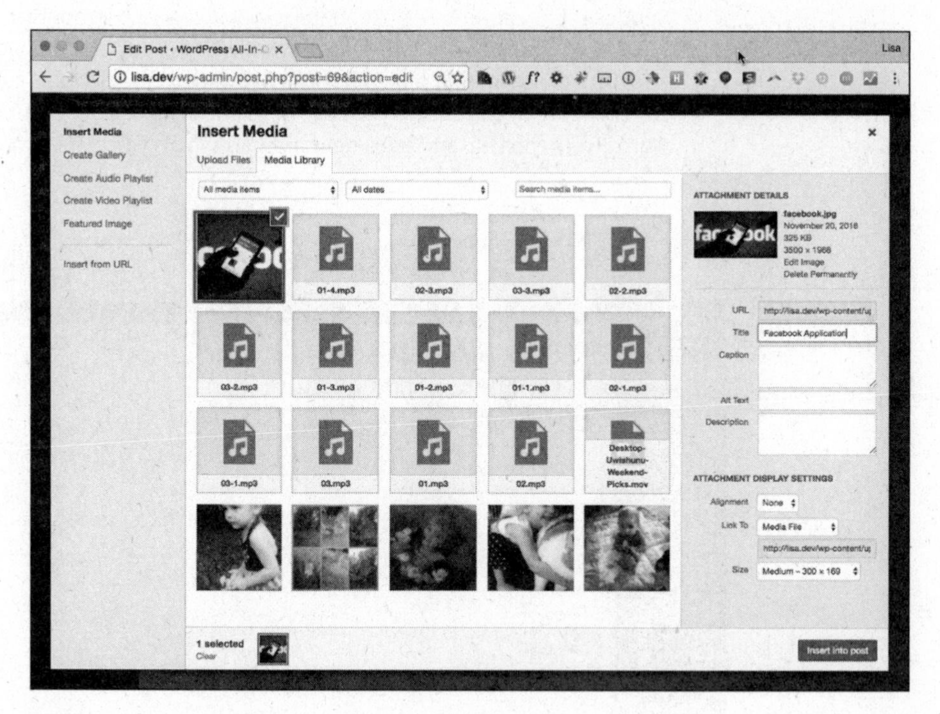

FIGURE 4-3: The Insert Media window.

Using links as currency

If content is king, links are the currency that keeps the king in power. No matter how good a site you have, how great your content is, and how well you optimize that content, you need links. Search engines assess the links flowing into your site for number and quality, and they evaluate your website accordingly.

If a high-quality site that has a high Google Page Rank (a value from 0 to 10) features a link to your page, search engines take notice and assume that you have authority on a subject. Search engines consider these high-quality links to be more important than low-quality links. Having a good amount of mid-quality links, however, can help as well. (This tactic, like many well-known approaches to improving site rank, is based only on trial and error. Google keeps its algorithm a secret, so no one knows for sure.)

Being included in a listing of links on a site, having a pingback or trackback when other authors mention your content in their articles, or even leaving a comment on someone's site can provide links back to your site. If you want to check out how many links you currently have coming to your site, go to Google, type **link:www.*yoursite*.com** in the search text box, and click Google Search. You can also search for competitors' sites to see where they're listed and to what sites they're linked.

Although you do need to try to get other sites to link to your site (called outside links) because outside links factor into search engine algorithms, you can help your own ranking by adding internal links. If you have an authoritative post or page on a particular subject, you should link internally to it within your site. Take ESPN.com (`http://www.espn.com/`), for example. The first time this site mentions an athlete in an article, it links to the profile of that athlete on the site. It essentially tells the search engines each time they visit ESPN.com that the player profile has relevancy, and the search engine indexes it. If you repeatedly link some of your internal pages that are gaining page rank to a profile page over a period of time, that profile page is going to garner a higher search engine ranking (especially if external sites are linking to it too).

This internal and external linking strategy uses the concept of pillar posts (authoritative or popular), in which you have a few pages of content that you consider to have high value and try to build external and internal links to them so that you can get these posts ranked highly in search results.

Submitting to search engines and directories

After you get some content on your website (ten posts or so), submit your website to some search engines. Plenty of sites out there charge you to submit your site to search engines, but honestly, you can submit your site easily yourself. Also, with the help of some plugins (described in Book 5, Chapter 5), you can get your information to search engines even more easily than you may think.

After you submit your website or site map, a search engine reviews it for search engine crawling errors; if everything checks out, you're on your way to having your site crawled and indexed. This process — from the submission of your site through its first appearance in search engine results — can easily take four to six weeks. So be patient. Don't resubmit, and don't freak out that search engines are never going to list your site. Give the process time.

Not to be confused with search engines are website and blog directories. Directories can lead to a small amount of traffic, and some directories, such as DMOZ (`www.dmoz.org`), actually supply information to search engines and other directories. The main benefit of getting listed in directories isn't really traffic but the amount of backlinks (links to your site from other websites) that you can build into your site.

TIP

Although submitting your site to directories may not be as important as submitting to search engines, you may still want to do it. Because filling out 40 or more forms is pretty monotonous, create a single document in which you prewrite all the necessary information: site title, URL, description, contact information, and your registration information. This template helps speed the submission process to these sites.

Optimizing Your Site under the Hood

Some optimization concepts really happen under the hood. You can't readily see these adjustments on your page, but they have an effect on how search engines deal with your content.

Metadata

The metadata on a website contains the information that describes to search engines what your site is about. Additionally, the information often contained in the metadata shows up as the actual search engine results in Google. The search engine pulls the page title and page description that appear in search results from the header of your site. If you do nothing to control this information, Google and other search engines often pull their description from the page title and the first few sentences of a post or page.

Although the title and the first few sentences sound good in principle, they probably don't represent what your site is actually about. You probably don't sum up your topic in the first two sentences of a post or page. Those first few lines likely aren't the best ad copy or the most enticing information. Fortunately, some plugins (such as the All in One SEO Pack plugin, which is in the WordPress Plugin Directory at `https://wordpress.org/plugins/all-in-one-seo-pack`) allow you to control these details on a post and page level. Additionally, theme frameworks (see Book 6, Chapter 7) such as Genesis offer you more control of your SEO information.

Include descriptive page titles, descriptions, and targeted keywords for each post via these plugins or frameworks. This information has an effect on your results and often helps people decide to click the link to your website.

The robots.txt file

When a search engine goes to your website, it first looks at your `robots.txt` file to get the information about what it should and shouldn't be looking for and where to look.

You can alter your `robots.txt` file to direct search engines to the information that they should crawl and to give specific content priority over other content. Several plugins allow you to configure your `robots.txt` file.

Researching Your Niche

Search Engine Optimization

When you're working to improve your SEO, you can use a lot of publicly available data. This data can help you determine where you should try to get links and what type of content you may want to target. These two sites can help you get a general picture of the niche you're working in:

- **Google (`www.google.com`):** You can find what types of links are flowing into a website by typing **link: `www.`*`yoursite`*`.com`** in the Google search text box and clicking Google Search. (Replace *yoursite.com* with the domain you want to target.) Google gives you a list of the sites linking to your site. By doing this search for other websites in your niche, you can find out the sources of their links — industry-specific directories you may not know about, places where they've guest-authored, or other resource sites that you may be able to get listed on.

  This data gives you information about what to target for a link-building campaign.
- **SEMrush (`https://www.semrush.com`):** SEMrush (see Figure 4-4) offers both paid and free versions, and spending a few dollars for a month's access to the light version of the product can be a good investment. (The free version lets

you look up only ten results at a time.) SEMrush allows you to see the terms for which other websites rank. Use this information to judge the health of the competitor's domain, the number of terms for which it ranks in Google's top 20, and the terms themselves.

You can use this information in a lot of different ways. You can see what terms you may want to work into your content, for example. SEMrush provides not only information about what terms search engines use to rank these sites, but also information about how competitive some of those keywords are with other websites using the same keywords as yours.

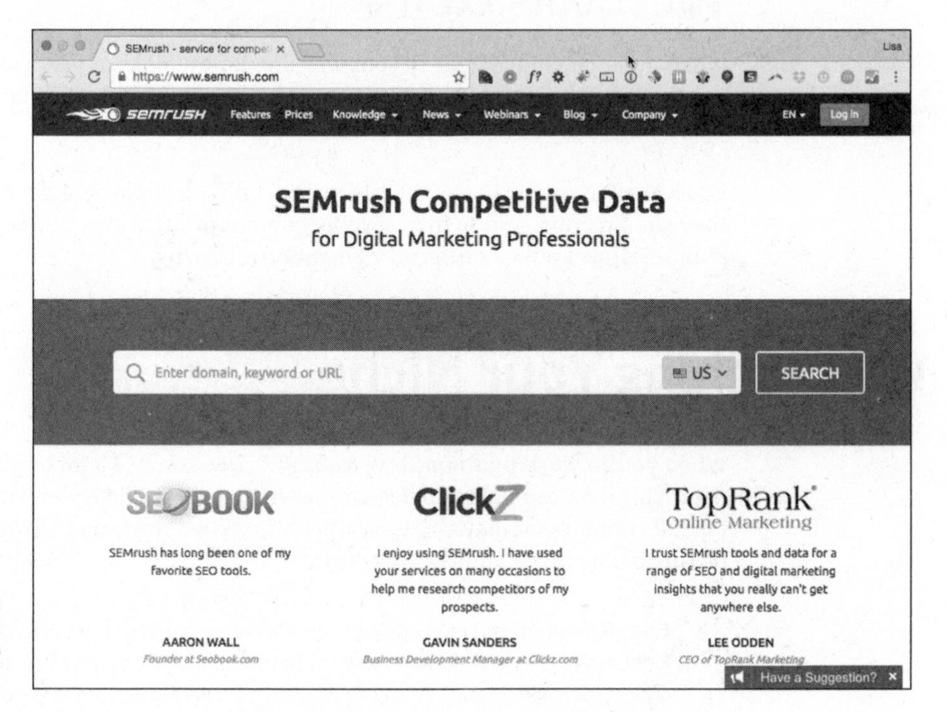

FIGURE 4-4: SEMrush helps you evaluate your competition.

Creating Search Engine Strategies

You can use the techniques discussed in previous sections of this chapter when you set up your site, write strategic content, and begin to build links into your website. The next section deals with setting up your website so that it's optimized for search engines.

Setting up your site

When setting up your site, you're going to want to follow some best practices to make sure that your site is optimized for search engines. Some of these best practices include

- **Permalinks:** First, set up your permalink structure. Log in to your WordPress account, and on the sidebar, select Permalinks in the Settings section. The Permalink Settings page appears. (See Figure 4-5.) Select the Post Name radio button.

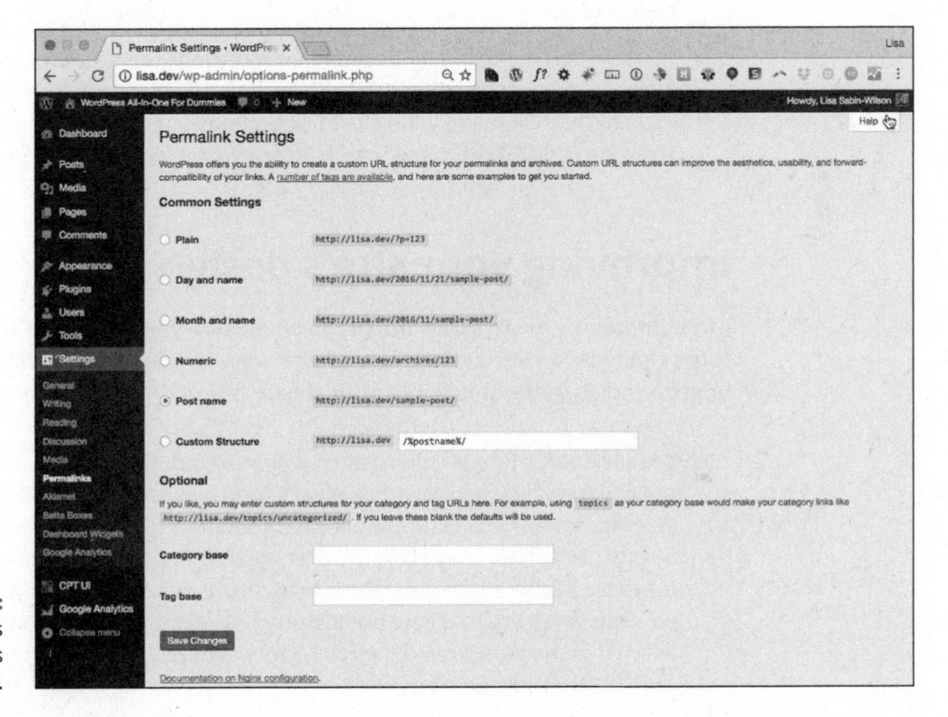

FIGURE 4-5: The WordPress Permalinks Settings page.

Making this change gives you a URL that contains just your domain and the title of your blog post. If you use a focused category structure in which you've carefully picked out keywords, you may want to add the category to the URL. In that case, you enter ***/%category%/%postname%/*** in the text box.

Avoid using the default URL structure, which includes just the number of your post, and don't use dates in the URL. These numbers have no real value when doing SEO. WordPress by default numbers all your posts and pages with specific ID numbers. If you haven't set up a custom permalink structure in WordPress, permalinks for your posts end up looking something like this:

http://*yourdomain*.com/?p=12 (where 12 is the specific post ID number). Although these numbers are used for many WordPress features, including exclusions of data and customized RSS feeds, you don't want these numbers in your URLs because they don't contain any keywords that describe the post.

Also, if you already have an established site and are just now setting up these permalinks, you must take the time to install a redirection plugin. You can find several of these plugins available in the Plugin Directory on WordPress.org. You must establish redirection for your older content so that you don't lose the links that search engines, such as Google and Yahoo!, have already indexed for your site. One good redirection plugin to use is simply called Redirection; you can find it at `https://wordpress.org/plugins/redirection`.

» **Privacy:** You don't want your site to fail to be indexed because you didn't set the correct privacy settings. On the WordPress Settings menu, click the Reading link. On the resulting Reading Settings screen, make sure that the check box titled Discourage Search Engines from Indexing This Site is cleared.

Improving your site's design

After improving your setup on the back end of your site, you'll want to make some changes in your design so your site works better with search engines. Some improvements you can make to your theme templates include

» **Breadcrumbs:** Breadcrumbs, often overlooked during website creation, provide the valuable navigation usually seen above on a page above the title. (See Figure 4-6.) Breadcrumbs are pretty valuable for usability and search engine navigation. They allow the average user to navigate the site easily, and they help search engines determine the structure and layout of your site. A good plugin to use to create breadcrumb navigation is Breadcrumb NavXT, which is in the WordPress Plugin Directory at `https://wordpress.org/plugins/breadcrumb-navxt`.

» **Validated code and speed:** If you're not a professional web designer, you probably don't do a lot of coding on your site. But if you make some small edits to your WordPress installation or add a lot of code through widgets, do it properly by putting it directly in your CSS rather than coding into your site. Coding these features properly helps improve the speed of your site and the way search engines crawl it. Book 6 contains a great deal of information about coding the templates in your theme; check out that minibook for more information about correct coding.

Breadcrumbs

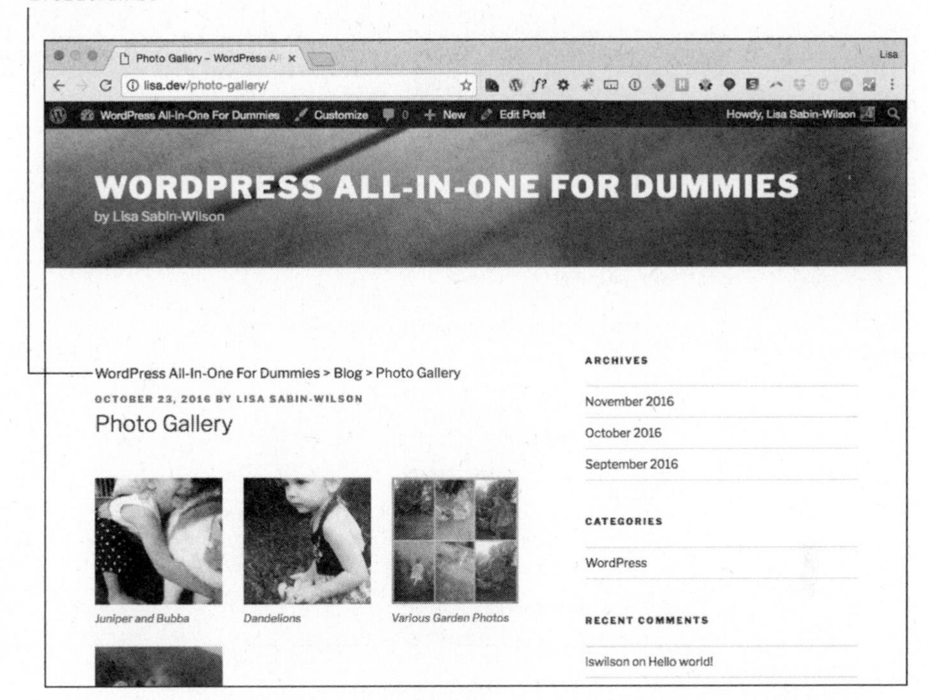

FIGURE 4-6: Users and search engines can follow the breadcrumbs.

When it comes to improving site speed, proper code has a lot to do with performance. You can take other steps to help improve the speed of your site, such as installing caching plugins, including the W3 Total Cache plugin (`https://wordpress.org/plugins/w3-total-cache`). The quality of your hosting (Book 2, Chapter 1), the size of your image files (make sure that you set image-file quality to web standards), the number of images you're using, and third-party widgets or scripts (such as installing a widget provided by Twitter or Facebook) can all affect the speed and performance of your site.

» **Pagination:** Another basic design feature that's often overlooked during site setup, *pagination* creates bottom navigation that allows people and search engines to navigate to other pages. (See Figure 4-7.) Pagination can really help both people and search engines navigate your category pages.

Some themes don't have built-in pagination, so you may need to add a plugin to accomplish this effect. A few of these kinds of plugins are on the market; check out Book 5, Chapter 5.

REMEMBER

Links pass on authority. When you link to a site or a site links to you, the link is saying that your site has value for the keyword in the link. So evaluate the links that you have, and think about whether you really want to link to those websites.

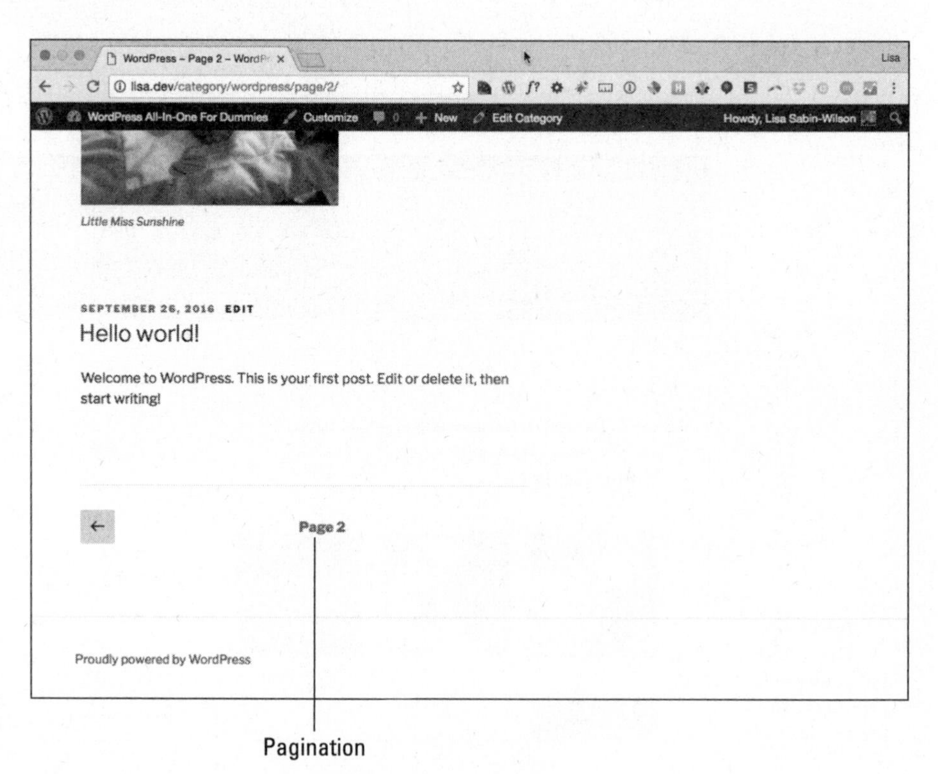

FIGURE 4-7: Pagination in action.

Dealing with duplicate content

WordPress does have one major problem when it comes to SEO: It creates so many places for your content to live that duplicate content can confuse search engines. Fortunately, plugins and some basic editing easily take care of these issues. Here's what to do:

- **Take care of your archive page on your site.** This page displays archives such as category, date-based archives, and so on. You don't want your archive page to present full posts — only truncated versions (short excerpts) of your posts. Check your theme to see how your archive is presented. If your archive shows complete posts, see whether your theme has instructions about how to change your archive presentation. (Each theme is unique, but check out the information in Book 6; it's full of great information about tweaking and altering theme template files.)
- **Make sure that search engines aren't indexing all your archives by using a robots plugin.** You want robots going through only your category archive, not the author index and other archives.

Creating an editorial SEO list/calendar

Planning your posts from now until the end of time can take some of the fun out of publishing. Still, it doesn't hurt to create a list of keywords that your competitors rank for and some of the content they've discussed. Take that list and apply it to new posts, or write *evergreen content* (topics that aren't time-sensitive) centered on what you want to say. Planning your content can really help in figuring out what keywords you want to target when you want to write content to improve for ranking for targeted keywords.

TIP

If you feel that your site content is more news- or current events–oriented, create a reference list of keywords to incorporate into your newer posts so that you can rank for these targeted terms.

Establishing a routine for publishing on your site

Although you can't really call this high strategy, getting into the habit of posting content regularly on your site helps you get the basics down. Here are some things to keep in mind:

- **Properly title your post.** Make sure that your post includes the keyword or phrase for which you're trying to rank.
- **Fix your URL.** Get rid of stop words or useless words from your URL, and make sure that the keywords you want to target appear in the URL of your post. *Stop words* are filler words such as *a, so, in, an,* and so on. For a comprehensive list of stop words, check out `www.link-assistant.com/seo-stop-words.html`.
- **Choose a category.** Make sure that you have your categories set up and that you place your posts in the proper categories. Whatever you do, don't use the uncategorized category; it brings no SEO value to the table.
- **Fill out metadata.** If you're using a theme framework, the form for metadata often appears right below the post box. If you aren't using a theme framework, you can use the All in One SEO Pack plugin (see Figure 4-8). When activated, this plugin usually appears toward the bottom of your posting page. Make sure that you completely fill out the title, description, keywords, and other information that the plugin or theme framework asks for.
- **Tag posts properly.** You may want to get into the habit of taking the keywords from the All in One SEO Pack plugin and pasting them into the tags section of the post.

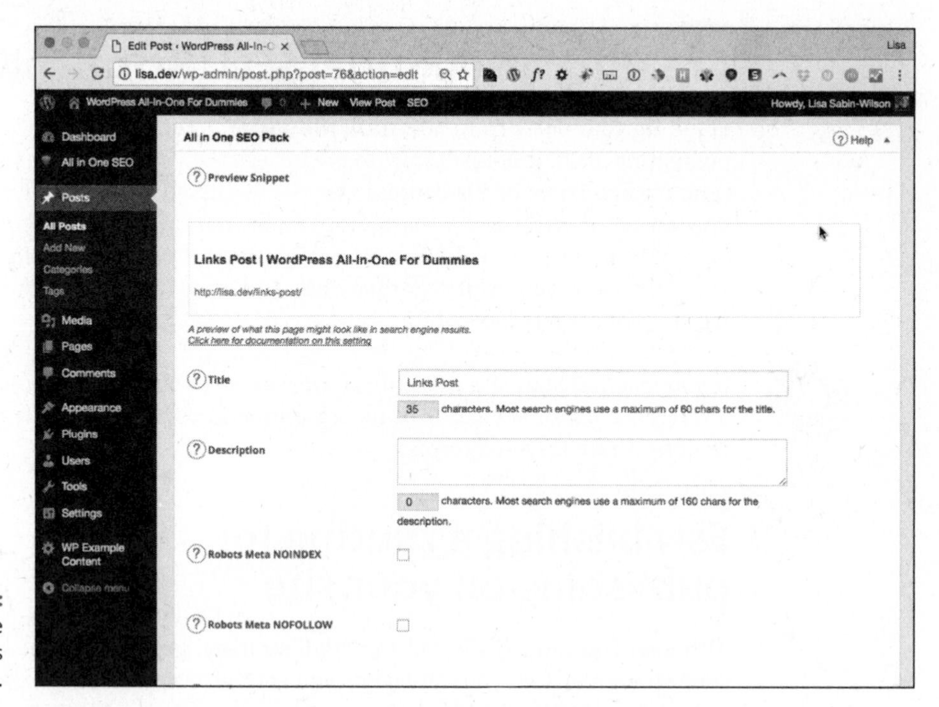

FIGURE 4-8: The All in One SEO Pack's metadata form.

» **Fill out image info.** Take the time to completely fill out your image info whenever you upload pictures to your posts. Every time you upload an image to WordPress, you see a screen in which you can fill in the URL slug, description, and alt text for the image you've uploaded.

Creating a link-building strategy

In previous sections of this chapter, I tackle most of the onsite SEO strategy and concepts. In this section, I explain how you can start working on your off-page strategy. Here are some things to keep in mind:

» **Fill out your social media profiles.** As I discuss in Book 5, Chapter 1, a lot of social media sites pass on page rank through their profiles. Take the time to fill out your social media profile properly and list your site in these profiles.

Social media sites allow you to link to your site with a descriptive word. Industry professionals say that this link has value to search engines, which is debatable, but adding it can never hurt.

- **Use forum signatures.** If you participate in forums, you can easily generate traffic and earn some links to your website from other websites by including your site URL in your forum signature.
- **Examine your competitor's links.** See where your competitors or other people in your niche are getting links — such as directories, lists, guest blogs, and friends' sites — and then try to get links on those sites. Try to determine the relationships, and figure out whether you can establish a relationship with those sites as well.
- **Guest-author.** Find some of the top sites in your niche and then ask them whether you can guest-author. Guest-authoring gives you a link from a respected source and builds a relationship with other content creators. Also, guest-authoring can't hurt your subscriber numbers; often, you see a bump after you guest-author on a large site.
- **Use website directory registration.** Directory registration, albeit a time-consuming affair, can often provide a large number of backlinks to your site from respected sources.
- **Comment on other sites.** A lot of sites pass on page rank because the links in their comment section are live. Make sure that when you engage other people, you properly fill out your information before you post, including the URL to your site. Don't start posting inane comments on random sites to get links. Doing so is considered to be rude and can lead to your site being marked as spam in various commenting systems.
- **Participate in social bookmarking.** Getting involved in Reddit, Digg, and other social-bookmarking communities allows you to participate in social media with people who have similar interests, and you can build links to your site by submitting content to social news and bookmarking sites.

IN THIS CHAPTER

» Using plugins for SEO best practices

» Breaking down your SEO configuration options

» Generating site maps

» Using redirect plugins

» Adding pagination

Chapter 5 Exploring Popular SEO Plugins

When you have the concepts of search engine optimization (SEO) down and the beginnings of your strategy properly mapped out, you can install the tools you need. In this chapter, I go through some of the most popular SEO-related plugins. All these plugins have good developers behind them and good track records.

Several plugins in the WordPress Plugin Directory assist with SEO, so it's hard to decide which ones to use. In Book 7, I cover plugins in detail, but in this chapter, I discuss the most common plugins, as well as the ones that I use myself, because they're solid, reliable plugins that deliver good SEO results.

Exploring Must-Use Plugins for SEO Best Practices

Here are the plugins that this chapter covers:

» **Yoast SEO:** Gives you complete control of the search-engine optimization of your site.

- **Google XML Sitemaps:** Generates an XML site map that's sent to Google, Yahoo!, Bing, and Ask.com. When your site has a site map, site crawlers can more efficiently crawl your site. One of the bonuses of the site map is that it notifies search engines every time you post.
- **Redirection:** Helps when you move from an old site to WordPress or when you want to change the URL structure of an established site. It allows you to manage 301 redirections (when the web address of a page has changed, a 301 redirect tells search engines where they can find the new web address of the page), track any 404 errors (errors that are displayed when you try to load a page that does not exist) that occur on the site, and manage any possible incorrect web address (URL) issues with your website.
- **WP-PageNavi:** Helps you achieve pagination for your WordPress site by allowing you to display page links at the bottom of each archive page and/or category page.

Check out Book 7, Chapter 2 for information on plugin installation.

Yoast SEO

The Yoast SEO plugin (`https://wordpress.org/plugins/wordpress-seo`) makes your life so much easier because it automates many SEO tasks for you. Of all the plugins I cover, this one is an absolute must for your site. It gives you a lot of control of your SEO, and it's very flexible.

This plugin breaks down each option on the configuration page, which allows you to preselect options right off the bat or make some changes to the plugin.

After you install this plugin, click the SEO link on the Dashboard to open the Yoast SEO Dashboard page, shown in Figure 5-1. Then click the General tab and follow these steps:

1. **Click the Open the Configuration Wizard button.**

   This step opens the welcome page (see Figure 5-1) and starts the configuration process.

2. **(Optional) Enter your name and email address to sign up for the Yoast email newsletter.**

   If you do not wish to subscribe to the newsletter, leave the text fields blank.

3. **Click the Next button.**

   This takes you to the Environment screen.

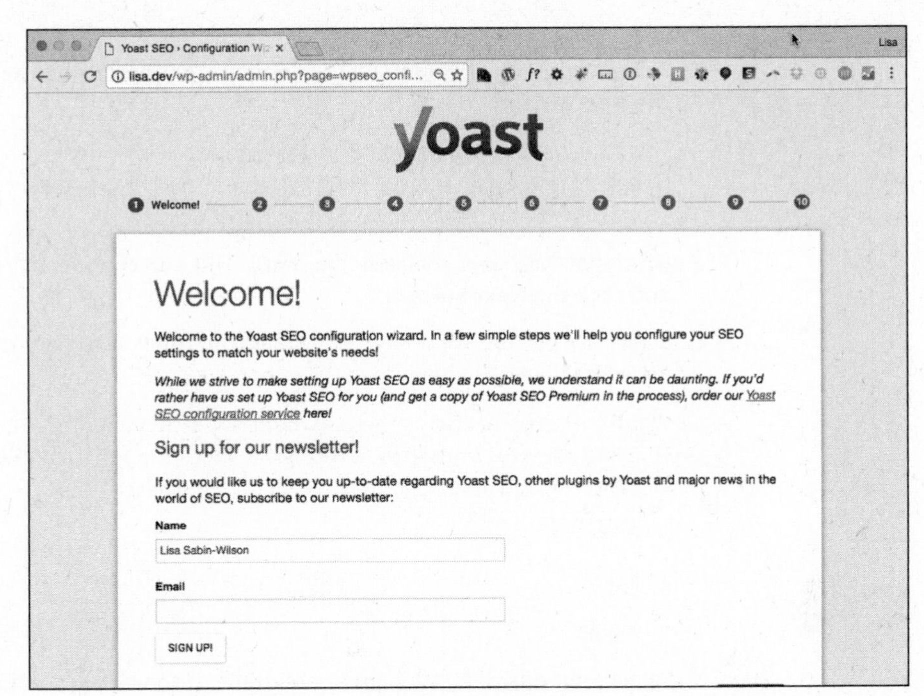

FIGURE 5-1: The Yoast SEO Welcome page.

4. **Select Production on the Environment page, and then click the Next button.**

   The options on this page include

   - *Production:* This site is a live site with real traffic.
   - *Staging:* This site is a copy of a live site used only for testing purposes.
   - *Development:* This site is running on a local computer and used only for development purposes.

5. **Select the type of site you're building on the Site Type page, and then click the Next button.**

   Available options on this page include

   - Blog
   - Webshop
   - News Site
   - Small Business Site
   - Other Corporate Site
   - Other Personal Site

6. **Select Company or Person on the Company or Person page and then click the Next button.**

   This data is shown as metadata on your site and is intended to appear in Google's Knowledge Graph, a knowledgebase used by Google to enhance its search results. Your site can be a company or a person.

7. **(Optional) Add your social media profile URLs on the Social Profiles page, and click the Next button.**
8. **Select the options for your post visibility on the Post Type Visibility page and click the Next button.**

   WordPress automatically generates a URL for each media item in the library. Enabling visibility here allows for Google to index the generated URL. Select Visible or Hidden options for the following post types:

   - Posts
   - Pages
   - Media

9. **Select your desired option on the Multiple Authors page, and click the Next button.**

   Does your site have (or will it have) multiple authors? Select Yes or No.

10. **(Optional) Enter your authentication code on the Google Search Console page, and click the Next button.**

    This step allows Yoast SEO to fetch your Google Search Console information. You can skip this section if you don't use Google Search Console.

11. **Enter your website name in the Title Settings page, select the character you'd like to use to separate your site title and content title in the Title Separator section, and click the Next button.**

    Type the title of your website in the Website Name text box. (This title is the same one that you entered in the General Settings page, as I discuss in Book 3, Chapter 2.)

    The Title Separator displays, for example, between your post title and site name. Symbols, such as a dash or an arrow, are shown in the size in which they'll appear in the search results.

12. **Click the Close button on the Success page.**

If you find that you need to change any of the settings in the configuration wizard, click any of the links on the SEO menu in your WordPress Dashboard to change the Title tags, Meta tags, and Social Media profile URLs.

Most of the remaining options that are selected by default should work fine for your site. Make sure, however, that Noindex is selected for the Archives pages (to find this setting, choose SEO⇨Titles & Metas⇨Archives) to make sure that the search engines are not indexing your archive pages. Indexing archive pages would provide the search engines with duplicate content that they have already indexed and that is one of the top ways of getting penalized by Google, which means you run the risk of having your site removed from Google's search engine results completely.

After you make all your selections, click the Save Changes button at the bottom of any of the Yoast SEO settings pages where you make any changes.

You can use the Yoast SEO plugin right out of the box without changing any of the default options. If you aren't confident in fine-tuning it, you don't have to. But don't forget to put in the proper information for your home page, including title, description, and keywords.

The Yoast SEO plugin also gives you full control of SEO settings for each individual page and post on your site by allowing you to configure the following:

- Facebook and Twitter share title, description, and image
- Title
- Slug (or permalink)
- Meta description (the short snippet of text that appears in search engine listings)
- Preferred focus keywords

Another nice feature of the Yoast SEO plugin is that it gives you real-time analysis of your content by rating its readability and keyword analysis. It also makes recommendations for improving both before you publish your content.

Yoast SEO creates Google XML site maps for you, which you can configure by choosing Dashboard⇨SEO⇨XML Sitemaps. Don't worry, though — if you decide not to use the Yoast SEO plugin, the next section has you covered with a plugin that creates XML site maps for you as well.

Google XML Sitemaps for WordPress

You can use Google XML Sitemaps (`https://wordpress.org/plugins/google-sitemap-generator`) right out of the box with very little configuration. After you install it, you need to tell the plugin to create your site map for the first time. You can accomplish this easy task by following these steps:

1. **Click the XML-Site map link on the Settings menu on your Dashboard.**

   The XML Sitemap Generator for WordPress options page appears in your browser window. (See Figure 5-2.)

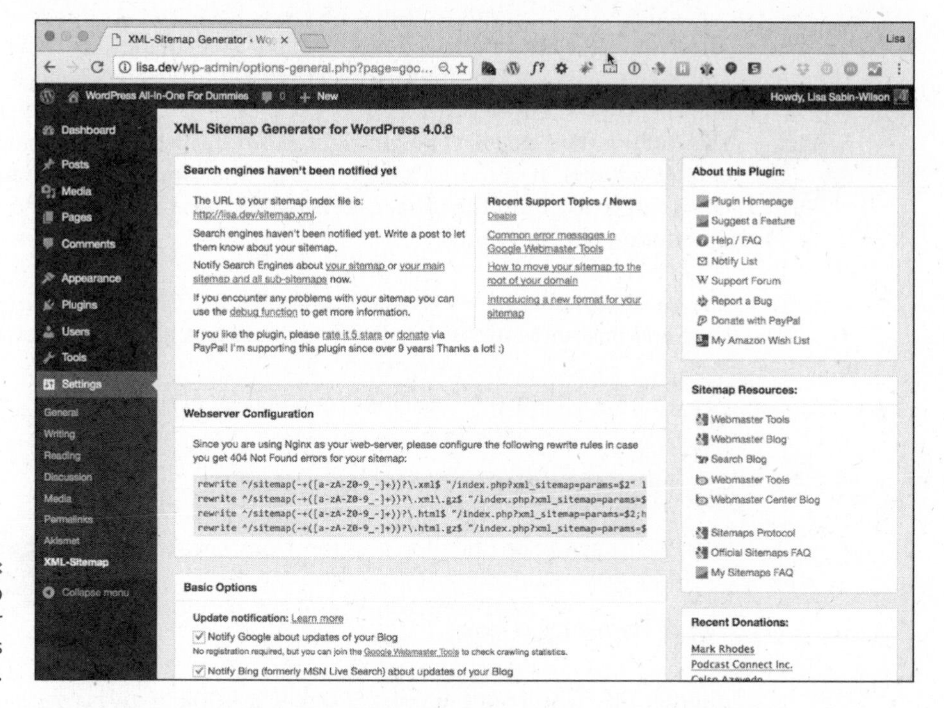

FIGURE 5-2: XML Sitemap Generator for WordPress settings.

2. **In the top module, titled Search Engines Haven't Been Notified Yet, click the Your Sitemap link in the option that begins with the words *Notify Search Engines*.**

   The XML Sitemap Generator for WordPress page refreshes, and the Search Engines Haven't Been Notified Yet module is replaced by the Result of the Last Ping module, showing the date when your site map was last generated.

3. **(Optional) View your site map in your browser.**

   Click the first site-map link in the top module or visit `http://`*yourdomain*.com/sitemap.xml (where *yourdomain*.com is your actual domain).

You never need to visit your site map or maintain it. The XML Sitemap Generator maintains the file for you. Every time you publish a new post or page on your website, the plugin automatically updates your site map with the information and notifies major search engines — such as Google, Bing, and Ask.com — that you've updated your site with new content. Basically, the plugin sends an invitation to the search engines to come to your site and index your new content in their search engines.

TIP

Having a Google Webmaster account can further assist Google in finding and indexing new content on your site. If you don't already have one of these accounts, visit `www.google.com`, click the Sign In link in the top-right corner, click the Create an Account Now link, and follow the onscreen steps to create a new Google account. After you sign in to the account, you can set up the Google Webmaster tools and add your site map to Google.

In the Basic Options section of the XML Sitemap Generator for WordPress plugin page (refer to Figure 5-2), select every check box you see.

All the other default settings are fine for you to use, so leave those as they are. In the Sitemap Content section, which is in the middle of the XML Sitemap Generator for WordPress page, select the following check boxes: Include Homepage, Include Posts, Include Static Pages, Include Categories, and Include the Last Modification Time. Making these selections allows search engines to crawl your site in the most efficient way.

Redirection

If you're redoing the URL (permalink) structure of your site or moving a site to WordPress from another blogging platform, such as Blogger or Tumblr, you really need to use the Redirection plugin (`https://wordpress.org/plugins/redirection`). Redirection allows you to maintain the links that are currently coming into your site by rerouting (or redirecting) people coming in through search engines and other existing links going to the new permalink. If you change URLs, you need to reroute/redirect old links to maintain the integrity of incoming traffic from websites and search engines that are still using the old URL.

Using Redirection is a pretty simple process. After you visit the Redirections page in your Dashboard, you find the Redirection link in the Tools menu: Once there, enter the old URL in the Source URL text box, enter the new URL in the Target URL text box, and then click the Add Redirection button. (See Figure 5-3.)

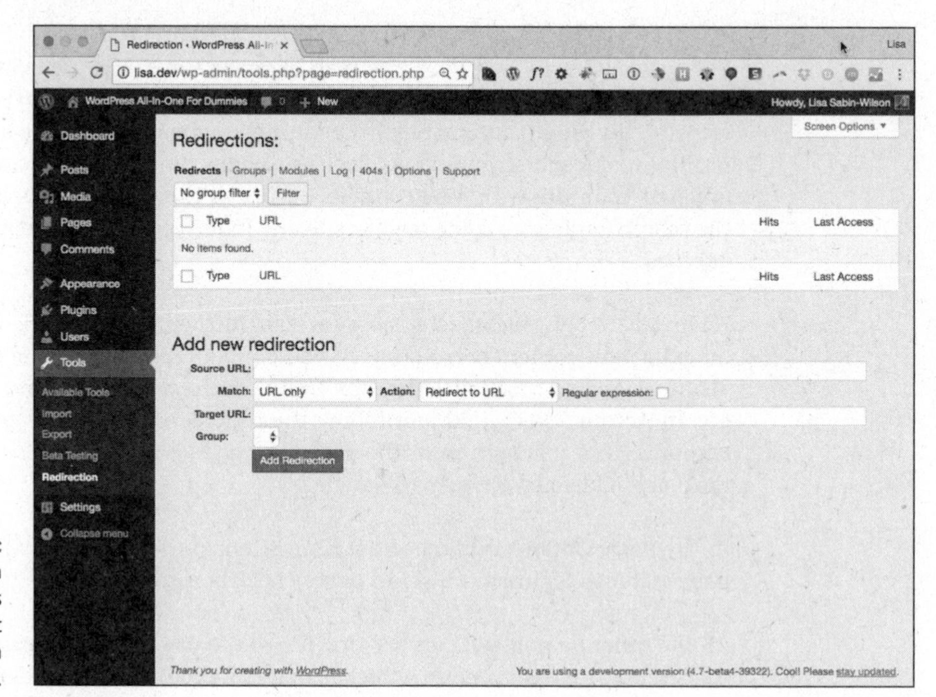

FIGURE 5-3: The Redirection plugin allows you to redirect traffic from your old URL.

WP-PageNavi

To create page navigation links below your site posts and archive listings for sites that have numbered pages, you need to install the WP-PageNavi plugin (`https://wordpress.org/plugins/wp-pagenavi`). Adjust the WP-PageNavi plugin settings to your liking. The default settings are shown in Figure 5-4.

This plugin provides a better user experience for your readers by making it easier for them to navigate through your content; it also allows search engines to go through your web page easily to index your pages and posts. After you install and activate the plugin, you need to insert the following code into your Main Index template (`index.php`) or any template your theme uses to display archives (such as a blog page, category page, or search page):

```
<?php wp_pagenavi(); ?>
```

The `wp_pagenavi();` template tag needs to be added on a line directly after The Loop. See Book 6, Chapter 3 for extensive discussion of The Loop in the Main Index template file to find out where to add this line of code.

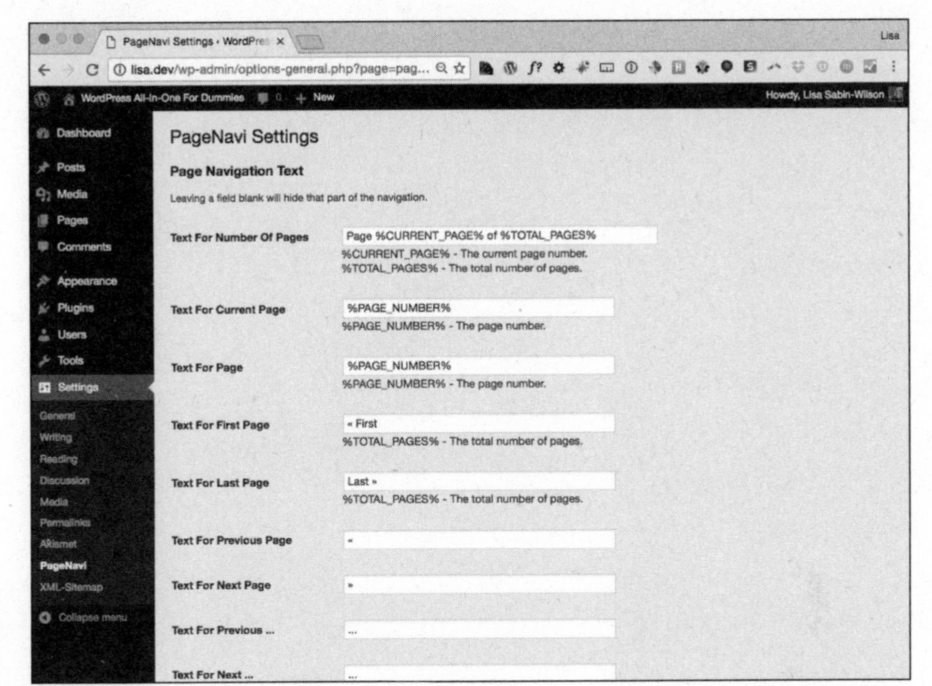

FIGURE 5-4: Adjust the WP-PageNavi settings.

You can experiment with where you want to place the `wp_pagenavi()` code in your template file to give you the look and feel you want. Additionally, you can control the look of the plugin by providing styling in your CSS (`style.css`) theme file for the WP-PageNavi plugin display, or you can have the plugin insert its default CSS into your regular CSS by deselecting the Use Pagenavi.css? option.

6

Customizing the Look of Your Site

Contents at a Glance

» Tweaking your header media

» Customizing color options

» Installing custom navigation menus

» Exploring widgets on your website

Chapter 1

Examining the Default Theme: Twenty Seventeen

Bundled with the release of WordPress 4.7 in December 2016 is the new default theme, Twenty Seventeen. The goal of the core development team for WordPress is to release a new default theme every year, which is why, with every new installation of WordPress, you find themes called Twenty Ten through Seventeen (each named to correspond with the year for which it was created).

With the release of WordPress version 4.7 at the end of 2016, the resulting community effort was Twenty Seventeen, a powerful theme with drop-down menu navigation, header and background image uploaders, multiple-page templates, widget-ready areas, parent-child theme support, and built-in mobile and tablet support.

The members of the WordPress team who worked on Twenty Seventeen describe it as follows:

"Twenty Seventeen brings your site to life with immersive featured images and subtle animations. With a focus on business sites, it features multiple

sections on the front page as well as widgets, navigation and social menus, a logo, and more. Personalize its asymmetrical grid with a custom color scheme and showcase your multimedia content with post formats. Our default theme for 2017 works great in many languages, for any abilities, and on any device."

These features make Twenty Seventeen an excellent base for many of your theme customization projects. This chapter shows you how to manage all the features of the default Twenty Seventeen theme, such as handling layouts, editing the header graphic and background colors, installing and using custom navigation menus, and using widgets on your site to add some great features.

Exploring the Layout and Structure

If you just want a simple look for your site, look no further than Twenty Seventeen. This theme offers a clean design style that's highly customizable. As such, the font treatments are sharp and easy to read. Many of the new built-in theme features allow you to make simple yet elegant tweaks to the theme, including uploading new feature images and adjusting the background colors. Figure 1-1 shows the Twenty Seventeen WordPress default theme, out of the box — that is to say, without any customizations.

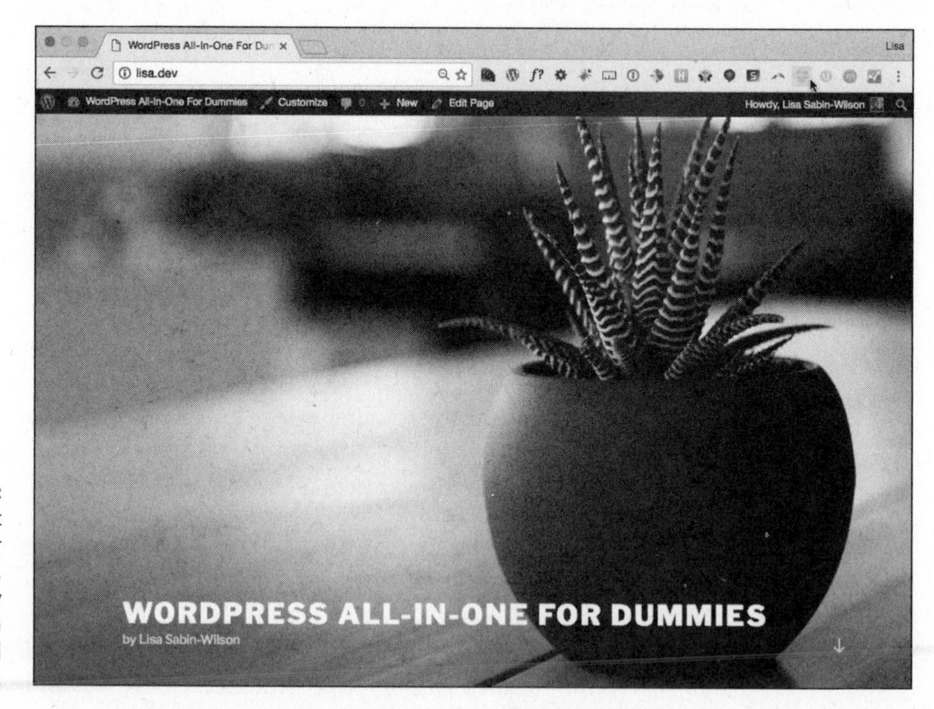

FIGURE 1-1: The default theme for WordPress, Twenty Seventeen, with only the title and byline changed.

The Twenty Seventeen theme's distinctive layout features include

- **Header media:** Twenty Seventeen uses trendy design for the header of your site by allowing you to upload a large feature image to display as a full-screen image with the site title displayed at the top (refer to Figure 1-1). Alternatively, you can use a video for the header area of the site with a fall-back image while the video loads.
- **Theme options:** The Theme Options section allows you to use published static page content to display modular content on your home page, making it easier than ever to build small content boxes that contain text, images, or any media that enables you to call attention to your site's offerings and content.
- **One-column page layout:** Twenty Seventeen's one-column layout for WordPress , shown in Figure 1-2, comes in very handy for such pages as product sales pages, email subscription form pages, photography or portfolio pages, and other content that you don't want to be upstaged by distractions on the sidebar.

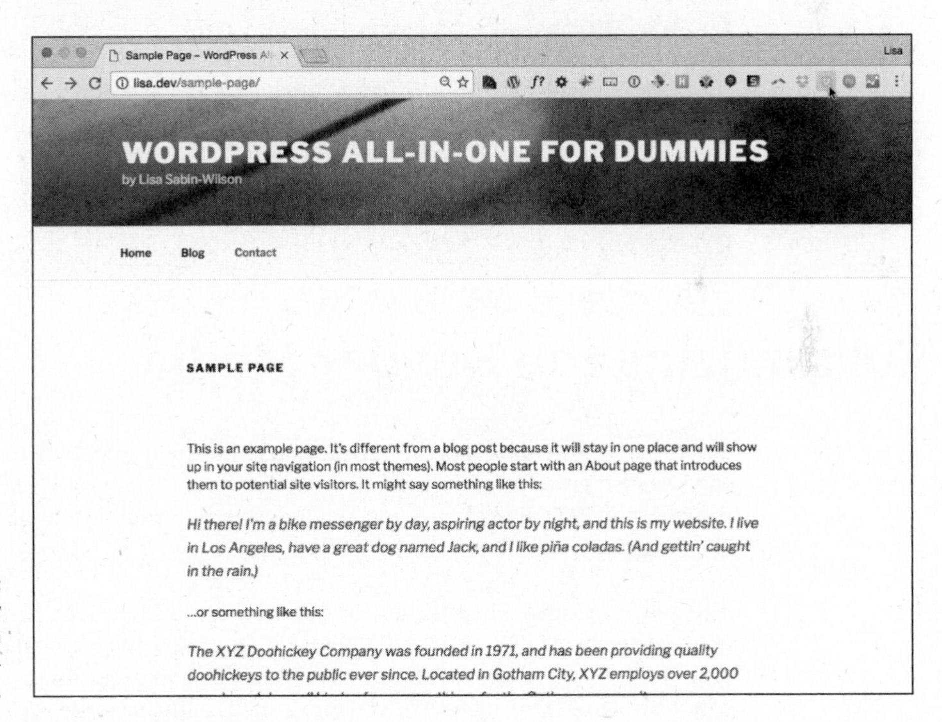

FIGURE 1-2: The Twenty Seventeen theme's one-column layout.

- **Color adjustment:** By default, the color scheme for the Twenty Seventeen theme is a white background with black text and content. You can easily change the colors of the header text and site background by using the theme customizer to select the light or dark theme, or you can even use a custom color by using a color picker. Figure 1-3 shows the Twenty Seventeen theme with the dark color scheme.

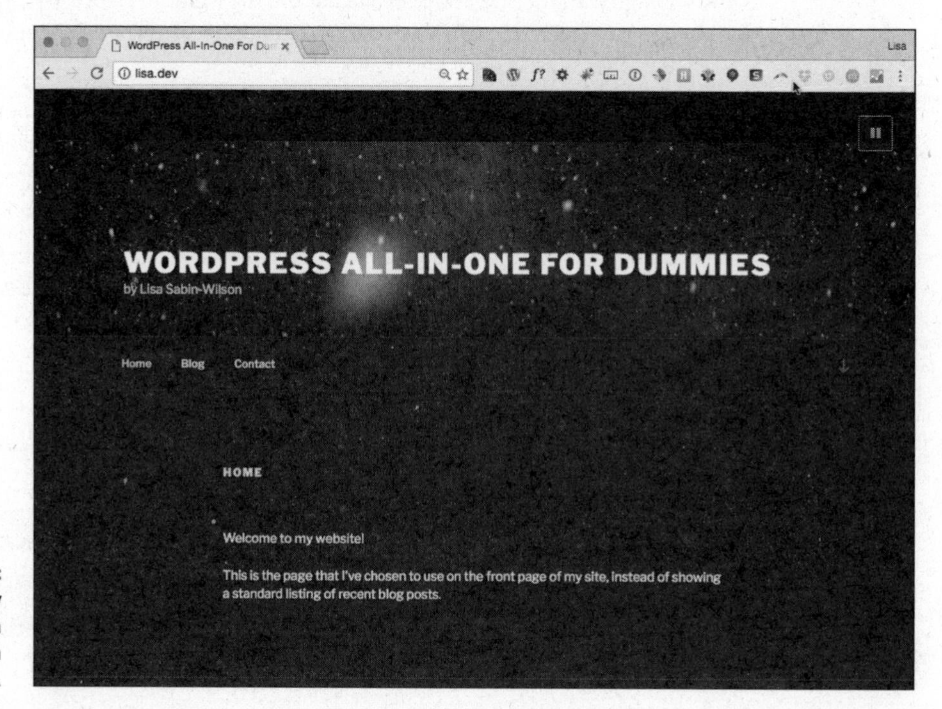

FIGURE 1-3: The Twenty Seventeen theme with custom colors.

Customizing the Header Media

Twenty Seventeen's Header Media options allow you to upload new, unique custom header graphics or a header video for your WordPress site. The option is called Header Media because you can configure an image or a video (or both) for the design of your site.

The recommended dimensions for your customized header media are at least 2,000 pixels wide by 1,200 high. If your photo or video is larger than that, you can crop it after you've uploaded it to WordPress, although for images, cropping with a graphics program (such as Adobe Photoshop) is the best way to get exact results.

Uploading a header image

Twenty Seventeen comes preloaded with one default header image, but WordPress allows you to upload one of your own. To install a custom header image, follow these steps with the Twenty Seventeen theme activated (read Book 6, Chapter 2 for information on activating themes):

1. **On the WordPress Dashboard, click the Header link on the Appearance menu.**

   The Custom Header page appears, with the Customizing Header Media panel displayed on the left side of the screen.

2. **In the Header Image section, click the Add New Image button.**

   This opens the Choose Image screen shown in Figure 1-4.

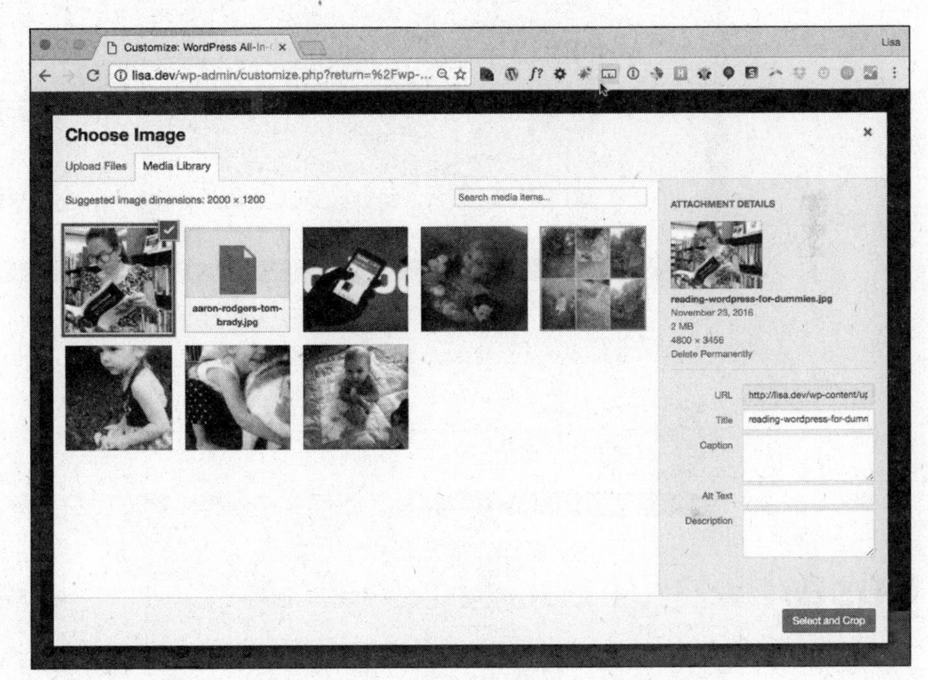

FIGURE 1-4: Selecting an image to use as a header image.

3. **Click the Select and Crop button.**

   This step opens the Crop Image screen, shown in Figure 1-5.

4. **(Optional) Crop the image to your liking.**

   To resize and crop your image, drag one of the eight tiny boxes located at the corners and the middle of the image. You can also click within the image and move the entire image up or down to get the placement and cropping effect that you want.

5. **Click the Crop Image button to crop your header image.**

6. **Click the Save & Publish button in the top-right corner of the Customizing Header Media screen to save your changes.**

FIGURE 1-5: Using the crop tool.

Figure 1-6 shows the Twenty Seventeen theme with a custom header image.

FIGURE 1-6: The Twenty Seventeen theme with new header image.

TIP

You can upload multiple images to use in the header image area and click the Randomize Uploaded Headers button in the Customizing Header Media panel to tell the Twenty Seventeen theme that you'd like to see a different header on each new page load. This option adds a little variety to your site!

Including a header video

The Twenty Seventeen theme comes with a default header image but not with a video for you to use as a header. One small text field, however, allows you to include a video from YouTube (`http://youtube.com`) so that visitors to your site see a video in the background of your header.

Follow these steps to configure a header video:

1. **On the WordPress Dashboard, click the Header link on the Appearance menu.**

   The Custom Header page appears, with the Customizing Header Media panel displayed on the left side of your screen.

2. **To upload your video file, click the Select Video button in the Header Video section.**

   The Select Video screen opens so you can select a video from the WordPress Media Library or upload a video file from your computer's hard drive. The Header Video supports the .mp4 file format for videos.

   If you would rather use a video from a service like YouTube, skip this step and proceed to Step 3.

3. **To use a video from YouTube, copy and then paste the link to that video in the Enter a YouTube video URL text field.**

4. **Click the Save & Publish button in the top-right corner of the Customizing Header Media panel.**

   This action saves and publishes any changes you made in the header media.

Customizing the Colors

After you explore the header image settings, you may want to pick a background color or change the color of the header text. The default background color in the Twenty Seventeen theme is white, but you can change the color scheme. Here's how:

1. **On the WordPress Dashboard, click the Customize link on the Appearance menu.**

   The Customizer panel opens on the left side of your screen.

2. **Click the Colors link.**

   The Customizing Colors screen opens.

3. **Select the Dark option to change your color scheme to a prebuilt dark palette.**

4. **Select the Custom option to configure your own color.**

   Selecting this option displays a color-palette bar, shown in Figure 1-7.

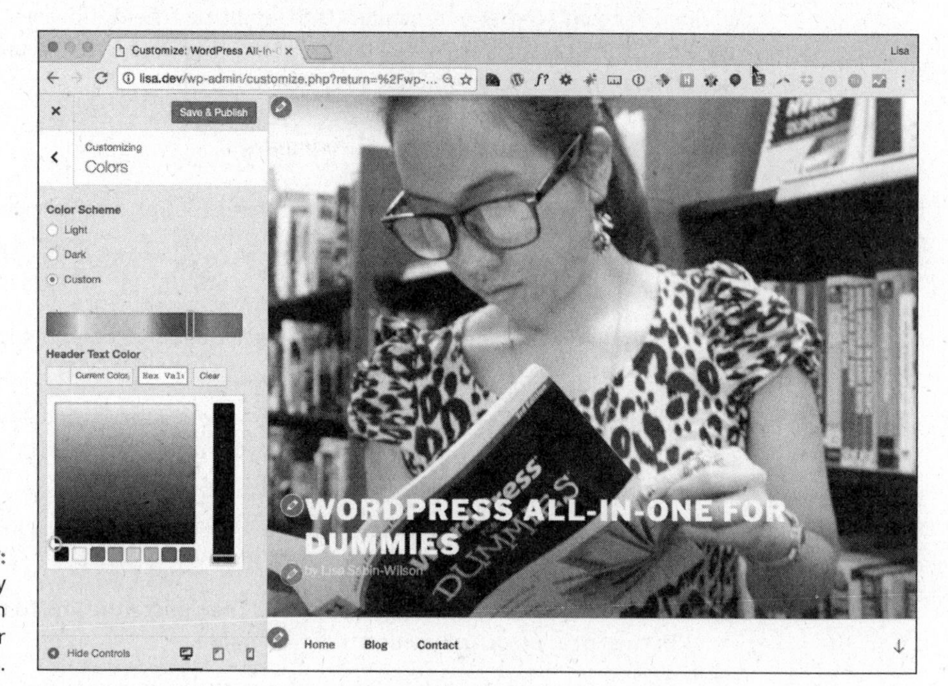

FIGURE 1-7: The Twenty Seventeen theme's color options.

5. **On the color bar, select your preferred color.**

   Click anywhere on the color bar to set your preferred background color.

6. **Click the Select Color button in the Header Text Color section to change the color of the text that appears in your header.**

   The Select Color button changes to Current Color after you click it to set the color value. You can click anywhere on the color palette that appears (refer to Figure 1-7) or you can type a six-digit hexadecimal code (*hex code,* for short) if you already know your preferred color.

TIP

Color values are defined in HTML and CSS by six-digit hexadecimal codes starting with the # sign, such as #000000 for black or #FFFFFF for white. (As noted in Book 6, Chapter 4, adjusting hexadecimal colors is one of the easiest ways to tweak the colors in your theme for a new look.)

7. **When you finalize your selections, click the Save & Publish button.**

TIP

The WordPress Customizer displays on the left side of your computer screen while a preview of your site displays on the right side. As you're making changes in the Customizer, you see a preview of what changes those options make on your site so you can preview before you publish.

Including Custom Navigation Menus

Navigational menus are vital parts of your site's design. They tell your site visitors where to go and how to access important information or areas of your site.

Similar to the WordPress Widgets feature, which lets you drag and drop widgets, the Menus feature offers an easy way to add and reorder a variety of navigational links to your site, as well as create secondary menu bars (if your theme offers multiple menu areas).

TIP

Additionally, the Menus feature improves WordPress by allowing you to easily create more traditional websites, which sometimes need multiple and more diverse navigational areas than a typical website layout uses or needs.

Twenty Seventeen comes with the appropriate code in the navigation menus to make use of this robust feature. (By default, Twenty Seventeen offers two menu navigation areas to include custom menus.)

To create a new navigation menu in Twenty Seventeen, follow these steps:

1. **On the WordPress Dashboard, click the Menus link on the Appearance menu.**

   The Menus screen loads, as shown in Figure 1-8.

2. **Enter a menu name in the Menu Name field and then click Create Menu.**

   After you create your new custom menu, the gray modules to the left become active, allowing you to add new links to your custom menu.

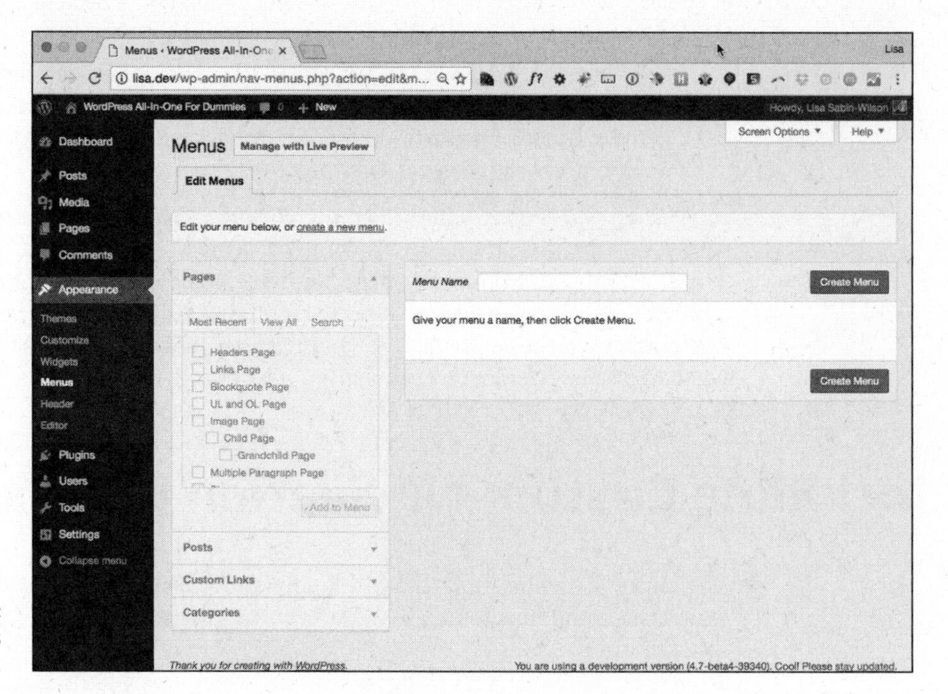

FIGURE 1-8: The WordPress Menus page.

3. **In the Manage Locations module at the top of the Menus screen, choose your new menu from the Top Menu drop-down menu and then click Save.**

   Your new menu is activated for display in your site's header.

4. **Repeat Steps 1 through 3 to configure a second menu for the Social Links Menu location.**

   The Twenty Seventeen theme is programmed to recognize links to Facebook, Twitter, and other social media services, so when you add social media links to the Social Links menu, Twenty Seventeen automatically adds social media icons in the footer area of your site. These icons are linked to the social media profile links you add to the menu.

5. **Click the Edit Menus tab of the Menus screen.**

6. **From the drop-down menu at the top of the Menus screen, choose the menu you want to edit.**

   For this section of the chapter, I created a menu called Main and assigned it to the Top Menu location. The following steps and images reflect those options.

7. **Add items to your new menu, such as custom links, pages, and categories.**

   Items that you can add to your menu include the following:

- *Pages:* To include existing pages on your menu, locate the Pages module (shown in Figure 1-9) and click the pages you want to include. After you do that, click the Add to Menu button and then click the Save Menu button.

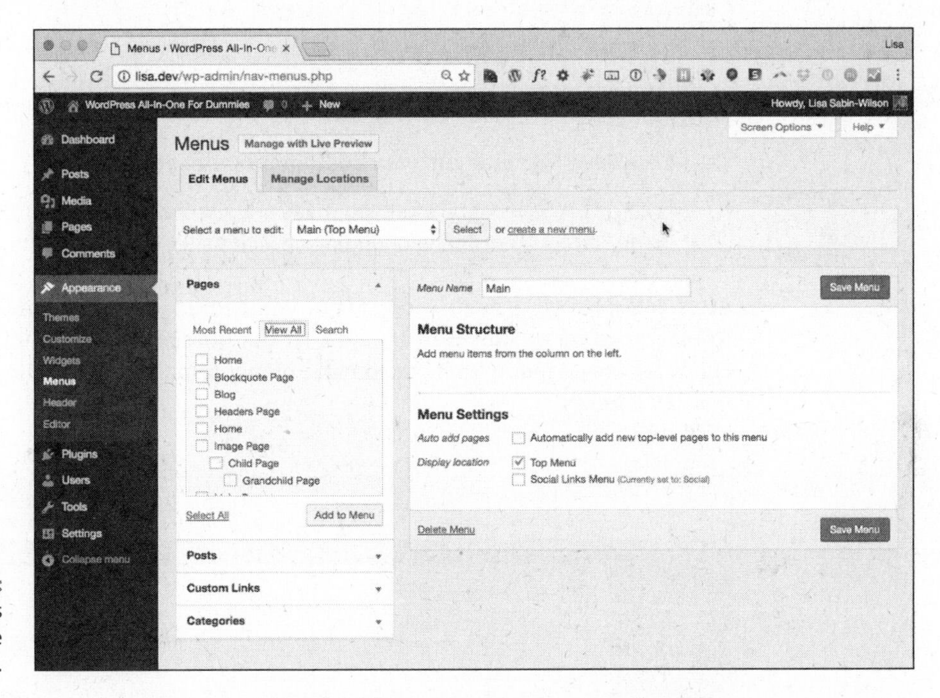

FIGURE 1-9: Selecting pages to add to the custom menu.

- *Categories:* To include existing categories on your menu, click the Categories link to open the Categories menu module; then click the categories you want to include. After you do that, click the Add to Menu button and then click the Save Menu button.
- *Custom Links:* You can add links to other websites, such as your Twitter or Facebook profile pages. Scroll to the Custom Links module (shown in Figure 1-10). In the URL field, type the web address you want to direct people to. In the Label field, add the word or phrase the menu displays for people to click. Then click the Add to Menu button, and then click the Save Menu button.

TIP

If you're using custom post types (covered in Book 6, Chapter 6), click the Screen Options tab in the top-right corner of the Dashboard screen (refer to Figure 1-9) and choose Custom Post Types under the Boxes heading. This option makes custom post type links available for addition to the menu.

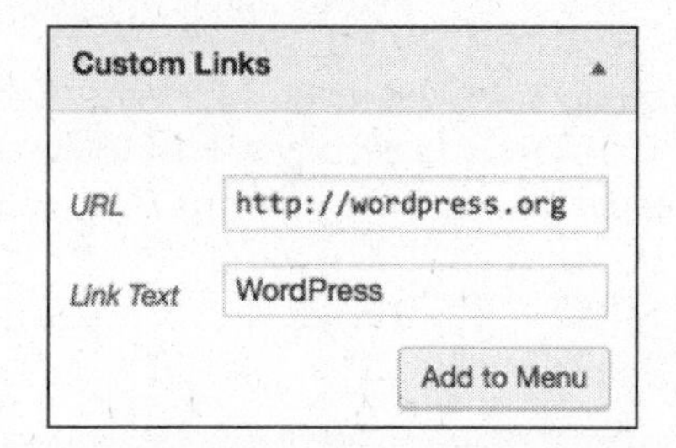

FIGURE 1-10: Adding links to a custom menu.

8. **Click Save Menu to add your custom menu to your theme.**

REMEMBER

Always click Save Menu after you make any significant change in a custom menu, such as reordering or adding new items.

After you save your navigation menu, you can use the drag-and-drop interface to rearrange it, as shown in Figure 1-11. Additionally, you can create submenus below top-level menu items by moving menu items slightly to the right below the top-level items.

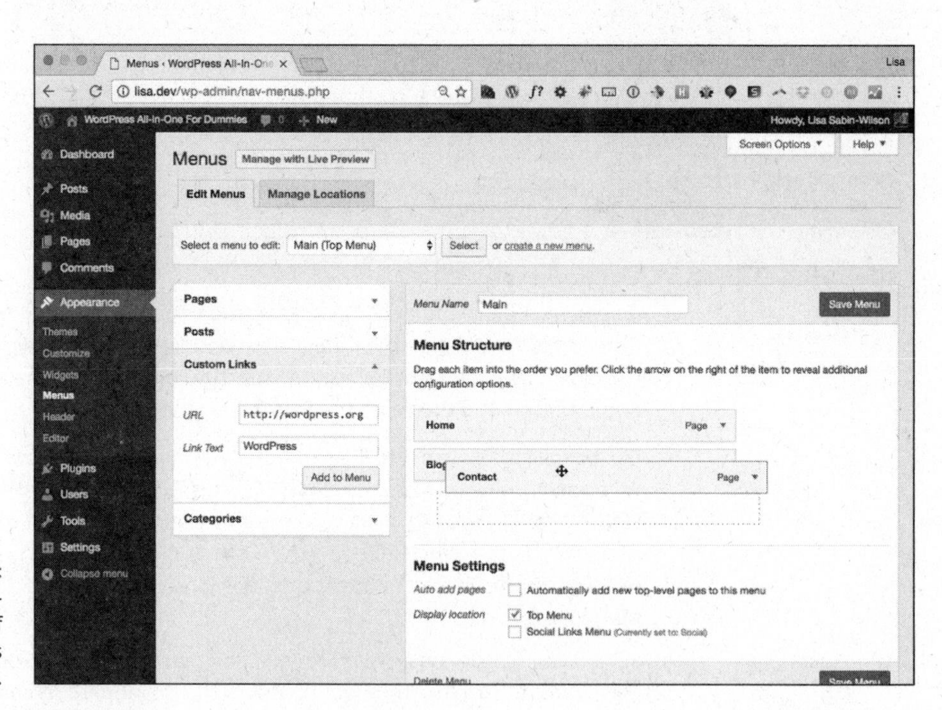

FIGURE 1-11: The drag-and-drop interface of the WordPress Menus screen.

TIP

Use the submenus feature to avoid cluttering the navigation bar. By organizing content logically, you can help readers find what they want faster even if you have lots of content for them to look through.

You can also create multiple custom menus and add them to your theme through widget areas by using any one of the following:

- Custom Menu widget (using widgets is covered in Book 6, Chapter 3)
- Multiple menu areas, providing the theme you are using supports the use of multiple menus (see Book 6, Chapter 6)
- Inserting the menu template tag in one of your theme's template files (information on working with theme template tags is found in Book 6, Chapter 6)

Enhancing Your Website with Widgets

WordPress widgets are helpful tools built into the WordPress application. They allow you to arrange the display of content on your blog sidebar, such as recent posts and archive lists. With widgets, you can arrange and display the content of the sidebar of your site without having to know a single bit of PHP or HTML.

Widget areas are the regions of your theme where you can insert and arrange content (such as a list of your recent blog posts or links to your favorite sites) or custom menus by dragging and dropping (and editing) available widgets (shown on the Dashboard's Widget page) into those corresponding areas. You can find more information on widget areas in Book 6, Chapter 3.

Many widgets offered by WordPress (and those added by some WordPress themes and plugins) provide drag-and-drop installation of more advanced functions that normally are available only if you write code directly into your theme files.

Click the Widgets link on the Appearance menu on the Dashboard to see the available widgets on the Widgets screen. This feature is a big draw because it lets you control what features you use and where you place them without having to know a lick of code.

To explore the Twenty Seventeen theme's widget-ready areas, include Sidebar, Footer 1 and Footer 2, as shown in Figure 1-12.

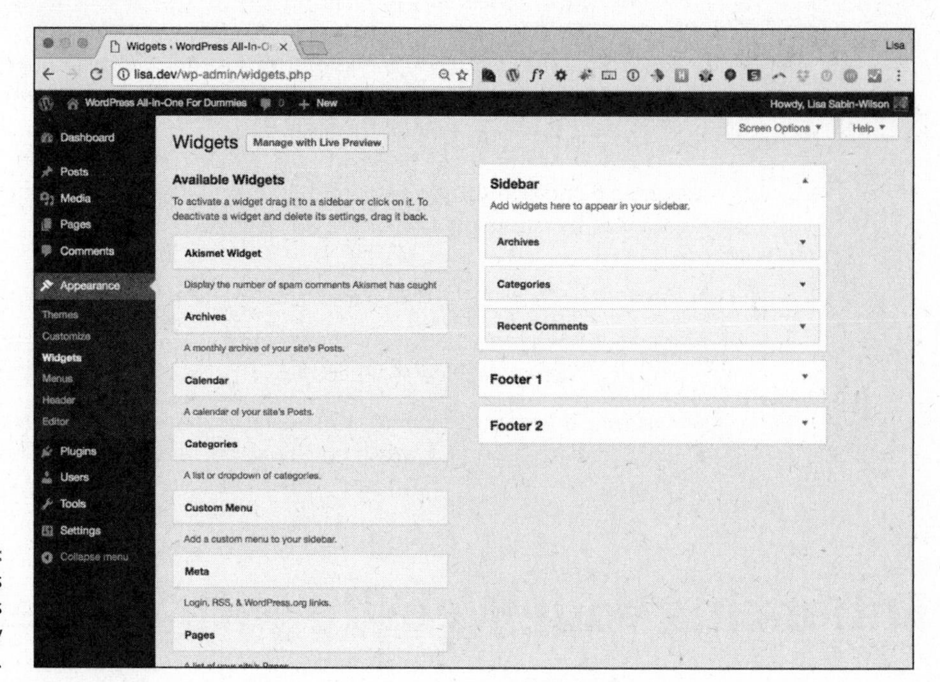

FIGURE 1-12: This page displays available widgets and widget-ready areas.

Adding widgets to your sidebar

The Widgets screen lists all the widgets that are available for your WordPress site. On the right side of the Widgets screen is the sidebar area designated in your theme. You drag your selected widget from the Available Widgets section to your chosen widget area on the right. To add a Search box to the right sidebar of the default layout, for example, drag the Search widget from the Available Widgets section to the Sidebar widget area.

To add a new widget to your sidebar, follow these steps:

1. **Find the widget you want to use.**

   The widgets are listed in the Available Widgets section. For the purpose of these steps, choose the Recent Posts widget.

2. **Drag and drop the widget to the Sidebar widget section on the right side of the page.**

   The widget is now located in the Sidebar widget section, and the content of the widget now appears on your site's sidebar.

3. **Click the arrow to the right of the widget title.**

   Options for the widget appear. Each widget has different options that you can configure. The Recent Posts widget, for example, lets you configure the title, the number of recent posts you want to display (the default is 5; the maximum is 15), and the date. (See Figure 1-13.)

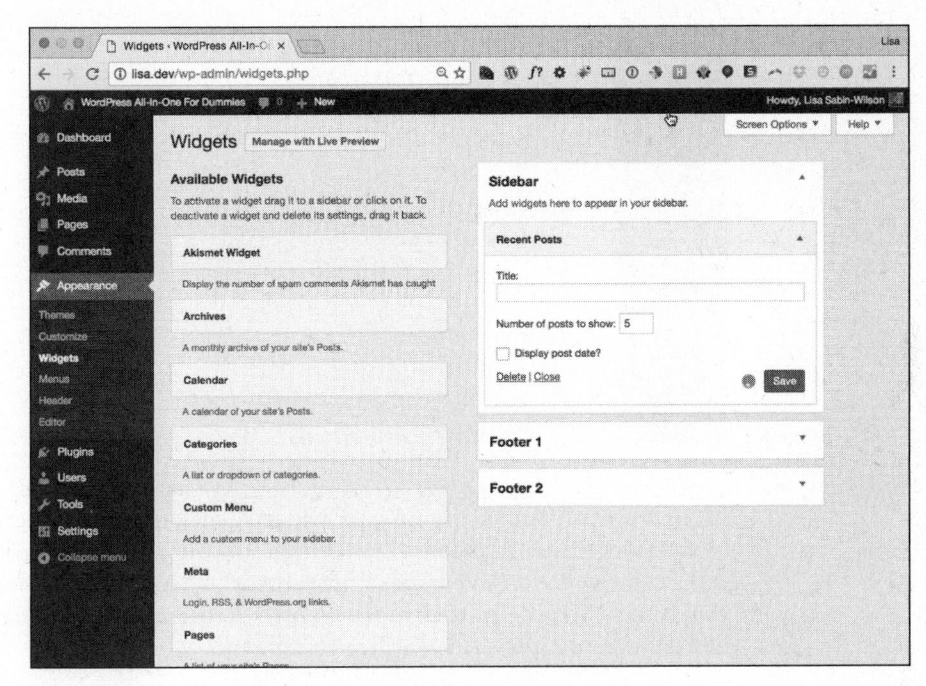

FIGURE 1-13: Editing the Recent Posts widget.

4. **Select your options, and click the Save button.**

   The options you've set are saved.

5. **Arrange your widgets in the order in which you want them to appear on your site by dragging and dropping them in the list.**

   Repeat this step until your widgets are arranged the way you want them.

TIP

To remove a widget from your sidebar, click the arrow to the right of the widget title to open the widget options; then click the Delete link. WordPress removes the widget from the right side of the page and places it back in the Available Widgets list. If you want to remove a widget but want WordPress to remember the settings that you configured for it, instead of clicking the Delete link, simply drag the widget into the Inactive Widgets area, shown in Figure 1-14, on the bottom of the Widgets screen. The widget and all your settings are stored for future use.

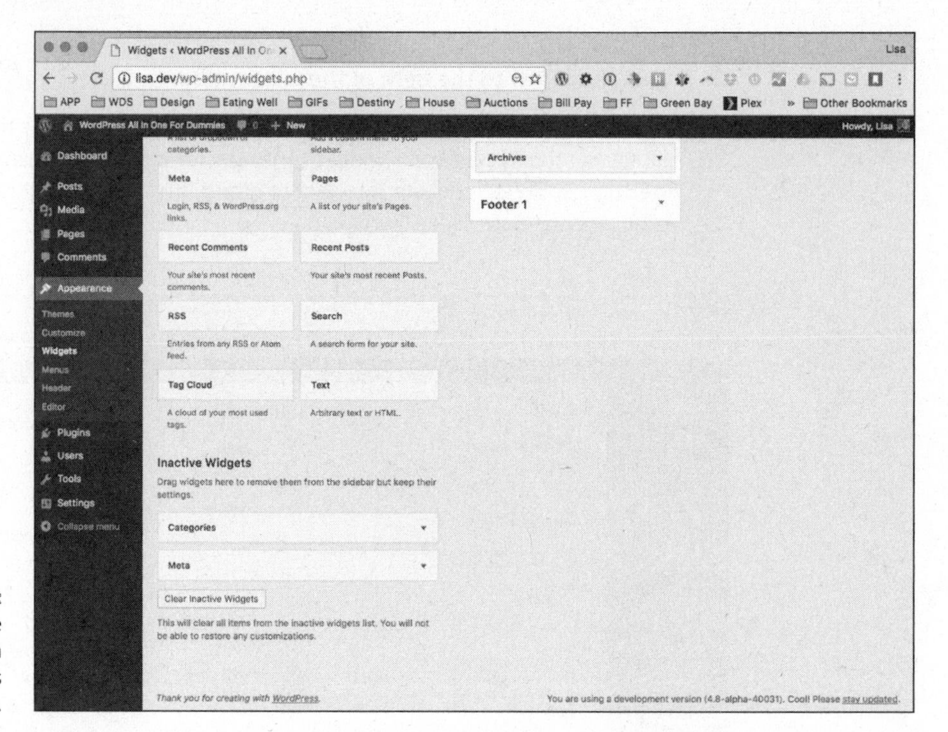

FIGURE 1-14: The Inactive Widgets area on the Widgets screen.

After you select and configure your widgets, click the Visit Site button at the top of your WordPress Dashboard (to the right of your site name). Your site's sidebar reflects the content (and order of the content) of the Widgets screen sidebar. How cool is that? You can go back to the Widgets screen and rearrange items, as well as add and remove items, to your heart's content.

REMEMBER

The number of options available for editing a widget depend on the widget. Some widgets have several editable options; others simply let you write a title. The Recent Posts widget (refer to Figure 1-13), for example, has three options: one for editing the title of the widget, one for setting how many recent posts to display, and one for specifying whether to display the post date.

Using the Text widget

The Text widget is one of the most popular and useful WordPress widgets because it enables you to add text and even HTML code to widget areas without editing the theme's template files. Therefore, you can designate several types of information on your site by including your desired text within it.

Here are some examples of how you can use the Text widget:

- **Add an email newsletter subscription form.** Add a form that allows site visitors to sign up for your email newsletter. Because adding an email newsletter subscription form often involves HTML, the Text widget is especially helpful.
- **Display business hours of operation.** Display the days and hours of your business operation where everyone can easily see them.
- **Post your updates from social networks.** Many social networking sites, such as Twitter and Facebook, offer embed codes that let you display your updates on those sites directly on your website. Social networking embed codes often include JavaScript, HTML, and CSS, which you can easily embed with the Text widget.
- **Announce special events and notices.** If your organization has a special sale, an announcement about a staff member, or an important notice about weather closings, for example, you can use the Text widget to post this information to your site in a few seconds.

WARNING

The WordPress Text widget doesn't allow you to include PHP code of any kind. This widget doesn't execute PHP code, such as special WordPress template tags or functions (like the ones you find in Book 6, Chapter 3). A great plugin called Advanced Text Widget, however, allows you to insert PHP code within it. You can download the Advanced Text Widget from the WordPress Plugin Directory at `https://wordpress.org/plugins/advanced-text-widget`. (You can find more information about using and installing WordPress plugins in Book 7.)

To add the Text widget, follow these steps:

1. **On the WordPress Dashboard, click the Widgets link on the Appearance menu.**
2. **Find the Text widget in the Available Widgets section.**
3. **Drag the Text widget to the desired widget area.**

   The Text widget opens, as shown in Figure 1-15.
4. **Add a widget headline in the Title field and any desired text in the Content box.**
5. **Click the Save button.**
6. **Click the Close link at the bottom of the Text widget box.**

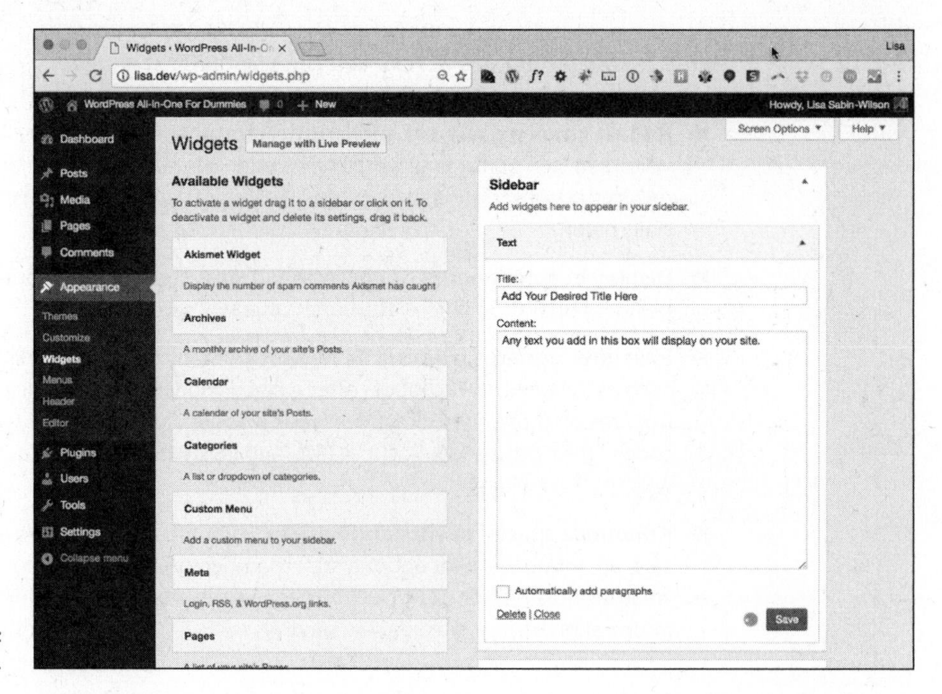

FIGURE 1-15: The Text widget.

Using the RSS widget

The RSS widget allows you to pull headlines from almost any RSS feed, including recent headlines from your other WordPress sites. You can also use it to pull in headlines from news sites or other sources that offer RSS feeds. This practice is commonly referred to as *aggregation*, which means that you're gathering information from a syndicated RSS feed source to display on your site.

After you drag the RSS widget to the appropriate widget area, the widget opens, and you can enter the RSS Feed URL you want to display. Additionally, you can easily tweak other settings, as shown in Figure 1-16, to add information to the widget area for your readers.

Follow these steps to add the RSS widget to your blog:

1. **Add the RSS widget to your sidebar on the Widgets screen.**

   Follow the steps in "Adding widgets to your sidebar" earlier in this chapter to add the widget.

2. **Click the arrow to the right of the RSS widget's name.**

   The widget opens, displaying options you can configure.

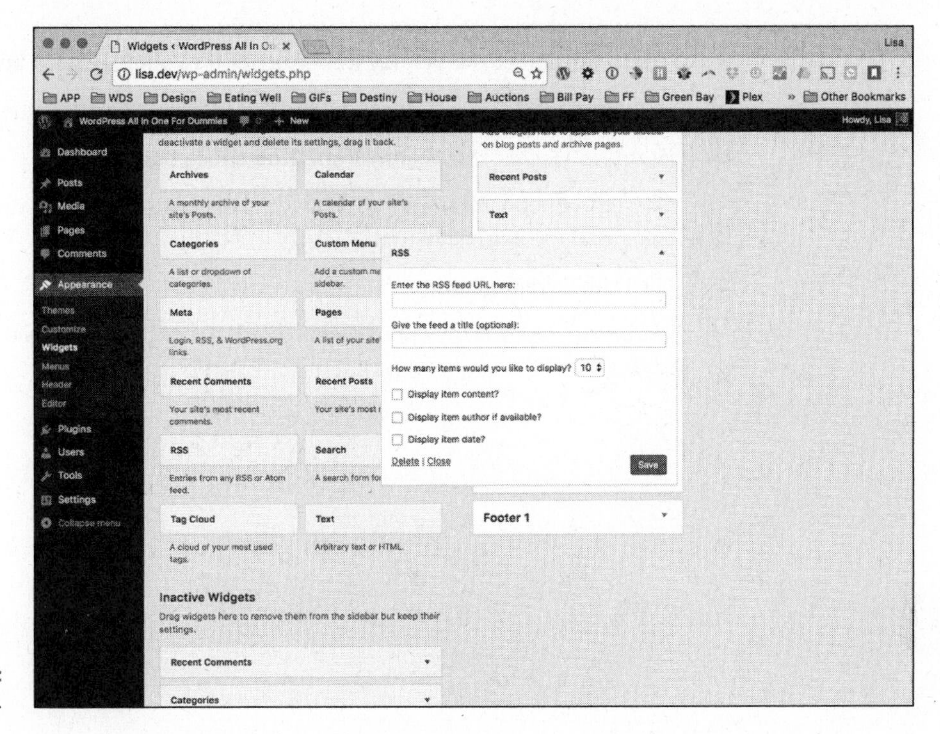

FIGURE 1-16: The RSS widget.

3. **In the Enter the RSS Feed URL Here text box, type the RSS URL of the site you want to add.**

   You can usually find the RSS Feed URL of a site listed on the sidebar.

4. **Type the title of the RSS widget.**

   This title appears in your site above the links from this site. If I wanted to add the RSS feed from my business site, for example, I'd type **WebDevStudios Feed**.

5. **Select the number of items from the RSS feed to display on your site.**

   The drop-down menu gives you a choice of 1–20.

6. **(Optional) Select the Display the Item Content check box.**

   Selecting this check box tells WordPress that you also want to display the content of the feed (usually, the content of the post from the feed URL). If you want to display only the title, leave the check box deselected.

7. **(Optional) Select the Display Item Author If Available check box.**

   Select this option if you want to display the author's name with the item's title.

8. **(Optional) Select the Display Item Date check box.**

   Select this option if you want to display the date when the item was published along with the item's title.

9. **Click the Save button.**

   WordPress saves all the options and reloads the Widgets page with your RSS widget intact.

» Exploring things to avoid with free themes

» Installing, previewing, and activating your new theme

» Discovering premium theme options

Chapter 2
Finding and Installing WordPress Themes

WordPress themes are simply a group of bundled files called *templates*, which, when activated in WordPress, determine the look and basic function of your site. (See Book 6, Chapter 3 for more about template files.)

Because themes set the design style of your site, including how content displays on it, they're the first and most basic tools you can use to customize your site to fit your unique needs. One of the most amazing benefits of the WordPress community is the thousands of free themes that are available — and the new ones released each week.

Although finding one WordPress theme among thousands of options can be challenging, it's a fun adventure. You can explore the various designs and features to ultimately find the right theme for yourself and your site. In this chapter, you discover the options for finding and installing free themes on your WordPress site. I also discuss premium theme options and tell you a few things to avoid.

Getting Started with Free Themes

With thousands of free WordPress themes available and new ones appearing all the time, your challenge is to find the right one for your site. Here are a few things to remember while you explore. (Also see the nearby sidebar "Are all WordPress themes free?" for information about free versus commercial themes.)

- **Free themes are excellent starting places.** Find a couple of free themes and use them as starting points for understanding how themes work and what you can do with them. Testing free themes, their layouts, and their options helps you identify what you want in a theme.
- **You'll switch themes frequently.** Typically, you'll find a WordPress theme that you adore, and then, a week or two later, you'll find another theme that fits you or your site better. Don't expect to stay with your initial choice. Something new will pop up on your radar screen. Eventually, you'll want to stick with one that fits your needs best and doesn't aggravate visitors because of continual changes.
- **You get what you pay for.** Although a plethora of free WordPress themes exists, you receive limited or no support for them. Free themes are often labors of love. The designers have full-time jobs and responsibilities, and they release these free projects for fun, passion, and a desire to contribute to the WordPress community. Therefore, you shouldn't expect (or demand) support for these themes. Some designers maintain very active and helpful forums to help users, but those forums are rare. Just be aware that with free themes, you're on your own.
- **Download themes from reputable sources.** Themes are essentially pieces of software. Therefore, they can contain things that could be scammy, spammy, or potentially harmful to your site or computer. Therefore, it's vital that you do your homework by reading online reviews and downloading themes from credible, trusted sources. The best place to find free WordPress themes is the WordPress Theme Directory (see Figure 2-1) at `https://wordpress.org/themes`.

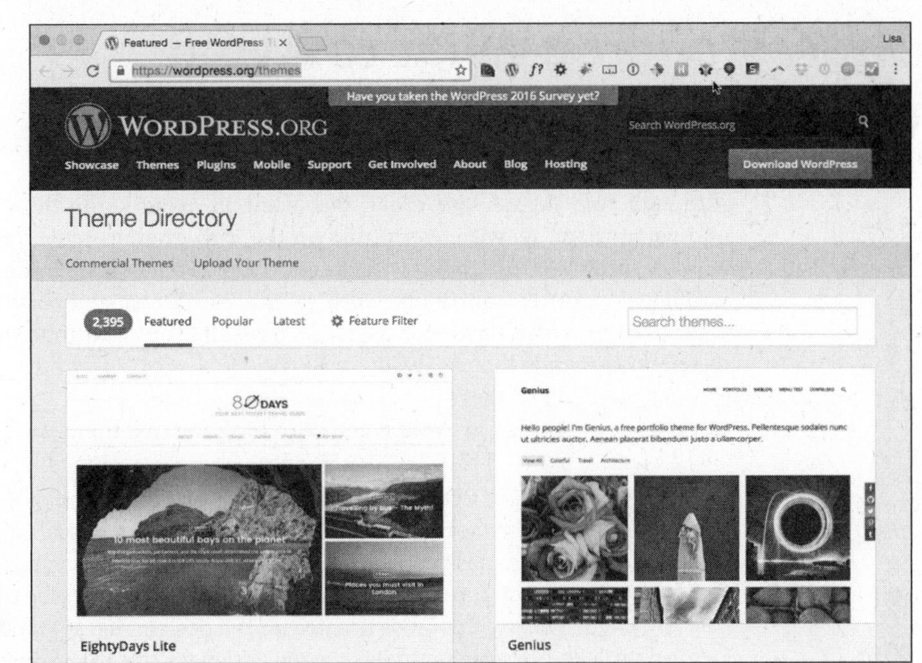

FIGURE 2-1: The most trusted resource for free themes.

ARE ALL WORDPRESS THEMES FREE?

Not all WordPress themes are created equal, and it's important for you, the user, to know the difference between free and commercial themes:

- **Free:** These themes are free, period. You can download and use them on your website at absolutely no cost. It's a courtesy to include a link to the designer in the footer of your website, but you can remove that link if you want to.
- **Commercial:** These themes cost money. Commercial themes usually are available for download only after you've paid anywhere from $10 to $500. The designer feels that these themes are a cut above the rest and, therefore, are worth the money you spend for them. Generally, you aren't allowed to remove any designer credits that appear in these themes, and you aren't allowed to redistribute the themes. ***Note:*** You *don't* find premium themes in the WordPress Theme Directory. I provide information on where to find premium themes at the end of this chapter.

Understanding What to Avoid with Free Themes

Although free themes are great, you want to avoid some things when finding and using them. As with everything on the web, themes have the potential to be abused. Although free themes were conceived to allow people (namely, designers and developers) to contribute work to the WordPress community, they've also been used to wreak havoc for users. As a result, you need to understand what to watch out for and what to avoid.

Here are some things to avoid when searching for free themes:

- **Spam links:** Many free themes outside the WordPress Theme Directory include links in the footer or sidebars that can be good or bad. The good uses of these links are designed to credit the original designer and possibly link to her website or portfolio. This practice — a nice reward to the creators — should be observed because it increases the designer's traffic and clients. Spam links, however, aren't links to the designer's site; they're links to sites you may not ordinarily associate with or endorse on your site. The best example is a link in the footer that links to odd, off-topic, and uncharacteristic keywords or phrases, such as *weight loss supplement* or *best flower deals.* Mostly, this spam technique is used to increase the advertised site's search engine ranking for that particular keyword by adding another link from your site — or, worse, to take a site visitor who clicks it to a site unrelated to the linked phrase.
- **Hidden and malicious code:** Unfortunately, the WordPress community has received reports of hidden, malicious code within a theme. This hidden code can produce spam links, security exploits, and abuses on your WordPress site. Hackers install code in various places that run this type of malware. Unscrupulous theme designers can, and do, place code in theme files that inserts hidden malware, virus links, and spam. Sometimes, you see a line or two of encrypted code that looks like it's just part of the theme code. Unless you have a great deal of knowledge of PHP, you may not know that the theme is infected with dangerous code.
- **Lack of continued development:** WordPress software continues to improve with each new update. Two or three times a year, WordPress releases new software versions, adding new features, security patches, and numerous other updates. Sometimes, a code function is superseded or replaced, causing a theme to break because it hasn't been updated for the new WordPress version. Additionally, because software updates add new features, the theme needs to be updated accordingly. Because free themes typically come without any warranty or support, one thing you should look for — especially if a

theme has many advanced back-end options — is whether the developer is actively maintaining the theme for current versions of WordPress. Active maintenance typically is more an issue with plugins than it is with themes, but the topic is worth noting.

- **Endlessly searching for free themes:** Avoid searching endlessly for the perfect theme. Trust me — you won't find it. You may find a great theme and then see another with a feature or design style you wish the previous theme had, but the new theme may lack certain other features. Infinite options can hinder you from making a final decision. Peruse the most popular themes in the WordPress Theme Directory, choose five that fit your criteria, and then move on. You always have the option to change a theme later, especially if you find the vast amount of choices in the directory to be overwhelming.

The results of these unsafe theme elements can range from simply annoying to downright dangerous, affecting the integrity and security of your computer and/or hosting account. For this reason, the WordPress Theme Directory is considered to be a safe place from which to download free themes. WordPress designers develop these themes and upload them to the directory, and the folks behind the WordPress platform vet each theme. In the official directory, themes that contain unsafe elements simply aren't allowed.

REMEMBER

The WordPress Theme Directory isn't the only place on the web to find free WordPress themes, but it's the place to find the most functional and *safest* themes available. Safe themes contain clean code and fundamental WordPress functions to ensure that your WordPress blog functions with the minimum requirements. The WordPress.org website lists the basic requirements that theme designers have to meet before their themes are accepted into the directory; you can find that listing of requirements at `https://wordpress.org/themes/about`. I highly recommend that you stick to the WordPress Theme Directory for free themes to use on your site; you can be certain that those themes don't contain any unsafe elements or malicious code.

Installing a Theme

After you find a WordPress theme, you can install the theme on your WordPress site via SFTP or the WordPress Dashboard.

To install a theme via SFTP, follow these steps:

1. **Download the theme file from the Theme Directory.**

   Typically, theme files are provided in a compressed format (`.zip` file).

I discuss how you can peruse the WordPress Theme Directory from your WordPress installation in the next section.

2. **Unzip or extract the theme's `.zip` file.**

   You see a new folder on your desktop, typically labeled with the corresponding theme name. (Visit Book 2, Chapter 2 if you need to refresh yourself on how to use SFTP.)

3. **Upload the theme folder to your web server.**

   Connect to your hosting server via SFTP, and upload the extracted theme folder to the `/wp-content/themes` folder on your server. (See Figure 2-2.)

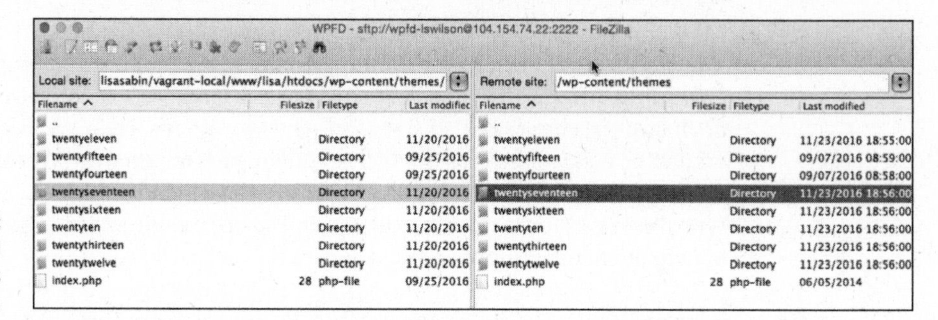

FIGURE 2-2: Upload and download panels in SFTP.

To install a theme via the Dashboard's theme installer, follow these steps:

1. **Download the theme file from the Theme Directory to your desktop.**

   Typically, theme files are provided in a compressed format (`.zip` file). When you use this method, you don't extract the `.zip` file, because the theme installer does that for you.

2. **Log in to your WordPress Dashboard.**
3. **Click the Themes link on the Appearance menu.**

   The Themes screen appears.

4. **Click the Add New button.**

   The Add Themes screen appears, displaying a submenu of links.

5. **Click the Upload Theme button.**

   The panel displays a utility to upload a theme in `.zip` format.

6. **Upload the `.zip` file you downloaded in Step 1.**

   Click the Choose File button and then locate and select the `.zip` file you stored on your computer.

7. **Click the Install Now button.**

   WordPress unpacks and installs the theme in the appropriate directory for you. Figure 2-3 shows the result of installing a theme via this method.

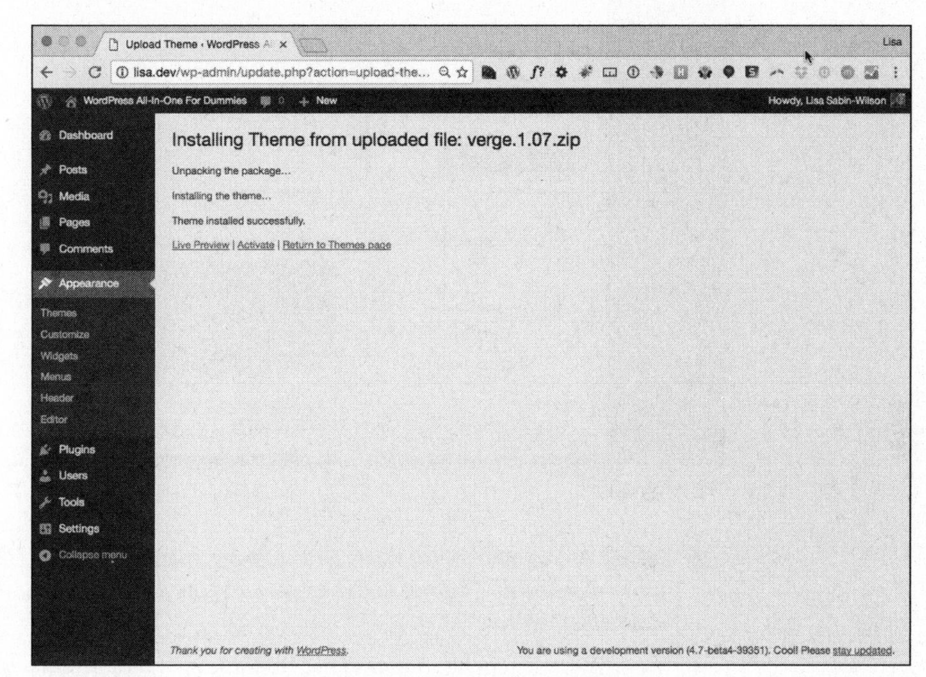

FIGURE 2-3: Installing a theme via the Dashboard's theme installer.

Browsing the free themes

Finding free themes via the Add Themes screen is extremely convenient because it lets you search the Theme Directory from your WordPress site. Start by clicking the Themes link on the Appearance menu of the WordPress Dashboard and then click the Add New button to open the Add Themes screen. (See Figure 2-4.)

After you navigate to the Add Themes screen, you see the following menu links:

- **Featured:** The themes shown in the Featured link have been selected by WordPress.org for you. The themes in this section are favorites of the members of the Themes team.
- **Popular:** If you don't have a theme in mind, the themes in this section are some of the most popular themes. I recommend that you install and test-drive one of these for your site's first theme.

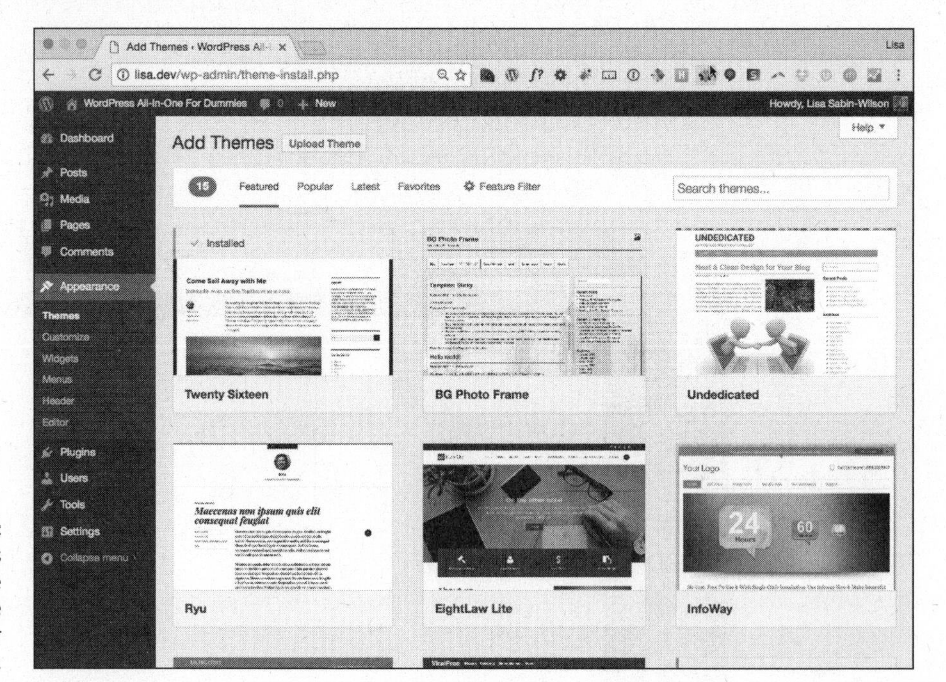

FIGURE 2-4: The Add Themes screen, where you can find free themes from your Dashboard.

- **Latest:** As WordPress improves and changes, many themes need updating to add new features. Themes in the Latest category are themes that have been updated recently.
- **Favorites:** If you marked themes as favorites on the WordPress website, you can find those themes on the Add Themes screen. Type your WordPress.org username in the text field, and click the Get Favorites button.
- **Feature Filter:** This link gives you a variety of filters to choose among to find a theme you're looking for. You can filter by Layout, Features, and Subject. After you select your desired filters, click the Apply Filters button to view the themes that match your set filters.
- **Search:** If you know the name of a free theme, you can easily search for it here by keyword, author, or tag. You can also refine your search based on specific features within the themes, including color, layout, and subject (such as Holiday).

After you find the theme that you want, click the Install button that appears when you hover your cursor over the theme thumbnail.

Previewing and activating a theme

After you upload a theme via SFTP or the theme installer, you can preview and activate your desired theme.

TIP

The WordPress Theme Preview option allows you to see how the theme would look on your site without actually activating it. If you have a site that's receiving traffic, it's best to preview any new theme before activating it to ensure that you'll be happy with its look and functionality. If you're trying to decide among several new theme options, you can preview them all before changing your live site.

To preview your new theme, follow these steps:

1. **Log in to your WordPress Dashboard.**
2. **Click the Themes link on the Appearance menu.**

   The Themes screen appears, displaying your current (activated) theme and any themes that are installed in the `/wp-content/themes` directory on your web server.
3. **Preview the theme you want to use.**

   Click the Live Preview button that appears when you hover your cursor over the theme thumbnail. A preview of your site with the theme appears in a pop-up window, as shown in Figure 2-5.

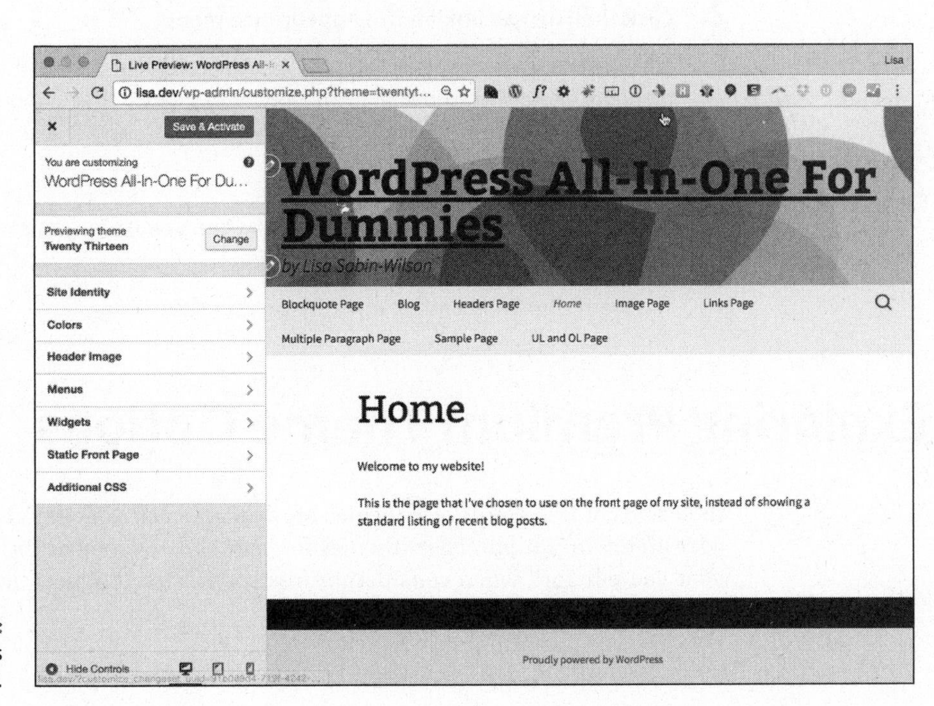

FIGURE 2-5: A WordPress theme preview.

4. **(Optional) Configure theme customization features.**

   Some, but not all, themes are developed to provide customization features. Figure 2-5 shows these customization options for the Twenty Thirteen theme:

   - Site Identity
   - Colors
   - Header Image
   - Menus
   - Widgets
   - Static Front Page
   - Additional CSS

5. **Choose whether to activate the theme.**

   Click the Save & Activate button in the top-right corner of the configuration panel to activate your new theme with the options you set in Step 4, or close the preview by clicking the Cancel (X) button in the top-left corner of the panel.

To activate a new theme without previewing it, follow these steps:

1. **Log in to your WordPress Dashboard.**
2. **Click the Themes link on the Appearance menu.**

   The Themes screen appears, displaying your current (activated) theme and any themes that are installed in the `/wp-content/themes` directory on your web server.

3. **Find the theme you want to use.**
4. **Click the Activate button that appears when you hover your mouse over the theme thumbnail.**

   The theme immediately becomes live on your site.

Exploring Premium Theme Options

Thousands of free WordPress themes are available, but you may also want to consider premium (for purchase) themes for your site. Remember the adage "You get what you pay for" when considering free services or products, including WordPress and free themes.

Typically, when you download and use something free, you get no assistance with the product or service. Requests for help generally go unanswered. Therefore, your expectations should be lower because you aren't paying anything. When you pay for something, you usually assume that you have support or service for your purchase and that the product is of high (or acceptable) quality.

WordPress, for example, is available free. Except for the active WordPress support forum, however, you have no guarantee of support while using the software. Moreover, you have no right to demand service.

Here are some things to consider when contemplating a premium theme. (I selected the commercial companies listed later in this chapter based on these criteria.)

- **Selection:** Many theme developers offer a rich, diverse theme selection, including themes designed for specific niche industries, topics, or uses (such as video, blogging, real estate, or magazine themes). Generally, you can find a good, solid theme to use for your site from one source.
- **Innovation:** To differentiate them from their free counterparts, premium themes include innovative features, such as theme settings or advanced options that extend WordPress to help you do more.
- **Great design with solid code:** Although many beautiful free themes are available, premium themes are professionally coded and beautifully designed, cost thousands of dollars, and require dozens of hours to build, which simply isn't feasible for many free theme developers.
- **Support:** Most commercial companies have full-time support staff to answer questions, troubleshoot issues, and point you to resources beyond their support. Often, premium theme developers spend more time helping customers troubleshoot issues outside the theme products. Therefore, purchasing a premium theme often provides a dedicated support community to question about advanced issues and upcoming WordPress features; otherwise, you're on your own.
- **Stability:** No doubt you've purchased a product or service from a company only to find later that the company has gone out of business. If you choose to use a premium theme, purchase a theme from an established company with a solid business model, a record of accomplishment, and a dedicated team devoted to building and supporting quality products.

REMEMBER

Although some free themes have some or all of the features in the preceding list, for the most part, they don't. Keep in mind that just because a designer calls a theme "premium" doesn't mean that the theme has passed through any kind of quality review. The view of what constitutes a premium theme can, and will, differ from one designer to the next.

Fully investigate any theme before you spend your money on it. Here are some things to check out before you pay:

- Email the designer who's selling the premium theme, and ask about a support policy.
- Find people who've purchased the theme, and contact them to ask about their experiences with the theme and the designer.
- Carefully read any terms that the designer has published on his site to find any licensing restrictions that exist.
- If the premium theme designer has a support forum, ask whether you can browse the forum to find out how actively the designer answers questions and provides support. Are users waiting weeks to get their questions answered, for example, or does the designer seem to be on top of support requests?
- Search online for the theme and the designer. Often, users of premium themes post about their experiences with the theme and the designer. You can find both positive and negative information about the theme and the designer before you buy.

These developers are doing some amazingly innovative things with WordPress themes, and I highly recommend that you explore their offerings:

- **iThemes (`https://ithemes.com`):** iThemes (see Figure 2-6) emphasizes business WordPress themes that use WordPress as a full-fledged, powerful content management system. The site's pride and joy is iThemes Builder, which is more a build-a-WordPress website tool than a typical theme.
- **Organic Themes (`https://organicthemes.com`):** Organic Themes (see Figure 2-7) has a great team, paid support moderators, and WordPress themes that are as solid, from a code standpoint, as they are beautiful.
- **WooThemes (`https://woocommerce.com/woothemes`):** WooThemes (see Figure 2-8) has a wide selection of high-quality themes with excellent options and support. Their most popular theme is Canvas, a highly customizable theme with more than 100 options you can use to personalize your site via a theme options panel.
- **Press75 (`http://press75.com`):** Press75 (see Figure 2-9) offers niche themes for photography, portfolios, and video. Check out the Video Elements theme for a great example.

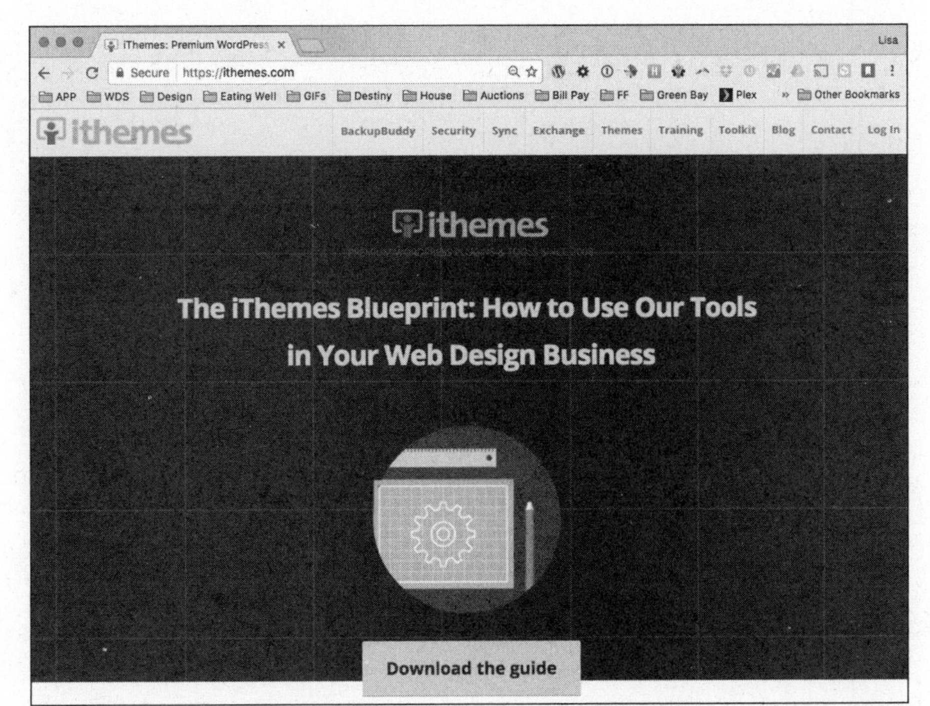

FIGURE 2-6: iThemes.com, provider of commercial WordPress themes.

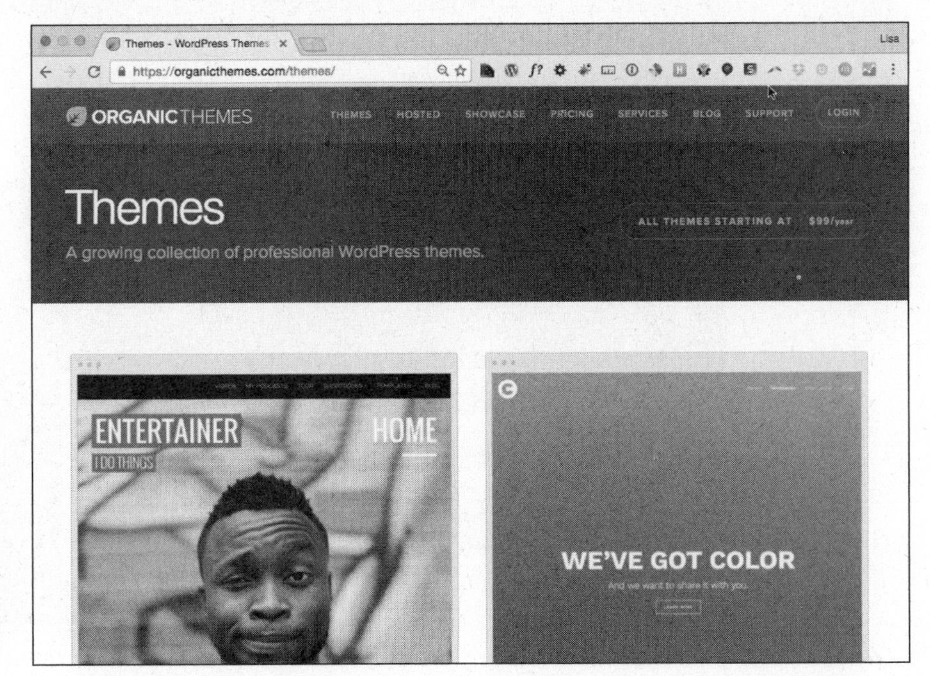

FIGURE 2-7: Organic Themes, another provider of commercial WordPress themes.

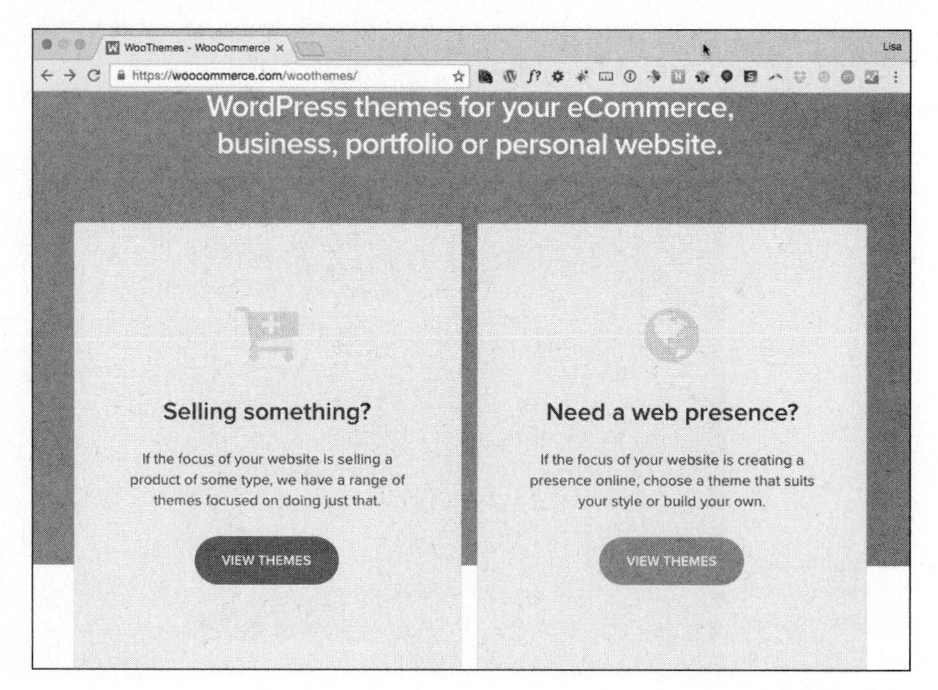

FIGURE 2-8: WooThemes has premium themes, community, and support.

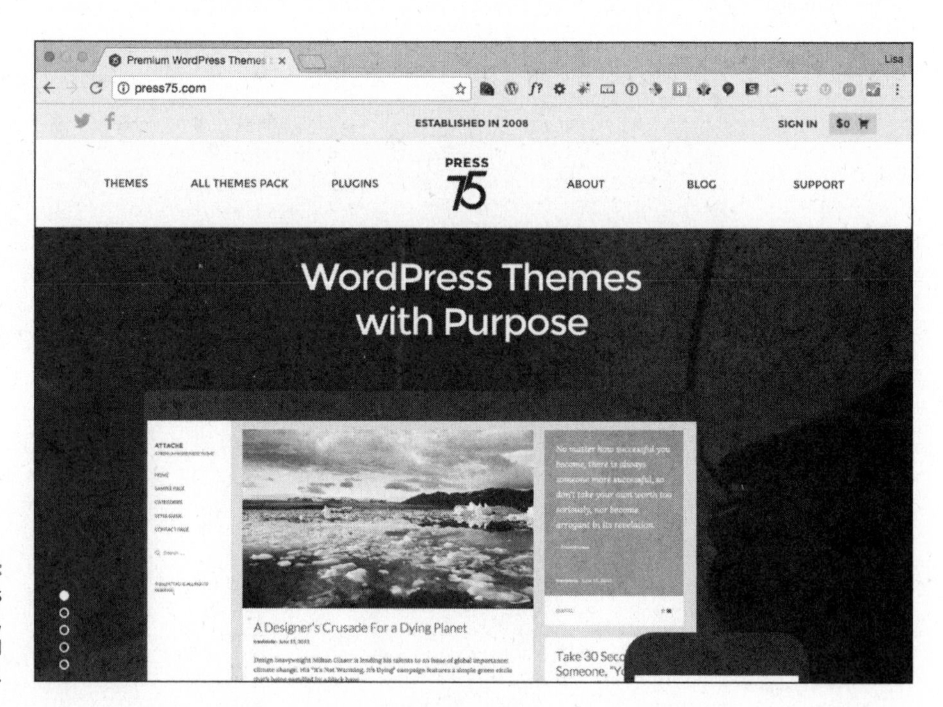

FIGURE 2-9: Press75 offers premium themes, demos, and theme packages.

TIP

You can't find, preview, or install premium themes by using the Add Themes feature on your WordPress Dashboard (covered in an earlier section of this chapter). You can find, purchase, and download premium themes only from third-party websites. After you find a premium theme you like, you need to install it via the SFTP method or by using the Dashboard upload feature. (See the earlier "Installing a Theme" section.) You can find a very nice selection of premium themes on the WordPress website at `https://wordpress.org/themes/commercial`.

» Exploring template tags

» Making widget areas

» Understanding the main template files

Chapter 3
Exploring the Anatomy of a Theme

This chapter breaks down the parts that make up your WordPress theme. Understanding your theme allows you greater flexibility when you customize it. Many of the problems people encounter with themes, such as not knowing which files edit certain functions of a site, comes from lack of understanding all the pieces.

There are those who like to get their hands dirty (present company included!). If you're one of them, you need to read this chapter. WordPress users who create their own themes do so in the interest of

- **Individuality:** Having a theme that no one else has. (If you use one of the free themes, you can pretty much count on the fact that at *least* a dozen other WordPress blogs have the same look as yours.)
- **Creativity:** Displaying your own personal flair and style.
- **Control:** Having full control of how the blog looks, acts, and delivers your content.

Many of you aren't at all interested in creating your own theme for your WordPress blog, however. Sometimes, it's just easier to leave matters to the professionals and hire an experienced WordPress theme developer to create a custom look for your WordPress website or to use one of the thousands of free themes provided by WordPress designers. (See Book 6, Chapter 2 for more on free themes.)

Creating themes does require you to step into the code of the templates, which can be a scary place sometimes — especially if you don't really know what you're looking at. A good place to start is understanding the structure of a WordPress website. Separately, the parts won't do you any good, but when you put them together, the real magic begins! This chapter covers the basics of doing just that, and near the end of the chapter, you find specific steps for putting your own theme together.

TIP

You don't need to know HTML to use WordPress. If you plan to create and design WordPress themes, however, you need some basic knowledge of HTML and Cascading Style Sheets (CSS). For assistance with HTML, check out *HTML, XHTML, and CSS For Dummies*, 7th Edition by Ed Tittel and Jeff Noble (published by John Wiley & Sons, Inc.).

Starting with the Basics

A WordPress theme is a collection of WordPress templates made up of WordPress template tags. When I refer to a WordPress *theme*, I'm talking about the group of templates that makes up the theme. When I talk about a WordPress *template*, I'm referring to only one of the template files that contain WordPress template tags. WordPress template tags make all the templates work together as a theme (more about this topic later in the chapter). These files include

- **The theme's stylesheet (`style.css`):** The stylesheet provides the theme's name, as well as the CSS rules that apply to the theme. (Later in this chapter, I go into detail about how stylesheets work.)
- **The Main Index template (`index.php`):** The index file is the first file that loads when a visitor comes to your site. It contains the HTML as well as any PHP code needed on your home page.
- **An optional functions file (`functions.php`):** This optional file is a place where you can add additional functionality to your site via PHP functions.

Template and functions files end with the `.php` extension. *PHP* is the scripting language used in WordPress, which your web server recognizes and interprets as such. (Book 2, Chapter 3 covers additional details on the PHP language that you'll find helpful.) These files contain more than just scripts, though. The PHP files also contain HTML, which is the basic markup language of web pages.

Within this set of PHP files is all the information your browser and web server need to make your website. Everything from the color of the background to the layout of the content is contained in this set of files.

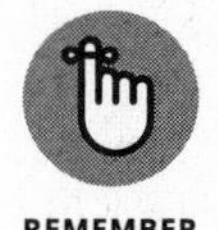
REMEMBER

The difference between a template and a theme can cause confusion. *Templates* are individual files. Each template file provides the structure in which your content will display. A *theme* is a set of templates. The theme uses the templates to make the whole site.

Understanding where the WordPress theme files are located on your web server gives you the ability to find and edit them, as needed. You can use two different methods to view and edit WordPress theme files by following these steps:

1. **Connect to your web server via SFTP, and have a look at the existing WordPress themes on your server.**

   The correct location is `/wp-content/themes/`. When you open this folder, you find the `/twentyseventeen` theme folder.

   REMEMBER

   If a theme is uploaded to any folder other than `/wp-content/themes`, it won't work.

2. **Open the folder for the Twenty Seventeen theme (`/wp-content/themes/twentyseventeen`), and look at the template files inside.**

   When you open the Twenty Seventeen theme folder (see Figure 3-1), you see several files. At *minimum*, you find these five templates in the default theme:

   - *Stylesheet* (`style.css`)
   - *Header* (`header.php`)
   - *Main Index* (`index.php`)
   - *Sidebar* (`sidebar.php`)
   - *Footer* (`footer.php`)

   These files are the main WordPress template files, which I discuss in more detail in this chapter. There are several template files, however, and you should try to explore all of them if you can. Take a peek inside to see the template functions they contain. These filenames are the same in every WordPress theme.

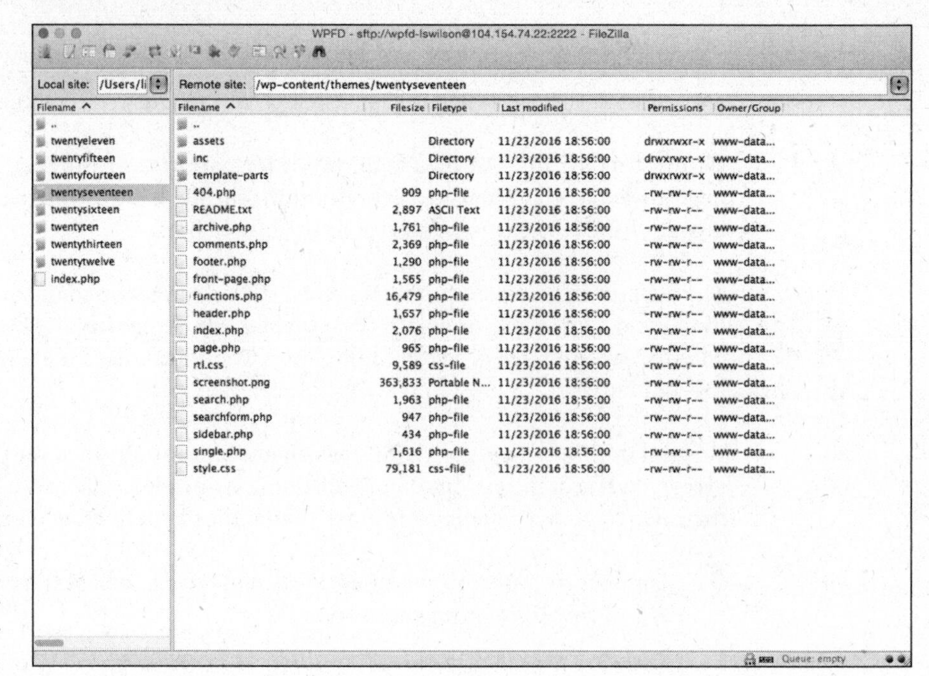

FIGURE 3-1: Twenty Seventeen theme in the /wp-content/themes/twentyseventeen folder on your web server.

3. **Log in to your WordPress Dashboard in your web browser window, and click the Editor link on the Appearance menu to look at the template files within a theme.**

   This page lists the various templates available within the active theme. (Figure 3-2 shows the templates in the default Twenty Seventeen theme.) A text box on the left side of the screen displays the contents of each template, and this box is also where you can edit the template file(s). To view and edit a template file, click the template name in the list on the right side of the page.

The Edit Themes screen also shows the template tags within the template file. These tags make all the magic happen on your site; they connect all the templates to form a theme. The "Exploring Template Tags, Values, and Parameters" section of this chapter discusses these template tags in detail, showing you what they mean and how they function.

TIP

Click the Documentation drop-down menu on the Themes screen to see all the template tags used in the template you're currently viewing. This list is helpful when you edit templates, and it gives you some insight into some of the different template tags used to create functions and features within your WordPress theme. (*Note:* The Documentation drop-down menu on the Themes screen doesn't appear when you view the stylesheet because no template tags are used in the `style.css` template — only CSS.)

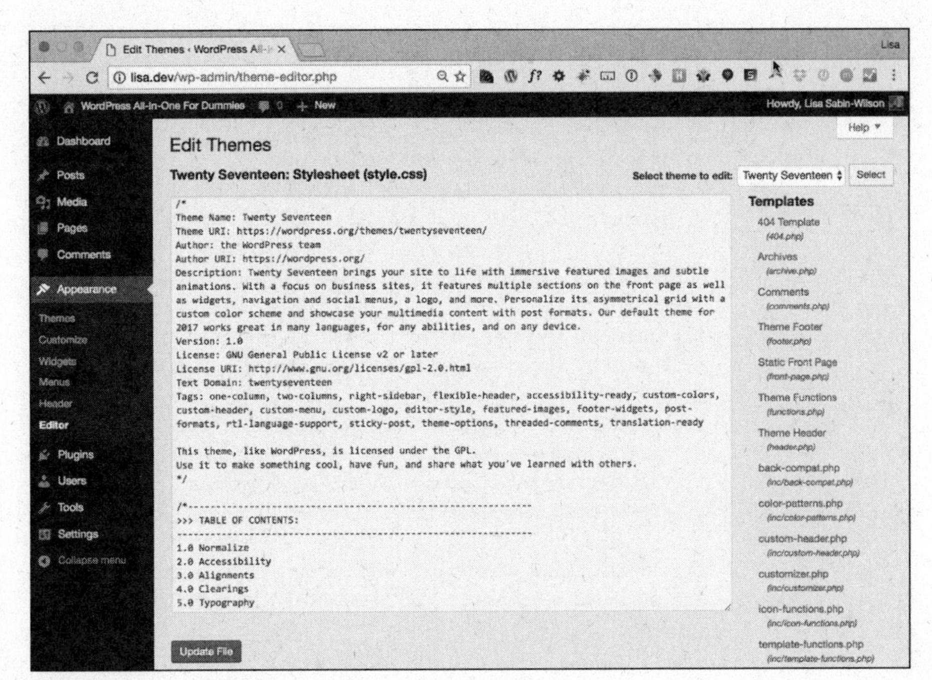

FIGURE 3-2: A list of templates available in the default Twenty Seventeen WordPress theme.

Understanding the Stylesheet

Every WordPress theme includes a `style.css` file. A browser uses this file, commonly known as the *stylesheet*, to style the theme. Style can include text colors, background images, and spacing between elements on the site. The stylesheet targets areas of the site to style by using CSS IDs and classes. *CSS IDs* and *classes* are simply means of naming a particular element of the site. IDs are used for elements that appear only once on a page, but classes can be used as many times as necessary. Although this file references *style*, it contains much more information about the theme.

At the beginning of the `style.css` file, a comment block known as the *stylesheet header* passes information about your theme to WordPress. *Comments* are code statements included only for programmers, developers, and any others who read the code. Computers ignore comment statements, but WordPress uses the stylesheet header to get information about your theme. In CSS, comments always begin with a forward slash followed by a star (`/*`) and end with a star followed by a forward slash (`*/`). The following code shows an example of the stylesheet header for the Twenty Seventeen theme:

```
./*
Theme Name: Twenty Seventeen
Theme URI: https://wordpress.org/themes/twentyseventeen/
Author: the WordPress team
```

```
Author URI: https://wordpress.org/
Description: Twenty Seventeen brings your site to life with immersive
   featured images and subtle animations. With a focus on business sites,
   it features multiple sections on the front page as well as widgets,
   navigation and social menus, a logo, and more. Personalize its
   asymmetrical grid with a custom color scheme and showcase your multimedia
   content with post formats. Our default theme for 2017 works great in many
   languages, for any abilities, and on any device.
Version: 1.0
License: GNU General Public License v2 or later
License URI: http://www.gnu.org/licenses/gpl-2.0.html
Text Domain: twentyseventeen
Tags: one-column, two-columns, right-sidebar, flexible-header, accessibility-
   ready, custom-colors, custom-header, custom-menu, custom-logo, editor-style,
   featured-images, footer-widgets, post-formats, rtl-language-support,
   sticky-post, theme-options, threaded-comments, translation-ready

This theme, like WordPress, is licensed under the GPL.
Use it to make something cool, have fun, and share what you've learned with
   others.
*/
```

Figure 3-3 shows how the Themes page of the Dashboard looks with the Twenty Seventeen theme activated. The title and information are taken directly from the `style.css` header.

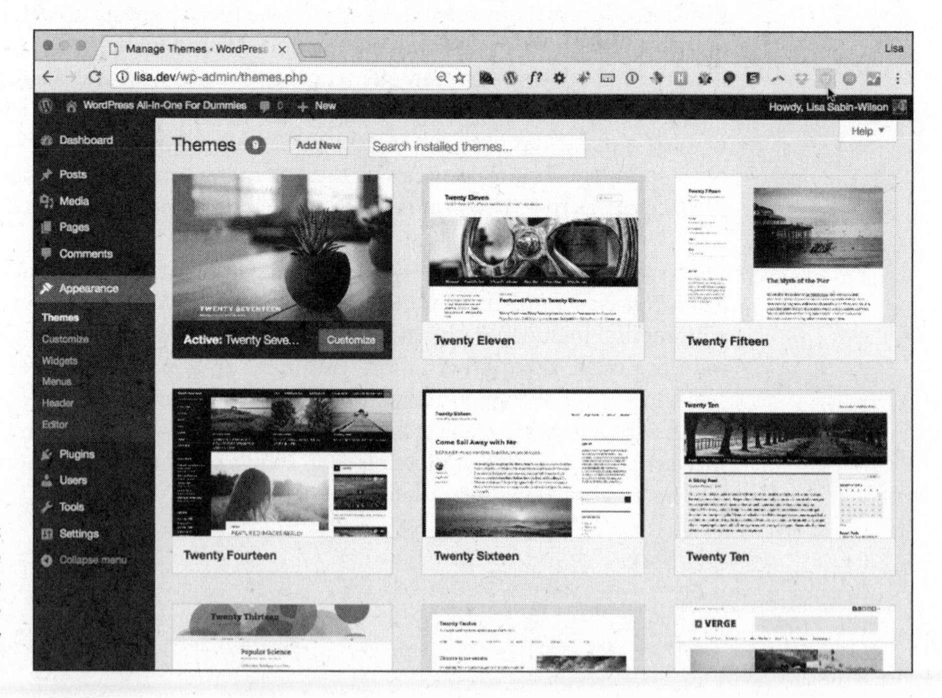

FIGURE 3-3: This page shows the currently active theme, Twenty Seventeen.

If you make modifications in the stylesheet header, the changes are reflected in the WordPress Dashboard on the Themes screen.

WARNING

Themes must provide this information in the stylesheet header, and no two themes can have the same information. Two themes with the same name and details would conflict in the theme-selection page. If you create your own theme based on another theme, make sure that you change this information first.

Below the stylesheet header are the CSS styles that drive the formatting and styling of your theme.

TIP

Book 6, Chapter 4 goes into detail about CSS, including some examples that you can use to tweak the style of your existing WordPress theme. Check it out!

Exploring Template Tags, Values, and Parameters

Some people are intimidated when they look at template tags. Really, template tags are just simple bits of PHP code that you can use inside a template file to display information dynamically. Before starting to play around with template tags in your WordPress templates, it's important to understand what makes up a template tag and why.

WordPress is based in PHP (a scripting language for creating web pages) and uses PHP commands to pull information from the MySQL database. Every tag begins with the function to start PHP and ends with a function to stop PHP. In the middle of those two commands lives the request to the database that tells WordPress to grab the data and display it.

A typical template tag looks like this:

```
<?php get_info(); ?>
```

This example tells WordPress to do three things:

- Start PHP (`<?php`).
- Use PHP to get information from the MySQL database and deliver it to your blog (`get_info();`).
- Stop PHP (`?>`).

In this case, `get_info` represents the tag function, which grabs information from the database to deliver it to your site. What information is retrieved depends on what tag function appears between the two PHP commands. As you may notice, there's a lot of starting and stopping of PHP throughout the WordPress templates. The process seems as though it would be resource-intensive, if not exhaustive — but it really isn't.

REMEMBER

For every PHP command you start, you need a stop command. Every time a command begins with `<?php`, somewhere later in the code is the closing `?>` command. PHP commands that aren't structured properly cause really ugly errors on your site, and they've been known to send programmers, developers, and hosting providers into loud screaming fits.

Understanding the basics

If every piece of content on your site were *hard-coded* (manually added to the template files), the content wouldn't be easy to use and modify. Template tags allow you to add information and content to your site dynamically. One example of adding information by using a template tag involves the `the_category` tag. Instead of typing all the categories and links that each post belongs in, you can use the `the_category()` tag in your template to automatically display all the categories as links.

Using template tags prevents duplication of effort by automating the process of adding content to your website.

When you use a template tag, you're really telling WordPress to do something or retrieve some information. Often, template tags are used to fetch data from the server and even display it on the front end. More than 100 template tags are built into WordPress, and the tags vary greatly in terms of what they can accomplish. You can find a complete list of template tags in the WordPress Codex (documentation for WordPress) at `https://codex.wordpress.org/Template_Tags`.

Template tags can be used only inside PHP blocks. The PHP blocks can be opened and closed as many times as necessary in a template file. When a PHP block is opened, the server knows that anything contained in the block is to be translated as PHP. The opening tag (`<?php`) must be followed at some point by the closing tag (`?>`). All blocks must contain these tags. A template tag is used in the same way that PHP functions are. The tag is always text with no spaces (may be separated by underscores or dashes), opening and closing brackets, and a semicolon. The following line of code shows you how it all looks:

```
<?php template_tag_name(); ?>
```

PHP is a fairly advanced coding language and has many built-in functions for you to use. If you aren't a PHP developer, keep it simple when you're attempting to add custom PHP. All code must be semantically perfect; otherwise, it won't work. Always read your code to make sure that you entered it correctly.

Some template tags can be used only inside The Loop, so check the codex for details. You can find out more about The Loop in "Examining the Main Index and The Loop" later in this chapter.

Using parameters

Because a template tag is a PHP function, you can pass parameters to the tag. A *parameter* is a variable that allows you to change or filter the output of a template tag. WordPress has three types of template tags:

- **Tags without parameters:** Some template tags don't require any options, so they don't need any parameters passed to them. The `is_user_logged_in()` tag, for example, doesn't accept any parameters because it returns only `true` or `false`.
- **Tags with PHP function–style parameters:** Template tags with PHP function–style parameters accept parameters that you pass to them by placing one or more values inside the function's parentheses. If you're using the `bloginfo()` tag, for example, you can filter the output to just the description by using

  ```
  <?php bloginfo('description'); ?>
  ```

  If there are multiple parameters, the order in which you list them is very important. Each function sets the necessary order of its variables, so double-check the order of your parameters.

  Always place the value in single quotes, and separate multiple parameters with commas.
- **Tags with query string–style parameters:** Template tags with query string–style parameters allow you to change the values of just the parameters you require. This ability to cherry-pick only those parameters you need to change is useful for template tags that have a large number of options. The `wp_list_pages()` tag, for example, has 18 parameters. Instead of using the PHP function–style parameters, you can use this function to get to the source of what you need and give it a value. If you want to list all your WordPress pages except page 24, for example, you use

  ```
  <?php wp_list_pages('exclude=24'); ?>
  ```

Query string–style parameters can be the most difficult to work with because they generally deal with the template tags that have the most possible parameters.

Table 3-1 helps you understand the three variations of parameters that WordPress uses.

TABLE 3-1 **Three Variations of Template Parameters**

Variation	Description	Example
Tags without parameters	These tags have no additional options available. Tags without parameters have nothing within the parentheses.	`the_tag();`
Tags with PHP function–style parameters	These tags have a comma-separated list of values placed within the tag parentheses.	`the_tag('1,2,3');`
Tags with query string–style parameters	These types of tags generally have several available parameters. This tag style enables you to change the value for each parameter without being required to provide values for all available parameters for the tag.	`the_tag('parameter=true');`

REMEMBER

The WordPress Codex, located at `https://codex.wordpress.org`, has every conceivable template tag and possible parameter known to the WordPress software. The tags and parameters in this chapter are the ones used most often by WordPress users.

Customizing common tags

Because template tags must be used inside the PHP template files, they can easily be customized with HTML. If you're using the PHP tag `wp_list_pages()`, for example, you could display it in an HTML unordered list so that the pages are easily accessible to users, like this:

```
<ul>
<?php wp_list_pages(); ?>
</ul>
```

This code displays all the pages that you created in WordPress as an unordered list. If you had the pages About, Blog, and Content, the list of pages would be displayed like this:

- About
- Blog
- Contact

Another example is titles. For proper search engine optimization (SEO), you should always put page titles in H1 HTML tags, like this:

```
<h1 class="pagetitle">
<?php the_title(); ?>
</h1>
```

Creating New Widget Areas

Many themes are *widget-ready,* meaning that you can insert widgets into them easily. Widgets allow you to add functionality to your sidebar without having to use code. Some common widget functionalities include displaying recent posts, displaying recent comments, adding a search box for searching content on a site, and adding static text. Even widget-ready themes have their limitations, however. You may find that the theme you chose doesn't have widget-ready areas in all the places you want them. You can make your own, however.

Registering your widget

To add a widget-ready area to the WordPress Dashboard Widget interface, first register the widget in your theme's `functions.php` file as follows:

```
function my_widgets_init() {
   register_sidebar( array (
    'name' => __( 'Widget Name'),
    'id' => 'widget-name',
    'description' => __( 'The primary widget area'),
    'before_widget' => '<li id="%1$s" class="widget-container %2$s">',
    'after_widget' => "</li>",
    'before_title' => '<h3 class="widget-title">',
    'after_title' => '</h3>',
    ) );
  }
add_action('widgets_init', 'my_widgets-init');
```

You can insert this code directly below the first opening PHP tag (<?php). Sometimes, it's to add a few extra lines when you're adding code. The extra empty lines around your code are ignored by the browser but can greatly increase readability of the code.

Within that code, you see seven *arrays* (sets of values that tell WordPress how you want your widgets to be handled and displayed):

- `name`: This name is unique to the widget and is displayed on the Widgets page on the Dashboard. It's helpful to register several widgetized areas on your site.
- `id`: This array is the unique ID given to the widget.
- `description`: This array is a text description of the widget. The text that gets placed here displays on the Widgets page on the Dashboard.
- `before_widget`: This array is the HTML markup that gets inserted directly before the widget. It's helpful for CSS styling purposes.
- `after_widget`: This array is the HTML markup that gets inserted directly after the widget.
- `before_title`: This array is the HTML markup that gets inserted directly before the widget title.
- `after_title`: This array is the HTML markup that gets inserted directly after the widget title.

REMEMBER

Even though you use `register_sidebar` to register a widget, widgets don't have to appear on a sidebar. Widgets can appear anywhere you want them to. The example code snippet earlier in this section registers a widget named Widget Name on the WordPress Dashboard. Additionally, it places the widget's content in an element that has the CSS class `widget` and puts `<h4>` tags around the widget's title.

Widgets that have been registered on the WordPress Dashboard are ready to be populated with content. On the Appearance menu of your site's Dashboard, you see a link titled Widgets. When you click the Widgets link, you see the new widget area you registered.

Displaying new widgets on your site

When a widget-ready area is registered with the WordPress Dashboard, you can display the area somewhere on your site. A common place for widget-ready areas is the sidebar.

To add a widget-ready area to your sidebar, pick a location within the sidebar and then locate that area in the HTML, which can vary from theme to theme. Many times, theme authors create their own `sidebar.php` file, and you can add this code there. After you find the area in the HTML, add the following code to the template:

```
<?php dynamic_sidebar('Widget Name'); ?>
```

This code displays the contents of the widget that you previously registered in the admin area.

Simplifying customization with functions

You may find that the simple code doesn't accomplish all the functionality that you need. You may want to style the widget's title separately from the content, for example. One solution is to create a custom PHP function that gives you a few more options. Open `functions.php`, and insert the following code directly below the opening `<?php` tag to create a function:

```
function add_new_widget_location( $name ) {
if ( ! function_exists( 'dynamic_sidebar' ) || ! dynamic_sidebar(
$name ) ) : ?>
<div class="widget">
<h4><?php echo $name; ?></h4>
<div class="widget">
<p>This section is widgetized. If you would like to
add content to this section, you may do so by using the Widgets
panel from within your WordPress Admin Dashboard. This Widget
Section is called "<strong><?php echo $name; ?></strong>"</p>
</div>
</div>
<?php endif; ?>
<?php
}
```

In this function, the first part checks to see whether a widget is assigned to this area. If so, the widget displays. If not, a message with the name of the widget area displays, which allows users to distinguish the widget area they want to add widgets to. Now if you want to display a widget by using this method, you go to the desired template file and insert the following code where you want the widget to appear:

```
<?php add_new_widget_location('Widget-Name'); ?>
```

Exploring common problems

A common problem in creating widget areas is forgetting the admin side. Although people successfully create a widget in the PHP template where they want it, they often fail to make it to the `functions.php` to register the new widget area.

Another common problem is omitting the widget code from the `functions.php` file. If you're adding widget areas to an existing site, you need to add the widget code to the bottom of the list of widgets in the `functions.php` file. Failure to do so causes the widget areas to shift their contents, which places your widgets out of order, causing you to have to redo them on the Widgets page of the WordPress Dashboard.

Examining the Main Index and The Loop

Your theme is required to have only two files: `style.css` and a main index file, known in WordPress as `index.php`. The `index.php` file is the first file WordPress tries to load when someone visits your site. Extremely flexible, `index.php` can be used as a stand-alone file, or it can include other templates. The Main Index template drags your blog posts out of the MySQL database and inserts them into your blog. This template is to your blog what the dance floor is to a nightclub: It's where all the action happens.

The filename of the Main Index template is `index.php`. You can find it in the `/wp-content/themes/twentyseventeen/` folder.

The first template tag in the Main Index template calls in the Header template, meaning that it pulls the information from the Header template into the Main Index template, as follows:

```
<?php get_header(); ?>
```

Your theme can work without calling in the Header template, but it will be missing several essential pieces — the CSS and the blog name and tagline, for starters.

The Main Index template in the Twenty Seventeen theme calls in three other files in a similar fashion:

```
get_template_part( 'template-parts/post/content',
get_post_format() );
```

This function includes the Post-Format-specific template for the content. If you want to override this function in a child theme, include a file called content-XXX.php (where XXX is the Post Format name, for example: content-audio.php or content-video.php) instead.

`get_sidebar();`

This function calls in the template file named `sidebar.php`.

`get_footer();`

This function calls in the template file named `footer.php`.

I cover each of these three functions and template files in upcoming sections of this chapter.

REMEMBER

The concept of *calling in* a template file by using a function or template tag is exactly what the Main Index template does with the four functions for the header, loop, sidebar, and footer templates explained later in this section.

Generally, one of the important functions of the main index is to contain The Loop. In WordPress, *The Loop* displays posts and pages on your site. Any PHP or HTML that you include in The Loop repeats for each of your posts that it displays. The Loop has a starting point and an ending point; anything placed between them is used to display each post, including any HTML, PHP, or CSS tags and codes.

Here's a look at what the WordPress Codex calls "The World's Simplest Index":

```
<?php
get_header();
if (have_posts()) :
  while (have_posts()) :
    the_post();
    the_content();
  endwhile;
endif;
get_sidebar();
get_footer();
?>
```

Here's how it works:

1. The template opens the php tag.
2. The Loop includes the header, meaning that it retrieves anything contained in the `header.php` file and displays it.

Exploring the Anatomy of a Theme

3. The Loop begins with the `while (have_posts()) :` bit.
4. Anything between the `while` and the `endwhile` repeats for each post that displays.

   The number of posts that displays is determined in the settings section of the WordPress Dashboard.
5. If your site has posts (and most blogs do, even when you first install them), WordPress proceeds with The Loop, starting with the piece of code that looks like this:

   ```
   if (have_posts()) :
     while (have_posts()) :
   ```

   This code tells WordPress to grab the posts from the MySQL database and display them on your page.
6. The Loop closes with this tag:

   ```
     endwhile;
   endif;
   ```

   Near the beginning of The Loop template is a template tag that looks like this:

   ```
   if (have_posts()) :
   ```

   To read that template tag in plain English, it says `If [this blog] has posts`.
7. If your site meets that condition (that is, if it has posts), WordPress proceeds with The Loop and displays your posts. If your site doesn't meet that condition (that is, doesn't have posts), WordPress displays a message that no posts exist.
8. When The Loop ends (at the `endwhile`), the index template goes on to execute the files for the sidebar and footer.

Although it's simple, The Loop is one of the core functions of WordPress.

WARNING

Misplacement of the `while` or `endwhile` statements causes The Loop to break. If you're having trouble with The Loop in an existing template, check your version against the original to see whether the `while` statements are misplaced.

REMEMBER

In your travels as a WordPress user, you may run across plugins or scripts with instructions that say something like this: "This must be placed within The Loop." That's The Loop that I discuss in this section, so pay particular attention. Understanding The Loop arms you with the knowledge you need to tackle and understand your WordPress themes.

The Loop is no different from any other template tag, in that it must begin with a function to start PHP and end with a function to stop PHP. The Loop begins with PHP and then makes a request: "While there are posts in my blog, display them on this page." This PHP function tells WordPress to grab the blog post information from the database and return it to the blog page. The end of The Loop is like a traffic cop with a big red stop sign telling WordPress to stop the function.

REMEMBER

You can set the number of posts displayed per page on the Reading Settings page of the WordPress Dashboard. The Loop abides by this rule and displays only the number of posts per page that you've set.

WordPress uses other template files besides the main index, such as the header, sidebar, and footer templates. The next sections give you a closer look at a few of them.

Header template

The Header template for your WordPress themes is the starting point for every WordPress theme because it tells web browsers the following:

- The title of your site
- The location of the CSS
- The RSS-feed URL
- The site URL
- The tagline (or description) of the site

In many themes, the first elements in the header are a main image and the navigation. These two elements are usually in the `header.php` because they load on every page and rarely change. The following statement is the built-in WordPress function to call the header template:

```
<?php get_header(); ?>
```

TIP

Every page on the web has to start with a few pieces of code. In every `header.php` file in any WordPress theme, you can find these bits of code at the top:

- The `DOCTYPE` (which stands for *document type declaration*) tells the browser which type of XHTML standards you're using. The Twenty Seventeen theme uses `<!DOCTYPE html>`, which is a declaration for W3C standards compliance mode and covers all major browser systems.

Exploring the Anatomy of a Theme

- The <html> tag (HTML stands for *Hypertext Markup Language*) tells the browser which language you're using to write your web pages.
- The <head> tag tells the browser that the information contained within the tag shouldn't be displayed on the site; rather, it's information about the document.

In the header template of the Twenty Seventeen theme, these bits of code look like this, and you should leave them intact:

```
<!DOCTYPE html>
<html <?php language_attributes(); ?> class="no-js no-svg">
<head>
```

TIP

On the Edit Themes page, click the Header template link to display the template code in the text box. Look closely, and you see that the <!DOCTYPE html> declaration, <html> tag, and <head> tag show up in the template.

The <head> tag needs to be closed at the end of the Header template, which looks like this: </head>. You also need to include a fourth tag, the <body> tag, which tells the browser where the information you want to display begins. Both the <body> and <html> tags need to be closed at the end of the template files (in footer.php), like this: </body></html>.

Using bloginfo parameters

The Header template makes much use of one WordPress template tag in particular: bloginfo();.

What differentiates the type of information that a tag pulls in is a *parameter*. Parameters are placed inside the parentheses of the tag, enclosed in single quotes. For the most part, these parameters pull information from the settings on your WordPress Dashboard. The template tag to get your site title, for example, looks like this:

```
<?php bloginfo('name'); ?>
```

Table 3-2 lists the various parameters you need for the bloginfo(); tag and shows you what the template tag looks like. The parameters in Table 3-2 are listed in the order of their appearance in the Twenty Seventeen header.php and template-parts/header/site/site-branding.php template file (called into the header.php file through the use of the get_template_part(); function), and pertain only to the bloginfo(); template tag.

TABLE 3-2 **Tag Values for bloginfo();**

Parameter	Information	Tag
`charset`	Character settings set in the General Settings screen	`<?php bloginfo('charset'); ?>`
`name`	Blog title, set in Settings/General	`<?php bloginfo('name'); ?>`
`description`	Tagline for your blog, set in Settings/General	`<?php bloginfo ('description'); ?>`

Creating title tags

Here's a useful tip about your blog's `<title>` tag: Search engines pick up the words used in the `<title>` tag as keywords to categorize your site in their directories.

The `<title></title>` tags are HTML tags that tell the browser to display the title of your website on the title bar of a visitor's browser. Figure 3-4 shows how the title of my business website sits on the title bar of the browser window.

FIGURE 3-4: The title displayed on the browser.

Search engines love the title bar. The more you can tweak that title to provide detailed descriptions of your site (otherwise SEO), the more the search engines love your site. Browsers show that love by giving your site higher rankings in their results.

The <title> tag is the code that lives in the Header template between these two tag markers: <title></title>. In the default Twenty Seventeen theme, this bit of code is located in the functions.php template file of the theme and looks like this:

```
add_theme_support( 'title-tag' );
```

The add_theme_support('title-tag'); function in the functions.php template tells WordPress to place the <title> tag in the <head> section of the website.

It may help for me to put this example in plain English. The way that the add_theme_support('title-tag'); function displays the title is based on the type of page that's being displayed — and it shrewdly uses SEO to help you with the browser powers that be.

The title bar of the browser window always displays your site's name unless you're on a single post page. In that case, it displays your site's title plus the title of the post on that page.

Displaying your blog name and tagline

Most WordPress themes show your site name and tagline in the header of the site, which means that this information is displayed in easy, readable text for all visitors (not just search engines) to see. My site name and tagline, for example, are

- **Site name:** Lisa Sabin-Wilson
- **Site tagline:** Designer, Author: WordPress For Dummies

You can use the bloginfo(); tag, found in the header.php template, plus a little HTML code to display your site name and tagline. Most sites have a clickable title that takes you back to the main page when the visitor clicks it. No matter where your visitors are on your site, they can always go back home by clicking the title of your site on the header.

To create a clickable title, use the following code:

```
<a href="<?php bloginfo('url'); ?>"><?php bloginfo('name'); ?></a>
```

The `bloginfo('url');` tag is your main Internet address, and the `bloginfo('name');` tag is the name of your site (refer to Table 3-2). So the code creates a link that looks something like this:

```
<a href="http://yourdomain.com">Your Site Name</a>
```

The tagline generally isn't linked back home. You can display it by using the following tag:

```
<?php bloginfo('description'); ?>
```

This tag pulls the tagline directly from the one that you set up on the General Settings screen of your WordPress Dashboard.

This example shows that WordPress is intuitive and user-friendly; you can do things such as change the blog name and tagline with a few keystrokes on the Dashboard. Changing your options on the Dashboard creates the change on every page of your site, with no coding experience required. Beautiful, isn't it?

In the Twenty Seventeen templates, these tags are surrounded by tags that look like these: <h1 class-"site-title"></h1> or <p class="site-description"></p>. These tags are HTML tags, which define the look and layout of the blog name and tagline in the CSS of your theme. Book 6, Chapter 4 covers CSS.

Sidebar template

The Sidebar template in WordPress has the filename `sidebar.php`. The sidebar is usually located on the left or right side of the main content area of your WordPress theme. (In the Twenty Seventeen theme, the sidebar is displayed to the right of the main content area.) The sidebar is a good place to put useful information about your site, such as a summary, advertisements, or testimonials.

Many themes use widget areas in the sidebar template. This practice allows you to display content easily on your WordPress pages and posts. In the default Twenty Seventeen theme, the `index.php` template file also includes the `sidebar.php` template file, which means that it tells WordPress to execute and display all the template functions included in the Sidebar template (`sidebar.php`). The line of code from the Main Index template (`index.php`) that performs this task looks like this:

```
<?php get_sidebar(); ?>
```

This code calls the Sidebar template and all the information it contains into your page.

Footer template

The Footer template in WordPress has the filename `footer.php`. The footer is generally at the bottom of the page and contains brief reference information about the site, such as copyright information, template design credits, and a mention of WordPress. Similarly to the Header and Sidebar templates, the Footer template gets called into the Main Index template through this bit of code:

```
<?php get_footer(); ?>
```

This code calls the Footer template and all the information it contains into your website.

The default Twenty Seventeen theme shows the site title and the statement `Proudly powered by WordPress`. You can use the footer to include all sorts of information about your site; for example, you can display a photograph or a list of links to other pages in your site — you don't have to restrict the footer to small bits of information.

Examining Other Template Files

To make your website work properly, WordPress uses all the theme files together. Some files, such as the header and footer, are used on every page; others, such as the Comments template (`comments.php`), are used only at specific times to pull in specific functions.

When someone visits your site, WordPress uses a series of queries to determine which templates to use.

Many more theme templates can be included in your theme. Here are some of the other template files you may want to use:

- **Comments template (`comments.php`):** The Comments template is required if you plan to host comments on your site; it provides all the template tags you need to display those comments. The template tag used to call the comments into the template is `<?php comments_template(); ?>`.
- **Single Post template (`single.php`):** When your visitors click the title or permalink of a post you published to your site, they're taken to that post's individual page. There, they can read the entire post, and if you have comments enabled, they see the comments form and can leave comments.

- **Page template (page.php):** You can use a Page template for static pages of your WordPress site.
- **Search Results (search.php):** You can use this template to create a custom display of search results on your site. When someone uses the search feature to search your site for specific keywords, this template formats the return of those results.
- **404 template (404.php):** Use this template to create a custom 404 page, which is the page visitors get when the browser can't find the page requested and returns that ugly 404 Page Cannot Be Found error.

The templates in the preceding list are optional. If these templates don't exist in your WordPress themes folder, nothing breaks. The Main Index template handles the display of these items (the single post page, the search results page, and so on). The only exception is the Comments template. If you want to display comments on your site, you must include that template in your theme.

Customizing Your Posts with Template Tags

This section covers the template tags that you use to display the body of each post you publish. The body of a post includes information such as the post date and time, title, author name, category, and content. Table 3-3 lists the common template tags you can use for posts, available in any WordPress theme template. The tags in Table 3-3 work only if you place them within The Loop (covered in "Examining the Main Index and The Loop" earlier in this chapter and located in the loop.php template file).

The last two tags in Table 3-3 aren't like the others. You don't place these tags in The Loop; instead, you insert them after The Loop but before the if statement ends. Here's an example:

```
<?php endwhile; ?>
<?php next_posts_link('&laquo; Previous Entries') ?>
<?php previous_posts_link('Next Entries &raquo;') ?>
<?php endif; ?>
```

TABLE 3-3 Template Tags for Posts

Tag	Function
`get_the_date();`	Displays the date of the post.
`get_the_time();`	Displays the time of the post.
`the_title();`	Displays the title of the post.
`the_permalink();`	Displays the permalink (URL) of the post.
`get_the_author();`	Displays the post author's name.
`the_author_link();`	Displays the URL of the post author's site.
`the_content('Read More...');`	Displays the content of the post. (If you use an excerpt [see next entry], the words *Read More* appear and are linked to the individual post page.)
`the_excerpt();`	Displays an excerpt (snippet) of the post.
`the_category();`	Displays the category (or categories) assigned to the post. If the post is assigned to multiple categories, commas separate them.
`comments_popup_link('No Comments', 'Comment (1)', 'Comments(%)');`	Displays a link to the comments, along with the comment count for the post in parentheses. (If no comments exist, the tag displays a *No Comments* message.)
`next_posts_link('« Previous Entries')`	Displays the words *Previous Entries* linked to the preceding page of post entries.
`previous_posts_link('Next Entries »')`	Displays the words *Next Entries* linked to the next page of post entries.

Putting It All Together

Template files can't do a whole lot by themselves. The real power comes when they're put together.

Connecting the templates

WordPress has built-in functions to include the main template files (such as `header.php`, `sidebar.php`, and `footer.php`) in other templates. An `include` function is a custom PHP function that is built into WordPress, allowing you to retrieve the content of another template file and display it along with the content of another template file. Table 3-4 shows the templates and the functions you use to include them.

TABLE 3-4 **Template Files and Include Functions**

Template Name	Include Function
header.php	<?php get_header(); ?>
sidebar.php	<?php get_sidebar(); ?>
footer.php	<?php get_footer(); ?>
search.php	<?php get_search_form(); ?>
comments.php	<?php comments_template(); ?>

If you want to include a file that doesn't have a built-in include function, you need a different piece of code. To add a unique sidebar to a certain page template, for example, you could name the sidebar file sidebar-page.php. To include that sidebar in another template, you use the following code:

```
<?php get_template_part('sidebar', 'page'); ?>
```

In this statement, the PHP get_template_part function looks through the main theme folder for the sidebar_page.php file and displays the sidebar.

In this section, you put together the guts of a basic Main Index template by using the information on templates and tags provided so far in this chapter. There seem to be endless lines of code when you view the index.php template file in the Twenty Seventeen theme, so I've simplified it for you with the following steps. These steps should give you a basic understanding of The Loop and common template tags and functions that you can use to create your own.

You create a new WordPress theme by using some of the basic WordPress templates. The first steps in pulling everything together are as follows:

1. **Connect to your web server via SFTP, click the wp-content folder, and then click the themes folder.**

   This folder contains the themes that are currently installed on your WordPress site. (See Book 2 if you need more information on SFTP.)

2. **Create a new folder, and call it mytheme.**

   In most SFTP programs, you can right-click and choose New Folder from the shortcut menu. (If you aren't sure how to create a folder, refer to your SFTP program's help files.)

3. **In your favored text editor (such as Notepad for the PC or TextEdit for the Mac), create and save the following files with the lines of code I've provided for each:**

   - *Header template:* Create the file with the following lines of code and then save with the filename `header.php`:

```
<!DOCTYPE html>
  <html <?php language_attributes(); ?> class="no-js">
   <head>
     <meta charset="<?php bloginfo( 'charset' ); ?>">
    <link rel="stylesheet" type="text/css" media="all" href="<?php
    bloginfo( 'stylesheet_url' ); ?>"/>

     <?php wp_head(); ?>
   </head>
   <body <?php body_class() ?>>
     <header class="masthead">
       <h1><a href="<?php bloginfo( 'url' ); ?>"><?php bloginfo(  'name' );
       ?></a></h1>
       <h2><?php bloginfo( 'description' ); ?></h2>
     </header>
     <div id="main">
```

   - *Theme Functions:* Create the file with the following lines of code and then save it using the filename `functions.php`:

```
<?php
add_theme_support( 'title-tag' );

if ( function_exists( 'register_sidebar' ) ) register_sidebar( array(
'name'=>'Sidebar',
));
?>
```

     The Theme Functions file registers the widget area for your site so that you're able to add widgets to your sidebar by using the WordPress widgets available on the Widgets page of the Dashboard.

   - *Sidebar template:* Create the file with the following lines of code and then save it with the filename `sidebar.php`:

```
<aside class="sidebar">
<ul>
<?php if ( !function_exists( 'dynamic_sidebar ') || !dynamic_sidebar(
'Sidebar' ) ) : ?>
```

```
<?php endif; ?>
</ul>
</aside>
```

The code tells WordPress where you want the WordPress widgets to display in your theme. In this case, widgets are displayed on the sidebar of your site.

- *Footer template:* Create the file with the following lines of code and then save with the filename `footer.php`:

```
<footer>
<p>&copy; Copyright <a href="<?php bloginfo( 'url' ); ?>"><?php
bloginfo( 'name' ); ?></a>. All Rights Reserved</p>
</footer>
<?php wp_footer(); ?>
</body>
</html>
```

- *Stylesheet:* Create the file with the following lines of code and then save it with the filename `style.css`:

```
/*
Theme Name: My Theme
Description: Basic Theme from WordPress For Dummies example
Author: Lisa Sabin-Wilson
Author URI: http://lisasabin-wilson.com
*/

body {
font-family: verdana, arial, helvetica, sans-serif;
font-size: 16px;
color: #555;
background: #ffffff;
}

header.masthead {
width: 950px;
margin: 0 auto;
background: black;
color: white;
padding: 5px;
text-align:center;
}
header.masthead h1 a {
color: white;
font-size: 28px;
```

```
font-family: Georgia;
text-decoration: none;
}

header.masthead h2 {
font-size: 16px;
font-family: Georgia;
color: #eee;
}

header.masthead nav {
background: #ffffff;
text-align: left;
height: 25px;
padding: 4px;
}


header.masthead nav ul {
list-style:none;
margin:0;
}

#main {
width: 950px;
margin: 0 auto;
padding: 20px ;
}

#main section {
width: 500px;
float:left;
}

#main .hentry {
margin: 10px 0;
}

aside.sidebar {
width: 290px;
margin: 0 15px;
float:right;
}
```

```
aside.sidebar ul {
list-style:none;
}

footer {
clear:both;
width: 960px;
height: 50px;
background: black;
color: white;
margin: 0 auto;
}

footer p {
text-align:center;
padding: 15px 0;
-}

footer a {
color:white;
}
```

(More CSS is covered in Chapter 4 of this book; this example gives you just some *very* basic styling to create your sample theme.)

Using the tags provided in Table 3-3, along with the information on The Loop and the calls to the Header, Sidebar, and Footer templates provided in earlier sections, you can follow the next steps for a bare-bones example of what the Main Index template looks like when you put the tags together.

WARNING

When typing templates, be sure to use a text editor such as Notepad or TextEdit. Using a word processing program such as Microsoft Word opens a whole slew of problems in your code. Word processing programs insert hidden characters and format quotation marks in a way that WordPress can't read.

Now that you have the basic theme foundation, the last template file you need to create is the Main Index template. To create a Main Index template to work with the other templates in your WordPress theme, open a new window in a text-editor program and then follow these steps. (Type the text in each of these steps on its own line. Press the Enter key after typing each line so that each tag starts on a new line.)

1. **Type** `<?php get_header(); ?>`.

   This template tag pulls the information in the Header template of your WordPress theme.

2. **Type** <section>.

   This tag is HTML markup that tells the browser that it's a grouping of content (in this case, blog posts).

3. **Type** <?php if (have_posts()) : ?>.

   This template tag is an if statement that asks, "Does this blog have posts?" If the answer is yes, the tag grabs the post content information from your MySQL database and displays the posts in your blog.

4. **Type** <?php while (have_posts()) : the_post(); ?>.

   This template tag starts The Loop.

5. **Type** <article <?php post_class() ?> id="post-<?php the_ID(); ?>">.

   This tag is HTML markup that tells the browser that it's the start of a new, single article, along with the post_class CSS designation.

6. **Type** <h1><a href="<?php the_permalink(); ?>"><?php the_title(); ?></a></h1>.

   This tag tells your blog to display the title of a post that's clickable (linked) to the URL of the post, surrounded by HTML Header tags.

7. **Type** Posted on <?php the_date(); ?> at <?php the_time(); ?>.

   This template tag displays the date and time when the post was made. With these template tags, the date and time format are determined by the format you set on the Dashboard.

8. **Type** Posted in <?php the_category(','); ?>.

   This template tag displays a comma-separated list of the categories to which you've assigned the post, such as *Posted in: category 1, category 2*.

9. **Type** <?php the_content('Read More..'); ?>.

   This template tag displays the actual content of the blog post. The 'Read More..' portion of this tag tells WordPress to display the words *ReadMore,* which are clickable (hyperlinked) to the post's permalink, where the reader can read the rest of the post in its entirety. This tag applies when you're displaying a post excerpt, as determined by the actual post configuration on the Dashboard.

10. **Type** Posted by: <?php the_author(); ?>.

    This template tag displays the author of the post in this manner: *Posted by: Lisa Sabin-Wilson.*

11. **Type** </article>.

    This tag is HTML markup that tells the browser that the article has ended.

12. Type <?php endwhile; ?>.

This template tag ends The Loop and tells WordPress to stop displaying posts here. WordPress knows exactly how many times The Loop needs to work, based on the setting on the WordPress Dashboard. That's exactly how many times WordPress will execute The Loop.

13. Type <?php next_posts_link('« Previous Entries'); ?>.

This template tag displays a clickable link to the preceding page of entries, if any.

14. Type <?php previous posts link('» Next Entries'); ?>.

This template tag displays a clickable link to the next page of entries, if any.

15. Type <?php else : ?>.

This template tag refers to the if question asked in Step 3. If the answer to that question is no, this step provides the else statement: IF this blog has posts, THEN list them here (Steps 3 and 4), or ELSE display the following message.

16. Type Not Found. Sorry, but you are looking for something that isn't here.

This tag is the message followed by the template tag displayed after the else statement from Step 15. You can reword this statement to have it say whatever you want.

17. Type <?php endif; ?>.

This template tag ends the if statement from Step 3.

18. Type </section>.

This tag is HTML5 markup that closes the <section> tag opened in Step 2 and tells the browser that this grouping of content has ended.

19. Type <?php get_sidebar(); ?>.

This template tag calls in the Sidebar template and pulls that information into the Main Index template.

20. Type </div>.

This tag is HTML markup closing the <div id="main"> that was opened in the header.php file.

21. Type <?php get_footer(); ?>.

This template tag calls in the Footer template and pulls that information into the Main Index template. ***Note:*** The code in the footer.php template ends the <body> and <html> tags that were started in the Header template (header.php).

When you're done, the display of the Main Index template code looks like this:

```
<?php get_header(); ?>
<section>
<?php if (have_posts()) : ?>

<?php while (have_posts()) : the_post(); ?>
 <article <?php post_class() ?> id="post-<?php the_ID(); ?>">
   <h1><a href="<?php the_permalink(); ?>"><?php the_title(); ?></a></h1>
   Posted on: <?php the_date(); ?> at <?php the_time(); ?>
   Posted in: <?php the_category(','); ?>
   <?php the_content('Read More..'); ?>
   Posted by: <?php the_author(); ?>
</article>

<?php endwhile; ?>
<?php next_posts_link( '&laquo; Previous Entries' ) ?>
<?php previous_posts_link( 'Next Entries &raquo;' ) ?>
<?php else : ?>
<p>Not Found
Sorry, but you are looking for something that isn't here.</p>
<?php endif; ?>
</section>
<?php get_sidebar(); ?>
</div>
<?php get_footer(); ?>
```

22. Save this file as `index.php` **, and upload it to the** `mythemes` **folder.**

In Notepad, you can save the file by choosing File➪Save As. Type the name of the file in the File Name text box and then click Save.

23. Activate the theme on the WordPress Dashboard, and view your site to see your handiwork in action!

TIP

My Main Index template code has one template tag that I explain in Chapter 4 of this book; that template tag is `<article <?php post_class() ?> id="post-<?php the_ID(); ?>">`. This tag helps you create some interesting styles in your template by using CSS.

The simple, basic Main Index template that you just built doesn't have the standard HTML markup in it, so you'll find that the visual display of your site differs from the default Twenty Seventeen theme. This example gives you the bare-bones basics of the Main Index template and The Loop in action.

TIP

If you're having a hard time typing the code provided in this section, I've made this sample theme available for download on my website. The `.zip` file contains the files discussed in this chapter so you can compare your efforts with mine, electronically. You can download the theme `.zip` file here: `http://lisasabin-wilson.com/wpfd/my-theme.zip`.

Using additional stylesheets

Often, a theme uses multiple stylesheets for browser compatibility or consistent organization. If you use multiple stylesheets, the process for including them in the template is the same as for any other stylesheet.

To add a new stylesheet, create a directory in the root theme folder called `css`. Next, create a new file called `mystyle.css` within the `css` folder. To include the file, you must edit the `header.php` file. This example shows the code you need to include in the new CSS file:

```
<link rel="stylesheet" href="<?php bloginfo('stylesheet_directory');
?>/css/mystyle.css" type="text/css" media="screen"/>
```

» Customizing your site with CSS

» Modifying your theme with CSS and HTML

» Customizing basic elements for uniqueness

» Exploring additional resources

Chapter 4

Customizing Your Theme

Customizing your WordPress theme's overall look with unique graphics and colors is one of the most fun and exciting aspects of using WordPress themes. You can take one of your favorite, easily customizable themes and personalize it with some simple changes to make it unique. (For more information on finding an existing theme, read Book 6, Chapter 2.)

After you find an existing free (or premium) WordPress theme that suits your needs, the next step is personalizing the theme through some of the following techniques:

- **Plugging in your own graphics:** The easiest way to make a theme your own is to add a graphical header that includes your logo and matching background graphics.
- **Adjusting colors:** You may like the structure and design of your theme but want to adjust the colors to match your own tastes or brand look. You can do this in the CSS (Cascading Style Sheets), too.
- **Adding/changing fonts:** You may want to change the font, or *typography,* on your site by using different font types, sizes, or colors. You can edit these display properties in the CSS.

Often, the customization process is one of trial and error. You have to mix and match different elements, tweaking and tinkering with graphics and CSS until you achieve design perfection. In this chapter, you explore the easiest ways to customize your WordPress theme through graphics and CSS.

Changing Your Background Graphic

Using background graphics is an easy way to set your site apart from others that use the same theme. Finding a background graphic for your site is much like finding just the right desktop background for your computer. You can choose among a variety of background graphics for your site, such as photography, abstract art, and repeatable patterns.

You can find ideas for new and different background graphics by checking out some of the CSS galleries on the web, such as `www.cssdrive.com` and `http://csselite.com`. Sites like these should be used only for inspiration, not theft. Be careful when using images from outside sources.

WARNING

You want to use only graphics and images that you have been given the right (through express permission or licenses that allow you to reuse) to use on your site). For this reason, always purchase graphics from reputable sources, such as these three online graphic sites:

- **iStockphoto (`www.istockphoto.com`):** iStockphoto has an extensive library of stock photography, vector illustrations, video and audio clips, and Adobe Flash media. You can sign up for an account and search libraries of image files to find the one that suits you or your client best. The files that you use from iStockphoto aren't free; you do have to pay for them. Be sure that you read the license for each image you use. The site has several licenses. The cheapest one is the Standard License, which has some limitations. You can use an illustration from iStockphoto in one website design, for example, but you can't use that same illustration in a theme design that you intend to sell multiple times (say, in a premium theme marketplace). Be sure to read the fine print!
- **Dreamstime (`https://www.dreamstime.com`):** Dreamstime is a major supplier of stock photography and digital images. Sign up for an account and search the huge library of digital image offerings. Dreamstime does offer free images at times, so keep your eyes out for those! Also, Dreamstime has different licenses for its image files, and you need to pay close attention to them. One nice feature is the Royalty Free licensing option, which allows you to pay for the image one time and then use the image as many times as you like. You can't redistribute the image in the same website theme repeatedly, however, such as in a template that you sell to the public.

- **GraphicRiver (`https://graphicriver.net`):** GraphicRiver offers stock graphic files such as Adobe Photoshop images, design templates, textures, vector graphics, and icons, to name just a few. The selection is vast, and the cost to download and use the graphic files is minimal. As with all graphic and image libraries, be sure to read the terms of use or any licensing attached to each of the files to make sure you're legally abiding by the terms.

TIP

Another great resource for free graphics and more is *Smashing Magazine* at `https://www.smashingmagazine.com` (click the Freebies link under Graphics in the left navigation menu). You can find hundreds of links and resources to free and often reusable graphics, such as textures and wallpapers for your site.

To best use background graphics, you must answer a few simple questions:

- **What type of background graphic do you want to use?** Do you want a repeatable pattern or texture, for example, or a black-and-white photograph of something in your business?
- **How do you want the background graphic to display in your browser?** Do you want to tile or repeat your background image in the browser window or pin it to a certain position no matter what size your guest's browser is?

The answers to those questions determine how you install a background graphic in your theme design.

REMEMBER

When working with graphics on the web, use GIF, JPG, or PNG image formats. For images with a small number of colors (such as charts, line art, and logos), GIF format works best. For other image types (screen shots with text and images, blended transparency, and so on), use JPG (also called JPEG) or PNG.

For web design, the characteristics of each image file format can help you decide which file format you need to use for your site. The most common image file formats and characteristics include the following:

- **JPG:** Suited for use with photographs and smaller images used in your web design projects. Although the JPG format compresses with lossy compression, you can adjust compression when you save a file in JPG a format. That is, you can choose the degree, or amount, of compression that occurs, from 1 to 100. Usually, you won't see a great deal of image quality loss with compression levels 1 through 20.
- **PNG:** Suited for larger graphics used in web design, such as the logo or main header graphic that helps brand the overall look of the website. A .png file uses lossless image compression; therefore, no data loss occurs during compression, so you get a cleaner, sharper image. You can also create and

save a .png file on a transparent canvas; .jpg files must have a white canvas or some other color that you designate.

» **GIF:** Compression of a .gif file is lossless; therefore, the image renders exactly the way you design it, without loss of quality. These files compress with lossless quality when the image uses 256 colors or fewer, however. For images that use more colors (higher quality), GIF isn't the greatest format to use. For images with a lot of colors, go with PNG format instead.

Uploading an image for background use

If you want to change the background graphic in your theme, follow these steps:

1. **Upload your new background graphic via SFTP to the images folder in your theme directory.**

   Typically, the images folder is at wp-content/themes/themename/images.

2. **On the WordPress Dashboard, click the Editor link on the Appearance menu.**

   The Edit Themes screen displays.

3. **Click the Stylesheet (style.css) link on the right side of the page.**

   The style.css template opens in the text-editor box on the left side of the Edit Themes screen.

4. **Scroll down to find the body CSS selector.**

   I discuss CSS selectors later in this chapter. The following code segment is a sample CSS snippet you can use to define the background color of your site:

```
body {
    background: #f1f1f1;
}
```

5. **(Optional) Modify the background property values from the code in Step 4.**

   Change

```
background: #f1f1f1;
```

   to

```
background #FFFFFF url('images/newbackground.gif');
```

   With this example, you added a new background image (newbackground.gif) to the existing code and changed the color code to white (#FFFFFF).

6. **Click the Update File button to save the stylesheet changes you made.**

   Your changes are saved and applied to your theme.

Positioning, repeating, and attaching images

After you upload a background graphic, you can use CSS background properties to position it how you want it. The main CSS properties — `background-position`, `background-repeat`, and `background-attachment` — help you achieve the desired effect. Table 4-1 describes the CSS background properties and the available values for changing them in your theme stylesheet. If you're a visual person, you'll enjoy testing and tweaking values to see the effects on your site.

TABLE 4-1 CSS Background Properties

Property	Description	Values	Example

`background-position`	Determines the starting point of your background image on your web page	`bottom center`
		`bottom right`
		`left center`
		`right center`
		`center center`
			`background-position: bottom center;`

`background-repeat`	Determines whether your background image will repeat or tile	`repeat` (repeats infinitely) `repeat-y` (repeats vertically) `repeat-x` (repeats horizontally) `no-repeat` (doesn't repeat)	`background-repeat: repeat-y;`

`background-attachment`
`background-origin`
`background-clip`
	Determines whether your background image is fixed or scrolls with the browser window
	Specifies the positioning area of the background images
	Specifies the painting area of the background images
		fixed
		scroll
		padding-box
		border-box
		content-box
		initial
		inherit
		border-box
		padding-box
		content-box
		initial
		inherit
			`background-attachment: scroll;`
			`background-origin: content-box;`
			`background-clip: padding-box;`

Suppose that your goal is to *tile,* or repeat, the background image so that it scales with the width of the browser on any computer. To achieve this, open the stylesheet again, and change

```
background: #f1f1f1;
```

to

```
background: #FFFFFF;
background-image: url(images/newbackground.gif);
background-repeat: repeat;
```

If your goal is to display a fixed image that doesn't scroll or move when your site visitor moves the browser, you can use the `background-position`, `background-repeat`, and `background-attachment` properties to display it exactly how you want it to appear.

You can tile (repeat) an image so that it covers the entire background space no matter how wide or long the website display is. To achieve this look, change `background: #f1f1f1` in your stylesheet to

```
background: #FFFFFF;
background-image: url(images/newbackground.gif);
background-position: top left;
background-repeat: repeat;
```

TIP

As you become more comfortable with CSS properties, you can start using shortening methods to make your CSS coding practice more efficient. The preceding block of code, for example, looks like this with CSS shortcode practice:

```
background: #fff url(images/newbackground.gif) repeat top left;
```

As you can see from these examples, changing the background graphic by using CSS involves options that depend on your creativity and design style more than anything else. But when you leverage these options properly, you can use CSS to take your design to the next level.

Changing Your Header Graphic

Creating a header graphic is one of the fastest ways to personalize a site and make it unique. The header graphic typically is the strongest graphic design element. Positioned at the top of your theme, a header graphic often includes a logo or other information about your site or business.

Here are some elements you might include in your header graphic:

- **Business name or logo:** This sounds obvious, but the header graphic is the prime way to identify the site. If you don't have a logo, you can simply stylize your business name for your header graphic, but your brand identity needs to be prominent and polished in the header graphic.
- **Profile photos:** If it's for a blog or an independent professional's site (say, for a real estate agent), you may want to include a studio-quality profile photo of the person to help your site guests know whom they're dealing with and to add a touch of warmth.
- **Taglines, important slogans, and keywords:** Use the header area to tell your visitors something about your site or business.
- **Contact information:** If you're doing a small-business website, including phone and address information is vital.
- **Background images:** Be creative with the header image behind all this information. Use a pattern or graphic that matches your brand colors and doesn't distract attention from the vital information you want to communicate.

REMEMBER

Most WordPress themes, particularly premium themes, allow you to upload new header graphics over existing ones from the WordPress Dashboard. Sometimes, the ability to upload a new header graphic is called a Custom Header Uploader script or feature. This feature allows you to turn off HTML overlay text and use only graphics for your header, too.

You can personalize your header graphic in the following ways:

- Replace or overwrite the theme's existing header image with an appropriate image of your choosing.
- Use a repeating graphic pattern.

Using a repeating graphic pattern is similar to using a repeating background image, which I discuss in "Positioning, repeating, and attaching images" earlier in this chapter. In the following sections, you replace your existing header image (in the free Quick-Vid theme from iThemes: `https://wordpress.org/themes/quick-vid/`) by using the Custom Header feature of many WordPress themes. Figure 4-1 shows the Quick-Vid theme's default header image.

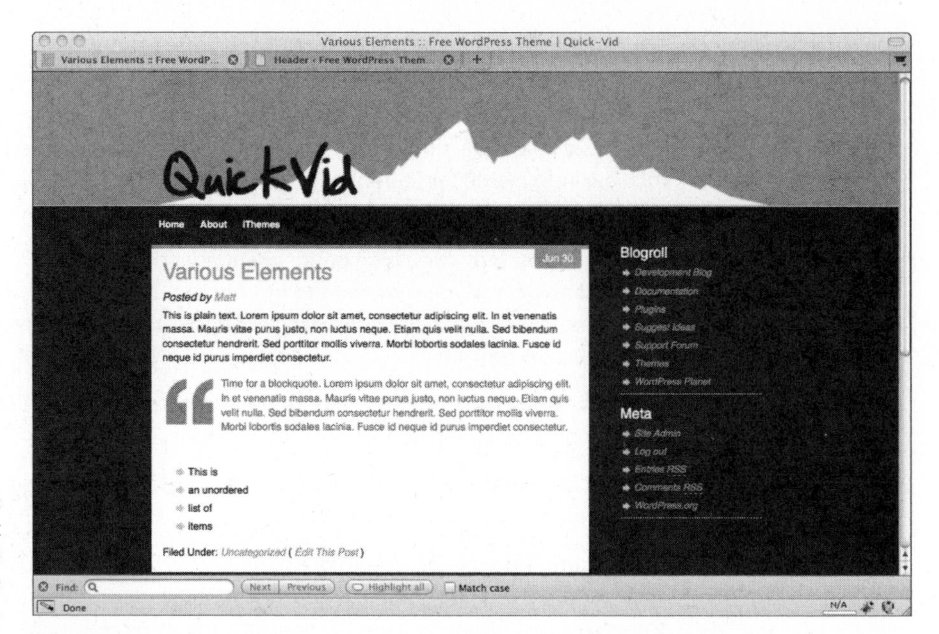

FIGURE 4-1: The default header of the free Quick-Vid theme from iThemes.

Considering the image dimensions

Generally, you want to replace the existing header image with an image that has exactly the same dimensions. To determine the dimensions of the existing image, find the default header graphic, and open it in an image-editing program, such as Photoshop. Create (or crop) your new header graphic to the same dimensions (in pixels) to minimize problems when adding the image to your theme.

Adobe Photoshop Elements is a handy design software tool for basic image editing. It has significantly fewer features than its bigger and older brother, Photoshop, but for most image editing jobs, it does great work for a fraction of the price.

Uploading a header image

Depending on your theme, replacing an existing header image is a fast and efficient way to make changes; you simply upload the graphic and refresh your site.

The Header feature is included in many popular themes. To add a new header graphic in your theme with the Custom Header feature, follow these steps:

1. **On the WordPress Dashboard, click the Header link on the Appearance menu.**

   The Customizing Header Image screen appears, as shown in Figure 4-2. Here, you can adjust your header area, add or remove text, and upload new graphics.

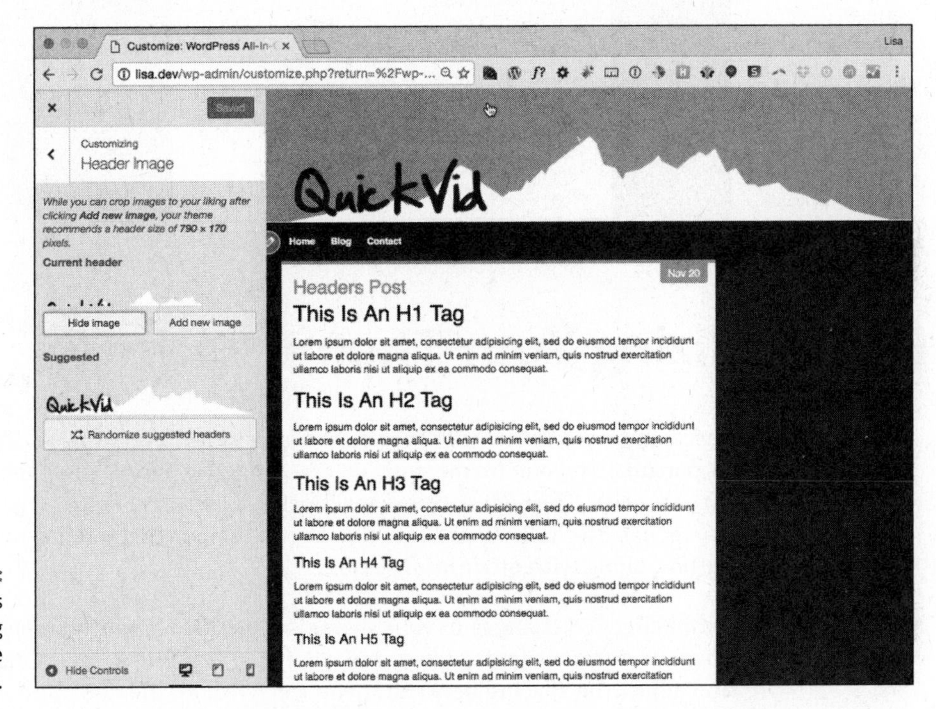

FIGURE 4-2: The WordPress Customizing Header Image screen.

2. **Upload your new header graphic by clicking the Add New Image button in the Current Header section.**

   If your image isn't sized to the specifications given, you'll be asked to crop it to fit.

3. **Click the Save & Publish button to save your changes.**

   The Save & Publish button changes to Saved once your changes have been updated. Figure 4-3 shows a new header graphic.

FIGURE 4-3: The Quick-Vid theme with a new header image.

Personalizing Your Theme with CSS

Cascading Style Sheets (CSS) are part of every WordPress theme. The primary way to personalize your theme with CSS is through your theme's default stylesheet (`style.css`). Through a comment block (shown in Figure 4-4), your theme's `style.css` file tells WordPress the theme name, the version number, and the author, along with other information.

By making CSS changes in your theme's stylesheet, you can apply unique styling (such as different fonts, sizes, and colors) to headlines, text, links, and borders, and adjust the spacing between them, too. With all the CSS options available, you can fine-tune the look and feel of different elements with simple tweaks.

To explore your theme's stylesheet, click the Editor link on the Appearance menu of the WordPress Dashboard. By default, your theme's main stylesheet (`style.css`) should appear. If not, look at the far-right side of the WordPress Dashboard below the Templates heading; scroll down to find the Styles heading; and click the Stylesheet file, shown in Figure 4-5.

WARNING

Making changes in the stylesheet or any other theme file can cause your site to load the theme improperly. Be careful what you change. When you make changes, ensure that you're on a playground or sandbox site so that you can easily restore your original file and don't permanently affect a live or important site. Always save an original copy of the stylesheet in a text program, such as Notepad (for the PC) or TextEdit (for the Mac), so you can find the original CSS and copy and paste it back into your stylesheet if necessary.

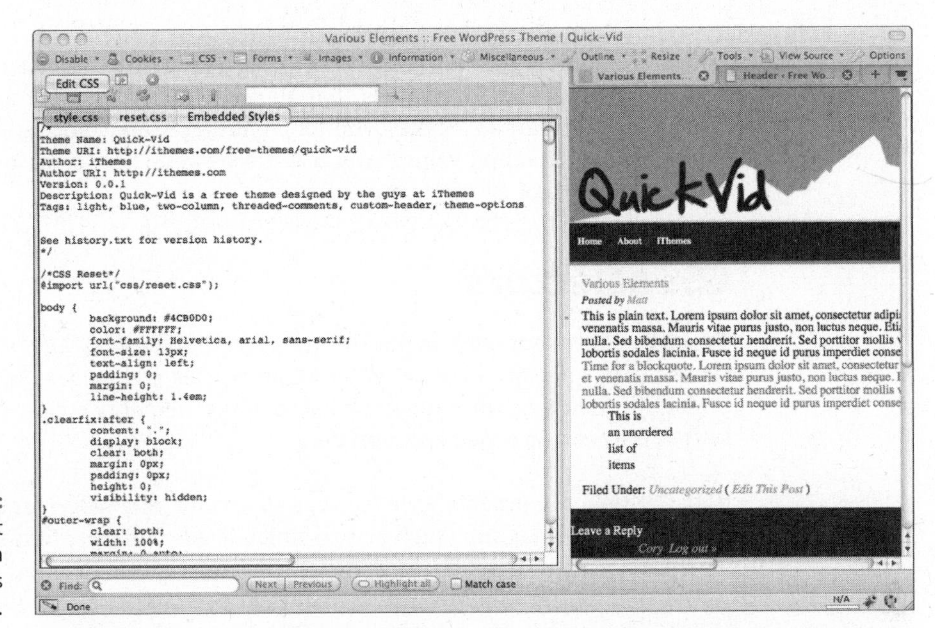

FIGURE 4-4: The comment block of a typical WordPress stylesheet.

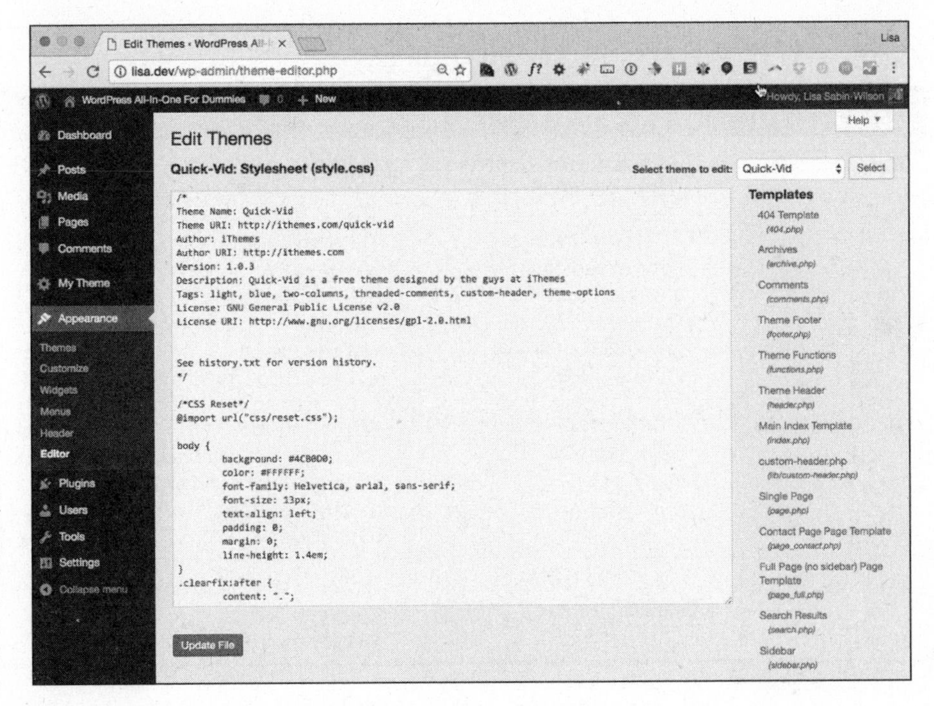

FIGURE 4-5: The list of template files contained in the Quick-Vid theme. Scroll down to see the Styles heading.

Knowing some key CSS concepts can help you personalize your theme's stylesheet. CSS is simply a set of commands that allows you to customize the look and feel of your HTML markup. Some common commands and tools are selectors, IDs, classes, properties, and values. You use these commands to customize HTML to display your design customizations.

CSS selectors

Typically, CSS *selectors* are named for the corresponding HTML elements, IDs, and classes that you want to style with CSS properties and values. Selectors are very important in CSS because they're used to select elements on an HTML/PHP page so that they can be styled appropriately.

With CSS, you can provide style (such as size, color, and placement) to the display of elements on your site (such as text links, header images, font size and colors, paragraph margins, and line spacing). *CSS selectors* contain names, properties, and values that define which HTML elements in the templates are to be styled with CSS. Table 4-2 lists some basic global CSS selectors.

TABLE 4-2 **Basic Global CSS Selectors**

CSS Selector	Description	HTML	CSS Example
body	Contains the elements of the overall site style	<body>	body {font-family: Georgia}
a	Sets the attributes for hyperlinks within your site	<a href=”http://wordpress.org”>WordPress</a>	a {color:blue}
h1, h2, h3, h4, h5, h6	Contains headings or headlines	<h1>My main title</h1>	h1 {color:black}
blockquote	Defines how indented text is styled	<blockquote> "A journey of a thousand miles begins with a single step."</blockquote>	blockquote {font-style: italic}
p	Sets formatting for paragraphs	<p>My first paragraph says to keep writing</p>	p {color: #000}

If you assign a style to the h1 selector, for example, it affects all <h1> tags in your HTML. Sometimes you want this behavior, but at other times, you want to affect a smaller subset of elements.

CSS IDs and classes

With CSS IDs and classes, you can define more elements to style. Generally, you use IDs to style one broader specific element (such as your header section) on your page. Classes style, define, and categorize more specifically grouped items (such as images and text alignment, widgets, or links to posts). The difference between CSS IDs and classes are

- **CSS IDs** are identified with the hash mark (#). `#header`, for example, indicates the header ID. Only one element can be identified with an ID.
- **CSS classes** are identified with a period (`.`); `.alignleft`, for example, indicates aligning an element to the left.

Table 4-3 lists some CSS IDs and classes.

TABLE 4-3 CSS IDs and Classes Examples

CSS IDs and Classes	Description	HTML	CSS Example
`#header`	Identifies the header section of your theme	`<div id="header">`	`#header {background: #000}`
`#footer`	Identifies the footer section of your theme	`<div id="footer">`	`#footer {background: #ccc}`
`.wp-caption-text`	Identifies the WordPress image caption	`<p class="wp-caption-text">This is a caption</p>`	`.wp-caption-text {color: #000}`
`.alignleft`	Identifies the left alignment feature in WordPress	`<img class="alignleft" src="http://yourdomain.com/photo.jpg" />`	`.alignleft {float:left}`

CSS properties and values

CSS properties are assigned to the CSS selector name. You also need to provide values for the CSS properties to define the style elements for the particular CSS selector you're working with.

The body selector that follows defines the overall look of your web page. background is a property and #DDDDDD is its value, and color is a property and #222222 is its value.

```
body {
background: #DDDDDD;
color: #222222;
}
```

REMEMBER

Every CSS property needs to be followed by a colon (:), and each CSS value needs to be followed by a semicolon (;).

Understanding that properties are assigned to selectors, as well as your options for the values, makes CSS a fun playground for personalizing your site. You can experiment with colors, fonts, font sizes, and more to tweak the appearance of your theme.

Understanding Basic HTML Techniques

HTML can help you customize and organize your theme. To understand how HTML and CSS work together, consider this example: If a website were a building, HTML would be the structure (the studs and foundation), and CSS would be the paint.

HTML contains the elements that CSS provides the styles for. All you have to do to apply a CSS style is use the right HTML element. Here's a very basic block of HTML:

```
<body>
<div id="content">
<h1>Headline Goes Here</h1>
<p>This is a sample sentence of body text. <blockquote>The journey
of a thousand miles starts with the first step.</blockquote> I'm
going to continue on this sentence and end it here.</p>
<p>Click <a href="http://corymiller.com">here</a> to visit my
website.</p>
</div>
</body>
```

All HTML elements must have opening and closing tags. Opening tags are contained in less-than (<) and greater-than (>) symbols. Closing tags are the same except that they're preceded by a forward slash (/).

Here is an example:

```
<h1>Headline Goes Here</h1>
```

Note that the HTML elements must be properly nested. In line 4 of the example HTML block, a paragraph tag is opened (`<p>`). Later in that line, a block quote is opened (`<blockquote>`) and nesting inside the paragraph tag. When you're editing this line, you can't end the paragraph (`</p>`) before you end the block quote (`</blockquote>`). Nested elements must close before the elements they're nested within close.

REMEMBER

Proper *tabbing*, or indenting, is important when you're writing HTML, mainly for readability so you can quickly scan code to find what you're looking for. A good rule is that if you didn't close a tag in the preceding line, indent one tab over. This practice allows you to see where each element begins and ends, and it can be very helpful in diagnosing problems.

For more in-depth tutorials on HTML, see the HTML section of w3schools.com at `http://www.w3schools.com/html/default.asp`.

Changing Basic Elements for a Unique Look

When you understand the basic concepts of personalizing your site with graphics and CSS, you begin to see how easy changing the look and feel of your site is with these tools. The next few sections explore some ways to accomplish an interesting design presentation or a unique and creative look.

Background colors and images

Changing the background image can completely change the feel of your site. You can also use background colors and images for other elements in your theme.

Background techniques include using solid colors and repeating gradients or patterns to achieve a subtle yet polished effect. (***Note:*** Use colors that accent the colors of your logo and don't hamper text readability.)

You can add CSS background colors and image effects to the following areas of your theme:

- Post and page content sections
- Sidebar widgets
- Comment blocks
- Footer area

Font family, color, and size

You can change the fonts in your theme for style or for readability purposes. Typographic (or font) design experts use simple font variations to achieve amazing design results. You can use fonts to separate headlines from body text (or widget headlines and text from the main content) to be less distracting. Table 4-4 lists some examples of often-used font properties.

TABLE 4-4 Fonts

Font Properties	Common Values	CSS Examples
`font-family`	`Georgia, Times, serif`	`body {font-family: Georgia; serif;}`
`font-size`	`px, %, em`	`body {font-size: 14px;}`
`font-style`	`Italic, underline`	`body {font-style: italic;}`
`font-weight`	`bold, bolder, normal`	`body {font-weight: normal}`

The web is actually kind of picky about how it displays fonts, as well as what kind of fonts you can use in the `font-family` property. Not all fonts display correctly. To be safe, here are some commonly used font families that display correctly in most browsers:

- **Serif fonts:** Times New Roman, Georgia, Garamond, Bookman Old Style
- **Sans-serif fonts:** Verdana, Arial, Tahoma, Trebuchet MS

REMEMBER

Serif fonts have little tails, or curlicues, at the edges of letters. (This book's text is in a serif font.) Sans-serif fonts have straight edges and no fancy styling. (The heading in Table 4-4 uses a sans-serif font. Look, Ma, no tails!)

Font color

With more than 16 million HTML color combinations available, you can find just the right shade of color for your project. After some time, you'll memorize your favorite color codes. Knowing codes for different shades of gray can help you quickly add an extra design touch. You can use the shades of gray listed in Table 4-5 for backgrounds, borders of design elements, and widget headers.

TABLE 4-5 **My Favorite CSS Colors**

Color	Value
White	#FFFFFF or #FFF
Black	#000000 or #000
Gray	#CCCCCC or #CCC #DDDDDD or #DDD #333333 or #333 #E0E0E0

You can easily change the color of your font by changing the `color` property of the CSS selector you want to tweak. You can use hexadecimal codes to define the colors.

You can define the overall font color in your site by defining it in the body CSS selector, like this:

```
body {
color: #333;
}
```

Font size

To tweak the size of your font, change the `font-size` property of the CSS selector you want to tweak. Generally, the following units of measurement determine font sizes:

- **px (pixel):** Increasing or decreasing the number of pixels increases or decreases the font size. `12px` is larger than `10px`.
- **pt (point):** As with pixels, increasing or decreasing the number of points affects the font size. `12pt` is larger than `10pt`.

» `em`: An em is a scalable unit of measurement that's equal to the current font size. If the font size of the body of the site is defined as `12px`, `1em` is equal to `12px`; likewise, `2em` is equal to `24px`.

» **% (percentage):** Increasing or decreasing the percentage number affects the font size. `50%` is equivalent to 7 pixels, and `100%` is equivalent to 14 pixels.

In the default template CSS, the font size is defined in the `<body>` tag in pixels, like this:

```
font-size: 12px;
```

Putting all three elements (`font-family`, `color`, and `font-size`) together in the `<body>` tag styles the font for the entire body of your site. Here's how the elements work together in the `<body>` tag of the default template CSS:

```
body {
font-size: 12px;
font-family: Georgia, "Bitstream Charter", serif;
color: #666;
}
```

When you want to change a font family in your CSS, open the stylesheet (`style.css`), search for `property: font-family`, change the values for that property, and then save your changes.

In the default template CSS, the font is defined in the `<body>` tag, like this:

```
font-family: Georgia, "Bitstream Charter", serif;
```

Borders

CSS borders can add flair to elements of your theme design. Table 4-6 illustrates common properties and CSS examples for borders in your theme design.

TABLE 4-6 Common Border Properties

Border Properties	Common Values	CSS Examples
`border-size`	px, em	`body {border-size: 1px;}`
`border-style`	solid, dotted, dashed	`body {border-style: solid}`
`border-color`	Hexadecimal values	`body {border-color: #CCCCCC}`

Finding Additional Resources

There may come a time when you want to explore customizing your theme further. Here are some recommended resources:

- **WordPress Codex (`https://codex.wordpress.org`):** Official WordPress documentation
- **W3Schools (`www.w3schools.com`):** Free, comprehensive HTML and CSS reference
- ***Smashing Magazine* (`https://www.smashingmagazine.com`):** Numerous tips and tricks for customizing a WordPress theme

IN THIS CHAPTER

» Defining the relationship between parent and child themes

» Tweaking child themes with styles

» Customizing child themes with images

» Modifying child themes with template files

Chapter 5

Understanding Parent and Child Themes

Using a theme exactly how the theme author released it is great. If a new version is released that fixes a browser compatibility issue or adds features offered by a new version of WordPress, a quick theme upgrade is very easy to do.

There's a good chance, however, that you'll want to tinker with the design, add features, or modify the theme structure. If you modify the theme, you won't be able to upgrade to a newly released version without modifying the theme again.

If only you could upgrade customized versions of themes with new features when they're released! Fortunately, child themes give you this best-of-both-worlds theme solution.

This chapter explores what child themes are, how to create a child theme–ready parent theme, and how to get the most out of using child themes.

Customizing Theme Style with Child Themes

A WordPress theme consists of a collection of template files, stylesheets, images, and JavaScript files. The theme controls the layout and design that your visitors see on the site. When such a theme is properly set up as a parent theme, it allows a *child theme*, or a subset of instructions, to override its files. This ensures that a child theme can selectively modify the layout, styling, and functionality of the parent theme.

The quickest way to understand child themes is by example. In this section, you create a simple child theme that modifies the style of the parent theme.

Currently, the default WordPress theme is Twenty Seventeen. Figure 5-1 shows how the Twenty Seventeen theme appears on a sample site.

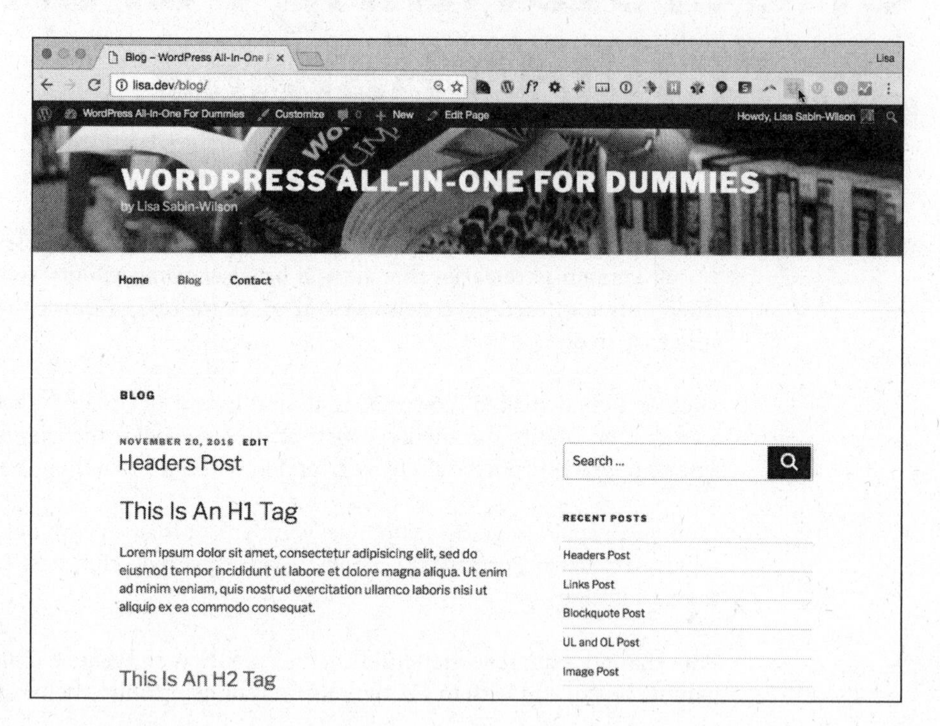

FIGURE 5-1: The Twenty Seventeen theme.

You likely have Twenty Seventeen on your WordPress site, and Twenty Seventeen is child theme–ready; therefore, it's a great candidate for creating a child theme.

Creating a child theme

Like regular themes, a child theme needs to reside in a directory inside the `/wp-content/themes` directory. The first step in creating a child theme is adding the directory that will hold it. For this example, connect to your hosting account via SFTP, and create a new directory called `twentyseventeen-child` inside the `/wp-content/themes` directory.

To register the `twentyseventeen-child` directory as a theme and to make it a child of the Twenty Seventeen theme, create a `style.css` file, and add the appropriate theme headers. To do this, type the following code in your favorite code or plain-text editor (such as Notepad for the PC or TextEdit for the Mac), and save the file as `style.css`:

```
/*
Theme Name: Twenty Seventeen Child
Description: My magnificent child theme
Author: Lisa Sabin-Wilson
Version: 1.0
Template: twentyseventeen
*/
```

Typically, you find the following headers in a WordPress theme:

- **`Theme Name`:** The theme user sees this name in the back end of WordPress.
- **`Description`:** This header provides the user any additional information about the theme. Currently, it appears only on the Manage Themes page (accessed by clicking the Themes link on the Appearance menu).
- **`Author`:** This header lists one or more theme authors. Currently, it appears only on the Manage Themes page (accessed by clicking the Themes link on the Appearance menu).
- **`Version`:** The version number is very useful for keeping track of outdated versions of the theme. It's always a good idea to update the version number when modifying a theme.
- **`Template`:** This header changes a theme into a child theme. The value of this header tells WordPress the directory name of the parent theme. Because your child theme uses Twenty Seventeen as the parent, your `style.css` needs to have a `Template` header with a value of `twentyseventeen` (the directory/folder name of the Twenty Seventeen theme).

Now activate the new Twenty Seventeen Child theme as your active theme. (If you need a reminder on how to activate a theme on your site, check out Book 6, Chapter 2.) You should see a site layout similar to the one shown in Figure 5-2.

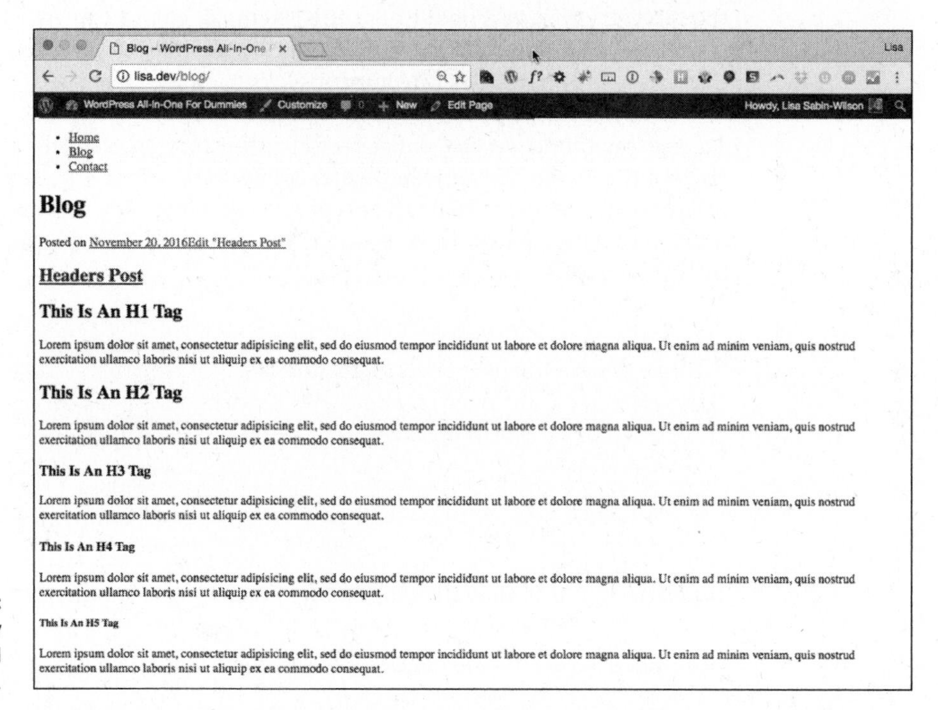

FIGURE 5-2: The Twenty Seventeen Child theme.

Figure 5-2 shows that the new theme doesn't look quite right. The problem is that the new child theme replaced the `style.css` file of the parent theme, yet the new child theme's `style.css` file is empty.

You could just copy and paste the contents of the parent theme's `style.css` file, but that would waste some of the potential of child themes.

Loading a parent theme's style

REMEMBER

One of the great things about CSS is how rules can override one another. If you list the same rule twice in your CSS, the rule that comes last takes precedence.

For example:

```
a {
color: blue;
}

a {
color: red;
}
```

This example is overly simple, but it nicely shows what I'm talking about. The first rule says that all links (`a` tags) should be blue, whereas the second rule says that links should be red. With CSS, the last instruction takes precedence, so the links will be red.

Using this feature of CSS, you can inherit all the styling of the parent theme and selectively modify it by overriding the rules of the parent theme. But how can you load the parent theme's `style.css` file so that it inherits the parent theme's styling?

Fortunately, CSS has another great feature that helps you do this with ease. Just add one line to the Twenty Seventeen Child theme's `style.css` file (in bold in the following example):

```
/*
Theme Name: Twenty Seventeen Child
Description: My magnificent child theme
Author: Lisa Sabin-Wilson
Version: 1.0
Template: twentyseventeen
*/
@import url('../twentyseventeen/style.css');
```

Several things are going on here, so let me break the code down piece by piece:

- **`@import`:** This code tells the browser to load another stylesheet. Using this code allows you to pull in the parent stylesheet quickly and easily.
- **`url`:** This code indicates that the value is a location, not a normal value.
- **`('../twentyseventeen/style.css');`:** This code is the location of the parent stylesheet. Notice the `/twentyseventeen` directory name. This name needs to be changed to match the `Template` value in the header so that the appropriate stylesheet is loaded.

Figure 5-3 shows how the site appears after the child theme's `style.css` file is updated to match the listing.

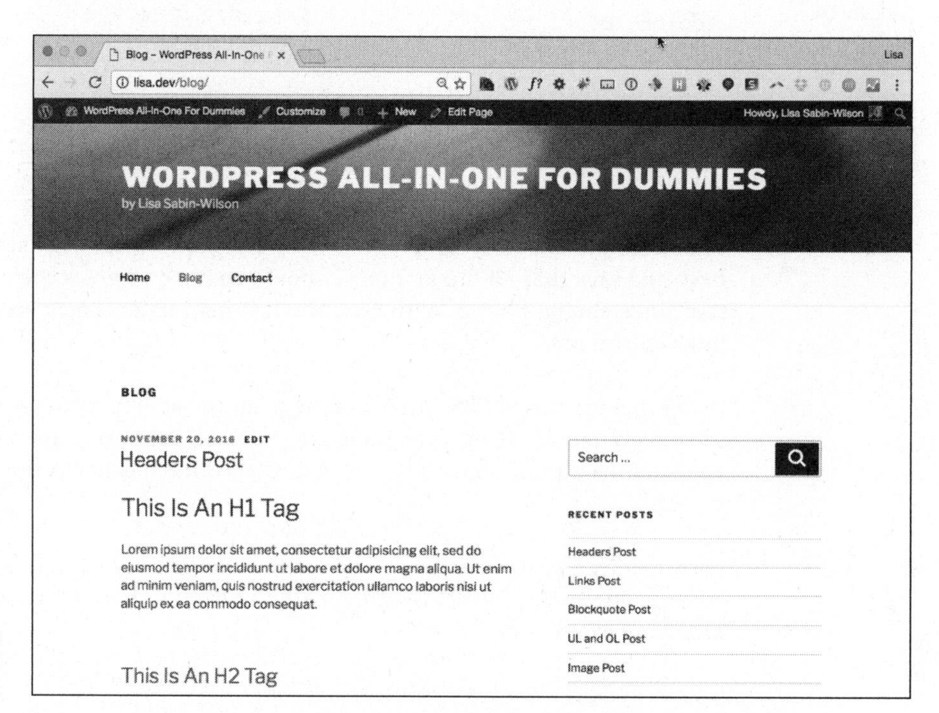

FIGURE 5-3: The updated child theme.

Customizing the parent theme's styling

Your Twenty Seventeen Child theme is set up to match the parent Twenty Seventeen theme. Now you can add new styling to the Twenty Seventeen Child theme's `style.css` file. To see a simple example of how customizing works, add a style that converts all h1, h2, and h3 headings to uppercase, like so:

```
/*
Theme Name: Twenty Seventeen Child
Description: My magnificent child theme
Author: Lisa Sabin-Wilson
Version: 1.0
Template: twentyseventeen
*/
@import url('../twentyseventeen/style.css');

h1, h2, h3 {
text-transform: uppercase;
}
```

Figure 5-4 shows how the child theme looks with the code additions applied. Getting better, isn't it? (*Hint:* The site title and post titles are all uppercase, unlike in Figure 5-3.)

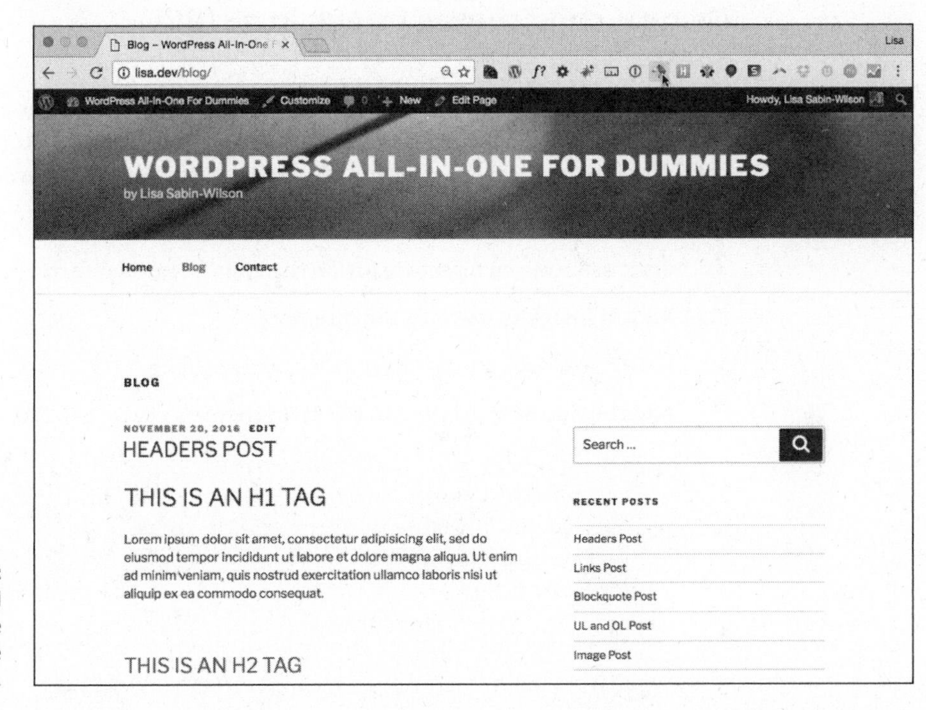

FIGURE 5-4: The updated child theme with uppercase headings.

As you can see, with just a few lines in a `style.css` file, you can create a new child theme that adds specific customizations to an existing theme. The change was quick and easy to make, and you didn't have to modify anything in the parent theme to make it work.

TIP

When upgrades to the parent theme are available, you can upgrade the parent to get the additional features, but you don't have to make your modifications again because you made your modifications in the child theme, not the parent theme.

Customizations that are more complex work the same way. Simply add the new rules after the import rule that adds the parent stylesheet.

Using images in child theme designs

Many themes use images to add nice touches to the design. Typically, these images are added to a directory named `images` inside the theme.

Just as a parent theme may refer to images in its `style.css` file, your child themes can have their own images directory. The following sections show how you can use these images.

Using a child theme image in a child theme stylesheet

Including a child theme image in a child theme stylesheet is common. To do so, you simply add the new image to the child theme's `images` directory and refer to it in the child theme's `style.css` file. To get a feel for the mechanics of this process, follow these steps:

1. **Create an `images` directory inside the child theme's directory.**
2. **Add an image to use into the directory.**

   For this example, add an image called body-bg.png.
3. **Add the necessary styling to the child theme's `style.css` file, as follows:**

```
/*
Theme Name: Twenty Seventeen Child
Description: My magnificent child theme
Author: Lisa Sabin-Wilson
Version: 1.0
Template: twentyseventeen
*/
@import url('../twentyseventeen/style.css');
body {
  background: url('images/body-bg.jpg');
}
.site-content-contain {
  background-color: transparent;
}
```

With a quick refresh of the site, you see that the site now has a new brick background. Figure 5-5 shows the results clearly.

Using images in a child theme

Child theme images are acceptable for most purposes. You can add your own images to the child theme even if the image doesn't exist in the parent theme folder — and you can accomplish that task without changing the parent theme at all.

In the footer of the Twenty Seventeen Child theme, I added a WordPress logo to the left of the phrase *Proudly powered by WordPress,* as shown in Figure 5-6. The logo doesn't appear in the footer of the Twenty Seventeen theme by default.

Blog – WordPress All-In-One
lisa.dev/blog/
WordPress All-In-One For Dummies
Customize
New
Edit Page
Howdy, Lisa Sabin-Wilson
WORDPRESS ALL-IN-ONE FOR DUMMIES
by Lisa Sabin-Wilson
Home
Blog
Contact
BLOG
NOVEMBER 20, 2016 EDIT
HEADERS POST
THIS IS AN H1 TAG
Lorem ipsum dolor sit amet, consectetur adipisicing elit, sed do eiusmod tempor incididunt ut labore et dolore magna aliqua. Ut enim ad minim veniam, quis nostrud exercitation ullamco laboris nisi ut aliquip ex ea commodo consequat.
Search ...
RECENT POSTS
Headers Post
Links Post
Blockquote Post
UL and OL Post

FIGURE 5-5: The Twenty Seventeen Child theme after the background image is edited.

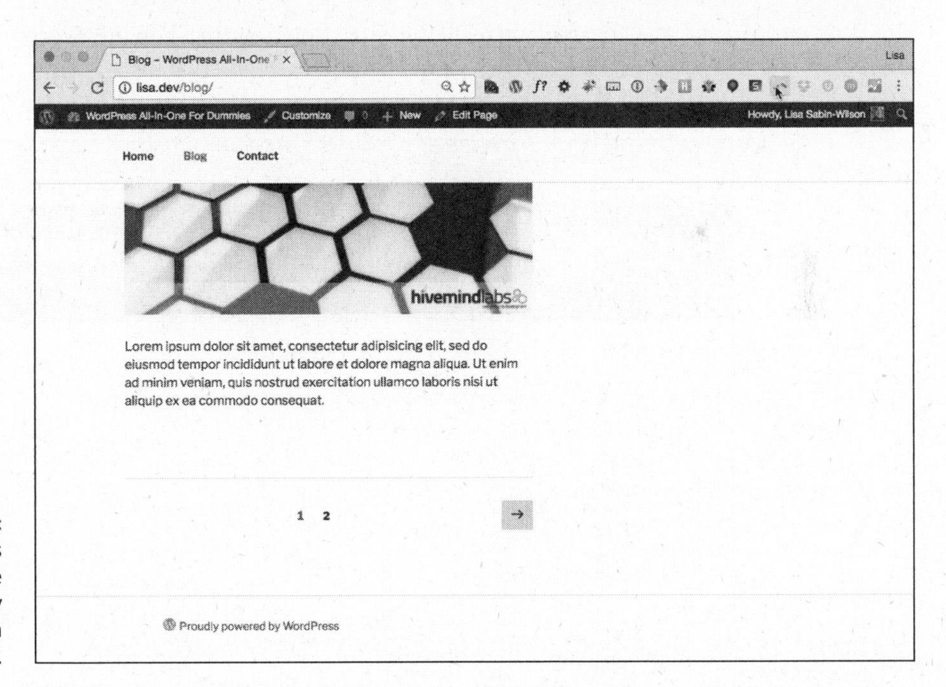

FIGURE 5-6: The WordPress logo in the Twenty Seventeen Child footer.

Create a folder in your child theme called `/images`, and add your selected images to that folder. Then you can call those images into your child theme by using the stylesheet (`style.css`) file in your child theme folder.

In this next example, I add the same WordPress logo in front of each widget title in the sidebar. Because the logo image already exists inside the child theme images folder (from the preceding example), I can simply add a customization to the child theme's `style.css` file to make this change, as follows:

```
/*
Theme Name: Twenty Seventeen Child
Description: My magnificent child theme
Author: Lisa Sabin-Wilson
Version: 1.0
Template: twentyseventeen
*/

@import url('../twentyseventeen/style.css');

.widget-title {
background: url(images/wp-blue.png) no-repeat left top;
padding: 20px 30px;
}
```

Save the file and refresh the site. Now you're showing WordPress pride. (See Figure 5-7.)

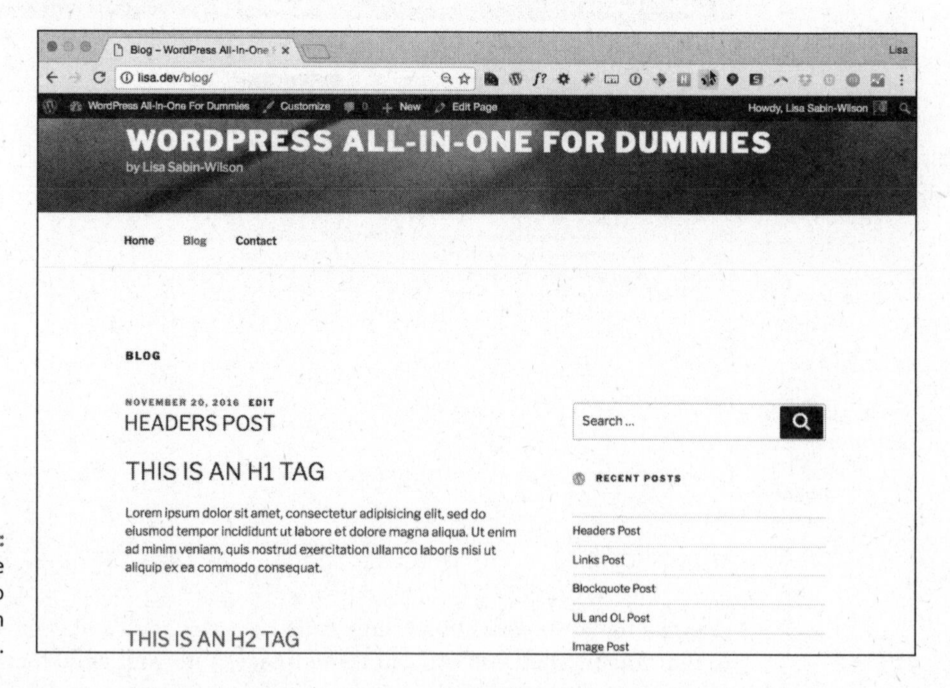

FIGURE 5-7: Showing the WordPress logo before each widget title.

Modifying Theme Structure with Child Themes

The preceding section shows how to use a child theme to modify the stylesheet of an existing theme. This feature is tremendously powerful. A talented CSS developer can use this technique to create an amazing variety of layouts and designs.

This example is just the beginning of the power of child themes. Although every child theme overrides the parent theme's `style.css` file, the child theme can override the parent theme's template files, too. Child themes aren't limited to overriding just template files; when necessary, child themes can also supply their own template files.

Template files are PHP files that WordPress runs to render different views of your site. A *site view* is the type of content being looked at. Examples of views are home, category archive, individual post, and page content.

Some examples of common template files are `index.php`, `archive.php`, `single.php`, `page.php`, `attachment.php`, and `search.php`. (You can read more about available template files, including how to use them, in Book 6, Chapter 3.)

You may wonder what purpose modifying template files of a parent theme serves. Although modifying the stylesheet of a parent theme can give you very powerful control of the design, it can't add new content, modify the underlying site structure, or change how the theme functions. To get that level of control, you need to modify the template files.

Overriding parent template files

When both the child theme and parent theme supply the same template file, the child theme file is used. The process of replacing the original parent template file is referred to as *overriding.*

REMEMBER

Although overriding each of the theme's template files can defeat the purpose of using a child theme (updates to those template files won't enhance the child theme), sometimes, producing a needed result makes doing so necessary.

The easiest way to customize a specific template file in a child theme is to copy the template file from the parent theme folder to the child theme folder. After the file is copied, it can be customized as needed, and the child theme reflects the changes.

A good example of a template file that can be overridden is the `footer.php` file. Customizing the footer allows for adding site-specific branding.

Adding new template files

A child theme can override existing parent template files, but it can supply template files that don't exist in the parent. Although you may never need your child themes to do this, this option can open possibilities for your designs.

This technique proves most valuable with page templates. The Twenty Seventeen theme has a page template named `page.php`. This page template creates a full-width layout for the content and removes the sidebar, as shown in Figure 5-8.

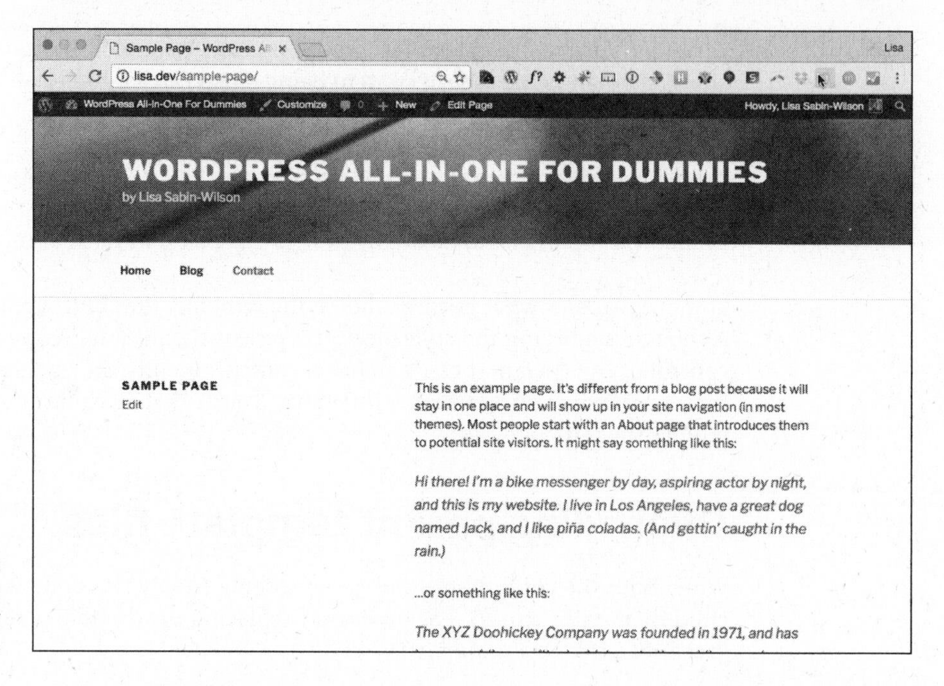

FIGURE 5-8: Page Template in the Twenty Seventeen theme.

The layout was intentionally set up this way to improve readability on pages where you may not want the distraction of other content in a sidebar. Sometimes, I like to have a full-width layout option so that I can embed a video, add a forum, or add other content that works well at full width; at other times, I want a static page that displays the sidebar. If you want to customize that template and override what the Twenty Seventeen theme currently has available, simply create a new page template with the same filename as the one you're replacing (in this case, `page.php`), and add the code to display the sidebar (`get_sidebar();`). Thereafter, WordPress uses the `page.php` template file in your child theme by default, ignoring the one in the Twenty Seventeen parent theme folder.

Removing template files

You may be asking why you'd want to remove a parent's template file. That's a good question. Unfortunately, the Twenty Seventeen theme doesn't provide a good example of why you'd want to do this. Therefore, you must use your imagination a bit.

Imagine that you're creating a child theme built from a parent theme called Example Parent. Example Parent is well designed, and a great child theme was quickly built from it. The child theme looks and works exactly the way you want it to, but there's a problem.

The Example Parent theme has a `home.php` template file that provides a highly customized non-blog home page. However, what you want is a standard blog home page. If the `home.php` file didn't exist in Example Parent, everything would work perfectly.

You can't remove the `home.php` file from Example Parent without modifying the theme, so you have to use a trick. Instead of removing the file, override the `home.php` file, and have it emulate `index.php`.

You may think that simply copying and pasting the Example Parent `index.php` code into the child theme's `home.php` file is a good approach. Although this technique works, there's a better way: You can tell WordPress to run the `index.php` file so that to `index.php` are respected. This single line of code inside the child theme's `home.php` is all that's needed to replace `home.php` with `index.php`:

```
<?php locate_template( array( 'index.php' ), true ); ?>
```

The `locate_template` function does a bit of magic. If the child theme supplies an `index.php` file, that file is used. If not, the parent `index.php` file is used.

This technique produces the same result that removing the parent theme's `home.php` file does. The `home.php` code is ignored, and the changes to `index.php` are respected.

Modifying the functions.php file

Like template files, child themes can provide a Theme Functions template, or `functions.php` file. Unlike template files, the `functions.php` file of a child theme doesn't override the file of the parent theme.

When a parent theme and a child theme each have a `functions.php` file, both the parent and child `functions.php` files run. The child theme's `functions.php` file runs first, and then the parent theme's `functions.php` file runs. This sequence is intentional, because it allows the child theme to replace functions defined in the parent theme.

The Twenty Seventeen `functions.php` file defines a function called `twentyseventeen_setup`. This function handles the configuration of many theme options and activates some additional features. Child themes can replace this function to change the default configuration and features of the theme, too.

The following lines of code summarize how the `functions.php` file allows the child theme to change the default configuration and features of the theme:

```
if ( ! function_exists( 'twentyseventeen_setup' ) ):
function twentyseventeen_setup() {
// removed code
}
endif;
```

Wrapping the function declaration in the `if` statement protects the site from breaking in the event of a code conflict and allows a child theme to define its own version of the function.

In the Twenty Seventeen Child theme, you can see how modifying this function affects the theme. Add a new `twentyseventeen_setup` function that adds post thumbnails support to the Twenty Seventeen Child theme's `functions.php` file, as follows:

```
<?php
function twentyseventeen_setup() {
add_theme_support( 'post-thumbnails' );
}
```

As a result of this change, the child theme no longer supports other special WordPress features, such as custom editor styling, automatic feed link generation, internationalization and location, and so on.

The takeaway from this example is that a child theme can provide its own custom version of the function, because the parent theme wraps the function declaration in an `if` block that checks for the function first.

Preparing a Parent Theme

WordPress makes it very easy for you to make parent themes. WordPress does most of the hard work, but you must follow some rules for a parent theme to function properly.

The words *stylesheet* and *template* are used numerous times in many contexts. Typically, *stylesheet* refers to a CSS file in a theme, and *template* refers to a template file in the theme. But these words also have specific meaning in the context of parent and child themes. You must understand the difference between a stylesheet and a template when working with parent and child themes.

REMEMBER

In WordPress, the active theme is the stylesheet, and the active theme's parent is the template. If the theme doesn't have a parent, the active theme is both the stylesheet and the template.

TECHNICAL STUFF

Originally, child themes could replace only the `style.css` file of a theme. The parent provided all the template files and `functions.php` code. Thus, the child theme provided style, and the parent theme provided the template files. The capabilities of child themes expanded in subsequent versions of WordPress, making the use of these terms for parent and child themes somewhat confusing.

Imagine two themes: Parent and Child. The following code is in the Parent theme's `header.php` file and loads an additional stylesheet provided by the theme:

```
<link type="text/css" rel="stylesheet" media="all" href="<?php
bloginfo('stylesheet_directory') ?>/reset.css"/>
```

The `bloginfo` function prints information about the blog configuration or settings. This example uses the function to print the URL location of the stylesheet directory. The site is hosted at `http://example.com`, and the Parent is the active theme. It produces the following output:

```
<link type="text/css" rel="stylesheet" media="all"
href="http://example.com/wp-content/themes/Parent/reset.css"/>
```

If the Child theme is activated, the output would be

```
<link type="text/css" rel="stylesheet" media="all"
href="http://example.com/wp-content/themes/Child/reset.css"/>
```

The location now refers to the reset.css file in the Child theme. This code could work if every child theme copies the reset.css file of the Parent theme, but requiring child themes to add files to function isn't good design. The solution is simple, however. Instead of using the stylesheet_directory in the bloginfo call, use template_directory. The code looks like this:

```
<link type="text/css" rel="stylesheet" media="all" href="<?php
bloginfo('template_directory') ?>/reset.css"/>
```

Now all child themes properly load the parent reset.css file.

When developing, use template_directory in stand-alone parent themes and stylesheet_directory in child themes.

- » Creating new templates
- » Activating custom menus
- » Exploring custom post types
- » Understanding post formats
- » Using post thumbnails for feature images
- » Building a theme options page

Chapter 6
Digging Into Advanced Theme Development

The previous chapters of this minibook describe WordPress themes and discuss using their structure to build your site. Delving into deeper topics can help you create flexible themes that offer users options for controlling the theme.

Whether you're building a theme for a client, the WordPress.org theme directory, or yourself, adding advanced theme features can make theme development easier and faster with a high-quality result. With these advanced theme concepts and tools, you can build robust, dynamic themes that allow for easier design customization and offer a variety of layout options.

Beyond just tools and methods of advanced theme development, this chapter provides some development practices that help projects succeed.

Getting Started with Advanced Theming

Before themes were added to WordPress, customizing the design of a site meant modifying the main WordPress `index.php` file and the default `print.css` file. Version 1.5 added the first theme support and rudimentary child theme support. Over time, WordPress began to support other features, such as custom headers, custom backgrounds, and featured images.

Additionally, the capabilities of themes have grown steadily. Incremental improvement — beginning with a small, simple starting point and improving it over time — works very well in theme development. By developing incrementally, you can build a theme from start to completion from an existing, well-tested theme (most themes are part of a larger incremental improvement process) and maximize your development time. I can't think of a single theme I've developed that wasn't built on another theme.

TIP

You don't need to develop each theme from scratch. Choosing a good starting point makes a big difference in how quickly you can get your project off the ground.

Finding a good starting point

Choosing a solid starting point on which to build your latest and greatest theme design can be time-consuming. Although exploring all the available themes in detail is tempting, I find that exhaustive searches waste more time than they save.

Begin with the most current theme unless you find a more suitable one. Because the design and capabilities of the theme were recently implemented, modifying it to meet your current project's needs is faster than rediscovering all the nuances of an older, unfamiliar theme.

You may wonder whether I ever build themes off other designers' themes. I do. These days, if a new theme comes out that shows how to integrate some new feature, I play around with the theme to understand the concept but always go back to one of my themes to implement the modification. The reason for this practice is simple: If I can implement the feature into my own design, I have a much better appreciation of how it works. Allowing someone else's code or design to do the heavy lifting can place a limitation on how I use that feature.

TIP

If you're new to theme development and haven't produced a theme of your own, start with the WordPress default theme, Twenty Seventeen. (See Book 6, Chapter 1 for a full analysis of the Twenty Seventeen theme.) This theme was developed to help new theme developers discover how themes work.

Note: All the examples in this chapter are built off the WordPress default Twenty Seventeen theme unless noted otherwise.

Customizing the theme to your needs

After you select a theme for your project, you should create a copy of the theme. This way, you can look at the unmodified version in case you accidentally remove something that causes the theme or design to break.

When you find code and styling that you don't need anymore, comment it out rather than delete it. This practice removes the functionality but still allows you to add it back if you change your mind.

You can comment out a line of PHP code by adding // in front of it, as in this example:

```
// add_editor_style();
```

To comment out CSS (Cascading Style Sheets) code, wrap a section in /* and */, as follows:

```
/**#content {
 margin: 0 280px 0 20px;
}
*/
#primary,
#secondary {
 float: right;
/* overflow: hidden;
*/
 width: 220px;
}
```

You comment out HTML code by using brackets starting with `<!--` and ending with `-->` surrounding the code, like this:

```
<!--<div id="content">this is a content area</div>-->
```

TIP

When you start finalizing the theme, go through the files and remove any blocks of commented styling and code to clean up your files.

Digging Into Advanced Theme Development

Adding New Template Files

Book 6, Chapter 3 introduces the concept of template files and gives you an overview of the template files available to you. Book 6, Chapter 5 explains the idea of overriding template files with child themes. The following sections explore some advanced uses of template files.

Although you rarely need to use all these techniques, being fluent in your options gives you flexibility to address specific needs quickly when they come up.

Creating named templates

WordPress recognizes three special areas of a theme: header, footer, and sidebar. The `get_header`, `get_footer`, and `get_sidebar` functions default to loading `header.php`, `footer.php`, and `sidebar.php`, respectively. Each of these functions also supports a name argument to allow you to load an alternative version of the file. Running `get_header('main')`, for example, causes WordPress to load `header-main.php`.

You may wonder why you'd use a name argument when you could just create a template file named whatever you like and load it directly. The reasons for using the `get_header`, `get_footer`, or `get_sidebar` functions with a name argument are

- Holding to a standard naming convention that other WordPress developers can easily understand
- Automatically providing support for child themes to override the parent theme's template file
- Offering a fallback that loads the unnamed template file if the named one doesn't exist

REMEMBER

In short, use the name argument feature if you have multiple, specialized header, footer, or sidebar template files.

You can use this named template feature along with theme options (discussed in "Exploring Theme Options" later in this chapter) to allow users to easily switch among different header, footer, and sidebar styles. On the Theme Options page, you can give users the ability to choose the specific header, footer, or sidebar template file they want, which is an easy way to change the layout or design of the site. For a good example of content you could add to a different sidebar file, see the nearby sidebar "WP_Query posts for category content," which discusses displaying a list of recent posts and filing them in a specific category on the sidebar of your site.

WP_QUERY POSTS FOR CATEGORY CONTENT

WordPress makes it possible to pull in very specific types of content on your website through the `WP_Query();` template class. You place this template tag before The Loop (see Book 6, Chapter 3), and it lets you specify which category you want to pull information from. If you have a category called WordPress, and you want to display the last three posts from that category on your front page, on your blog sidebar, or somewhere else on your site, you can use this template tag.

The `WP_Query();` template class has several parameters that let you display different types of content, such as posts in specific categories, content from specific pages/posts, or dates in your blog archives. The `WP_Query();` class lets you pass many variables and parameters. It's not limited to categories, either; you can also use it for pages, posts, tags, and more. Visit the WordPress Codex at `https://codex.wordpress.org/Class_Reference/WP_Query` to read about this feature.

To query the posts on your blog to pull out posts from just one specific category, you can use the following tag with the associated arguments for the available parameters. This example tells WordPress to query all posts that exist on your site and list the last five posts in the Books category:

```
<?php $the_query = new WP_Query('posts_per_page=5&category_name=books'); ?>
```

Simply place this code on a line above the start of The Loop; you can use it on a sidebar to display clickable titles of the last five posts in the Books category. (When the reader clicks a title, he or she is taken to the individual post page to read the full post.)

```
<?php $the_query = WP_Query('posts_per_page=5&category_name=books'); ?>
<?php while ($the_query->have_posts()) : $the_query->the_post(); ?>
<strong><a href="<?php the_permalink() ?>" rel="bookmark" title="Permanent
    Link to
<?php the_title_attribute(); ?>"><?php the_title(); ?></a></strong>
<?php the_excerpt(); endwhile; ?>
```


Creating and using template parts

A template part is very similar to the header, footer, and sidebar templates except that its use isn't limited to header, footer, and sidebar.

The `get_header`, `get_footer`, and `get_sidebar` functions allow for code that's duplicated in many of the template files to be placed in a single file and loaded by means of a standard process. The purpose of template parts is to offer a standardized function that can be used to load sections of code specific to an individual theme. Sections of code that add a specialized section of header widgets or display a block of ads can be placed in individual files and easily loaded as a template part.

You load template parts by using the `get_template_part` function. The `get_template_part` function accepts two arguments: `slug` and `name`. The `slug` argument is required and describes the generic type of template part to be loaded, such as `loop`. The `name` argument is optional and selects a specialized template part, such as `post`.

A call to `get_template_part` with just the `slug` argument tries to load a template file with the filename *`slug`*`.php`. Thus, a call to `get_template_part('loop')` tries to load `loop.php`, and a call to `get_template_part('header-widgets')` tries to load `header-widgets.php`. See a pattern here? The *`slug`* part of the filename refers to the name of the template file, minus the `.php` extension, because WordPress already assumes that the file is a `.php` file.

A call to `get_template_part` with both the `slug` and `name` arguments tries to load a template file with the filename *`slug-name`*`.php`. If a template file with the filename *`slug-name`*`.php` doesn't exist, WordPress tries to load a template file with the filename *`slug`*`.php`. Thus, a call to `get_template_part('loop', 'post')` first tries to load `loop-post.php` followed by `loop.php` if `loop-post.php` doesn't exist; a call to `get_template_part('header-widgets', 'post')` first tries to load `header-widgets-post.php` followed by `header-widgets.php` if `header-widgets-post.php` doesn't exist.

The Twenty Seventeen theme offers a good example of the template part feature in use. It uses a template part called `content` to allow the page or post content, within The Loop, to get pulled into individual files.

REMEMBER

The Loop is the section of code in most theme template files that uses a PHP `while` loop to loop through the set of post, page, and archive content (to name a few content examples) and display it. The presence of The Loop in a template file is crucial for a theme to function properly. Book 6, Chapter 3 examines The Loop in detail.

Twenty Seventeen's `index.php` template file shows a template part for the `content` template part in action:

```
<?php
  if ( have_posts() ) :

  /* Start the Loop */
  while ( have_posts() ) : the_post();
```

```
  /*
   * Include the Post-Format-specific template for the content.

   * If you want to override this in a child theme, include a
   file

   * called content-___.php (where ___ is the Post Format name)
   and that will be used instead.

   */
    get_template_part( 'template-parts/post/content', get_post_
   format() );

  endwhile;

the_posts_pagination( array(
   'prev_text' => twentyseventeen_get_svg( array( 'icon' =>
    'arrow-left' ) ) . '<span class="screen-reader-text">' .
    __( 'Previous page', 'twentyseventeen' ) . '</span>',
   'next_text' => '<span class="screen-reader-text">' . __(
    'Next page', 'twentyseventeen' ) . '</span>' .
    twentyseventeen_get_svg( array( 'icon' => 'arrow-right' ) ),
   'before_page_number' => '<span class="meta-nav screen-reader-
    text">' . __( 'Page', 'twentyseventeen' ) . ' </span>',

) );

else :

   get_template_part( 'template-parts/post/content', 'none' );

endif;
?>
```

Loading the content by using a template part, Twenty Seventeen cleans up the `index.php` code considerably compared with other themes. This cleanup of the template file code is just the icing on the cake; the true benefits are the improvements in theme development.

Twenty Seventeen's `index.php` template file calls for a template part with a `slug` of `content` and a `name` of `get_post_format()`. The `get_post_format();` tag pulls in the defined format for the post (check out "Adding support for post formats" later in this chapter), such as asides, image, and link. If a post format exists, the `get_template_part();` calls it in. If the post format is defined as aside, the `get_template_part();` pulls in the `content-aside.php` template. If no post format has been defined, Twenty Seventeen simply uses the content.php template. A child theme (child themes are discussed at length in Book 6, Chapter 5) could supply a content.php file to customize just The Loop for index.php. Because of Twenty Seventeen's use of template parts and using both arguments of the get_template_part function, the child theme can do this without having to supply a customized index.php file.

With Twenty Seventeen's code for the header, The Loop, sidebar, and footer placed in separate files, the template files become much easier to customize for specific uses. You can see the difference by comparing index.php with the page.php template files.

The index.php listing:

```
<?php get_header(); ?>

<div class="wrap">
<?php if ( is_home() && ! is_front_page() ) : ?>
<header class="page-header">
<h1 class="page-title"><?php single_post_title(); ?></h1>
</header>
<?php else : ?>
<header class="page-header">
<h2 class="page-title"><?php _e( 'Posts', 'twentyseventeen' ); ?></h2>
</header>
<?php endif; ?>

<div id="primary" class="content-area">
<main id="main" class="site-main" role="main">

<?php
if ( have_posts() ) :

/* Start the Loop */
while ( have_posts() ) : the_post();

/*
 * Include the Post-Format-specific template for the content.
 * If you want to override this in a child theme, include a file
 * called content-___.php (where ___ is the Post Format name), and that will
    be used instead.
 */
get_template_part( 'template-parts/post/content', get_post_format() );

endwhile;
```

```
the_posts_pagination( array(
  'prev_text' => twentyseventeen_get_svg( array( 'icon' => 'arrow-left' ) ) .
     '<span class="screen-reader-text">' . __( 'Previous page',
     'twentyseventeen' ) . '</span>',
   'next_text' => '<span class="screen-reader-text">' . __( 'Next page',
     'twentyseventeen' ) . '</span>' . twentyseventeen_get_svg( array(
     'icon' => 'arrow-right' ) ),
  'before_page_number' => '<span class="meta-nav screen-reader-text">' . __(
     'Page', 'twentyseventeen' ) . ' </span>',
) );

else :

get_template_part( 'template-parts/post/content', 'none' );

endif;
?>

</main><!-- #main -->
</div><!-- #primary -->
<?php get_sidebar(); ?>
</div><!-- .wrap -->

<?php get_footer(); ?>
```

The page.php listing:

```
<?php get_header(); ?>
<div class="wrap">
<div id="primary" class="content-area">
<main id="main" class="site-main" role="main">

<?php
while ( have_posts() ) : the_post();

get_template_part( 'template-parts/page/content', 'page' );

// If comments are open or we have at least one comment, load up the comment
    template.
if ( comments_open() || get_comments_number() ) :
comments_template();
endif;

endwhile; // End of the loop.
?>
```

```
</main><!-- #main -->
</div><!-- #primary -->
</div><!-- .wrap -->

<?php get_footer();?>
```

Before template parts, the full Loop code was duplicated in the index.php and page.php files. As a result, a modification in the index.php file's Loop code required the same modification in the page.php file. Imagine having to make the same modification in five template files. Repeatedly making the same modifications quickly becomes tiring, and each modification increases the chance of making mistakes. Using a template part means that the modification needs to be made only one time.

In the index.php and page.php example, the get_template_part call allows you to easily create as many customized page templates as needed without having to duplicate The Loop code. Without the duplicate code, the code for The Loop can be easily modified in one place.

When you start duplicating sections of code in numerous template files, place the code in a separate file and use the get_template_part function to load it where needed.

Exploring content-specific standard templates

The template files discussed so far span a wide scope of site views specific to the view and not the content. The category.php template file, for example, applies to all category archive views but not to a specific category, and the page.php template file applies to all page views but not to a specific page. You can create template files for specific content and not just the view, however.

Four content-specific template types are available: author, category, page, and tag. Each one allows you to refer to specific content by the term's ID (an individual author's ID, for example) or by the slug.

The slug discussed in this section differs from the slug argument of the get_template_part function described in the preceding section. For this section, *slug* refers to a post, page, or category slug (to name a few types of slugs), such as a Press Releases category with a slug of press-releases or a post titled "Hello World" with a slug of hello-world.

Suppose that you have an About Us page with an id of 138 and a slug of about-us. You can create a template for this specific page by creating a file named either page-138.php or page-about-us.php. In the same way, if you want to create a template specific to an awesome author named Lisa with an id of 7 and a slug of lisa, you can create a file named author-7.php or author-lisa.php.

Creating a template by using the slug can be extremely helpful for making templates for sites that you don't manage. If you want to share a theme that you created, you could create a category-featured.php template, and this template would automatically apply to any category view that has a slug of featured.

When you use categories as the example, the file-naming convention is as follows:

- A template with the filename category.php is a catch-all (default) for the display for all categories. (Alternatively, a template with the filename archives.php will display categories if a category.php doesn't exist.)
- Add a dash and the category ID number to the end of the filename (as shown in Table 6-1) to specify a template for an individual category.
- Alternatively, add a dash and the category slug to the end of the filename (as shown in Table 6-1) to define it as a template for that particular category. If you have a category called Books, for example, the category slug is books; the individual category template file would be named category-books.php.
- If you don't have a category.php, an archives.php, or category-#.php file, the category display pulls from the Main Index template (index.php).

Table 6-1 gives you some examples of file-naming conventions for category templates.

TABLE 6-1 Category Template File-Naming Conventions

If the Category ID or Slug Is . . .	The Category Template Filename Is . . .
1	category-1.php
2	category-2.php
3	category-3.php
books	category-books.php
movies	category-movies.php
music	category-music.php

REMEMBER

Because creating a template by using slugs is so useful (and because an ID is relevant only to a specific site), you may wonder why the id option exists. The short answer is that the id option existed before the slug option did, but it's still valuable in specific instances. You can use the id option for a content-specific template without worrying about the customization breaking when the slug changes, for example. This option is especially helpful if you set up the site for someone else and can't trust him or her to leave the slugs alone (such as changing a category with a slug of news to press-releases).

Using page templates

Although the page-*slug*.php feature is very helpful, sometimes, requiring the theme's user to use the name you choose for a specific feature is too difficult or unnecessary. Page templates allow you to create a stand-alone template (just like page.php or single.php) that the user can selectively use on any specific page he or she chooses. As opposed to the page-*slug*.php feature, a page template can be used on more than one page. The combined features of user selection and multiple uses make page templates much more powerful theme tools than page-*slug*.php templates.

For more on page templates, see Chapters 1, 3, and 5 of this minibook (Book 6).

To make a template a page template, simply add Template Name: *Descriptive Name* to a comment section at the top of the template file. The following code is a page template created with the name "Full-width Page Template, No Sidebar" in the Twenty Seventeen theme:

```
<?php
/**
 * Template Name: Full-Width Page Template, No Sidebar
 *
 * Description: Twenty Seventeen loves the no-sidebar look as much as
 * you do. Use this page template to remove the sidebar from any page.
 *
 * Tip: To remove the sidebar from all posts and pages simply remove
 * any active widgets from the Main Sidebar area, and the sidebar will
 * disappear everywhere.
 *
 * @package WordPress
 * @subpackage Twenty_Twelve
 * @since Twenty Seventeen 1.0
 */
```

This registers the template file as a page template and adds Full-width Page Template, No Sidebar to the Page Attributes module's Template drop-down list, as shown in Figure 6-1. (Check out Book 4, Chapters 1 and 2 for information on publishing pages.) Using a template on a static page is a two-step process: Upload the template, and tell WordPress to use the template by tweaking the page's code.

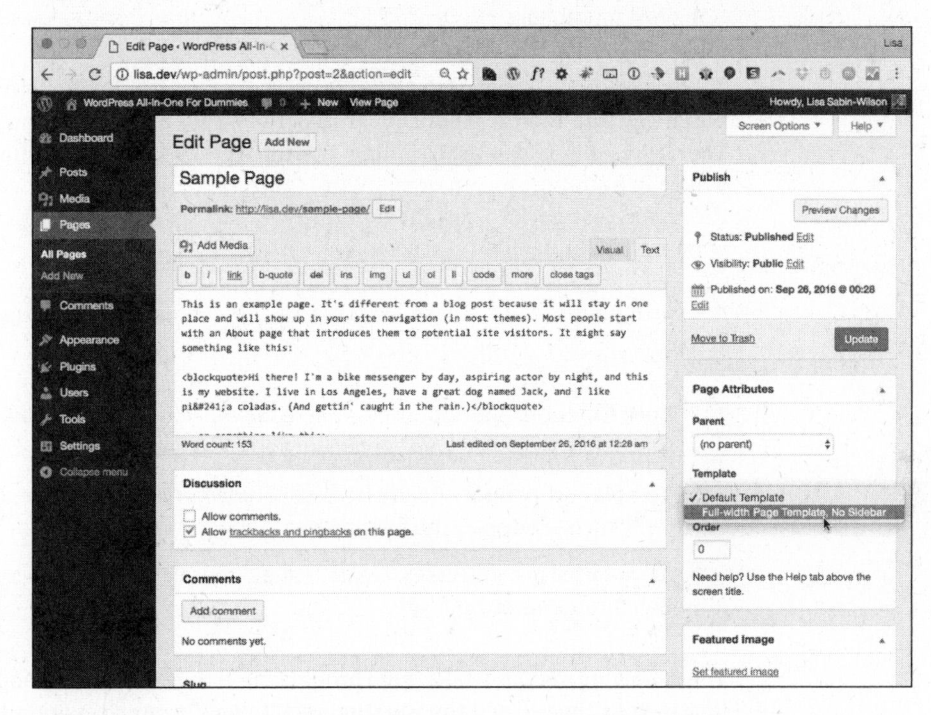

FIGURE 6-1: The Dashboard showing page attributes.

By providing a robust set of page templates, you can offer users of your theme an easy-to-use set of options for formatting their pages. These options can be used only for pages, but named header, footer, sidebar, and template parts can be used to offer users options on other site views.

Adding Theme Support for Built-In Features

The WordPress core offers several great tools that you can easily add to a theme to give the it more customization options. WordPress provides several built-in features that enable you to enhance your site and theme. This section covers five of the most popular features:

- Custom navigation menus
- Custom post types

- Custom taxonomies
- Post formats
- Post thumbnails

These features are part of the WordPress core, but they aren't activated by default. When you add theme support, you're activating a built-in feature of your theme. Therefore, when you're traveling around the WordPress community, whether you're in a support forum or at a WordCamp event, and you hear someone say that the theme supports a certain feature, you can smile, because you know exactly what he's talking about.

Activating support for these features in the theme you're using involves a few steps:

- **Core function:** Add support for the feature in your theme by including the core function in your theme's Theme Functions template file (`functions.php`).
- **Template function:** Add the necessary function tags in your theme template(s) to display the features on your website.
- **Templates:** In some cases, create feature-specific templates to create enhancements on your site.

The following sections take you through the five features. You add the core function to your theme, add the function tags to your templates and, if indicated, create a feature-specific template in your theme to handle the added options.

Adding support for custom menus

The WordPress menu-building feature is a great tool for theme developers. Before the addition of this tool, theme developers implemented their own menu solutions, creating a huge number of themes with navigation customization requiring coding and a small set of themes with very different ways of handling navigation. Creating complex, multiple-level menus on your WordPress site takes just a few steps, as outlined in this section.

A *navigation menu* is a listing of links that displays on your site. These links can be links to pages, posts, or categories within your site, or they can be links to other sites. You can define navigation menus on your site with the built-in Custom Menus feature in WordPress.

It's to your advantage to provide at least one navigation menu on your site so that readers can see everything your site has to offer. Providing visitors a link — or several links — is in keeping with the point-and-click spirit of the web.

The Twenty Seventeen theme already supports menus. Looking at Twenty Seventeen's `functions.php` file, you can see that the following lines of code handle registering the theme's menu:

```
// This theme uses wp_nav_menu() in two locations.
    register_nav_menus( array(
    'top'    => __( 'Top Menu', 'twentyseventeen' ),
    'social' => __( 'Social Links Menu', 'twentyseventeen' ),
    ) );
```

This code registers two navigation areas, called Top Menu and Social Links Menu, with a theme location name of `top` and `social`. With the Twenty Seventeen theme active, click the Menus link on the Appearance menu to load the Menus screen on the Dashboard and view the Top Menu location.

Digging Into Advanced Theme Development

Core menu function and template tags

The Custom Menu feature is already built into the default Twenty Seventeen WordPress theme, so you don't have to worry about preparing your theme for it. If you're using a different theme, however, adding this functionality is easy:

1. **Click the Editor link on the Appearance menu and then click the Theme Functions template file (`functions.php`).**

   The Theme Functions template opens in the text editor on the left side of the Edit Themes page.

2. **Add a new line to the Theme Functions template file and then type the following function:**

   ```
   // ADD MENU SUPPORT
   add_theme_support( 'nav-menus' );
   ```

3. **Click the Update File button to save the changes to the template.**

   This template tag tells WordPress that your theme can use the Custom Menu feature, and a Menus link now appears on the Appearance menu of the Dashboard.

4. **Open the Header template (`header.php`).**

   Click the Header link on the Edit Themes page to open the Header template in the text editor on the left side of the Edit Themes page.

5. **Add the following template tag by typing it on a new line in the Header template (`header.php`):**

   ```
   <?php wp_nav_menu(); ?>
   ```

This template tag is needed so the menu you build by using the Custom Menu feature displays at the top of your website. Table 6-2 gives the details on the different parameters you can use with the `wp_nav_menu();` template tag to further customize the display to suit your needs.

6. **Click the Update File button at the bottom of the page to save the changes you made in the Header template.**

TABLE 6-2 **Common Tag Parameters for wp_nav_menu();**

Parameter	Information	Default	Tag Example
`id`	The unique ID of the menu (because you can create several menus, each has a unique ID number)	Blank	`wp_nav_menu( array ( 'id' => '1' ) );`
`slug`	The menu name in slug form (for example, nav-menu)	Blank	`wp_nav_menu( array ( 'slug' => 'nav-menu' ) );`
`menu`	The menu name	Blank	`wp_nav_menu( array ( 'menu' => 'Nav Menu' ) );` or `wp_nav_menu('Nav Menu');`
`menu_class`	The CSS class used to style the menu list	Menu	`wp_nav_menu( array ( 'menu_class' => 'mymenu' ) );`
`format`	The HTML markup used to style the list (an unordered list [`ul/li`] or a `div` class)	`Div`	`wp_nav_menu( array ( 'format' => 'ul' ) );`
`fallback_cb`	The parameter that creates a fallback if a custom menu doesn't exist	`wp_page_menu` (the default list of page links)	`wp_nav_menu( array ( ' fallback_cb' => 'wp_page_menu' ) );`
`before`	The text that displays before the link text	None	`wp_nav_menu( array ( 'before' => 'Click Here' ) );`
`after`	The text that displays after the link text	None	`wp_nav_menu( array ( 'after' => '»' ) );`

Figure 6-2 shows the default Twenty Seventeen theme with a navigation menu (Home, Blog, and Contact) below the theme's header graphic.

FIGURE 6-2: The Twenty Seventeen theme with a navigation menu below the header.

A menu called Main was created in the WordPress Dashboard. (See Book 6, Chapter 1 for details on creating menus in the WordPress Dashboard.) The template tag used in the theme to display the menu looks like this:

```
<?php wp_nav_menu('Main'); ?>
```

The HTML markup for the menu is generated as an unordered list, by default, and looks like this:

```
<ul id="top-menu" class="menu">
<li id="menu-item-110" class="menu-item menu-item-type-post_type menu-item-
   object-page menu-item-home current-menu-item page_item page-item-15
   current_page_item menu-item-110"><a href="http://lisa.dev/">Home</a></li>
<li id="menu-item-111" class="menu-item menu-item-type-post_type menu-item-
   object-page menu-item-111"><a href="http://lisa.dev/blog/">Blog</a></li>
<li id="menu-item-112" class="menu-item menu-item-type-post_type menu-item-
   object-page menu-item-112"><a href="http://lisa.dev/sample-page/">
   Contact</a></li>
</ul>
```

Notice that in the HTML markup, the `<ul id="top-menu" class="menu">` line defines the CSS ID and class.

The ID reflects the name that you give your menu. Because the menu is named Top Menu, the CSS ID is `top-menu`. If the menu were named Foo, the ID would be `foo`. WordPress allows you to use CSS to create different styles and formats for your different menus by assigning menu names in the CSS and HTML markup.

When developing themes for yourself or others to use, make sure that the CSS you define for the menus can do things like account for subpages by creating drop-down menus. You can accomplish this task in several ways. Listing 6-1 gives you just one example: a block of CSS that you can use to create a nice style for your menu. (This CSS example assumes that you have a menu named Main; therefore, the HTML and CSS markups use `menu-main`.)

LISTING 6-1: Sample CSS for Drop-Down Menu Navigation

```
#menu-main {
....width: 960px;
....font-family: Georgia, Times New Roman, Trebuchet MS;
....font-size: 16px;
....color: #FFFFFF;
....margin: 0 auto 0;
....clear: both;
....overflow: hidden;
....}

#menu-main ul {
....width: 100%;
....float: left;
....list-style: none;
....margin: 0;
....padding: 0;
....}

#menu-main li {
....float: left;
....list-style: none;
....}

#menu-main li a {
....color: #FFFFFF;
....display: block;
....font-size: 16px;
```

```
....margin: 0;
....padding: 12px 15px 12px 15px;
....text-decoration: none;
....position: relative;
....}

#menu-main li a:hover, #menu-main li a:active, #menu-main .current_page_item a,
   #menu-main .current-cat a, #menu-main .current-menu-item {
....color: #CCCCCC;
....}

#menu-main li li a, #menu-main li li a:link, #menu-main li li a:visited {
....background: #555555;
....color: #FFFFFF;
....width: 138px;
....font-size: 12px;
....margin: 0;
....padding: 5px 10px 5px 10px;
....border-left: 1px solid #FFFFFF;
....border-right: 1px solid #FFFFFF;
....border-bottom: 1px solid #FFFFFF;
....position: relative;
....}

#menu-main li li a:hover, #menu-main li li a:active {
....background: #333333;
....color: #FFFFFF;
....}

#menu-main li ul {
....z-index: 9999;
....position: absolute;
....left: -999em;
....height: auto;
....width: 160px;
....}

#menu-main li ul a {
....width: 140px;
....}
```

(continued)

LISTING 6-1: *(continued)*

```
#menu-main li ul ul {
....margin: -31px 0 0 159px;
....}

#menu-main li:hover ul ul, #menu-main li:hover ul ul ul {
....left: -999em;
....}


#menu-main li:hover ul, #menu-main li li:hover ul, #menu-main li li li:hover ul
   {
....left: auto;
....}

#menu-main li:hover {
....position: static;
....}
```

REMEMBER

The CSS you use to customize the display of your menus will differ. The example in Listing 6-1 is just that: an example. After you get the hang of using CSS, you can try different methods, colors, and styling to create a custom look. (You can find additional information about basic HTML and CSS in Book 6, Chapter 4.)

Custom menus using widgets

You don't have to use the `wp_nav_menu();` template tag to display the menus on your site because WordPress also provides a Custom Menu widget that you can add to your theme. As a result, you can use widgets instead of template tags to display the navigation menus on your site. This widget is especially helpful if you've created multiple menus for use on your site in various different places. Have a look at Book 6, Chapter 4 for more information on using WordPress widgets.

Your first step is to register a special widget area for your theme to handle the Custom Menu widget display. Open your theme's `functions.php` file, and add the following lines of code:

```
// ADD MENU WIDGET
if ( function_exists('register_sidebar') )
    register_sidebar(array('name'=>'Menu',));
```

These few lines of code create a new Menu widget area on the Widgets page of your Dashboard. You can drag the Custom Menu widget into the Menu widget to indicate that you want to display a custom menu in that area. Figure 6-3 shows the Menu widget area with the Custom Menu widget added.

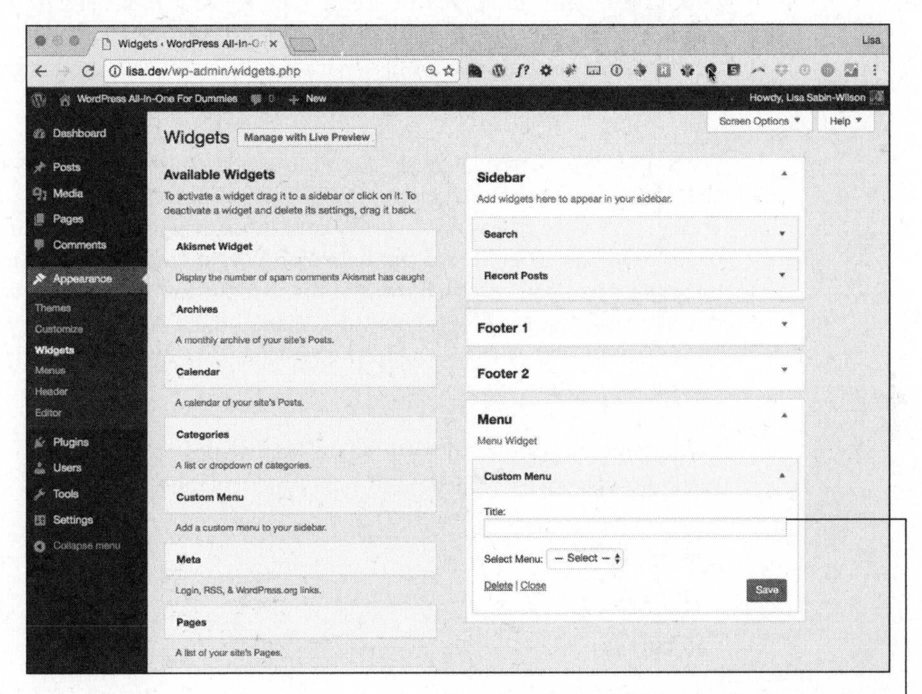

Custom Menu widget

FIGURE 6-3: Widgets page displaying a Menu widget area with a Custom Menu widget.

To add the widget area to your theme, open the Theme Editor (click the Editor link on the Appearance menu), open the `header.php` file, and add these lines of code in the area in which you want to display the Menu widget:

```
<ul>
<?php if ( !function_exists('dynamic_sidebar') || !dynamic_sidebar('Menu') ) : ?>
<?php endif; ?>
</ul>
```

These lines of code tell WordPress that you want information contained in the Menu widget area to display on your site.

Adding support for custom post types

Custom post types and custom taxonomies have expanded the content management system (CMS) capabilities of WordPress and are likely to become a big part of plugin and theme features as more developers become familiar with their use. *Custom post types* allow you to create new content types separate from posts and pages, such as movie reviews or recipes. *Custom taxonomies* allow you to create new types of content grouping separate from categories and tags, such as genres for movie reviews or seasons for recipes.

Posts and pages are nice generic containers of content. A *page* is timeless content that has a hierarchal structure; it can have a parent, forming a nested, or hierarchal, structure of pages. A *post* is content that's listed in linear (not hierarchal) order based on when it was published and organized in categories and tags. What happens when you want a hybrid of these features? What if you want content that doesn't show up in the post listings, displays the posting date, and doesn't have either categories or tags? Custom post types are created to satisfy this desire to customize content types.

By default, WordPress already has different post types built in to the software, ready for you to use. The default post types include

- Blog posts
- Pages
- Menus
- Attachments
- Revisions

Custom post types give you the ability to create new and useful types of content on your website, including a smart and easy way to publish those content types on your site.

The possibilities for the use of custom post types are endless. To kick-start your imagination, here are some of the most popular, useful ideas that others have implemented on sites:

- Photo gallery
- Podcast or video
- Book reviews
- Coupons and special offers
- Events calendar

Core custom post type function

To create and use custom post types on your site, you need to be sure that your WordPress theme contains the correct code and functions. This section shows you how to create a very basic, generic custom post type called Generic Content. Follow these steps to create the same basic custom post type:

1. **Open the Theme Functions template file (`functions.php`).**

   Click the Editor link on the Appearance menu to open the Theme Editor screen. Then click the Theme Functions template link to open the `functions.php` file in the text editor.

2. **Add the custom post types code to the bottom of the Theme Functions template file.**

   Scroll to the bottom of the `functions.php` file, and include the following code to add a Generic Content custom post type to your site:

```
// ADD CUSTOM POST TYPE
add_action( 'init', 'create_post_type' );
function create_post_type() {
  register_post_type( 'generic-content',
    array(
      'labels' => array(
        'name' => __( 'Generic Content' ),
        'singular_name' => __( 'Generic Content' )
      ),
      'public' => true
    )
  );
}
```

3. **Click the Update File button to save the changes you made in the `functions.php` file.**

The `register_post_type` function can accept several arguments and parameters, which are detailed in Table 6-3. You can use a variety and combination of different arguments and parameters to create a specific post type. You can find more information on Custom Post Types and using the `register_post_type` function in the WordPress codex at `http://codex.wordpress.org/Function_Reference/register_post_type`.

TIP

If you really don't feel up to writing this code in the Theme Functions template file, check out a nifty plugin developed for WordPress called Custom Post Type UI by WebDevStudios (`https://webdevstudios.com`). This plugin provides an interface for your WordPress Dashboard that simplifies the creation of custom post types on your site and bypasses the need to create the code in the Theme Functions template file (`functions.php`). You can find the plugin at `https://wordpress.org/plugins/custom-post-type-ui`.

After you complete the steps to add the generic content post type to your site, the Generic Content post type appears on the left navigation menu of the Dashboard, as shown in Figure 6-4.

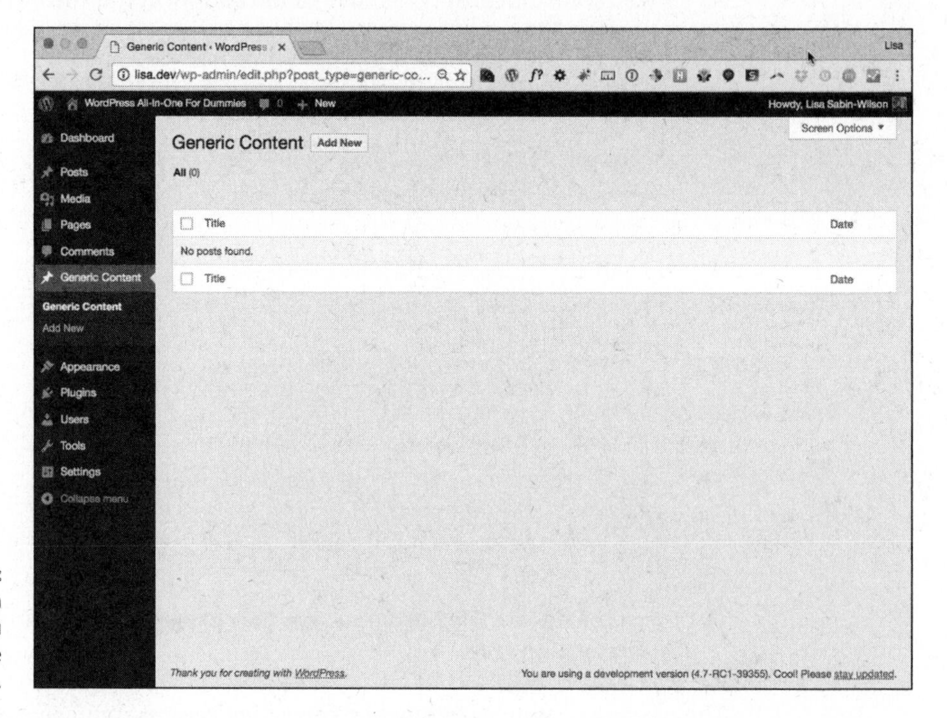

FIGURE 6-4: A new custom post type menu appears on the Dashboard.

You add and publish new content by using the new custom post type the same way you do when you write and publish blog posts. The published content isn't added to the chronological listing of blog posts; it's treated as separate content, just like static pages.

Generic Content is part of the permalink structure, and the permalink looks similar to `http://`*`yourdomain`*`.com/generic-content/new-article`.

TABLE 6-3 **Arguments and Parameters for register_post_type();**

Parameter	Information	Parameters	Example
label	A plural descriptive name for the post type.	None.	'label' => __('Generic Content'),
labels	An array of descriptive labels for the post type.	Default is empty and set to the label value. name singular_name add_new add_new_item edit_item new_item view_item view_items search_items not_found not_found_in_trash parent_item_colon all_items archives attributes insert_into_item uploaded_into_item uploaded_to_this_item featured_image set_featured_image remove_featured_image use_featured_image menu_name filter_items_list	'name' => __('Generic Content'),

(continued)

TABLE 6-3 *(continued)*

Parameter	Information	Parameters	Example

		`items_list_navigation`
		`items_list`
		`name_admin_bar`	

`description`	The description of the post type; displayed on the Dashboard to represent the post type.	None.	`'description' => __( 'This is a description of the Generic Content type' ),`

`public`
`show_ui`
`publicly_queryable`
`exclude_from_search`
`show_in_nav_menus`
	Sets whether the post type is public.
	The three other arguments are
	`show_ui`: Show admin screens
	`publicly_queryable`: Query for this post type from the front end
	`exclude_from_search`: Show post type in search results
		`true` or `false`: Default is false.	`'public' => true,`
			`'show_ui' => true,`
			`'publicly_queryable' => true,`
			`'exclude_from_search' => false,`
			`'show_in_nav_menus' => true`

`menu_position`	Sets the position of the post type menu item on the Dashboard navigation menu.	Default: 20; By default, custom post types appear after the Comments menu of the Dashboard. Set integer in intervals of five (5, 10, 15, 20, and so on).	`'menu_position' => 25,`
`menu_icon`	Defines a custom icon (graphic) to the post type menu item on the Dashboard navigation menu. Creates and uploads the image into the `images` directory of your theme folder.	None.	`'menu_icon' => get_stylesheet_directory_uri() . '/images/generic-content.png',`

Parameter	Information	Parameters	Example
hierarchical	Tells WordPress whether to display the post type content list in a hierarchical manner.	true or false: Default is true.	'hierarchical' => true,
query_var	Controls whether this post type can be used with a query variable such as query_posts (see preceding section) or WP_Query.	true or false: Default is false.	'query_var' => true,
capability_type	Defines permissions for users to edit, create, or read the custom post type.	post (default): Gives the same capabilities to those who can edit, create, and read blog posts.	'query_var' => post,
capabilities	Tells WordPress what capabilities are accepted for this post type.	Default: empty, the capability_type value is used. edit_post: allows post type to be edited read_post: allows post type to be read delete_post: allows post type to be deleted	'capabilities' => edit_post,
map_meta_cap	Tells WordPress whether to use the default internal meta capabilities.	true or false: Default is false.	'map_meta_cap' => true,
supports	Defines what meta boxes, or modules, are available for this post type in the Dashboard.	title: Text box for the post title editor: Text box for the post content comments: Check boxes to toggle comments on/off trackbacks: Check boxes to toggle trackbacks and pingbacks on/off	'supports' => array('title', 'editor', 'excerpt', 'custom-fields', 'thumbnail'),

(continued)

TABLE 6-3 *(continued)*

Parameter	Information	Parameters	Example
		revisions: Allows post revisions author: Drop-down box to define post author excerpt: Text box for the post excerpt thumbnail: The featured image selection custom-fields: Custom fields input area page-attributes: The page parent and page template drop-down menus	
rewrite	Rewrites the permalink structure for the post type.	true or false; Two other arguments are available: slug: Permalink slug to prepend to your custom post types with_front: If you've set your permalink structure with a specific prefix such as /blog	'rewrite' => array('slug' => 'my-content', 'with_front' => false),
has_archive	Tells WordPress whether to enable post archives for this post type.	true or false: Default is false. (If true, WordPress uses the post type name as its slug in the permalink URL.)	'has_archive' => true,
can_export	Tells WordPress whether this post type can be exported by using the built-in content exporter.	true or false: Default is true.	'can_export' => false,
taxonomies	Uses existing WordPress taxonomies (category and tag).	Category post_tag	'taxonomies' => array('post_tag', 'category'),

TIP

The three modules WordPress gives you to add navigation menus are Custom Links, Pages, and Categories. On the Menus screen of the WordPress Dashboard, click the Screen Options tab in the top-right corner. The check box next to the post types you've created enables your custom post types on the menus you create.

Custom post type templates

By default, custom post types use the `single.php` template in your theme — that is, they do unless you create a specific template for your custom post type if you find the regular WordPress `single.php` template to be too limiting for your post type.

The preceding section has the code for building a simple Generic Content custom post. After that code is added, a Generic Content menu appears on the WordPress Dashboard. Click the Add New link on the Generic Content menu and publish a new post to add some content for testing. In this example, a new Generic Content type with a title of Test and a slug of `test` is added. Because the Generic Content type doesn't have a specific template, it uses the `single.php` template, and resulting posts look no different from a standard one.

TIP

If you get a Not Found page when you try to go to a new custom post type entry, reset your permalink settings. Click the Permalinks link on the Settings menu of the WordPress Dashboard and then click the Save Changes button. WordPress resets the permalinks, which adds the new custom post type link formats in the process.

To build a template specific for the Generic Content post type, add a new template named `single-`*`posttype`*`.php`, where *`posttype`* is the first argument passed to the `register_post_type` function from the preceding section. For this example, the single template file specific to Sample Post Type is `single-generic-content.php`. Any modifications made in this template file appear only for instances of the Generic Content post type.

A basic structure for `single-generic-content.php` looks like this:

```
<?php get_header(); ?>
<div id="container">
    <div id="content" role="main">
        <?php get_template_part( 'template-parts/post/content', single-
    generic-content );?>
    </div><!-- #content -->
</div><!-- #container -->
<?php get_sidebar(); ?>
<?php get_footer(); ?>
```

When you use the template part, creating a file called `loop-generic-content.php` allows for easy customization of The Loop for the Generic Content post type entry.

Adding support for custom taxonomies

There are times when being limited to categories and tags just is not enough for a site. A movie review custom post type, for example, might need a variety of new taxonomies or grouping options. Organizing movie reviews by director, star, review rating, film genre, and Motion Picture Association of America (MPAA) rating allows visitors to the site to view different groupings of reviews that may interest them. Like the custom post type example, this example creates a very simple taxonomy to test custom taxonomy–specific templates. For this example, a new post taxonomy called Sample Taxonomy is created.

To register this new taxonomy, use the `register_taxonomy` function. Adding the following code to the bottom of your theme's `functions.php` file registers the new sample taxonomy custom taxonomy specifically for WordPress built-in posts, adds a Sample Taxonomy link to the Posts menu entry to manage the Sample Taxonomy entries, and adds sample taxonomy options to the editor for posts.

```
register_taxonomy( 'sample-taxonomy', 'post', array( 'label' => 'Sample
    Taxonomy' ) );
```

This function call gives the new custom taxonomy an internal name of `sample-taxonomy`, assigns the new taxonomy to Posts, and gives the taxonomy a human-readable name of Sample Taxonomy.

After adding this code to your theme, you can create and assign Sample Taxonomies when creating a new post or editing an existing post. For this example, you could add a sample taxonomy with a name of Testing to an existing post and update the post.

With the Testing taxonomy added, you can visit `example.com/sample-taxonomy/testing` to get the archive page for the new sample taxonomy.

TIP

If you get a Not Found page or don't get an archive listing when you try to go to a specific sample taxonomy entry's archive, resave your permalink settings. Click the Permalinks link on the Settings menu of the WordPress Dashboard and then click Save Changes. This step forces WordPress to reset the permalinks, which adds the new custom taxonomy link formats in the process.

Adding a new template file called `taxonomy-sample-taxonomy.php` allows you to add a template that's specific to this new custom taxonomy. As you can with categories and tags, you can add a template that's specific to a single custom taxonomy entry. Therefore, a template specific to a sample taxonomy with a slug of `testing` would have the filename `taxonomy-sample-taxonomy-testing.php`.

REMEMBER

Custom taxonomies appeal only to owners of specific types of sites that deal mainly in niche areas of content: owners who want to really drill down in navigation and grouping options for their content. You can find more about custom taxonomies in the WordPress Codex at `https://codex.wordpress.org/Function_Reference/register_taxonomy`.

Adding support for post formats

Introduced in Version 3.1 of WordPress, the Post Formats feature allows you to designate a different content display and style for certain types of designated posts. Unlike with custom post types, you can't create different post formats because WordPress has already assigned them for you. It's up to you what post format, if any, you want to use in your theme.

Here are the nine types of WordPress post formats:

- **Aside:** A very short post that shares a random thought or idea. Typically, an aside is shared without a post title or any category/tag designations. It's simply a random, one-off thought — not a full post — shared on your site.
- **Audio:** A post that shares audio files or podcasts. Usually, audio posts have very little text and include a built-in audio player that visitors can click to listen to the content.
- **Chat:** A transcript of an online conversation that can be formatted to look like a chat (or instant-message) window.
- **Gallery:** A gallery of clickable images, in which clicking an image opens a larger version of the photo. Often, galleries don't contain text (but may have titles) and are used only for the display of images.
- **Image:** A post that shares a single image. The image may or may not have text or a caption.
- **Link:** A post that provides a link you find useful and want to share with your readers. These post formats often contain a title and sometimes a short bit of text that describes the link.
- **Quote:** A post that displays a quotation on your site. Often, users include a byline or the quote's source.
- **Status:** A short status update, usually limited to 140 characters or less. (Think Twitter!)
- **Video:** A post that displays a video, usually embedded within a video player (such as YouTube) so that your readers can play the video without leaving your site.

You have only these nine designated post formats to work with. You can use one or all of them in your theme, depending on your specific needs.

TIP

If you find that your site needs a type of post format that currently isn't available, consider adding it as a custom post type.

Core post format function

To add support for post formats to your theme, you need to add the core function call to your Theme Functions template file (functions.php). After you do, you'll see the magic that occurs on the Add New Post page of your WordPress Dashboard! Here's how to add post formats support to your theme:

1. **Click the Editor link on the Appearance menu of your Dashboard.**

   The Edit Themes screen appears.

2. **Open the Themes Function file in the text editor.**

   The link for the Theme Functions template file is on the right side of the Edit Themes page. Clicking this link opens the Theme Functions template file (functions.php) in the text editor on the left side of the Edit Themes page.

3. **Add the following function on a new line:**

```
add_theme_support( 'post-formats', array( 'aside', 'chat', 'gallery', 'image',
    'link', 'quote', 'status', 'video', 'audio' ) );
```

TIP

   This code sample adds all nine post formats to the theme. You don't have to use all nine; you can include only the formats that you need in your theme and leave the rest alone.

4. **Click the Update File button to save the changes you made in the functions.php file.**

You won't notice an immediate change on your site when you save your new Theme Functions template file with the Post Formats support added. To see what WordPress added to your site, you need to visit the Add New Post page (which you access by clicking the Add New link on the Posts menu).

The change is subtle, but if you follow the steps to add post format support, you see a Format module on the right side of the page, as shown in Figure 6-5. Choose your desired format to designate a format for your post. In Figure 6-5, all nine post format options are listed. You also see a tenth format option, Standard (or Default), which is used when you don't select a specific format for your post.

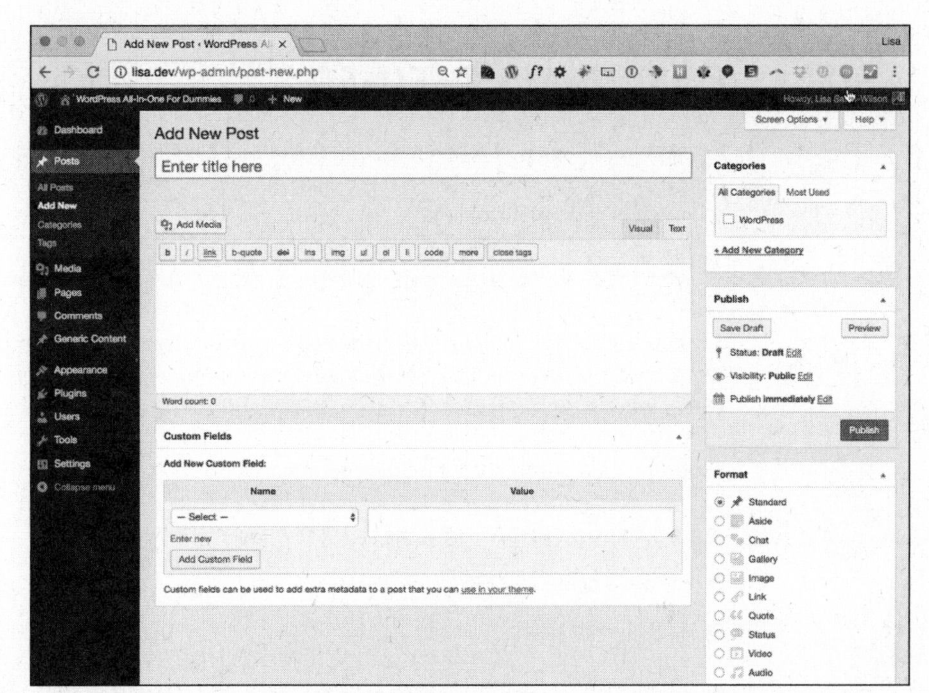

FIGURE 6-5: The Format module of the Add New Post screen.

Template tags for post formats

Adding Post Format support to your theme isn't enough. If you're going to add post format support, you really should provide some unique styling for each type of format; otherwise, your post formats will look like your blog posts, and the point of adding them to your theme will be lost.

You can display your post format in two ways:

- **Content:** For each format, you can designate what content you want to display. If you don't want to display a title for your Asides format, for example, leave out the template tag that calls it, but leave the tag in for your Video post format.
- **Style:** When you use the HTML markup that's provided by the `post_class();` tag, each of your formats has a CSS class assigned to it. Use those CSS classes to provide unique styles for fonts, colors, backgrounds, and borders to each of your post formats. The nearby sidebar "Post class defined" discusses how to use HTML and CSS to create custom styles in your template.

POST CLASS DEFINED

In the default Twenty Seventeen theme, examine the code for the content.php template (in the folder /template-parts/post/). About three fourths of the way in, you see a line of code that looks like this:

```
<article id="post-<?php the_ID(); ?>" <?php post_class(); ?>>
```

The cool part of that template tag is the post_class() section. This template tag tells WordPress to insert specific HTML markup into your template that allows you to use CSS to make custom styles for sticky posts, categories, tags, and post formats.

Suppose that a post has the following options set:

- Stick this post to the front page
- Filed in a category called WordPress
- Tagged with News

When the post_class() tag is used in the template, WordPress inserts HTML markup that allows the use of CSS to style posts assigned to specific tags, categories, logged-in status, or post formats, differently. WordPress inserts the following HTML markup for the post:

```
<article class="post sticky category-wordpress tag-news">
```

Likewise, for post formats, if a post is published via the Images post format, the post_class() tag in the template contains the following HTML markup, indicating that this post should be formatted for an image display:

```
<article class="post type-post format-image">
```

Add this information to the CSS and HTML information provided in Book 6, Chapter 3, and you see how you can use CSS along with the post_class(); tag to provide custom styles for each post type on your site and unique styles for the categories and tags you use in your posts.

Adding unique styles for your post formats starts with creating the content designations you want to display for each format. Earlier in this section, I provide a list of nine post formats and some ideas about what you can do to display them on your site. The possibilities are endless; the use of post formats is really up to you. See Book 6, Chapter 3 for more information on the content-related template tags you can use in these areas. The following steps take you through the creation of a very simple, stripped-down Main Index file (index.php):

1. **Open your favorite text editor, such as Notepad (for PC) or TextEdit (for Mac).**

2. **Type** <?php get_header(); ?>.

   This function includes all the code from your theme's header.php file.

3. **Type the following two lines:**

```
<?php if (have_posts()) : ?>
<?php while (have_posts()) : the_post(); ?>
```

   These two lines of code indicate the beginning of The Loop (discussed in Book 6, Chapter 3).

4. **Type** <div id="post-<?php the_ID(); ?>" <?php post_class(); ?>>.

   This line provides HTML and CSS markup, using the post_class(); function that provides unique CSS classes for each of your post formats. (See the nearby sidebar "Post class defined.")

5. **Type** <?php.

   This part initiates the start of a PHP function.

6. **Type the following lines to provide content for the Asides post format:**

```
if ( has_post_format( 'aside' )) {
echo the_content();
}
```

7. **Type the following lines to provide content for the Gallery post format:**

```
elseif ( has_post_format( 'gallery' )) {
echo '<h3>';
echo the_title();
echo '</h3>';
echo the_content();
}
```

8. **Type the following lines to provide content for the Image post format:**

```
elseif ( has_post_format( 'image' )) {
echo '<h3>';
echo the_title();
echo '</h3>';
echo the_post_thumbnail('image-format');
echo the_content();
}
```

9. **Type the following lines to provide content for the Link post format:**

```
elseif ( has_post_format( 'link' )) {
echo '<h3>';
echo the_title();
echo '</h3>';
echo the_content();
}
```

10. **Type the following lines to provide content for the Quote post format:**

```
elseif ( has_post_format( 'quote' )) {
echo the_content();
}
```

11. **Type the following lines to provide content for the Status post format:**

```
elseif ( has_post_format( 'status' )) {
echo the_content();
}
```

12. **Type the following lines to provide content for the Video post format:**

```
elseif ( has_post_format( 'video' )) {
echo '<h3>';
echo the_title();
echo '</h3>';
echo the_content();
}
```

13. **Type the following lines to provide content for the Audio post format:**

```
elseif ( has_post_format( 'audio' )) {
echo '<h3>';
echo the_title();
echo '</h3>';
echo the_content();
}
```

14. **Type the following lines to provide content for all other (Default) posts:**

```
else {
echo '<h3>';
echo the_title();
echo '</h3>';
echo the_content();
}
```


15. **Type** `?>`**.**

    This line ends the PHP function.

16. **Type** `</div>`**.**

    This line closes the HTML `div` tag you opened in Step 4.

17. **Type** `<?php endwhile; else: ?> <?php endif; ?>`**.**

    This line closes the `while` and `if` statements that were opened in Step 3.

18. **Type** `<?php get_sidebar(); ?>`**.**

    This function calls in the code included in the `sidebar.php` file of your theme.

19. **Type** `<?php get_footer(); ?>`**.**

    This function calls in the code included in the `footer.php` file of your theme.

20. **Save your file as** `index.php`**.**

    Upload it to your theme folder, replacing your existing `index.php` file.

Listing 6-2 displays the full code for your new `index.php` file.

LISTING 6-2: A Simple Post Formats Template

```
<?php get_header(); ?>
<?php if (have_posts()) : ?>
<?php while (have_posts()) : the_post(); ?>
<div id="post-<?php the_ID(); ?>" <?php post_class(); ?>>
<?php

if ( has_post_format( 'aside' )) {
....echo the_content();
}

elseif ( has_post_format( 'gallery' )) {
....echo '<h3>';
....echo the_title();
....echo '</h3>';
....echo the_content();
}

elseif ( has_post_format( 'gallery' )) {
....echo '<h3>';
....echo the_title();
....echo '</h3>';
....echo the_content();
}

elseif ( has_post_format( 'image' )) {
....echo '<h3>';
....echo the_title();
....echo '</h3>';
....echo the_post_thumbnail('image-format');
....echo the_content();
}

elseif ( has_post_format( 'link' )) {
....echo '<h3>';
....echo the_title();
....echo '</h3>';
....echo the_content();
}

elseif ( has_post_format( 'quote' )) {
....echo the_content();
}

elseif ( has_post_format( 'status' )) {
....echo the_content();
}
```

```
elseif ( has_post_format( 'video' )) {
....echo '<h3>';
....echo the_title();
....echo '</h3>';
....echo the_content();
}

elseif ( has_post_format( 'audio' )) {
....echo '<h3>';
....echo the_title();
....echo '</h3>';
....echo the_content();
}

else {
....echo '<h3>';
....echo the_title();
....echo '</h3>';
....echo the_content();
}
?>
</div>
<?php endwhile; else: ?>
<?php endif; ?>
<?php get_sidebar(); ?>
<?php get_footer(); ?>
```

The example in Listing 6-2 is a very simple one that doesn't include a whole lot of HTML markup or CSS classes. Therefore, you can focus on the code bits that are required to designate and define different content displays for your post formats. You can see in Listing 6-2 that some of the formats contain the template tag to display the title, `the_title();`, and others don't, but they all contain the template tag to display the content of the post: `the_content();`. As I mention previously, you can play with different content types and markup that you want to add to your post formats.

By coupling your template additions for post formats with the `post_class();` tag, which adds special CSS classes and markup for each post format type, you can customize the display of each post format to your heart's content.

Adding support for post thumbnails

The WordPress feature called Post Thumbnails (also known as Featured Images) takes a lot of the work out of associating an image with a post and using the correct size each time. A popular way to display content in WordPress themes involves using thumbnail images with snippets (excerpts) of text; the thumbnail images are consistent in size and placement within your theme. Before the inclusion of post thumbnails in WordPress, users had to open their images in an image-editing program (such as Adobe Photoshop) and crop and resize their images to the desired sizes, or they had to use fancy scripts that resized images on the fly, which tended to be resource-intensive on web servers and weren't optimal solutions. How about using a CMS that crops and resizes your images to the exact dimensions that you specify? Yep, WordPress does that for you with just a few adjustments.

By default, when you upload an image, WordPress creates three versions of your image based on dimensions that are set on your Dashboard (click the Media link on the Settings menu):

- **Thumbnail size:** Default dimensions are 150px x 150px.
- **Medium size:** Default dimensions are 300px x 300px.
- **Large size:** Default dimensions are 1024px x 1024px.

Therefore, when you upload an image, you actually end up with four sizes of that image stored on your web server: thumbnail, medium, large, and the original image. Images are cropped and resized proportionally, and when you use them in your posts, you can designate which size to use in the image options of the Add Image page. (See Book 4, Chapter 3 for details on uploading images in WordPress.)

In the Twenty Seventeen theme, specifically, you see the following in the `functions.php` file, which define image sizes:

- `add_image_size( 'twentyseventeen-featured-image', 2000, 1200, true );`
- `add_image_size( 'twentyseventeen-thumbnail-avatar', 100, 100, true );`

Within the WordPress image uploader, you can designate a particular image as the featured image of the post (see Figure 6-6), and then — using the Featured Images function that you add to your theme — you can include template tags to display your chosen featured image with your post. This technique is helpful for creating the magazine- or news-style themes that are popular on WordPress sites.

In the following sections, I also cover adding support for image sizes other than the default image sizes set on the Media Settings page of your Dashboard. This technique is helpful on sections of your site where you want to display much smaller thumbnails or larger versions of medium-size thumbnails.

FIGURE 6-6: Featured Image in use with the Twenty Seventeen theme.

Core post thumbnails function and template tags

Adding support for post thumbnails includes adding one line of code to your Theme Functions template file (`functions.php`):

```
add_theme_support( 'post-thumbnails' );
```

After you add this line of code to your Theme Functions template file, you can use the Featured Image feature for your posts. You can designate featured images by using the function in the WordPress image uploader. The function is also an option on the Add New Post page, where you write and publish your posts.

After you add featured images to your posts, make sure that you add the correct tag in your template(s) so that the featured images display on your site in the areas where you want them to display. Open your `index.php` template, and include the following line of code to include the default thumbnail-size version of your chosen featured image in your posts:

```
<?php if ( has_post_thumbnail() ) { the_post_thumbnail('thumbnail'); ?>
```

The first part of that line of code checks whether a featured image is associated with the post. If so, the image displays; if not, the code returns nothing. You can also include the other default image sizes (set in the Media Settings screen of the

Dashboard, shown in Figure 6-7) for medium-, large-, and full-size images by using these tags:

```
<?php if ( has_post_thumbnail() ) { the_post_thumbnail('medium'); ?>
```

```
<?php if ( has_post_thumbnail() ) { the_post_thumbnail('large'); ?>
```

```
<?php if ( has_post_thumbnail() ) { the_post_thumbnail('full'); ?>
```

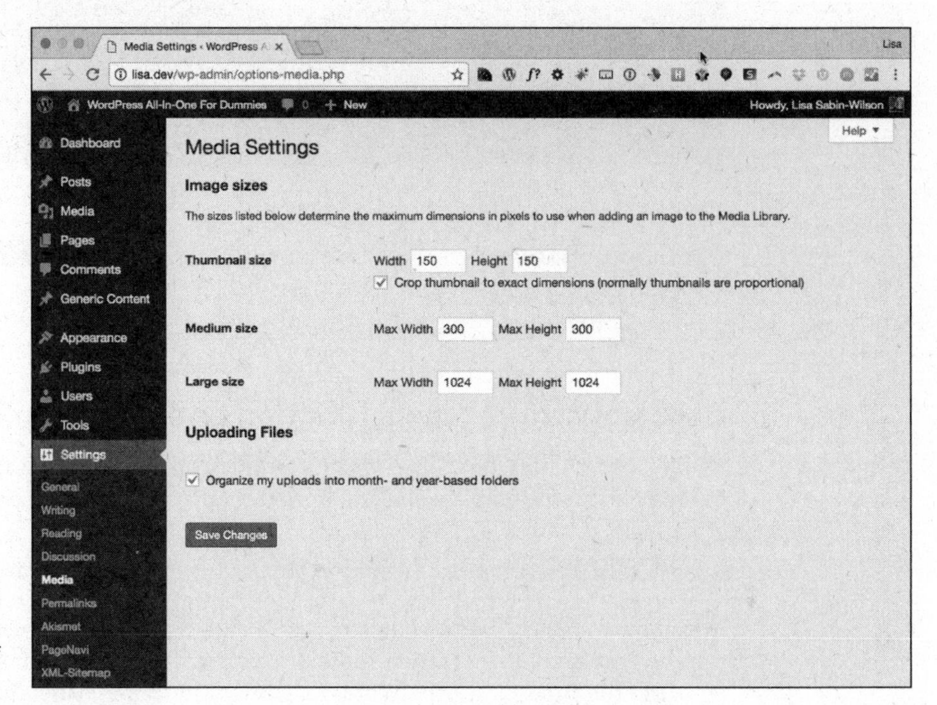

FIGURE 6-7: The Media Settings page of the Dashboard.

Custom image sizes for post thumbnails

If the default image sizes in WordPress (thumbnail, medium, large, and full) don't satisfy you, and you want to display images with different dimensions, WordPress makes it relatively easy to add custom image sizes to your Theme Functions template file. Then you use the `the_post_thumbnail` function to display the post thumbnail in your theme.

You can add unlimited sizes for your images. The following example shows how to add a new image size of 600px x 300px. Add this line to your Theme

Functions template file (`functions.php`) below the `add_theme_support('post-thumbnails')` function you added:

```
add_image_size( 'custom', 600, 300, true);
```

This code tells WordPress to create an additional version of the images you upload and to crop and resize them to 600px x 300px. Notice the four parameters in the `add_image_size` function:

- **Name (`$name`):** Gives the image size a unique name that you can use in your template tag. The image size in this example uses the name '`custom`'.
- **Width (`$width`):** Gives the image size a width dimension in numbers. In this example, the width is defined as `600`.
- **Height (`$height`):** Gives the image size a height dimension in numbers. In this example, the height is defined as `300`.
- **Crop (`$crop`):** This parameter (which is optional) tells WordPress whether to crop the image to exact dimensions or do a soft proportional resizing of the image. In this example, the parameter is set to `true` (accepted arguments: `true` or `false`).

Digging Into Advanced Theme Development

Adding the custom image size to your template to display the featured image is the same as adding default image sizes. The only difference is the name of the image set in the parentheses of the template tag. The custom image size in this example uses the following tag:

```
<?php if ( has_post_thumbnail() ) { the_post_thumbnail('custom'); ?>
```

Exploring Theme Options

One of the key features of an advanced theme is a Theme Options page. A Theme Options page allows the theme user to supply information to the theme without having to modify the theme files. Although a single-use theme could have this information hard-coded into the theme, that solution is an inelegant one. If the theme is used more than once or is managed by a nondeveloper, having an easy-to-change setting on the back end allows changes to be made quickly and easily.

Use a Theme Options page when the information is specific to the user and not to the theme design. Web analytics code (such as visitor-tracking JavaScript from Google Analytics or Woopra) is a good example of this user-specific information.

Because hundreds of analytics providers exist, most analytics providers require the JavaScript code to be customized for the specific site. The theme could have several different header and footer files, providing easy-to-use theme options. Adding JavaScript code to the header and the footer rather than requiring theme file modifications can make using your theme much easier.

TIP

Early in the design process, consider what a user may want to modify. Advanced uses of a Theme Options page vary widely and include design editors, color pickers, font options, and settings that modify the theme layout (switch a sidebar from one side of the theme to another, for example). The options offered depend on the project and the design.

Understanding theme-options basics

Before jumping into the code, you should understand the basic concepts of how theme options work.

Before WordPress Version 2.8, adding options to your theme required the developer to code the entire process, including providing an input form to accept the options, storing the options in the database, and retrieving the options from the database to use them. Fortunately, things have gotten much better. Some work is still required, but adding options is much easier now.

To let the user access the theme options, an input form is required. This process requires the most work because the form still needs to be manually created and managed. The form will need to be added to the back end so that the user can access it. Adding a new option to the Appearance menu allows the user to find the Theme Options page. Fortunately, WordPress offers an easy-to-use function called `add_theme_page`. To have WordPress manage as much as possible for you, the code needs to tell WordPress to store the data. The `register_setting` function can handle this task.

Building a simple theme-options page

Now that you know what pieces you need to build the Theme Options page, you can jump into the code. Open a plain-text editor, and enter the code shown in Listing 6-3.

LISTING 6-3: The Theme Options Page

```
<?php
function cm_theme_options_init()                                          →2
 register_setting( 'cm_theme_options', 'theme_options' );                 →3
}
add_action( 'admin_init', 'cm_theme_options_init' );                      →5
function cm_theme_options_menu() {                                        →6
 add_theme_page( 'Theme Options', 'Theme Options', 'manage_options',
   'cm_theme_options', 'cm_theme_options_page' );                         →7
}
add_action( 'admin_menu', 'cm_theme_options_menu' );                      →9
function cm_theme_options_page() {                                        →10
 ?>
 <div class="wrap">                                                       →12
 <?php screen_icon(); ?>                                                  →13
 <h2>Theme Options</h2>                                                   →14
 <form method="post" action="options.php">                                →15
 <?php settings_fields( 'cm_theme_options' ); ?>                          →16
 <?php $options = get_option( 'theme_options' ); ?>                       →17
 <Table class="form-table">                                               →18
 <tr valign="top">                                                        →19
<th scope="row">Checkbox</th>                                             →20
 <td><input name="theme_options[checkbox]" type="checkbox"
   value="1" <?php checked('1', $options['checkbox']); ?>/></td>          →21
 </tr>
 <tr valign="top"><th scope="row">Text</th>                               →23
 <td><input type="text" name="theme_options[text]" value="<?php echo
   $options['text']; ?>"/></td>                                           →24
 </tr>
 <tr valign="top">                                                        →26
<th scope="row">Text Area</th>                                            →27
 <td><textarea name="theme_options[text_area]"><?php echo $options
   ['text_area']; ?></textarea></td>                                      →28
 </tr>
 </table>
 <p class="submit"><input type="submit" class="button-primary"
   value="<?php _e( 'Save Changes' ); ?>"/></p>                           →31
 </form>
 </div>
 <?php
}
?>
```

Digging Into Advanced Theme Development

Here's a brief explanation of what the various lines do:

→2 This line creates a new function that calls register_setting, the function that tells WordPress about the need to store data.

→3 This line tells WordPress that you're creating a new settings group called cm_theme_options. The theme_options argument sets the WordPress options name used to store and retrieve the theme options. You'll want to change these option to be unique to your theme so that you don't accidentally load or save over settings from other themes or plugins.

→5 The new cm_theme_options_init function needs to be called to work. This line causes the function to be called during the admin_init action, which is a good action to use to run functions that need to be called on each admin page load.

→6 This new function handles registering the new menu entry that will show your form.

→7 The add_theme_page function adds a new menu entry on the Appearance menu. In order, the arguments are page title (appears on the title bar of the browser), menu entry name, required access level to visit the page, the variable name of the page (this needs to be unique for the page to work), and the function that should be run when visiting the menu location. This last argument (cm_theme_options_page in this example) is the name of the function that holds the options form.

→9 The new cm_theme_options_menu function needs to be called to work. This line causes the function to be called during the admin_menu action, which is when new menu entries should be added.

→10 This new function produces the input form for editing the theme options.

→12 Wrapping a form in the wrap class applies WordPress's default formatting.

→13 This line outputs the Appearance icon in front of the heading that follows.

→14 This line adds a title to the Theme Options page.

→15 This line starts the HTML form with an action that points to options.php and handles saving the data.

→16 The settings_fields is a function that adds some hidden inputs that allow the options to save properly. The cm_theme_options argument must match the first argument passed to the register_setting function. If this function is missing or if the argument doesn't match the first argument of the register_setting function, the options won't save properly.

→17 This line loads the saved theme options into the $options variable. The theme_options argument must match the second argument passed to the register_setting function.

→18 This line gives the table a class of form-table that applies WordPress's default form styling.

→19–20 These lines start a new row to hold the first option and add a description row header (the content inside the th tag). As indicated by the description, this option is a generic check-box input.

→21 This line adds the check-box input. The checked function from WordPress handles outputting the required HTML if a checked state was previously saved. The theme_options[checkbox] portion matches the second argument passed to the register_setting function, followed by the name of the specific option (in this case, checkbox). The $options['checkbox'] loads the specific option from the $options array.

→23 This line starts a new row to hold another option and adds a description row header (the content inside the th tag). As indicated by the description, this option is a generic text input.

→24 This line adds the text input. The echo outputs the existing value so that it prepopulates the input. The theme_options[text] portion matches the second argument passed to the register_setting function, followed by the name of the specific option (in this case, text). The $options['text'] loads the specific option from the $options array.

→26–27 These lines start a new row to hold another option and add a description row header (the content inside the th tag). As indicated by the description, this option is a generic text-area input.

→28 This line adds the text area input. The echo outputs the existing value so that it prepopulates the input. The theme_options[text_area] portion matches the second argument passed to the register_setting function, followed by the name of the specific option (in this case, text_area). The $options['text_area'] loads the specific option from the $options array.

→31 This line adds a button with a label of Save Changes. Giving the input a class of button-primary and wrapping it in a p tag with a class of submit applies WordPress's default button styling.

To load this file, you need to add a line of code to the theme's `functions.php` file. Edit the `functions.php` file, and add the following line at the bottom of the file:

```
require_once( 'theme-options.php' );
```

Click the Themes Options link on the Appearance menu. The Theme Options screen appears, as shown in Figure 6-8.

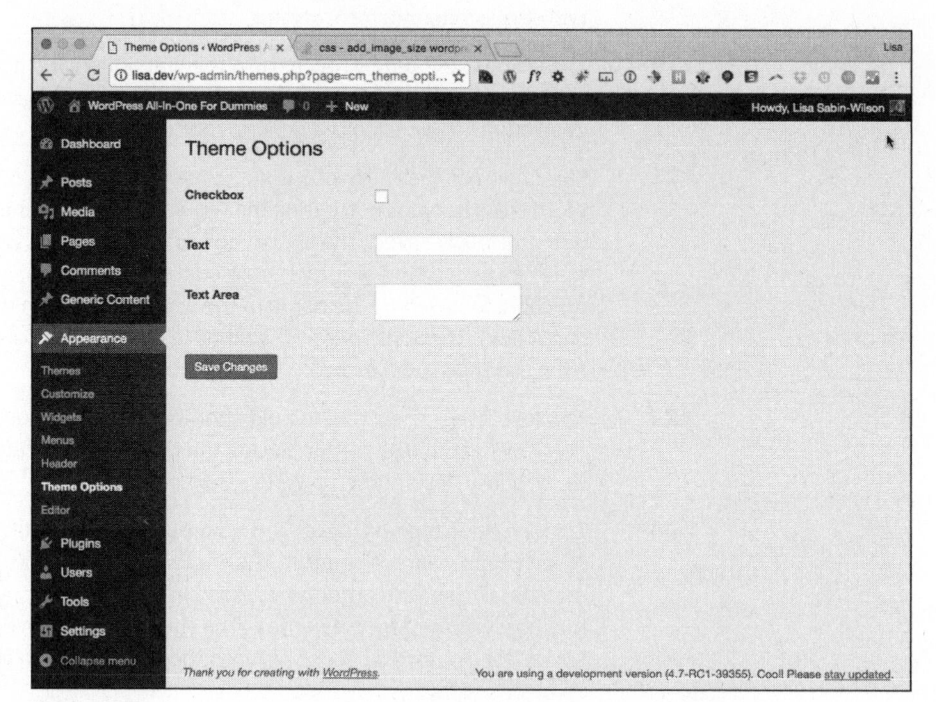

FIGURE 6-8: The new Theme Options page in the WordPress back end.

Using theme options in the theme

Compared with setting up a Theme Options page, using the stored options is very easy. To make it easier, add the following code (a quick function that makes using the options as simple as making a single function call) to your theme's `functions.php` file:

```
<?php
function get_theme_option( $option_name ) {
    global $theme_options;
        if ( ! isset( $theme_options ) )
            $theme_options = get_option( 'theme_options' );
```

```
        if ( isset( $theme_options[$option_name] ) )
            return $theme_options[$option_name];
    return '';
}
?>
```

The get_theme_option function takes an option name as its only argument and returns that option's value. To get the check-box option value, for example, simply call get_theme_option('checkbox').

If your theme has a section that can be enabled and disabled by a theme option check box, a section of code such as the following works very well:

```
<?php if ( get_theme_option('checkbox') ) : ?>
    <!-- example code -->
<?php endif; ?>
```

Typically, text or text area options output user-provided content. By using a check to see whether an option has a value, your theme can offer a default set of text that can be overridden by text entered in a theme option:

```
<div class="footer-right">
    <?php if ( get_theme_option('text') ) : ?>
        <?php echo get_theme_option('text'); ?>
    <?php else : ?>
        <p>Sample Theme by Lisa Sabin-Wilson</p>
    <?php endif; ?>
    <p>Powered by <a
href="http://wordpress.org">WordPress</a></p>
</div>
```

» Exploring popular theme frameworks

» Recognizing common framework features

Chapter 7

Using Theme Frameworks to Simplify Customization

As theme development for WordPress became a more complex task, theme designers began to realize that they were using the same snippets of code and functions repeatedly to accomplish the same tasks in every theme they built. When it came time for them to upgrade their themes (such as when WordPress released a new version with new features), they found themselves updating the same functions and adding the same features over and over to several themes they had developed. This situation is why theme frameworks were born. Essentially, a *theme framework* is a single theme that's a foundation for other themes to be built on.

Book 6, Chapter 5 discusses child themes, including how to build them. With theme frameworks, the parent theme (the framework) contains all the WordPress functions and template tags, and you can build child themes on top of them. The nice thing about this setup is that the original theme developer has to update only one theme — the framework — to upgrade all his theme offerings (the child themes).

Frameworks come with the tools developers can use to make a custom theme with great efficiency. Using a framework that provides these tools is much faster than building your own tools every time you want to modify a standard theme.

In this chapter, I explore some popular theme frameworks and the tools these frameworks contain, which make them appealing to developers who want to create custom themes.

Understanding Theme Frameworks

Many theme frameworks are available on the WordPress market. The goal of these frameworks is to allow you to create custom websites and themes without requiring you to be an expert programmer. Creating custom layouts, designs, and functionality can be difficult, and theme frameworks bridge the gap.

At its core, a theme framework is still just a WordPress theme. You install it and activate it just as you do any other theme. The real power of theme frameworks usually comes in theme options, child themes, and layout customization. One of the most important aspects of using theme frameworks is starting with the right one for your project.

When you install a theme framework, you may be surprised to find limited or no styling in the theme. Generally, theme frameworks are meant to be blank canvases that you fill with your own styles. The goal for a framework is to get out of your way while you're developing. By doing so, it allows you to use the provided tools instead of having to remove a lot of unnecessary elements and styling.

Think of a framework as being like a toolbox. All the tools you need are packaged nicely inside. You take out only the tools that you need for a given project.

Discovering Popular Frameworks

Many theme frameworks are available from a variety of sources. Here's a look at a few of the most popular theme frameworks.

Hybrid Core

The Hybrid Core framework features 15 custom page templates and 8 widget-ready areas. Additionally, six child themes are available from the Hybrid Core website at http://themehybrid.com/hybrid-core. Figure 7-1 displays the Spring Song theme, which is available as a Hybrid Core child theme. Hybrid Core also supports a series of add-on plugins specific to this theme. These add-ons include such features as a Tabs plugin, Hooks plugin, and Page Template packs.

FIGURE 7-1: Spring Song child theme for Hybrid Core.

TIP

The Hooks plugin in particular can be very handy if you're unfamiliar with PHP programming because it provides a graphical interface to latch into hooks, which I explore later in this chapter.

Hybrid Core, its child themes, and all its add-on plugins are available free. You can download them from http://themehybrid.com.

Key features of Hybrid Core include

- Theme Options menu
- Support for child themes
- Add-on plugins that extend functionality

Genesis

Genesis (see Figure 7-2) includes six default layout options, a prerelease security audit from WordPress lead developer Mark Jaquith, and a comprehensive array of search engine optimization (SEO) settings. Another great feature of Genesis is a built-in theme store on the WordPress Dashboard that allows you to easily choose, purchase, and activate more than 18 child themes. Like other frameworks, Genesis has some theme-specific plugins that add functionality. You can purchase Genesis from StudioPress (`www.studiopress.com`) for $59.95; it includes one child theme. Additional child themes are available for anywhere from $99.95 to $129.95 each.

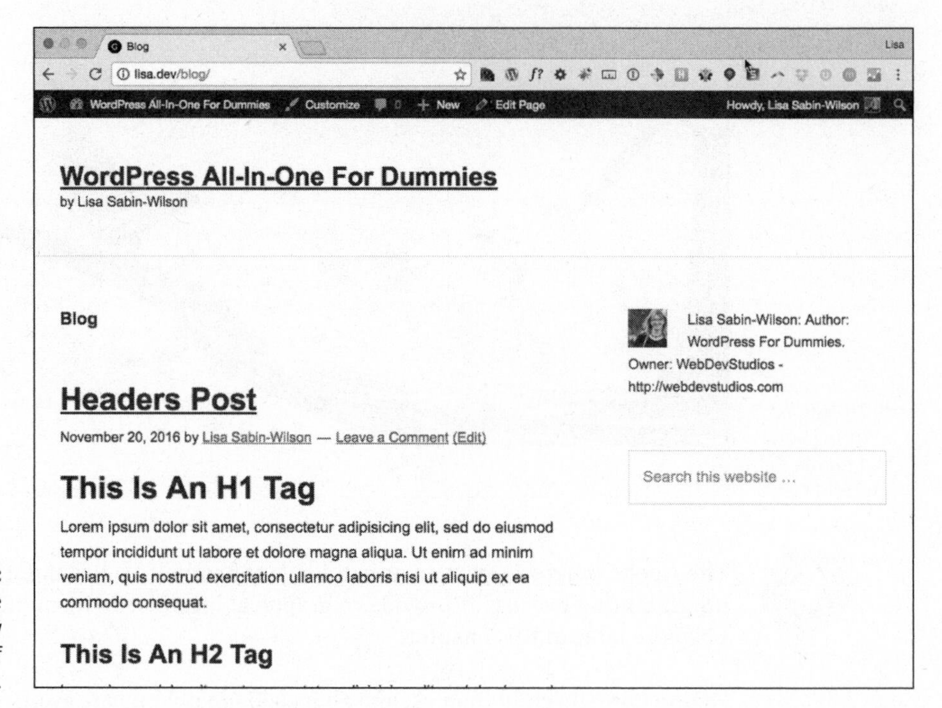

FIGURE 7-2: A home page with a new installation of iThemes Builder.

Key features are

- Theme Options menu
- Support for child themes
- Style manager for customization
- SEO Options
- Add-on plugins that extend functionality

Using Common Framework Features

Theme frameworks offer a host of features that make your life easier when it comes to building a website. Individual frameworks offer many unique features but also a common set of features. These common features generally allow for faster and easier development of your WordPress website.

Theme functions

Most themes include a `functions.php` file that contains functions for the theme, but some theme frameworks take this file to the next level by offering customization options with these functions that rival many plugins.

In the Genesis theme, a custom function allows you to create new widget areas. The `genesis_register_sidebar()` function, which takes care of the heavy lifting for widgetizing a new area, gives you some options that let you customize it easily. The following example shows how you might use this function in your theme's `functions.php` file:

```
genesis_register_sidebar(array(
    'name'=>'My New Widget',
    'description' => 'This widget is new.',
    'before_title'=>'<h4 class="mywidget">',
    'after_title'=>'</h4>'
));
```

This function allows you to enter a few customizations into the function, such as a name for your widget, a description, and any HTML that you want to appear before and after your widget.

Many standard themes also provide functions that are used in the theme, but theme frameworks offer many additional custom functions that the theme doesn't. The custom functions help you make the theme do exactly what you need for a specific site.

REMEMBER

Theme functions can vary greatly, so most theme authors have ample documentation through their site or forums where you can get more information about available functions.

Hooks

Many theme frameworks provide *hooks* to allow you to access and modify features of the theme. Hooks are small functions in WordPress that allow you to "hook into" other functions within WordPress and its themes or plugins. Hooks may seem to be a little advanced, but with a little practice, hooks are quite efficient at modifying a theme. Theme frameworks provide hooks that allow you to latch into functions of the theme and then call or modify them at a specific time.

The two types of hooks are

- **Action:** Performs during the loading of the theme when you can latch in a specific function. If you want one of your functions to execute at the same time a file in the theme is loaded, you can use an action to hook the function to that file's load.
- **Filter:** Modifies data while it passes to the theme or to the browser screen. Some theme frameworks allow you to filter the classes that the theme applies to elements.

You can find an example of modifying your theme with hooks in iThemes Builder. Consider the hook to add meta data: `builder_add_meta_data`. This hook can be useful for SEO in your theme.

To add custom meta output, you can replace the default function with a custom one. To do that, remove the existing action and add your own, like this:

```
<?php
remove_action('builder_add_meta_data','builder_seo_options');
add_action('builder_add_meta_data','my_custom_builder_seo_options');
?>
```

The default SEO options were removed with `remove_action` and replaced with a new function called `my_custom_builder_seo_options`. Next, you need to define what `my_custom_builder_seo_options` will do, as follows:

```
<?php
function my_custom_builder_seo_options() {
   /*Add custom seo options here.*/
}
?>
```

This code creates a basic PHP function where you can define what custom SEO options you want to use in the theme.

Because many frameworks have 100 or more hooks, most frameworks provide documentation (through their websites or forums) stating what hooks are available, what each hook does, and what parameters are available to modify how each hook works.

Child themes

Some frameworks allow you to modify the theme by using child themes. (Find out more about the parent/child theme relationship in Book 6, Chapter 5.) Child themes can be as simple as stylesheets, but they derive their power from the parent theme's template and function files.

A huge advantage of using a child theme is that it protects any customizations you make from being overwritten if a newer version of the theme comes out. For frameworks, this protection is especially important because changes in the core theme may be more frequent than regular themes due to the need to add more hooks, functions, or options over time.

Hybrid Core, Genesis, and iThemes Builder all extend their frameworks through child themes. Many frameworks provide child themes for free; others build child themes to sell.

Layout options

The ability to change the layout of a framework is important for many users. Different frameworks use different methods for this purpose. Some frameworks use template files to allow layouts to be reorganized; others provide an interface to allow layouts to be created from scratch.

Styling

Many theme frameworks incorporate methods that let you customize the style of the website without needing to know CSS (Cascading Style Sheets). (See Book 6, Chapter 4.) Frameworks that use a what-you-see-is-what-you-get (or WYSIWYG) style editor, allow you to easily match your theme's colors and design to your branding. Additionally, many editors include color pickers so that you don't have to use hexadecimal values to choose colors.

Other common elements that can be styled are borders, fonts, and headers.

Customizing Theme Frameworks

You may find that you're changing the same elements every time you set up a website by using a theme framework. Here is a list of where to start when customizing a theme framework for a website:

TIP

- **Add a custom header or logo image.** Adding a nice header graphic to a site makes it look unique from the beginning. If graphics aren't your specialty, use a good designer for your header graphic. This crucial element will catch your visitors' eyes as soon as they hit the page.
- **Change the colors of the background and links to match the header and branding of the site.** Many frameworks provide a simple interface for this task; others require you to open the `style.css` file to change the color. Either way, changing the background and link colors to match the site adds cohesiveness throughout the branding of the site.
- **Consider the home page layout.** This consideration is based on what the site is trying to achieve. If you want the site to focus on the blog, for example, you might place it on the front page with a left or right sidebar. If your site is more static, you might create a layout with many widget areas on the home page that display things from around the site. The home page is the landing point for many of your site's visitors, so it's important to consider its layout early in the process.
- **Decide how to lay out the inside pages and blog post pages.** These pages are just as important as the home page. The form and function of the page and blog post layouts need to be well planned to accommodate the parts of the site that you want every user to see, such as ads on the sidebar or an email newsletter sign-up form at the bottom of every post. In the case of pages, you might include information about your products and services in the sidebars and feature areas.
- **Add a contact form to the site.** Don't overlook installing one of those vital items on almost every site. Some frameworks offer built-in contact forms that you can add to any page.

If your theme framework doesn't offer a built-in contact form, many free plugins include this functionality, including Contact Form 7 from `http://contactform7.com`. If you're looking for a more robust form plugin, Gravity Forms by Rocket Genius (`www.gravityforms.com`) is one of the best form-creation plugins. It's a premium plugin but, at $39 per year, well worth the price.

7

Using and Developing Plugins

Contents at a Glance

IN THIS CHAPTER

» **Extending WordPress with plugins**

» **Comparing plugins to themes**

» **Finding plugins**

» **Evaluating the trustworthiness of plugins**

Chapter 1
Introducing WordPress Plugins

Half the fun of running a WordPress-powered website is playing with the hundreds of plugins that you can install to extend your site's functions and options. WordPress plugins are like those really cool custom rims you put on your car: Although they don't come with the car, they're awesome accessories that make your car better than all the rest.

By itself, WordPress is a very powerful program for web publishing, but by customizing WordPress with *plugins* — add-on programs that give WordPress almost limitless ways to handle web content — you can make it even more powerful. You can choose any plugins you need to expand your online possibilities. Plugins can turn your WordPress installation into a full-featured gallery for posting images on the web, an online store to sell your products, a user forum, or a social networking site. WordPress plugins can be simple, adding a few minor features, or complex enough to change your entire WordPress site's functionality.

To help you use plugins to customize your site, this chapter introduces you to plugins: what they are, how to find and install them, and how to use them to enhance your site and make it unique. Using plugins can also greatly improve your visitors' experiences by providing them various tools to interact and participate — just the way you want them to!

DEVELOPING PLUGINS — A COMMUNITY ACTIVITY

Although plugins are written and developed by people who have the skills required to do so, the WordPress user community is also largely responsible for the ongoing development of plugins. Ultimately, the end users are the ones who put those plugins to the true test on their own sites. Those same users are also the first to speak up and let the developers know when something isn't working right, helping the developers troubleshoot and fine-tune their plugins.

The most popular plugins are created by developers who encourage open communication with the user base. Overall, WordPress is one of those great open-source projects in which the relationship between developers and users fosters a creative environment that keeps the project fresh and exciting every step of the way.

Finding Out What Plugins Are

A *plugin* is a small program that, when added to WordPress, interacts with the software to provide some extensibility to the software. Plugins aren't part of the core software; neither are they software programs themselves. They typically don't function as stand-alone software. They do require the host program (WordPress, in this case) to function.

Plugin developers are the people who write these gems and share them with the rest of us — usually for free. As is WordPress, many plugins are free to anyone who wants to further tailor and customize a site to meet specific needs.

Thousands of plugins are available for WordPress — certainly way too many for me to list in this chapter alone. I could, but then you'd need heavy machinery to lift this book off the shelf! Here are just a few examples of things that plugins let you add to your WordPress site:

- **Email notification:** Your biggest fans can sign up to have an email notification sent to them every time you update your website.
- **Social media integration:** Allow your readers to submit your content to some of the most popular social networking services, such as Digg, Twitter, Facebook, and Reddit.

- **Stats program:** Keep track of where your traffic is coming from; which posts on your site are most popular; and how much traffic is coming through your website on a daily, monthly, and yearly basis.

This chapter takes you through the process of finding plugins, installing them on your WordPress site, and managing and troubleshooting them.

Extending WordPress with Plugins

WordPress by itself is an amazing tool. The features built into WordPress are meant to be the ones that you'll benefit from most. All the desired site features that fall *outside* what is built into WordPress are considered to be the territory of plugins.

There's a popular saying among WordPress users: "There's a plugin for that." The idea is that if you want WordPress to do something new, you have a good chance of finding an existing plugin that can help you do what you want. Currently, more than 48,000 plugins are available in the WordPress Plugin Directory (`https://wordpress.org/plugins`), and this number is growing at a rate of a few new plugins each day. In addition, thousands of additional plugins that are outside the Plugin Directory are available for free or for a fee. So if you have an idea for a new feature for your site, you just may find a plugin for that feature.

Suppose that you want to easily add recipes to your site. A Google search for *wordpress plugin recipes* results in links to the EasyRecipe plugin (`https://wordpress.org/plugins/easyrecipe`) and the ZipList Recipe Plugin (`https://wordpress.org/plugins/ziplist-recipe-plugin`). You can find even more recipe-related plugins with a more in-depth search.

Identifying Core Plugins

Some plugins hold a very special place in WordPress, in that they ship with the WordPress software and are included by default in every WordPress installation.

For the past few years, two plugins have held this special position:

- **Akismet:** The Akismet plugin has the sole purpose of protecting your blog from comment spam. Although other plugins address the issue of comment spam, the fact that Akismet is packaged with WordPress and works quite well means that most WordPress users rely on Akismet for their needs. Book 3, Chapter 4 covers how to activate and configure Akismet on your site.

- **Hello Dolly:** The Hello Dolly plugin helps you get your feet wet in plugin development, if you're interested. It was first released with WordPress version 1.2 and is considered to be the oldest WordPress plugin. When the plugin is active, the tops of your Dashboard pages show a random lyric from the song "Hello, Dolly!"

Figure 1-1 shows the core plugins you find in WordPress when you first install it.

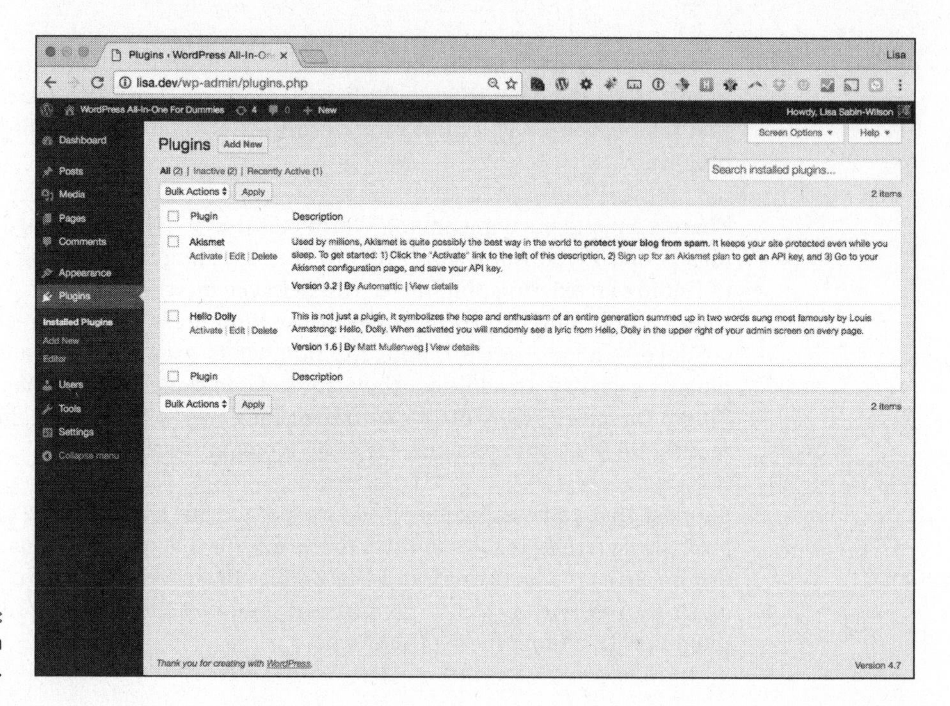

FIGURE 1-1: Core plugins in WordPress.

The idea of core plugins is to offer a base set of plugins to introduce you to the concept of plugins while also providing a benefit. The Akismet plugin is useful because comment spam is a big issue for websites. The Hello Dolly plugin is useful as a nice starting point for understanding what plugins are and how they're coded.

Although WordPress automatically includes these plugins, your site doesn't have to run them. Plugins are disabled by default and must be manually activated to be used. Core plugins can be deleted, just as any other plugins can, and they won't be replaced when you upgrade WordPress. (If you need to install or delete a plugin, turn to Book 7, Chapter 2.)

Future versions of WordPress may offer different sets of core plugins. It's possible that one or both of the current core plugins will cease being core plugins and that other plugins will be included. Although this topic has been much discussed in WordPress development circles over the past few years, at this writing, no definitive decisions have been made. So the current set of core plugins is likely to stay for a while longer.

Distinguishing between Plugins and Themes

Because themes can contain large amounts of code and add new features or other modifications to WordPress, you may wonder how plugins are different from themes. In reality, only a few technical differences exist between plugins and themes, but the ideas of what plugins and themes are supposed to be are quite different. (For more about themes, see Book 6.)

At the most basic level, the difference between plugins and themes is that they reside in different directories. Plugins are in the `wp-content/plugins` directory of your WordPress site. Themes are in the `wp-content/themes` directory.

TECHNICAL STUFF

The `wp-content/plugins` and `wp-content/themes` directories are set up this way by default. You can change both of these locations, but this change is very rarely made. The possibility is something to be aware of if you're working on a WordPress site and are having a hard time locating a specific plugin or theme directory.

The most important difference that separates plugins from themes is that a WordPress site always has one, and only one, active theme, but it can have as many active plugins as you want — even none. This difference is important because it means that switching from one theme to another prevents you from using the features of the old theme. By contrast, activating a new plugin doesn't prevent you from making use of the other active plugins.

REMEMBER

Plugins are capable of changing nearly every aspect of WordPress. The Multiple Themes plugin, for example (available at `https://wordpress.org/plugins/jonradio-multiple-themes`), allows you to use different themes for specific parts of your WordPress site. Thus, you can overcome even the limitation of having only one active theme on a site by using a plugin.

Because WordPress can have only one theme but many plugins activated at one time, it's important that the features that modify WordPress are limited to just plugins, whereas themes should remain focused on the appearance of the site.

For you, this separation of functionality and appearance is the most important difference between plugins and themes.

This separation of functionality into plugins and appearance into themes isn't enforced by WordPress, but it's a good practice to follow. You can build a theme that includes too much functionality, and you may start to rely on those functions to make your site work, which ultimately makes switching to another theme difficult.

The functionality role of plugins doesn't mean that control of the appearance of a WordPress site is limited to themes. Plugins are just as capable of modifying the site's appearance as a theme is. The WPtouch Mobile Plugin, for example (available at `https://wordpress.org/plugins/wptouch`), can provide a completely different version of your site to mobile devices such as smartphones by replacing the functionality of the theme when the user visits the site on a mobile device.

Other technical differences separate plugins and themes. The differences matter mostly to developers, but it could be important for you to know these differences as a nondeveloper WordPress user. Plugins load before the theme, which gives plugins some special privileges over themes and can even result in one or more plugins preventing the theme from loading. The built-in WordPress functions in the `wp-includes/pluggable.php` file can be overridden with customized functions, and only plugins load early enough to override these functions. Themes support a series of structured template files and require a minimum set of files to be valid. By comparison, plugins have no structured set of files and require only a single `.php` file with a comment block at the top to tell WordPress that the file is a plugin. One of the final technical differences is that themes support a concept called child themes, wherein one theme can require another theme to be present to function; no such feature is available to plugins.

Finding Plugins in the WordPress Plugin Directory

The largest and most widely used source of free WordPress plugins is the WordPress Plugin Directory (`http://wordpress.org/extend/plugins`), shown in Figure 1-2. This directory is filled with more than 48,000 plugins that cover an extremely broad range of features. Due to the large number of plugins that are freely available, as well as the fact that each plugin listing includes ratings and details such as user-reported compatibility with WordPress versions, the Plugin Directory should be your first stop when you're looking for a new plugin to fill a specific need.

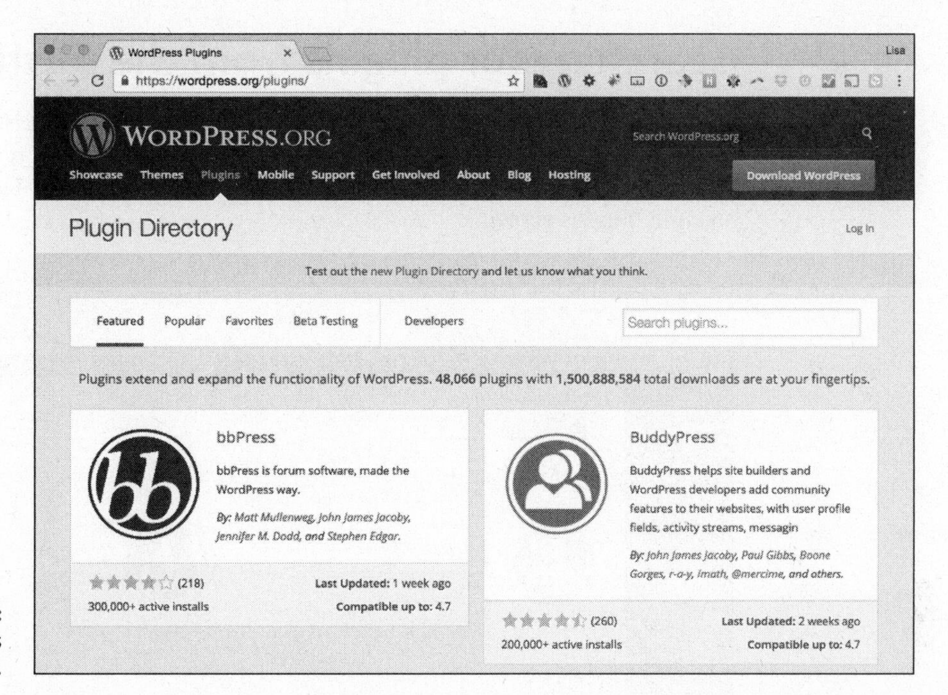

FIGURE 1-2: The WordPress Plugin Directory.

Plugins in the Plugin Directory should not be considered to be "official" or "supported" by WordPress. Anyone can submit plugins to the directory. Some restrictions exist on what can be listed in the Plugin Directory, but these restrictions mainly focus on licensing guidelines and blatant attempts to exploit users.

WARNING

Be critical of anything that you add to your site. Plugins receive very little code review after they're added to the Plugin Directory, and adding buggy code to your site can cause your site to crash. Also, adding malicious code to your site can enable other people to gain access to your site without your authorization. This isn't to say that plugins in the Plugin Directory shouldn't be trusted; rather, you should never add anything to your site without doing some checking-up on the plugin, theme, or code.

Although you can search for plugins on the Plugin Directory site, WordPress has a built-in feature for searching the Plugin Directory. This feature even allows you to install the Plugin Directory from WordPress without having to download the plugin and upload it to your site.

The following sections show you how to find plugins.

Searching for plugins from the Dashboard

After logging in to your WordPress Dashboard, click the Add New link on the Plugins menu. You see the Add Plugins screen, which you use to install plugins from the Dashboard; it's also where you can search for plugins. Figure 1-3 shows the Add Plugins screen.

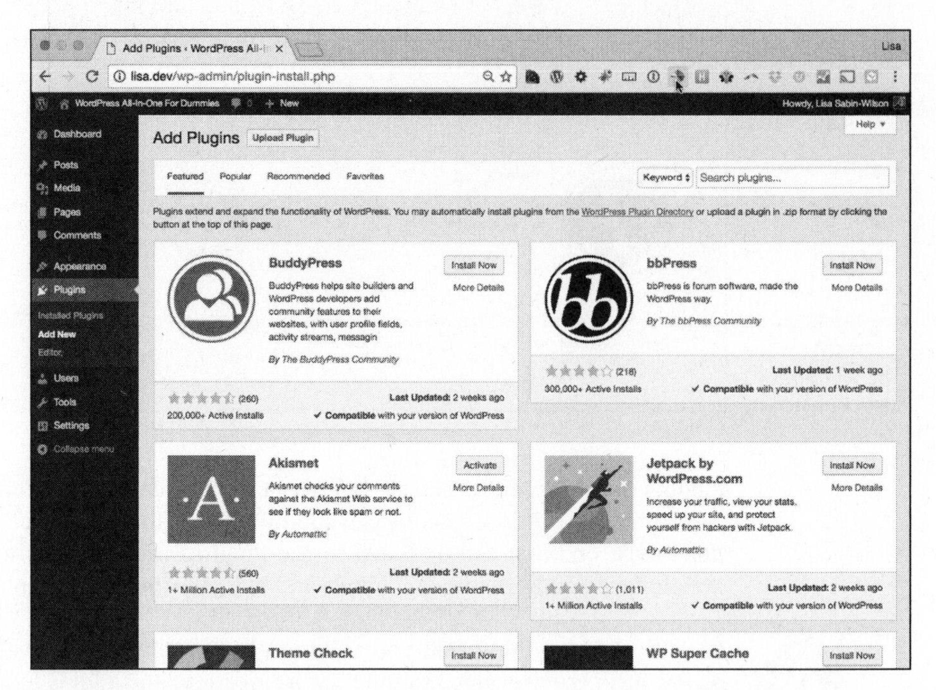

FIGURE 1-3: The Add Plugins screen.

At the top of the Add Plugins page is a series of links that provide ways to find plugins. (If you're looking to install a plugin, turn to Book 7, Chapter 2.)

Add Plugin

Figure 1-3 shows the Add Plugin screen, which allows you to search the WordPress Plugin Directory by typing terms in the Search box or by clicking the Featured, Popular, or Recommended tag links to narrow the list of plugins.

After you use either of the search options, the screen changes to a search-results page, which lists plugins that match the search query. As shown in Figure 1-4, the search results provide a wealth of information about each found plugin.

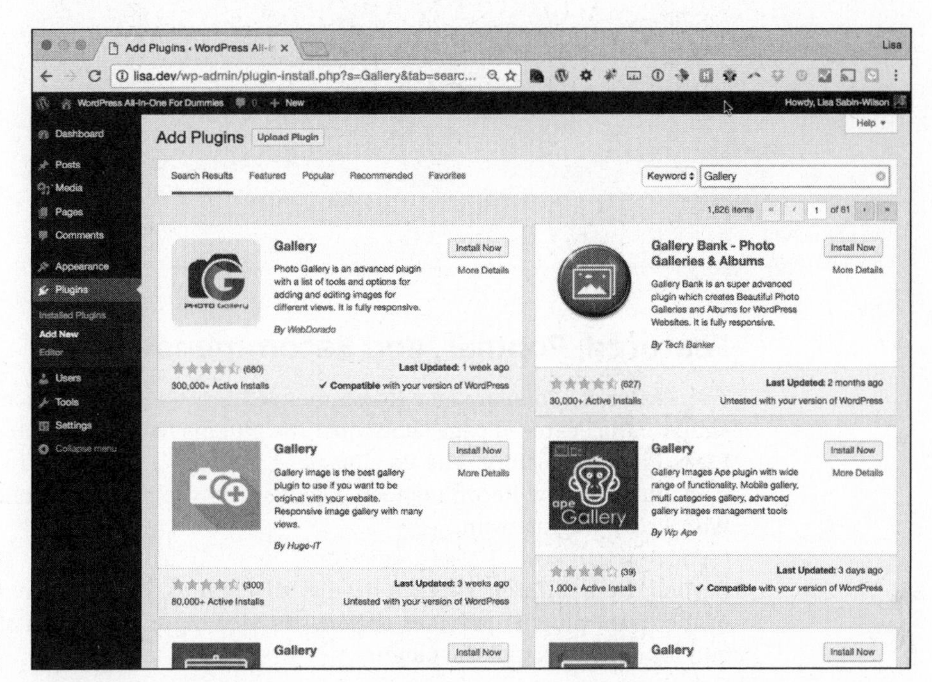

FIGURE 1-4: Plugin search results after a search for the term *Gallery*.

For each plugin listed on the page, the search results show the plugin's name, description, rating, time of last update, and compatibility with the version of WordPress you're using. Click the More Details link to read and discover more information about the plugin or the Install Now button to install the plugin on your site. (I describe these options in detail in "Evaluating plugins before installing" later in this chapter.)

Upload Plugin

Click the Upload Plugin button at the top of the Add Plugins screen (refer to Figure 1-3) to display the Upload Plugin screen, shown in Figure 1-5. The Upload section allows for easy installation of downloaded plugin `.zip` files without using SFTP or some other method to upload the files to the server. This feature makes it very quick and easy to install downloaded plugin `.zip` files. Although you can do this with plugins you find in the WordPress Plugin Directory, this feature is mostly used to install plugins that aren't available in the Plugin Directory because they can't be installed via a search in the Add Plugins screen.

After you upload your desired plugin file, the Add Plugins screen appears, and the options that activate the newly installed plugin become available.

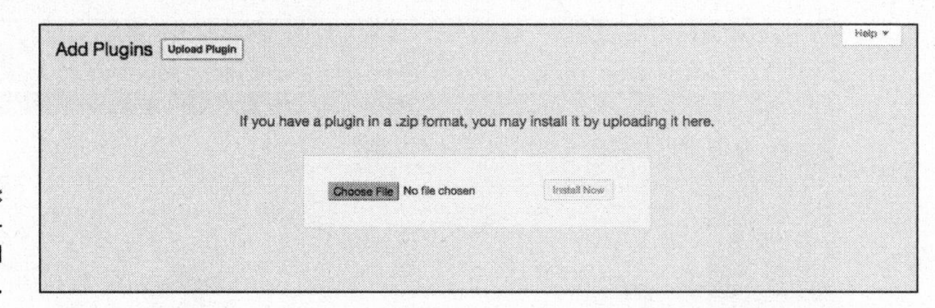

FIGURE 1-5: The Upload section of the Add Plugins page.

Featured, Popular, and Recommended

The Featured, Popular, and Recommended screens of the Dashboard are very similar. The Featured page shows just the plugins listed as Featured in the WordPress Plugin Directory. The Popular page includes a list of all the plugins sorted by popularity. The Recommended page lists all the plugins recommended by the WordPress plugins team.

Beyond these differences, each page is identical to the search-results page. Each of the listed plugins provides options for viewing more details about the plugin and for quickly install the plugin.

Finding plugins through WordPress.org

In the following sections, I show you how to find, upload, and install the popular Twitter plugin, developed by Twitter. I'm using the Twitter plugin as a real-world example to take you through the mechanics involved in downloading, unpacking, uploading, activating, and using a plugin in WordPress.

The Twitter plugin gives your readers the opportunity to share your content on the Twitter social networking site.

REMEMBER

Installing the Twitter plugin takes you through the process, but keep in mind that every plugin is different. Reading the description and installation instructions for each plugin you want to install is very important.

Finding and downloading the files

The first step in using plugins is locating the one you want to install. The absolutely best place to find WordPress plugins is the official WordPress Plugins Directory at `https://wordpress.org/plugins`, where, at this writing, you can find more than 48,000 plugins available for download.

To find the Twitter plugin, follow these steps:

1. **Go to the official WordPress Plugin Directory, located at `http://wordpress.org/plugins`.**
2. **In the Search box at the top of the Plugin Directory home page, enter the keyword** Twitter **and then click the Search button or press Enter.**
3. **Locate the Twitter plugin on the search-results page (see Figure 1-6), and click the plugin's name.**

   The Twitter plugin page opens in the WordPress Plugin Directory, where you find a description of the plugin as well as other information about the plugin (see Figure 1-7). In Figure 1-7, take note of the important information on the right side of the page:

   - *Download Version xx:* This button is the button that you click to download the plugin. The number represents the most recent version number of the plugin.
   - *Requires:* This line tells you what version of WordPress you need to use this plugin successfully. Figure 1-7 shows that the Twitter plugin requires WordPress version 3.9 or later, which means that this plugin doesn't work with WordPress versions earlier than 3.9. Helpful!
   - *Compatible Up To:* This line tells you what version of WordPress this plugin is compatible up to. If this section tells you that the plugin is compatible up to version 4.2, for example, you usually can't use the plugin with versions later than 4.2. I say *usually* because the plugin developer may not update the information in this section — especially if the plugin files themselves haven't changed. The best way to check is to download the plugin, install it, and see whether it works! (Figure 1-7 shows that the Twitter plugin is compatible up to WordPress version 4.7.)
   - *Last Updated:* This line displays the date when the plugin was last updated by the author.
   - *Active Installs:* This number tells you how many times this plugin has been downloaded and installed by other WordPress users.
   - *Ratings:* With a rating system of 1 to 5 stars (1 being the lowest and 5 being the highest), you can see how other WordPress users have rated this plugin.

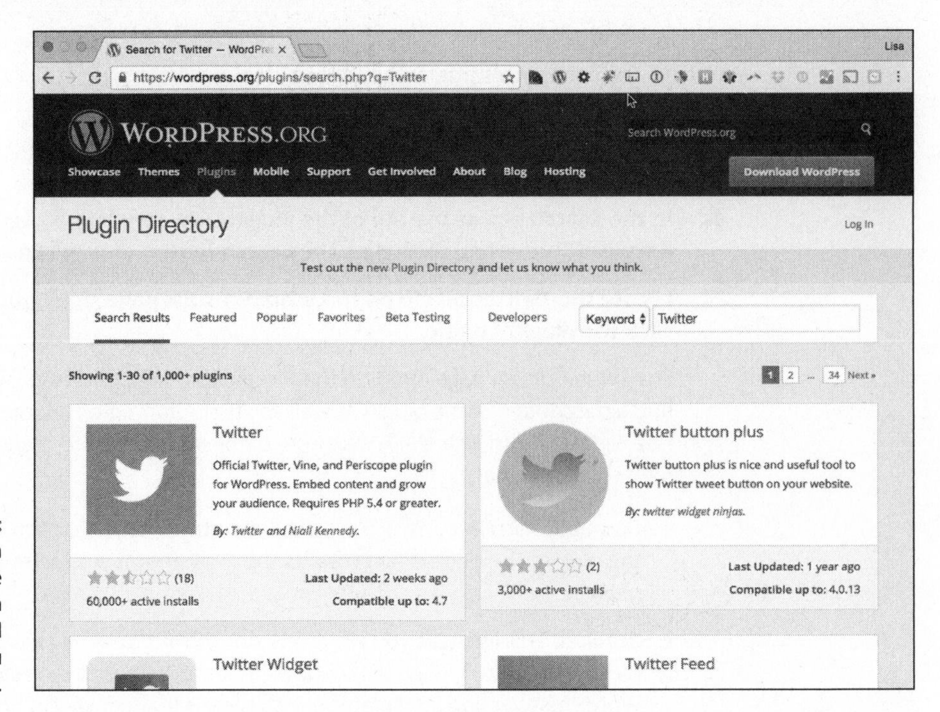

FIGURE 1-6: Use the search feature of the WordPress Plugin Directory to find the plugin you need.

FIGURE 1-7: The download page for the Twitter plugin.

4. **Click the Download button for the plugin version you want to download.**

   If you're using Internet Explorer, click the Download button, and a dialog box opens, asking whether you want to open or save the file. Click Save to save the `.zip` file to your hard drive, and *remember where you saved it.*

   If you're using Mozilla Firefox, click the Download button, and a dialog box opens, asking what Firefox should do with the file. Select the Save File radio button and then click OK to save the file to your hard drive. Again, *remember where you saved it.*

   For other browsers, follow the download instructions in the resulting dialog box.

5. **Locate the file on your hard drive and open it with your favorite decompression program.**

   If you're unsure how to use your decompression program, refer to the documentation available with the program.

6. **Unpack (decompress) the plugin files you downloaded for the Twitter plugin.**

Evaluating plugins before installing

When you've found a plugin via the Dashboard's Install Plugins page, you can find a wealth of information about that plugin to help you decide whether to download it or go on to the next one.

The methods for evaluating plugins described in this section are no substitute for thoroughly testing a plugin. Testing a plugin is good practice unless you're familiar enough with the code and the developers that bugs or security issues seem to be unlikely. To test a plugin, set up a stand-alone site to be used just for testing, install the plugin, and check for any issues before trusting the plugin on your main site.

TIP

Look at the version number of the plugin. If the version number includes *Alpha* or *Beta*, the plugin is being tested and may have bugs that could affect your site. You may want to wait until the plugin has been thoroughly tested and released as a full version. Generally, the higher the version number, the more *mature* (that is, tested and stable) the plugin is.

Don't rely on just one of these methods to assess the trustworthiness of a plugin. Combine them to get a sense of what other users think about the plugin. If the plugin has a five-star rating from 500 users but also has dozens of negative feedback comments with little positive commentary, don't trust the plugin very much. If a plugin has a three-star rating from 10 users but has nothing but positive comments, however, the plugin may have some issues yet may still work very well for some users.

As with many things in life, you have no guarantees with plugins. The best thing you can do is find information about the plugin to determine whether it is trustworthy.

Details

In the WordPress Dashboard Add Plugins screen (refer to Figure 1-3), click a plugin's More Details link to find information taken from the plugin's page in the WordPress Plugin Directory. Figure 1-8 shows the details that are available for the Twitter plugin, including tabs of information about the plugin, such as Description, Installation, Screenshots, and Changelog.

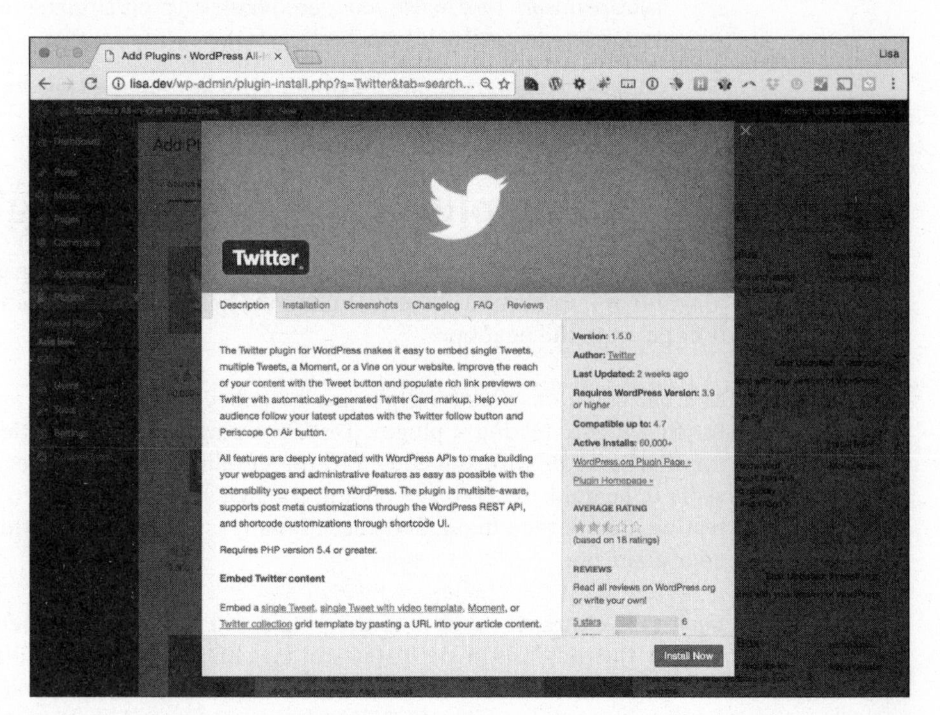

FIGURE 1-8: Details for the Twitter plugin.

Make sure to check out each plugin's Description page. You can find important information on this page that isn't present on the search-results page. When you're considering a plugin that you don't have experience with, this information can help you determine how reliable and trustworthy the plugin is.

Ratings

Consider the plugin's rating and the number of people who submitted a rating. The more people who rated the plugin, the more you can trust the rating; the fewer people who rated the plugin, the less you can trust the rating. A plugin that has fewer than 10 ratings probably isn't very trustworthy, or it's relatively new. A plugin that has more than 100 ratings is very trustworthy. Any plugin rated between 15 and 100 times is acceptably trustworthy.

If a plugin has a large percentage of one- or two-star ratings, treat the plugin very suspiciously. Take the extra step of visiting the plugin's page in the Plugin Directory to see what other people are saying about the plugin. You can do this by clicking the WordPress.org Plugin Page link on the right side of the Description page. On the plugin's page, as shown in Figure 1-9, click the View Support Forum link or the Support link at the top of the page to see the information posted by users, both positive and negative. You can determine whether the issues that other people experienced are likely to affect you.

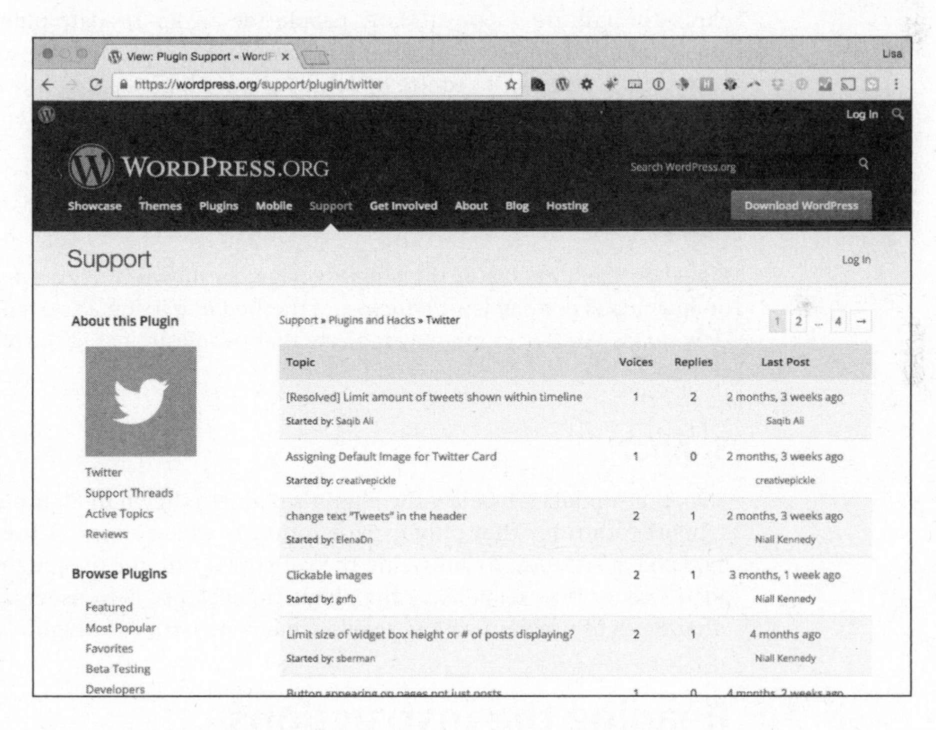

FIGURE 1-9: The Support page for the Twitter plugin.

Downloads

The next detail to consider is the number of downloads. The higher the number of downloads, the more likely the plugin is to work well; plugins that don't work very well typically don't pick up enough popularity to get many downloads. If a plugin has hundreds of thousands of downloads or more, it's extremely popular. Plugins with tens of thousands of downloads are popular and may grow even more popular. If the plugin has fewer than 10,000 downloads, you can't determine anything just by looking at the download count.

REMEMBER

A low download count doesn't necessarily count against a plugin. Some plugins simply provide a feature that has a very limited audience. Thus, the download count is an indicator but not proof of quality (or lack thereof).

You should take the Compatible Up To and Last Updated information very lightly. If a plugin indicates that the compatible-up-to version is for a very old WordPress version, it may have issues in the latest versions of WordPress, but plenty of plugins work just fine with current versions of WordPress even though they don't explicitly indicate support. Many people see an up-to-date plugin as a sign of quality and upkeep. This reasoning is flawed, however, because some plugins are very simple and don't require updating often. Plugins shouldn't be updated just to bump that number; thus, a plugin that hasn't been updated for a while may be functioning perfectly well without any updates.

Stats

Listed on the Stats tab of the plugin's page, as shown in Figure 1-10, the number of downloads per day isn't a foolproof method of getting a trusted plugin, but the Downloads Per Day graph may indicate that people are using the plugin with some success.

Support

Click the Support tab below the plugin's banner (shown in Figure 1-9) to view the support forum for that plugin. This forum is where users of the plugin request help and assistance. By browsing the support forum for the plugin, you can get a good feel for how responsive the plugin's developer is to users, and you can see what types of problems other people are having with the plugin.

Reading the instructions

Frequently, the plugin developer includes a `readme` file inside the `.zip` file. Do what the title of the file says: Read it. Often, it contains the exact documentation and instructions that you can find on the plugin developer's website.

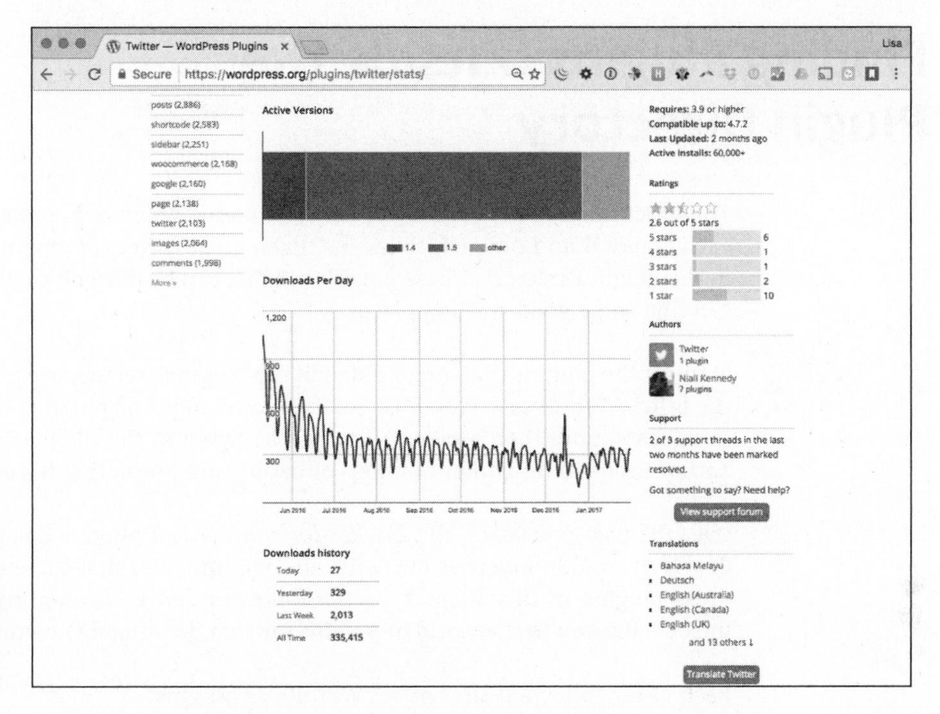

FIGURE 1-10: The download history for a plugin within the Plugin Directory.

Make sure that you read the instructions carefully and follow them correctly. Ninety-nine percent of WordPress plugins have great documentation and instructions from the plugin developers. If you don't follow the instructions correctly, the best scenario is that the plugin just won't work on your site. At worst, the plugin may create all sorts of ugly errors, requiring you to restart plugin installation from step one.

TIP

You can open `readme.txt` files in any text-editor program, such as Notepad or WordPad on a PC, or TextEdit on a Mac.

In the case of the Twitter plugin, the `readme.txt` file contains information regarding the requirements of the plugin and useful information on how to use the plugin after you have it installed and activated on your site.

REMEMBER

Every plugin is different in terms of where the plugin files are uploaded and what configurations and setup are necessary to make the plugin work on your site. Read the installation instructions very carefully, and follow those instructions to the letter to install the plugin correctly on your site.

Finding Plugins outside the Plugin Directory

The exact number of plugins outside the Plugin Directory is unknown. There are easily more than 1,000, which means that a great variety of plugins exists outside in the Plugin Directory. These outside plugins can be difficult to discover, but you can find some good starting places.

Many of the plugins that aren't listed in the Plugin Directory are paid plugins, and the official WordPress Plugin Directory allows only free plugins. If a plugin is for sale (costs more than a penny), it can't be listed in the Plugin Directory, so the author needs to find other methods of listing and promoting his or her products.

Over the past few years, the market for commercial plugins has grown tremendously. It wouldn't be possible to list all the companies that currently offer WordPress plugins in this chapter, so the following list is a sampling. It's a way of introducing you to the world of plugins outside the Plugin Directory.

Each of the following sites offers WordPress plugins:

- **CodeCanyon (`https://codecanyon.net`):** With thousands of plugins, this online marketplace is the paid plugin version of the Plugin Directory. Just as the Plugin Directory contains plugins from a large number of developers, CodeCanyon is a collection of plugins from various developers rather than plugins created by a single company.
- **Gravity Forms (`www.gravityforms.com`):** For many WordPress users, Gravity Forms is the plugin to pay for. Typically, it's the first and last recommendation people give to someone who wants to create forms in WordPress.
- **iThemes (`https://ithemes.com`):** Starting as a theme developer, iThemes branched out into developing plugins as well. The most popular offering is BackupBuddy, a plugin for backing up and restoring your sites.
- **WooCommerce (`https://woocommerce.com`):** WooCommerce is a popular plugin with a full-featured e-commerce solution that you can integrate into your WordPress site. Essentially, this plugin turns your WordPress site into an online store.

Although these sites give you a taste of what commercial plugin sites have to offer, having other sources of information about new, exciting plugins can also be very helpful. Many popular WordPress news sites talk about all things WordPress, including reviews and discussions of specific plugins. Check out the following sites if you want to know more about what plugins are being talked about:

- **WPBeginner (`www.wpbeginner.com`):** This site is dedicated to helping new WordPress users get up and running quickly. It features a very active blog on a variety of topics. The site often features posts about how to use plugins to create specific types of solutions for your site.
- **Post Status (`https://poststatus.com`):** Post Status is an all-things-WordPress news site. If there's buzz on a topic in the WordPress world, you're likely to find discussions about it here.

One of the great things about using a community or news site to discover new plugins is that you aren't alone in deciding whether to trust a plugin. You can get some outside opinions before you take a chance on a plugin.

If you aren't finding what you want in the Plugin Directory, don't know anyone who offers the solution you're looking for, and aren't seeing anything on community sites, it's time to go to a trusty search engine (such as Google) and see what you can find.

A good way to start is to search for the words *wordpress* and *plugin* along with one to a few words describing the feature you want. If you want a plugin that provides advanced image-gallery features, for example, search for *wordpress plugin image gallery.* As long as your search isn't too specific, you're likely to get many results. The results often contain blog posts that review specific plugins or list recommended plugins.

Some developers include malware, viruses, and other unwanted executables in their plugin code. Your best bet is to use plugins from the official WordPress Plugin Directory or to purchase plugins from a reputable seller. Do your research first. Read up on plugin security in Book 2, Chapter 5.

Comparing Free and Commercial Plugins

Thousands of plugins are available for free, and thousands of plugins have a price. What are the benefits of a free plugin versus a paid plugin? This question is a tough one to answer.

It's tempting to think that some plugins are better than others, which is why they cost money. Unfortunately, things aren't that simple. Some amazing plugins that I would gladly pay for are free, and some terrible plugins that I wouldn't pay for have a cost.

Often, a paid plugin includes support, which means that the company or individual selling the plugin is offering assurance that if you have problems, you'll receive support and updates to address bugs and other issues.

Free plugins typically list places to make support requests or to ask questions, but nothing ensures that the developer will respond to your requests within a certain period — or at all. Even though developers have no obligation to help with support requests from their plugin's users, many developers work hard to help users with reported issues and other problems. Fortunately, because many free plugins are in the Plugin Directory, and the Plugin Directory includes a built-in support forum and rating score, you can easily see how responsive a plugin author is to support issues.

Personally, I believe that the commercial plugin model works in an environment of tens of thousands of free plugins because many WordPress users want the assurance that when they have problems, they have a place to ask questions and get help.

So if people can get paid to produce plugins, why are so many plugins free? This question is another great one.

One reason why so many plugins are available for free is that many WordPress developers are generous people who believe in sharing their plugins with the community. Other developers feel that having their plugins available to the millions of WordPress users via the Plugin Directory is a great way to market their talents, which can lead to contract work and employment. Buzzwords on a résumé are far less valuable than a plugin you wrote that was downloaded thousands or millions of times.

Another reason to release a free plugin is to entice people to pay for upgrades — a model often referred to as *freemium*. Freemium plugins often have paid plugins that add features to the free plugin. Thus, the freemium model is a mix of the free and paid plugin models, and gets the best of both worlds. The free plugin can be listed in the Plugin Directory, giving the plugin a large amount of exposure. You can get a feel for how the plugin functions, and if you want the additional features, you can purchase the paid plugin.

An example of the freemium model is the WooCommerce plugin (`https://woocommerce.com`). The main plugin is available for free in the Plugin Directory (`https://wordpress.org/plugins/woocommerce`), yet it supports a large number of paid plugins to add more features. By itself, the WooCommerce plugin turns the site into a shopping cart. To extend this functionality, paid plugins are available to add payment processing for specific credit-card processors, shipping, download managers, and many other features.

The biggest difference between free and paid plugins is that sometimes, you won't find what you need in a free plugin and have to go with a paid plugin. In the end, what you download is up to you. Many great free plugins are available, and so are many great paid plugins. If you want the features offered by a paid plugin and are willing to pay the price, paid plugins can be very good investments for your site.

TIP

Many free plugins have links to donation pages. If you find a free plugin to be valuable, please send a donation to the developer. Most developers of free plugins say that they rarely, if ever, receive donations. Even a few dollars can really encourage the developer to keep updating old plugins and releasing new free plugins.

» Manually installing WordPress plugins

» Updating plugins

» Activating and deactivating plugins

» Taking plugins out of WordPress

» Using the core plugins

Chapter 2
Installing and Managing Plugins

With more than 48,000 plugins available, you have a huge number of options for customizing your site. Book 7, Chapter 1 details the types of plugins available and where to find them. In this chapter, you start putting these plugins to use. This chapter is dedicated to helping you install, activate, deactivate, update, and delete plugins.

Installing Plugins within the WordPress Dashboard

When you've found a plugin in the WordPress Plugin Directory (see Book 7, Chapter 1) that you want to install, you can install it directly from the Dashboard. (If you found a plugin that isn't in the Directory, you have to install it manually. See the later section "Installing Plugins Manually.")

WordPress makes it super-easy to find, install, and then activate plugins for use on your site. Just follow these simple steps:

1. **Click the Add New link on the Plugins menu.**

   The Install Plugins page opens, allowing you to browse the official WordPress Plugin Directory from your WordPress Dashboard.

2. **Search for a plugin to install on your site.**

   Enter a keyword for a plugin you'd like to search for. If you want to search for plugins that allow you to add features for integration with Twitter on your site, for example, enter the word **Twitter** in the Search Plugins text box to return a list of plugins that deal specifically with the social network called Twitter.

TIP

   You can also discover new plugins by clicking any of the provided categories at the top of the Add Plugins screen: Featured, Popular, and Recommended, to name a few.

   For this example, you want to install a plugin that will integrate your site with the social network Twitter. To find it, enter **Twitter** in the Search text box on the Install Plugins page; then click the Search button.

   Figure 2-1 shows the results page for the Twitter search phrase. The first plugin listed, called simply Twitter, is a plugin developed by the developers at Twitter. This plugin the one you want to install.

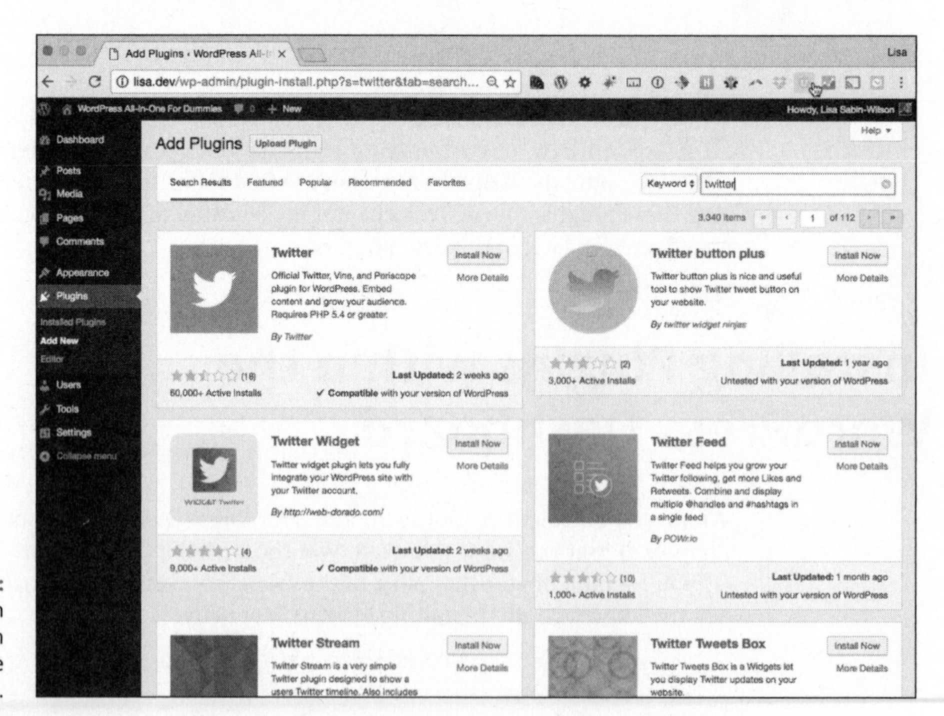

FIGURE 2-1: The Add Plugin screen's search results for the Twitter plugin.

3. **Click the More Details link.**

   A Description window opens, giving you information about the Twitter plugin, including a description, version number, author name, and Install Now button.

4. **Click the Install Now button.**

   You go to the Installing Plugins page within your WordPress Dashboard, where you find a confirmation message that the plugin has been downloaded, unpacked, and successfully installed.

5. **Specify whether to activate the plugin or proceed to the Plugins page.**

   Two links are shown below the confirmation message:

   - *Activate Plugin:* Click this link to activate the plugin you just installed on your blog.
   - *Return to Plugin Installer:* Click this link to go to the Install Plugins page without activating the plugin.

WARNING

Autoinstallation of plugins from your WordPress Dashboard works in most web-hosting configurations. Some web-hosting services, however, don't allow the kind of access that WordPress needs to complete the autoinstallation. If you get any errors or find that you're unable to use the plugin autoinstallation feature, get in touch with your web-hosting provider to find out whether it can assist you.

TIP

If the Dashboard displays any kind of error message after you install the plugin, copy the message and paste it into a support ticket in the WordPress.org support forum (`http://wordpress.org/support`) to elicit help from other WordPress users about the source of the problem and the possible solution. When you post about the issue, provide as much information about the issue as possible, including a screen shot or pasted details.

Installing Plugins Manually

Installing plugins from the Dashboard is so easy that you'll probably never need to know how to install a plugin manually via SFTP. (Book 2, Chapter 2 explains how to use SFTP.) But the technique is still helpful to know in case the WordPress Plugin Directory is down or unavailable.

Follow these steps to install a plugin by using SFTP:

1. **Go to the plugin's page in the WordPress Plugin Directory.**

   For this example, go to the Twitter plugin's page at `https://wordpress.org/plugins/twitter`.

2. **Click the red Download button to transfer the plugin's .zip file to your computer.**

3. **Unzip the plugin files.**

   All plugins that you download from the Plugin Directory are in `.zip` format. Most operating systems (Windows, Mac, and so on) have built-in tools to open `.zip` files. After opening the `.zip` file, extract the directory contained inside, and put it in a directory on your computer that's easily accessible.

4. **Connect to your site's server by using SFTP.**

   For details on how to use SFTP, see Book 2, Chapter 2. If you have any difficulty connecting to your server, contact your hosting provider and ask for assistance in connecting to your server via SFTP.

5. **Navigate to the `wp-content` folder within the WordPress installation for your website or blog.**

REMEMBER

   The location of your WordPress installation can differ with every hosting provider. Make sure that you know the location before you proceed. Check out Book 2, Chapters 2 and 4 for information on where the WordPress installation is located on your web server.

6. **Navigate to the `/wp-content/plugins` directory.**

   First, navigate to `wp-content`. Inside this directory are the `plugins` and `themes` directories, along with a few others. Navigate to the `plugins` directory, which is where all plugins reside.

7. **Upload the plugin's folder to the `/wp-content/plugins` directory on your web server.**

   The plugin folder is named for the plugin (if you're uploading the Twitter plugin, for example, the folder is `/twitter`) and contains all the files for that plugin.

Go to the Dashboard's Plugins page, and you see the new plugin listed. If you make a mistake, delete all the newly uploaded files and begin again.

Upgrading Plugins

For a lot of reasons, mainly related to security and feature updates, always use the most up-to-date versions of the plugins in your blog. But with everything you have to do every day, how can you possibly keep up with knowing whether the plugins you're using have been updated?

You don't have to. WordPress does it for you.

Figure 2-2 shows that an out-of-date version (3.1.1) of Akismet is installed. WordPress notifies you when a new update is available for a plugin in four ways, as shown in Figure 2-2 and Figure 2-3:

- **Dashboard Updates link:** The Updates link below the Dashboard menu displays a red circle with a white number. The number indicates how many plugins have updates available. (In Figure 2-2, one plugin on my site has an update available.) Click the Updates link to see which plugins have updates available.
- **Toolbar:** When a new update is available, a small icon appears on the toolbar at the top of your Dashboard, to the right of your site title, as shown in Figure 2-2.
- **Plugins menu's title:** The Plugins menu's title also displays a red circle with a white number. As with the Updates link, the number indicates how many plugins have updates available, as shown in Figure 2-2.

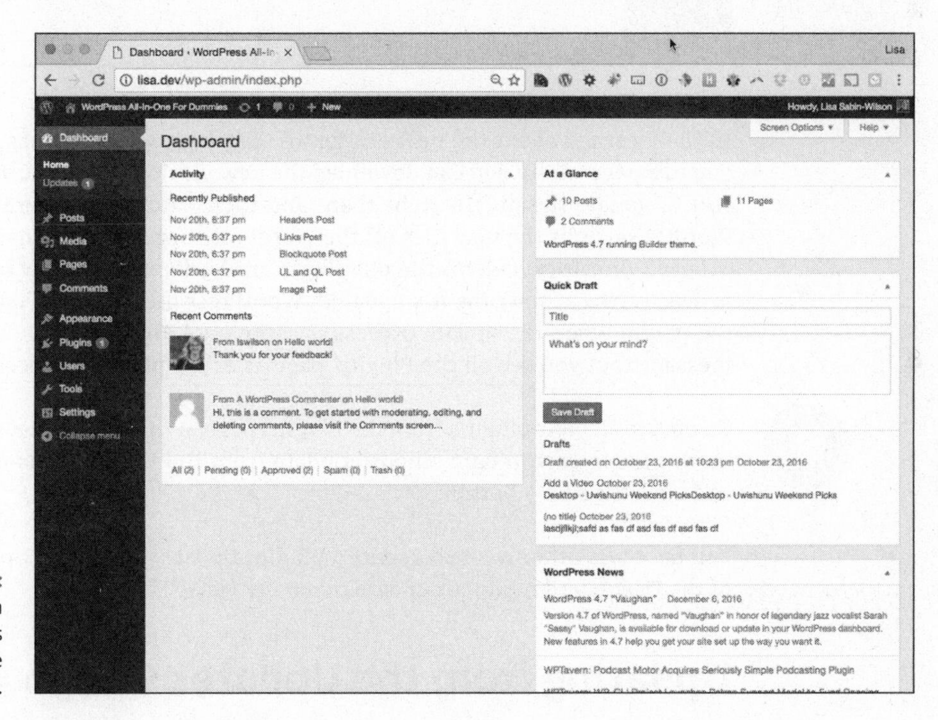

FIGURE 2-2: Various plugin update indicators displayed on the Dashboard.

- **Plugins page:** Figure 2-3 shows the Plugins page. Below the Akismet plugin is a message that says `There is a new version of Akismet available. View version 3.2 details or update now.`

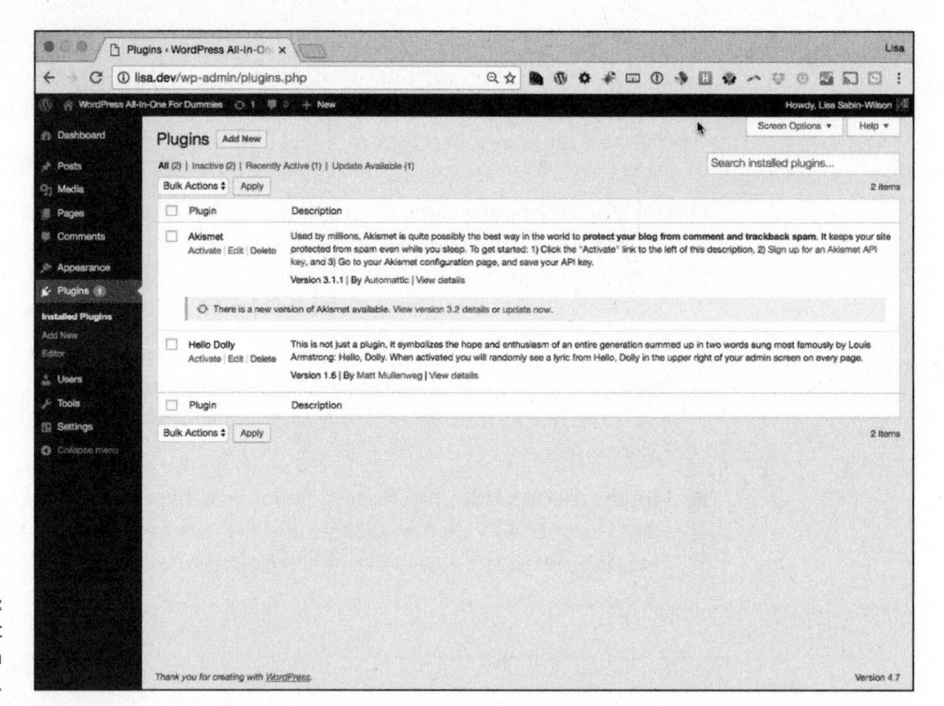

FIGURE 2-3: The Akismet plugin with an available update.

In the message about the new version of the plugin are two links. One link takes you to a page where you can download the new version, and the other link enables you to update the plugin right there and then. Click the Update Now link, and WordPress grabs the new files off the WordPress.org server, uploads them to your `plugins` directory, deletes the old plugin, and activates the new one. (If a plugin is deactivated at the time it's updated, WordPress gives you the option to activate the plugin when the update process is complete.) Figure 2-4 shows the `Updated` message that you see on the Plugins page after the plugin has been upgraded.

TIP

To update all your plugins from the Plugins page at the same time, select the check mark next to each plugin, choose Update from the Bulk Actions drop-down list, and click the Apply button.

You have two other ways to update a plugin: Update it from the Updates submenu of the Dashboard menu, or update manually via SFTP.

Updating from the Updates screen

The Updates screen, accessible as a submenu of the Dashboard menu, provides a quick way to update WordPress, plugins, and themes in one place. As shown in Figure 2-5, the Updates screen is a one-stop shop for all the updates across your site.

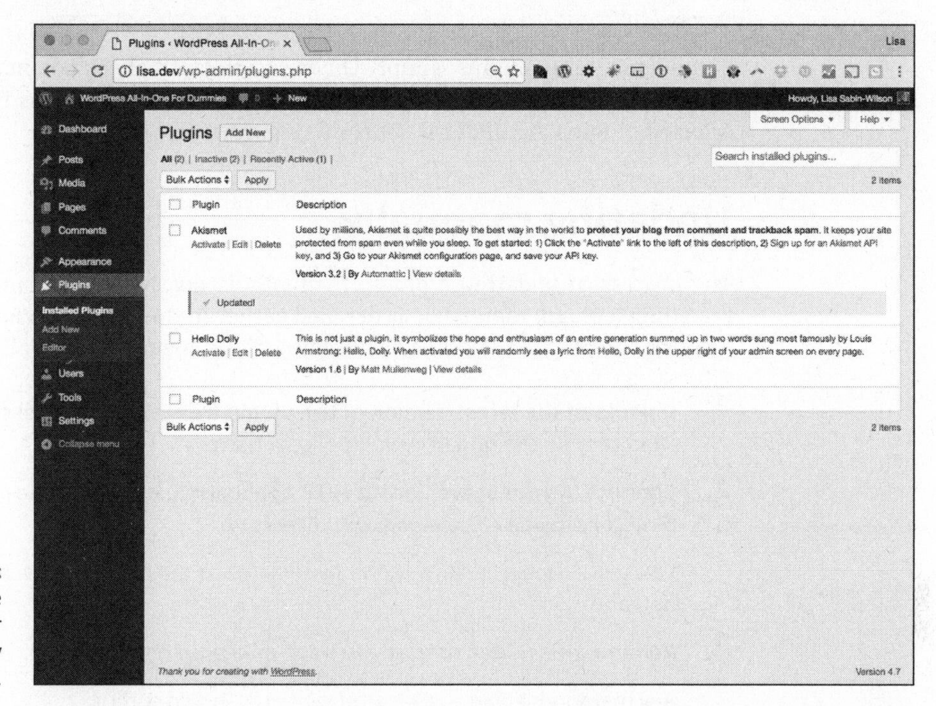

FIGURE 2-4: Updated message appears after you successfully update a plugin.

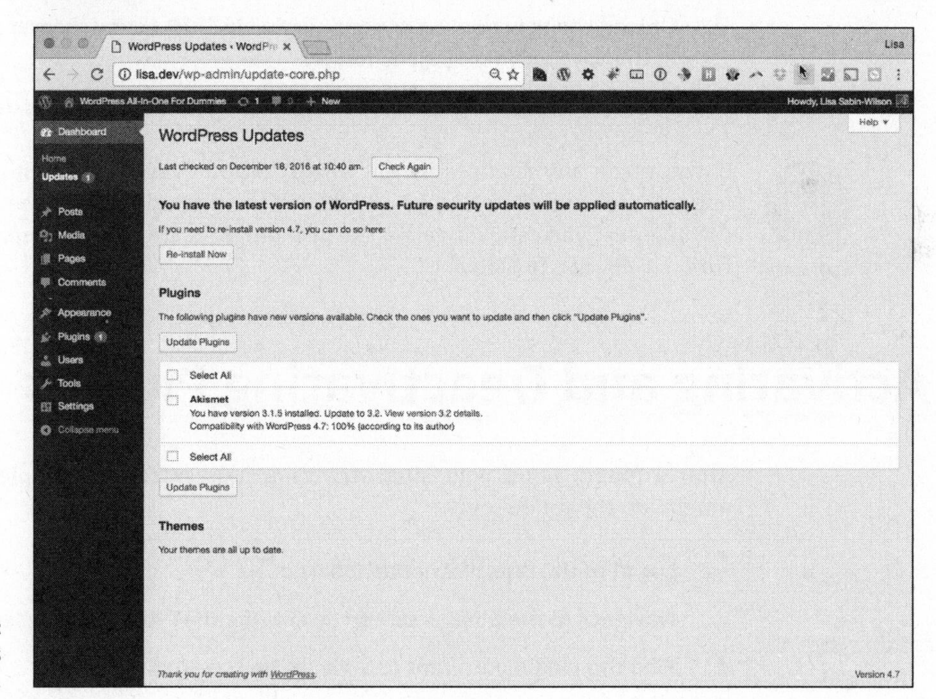

FIGURE 2-5: The WordPress Updates screen.

To update all the plugins, simply choose a Select All check box next to one of the headers (as shown in Figure 2-5), and click the Update Plugins button. All your selected plugins are updated. (Refer to Figure 2-4.)

Updating manually

The process of updating a plugin manually is nearly identical to the process of installing a plugin manually, as detailed earlier in this chapter. The only change is deleting the plugin's directory before uploading the new files. Follow these steps:

1. **Download the latest version of the plugin from the WordPress Plugin Directory or the plugin developer's website.**
2. **Connect to your server via an SFTP application, and go to the `plugins` directory inside the `wp-content` directory.**

   You should see a folder with the same name as the plugin you want to upgrade.
3. **Rename this folder so that you have a backup if you need it.**

   Any memorable name, such as *plugin*-`old`, should suffice.
4. **Upload the new version of your plugin via SFTP to your server so that it's in the `wp-content/plugins` folder.**
5. **Log in to the WordPress Dashboard, and activate your upgraded plugin.**

REMEMBER

If you made any changes in the configuration files of your plugin before your upgrade, make those changes again after the upgrade. If you need to back out of the upgrade, you can just delete the new plugin's directory and rename the folder from *plugin*-`old` to *plugin.*

Activating and Deactivating Plugins

After a plugin is on your site, activating it is extremely simple. To activate a plugin, do the following:

1. **Log in to the WordPress Dashboard.**
2. **Navigate to the Plugins screen by clicking the Plugins menu link.**
3. **Find the plugin you want to activate on the Plugins screen.**
4. **Click the Activate link just below the plugin's name.**

If everything goes well, the links below the plugin change from Activate, Edit, and Delete to Settings, Deactivate, and Edit, as shown in Figure 2-6. These changes indicate that the plugin has been activated successfully.

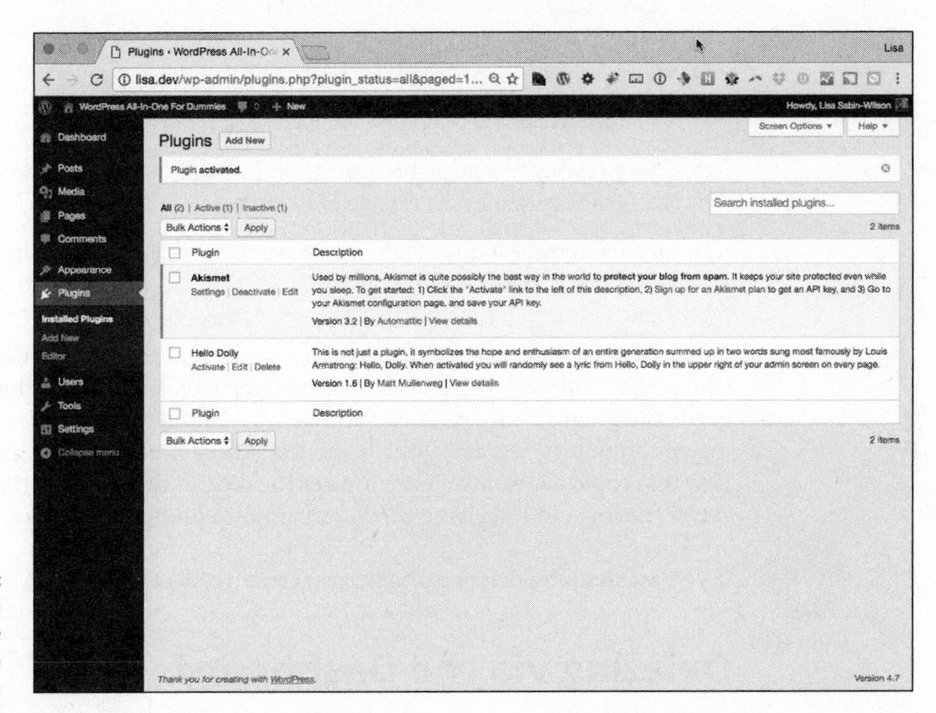

FIGURE 2-6: An activated plugin on the Plugins screen in the Dashboard.

TIP

If a long error message appears in the activation notice, the plugin has an issue that's preventing it from activating. Copy the message that appears in the notice, and send the details to the plugin's author for help fixing the issue.

Deactivating a plugin involves the same process as activating a plugin. Simply follow the same steps, but click the Deactivate link for the plugin that should be deactivated. You get a message at the top of your Dashboard telling you that the plugin has been successfully deactivated.

WARNING

Whatever you do, do *not* ignore the plugin update messages that WordPress gives you. Plugin developers usually release new versions because of security problems or vulnerabilities that require an upgrade. If you notice that an upgrade is available for a plugin you're using, stop what you're doing and upgrade it. The process takes only a few seconds.

Deleting Plugins

Sometimes, it's simply time to let go of a plugin and remove it from the site. You could have many reasons for deleting a plugin:

- You no longer need the feature offered by the plugin.
- You want to replace the plugin with a different one.
- You're retiring the plugin because its functionality has been replaced by features built into a new version of WordPress.
- You're removing it due to performance issues because the plugin simply requires too many resources to run.

It may be tempting to simply deactivate undesired plugins and leave them sitting in your `plugins` directory, but take the extra step to delete plugins that you no longer need. The plugin's `.php` files can still be run manually if someone (or some automated computer program) directly requests those files. If the plugin had a security flaw that could allow direct execution of the code to compromise the security of the server, having old code lying around is simply a problem waiting to happen.

REMEMBER

If you accidentally delete a plugin, you can always reinstall it.

Deleting via the Dashboard

You handle deleting plugins from the Plugins page. Before you delete a plugin, however, you must deactivate it. (See the earlier section "Activating and Deactivating Plugins.")

Ready to delete the plugin? Click the Delete link just below the plugin's name. As shown in Figure 2-7, you have to confirm that you want to delete a plugin before that action takes place.

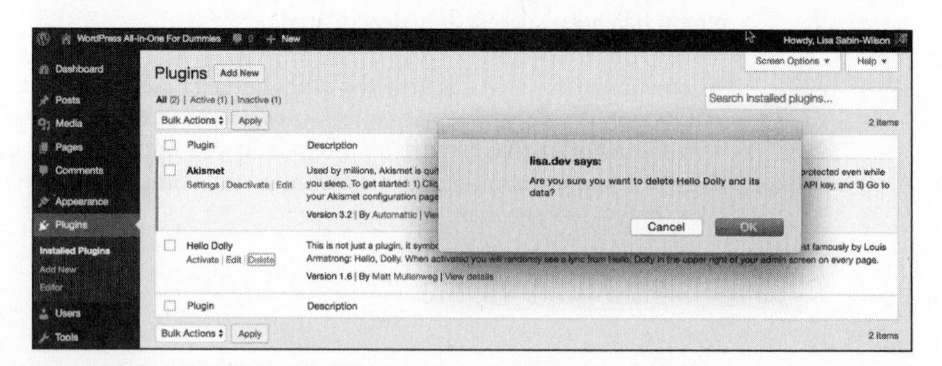

FIGURE 2-7: The confirmation message for deleting a plugin.

After confirming the deletion of a plugin, you return to the Plugins page, which displays a notice confirming that the plugin was deleted, as shown in Figure 2-8.

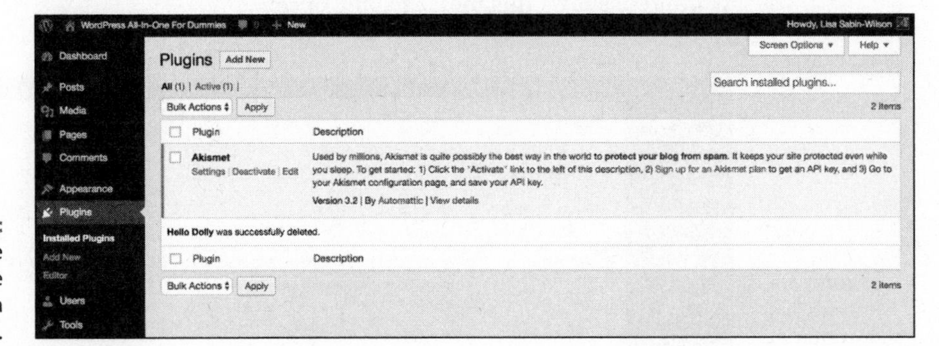

FIGURE 2-8: A notice confirming the deletion of a plugin.

Deleting manually

You can delete a plugin manually by removing the plugin's directory from the `wp-content/plugins` directory. Because the files no longer exist, the plugin simply stops running. It may be helpful to deactivate the plugin first, but doing so isn't required.

Deleting a plugin manually can be very helpful when a plugin has a fatal error that causes the site to crash. If you can't gain control of the site again, deleting the plugin's directory manually could quickly return control of your site.

Because this process involves using SFTP, as upgrading a plugin manually does, this process is very similar to installing a plugin manually. The main difference is that rather than uploading the plugin's directory, you're deleting it.

TIP

Before deleting a plugin, download the directory to a local system first, just so you don't lose any data that would be difficult to get back later. If the goal is to force the plugin to deactivate, you can rename the plugin's directory rather than delete it. Renaming prevents WordPress from locating the plugin, thereby disabling it.

The process works as follows:

1. **Connect to your site's server by using SFTP.**
2. **Navigate to the site's directory.**
3. **Navigate to the `wp-content/plugins` directory.**
4. **Delete the plugin's directory.**

The Plugins screen shows a message confirming that the plugin is deactivated due to an error (missing files). (See Figure 2-9.) Note that this message is shown only when you go to the Plugins screen after deleting the plugin manually and that WordPress shows this message only once.

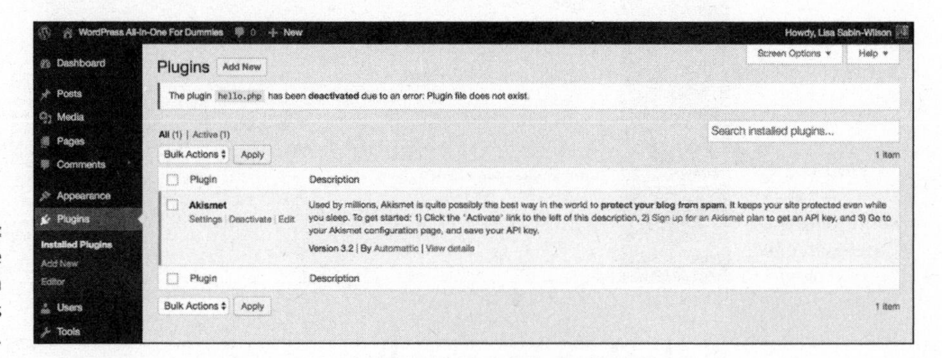

FIGURE 2-9: The message shown after an active plugin is deleted manually.

Exploring the Plugins Included with WordPress

At this writing, WordPress packages two plugins with the installation files, as mentioned in Book 7, Chapter 1:

- **Akismet:** This plugin is essential.
- **Hello Dolly:** This plugin isn't necessary to make your site run smoothly, but it adds some fun.

Incorporating Akismet

It's my humble opinion that Akismet is the mother of all plugins and that no WordPress site is complete without a fully activated version. Apparently, WordPress agrees, because the plugin has been packaged in every WordPress software release since version 2.0. Akismet was created by the folks at Automattic — the same folks who bring you the Jetpack plugin. Akismet is the answer to comment and trackback spam.

Matt Mullenweg of Automattic says that Akismet is a "collaborative effort to make comment and trackback spam a nonissue and restore innocence to blogging, so you never have to worry about spam again."

To use the plugin, follow these steps:

1. **On the Plugins page, click the Activate link below the Akismet plugin's name.**

   A message appears at the top of the page, saying `Activate your Akismet account. Almost done — activate your account and say goodbye to comment spam`. (See Figure 2-10.)

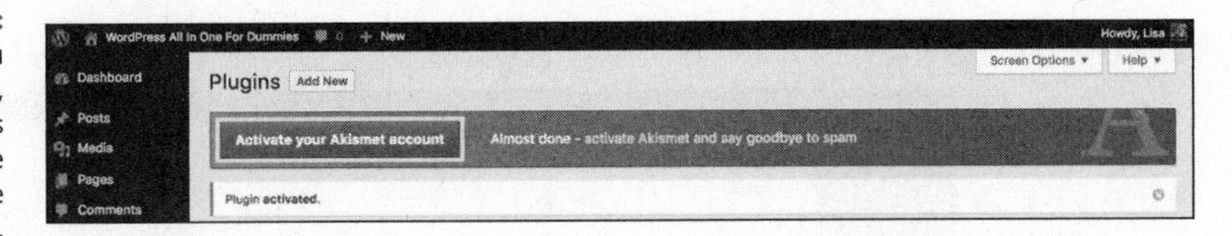

FIGURE 2-10: After you activate Akismet, WordPress tells you that the plugin isn't quite ready to use yet.

2. **Click the Activate Your Akismet Account button.**

   Clicking this link takes you to the Akismet screen of your WordPress Dashboard, where you can create an API key, which is required for the Akismet plugin.

3. **Click the Get Your API Key button.**

   This step opens the Akismet website (at `https://akismet.com/wordpress`) in your browser window

4. **Click the Get an Akismet API Key button.**

   This step opens the sign-up page on the Akismet website. Because Akismet is hooked to the WordPress.com service, if you already have a WordPress.com account, click the I Already Have a WordPress.com Account! link.

   Otherwise, fill in the text fields, and provide your email address, desired username, and password. After you have do this, Akismet requests that you authorize your account. Click OK to authorize.

5. **Enter the URL of the site where you'll use Akismet, and select a plan:**

   - *Basic:* Free for people who own one small, personal, WordPress-powered site. You can choose to pay nothing ($0), or if you'd like to contribute a little cash toward the cause of combatting spam, you can opt to spend up to $48 per year for your Akismet key subscription.
   - *Plus:* $5 per month for people who own one small, nonpersonal (or business) WordPress-powered site.
   - *Premium:* $9 per month per site for people who run multiple sites and want to cover each site under one Akismet account. This plan offers everything that the Basic and Plus plans offer, plus additional security protection.

6. **Select and pay for (if needed) your Akismet key.**

   After you've gone through the sign-up process, Akismet provides your API key. Copy that key by right-clicking it and choosing Copy from the shortcut menu.

7. **When you have your API key, go to the Akismet screen of your Dashboard (see Figure 2-11) by clicking the Akismet link on the Settings menu.**

8. **Enter your Akismet API key in the text box labeled Manually Enter an API Key and then click the Use This Key button to fully activate the Akismet plugin.**

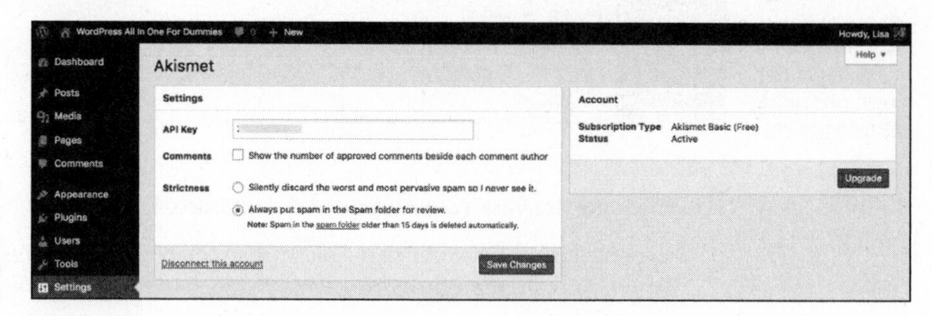

FIGURE 2-11: The Akismet screen of the WordPress Dashboard.

On the Akismet configuration page, after you've entered and saved your key, you can select two options to further manage your spam protection:

- **Comments:** Select the check box titled Show the Number of Approved Comments Beside Each Comment Author to tell Akismet to display the number of approved comments each comment author has on your blog.
- **Strictness:** By default, Akismet always puts detected spam email in the spam folder for your review. You can, however, select the radio button titled Silently Discard the Worst and Most Pervasive Spam So I Never See It.

Akismet catches spam and throws it into a queue, holding the spam for 15 days and then deleting it from your database. It's probably worth your while to check the Akismet Spam page once a week to make sure that the plugin hasn't captured any legitimate comments or trackbacks.

You can rescue those nonspam-captured comments and trackbacks by following these steps (after you've logged in to your WordPress Dashboard):

1. **Click the Comments menu.**

   The Comments page appears, displaying a list of the most recent comments on your site.

2. **Click the Spam link.**

   The Comments page displays all comments that the plugin caught.

3. **Browse the list of spam comments, looking for any legitimate comments or trackbacks.**

4. **If you locate a comment or trackback that's legitimate, click the Not Spam link that appears directly below the entry when you hover your cursor over it.**

   The comment is marked as legitimate. In other words, you don't consider this comment to be spam. The comment is then approved and published on your site.

REMEMBER

Check your spam filter often. While writing this chapter, I found four legitimate comments caught in my spam filter and was able to de-spam them, releasing them from the binds of Akismet and unleashing them upon the world.

The folks at Automattic did a fine thing with Akismet. Since the emergence of Akismet, I've barely had to think about comment or trackback spam, except for the few times a month I check my Akismet spam queue.

Saying Hello Dolly

Matt Mullenweg, co-founder of WordPress, developed the Hello Dolly plugin. Anyone who follows the development of WordPress knows that Mullenweg is a huge jazz fan. How do we know this? Every single release of WordPress is named after some jazz great. One of the most recent releases of the software, for example, is named Vaughn, after legendary jazz vocalist Sarah Vaughn; another, earlier, release was named Coltrane, after the late American jazz saxophonist and composer John Coltrane.

So it isn't surprising that Mullenweg developed a plugin named Hello Dolly. Here's the description of the plugin that you see on the Plugins screen of your Dashboard:

> This is not just a plugin, it symbolizes the hope and enthusiasm of an entire generation summed up in two words sung most famously by Louis Armstrong: "Hello, Dolly." When activated, you will randomly see a lyric from "Hello, Dolly" in the upper right of your admin screen on every page.

Is it necessary? No. Is it fun? Sure!

Activate the Hello Dolly plugin from the Plugins screen of your WordPress Dashboard. When you've activated it, your WordPress blog greets you with a different lyric from the song "Hello, Dolly!" each time.

If you want to change the lyrics in this plugin, you can edit them by clicking the Edit link to the right of the Hello Dolly plugin on the Plugins screen. The Plugin Editor opens, allowing you to edit the file in a text editor. Make sure that each line of the lyric has its own line in the plugin file. This plugin may not seem to be very useful to you — in fact, it may not be useful to the majority of WordPress users — but the real purpose of the plugin is to provide WordPress plugin developers a simple example of how to write a plugin.

» Discovering settings pages

» Using widgets and shortcodes

Chapter 3
Configuring and Using Plugins

The types of features offered by WordPress plugins are extremely diverse. Similarly, the ways of interacting with plugins are also extremely diverse. Some plugins don't have an interface and can be activated or deactivated only, whereas others provide one or more settings screens to control how the plugins behaves. Other plugins offer widgets and *shortcodes* (short, easy-to-remember codes used to execute PHP functions) to add new features to sidebars and content.

This chapter digs into the topic of how to interact with plugins. Although this topic is a big one, the examples in this chapter prepare you for the different ways of interacting with plugins.

Exploring Activate-and-Go Plugins

Certain plugins are easy to use because they don't have any settings or features to interact with; I call them *activate-and-go* plugins. You simply activate them, and they do what they're intended to do.

Like WordPress plugins as a whole, activate-and-go plugins offer a wide variety of features. The following list offers a sampling of activate-and-go plugins that you'll find useful on your website:

- **JC Ajax Comments (`https://wordpress.org/plugins/jc-ajax-comment/`):** When pages and posts get large numbers of comments, they sometimes load more slowly, making the site seem sluggish. The JC Ajax Comments plugin makes such pages and posts load much more quickly by having the content load first without the comments and then quietly pulling down the comments separately.
- **BBQ: Block Bad Queries (`https://wordpress.org/plugins/block-bad-queries`):** The BBQ plugin helps protect your site against attackers trying to exploit specific security vulnerabilities. This plugin doesn't require any configuration; it automatically scans all requests coming to the site and protects against bad ones.
- **Disable WordPress Core Updates (`https://wordpress.org/plugins/disable-wordpress-core-update`):** The capability of WordPress to update itself automatically has been a tremendous help to WordPress users. Keeping WordPress updated not only offers new features and enhancements, but also helps keep your site safe from attackers. For some users, notifications to update WordPress can become distractions, especially if the site is run by many people but a single person is responsible for handling site updates. The Disable WordPress Core Updates plugin disables automatic checks for new WordPress versions and also disables any notifications that a new version is available. WordPress can still be updated from the Dashboard (as described in Book 2, Chapter 6), but the notifications no longer appear.

To use any of these plugins, simply install and activate them as discussed in Book 7, Chapter 2. When a plugin is activated, it starts doing its job.

REMEMBER

For some plugins, such as BBQ or JC Ajax Comments, the result of activating the plugin may be underwhelming; the plugin simply does its work behind the scenes and doesn't change anything that's visible to you. But just because you don't see any immediate change doesn't mean that plugins aren't doing their jobs.

Discovering Settings Pages

Many popular plugins have settings pages where you tweak the functionality of the plugin and tailor it to the specific needs of your site. Often, these settings need to be configured once and updated only when the plugin changes.

The following sections explore a selection of the most popular WordPress plugins, show you how to access settings pages, and describe what you can expect from them.

Typically, you access settings pages from submenus of the Dashboard's Settings page. Another common place to access plugin settings pages — especially for plugins that provide advanced features such as site caching — is the Dashboard's Tools menu.

TIP

If you have a hard time finding the settings page for a plugin, check the plugin's page for details on how to access the settings. For plugins in the WordPress Plugin Directory, check the installation and FAQ tabs. If the Plugin Directory page has a screen-shots tab, one of the screen shots usually shows the settings page.

Akismet

Akismet is bundled with WordPress and likely is already installed on your WordPress site. After you activate Akismet, a notice appears, saying that the plugin requires additional configuration before it will function.

TIP

Check for an activation notice after you install any plugin. Although most plugins don't offer such a notice, if one is available, it lets you know how to get started with the plugin. The Akismet plugin always has an activation notice after installation. (Information about the Akismet plugin, including installation instructions, is in Chapter 2 of this minibook.)

Google XML Sitemaps

The Google XML Sitemaps plugin is a good next step for diving into plugin settings pages. Google XML Sitemaps has several options and shows you just how intricate settings pages can get.

Google XML Sitemaps is one of WordPress's most popular plugins, with more than 9 million downloads. You can find it in the Plugin Directory at `https://wordpress.org/plugins/google-sitemap-generator`.

Google XML Sitemaps makes it easy to automatically add support for site maps to your WordPress site. Although most WordPress sites can be scanned easily by search engines, adding site maps adds a level of safety to ensure that all the content on the site can be found.

With default settings, the plugin automatically generates site maps as content is added to or modified on the site. In addition, it notifies Google and Bing of these updates so that it can update the search engine cache with this new data. (Book 5, Chapter 5 covers Google XML Sitemaps in depth.)

The Google XML Sitemaps plugin settings page is available from the XML-Sitemap submenu of the Dashboard's Settings menu. Notice that the menu name is different from the plugin name.

REMEMBER

Submenu names are limited in length, which means that longer plugin names are shortened to fit properly.

Figure 3-1 shows a portion of the Google XML Sitemaps plugin's settings page.

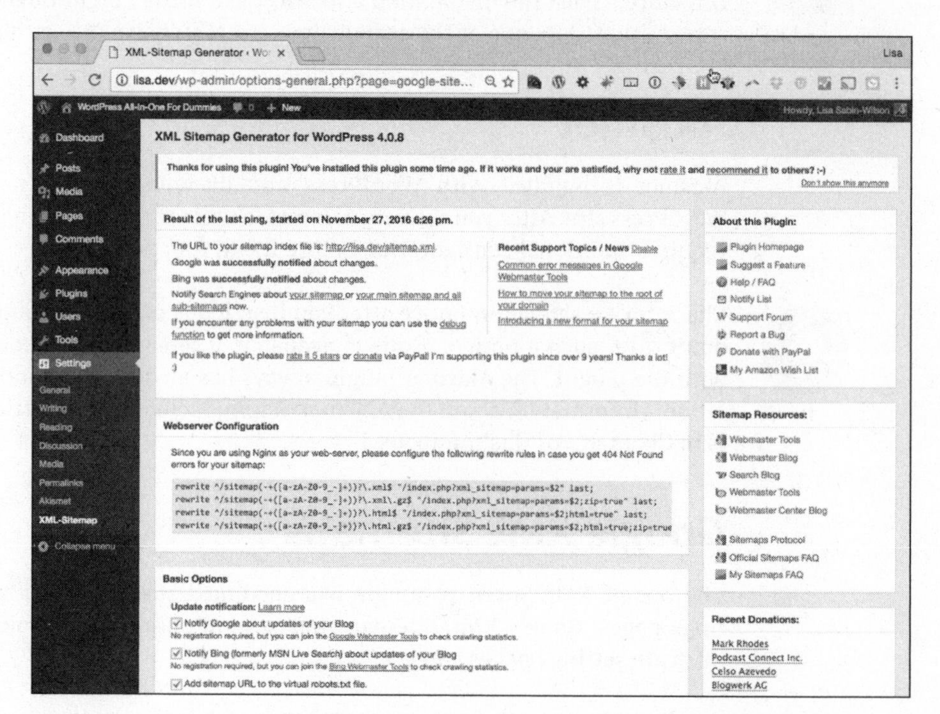

FIGURE 3-1: The settings page for Google XML Sitemaps.

Like the Akismet plugin, Google XML Sitemaps requires an additional step to be fully functional. Unlike Akismet, Google XML Sitemaps is very quiet about how to get set up; it doesn't provide that Dashboard-wide notification message. For this reason, it's very important to carefully read settings pages and plugin documentation. You can easily miss something extremely important.

Below the information box used to generate and regenerate site maps are the settings for the plugin. Along the right side are resources about the plugin and about site maps. This type of format isn't uncommon for plugin settings pages.

Scrolling through the page reveals just how exhaustive the available settings are. The settings range from basic options (such as enabling or disabling automatic site map generation when the site's content is changed) to options that control the information in the generated site map to advanced options that control how many server resources the plugin can consume when generating the site map. Nearly every aspect of the plugin's functionality is represented as an option on the settings screen, offering a large amount of flexibility in how the plugin functions on the site.

Many popular plugins have this type of settings page. Although the settings can be excessive for some people, most users can get very good results simply by using the default settings. The settings of plugins such as Google XML Sitemaps are available for people who desire extra control of functionality. I recommend reading the settings to get an idea of what options are available.

REMEMBER

If you don't understand a setting, leave it in its default state.

All in One SEO Pack

The All in One SEO Pack plugin, also known as AIOSEOP or AIO SEO, focuses on improving the SEO (search engine optimization) of your WordPress site. If you're unfamiliar with SEO, see Book 5, Chapter 4.

With more than 13 million downloads, All in One SEO Pack is one of the most-downloaded plugins on the WordPress Plugin Directory. You can find it in the Plugin Directory at `https://wordpress.org/plugins/all-in-one-seo-pack`.

After activation, the All In One SEO plugin inserts links into the left navigation menu of your WordPress Dashboard. Click the General Settings link to load the All in One SEO Pack Plugin Options screen, which is shown in Figure 3-2.

This settings page has a variety of settings to control many features of the plugin. The portion of the page shown in Figure 3-2 doesn't display title settings; you have to scroll down the screen to find those settings. Typically, titles are controlled by the theme and are modifiable only through code changes. Opening control of titles without requiring code modifications is one of the primary reasons why plugins such as All in One SEO are so popular.

Many other settings go beyond control of titles. Some of the most-often-used settings on this page are the ones that control automatic generation of keywords and description metadata, integrate the site with Google+ and Google Analytics, and determine what content is marked as `noindex`.

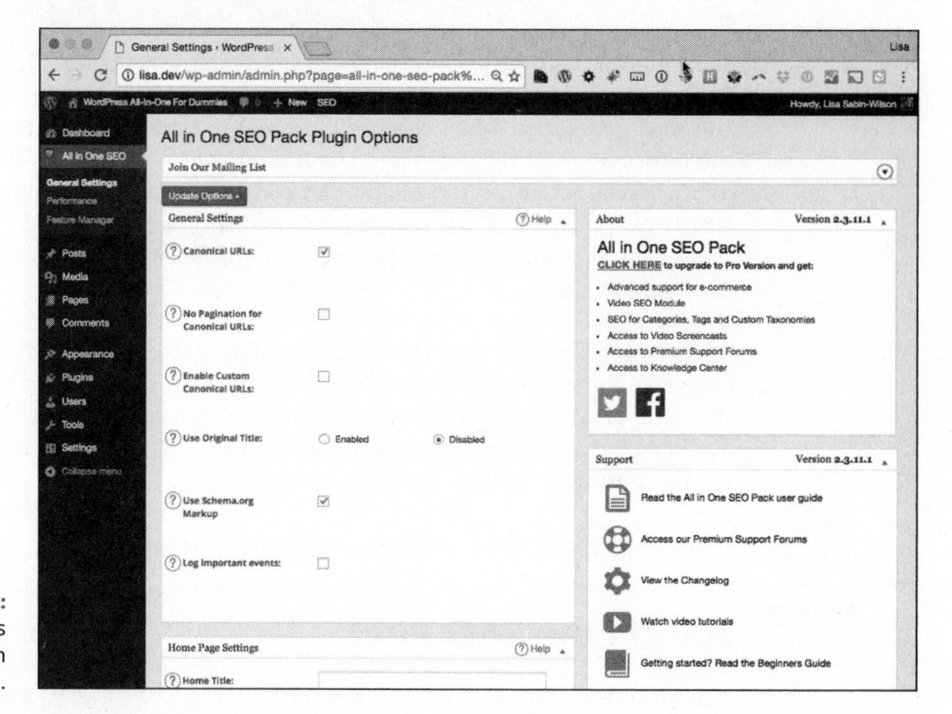

FIGURE 3-2: The settings page for All in One SEO.

When you mark a specific page with `noindex`, search engines ignore the content of the page and don't return search results that link to it.

Like the settings for Akismet and Google XML Sitemaps, the settings for the All in One SEO Pack plugin affect the whole site. Although some of the settings apply only to specific parts of the site, the settings page as a whole focuses on the entire site (which is true of most plugin settings pages). If the plugin creates a stand-alone settings page, the settings on that page typically apply to the whole site unless the setting specifies otherwise.

Being able to customize the title, description, and keywords on each page or post is very helpful. Because managing such customizations for the site's content would be difficult to control in one settings page, All in One SEO Pack provides additional settings in the editors for posts and pages. Figure 3-3 shows the settings box added to the editor by All in One SEO Pack.

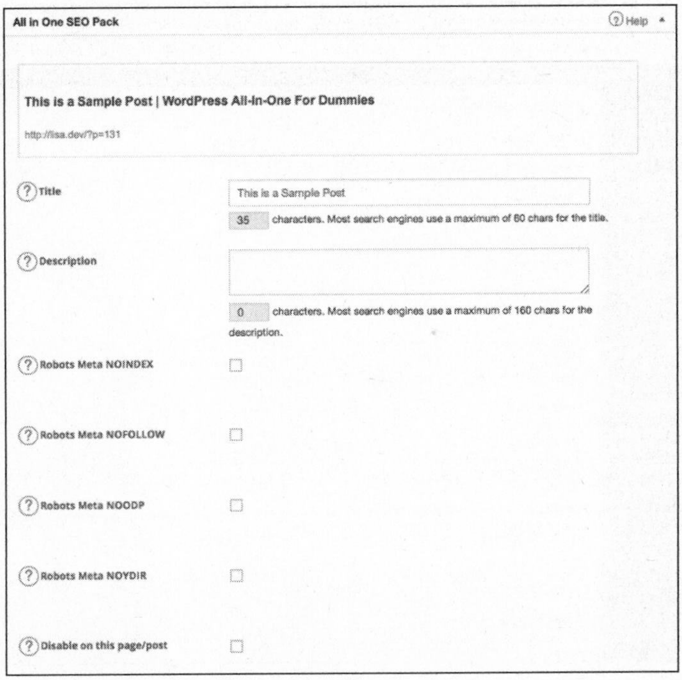

All in One SEO Pack
Help

This is a Sample Post | WordPress All-In-One For Dummies
http://lisa.dev/?p=131

Title
This is a Sample Post
35 characters. Most search engines use a maximum of 60 chars for the title.

Description
0 characters. Most search engines use a maximum of 160 chars for the description.

Robots Meta NOINDEX

Robots Meta NOFOLLOW

Robots Meta NOODP

Robots Meta NOYDIR

Disable on this page/post

FIGURE 3-3: All in One SEO Pack settings that control SEO features for a specific post.

Using Widgets

Widgets offer a very powerful, flexible way to add specific kinds of content to your site's sidebars. WordPress comes with several widgets, such as a calendar, a list of pages on the site, a list of recent comments, and a site-search tool. Plugins can expand this set of default widgets by adding their own. The following sections review plugins that add their own widgets to show you how plugin-provided widgets offer new options to enhance your site.

TIP

You manage widgets on the Widgets screen in the Dashboard. After logging in to your site's Dashboard, hover your mouse over the Appearance menu and then click the Widgets link to access the Widgets page.

Akismet

After you activate and set up a valid application programming interface (API) key for the Akismet Widget, a new widget named Akismet Widget appears on your Widgets page, as shown in Figure 3-4. (You can find more about the Akismet API key in Book 7, Chapter 2.)

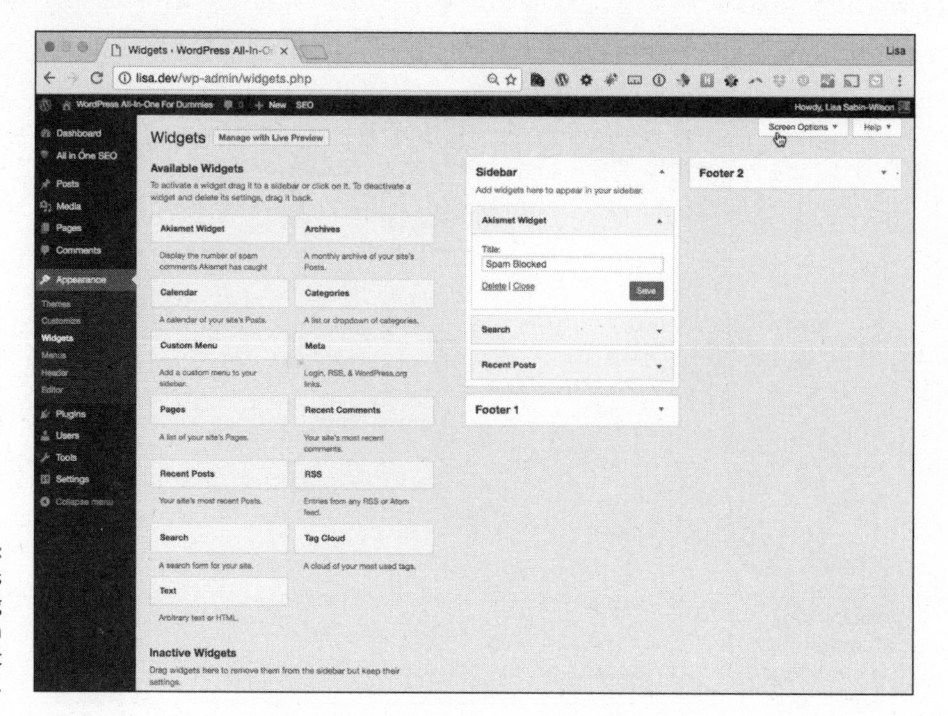

FIGURE 3-4: The Widgets page, showing the addition of the Akismet Widget.

To use the Akismet Widget, drag it from the Available Widgets section and drop it into one of the sidebars on the right side of the Widgets page. As shown in Figure 3-4, after you drop the widget into a sidebar, the settings for the widget become available. The settings are quite simple because only the title of the widget can be modified. At a minimum, most widgets offer a title setting. Although some exceptions exist, most widgets treat the title as optional and simply don't show a title if the setting is empty.

Now the Akismet Widget appears on your site, displaying a counter that shows how many spam comments the Akismet plugin has blocked on the site. Figure 3-4 shows that the widget has blocked no comments.

Twitter Widget Pro

The Twitter Widget Pro plugin serves a single purpose: making it easy to add a Twitter stream to your site. This feature takes the form of a widget, meaning that you can add the Twitter stream to any sidebar on your site. Twitter Widget Pro is available from the Plugin Directory at `https://wordpress.org/plugins/twitter-widget-pro`.

As shown in Figure 3-6, the Twitter Widget Pro Widget provides a large number of settings that control its output.

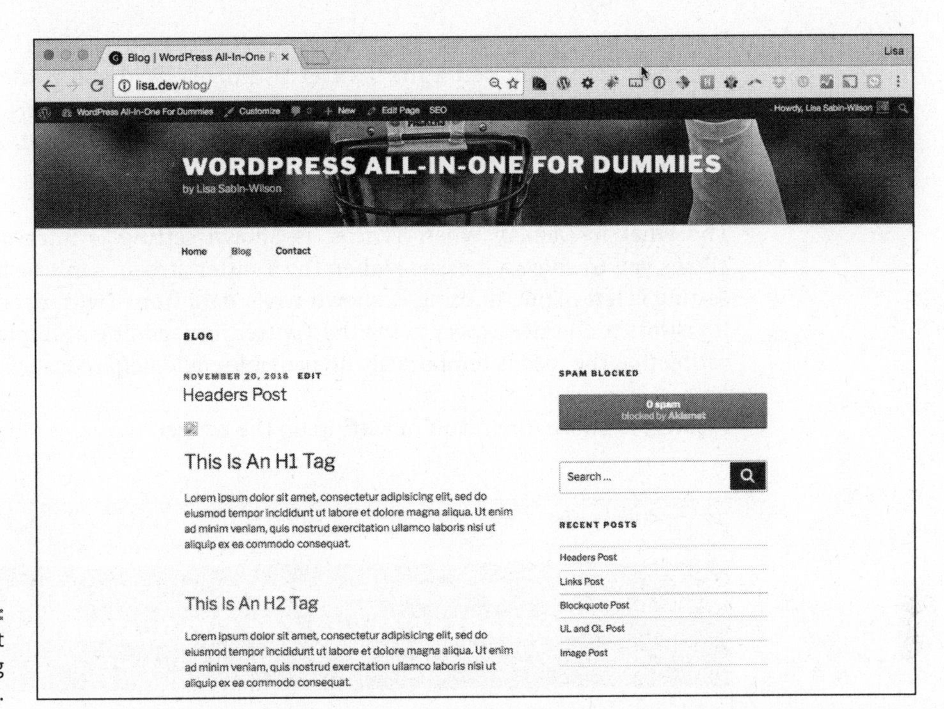

FIGURE 3-5: The Akismet Widget running on the site.

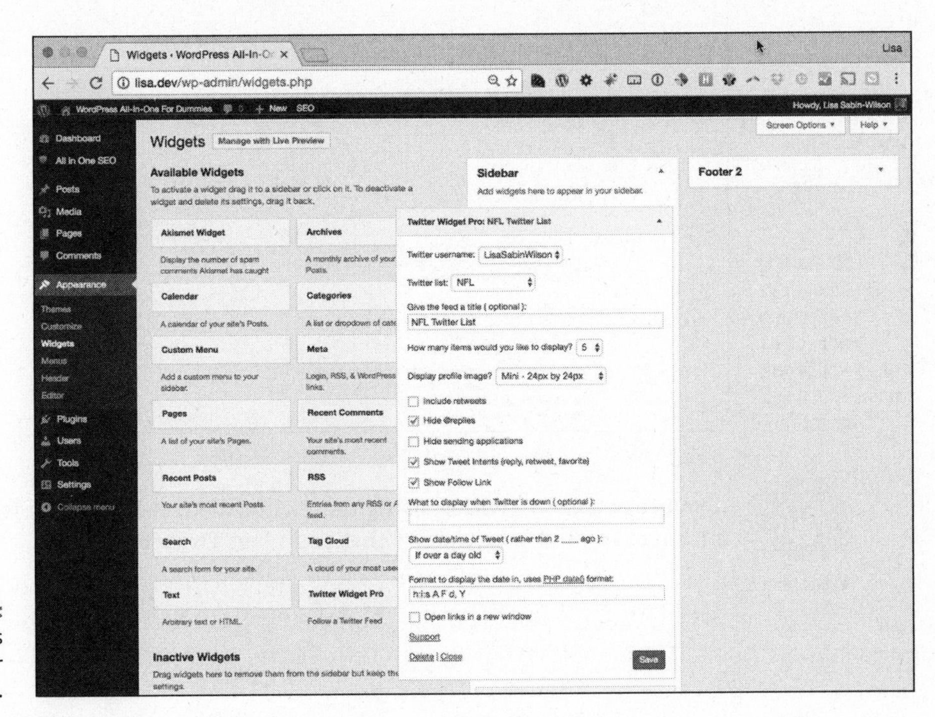

FIGURE 3-6: The settings for Twitter Widget Pro.

The most important setting is Twitter Username. Without a valid Twitter username, the widget won't produce any output. If the widget fails to render anything on your site, double-check the username to ensure that it's a valid Twitter username.

The What to Display When Twitter Is Shown setting is interesting because it allows you to show a message when the Twitter stream can't be accessed. If this setting is left blank, nothing is shown when data from Twitter can't be retrieved. If visitors to the site expect to see the Twitter feed, adding a simple message indicating that the feed is temporarily unavailable could help reduce visitor confusion.

Figure 3-7 shows the result of setting up the widget.

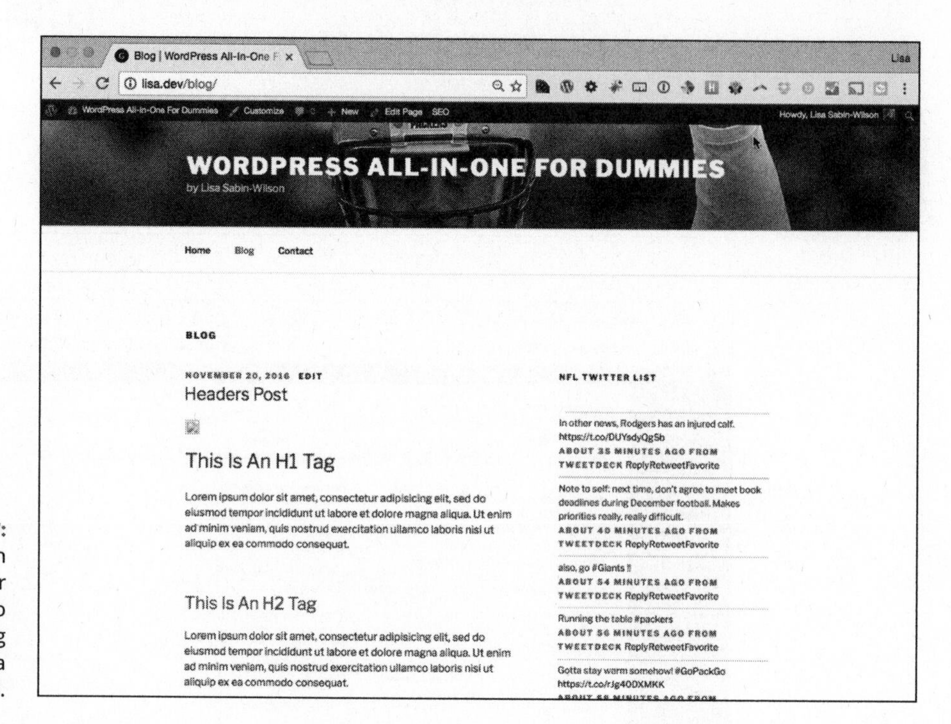

FIGURE 3-7: The widget in the Twitter Widget Pro plugin showing the latest from a Twitter feed.

REMEMBER

If you're active on Twitter, using the Twitter Widget Pro plugin is an easy way to inform or remind your visitors that you're on Twitter.

Additional widgets to try

Akismet and Twitter Widget Pro just scratch the surface of what's possible with widgets offered by plugins. Following are additional widgets that can help fill in your sidebars:

- **List Custom Taxonomy Widget:** This plugin is available in the Plugin Directory at `https://wordpress.org/plugins/list-custom-taxonomy-widget/`. The widget in this plugin provides you with a way to display custom taxonomies on your site. (Book 6, Chapter 6 covers custom taxonomies.)
- **Image Widget:** Sometimes, you just want to add an image to a sidebar. Although the built-in Text Widget and some HTML can provide this functionality, some users don't know how to create the HTML for the image or want a simpler solution for adding images. The Image Widget plugin provides a widget that allows you to easily upload an image and display it on the site without writing or copying and pasting HTML markup. You can find the plugin in the Plugin Directory at `https://wordpress.org/plugins/image-widget`.
- **Yet Another Related Posts Plugin:** This plugin (also known as YARPP) is available in the Plugin Directory at `https://wordpress.org/plugins/yet-another-related-posts-plugin`. The plugin provides a variety of methods to display links to content on your site that are related to the current page, post, or other type of content that is currently being viewed. One of these methods is to use the provided widget so you can easily see a list of links to similar content on your site's sidebar. Because these lists are automatically generated and updated, the site's related content starts to cross-link with other related content without requiring you to manage those lists manually.

With more than 3,000 plugins currently listed with the *widget* tag in the Plugin Directory (`https://wordpress.org/plugins/tags/widget`), a wealth of new widgets for use in your sidebars is at your fingertips. One of those plugins may offer the perfect widget for adding value to your sidebars.

Enhancing Content with Shortcodes

Widgets can add functionality, navigational aids, and other useful bits of information to your sidebars. What if you want to add dynamic elements (such as automatically generated lists of related content or embedded videos) without having to switch to the HTML editor and deal with complex embed codes? In this situation, shortcodes come to the rescue.

Just as widgets allow code to generate content for use on a sidebar, shortcodes allow code to generate additional content inside a post, page, or other content type. Shortcodes always appear between brackets, like the shortcode used for the insertion of a photo gallery on your website (explained in the "Gallery shortcode" section in this chapter): `[gallery]`. In the following sections, you find out about a few useful shortcodes.

Gallery shortcode

One of the shortcodes built into WordPress is `gallery`. (See Book 4, Chapter 3 for more about the `[gallery]` shortcode.)

The most basic gallery shortcode is `[gallery]`. In this form, all the default arguments are used. (Shortcodes can also support optional arguments that allow customization.) By default, a gallery is arranged into three columns and uses thumbnail-size images. The following shortcode displays the gallery in two columns and uses medium-size images:

```
[gallery columns="2" size="medium"]
```

In many ways, shortcodes are similar to HTML tags. The `gallery` shortcode looks like an opening HTML tag that swaps the < and > characters with `[` and `]`.

Embed shortcode

Shortcodes can also surround text by using an opening and closing shortcode. The `embed` shortcode, another shortcode provided by WordPress, is one example of a shortcode that is used before and after a video link to tell WordPress to embed it in a video player.

WordPress, by default, automatically changes video links to embedded videos in a player for videos from a defined list of allowed video sites. (See Book 4, Chapter 4 for details on which sites are supported.) Although the automatic embedding happens when supported video links are left on a line on their own, supported video links can be surrounded by the `embed` shortcode to explicitly indicate that the link is to be changed into an embedded video, as in this example:

```
[embed] http://wordpress.tv/2016/12/07/matt-mullenweg-state-of-the-
    word-2016/[/embed]
```

When you add this shortcode to your post or page content, Matt Mullenweg's 2016 State of the Word video displays in place of the shortcode, as shown in Figure 3-8.

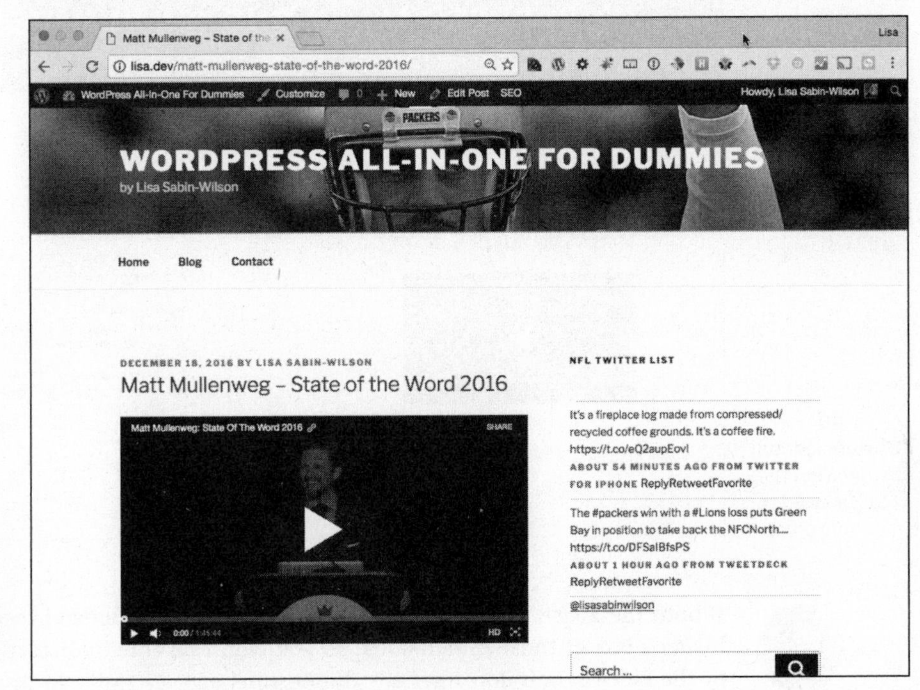

FIGURE 3-8: An embedded video replacing embed shortcode.

You may wonder why you'd want to use the `embed` shortcode instead of simply putting the link on its own line. The reason is that like the `gallery` shortcode, the `embed` shortcode supports arguments that allow you to customize the display of the video. The supported arguments are `width` and `height`. The following shortcode modifies the embedded video to have a width of 400 pixels:

```
[embed width="400"] http://wordpress.tv/2016/12/07/matt-mullenweg-state-of-
    the-word-2016/[/embed]
```

Figure 3-9 shows the result of this change. Notice that the entire video is smaller, is because reducing the width to 400 pixels automatically scales down the height as well.

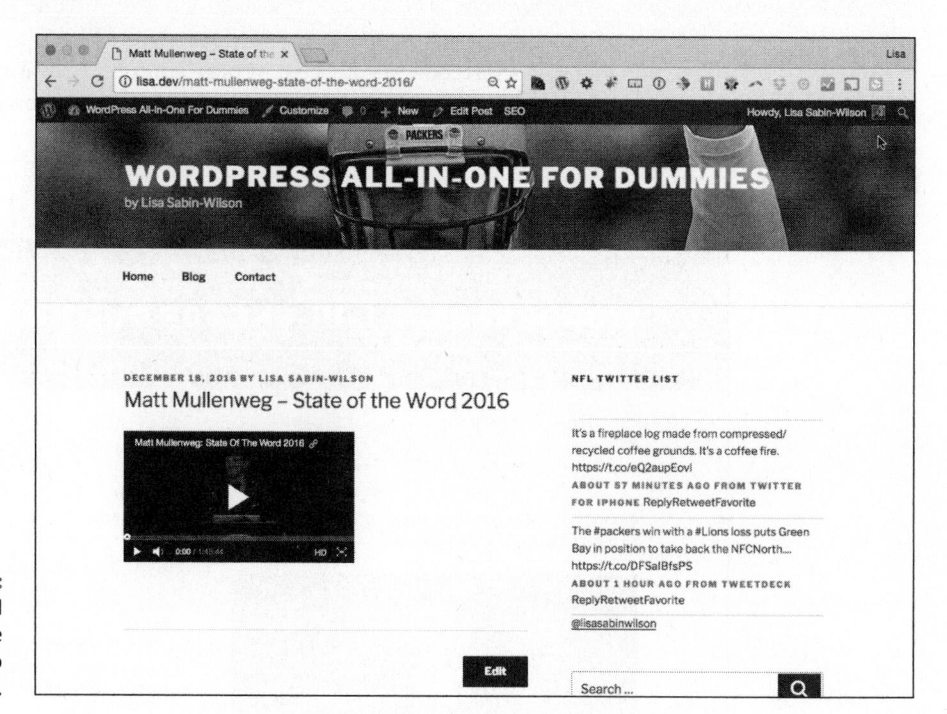

FIGURE 3-9: The embedded video with the width reduced to 400 pixels.

TIP

If both the `width` and `height` arguments are used, the video is scaled down to fit inside a box of those dimensions, so you won't be able to distort the aspect ratio of the video if you don't get the dimensions exactly right. In practice, it's often easiest to supply the `width` argument and not the `height` argument.

Twitter Widget Pro shortcode

The Twitter Widget Pro plugin provides more than just a widget; it also adds support for the `twitter-widget` shortcode. This shortcode provides the same functionality as the widget, but it can be added to content and uses the shortcode method of controlling arguments rather than a widget editor.

At its most basic, the `twitter-widget` shortcode looks like this:

```
[twitter-widget username="lisasabinwilson"]
```

Notice that the `username` argument is set to my Twitter username, `lisasabinwilson`. When you use this shortcode, replace the `lisasabinwilson` username with the Twitter username that you want to use.

The FAQ page for Twitter Widgets Pro in the Plugin Directory includes a list of the arguments that are available for the shortcode. For example, you can display a title for the widget included in the Twitter Widgets Pro plugin by using this shortcode:

```
[twitter-widget username="lisasabinwilson" title=
    "Twitter Feed"]
```

The title is modified, and replies are hidden, which means that you can easily produce the same results whether you decide to use Twitter Widget Pro's widget or the shortcode feature.

WP Google Maps shortcode

The WP Google Maps plugin offers both a widget and shortcode that allows to add a Google map to your site. You can find the plugin in the Plugin Directory at `https://wordpress.org/plugins/wp-google-maps`.

The WP Google Maps Plugin is noteworthy because its shortcodes can quickly become very complex and long. Consider the following embed code provided by Google Maps, which displays a map of Lambeau Field in Green Bay, Wisconsin:

```
<iframe src="https://www.google.com/maps/d/embed?mid=1UC_qvy-WGYOWOtv-
    mZ_3Jg7IYXQ&hl=en_US" width="640" height="480"></iframe>
```

That code would be quite difficult to type without making any errors. Fortunately, the WP Google Maps plugin provides a shortcode tool that makes it easy to have the code generate such complex shortcodes.

After activating the plugin, follow these steps to use the shortcode:

1. **Click the Maps link on the Dashboard menu.**

   The My Maps screen loads on your Dashboard.

2. **Click the Edit link below My First Map.**

   The Create Your Map screen loads on your Dashboard.

3. **In the Address/GPS text field, type the address of the location that you want to display on your map.**

   This location can be a city, state, country, or street address.

4. **Click the Add Marker button to add the location to the map.**
5. **Click the Save Map button.**
6. **Copy the text displayed in the Short Code text box.**

   This code looks something like `[wpgmza id="1"]`, as shown in Figure 3-10.

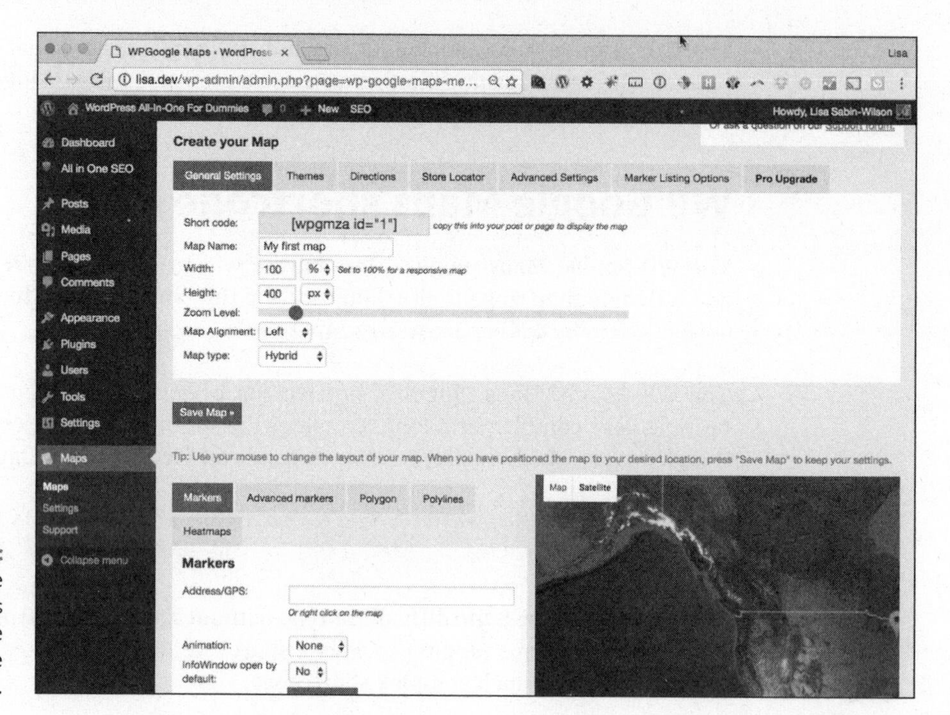

FIGURE 3-10: The WP Google Maps plugin's settings page displays the shortcode.

7. **Paste the copied code into a new page or post.**

   When you publish the post, you see the map embedded in that page on your live website. As shown in Figure 3-11, the shortcode produces an interactive map that visitors can easily navigate. By clicking the marker, visitors can access options to get directions to or from the location. Although you should always give an address, the map gives visitors a much better understanding of where a location is, which means that they don't have to leave the site to look up the address elsewhere.

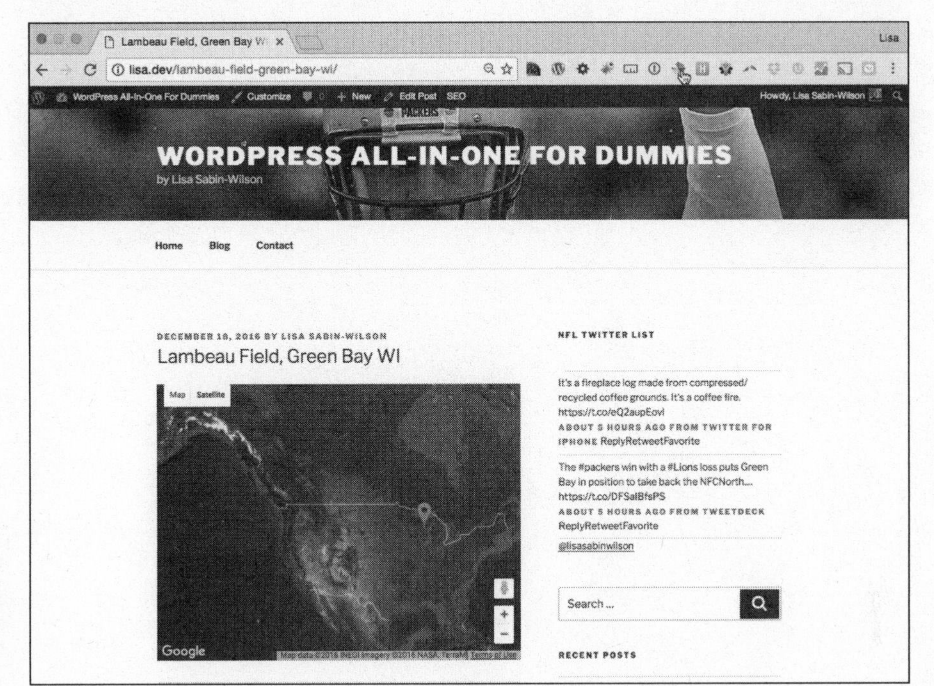

FIGURE 3-11: The map generated by the WP Google Maps plugin's shortcode.

» Removing plugin features

» Modifying plugin output

» Changing shortcode names

Chapter 4
Modifying Existing Plugin Code

WordPress has more than 48,000 plugins in the Plugin Directory, so you'd think that you could find a plugin to do everything you could possibly need on your WordPress website. But not even the best plugins can meet the needs of every user.

This chapter explores the idea of taking an existing plugin and tweaking it to meet your specific needs. With a little bit of programming knowledge and some determination, modifying existing plugin code is very possible. Although you won't become a full-fledged developer overnight, making changes to existing plugins can definitely get the mental gears spinning, figuring out how to do more and more with your programming knowledge.

The examples in this chapter are simple, offering a basic introduction to modifying plugins. Chapter 5 of this minibook goes into much more depth with regard to plugin development.

Setting the Foundation for Modifying Plugins

Before you start modifying plugins, you should do the following:

- » **Set up a development site.** That way, if you accidentally break your site, no harm is done.
- » **Display error messages.** By default, most WordPress sites hide error messages. You want to see those messages when developing plugins, however, because they can provide valuable feedback.

  Find the `wp-config.php` file (it's in the main WordPress installation on your web server), and change `WP_DEBUG define` (scroll to the bottom of the file) from `false` to `true`. Now your error-message information will display.

  TIP

  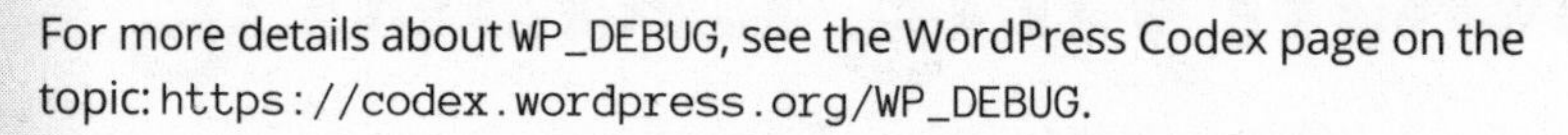

  For more details about `WP_DEBUG`, see the WordPress Codex page on the topic: `https://codex.wordpress.org/WP_DEBUG`.

  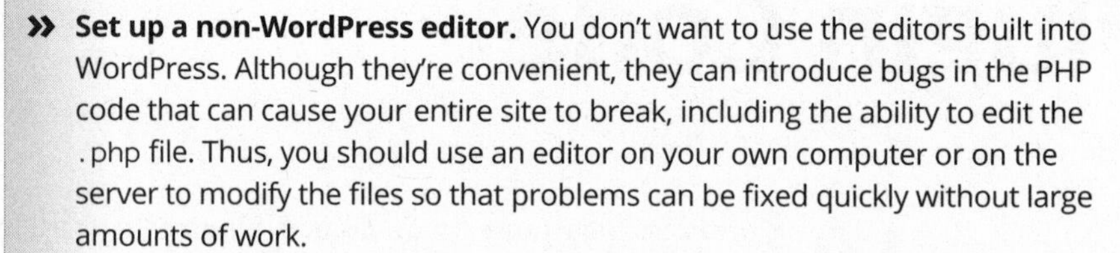

- » **Set up a non-WordPress editor.** You don't want to use the editors built into WordPress. Although they're convenient, they can introduce bugs in the PHP code that can cause your entire site to break, including the ability to edit the `.php` file. Thus, you should use an editor on your own computer or on the server to modify the files so that problems can be fixed quickly without large amounts of work.

REMEMBER

All code is covered by copyright. For a plugin to be added to the Plugin Directory, the code needs to use the GPLv2 license or above. Any code in the Plugin Directory is available for you to use, modify, and even redistribute; the only limitation is that your modified code must use the same licensing as the original code. Book 1, Chapter 2 covers licensing and the General Public License (GPL).

It's important to know that when you start making changes, you're on your own. You can no longer just update the plugin to gain access to bug fixes or new features. To update to a newer version of the plugin, you'd have to download the latest version of the code and modify it again to make your desired changes.

Because WordPress supports updating plugins automatically, you want to ensure that you don't accidentally update a modified plugin and lose your modifications. To prevent such a situation from happening, do the following:

» **Change the name of the plugin's directory.** The directory name is used as the basis for determining what to sync the plugin to for update purposes. If you change the directory name, the plugin is no longer a candidate for automatic updates. Make sure that you pick a name that doesn't exist in the Plugin Directory. You can verify that you have a unique name by trying to go to

```
https://wordpress.org/plugins/new-directory-name
```

where *new-directory-name* is replaced by the name you want to use for your directory. Note that after you change the name of the directory, the plugin needs to be activated again.

» **Modify the name of the plugin.** Although this step isn't strictly necessary, it does help ensure that the plugin stands out on its own. Even simply adding `(modified)` to the name helps ensure that the modified version is kept separate from the normal version of the plugin. It also serves as a reminder to you that the plugin is modified and shouldn't be treated as a normal plugin. In other words, it's a reminder to take care when deciding to update or delete the plugin.

» **Add your name to the listing of plugin authors.** This step tells people where to send questions about the plugin — especially important if the modified plugin is redistributed or used in an environment with many users.

Removing Part of a Form

One of the easiest modifications to make is to remove something from an existing plugin.

The All in One SEO Pack plugin, available at `https://wordpress.org/plugins/all-in-one-seo-pack`, provides a large number of settings that control search engine optimization (SEO) features for each post and page. Suppose that you manage a site, with several editors and authors who frequently ask questions about how to use the keywords input fields that are on all post and page editors (as shown in Figure 4-1). You've decided that the keywords aren't important to the site, and you'd like to avoid all the questions from the editors and authors. So you want to remove the input that appears in the page editors.

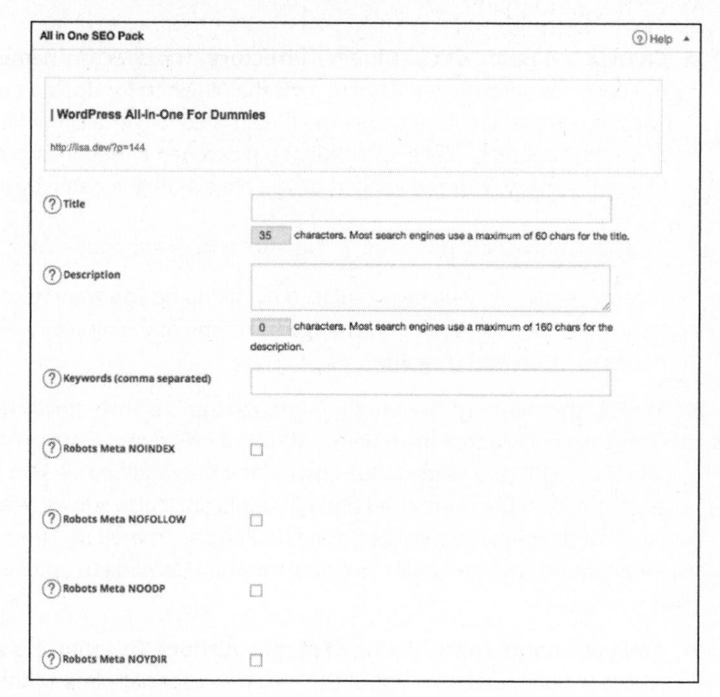

FIGURE 4-1: The All in One SEO Pack's post-/page-specific settings before modifications.

Follow these steps:

1. **Search the plugin's files for a match for *Keywords (comma separated)*.**

   I searched for "Keywords (comma separated)" because this string (a portion of code contained inside quotes) is a unique string inside the plugin's code. After digging around in the files, I found the section responsible for this form on lines 987 to 990 of the plugin's `aioseop_class.php` file.

2. **Remove the necessary lines.**

   The section of code that adds that setting looks like this:

```
'keywords' => array(
  'name' => __( 'Keywords (comma separated)', 'all-in-one-
  seo-pack' ),
  'type' => 'text',
),
```

   To make the modification properly, you can comment out those four lines with double slashes in front of each line, like this:

```
// 'keywords' => array(
// 'name' => __( 'Keywords (comma separated)', 'all-in-
  one-seo-pack' ),
```

```
    // 'type' => 'text',
    // ),
```

3. **Save the modification, and upload the change to the server.**

   The editor page looks like Figure 4-2.

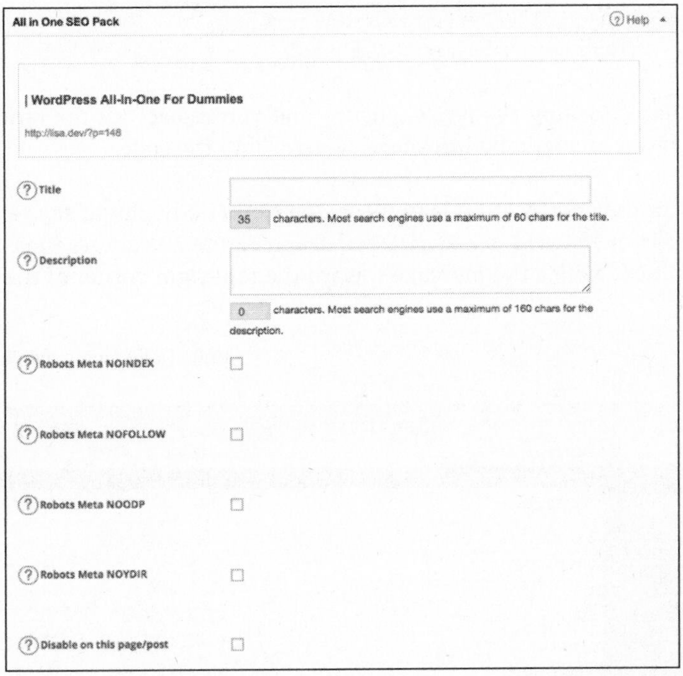

FIGURE 4-2: The All in One SEO Pack's post-/page-specific settings after the Keywords setting is removed.

After making this change, load your website into your browser to make sure that your site still loads with no error messages displayed, to ensure that nothing has broken due to this modification. Save a variety of settings for the SEO feature, and ensure that the modifications still take effect after the settings are saved.

Modifying the Hello Dolly Plugin's Lyrics

The Hello Dolly plugin is included with WordPress and is an easy plugin to modify. If you don't have this plugin installed on your site, you can find it in the Plugin Directory at `https://wordpress.org/plugins/hello-dolly`.

In the `hello.php` file of the plugin is a variable named `$lyrics` that stores the lyric lines of the "Hello, Dolly" song. By replacing this text with your own text, you can change the random selection of a "Hello, Dolly" lyric line to anything you desire.

The `$lyrics` variable can be replaced in its entirety by new text, such as this:

```
$lyrics = "I love WordPress.
   There's a plugin for that.";
```

REMEMBER

When changing the lyrics, ensure that you replace all the lyrics text; otherwise, you can accidentally introduce an error into the code.

After the modification is in place, the Hello Dolly plugin says either `I love WordPress` or `There's a plugin for that`. Figure 4-3 shows the result of this modification. Notice the message toward the top-right corner of the screen.

Hello Dolly plugin displays a custom quote.

FIGURE 4-3: The modified Hello Dolly plugin now declares "I love WordPress."

When a *string* is used just for output, it's typically safe to modify the text without creating any bugs or other issues with the plugin. By changing a string's text, you can change the output of the plugin without much effort.

TIP

WordPress has a built-in mechanism to replace strings in code with new strings. This feature is called *localization* and is typically used to change text into another language. As long as the plugin properly uses the localization functions for all its output strings, you can use this feature to change specific strings as desired, even if you simply change the words used rather than the language. This feature is discussed in Book 7, Chapter 6.

Changing a Shortcode's Name

Sometimes, shortcodes have hard-to-remember names. This problem typically comes up because the plugin author is trying to avoid creating conflicts with other plugins by using the same shortcode name. Although doing so is good practice because it helps prevent code conflicts, looking up a shortcode's name each time you want to use it can be annoying.

The Posts in Page plugin (available at `https://wordpress.org/plugins/posts-in-page/installation`) provides an example of how shortcodes can be hard to remember. It comes with two shortcodes: `ic_add_post` and `ic_add_posts`. From a developer standpoint, these names make sense, because the plugin author is IvyCat. Thus, the initials `ic` prefix each shortcode name, helping ensure that the shortcode names are unique. From a user standpoint, however, the names just cause frustration.

A nice feature of shortcodes is that the code that handles the shortcode can be connected to multiple names, which means that you can add extra names for a shortcode rather than simply change the old name to a new name. Adding a second name is helpful because it prevents any existing uses of the old shortcode name from breaking after the change.

Searching the plugin's files for `add_shortcode` — the function that creates new shortcodes — shows that the shortcodes are created in the `posts_in_page.php` file. The two lines of code that create the current shortcodes are on lines 45 and 46 and look like this:

```
add_shortcode( 'ic_add_posts', array( &$this, 'posts_in_page' ) );
add_shortcode( 'ic_add_post', array( &$this, 'post_in_page' ) );
```

Modifying Existing Plugin Code

The new name for `ic_add_posts` will be `show-posts`. The new name for `ic_add_post` will be `show-post`. To add these shortcodes, simply copy and paste the original two `add_shortcode` function calls and then modify each shortcode name to be the new name. After you make the changes, lines 45 and 46 look like the following:

```
add_shortcode( 'show-posts', array( &$this, 'posts_in_page' ) );
add_shortcode( 'show-post', array( &$this, 'post_in_page' ) );
```

The functionality is exactly the same as the `ic_add_posts` shortcode; it simply has a name that you can remember more easily.

TIP

It's possible to register additional names for a specific shortcode without modifying the shortcode's plugin code. You can accomplish this registration by creating a custom plugin that simply has the `add_shortcode` function call that connects the new shortcode name with the callback function name (the second argument of the function). Creating new plugins is discussed in Book 7, Chapter 5.

» **Filtering content**

» **Adding shortcodes and widgets**

» **Creating a plugin settings page**

Chapter **5**

Creating Simple Plugins from Scratch

You can extend WordPress functionality through plugins and themes without modifying any WordPress core files. This method allows customizing WordPress while still permitting easy upgrades when new versions of WordPress are released. By using the WordPress software's built-in *action hooks* (placeholder functions that allow plugin developers to execute code hooked into them) and *filter hooks* (other placeholder functions that you can use to apply parameters to filter results), you can create just about any functionality you can imagine.

This chapter takes you on a crash course in creating plugins. The plugins I show you how to build start simple and iteratively introduce new concepts as the functionality gets deeper and more complex. Having a foundational knowledge of PHP is helpful for getting the most out of this chapter, but even beginning PHP developers should be able to get value out of each project.

REMEMBER

This book doesn't turn you into a PHP programmer or MySQL database administrator; Book 2, Chapter 3 gives you a glimpse of how PHP and MySQL work together to help WordPress build your website. If you're interested in finding out how to program PHP or becoming a MySQL database administrator, check out *PHP, MySQL, JavaScript & HTML5 All-in-One For Dummies* by Steve Suehring and Janet Valade (John Wiley & Sons, Inc.).

You may be tempted to edit the core code of WordPress rather than write a plugin to achieve the desired functionality. This method isn't recommended; it makes upgrading difficult and can cause various problems, including serious security issues.

To make plugin development safer, use a test site so that you don't introduce bugs that can break your site. Breaking an active site while developing a plugin is an easy way to annoy visitors.

When writing a plugin, use a simple text editor such as Notepad (Windows) or TextEdit (Mac). Don't use the editors built into WordPress to edit code; they can introduce bugs that can break the site.

Understanding Plugin Structure

All that's required for WordPress to see a plugin is a PHP file in the `wp-content/plugins` directory of the site, with some special information at the top of the file. This information at the top of a plugin file, typically referred to as the plugin's *file header,* is what WordPress looks for when determining which plugins are installed on the site. A freshly installed WordPress site makes a good starting point to understand how this structure works in practice.

Inspecting WordPress's core plugins

As discussed in Book 7, Chapter 1, WordPress includes two core plugins: Akismet and Hello Dolly. Looking at the files for each of these plugins helps you understand how you can structure your own plugins.

Inside a fresh WordPress site's `wp-content/plugins` directory, you find a directory named `/akismet` and two files named `hello.php` and `index.php`. The `hello.php` file is for the Hello Dolly plugin and has the following text at the top of the file:

```
<?php
/**
 * @package Hello_Dolly
 * @version 1.6
 */
/*
Plugin Name: Hello Dolly
Plugin URI: https://wordpress.org/plugins/hello-dolly/
```

```
Description: This is not just a plugin, it symbolizes the hope and enthusiasm
    of an entire generation summed up in two words sung most famously by
    Louis Armstrong: Hello, Dolly. When activated you will randomly see a
    lyric from <cite>Hello, Dolly</cite> in the upper right of your admin
    screen on every page.
Author: Matt Mullenweg
Version: 1.6
Author URI: http://ma.tt/
*/
```

This section is the file header, which tells WordPress about the plugin. The Plugin Name, Plugin URI, and Description sections of the file header are referred to as *fields*. I discuss the fields and their uses in Book 7, Chapter 6.

WARNING

If you remove the file header, the Hello Dolly plugin becomes unavailable, because WordPress no longer recognizes it as a plugin.

Open the `index.php` file in the `/wp-content/plugins/` folder, and you see the following few lines of code:

```
<?php
// Silence is golden.
?>
```

Because this file doesn't have a file header, it isn't a plugin. The file is in the `plugins` directory to prevent people from going to *domain.com*`/wp-content/plugins` (where *domain.com* is your site's domain name) to get a full listing of all the plugins on your site. Because the `index.php` file doesn't output anything, people who are trying to get a list of your plugins simply see a blank screen.

All that remains in the `/wp-content/plugins` directory is the `/akismet` directory. Inside this directory are three files: `admin.php`, `akismet.php`, and `legacy.php`. If you open each file, you can see that only the `akismet.php` file contains the file header:

```
/**
 * @package Akismet
 */
/*
Plugin Name: Akismet
Plugin URI: https://akismet.com/
Description: Used by millions, Akismet is quite possibly the best way in the
    world to <strong>protect your blog from spam</strong>. It keeps your site
    protected even while you sleep. To get started: 1) Click the "Activate"
    link to the left of this description, 2) <a href="https://akismet.com/
    get/">Sign up for an Akismet plan</a> to get an API key, and 3) Go to
    your Akismet configuration page, and save your API key.
```

```
Version: 3.2
Author: Automattic
Author URI: https://automattic.com/wordpress-plugins/
License: GPLv2 or later
Text Domain: akismet
*/
```

Because the `akismet.php` file has the file header, WordPress recognizes the `/akismet` directory as a plugin. If the `akismet.php` file is removed, the Akismet plugin disappears from the listing of available plugins in your WordPress installation. (On the Dashboard, click the Plugins link to see the Plugins page.)

Knowing the requirements

Looking at the way the default plugins are set up gives you an idea of how to set up your plugins, but knowing all the requirements is nice so that you don't make mistakes. The reality is that WordPress has very few requirements for plugin setup.

Requirement 1: File header

The file header allows WordPress to recognize your plugin. Without this key piece of information, your plugin won't show up as an available plugin, and you won't be able to activate it.

Although the file header has many fields, only Plugin Name is required. The following is a valid file header, for example:

```
/*
Plugin Name: Example Plugin
*/
```

Providing additional information can be very helpful, of course, but if you're making a plugin for yourself quickly, the plugin name is all that's required. See Book 7, Chapter 6 for more information about the file header.

Requirement 2: Correct placement of main plugin file

The main plugin file (the one with the file header) must be in the `/wp-content/plugins` directory or inside a directory immediately inside the `/wp-content/plugins` directory.

Here are some examples of valid locations for the main plugin `.php` file:

- wp-content/plugins/example.php
- wp-content/plugins/example/example.php

Here are some examples of invalid locations for the main plugin .php file:

- wp-content/example.php
- wp-content/plugins/example/lib/example.php

WARNING

You can place the main plugin file too deep. WordPress looks only in the /wp-content/plugins directory and inside the first level of the directories contained in /wp-content/plugins — no deeper. If you place the file too deep within the plugin directory, it won't work.

Following best practices

WordPress's requirements for plugins are very lax, allowing you to set up your plugin any way you want. You can name the main plugin file and plugin directory anything you like. You can even put multiple main plugin files inside a single directory. Just because you can, however, doesn't mean that you should. Following are some best practices that create some consistency.

Best Practice 1: Always use a plugin directory

Hello Dolly doesn't reside in a directory because it's simple enough to need only one file. Each plugin should reside in its own directory, however, even if it needs only one file.

When you create a plugin, a single file may be enough to do what you need, but further development may require adding more files. It's better to put the plugin in a directory from the start than to restructure it later.

WARNING

Moving or renaming a main plugin file deactivates the plugin, because WordPress stores the plugin's activation state based on the path to the main plugin file.

Do yourself and any users of your plugin a favor: Always place your plugins inside a directory.

Best Practice 2: Use meaningful, unique names

When doing any WordPress development (whether for a plugin or theme), keep in mind that your code shares space with code written by other people (other plugin developers, WordPress core developers, theme developers, and so on). This fact means that you should never use simple names for anything; the names of your plugins should be unique.

You might think that naming a plugin Plugin allows you to move past the boring stuff and on to development, but it just makes everything difficult to keep track of. If your plugin produces a widget that displays a list of recent movie reviews, for example, Lisa Sabin-Wilson's Movie Reviews Widget Plugin is much more meaningful than Widget Plugin.

Best Practice 3: Match the plugin and plugin directory names

Make sure that your plugin's directory name makes it easy to find the plugin in the `/wp-content/plugins` directory.

Going with the preceding example, having Lisa Sabin-Wilson's Movie Reviews Widget in a `widget` directory makes finding the widget difficult. The directory name doesn't have to match, but it should make sense. Some good directory names for this example would be `/movie-reviews-widget`, `/lsw-movie-reviews-widget`, and `/movie-reviews`.

Best Practice 4: Don't use spaces in directory or filenames

Although modern desktop operating systems can handle directories and files that have spaces in the names, some web servers can't. A good practice is to use a hyphen (-) in place of a space when naming files and directories. In other words, use `movie-reviews-widget` rather than `movie reviews widget`.

Avoiding using spaces in file and directory names will save you many headaches.

Best Practice 5: Use consistent main plugin filenames

Although you can name the plugin's main file anything, coming up with a consistent naming scheme for plugins is a good idea.

The most popular naming scheme matches the main plugin `.php` filename with the plugin directory name. The main plugin file for a plugin directory called `/movie-reviews`, for example, might be is `movie-reviews.php`. The problem with this naming scheme is that it doesn't mean anything. A plugin filename should always indicate that file's purpose. The purpose of the `movie-reviews.php` file is clear only when you know that many developers name the main plugin file the same as the plugin directory.

Another naming scheme uses a consistent filename for all plugins. A main plugin file called `init.php`, for example, indicates that the file is used to initialize the plugin. (`init` is the abbreviation for *initialize.*) The name `init.php` makes the purpose of the file clear regardless of the plugin name or purpose.

Creating Your First Plugin

When you're developing something new, taking very small steps usually is the best approach. That way, if something breaks, the problem is clear. Doing multiple new things at the same time makes finding where something went wrong difficult.

Sticking with this concept, the first plugin you create in this chapter is a plugin that can be activated and deactivated but doesn't do anything. In other words, it's a fully functional plugin shell that's ready for code to be added.

Because this plugin is an example that doesn't do anything, I named it Example: Do Nothing.

Uploading the plugin file to a directory

For this plugin, all you need is a main plugin file. Follow these steps to upload the plugin file to its own directory:

1. **Connect to your web server via SFTP.**

   Check out Book 2, Chapter 2 for details on using SFTP.

2. **Browse to the `/wp-content/plugins` directory in your WordPress installation directory.**

   If you're unsure where your WordPress installation directory is located, see Book 2, Chapter 4, where I cover installing WordPress on your web server.

3. **Create a new directory within `/wp-content/plugins` called `/example-do-nothing`.**

   Most SFTP programs allow you to right-click within the folder and choose Add New Folder or Add New Directory from the contextual menu.

4. **Create an empty `.php` file with the filename `init.php`.**

   Use your favorite text editor, such as Notepad for PC or TextEdit for Mac, to open a new file; then save it with the filename `init.php`.

5. **Upload your blank `init.php` file to `/wp-content/plugins/example-do-nothing`.**

   Your plugin directory and plugin file are set up. In the next section, you add code to the `init.php` plugin file.

Adding the file header

Open the `init.php` file you created in "Uploading the plugin file to a directory" earlier in this chapter. (Most SFTP programs have built-in text editors that allow you to right-click the file and choose Edit from the contextual menu.) Add the following lines of code to create the file header:

```
<?php
/*
Plugin Name: Example: Do Nothing
Description: This plugin does nothing. It is an example of how to create a
   valid WordPress plugin.
*/
```

Adding the `?>` tag at the end of a `.php` file is optional at this point. Leaving the tag out is helpful because it prevents you from accidentally adding code after it, which may cause the PHP code to break.

Adding a plugin description isn't necessary, but a description makes the purpose of the plugin clear to anyone who reads your code. Additionally, the plugin description displays on the Plugins page on your Dashboard to give users a good idea of the purpose of your plugin. When developing, you wind up with many plugins that you used for simple tests or didn't finish. Having solid names and descriptions imposes order on the chaos so that you don't forget important code or accidentally delete it.

Be sure to save the `init.php` file and upload it to the `/wp-content/plugins/example-do-nothing` directory on your web server.

Testing the plugin

After modifying the `init.php` file and saving it in the `/wp-content/plugins/example-do-nothing` directory, visit your WordPress Dashboard, and click the Plugins link on the navigation menu to view the Plugins screen. Your new plugin is listed with the title Example: Do Nothing, as shown in Figure 5-1.

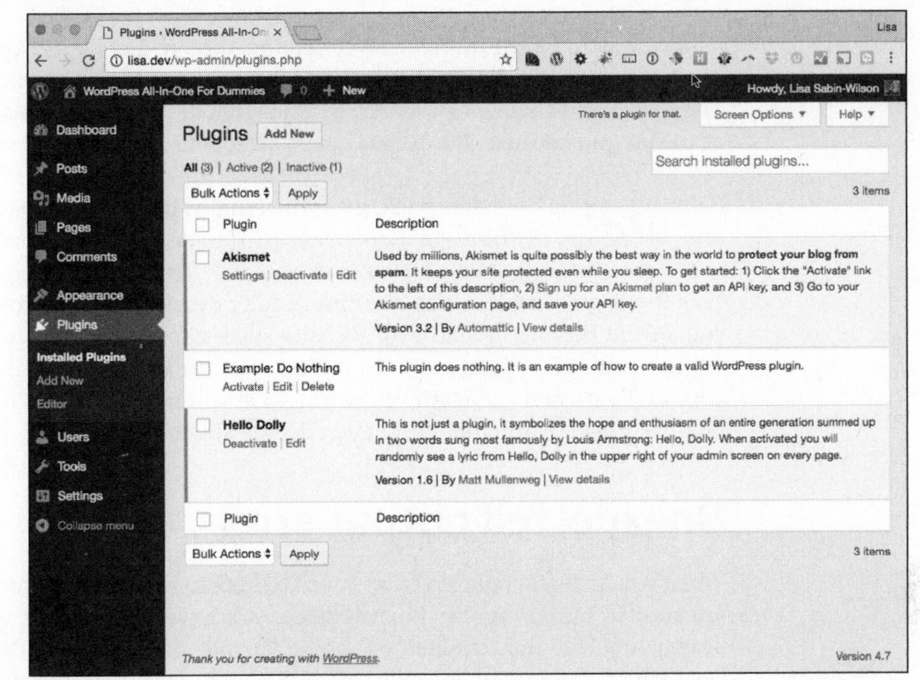

FIGURE 5-1: The Plugins page, showing the sample plugin in the list.

Click the Activate link directly below the title. The Plugins page displays a `Plugin activated` message, which indicates that the Example: Do Nothing plugin was activated in your WordPress installation. Although your new plugin doesn't do anything, you have a simple WordPress plugin with the correct file structure, naming conventions, and headers.

Fixing Problems

Potentially, several things could go wrong. If you're having problems, delete the plugin file from its directory and start over. If you still have problems, the following sections cover some common issues and give you possible solutions that you can try.

White screen of nothingness

A common problem in plugin development is making a change and finding that every attempt to load the site in your browser window results in a blank white screen. A code error is breaking WordPress when it tries to run your plugin code.

A quick way to fix this problem is to rename the /wp-content/plugins/example-do-nothing plugin directory on your web server to something like /wp-content/plugins/old.example-do-nothing. This change causes automatic deactivation of the plugin, because WordPress won't be able to locate it.

Before changing the name back, go to the Plugins page of your Dashboard. A message at the top of the page states The plugin example-do-nothing/init.php has been deactivated due to an error: Plugin file does not exist. This message confirms that WordPress fully deactivated the broken plugin; now you should be able to load your website successfully without seeing the dreaded white screen of nothingness. After that, you can change the filename back, fix your problem, and try again. If the plugin is still broken, WordPress prevents the plugin from activating and gives you details about the error.

Unexpected output error

When you activate a plugin from your Dashboard and see an error message about unexpected output on the Plugins page, you have code or text within the main plugin .php file that is outside a <?php ?> code block. Every PHP function must start with a command that tells your web server to initiate (or start) PHP. If your plugin file is missing the <?php line, an unexpected-output error occurs, and WordPress doesn't activate your plugin.

Have some fun; try to create this error so that you'll know it when you see it. You can intentionally create the error by following these steps:

1. **Connect to your web server via SFTP.**
2. **Browse to the /wp-content/plugins/example-do-nothing directory.**
3. **Open the init.php file in your text editor.**
4. **Remove the <?php line from the top of the init.php file.**
5. **Save the init.php file.**
6. **Upload the file to the /wp-content/plugins/example-do-nothing directory.**

When you try to activate the Example: Do Nothing plugin, the text of the plugin header displays at the top of the Plugins page, as shown in Figure 5-2:

```
/* Plugin Name: Example: Do Nothing Description: This plugin does
nothing. It is an example of how to create a valid WordPress
plugin. */.
```

All this fuss occurs because of a missing <?php line.

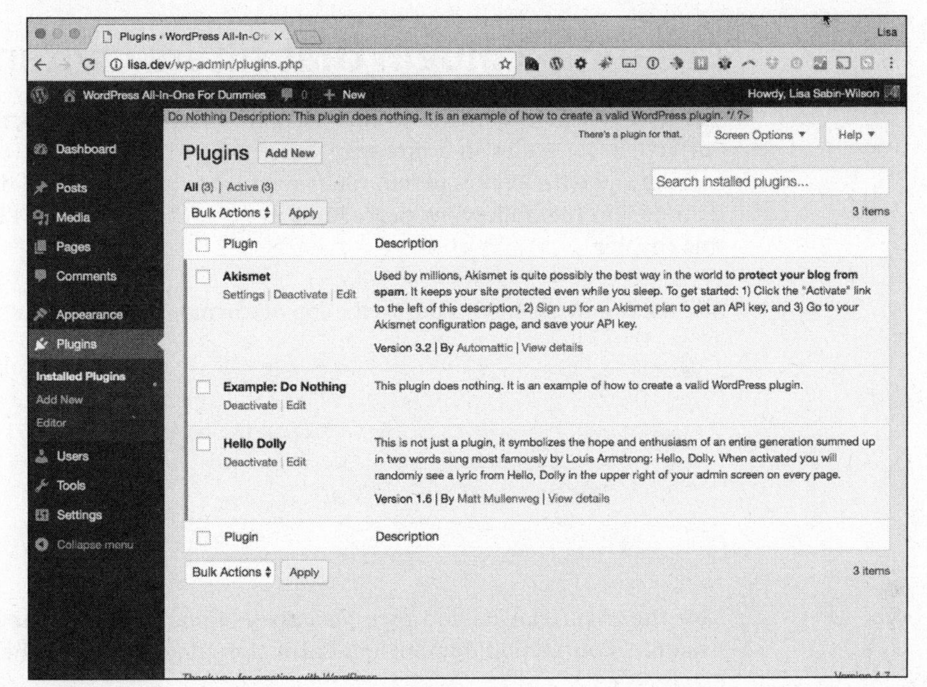

FIGURE 5-2: An output error message displayed on the Plugins page.

Filtering Content

It's time to create a WordPress plugin that actually does something and, in the process, discover more basics of WordPress plugin development.

A powerful feature of WordPress is its numerous filters. By latching code to a filter, you can modify information as it flows through WordPress; therefore, you can modify the information that WordPress displays or stores.

Suppose that you have a habit of using contractions far too often. Your readership mocks you and your penchant for the practice of merging words. At night, you worry about whether you missed an instance of *it's*, *we're*, or *I'll*.

This habit is causing you to lose sleep. You tried listening to the self-help tapes; you reviewed every word and went to therapy to find the deep-seated cause of your craving for contractions. Despite your best efforts and the continuous ridicule, you can't help but sound like an etiquette contrarian.

Fortunately, a cure is available. With a simple filter and a bit of code, you can create a simple WordPress plugin that disguises your grammatical ailment — as you do in the next section of this chapter, "Creating a plugin that filters content."

Creating a plugin that filters content

The plugin that you create in this section is Example: Contraction Compulsion Correction; it resides in a directory called `/example-contraction-compulsion-correction` with a main plugin file named `init.php`. To create the directory and main plugin file, follow the procedures in "Creating Your First Plugin" earlier in this chapter.

Add the following file header to the top of the main plugin (`init.php`) file:

```
<?php
/*
Plugin Name: Example: Contraction Compulsion Correction
Description: This plugin cannot solve your contraction issues, but it can
   hide them by fixing them on the fly.
*/
?>
```

Save the `init.php` file and then visit the Plugins page on your Dashboard. The Example: Contraction Compulsion Correction plugin appears there, as shown in Figure 5-3.

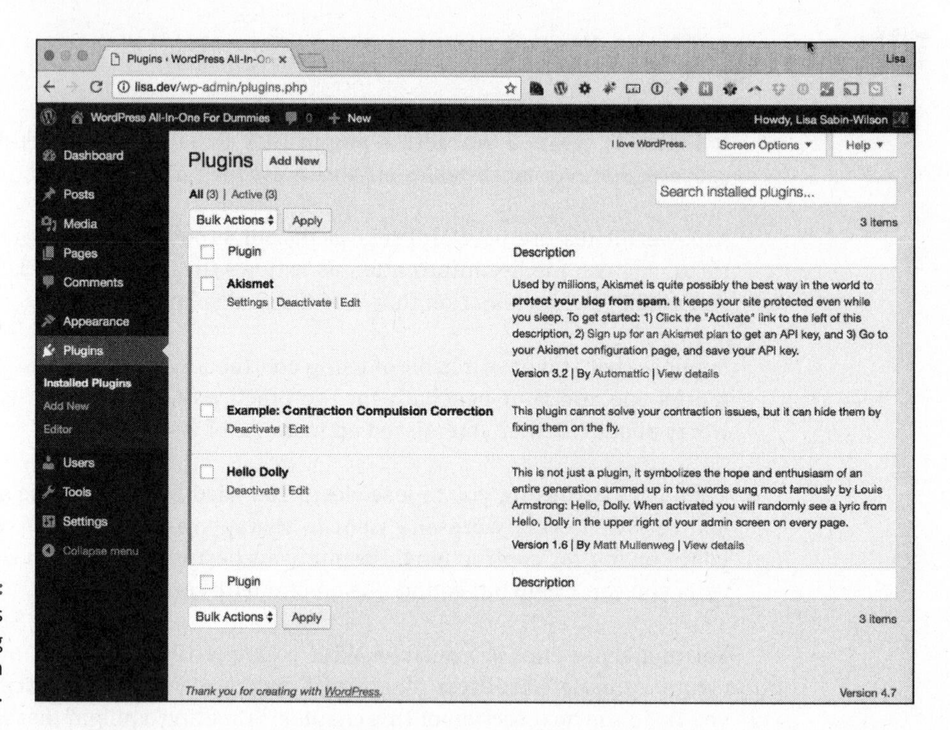

FIGURE 5-3: The Plugins page showing your new plugin in the list.

Testing the filter

The filter you use in this section is `the_content`, which replaces all the content in your site's posts and pages with a simple message. If the filter works as expected, you can expand it to hide the contractions that you published in your posts (or pages).

TIP

`the_content` is just one of hundreds of filters in WordPress. You can find information about filters in the WordPress.org Codex (`https://codex.wordpress.org/Plugin_API/Filter_Reference`).

Include the `the_content` filter in your plugin to replace all the content on your website (posts and pages) with a single phrase. (You change the content that gets filtered in the following section to filter the contractions out of your published content.) Follow these steps:

1. **Connect to your web server via SFTP.**
2. **Browse to this directory:**

```
/wp-content/plugins/example-contraction-compulsion-correction
```

3. **Open the `init.php` file in your text editor.**
4. **Type the following lines of code at the end of the file (after the file header) but before the closing `?>`:**

```
add_filter('the_content','my_filter_the_content');

function my_filter_the_content($content) {
  $content = "Test content replacement.";
  return $content;
}
```

5. **Save your `init.php` file, and upload it to the `/wp-content/plugins/example-contraction-compulsion-correction` folder.**

The first line of code in Step 4 tells WordPress to apply the filter after the plugin is activated. The next few lines of code define the function (`function my_filter_the_content ($content)`) with a variable (`$content`), define the `$content` variable (`$content = "Test content replacement.";`), and tell WordPress to return `$content` within the body of your published posts and pages. Check out the nearby sidebar "Using curly brackets (complex syntax)" for information about using correct PHP syntax.

USING CURLY BRACKETS (COMPLEX SYNTAX)

You see curly brackets within code. Curly brackets (referred to as *complex syntax* in the PHP Manual (http://www.php.net/manual/en/language.types.string.php) serve to open and then close the function definition, or expression. The code samples in "Testing the filter" name the function function my_filter_the_content ($content). An open curly bracket, indicating the start of the function expression, follows that line. Immediately after the two lines $content="Test content replacement."; and return $content, which are the expression for the function, you see the closing curly bracket that indicates the end of the function expression. Without these curly brackets, your code won't work correctly.

Check out the entire PHP manual online at http://php.net/manual/en/index.php to brush up on correct PHP code syntax, such as when to use single quotation marks instead of double quotation marks and how to use the semicolon (;).

With the the_content filter in place in your plugin, visit the Plugins page of your Dashboard, and activate the Example: Contraction Compulsion Correction plugin. After you activate the plugin, view any post or page on your website. The result: Test content replacement replaces the content of that entry. Your new plugin is filtering content on your website. (See Figure 5-4.) In the next section, "Replacing contractions in your content," you apply the real filter that fulfills the purpose of the plugin you're creating.

Replacing contractions in your content

To replace all the contractions in your content with full phrases or words, change the code in the init.php plugin file. Follow these steps:

1. **Connect to your web server via SFTP.**
2. **Browse to the /wp-content/plugins/example-contraction-compulsion-correction directory.**
3. **Open the init.php file in your text editor.**
4. **Remove the following lines of code:**

```
add_filter('the_content','my_filter_the_content');
function my_filter_the_content($content) {
  $content = "Test content replacement.";
  return $content;
}
```

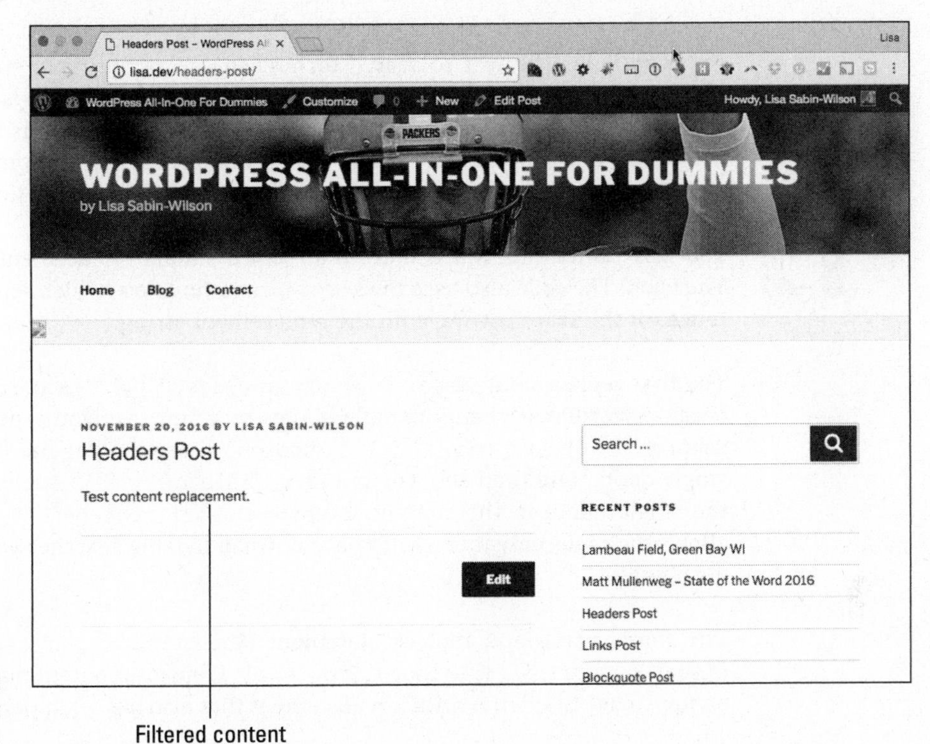

FIGURE 5-4: Filtered content on your website.

5. **Type the following lines of code at the end of the file (after the file header):**

```
add_filter('the_content','my_filter_the_content');
function my_filter_the_content($content) {
  $replacements = array(
    "isn't" => "is not",
    "we'll" => "we will",
    "you'll" => "you will",
    "can't" => "cannot",
    );
  foreach($replacements as $search => $replace) {
    $search = str_replace("'","’",$search);
    $content = str_replace(ucfirst($search),ucfirst($replace),$content);
    $content = str_ireplace($search,$replace,$content);
  }
  return $content;
}
```

6. **Save your `init.php` file, and upload it to the `/wp-content/plugins/-contraction-compulsion-correction` folder.**

An array holds the text to search for and to use as the replacement. The array defines the words you're replacing within your content and loops to make all the replacements. In this example, *isn't* is replaced by *is not*, *we'll* is replaced by *will not*, and so on. This example covers only a small subset of the contractions, however. You have to modify the example to fit your specific contraction compulsions.

TIP

You may notice that much more than just a simple replacement is going on in The Loop. The code also uses the `str_replace` function, which replaces all occurrences of the search string with the replacement string.

The first replacement (`$search = str_replace("'","’",$search);`) is needed because WordPress changes single quotes to a fancy version represented by "’". `$search = str_replace("'","’",$search);` searches for the instances of the single quote, and then `$content = str_replace("’","’",$content);` replaces the single quote in the content. `$search = str_replace("'","’",$search);` allows the replacements array to have normal-looking searches with regular single quotes.

The third search and replace statement (`$content = str_replace(ucfirst($search),ucfirst($replace),$content);`) replaces content matches that have an uppercase first letter with a replacement that also has an uppercase first letter.

The last search and replace statement: (`$content = str_ireplace($search,$replace,$content);`) does a non–case-sensitive search to replace all remaining matches with the lowercase version of the replacement.

To test your contraction replacement plugin, follow these steps:

1. **Log in to your Dashboard.**
2. **Visit the Add New Post page (by hovering your cursor over Posts and then clicking the Add New link).**

   The Add New Post page loads on your Dashboard so that you can write and publish a new post. (See Book 4, Chapter 1.)
3. **Type a title for your post in the Title text field.**
4. **Type the following text in the post editor:**

   Isn't it grand that we'll soon be sailing on the ocean blue? You'll see. We'll have a great time. I can't wait.

   Notice the contractions *Isn't, we'll, You'll, We'll,* and *can't.* Figure 5-5 shows an Add New Post page that contains these phrases.
5. **Publish your post by clicking the Publish button.**

   Figure 5-6 displays the post on a website with the contractions replaced by the appropriate words, as defined in the plugin function.

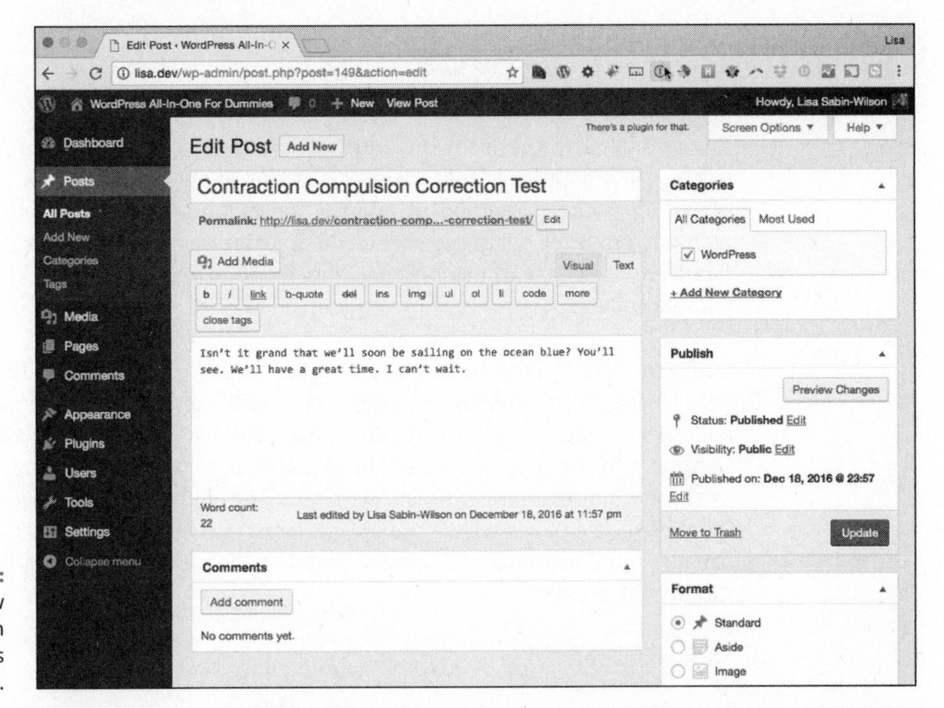

FIGURE 5-5: The Add New Post page with the contractions in place.

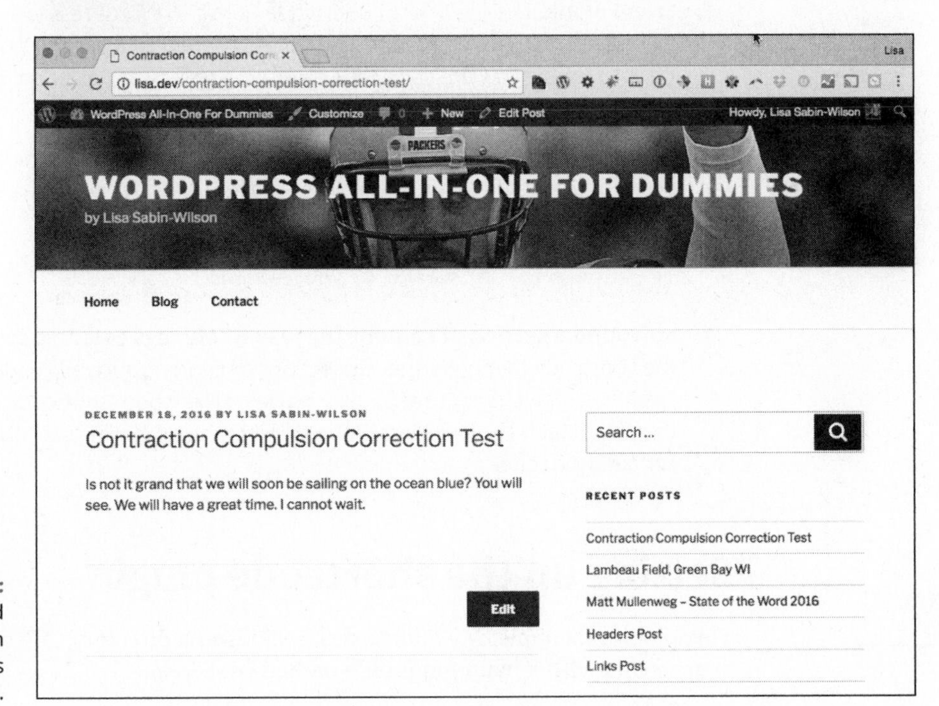

FIGURE 5-6: The published post with contractions replaced.

Creating Simple Plugins from Scratch

Creating Shortcodes

You can use the the_content filter to add new sections of content to posts and pages, such as a form that visitors can use to subscribe to new posts. This type of all-or-nothing method of adding content often adds content where it isn't wanted, however. Using a shortcode is often a much better solution; it offers flexibility and ease of use because shortcode is a shorthand version of fully executable code.

You can employ shortcodes for a wide variety of uses. WordPress includes a [gallery] shortcode, for example, that you can add to any post to display a gallery of images assigned to that post in place of the shortcode. (See Book 4, Chapter 3 for details on adding photo galleries to your posts.) Many plugins (such as those for a forum or contact form) use shortcodes to allow plugin users to designate specific pages where the plugin front end should appear. These shortcodes can surround sections of content, allowing the code that powers the shortcode to modify the content.

Shortcodes can

- **Stand alone:** The [gallery] shortcode built into WordPress is a good example. Simply adding [gallery] to the content of a page or post is all you need to do to allow the shortcode to insert a gallery of uploaded images and display them within the body of a post or page.
- **Support arguments that pass specific information to the shortcode:** This shortcode gives the default WordPress gallery a width of 400 pixels and the caption My Venice Vacation:

```
[gallery width="400" caption="My Venice Vacation"
```

- **Surround a section of content the way HTML tags can:** This shortcut allows the shortcode to modify specific sections of content, such as [code lang="php"]<?php the_title(); ?>[/code]. (The "code" shortcode doesn't exist by default in WordPress. You build it in the upcoming section "Building a simple shortcode.")

Setting up the shortcode plugin

This plugin, Example: My Shortcodes, resides in a directory called /example-my-shortcodes with a main plugin file named init.php.

The reason for the relatively generic name is that you can use this plugin to create multiple shortcodes. To start, create the directory and main plugin file (see "Creating Your First Plugin" earlier in this chapter for instructions), and add the following file header to the init.php file:

```
<?php
/*
Plugin Name: Example: My Shortcodes
Description: This plugin provides the digg and code shortcodes.
*/
?>
```

Building a simple shortcode

In many ways, shortcodes are coded like filters (similar to the the_content filter) except that with shortcodes, the content to be filtered is optional. For this shortcode, you won't worry about content filtering or shortcode attributes; the shortcode is simple enough that it doesn't need either.

The shortcode you create in this section is named [digg]. Adding this shortcode to a post displays a Digg This Post link, which allows that post to be submitted to http://digg.com (a site for keeping track of interesting links).

Creating a shortcode requires two things:

- **Shortcode function:** This function handles the creation of the shortcode and a call to the add_shortcode function.
- **Shortcode arguments:** The add_shortcode function accepts two arguments: the name of the shortcode and the function used by the shortcode.

To get started with your shortcode, add the following code to the end of the Example: My Shortcodes plugin's init.php file, before the closing ?>:

```
add_shortcode('digg','my_digg_shortcode');

function my_digg_shortcode() {
return "<p><a href='http://digg.com/submit?url=".urlencode(get_permalink()).
    "&bodytext=".urlencode(get_the_title())."'>Digg This Post</a></p>";
}
```

REMEMBER

HTML links have to follow some rules because only certain characters are permitted. The urlencode function used for both get_permalink and get_the_title, for example, ensures that the information added to the link results in the creation of a valid link.

Save the changes, make sure that the plugin is active, and then add a post that has `[digg]` in the content. If everything works properly, you should see a Digg This Post link in place of the shortcode when you view the post on your website. (See Figure 5-7.)

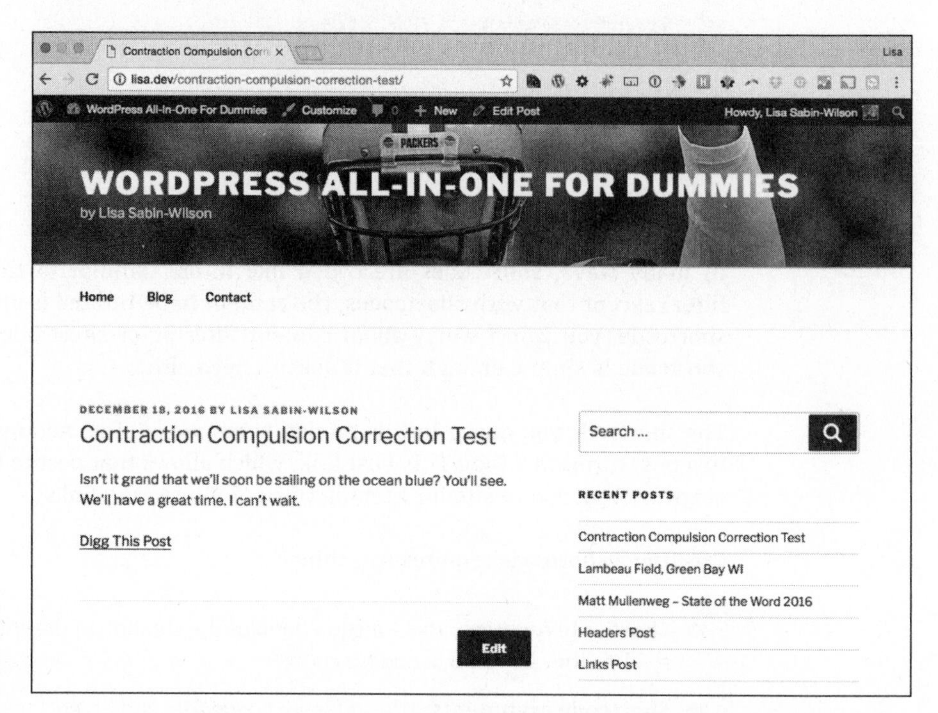

FIGURE 5-7: Digg This Post link added to a post.

Shortcodes can display depending on specific criteria. It's easy to modify the `[digg]` shortcode to display the Digg This Post link if an individual post is being viewed rather than a list, such as the home page or a category archive. The following code shows an updated `my_digg_shortcode` function that uses the `is_single` template tag function to return an empty string if an individual post isn't being viewed:

```
add_shortcode('digg','my_digg_shortcode');

function my_digg_shortcode() {
  if(!is_single()) return '';

  return "<p><a href='http://digg.com/submit?url=".urlencode(get_
    permalink()). "&bodytext=".urlencode(get_the_title())."'>Digg This Post
    </a></p>";
  }
```

Notice the exclamation point (`!`) in front of the `is_single` function call. In PHP language, the exclamation point means *not.* Therefore, the `if` statement translates to "If the current view is not a single post, return an empty string." Because the `if` statement fails when the view is a single post, the original functionality of returning the Digg This Post link is used. The function fires only when a visitor is viewing a single, individual post page, not when the visitor is viewing any other type of page (such as a static page or a category archive page).

Using shortcode attributes

By using attributes, you can customize the shortcode output to meet specific needs without having to rewrite an existing shortcode or create a new one.

The `[digg]` shortcode is a good example of how you can use an attribute to customize the shortcode output. Notice that the generated link to Digg includes `bodytext=`. The text added after `bodytext=` is the default description text for the submitted link, which is the text that displays on your site. The shortcode sends the title of the post (`get_the_title`). Via an attribute, this behavior can be made default behavior while allowing the user to supply a customized description.

To add the attribute support, you need to update the `my_digg_shortcode` function again. Update the function to match the following:

```
add_shortcode('digg','my_digg_shortcode');

function my_digg_shortcode($attributes=array()) {
if(!is_single()) return '';

$attributes=shortcode_atts(
array('description'=>get_the_title()),
$attributes
);
extract($attributes);

return "<p><a href='http://digg.com/submit?url=".urlencode(get_permalink()).
   "&bodytext=".urlencode($description)."'>Digg This Post</a></p>";
}
```

Note the following:

» The `$attributes` argument is added to the `my_digg_shortcode` function declaration. Without this argument, the shortcode function can't receive any of the attributes set on the shortcode. The `=array()` ensures that the

$attributes variable is set to an empty array if the shortcode doesn't have any attributes set.

- The call to the shortcode_atts function passes in an array of default attribute values and merges these defaults with the attributes used in the actual shortcode. Then the code stores this resulting array back in the $attributes variable. Without this section, the attributes won't have a default value.

TIP

It's a good idea to set the defaults even if the default is an empty string.

- The extract function takes the array of attributes and breaks the information into individual plugins. In this example, the function creates the $description variable because that variable is the used attribute. If a shortcode uses title and id attributes, using extract creates $title and $id variables.
- Because the $description variable holds the description that should be used, the get_the_title function call in the returned string is replaced by the $description variable.

Update your shortcode in the body of your test post to use this new description attribute:

```
[digg description="Shortcodes are awesome!"]
```

After saving the post changes, view the updated post, and hover your mouse over the Digg This Post link. You should see a link with the following format:

```
http://digg.com/submit?url=http://domain.com/
    testing-shortcodes/&bodytext=Shortcodes+are+awesome!
```

Adding content to shortcodes

The final piece of the shortcodes puzzle is content. When you wrap a shortcode around a section of content, the shortcode function can modify the content in creative ways.

The example in this section creates a new shortcode called [code], which allows designated sections of content to be formatted as code.

To get the new [code] shortcode running, add the code shown in Listing 5-1 to the bottom of your plugin init.php file. (***Note:*** The arrows and numbers aren't part of the actual code; I include them for purposes of explanation.)

LISTING 5-1: The Code Shortcode

```
function my_code_shortcode($attributes=array(),$content='') {        →1
if(empty($content)) return '';                                       →2

$attributes=shortcode_atts(                                          →3
array('lang'=>''),                                                   →4
$attributes                                                          →5
);                                                                   →6
extract($attributes);                                                →7

$content=str_replace("</p>\n<p>","\n\n",$content);                   →8
$content=str_replace('<p>','',$content);                             →9
$content=str_replace('</p>','',$content);                            →10
$content=str_replace('<br />','',$content);                          →11

$style='white-space:pre;overflow:auto;';                             →12
$style.='font:"Courier New",Courier,Fixed;';                         →13

if('php'==$lang) {                                                   →14
$style.='background-color:#8BD2FF;color:#FFF;';                      →15
}                                                                    →16
else if('css'==$lang) {                                              →17
$style.='background-color:#DFE0B0;color:#333;';                      →18
}                                                                    →18
else {                                                               →19
$style.='background-color:#EEE;color:#000;';                         →20
}                                                                    →21

return "<pre class='$lang' style='$style'>$content</pre>";           →22
}                                                                    →23
add_shortcode('code','my_code_shortcode');                           →24
```

Before digging into how everything works, save the changes to the plugin, and add the following shortcodes to a post:

```
[code]This is a basic code test[/code]

[code lang="php"]echo "This is PHP code.";[/code]

[code lang="css"]p { color:#FFF; }[/code]
```

Figure 5-8 shows that these shortcodes produce some fixed-space boxes with different background colors and styling to contain the code.

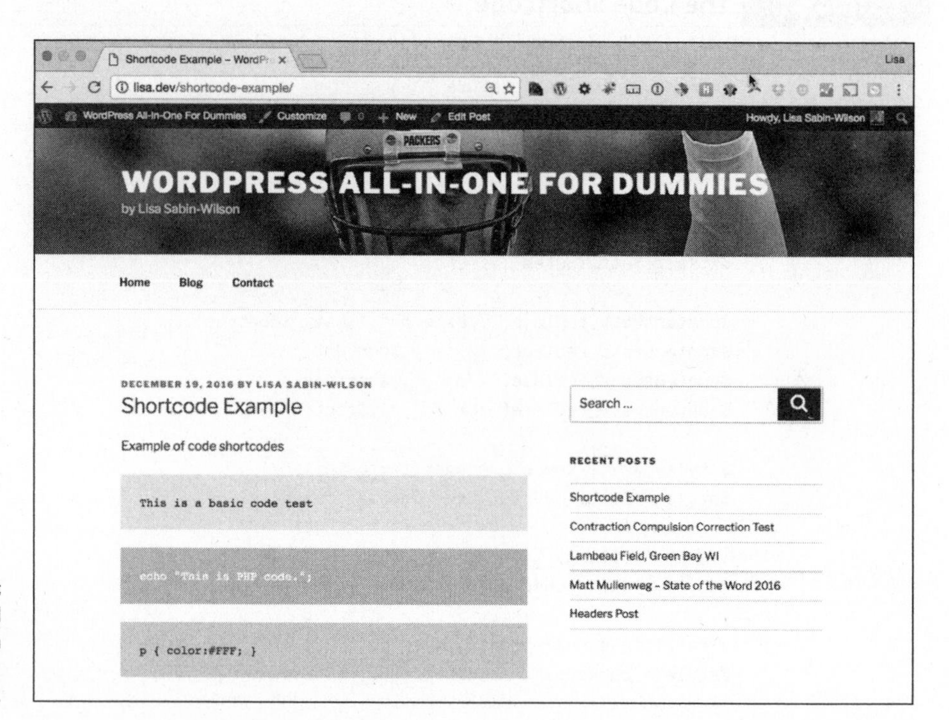

FIGURE 5-8: Code added to a post via the [code] shortcode.

Now that it's clear what the shortcode is doing, you can dissect the way it works:

→1 Just as it's important to add the `$attributes` variable to the function declaration to get access to the shortcode attributes, the `$content` variable is needed to access the content of the shortcode.

→2 If the `$content` variable is empty, the function returns an empty string because going any further with empty content is unnecessary.

→3–7 As with the `[digg]` shortcode from the preceding section, the `shortcode_atts` function establishes base defaults. By default, the `lang` attribute is an empty string. The extract function fills out the `$lang` variable.

→8–11 This set of `str_replace` function calls allows for proper handling of multiline content by the `[code]` shortcode. The problem is that WordPress always tries to add the `<p>` (paragraph) and `<br />` (line break) HTML markup tags even when it shouldn't. The `str_replace` replaces the separation of two `<p>` tag sections with two new lines (a new line is represented by the `\n` code) and then removes all the remaining `<p>` and `<br />` tags inserted by WordPress, allowing the content to display properly when it's wrapped in a `<pre>` (preformat for code) tag.

→12–13 The `$style` variable stores the basic CSS (Cascading Style Sheet; see Book 6, Chapter 4 for more information) styling for the [code] shortcode output.

→14–21 This set of conditional code determines the background and text color based on the `lang` attribute. The php receives a blue background with white text, the `css` receives a khaki background with dark text, and the default is a gray background with black text.

→22 The last line of the function returns the shortcode output by wrapping the content in a `<pre>` HTML markup tag. The `<pre>` tag uses the generated style and adds the `$lang` as a class. The addition of the class allows for more customization through a stylesheet.

→24 This line adds the shortcode to the WordPress code action hook and is required for the shortcode to *fire* (execute) properly.

Adding Widgets

Widgets are individual features you can add to theme sidebars. Widgets can add a simple site search, display a calendar, list the most recent posts, and show RSS feed updates. These features just scratch the surface of what widgets offer and what widgets are capable of doing. WordPress has a handy widget API that makes widget creation very easy.

Coding a simple widget

Widgets are a bit different from filters and shortcodes. Instead of creating one function and registering it, widgets are collections of functions packaged in a container called a *class*. A class is registered as a widget by means of the `register_widget` function.

Like shortcodes, multiple plugins can be housed in a single plugin file. Because the code for some of these widgets gets lengthy, however, each widget is its own plugin.

The widget plugin that you build in this section creates a widget that you can use on the Widgets page of your Dashboard (hover your mouse over Appearance and then click the Widgets link). This plugin, called Example: My User Widget, places a widget on your sidebar that displays `You are logged in as` *Name* or `Welcome Guest` depending on whether you're logged in. This widget is called My User Widget.

As you did in the preceding examples, create a plugin directory in the /wp-content/plugins directory by adding a new folder called /example-my-user-widget, and add an empty init.php file that will serve as your main plugin .php file. Then follow these steps to create the sample plugin:

1. **Open the init.php file in your text editor.**
2. **Add the file header to the top of the init.php file:**

```
<?php
/*
Plugin Name: Example: My User Widget
Description: This plugin provides a simple widget that shows the name of the logged
    in user
*/
```

3. **Press the Enter key, and type the next line of code:**

```
class My_User_Widget extends WP_Widget {
```

This code creates a new class called My_User_Widget, which is based on (and extends) the structure of the existing WP_Widget class.

A class is a way of collecting a set of functions and a set of data in a logical group. In this case, everything that's in the My_User_Widget class is specific to the My User Widget widget. Modifying this class doesn't affect any other widgets on your site.

The WP_Widget class is the central feature of WordPress's Widget API. This class provides the structure and most of the code that powers widgets. When the My_User_Widget class extends the WP_Widget class, the My_User_Widget class automatically gains all the features of the WP_Widget class. This situation means that only the code that needs to be customized for the specific widget needs to be defined, because the WP_Widget class handles everything else.

4. **Press Enter, and type the following lines of code in the init.php file:**

```
function My_User_Widget() {
  parent::WP_Widget(false,'My User Widget');
}
```

The My_User_Widget function has the same name as the My_User_Widget class. A PHP class function that has the same name as the class is a *constructor.* This function is called automatically when the class code is run. (For widgets, WordPress automatically runs the registered widget code behind the scenes.) Constructors run necessary initialization code so that the class behaves properly.

By calling parent::WP_Widget, the My_User_Widget class can tell WordPress about the new widget. In this instance, the My_User_Widget says that the default base ID should be used (this is the false argument you see in the function code, and it defaults to the lowercase version of the class name, my_user_widget, in this case) and that the widget's name is My User Widget.

5. **Press Enter, and type the following line of code in the init.php file:**

```
function widget($args) {
```

The widget function displays the widget content. This function accepts two parameters: $args and $instance. For this example, only $args is needed. You use $instance when you create the next widget.

6. **Press Enter, and type the following lines of code in the init.php file:**

```
$user=wp_get_current_user();
  if(!isset($user->user_nicename)) {
  $message='Welcome Guest';
  }
else {
  $message="You are logged in as {$user->user_nicename}";
}
```

This code finds information about the current user and sets a message to display that depends on whether the user is logged in. This status is determined by checking the $user variable. If the $user->user_nicename variable isn't set, the user isn't logged in.

7. **Press Enter, and type the following lines of code in the init.php file:**

```
extract($args);
  echo $before_widget;
  echo "<p>$message</p>
";
  echo $after_widget;
}
}
```

The $args variable contains several important details about how the sidebar wants widgets to be formatted. The variable is passed into the extract function to pull the settings into stand-alone variables. The four main variables used are $before_widget, $after_widget, $before_title, and $after_title. Because this widget doesn't have a title, the title variables aren't used. The $before_widget variable should always be included before any widget content, and the $after_widget variable should always be included after the widget content.

8. **Press Enter, and type the following lines of code in the `init.php` file:**

```
function register_my_user_widget() {
  register_widget('My_User_Widget');
}
add_action('widgets_init','register_my_user_widget');
```

To register a widget with WordPress (so that WordPress recognizes it as a widget), use the `register_widget` function, and pass it the name of the `widget` class.

Although calling the `register_widget` immediately after the class definition would be nice, the process isn't that simple. The code of the widget that includes the different functions must run before the widget can be registered. When code needs to run at specific times, the code is placed in a function, and the `add_action` function is used to have WordPress run the function at a specific time.

These specific points in time are *actions.* For widget registration, you want to use the `widget_init` action. This action happens after WordPress finishes setting up the code necessary to handle widget registrations, yet occurs early enough for the widget to be registered in time.

REMEMBER

Don't forget to add the closing `?>` tag, which tells WordPress that the PHP execution in this plugin has come to an end.

When this final piece is in place, the widget is ready for use. When you're done, the entire code block looks like this:

```
<?php
/*
Plugin Name: Example: My User Widget
Description: This plugin provides a simple widget that shows the name of the
    logged in user
*/
class My_User_Widget extends WP_Widget {
  function My_User_Widget() {
    parent::WP_Widget(false,'My User Widget');
  }
  function widget($args) {
    $user=wp_get_current_user();
    if(!isset($user->user_nicename)) {
      $message='Welcome Guest';
  }
    else {
      $message="You are logged in as {$user->user_nicename}";
    }
```

```
  extract($args);
    echo $before_widget;
    echo "<p>$message</p>";
    echo $after_widget;
  }
}

function register_my_user_widget() {
  register_widget('My_User_Widget');
}

add_action('widgets_init','register_my_user_widget');
?>
```

Before WordPress 2.8, widget code had to manage everything by itself. For a widget to be used more than once, complex and bug-prone code needed to be produced and maintained. Fortunately, the `WP_Widget` class handles this task seamlessly. You no longer need to worry about single-use or multiple-use widgets, because all widgets coded to use the `WP_Widget` class automatically become multiple-use widgets.

Open the Widgets page on your Dashboard (hover your cursor over Appearance and click the Widgets link). You see a new widget called My User Widget.

Adding an options editor to a widget

Although some widgets work properly without any type of customization, most widgets need to allow the user to supply a title. Thanks to the `WP_Widget` class, adding options to a widget is easy.

The My User Widget example uses the `widget` function to display the widget's content. This addition to the widget code uses two additional functions to handle the widget options: `form` and `update`. The `form` function displays the HTML form inputs that allow the user to configure the widgets options. The `update` function allows the widget code to process the submitted data to ensure that only valid input is saved.

In this section, you create a basic clone of WordPress's Text Widget. Although the process is a bit simple, this cloning allows you to focus on using widget options without getting caught up in the details of a complex widget concept.

To start coding, set up the plugin environment by creating a new directory inside your `/wp-content/plugins` directory, call it `/example-my-text-widget`, and include a blank `init.php` file as your main plugin `.php` file. The plugin you're creating is Example: My Text Widget, which creates a widget called My Text Widget.

Follow these steps to create the `init.php` file for your Example: My Text Widget plugin:

1. **Open the `init.php` file in your text editor.**
2. **Create the file header by adding this code to the top of the `init.php` file:**

```
<?php
/*
Plugin Name: Example: My Text Widget
Description: This plugin provides a basic Text Widget clone complete with widget
   options
*/
```

3. **Press Enter, and type the following line of code in the `init.php` file:**

```
class My_Text_Widget extends WP_Widget {
```

   As with the My User Widget in the preceding section, this widget is created by extending WordPress's `WP_Widget` class. This new widget's class is `My_Text_Widget`.

4. **Press Enter, and type the following lines of code in the `init.php` file:**

```
function My_Text_Widget() {
  $widget_ops=array('description'=>'Simple Text Widget clone');
  $control_ops=array('width'=>400);
  parent::WP_Widget(false,'My Text Widget',$widget_ops,$control_ops);
}
```

   The constructor for this plugin is a bit different this time. The `parent::WP_Widget` function is still called to set up the widget, but two more arguments are given: `$widget_ops` and `$control_ops`. These two arguments allow a variety of widget options to be set. The `$widget_ops` argument can set two options: `description` and `classname`. The description appears below the name of the widget in the widgets listing. The `classname` option sets the class that the rendered widget uses.

5. **Press Enter, and type the following line of code in the `init.php` file:**

```
function form($instance) {
```

   The `form` class function displays an HTML form that allows the user to set the options used by the widget. The `$instance` variable is an array containing the current widget options. When the widget is new, this `$instance` variable is an empty array.

6. **Press Enter, and type the following line of code in the `init.php` file:**

```
$instance=wp_parse_args($instance,array('title'=>'','text'=>''));
```

Shortcode plugins used the `shortcode_atts` function to merge default options with ones from the shortcode, as explained previously in this chapter in the "Using shortcode attributes" section. In this example, `wp_parse_args` performs the same task by merging the existing `$instance` options with default option values.

7. **Press Enter, and type the following line of code in the `init.php` file:**

```
extract($instance);
```

The `extract` function pulls the title and text options in the `$instance` variable into the stand-alone variables `$title` and `$text`.

8. **Press Enter, and type ?>.**

By using the close tag, `?>`, you can more easily display a large amount of HTML without having to echo out each line.

9. **Press Enter, and type the following lines of code in the `init.php` file:**

```
<p>
  <label for="<?php echo $this->get_field_id('title'); ?>">
    <?php _e('Title:'); ?>
    <input
      class="widefat"
      type="text"
      id="<?php echo $this->get_field_id('title'); ?>"
      name="<?php echo $this->get_field_name('title'); ?>"
      value="<?php echo esc_attr($title); ?>"
/>
</label>
</p>
```

The block of HTML displays the title input.

Notice the `$this->get_field_id` and `$this->get_field_name` function calls. These functions, provided by the `WP_Widget` class, produce the needed `id` and `name` values specific to this widget instance. To use these functions, simply pass in the name of the option that is used — `title`, in this case.

The `_e('Title:');` section simply prints `Title:`. This section is wrapped in a call to the `_e` function because the `_e` function allows the text to be translated into other languages.

The value attribute of the `input` tag sets the default value of the field. Because this widget will be populated with the current title, the `$title` variable is included in this attribute. First, however, the `$title` variable is passed through the `esc_attr` function, which allows the text to be formatted properly for use as an attribute value. If the `esc_attr` isn't used, some values in the title, such as double quotation marks, could break the HTML.

10. Press Enter, and type the following lines of code in the `init.php` file:

```
<textarea
  class="widefat"
  rows="16"
  id="<?php echo $this->get_field_id('text'); ?>"
  name="<?php echo $this->get_field_name('text'); ?>"
>
<?php echo esc_attr($text); ?>
</textarea>
```

This block of HTML displays the `textarea` input that allows the user to input the text that she wants to display. There are only two differences between this input and the preceding one: The `textarea` and `text` inputs have a different format, and this input doesn't have a description.

11. Press Enter, and type `<?php`**.**

The form HTML is complete, so the open tag, `<?php`, is used to switch back to PHP code.

12. Press Enter, and type `}`**.**

The `form` function closes.

13. Press Enter, and type the following line of code in the `init.php` file:

```
function update($new_instance,$old_instance) {
```

The `update` class function processes the submitted form data. The `$new_instance` argument provides the data submitted by the form. The `$old_instance` argument provides the widget's old options.

14. Press Enter, and type the following line of code in the `init.php` file:

```
$instance=array();
```

A new empty array variable, `$instance`, is created. This variable stores the final options values.

15. **Press Enter, and type the following lines of code in the `init.php` file:**

```
$instance['title']=strip_tags($new_instance['title']);
$instance['text']=$new_instance['text'];
```

Store the title and text options from the `$new_instance` variable in the `$instance` variable. The title option is run through the `strip_tags` function so that no HTML tags are stored in the title option.

16. **Press Enter, and type the following line of code in the `init.php` file:**

```
return $instance;
```

The `update` function works like a `filter` function. The data is passed in and is manipulated as desired; then the final value is returned.

After seeing how this function works, you may wonder why the `$instance` variable was needed. Even though using `$new_instance` directly would be simpler, it also could produce unexpected results. It's possible for unexpected data to come through as part of the `$new_instance` variable. By creating the `$instance` variable and assigning only known options to it, you can be assured that you know exactly what data is stored for the widget and that your code has had a chance to clean up that data.

17. **Press Enter, and type }.**

Close the `update` function.

18. **Press Enter, and type the following line of code in the `init.php` file:**

```
function widget($args,$instance) {
```

The `widget` class function is the same as before but now has the `$instance` argument. The `$instance` argument stores the options set for the widget.

19. **Press Enter, and type the following lines of code in the `init.php` file:**

```
extract($args);
extract($instance);
```

Use the `extract` function on both `$args` and `$instance` to populate easy-to-use variables for each.

20. **Press Enter, and type the following line of code in the `init.php` file:**

```
$title=apply_filters('widget_title',$title,$instance,$this->id_base);
```

In "Creating a plugin that filters content" earlier in this chapter, which discusses the Example: Contraction Compulsion Correction plugin, the `add_filter` function adds a function to be used as a filter. The `apply_filters` in this example function are how those filter functions are used. This line of code translates to "Store the result of the `widget_title` filters in the `$title` variable." Each filter is passed the `$title`, `$instance`, and `$this->id_base` variables. Every widget that has a title should have this line of code so that filters have a chance to filter all widget titles.

21. **Press Enter, and type the following lines of code in the `init.php` file:**

```
echo $before_widget;
if(!empty($title)) echo $before_title . $title . $after_title;
echo $text;
echo $after_widget;
```

As with the previous Example: My User widget, the `$before_widget` variable is included before the rest of the widget content ,and the `$after_widget` variable is included after all the other widget content. Because this widget supports a title, the code adds the `$before_title` and `$after_title` variables, which, like the `$before_widget` and `$after_widget` variables, come from the `$args` argument passed to the function. The `if` statement ensures that the title appears only if the title isn't empty.

REMEMBER

! means *not.*

22. **Press Enter, and type }.**

The widget function class closes.

23. **Press Enter, and type }.**

The `My_Text_Widget` class closes.

24. **Press Enter, and type the following lines of code in the `init.php` file:**

```
function register_my_text_widget() {
  register_widget('My_Text_Widget');
}

add_action('widgets_init','register_my_text_widget');
```

Register the widget. Notice that the `My_Text_Widget` argument of the `register_widget` function matches the name of this widget's class.

When you finish the preceding steps, the entire block of code in your `init.php` file looks like Listing 5-2.

LISTING 5-2: **The init.php File for the Example: My Text Widget Plugin**

```
<?php
/*
Plugin Name: Example: My Text Widget
Description: This plugin provides a basic Text Widget clone complete with widget
   options
*/

class My_Text_Widget extends WP_Widget {
  function My_Text_Widget() {
    $widget_ops=array('description'=>'Simple Text Widget clone');
    $control_ops=array('width'=>400);
    parent::WP_Widget(false,'My Text Widget',$widget_ops,$control_ops);
  }
  function form($instance) {
    $instance=wp_parse_args($instance,array('title'=>'','text'=>''));
    extract($instance);
?>
<label for="<?php echo $this->get_field_id('title'); ?>">
<?php _e('Title:'); ?>
<input
  class="widefat"
  type="text"
  id="<?php echo $this->get_field_id('title'); ?>"
  name="<?php echo $this->get_field_name('title'); ?>"
  value="<?php echo esc_attr($title); ?>"
/>
</label>
</p>

<textarea
  class="widefat"
  rows="16"
  id="<?php echo $this->get_field_id('text'); ?>"
  name="<?php echo $this->get_field_name('text'); ?>"
>
<?php echo esc_attr($text); ?>
</textarea>

<?php
}
  function update($new_instance,$old_instance) {
    $instance=array();
    $instance['title']=strip_tags($new_instance['title']);
    $instance['text']=$new_instance['text'];
    return $instance;
  }
```

(continued)

LISTING 5-2: *(continued)*

```
  function widget($args,$instance) {
    extract($args);
    extract($instance);
    $title=apply_filters('widget_title',$title,$instance,$this->id_base);
      echo $before_widget;
      if(!empty($title)) echo $before_title . $title . $after_title;
      echo $text;
      echo $after_widget;
  }
}
  function register_my_text_widget() {
    register_widget('My_Text_Widget');
  }

add_action('widgets_init','register_my_text_widget');}
?>
```

Now the widget is ready for use. Open the Widgets page of your Dashboard (hover your cursor over Appearance and click the Widgets link). You see a new widget called My Text Widget. When expanded, the widget has a Title field and text box for the user to configure and add content to, as shown in Figure 5-9.

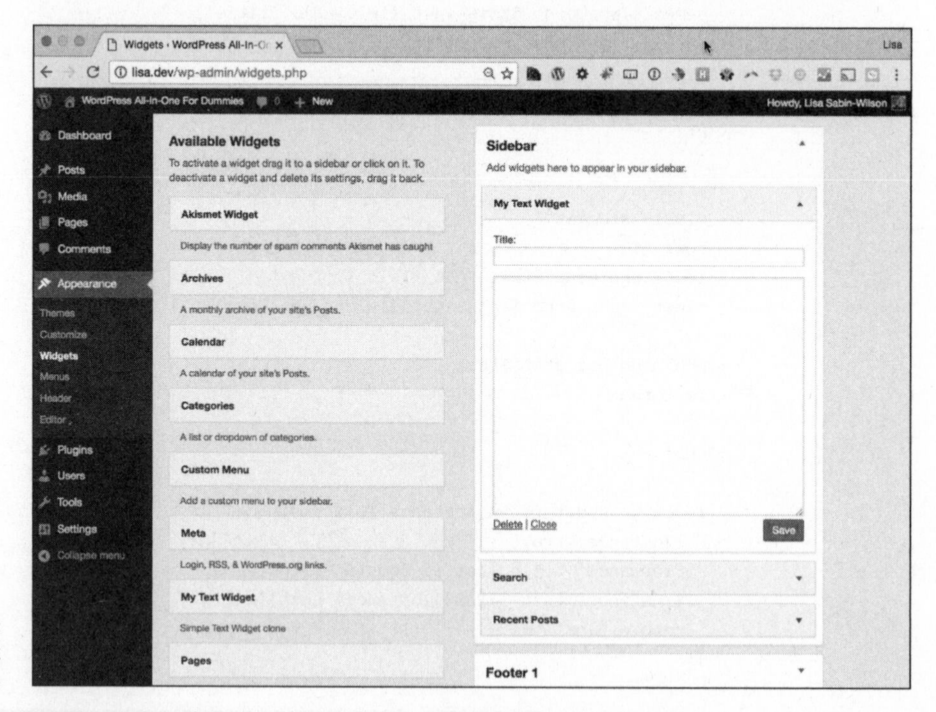

FIGURE 5-9: The My Text Widget.

Building a Settings Page

Many plugins offer a settings page that allows the user to customize plugin options. The options offered by a settings page vary from a few check boxes, drop-down lists, or text inputs to multiple advanced editors that allow the user to build data sets, set up forums, or do advanced content management. Although the following sections focus on building a simple settings page, you can expand the concept to fill any type of plugin need.

Reduced to a bare minimum, a basic settings page consists of code that displays the page, stores the settings, and adds the page to the WordPress admin menu. The plugin you create in this section gives you a solid foundation that you can use to build your own plugin settings pages.

Setting up a plugin with a settings page

The plugin that you create in this section is Example: Settings Page. To get started with this plugin, create a new plugins directory named `/example-settings-page`. All the files for this plugin go in this new directory. That's right: *files*. You add multiple plugin files to the plugin directory.

TIP

Separate different functionality into separate files because as plugins get larger, having everything in one file quickly becomes hard to manage. Similar to the way that you name the `init.php` file (because it *initializes* the plugin), you should name the other files that the plugin uses so that you can easily discern the purpose of each file by looking at the filenames.

This new plugin has five files, so create five blank files and name them as follows:

- `init.php`: Contains the file header and loads the other needed plugin files
- `settings-page.php`: Holds the settings page code
- `default-settings.php`: Sets the default values used for the settings
- `msp-form-class.php`: Provides a class that makes adding form inputs easy
- `settings-functions.php`: Provides functions that load and save the settings

By dividing the plugin code into logical groups, each of which has its own file, you make the code much easier to maintain. With this type of setup, as you add new features to the plugin, you can easily isolate the features in their own files.

You must create all the files before the plugin can function. To start, add the following code to your plugin's init.php file:

```
<?php
/*
Plugin Name: Example: Settings Page
Description: This plugin offers a solid starting point for building settings
   pages
*/

require_once(dirname(__FILE__).'/default-settings.php');
require_once(dirname(__FILE__).'/msp-form-class.php');
require_once(dirname(__FILE__).'/settings-functions.php');
require_once(dirname(__FILE__).'/settings-page.php');
```

The require_once function loads another .php file. If the file isn't found, a fatal error occurs, and the plugin doesn't function. Therefore, trying to activate the plugin at this point causes the plugin to fail. This situation is good, because it prevents the plugin from activating when needed files are missing, which can happen if an incomplete upload of the plugin files occurs.

Four primary functions load other PHP code: include, include_once, require, and require_once. The include functions don't cause a fatal error if a problem occurs in loading the requested file, but the require functions do. The functions that add _once don't load the file if the file has already been loaded. This situation helps prevent the code from accidentally loading a file multiple times, which can break the code or cause unexpected behavior.

Directly below the require_once functions, add the following line:

```
dirname(__FILE__).'/default-settings.php'
```

This section of code finds the full path to the default-settings.php file. Refer to the file's full path to prevent issues with some server setups.

That's all the code that the init.php file needs for this plugin. Save and upload the file to your plugin directory at /wp-content/plugins/example-settings-page. When you add new features to the plugin, you need to add new files to hold these features and use a require_once call to load each one.

Adding a new Admin menu entry

Add the following lines of code to the settings-page.php file from the preceding section:

```
<?php

function msp_add_admin_menu() {
  add_options_page(
    'Example Settings Page',
    'Example Settings',
    'manage_options',
    'msp-example-settings-page',
    'msp_display_settings_page'
  );
  if(isset($_POST['msp_save'])) {
    add_action("admin_head-$page",'msp_save_settings_handler');
  }
}
add_action('admin_menu','msp_add_admin_menu');
```

This code adds a new Example Setting link to the Dashboard's Settings menu. The `add_options_page` function is the key to adding this new menu. The following list describes what each argument in this function does:

- **Example Settings Page:** The first argument sets the title of the page. The browser, not the settings page, displays this title.
- **Example Settings:** The second argument is the name of the menu. The space is limited, so keep the name short.
- **manage_options:** The third argument is the capability that the user must have to see and access the menu. For settings pages, the `manage_options` capability typically is the best choice, as only administrators have this capability by default. You can use `edit_others_posts` to give access to administrators and editors, and you can use `edit_published_posts` for administrators, editors, and authors.
- **msp-example-settings-page:** The fourth argument is a name that WordPress uses internally to navigate to the page. Make sure that this name is unique; otherwise, you may run into problems with other plugins. The `msp` at the front of the name stands for *My Settings Page*. Consistently using a prefix in this manner helps prevent duplicate name issues.
- **msp_display_settings_page:** The fifth argument is the name of the function that's called when the page is viewed. As with the fourth argument, make sure that this name is unique to your plugin. Notice that the `msp` prefix is used here, too.

Just as the `register_widget` function can't be called when a plugin first loads, the `add_options_page` function must wait until WordPress is ready before it can be called. To call the function when WordPress is ready, the `add_options_page` function is wrapped in a function that's called when the `admin_menu` action is run.

Other than registering the new admin menu, this code adds a page-specific action if form data with a msp_save variable has been submitted. When this action fires, the msp_save_settings_handler function processes and saves the submitted form data.

Creating a settings form

There are as many ways to set up a settings form as there are hot dogs in Chicago. Well, maybe not quite that many, but I'm sure that the numbers are close. Most of these methods use large sections of repeated code for each input.

This form shows how you can integrate each HTML form input type into your settings form. Instead of coding each input manually — a time-consuming, error-prone process — you can use a set of functions that WordPress provides to make adding new inputs easy.

Open your plugin settings-page.php file, and follow these steps to add the code that handles the HTML form input for your settings page:

1. **Open the settings-page.php file in your text editor.**
2. **Add the following line of code at the bottom of the settings-php file:**

   ```
   function msp_display_settings_page() {
   $form=new MSP_Form(msp_get_settings());
   ?>
   ```

   The msp-form-class.php file, which you create later in the "Creating the MSP_Form class" section of this chapter, contains a new class called MSP_Form. This class accepts an array of settings to assign the initial values of the form inputs. The settings come from the msp_get_settings function, which is defined in the settings-functions.php file (covered in the next section, "Configuring default settings").

   When you use a class like this one, a variable stores an *object*. In this case, the object is stored in a variable called $form. Later in this form code, the $form variable adds new form inputs by using the functions provided by MSP_Form.
3. **Press Enter, and type the following lines of code:**

   ```
   <?php if(isset($_GET['updated'])) : ?>
   <div id="message" class="updated fade">
     <p><strong>Settings saved.</strong></p>
   </div>
   <?php endif; ?>
   ```

When the form data is saved, the page redirects to a new settings-page link that contains a variable named `updated`. This section of code displays a `Settings saved` message when this redirect happens, confirming to the user that the settings changes are saved.

4. **Press Enter, and type the following lines of code:**

```
<div class="wrap">
<?php screen_icon(); ?>
<h2>Example Settings Page</h2>
```

This section simply follows the page structure used by WordPress's built-in editors so that your settings page fits the style of WordPress.

5. **Press Enter, and type the following line of code:**

```
<form method="post" action="<?php echo $_SERVER['REQUEST_URI']; ?>">
```

Many settings pages use code that submits the form data to the `options.php` page built into WordPress. This settings-page form sends the data back to itself, allowing this plugin's code to have full control of how the data is stored.

6. **Press Enter, and type the following line of code:**

```
<?php wp_nonce_field('msp-update-settings'); ?>
```

The `wp_nonce_field` function is part of the nonce system of WordPress. nonce — which stands for *number used once* or *number once* — protects a site from attackers and increases security by validating the contents of a form field. All your WordPress-managed forms should use nonces — and use them properly. The key is to have the `wp_nonce_field` function call (with a unique attribute to identify the form's purpose) inside the HTML form tags and a call to `check_admin_referer` (that uses a matching attribute) before any action is taken with the form data. The `check_admin_referer` call is in the `msp_save_settings_handler` function at the end of this file's code.

7. **Press Enter, and type the following code:**

```
<table class="form-table">
<tr valign="top">
<th scope="row"><label for="text">Text</label></th>
<td>
<?php $form->add_text('text'); ?>
</td>
</tr>
<tr valign="top">
<th scope="row"><label for="textarea">Text Area</label></th>
<td>
```

```
<?php $form->add_textarea('textarea'); ?>
</td>
</tr>
```

This code is the table structure commonly used in editors throughout WordPress, employing the HTML markup to create tables. This structure is simple and easy to expand by adding options.

TIP

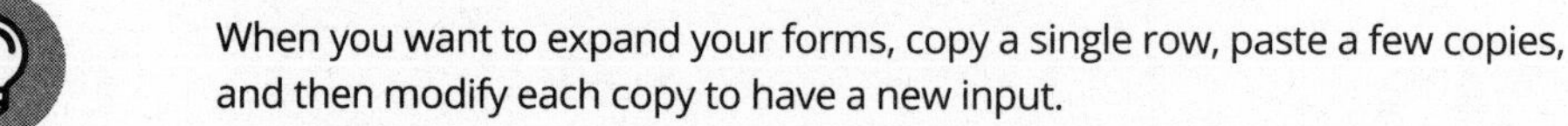

When you want to expand your forms, copy a single row, paste a few copies, and then modify each copy to have a new input.

The call to the `$form->add_text` function adds the text input to the form. This function accepts an argument that defines the name of the setting (`text`, in this case). The `textarea` section of the code follows the same format as this section.

8. **Press Enter, and type the following lines of code:**

```
<tr valign="top">
<th scope="row">Multi-Checkbox</th>
<td>
<label title="1">
<?php $form->add_multicheckbox('multicheckbox','1'); ?> 1
</label><br/>
<label title="2">
<?php $form->add_multicheckbox('multicheckbox','2'); ?> 2
</label><br/>
<label title="3">
<?php $form->add_multicheckbox('multicheckbox','3'); ?> 3
</label><br/>
<label title="4">
<?php $form->add_multicheckbox('multicheckbox','4'); ?> 4
</label>
</td>
</tr>
```

This section uses the `MSP_Form` class's `add_multicheckbox` function to create a form input that allows multiple values to be stored in a single setting. The second argument used for this function sets the value of the specific check box. When data is loaded from this type of setting, the values of the checked inputs are stored in an array.

9. **Press Enter, and type the following lines of code:**

```
<tr valign="top">
<th scope="row">Radio</th>
<td>
<label title="Option 1">
<?php $form->add_radio('radio','1'); ?>
Option 1
</label>
<br/>
<label title="Option 2">
<?php $form->add_radio('radio','2'); ?>
Option 2
</label>
<br/>
<label title="Option 3">
<?php $form->add_radio('radio','3'); ?>
Option 3
</label>
</td>
</tr>
```

This section uses the `MSP_Form` class's `add_radio` function to create a form input by using radio buttons. The radio button uses the same format as the `multicheckbox` inputs in Step 8. The second argument sets the value of the individual input. The data loaded for a radio setting is a string, not an array.

10. **Press Enter, and type the following lines of code:**

```
<tr valign="top">
<th scope="row"><label for="select">Select</label></th>
<td>
<?php
$options = array(
  '1'=>'Option 1',
  '2'=>'Option 2',
  '3'=>'Option 3',
);
?>
<?php $form->add_select('select',$options); ?>
</td>
</tr>
</table>
```

Select inputs can be difficult to code by hand, but the `MSP_Form` class makes adding a select input easy. First, create an associative, or *named,* array. A named array sets the key to be used for each value. In other words, each array input looks like `'1'=>'Option 1'`, where `1` is the value to be stored for that setting and `Option 1` is the description to show for that value. Then the named array is passed in as the second argument to the `add_select` function.

11. **Press Enter, and type the following lines of code:**

```
<p class="submit">
<input type="submit" name="msp_save" class="button-primary" value="Save Changes"/>
</p>
```

Note the `msp_save` value used for the Submit button's name. WordPress uses this name to check for a settings-form submission toward the top of this file.

12. **Press Enter, and type the following lines of code:**

```
<?php $form->add_used_inputs(); ?>
</form>
</div>
```

This call to `add_used_inputs` is important for handling the submitted form data.

13. **Press Enter, and type the following lines of code:**

```
<?php
}
function msp_save_settings_handler() {
check_admin_referer('msp-update-settings');
$settings=MSP_Form::get_post_data();
msp_update_settings($settings);
}
```

The `msp_save_settings_handler` completes the `settings-page.php` code. First, the nonce set inside the form is checked with the `check_admin_referer` to ensure that the nonces are used properly. The `MSP_Form` class function `get_post_data` is used to store the submitted settings in the `$settings` variable. Then these settings are passed to the `msp_update_settings` function (defined in the `settings-functions.php` file) to save the settings.

When you finish, the code in your entire `settings-page.php` file looks like Listing 5-3.

LISTING 5-3: HTML Form Input Code for the Settings Page

```
<?php

function msp_add_admin_menu() {
add_options_page(
  'Example Settings Page',
  'Example Settings',
  'manage_options',
  'msp-example-settings-page',
  'msp_display_settings_page'
);
if(isset($_POST['msp_save'])) {
  add_action("admin_head-$page",'msp_save_settings_handler');
}
}
add_action('admin_menu','msp_add_admin_menu');

function msp_display_settings_page() {
$form=new MSP_Form(msp_get_settings());
?>
<?php if(isset($_GET['updated'])) : ?>
<div id="message" class="updated fade">
  <p><strong>Settings saved.</strong></p>
</div>
<?php endif; ?>
<div class="wrap">
<?php screen_icon(); ?>
<h2>Example Settings Page</h2>
<form method="post" action="<?php echo $_SERVER['REQUEST_URI']; ?>">
<?php wp_nonce_field('msp-update-settings'); ?>
<table class="form-table">
<tr valign="top">
<th scope="row"><label for="text">Text</label></th>
<td>
<?php $form->add_text('text'); ?>
</td>
</tr>
<tr valign="top">
<th scope="row"><label for="textarea">Text Area</label></th>
<td>
<?php $form->add_textarea('textarea'); ?>
</td>
</tr>
<tr valign="top">
<th scope="row"><label for="checkbox">Checkbox</label></th>
<td>
```

(continued)

LISTING 5-3: *(continued)*

```
<?php $form->add_checkbox('checkbox'); ?>
</td>
</tr>
<tr valign="top">
<th scope="row">Multi-Checkbox</th>
<td>
<label title="1">
<?php $form->add_multicheckbox('multicheckbox','1'); ?> 1
</label><br/>
<label title="2">
<?php $form->add_multicheckbox('multicheckbox','2'); ?> 2
</label><br/>
<label title="3">
<?php $form->add_multicheckbox('multicheckbox','3'); ?> 3
</label><br/>
<label title="4">
<?php $form->add_multicheckbox('multicheckbox','4'); ?> 4
</label>
</td>
</tr>
<tr valign="top">
<th scope="row">Radio</th>
<td>
<label title="Option 1">
<?php $form->add_radio('radio','1'); ?>
Option 1
</label>
<br/>
<label title="Option 2">
<?php $form->add_radio('radio','2'); ?>
Option 2
</label>
<br/>
<label title="Option 3">
<?php $form->add_radio('radio','3'); ?>
Option 3
</label>
</td>
</tr>
<tr valign="top">
<th scope="row"><label for="select">Select?> </th>
<td>
<?php
$options = array(
  '1'=>'Option 1',
  '2'=>'Option 2',
  '3'=>'Option 3',
```

```
);
?>
<?php $form->add_select('select',$options); ?>
</td>
</tr>
</table>
<p class="submit">
<input type="submit" name="msp_save" class="button-primary" value="Save Changes"/>
</p>
<?php $form->add_used_inputs(); ?>
</form>
</div>
<?php
}
function msp_save_settings_handler() {
check_admin_referer('msp-update-settings');
$settings=MSP_Form::get_post_data();
msp_update_settings($settings);
}
```

Configuring default settings

A nice feature of this settings-form code allows you to make default settings easily. These default settings are loaded even if the settings form hasn't been saved, allowing for reliable use of the defaults.

To set up the defaults, add the following code to a new file, and save it as `default-settings.php` in your `/example-settings page` plugin directory:

```
<?php
function msp_get_default_settings() {
  $defaults=array(
   'text'  => 'Sample Text',
   'textarea' => 'Sample Textarea Text',
   'checkbox' => '1',
   'multicheckbox' => array('2','3'),
   'radio'    => '2',
   'select'   => '3',
  );
  return $defaults;
}
```

Creating Simple Plugins from Scratch

By modifying the `$defaults` array, you can add new default values, remove them, or modify them easily. The format for this type of array is '*setting_name*'=>'*setting_value*', where *setting_name* is the name of the setting and *setting_value* is the default value you want to use for that setting.

Adding settings functions

The settings functions manage loading and saving settings. Both functions are optimized for performance and reliability. Add the following code to a new file, and save it as `settings-functions.php` in your `/example-settings-page` plugin directory to add the settings function:

```
<?php
function msp_get_settings($name=null) {
static $settings=null;

if(is_null($settings)) {
$settings=get_option('msp-example-settings');
if(!is_array($settings)) $settings=array();

$defaults=msp_get_default_settings();
$settings=array_merge($defaults,$settings);
}

if(is_null($name)) return $settings;
if(isset($settings[$name])) return $settings[$name];
return '';
}

function msp_update_settings($settings) {
update_option('msp-example-settings',$settings);

$redirect_url=array_shift(explode('?',$_SERVER['REQUEST_URI']));
$redirect_url.='?page='.$_REQUEST['page'].'&updated=true';

wp_redirect($redirect_url);
}

$all_settings=msp_get_settings();
$text=msp_get_settings('text');
```

Because you call the function with no argument, WordPress returns all the settings. When you pass a setting name, WordPress returns the value of just that setting.

The msp_update_settings function accepts a new array of settings to be saved. When the settings are saved, a new redirect link with the updated variable is built and displays the Settings saved message. Then WordPress's wp_redirect function redirects the site to this new link.

Creating the MSP_Form class

The MSP_Form class code is quite lengthy but well worth the effort of creating because it makes form building so easy.

Add the code in Listing 5-4 to a new file, and save it as msp-form-class.php in your /example-settings-page plugin directory.

LISTING 5-4: The MSP_Form Class Code

```
<?php
class MSP_Form {
var $inputs=array();
var $settings=array();
function MSP_Form($settings=array()) {
if(is_array($settings)) $this->settings=$settings;
}
function add_used_inputs() {
$value=implode(',',array_unique($this->inputs));
$this->add_hidden('___msp_form_used_inputs',$value);
}
function get_post_data() {
if(!isset($_POST['___msp_form_used_inputs'])) {
return $_POST;
}
$data=array();
$inputs=explode(',',$_POST['___msp_form_used_inputs']);
foreach((array)$inputs as $var) {
$real_var=str_replace('[]','',$var);
if(isset($_POST[$real_var])) {
$data[$real_var]=stripslashes_deep($_POST[$real_var]);
}
else if ($var!=$real_var) {
$data[$real_var]=array();
}
else {
$data[$real_var]='';
}
}
return $data;
}
```

(continued)

LISTING 5-4: (continued)

```
// $form->add_text('name');
function add_text($name,$options=array()) {
if(!isset($options['class'])) $options['class']='regular-text';
$this->_add_input('text',$name,$options);
}
function add_textarea($name,$options=array()) {
$this->_add_input('textarea',$name,$options);
}
function add_checkbox($name,$options=array()) {
if($this->_get_setting($name)) $options['checked']='checked';
$this->_add_input('checkbox',$name,$options);
}
function add_file($name,$options=array()) {
$this->_add_input('file',$name,$options);
}
function add_password($name,$options=array()) {
$this->_add_input('password',$name,$options);
}

// $form->add_select('num',array('1'=>'One','2'=>'Two'));
function add_select($name,$values,$options=array()) {
$options['values']=$values;
$this->_add_input('select',$name,$options);
}

// $form->add_radio('type','extended');
function add_radio($name,$value,$options=array()) {
if($this->_get_setting($name)==$value) $options['checked']='checked';
$options['value']=$value;
$this->_add_input('radio',$name,$options);
}
function add_multicheckbox($name,$value,$options=array()) {
$setting=$this->_get_setting($name);
if(is_array($setting) && in_array($value,$setting)) {
$options['checked']='checked';
}
$options['value']=$value;
$this->_add_input('checkbox',"{$name}[]",$options);
}
function add_hidden($name,$value,$options=array()) {
$options['value']=$value;
$this->_add_input('hidden',$name,$options);
}

// $form->add_submit('save','Save');
function add_submit($name,$description,$options=array()) {
```

```
$options['value']=$description;
$this->_add_input('submit',$name,$options);
}
function add_button($name,$description,$options=array()) {
$options['value']=$description;
$this->_add_input('button',$name,$options);
}
function add_reset($name,$description,$options=array()) {
$options['value']=$description;
$this->_add_input('reset',$name,$options);
}

// $form->add_image('imagemap','http://domain.com/img.gif');
function add_image($name,$image_url,$options=array()) {
$options['src']=$image_url;
$this->_add_input('image',$name,$options);
}

// This function should not be called directly.
function _add_input($type,$name,$options=array()) {
$this->inputs[]=$name;
$settings_var=str_replace('[]','',$name);
$css_var=str_replace('[','-',str_replace(']','',$settings_var));

if(!is_array($options)) $options=array();
$options['type']=$type;
$options['name']=$name;
if(!isset($options['value']) && 'checkbox'!=$type) {
$options['value']=$this->_get_setting($settings_var);
}

if('radio'==$type || $settings_var!=$name) {
if(empty($options['class'])) $options['class']=$css_var;
}
else {
if(empty($options['id'])) $options['id']=$css_var;
}

$scrublist=array($type=>array());
$scrublist['textarea']=array('value','type');
$scrublist['file']=array('value');
$scrublist['dropdown']=array('value','values','type');

$attributes = '';
foreach($options as $var => $val) {
  if(!is_array($val) && !in_array($var,$scrublist[$type]))
  $attributes.="$var='".esc_attr($val)."' ";
}
```

(continued)

Creating Simple Plugins from Scratch

LISTING 5-4: (continued)

```
if('textarea'==$options['type']) {
  echo "<textarea $attributes>";
  echo format_to_edit($options['value']);
  echo "</textarea>\n";
}
else if('select'==$options['type']) {
  echo "<select $attributes>\n";
  if(is_array($options['values'])) {
  foreach($options['values'] as $val=>$name) {
  $selected=($options['value']==$val)?' selected="selected"':'';
  echo "<option value=\"$val\"$selected>$name</option>\n";
  }
  }
  echo "</select>\n";
}
else {
  echo "<input $attributes/>\n";
}

}

// This function should not be called directly.
function _get_setting($name) {
if(isset($this->settings[$name])) {
return $this->settings[$name];
}
return '';
}

}
```

You can use 13 functions to add inputs to a form. Following are examples of how to use them:

```
$form->add_text('setting_name');
$form->add_textarea('setting_name');
$form->add_checkbox('setting_name');
$form->add_file('setting_name');
$form->add_password('setting_name');
$form->add_select('setting_name',array('1'=>'One','2'=>'Two'));
$form->add_radio('setting_name','value_1');
$form->add_radio('setting_name','value_2');
$form->add_multicheckbox('setting_name','value_1');
$form->add_multicheckbox('setting_name','value_2');
$form->add_hidden('setting_name','value');
```

```
$form->add_submit('setting_name','value');
$form->add_button('setting_name','value');
$form->add_reset('setting_name','value');
$form->add_image('imagemap','http://domain.com/img.gif');
```

A feature that all the form input functions support is an optional last parameter to set additional HTML tag attributes in the final output. Here's an example block of code that sets additional HTML markup in the final output:

```
$form->add_text('setting_name',array('class'=>'regular-text code'));
$form->add_text('setting_name',array('style'=>'width:250px;font-size:2em;'));
$form->add_textarea('setting_name',array('rows'=>'10','cols'=>'30'));
$form->add_checkbox('setting_name',array('checked'=>'checked'));
```

This code adds the class `regular-text` and `code` to the first text input, adds inline styling to control the width and font size of the second text input, uses 10 rows and 30 columns for the `textarea` input, and forces the `checkbox` to always be selected on page load for the last input.

You can use the last optional argument on any of the form inputs to set needed tag attributes. By using this argument in creative ways, you can always use the input form functions rather than code inputs from scratch.

Making sure the settings page works

After you have the files entered, you're ready to activate the Example: Settings Page plugin.

TIP

Because the code is divided, if WordPress displays an error message saying that the plugin could not be activated, note which file and line number had the error. This information help you find and fix your error quickly.

After activating the plugin, visit your Dashboard, hover your cursor over Settings, and click the Example Settings link to see the settings form in action, as shown in Figure 5-10. Notice that the form is populated by the default values you set. These default values are set the first time the `msp_get_settings` function is called. Therefore, even if the user has never gone to your settings page, the defaults are available. In addition, when you add new settings and update the defaults to the new settings, those new defaults also become available automatically.

After testing the settings options, go into the `settings-page.php` file, and add new settings. When you have the new settings, make sure to update the `$defaults` array in `default-settings.php`.

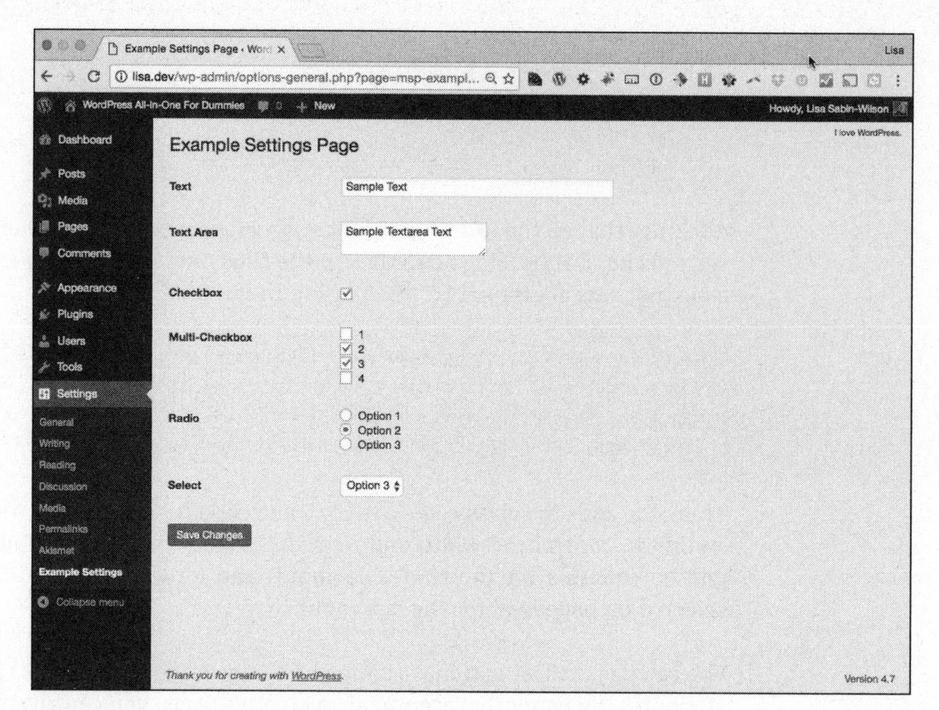

FIGURE 5-10: The Example Settings page.

You can extend this plugin with code from another example plugin. To load the settings, simply use the `msp_get_settings` function.

You could update the `widget` function from the Example: My User Widget plugin (see " Coding a simple widget" earlier in this chapter) to use two settings: `guest_message` and `user_message`. Enter the following code:

```
function widget($args) {
$user=wp_get_current_user();
if(!isset($user->user_nicename))
$message=msp_get_settings('guest_message');
else
$message=msp_get_settings('user_message').$user->user_nicename;
extract($args);
echo $before_widget;
echo "<p>$message</p>";
echo $after_widget;
}
```

» **Naming, filing, and locating your plugin**

» **Internationalizing your plugin**

Chapter **6**

Exploring Plugin Development Best Practices

Starting to develop WordPress plugins is relatively easy. Developing WordPress plugins well is much more difficult.

The key to doing development well is sticking with a set of standards to ensure that your plugin is well designed and implemented. A set of standards that many people can agree upon is typically referred to as *best practices.* By adopting best practices as your own personal development standard, you ensure that other developers can easily understand your plugin's structure and code. Doing so makes collaboration much smoother. In other words, if all WordPress plugin developers followed best practices, the WordPress development world would be a happier, more productive place.

This chapter delves deeper into best practices and is dedicated to taking your plugin's quality to the next level.

Adding a File Header

The most fundamental best practice in creating WordPress plugins is ensuring that your plugin has a *file header* at the top of the main plugin file. As discussed in Book 7, Chapter 5, the file header is the part of a plugin file that identifies the file as a plugin. Without the file header, the plugin isn't actually a plugin and can't be enabled on the WordPress Dashboard.

TIP

Even if you don't distribute your plugin in the WordPress Plugin Directory (`https://wordpress.org/plugins`), take the time to fill in the file header with the name, description, author, version, and license. This info is helpful to all your plugin users.

Use the following header names to supply information about your plugin:

» **Plugin Name:** The value of this entry is listed as the name of the plugin on the Dashboard's Plugins page. The Plugin Name is the only required entry in a plugin's file header. If this entry isn't present, WordPress ignores the plugin.

When giving your plugin a name, make sure that you choose a name that is

- *Unique:* WordPress uses the name to check for plugin updates. If you name the plugin *Akismet,* for example, WordPress could offer to let the user automatically upgrade the plugin, resulting in your plugin code's being replaced by the actual Akismet plugin. Starting all your plugin names with your own name is a good way to achieve unique names. Following are some unique plugin names: Lisa's Twitter Widget, Lisa's Amazon Affiliate Shortcodes, and Lisa's Really Cool Plugin.
- *Descriptive:* Use a name that describes the plugin's purpose. Lisa's Twitter Widget and Lisa's Amazon Affiliate Shortcodes both describe what the plugin offers quite well, but Lisa's Really Cool Plugin doesn't identify its purpose.

REMEMBER

Even if you use the plugin only on your own site, by using a nondescriptive name, you may end up having to look at the plugin's code just to remember what the plugin actually does.

» **Description:** This entry is meant to be a brief explanation of what features the plugin offers. The description entry appears next to each plugin listed on the Dashboard's Plugins page. Because the plugin's listing shares space with other plugins, don't add a large description. Limit the description to one to three sentences.

You can put HTML in the description, as well as add links for plugin documentation and other resources.

WARNING

Don't abuse this feature to make your plugin stand out from all the others. If your plugin is installed, you've already won over the user and motivated him to install the plugin. Don't lose a user by spamming his plugin listing.

- **Version:** If you share the code for your plugin, the version entry of the plugin could be one of the most important entries in the plugin's file header. If the version number is updated properly, when a user reports an issue, you quickly know exactly what code the user is running, whether the plugin is outdated, and whether the current code has a bug. The version entry is simple but very powerful.
- **Plugin URI:** Enter the address of the website where you talk in depth about the plugin. At minimum, provide information about what the plugin is and any necessary instructions on using it. Also, it's a good idea to allow comments so that people can provide feedback, both good and bad. Feedback works as a simple support system for your plugin.
- **Author:** List the name(s) of the plugin author(s). Sometimes, a company's name is used instead of a specific developer's.
- **Author URI:** List the address of the plugin author's website.
- **License:** This entry is the name of the license under which your plugin is released. For most plugins, this entry should be GPLv2, because it matches the license under which WordPress is released. When you submit a plugin to the Plugin Directory, that plugin must be licensed under GPL version 2 or later. For more information on GPL licensing, including how it pertains to your plugin development practices, see Book 1, Chapter 2.
- **Text Domain:** Creates a unique identifier that later allows the plugin to accept translations for other languages (other than English). The value of this entry is used as the domain for translating the other entries and should match the domain used in the `load_plugin_textdomain` function. For details about translating, see "Internationalizing or Localizing Your Plugin" later in this chapter.

  This entry is rarely used in plugins, but it's available to allow translators to supply translations for the details in the file header.
- **Domain Path:** This entry is used with the Text Domain entry to offer file-header translations. The value of this entry is the name of the directory inside the plugin's directory (such as `/language/` or `/translations/`) where the translation files are located. The directory must begin with a forward slash for the translation files to be found. If this entry isn't used, the plugin's directory is searched for the translation files.

  Like Text Domain, the Domain Path entry is rarely used.

» **Network:** When WordPress is running as a network (as discussed in Book 8), this entry allows plugins to indicate that they must be active for the entire network rather than a single site. The only accepted value for this entry is `true`; any other value is treated the same as a blank or missing entry. If WordPress isn't running as a network, this entry is ignored.

  Using this entry is helpful if your plugin provides very low-level features, such as advanced caching. Because such a feature could create problems if it's active on only some sites in the network, this entry forces an all-or-nothing activation of the plugin. Either all the sites in the network run the plugin, or no site does.

» **Site Wide Only:** This deprecated entry is superseded by the Network entry and is supported only for backward compatibility. If the Network entry is supplied, this entry is ignored. As with Network, the only recognized value is `true`. You shouldn't use this entry as support, however, because it may be removed in future versions.

Although you may not need all the entry options, a file header that uses all the options looks like the following. (The Site Wide Only option is left out because it is replaced by the Network option.)

```
<?php
/*
  Plugin Name: Lisa's Twitter Widget
  Description: Display Twitter feeds in any sidebar on your site.
  Version: 1.0.0
  Plugin URI: http://example.com/twitter-widget
  Author: Lisa Sabin-Wilson
  Author URI: http://example.com
  License: GPLv2
  Text Domain: lsw-twitter-widget
  Domain Path: /language/
  Network: true
*/
?>
```

TIP

It's customary to include a licensing statement in the header of your plugin to indicate adherence to the GPLv2 license. This statement is easy to include and is formatted as follows:

```
<?php
/*

This program is free software; you can redistribute it and/or
modify it under the terms of the GNU General Public License
as published by the Free Software Foundation; either version 2
of the License, or (at your option) any later version.
```

```
This program is distributed in the hope that it will be useful,
but WITHOUT ANY WARRANTY; without even the implied warranty of
MERCHANTABILITY or FITNESS FOR A PARTICULAR PURPOSE. See the
GNU General Public License for more details.

You should have received a copy of the GNU General Public License
along with this program; if not, write to the Free Software
Foundation, Inc., 51 Franklin Street, Fifth Floor, Boston, MA 02110-1301, USA.
*/
?>
```

Figure 6-1 shows the Plugins page, displaying information pulled from each plugin's file header.

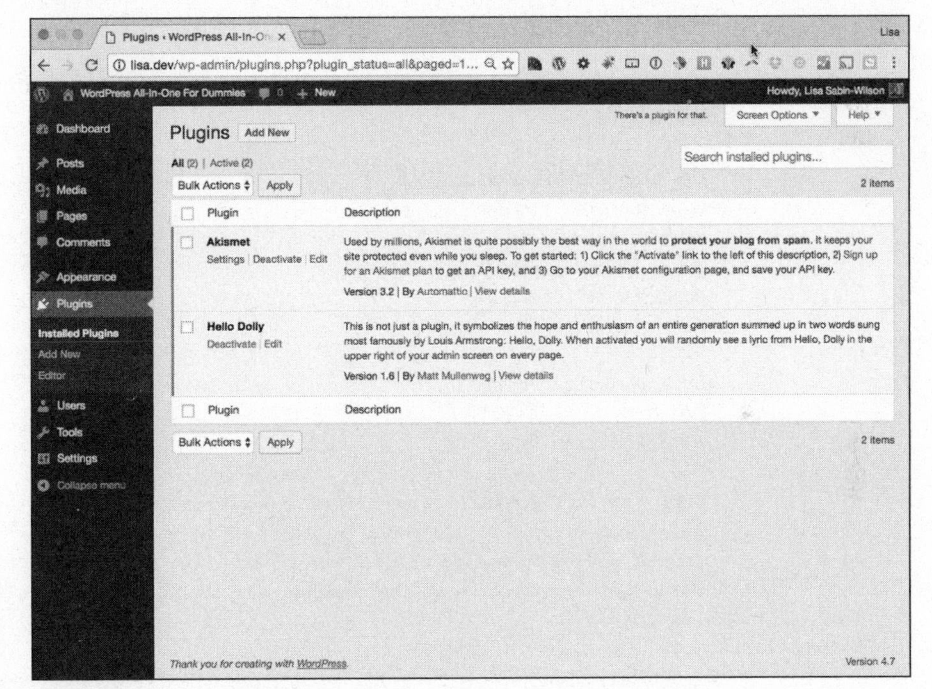

FIGURE 6-1: The WordPress Plugins page displays plugin information.

Creating a readme.txt File

Before you submit a plugin to the WordPress Plugin Directory, make sure that your plugin includes a `readme.txt` file. WordPress won't accept your plugin without a `readme.txt` file, which is included with the rest of your plugin's files when a user downloads the plugin. The file should contain the information the user needs to use the plugin properly.

The information contained in the readme.txt file is much more elaborate than the information in the file header. All the plugin pages in the Plugin Directory are generated from information in the readme.txt files of the plugins. Thus, this information is very important if you want people to take your plugin seriously.

TECHNICAL STUFF

The file is formatted in a slightly modified version of Markdown format. *Markdown* is a simple syntax for formatting content to contain headings, text styling (such as underline and bold), links, and so on. The main difference between the format used for WordPress plugin readme.txt files and regular Markdown is that the readme.txt files use a different format for headings (as you see in the following example), and a WordPress-specific section similar to the file header of the main plugin file appears at the top of readme.txt files.

Before digging into the details, look through the following example readme.txt file contents:

```
=== Lisa's Twitter Widget ===
Contributors: lisasabinwilson
Stable tag: 1.0.0
Tags: twitter, widget, social media
Requires at least: 3.3
Tested up to: 4.7

Display Twitter feeds in any sidebar on your site.

== Description ==

Lisa's Twitter Widget uses the power of WordPress's widget system to allow
you to quickly and easily add a Twitter feed to your site. After adding the
Twitter Widget to one of your sidebars, enter the desired Twitter username in
the widget's options, save the changes, and the Twitter feed will show up on
your site.

== Installation ==

Extract the zip file and just drop the contents in the wp-content/plugins/
directory of your WordPress installation and then activate the Plugin from
the Plugins page in your WordPress Dashboard.
```

Notice that the lines with the equal signs separate the file into sections. When the plugin's page is generated, the text on the topmost line with the three equal signs on each side (which I refer to as the *header section*) is used as the plugin's name, and the text after the Description and Installation sections is used to populate the Description and Installation tabs on the plugin's page.

You can do much more with a `readme.txt` file than I show here. This example is a small, simplified version of what you see in active plugins with well-crafted `readme.txt` files.

TIP

If you're concerned about whether your plugin's `readme.txt` file adheres to the expectations of the WordPress Plugin Directory, use the WordPress validator tool (`https://wordpress.org/plugins/about/validator`). This tool tells you whether your `readme.txt` file contains all the necessary components and information. If you're still having problems, the handy `readme.txt` file generator at `http://sudarmuthu.com/wordpress/wp-readme` helps you generate a valid `readme.txt` file for your plugin by making sure that the file meets the basic requirements.

Setting up the header section

The header section at the top of the file is similar to the file header in the actual plugin file because like the file header, the header section is parsed by code to format the data in specific ways. The format of the header section is as follows:

```
=== Plugin Name ===
entry: value
entry: value
entry: value

Brief description of the plugin.
```

The plugin name and short description are required and should be unique to the plugin and description. Like the description of the file header, this description should be brief (one to three sentences).

You can add several entries to the header section. Table 6-1 lists the available entries for the header section. Unless otherwise noted, each entry is required and should be included in the header section.

TABLE 6-1 **Entries for the readme.txt File Header Section**

Component	Description
Contributors	Comma-separated list of contributors' wordpress.org usernames. If valid usernames aren't used, the plugin page won't link properly to each contributor's other plugins.
Donation Link	Web address for a page that accepts donations for the plugin. This entry is optional.
License	How the plugin is licensed — typically, GPLv2.
License URI	The web address that contains the full details of the license — typically, `www.gnu.org/licenses/gpl-2.0.html`.
Requires At Least	The earliest version of WordPress with which your plugin is known to be compatible. If you haven't tested the plugin with older versions of WordPress or don't test updated versions of the plugin with older versions of WordPress, set this entry to the latest version of WordPress to ensure that you don't indicate nonexistent compatibility.
Stable Tag	Indicates the subversion tag of the latest stable version of your plugin. If no stable tag is provided, trunk is assumed to be stable.
Tags	A comma-separated list of descriptive tags related to the plugin. You can find a list of existing tags at `https://wordpress.org/plugins/tags`.
Tested Up To	The latest version of WordPress in which you've successfully tested your plugin. You can use prerelease versions of WordPress (such as 3.5-beta3) if the plugin is compatible with the development version.

Adding other sections

After the header section, the rest of the `readme.txt` file is dedicated to providing information used to create the plugin's Plugin Directory page. This information is divided into sections that display on the plugin's page as tabs. The following list describes the tabs and what they're meant to be used for, and also provides any special details:

- **Description:** This tab appears by default, so put the information that you want people to see when they first visit the plugin's page in this section. Focus on providing information about what features the plugin offers, what makes the plugin different from other plugins with similar features, and what requirements may prevent people from using the plugin successfully.
- **Installation:** In this section, you should include any special instructions on how to install and configure the plugin properly.
- **Frequently Asked Questions:** Are users asking you the same questions over and over? Do you expect certain questions? This tab is the place for you to answer your users' questions.
- **Changelog:** This section is helpful for listing the changes in each version. Be sure to include details about new features, enhancements, and bug fixes so that

current users who are curious about the reason for the new release can easily find what to expect after the upgrade. It's common to remove older listings, because this section gets long. The number of listings to keep is up to you.

- **Screenshots:** This section is for listing pictures and videos that show the plugin in action. The screen shots follow a specific format:
  - They must be added to the plugin's subversion repository in the `/assets` directory.
  - The names should be `screenshot-1.png`, `screenshot-2.png`, `screenshot-3.png`, and so on.
  - The supported formats are PNG, JPEG, and GIF.

TIP

  Add a description for each screen shot by using a numbered list. The description is displayed below the image with the matching number.

- **Other Notes:** The Other Notes tab is special in that it's created by merging all other sections, so you can create as many custom sections as you like and have each of the custom sections appear on the Other Notes tab.

The following example shows a `readme.txt` with each of these sections:

```
=== Lisa's Twitter Widget ===
Contributors: lisasabinwilson
Stable tag: 1.0.0
Donation link: http://example.com
Tags: twitter, widget, social media
Requires at least: 3.3
Tested up to: 4.7
License: GPLv2
License URI: http://www.gnu.org/licenses/gpl-2.0.html

Display Twitter feeds in any sidebar on your site.

== Description ==

Lisa's Twitter Widget uses the power of WordPress's widget system to allow
you to quickly and easily add a Twitter feed to your site. After adding the
Twitter Widget to one of your sidebars, enter the desired Twitter username in
the widget's options, save the changes, and the Twitter feed will show up on
your site.

== Installation ==

Extract the zip file and just drop the contents in the wp-content/plugins/
directory of your WordPress installation and then activate the Plugin from
the Plugins page in your WordPress Dashboard.
```

```
== Frequently Asked Questions ==

= Can the widget combine the feeds from more than one username? =

Not at this time. Such a feature may be added in a future version. For now,
you will simply have to use more than one widget.

= Can I load a feed for a hashtag? =

Yes. Rather than supplying a username, a hashtag can be used (ensure that you
add the # before the hashtag) to show the latest feed for that hashtag.

== Changelog ==

= 1.0.0 =
* Release-ready version
* Fixed issue with feed caching not working properly, causing site slowdowns.

= 0.0.1 =
* Development version. It still has some bugs.

== Screenshots ==
1. Configuration options for the Twitter Widget.
2. The Twitter Widget showing a feed on the site.

== Thanks ==

My thanks to Twitter and its fantastic API. If it wasn't for the API,
     this plugin would not be possible.
```

TIP

The Markdown format has many options for formatting the content. Go to `http://daringfireball.net/projects/markdown/syntax` for full details on the options available in Markdown.

Besides setting up plugin pages in the Plugin Directory, the `readme.txt` file serves two important purposes: controlling the released version of the plugin and offering content to the search function in the Plugin Directory.

The `readme.txt` file in your plugin's repository controls the released version of the plugin. After you update the plugin with a new version number, update the stable version value in the `/trunk/readme.txt` file to reflect the new version number. If you don't update this value, the new version won't be released.

Internationalizing or Localizing Your Plugin

WordPress users exist across the United States, Russia, Japan, Germany, and all points between. Therefore, the next person to download and use your plugin may not speak the same language that you do. If you write and distribute your plugin only in English, it may be useless to the next person if he speaks only German. The WordPress software has internationalization built into it, however, which means it can be *localized*, or translated, into different languages.

REMEMBER

You aren't translating the file into different languages (unless you want to). Rather, you're providing a mechanism of support for people who want to provide translation for your plugin through the creation of `.mo` (machine object) files. People in many countries have created `.mo` files that translate WordPress into different languages; by providing localization for your plugin, you're enabling them to translate your plugin text as well. If you're interested in translating WordPress into a different language, check out the resource page at `https://make.wordpress.org/polyglots/handbook`.

Using GetText functions for text strings

WordPress provides two main localization functions: `__` and `_e`. These functions use the `GetText` translation utility installed on your web server. These two functions let you wrap plain text into strings of text to be translated. You need to account for two types of text strings in your plugin file:

» **HTML:** Example: `<h1>Plugin Name</h1>`

To wrap HTML text strings within the `GetText` function call, wrap it by using the `_e` function like this:

```
<h1><?php _e('Plugin Name', 'plugin-name'); ?></h1>
```

This function tells PHP to echo (`_e`) or display the string of text on your web browser screen but adds the benefit of using the `GetText` function, which allows that string of text to be translated.

» **PHP:** Example: `<?php comments_number( 'No Responses', 'One Response', '% Responses' );?>`

To wrap PHP text strings with the GetText function, wrap it by using the __ function, like this:

```
<?php comments_number(__('No Responses',plugin-name),('One Response', 'plugin-
    name'),( '% Responses', 'plugin-name') );?>
```

Unlike the echo function (_e), the __ function is used when you need to add a string of text to an existing function call (in this case, comments_number()).

REMEMBER

Avoid slang when writing your text strings in your plugin file. Slang is significant to only a certain demographic (age, geographic location, and so on) and may not translate well in other languages.

The second argument within the GetText string for the PHP text string example is plugin-name. This argument defines the domain of the text and tells GetText to return the translations only from the dictionary supplied with that domain name. This domain is the *plugin text domain,* and most plugin authors use the name of the plugin (separated by hyphens) as the definer. Because some of the text you provide in your plugin is unique and, most likely, doesn't exist within the WordPress core language files, use the text domain in your GetText functions to ensure that the GetText function pulls the language dictionary in your plugin, instead of attempting to pull the text from the core WordPress language files.

In a plugin file, you define the text domain like this:

```
$my_translator_domain = PLUGIN-NAME;
$my_translator_is_setup = 0;
function fabfunc_setup(){
  global $my_translator_domain, $my_translator_is_setup;
  if($my_translator_is_setup) {
    return;
  }
  load_plugin_textdomain($my_translator_domain,
      PLUGINDIR.'/'.dirname(plugin_basename(__FILE__)),

      dirname(plugin_basename(__FILE__)));
}
```

TIP

If you do not want to add the load_plugin_textdomain(); to your plugin and the plugin requires at least version 4.6 of the WordPress software, you do have the option of using translate.wordpress.org to translate your plugin into different languages. To do this, make sure your Readme.txt file contains this line in the header Requires at least: 4.6.

These lines of code simply tell WordPress where your plugin file is located (the text domain), which in turn informs WordPress where to find the .pot, or translation file, for your plugin.

Creating the POT file

After you create your plugin and include all text strings within the GetText functions __ and _e, you need to create a .pot (portable object template) file, which contains translations for all the strings of text that you wrapped in the GetText functions. Typically, you create the .pot file in your own language, in a special format, thereby allowing other translators to create their own .po (portable object) file or .mo (machine object) file in their language, using yours as the guide to translate by.

The .pot file is the original translation file, and the .po file is a text file that includes your original text (from the .pot) along with the translation for the text. Alternatively, you can use an .mo file, which is basically the same as a .po file. Although .po files are written in plain text meant to be human-readable, .mo files are compiled to make it easy for computers to read. Most web servers use .mo files to provide translations for .pot files.

TIP

WordPress has an extensive .pot file that you can use as a template for your own. Download it at http://svn.automattic.com/wordpress-i18n/pot/trunk/wordpress.pot.

Additionally, you can translate .pot files into .mo files by using translation tools such as Poedit (https://poedit.net), a free tool that takes the original .pot file and the provided translations in a .po file and merges them into a compiled .mo file for your web server to deliver the translated text.

The .pot file begins with a *header* section, which contains required information about what your translation is for. The .pot header section looks like this:

```
# Copyright (C) 2017 WordPress
# This file is distributed under the same license as the WordPress package.
msgid ""
msgstr ""
"Project-Id-Version: WordPress 4.8-alpha-40038\n"
"Report-Msgid-Bugs-To: https://make.wordpress.org/polyglots/\n"
"POT-Creation-Date: 2017-02-01 22:13:18+00:00\n"
"MIME-Version: 1.0\n"
"Content-Type: text/plain; charset=UTF-8\n"
```

```
"Content-Transfer-Encoding: 8bit\n"
"PO-Revision-Date: 2017-MO-DA HO:MI+ZONE\n"
"Last-Translator: FULL NAME <EMAIL@ADDRESS>\n"
"Language-Team: LANGUAGE <LL@li.org>\n"
```

All the capitalized, italicized terms in this code example are placeholders. Replace these terms with your own information.

The format of the .pot file is specific and needs to contain the following information:

- **Filename:** The name of the file in which the text string exists. If the plugin file is wordpress-twitter-connect.php, for example, you need to include that filename in this section.
- **Line of code:** The line number of the text string in question.
- **msgid:** The source of the message, or the exact string of text that you included within one of the GetText functions: __ or _e.
- **msgstr:** A blank string where the translation (in the subsequent .pot files) is inserted.

For your default .pot file, to format a text string by using the GetText function (`<h1><?php _e('WordPress Twitter Connect'); ?></h1>`), which exists on the second line of the wordpress-twitter-connect.php plugin file, you include three lines in the .pot file that look like this:

```
#: wordpress-twitter-connect.php:2
msgid: "WordPress Twitter Connect"
msgstr: ""
```

You need to go through all the text strings in your plugin file that you wrapped in the GetText functions and define them in the .pot file in the format provided. Now if anyone wants to create a .po file for your plugin in a different language, he or she simply copies his or her language translation of your .pot file between the quotation marks for the msgstr: section for each text string included in the original .pot file.

You need to include all .pot and .po (or .mo) files need in your plugin folder for the translations to be delivered to your website. Have a look at the directory structure of the popular WordPress All in One SEO Pack plugin in Figure 6-2. You see the original .pot file along with the translated .mo files listed within the /wp-content/plugins/all-in-one-seo-pack/ plugin folder.

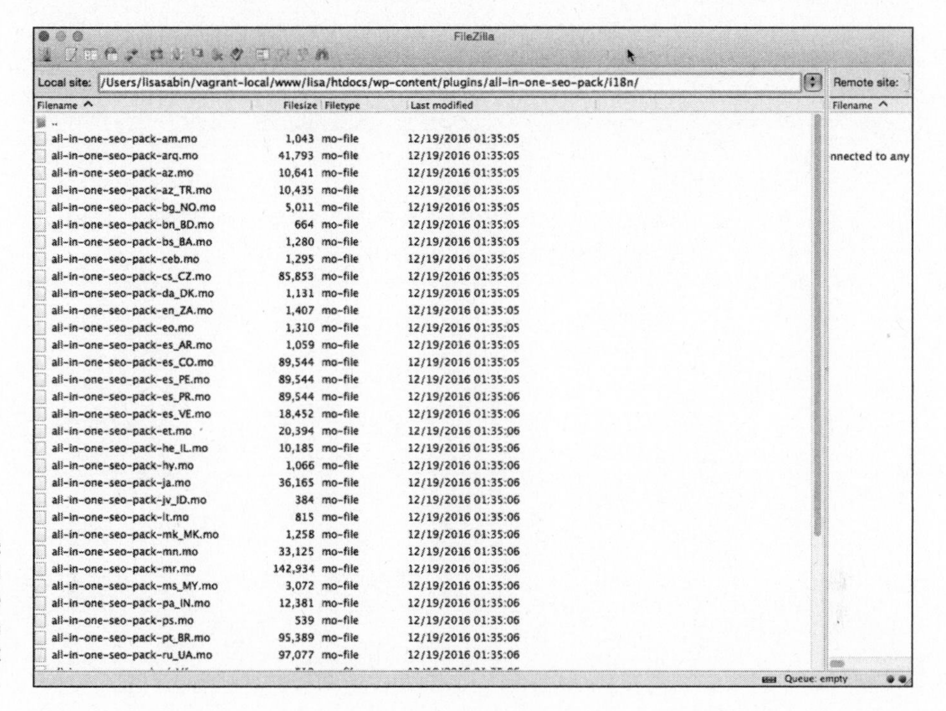

FIGURE 6-2: The .pot and .mo files for the WordPress All in One SEO Pack plugin.

You, or other translators, can create unlimited `.mo` files for several languages. Make sure that you name the language file according to the standardized naming conventions for the language. The naming convention is `language_`*`COUNTRY`*`.mo`. The French `.mo` file for the `wordpress-twitter-connect.php` plugin, for example, is `wordpress-twitter-connect-fr_FR.mo`.

TIP

You can find a full list of language codes at `www.w3schools.com/tags/ref_language_codes.asp`. For a full list of country codes, visit the Online Browsing Platform, where you can search for specific country codes: `www.iso.org/obp/ui`.

» Plugifying your plugins

» Enhancing plugins

» Using custom post types and shortcodes

» Enhancing editors

» Fixing links

Chapter 7
Plugin Tips and Tricks

When you have a WordPress plugin or two under your belt, you'll discover that you want to interact with many more parts of WordPress. WordPress is always coming out with new functionality and, along with it, new application programming interface (API) hooks, known as *action* and *filter hooks*, covered in Book 7, Chapter 5. This chapter discusses some of this functionality and offers you ways to extend your use of WordPress plugins. Because this functionality involves some simple programming skills, I assume (for the purposes of this chapter) that you have some basic PHP and WordPress plugin development knowledge.

Using a Plugin Template

When you start writing WordPress plugins, you find that you spend a significant amount of time rewriting the same things. Most plugins have the same basic structure and are set up the same way, meaning that they all deal with creating settings pages, storing options, and interacting with particular plugins, among other things. You can save hours of work each time you start a new plugin if you create a template.

Such a template varies from person to person, depending on programming styles, preferences, and the types of plugins to include. If you often write plugins that use your own database tables, for example, you should include tables in your template. Similarly, if your plugins almost never require options pages, leave those pages out of your template.

To create your own template, determine what functionality and structure your plugins usually contain. Then follow these steps:

1. **Create your file structure.**

   As you write more plugins, you'll find yourself repeating the same general filenames. If you find that you're including enough JavaScript and Cascading Style Sheets (CSS) in your plugins to necessitate their own files or directories, include them in your template. If you're using a lot of JavaScript or CSS, you could modify the file structure of your plugin template to look something like the one shown in Figure 7-1.

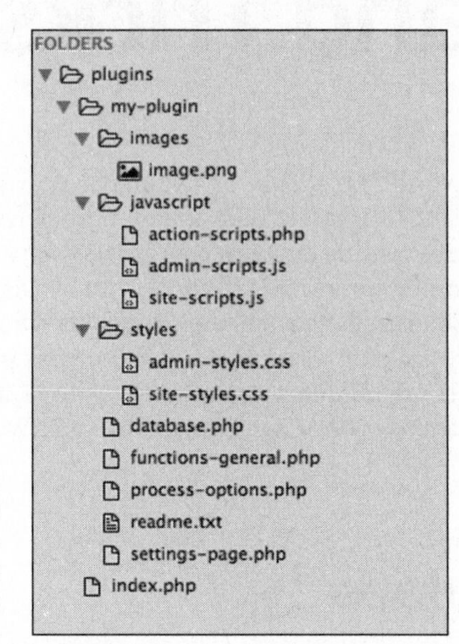

FIGURE 7-1: The recommended file structure for a plugin.

2. **Determine what functionality you generally have in your plugins.**

   If your plugins usually contain masses of code in a class, you can set up a basic class for your plugin template. Likewise, if your plugins typically have a single options page or a system of top-level and submenu pages, you can set up a general template.

3. **Create your primary plugin .php file.**

Usually, this file contains some general add_action calls, includes to other files, and other general initializations. (Check out Book 7, Chapter 5 for information on add_action calls and other plugin functions.) If you always call certain actions, set them up in your primary plugin .php file template. If you always register a plugin function to be run when the plugin is activated (register_activation_hook), for example, and add a menu item for the plugin on the Dashboard (admin_menu), add those calls to your primary template, as follows:

```
<?php
$myInstance = new myPlugin();

add_action('register_activation_hook','my_activation_plugin');
add_action('admin_menu',array($myInstance, 'admin_menu'));

?>
```

4. **Set up the functions you use most often in the body of your primary plugin .php file.**

The line of code used here — function my_activation_plugin — was added in Step 3 through the add_action hook. In your plugin template, you define any scripts your plugin uses by adding this function, which fires when a user activates the plugin:

```
<?php

function my_activation_plugin(){
//plugin activation scripts here
}
?>
```

5. **(Optional) Create your basic class structure.**

You might add a few lines of code that resemble the following:

```
<?php
class myPlugin {
     var $options = ;
     var $db_version = '1';
     function myPlugin() {
     add_action('admin_init',array($this,'admin_init');
     }
     function admin_init(){
          //admin initializations
     }
```

```
        function process_options($args,$data){
            //process our options here
        }
        function admin_menu(){
            //code for admin menu
        }
        function __construct(){
            //PHP 5 Constructor here
        }
    } //end class
    ?>
```

Obviously, your class template may be more detailed than this example, depending on your particular coding styles and the types of plugins you like to write.

In addition, you may want to set up a basic plugin options page along with plugin options management scripts. Everyone uses different techniques for such things as processing plugin options. When you determine your particular type, include the basic format in your template.

REMEMBER

As your programming style, WordPress, or your interest in different types of plugins changes over time, you'll find that your template needs change, too. Make sure that you update your templates.

Making Your Plugin Pluggable

The WordPress API provides a great solution for hooking in and extending or modifying functionality. Sometimes, however, you may find it useful to hook into another plugin rather than WordPress itself and to extend or modify that plugin's functionality instead.

Here's an example of making your plugin pluggable: interaction (or lack of) between WooCommerce (`https://woocommerce.com`) and the All in One SEO Pack plugin. All in One SEO Pack, among other things, adds a custom document title, description, keywords, and canonical URL to each page of a WordPress site. People often ask how to have their plugins modify this information before it appears onscreen. Forum plugins and shopping-cart plugins particularly need this function, which All in One SEO Pack could easily satisfy if it would just provide a method of accessing its functionality. It has. Read on.

WooCommerce (and plugins with similar functionality) creates its own virtual pages for things like product listings, all contained on a single WordPress page. If you define `http://mywebstore.com/shop` as the WordPress page WooCommerce uses, for example, `http://mywebstore.com/shop/product-name` and all other product pages are created dynamically by the WooCommerce plugin and are outside the reach of other WordPress plugins. So WordPress and the All in One SEO Pack plugin don't know about them. You can see the problem for purposes of search engine optimization (SEO): WordPress and All in One SEO Pack think that all these product pages are the same as the shop page, with the same titles, canonical URL, and so on. Code could have been written into the plugin to compensate for the needs of the WooCommerce plugin, but that coding would be never-ending; an infinite number of other plugins might need to hook into the All in One SEO Pack functions. To solve this problem, the All in One SEO Pack API was born.

The WooCommerce plugin has an important need to hook into the document title, meta description, meta keywords, and canonical URL that the All in One SEO Pack plugin produces. Because WordPress and All in One SEO Pack weren't aware of the generated product pages, they had the same canonical URL, which is detrimental for SEO purposes. The fix was simple. In the All in One SEO Pack plugin, after the canonical URL is generated and immediately before printing it to the screen, the `apply_filters` function is used on the variable. This solution allows WooCommerce to use `add_filter` to hook in and filter the canonical URL, returning the appropriate URL for that page. That is, developers added the following code to the All in One SEO Pack plugin:

```
function prepare_canonical_url(){
     $canonical_url = my_determine_canonical_url_function();
     $new_canonical_url = apply_filters( 'aioseop_canonical_url',
   $canonical_url);

     return $new_canonical_url;
}
```

This code returns the value of the canonical URL and lets another plugin filter it, if desired, before returning the final value.

The part you can add in the primary WooCommerce plugin file (at the end of the file before the closing `?>`) is just as simple. It works the same way `add_filter` does for hooking into the WordPress API:

```
add_filter('aioseop_canonical_url','wooc_change_canonical_url');

function wooc_change_canonical_url($old_url){
     $new_url = determine_current_product_page_url($old_url);
     return $new_url;
}
```

This filter provides a simple solution that allows other plugins (and themes) to hook into a plugin without the need to add specific code for any plugin.

Enhancing Plugins with CSS and JavaScript

You can add functionality to a plugin in many ways. The following sections look at two methods: CSS styling and JavaScript. You may never develop a plugin that uses either method, but chances are good that you, as a budding plugin developer, may need this information at some point.

Calling stylesheets within a plugin

Controlling how your plugin's output looks onscreen (on the WordPress Dashboard or the front end of the website or blog) is best done through a stylesheet. If you've been around web design and HTML, you're probably familiar with CSS. Nearly every styling aspect of a website is controlled by a stylesheet, and WordPress websites are no exceptions. If you want to read the authoritative guide to stylesheets, visit the W3C.org website at `www.w3.org/Style/CSS`. (For more on CSS, see Book 6.)

You can use a single stylesheet to control how your Plugin Options page looks on the Dashboard, how your plugin widget looks on the Dashboard, or how your plugin displays information on the front-end website.

Create and use separate stylesheets for the plugin on the Dashboard and the plugin's display on the front end, because the stylesheets are called at different times. The back-end stylesheet is called when you're administering your site on the WordPress Dashboard, whereas the front-end stylesheet is called when a user visits the website. Additionally, using separate stylesheets makes style management easier and cleaner.

The best practice for adding stylesheets within your plugin is to create a `/styles` directory, such as `/my-plugin/styles`. Place your stylesheets for the back end and front end inside this directory (refer to Figure 7-1 earlier in this chapter). To call a stylesheet from your plugin, you should use the built-in WordPress `wp_enqueue_style` function, which creates a queuing system in WordPress for loading stylesheets only when they're needed instead of on every page. Additionally, this function supports dependencies, so you can specify whether your stylesheet depends on another that should be called first. This queuing system is used for scripts, too. The `wp_enqueue_scripts` function does the same for scripts as I discuss a little later in this section.

Suppose that you're creating a gallery plugin to display images on your website. You want your gallery to look nice, so you want to create a stylesheet that controls how the images display. You can call that stylesheet in your plugin by using a simple function and action hook. These lines of code get added to your primary plugin .php file at the end, just before the closing ?> tag. Follow these steps:

1. **Create a function in your primary plugin .php file to register your stylesheet and invoke wp_enqueue_style, as follows:**

```
function add_my_plugin_stylesheet() {
   wp_register_style('mypluginstylesheet', '/wp-content/plugins/ my-plugin/
    styles/site-style.css');
   wp_enqueue_style('mypluginstylesheet');
}
```

2. **Use the wp_print_styles action hook, and call your function:**

```
add_action( 'wp_print_styles', 'add_my_plugin_stylesheet' );
```

Here's a breakdown of the hooks in the function:

- **The wp_register_style function registers your stylesheet for later use by wp_enqueue_style, as follows:**

  ```
  wp_register_style( $handle, $src, $deps, $ver, $media )
  ```

  The function has several parameters; the first is $handle, which is the name of your stylesheet.

  WARNING

  $handle must be unique. You can't have multiple stylesheets with the same name in the same directory.

  The second parameter is $src, the path to your stylesheet from the root of WordPress — in this case, the full path to the file within the plugin's styles directory.

  The remaining parameters are optional. To find out more about them, read the WordPress documentation for this function at https://codex.wordpress.org/Function_Reference/wp_register_style.

- **The wp_enqueue_style function queues the stylesheet, as follows:**

  ```
  wp_enqueue_style( $handle, $src, $deps, $ver, $media )
  ```

  The $handle parameter is the name of your stylesheet as registered with wp_register_style. The $src parameter is the path, but you don't need this parameter because you already registered the stylesheet path. The remaining parameters are optional and explained in the WordPress documentation for this function at https://developer.wordpress.org/reference/functions/wp_enqueue_style.

» **The action hook that calls the function uses `wp_print_styles` to output the stylesheet to the browser.**

In the source code, the line of code that shows the stylesheet being included looks like this:

```
<link rel='stylesheet' id='twentyseventeen-style-css'  href='http://lisa.dev/
    wp-content/themes/twentyseventeen/style.css?ver=4.8-alpha-40031' type='text/css'
    media='all'/>
```

Another example uses a stylesheet for the plugin's admin interface, which controls how your plugin option page within the Dashboard appear. These lines of code also get added to your plugin's primary .php file just before the closing ?> tag:

```
add_action('admin_init', 'myplugin_admin_init');

function myplugin_admin_init() {
   wp_register_style('mypluginadminstylesheet', '/wp-content/plugins/
   my-plugin/admin-styles.css');
    add_action('admin_print_styles' 'myplugin_admin_style');
    function myplugin_admin_style() {
      wp_enqueue_style('mypluginadminstylesheet');
   }
}
```

This example uses some hooks that are specific to the WordPress Dashboard:

» The action hook calls `admin_init`. This hook makes sure that the function is called when the Dashboard is accessed. The callback function is `myplugin_admin_init`.

» The function registers the stylesheet, using `wp_register_style`.

» An action hook calls the `myplugin_admin_style` function. The `admin_print_styles` hook is used because it's specific to the WordPress Dashboard display.

» The function queues the stylesheet, using `wp_enqueue_style`.

In the source code, the line of code that shows the stylesheet being included looks like this:

```
<link rel='stylesheet' id='aioseop-module-style-css'  href='http://lisa.dev/
    wp-content/plugins/all-in-one-seo-pack/css/modules/aioseop-module.css'
    type='text/css' media='all'/>
```

Calling JavaScript within a plugin

After using the wp_register_style and wp_enqueue_style functions to call stylesheets within a plugin, you see how similar functions can call JavaScript, which has many uses within a plugin.

JavaScript can control functionality within a form or display something with an effect. WordPress comes with some JavaScript in the core that you can call in your plugin, or you can write your own. As with stylesheets, it's best to store JavaScript in a separate subdirectory within your plugin, such as /my-plugin/javascript.

Instead of using wp_register_style and wp_enqueue_style to register and queue JavaScript, you must use wp_register_script and wp_enqueue_script. These functions work in much the same way and have much the same parameters. Here's an example to add to your plugin's primary .php file, before the closing ?> tag:

```
if ( !is_admin() ) {
   wp_register_script('custom_script','/wp-content/plugins/ my-plugin/
   javascript/custom-script.js',);
   wp_enqueue_script('custom_script');
}
```

Immediately, you notice that the wp_enqueue_script function loads scripts in the front end of your website and on the Dashboard. Because this situation can cause conflicts with other scripts that WordPress uses on the Dashboard display, the "if is not" (!is_admin) instruction tells the plugin to load JavaScript only if it's not being loaded on the Dashboard. This code loads custom-script.js only on the front end of the website (that is, what your site visitors see). You could add a more specific conditional if instruction to load JavaScript only on a certain page.

If you want to load the JavaScript in wp-admin, the action hook admin_init loads your callback function when wp-admin is accessed and the admin_print_script function outputs the script to the browser, just like the stylesheet example.

Working with Custom Post Types

One of the most confusing features of WordPress is custom post types. The feature is also useful, powerful, and easy to implement and use when you understand how it works. WordPress has five default post types:

- **Post:** The most commonly used post type. Content appears in a blog in reverse sequential time order.

» **Page:** Similar to posts except for not using the time-based structure of posts. Pages can be organized in a hierarchy and have their own URLs off the main site URL.

» **Attachment:** A special post type that holds information about files uploaded through the WordPress Media upload system.

» **Revisions:** A post type that holds past revisions of posts and pages as well as drafts.

» **Nav Menus:** A post type that holds information about each item on a navigation menu.

A post type is really a type of content stored in the `wp_posts` table inside the WordPress database. The post type is stored in the `wp_posts` table's `post_type` column. The information in the `post_type` column differentiates each type of content so that WordPress, a theme, or a plugin can treat the specific content types differently.

When you understand that a post type is just a method to distinguish how different content types are used, you can investigate custom post types.

Suppose that you have a website about movies. Movies have common attributes, such as actors, directors, writers, and producers. But perhaps you don't want to store your movie information in a post or a page because it doesn't fit any content type. In this situation, custom post types are useful. You can create a custom post type for movies and apply the common attributes (actors, directors, and so on). You can use a theme to handle movies differently from a post or a page by using a custom template for the `movies` post type and then creating different styling attributes and templates for the `movies` post type. You can search and archive movies differently with a custom post type.

One method of creating a simple custom post type in WordPress is to add the following lines of code to the Theme Functions template file (located in the theme file and called `functions.php`):

```
add_action('init','create_post_type');                              →1
function create_post_type() {
  register_post_type( 'movies',                                     →3
    array(                                                          →4
      'labels' => array(                                            →5
        'name' => ('Movies'),
        'singular_name' => ('Movie'),
        'rewrite' => array('slug' => 'movies'),
      ),
```

```
        'public' => true,                                         →10
    )
  );
}
```

Here's what's going on in the code:

- →1 The first line is the action hook, which uses `'init'` so that it's called on the front end and on the Dashboard to display the custom post type in both.
- →3 The callback function starts with the `register_post_type` function and the custom post type name. This function creates the custom post type and gives it properties.
- →4 Next is an array of arguments that are the custom post type properties.
- →5 The `'labels'` arguments include the name that appears on the Dashboard menu, the name to be used (Movies), and the slug in the URL of the posts (`http://yourdomain.com/movies`, for example) in this custom post type.
- →10 The `'public'` argument controls whether the custom post type displays on the Dashboard.

Figure 7-2 shows how the Custom Post Type page and menu item look on the Dashboard. Figure 7-3 shows a custom post type on a website.

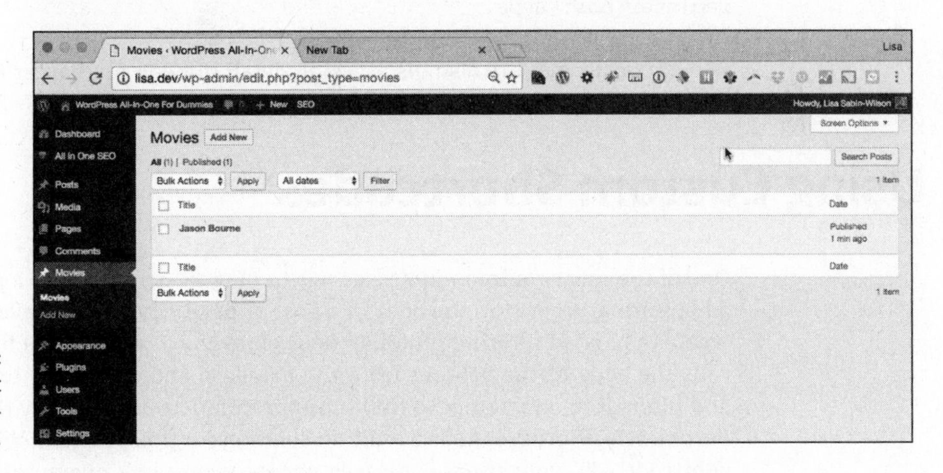

FIGURE 7-2: A custom post type on the WordPress Dashboard.

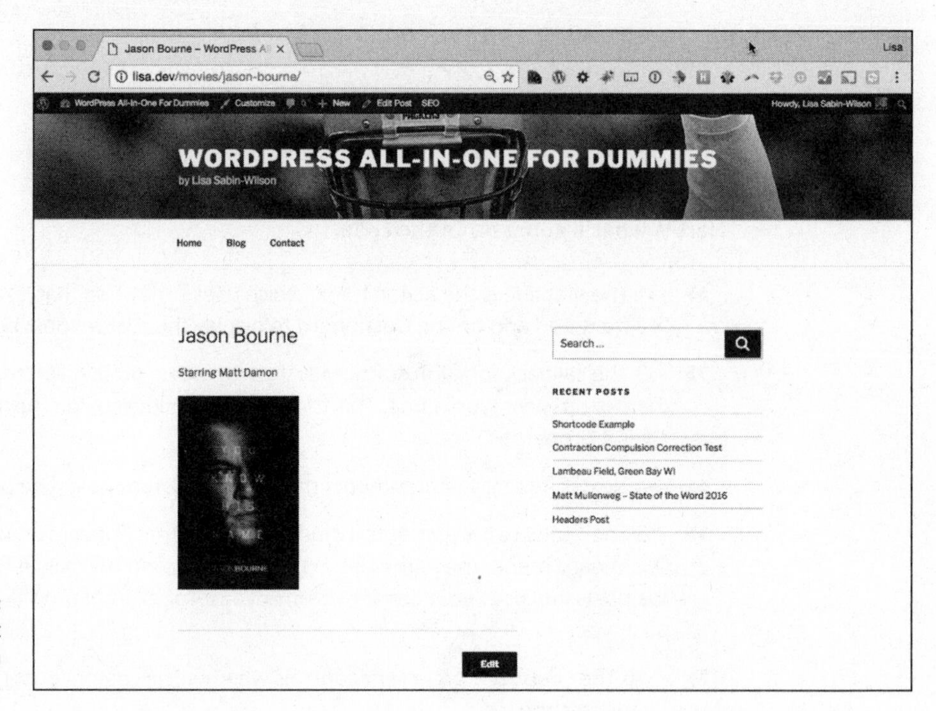

FIGURE 7-3: A custom post type shown on a website.

Many other arguments associated with `register_post_type` give this function its real power. For full documentation of all the arguments and the use of this function, check out `https://codex.wordpress.org/Function_Reference/register_post_type`.

Custom post types are also discussed in detail in Book 6, Chapter 6.

Using Custom Shortcodes

One of the most common inefficiencies in plugins occurs when a plugin wants to add information within the body of a post or page. The plugin developer manually creates a bloated filtering function, hooks into `the_content` (the function tag that calls the body of the content from the database and delivers it to your website), and filters it in an attempt to find the appropriate spot to display the information. Fortunately, WordPress has a built-in solution for this problem. When you use the shortcode API, your users can easily choose where in a given post to display the information that your plugin provides.

Suppose that you have a string of data that your plugin generates dynamically, and you want your users to determine where in each post and page that string displays. Users type a shortcode like this within the body of their content to display information from a plugin:

```
[myshortcode]
```

On the developer side, you use the `add_shortcode` function and add it to your primary plugin `.php` file, as follows:

```
<?php add_shortcode($tag, $func); ?>
```

The `add_shortcode` function accepts two parameters:

- The `$tag` parameter is the string that users type within the body of their content to make a call to the plugin shortcode. (In the preceding example, `[myshortcode]` is what users type, so your `$tag` parameter would be `myshortcode`.)
- The `$func` parameter is your callback function (a function that you still need to define in the body of your primary plugin `.php` file, as covered in the next section) that returns the output of the called shortcode.

The shortcode function gets added to your primary plugin `.php` code before the closing `?>` tag:

```
add_shortcode('myshortcode','my_shortcode');

function my_shortcode(){
     return "this is the text displayed by the shortcode";
}
```

In this example, you add the shortcode hook `add_shortcode('myshortcode','my_shortcode');` and then give definition to the function (`$func`) called `my_shortcode` by telling WordPress to output the text `this is the text displayed by the shortcode`.

All your user has to do is type **[myshortcode]** somewhere in the body of his post/page editor (on the Dashboard, hover the cursor over Post and then click the Add New link), as shown in Figure 7-4. When users view the site, the shortcode that the user entered in the body of his post is translated by WordPress to display the returned value, or *output*, of the shortcode function, as shown in Figure 7-5.

WARNING

Shortcode names must be unique to your own plugin, so you should use a name that is specific to your plugin. If your plugin is called Super SEO Plugin, you could name your shortcode `[superseoplugincode]` in an attempt to make sure that no other plugin uses your shortcode. Another plugin that uses a shortcode with the same name will cause a conflict.

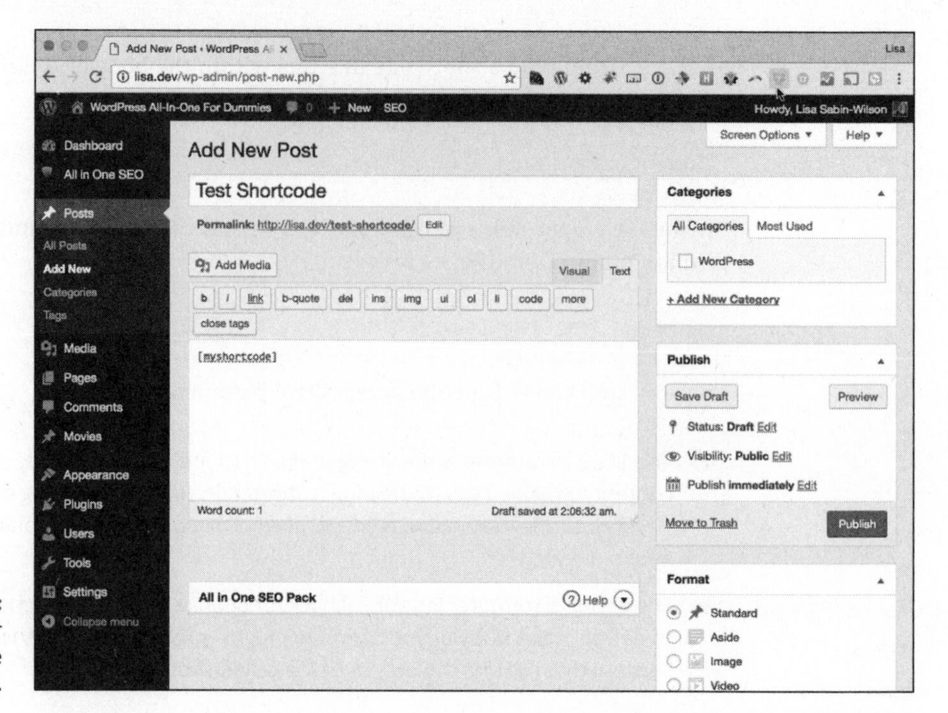

FIGURE 7-4: The post editor showing a simple shortcode.

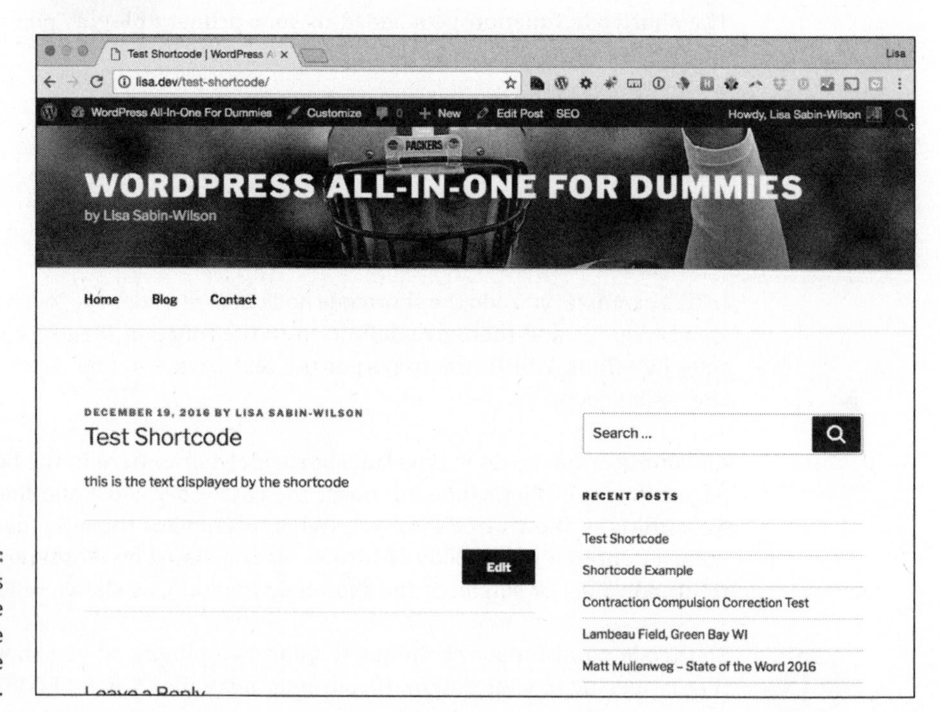

FIGURE 7-5: The shortcode is replaced by the returned value of the shortcode function.

Shortcodes can include arguments to be passed into the shortcode function:

```
<?php
add shortcode('myshortcode','my_shortcode');

function my_shortcode($attr, $content){
    return 'My name is ' . $attr['first'] . $attr['last'];
}
```

Calling this code with

```
[myshortcode first="John" last="Smith"]
```

outputs `My name is John Smith.`

Adding Functionality to Profile Filters

WordPress provides three contact settings in the Profile screen of the WordPress Dashboard by default: Email, Website, and Google+. These settings are extensible, of course, which means that you can easily add new contact methods by using filters. Adding more settings is painless, and you can even add a little extra functionality while you're at it.

Users fill out their profile data on the WordPress Dashboard by hovering the mouse over Users and then clicking the Your Profile link. (See Book 3, Chapter 2.) User profile fields are stored in the WordPress database inside the `user_metadata` table; you can easily fetch them by using `get_the_author_meta('url')` and print them with `the_author_meta('url')`. If you add a Twitter Contact Info field, it appears in profiles, and you can use `the_author_meta('twitter')` template tags in your theme to print the account name.

Figure 7-6 shows the Twitter Contact Info field in a profile within the WordPress Dashboard.

`the_author_meta()` template tag has a hook called `the_author_{$field}`, where the PHP variable `$field` is the requested meta field assigned to each contact type in the user profile files, such as `aim` in the preceding example. These dynamic hooks are powerful because they allow you to narrow your target.

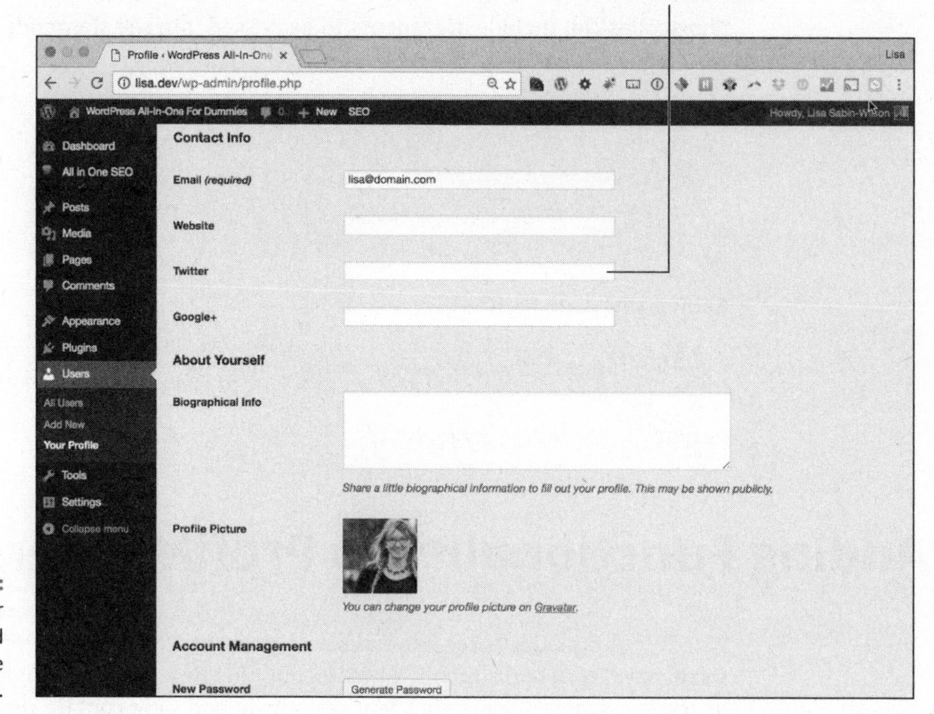

FIGURE 7-6: Custom Twitter profile field shown on the Dashboard.

In this example, I use the dynamic `the_author_twitter` hook to change the result from `"lisasabinwilson"` to `<a href="http://twitter.com/lisasabinwilson">@lisasabinwilson</a>`. When you call `the_author_meta('twitter')` in your theme, you get a clickable link to my Twitter profile. Start by entering the following lines of code in your Theme Functions file (`functions.php`) in your active theme folder (add this code toward the bottom of the file, before the closing `?>` tag):

```
/**
 * Add Twitter to the list of contact methods captured via profiles.
 */
function my_add_twitter_author_meta( $contact_methods ) {
  $contact_methods['twitter'] = 'Twitter';
return $contact_methods;
}

add_filter( 'user_contactmethods', 'my_add_twitter_author_meta' );

/**
 * Convert staff Twitter accounts to links to twitter.com.
 */
```

```
function my_link_author_twitter_accounts( $value ) {
  if ( strlen( $value ) ) {
    $url = esc_url( 'http://twitter.com/' . $value );
    $value = '<a href="' . $url . '">@' . esc_html( $value ) . '</a>';
  }
return $value;
}
add_filter( 'the_author_twitter', 'my_link_author_twitter_accounts' );
```


Correcting Hyperlink Problems

Most websites use underline to style hyperlinks. When you're producing content in WordPress, highlighting words and phrases quickly to add hyperlinks can lead to hyperlinking (and underlining) the spaces before and after your anchor text.

For some people, this situation is enough to convince them to hide underlines for hyperlinks, even though that practice may not be desirable.

Here's a snippet that filters blog post content to ensure that you don't have any spaces on the wrong sides of a tag or between a closing tag and punctuation. Add this code to your Theme Functions (`functions.php`) file:

```
/**
 * Prevents underlined spaces to the left and right of links.
 *
 * @param string $content Content
 * @return string Content
 */

function my_anchor_text_trim_spaces( $content ) {
  // Remove spaces immediately after an <a> tag.
  $content = preg_replace( '#<a([^>]+)>\s+#', ' <a$1>', $content );
  // Remove spaces immediately before an </a> tag.
  $content = preg_replace( '#\s+</a>#', '</a> ', $content );
  // Remove single spaces between an </a> tag and punctuation.
  $content = preg_replace( '#</a>\s([.,!?;])#', '</a>$1', $content );
  return $content;
}

add_filter( 'the_content', 'my_anchor_text_trim_spaces' );
```

REMEMBER

HTML ignores more than one space in a row (also more than one tab character and line break) unless you're using the `pre` element or nonbreaking space entities (` `). Therefore, even if your converted text contains two consecutive spaces, the browser won't show the text any differently.

8

Running Multiple Sites with WordPress

Contents at a Glance

IN THIS CHAPTER

» Discovering where multiple sites began

» Exploring what you can do with multiple sites

» Configuring your web server

Chapter 1
An Introduction to Multiple Sites

This chapter introduces you to the network feature that is built into the WordPress software. The network feature allows you, the site owner, to add and maintain multiple sites within one installation of WordPress. In this chapter, you discover how to set up the WordPress network feature, explore settings and configurations, gain an understanding of the Network Administrator role, determine which configuration is right for you, and find some great resources to help you on your way.

When you enable the network features, users of your network can run their own sites within your installation of WordPress. They also have access to their own Dashboards, which have the same options and features that I discuss in the past several chapters. Heck, it probably would be a great idea to buy a copy of this book for every member of your network so that everyone can become familiar with the WordPress Dashboard and features, too. At least have a copy on hand so people can borrow yours!

Deciding When to Use the Multisite Feature

Usually, for multiple users to post to one site, the default WordPress set up, is sufficient. The *Multi* part of the WordPress Multisite feature's name doesn't refer to how many users were added to your WordPress website; it refers to the ability to run multiple sites in one installation of the WordPress software. *Multisite* is a bit of a misnomer and an inaccurate depiction of what the software actually does. A *network of sites* is a much closer description.

Determining whether to use the Multisite feature depends on user access and publishing activity. Each site in the network shares a codebase and users but is a self-contained unit. Users still have to access the back end of each site to manage options or post to that site. A limited number of general options is available networkwide, and posting is not one of those options.

You can use multiple sites in a network to give the appearance that only one site exists. Put the same theme on each site, and the visitor doesn't realize that the sites are separate. This technique is a good way to separate sections of a magazine site, using editors for complete sections (sites) but not letting them access other parts of the network or the back ends of other sites.

Another factor to consider is how comfortable you are with editing files directly on the server. Setting up the network involves accessing the server directly, and ongoing maintenance and support for your users often leads to the network owner doing the necessary maintenance, which is not for the faint of heart.

Generally, you should use a network of sites in the following cases:

- **You want multiple sites and one installation.** You're a blogger or site owner who wants to maintain another site, possibly with a subdomain or a separate domain, all on one web host. You're comfortable with editing files, you want to work with one codebase to make site maintenance easier, and most of your plugins and themes are accessible to all the sites. You can have one login across the sites and manage each site individually.
- **You want to host blogs or sites for others.** This process is a little more involved. You want to set up a network in which users sign up for their own sites or blogs below (or on) your main site and you maintain the technical aspects.

Because all files are shared, some aspects are locked down for security purposes. One of the most puzzling security measures for new users is suppression of errors. Most PHP errors (such as those that occur when you install a faulty plugin or incorrectly edit a file) don't output messages to the screen. Instead, WordPress displays what I like to call the White Screen of Death.

Finding and using error logs and doing general debugging are necessary skills for managing your own network. Even if your web host sets up the ongoing daily or weekly tasks for you, managing a network can involve a steep learning curve.

REMEMBER

When you enable the Multisite feature, the existing WordPress site becomes the main site in the installation.

Although WordPress can be quite powerful, in the following situations, managing multiple sites has limitations:

- **One web account is used for the installation.** You can't use multiple hosting accounts.
- **You want to post to multiple blogs at the same time.** WordPress doesn't allow this practice by default.
- **If you choose subdirectory sites, the main site regenerates permalinks with `/blog/` in them to prevent collisions with subsites.** Plugins are available that prevent this regeneration.

The best example of a large blog network with millions of blogs and users is the hosted service at WordPress.com (`https://wordpress.com`). At WordPress.com, people are invited to sign up for an account and start a blog by using the Multisite feature within the WordPress platform on the WordPress server. When you enable this feature in your own domain and enable the user registration feature, you're inviting users to do the following:

- Create an account
- Create a blog in your WordPress installation (on your domain)
- Create content by publishing blog posts
- Upload media files such as photos, audio, and video
- Invite friends to view their blog or to sign up for their own accounts

Understanding the Difference between Sites and Blogs

Each additional blog in a WordPress Multisite network is a *site* instead of a *blog.* What's the difference?

Largely, the difference is one of perception. Everything functions the same way, but people see greater possibilities when they no longer think of each site as being "just" a blog. WordPress can be much more:

- With the addition of the Domain Mapping plugin (see Book 8, Chapter 6), you can manage multiple sites that have different, unique domain names. None of these sites even has to be a blog. The sites can have blog elements, or they can be static sites that use only pages.
- The built-in options let you choose between subdomains and subfolder sites when you install the network. If you install WordPress in the root of your web space, you get `subdomain.`*`yourdomain.com`* (if you choose subdomains) or *`yourdomain.com`*`/subfolder` (if you choose subfolders). Book 8, Chapter 2 discusses the differences and advantages.

  REMEMBER

  After you choose the kind of sites you want to host and create those sites, you can't change them. These sites are served virtually, meaning that they don't exist as files or folders anywhere on the server; they exist only in the database. The correct location is served to the browser by means of rewrite rules in the `.htaccess` file. (See Book 2, Chapter 5.)
- The main, or parent, site in the network can also be a landing page for the entire network of sites, showcasing content from other sites in the network and drawing in visitors further.

Setting Up the Optimal Hosting Environment

This chapter assumes that you already have the WordPress software installed and running correctly on your web server and that your web server meets the minimum requirements to run WordPress. (See Chapter 2 in this minibook.)

Before you enable the WordPress network feature, you need to determine how you're going to use the feature. You have a couple of options:

» Manage just a few of your own WordPress websites.

» Run a full-blown content network with several hundred sites and multiple users.

If you're planning to run just a few of your own sites with the WordPress network feature, your current hosting situation probably is well suited to the task. (See Chapter 3 of this minibook for information on web-hosting services.) If you plan to host a large network with hundreds of blogs and multiple users, however, you should consider contacting your host and increasing your bandwidth.

WARNING

In addition to the security measures, time, and administrative tasks that go into running a community of websites, you've got a few things to worry about. Creating a community increases the resource use, bandwidth, and disk space on your web server. In many cases, if you go over the limits allotted to you by your web host, you incur great cost. Make sure that you anticipate your bandwidth and disk-space needs before running a large network on your website! (Don't say that I didn't warn you.)

Checking out shared versus dedicated hosting

Many WordPress network communities start with grand dreams of being large, active communities. Be realistic about how your community will operate so that you can make the right hosting choice for yourself and your community.

A small multisite community is easy to handle with a shared-server solution, whereas a large, active community should operate on a dedicated server. The difference between the two types of communities lies in the servers:

» **Shared-server solution:** You have one account on one server that has several other accounts on it. Think of this arrangement as apartment living. One apartment building has several apartments where multiple people live under one roof.

» **Dedicated server:** You have one account. You have one server. That server is dedicated to your account, and your account is dedicated to the server. Think of this arrangement as owning a home and not sharing your living space with anyone else.

A dedicated server solution is a more expensive investment for your community, whereas a shared-server solution is more economical. Base your decision on your realistic estimates of how big and how active your community will be. You can move from a shared-server solution to a dedicated server solution if your community gets larger than you expected, but it's easier to start with the right solution for your community from Day One.

Exploring subdomains versus subdirectories

The WordPress Multisite feature gives you two ways to run a network of sites on your domain: the subdomain option and the subdirectory option. The more popular option (and recommended structure) sets up subdomains for the sites created in a WordPress network. In the subdomain option, the username of the site appears first, followed by your domain name. In the subdirectory option, your domain name appears first, followed by the username of the site.

Which option should you choose? You can see the differences in the URLs of these two options:

- A subdomain looks like this: `http://`*`username.yourdomain.com`*.
- A subdirectory looks like this: `http://`*`yourdomain.com/username`*.

While the network is being set up, tables of information about the network — including the main site URL — are added to the database. If you're developing a site or want to change the domain, you need to change every reference to the domain name in the database. See Book 2, Chapter 3 for information about the WordPress database structure, including how data is stored in tables, as well as how to use a popular database administration tool called phpMyAdmin to manage, view, and edit database tables.

Choosing Linux, Apache, MySQL, and PHP server environments

A network of sites works best on a LAMP (Linux, Apache, MySQL, and PHP) server with the `mod_rewrite` Apache module enabled. `mod_rewrite` is an Apache module that builds URLs that are easier to read than standard URLs. (See the nearby "Apache `mod_rewrite`" sidebar for more information.) In WordPress, this Apache module is used for permalinks. If your installation uses any permalink other than the default, `?p=123`, you're okay. Your web host can help you determine whether your web server allows Apache `mod_rewrite`, which is a requirement for setting up the WordPress Multisite feature. (You can find more information on permalink structure in Book 3, Chapter 2.)

APACHE MOD_REWRITE

The Apache Software Foundation supplies th3e Apache software for free at `https://www.apache.org/free/`. Apache is loaded and running on your web server. Usually, the only person who has access to Apache files is the web server administrator (generally, your web host). Depending on your own web server account and configuration, you may not have access to the Apache software files.

The Apache module that's necessary for the WordPress network to create nice permalink URLs is called `mod_rewrite`. This module must be configured so that it's active and installed on your server.

You (or your web host) can make sure that the Apache `mod_rewrite` is activated on your server. To do so, open the `httpd.conf` file, and verify that it includes the following line:

```
LoadModule rewrite_module /libexec/mod_rewrite.so
```

If not, type this code on its own line in the `httpd.conf` file, and save the file. You probably need to restart Apache before the change takes effect.

For the purposes of this chapter, I stick to the LAMP server setup, because it's most similar to the average web host and is most widely used.

WARNING

The Apache `mod_rewrite` module is required for WordPress multisites. If you don't know whether your current hosting environment has this module in place, drop an email to your hosting provider and ask. The provider can answer that question for you (in addition to installing the module for you in the event that your server doesn't have it).

TECHNICAL STUFF

Networks also work well on Nginx and Lightspeed servers, but many users have reported having great difficulty on IIS (Windows) servers. Therefore, I don't recommend setting up WordPress with multisite features in a Windows server environment.

Subdomain sites work by way of a virtual host entry in Apache, also known as a wildcard subdomain. On shared hosts, your web hosting provider support team has to enable this entry for you, or the provider may already have done so for all accounts. It's best to ask your hosting provider before you begin. In these situations, the domain you use for your installation must be the default domain in your account. Otherwise, the URLs of your subsites won't work properly or will contain a folder name.

WARNING

Some hosts may require you to have a dedicated IP address, but this address isn't a software requirement for a WordPress network to function.

Before proceeding with the final steps in enabling the WordPress multisite feature, you need to get a few items in order on your web server. You also need to decide how to handle the multiple blogs in your network. The configurations in the next section, *"Adding a virtual host to the Apache configuration,"* need to be in place to run the WordPress network successfully.

Adding a virtual host to the Apache configuration

You need to add a hostname record that points to your web server. To add this record, you use the domain name server (DNS) configuration tool available in your web server administration software, such as Web Host Manager (`https://cpanel.com/products/`).

WARNING

In this section, you edit and configure Apache server files. If you can perform the configurations in this section yourself (and if you have access to the Apache configuration files), this section is for you. If you don't know how, are uncomfortable with adjusting these settings, or don't have access to the configurations in your web server software, you need to ask your hosting provider for help or hire a consultant to perform the configurations for you. I can't stress enough that you shouldn't edit the Apache server files yourself if you aren't comfortable with it or don't fully understand what you're doing. Web hosting providers have support staff to help you with these things; take advantage of the help if you need it!

The hostname record looks like this: `*.yourdomain.com` (where `yourdomain.com` is your actual domain name). Follow these steps to enable the wildcard subdomains in Apache:

1. **Log in to your server as the root user.**
2. **Open the `httpd.conf` file or the `vhost include` file for your current web account.**
3. **Find the virtual host section for your domain.**
4. **Add the wildcard subdomain record next to the domain name.**

   The record looks like this:

   ```
   ServerAlias yourdomain.com *.yourdomain.com
   ```

5. **Save the file.**
6. **Restart Apache.**

You also need to add a wildcard subdomain DNS record. Depending on how your domain is set up, you can add this record in your registrar or your web hosting account. If you simply point the domain record to your web host's name servers, you can add more DNS records at your web host in the web server administration interface, such as Web Host Manager.

You also should add a `CNAME` record with a value of *. CNAME (which stands for *canonical name*) is a record stored in the DNS settings of your Apache web server that tells Apache to associate a new subdomain with the main account domain. Applying the value * tells Apache to send any subdomain requests to your main domain; from there, WordPress looks up that subdomain in the database to see whether it exists.

Networks require a great deal more server memory (RAM) than typical WordPress sites do (those that don't use the Multisite feature), simply because multisites are bigger, have more traffic, and use more database space and resources because multiple sites are running. You aren't simply adding instances of WordPress; you're multiplying the processing and resource use of the server when you run the WordPress multisite feature. Although smaller instances of a network run well enough on most web hosts, you may find that when your network grows, you need more memory. I generally recommend starting with a hosting account that has access to at least 256MB of RAM.

For each site created, nine tables are added to the single database. Each table has a prefix similar to `wp_ ID_tablename` (where `ID` is a unique ID number assigned to the site).

The only exception is the main site; its tables remain untouched and remain the same. (Book 2, Chapter 3 shows how the main site's tables look.) With WordPress multisites, all new installations leave the main blog tables untouched and number additional site tables sequentially for every site that's added to the network.

Much discussion about the database layout has occurred in Trac, the WordPress bug tracking system, and in the WordPress.org forums. Although Trac may seem to be unwieldy, it scales appropriately. The average user builds a small to medium-size network, which usually needs no more than a virtual private server (VPS) account.

» Configuring the multisite installation

» Disabling multisite

» Correcting errors

Chapter 2

Setting Up and Configuring Multisite Features

This chapter covers how to find the files you need to edit multiple sites, how to enable multiple sites in the network, and how to remove multiple sites when you no longer want to have multiple sites in your WordPress install.

By default, access to multisite settings is disabled to ensure that users don't set up their networks without researching all that the setup entails. Setting up the Multisite feature is more than configuring options or turning on a feature. Before enabling and setting up a network, be sure that you read Book 8, Chapter 1.

Here's what you need:

- Backups of your site (explained in Book 2, Chapter 7)
- Access to the `wp-config.php` file for editing (see Book 2, Chapter 5)

» Enabled wildcard subdomains (covered in Book 8, Chapter 1) if you're using subdomains

Enabling the Multisite Feature

You need to enable access to the Multisite menu so that you can set up the network and allow the creation of multiple sites.

TIP

Before you begin your edits, make a copy of your `wp-config.php` file, and keep it in a safe place on your computer.

Follow these steps:

1. **Connect to your web server via SFTP.**

   For details on SFTP, see Book 2, Chapter 2.

2. **Locate the `wp-config.php` file in the root or `/public_html` directory of your website.**

   This file is with the main WordPress files.

3. **Open the `wp-config.php` file for editing in your favorite text editor.**

   For Windows users, Notepad does the trick. For Macs, use TextEdit.

4. **Click at the end of the line that reads `define('DB_COLLATE', '');` and press Enter to create a new blank line.**

   TIP

   Some SFTP clients let you right-click the filename on the server and choose Edit from the contextual menu to edit the file within your SFTP program.

5. **Type** define('WP_ALLOW_MULTISITE', true);.

   This line of code tells WordPress that you intend to use the multisite feature; additionally, it activates the Network option on the Tools menu of your WordPress Dashboard (covered later in this chapter).

6. **Save the `wp-config.php` file, and upload it to your website.**

When you log in to the Dashboard of WordPress, you see a Network Setup link on the Tools menu, as shown in Figure 2-1.

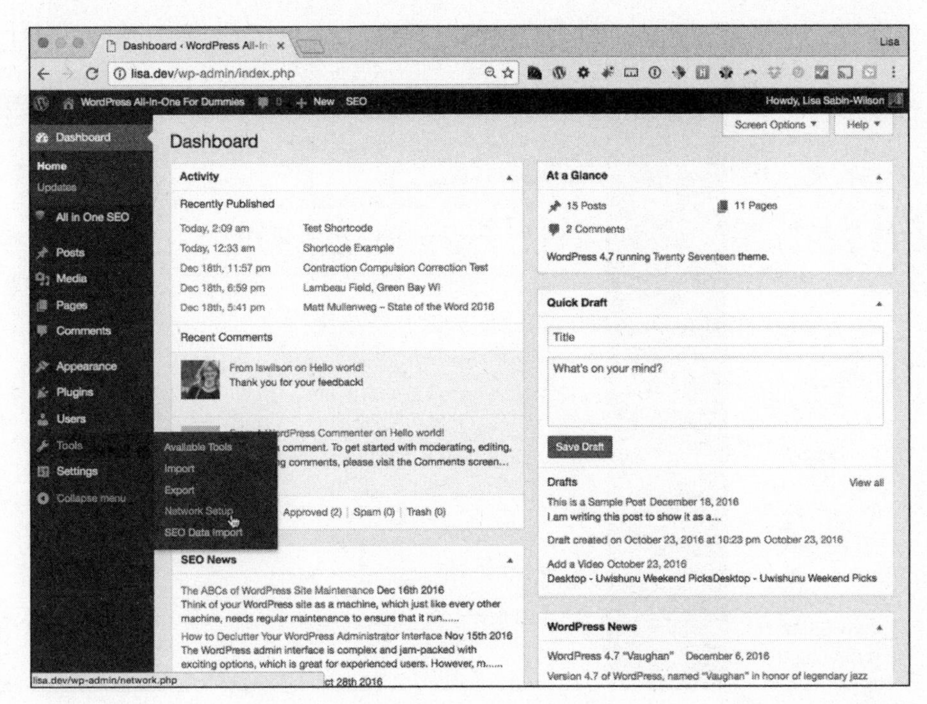

FIGURE 2-1: The Network Setup link on the Tools menu.

Clicking this menu item displays the Create a Network of WordPress Sites page, which is covered in the next section, "Installing the Network on Your Site."

TIP

If you have any plugins installed and activated on your WordPress installation, deactivate them before you proceed with the network setup. WordPress won't allow you to continue until you deactivate all your plugins.

Installing the Network on Your Site

The Network Details heading on the Create a Network of WordPress Sites page has options filled in automatically. The server address, for example, is pulled from your installation and can't be edited. The Network Title and Administrator E-Mail Address are pulled from your installation database, too, because your initial WordPress site is the main site in the network. Click the Install button, and WordPress creates new network tables in the database. The page refreshes, and the Enabling the Network page appears, as shown in Figure 2-2.

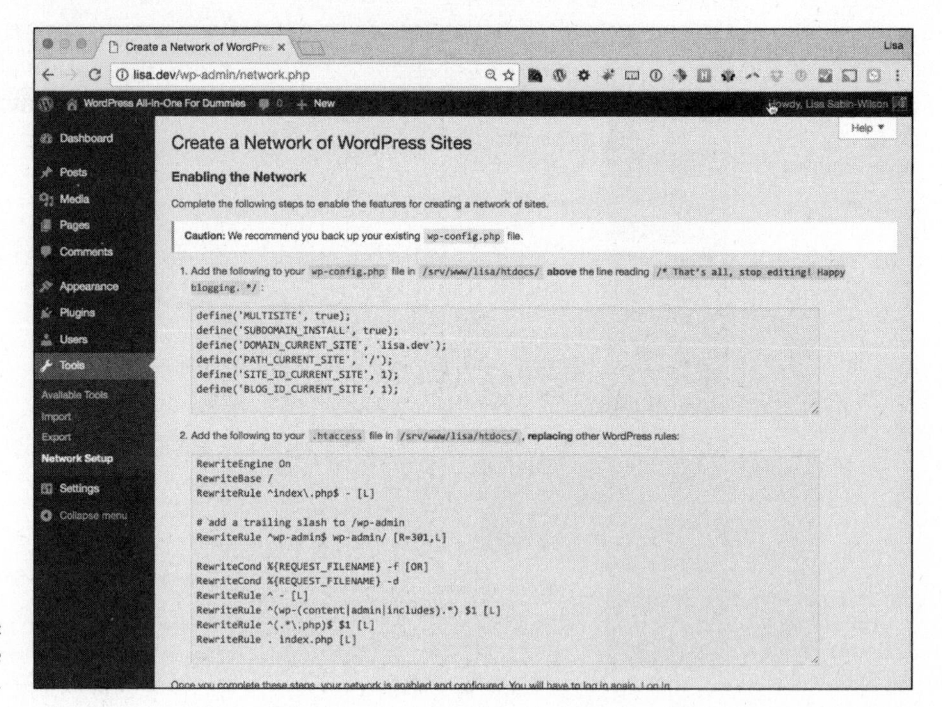

FIGURE 2-2: The Enabling the Network page.

From the Enabling the Network page, follow these steps to install the multisite feature after you install the network. *Note:* These steps require you to edit web server files, so be sure to have your text-editor program handy.

1. **Add the network-related configuration lines to the `wp-config.php` file.**

   On the Create a Network of WordPress Sites page, WordPress lists up to seven lines of configuration rules that need to be added to the `wp-config.php` file. The first line includes the line discussed in "Enabling the Multisite Feature" earlier in this chapter: `define('multisite', true );`. You can skip that line, copy the rest of the lines, and then paste them below the `define('multisite', true);` line in your `wp-config.php` file. The lines of code you add look like this:

   ```
   define('MULTISITE', true);
   define('SUBDOMAIN_INSTALL', true);
   define('DOMAIN_CURRENT_SITE', 'domain.com');
   define('PATH_CURRENT_SITE', '/');
   define('SITE_ID_CURRENT_SITE', 1);
   define('BLOG_ID_CURRENT_SITE', 1);
   ```

   These lines of code provide configuration settings for WordPress by telling it whether it's using subdomains, what the base URL of your website is, and what your site's current path is. This code also assigns a unique ID of `1` to your website and blog for the main installation site of your WordPress Multisite network. By default, WordPress sets up your network to use subdomains instead of subdirectories. If you want to use subdirectories, make sure that you have the VHost and Apache `mod_rewrite` configurations in place on your server. (See Book 8, Chapter 1 for details on Apache mod_rewrite.) Then change this line of code

```
define('SUBDOMAIN_INSTALL', true);
```

to

```
define('SUBDOMAIN_INSTALL', false);
```

WARNING

The lines of code that appear on the Create a Network of WordPress Sites page are unique to *your* installation of WordPress. Make sure that you copy the lines of code on this page to *your* installation, because this code is specific to your site's setup.

2. **Add the rewrite rules to the `.htaccess` file on your web server.**

WordPress lists up to 13 lines of code that you need to add to the `.htaccess` file on your web server in the WordPress installation directory. The lines look something like this:

```
RewriteEngine On
RewriteBase /
RewriteRule ^index\.php$ - [L]

# add a trailing slash to /wp-admin
RewriteRule ^wp-admin$ wp-admin/ [R=301,L]

RewriteCond %{REQUEST_FILENAME} -f [OR]
RewriteCond %{REQUEST_FILENAME} -d
RewriteRule ^ - [L]
RewriteRule ^(wp-(content|admin|includes).*) $1 [L]
RewriteRule ^(.*\.php)$ $1 [L]
RewriteRule . index.php [L]
RewriteBase /
RewriteRule ^index\.php$ - [L]
```

REMEMBER

Book 8, Chapter 1 discusses the Apache `mod_rewrite` module. You must have this module installed on your web server to run the WordPress multisite feature. The rules you add to the `.htaccess` file on your web server are `mod_rewrite` rules, and they need to be in place so that your web server tells WordPress how to handle things, such as permalinks for blog posts and pages, media, and other uploaded files. If these rules aren't in place, the WordPress multisite feature won't work correctly.

3. **Copy the lines of code from the Enabling the Network page, open the `.htaccess` file, and paste the lines of code there.**

   Replace the rules that already exist in that file.

4. **Save the `.htaccess` file, and upload it to your web server.**

5. **Click the login link at the bottom of the Enabling the Network page.**

   You're logged out of WordPress because by following these steps, you changed some of the browser cookie-handling rules in the `wp-config.php` and `.htaccess` files.

Completing the installation steps activates a Network Admin menu item on the top-right menu of links on your WordPress Dashboard. The Network Admin Dashboard is where you, as the site owner, administer and manage your multisite WordPress network. (See Book 8, Chapter 3.)

Disabling the Network

At some point, you may decide that running a network of sites isn't for you, and you want to disable the multisite feature. Before disabling the network, save any content from the other sites by making a full backup of your site. Book 2, Chapter 7 has detailed information about backing up your site.

The first step is restoring the original `wp-config.php` file and `.htaccess` files. This step causes your WordPress installation to stop displaying the Network Admin menu and the extra sites.

You may also want to delete the tables that were added, permanently removing the extra sites from your installation. Book 2, Chapter 3 takes you through the WordPress database, including the use of a popular database administration tool called phpMyAdmin. You can use that tool to delete the multisite tables from your WordPress database when you want to deactivate the feature. The extra database tables that are no longer required when you aren't running the WordPress multisite feature include

- **`wp_blogs`:** This database table contains one record per site and is used for site lookup.
- **`wp_blog_versions`:** This database table is used internally for upgrades.
- **`wp_registration_log`:** This database table contains information on sites created when a user signs up, if the user chose to create a site at the same time.

- **wp_signups:** This database table contains information on users who signed up for the network.
- **wp_site:** This database table contains one record per WordPress network.
- **wp_sitemeta:** This database table contains network settings.

Additionally, you can delete any database tables that have blog IDs associated with them. These tables start with prefixes that look like wp_1_, wp_2_, wp_3_, and so on.

WordPress adds new tables each time you add a site to your network. Those database tables are assigned unique numbers incrementally.

Dealing with Common Errors

Occasionally, you may enter a configuration setting incorrectly or change your mind about the kind of network you require. If you installed WordPress, enabled the network, and then decide to move the network to a new location, you encounter errors when changing the URL. The proper method is to move WordPress first, disable the network if you installed it, and then enable the network at the new location.

To change from subdomains to subdirectories, or vice versa, follow these steps:

1. **Delete the extra sites, if any were created.**
2. **Edit `wp-config.php`, changing the value of `define( 'SUBDOMAIN_INSTALL', true );` to `define( 'SUBDOMAIN_INSTALL', false );`.**

   To switch from subdomains to subdirectories, change `false` to `true`:

   ```
   define ('SUBDOMAIN_INSTALL', true);
   ```
3. **Save the `wp-config.php` file, and upload it to your website.**
4. **Visit the Dashboard of WordPress, click the Permalink link on the Settings menu, and click the Save Changes button.**

   This step saves and resets your permalink structure settings and flushes the internal rewrite rules, which are slightly different in subdomains than they are in subdirectories.

You can't complete this process if you want to keep extra sites.

» Handling network management

» Preventing spam and splogs

Chapter 3

Becoming a Network Admin

When you enable the WordPress network option and become a network admin, you can examine the various settings that are available to you and review the responsibilities of running a network.

As a network admin, you can access the Network Admin Dashboard, which includes several submenus, as well as the overall settings for your network. This chapter discusses the menu items and options on the Network Admin page, guides you in setting network options, and discusses the best ways to prevent spam and spam blogs (splogs).

Exploring the Network Admin Dashboard

When the WordPress network is fully enabled and configured, you see a new My Sites link on the menu in the top-right corner of your WordPress Dashboard. In this section, I show you how to become a Network Admin and explain everything you need to know about being Network Admin of your new WordPress network.

If you hover your cursor over that link, the Network Admin link appears in the drop-down menu in the top-left section of the Dashboard, as shown in Figure 3-1.

WordPress separated the Network Admin menu features from the regular (Site Admin) Dashboard menu features to make it easier for you to see which part of your site you're managing. If you're performing actions that maintain your main website — such as publishing posts or pages, creating or editing categories, and so on — you work on the regular Dashboard (Site Admin). If, however, you're managing any of the network sites, plugins, and themes for the network sites or registered users, you work in the Network Admin section of the Dashboard.

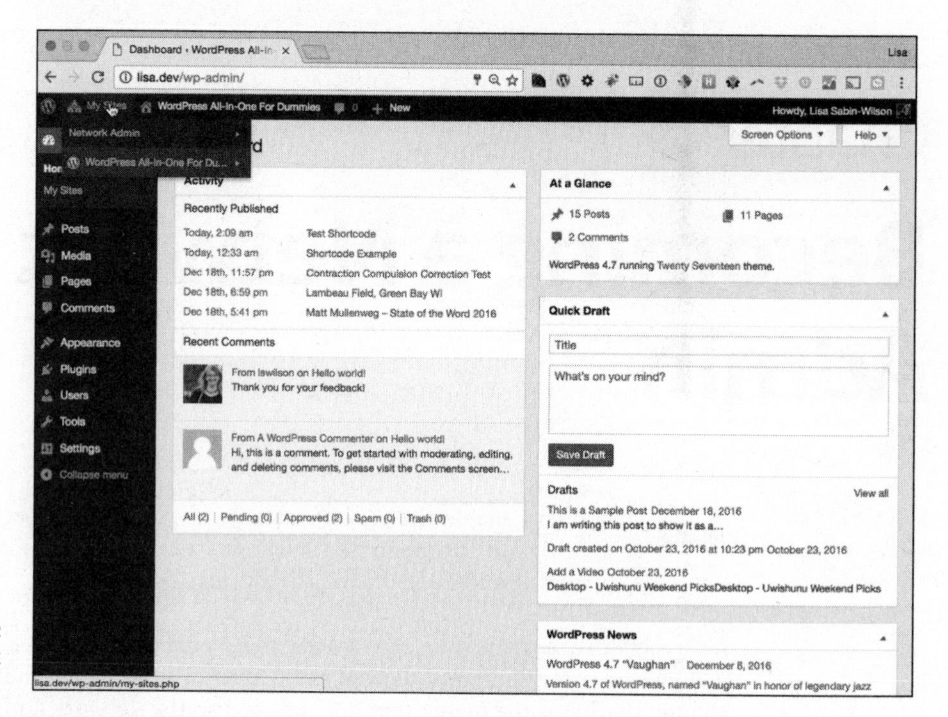

FIGURE 3-1: The Network Admin link.

REMEMBER

Keep in mind the distinct difference between the Site Admin and Network Admin Dashboards, as well as their menu features. WordPress does its best to know which features you're attempting to work with, but if you find yourself getting lost on the Dashboard or not finding a menu or feature that you're used to seeing, make sure you're working in the correct section of the Dashboard.

The Network Admin Dashboard (see Figure 3-2) is similar to the regular Word-Press Dashboard, but as you may notice, the modules pertain to the network of sites. Options include creating a site, creating a user, and searching existing sites and users. Obviously, you won't perform this search if you don't have any users or sites yet. This function is extremely useful when you have a community of users and sites within your network, however.

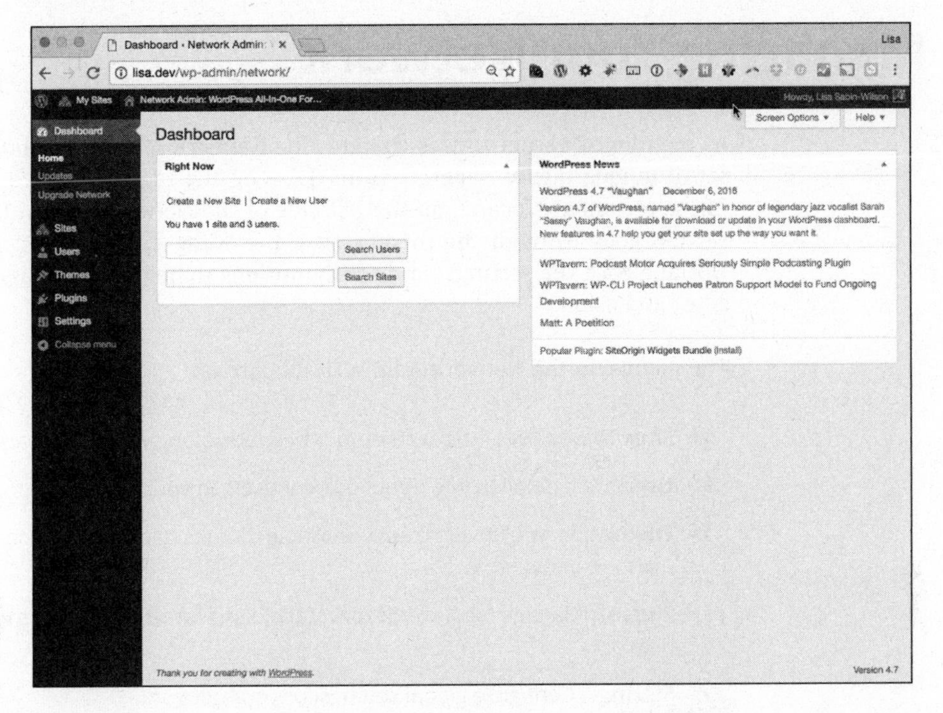

FIGURE 3-2: The Network Admin Dashboard.

TIP

The Network Admin Dashboard is configurable, just like the regular Dashboard; you can move the modules around and edit their settings. See Book 3, Chapter 1 for more information about arranging the Dashboard modules to suit your tastes.

The Search Users feature allows you to search usernames and user email addresses. If you search for the user *Lisa*, for example, your results include any user whose username or email address contains *Lisa*, so you can receive multiple results when using just one search word or phrase. The Search Sites feature returns any blog content within your community that contains your search term, too.

The Network Admin Dashboard has two useful links near the top-left corner:

- **Create a New Site:** Click this link to create a site within your network. The Add New Site page appears, allowing you to add the site. Find out how to add a site in the upcoming "Sites" section.
- **Create a New User:** Click this link to create a new user account within your community. The Add New User page appears, allowing you to add a user to your community. Find out how to add a new user in the "Users" section later in the chapter.

Additionally, the Network Admin Dashboard gives you a real-time count of how many sites and users you have in your network, which is nice-to-know information for any network admin.

Managing Your Network

As mentioned earlier in this chapter, the Network Admin Dashboard has its own set of unique menus separate from those on the regular Site Admin Dashboard. Those menus are located on the left side of the Network Admin Dashboard. This section goes through the menu items, providing explanations; instructions on working with the settings; and configurations to help you manage your network, sites, and users.

The menus on the Network Admin Dashboard are

- **Sites:** View a list of the sites in your network, along with details about them.
- **Users:** See detailed info about current users in your network.
- **Themes:** View all the currently available themes to enable or disable them for use in your network.
- **Plugins:** Manage (activate/deactivate) plugins for use on all sites within your network.
- **Settings:** Configure global settings for your network.
- **Updates:** Upgrade all sites in your network with one click.

All the items on the Network Admin Dashboard are important, and you'll use them frequently throughout the life of your network. Normally, I'd take you through each of the menu items in order so that you can follow along on your Dashboard, but it's important to perform some preliminary configurations on your network. Therefore, I start the following sections with the Settings menu; then I take you through the other menu items in order of their appearance on the Network Admin Dashboard.

Settings

When you click the Settings menu link on the Network Admin Dashboard, the Network Settings page appears. The Settings page contains several sections of options for you to configure to set up your network the way you want to.

REMEMBER

When you finish configuring the settings on the Network Settings page, don't forget to click the Save Changes button at the bottom of the page, below the final Menu Settings section. (See "Menu Settings" later in the chapter.) If you navigate away from the Network Settings page without clicking the Save Changes button, your configurations aren't saved, and you need to go through the entire process again.

Operational Settings

The Operational Settings section, shown in Figure 3-3, has Network Title and Network Admin Email settings:

- **Network Title:** This setting is the title of your overall network of sites. This name is included in all communications regarding your network, including emails that new users receive when they register a new site within your network. Type your desired network title in the text box.
- **Network Admin Email:** This setting is the email address that all correspondence from your website is addressed from, including all registration and sign-up emails that new users receive when they register a new site and/or user account within your network. In the text box, type the email that you want to use for these purposes.

Network Title — WordPress All-In-One For Dummies Sites

Network Admin Email — lisa@domain.com

This email address will receive notifications. Registration and support emails will also come from this address.

FIGURE 3-3: The Operational Settings section of the Network Settings page.

Registration Settings

The Registration Settings section (see Figure 3-4) allows you to control various aspects of allowing users to sign up to your network. The most important option specifies whether to allow open registration.

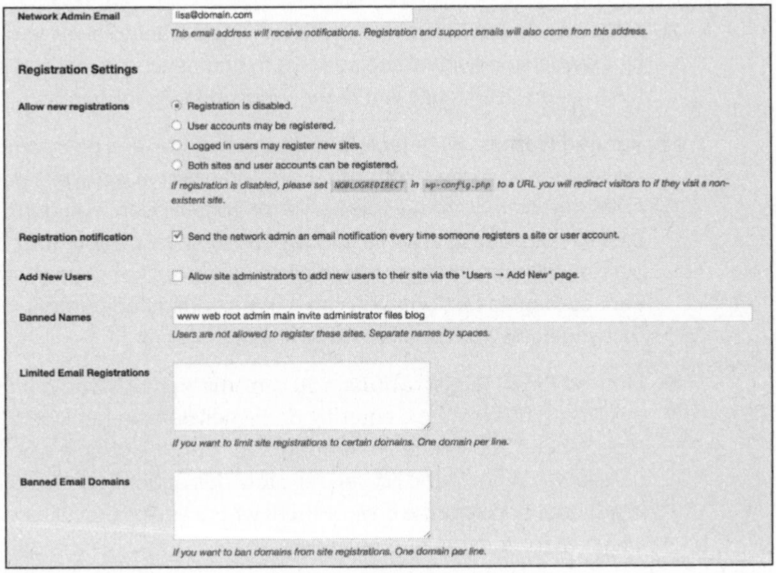

Network Admin Email — lisa@domain.com

This email address will receive notifications. Registration and support emails will also come from this address.

Registration Settings

Allow new registrations
- Registration is disabled.
- User accounts may be registered.
- Logged in users may register new sites.
- Both sites and user accounts can be registered.

If registration is disabled, please set `NOBLOGREDIRECT` in `wp-config.php` to a URL you will redirect visitors to if they visit a non-existent site.

Registration notification — Send the network admin an email notification every time someone registers a site or user account.

Add New Users — Allow site administrators to add new users to their site via the "Users → Add New" page.

Banned Names — www web root admin main invite administrator files blog

Users are not allowed to register these sites. Separate names by spaces.

Limited Email Registrations

If you want to limit site registrations to certain domains. One domain per line.

Banned Email Domains

If you want to ban domains from site registrations. One domain per line.

FIGURE 3-4: The Registration Settings section of the Network Settings page.

Decide how you want to handle registration on your network, and set one of the following options by selecting its radio button:

- **Registration Is Disabled:** Disallows new user registration. When selected, this option prevents people who visit your site from registering for a user account. Registration is disabled by default.
- **User Accounts May Be Registered:** Allows people to create only user accounts; they won't be able to create blogs within your network.
- **Logged In Users May Register New Sites:** Allows only existing users — those who are already logged in — to create new blogs within your network. This option also disables new user registration. Select this option if you don't want just anyone to register for an account. Instead, you (as the site administrator) can add users at your discretion.
- **Both Sites and User Accounts Can Be Registered:** Allows users to register an account and a site on your network during the registration process.

These options apply only to outside users. As a network admin, you can create new sites and users any time you want by setting the necessary options on the Network Admin Dashboard. (For information about creating users, see the upcoming "Users" section.)

The remaining options in the Registration Settings section are as follows:

- **Registration Notification:** When this option is selected, an email is sent to the network admin every time a user or a site is created on the system, even if the network admin creates the new site.
- **Add New Users:** Select this check box if you want to allow your community blog owners (individual site admins) to add new users to their own community blog via the Users page within their individual Dashboards.
- **Banned Names:** By default, WordPress bans several usernames from being registered within your community, including *www, web, root, admin, main, invite,* and *administrator*. This ban is for good reason. You don't want a random user to register a username such as *admin,* because you don't want that person misrepresenting himself as an administrator of your site. In the Banned Names text box, you can enter an unlimited number of usernames that you don't want to allow on your site.
- **Limited Email Registrations:** You can limit sign-ups based on email domains by filling in this text box, entering one email domain per line. If you have open registrations but limited email addresses, only the people who have email domains that are in the list can register. This option is an excellent one to use in a school or corporate environment where you're providing email addresses and sites to students or employees.

- **Banned Email Domains:** This feature, which is the reverse of Limited Email Registration, blocks all sign-ups from a particular domain and can be useful in stopping spammers. You can enter **gmail.com** in the text box, for example, to ban anyone who tries to sign up with a Gmail address.

New Site Settings

The New Site Settings section (you need to scroll down to see it) is a configurable list of items that populates default values when a new site is created. The list includes the values that appear in welcome emails, on a user's first post page, and on a new site's first page, as shown in Figure 3-5:

- **Welcome Email:** The email text that owners of newly registered sites in your network receive when their registration is complete. You can leave the default message in place, if you like, or you can type the text of a email you want new site owners to receive when they register a site within your network.

  A few variables you can use in this email aren't explained entirely on the Site Options page, including the following:

  - `SITE_NAME`: Inserts the name of your WordPress site
  - `BLOG_URL`: Inserts the URL of the new member's blog
  - `USERNAME`: Inserts the new member's username
  - `PASSWORD`: Inserts the new member's password
  - `BLOG_URLwp-login.php`: Inserts the hyperlinked login URL for the new member's blog
  - `SITE_URL`: Inserts the hyperlinked URL for your WordPress site
- **Welcome User Email:** The email text that is automatically sent to a new user when she registers for an account in your network. You can use the variables from the preceding item to personalize the email a bit rather than use the default text shown in Figure 3-5.
- **First Post:** The first, default post that appears on every newly created site in your network. WordPress provides some default text that you can leave in place, or you can type your desired text in the text box.

  You can use this area to provide useful information about your site and services. This post also serves as a nice guide for new users; they can view it on the Dashboard's Edit Post page to see how it was entered and formatted, and then use that page as a guide for creating their own blog posts. You can also use the variables mentioned for the Welcome Email item to have WordPress add some information for you automatically.

- **First Page:** Like the First Post setting, displays default text on every newly created site in your network. The First Page text box doesn't include default text, however; if you leave it blank, no default page is created.
- **First Comment:** Displays on the first default post on every newly created site within your network. Type the text that you want to appear in the first comment on every site that's created in your community.
- **First Comment Author:** The name of the author of the first comment on new sites in your network. This option isn't shown in Figure 3-5; you need to scroll down to see it.
- **First Comment URL:** The web address (URL) for the author of the first comment. This URL links the first comment author's name to the URL you type here. (This option doesn't appear in Figure 3-5.)

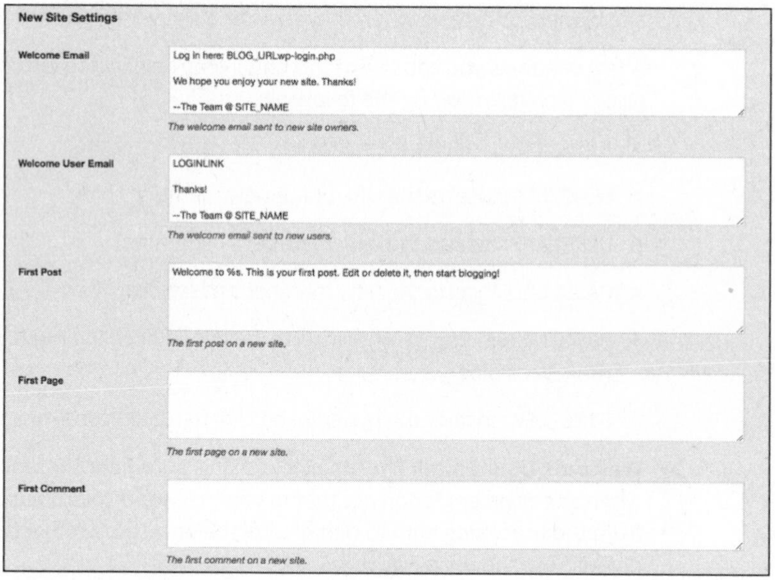

New Site Settings

Welcome Email
Log in here: BLOG_URLwp-login.php
We hope you enjoy your new site. Thanks!
--The Team @ SITE_NAME
The welcome email sent to new site owners.

Welcome User Email
LOGINLINK
Thanks!
--The Team @ SITE_NAME
The welcome email sent to new users.

First Post
Welcome to %s. This is your first post. Edit or delete it, then start blogging!
The first post on a new site.

First Page
The first page on a new site.

First Comment
The first comment on a new site.

FIGURE 3-5: New Site Settings section of the Network Settings page.

Upload Settings

Scrolling down the Network Settings page, you get to the Upload Settings section (see Figure 3-6). These settings define the types of files that site owners within your network can upload by using the media upload feature on the WordPress Write Posts and Write Page areas. (See Book 3, Chapter 3.) The options in the Upload Settings section have default settings that are already filled in for you:

- **Site Upload Space:** If you leave this check box deselected, users are allowed to use all the space they want for uploads; no limits apply. Select the check box to limit the available space per site and then enter the amount in megabytes; default storage space is 100MB. You give users this amount of hard-drive space to store the files that they upload to their blogs. If you want to change the default storage space, type a new number in the text box.

- **Upload File Types:** This setting defines the types of files that site owners can upload to their sites from their Dashboards. Users can't upload any file types that don't appear in this text box. By default, WordPress allows the following file types: `.jpg`, `.jpeg`, `.png`, `.gif`, `.mp3`, `.mov`, `.avi`, `.wmv`, `.midi`, `.mid`, `.mt2s`, and `.pdf`. You can remove any default file types and replace them with new ones.
- **Max Upload File Size:** This amount is in kilobytes (KB), and the default setting is 1500KB, which means that a user can't upload a file larger than 1500KB. Adjust this number as you see fit by typing a new number in the text box.

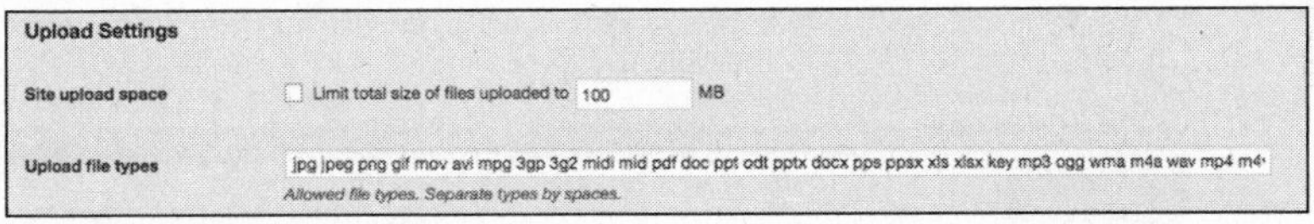

FIGURE 3-6: Upload Settings section of the Network Settings page.

Menu Settings

The Plugins administration menu is disabled on the Dashboard for all network sites (except for the network admin). The network admin always has access to the Plugins menu. If you leave this option unselected, the Plugins page is visible to users on their own site's Dashboards. Select the check box to enable the Plugins administration menu for your network users (see Figure 3-7). For more information about using plugins with WordPress, see Book 7.

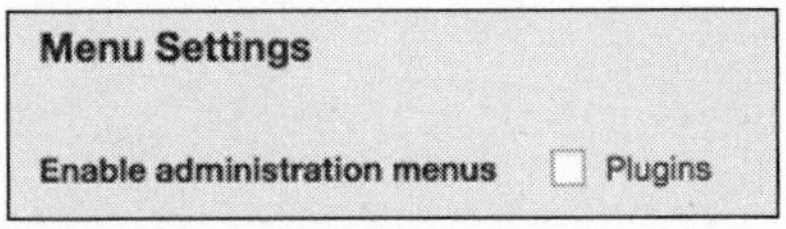

FIGURE 3-7: The Network Settings page's Menu Settings section.

Sites

Clicking the Sites link on the Network Admin Dashboard takes you to the Sites page, where you can manage your individual sites. Although each site in the network has its own Dashboard for basic tasks — posting, changing themes, and so on — the Sites page is where you create sites, delete sites, and edit the properties of the sites within your network. Editing information from this page is handy when you have problems accessing a site's back-end Dashboard.

The Sites page also lists all the sites within your network. The list shows the following statistics about each community site:

- **Path:** The site's path in your network. This setting means that the site's domain is `newsite`.*yourdomain.com* if you're using a subdomain setup or *yoursite.com*/`newsite` if you're using a subdirectory setup.
- **Last Updated:** The date when the site was last updated (or published to).
- **Registered:** The date when the site was registered in your network.
- **Users:** The username and email address associated with the user(s) of that site.

When you hover your pointer over the pathname of a site in your network, you see a handy list of links that helps you manage the site. Figure 3-8 shows the options that appear below a site listing when you hover your cursor over a site name.

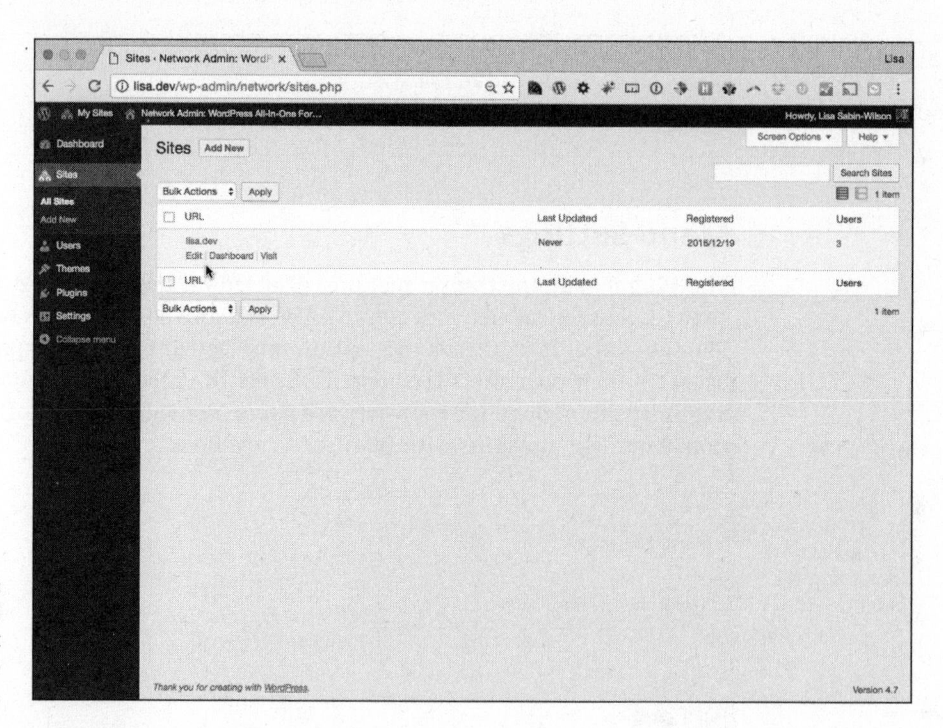

FIGURE 3-8: Individual site-management options on the Sites page.

The management options for network sites, some of which appear in Figure 3-8, are as follows:

- **Edit:** Click this link to go to the Edit Site page (see Figure 3-9), where you can change aspects of each site.
- **Dashboard:** Click this link to go to the site's Dashboard.

- **Visit:** Click this link to visit the live site in your web browser.
- **Deactivate:** Click this link to mark the site for deletion from your network. A pop-up window asks you to confirm your intention to deactivate the site. Click the Yes button to confirm. The user's site displays a message stating that the site has been deleted. You can reverse this action by revisiting the Sites page and clicking the Activate link that appears below the site pathname. (The Activate link appears only for sites that are marked as Deactivated.)
- **Archive:** Click this link to archive the site on your network, which prevents visitors from viewing it. The user's site displays the message `This site has been archived or suspended`. You can reverse this action by revisiting the Sites page and clicking the Unarchive link that appears below the site's pathname. (The Unarchive link appears only for sites that are marked as Archived.)
- **Spam:** Click this link to mark the site as spam and block users from accessing it via the Dashboard. Users see the message `This site has been archived or suspended`. You can reverse this action by revisiting the Sites page and clicking the Not Spam link that appears below the site's pathname. (The Not Spam link appears only for sites that are marked as Spam.)
- **Delete:** Click this link to delete the site from your network of sites. Although you see a confirmation screen that asks you to confirm your intention to delete the site, after you've done it, you can't reverse this decision.

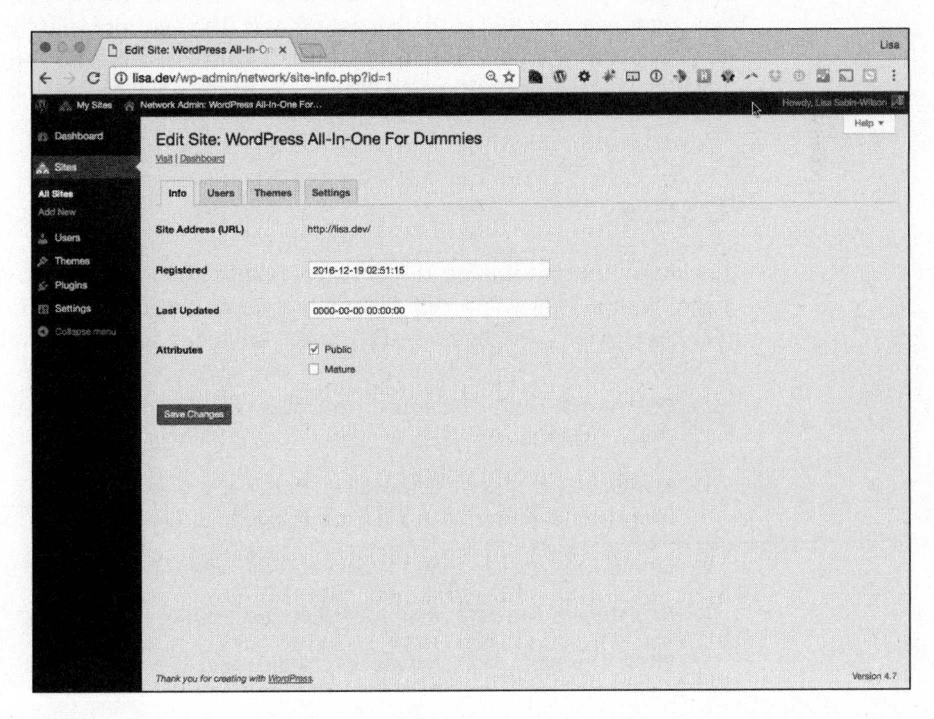

FIGURE 3-9: The Edit Site page.

Generally, you use the Edit Site page only when the settings are unavailable on the Dashboard of that particular site. Configure the options that appear on the four tabs of the Edit Sites page:

- **Info:** You can edit the site's domain, path, registered date, updated date, and attributes (Public, Archived, Spam, Deleted, Mature).
- **Users:** You can manage the users who are assigned to the site, as well as add users to the site in the Add New User section.
- **Themes:** You can enable themes for this site — particularly useful if you have themes that aren't network-enabled (see the upcoming "Themes" section). All the themes that aren't enabled within your network are listed on the Themes tab, which allows you to enable themes on a per-site basis.
- **Settings:** The settings on this tab cover all the database settings for the site that you're editing. You rarely need to edit these settings because you, as the network admin, have access to each user's Dashboard and can make any changes in the site's configuration settings there.

Also on the Sites menu of the Network Admin Dashboard, you see an Add New link; click it to load the Add New Site page on your Network Admin Dashboard. You can create a new site from the Add New Site page, shown in Figure 3-10. Fill in the Site Address (URL), Site Title, Site Language and Admin Email text boxes, and click the Add Site button to add the new site to your network. If the admin email address you enter is associated with an existing user, the new site is assigned to that user in your network. If the user doesn't exist, WordPress creates a new user, and an email is sent with a notification. The site is immediately accessible. The email that the user receives contains a link to the site, a login link, username, and password.

Users

Clicking the Users link on the Network Admin Dashboard takes you to the Users page, where you see a full list of members, or users, within your network. The Users page (see Figure 3-11) lists the following information about each user:

- **Username:** The login name the member uses when she logs in to her account in your community.
- **Name:** The user's real name, taken from her profile. If the user hasn't provided her name in her profile, this column is blank.
- **Email:** The email address the user entered when she registered on your site.
- **Registered:** The date when the user registered.
- **Sites:** Any sites of which the user is a member.

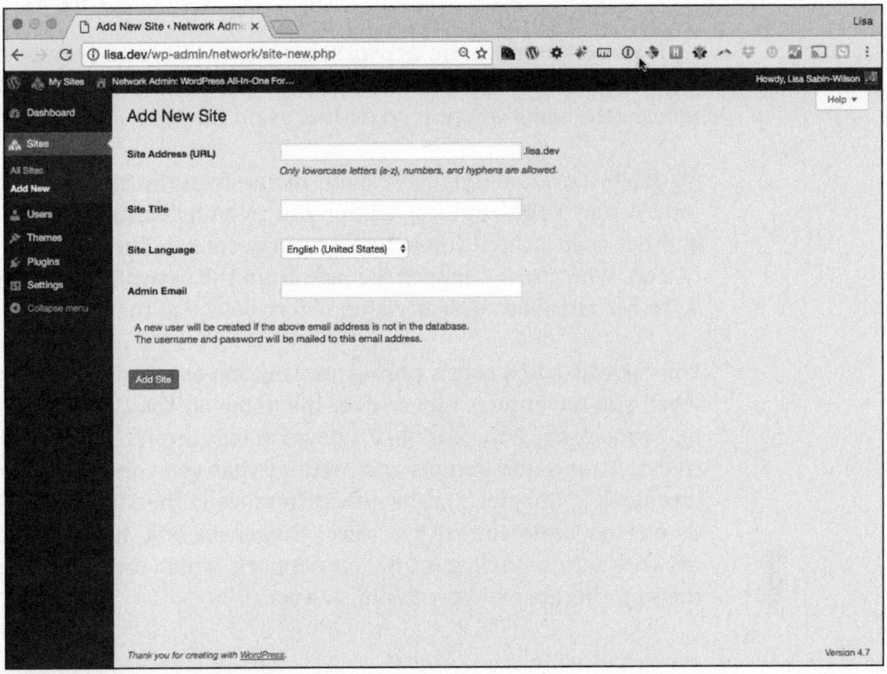

FIGURE 3-10: The Add New Site page of the Network Admin Dashboard.

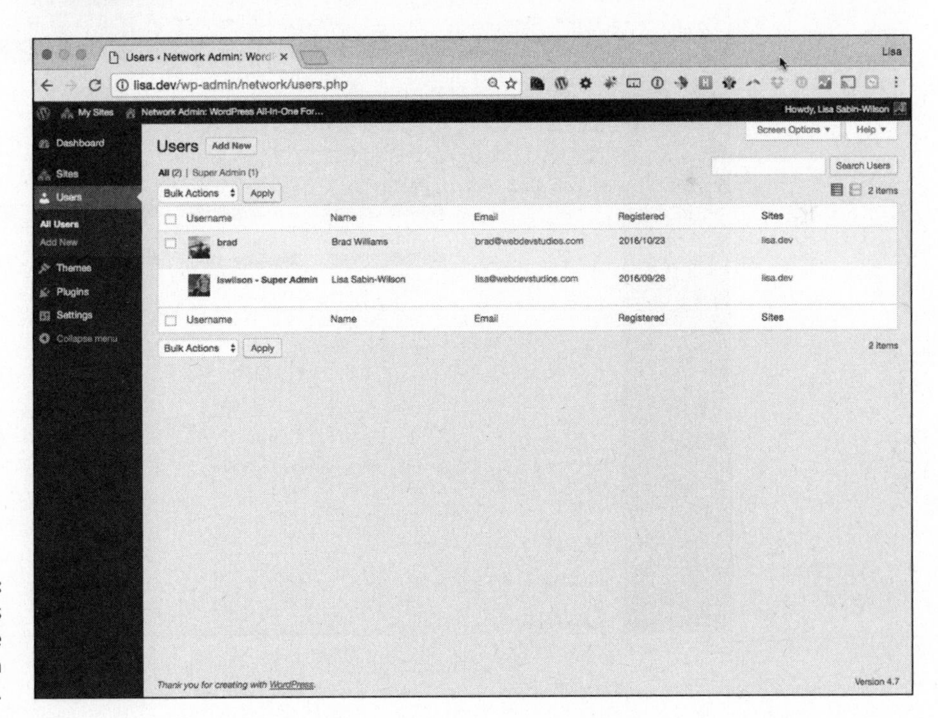

FIGURE 3-11: The Users page of the Network Admin Dashboard.

You can add users to the network, manage users, and even delete users by clicking the links that appear below their names when you hover over them with your mouse (the same way you do with sites on the Sites page).

To delete a user, simply hover your mouse over the username and click the Delete link. A new page appears, telling you to transfer this user's posts and links to another user account (most likely, your account). Then click the Confirm Deletion button. WordPress removes the user from the network. This action is irreversible, so be certain about your decision before you click that button!

You can also edit a user's profile information by clicking the Edit link that appears when you hover your mouse over his name on the Users page. Clicking that link takes you to the Edit User page, shown in Figure 3-12. The options on this page are (mostly) the same options and settings that you configured for your own profile (see Book 3, Chapter 2). The only difference is the Super Admin setting, which is deselected by default. If you select this check box, however, you grant this user network admin privileges for your network, which means that the user has exactly the same access and permissions as you.

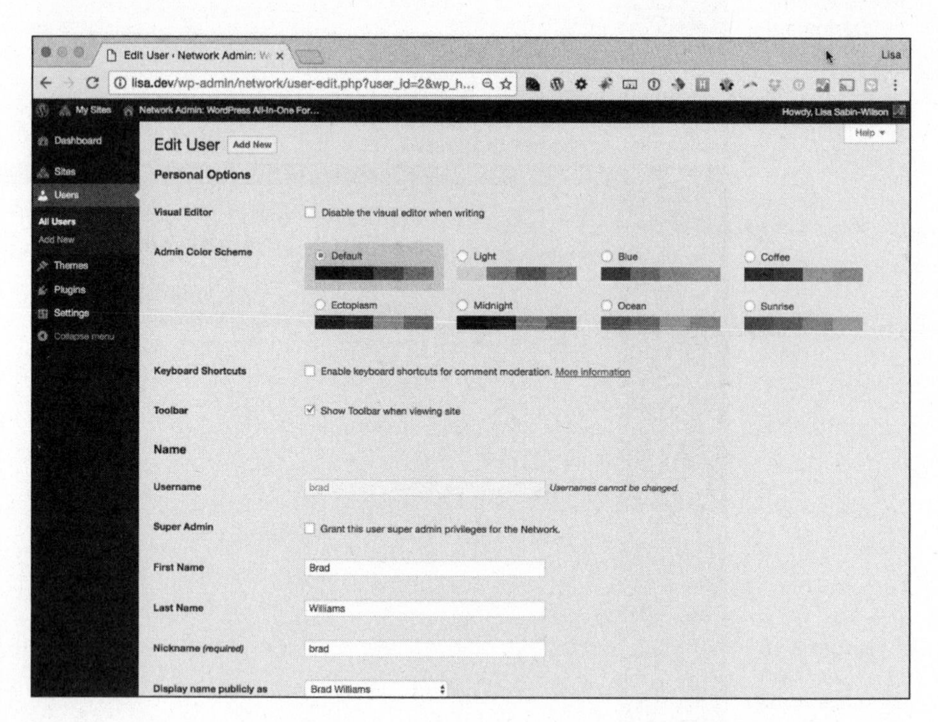

FIGURE 3-12: The Edit User page.

TIP

At this writing, the terms *Super Admin* and *Network Admin* are interchangeable. When WordPress first merged the WordPress MU (multi-user) code base with the regular WordPress software, the term used to describe the Network Admin was Super Admin. Right now, Network Admin is the standard term, but Super Admin is still used in some areas of both the Network Admin Dashboard and the regular Dashboard. This situation will mostly likely change in the near future, when the folks at WordPress realize the discrepancy and make the updates in later versions of the software.

Also on the Users menu of the Network Admin Dashboard, you see a link called Add New. Click that link to load the Add New User page on your Network Admin Dashboard (shown in Figure 3-13).

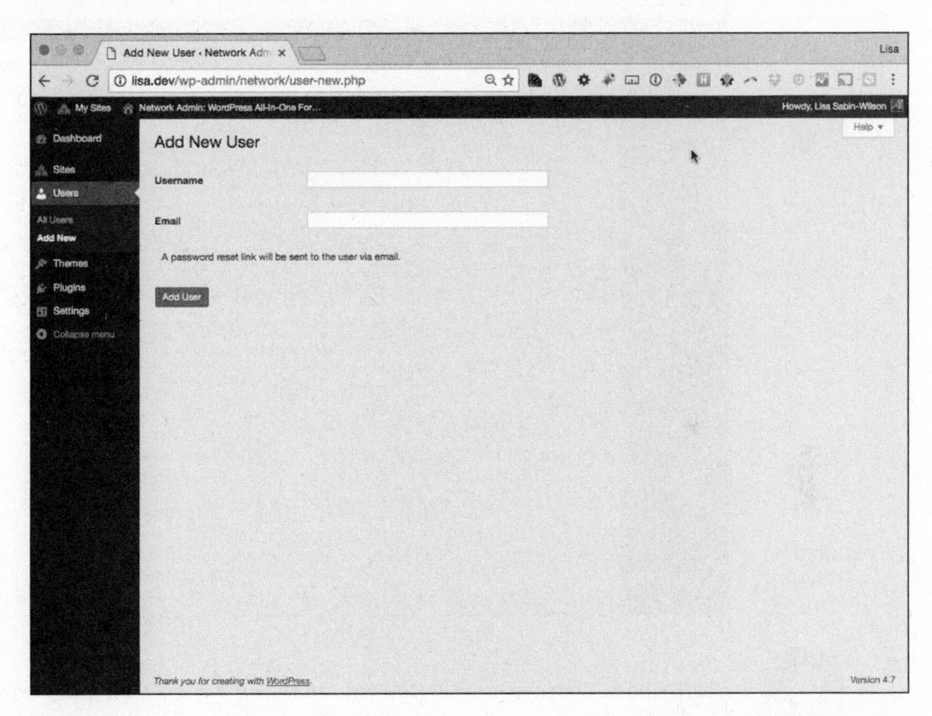

FIGURE 3-13: The Add New User page of the Network Admin Dashboard.

You can add a new user from the Add New User page by typing the username and email address of the user you want to add and then clicking the Add User button. The new user is sent an email alerting him to the new account, along with the site URL, his username, and his password. (WordPress randomly generates the password when the user account is created.)

Themes

When the multisite feature is enabled, only users who have network admin access have permission to install themes, which are shared across the network. For details on how to find, install, and activate new themes in your WordPress installation, see Book 6, Chapters 1 and 2. After you install a theme, you must enable it in your network; then the theme appears on the Appearance menu of each site, and users in your network can activate it on their sites. To access the Themes page (shown in Figure 3-14), click the Installed Themes link on the Themes menu of the Network Admin Dashboard.

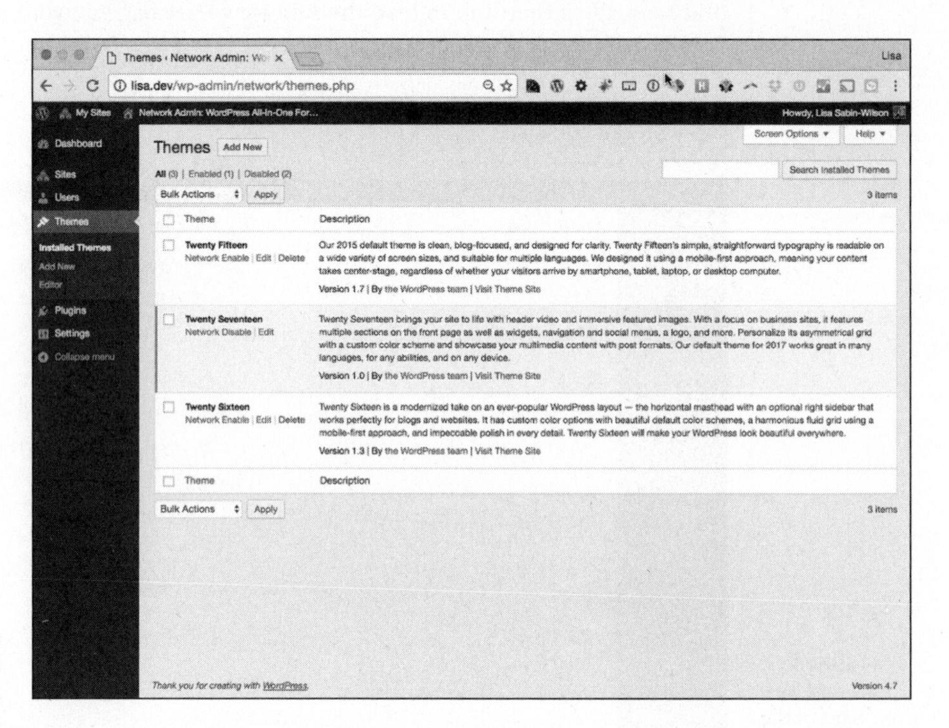

FIGURE 3-14: The Themes page.

TIP

See Book 8, Chapter 5 for details on enabling a theme on a per-site basis.

Plugins

Most WordPress plugins can work on your network. There are, however, some special plugins and some special considerations for using plugins in a network. (For details on finding, installing, and activating plugins in WordPress, see Book 7, Chapters 1 and 2.)

Browse to the Plugins page of your Network Admin Dashboard by clicking the Plugins link on the Plugins menu. The Plugins page is almost the same as described in Book 7, but if you don't know where to look, you can easily miss one very small, subtle difference. Check out Figure 3-15, and look below the name of the plugin. Do you see the Network Activate link? That link is the big difference between plugins listed on the regular Dashboard and those listed on the Network Admin Dashboard. As the Network Admin, you can enable certain plugins to be activated globally (across your entire network). All sites in your network can use the network-activated plugin's features. By contrast, plugins that you activate on the regular Dashboard (Site Admin) are activated and available only for *your* main website.

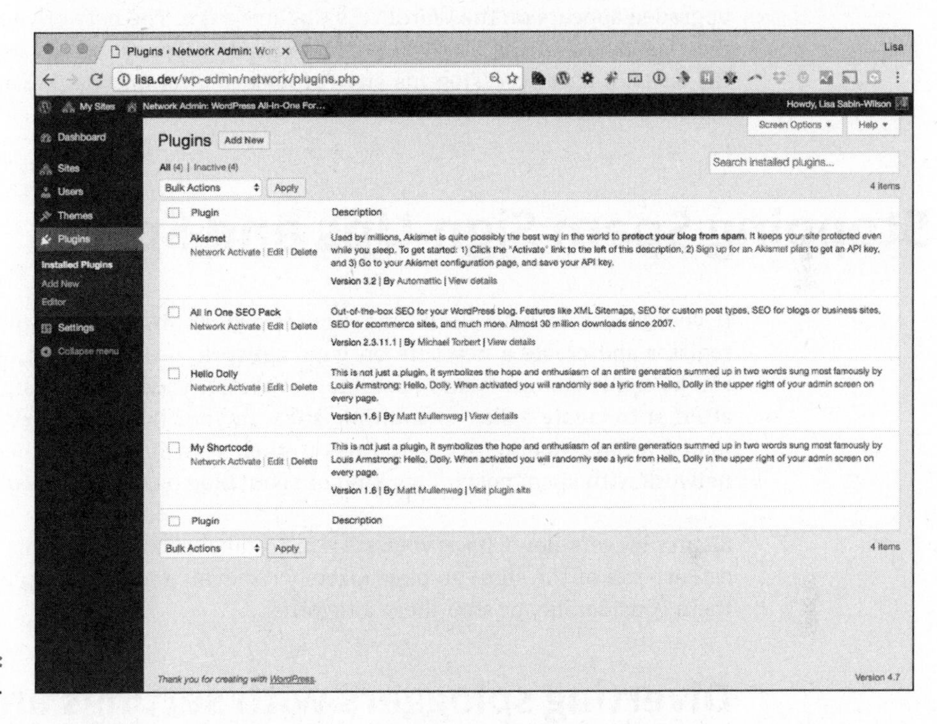

FIGURE 3-15: The Plugins page.

TIP

In the list of plugins, you see the plugins that you have network activated. From here you can activate and deactivate those plugins as you desire.

Network admins are the only people who can install new plugins on the site; regular users don't have that kind of access (unless you made them network admins in their User settings).

Also located on the Plugins menu of the Network Admin Dashboard are two other links: Add New and Editor. The Add New link takes you to the Add New Plugins page, where you can add and install new plugins by searching the WordPress Plugin Directory within your Dashboard (see Book 7, Chapters 1 and 2), and the Editor link gives you access to the Plugin Editor.

Updates

Clicking the Updates link on the Network Admin Dashboard gives you access to the WordPress Updates page, which takes you through the process of upgrading your WordPress installation (see Book 2, Chapter 6). For a network site, however, WordPress takes the extra step of upgrading all sites in your network so that they all use the upgraded feature sets.

If the process of upgrading network sites stalls or stops, the URL of the last site upgraded appears on the WordPress Updates page. The network admin can access the Dashboard of the site where the upgrade stopped, which usually clears up the issue. A user accessing his site Dashboard after an upgrade also triggers the update process.

Stopping Spam Sign-Ups and Splogs

If you choose to have open sign-ups, in which any member of the public can register and create a new site on your network, at some point, automated bots run by malicious users and spammers will visit your network sign-up page and attempt to create a site, or multiple sites, in your network. They do so by automated means, hoping to create links to their own sites or fill their sites on your network with spam posts. This kind of spam blog or site is a *splog.*

Spam bloggers don't hack your system to take advantage of this situation; they call aspects of the sign-up page directly. You can do a few simple things to slow them considerably or stop them altogether.

Diverting sploggers with settings and code

In the Registration Settings section of the Network Settings page (refer to Figure 3-4 earlier in this chapter), deselect the check box for the Add New Users setting to stop many spammers. When spammers access the system to set up a spam site, they often use the Add New Users feature to create many other blogs via programs built into the bots.

Spammers often find your site via Google Search for the link to the sign-up page. You can stop Google and other search engines from crawling your sign-up page by adding `rel=nofollow,noindex` to the sign-up page link. Wherever you add a link to your sign-up page, inviting new users to sign up, the HTML code you use to add the `nofollow,noindex` looks like this:

```
<a href="http://yoursite.com/wp-signup.php" rel="nofollow,noindex ">Get your
   own site here</a>
```

TIP

Add a link to any page or widget area to instruct legitimate visitors to sign up for a site in your network.

Preventing spam with plugins

Plugins can help stop spam blogs, too. The Moderate New Blogs plugin interrupts the user sign-up process and sends you (the network admin) an email notification that a user has signed up for a site. Then you can determine whether the site is legitimate. Download the plugin at `https://wordpress.org/plugins/moderate-new-blogs`.

The WangGuard/Sploghunter plugin was written mainly to stop sploggers and prevent spam signups on a WordPress site — with or without the network feature activated. You can get the plugin at `https://wordpress.org/plugins/wangguard`. This plugin is free for personal use or if you have less than 5,000 registrations or make less than $200 per month of income on your site.

The Cookies for Comments plugin (available at `https://wordpress.org/plugins/cookies-for-comments`) leaves a cookie in a visitor's browser. If the sign-up page is visited, the plugin checks for the cookie. If no cookie exists, the sign-up fails. Be sure to check the installation directions for this plugin, because it requires an `.htaccess` file edit.

IN THIS CHAPTER

» Understanding default user management

» Changing default user behavior

» Exploring user access to site features

Chapter 4

Managing Users and Access Control

In Book 8, Chapter 3, I discuss the Network Admin menu you have access to on your Dashboard to manage aspects of your network. In this chapter, I explain how to manage users across the network, including how to change some of the default management options to suit your needs.

One of the hardest things for new network admins to understand is that although each site is managed separately, users are global. That is, after a user logs in, he is logged in across the entire network and has the ability to comment on any site that has commenting enabled. (See Book 3, Chapter 2.) The user can visit the Dashboard of the main site to manage his profile information and access the Dashboard's My Sites menu to reach sites that he administers. The user also registers at the main site — not at individual sites in the network.

Setting Default User Permissions

When you enable the multisite feature, new site and new user registrations are turned off by default. You can add new sites and users from the Network Admin Dashboard, however. To let users sign up for your network, follow these steps:

1. **Log in to the Network Admin Dashboard and then click the Settings menu link.**

   The Settings page loads in your browser window.

2. **In the Registration Settings section, select the User Accounts May Be Registered option (as shown in Figure 4-1).**

   This setting allows users to register on your network. It also assigns them to the main site as Subscribers but doesn't allow them to create new sites.

3. **Click the Save Changes button at the bottom of the page (not shown in Figure 4-1).**

TIP

By selecting the User Accounts May Be Registered option on the Network Admin Settings page, you not only allow users to register a new account, but also give them the option to create a new site on your network.

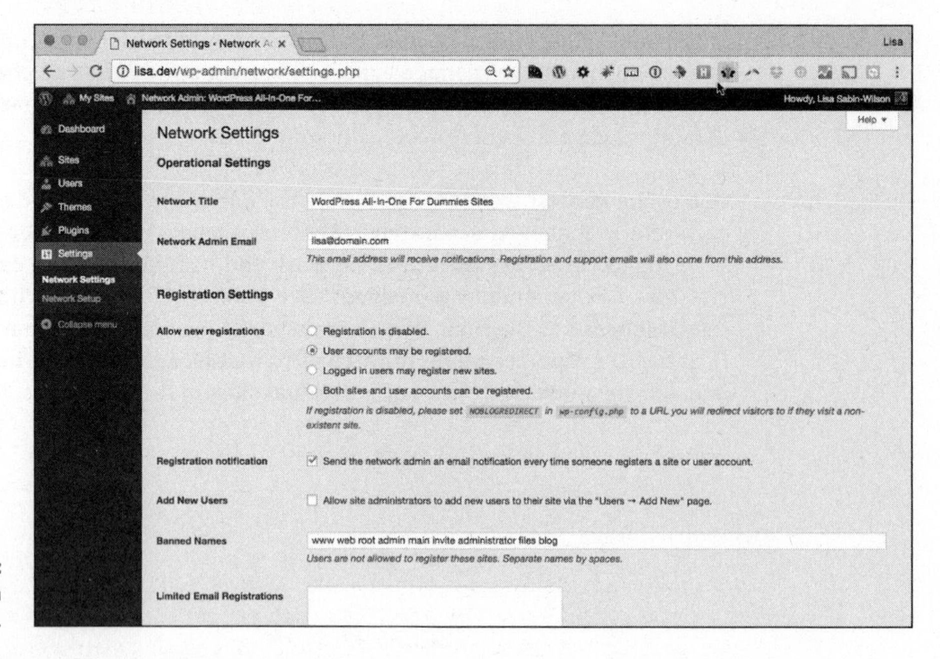

FIGURE 4-1: User registration options.

Registering users

When signing up, the user is directed to the main site of the installation and then added to one of the subsites. The subsite may be her site (if she chose to have a site when registering) or an existing site. If the site is any existing site other than the main site, you, as the network admin, must manually add the user to that site. The user who owns the site can manually add users as well if you enabled the option in Network Admin settings that allows site admins to add new users to their sites.

The registration page (see Figure 4-2) is located at `http://yourdomain.com/wp-signup.php`. This sign-up page bypasses the regular WordPress registration page. (See Book 3, Chapter 3.)

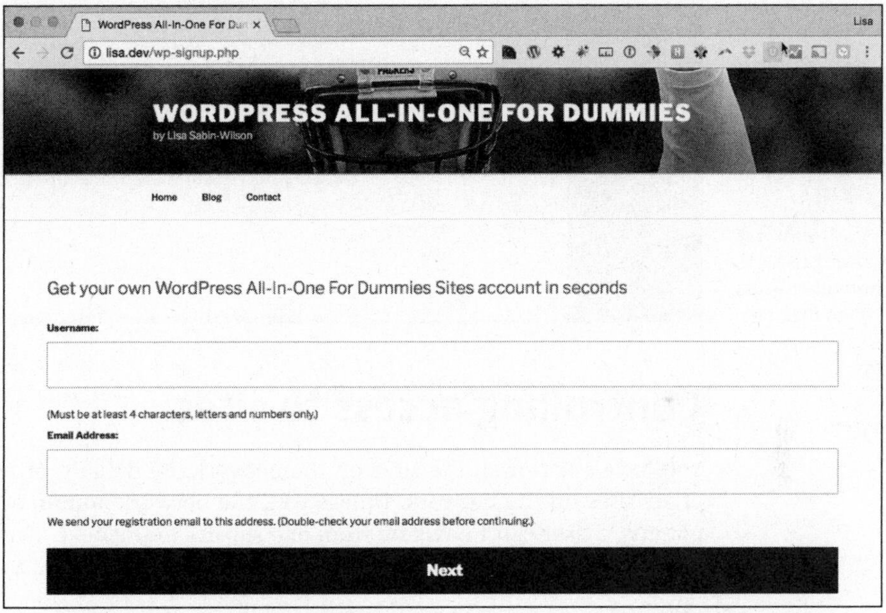

FIGURE 4-2: The network sign-up page.

After filling out the form, the user receives an email with a link to activate her account. When she does, she can immediately log in and manage her details; she's directed to her primary site, which is the main site or Dashboard site if she has no site to administer.

Users can also be added to existing sites in the network. You can always assign users to specific sites on a per-case basis. When you set up a network and enable the option titled Allow Site Administrators to Add New Users to Their Site via the Users → Add New Page (shown in Figure 4-3), you allow site admins to add other users in the network to their sites. Although the Add New Users setting is turned off by default, you can enable it by selecting the Allow Site Administrators option on the Settings page of the Network Admin Dashboard.

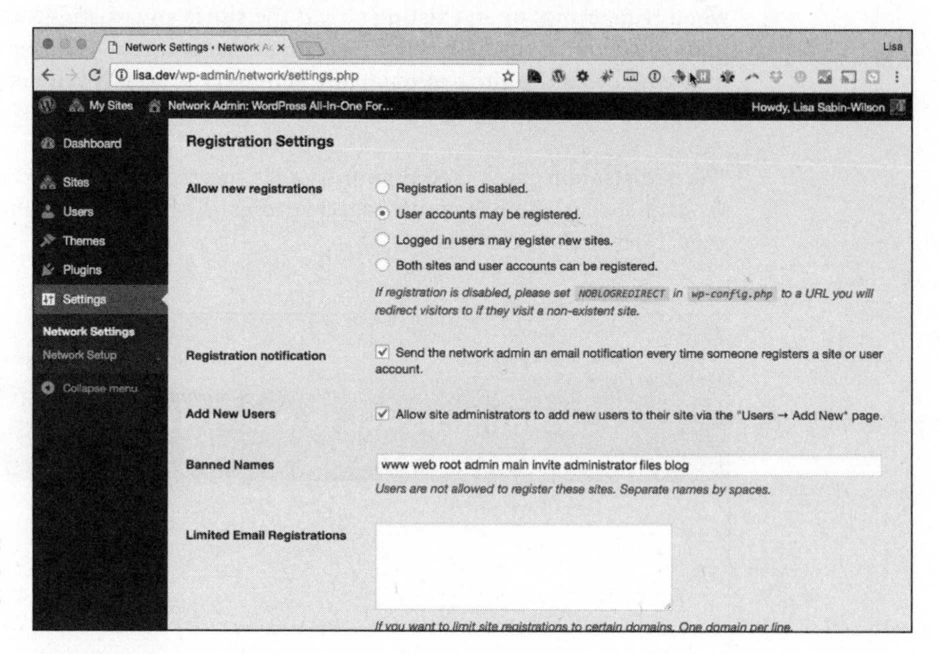

FIGURE 4-3: Allowing users to add other users to their site.

Controlling access to sites

You have a list of all the sites on the network; by default, other users can't find other sites on the network. Unless you, the network admin, add such ability via plugins, a user can't navigate from one subsite to the next. The only list provided to a user is the Dashboard's My Sites menu, shown in Figure 4-4.

The My Sites page lists only sites for which the user is administrator, not sites on which the user has a lesser role. Additionally, the My Sites menu has a link that permits the user to create more sites (if the network admin has allowed that function via the Settings menu of the Network Admin Dashboard).

By default, users can create no sites or an unlimited number of sites. You can limit the number of sites that a user can create by installing the Limit Blogs Per User plugin (`https://github.com/toddnestor/wp_limit_blogs_per_user`).

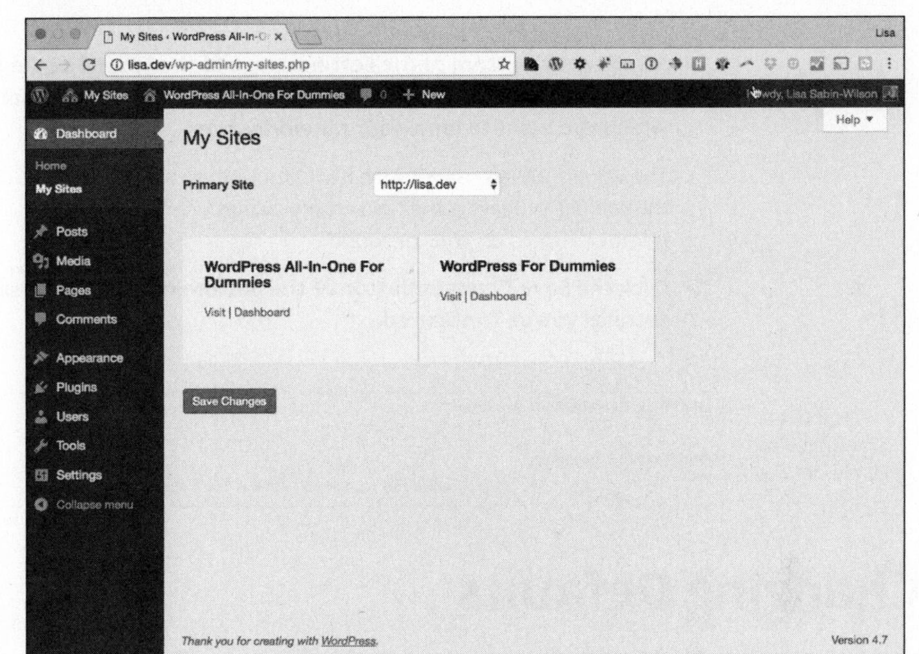

FIGURE 4-4: The My Sites page shows sites that the user administers.

Follow these steps to limit the number of sites that your users can create:

1. **Click the Network Admin link located in the top-left corner of your Dashboard, below My Sites.**
2. **Hover your cursor over the Plugins menu, and click the Add New link.**

   The Add Plugins page of your Network Admin Dashboard opens.
3. **In the Search field, type the name of the plugin:** Limit Blogs Per User.
4. **Be sure that the Keyword option is selected in the drop-down menu next to the search field so that WordPress knows to search by keywords (not by Author or Tag).**
5. **Click the Search Plugins button.**
6. **On the search results page, click the Install Now button for the Limit Blogs Per User plugin.**
7. **Click the Network Activate button on the Add Plugin page.**

   The Limit Blogs Per User plugin is now active on your network.
8. **Click the Settings link of the Network Admin Dashboard.**

9. **Scroll to the bottom of the Settings page to the Limit Blog Registrations Per User heading (shown in Figure 4-5), and enter the number of sites to which you want to limit your network users.**

   The value 1 allows users to create no more than one site, and so on. Entering the value 0, or leaving the field empty, allows users to create an unlimited number of sites on your network.

10. **Click the Save Changes button at the bottom of the page to save all the settings you've configured.**

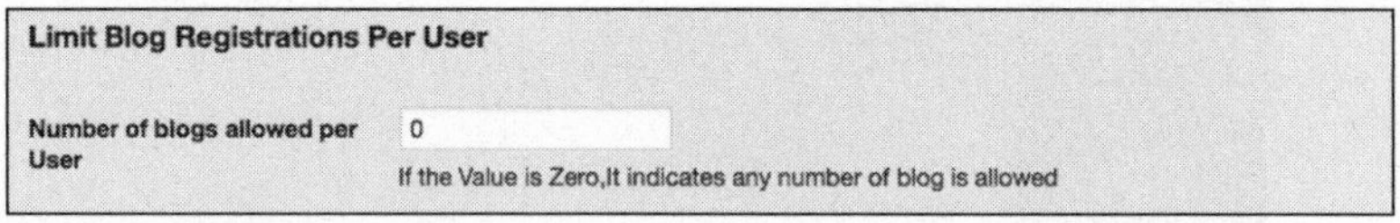

FIGURE 4-5: The Limit Blog Registrations Per User option.

Limit Blog Registrations Per User

Number of blogs allowed per User 0

If the Value is Zero,It indicates any number of blog is allowed

Changing Defaults

Depending on your specific needs, you may find yourself wanting to change how users are added to sites, as subscribers, within your network. By default, for example, users can't add themselves to a random network site without making the request to the network admin or the administrator of the site to which they want to be added. This setup may work fine for most sites, but if you want your users to be able to register with existing sites within your network, read on; these sections are for you.

Making sign-up site-specific

For many people, signing up on the main site and then needing to be added to a child site can be confusing. Plugins, however, can make the process easier and less confusing for everyone.

If you want existing users to add themselves to existing sites on the network, the Join My Multisite widget plugin (available in the WordPress Plugin Directory at `https://wordpress.org/plugins/join-my-multisite`) allows them to do so. Install this plugin as a regular plugin, as I outline in Book 7, Chapters 1 and 2.

When the network is activated, the plugin adds a widget on the Widgets page of every user's Dashboard. (Hover your mouse over Appearance and then click the Widgets link.) The new widget is called Join My Site. The user must drag the widget to the appropriate sidebar to display the Join My Multisite widget in the sidebar. (For details on using widgets, see Book 6, Chapter 1.) The site displays a welcome message on the sidebar and a Register for an Account button that users can click to register on the site. (See Figure 4-6.)

REMEMBER

If the user isn't logged in to the network, the welcome message is `If you want to add yourself to this site, please log in.`, so only users who are already network members and are logged in can add themselves to network sites by using the Join My Site widget.

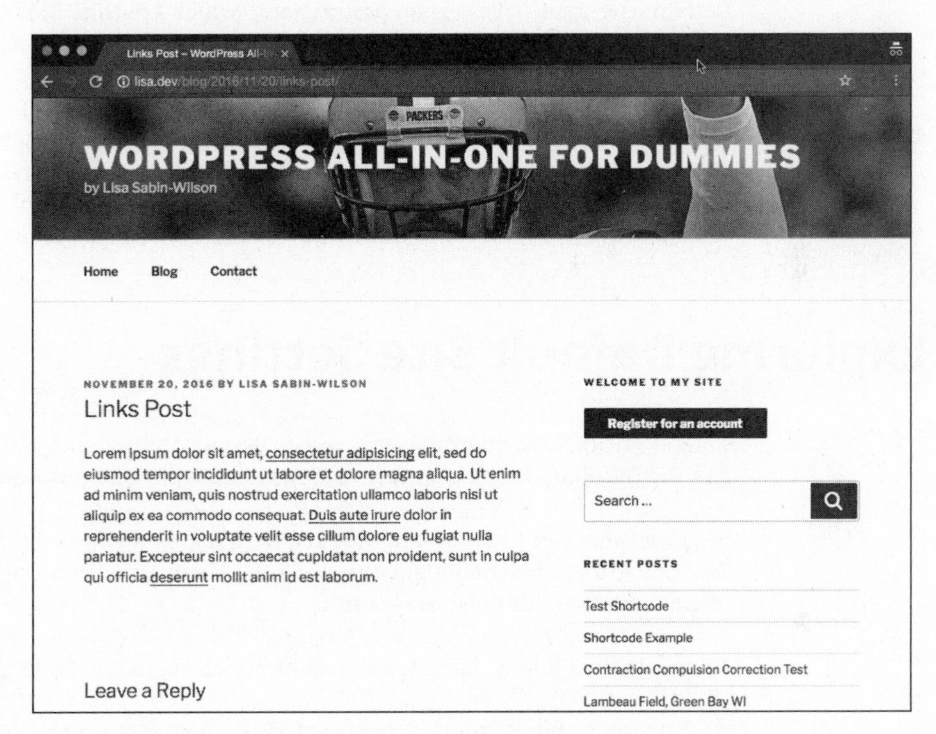

FIGURE 4-6: The Add Users widget on a site.

Changing roles on sign-up

When he's added to a network or a site, a user is assigned the role of Subscriber by default. You may want to assign a different role to the user and automatically add him to your other sites on the network. (Book 3, Chapter 3 explains roles and permissions.)

When a user signs up for his own site, for example, you may want to assign him a nonadministrator role. You may want to set his role to Editor to restrict the Dashboard menus he can access and to prevent him from using some of the functionality of WordPress. You may want to have new site owners sign up as Editors of the sites, a role that gives them fewer permissions on the Dashboard.

The Multisite User Role Manager plugin at https://wordpress.org/plugins/multisite-user-role-manager/ allows you to set a role other than administrator for new users who choose to have sites of their own. This plugin also allows you to set new user roles on other sites on your network (such as the default Subscriber role).

Locking down menus

Certain user roles have certain permissions (which I outline in Book 3, Chapter 3) that give users access to various menus of the Dashboard. You may want to close areas that you don't want users to access, however.

You can limit access to menus via the Menus plugin, available at https://wordpress.org/plugins/menus. Once installed, this plugin gives you a settings page to control which menus can be accessed by specific user roles. You find this settings page in the Network Admin Dashboard at Settings⇨Menu Settings.

Exploring Default Site Settings

Default settings can control user access to various things, such as menus, themes, and the Dashboard. The next few sections discuss the network settings in detail.

Because users can't add or edit plugins, the Plugins menu is disabled by default. You can still access the Plugins page via the Network Admin Dashboard Plugins menu, but other administrators can't.

To enable the Plugins menu for site administrators, follow these steps:

1. **On the Network Admin Dashboard, click the Settings menu link.**
2. **Scroll down to the Menu Settings section.**

   The check box for the Plugins menu is deselected, which means that users can't see the menu regardless of their roles.
3. **Select the Plugins check box to make the Plugins menu available to site administrators, as shown in Figure 4-7.**
4. **Save your selection by clicking the Save Changes button.**

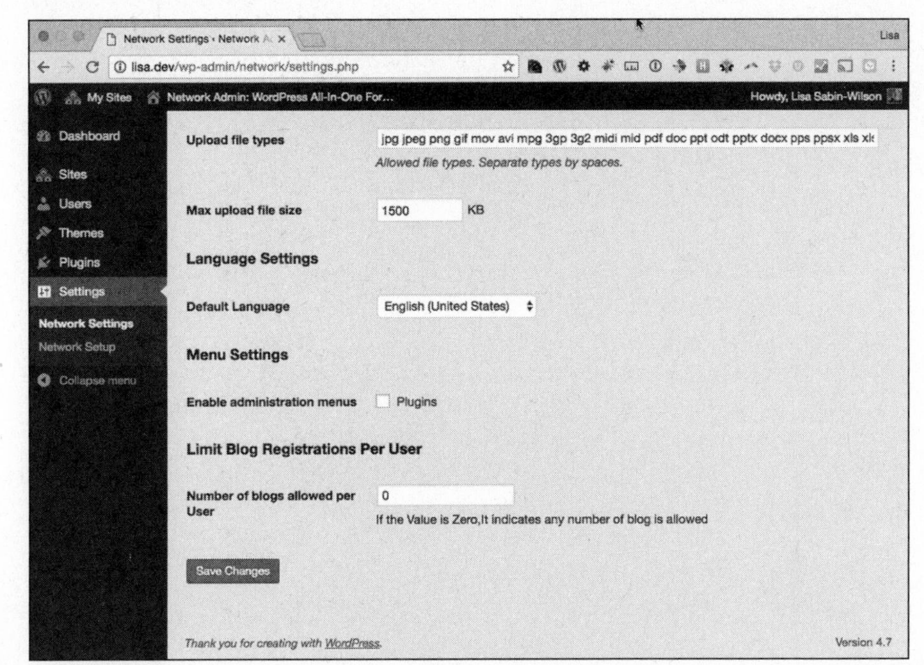

FIGURE 4-7: Enabling the Plugins menu for site administrators.

REMEMBER

Similarly, you must enable any themes installed on the network before a site administrator can choose the theme from the Appearance menu. I explain how to do so in Book 8, Chapter 3.

» Understanding networkwide plugins

» Finding networkwide and Must-Use plugins

Chapter 5 Using Network Plugins and Themes

When you add new plugins and themes to your WordPress installation, you add new functionality and aesthetics. You don't just multiply your choices, however; the possibilities become endless. You can gather and display information from across the network, for example, or have the same features available to everyone. You can choose to have the same theme on all sites or different themes. You can not only manage plugins and themes on a global level, but also have site-specific control.

In this chapter, I show you how certain functionality appears across the network and how certain plugins look by default on all sites for all users. I also cover controlling access to different themes for different sites.

One of the interesting features of a network is the extensive use of the `mu-plugins` folder. In this chapter, I describe exactly how this folder processes plugin code. I also cover the Network Activate link on the Plugins page, which is similar to the Activate link but has important differences.

REMEMBER

This chapter doesn't cover installing plugins and themes. I cover plugins in Book 7, Chapter 2 and themes in Book 6, Chapter 2.

Using One Theme on Multiple Sites

In certain situations (when you want consistent branding and design across your entire network, for example), each site in a network is used as a subsection of the main site. You could set up WordPress networks as a magazine-style design on your main site and populate the content with different posts from sites within your network, aggregating all the content to the main site. You can see an example on a site on the New York Times network site (`www.nytimes.com/interactive/blogs/directory.html`), which is shown in Figure 5-1.

The New York Times Blogs site is run by the New York Times, a media company in the United States. The goal is create a directory of blogs in the New York Times network, organized by topic. On the New York Times Blog site, all the titles link to different blogs within the site's network. The New York Times allows the administrators of its network sites to use different logos in the header of each site; however, each site uses the same basic theme so the branding across sites is consistent with the main Blogs site.

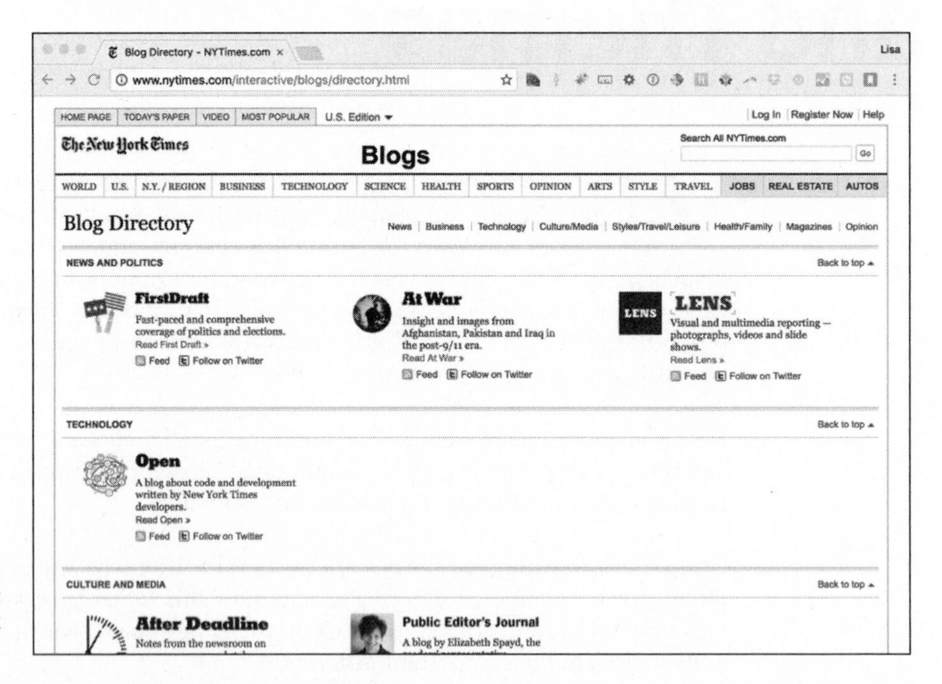

FIGURE 5-1: The New York Times Blogs site.

Although each site on the network operates separately from the main (network admin) site, you may want each site to look the same as the main site to tie into the main site visually, through design and branding, and to provide a consistent experience for visitors to all sites within a network. You may have a custom theme specially made for the main site, with added features to display networkwide content. If consistency and network branding are your goals, you may want to create a single theme that's used on all sites within your network (other than your main site).

REMEMBER

Book 6 discusses how WordPress accesses themes stored on the web server. When the network is enabled, these themes are shared among all sites and are available on the site administrator's Dashboard. If a change is made in a theme file, every site on the network using that theme experiences the change, because only one copy of the theme is being served. When a theme is enabled, it appears on the Manage Themes page of the administrator's Dashboard (which the users access by hovering the pointer over the Appearance menu on the Network Admin Dashboard and clicking the Themes link). Users can choose to activate this theme so that it displays on the front side of the site. You must activate a theme for use across all sites on a network by clicking the Network Enable link below the Theme name on the Network Admin Dashboard. (Access this Themes page by clicking the Themes menu link on the Network Admin Dashboard.)

The main network site could have 20 themes installed in the main WordPress installation, but if you haven't enabled them for use across the entire network, site administrators can't see network-disabled themes on their Dashboards and, therefore, can't use them on their sites.

DEFAULT THEME PLUGIN

A wonderful plugin called Default Theme adds a simple item to the Network Admin Dashboard, which you can access by clicking the Settings menu link. Simply, the Default Theme setting gives you the option to assign a default theme for new sites.

The Default Theme plugin isn't free, unfortunately. It's available from the development group at WPMU DEV at `https://premium.wpmudev.org/project/default-theme`. To access the plugin, you need to purchase membership on the site, but don't let that requirement deter you.

The WPMU DEV membership gives you access to hundreds of WordPress network-related plugins and themes for one monthly membership. I recommend it highly and feel that it's worth every penny it costs. After you have your WPMU DEV membership, you can begin downloading hundreds of plugins and themes for your WordPress network. You can purchase a membership by visiting `https://premium.wpmudev.org`.

If a consistent network design is what you're after, you'll run into a few troubles with the WordPress network because, by default, no matter what themes you've activated, the default WordPress Twenty Seventeen theme gets activated whenever new sites are created within your network. It would be nice for WordPress to provide a global setting in the Network Admin Dashboard that allows you to assign the default theme to every site on your network, but that currently isn't the case unless you want to edit some code in the WordPress configuration file (which I cover in "Setting the default theme for sites" later in this chapter).

Enabling themes for individual sites

You may have a customized theme for one member site that you don't want other sites within the network to use or have access to. As the network admin, you can edit each site on the network. You can perform some basic tasks, such as enabling or disabling themes, or add new themes to the network without leaving your own Dashboard. If you want to have a theme available for use on only one site, not available for other sites to choose, follow these steps:

1. **Click the Network Admin link on the My Sites menu (in the top-left corner of your Dashboard) and then click the Sites menu link.**

   The Sites screen appears, showing a list of all sites across the network, sorted by creation date, as shown in Figure 5-2.

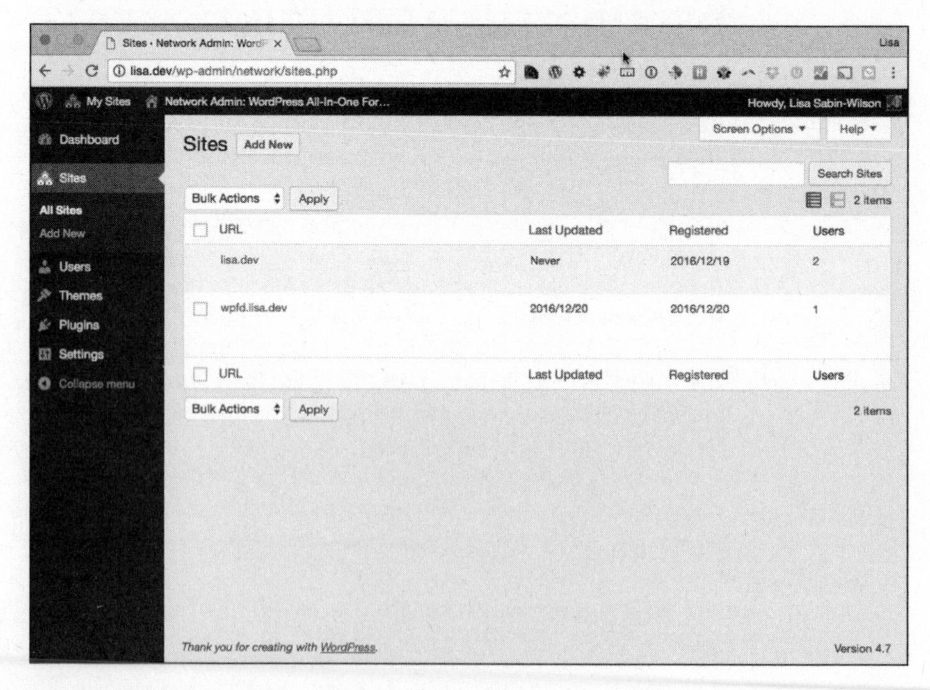

FIGURE 5-2: A list of sites on the network.

2. **Hover your cursor over the site you want to enable a theme for and then click the Edit link.**

   The Edit Site screen appears.

3. **Click the Themes tab.**

   The Edit Sites screen changes to show a list of themes that you can enable for the site you're editing. (See Figure 5-3.)

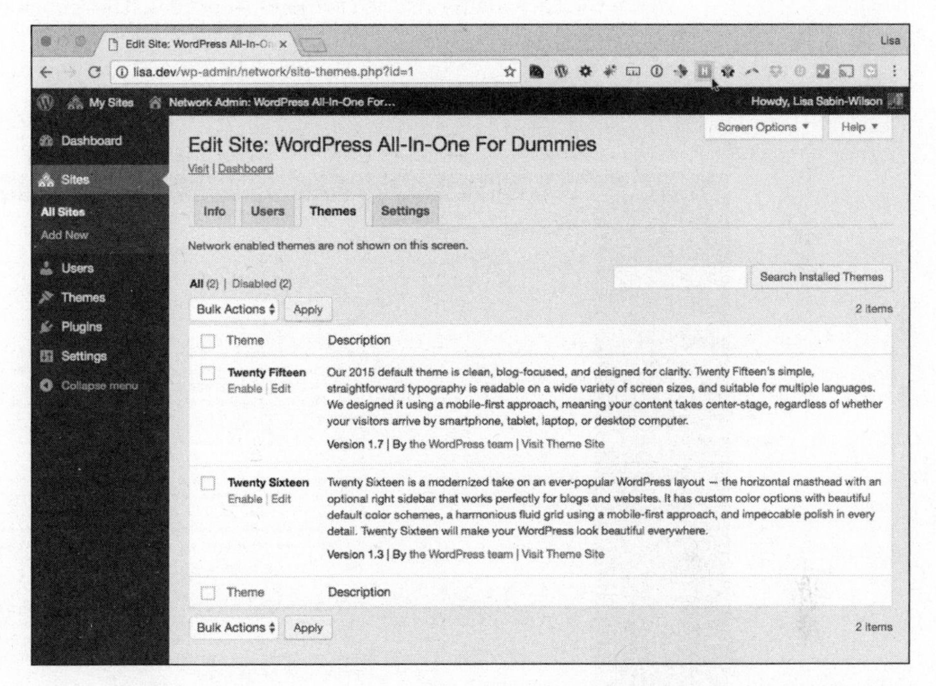

FIGURE 5-3: The Edit Site screen with the Themes tab active.

4. **Click the Enable link for the theme you want to enable on the site you're editing.**

   The Edit Site screen refreshes with the Theme tab still active and displays a message stating that the theme has been enabled. Your selected theme is now enabled on the site.

5. **Repeat these steps for any other sites on which you want to enable a theme.**

Installing themes for network use

Installing a theme for use on your network involves the same process you use to install a theme on your individual site (see Book 6, Chapter 2), but with an extra step: You have to enable each theme on the Network Admin Dashboard to activate it on the Appearance menu of the individual site administrators' Dashboards for sites within your network. Here's how to enable a theme so that all your site owners can use it on their sites:

1. **Click the Network Admin link on the My Sites menu (in the top-left corner of your Dashboard) and then click the Themes menu link.**

   The Themes screen appears, displaying a list of installed themes, as shown in Figure 5-4. Each theme installed in the `/wp-content/themes` folder is listed on this screen.

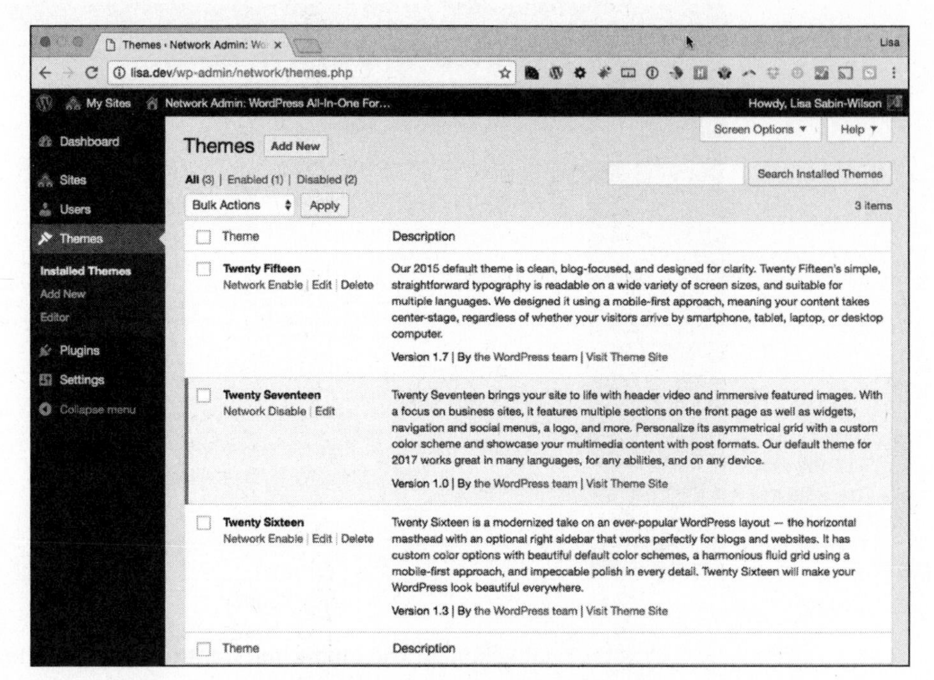

FIGURE 5-4: A list of themes on the network.

2. **Click the Network Enable link for the theme you want to use.**

   Enabling a theme on the Themes screen causes it to appear in the list of available themes on each network site's Dashboard (but doesn't change any user's active theme; it merely makes the theme available for use).

3. **Repeat these steps to enable more themes on your network.**

REMEMBER

Installing a new theme in your main WordPress installation doesn't mean that it's available for use networkwide. As the network admin, you always have to enable the theme before your site owners can use it.

Setting the default theme for sites

When a new site is created on the network, by default, it displays the Twenty Seventeen theme provided within WordPress. If you want to use a different theme as the default for all new sites created, add a `define` statement to the `wp-config.php` file of your WordPress installation. (Check out Book 2, Chapter 5 to familiarize yourself with the `wp-config.php` file you're modifying in this section.)

Install your theme on the server, as I outline in Book 6, Chapters 1 and 2. You may also want to enable the theme networkwide, as outlined in the preceding section. This step isn't necessary, but if you have other themes available, and if the active theme is disabled, a user who switches away from that theme won't be able to switch back to it.

Because the Twenty Seventeen WordPress theme is already the default, use the theme from 2016 called (funnily enough) Twenty Sixteen as the theme you want to set as the default for all sites within the network. Follow these steps:

1. **Log in to your web server via SFTP.**

   See Book 2, Chapter 2 for details on using SFTP.

2. **Open the `wp-config.php` file in your favorite text editor.**

   See Book 2, Chapter 5 for details about where you can find the `wp-config.php` file on your web server.

TIP

   Save a copy of your original `wp-config.php` file to your desktop before editing it in case you make any mistakes or typos in the next few steps.

3. **Locate the following line of code in the `wp-config.php` file:**

```
define ('WPLANG', '');
```

   You can find this line toward the bottom of the file; scroll until you locate it.

4. **Add a new blank line below it.**
5. **Type** define('WP_DEFAULT_THEME', 'twentysixteen');.

   This one line of code tells WordPress to use the Twenty Sixteen theme as the default theme for all new sites within your network.

6. **Save the `wp-config.php` file, and upload it to your web server.**

   The `twentysixteen` in quotes refers to the theme's folder name on the web server. The name within the quotes should be identical to the name of the folder where the theme files reside. All new sites created now display the Twenty Sixteen theme.

Gathering and Displaying Networkwide Content

Depending on your needs, you may want to gather content from sites across your network to display on the front page of the main site (as the New York Times Blogs site does). Although some plugins can do this for you, you can accomplish the same thing by placing a few lines of code in your theme template file.

The main page of your network is controlled by the theme that's active on the Themes page of your regular Dashboard (which you access by hovering your pointer over Appearance and clicking the Themes link). You can customize this theme with some code samples in the next section to suit your particular needs.

Adding posts from network sites

One of the best ways to pull visitors to your site is to display a short list of headlines from posts made on other sites within your network. For a single WordPress site, the Recent Posts widget can handle this task. When you're running a network, you have no built-in way to pull a list of posts from across all the sites in your network. But the Network Posts Extended plugin, available in the WordPress Plugin Directory at `https://wordpress.org/plugins/network-posts-extended`, can do this job for you quickly and efficiently. The plugin includes a handy widget that makes it easy for you to add recent posts from across your network of sites to your main website.

Listing network sites

To list all the sites in the network, use the Multisite Directory plugin, available free from the WordPress Plugin Directory at `https://wordpress.org/plugins/multisite-directory`. You install this plugin just as you do any other plugin in WordPress; see Book 7, Chapters 1 and 2 for information on installing WordPress plugins.

To use the features that this plugin provides, create Categories in the Sites menu in the Network Admin Dashboard under Sites⇨Categories; then assign each site within your network to a category. If you're running a network of sites about books, you might create categories of genres such as Mystery, Romance, and Biography. Then you can assign each site in your network to the appropriate genre, or category. When you use the Multisite Directory widget, included in the plugin, you can display a directory of sites in your book network by genre to help your visitors find the sites that interest them most.

Displaying most-commented posts

When you're running multiple sites, you may want to display a listing of the most-commented posts across the network. The WP Multisite Most Commented Posts widgets plugin (available at `https://wordpress.org/plugins/wp-multisite-popular-posts`) lets you do just that.

Install the plugin as outlined in Book 7, and click the Network Activate link. Hover your pointer over Appearance, and click the Widgets link on the Dashboard of your site to load the Widgets page. A new widget is added, called WP Multisite Popular Posts.

If you drag this widget to the sidebar of your choosing, it displays a list of the most popular posts on every site across the network. With this plugin, popular posts are determined by the posts that have the greatest number of comments. If you expand the WP Multisite Popular Posts widget, you see the following configurable options (shown in Figure 5-5):

- **Widget Title:** This title is displayed on your site, above the widget information.
- **Select Max Comments:** This option lets you set 1 to 10 comments.
- **Select Time Frame:** This option sets the time frame for the posts. You can display the most popular posts of all time, for example, or just the ones posted over the past month.
- **Show Number of Comments:** Select this box to tell the widget to show the number of comments left for each post displayed.
- **Show Total Posts:** Select this box to tell the widget to display the total number of posts on your network.

REMEMBER

Be sure to click the Save button within the widget to save all the settings you create here.

Using sitewide tags and categories

The WordPress MU Sitewide Tags plugin pulls information from every new post on each site and reproduces it on a site that the plugin creates on your network. This site, called Tags Blog, aggregates the posts from every site. You may also set Tags Blog as the main site of your network so that all new posts appear on the front page.

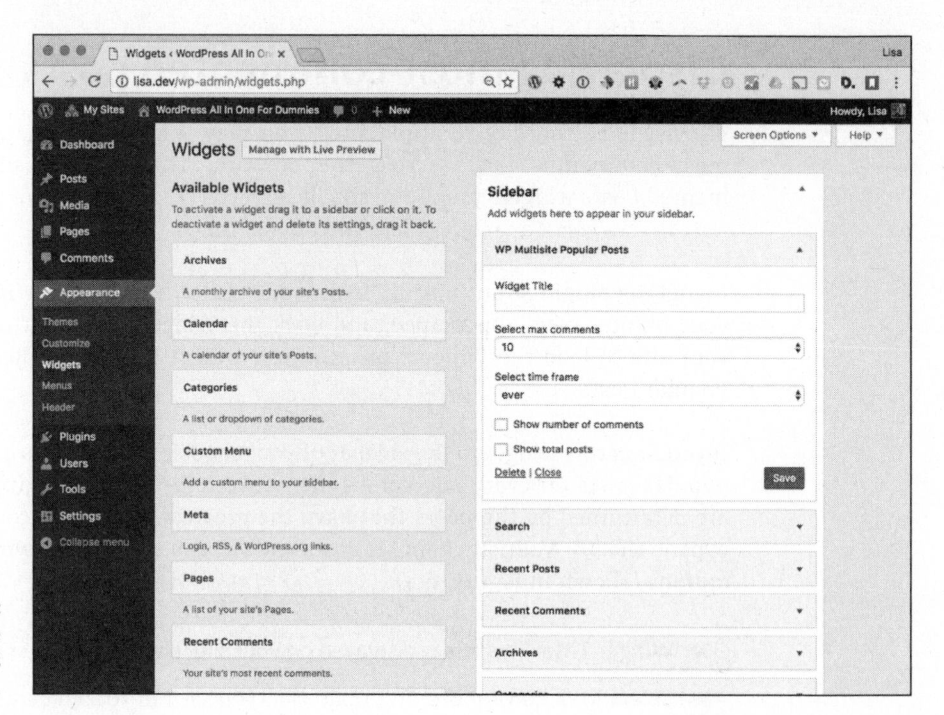

FIGURE 5-5: The WP Multisite Popular Posts widget options.

The plugin pulls and reposts almost all content from a post on another network site, including the title, content, tags, categories, and author information. At this time, however, it doesn't pull the post thumbnail or the comments.

By default, this plugin is set to create a new site called Tags Blog and saves up to 5,000 posts before it starts to remove older ones. Each post it aggregates retains its original permalink (the full URL to the post) and all post meta information (the post's original author, the date and time it was published, and any categories and tags assigned to it).

All the aggregated posts from across the network are saved and published on one site; the posts on the Tags Blog get displayed on one page as if the network posts were posted to a single site. Each new post is saved to the Tags Blog site when it's created. Then users or visitors can search the Tags Blog site and see tags from across the network, search aggregated posts across the network, and see posts by all network authors, among other things. The possibilities for networkwide aggregation and display are endless.

REMEMBER

Because the original permalink of the post is retained, search engines don't read these aggregated posts as duplicate content; therefore, each site retains any page rankings or search engine optimization (SEO) juice.

In order to install the plugin, you can can find and download the Sitewide Tags plugin at `https://wordpress.org/plugins/wordpress-mu-sitewide-tags` to install manually on your site, or you may install it by using the built-in plugin installer in your WordPress Dashboard.

After you install the plugin, activate it on the network by moving it to the `mu-plugins` folder (discussed later in this chapter in "Using and Installing Networkwide Plugins"). When you move the plugin to this folder on the server, it's activated automatically and starts working.

When the plugin is active, you need to enable it. Visit the Global Tags page on your Network Admin Dashboard (click the Sitewide Tags link on the Settings menu). Figure 5-6 shows the Global Tags page, where you enable the Tags Blog on your network by selecting the Enabled check box on the Global Tags page.

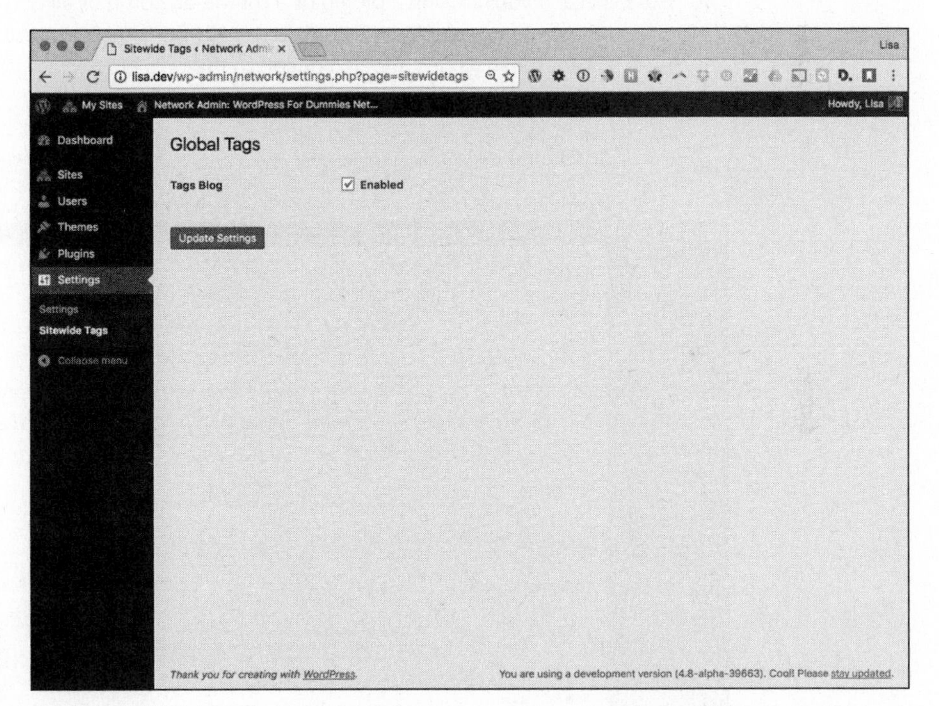

FIGURE 5-6: Enabling Tags Blog.

After you click the Update Settings button on the Global Tags page, the page refreshes. You see that the Global Tags section has changed and provides new options (see Figure 5-7), including the following:

- **Tags Blog:** In the text field, type the name of the blog to which you want to aggregate all posts on the network, or select the option to aggregate all the posts to the main blog.

» **Max Posts:** This option is set to 5,000 by default. Beyond that number, the plugin automatically deletes older posts from the Tags Blog site.

» **Include Pages:** Select this box to enable the option, which includes any pages that users create on their network sites and pushes those pages to the assigned tags blog.

» **Include Post Thumbnails:** Select this option to have WordPress include thumbnail images from across the network.

» **Privacy:** This option determines whether the Tags Blog site can be indexed by search engines.

» **Non-Public Blogs:** When enabled, this feature aggregates posts from sites that have changed their privacy settings (hover the pointer over Settings, and click the Privacy link) to nonpublic.

» **Post Meta:** If you're using a plugin or a theme on some or all of the network sites that create custom files, enter the specific field names in use on your site here so that those values are also pulled to the tags blog.

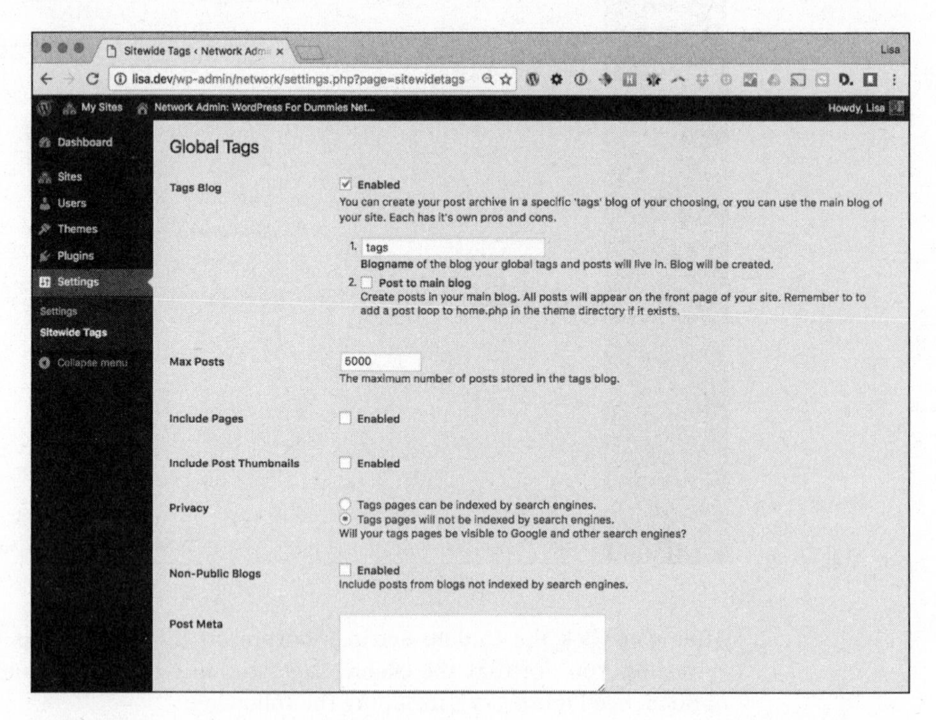

FIGURE 5-7: Global Tags options.

Using and Installing Networkwide Plugins

Networkwide plugins perform an action globally on all sites on the network. Because you're working with one code base, you need only one copy of a plugin. All sites within the network use the same copy.

When you have a single installation of WordPress on a single site, the Activate link on the Plugins page turns on that plugin for that site. (See Book 7, Chapters 1 and 2.) When you have multiple sites on a network using the Multisite feature, the Activate link works the same way for the site on which you activate the plugin.

You see a Network Activate link on the Plugins page of the Network Admin Dashboard, which activates the plugin on all sites on the network. This option simply turns on the plugin for all network sites; it doesn't allow you to manage plugin options globally unless the plugin itself is coded to do so. You can see a list of network-activated plugins on the Plugins page of the Network Admin Dashboard (which you access by clicking the Plugins menu link).

REMEMBER

Any changes made in this copy affect every site within your network.

A special breed of plugins — the Must-Use plugins — gets installed in the `/wp-content/mu-plugins` folder on your web server. Any plugin file placed inside this folder runs as though it were part of WordPress. The plugins in this folder execute automatically, without the need for activation from your Dashboard.

WARNING

In fact, you *can't* access the files in this folder from the WordPress Dashboard. If you use the Install Plugins page (hover your pointer over Plugins, and click the Add New link) to find and install a Must-Use plugin, you may be required to move the plugin files from the `plugins` folder to the `mu-plugins` folder. The plugin's `readme.txt` file always states whether the plugin needs to be moved into the Must-Use (`mu-plugins`) folder.

Generally, plugins placed in the `/wp-content/mu-plugins` folder are for networkwide features or customizations that users can't disable. An example is a custom-branded login page on each site on your network. If a plugin design adds a new menu item, the menu item appears as soon as the plugin is placed in the `/wp-content/mu-plugins` folder, without further need for activation on the Dashboard.

Not all plugins placed in the `/wp-content/mu-plugins` folder appear in the plugins list (hover your pointer over Plugins, and click the Plugins link), because not all of them require activation.

Place the main Must-Use plugin file in the /wp-content/mu-plugins folder, not in a subfolder. If multiple files exist within the plugin, some plugins use a file with a command to include the subfolder so that the code executes.

REMEMBER

You still control plugin settings on a per-site basis; you must visit the back end of each site if you want to alter any settings provided by the plugin.

Here's how to create the /wp-content/mu-plugins folder and install a network-wide plugin:

1. **Connect to your web server via SFTP.**
2. **Navigate to the /wp-content folder.**

   You see the subdirectories plugins and themes.
3. **Using your SFTP program, create a mu-plugins subdirectory.**

   Most SFTP programs allow you to right-click and choose to add a new folder.
4. **Upload the plugin file — not the plugin folder — to the /wp-content/mu-plugins folder on your web server.**

   The plugin immediately runs on your installation. Generally speaking, the only plugins that go in this folder are ones in which the plugin's instructions (typically located in the readme.txt file) explicitly state to put it in the /mu-plugins folder.

Discovering Handy Multisite Plugins

You can find multisite plugins that take advantage of WordPress's multisite functionality in the WordPress Plugin Directory at https://wordpress.org/plugins. Usually, multisite plugins are tagged with certain keywords that help you find them, such as *wpmu, wordpressmu, multisite,* and *network.*

When you click the linked tags for these terms in the directory, you're taken to a page that lists the related plugins, such as https://wordpress.org/plugins/tags/multisite or https://wordpress.org/plugins/tags/network.

Additionally, you can find more plugins by searching for them in search engines and by asking about them on the WordPress Forums page (https://wordpress.org/support).

» Understanding how to map domains

» Mapping many domains

» Establishing multiple domains

Chapter 6
Using Multiple Domains within Your Network

With a network of multiple sites readily available in WordPress, many people have expressed the desire to run multiple sites on their own separate domain names through one installation. Before the network feature was added to the WordPress software, you could run only one site per installation of the software. Now it's possible to run several sites under one installation of WordPress by activating the network feature, which I discuss in Book 8, Chapters 4 and 5.

In this chapter, I discuss using multiple domains and a feature called *domain mapping,* which enables you to run not only multiple sites, but also multiple sites with unique domain names that are not tied to the main.

Note: To tackle this chapter, you need to understand domains (Book 2, Chapter 1) and domain name server (DNS) records.

Finding Your Way with Domain Mapping

Domain mapping means telling your web server which domains you want WordPress to answer to and which site you want visitors to see when they request that domain. This process is more than domain forwarding or masking, because the

URLs for your posts have the full domain name in them. Instead of the child site's being in `secondsite.`*yourdomain*`.com` format, it can be *myotherdomain*`.com`.

Domain mapping isn't possible in certain instances, however. If your WordPress install is in a subfolder, and this folder is part of the URL, any mapped domain also contains this folder name. In that case, it would be better to move the installation so that it isn't in a subfolder.

You also need to access your web host's control panel (where you manage DNS records on your web server) and the control panel for your domain name registrar, which may be a different company.

REMEMBER

The network install, by default, lets you choose between a subfolder setup and a subdomain setup. This step is still required before you can specify a domain for that site. I cover how to enable the network in Book 8, Chapter 2. Be sure to set up the network and ensure that it's functioning properly before you attempt to map domains.

Adding domains

You need to set up your web server to accept any incoming requests for the domain you want to map and the location to send them to. I use the cPanel control panel (Book 2, Chapter 2) in this section (with the HostGator hosting provider; your cPanel may look a little different) because it's quite popular and available on many web hosts. On cPanel-based web hosts, this task is referred to as *domain parking.*

Follow these steps to add on a domain on your web hosting account via cPanel:

1. **Log in to your website's cPanel.**

   The address is provided by your web host and usually available at `http://yourdomain.com/cpanel`.

2. **In the Domains section, click the Addon Domains icon, shown in Figure 6-1.**

   The Addon Domains page displays in your browser window and lists any domains you have parked (if you previously parked any) and provides a form where you can enter a new domain.

3. **In the Create an Addon Domain section, enter the domain name you want to map.**

   The domain is directed to the root folder of your website, which is where your WordPress installation should be located. If it isn't, follow the steps in the next section.

4. **Click the Add Domain button.**

   The screen refreshes and shows a message saying that your add-on domain has been created. (See Figure 6-2.)

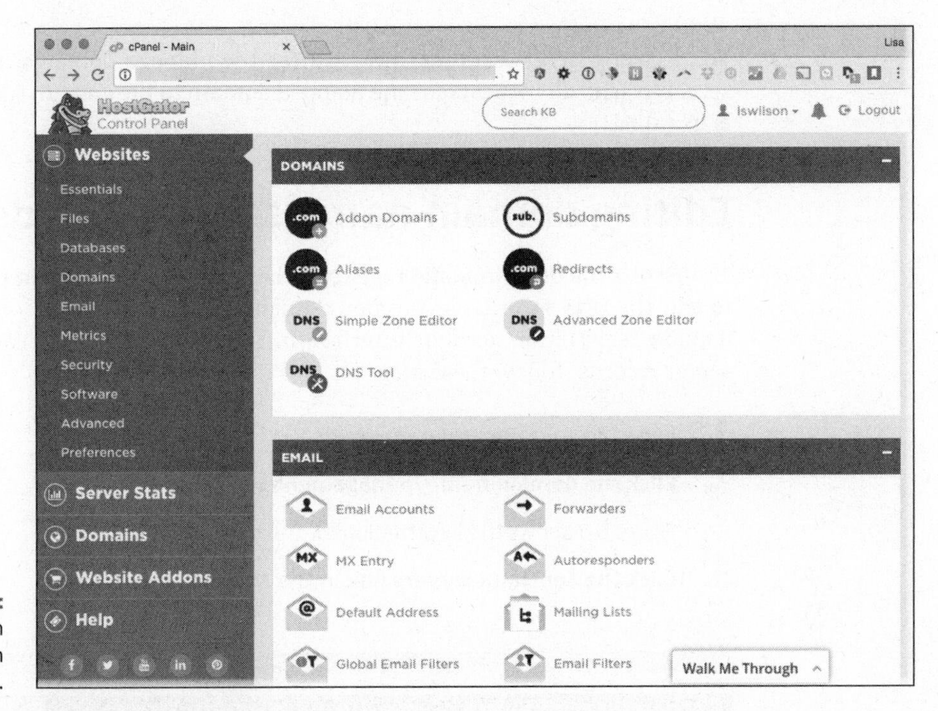

FIGURE 6-1: The Addon Domains icon in cPanel.

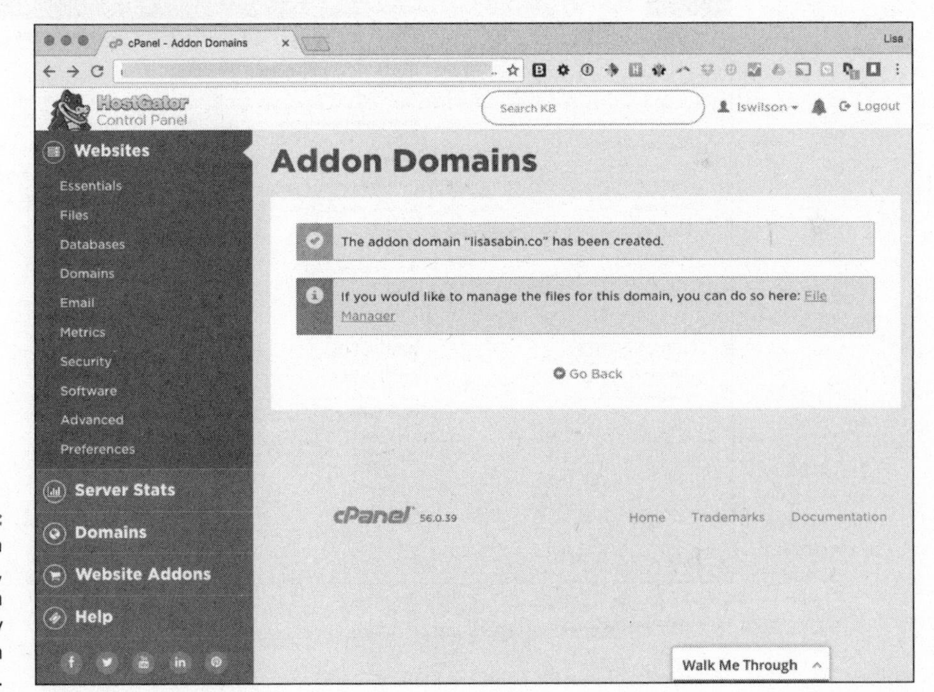

FIGURE 6-2: The Addon Domains page, showing a successfully created add-on domain.

You're using a `ServerAlias` directive for the mapped domains, telling the web server to send all requests for the mapped domain to the domain where WordPress is installed.

Editing domain name server records

To instruct the domain name registrar where to send the domain name, you need to edit the DNS records. A common domain name registrar is GoDaddy; I use its domain registration account interface in the following steps. To edit the name server records, follow these steps:

1. **Log in to your domain name registrar.**
2. **Click the domain name management tools.**

   Figure 6-3 shows the information for the domain to map.
3. **Click the Set Nameservers link in the Nameservers section.**

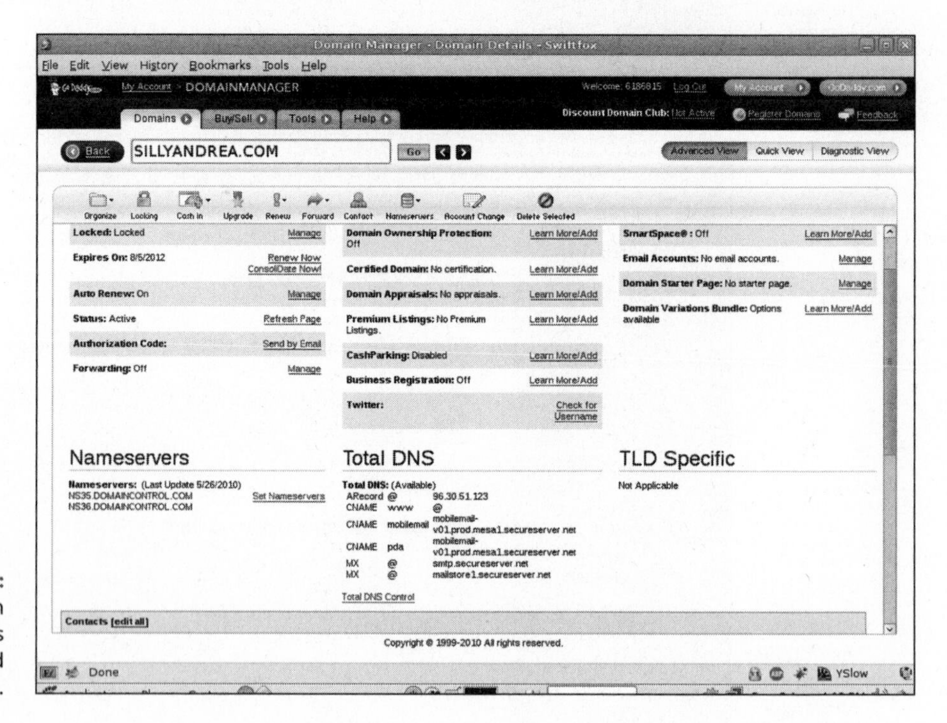

FIGURE 6-3: The domain name records of a mapped domain.

4. **Type the name servers for your web host where your WordPress install is located and then save your changes by clicking the Save Changes button.**

   The servers around the world now know that your domain "lives" at this web server location. Name server changes may take up to 24 hours to propagate across the Internet.

Installing the Domain Mapping Plugin

Before you can add your mapped domains to WordPress, you need to install the WordPress MU Domain Mapping plugin to help handle this task in WordPress. The Domain Mapping plugin doesn't do any setup on the server side; it helps rename the site and takes care of any login issues. To install this plugin, follow these steps:

1. **Download the plugin from `https://wordpress.org/plugins/wordpress-mu-domain-mapping`.**
2. **Unzip the plugin on your local computer.**

   Inside are two PHP files: `domain-mapping.php` and `sunrise.php`.
3. **Open your SFTP program, and navigate to your website's `wp-content` folder.**

   For details on how to do this, see Book 2, Chapter 2.
4. **Upload the `sunrise.php` file directly into the `/wp-content` folder.**
5. **Download a copy of your `wp-config.php` file by using your SFTP program.**
6. **Open `wp-config.php` on your computer with a text editor, and add the following line below the `define( 'MULTISITE', true );` line.**

```
define( 'SUNRISE', 'on' );
```

7. **Save the file, and upload it to your website.**

   The plugin is immediately available (and running) on your network. All you need to do is set up the options and map a domain to a site. Two new items are added to the Super Admin menu: Domain Mapping and Domains. On the user administrator side, a new Domain Mapping item appears on the Tools menu.

REMEMBER

The network admin needs to activate domain mapping on the Domain Mapping page (by choosing Network Admin⇨Domain Mapping) before a user can map a domain by enabling the Domain Mapping feature.

Obtaining your IP address

An *IP address* is a number assigned to every website and computer connected to the Internet. This number is used in domain mapping to help direct Internet traffic to the appropriate site on your network. You can find the IP address of your website three ways:

- Your web host provider can tell you.
- The address may appear within the web host's control panel.
- You can visit an IP lookup website. Such websites can tell you the IP of your website when you provide your domain name.

To find your address on an IP lookup website, follow these steps:

1. **Visit the handy online IP checker located at `http://ipinfo.info/html/ip_checker.php`.**
2. **Enter the domain name of your website and then click Go.**
3. **Write down the IP address that appears.**

   Figure 6-4 displays the IP address of my business site, `https://webdevstudios.com`.

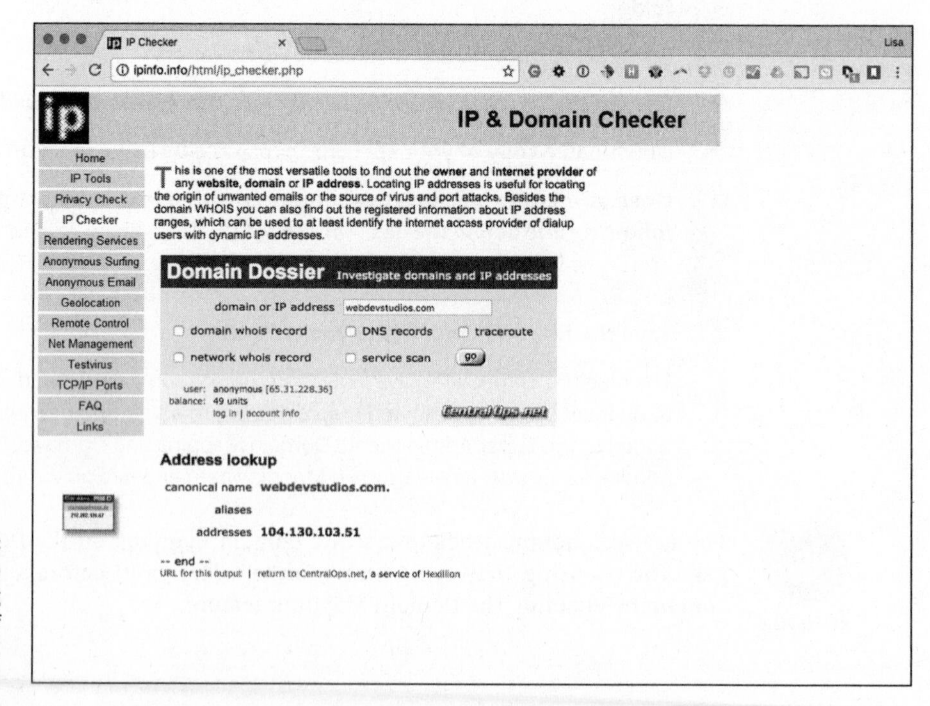

FIGURE 6-4: IP address lookup record revealing the IP address of webdevstudios.com.

4. **On your WordPress Dashboard, choose Network Admin➪Domain Mapping, enter your IP address, and click Save.**

Mapping a domain to a site

To map a domain to a site in your network, here's what you need to do:

1. **Navigate to the Network Admin Dashboard.**
2. **Click the Domain Mapping link on the Settings menu.**

   The Domain Mapping Configuration screen appears, as shown in Figure 6-5.

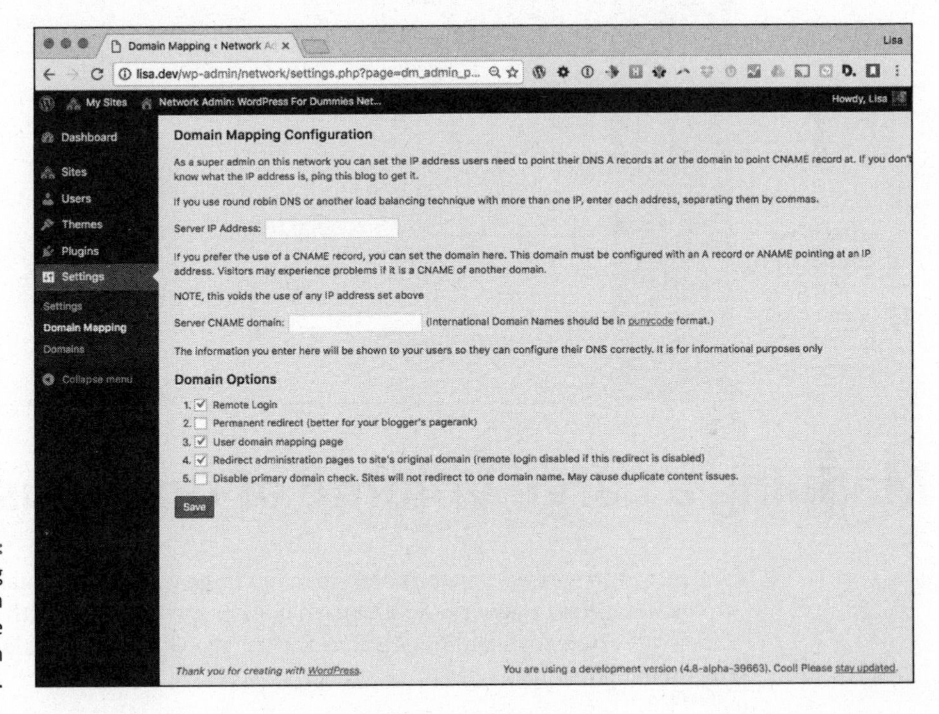

FIGURE 6-5: Domain Mapping Configuration screen of the Network Admin Dashboard.

3. **Enter the IP address of your web server.**
4. **Leave the rest of the default settings unchanged.**
5. **Click the Save button to save your changes.**

   The site now appears when you enter the mapped domain URL in your web browser's address bar.

With this plugin configured, you can map a domain to a site by choosing Network Admin➪Domains. (See Figure 6-6.) ***Note:*** You need to know the ID number of the site you want to map.

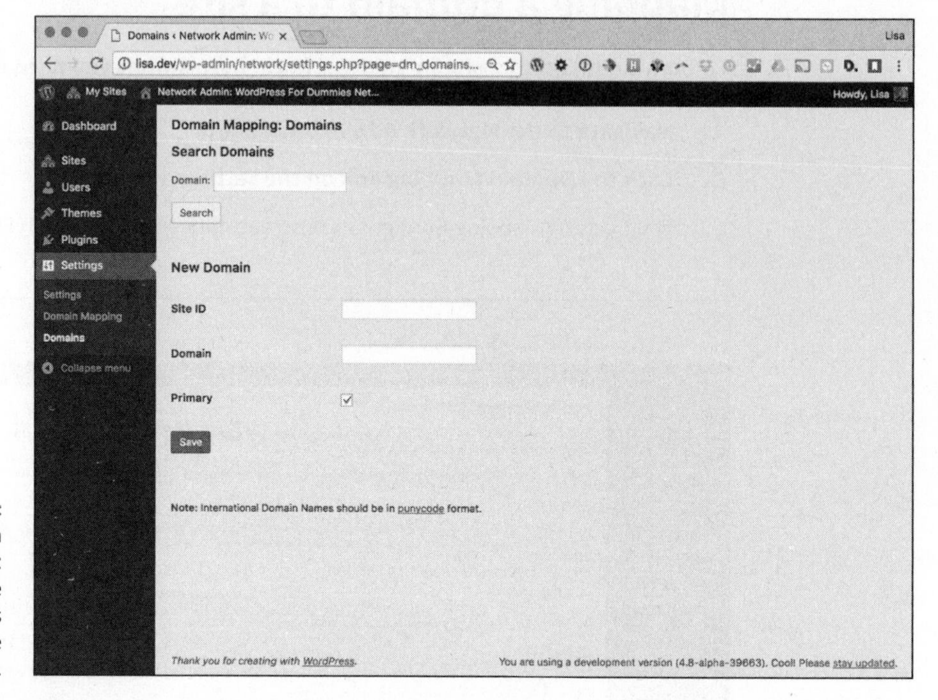

FIGURE 6-6: The Domain Mapping: Domains page maps domains from a single location.

Mapping a Large Volume of Domains

For some enterprises, you may need to map a large volume of domains (10 or more) to the WordPress network. Adding each domain to the server with a `ServerAlias` directive is time-consuming. Also, as the list grows, the server slows while reading all the domains.

The time necessary to add these domains can be shortened considerably by using a wildcard host. To use a wildcard host, you need to access your website via a terminal or via Secure Shell (SSH) with the root user. The Secure Shell is available only on a virtual private server (VPS) or a dedicated host. The ideal situation for using a wildcard host is when the main installation of WordPress is the default domain on the server. A quick way to check whether your WordPress main installation domain is the default domain on your web server is to type your IP address in your browser's address bar. If your main WordPress site displays in your browser, you can proceed with using a wildcard host. If not, you need to obtain a dedicated IP address from your web hosting provider; contact your provider to set one up for you.

Configuring Apache

Adding a wildcard host to your web server requires that you access the Apache configuration files on your web server. This section assumes that you have access to those files; if not, ask your web hosting provider either to provide the access you need or to complete the steps for you to add the wildcard host to your account.

Here's how you set this feature up:

1. **Log in to your website with the root user via a terminal.**
2. **Navigate to the configuration files in the folder located at `/etc/httpd/` by typing**

   ```
   cd /etc/httpd/
   ```

3. **Open the `httpd.conf` file by typing**

   ```
   vi httpd.conf
   ```

   Page down in the file until you see the vhost section. Find the vhost section that contains the information about your WordPress installation and the main domain of your network. (Depending on the number of domains hosted on your server, the `httpd.conf` file may contain several vhost entries; be sure that you're editing the vhost that contains the main domain of your WordPress install.)

4. **Press the Insert key to begin editing the file.**
5. **Comment out the lines, and place the wildcard as follows:**

   ```
   <VirtualHost *:80>
   ```

6. **Save the changes by pressing the Esc key, typing** :wq, **and then pressing Enter.**
7. **On the command line, restart Apache by typing**

   ```
   /etc/init.d/apache restart
   ```

Now you can map domains in volume by following these steps:

1. **Log in to your domain name registrar.**
2. **Click the domain name management tools for the domain you want to map.**

3. **Click Total DNS records.**
4. **Locate the A records at the top of the page, and insert the IP address of your WordPress network.**

   (I show you how to obtain this address in "Obtaining your IP address" earlier in this chapter.)

   Figure 6-7 shows an A record and the web-server IP address it points to. The domain is sent to that IP address regardless of name server.

5. **Choose Network Admin⇨Domains from your WordPress Dashboard.**

   The Domain Mapping: Domains page appears, as shown in Figure 6-7.

6. **Enter the ID of the site you want to map.**

   You can get the ID number from the Sites page (choose Super Admin⇨Sites).

7. **Enter the domain name you want to map to this site.**
8. **Click Save.**

   The page refreshes and shows you a list of mapped domains.

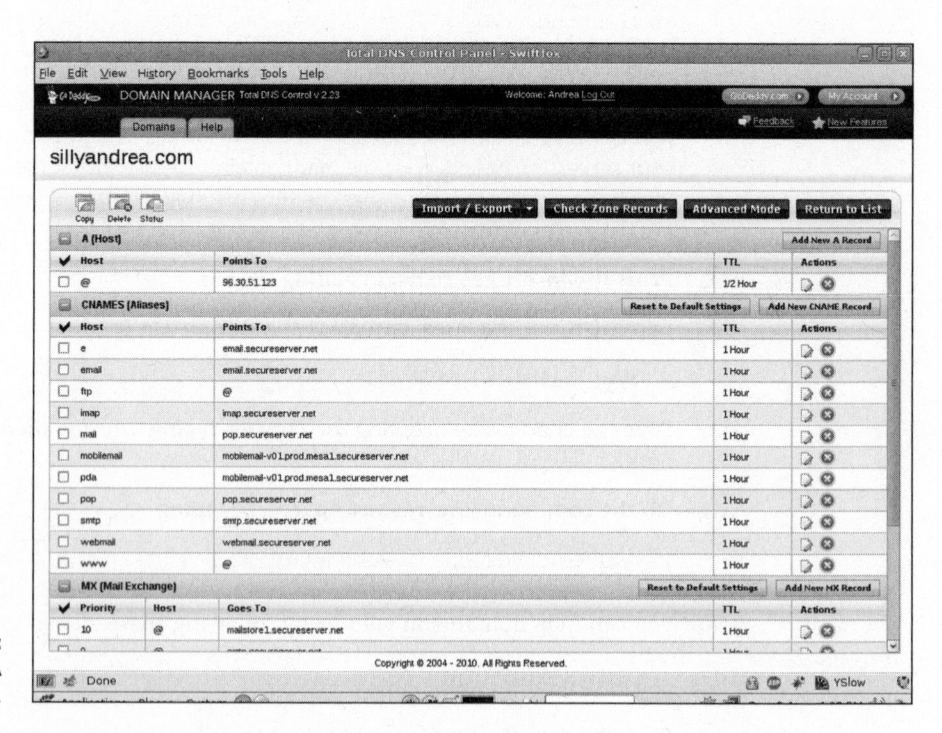

FIGURE 6-7: Domain A records.

There is no longer any need to add or point domains at the web host. The server is instructed to take any domain name request and send it to the WordPress network. WordPress associates the mapped domain with the correct site.

Hiding the original installation domain

The domain mapping plugin, mentioned in "Installing the Domain Mapping Plugin" earlier in this chapter, lets you access the subsite by the original location regardless of whether it's a subdomain site or a subfolder site, so you can use domain mapping no matter which setup you chose for your network (subdomains or subdirectories). The domain mapped for the child site is also the domain used on all uploaded media files, which maintains consistency for the site.

In some cases, you may want to hide the original installation domain. If your main installation domain is an obscure-looking domain such as `http://00954-yourvpsdomain-ba.com`, for example, you want to hide that domain because your site visitors can't easily remember or use it. If you want to hide the original installation domain, here's how you can do so:

1. **Choose Network Admin⇨Sites.**

   The Edit Site screen appears on your Network Admin Dashboard.

2. **Hover your pointer over the name of the site you want to edit and click the Edit link that appears, as shown in Figure 6-8.**

   The Edit Site page displays in your browser window.

3. **Find all instances of the original domain name, and change them to the new mapped domain.**

   Be sure to click each tab on the Edit Site page (Info, Users, Themes, and Settings) to change the original domain name to the new mapped domain wherever it appears on the Edit Site page. Keep any folder names intact.

4. **Save your changes by clicking the Save Changes button, as shown in Figure 6-9.**

Your mapped site is now inaccessible at the original subsite name (the subdomain or subfolder), and any references to it have been changed. Previous links within the body of posts, however, aren't updated automatically, so you need to edit the posts manually to change the links to reflect your newly mapped domain.

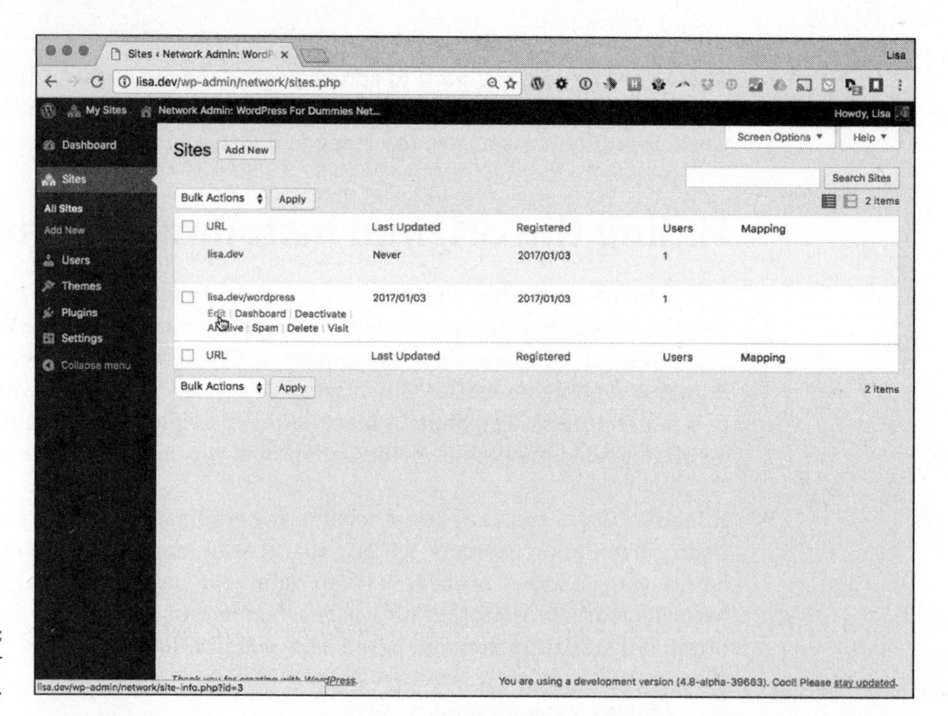

FIGURE 6-8: The Edit link for individual sites.

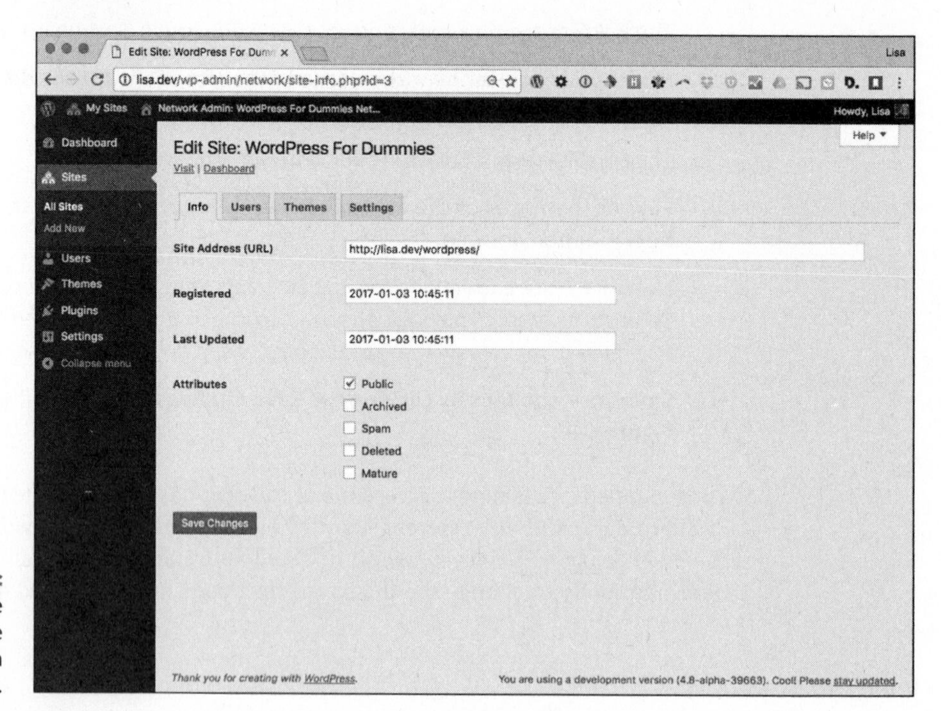

FIGURE 6-9: The Edit Site page of the Network Admin Dashboard.

Setting Up Multiple Networks

Multiple networks are supported in the WordPress code base, but WordPress has no built-in menu or interface for them on the Dashboard. Running multiple networks in one installation is an advanced feature that allows you to have another network in the same installation acting as a second independent network of sites. The new network can use its fully qualified domain name or a subdomain. The extra networks inherit the same type of sites. If you installed your original network by using subdomain sites, the extra network will also have subdomain sites. The network admin carries over to the new network, too. Additionally, you can add other network admins to the second network who will not have network admin access on the original network.

The plugin that helps you do all these things is WP Multi Network (available at `https://wordpress.org/plugins/wp-multi-network`). You install and manage the WP Multi Network plugin in a way that's similar to how you install and manage the Domain Mapping plugin. The domain for the new network still needs to be parked on the install, but the creation of the network is done on the Network options page after you install the WP Multi Network plugin. You can't take an existing site on the network and turn it into a second network; you must set up a new site when the new network is created.

To create a new network, fill in the fields of the Add New Network page in Figure 6-10:

- **Network title:** The name of the network you're creating (example: My New Network)
- **Domain:** The domain name you'll use for this new network (example: *mynewnetwork*`.com`)
- **Path:** The server path your new network will use (example: `/home/`*mynewnetwork*`/public_html/`)
- **Site Name:** The name of the site that will serve as the main site in this network (example: Network Main Site)

When you're done, click the Create Network button at the bottom of the Add New Network page. WordPress creates your new network, and you can assign child sites to it.

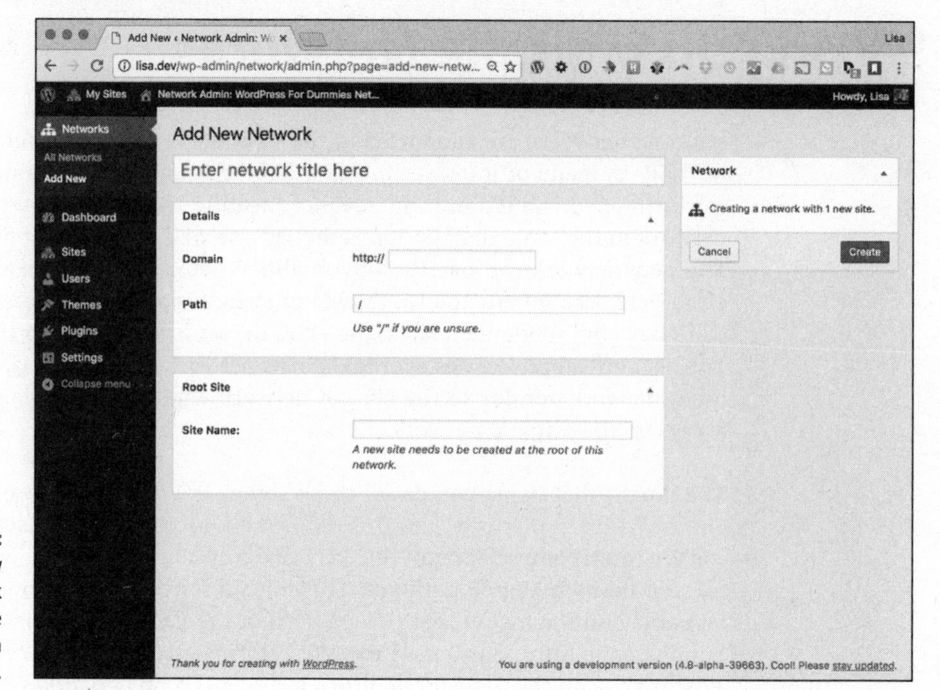

FIGURE 6-10: The Add New Network page in the Network Admin Dashboard.

Index

A

B

C

D

E

F

G

H

I

J

K

L

M

N

O

P

Q

R

S

T

U

V

W

Z

About the Author

Lisa Sabin-Wilson *(WordPress For Dummies, WordPress Web Design For Dummies)* has 14 years' experience working with the WordPress platform, having adopted it early in its first year of release in 2003. Lisa is the owner of a successful WordPress design and development agency, WebDevStudios (`http://webdevstudios.com`) and is a regular speaker on topics related to design and WordPress at several national conferences. Additionally, she hosts WordPress workshops around the country, teaching people how to use the WordPress platform to publish their own sites on the World Wide Web. You can find Lisa online on Twitter at `@LisaSabinWilson`.

Dedication

To WordPress . . . and all that entails from the developers, designers, forum helpers, bug testers, educators, consultants, plugin makers, and theme bakers.

Author's Acknowledgments

Every person involved in the WordPress community plays a vital role in making this whole thing work, and work well. Kudos to all of you! Also, big thanks to my wonderful husband, Chris Wilson, for his incredible support, backbone, and ability to put up with my crazy days of writing — I could *not* have done it without you!

Special thanks to the co-authors of the first edition of this book who helped form the framework of the publication and ensured its initial success: Cory Miller, Kevin Palmer, Andrea Rennick, and Michael Torbert.

Publisher's Acknowledgments

Acquisitions Editor: Amy Fandrei
Project Editor: Charlotte Kughen
Copy Editor: Kathy Simpson
Technical Editor: Donna Baker
Editorial Assistant: Serena Novosel
Sr. Editorial Assistant: Cherie Case
Production Editor: Antony Sami
Cover Image: ©burakpekakcan/Getty Images

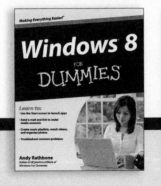

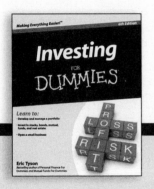

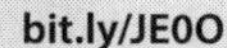